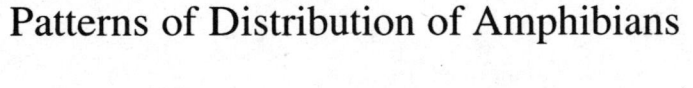

Patterns of Distribution of Amphibians

Patterns of Distribution
of Amphibians

A GLOBAL PERSPECTIVE

EDITED BY

William E. Duellman

The Johns Hopkins University Press

Baltimore and London

The Johns Hopkins University Press
2715 North Charles Street
Baltimore, MD 21218-4363
www.press.jhu.edu

A catalog record for this book is available from the British Library.

Library of Congress Cataloging-in-Publication Data

Patterns of distribution of amphibians : a global perspective /
edited by William E. Duellman.
p.   cm.
Includes bibliographical references (p.   ) and index.
ISBN 0-8018-6115-2 (alk. paper)
1. Amphibians—Geographical distribution.
I. Duellman, William Edward, 1930–
QL648.P37   1999
597.8'09—dc21

98-49449
CIP

Contents

Preface

The frogs, salamanders, and caecilians that make up the class Amphibia have been a part of folklore and have fascinated naturalists since ancient times. During the past two and a half centuries, our knowledge of these animals has increased enormously. More than 4700 species are now recognized, and although much remains to be learned and many new species await discovery, herpetologists now have a reasonably good idea of the diversity and general patterns of distribution of amphibians. Surprisingly, no modern synthesis of amphibian distributions exists. The need for such a synthesis is especially critical now because of declining amphibian populations throughout the world and the necessity for urgent conservation measures.

Material in the chapters in this book were organized for presentation in a symposium at the Second World Congress of Herpetology in Adelaide, Australia, in December 1993. This project has evolved since the World Congress, as have the amphibians of the world and, indeed, even the position of the major tectonic plates. For various reasons, several of the participants were unable to attend the congress, and subsequently one of the intended contributors withdrew from the project. As is too often the case in undertakings such as this, some authors provided manuscripts long before the last ones were received. To the former, I extend my apologies. To the latter, I am thankful that my dogged persistence resulted in manuscripts, some of which were prepared under trying conditions. All authors had the opportunity to provide addenda to their chapters; if an addendum exists, it is appended to the final page of each of their chapters.

This volume is not an atlas of distribution maps. Instead, the authors have sought to determine patterns of distribution and to correlate these with physical, biotic, and historical factors. Although the same basic material is covered in each of the regional treatments, the detail of coverage is variable. This is owing principally to the extent of knowledge of the amphibians in different parts of the world; for example, amphibians are much better known in Europe and North America than in Africa and South America. Likewise, conservation efforts are much more aggressive in Australia, Europe, and North America than in other parts of the world. In Chapter 1, I have endeavored to present an overview of patterns of distributions and to summarize threats to amphibian diversity and measures to protect this diversity and the distributions of taxa. Furthermore, I present my ideas pertaining to future research and conservation measures.

Except for Chapter 1, this volume contains syntheses of amphibian distributions in the various regions of the world. Each of these chapters concludes with one or more appendices in which the distributions of amphibians in the areas under consideration are listed; consequently, the index is nearly a checklist of amphibian species of the world (exclusive of species in Madagascar and Seychelles). In some cases, determining limits of the regions necessitated some degree of compromise. Thus, the coverage of North America is restricted to the Nearctic Realm, which includes northern Mexico and the Peninsula of Baja California (Chapters 2 and 3). A more extensive discussion on the southern boundary of the Palearctic Realm is presented by Borkin (Chapter 6), who concludes that the amphibian fauna of the Arabian Peninsula is part of the Ethiopian Realm and not the Palearctic. Some chapters contain special features. In Chapter 2, Duellman and Sweet incorporate previously published phylogenetic analyses of large taxa with respect to biogeography. Hedges (Chapter 4) provides maps of species densities and correlations of sizes of frogs and elevations in the West Indies. Campbell (Chapter 3) and Borkin (Chapter 6) tabulate distributions of amphibians by countries in Middle America and Europe and temperate Asia (exclusive of China), respectively; moreover, Campbell provides a taxonomic guide to the literature on Middle American amphibians, and Borkin provides a great service to western herpetologists by thoroughly reviewing the literature on amphibians in the Palearctic Realm. Likewise, Zhao (Chapter 7) gives an in-depth analysis of patterns of distribution in continental temperate Asia and the fringing archipelagos and islands. Inger (Chapter 8) and Poynton (Chapter 9) include essays on biogeographical analyses, and Tyler (Chapter 10) investigates the sizes of anurans with respect to aridity.

One conclusion that is evident from the analyses of patterns of distribution of amphibians is that the greatest diversity is in the highlands in the tropical regions, whether these be in Africa, New Guinea, or South America. Thus, the preservation of amphibian diversity requires suitable reserves in these montane regions. Also, there is a global imbalance in the diversity of amphibians; more than half of the species occur in the Neotropical Realm (Chapter 1).

I am indebted to the following colleagues for thoroughly reviewing manuscripts: Kraig Adler, Jonathan A. Campbell, Alan Channing, Brian I. Crother, Hélio da Silva, Ignacio de la Riva, W. Ronald Heyer, Robert F. Inger, John D. Lynch, Masafumi Matsui, Joseph R. Mendelson III, Jean-Luc Perret, Jay M. Savage, John E. Simmons, Linda Trueb, and David B. Wake. Their efforts greatly increased the quality of the chapters contained herein. Finally, I am grateful to Ginger Berman, Science Editor, the Johns Hopkins University Press, for her eagerness in soliciting the manuscript. Likewise, I am indebted to Lee Campbell Sioles, Martha Farlow, Michael Baker, and Tom Roche of the press for their expertise in shepherding the work through the publication process.

William E. Duellman

Patterns of Distribution of Amphibians

1. Global Distribution of Amphibians: Patterns, Conservation, and Future Challenges

WILLIAM E. DUELLMAN

Natural History Museum and
Department of Ecology and Evolutionary Biology
The University of Kansas
Lawrence, Kansas 66045-2454, USA

ABSTRACT   Globally, the greatest taxonomic diversity of amphibians is in the Gondwanan continents. Comparison of realms reveals the highest diversity in anurans to be in the Neotropics, followed in decreasing order by the Oriental and the Ethiopian realms. The Neotropical Realm also has the greatest diversity of caecilians and salamanders (bolitoglossine plethodontids); salamander diversity also is high in the Nearctic Realm, followed in decreasing order by the Palearctic and Oriental realms. Caecilians are more diverse in the Oriental than in the Ethiopian Realm. The amphibian species density per area of continents is not correlated with the size of the continent; the highest density of species is in South America, followed by North America. On islands, anuran diversity is greatest on large islands—Madagascar, followed by New Guinea and Borneo, but diversity is not correlated with sizes of the islands. In an overall global view, we observe that more than half of the species of amphibians occur in the New World. Continental regional species richness and endemism are not necessarily correlated; whereas richness may be higher in lowland regions, endemism is greater in montane regions. Species richness usually is lower on islands than comparably sized areas on continents, but endemism is higher.

Threats to amphibian diversity include habitat destruction, agricultural practices, pollution, road mortality, introduction of exotic species, harvesting, and collecting. Furthermore, populations of amphibians in seemingly pristine habitats are declining from for the most part unidentified causes. Among the primary conservation initiatives that have benefited amphibians are: habitat preservation; protection of individual species by facilitating breeding migrations and protecting or creating suitable breeding sites; and legislation regulating the collecting and trafficking of selected species. Changes in some agricultural and forestry practices may benefit populations of some species of amphibians.

Future challenges include more extensive and thorough inventories—especially in tropical regions of the world; results of inventories and existing museum records need to be in accessible electronic databases and the data associated with geographic information systems so as to determine patterns of distribution, species richness, and endemism more accurately. These data are essential documentation for establishment of reserves, which require effective management and programs for research. It is essential that conservation efforts emphasize education, and educated, knowledgeable policy makers are responsible for legislation. The great amount of funding necessary to meet these challenges must come from government agencies, private foundations, corporations, and ecotourism.

Key words: Amphibians; Global patterns of distribution; Species richness and endemism; Threats to diversity; Declining populations; Conservation; Proposed future initiatives.

INTRODUCTION

More than a century has passed since Boulenger's (1882) brief essay on the geographical distribution of the Batrachia; therein, he defined a northern zone of the world as being characterized by an abundance of salamanders and an absence of caecilians and an equatorial southern zone by the presence of caecilians and the absence of salamanders. Since Boulenger's work, major contributions from a global perspective have addressed distributions of family-groups with respect to earth history; of these, Savage's (1973) synthesis of anuran historical biogeography is the most detailed, and Duellman and Trueb's (1986) summary of amphibian distributions is the most comprehensive. Aside from these efforts, there have been syntheses of regional distribution patterns on continents—e.g., South America (numerous authors in Duellman, 1979), Middle America (Savage, 1982), the Palearctic Realm (Shcherbak, 1982), China (Zhao and Adler, 1993), and Europe (Gasc et al., 1997)—and archipelagos—e.g., Philippine Islands (Inger, 1954) and the West Indies (Hedges, 1996). There have been few attempts at intercontinental comparisons, and most of these have adopted an historical perspective (e.g., Tyler's [1979] comparison of Australia and South America, and Laurent's [1979] comparison of Africa and South America), but Duellman (1993a) made historical and ecological comparisons between Africa and South America.

Herein, I attempt to synthesize the global distribution patterns of amphibians. The differences and similarities among continents, major islands, and archipelagos are described. Patterns of distribution with respect to species diversity (richness) and endemism are addressed; these patterns provide a basis for an assessment of, and recommendations for, conservation measures. In preparing this synthesis, I have drawn shamelessly on data provided by authors of other chapters in this volume. Like them, I have used the most recent lists of amphibians compiled by Frost (1985) and Duellman (1993b) as the taxonomic basis. My sources for areas have been those provided by the authors herein and the third edition of Merriam Webster's Geographical Dictionary (1979).

SYNTHESIS OF GLOBAL DISTRIBUTION PATTERNS

TAXONOMIC COMPARISONS

The present familial classification does not necessarily reflect the phylogenetic relationships of many family-groups of amphibians. This is especially evident among the neobatrachian families (Ford and Cannatella, 1993). The classifications of families of salamanders (Larson and Dimmick, 1993) and caecilians (Nussbaum and Wilkinson, 1989) seem to be more reasonably resolved, except for the putative paraphyly of the Caeciliidae. Many families are restricted to one zoogeographic realm; some families are highly diverse in one realm and poorly represented elsewhere, and a few families are widely distributed on the planet (Table 1:1). Comparisons of families and some widespread genera follow.

Anura

The Neotropical Realm contains six endemic families of anurans with 17 genera and 313 species; of these, the largest endemic families by far are Centrolenidae (3 genera and 125 species) and Dendrobatidae (8 genera and 179 species). Although both families are primarily South American, both range into Middle America, where there are 13 species of centrolenids and 17 of dendrobatids. The other endemic families (Allophrynidae, Brachycephalidae, Pseudidae, and Rhinodermatidae) account for only six genera and nine species. Two other large families have their greatest diversity in the Neotropics. Of the 977 species of Leptodactylidae, only five range into the Nearctic, and only three are endemic to the Nearctic. Of 753 species of Hylidae, 575 species in 32 genera occur in the neotropics; the only other large radiation in this family is in the Australo-Papuan Realm, with three genera and 147 species. Also, Rhinophrynidae (1 species) essentially is endemic to the Middle American part of the Neotropics, with *Rhinophrynus* barely entering the Nearctic Realm.

Bufonidae is distributed worldwide except for the Australo-Papuan Realm and Madagascar, but it is represented only by *Bufo* in the Nearctic and Palearctic realms. In addition to *Bufo,* 13 genera occur in Africa, 11 in the Neotropics, and 6 in the Oriental Realm. Similar patterns of paucity in the northern hemisphere are apparent in two other widespread families. Microhylidae is represented by two gen-

Table 1:1. Distribution of living families of amphibians in zoogeographic realms. Numbers are genera/species. Some taxa are listed in more than one realm. Palearctic includes Mediterranean Africa; Ethiopian includes Madagascar and Seychelles.

Family	Nearctic	Neotropical	Palearctic	Ethiopian	Oriental	Australo-Papuan
ANURA:						
Allophrynidae	–	1/1	–	–	–	–
Arthroleptidae	–	–	–	7/73	–	–
Ascaphidae	1/1	–	–	–	–	–
Brachycephalidae	–	2/3	–	–	–	–
Bufonidae	1/21	12/188	1/24	14/82	7/76	–
Centrolenidae	–	3/125	–	–	–	–
Dendrobatidae	–	8/179	–	–	–	–
Discoglossidae	–	–	3/13	–	2/5	–
Heleophrynidae	–	–	–	1/5	–	–
Hemisotidae	–	–	–	1/8	–	–
Hylidae	4/25	32/575	1/7	–	1/7	3/147
Hyperoliidae	–	–	–	18/210	–	–
Leiopelmatidae	–	–	–	–	–	1/4
Leptodactylidae	2/5	49/972	–	–	–	–
Mantellidae	–	–	–	3/68	–	–
Megophryidae	–	–	1/5	–	6/95	–
Microhylidae	2/4	19/45	1/1	18/77	14/78	16/120
Myobatrachidae	–	–	–	–	–	24/122
Pelobatidae	2/7	2/2	1/4	–	–	–
Pelodytidae	–	–	1/2	–	–	–
Pipidae	–	1/6	–	4/21	–	–
Pseudidae	–	2/3	–	–	–	–
Ranidae	1/26	1/33	6/47	24/183	23/295	7/50
Rhacophoridae	–	–	2/3	3/31	6/189	–
Rhinodermatidae	–	1/2	–	–	–	–
Rhinophrynidae	1/1	1/1	–	–	–	–
Sooglossidae	–	–	–	2/3	–	–
CAUDATA:						
Ambystomatidae	1/17	1/18	–	–	–	–
Amphiumidae	1/3	–	–	–	–	–
Cryptobranchidae	1/1	–	1/2	–	–	–
Dicamptodontidae	1/4	–	–	–	–	–
Hynobiidae	–	–	5/27	–	5/8	–
Plethodontidae	15/107	12/188	1/7	–	–	–
Proteidae	1/6	–	1/1	–	–	–
Rhyacotritonidae	1/4	–	–	–	–	–
Salamandridae	2/6	1/1	9/32	–	5/21	–
Sirenidae	2/4	1/1	–	–	–	–
GYMNOPHIONA:						
Caeciliidae	–	10/58	–	9/24	2/4	–
Ichthyophiidae	–	–	–	–	2/36	–
Rhinatrematidae	–	2/9	–	–	–	–
Scolecomorphidae	–	–	–	2/5	–	–
Typhlonectidae	–	4/15	–	–	–	–
Uraeotyphlidae	–	–	–	–	1/4	–
Total Anura	15/90	154/2135	17/103	95/761	59/745	51/443
Total Caudata	25/151	15/248	17/89	–	10/29	–
Total Gymnophiona	–	16/82	–	11/29	5/44	–
Total Amphibia	40/241	185/2465	34/192	106/790	74/818	51/443

era and four species in North America and one species in the Palearctic, but numbers are vastly greater on the southern continents and associated islands—19 genera with 45 species in the Neotropics, 18 genera with 77 species in the Ethiopian Realm, 14 genera with 78 species in the Oriental Realm, and 16 genera with 120 species in the Australo-Papuan Realm. Although the genus *Rana* is represented by numerous species in the New World (26 in the Nearctic and 33 in the Neotropical realms) and 41 species in the Palearctic Realm, this is the only genus of this diverse family in those regions, except for a few Oriental genera that are in the southeastern periphery of the Palearctic Realm. However, Ranidae is most diverse in the southern lands of the Old World—50 species in 7 genera in the Australo-Papuan, 183 species in 24 genera in the Ethiopian, and 295 species in 23 genera in the Oriental realms.

Among anurans, the Nearctic has one endemic family, the monotypic Ascaphidae; the Palearctic also has only one endemic family, Pelodytidae, with two species. However, the greatest diversity of discoglossids is in the Palearctic Realm, which has three of the four genera and 13 of the 17 species; other discoglossids are in the Oriental Realm.

The Ethiopian Realm contains six endemic families of anurans—Arthroleptidae, Heleophrynidae, Hemisotidae, Hyperoliidae (Africa, Madagascar, and Seychelles), Mantellidae (Madagascar only), and Sooglossidae (Seychelles only). Only Myobatrachidae and Leiopelmatidae (the latter in New Zealand only) are endemic to the Australo-Papuan Realm. The Oriental Realm has no endemic families of anurans, but it contains by far the greatest diversity of rhacophorids (6 genera, 189 species); with the exception of three species of *Chiromantis* in Africa, other rhacophorid genera (2) and species (28) occur on Madagascar.

Three large genera, *Bufo, Hyla,* and *Rana,* have nearly cosmopolitan distributions, but the first two do not occur in the Australo-Papuan Realm. Most likely, all three genera are paraphyletic. Cannatella (1986) considered *Bufo* to be a derived lineage within Bufonidae, and Graybeal and Cannatella (1995) stated that there was no evidence for the monophyly of *Bufo*. In Graybeal's (1997) phylogenetic analysis, which revealed many homoplasies and has a low consistency index, *Bufo* is suggested to be paraphyletic with respect to several other genera of

Table 1:2. Distribution of three large genera in zoogeographic realms.

Realm	*Bufo*	*Hyla*	*Rana*
Nearctic	21[a]	10[b]	26[c]
Neotropical	74[a]	269[b]	33[c]
Palearctic	24[d,e]	7[f]	41[g]
Oriental	47[d]	7[f]	99[g]
Ethiopian	56[e]	0	9
Australo-Papuan	0	0	11

[a]Seven species shared by Nearctic and Neotropical.
[b]Two species shared by Nearctic and Neotropical.
[c]Five species shared by Nearctic and Neotropical.
[d]Nine species shared by Palearctic and Oriental.
[e]One species shared by Palearctic and Ethiopian.
[f]One species shared by Palearctic and Oriental.
[g]Sixteen species shared by Palearctic and Oriental.

bufonids, and some of the phenetic groups of *Bufo* recognized by Inger (1972), Tandy and Keith (1972), and Duellman and Schulte (1992) are not demonstrably monophyletic. The large genus *Hyla* undoubtedly is paraphyletic with respect to many genera—e.g., *Hyla eximia* Group and *Pseudacris* (Hedges, 1986), *Hyla bistincta* Group and *Plectrohyla* (Duellman and Campbell, 1992); likewise, there is no strong evidence for the monophyly of *Hyla*. The large genus *Rana* was divided into several genera by Dubois (1987, 1992). Inger (1996) soundly criticized the proposed classification and concluded (p. 245) "... acceptance of Dubois's classification, even on a provisional basis is likely to lead to confusion rather than understanding." These caveats notwithstanding, certain patterns seem to be evident in these large genera.

The lineage that includes *Bufo,* as now defined, seems to be western Gondwanan in origin. The genus is most speciose in the Neotropical (74) and Ethiopian (56) realms, followed by the Oriental Realm (47); 9 of the Oriental species are among the 24 species in the Palearctic Realm, and 7 of the Neotropical species are among the 21 species in the Nearctic Realm (Table 1:2). Hylidae primarily is distributed in the Neotropical and Australo-Papuan realms; 93% of the species of *Hyla* occur in the Neotropics. The contribution of this genus to the anuran faunas of the Nearctic, Palearctic, and Oriental realms is negligible, and the genus is absent in the Ethiopian and Australo-Papuan realms. In contrast to *Bufo* and *Hyla,* the greatest diversity of *Rana* is in the Oriental Realm (99 species, 16 of which are among the 41 species in the Palearctic Realm), followed by 33 species in the Neotropics, of which only three occur in South America.

Caudata

Most salamanders occur in the Northern Hemisphere; they are absent in the Australo-Papuan and Ethiopian realms. At the familial level, the greatest diversity of salamanders is in the Nearctic Realm, where all families (except Hynobiidae) occur and Amphiumidae, Dicamptodontidae, and Rhyacotritonidae are endemic. The largest distributions of the Ambystomatidae and Sirenidae also are in the Nearctic Realm; both families range into the northern part of the Middle American portion of the Neotropics. Salamanders of the family Proteidae are represented by one genus and six species in the Nearctic, and one genus and one species in the Palearctic realms. Likewise, cryptobranchids are represented by one monotypic genus in North America, and one genus with two species in the southeastern part of the Palearctic Realm. Salamandridae has the largest distribution with representatives in the Nearctic (*Notophthalmus meridionalis* peripherally in the Neotropics), Palearctic, and Oriental realms; the greatest diversity (9 genera and 32 species) is in the Palearctic. The only family of salamanders restricted to the Old World is Hynobiidae in the southern and eastern Palearctic and adjacent Oriental realms.

Plethodontidae has its greatest number of genera (25) in the Nearctic; one of these, *Hydromantes,* is represented by three species in western North America and seven species in the Mediterranean Region of the Palearctic. However, the greatest number of species (188) is in the Neotropics, especially in Middle America, where the tribe Bolitoglossini is represented by 12 genera, two of which (*Bolitoglossa* and *Oedipina*) also extend into South America.

Gymnophiona

Caecilians are pantropical in distribution; they are absent in the Nearctic, Palearctic, and Australo-Papuan realms, but they also are unknown from Madagascar. Rhinatrematidae and Typhlonectidae are restricted to the Neotropics; Ichthyophiidae and Uraeotyphlidae are limited to the Oriental Realm, and Scolecomorphidae is endemic to the Ethiopian Realm. The large, apparently paraphyletic Caeciliidae is most diverse in the Neotropics (10 genera

and 58 species), but the family is represented in Africa (6 genera and 17 species), the Seychelles (3 genera and 7 species), and the Oriental Realm (2 genera and 4 species).

REGIONAL COMPARISONS

Continents

The numbers of families, genera, and species are highly variable among zoogeographic realms (Fig. 1:1). Because of the absence of caecilians in Australia and most of Eurasia and North America, and the absence of salamanders in Australia and Africa, continental comparisons are meaningful only among anurans. It is evident from Table 1:3 that there is no correlation between the numbers of species of anurans and the sizes of the continents. Not only is the greatest number of species of amphibians in South America, but the highest density of species occurs there.

Among the three Gondwanan continents inhabited by anurans, Australia is the smallest, but it has a slightly higher density of species than Africa; combined, these two continents have fewer than half as many species as South America. In contrast to South America, the major parts of Australia and Africa are arid or semiarid. As pointed out by Duellman (1993a), approximately 75% of Sub-Saharan Africa is tropical dry forest and savanna-grassland, whereas only about 43% of South America supports those types of vegetation. About 35% of the species of anurans in Sub-Saharan Africa inhabit these semiarid environments, and the proportion is much lower (about 9%) in South America. Lowland tropical rainforest covers proportionally twice as much area in South America as it does in Africa, and the proportional number of species of anurans in rainforest is about the same on the two continents. In contrast, montane tropical forest accounts for only 1.6% of the area in Sub-Saharan Africa and 4.3% of the area in South America; in Africa, only 17% of the anurans inhabit this type of forest, as opposed to 38% in South America (Duellman, 1993a). Moreover, in the New World tropics, many species of anurans and plethodontid salamanders inhabit bromeliads; these water-holding plants are absent in the Old World.

Comparison of the tropical parts of Eurasia and North America (including associated islands in both regions) reveals that there are slightly more species in tropical Asia than in tropical North America, but

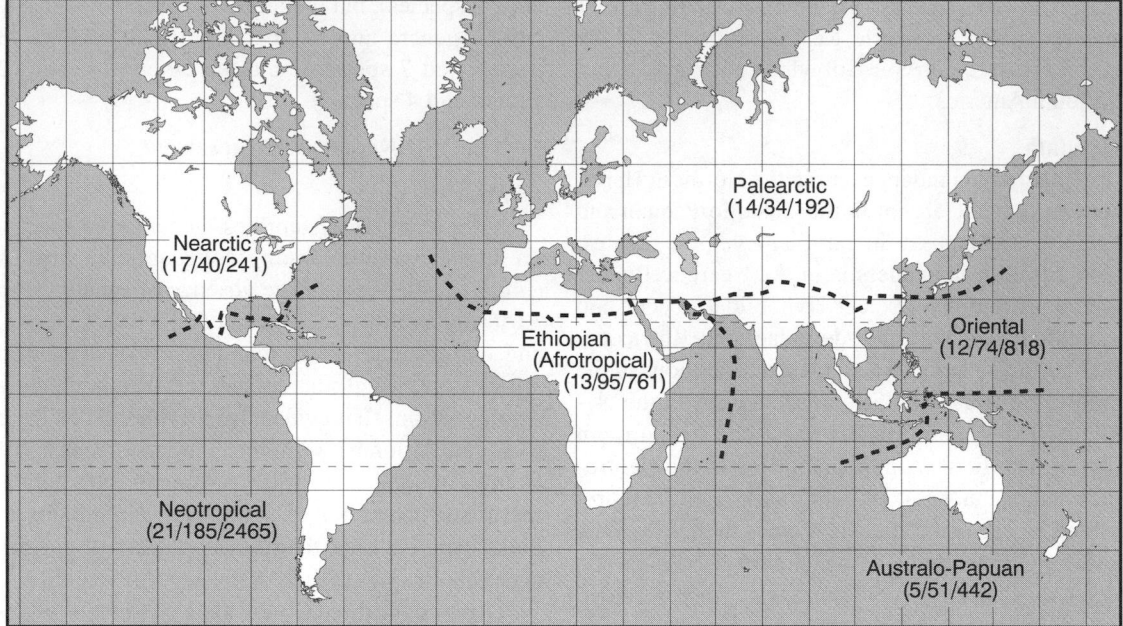

Fig. 1:1.   Numbers of taxa (families/genera/species) of amphibians in zoogeographic realms (demarcated by heavy dotted lines).

that the density of species per unit area is much higher in tropical North America than in tropical Asia (Table 1:4). As expected, the tropical parts of both continents have much greater numbers of species (and densities) than do the temperate parts of the continents (Table 1:4).

    Comparison of these continents and regions reveal that amphibians are far more diverse in tropical regions than in temperate ones. Furthermore, the Neotropical Realm has much greater diversity than any other part of the world, and the Nearctic is richer than the Palearctic.

Islands

    Comparison of anuran diversity on tropical islands reveals that the greatest diversity is on the largest islands—Madagascar, followed by New Guinea, and Borneo; however, in these three islands, diversity is not correlated with area (Table 1:5). The number of species per area on islands is higher in the West Indies than in the Greater Sundas, but the amphibian faunas on Sumatra and especially Sulawesi are poorly known (R. F. Inger, pers. comm.). The islands of Hainan and Taiwan on the continental shelf of southeastern China have relatively high species densities of anurans. On the islands and archipelagos bordering eastern Asia, there is a steady decline in the numbers of species of anurans from south to

Table 1:3. Anuran faunas on continents.

Continent	Area (km^2)	No. of Species	Species/ (10^6 km^2)
Eurasia[1]	54,745,798	940	17.2
Africa	30,244,049	632	20.9
North America[2]	22,071,442	979	44.4
South America	17,793,000	1742	97.9
Australia	7,686,884	210	27.3

[1] Including associated islands (Japan, Philippines, East Indies, etc.).
[2] Excluding Greenland but including West Indies.

north—59 species in the Philippines (299,536 km^2; density = 196.97 species/10^6 km^2), 21 species on Taiwan (35,760 km^2; density = 587.2 species/10^6 km^2), 17 species in the Ryukyu Archipelago (2196 km^2; density = 7741.3 species/10^6 km^2), and 17 species in the Japanese Archipelago (excluding Ryukyu Archipelago) (360,927 km^2; density = 47.1 species/ 10^6 km^2). The amphibian fauna of the Japanese Archipelago is enhanced by 20 species of salamanders; also two species of salamanders are known from the Ryukyu Archipelago, three salamanders from Taiwan, and two caecilians from the Philippines.

    Comparatively few species of anurans are known from the archipelagos to the east of New Guinea—17 species in the Bismarck Archipelago

Table 1:4. Numbers and densities of amphibians in tropical and temperate parts of Eurasia and North America.

	Eurasia				North America			
	Tropical		Temperate		Tropical		Temperate	
Order	No. of Species	Species/ (10^6 km^2)	No. of Species	Species/ (10^6 km^2)	No. of Species	Species/ (10^6 km^2)	No. of Species	Species/ (10^6 km^2)
Anura	745	100.07	103	2.17	553	298.75	90	4.56
Caudata	29	3.90	89	1.88	197	106.43	151	7.56
Gymnophiona	44	591	0	0.00	12	6.48	0	0.00
Total	818	109.87	192	4.05	762	411.67	241	12.23

(49,658 km^2; density = 342.3 species/10^6 km^2), and 26 species on the Solomon Islands (29,785 km^2; density = 872.9 species/10^6 km^2). When anuran diversity is viewed with respect to island (or archipelago) size (Fig. 1:2), it is evident that the reported numbers of species are low on the islands of Java and Sulawesi and in the Bismarck and Japanese archipelagos. The low number of species in the latter is correlated with the temperate location of that archipelago, whereas the low numbers elsewhere possibly are because the islands have been sampled insufficiently, either recently or before extensive habitat destruction (in the case of Java).

NEW WORLD VERSUS OLD WORLD DIVERSITY

Table 1:1 reveals that far more species of all three orders of amphibians occur in the New World than in the Old World. Combined, the Nearctic and Neotropical realms accommodate 42% of the gen-

era and 52% of the species of anurans, 58% of the genera and 71% of the species of salamanders, and 50% of the genera and 53% of the species of caecilians. Comparison of the families of amphibians inhabiting the six major terrestrial realms of the world reveals that by far the greatest diversity of families, genera, and species is in the Neotropical Realm, inhabited by 49% of the families, 47% of the genera, and 54% of the species.

Are these differences real or are they artifacts of sampling? Species-discovery rates have essentially the same slopes and are leveling off in temperate North America, temperate Europe, and the West Indies (Fig. 1:3). The rates in tropical Asia (Oriental), Middle America, Australo-Papuan region, and Sub-Saharan Africa have greater slopes, and only Africa

Table 1:5. Anuran diversity on islands.

Island	Area (km^2)	No. of Species	Species/ (10^6 km^2)
WEST INDIES:			
Cuba	105,007	50	476.2
Hispaniola	76,470	63	823.9
Jamaica	10,992	22	2001.5
Puerto Rico	8,768	20	2281.0
GREATER SUNDAS:			
Borneo	751,929	137	182.2
Sumatra	472,784	64	135.4
Sulawesi	188,487	21	111.4
Java	126,700	28	220.1
CHINESE ISLANDS:			
Hainan	33,991	27	798.3
Taiwan	35,760	21	587.2
OTHER ISLANDS:			
Madagascar	587,042	156	265.8
New Guinea	884,824	205	231.9

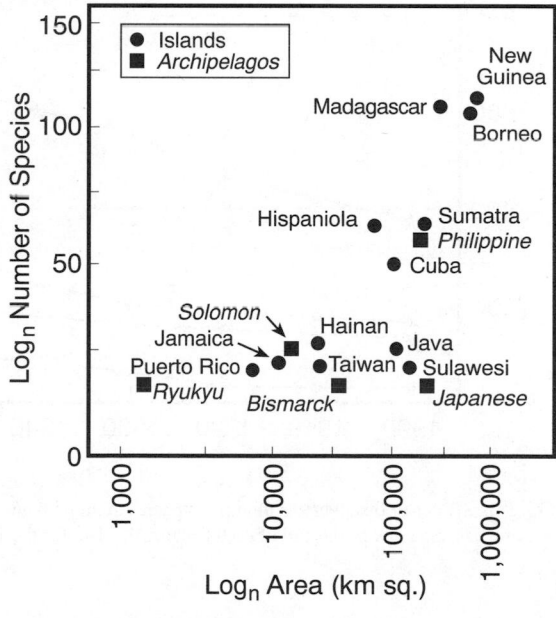

Fig. 1:2. Relationship of numbers of species of anurans and sizes of islands and archipelagos.

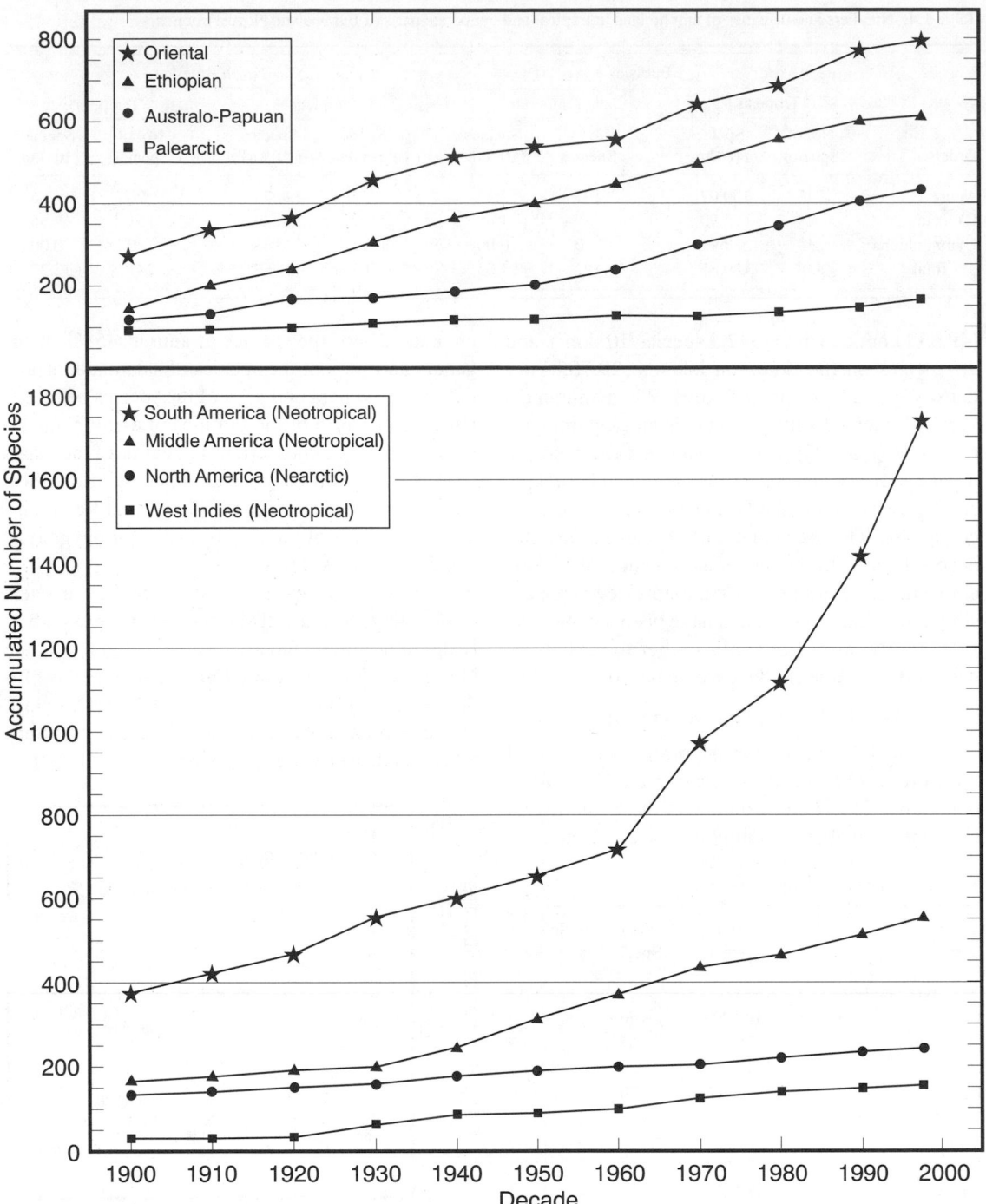

Fig. 1:3.   Rates of discovery (naming) of species of amphibians in different parts of the world during the Twentieth Century. Symbols are associated with the end of each decade; the data for the 1990s are incomplete.

seems to be asymptotic. In contrast, the species-discovery rate in South America shows a great increase over the past four decades, during which time more than 60% of the species have been named.

Expectations for a significant increase in the number of species yet to be discovered in Europe and North America are low (but several undescribed *Batrachoseps* and neotenic *Eurycea* are known); there probably are a few more species to be discovered in mountainous temperate regions of Asia, in Australia, and in the West Indies (especially Cuba and Hispaniola). Proportionately more species may be discovered in Middle America and Sub-Saharan Africa; many species from Madagascar await description. However, the regions of the world where the largest numbers of species of anurans remain to be discovered are primarily the mountainous regions of South America, the mountainous regions of southeast Asia ("Indochina" and southwestern China), and the islands in the Indo-Australian Archipelago (Sulawesi, Sumatra, and especially the Irian Jaya part of New Guinea).

Thus, the inequities in amphibian diversity and the numbers of species per area are not expected to change significantly with the acquisition of material from regions that are poorly known, although the comparative diversity within regions will be refined. It seems unlikely that the dominance of amphibian diversity in the New World, and especially in the Neotropics will be challenged by new discoveries in the Old World.

The high diversity of amphibians in the New World can be attributed to large radiations of taxa either absent in the Old World or poorly represented there. Plethodontids account for 67% of the species of salamanders worldwide; only seven of these occur in the Old World. Of the 151 species of salamanders in temperate North America, 107 (71%) are plethodontids; removal of plethodontids from the amphibian faunas of temperate North America and temperate Eurasia would result in Eurasia having more than twice as many salamanders as North America. Centrolenids, dendrobatids, and leptodactylids are large families only represented in the New World, and the majority of hylids are restricted to the New World (Table 1:1). Numerically, these four families of anurans and the plethodontids account for more than 80% of the New World amphibian fauna and about 97% of the Neotropical amphibian fauna. In contrast, the numbers of species in families endemic to the Old World are far fewer; the largest such families are Rhacophoridae with 223 species, Hyperoliidae with 210 species, Myobatrachidae with 122 species, and Megophryidae with 100 species. Eight other families of anurans endemic to the Old World contain collectively only 181 species. Thus, the 11 families of anurans endemic to the Old World contain fewer species (836) than one (Leptodactylidae with 977 species) of the nine families endemic to the New World. In part, this great discrepancy results from the highly speciose leptodactylid genus *Eleutherodactylus,* the more than 500 species of which were allocated to five subgenera by Lynch and Duellman (1997).

Only three families of amphibians shared by the New World and the Old World have greater diversity in the Old World. Microhylidae is represented in the Old World by 275 species in 48 genera and in the New World by 46 species in 19 genera. Ranidae is represented in the Old World by 574 species in 48 genera and in the New World by 54 species in one genus (*Rana*); Salamandridae is represented in the Old World by 51 species in 13 genera and in the New World by six species in two genera. The families of salamanders and caecilians not represented in the New World are Hynobiidae with 36 species in seven genera, Ichthyophiidae with 36 species in two genera, Scolecomorphidae with five species in two genera, and Uraeotyphlidae with four species in one genus.

SPECIES RICHNESS AND ENDEMISM

In the foregoing section, richness and endemism at the family level was addressed, and some major patterns were emphasized. These patterns are a reflection of major events in earth history, such as the fragmentation of Gondwanaland and the associated histories of familial lineages. The fossil records of most lineages of amphibians are poor or nonexistent for many large families (e.g., Centrolenidae, Dendrobatidae, Hyperoliidae, Rhacophoridae, and Plethodontidae). Although the fossil history of a few families (e.g., Pelobatidae, Pipidae, and Salamandridae) is extensive (Estes, 1981; Sanchíz, 1997), few attempts have been made to show the relationships among fossil and living taxa; a notable exception is the cladistic analysis of pipids and their relatives by Báez and Trueb (1997). Areal biogeography resulting from such phylogenetic analyses must be congruent with earth history, and it can be inferred that previous environmental conditions limited the distributions of species in the past by processes similar to those operating today to limit the distributions of modern species (Brown, 1995).

Although environmental factors may limit the kinds of amphibians that may be able to inhabit a given area, history plays an important role in the kinds of amphibians that are available to inhabit the area. For example, throughout most of the world, toads (*Bufo*) are diverse in semiarid habitats, but bufonids are not native to Australia, a continent that is largely arid and semiarid. In Australia, the "toad niche" is filled by some species of *Cyclorana* (Hylidae) and a radiation of fossorial myobatrachids—*Arenophryne, Neobatrachus,* and *Notaden.* Likewise, life-history traits are important in the diversity of species. Direct development of eggs (absence of free-living larvae) precludes the necessity of ponds for oviposition. This reproductive mode is characteristic of lineages of amphibians that are especially diverse in montane regions—bolitoglossine plethodontids in Middle America, *Eleutherodactylus* in the Neotropics, *Platymantis* on the islands in the southwestern Pacific, and asterophryine and genyophrynine microhylids in New Guinea. It has been shown that anurans having reproductive modes that involve nonaquatic eggs are much more diverse in regions in South America that have high atmospheric humidity (Duellman, 1988). Also, many lineages of anurans have larvae that develop in fast-flowing streams; such lineages are restricted to areas of high topographic relief and account for about 18% of the species (mostly megophryids and ranids) in southern Asia (Inger, this volume). Stream-adapted larvae are characteristic of 23% of the species (mostly hylids and leptodactylids) in a community in southeastern Brazil, and up to 53% of the species (bufonids, centrolenids, dendrobatids, and hylids) in some communities in the Andes (Duellman, 1988).

Physiological tolerances of species also play a role in species richness. As noted by Duellman and Sweet (this volume), a few North American amphibians are able to survive the long, cold winters and short summers in the northern part of the continent, and at least some of these species can withstand freezing of extracellular fluids. Optimum body temperatures generally are much lower in salamanders than in anurans (Duellman and Trueb, 1986); thus, in the Middle American highlands, plethodontid salamanders have proportionately greater species richness at higher elevations than do anurans (Campbell, this volume). Likewise, because of the problems

presented by evaporative water loss, amphibians are far more diverse and abundant in regions having high humidity throughout the year, whether this be the result of high rainfall and/or low evaporation rates.

Most terrestrial anurans and salamanders are generalist insectivores, although many salamanders include small gastropods in their diets, and some anurans, salamanders, and many caecilians feed on earthworms; the nature of the prey seems to depend mainly on its availability and the size of the gape of the amphibian. However, some species of anurans specialize on ants (especially microhylids and dendrobatids) or termites (e.g., hemisotids and some leptodactylids). Availability of prey may vary seasonally, such as in cool temperate or monsoonal climates.

Thus, the determinants of species richness include the history of the region, the history of the lineages inhabiting the region, past and present climatic and topographic conditions, the physiological and life-history adaptations of the lineages, and the availability (totally and temporally) of sufficient resources—food, shelter, and breeding sites. It is evident that within environments suitable for amphibian habitation the greatest richness of amphibians exists in areas of climatic stability, which in turn supports high habitat heterogeneity. Constancy of climate and vegetation creates a stable environment for animals and allows them to specialize on food and microhabitat, as was demonstrated among three types of habitat in Thailand by Inger and Colwell (1977). Thus, regions with stable environments permit the evolution of finer specializations and adaptations than do regions with erratic climates.

Because, theoretically, more species can occupy a given unit of habitat space in stable environments, niches are smaller (i.e., species are more specialized) than in unstable environments; the concept of "species packing" in aseasonal tropical environments formerly pervaded the design of competition and equilibrium models (Klopfer and MacArthur, 1961; MacArthur, 1969, 1970). However, such models have been scrutinized critically, for example by Conner and Simberloff (1986), who, together with Holt (1984) emphasized the importance of predation in structuring communities. Solid evidence for interspecific competition among amphibians is scarce. Interspecific competition has been demonstrated among plethodontid salamanders in the Appalachian Mountains in North America (Hairston,

1987, and papers cited therein). Patterns of food-size partitioning among species in an ant-eating terrestrial guild of anurans in the Amazon Basin in Peru show a considerable overlap in food, but this overlap diminishes during the dry season when food is less abundant; thus, the frogs may be avoiding competition by specializing on certain prey when food resources are limited (Toft, 1980). Likewise, the limited data on predation on amphibians have not been synthesized, except for Arnold's (1972) demonstration of a strong correlation between numbers of species of anurans and snakes that feed on them along a latitudinal gradient.

Last, species richness of amphibians may be influenced by richness of other organisms that utilize the same microhabitats and food resources. For example, Duellman and Pianka (1990) suggested that if the carrying capacity of the environment limits the number of nocturnal insectivores through limited food, perches, or shelter, some sort of balance between numbers of species of nocturnal, arboreal frogs and lizards might be expected, and they showed an inverse relationship between the numbers of species of nocturnal frogs and lizards in comparing communities of lizards and frogs in the Old and New World tropics.

Determination of endemism has some inherent problems that have been addressed by several workers (e.g., Axelius, 1991; Anderson, 1994; Harold and Mooi, 1994; Morrone, 1994). One of the problems is the definition of an area of endemism, which most simply can be defined as an area that encompasses the entire distributions of two or more species. A second problem concerns sizes of areas and the corresponding numbers of endemic species. Obviously, in a large area, the absolute number of endemic species will be greater than in any portion of that area. For example, 39 (43.3%) of the 90 species of amphibians occurring in the Appalachian Mountains of North America are endemic to those mountains, but the degree of endemism is variable within four natural regions in those highlands—0.0–17.3% ($\bar{x} =$ 13.2) (Duellman and Sweet, this volume). However, a large area does not necessarily indicate that there are many species or a large number of endemics. In the case of amphibians (ectotherms that require moisture) cold and dry regions have few species. For example, two of the largest regions in the northern hemisphere, Siberia and the Laurentian Uplands, have 9 and 14 species, respectively; neither has endemic species. Third, in an analysis of distribution and endemism in Australian anurans, Anderson (1994) questioned why there should be a correlation between an increase in numbers of species and in numbers of endemics. Although there is an overall correlation between the number of species in an area and the number of endemics in the area, considerable variation exists. For example, in the Austral forests of South America, all 32 species are endemic, but in the South American llanos, only 7 (22.6%) of 31 species are endemic (Duellman, this volume). Six species are endemic to the Great Plains of North America and six to a comparable region, Central Asia, but the former is inhabited by 37 species and the latter by only 17 (Borkin, and Duellman and Sweet, this volume).

Is there a pattern in relationships between numbers of species and numbers of endemics in various parts of the world? Comparison of these parameters for lowland and montane continental regions in the Afrotropical and Neotropical realms show an overall significant correlation ($r^2 = 0.96$) between numbers of species and numbers of endemics in 57 regions (Fig. 1:4). Comparison of 16 regions in the Nearctic and Palearctic realms show a much lower correlation ($r^2 = 0.76$; Fig. 1:5.), whereas comparisons of eight islands and five archipelagos show the same correlation ($r^2 = 0.96$) as among the continental tropical regions (Fig. 1:6).

Endemism is expected to be high on islands. This is especially true in the Greater Antilles, where endemism on each of the four major islands is 100%, except that two of 50 species on Cuba occur elsewhere, but not on the other Greater Antilles (Hedges, this volume). Also, high degrees of endemism might be expected in highland regions. For example, endemism is 92.8–95.0% ($\bar{x} = 93.8$) in three highland regions, compared to 22.5–100.0% ($\bar{x} = 57.5$) in 9 lowland regions, in South America. However, the degree of endemism is much lower in nine montane regions in Middle America (7.5–61.0%, $\bar{x} = 35.9$) and 13 montane regions in Africa (10.0–78.0%, $\bar{x} = 40.7$), and these values are comparably higher than those for 8 lowland regions in Middle America (0.0–14.4%, $\bar{x} = 7.1$) and 12 lowland regions in Africa (0.0–70.0%, $\bar{x} = 36.2$).

The notably lower correlation between numbers of species and numbers of endemics in the northern continents may be a reflection of the sizes of the areas that were analyzed. This is especially true for the At-

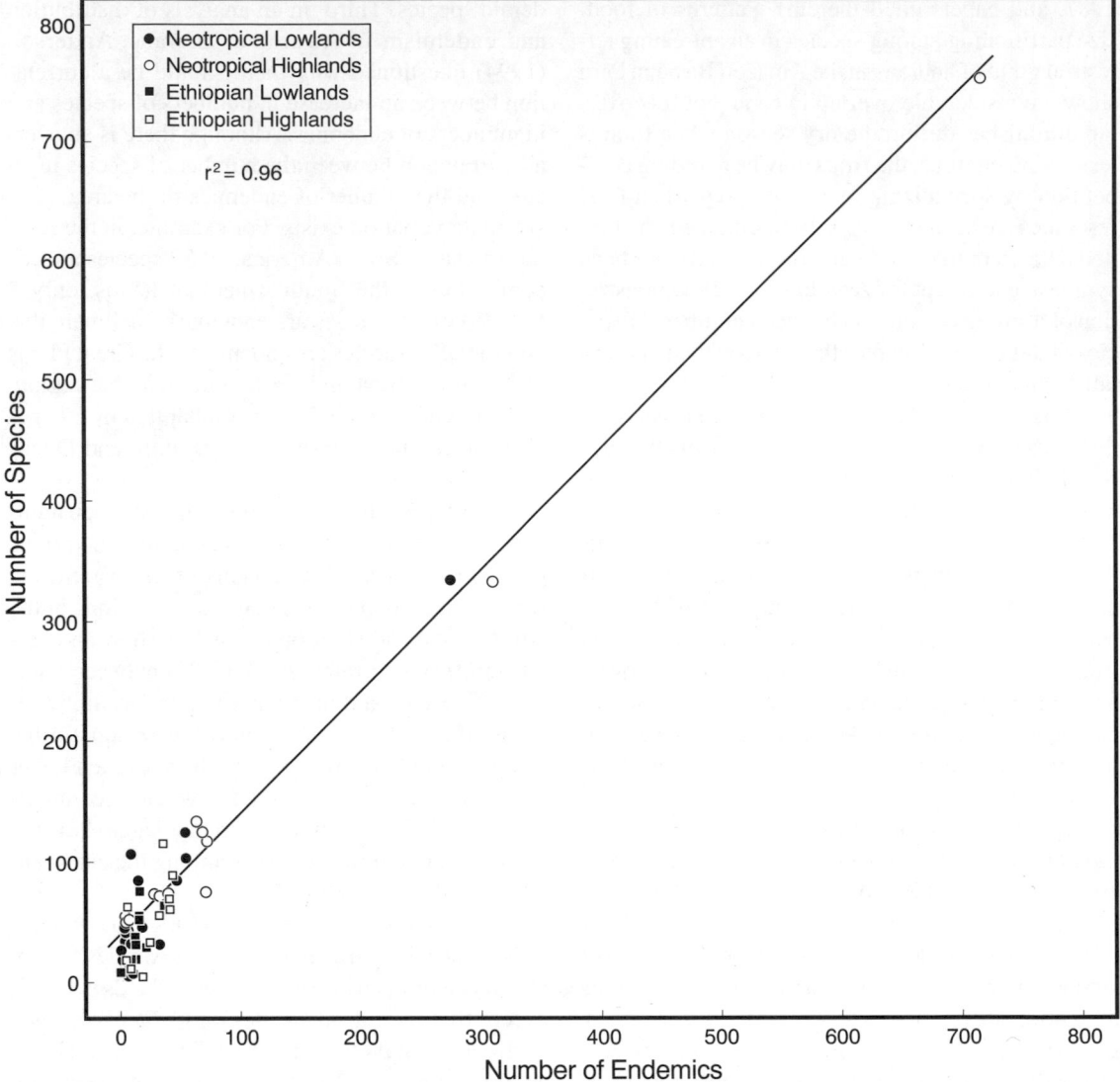

Fig. 1:4.   Correlation of species richness and endemism of amphibians in regions in the Neotropical and Afrotropical Realms; the 57 regions are those recognized by Campbell, Duellman, and Poynton (this volume).

lantic Coastal Plain and the Appalachian Highlands in North America and the Mediterranean Region in Eurasia. Realistically, area should be factored into any equation of richness versus endemism, and comparisons should be made among regions having similar climates and topography. For example, in South America, the area of the Guiana Highlands is only 1% of that of the Andes, but the Guiana Highlands have about 10% of the number of species as the Andes. Consequently, the species per area is 10 times greater in the Guiana Highlands than in the Andes, but the percentage of endemics is nearly equal (93.7% and 95.0%).

As noted by Raxworthy and Nussbaum (1996), patterns of endemism are the products of historical events associated with speciation and responses of species to changing environmental conditions. For example, the 100% endemism in the austral forests in South America mostly is a result of historical isolation of lineages of primitive telmatobiine leptodactylid frogs (Formas, 1979). Likewise, historically, Hispaniola was two islands; this history is reflected in the patterns of distribution and endemism (Hedges, this volume). The table-top mountains (*tepuis*) constituting the Guiana Highlands in north-

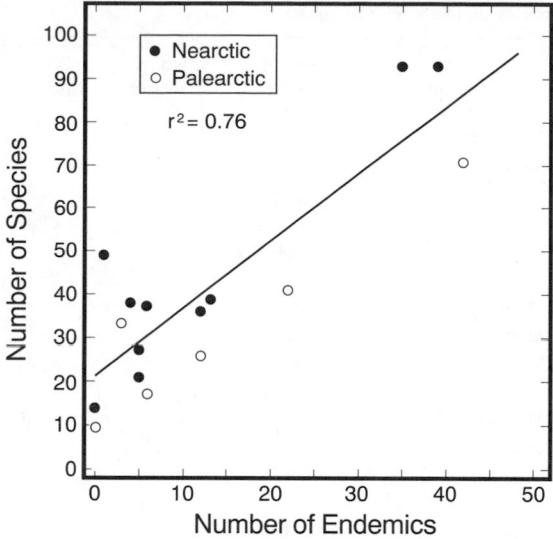

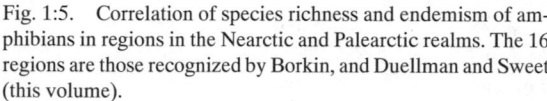

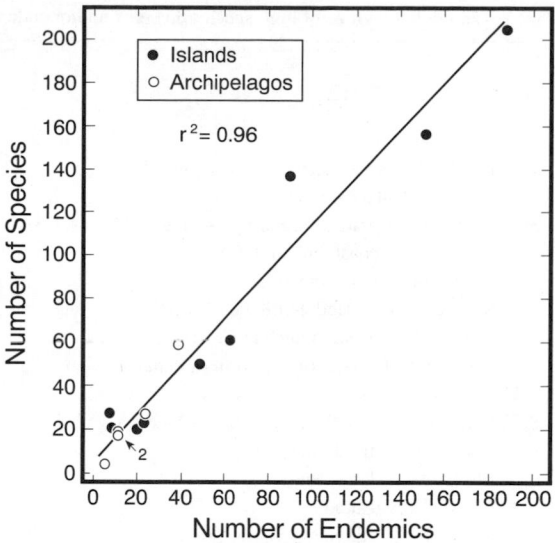

Fig. 1:5. Correlation of species richness and endemism of amphibians in regions in the Nearctic and Palearctic realms. The 16 regions are those recognized by Borkin, and Duellman and Sweet (this volume).

Fig. 1:6. Correlation of species richness and endemism of anurans on islands and archipelagos. Islands are Borneo, Cuba, Hainan, Hispaniola, Jamaica, Madagascar, New Guinea, Puerto Rico, and Taiwan. Archipelagos are Bismarck, Japan (excluding Ryukyu Islands), Philippine, Ryukyu, Seychelle, and Solomon.

eastern South America presumably are remnants of a Mesozoic plateau. The high degree of endemism in the Guiana Highlands results from endemism on individual *tepuis;* thus, the Guiana Highlands can be viewed as a continental archipelago. Habitat islands resulting from changing environmental conditions are exemplified by the supratreeline habitats in the northern Andes of South America, where endemism is high among groups of these islands that were continuous during the Pleistocene (Lynch et al., 1997).

Thus, in amphibians, we see overall high species richness in tropical regions, but also high richness in salamanders in humid, temperate mountains. Endemism to particular regions is generally higher in tropical regions than in temperate ones, and in montane regions than in lowland ones (Table 1:6; Fig. 1:7). Species richness is lower, but endemism is higher, on islands than in continental regions of the same size and having similar environments.

CONSERVATION OF AMPHIBIANS

More than a century ago, a few biologists (e.g., Wallace, 1869) noted the need for conservation. With the exception of national parks established primarily to preserve unique and fascinating landscapes (e.g., the geysers in Yellowstone National Park in the USA), most conservation efforts were directed toward protecting animals from commercial interests. These included laws passed in the late 19th Century in many states in the USA to protect game animals, the Lacey Act (USA) in 1900 to control commercial trade in bird feathers, and the Migratory Bird Act (USA) passed in 1913 (Graham, 1971; Matthiessen, 1959; Nash, 1990). Perhaps Carson's (1962) eloquent plea for environmental awareness was instrumental in initiating active conservation programs. Although some organizations with an

emphasis on conservation have existed for more than a century (e.g., Audubon Society founded in 1896), more recently, it has become fashionable for people, especially celebrities and politicians, to support a "cause." Many such persons enthusiastically associated their names with programs such as "save the redwoods," "save the whales," and "save the elephants." It seemed that in conservation, as in the national economy, skyscrapers, and passenger planes, the adage of "bigger is better" was the key to recognition. Large game preserves were established in Africa to protect the unique assemblages of large African mammals, the so-called "charismatic megafauna." Still, conservation efforts commonly emphasized certain taxa—especially animals that would entice public sympathy (e.g., American Bi-

Table 1:6. Global regions of high species diversity and/or endemism. Key numbers refer to Figure 1:7.

Key	Region	Topography	No. of Species	No. of Endemics	Percent Endemic
	NORTH AMERICA:				
1	Pacific–Cascade–Sierra Nevada ranges, USA	Montane	52	43	83
2	Interior Highlands, USA	Montane	36	12	33
3	Southern Appalachian Mts. & associated plateaus, USA	Montane	101	37	37
4	Southeastern coastal plain, USA	Lowland	68	27	40
	MIDDLE AMERICA:				
5	Southern Sierra Madre Oriental, Mexico	Montane	118	74	63
6	Highlands of western nuclear Central America	Montane	126	70	56
7	Highlands of Costa Rica & western Panama	Montane	133	68	51
	WEST INDIES:				
8	Macizo de Sagua-Baracoa & Sierra Maestra, Cuba	Montane & lowland	35	23	66
9	Cockpit Country, Jamaica	Montane	12	2	17
10	Massif de la Hotte, Haiti	Montane	32	16	50
	SOUTH AMERICA:				
11	Sierra Nevada de Santa Marta, Colombia	Montane	19	19	100
12	Cordillera Occidental, Colombia & Ecuador	Montane	200	156	78
13	Cordillera Oriental, southern Colombia & Ecuador	Montane	159	147	92
14	Tropical southern Andes, Bolivia & Peru	Montane	132	101	77
15	Upper Amazon Basin, Ecuador & northern Peru	Lowland	198	119	60
16	Upper Amazon Basin, southern Peru	Lowland	102	22	18
17	Guiana Highlands, Venezuela & Guianas	Montane	76	71	93
18	Atlantic Coastal Forest, Brazil	Montane & lowland	334	310	93
19	Austral Temperate Forest, Argentina, Chile	Montane & lowland	32	32	100
	EURASIA:				
20	Pyrenees Mountains, Spain, France	Montane	19	5	26
21	Northwestern Italy	Montane & lowland	20	13	41
22	Caucasus Mountains, Russia, Georgia	Montane & lowland	16	4	25
23	Hengduanshan Mountains, southwestern China	Montane	111	52	47
24	Honshu Island, Japan	Montane & lowland	30	8	27
25	Western Ghats, India	Montane	114	84	74
26	Southern Himalayan slopes, India	Montane	68	26	38
27	Highlands of Burma, Thailand, Vietnam	Montane	129	50	39
28	Highlands of Borneo,[1] Indonesia, Malaysia	Montane	137	90	66
	AFRICA:				
29	Cameroonian Highlands, Cameroon	Montane	62	39	63
30	Central Highlands, eastern rim of Zaire Basin	Montane	69	50	72
31	Ethiopian Highlands, Ethiopia	Montane	32	25	78
32	Eastern Highlands, Kenya	Montane	57	32	56
33	Cape Region, South Africa	Lowland	30	21	70
34	Central Highlands, Madagascar	Montane	66	61	92
35	Seychelles Islands	Montane & lowland	12	11	92
	AUSTRALO-PAPUAN REGION:				
36	Central Highlands, New Guinea[1]	Montane	205	193	94
37	Solomon Islands	Montane & lowland	26	24	92
38	Kimberley District, northwestern Australia	Lowland	36	12	33
39	Northeastern Queensland, Australia	Montane & lowland	56	29	52
40	Uplands, SE Queensland & N New South Wales, Australia	Montane & Lowland	48	18	38
41	Southwestern Australia	Lowland	25	19	76
42	Southeastern Australia	Montane & lowland	29	13	45
43	North Island and islands in Cook Strait, New Zealand	Montane & lowland	4	4	100

[1]Numbers for entire island.

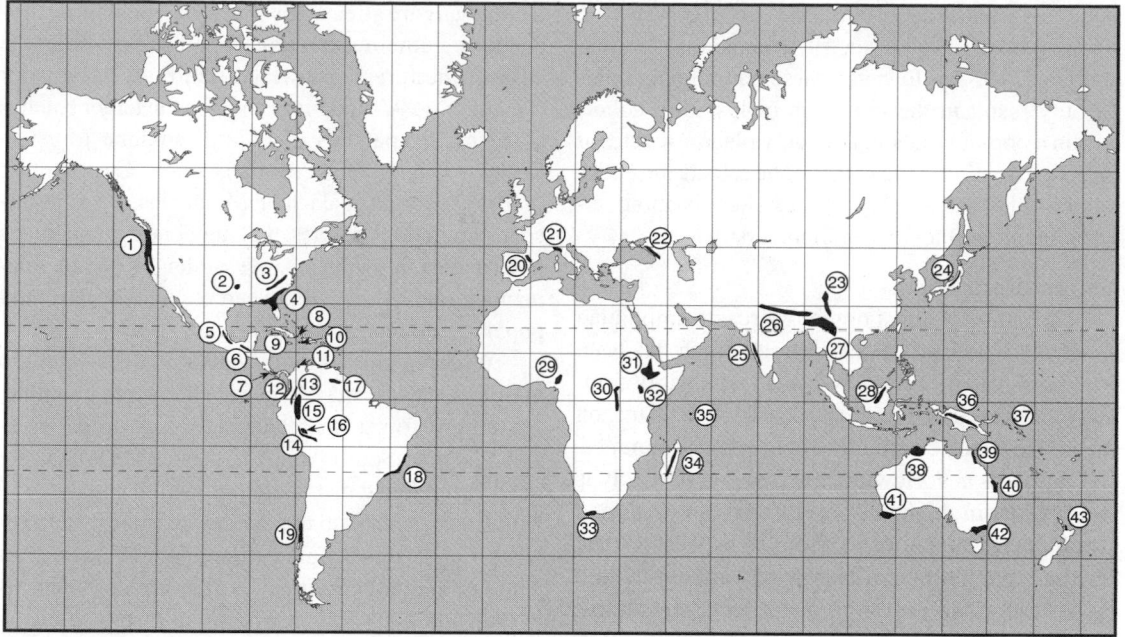

Fig. 1:7.   Regions of significant species richness and/or endemism. Numbers refer to key in Table 1:6.

son Society founded in 1905). This marketing strategy was used successfully by the World Wildlife Foundation, which featured the giant panda as its logo. This approach pervaded conservation organizations until the last decade or so; for example, in a workshop convened by the World Wildlife Fund in 1984 to identify and preserve Andean cloud forests, the assembled biologists were told that such a "flagship species" was necessary to "sell" the conservation efforts for a particular area.

Successful conservation efforts almost always were associated with large mammals (e.g., Bengal tiger) or gaudy birds (e.g., flamingos and parrots), but even large reptiles gained prominence—giant tortoises for the Galapagos National Park in Ecuador and the Komodo dragon for the Komodo National Park in Indonesia. However, celebrities, politicians, and conservation organizations neglected amphibians—apparently, they were not marketable. Although many biologists were aware of conservation needs for a few species of amphibians (e.g., Honegger, 1981), it was not until 1989 that biologists at the First World Congress of Herpetology noted a worldwide decline in amphibian populations, and this group of organisms became a major focus of conservationists and a subject of public interest. Declining amphibian populations even made the

cover story of *The New York Times Magazine* (Yoffe, 1992). Biologists and conservationists rallied to form the Declining Amphibian Populations Task Force (DAPTF) within the Species Survival Commission of the International Union for the Conservation of Nature. The DAPTF coordinates activities of many international working groups, sponsors selected research projects on specific suspected causes of declines, and publishes a quarterly newsletter, *Froglog*.

Herein, I attempt to review the threats to amphibian diversity and the efforts to preserve that diversity. Also, I endeavor to suggest conservation goals and the kinds of activities and legislation that may make those goals attainable.

THREATS TO AMPHIBIAN DIVERSITY

As noted previously, amphibians are most diverse in regions of stable, humid, and sufficiently warm climates. In variable climates and even in benign regimes, aperiodic, abiotic catastrophes can have disastrous effects on amphibian populations. Examples include effects of drought on survival of larvae (Bell and Lawton, 1975; Semlitsch, 1983) or adults (Jaeger, 1980; Tyler, 1991), effects of flooding on survival of larvae (Metter, 1968) or adults

(Duellman and Trueb, 1986), and exceptionally deep freezing on larvae (Heyer, 1979) or adults (Manion and Cory, 1952). However, these natural phenomena only result in the reduction or loss of a cohort within a population, and they only place a species at risk of extinction if the catastrophe affects the entire range of the species. The concern here is about anthropogenic influences on biodiversity.

Habitat Destruction

Certainly the most obvious threat to amphibian biodiversity is habitat destruction, which has been exemplified in recent years by the vast areas of tropical rainforest that have been cleared in the Amazon Basin. However, clearcutting of forests for commercial timber is a worldwide problem that probably is most serious in southern Asia, Indonesia, Russia, and North America, and most recently in West Africa. The age-old practice of slash-and-burn agriculture by squatters had only minimal effects on tropical forests, but in recent years, large commercial interests have cut and burned vast areas for seeding for cattle ranches and plantations of oil palms, bananas, and other commercial products. For decades, coffee was grown in the shade of native trees, but new strains of coffee trees were developed to survive in direct sunlight; thus, the species-diverse forest was reduced to a monoculture of coffee trees. Likewise, bulldozing of scrub forest for irrigated cotton plantations has been a common occurrence, especially on the Pacific lowlands of Mexico and Central America and in parts of Africa and the Near East. Overgrazing by cattle, sheep, and goats is nearly a worldwide phenomenon that has changed much grassland to semiarid desert. Add to these destructive practices the flooding of large regions by damming of rivers for hydroelectric plants (largest ever undertaken is now being built in the lower Yangtze River in China), and we witness an ever-increasing demand for space, food, clothing, and energy for an ever-increasing human population.

These destructive practices affect biodiversity as a whole, not solely amphibians. The distributions of species become restricted and fragmented. Little information is available on these effects on populations of amphibians. At the Minimum Critical Size Ecosystems Study Area in Amazonian Brazil, 90% of the species of amphibians in the area were found in forest fragments as small as 350 ha (Zimmerman and Bierregaard, 1986). Frogs of the genus *Eleuth-*

erodactylus are affected by forest fragmentation; they are more numerous at increasing distances from the edge of the forest in lowland rainforest in Ecuador (Pearman, 1997) and more abundant in larger patches than smaller patches of humid montane forest in Ecuador (Marsh and Pearman, 1997). Clear-cutting of forests in the Appalachian Mountains, USA, resulted in population declines of plethodontid salamanders to near zero in two years, but within 4–6 years, salamanders moved into the clearcut areas, where their populations increased rapidly (Ash, 1997).

Whereas habitat modification is detrimental to many species, it may benefit others. For example, clearing of forest for roads and farms in the Amazon Basin in South America creates areas that are quickly inhabited by two anurans (*Bufo marinus* and *Scinax rubra*) that do not occur in deep forest. Likewise, Inger (this volume) identified 11 species of anurans in southeastern Asia that are commensals of man and that rarely are found in forests. Lawson (1993) noted the abundance of eight species of rainforest anurans in farmbush and three species in palm plantations in Cameroon. The construction of farm ponds in the Great Plains of North America provides habitat especially utilized by *Acris crepitans* and *Ambystoma tigrinum* in an otherwise rather inhospitable environment; likewise, drainage canals and farm ponds have resulted in expansion of the range of *Xenopus laevis* in South Africa, and irrigated rice paddies provide habitat for many anurans in southeast Asia (Gans, 1994). All of these human commensals, which account for a small fraction of amphibians, have widespread geographic distributions, generally attain large populations, and often have negative impacts on native species. Thus, although habitat modification may benefit a few species, it is detrimental to most species and results in a net loss of biodiversity.

Agricultural Practices

In addition to the destruction of natural vegetation, agriculture has secondary effects on amphibians. Clearing of forests or plowing and overgrazing of grasslands results in erosion causing siltation of streams, rivers, and lakes—habitats for adults of some, and larvae of many, species. Siltation has potentially disastrous effects on eggs and larvae that develop in clear streams and potentially can greatly reduce anuran diversity in regions where this mode of reproduction is common—e.g., mountains in northeastern Australia, austral forests in South America, Atlantic forests in eastern Brazil, and on islands, such as Borneo. The same conditions affect plethodontid salamander diversity in the Appalachian Highlands in North America.

Changes in agricultural practices can endanger amphibian populations. For example, changing silvicultural practices in pine flatlands in Florida, USA, has had detrimental effects on a population of *Ambystoma cingulatum* (Means et al., 1996). Plantations of eucalyptus trees have had deleterious effects on amphibian populations in Spain and Portugal (Vences, 1993), Uruguay (Meneghel, 1992), and Ecuador (pers. obs.). Changing irrigation practices have affected populations of *Hyla japonica, Rana japonica,* and *Rana porosa* in Japan (Fujiokoa and Lane, 1997).

Humid slopes of mountains in southern Mexico and Central America accommodate many species of anurans and salamanders, and many of the latter have restricted geographic ranges. In these areas, coffee trees were grown in the shade of native species of trees, many laden with bromeliads inhabited by amphibians. Such coffee plantations have been important refuges of biodiversity, but the change to sun-grown coffee has resulted in the loss of the native trees and associated animal diversity (Perfecto et al., 1996). The relationships between different kinds of forest management and amphibians was reviewed by DeMaynadier and Hunter (1995).

Agrichemicals (fertilizers, herbicides, fungicides, and pesticides) are being used with increasing frequency, especially in developing countries. These chemicals not only have an effect on amphibians subjected to them directly, but some of the compounds pass through the soil and enter waterways, where they have serious effects on amphibian eggs and larvae (Hall and Henry, 1992; Harfenist et al., 1989; Vitt et al., 1990). Recent studies have demonstrated detrimental effects of specific agrichemicals on various amphibians—e.g., parathion on eggs and larvae of *Bufo arenarum* (Anguiano et al., 1994), ammonium nitrate on *Bufo americanus* and *Pseudacris triseriata* (Hecnar, 1995), fenitrothion on larvae of *Bufo americanus* and *Ambystoma maculatum* (Berrill et al., 1995), and polychlorinated biphenyl on amphibian abundance in South Carolina, USA (Foutenot et al., 1996). A worldwide survey of chemical effects was tabulated by Tyler (1989).

Although morphological anomalies, such as extra limbs, have been known in amphibians for a long time (Tyler, 1989), the alarmingly high number of such anomalies in north-central USA and Canada in recent years has caused concern. These anomalies have been correlated with pesticide runoff (Ouellet et al., 1997), but Sessions and Ruth (1990) reported that abnormalities in the hind limbs of *Ambystoma macrodactylum* and *Pseudacris regilla* can result from digenean parasites infesting the limb buds of larvae.

Pollution

The discharge of industrial wastes into waterways has been a common practice for more than a century. The eggs and larvae of amphibians are known to accumulate petroleum (Mahaney, 1994) and heavy metals in their tissues. See Tyler (1989) for a review.

Atmospheric acid pollution (acid rain) is an insidious form of environmental contamination, because the deposition may occur long distances from the source. The acidic pollution lowers the pH of water below tolerance levels of aquatic eggs and larvae. Wyman (1990) identified three major features of acid rain that affect amphibians. (1) Breeding in temporary ponds that fill with rain or runoff may occur before the water can buffer the low pH. (2) Breeding may occur in the early spring in temperate regions in ponds formed by snow melt; the snow has accumulated acid and therefore forms acidic ponds. (3) At high elevations, soils become acidic because the underlying bedrock has low acid-neutralizing capabilities. Low pH affects sperm motility and viability, thereby reducing success of fertilization, and sometimes resulting in embryos with developmental abnormalities (Pierce, 1985). Furthermore, low pH has been shown to reduce swimming ability in larvae of *Ambystoma laterale* and *A. maculatum* (Kutka, 1994). Acid precipitation has been implicated in population declines of *Bufo calamita* in Great Britain (Beebee et al., 1990) and *Ambystoma tigrinum* in the Rocky Mountains, USA.

Atmospheric pollution by chlorofluorocarbons that thin the ozone layer results in greater ultraviolet (UV) radiation reaching the earth (Tevini, 1993); this increased radiation is especially harmful to aquatic ecosystems (Hader, 1997). Amphibian eggs and larvae that develop in sunlight have melanin layers that protect tissues from normal amounts of UV light. Some adult amphibians living at high elevations actually bask in the sun, and these have melanin in the skin and/or on the gonads (Duellman and Trueb, 1986). Thus, amphibians have evolved measures for protection against normal amounts of UV radiation. However, increased levels of UV radiation may penetrate melanin layers and be detrimental to underlying tissues, especially embryos. It has been shown in field experiments that in clutches of eggs of *Ambystoma macrodactylum* developing in sunlight there is a much higher incidence of developmental anomalies and much lower survivorship than in eggs shielded from sunlight (Blaustein et al., 1997). Low survivorship of eggs of *Rana aurora* in response to ambient UV radiation was demonstrated by Blaustein et al. (1996), and UV radiation has been implicated in developmental abnormalities in *Bufo boreas* (Blaustein et al., 1997).

Road Mortality

Especially at times of migration to breeding sites, incredibly large numbers of amphibians are killed accidentally by motor vehicles on roads each year (Fahrig et al., 1995; Langton, 1989). Ehmann and Cogger (1985) estimated conservatively that more than 5 million such fatalities of amphibians and reptiles occur annually on paved roads in Australia.

Introduction of Exotic Species

Introduction of exotic fishes, especially game fish such as trout, in many parts of the world has been associated with local declines or extinctions of several species of anurans (Bradford, 1989; Kuzmin, 1996; Pough et al., 1997). Minnows of the genus *Gambusia* have been introduced in many places to control mosquito larvae, but these minnows also feed on amphibian eggs and larvae. Likewise, the introduction of *Rana catesbeiana* was implicated in declines in other species of anurans (Hays and Jennings, 1986), and rats have caused the extinction of some populations of *Leiopelma* in New Zealand (Pough et al., 1997).

Harvesting and Collecting

Indigenous peoples in various parts of the world eat frogs, but insofar as is known, their hunting and consumption of frogs has had no marked effects on populations of anurans. On the other hand, commercial exploitation is a serious matter. According to Pough et al. (1997), the most commercially important frogs for human consumption are ranids (*Rana*

catesbeiana, R. esculenta, Hoplobatrachus tigerinus, and *Pyxicephalus adspersus*); they reported that more than 6000 tons of frog legs were imported into the USA in 1976 (primarily from Japan and India) and that up to 4000 tons are imported annually by France (primarily from Bangladesh and Indonesia). Commercial "farming" of *Rana catesbeiana* has been marginally successful economically and has resulted in that species being introduced from North America to several parts of the world, principally Latin America.

Over the years, countless millions of frogs have been sacrificed for dissection and experimentation in biology classes and research laboratories around the world. In North America, the "laboratory frog" has been *Rana "pipiens."* Elsewhere, the most commonly used frogs have been *Rana esculenta* in Europe, *Leptodactylus ocellatus* in Argentina and Brazil, *Xenopus laevis* in South Africa, and *Bufo marinus* in Australia. Most of these frogs were taken from the wild.

Although amphibians have been common pets for many years, especially in Europe, their popularity has grown tremendously during the past two decades, especially in the USA; furthermore, terrarists are seeking a great variety of exotic species. Consequently, collecting and sale of amphibians in the pet trade has become a lucrative business, and has become a focus of international regulation. The number of amphibians collected worldwide for such purposes is unknown, but an idea of the magnitude of such commercial activities is illustrated by the documentation of the collecting of 41,493 anurans and 1050 salamanders in Florida in two years (June 1990–June 1992) (Pough et al., 1997), the documented exportation of 13,052 dendrobatid frogs from eight Central and South American countries in 1987–1993 (Gorzula, 1996), and an increase from 230 in 1980 to 20,000 in 1990 of frogs of the genus *Mantella* exported from Madagascar (Dayton, 1994). It has been estimated that more than 7.3 million amphibians and reptiles were kept as pets in the USA alone in 1994 (Ramus, 1995).

Many conservationists and governmental regulators decry scientific collecting, even for documenting biodiversity. Most field biologists are aware of threats to amphibian populations and collect only specimens that are necessary for research. I am unaware of any species of amphibian that has been threatened by professional scientific collecting. In viewing amphibians and reptiles together, Campbell and Frost (1993) noted that documented commercial trade in amphibians and reptiles and their products is responsible for more than five times as much collecting *per year* than scientific collecting has accomplished *in its entire history,* and that scientific collecting is responsible for less than 1% of the annual regulated take of amphibians and reptiles. All of the scientific specimens added to museum collections in the USA *in the past two centuries* amounts to less than one fourth of the numbers of individuals being kept as pets in one year in the USA!

DECLINING AMPHIBIAN POPULATIONS

It is obvious that anthropogenic effects, such as habitat destruction and modification, have been responsible for the reduction or elimination of many populations of amphibians (and other organisms). However, in the past two decades there have been notable declines in, and disappearances of, amphibian populations that do not seem to be associated with the usual explanations for the "biodiversity crisis." The explosion of literature on this subject was reviewed by Blaustein and Wake (1995) and Halliday and Heyer (1997); a highly readable narrative on the subject was prepared by Phillips (1994). The only global summary of declining amphibians was compiled by the Declining Amphibian Populations Task Force (Vial and Saylor, 1993). Based on data provided by more than 100 working groups of biologists, they tallied 159 species declining, 35 locally absent, and two extinct, worldwide. National efforts also are being made; of these, the most detailed and comprehensive is that for Australia (Tyler, 1997).

A major concern is the decline or disappearance of populations in pristine habitats that show no visible (to humans) environmental degradation. The most often cited example is the Monteverde Cloud Forest Reserve in Costa Rica, where what appeared to be a substantial population of the endemic *Bufo periglenes* disappeared after 1987. Of 36 species of aquatic-breeding frogs in Monteverde, 22 species disappeared after 1987, and only four have returned (Pounds et al., 1997). Populations in pristine, montane habitats seem to show the most noticeable declines, as noted in Australia by Tyler (1991) and Richards et al. (1993). The same phenomenon applies to the mountains in western North America (Stebbins and Cohen, 1995), the Venezuelan Andes (La Marca and Lötters, 1997), the Ecuadorian Andes (L. A. Coloma, pers. comm.), and highlands in Guatemala (Campbell, 1998). The argument that these declines may be natural fluctuations (Pechmann et al., 1991; Pechmann and Wilbur, 1994) was challenged by Blaustein (1994) and Blaustein and Wake (1995), among others. In elaborate tests of null models and data from Monteverde, Costa Rica, Pounds et al. (1997) demonstrated that the disappearances there were highly improbable in the context of normal demographic variability.

In contrast, there is no compelling evidence for comparable population declines of unknown causes in warm temperate and tropical lowlands. Results of studies of populations of four species of amphibians over a 12-yr period at the Savannah River Ecology Laboratory in southeastern USA showed that abundance was closely correlated with climatic variation; dry years, during which recruitment was low, were followed by one or more years of lower anuran abundance (Pechmann et al., 1991). Likewise, at a site in seasonal lowland rainforest in the Amazon Basin in South America, Duellman (1995) monitored population fluctuation of 20 species of anurans for six years; he concluded that anuran activity (as a measure of abundance) was closely correlated with periods of heavy rainfall and not total rainfall over a sampling period and that, although populations seemed to fluctuate, there was no evidence for declines. At Boracéia, a subtropical site in southeastern Brazil, Heyer et al. (1988, 1990) reported that seven species of anurans declined and five disappeared between 1979 and 1982, but they concluded that unusually hard frosts in June and July 1979 had been the major factor contributing to the declines and disappearances. Observations over a 15-yr period at the Reserva Atlântica in southeastern Brazil by Weygoldt (1989) revealed that between 1981 and 1987, 8 of 13 species had declined or disappeared; he believed that the major cause was a series of extremely dry winters.

Possible causes of population declines include (1) increased UV radiation that affects developing embryos (Blaustein et al., 1997), (2) biomagnification of toxic substances as their concentration increases up the food web (Stebbins and Cohen, 1995), (3) chemically induced alterations in sexual development by xenobiotics affecting endocrine systems

(Stebbins and Cohen, 1995), (4) epidemics of microparasites caused by climatic factors (Pounds and Crump, 1994), and (5) attainment of critical concentrations of atmospheric contaminants at times of abnormally warm and dry conditions (Pounds and Crump, 1994). The last two possible causes are associated with climatic changes, such as those imposed by the El Niño weather pattern on Central America in the late 1980s. The view that epidemic disease was the cause of catastrophic declines in amphibian populations in northeastern Australia (Laurance et al., 1996) was criticized for lack of evidence by Hero and Gillespie (1997) and Alford and Richards (1997).

With the exception of the experimental work on effects of UV radiation on amphibian embryos in northwestern USA (Blaustein et al., 1997), direct evidence on causes is lacking. The amphibians that have disappeared from high elevations are either diurnal species, such as *Atelopus, Colostethus,* and *Taudactylus* that lay their eggs in water, or are nocturnal species, such as *Ambystoma* and *Rana* that lay their eggs in water. Thus, these species, either as adults or embryos, possibly are subjected to increased UV radiation as a result of thinning of the ozone layer. The diel and reproductive habits of coexisting species are reflected in differential survival of populations. For example, in the high Andes of Ecuador, the diurnal species of anurans that deposit eggs in water (*Atelopus* and *Colostethus*) have disappeared, whereas the nocturnal species that deposit eggs under stones (*Eleutherodactylus*) or carry their eggs in a dorsal pouch (*Gastrotheca*) have survived. However, increased UV radiation seems unlikely to be responsible for declines in the montane forests at Monteverde, Costa Rica, where aquatic eggs are subjected to few, if any, direct rays of the sun.

CONSERVATION INITIATIVES

Some environmental problems have global implications and have resulted in well-publicized international conferences, such as the Earth Summit (for global protocols on the environment) in Rio de Janeiro, Brazil, in 1992 and the International Convention of Global Warming in Kyoto, Japan, in 1997. Unfortunately, little of substance has resulted from such conferences, which have been characterized by emphasis on national economies rather than potential global catastrophes. Thus, when we view con-

servation initiatives with respect to amphibians, we must dwell on a few effective international endeavors and more local situations for habitat preservation.

A major philosophical change in conservation biology has been the deemphasis of so-called flagship or umbrella species in conservation planning; this approach has been criticized in that it does not provide whole landscape or community solutions to conservation (Franklin, 1993; Hobbs, 1994; Walker, 1995). A multispecies approach for conservation of natural landscapes is coming into vogue (Lambeck, 1997). Furthermore, biologists knowledgeable about biodiversity are contributing more to conservation initiatives. For example, an assessment of biodiversity in terrestrial ecosystems in Latin America and the Caribbean by diverse field biologists formulated conservation priorities in Latin America (Dinerstein et al., 1995).

For any broad-scale conservation programs to maintain biodiversity, we must know the patterns of distribution of species. Therefore, distributions must be mapped. Although small-scale maps, such as those in the field guides for North America by Stebbins (1985) and Conant and Collins (1991) and for Australia by Barker et al. (1995), are useful for broad pictures, more detailed maps are required for accurate determination of ranges. This need has led to the production of herpetological atlases, the most complete of which so far contain dot maps in 50 X 50-km grids for Europe (Gasc et al., 1997) and 20 X 20-km grids for the Iberian Peninsula (Pleguezuelos, 1997).

Habitat Preservation

As acknowledgment of the need for conservation of natural areas and their biotas, governments throughout the world have established hundreds of national parks, nature preserves, and other protected areas. Some of these are international endeavors, such as La Amistad Reserve in Costa Rica and Panama and the Tambopata-Candamo Reserve in Peru and Bolivia; these were established for the expressed purpose of preserving continuous pristine environments.

Setting aside and managing protected areas is costly, and many developing nations are unable to bear the financial burdens. This problem has been ameliorated partially by the debt-for-nature swap program, whereby a large nongovernmental conservation agency purchases part of an international loan at a discount from a lending institution and establishes a protected area or conservation program in the debtor nation. The first area established by a debt-for-nature swap was the Beni Biosphere Reserve in Bolivia in 1987. Costa Rica, Madagascar, Mexico, Ecuador, and the Philippines are among countries that have established protected areas in this manner (Ayres, 1989).

Unfortunately, in most developing nations, many so-called protected areas are "paper parks." At the urging of governmental agencies or nongovernmental conservation organizations, governments decree a certain area as a national park or other nominal protected area. Often such areas have not been surveyed, nor are they actually protected. Although by decree hunting, logging, mining, and habitation is prohibited, enforcement is lacking. Again, economies cannot support the management of the parks.

In developing countries, tourism to natural areas (ecotourism or nature tourism) is becoming increasingly popular. Ecotourism can generate sorely needed revenue for management of reserves and for local economies, as well as heightened local awareness of the importance of conservation and incentives for governments and dwellers in and around potentially attractive areas to preserve them. Ecotourism has been a mainstay in the economies of some African countries, especially Kenya, and by the late 1960s, became a major source of revenue for the Galapagos Islands of Ecuador. The popularity of South American rainforests for ecotourism is shown by the existence of more than 30 parks in Ecuador, French Guiana, Peru, and Venezuela that have facilities for tourists (Castner, 1990). Tour companies in Europe and North America are offering an increasing variety of nature, birding, and even herpetological tours that range from low to demanding physical activity in primitive to luxurious accommodations. The majority of tourists who visit parks and reserves in developing countries are members of such tours (Boos, 1990).

Some tour companies have made contributions to conservation and research in developing nations. For example, with a highly successful trade in ecotourism, International Expeditions, Inc. (USA) and Explorama Tours (Peru) established the Amazon Center for Environmental Education and Research (ACEER), which now boasts a canopy walkway through the rainforest and a nearby research laboratory, as well as comfortable accommodations for tourists and investigators; these facilities were made possible by funds from tourism. Also, ecotourism has spawned the production of high-quality guides (with color photographs) to the amphibians of several regions visited by such tourists. Examples are guides to the amphibians of Madagascar (Glaw and Vences, 1994), Iquitos region of Peru (Rodríguez and Duellman, 1994), Sabah in Borneo (Inger and Steubing, 1989), and the Mayan region of Middle America (Campbell, 1998).

Because of the absence of inventories of the biotas of most protected areas, especially in developing nations, it is not possible to ascertain the variety of species included therein. All too often, protected areas are set aside without knowledge of their biodiversity. Inventories of biodiversity are seldom a part of the planning stages of parks and reserves. However, there are exceptions, such as the Korup National Park Project in Cameroon (Lawson, 1993) and the purchase of tracts of land by the Nature Conservancy after inventories have ascertained the biodiversity or at least the presence of species in need of protection.

Policies prohibiting or greatly restricting collecting in many protected areas seem to be consistent with preservation, but often the question is raised: What is being protected? For example, the policy prohibiting collecting in Manu National Park in Peru resulted in a published list of frogs from Cocha Cashu in the park (Rodríguez and Cadle, 1990) containing many unidentified species and others that were suspect. The absence of voucher specimens hinders accurate documentation. In this particular case, the park is situated within the Manu Biosphere Reserve, which includes a buffer zone, in which one site (Pakitza) has been subjected to a thorough biological inventory (Wilson and Sandoval, 1996), including amphibians (Morales and McDiarmid, 1996). Similar thorough biological investigations were carried out in the Cuzco Amazónico Reserve about 250 km to the east (Duellman and Koechlin, 1991; Duellman and Salas, 1991).

Agricultural and Forestry Practices

Conservation plays an increasingly important role in agricultural practices. For example, in some countries, the use of deleterious pesticides, such as DDT, has been outlawed, but their use persists in many developing countries. New concepts of sustainability are being investigated. In recognizing the importance of ecological constraints in resource management, Callicott and Mumford (1997:32) stated: "Neither the classic resource management concept of maximum sustainable yield nor the concept of sustainable development are useful to contemporary, nonanthropocentric, ecologically informed conservation biology." Forestry practices are implementing experimental procedures that avoid extensive clearcutting, such as retaining segments of interconnected natural stands (Franklin, 1992; Franklin et al., 1996). However, the effectiveness of forest fragments and interconnecting corridors in conserving ecosystems has not been evaluated fully (Bierregaard et al., 1992; Hobbs, 1992; Janzen, 1983), and little information is available regarding their effects on amphibians (Zimmerman and Bierregaard, 1986; Marsh and Pearman, 1997).

As noted in a foregoing section, the change from shade-grown to sun-grown coffee converted huge areas of managed forest to a monoculture. Now efforts are being made to go back to shade-grown coffee with an attendant increase in biodiversity; such efforts have been shown to be economically feasible in Guatemala and El Salvador (Rice and Ward, 1996).

Species Protection

In addition to preservation of habitat, efforts to save populations of amphibians involve decreasing mortality rates during migration to breeding sites, creation of suitable breeding sites, and the relocation, repatriation, and translocation of populations. Beginning in the 1960s, European conservationists initiated programs to reduce the mortality of migrating amphibians on roads by cautioning motorists to beware of such migrants and even setting up detours to avoid large numbers of amphibians on certain roads. Presumably these efforts may have reduced mortality on roads; nonetheless, large numbers of amphibians still were killed on the roads. Subsequently, drift fences and under-road culverts were constructed where some major migrations take place; generally these have met with success in Denmark (Graff, 1996), England (Cooke, 1988), Luxembourg (Engel and Bressanutti, 1993), Netherlands (Chardon et al., 1996), Spain (Yanes et al., 1995), and the USA (Piersan, 1987).

In the past decade, numerous efforts have been made to restore or create breeding sites for amphibians. Especially in Europe, ponds have been created; these have had positive influences on populations of frogs, such as *Bufo calamita* (Denton et al.,

1997) and *Hyla arborea* (Berninghausen, 1995; Meier, 1995), and salamanders, such as *Triturus alpestris* (Mikkelsen, 1993). Spawning sites for *Andrias japonicus* have been restored in the Ichi River, Japan (Tochimoto, 1995).

The problems associated with, and consequences of, translocations and reintroductions of amphibian populations were discussed in depth by Dodd and Seigel (1991) with lengthy comments by Burke (1991) and Reinert (1991), and by Andrén and Nilson (1995). In most cases, translocations and reintroductions have had limited success; for example, of six species translocated or reintroduced in the USA, England, Russia, and Puerto Rico and reported by Dodd and Seigel (1991), reproduction in the new populations only occurred in two—*Pelobates syriacus* and *Triturus vittatus*. Extensive knowledge of habitat requirements and population dynamics has been instrumental in the success of reintroductions of *Bufo calamita* in England, where only 30% of the reintroductions were unsuccessful and five of those were among the first six attempts (Denton et al., 1997).

Although the demands for amphibians, especially frogs, in teaching and research laboratories has diminished in recent years, tens of thousands of frogs are used annually. Laboratory-reared frogs, such as *Xenopus laevis,* can be bred economically and used for basic instruction (Bernhardt et al., 1991), thereby greatly relieving hunting pressure on native populations of *Rana.*

Regulations

In well-meaning efforts to protect species and certain populations, international organizations and governments throughout the world have enacted a bewildering array of regulations. Whereas most of this legislation was designed to regulate commercial exploitation of animals (and plants), the ever-changing regulations commonly restrict scientific activities that are essential to secure the very information that might be useful in determining how best to protect threatened species. The mere listing of a species on some list really does little to protect the species when its habitat is being destroyed. Furthermore, many inconsistencies exist because different criteria have been used to determine what species are to be protected.

As an example, I compare three up-to-date lists taken from the Internet of so-called threatened and endangered species of amphibians—Convention on

International Trade in Endangered Species of Wild Fauna and Flora (CITES), World Conservation Monitoring Centre of the International Union for the Conservation of Nature (IUCN), and the United States Fish and Wildlife Service (USFWS). On the CITES list, 81 anurans and 6 salamanders are categorized as threatened or endangered, whereas on the IUCN list, 61 anurans and 20 salamanders are so classified, and another 51 anurans and 25 salamanders are considered to be vulnerable. Only four species—*Bufo retiformis, Ambystoma lermaense, A. mexicanum,* and *Andrias japonicus*—are on both lists!

Within the USA (including Puerto Rico), four anurans and three salamanders are listed as threatened, and three anurans and six salamanders are listed as endangered. Of these 16 species, only the four species listed above also are placed in these categories by CITES and IUCN, but the latter also agrees with the USFWS in listing *Bufo houstonensis* as an endangered species. However, the IUCN classifies five additional anurans and three additional salamanders in North America as endangered and one more anuran and three more salamanders as threatened; none of these is listed by CITES.

Why the discrepancies? The IUCN lists species as endangered, threatened, or vulnerable based on stringent criteria and data provided by biologists. The listing of species by the USFWS also is based on biological data but restricted by political pressures from commercial and industrial interests, because the listing of a species as endangered might obstruct a multimillion-dollar development project. On the other hand, CITES is concerned with international trafficking in animals and plants and had its origins in the protection of species that provided valuable commercial products, such as ivory, furs, and tortoise shell. Of the 81 species of anurans listed as potentially threatened (CITES, Appendix II), 55 (68%) are species of brightly colored dendrobatids, many of which are bred by German and Dutch herpetoculturists, who sell the offspring at high prices

in the pet trade. Curtailing the exportation of wild dendrobatids makes the breeders' products all the more valuable. Did such commercial interests influence the listing of these dendrobatids by CITES, or were they listed because natural populations were threatened? There is no evidence for the latter (Mrosovsky, 1988; Myers and Daly, 1993).

A further complication is the increasing concern, especially among developing nations, about protecting their genetic resources. Legislation is guided less by concern for the protection of diversity than by visions of potential economic benefits. One large pharmaceutical corporation (Merck) has invested in biological inventories in Costa Rica with the provision that the corporation has the rights to drugs that are discovered through assays of the plants and animals inventoried. Numerous drugs have been developed from amphibians, and certainly many more await discovery (Tyler, 1995). However, a negative aspect of such investments is that governments tend to put a value (often inflated) on all species; such action restricts bonafide scientific research.

Because of the burgeoning amount of international, national, and state regulations, biologists and conservationists are frustrated in their efforts to obtain the kinds of documentation necessary for their research. Although permits to conduct biological research may be granted, presumably well-intentioned bureaucrats (often political appointees in ministries of commerce, tourism, or more commonly agriculture) who have no real knowledge of the biota or the kinds of efforts that are required to inventory biodiversity, require or request completely unreasonable documentation or fees. For example, prior to undertaking an inventory of the amphibians in a region that has not been inventoried, the biologist is asked to provide a list of species, and numbers of individuals of each, to be collected! I am aware of two major inventories of amphibians and reptiles in two South American countries that have been discontinued because of unrealistic restrictions.

FUTURE CHALLENGES

Biodiversity and its conservation face incredible challenges in the immediate future. Whereas our natural resources are dwindling and being degraded at a rate far greater than any time in human history,

new technology provides the means to assay the diversity of life on earth much more effectively than ever before. Below I present my perceptions of the major challenges, with emphasis on amphibians.

DIVERSITY

The amphibian faunas of many parts of the world remain essentially unknown, and the faunas in many other regions are far from completely documented. These regions must be surveyed thoroughly; the occurrence of species needs to be documented by preserved museum specimens that are accompanied by accurate locality, life-history, and ecological data, and have associated tissues, color photographs, and recordings of calls in the case of anurans. But, museum specimens and their associated data are only the beginning. Museums house a tremendous wealth of biological data, the effective use of which necessitates development of a global electronic network that can be accessed in a variety of ways to answer existing questions about biodiversity. Additionally, locality data should be incorporated into geographic information systems, so that new classes of questions can be asked. This need is being addressed by the Biological Informatics Working Group of the Organization for Economic Cooperation and Development (a consortium of 29 nations).

If these procedures are followed, future compilations of distributional patterns will be far more accurate than the rather crude analyses presented in this volume. Imagine being able to access electronically not only all records of occurrence of amphibians in a given region, but also climatic, vegetational, and soil data, all mapped with respect to topography. With such information, comparative species richness and areas of endemism will become clear, but more importantly, richness and endemism can be associated with other biotic and abiotic factors that will provide a basis for interpreting biodiversity data.

These suggestions are included in a document, "Systematics Agenda 2000, Charting the Biosphere" prepared by a committee of biologists under the auspices of the American Society of Plant Taxonomists, the Society of Systematic Biologists, and the Willi Hennig Society. This document (1994:1) emphasizes three interrelated scientific missions: "Mission 1: To discover, describe, and inventory species diversity. Mission 2: To analyze and synthesize the information derived from this global discovery effort into a predictive classification system that reflects the history of life. Mission 3: To organize the information derived from this global program in an efficiently retrievable form that best meets the needs of science and society."

CONSERVATION

The above kinds of knowledge are essential for establishing nature reserves for the protection of biodiversity, endemism, and distinctive evolutionary lineages. For amphibians, the global regions of high diversity and endemism (Table 1:6) can be refined and modified on the basis of more complete and more accurate data. There is an obvious need for large preserves in lowland regions of high species richness. Overall richness is highest, but local richness is comparatively low, in montane regions, but species in montane regions tend to have limited distributions; consequently, many small protected areas are needed to preserve the montane diversity. Protected areas should be bordered by buffer zones and be patrolled adequately so as to prevent human encroachment. Management of protected areas should include research facilities for biologists to carry out studies on the ecology and life histories of species and to monitor populations. At appropriate sites, ecotourism should be encouraged by the provision of adequate facilities and knowledgeable bilingual guides.

In addition to protection of areas having high species diversity and endemism, efforts must be intensified in several ways: (1) increased monitoring of populations, (2) more extensive efforts for protection during migrations, (3) more extensive provision of breeding sites for local populations, (3) more research on the causes of population declines and more efforts to reduce identifiable causes locally, (4) curtailment of the introduction of exotic species, especially predaceous fishes, and (5) greatly increased frog farming in order to meet commercial culinary and instructional demands. In order to address points 1 and 3, it is necessary to ease restrictions on scientific collecting and to use electronic databases of existing museum collections to identify declining and threatened species, such as that accomplished for mammals in southwestern Australia by McCarthy (1997).

EDUCATION

Education is the key to meeting future challenges effectively. Education of persons to be directly involved in biodiversity and conservation is only one facet of the problem. An informed and concerned populace is equally important. Persons in the vicinity of protected areas need to be educated about the

DUELLMAN: GLOBAL DISTRIBUTION OF AMPHIBIANS

economic value of such areas and shown that they can benefit from their existence. Moreover, the general populace, especially children, need to gain an appreciation of wildlife and its benefits; this can be accomplished by informed and dedicated teachers everywhere and by an increase in, and more effective orientation of, nature programs on television. All such programs should be made available to teachers.

At the university level, many more biologists must be trained in field techniques and in taxonomy in order to carry out necessary inventories and monitoring programs; training should be especially intensified in developing nations. Moreover, biologists need to be cross-trained in biological informatics and geographical information systems. Likewise, persons with biological training should be educated in appropriate disciplines in order to make them effective policy makers in government agencies and in nongovernmental organization. Legislation usually is promulgated by lawyers; there is a need to cross-train biologists in law.

At another level of interest, amphibians usually are ignored. Consider the worldwide interest in birds. There are tens of thousands of birders worldwide. They belong to clubs, many of which are parts of national organizations, such as the National Audubon Society, which have influence at local and national levels. These interests have resulted, at least in part, because ornithologists have encouraged amateur birders. In contrast, few herpetologists have encouraged amateurs, as is evident in that only in recent years have there been a large number of "popular" books on amphibians and reptiles, most of which have been written by nonscientists. The time is ripe for more organizations concerned with amphibians. Members of such organizations can aid in monitoring populations, create breeding sites, shepherd migrating individuals, lobby local governmental agencies, and maintain a species life-list. Fortunately, some regional herpetological societies are undertaking such activities.

LEGISLATION

The morass of national and international regulations proposed to protect living organisms is only moderately effective. This has resulted from many laws being made by individuals who may be unduly influenced by commercial interests, instead of being made by informed biologists. From available information on the numbers of animals exported and imported, it seems that it is much easier to import animals commercially than scientifically. In an effort to meet the challenges of the future, legislation needs to be changed so as to make the goals of biodiversity research and conservation attainable. The following should be considered.

1. Separation of regulations for collection and exportation of scientific specimens from commercial regulations with the implementation of scientific procedures placed in the hands of scientists. Scientific institutions could be certified by an international body (perhaps the United Nations or the Organization for Economic Cooperation and Development) so that personnel from those institutions could undertake necessary investigations, data resulting from which would be openly available electronically.

2. Establishment of an international protocol for the commercial development of products (e.g., drugs) from native populations in a given country with appropriate compensation in the form of royalties to the country of origin.

3. Strict regulation of the pet trade. Many persons are highly accomplished breeders of amphibians. Such persons should be licensed internationally and be allowed to import amphibians for breeding purposes; only captive-bred individuals should be sold to pet dealers. Furthermore, each individual used in breeding should be accompanied by locality data; captive-bred individuals should be accompanied by a certificate giving data on specific place of origin. Owners of pet amphibians should be encouraged to donate dead individuals (with accompanying data) to scientific institutions. Furthermore, breeders should be encouraged to collaborate with field biologists to further knowledge about behavior and life histories.

FUNDING

To meet these challenges of the future, vast sums of money are needed. Governmental granting agencies throughout the world must be persuaded to increase funding for biodiversity research, as do private foundations. Partnerships need to be established between biologists (and conservationists) with national and multinational corporations for funding of biological studies. The results of these studies can provide corporations with baseline data needed for projects, such as reclamation, or make available biological materials that a corporation can turn into a profitable product, such as drugs. On a smaller, local scale, tourism companies and national or state tourism agencies should work with, and support, biologists to study and maintain the biota that attracts the ecotourist.

Obviously, the thorough inventorying of amphibians (and other organisms) worldwide, monitoring populations, educating taxonomists and conservationists, establishing comprehensive electronic databases, incorporating biodiversity information into geographical information systems, and informed conservation and management require large sums of money. Therefore, it is terribly disconcerting that the National Aeronautic and Space Administration (USA) is contemplating expenditures of $4 billion over the next ten years for the exploration of Mars including the search for evidence of previous life there, but only a small pittance of that amount is likely to be available for studies of biodiversity on Earth and its conservation. There is no evidence that Mars has changed perceptibly during our lifetimes, whereas overwhelming evidence points to environmental degradation on our Earth. Is it not more reasonable for humans to invest in saving our own living planet before spending billions on a dead planet that will be there unchanged for eons of time?

Acknowledgments: First of all, I am indebted to the contributors of the chapters in this volume; they provided much of the data which I synthesized herein. Also, I thank Luis A. Coloma, Robert F. Inger, Meredith A. Lane, Jay M. Savage, and Michael J. Tyler for answering specific requests for information, José A. Gobbi for providing certain literature, Hélio da Silva for discovering electronic base maps, John E. Simmons for providing interesting literature of which I was unaware and for garnering information from the Internet, and Kraig Adler, Jonathan A. Campbell, John E. Simmons, and Linda Trueb for critical reviews of the manuscript.

LITERATURE CITED

ALFORD, R. A., AND S. J. RICHARDS. 1997. Lack of evidence for epidemic disease as an agent in the catastrophic decline of Australian rain forest frogs. Conserv. Biol. 11:1026–1029.

ANDERSON, S. 1994. Area and endemism. Quart. Rev. Biol. 69:451–471.

ANDRÉN, C., AND G. NILSON. 1995. Translocation of amphibians and reptiles. Consequence of introductions, re-introductions, and re-inforcement for conservation and definitions of concepts. Mem. Soc. Fauna Flora Fennica 71:84–87.

ANGUIANO, O. L., C. M. MONTAGNA, M. CHIFFLET DE LLAMAS, L. GAUNA, AND A. M. PECHEN DE ANGELO. 1994. Comparative toxicity of parathion in early embryos and larvae of the toad, *Bufo arenarum* Hensel. Bull. Environ. Contam. Toxicol. 52:649–655.

ARNOLD, S. J. 1972. Species densities of predators and their prey. Am. Nat. 106:220–236.

ASH, A. N. 1997. Disappearance and return of plethodontid salamanders to clearcut plots in the southern Blue Ridge Mountains. Conserv. Biol. 11:983–989.

AXELIUS, B. 1991. Areas of distribution and areas of endemism. Cladistics 7:197–199.

AYRES, J. M. 1989. Debt-for-equity swaps and the conservation of tropical rain forests. Trends Ecol. Evol. 4:331–332.

BÁEZ, A. M., AND L. TRUEB. 1997. Redescription of the Paleogene *Shelania pascuali* from Patagonia and its bearing on the relationships of fossil and Recent pipoid frogs. Sci. Pap. Nat. Hist. Mus. Univ. Kansas 4:1–41.

BARKER, J., G. C. GRIGG, AND M. J. TYLER. 1995. *A Guide to Australian Frogs.* Chipping Norton, Australia: Surrey Beatty & Sons.

BEEBEE, T. J., C. R. J. FLOWER, A. C. STEVENSON, S. T. PATRICK, P. G. APPLEBY, C. FLETCHER, C. MARSH, J. NATKANSKI, B. RIPPEY, AND R. W. BATTARBEE. 1990. Decline of the natterjack toad *Bufo calamita* in Britain: palaeoecological, documentary and experimental evidence for breeding site acidification. Biol. Conserv. 53:1–20.

BELL, G., AND J. H. LAWTON. 1975. The ecology of the eggs and larvae of the smooth newt, *Triturus vulgaris* (Linn.). J. Anim. Ecol. 44:393–424.

BERNHARDT, D. M., S. M. COOGAN, P. D. DANIELSON, A. DANNHAUER, A. DE MAJEWSKI, E. R. VANDER SCHAAF, AND S. J. ZOTTOLI. 1991. Conservation in the teaching laboratory - substitution of *Xenopus* for *Rana.* BioScience 41:578–580.

BERNINGHAUSEN, F. 1995. Erfolgreische Laubfroschenwiederansiedelung seit 1984 im Landkreis Roteenburg, Niedersachsen. Mertensiella 6:149–162.

BERRILL, M., S. BERTRAM, B. PAULI, D. COULSON, M. KOLOHON, AND D. OSTRANDER. 1995. Comparative sensitivity of amphibians tadpoles to single and pulsed exposures of the forest-use insecticide fenithrothrion. Environ. Toxicol. Chem. 14:1011–1018.

BIERREGAARD, R. O., T. J. LOVEJOY, V. KAPOS, A. SANTOS, AND R. W. HUTCHINGS. 1992. The biological dynamics of tropical forest fragments. BioScience 42:859–856.

BLAUSTEIN, A. R. 1994. Chicken Little or Nero's fiddle? A perspective on declining amphibian populations. Herpetologica 50:85–97.

BLAUSTEIN, A. R., P. D. HOFFMAN, J. M. KIESECKER, AND J. B. HAYS. 1996. DNA repair activity and and resistance to solar UV-B radiation in eggs of the red-legged frog. Conserv. Biol. 10:1398–1402.

BLAUSTEIN, A. R., J. M. KIESECKER, D. P. CHIVERS, AND R. G. ANTHONY. 1997. Ambient UV-B radiation causes deformities in amphibian embryos. Proc. Natl. Acad. Sci. USA 94:13735–13737.

BLAUSTEIN, A. R., AND D. B. WAKE. 1995. The puzzle of declining amphibian populations. Sci. Am. 272(4):52–57.

BOOS, E. 1990. *Ecotourism: The Potentials and Pitfalls.* 2 vols. Washington, D. C.: World Wildlife Fund—US.

BOULENGER, G. A. 1882. *Catalogue of the Batrachia Gradientia s. Caudata and Batrachia Apoda in the Collection of the British Museum,* 2nd Ed. London: British Museum.

BRADFORD, D. F. 1989. Allopatric distribution of native frogs and introduced fishes in high Sierra Nevada lakes of California: implications of the negative impact of fish introductions. Copeia 1989:775–778.

BROWN, J. H. 1995. *Macroecology.* Chicago: Univ. Chicago Press.

BURKE, R. L. 1991. Relocations, repatriations, and translocations of amphibians and reptiles: taking a broader view. Herpetologica 47:350–357.

CALLICOTT, J. B., AND K. MUMFORD. 1997. Ecological sustainability as a conservation concept. Conserv. Biol. 11:32–40.

CAMPBELL, J. A. 1998. *Amphibians and reptiles of northern Guatemala, the Yucatán, and Belize.* Norman, Oklahoma: Univ. Oklahoma Press.

CAMPBELL, J. A., AND D. R. FROST. 1993. Anguid lizards of the genus *Abronia:* revisionary notes, descriptions of four new species, a phylogenetic analysis, and key. Bull. Am. Mus. Nat. Hist. 216:1–121.

CANNATELLA, D. C. 1986. A new genus of bufonid (Anura) from South America, and phylogenetic relationships of the neotropical genera. Herpetologica 42:197–205.

CARSON, R. 1962. *Silent Spring.* Boston: Houghton Mifflin.

CASTNER, J. L. 1990. *Rainforests. A Guide to Research and Tourist Facilities at Selected Tropical Forest Sites in Central and South America.* Gainesville, Florida: Feline Press.

CHARDON, P., C. VOS, AND H. DE VRIES. 1996. The use of amphibian tunnels under roads. Levende Nat. 97:110–115.

CONANT, R., AND J. T. COLLINS. 1991. *A Field Guide to Reptiles and Amphibians. Eastern and Central North America,* 3rd Ed. Boston: Houghton Mifflin Co.

CONNER, E. F., AND D. SIMBERLOFF. 1986. Competition, scientific method, and null models in ecology. Am. Sci. 74:155–162.

COOKE, A. S. 1988. Mortality of toads (*Bufo bufo*) on roads near a Cambridgeshire breeding site. British Herpetol. Soc. Bull. 26:29–30.

DAYTON, K. 1994. A lizard in the bush. New Scientist 141:12–13.

DEMAYNADIER, P. G., AND M. L. HUNTER, JR. 1995. The relationship between forest management and amphibian ecology: a review of the North American literature. Environ. Rev. 3:230–261.

DENTON, J. S., S. P. HITCHINGS, T. J. C. BEEBEE, AND A. GENT. 1997. A recovery program for the natterjack toad (*Bufo calamita*) in Britain. Conserv. Biol. 11:1329–1338.

DINERSTEIN, E., D. M. OLSON, D. J. GRAHAM, A. L. WEBSTER, S. A. PRIMM, M. P. BOOKBINDER, AND G. LEDEK. 1995. *A Conservation Assessment of the Terrestrial Ecoregions of Latin America and the Caribbean.* Washington, D. C.: The World Bank.

DODD, C. K., JR., AND R. A. SEIGEL. 1991. Relocation, repatriation, and translocation of amphibians and reptiles: Are they conservation strategies that work? Herpetologica 47:336–350.

DUBOIS, A. 1987. Miscellanea taxinomica batrachologica (I). Alytes 5:7–95.

DUBOIS, A. 1992. Notes sur la classification des Ranidae (Amphibiens Anoures). Bull. Mens. Soc. Linn. Lyon 61:305–352.

DUELLMAN, W. E. (ed.). 1979. The South American herpetofauna; its origins, evolution, and dispersal. Monogr. Mus. Nat. Hist. Univ. Kansas 7:1–485.

DUELLMAN, W. E. 1988. Patterns of species diversity in anuran amphibians in the American tropics. Ann. Missouri Bot. Gard. 75:79–104.

DUELLMAN, W. E. 1993a. Amphibians in Africa and South America: evolutionary history and ecological comparisons. Pp. 200–243 in P. Goldblatt (ed.), *Biological Relationships between Africa and South America.* New Haven: Yale Univ. Press.

DUELLMAN, W. E. 1993b. Amphibian species of the world: additions and corrections. Spec. Publ. Mus. Nat. Hist. Univ. Kansas 21:1–372.

DUELLMAN, W. E. 1995. Temporal fluctuations in abundances of anuran amphibians in a seasonal Amazonian rainforest. J. Herpetol. 29:13–21.

DUELLMAN, W. E., AND J. A. CAMPBELL. 1992. Hylid frogs of the genus *Plectrohyla:* systematics and phylogenetic relationships. Misc. Publ. Mus. Zool. Univ. Michigan 181:1–32.

DUELLMAN, W. E., AND J. E. KOECHLIN. 1991. The Reserva Cuzco Amazonico, Peru: biological investigations, conservation, and ecotourism. Occas. Pap. Mus. Nat. Hist. Univ. Kansas 142:1–38.

DUELLMAN, W. E., AND E. R. PIANKA. 1990. Biogeography of nocturnal insectivores: historical events and ecological filters. Ann. Rev. Ecol. Syst. 21:57–68.

DUELLMAN, W. E., AND A. W. SALAS. 1991. Annotated checklist of the amphibians and reptiles of Cuzco Amazonico, Peru. Occas. Pap. Mus. Nat. Hist. Univ. Kansas 143:1–13.

DUELLMAN, W. E., AND R. SCHULTE. 1992. Description of a new species of *Bufo* from northern Peru and comments on phenetic groups of South American toads (Anura: Bufonidae). Copeia 1992:162–172.

DUELLMAN, W. E., AND L. TRUEB. 1986. *Biology of Amphibians.* New York: McGraw-Hill Book Co.

EHMANN, H., AND H. COGGER. 1985. Australia's endangered herpetofauna: a review of criteria and policies. Pp. 435–448 in G. Grigg, R. Shine, and H. Ehmann (eds.), *The Biology of Australasian Frogs and Reptiles.* Chipping Norton, Australia: Surrey Beatty and Sons.

ENGEL, E., AND C. BRESSANUTTI. 1993. Zur Funktionsfahigkeit einer neuen Amphibienschutzanlage in der Gemeinde Kehlen. Bull. Soc. Nat. Luxembourgeois 94:121–127.

ESTES, R. 1981. *Encyclopedia of Paleoherpetology. Part 2. Gymnophiona, Caudata.* New York: Gustav Fischer Verlag.

FAHRIG, L., J. H. PEDLAR, S. E. POPE, P. D. TAYLOR, AND J. F. WEGNER. 1995. Effects of road traffic on amphibian density. Biol. Conserv. 73:177–182.

FORD, L. S., AND D. C. CANNATELLA. 1993. The major clades of frogs. Herpetol. Monogr. 7:94–117.

FORMAS, J. R. 1979. La herpetofauna de los bosques temperados de Sudamérica. Pp. 341–369 in W. E. Duellman (ed.), The South American herpetofauna; its origins, evolution, and dispersal. Monogr. Mus. Nat. Hist. Univ. Kansas 7:1–485.

FOUTENOT, L. W., G. PITTMAN NOBLET, S. G. PLATT, AND J. M. AKINS. 1996. A survey of herpetofauna inhabiting polychlorinated biphenyl contaminated and reference watersheds in Pickens County, South Carolina. J. Elisha Mitchell Sci. Soc. 112:20–30.

FRANKLIN, J. F. 1992. Scientific basis for new perspectives in forests and streams. Pp. 25–72 in R. J. Naiman (ed.), *Watershed Management: Balancing Sustainability and Environmental Change.* New York: Springer-Verlag.

FRANKLIN, J. F. 1993. Preserving biodiversity: species, ecosystems, or landscapes? Ecol. Applica. 3:202–205.

FRANKLIN, J. F., D. E. BERG, D. A. THORNBURG, AND J. C. TAPPEINER. 1996. Alternative silviculture approaches to timber harvest: variable retention harvest systems. *Forestry in the 21st Century*. Covelo, California: Island Press.

FROST, D. R. 1985. *Amphibian Species of the World*. Lawrence, Kansas: Association of Systematics Collections and Allen Press, Inc.

FUJIOKOA, M., AND S. L. LANE. 1997. The impact of changing irrigation practices in rice fields on frog populations of the Kanto Plain, central Japan. Ecol. Res. 12:101–108.

GANS, C. 1994. Frogs and paddy: problems of management. J. Bombay Nat. Hist. Soc. 91:29–36.

GASC, J.-P., A. CABELA, J. CRNOBRNJA-ISILOVIC, D. DOLMEN, K. GROSSENBACHER, P. HAFFNER, J. LESCURE, H. MARTENS, J. P. MARTÍNEZ RICA, H. MAURIN, M. E. OLIVEIRA, T. S. SOFIANIDOU, M. VIETH, AND A. ZUIDERWIJK (eds.). 1997. *Atlas of Amphibians and Reptiles of Europe*. Paris: Muséum National d'Histoire Naturelle.

GLAW, F., AND M. VENCES. 1994. *A Fieldguide to the Amphibians and Reptiles of Madagascar*. 2nd Ed. Köln, Germany: M. Vences & F. Glaw Verlags, GbR.

GORZULA, S. 1996. The trade in dendrobatid frogs from 1987 to 1993. Herpetol. Rev. 2:116–123.

GRAFF, H. 1996. En paddetunnel til ingen verdens nytte? Nord. Herpetol. Foren. 39:37–40.

GRAHAM, F. 1971. *Man's Dominion. The Story of Conservation in America*. New York: J. P. Lippincott.

GRAYBEAL, A. 1997. Phylogenetic relationships of bufonid frogs and tests of alternate macroevolutionary hypotheses characterizing their radiation. Zool. J. Linnean Soc. 119:297–338.

GRAYBEAL, A., AND D. C. CANNATELLA. 1995. A new taxon of Bufonidae from Peru, with descriptions of two new species and a review of the phylogenetic status of supraspecific bufonid taxa. Herpetologica 51:105–131.

HADER, D.-P. 1997. *The Effects of Ozone Depletion on Aquatic Ecosystems*. Austin, Texas: R. G. Landes.

HAIRSTON, N. G. 1987. *Community Ecology and Salamander Guilds*. Cambridge: Cambridge Univ. Press.

HALL, R. J., AND P. F. P. HENRY. 1992. Assessing the effects of pesticides on amphibians and reptiles. Herpetol. J. 2:65–71.

HALLIDAY, T. R., AND W. R. HEYER. 1997. The case of the vanishing frogs. MIT's Techno. Rev. May/June 1997:56–63.

HARFENIST, A., T. POWER, K. L. CLARK, AND D. B. PEAKALL. 1989. *A Review and Evaluation of the Amphibian Toxicological Literature*. Ottawa: Canadian Wildlife Service Technical Report.

HAROLD, A. S., AND R. D. MOOI. 1994. Areas of endemism: definition and recognition criteria. Syst. Biol. 43:261–266.

HARTE, J., AND E. HOFFMAN. 1989. Possible effects of acidic deposition on a Rocky Mountain population of the tiger salamander *Ambystoma tigrinum*. Conserv. Biol. 3:149–158.

HAYS, M. P., AND M. R. JENNINGS. 1986. Decline of ranid frog species in western North America: are bullfrogs (*Rana catesbeiana*) responsible? J. Herpetol. 20:490–509.

HECNAR, S. J. 1995. Acute and chronic toxicity of ammonium nitrate to amphibians from southern Ontario. Environ. Toxicol. Chem. 14:2131–2137.

HEDGES, S. B. 1986. An electrophoretic analysis of holarctic hylid frog evolution. Syst. Zool. 35:1–21.

HEDGES, S. B. 1996. The origin of the West Indian amphibians and reptiles. Pp. 95–128 *in* R. Powell and R. W. Henderson (eds.), *Contributions to West Indian Herpetology: A Trib-*

ute to Albert Schwartz. Ithaca, NY: Society for the Study of Amphibians and Reptiles.

HERO, J.-M., AND G. R. GILLESPIE. 1997. Epidemic disease and amphibian declines in Australia. Conserv. Biol. 11:1023–1025.

HEYER, W. R. 1979. Annual variation in larval amphibian populations within a temporary pond. J. Washington Acad. Sci. 69:65–74.

HEYER, W. R., A. S. RAND, C. A. G. DA CRUZ, AND O. PEIXOTO. 1988. Decimations, extinctions, and colonizations of frog populations in southeast Brazil and their evolutionary implications. Biotropica 20:230–235.

HEYER, W. R., A. S. RAND, C. A. G. DA CRUZ, O. PEIXOTO, AND C. E. NELSON. 1990. Frogs of Boracéia. Arq. Zool. Mus. Zool. Univ. São Paulo 31:231–410.

HOBBS, R. J. 1992. The role of corridors in conservation: solution or bandwagon? Trends Ecol. Evol. 7:389–392.

HOBBS, R. J. 1994. Landscape ecology and conservation: moving from description to application. Pacific Conserv. Biol. 1:170–176.

HOLT, R. D. 1984. Spatial heterogeneity, indirect interactions, and the coexistence of prey species. Am. Nat. 124:377–406.

HONEGGER, R. E. 1981. *Threatened Amphibians and Reptiles in Europe*. Wiesbaden: Akademische Verlagsgesellschaft.

INGER, R. F. 1954. Systematics and zoogeography of Philippine Amphibia. Fieldiana: Zool. 33:183–531.

INGER, R. F. 1972. *Bufo* of Eurasia. Pp. 102–118 *in* W. F. Blair (ed.), *Evolution in the Genus* Bufo. Austin: Univ. Texas Press.

INGER, R. F. 1996. Commentary on a proposed classification of the family Ranidae. Herpetologica 52:241–246.

INGER, R. F., AND R. K. COLWELL. 1977. Organization of contiguous communities of amphibians and reptiles in Thailand. Ecol. Monogr. 47:229–253.

INGER, R. F., AND R. B. STEUBING. 1989. *Frogs of Sabah*. Sabah, Malaysia: Sabah Parks Trustees.

JAEGER, R. G. 1980. Density-dependent and density-independent causes of extinction of a salamander population. Evolution 34:617–621.

JANZEN, D. H. 1983. No park is an island: increase in interference from outside as park size decreases. Oikos 41:402–410.

KLOPFER, P. H., AND R. H. MACARTHUR. 1961. On the causes of tropical species diversity. Am. Nat. 95:223–226.

KUTKA, F. J. 1994. Low pH effects on swimming activity of *Ambystoma* alamander larvae. Environ. Toxicol. Chem. 13:1821–24.

KUZMIN, S. L. 1996. Threatened amphibians in the former Soviet Union: the current situation and main threats. Oryx 30:24–30.

LA MARCA, E., AND D. LÖTTERS. 1997. Monitoring of declines in Venezuelan *Atelopus* (Amphibia: Anura: Bufonidae). Pp. 207–213 *in* W. Böhme, W. Bischoff, and T. Ziegler (eds.). *Herpetologia Bonnensis*. Bonn, Germany: Museum Alexander Koenig.

LAMBECK, R. J. 1997. Focal species: a multi-species umbrella for nature conservation. Conserv. Biol. 11:849–856.

LANGTON, T. E. S. (ed.). 1989. *Amphibians and Roads*. Shefford, England: ACO Polymer Products.

LARSON, A., AND W. W. DIMMICK. 1993. Phylogenetic relationships of the salamander families: an analysis of congruence among morphological and molecular characters. Herpetol. Monogr. 7:77–93.

LAURANCE, W. F., K. R. MACDONALD, AND R. SPEARE. 1996. Epidemic disease and the catastrophic decline of Australian rain forest frogs. Conserv. Biol. 10:406–413.

LAURENT, R. F. 1979. Herpetofaunal relationships between Africa and South America. Pp. 55–71 *in* W. E. Duellman (ed.), The South American herpetofauna; its origins, evolution, and dispersal. Monogr. Mus. Nat. Hist. Univ. Kansas 7:1-485.

LAWSON, D. P. 1993. The reptiles and amphibians of the Korup National Park Project, Cameroon. Herpetol. Nat. Hist. 1:27–90.

LYNCH, J. D., AND W. E. DUELLMAN. 1997. Frogs of the genus *Eleutherodactylus* in western Ecuador: systematics, ecology, and biogeography. Spec. Publ. Nat. Hist. Mus. Univ. Kansas 23:1–236.

LYNCH, J. D., P. M. RUÍZ-CARRANZA, AND M. C. ARDILA-ROBALO. 1997. Biogeographic patterns of Colombian frogs and toads. Rev. Acad. Colombiana Cien. 21:237–248.

MACARTHUR, R. H. 1969. Species packing and what competition minimizes. Proc. Nat. Acad. Sci. USA 64:1369–1371.

MACARTHUR, R. H. 1970. Species packing and competitive equilibrium for many species. Theor. Pop. Biol. 1:1–11.

MAHANEY, P. A. 1994. Effects of freshwater petroleum contamination on amphibian hatching and metamorphosis. Environ. Taxicol. Chem. 13:259–265.

MANION, J. J., AND L. CORY. 1952. Winter kill of Rana pipiens in shallow ponds. Herpetologica 8:32.

MARSH, D. M., AND P. B. PEARMAN. 1997. Effects of habitat fragmentation on the abundance of two species of leptodactylid frogs in an Andean montane forest. Conserv. Biol. 11:1323–1328.

MATTHIESSEN, P. 1959. *Wildlife in America.* New York: Macmillan Co.

McCARTHY, M. A. 1997. Identifying declining and threatened species with museum data. Biol. Conserv. 83:9–17.

MEANS, D. B., J. G. PALIS, AND M. BAGGETT. 1996. Effects of slash pine silviculture on a Florida population of flatwoods salamander. Conserv. Biol. 10:426–437.

MEIER, E. 1995. Bestandsentwicklungen des Laubfrosches (*Hyla arborea* L.) in der westfalischen Bucht. Mertensiella 6:73–93.

MENEGHEL, M. D. 1992. Posibles efectos de plantaciones de eucaliptos en Río Negro y Paysandu sobre la herpetofauna residente. Bol. Soc. Zool. Uruguay 7:27–28.

METTER, D. E. 1968. The influence of floods on population structure of *Ascaphus truei* (Stejneger). J. Herpetol. 1:105–106.

MIKKELSEN, U. S. 1993. The alpine newt in Denmark. Flora og Fauna 99:3–9.

MORALES, V. R., AND R. W. McDIARMID. 1996. Annotated checklist of the amphibians and reptiles of Pakitza, Manu National Park Reserve Zone, with comments on the herpetofauna of Madre de Dios, Perú. Pp. 503–522 *in* D. E. Wilson and A. Sandoval (eds.), *Manu.* Washington, D. C.: Smithsonian Institution.

MORRONE, J. A. 1994. On the identification of areas of endemism. Syst. Biol. 43:438–441.

MROSOVSKY, N. 1988. The CITES conservation circus. Nature 331:563.

MYERS, C. W., AND J. W. DALY. 1993. Tropical poison frogs. Science 262:1193.

NASH, R. R. 1990. *American Environmentalism. Readings in Conservation History,* 3rd Ed. New York: McGraw-Hill Book Co.

NUSSBAUM, R. A., AND M. WILKINSON. 1989. On the classification and phylogeny of caecilians (Amphibia: Gymnophiona), a critical review. Herpetol. Monogr. 3:1–42.

OUELLET, M., J. BONIN, J. RODRIGUE, J. L. DESGRANGES, AND S. LAIR. 1997. Hindlimb deformities (ectromelia, ectrodactyly) in free-living anurans from agricultural habitats. J. Wildlife Diseases 33:95–104.

PEARMAN, P. B. 1997. Correlates of amphibian diversity in an altered landscape in Amazonian Ecuador. Conserv. Biol. 11:1211–1225.

PECHMANN, J. H. K., D. E. SCOTT, R. D. SEMLITSCH, J. P. CALDWELL, L. J. VITT, AND J. W. GIBBONS. 1991. Declining amphibian populations: the problem of separating human impacts from natural fluctuations. Science 253:892–895.

PECHMANN, J. H. K., AND H. M. WILBUR. 1994. Putting declining amphibian populations in perspective: natural fluctuations and human impacts. Herpetologica 50:65–84.

PERFECTO, I., R. A. RICE, R. GREENBERG, AND M. E. VAN DER VOORT. 1996. Shade coffee: a disappearing refuge for biodiversity. BioScience 46:598–608.

PHILLIPS, K. 1994. *Tracking the Vanishing Frogs.* New York: St. Martin's Press.

PIERCE, B. A. 1985. Acid tolerance in amphibians. BioScience 35:239–243.

PIERSAN D. 1987. Salamander tunnels in Amherst, Massachusetts, USA. Herpetofauna News 10:1–2.

PLEGUEZUELOS, J. M. 1997. *Distribución y Biogeografía de los Anfibios y Reptiles en España y Portugal.* Granada, Spain: Univ. Granada and Asoc. Herpetol. Española.

POUGH, F. H., R. M. ANDREWS, J. E. CADLE, M. L. CRUMP, A. H. SAVITZKY, K. D. WELLS. 1997. *Herpetology.* Upper Saddle River, New Jersey: Prentice Hall.

POUNDS, J. A., AND M. L. CRUMP. 1994. Amphibian declines and climate disturbance: the case of the golden toad and the harlequin frog. Conserv. Biol. 8:72–75.

POUNDS, J. A., M. P. L. FOGDEN, J. M. SAVAGE, AND G. C. GORMAN. 1997. Tests of null models for amphibian declines on a tropical mountain. Conserv. Biol. 11:1307–1322.

RAMUS, E. 1995. 7.3 million herps kept as pets in the U. S. Reptile & Amphibian Mag. Jan.-Feb. 1995:126.

RAXWORTHY, C. J., AND R. A. NUSSBAUM. 1996. Patterns of endemism for terrestrial vertebrates in eastern Madagascar. Pp. 369–383 *in* W. R. Lourenço (ed.), *Biogeographie de Madagascar.* Paris: ORSTROM.

REINERT, H. K. 1991. Translocation as a conservation strategy for amphibians and reptiles: some comments, concerns, and observations. Herpetologica 47:357–363.

RICE, R. A., AND J. R. WARD. 1996. *Coffee, Conservation, and Commerce in the Western Hemisphere.* Washington, D. C.: Natural Resources Defense Council.

RICHARDS, S. J., K. R. MACDONALD, AND R. A. ALFORD. 1993. Declines in populations of Australia's endemic tropical rainforest frogs. Pacific Conserv. Biol. 1:66–77.

RODRÍGUEZ, L. O., AND J. E. CADLE. 1990. A preliminary overview of the herpetofauna of Cocha Cashu, Manu National Park, Peru. Pp. 410–425 *in* A. H. Gentry (ed.), *Four Neotropical Rainforsts.* New Haven, Connecticut: Yale Univ. Press.

RODRÍGUEZ, L. O., AND W. E. DUELLMAN. 1994. *Guide to the Frogs of the Iquitos Region, Amazonian Peru.* Spec. Publ. Nat. Hist. Mus. Univ. Kansas 22:1–80, pls. 1–12.

SANCHIZ, B. 1997. *Encyclopedia of Paleoherpetology. Part 4. Salientia.* New York: Gustav Fischer Verlag.

SAVAGE, J. M. 1973. The geographic distribution of frogs: patterns and predictions. Pp. 351–445 *in* J. L. Vial (ed.), *Evolutionary Biology of the Anurans*. Columbia: Univ. Missouri Press.

SAVAGE, J. M. 1982. The enigma of the Central American herpetofauna: dispersal or vicariance? Ann. Missouri Bot. Gard. 69:464–547.

SEMLITSCH, R. D. 1983. Structure and dynamics of two breeding populations of the eastern tiger salamander, *Ambystoma tigrinum*. Copeia 1983:608–616.

SESSIONS, S. K., AND S. B. RUTH. 1990. Explanation for naturally occurring supernumerary limbs in amphibians. J. Exp. Zool. 254:38–47.

SHCHERBAK [SZCZERBAK], N. N. 1982. Grundzüge einer herpetogeographischen Gliederung der Paläarktis. Vertebrata Hungarica 21:227–239.

STEBBINS, R. C. 1985. *A Field Guide to Western Reptiles and Amphibians,* 2nd ed. Boston: Houghton Mifflin Co.

STEBBINS, R. C., AND N. W. COHEN. 1995. *A Natural History of Amphibians.* Princeton, New Jersey: Princeton University Press.

TANDY, M., AND R. KEITH. 1972. *Bufo* of Africa. Pp. 119–170 *in* W. F. Blair (ed.), *Evolution in the Genus* Bufo. Austin: Univ. Texas Press.

TEVINI, M. 1993. *UV-B Radiation and Ozone Depletion.* Boca Raton, Florida: Lewis Publishing Co.

TOCHIMOTO, T. 1995. Ecological studies on the Japanese giant salamander, *Andrias japonicus,* in the Ichi River in Hyogo Prefecture. 10. An attempt to rebuild spawning places along the river. J. Japanese Assoc. Zool. Gard. Aquar. 37:7–12.

TOFT, C. A. 1980. Feeding ecology of thirteen syntopic species of anurans in a seasonal tropical environment. Oecologia 45:131–141.

TYLER, M. J. 1979. Herpetofaunal relationships of South America with Australia. Pp. 73–106 *in* W. E. Duellman (ed.), The South American herpetofauna; its origins, evolution, and dispersal. Monogr. Mus. Nat. Hist. Univ. Kansas 7:1-485.

TYLER, M. J. 1989. *Australian Frogs.* Ringwood, Victoria: Viking O'Neil.

TYLER, M. J. 1991. Declining amphibian populations—a global phenomenon? An Australian perspective. Alytes 9:43–50.

TYLER, M. J. 1995. Frogs and drugs. Australian Nat. Hist. (Autumn) 1995:47–51.

TYLER, M. J. 1997. *The Action Plan for Australian Frogs.* Canberra: Wildlife Australia.

VENCES, M. 1993. Habitat choice of the salamander *Chioglossa lusitanica:* the effect of eucalypt plantations. Amphibia-Reptilia 14:201–212.

VIAL, J. L., AND L. SAYLOR. 1993. The status of amphibian populations. Corvalis, Oregon: Declining Amphibian Populations Task Force, Working Document 1:1–98.

VITT, L. J., J. P. CALDWELL, H. M. WILBUR, AND D. C. SMITH. 1990. Amphibians as harbingers of decay. BioScience 40:418.

WALKER, B. H. 1995. Conserving biological diversity through ecosystem resilience. Conserv. Biol. 9:747–752.

WALLACE, A. R. 1869. *The Malay Archipelago.* New York: Harper.

WEYGOLDT, P. 1989. Changes in composition of mountain stream frog communities in the Atlantic Mountains of Brasil: frogs as indicators of environmental deterioration? Stud. Neotrop. Fauna Environ. 243:249–255.

WILSON, D. E., AND A. SANDOVAL (eds.). 1996. *Manu.* Washington, D. C.: Smithsonian Institution.

WYMAN, R. L. 1990. What's happening to the amphibians? Conserv. Biol. 4:350–352.

YANES, M., J. M. VELASCO, AND F. SUÁREZ. 1995. Permeability of roads and railroads to vertebrates: the importance of culverts. Biol. Conserv. 71:217–222.

YOFFE, E. 1992. Silence of the frogs. New York Times Mag. Dec. 13, 1992:36–39, 64, 66, 76.

ZHAO, E.-M., AND K. ADLER. 1993. *Herpetology of China.* Oxford, Ohio: Society for the Study of Amphibians and Reptiles.

ZIMMERMAN, B. L., AND B. L. BIERREGAARD. 1986. Relevance of the equilibrium theory of island biogeography and species-area relations to conservation with a case from Amazonia. J. Biogeogr. 13:133–143.

2. Distribution Patterns of Amphibians in the Nearctic Region of North America

WILLIAM E. DUELLMAN AND SAMUEL S. SWEET

Natural History Museum and
Department of Ecology and Evolutionary Biology
The University of Kansas
Lawrence, Kansas 66045-2454, USA

Department of Ecology, Evolution
& Marine Biology
University of California
Santa Barbara, California 93106, USA

ABSTRACT   The Nearctic Region of North America has 241 native species of amphibians (90 anurans and 151 salamanders). Analysis of the patterns of distribution in five physiographic divisions reveals the occurrence of 83–93 species in each of the divisions, except the boreal Laurentian Uplands, where only 14 species occur. Percentage of endemism ranges from zero in the Laurentian Uplands to 81% in the North American Cordillera. Analysis of patterns of distribution in 45 natural regions shows that the greatest numbers of species occur in the Southern Appalachians, Allegheny Plateau, Gulf Coastal Plain, and Southern Coastal Plain, with 66, 56, 55, and 54 species, respectively. Salamanders are much more diverse than anurans in the Allegheny Plateau and Southern Appalachian regions than on the coastal plain, where the numbers are about equal. Also there are more species of salamanders than anurans in the mountains of western North America, whereas in the midcontinental, lowland regions and the basins between the mountains in western North America, anurans are more numerous. Areas of high diversity and endemism of salamanders (especially plethodontids) are in the Appalachian Highlands, Interior Highlands, and the Pacific–Cascade–Sierra Nevada ranges. Secondary areas, especially for aquatic salamanders (Amphiumidae, Sirenidae, and hemidactyliine Plethodontidae) are the Atlantic and Gulf coastal plains and the Edwards Plateau. The major area of diversity and endemism of anurans is the coastal plain in southeastern North America; a secondary area includes the northern parts of the Pacific–Cascade–Sierra Nevada ranges. The greatest diversities of ranid and hylid frogs, and ambystomatid salamanders also are on the coastal plains of southeastern North America, whereas the greatest diversity of *Bufo* is in the southwestern United States and adjacent Mexico. Phylogenetic analyses reveal that basal stocks of *Bufo* inhabit these same areas, which also are centers for the evolution of *Rana*. In contrast, basal stocks of plethodontine and desmognathine salamanders are in the Appalachian Highlands. Most generic lineages seem to have differentiated prior to the Miocene, when climatic changes resulted in the formation of grasslands and deserts in central and southwestern North America. The major threat to amphibians in North America is habitat degradation; species with small ranges or narrowly restricted habitat requirements are the most susceptible to extinction. Declining populations are most noticeable among species of *Bufo* and *Rana* at high elevations in western North America. Although many species are protected in various states and provinces, only 13 amphibians are listed as threatened or endangered under the United States Endangered Species Act.

Key words: Amphibians; North America; Patterns of distribution; Phylogenetics; Historical biogeography; Conservation.

INTRODUCTION

Surprisingly, an analysis of patterns of distribution of amphibians in North America is more problematic than it might seem. The existence of detailed and carefully verified distribution maps of all of the species (principally in the field guides by Stebbins [1985] and Conant and Collins [1991]) creates an illusion of simplicity. The tremendous amount of literature on North American amphibians is a blessing in disguise, for at one and the same time, the overwhelming quantity of information incorporates detailed information on many genera and species, but adequate information is missing for some taxa. Had we included all of the detailed information available, this chapter easily would have been thrice its length. Conversely, if the analyses had been reduced to the lowest common denominator, some important aspects would have been omitted entirely. Thus, our treatment of North American amphibians represents a compromise that hopefully will fulfill the basic needs of investigators and at the same time inspire them to seek additional pieces of the puzzle that will elucidate the biogeographic picture and historical scenarios.

The illusion of relative simplicity was one of the reasons that one of us (SSS) originally undertook the task of analyzing the North American amphibian fauna. He tabulated species occurrence in regions, compiled maps of species densities, and presented preliminary results at the Second World Congress of Herpetology in 1993. However, he was dissatisfied with the results and allowed other commitments to take precedence over completing the analyses and writing the manuscript. In May 1997, he turned over the material to WED, who updated the data and maps, expanded the analyses, incorporated historical data, and wrote the first draft of the manuscript.

Distributions of amphibians in the Nearctic are better known than elsewhere in the world except for Europe; therefore, it seems incomprehensible that no biogeographic synthesis of North American amphibians exists. Maps of vegetation and/or landforms have accompanied detailed distribution maps of species in several regional studies (e.g., Stebbins, 1951; Nussbaum et al., 1983), but no analyses of distributions were provided. The only previous attempt to synthesize distributional data on a broad scale was by Kiester (1971), who accumulated distributional data in quadrats (100 X 100 mi); he used the data to draw species-density maps and discuss latitudinal species densities.

Herein, we provide a classification of physiographic divisions and natural regions in North America. We analyze the distribution patterns of amphibians with respect to these divisions and regions and, in so doing, point out the biotic similarities and differences among regions and identify areas of high diversity and endemism. Furthermore, we analyze the distribution patterns taxonomically, point out major disjunctions, and interpret the historical biogeography based on available data from the fossil record and existing phylogenetic analyses. We interpret patterns of amphibian distributions with respect to topography, climate, and earth history, especially Pleistocene glaciation and associated climatic changes. Finally, we review conservation efforts with respect to amphibians in North America.

North America is a large continent extending from 72° N lat. to 16° S lat. North of the continuous landmass are many large islands, most of which are above the Arctic Circle; the northernmost of these, Ellesmere Island, terminates at 83° N lat. Thus, nearly half of the continent, and its associated islands, lies north of 50° N lat. To the south, the continent terminates at the Isthmus of Tehuantepec in southern Mexico. Although this definition is appropriate in a broad geographic sense, the boundary between the Nearctic and the Middle American (Mesoamerican or Neotropical) faunas is in northern Mexico (Eisenmann, 1955; Herskovitz, 1958; Savage, 1966). Therefore, in coordination with J. A. Campbell's analysis of distributions of Middle American amphibians (this volume), we include in the Nearctic the Peninsula of Baja California, the Sonoran and Chihuahuan deserts, and the coastal plain of northern Tamaulipas, all in Mexico. As so defined, the Nearctic is composed of all of Canada and continental United States of America (USA), and parts of northern Mexico (Fig. 2:1).

MATERIALS AND METHODS

The bases for nearly all of our distributional data are the maps in the field guides to North American amphibians and reptiles by Stebbins (1985) and Conant and Collins (1991). Some distributions were modified, based on new data published in the section of Geographic Distribution in *Herpetological*

Fig. 2:1.   North America with the provinces and territories of Canada and the states of the USA.

Review and subsequent accounts in the *Catalogue of American Amphibians and Reptiles* (both published by the Society for the Study of Amphibians and Reptiles). Furthermore, the field guides were updated by the addition of taxa and taxonomic changes by Hedges (1986), Good (1989), Good and Wake (1992), Chippindale et al. (1993), Hairston (1993), Mole and Kezer (1993), Platz (1993), Tilley and Mahoney (1996), Wake (1996), Green et al. (1997),

and Highton (1997). Distributions of the cryptic species *Hyla chrysoscelis* and *H. versicolor* were scored as the same, because this is the way they were mapped by Conant and Collins (1991), and because in many areas, these taxa have not been distinguished from one another. Classification follows that of Frost (1985), as updated by Duellman (1993).

The recognition of some populations at specific or subspecific levels is questionable. Collins (1991

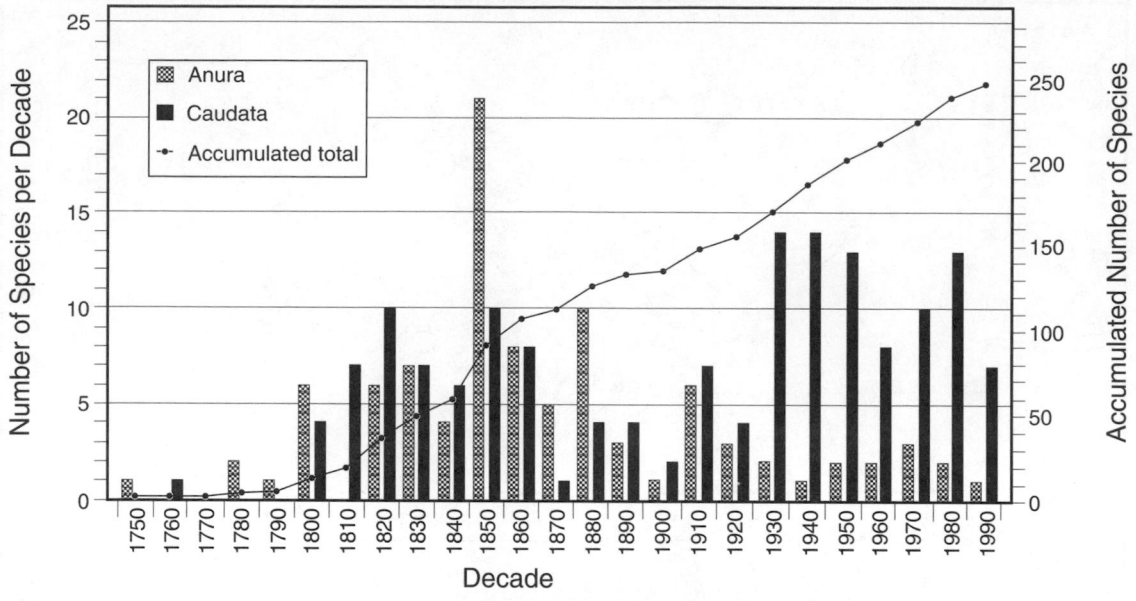

Fig. 2:2. Discovery rates of species of amphibians in North America.

[with a further attempt at justification, 1992]) proposed wholesale elevation of many subspecies of North American amphibians and reptiles to specific status; because these changes were suggested without justification, their elevation and his reasoning has been soundly criticized (Dowling, 1993; Frost et al., 1992; Montanucci, 1992; Van Devender et al., 1992). More recently, Collins (1997) based recognition of certain taxa at specific or subspecific levels on a consensus of specialists; in so doing, some subspecies proposed to be elevated to species by Collins (1991) were recognized as such, but others were retained at the subspecific level. Among subspecies that were elevated to the specific level by Collins (1991), molecular evidence (Highton, 1995, 1997) supports specific status for *Plethodon idahoensis* and *P. angusticlavius*. Collins's (1997) elevation of *Ambystoma tigrinum mavortium* to species status clearly is contradictory to evidence provided by Shaffer and McKnight (1996). We have been somewhat arbitrary in our recognition as species of other taxa generally placed at the subspecies level.

Using the distribution maps in Stebbins (1985) and Conant and Collins (1991), we scored the occurrence of species in physiographic and natural regions. Usually extremely peripheral occurrence was ignored because of the large scale of the maps. Areas of regions were obtained from the maps of the regions by use of a Micro-Plan II™ image analysis system (Laboratory Computer Systems, Inc., Cambridge, Massachusetts). Faunal similarities among regions were calculated by using the Coefficient of Biogeographic Resemblance (Duellman, 1990): $2C/(N_1 + N_2)$, where N_1 is the number of species in Region 1, N_2 is the number of species in Region 2, and C is the number of species in common to the two regions. Throughout, where grammatically appropriate, we abbreviate North America as NA and United States of America as USA.

THE AMPHIBIAN FAUNA

The discovery rate of amphibians in NA has progressed steadily, but distinct differences exist between the rates for anurans and salamanders (Fig. 2:2). By 1800, only four anurans (*Bufo marinus* [described from South America], *Bufo terrestris*, *Hyla cinerea*, *Rana pipiens*) and one salamander (*Siren lacertina*) had been named. By 1900, 77% of the currently recognized species of anurans, but only 41% of the salamanders, had been named. Nearly half of the species of salamanders have been recognized in the last half century. Many of these have been discovered by biochemical techniques (e.g., Highton et al., 1989; Good and Wake, 1992). We are aware of several studies now in progress that

Table 2:1. Taxonomic composition of the amphibian fauna of North America (excluding introduced taxa). Numbers in each column are total, endemics, and percent endemic.

Order	Families			Genera			Species		
Anura	8	1	12.5	15	3	20.0	90	73	81.1
Caudata	9	5	55.5	25	24	96.0	151	150	99.3
Total	17	6	35.3	40	27	67.5	241	217	90.0

will increase further the number of species of North American amphibians, especially salamanders.

The amphibian fauna of the Nearctic Region consists of 241 native species placed in 38 genera of 16 families. Salamanders make up 62.6% of the fauna (Table 2:1). The surface area of NA included in the Nearctic Region, exclusive of the islands in the Arctic Sea, is 19,717,189 km^2; the species density of amphibians is 12.23 (anurans 4.56; salamanders 7.56) per 10^6 km^2. North America is home to about 38% of the species of salamanders known worldwide, whereas only about 2.3% of the species of anurans known worldwide occur there. Four families are endemic to the Nearctic—Ascaphidae (1 genus, 1 species), Amphiumidae (1 genus, 3 species), Dicamptodontidae (1 genus, 4 species), and Rhyacotritonidae (1 genus, 4 species). Two other families are endemic to NA but range farther into Mexico; Sirenidae (2 genera, 4 species) has one species ranging into Mexico, and Ambystomatidae (1 genus, 35 species) has equal numbers of species in Mexico and the USA.

The only extra-Nearctic region with which native species are shared is Middle America; 17 species of anurans and one of salamanders also occur in non-Nearctic Mexico. Three West Indian species of anurans (*Eleutherodactylus coqui, E. planirostris,* and *Osteopilus septentrionalis*) are regarded as introduced species in NA. The only other extralimital anuran that has become established through introduction is the African pipid *Xenopus laevis.* The frog *Rana catesbeiana* and the salamander *Ambystoma tigrinum,* both native to NA east of the Rocky Moun-

tains, were introduced in many regions in western NA, where the species have spread. It is difficult to ascertain the original ranges of these species, so they are treated as native throughout their range. On the other hand, *Rana clamitans,* another eastern species, was introduced in northwestern USA and is treated as an introduced species there. Several amphibians (e.g., *Ambystoma tigrinum* and species of *Bufo* and *Rana*) have expanded their ranges in the southwestern USA in historic times as a consequence of large-scale water-transfer projects, such as in the Imperial Valley of California, or increased their ranges concomitant with the widespread creation of small impoundments in arid lowland regions.

Three globally widespread genera of anurans occur in NA. Of these, *Bufo* is represented worldwide by 212 species, of which 21 (9.9%) occur in NA. *Rana* contains 225 species, of which 26 (11.6%) occur in NA. In contrast, of the 186 species of *Hyla* worldwide, only 10 (5.3%) are found in the Nearctic Region. The most speciose family in NA is the Plethodontidae, represented by 15 genera (14 endemic) containing 106 species (all endemic); the genus *Plethodon* accounts for 43 of these species, followed by *Desmognathus* and *Eurycea* with 17 and 13 species, respectively. Other large genera of North American amphibians are *Ambystoma* (17 species), *Pseudacris* (11), and *Batrachoseps* (9). Of the 15 genera of anurans, the four most speciose genera (*Bufo, Hyla, Pseudacris,* and *Rana*) contain 72.3% of the species, whereas of the 25 genera of salamanders, the four most speciose genera (*Ambystoma, Desmognathus, Eurycea,* and *Plethodon*) contain 60.3% of the species.

NATURAL REGIONS OF NORTH AMERICA

In the broadest structural terms, NA consists of a great northeastern craton, the Laurentian Shield, flanked by eastern and western mountain systems that converge southward and enclose a broad mid-continental plain. The Appalachian system bordering the Atlantic and Gulf coasts resulted from ac-

cretion of exotic terranes in Ordovician time and a succession of orogenic episodes (Taconian, Acadian, and Appalachian) extending from the Devonian to the late Triassic, as the Iapetus Ocean closed in the formation of Pangaea. The western margin of NA incorporates a larger series of exotic terranes

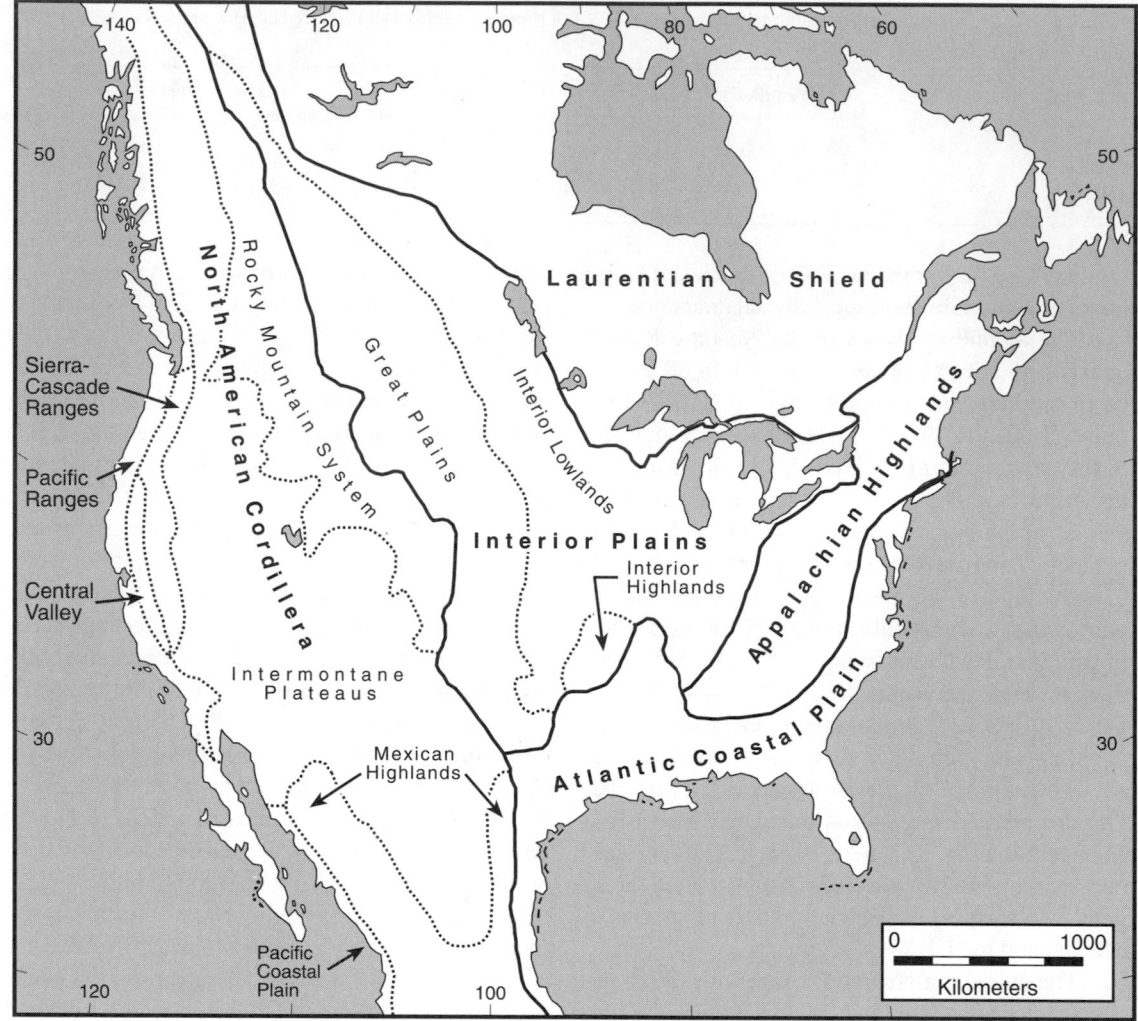

Fig. 2:3. Physiographic divisions (heavy lines) and pertinent provinces (dotted lines) of North America. Adapted from Lobeck (1948).

swept from subducting plates in the late Paleozoic, and a series of mountain systems (together termed the North American Cordillera) resulting from the subduction, compression, and translation of the region since middle Mesozoic times. Apart from marine transgressions (notably in the Cretaceous), the continent east of the Cordillera has undergone few structural changes since the middle Jurassic, whereas the Cordillera and western coastal region have experienced nearly continuous (and continuing) structural deformation. For example, since Miocene time, the peninsula of Baja California has been torn free of mainland Mexico and translated northwestward up to 300 km, with substantial effects on regional topography and climate.

Lobeck (1948) updated and refined Fenneman's (1917) classic divisions of the USA into physiographic regions and expanded the coverage to include all of NA. Lobeck (1948) recognized six major divisions, one of which, the Antillean Mountain System, is extralimital to the Nearctic. Each division was divided into provinces, which were subdivided into smaller units. For the purposes of this study, we utilize the five remaining divisions (Fig. 2:3).

However, analysis of distribution patterns based solely on these five physiographic divisions results in only general patterns and eliminates many important factors, such as temperature, precipitation, and dominant vegetation. In any region as large as NA, the geographic patterns of temperature and precipitation, which are influenced by wind currents and topography, have profound effects on the nature of the vegetation and concomitantly on the habitats available to amphibians and other organisms.

We have attempted only to examine broad patterns of temperature and precipitation and, in so doing, have excluded evapotranspiration. As expected, the latitudinal gradient in temperature is great, ranging from subtropical to Arctic conditions. Comparison of mean monthly temperatures in midwinter (January) and midsummer (July) reveals that the decline in temperatures is far more evenly graded latitudinally in the eastern part of the continent than in the west, where the much greater relief results in more complex patterns (Figs. 2:4, 2:5). The effect of the Pacific Ocean is appreciable along the west coast, where temperatures are warmer in the winter and cooler in the summer than at comparable latitudes on the east coast.

Annual precipitation ranges from more than 1500 mm annually in parts of the southeast to less than 100 mm annually in places in the southwest. East of the Rocky Mountains, the amount of precipitation gradually declines from the southeastern part of the coastal plain to the western part of the Great Plains. Likewise, there is a decline northwestward of about 40° N lat. (Fig. 2:6). The combination of prevailing winds and differences in temperature of ocean currents and the land (and its relief) results in far different patterns of precipitation in the western part of NA. Southern California and Baja California receive less than 500 mm of precipitation, nearly all of which falls in the winter months. In contrast, northward along the west coast, precipitation exceeds 2000 mm annually along the coast and lesser amounts inland at higher elevations. High amounts of precipitation grace the western (windward) slopes of the high mountains paralleling the west coast; the eastern slopes of the Sierra-Cascade ranges are much drier. These mountains create a major rain shadow to the east.

Our major dilemma has been the definition of regions for purposes of analysis of amphibian distributions. North America has been the focus of diverse attempts to define biogeographic regions (Kendeigh, 1954). The earliest was the life-zone concept of Merriam (1890, 1894). Although temperatures (isotherms) are the theoretical basis for the classification of zones, in practice, floristic components have been used. Life-zones are discontinuous areas

that have been used more satisfactorily in mountainous regions in western NA, where floristic and temperature breaks are sharp, than in eastern NA, where much more gradual changes in elevation result in less sharp climatic, floristic, or faunal changes. We reject the life-zone approach as being ineffective in assaying continental patterns of distribution.

A second method of faunal analysis is the biotic province concept proposed by Vestal (1914) and refined for NA by Dice (1943). A biotic province was defined by Dice (1943) as a considerable and continuous geographic area that is characterized by the occurrence of one or more ecological associations that differ, at least in proportional area covered, from the associations of adjacent provinces. Because this concept was so commonly misused, it has been rejected by most biogeographers (Peters, 1955).

A third approach is the ecological concept of the biome (Clements and Shelford, 1939). This concept was reviewed thoroughly by Kendeigh (1961), who defined a biome as a biotic community characterized by distinctiveness in life-forms of the important climax species; he also recognized bioassociations, subdivisions of a biome distinguished by uniformity and distinctiveness in the species composition of the climax community. Terrestrial biomes include categories such as coniferous forest, grassland, etc. In many respects, biomes and bioassociations are like life-zones and equally difficult to use in biogeographical analyses.

More recently, Bailey (1976, 1978) mapped ecoregions of the USA, and Crowley (1967) did essentially the same thing for Canada. Each of these regions has geographical continuity on the continent, and each is characterized by dominant types of vegetation. However, some of Bailey's ecoregions are too general to be meaningful at the level of analysis that we have attempted. For example, his Southern Deciduous Forest Region includes the Piedmont, the Atlantic Coastal Plain, the eastern part of the Gulf Coastal Plain, and peninsular Florida, whereas the Everglades in southern Florida is recognized as a separate region, Tropical Savanna. A cursory glance at distribution maps of amphibians in southeastern NA reveals that there are distinct patterns within the so-called Southern Deciduous Forest. Furthermore, the inclusion of the Piedmont and coastal plain in the same region combines parts of two distinct physiographic divisions.

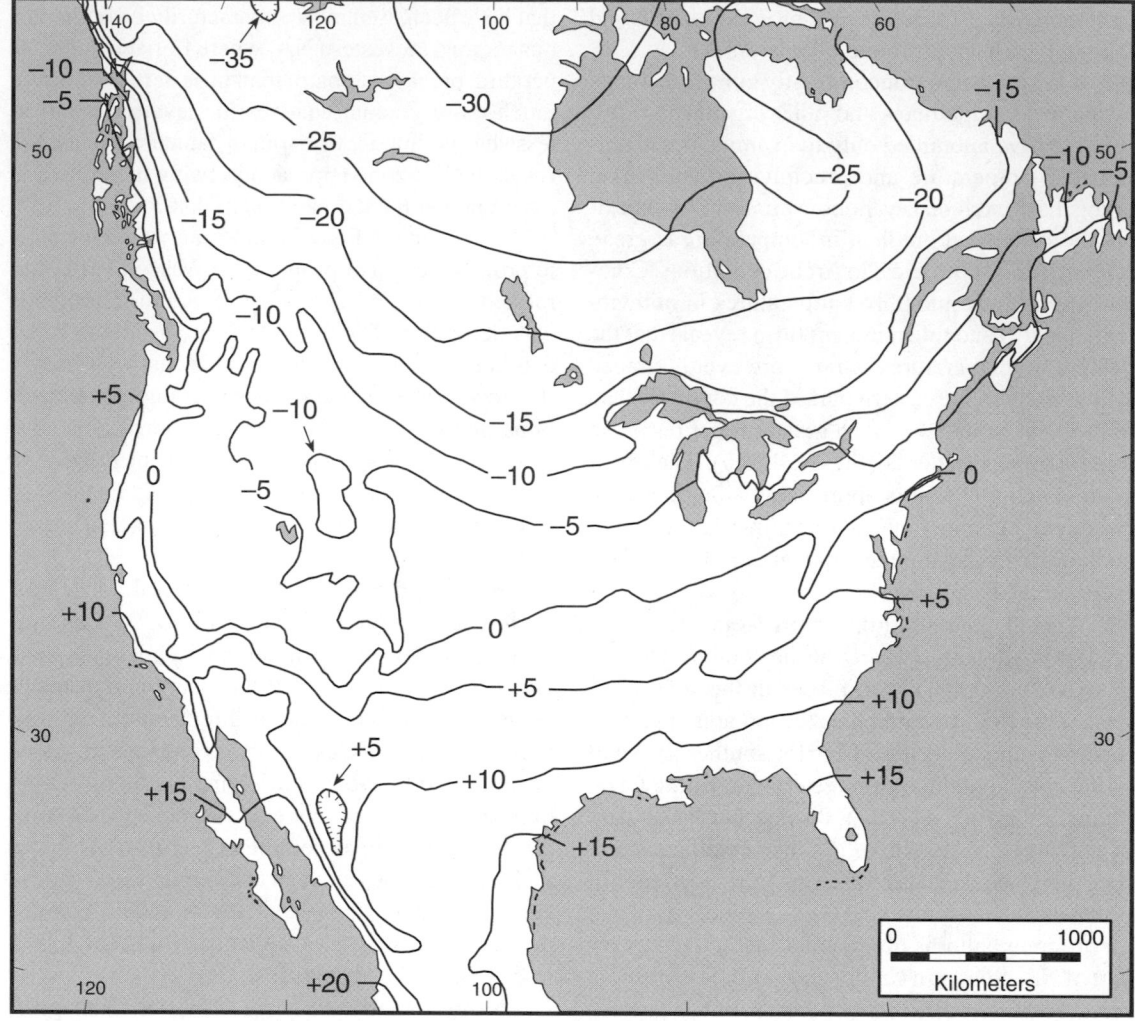

Fig. 2:4.   Mean isotherms (°C) for the month of January. Based on World Meteorological Organization (1979).

In our attempt to define regions having both physiographic and ecological bases, we have hybridized and/or partitioned many of Bailey's ecoregions. This resulted in the recognition of 45 regions, which we term Natural Regions (Figs. 2:7, 2:8). A Natural Region can be defined as being a continuous area within a physiographic division having climatic conditions that result in a characteristic type of vegetation. Thus, these regions can be viewed as geographic units that have certain kinds of animal habitats.

Most of these regions could be divided into ecological subregions for a finer scale of analysis than we have done here. For example, the Southern Coastal Plain includes such diverse habitats as hardwood forest, pine forest, and cypress swamps. Furthermore, similar ecological conditions prevail in widely separated regions (e.g., cool, moist fir forests in the Cascade Range in northwestern USA and the Southern Appalachians in southeastern NA). We group the Natural Regions in the appropriate Physiographic Divisions, some of which are divided into Physiographic Provinces. In many cases, especially in the lowlands, the boundaries of these regions are not distinct, because one region gradually merges with the other. Thus, the regions as mapped are generalized; the reader must keep in mind that transitional areas exist.

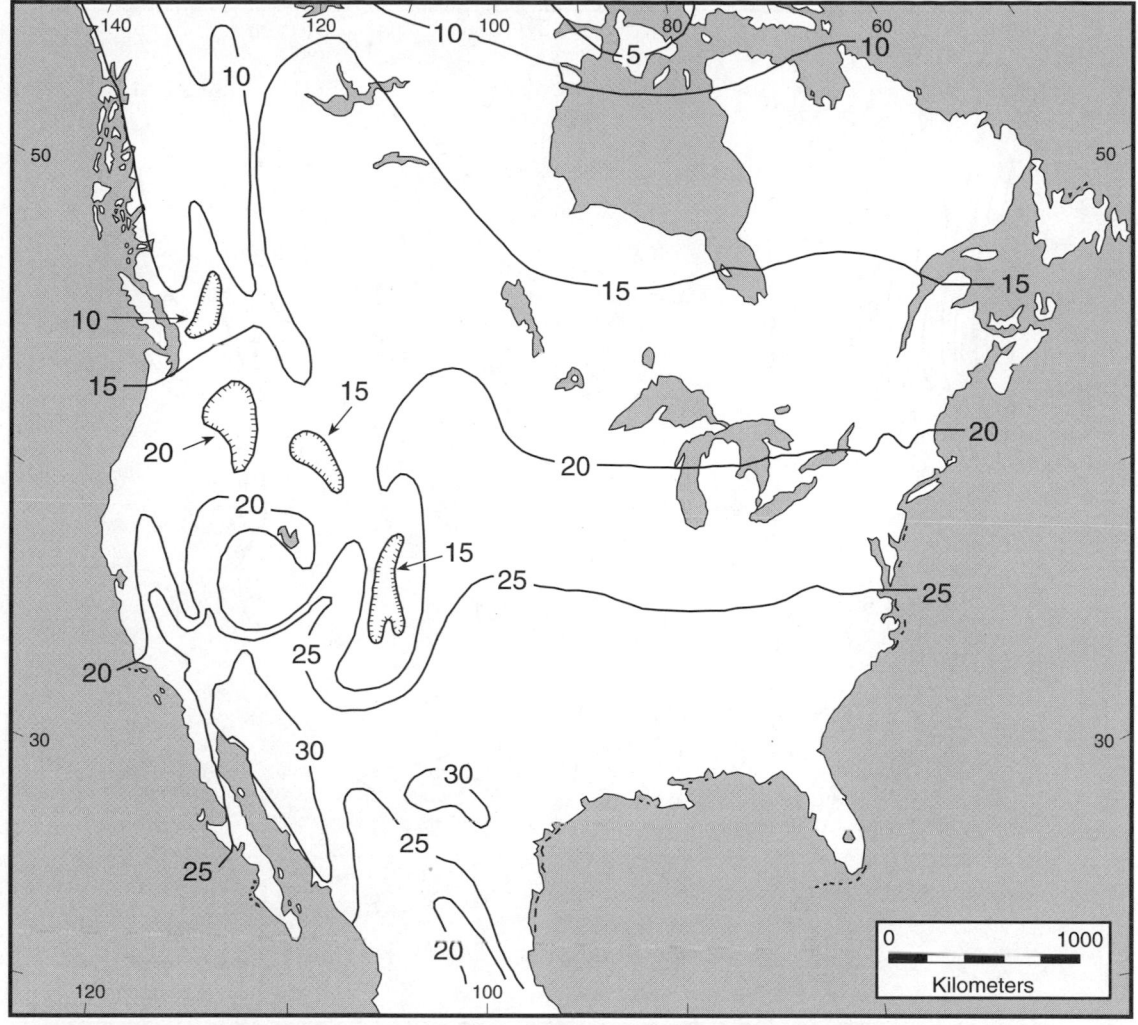

Fig. 2:5. Mean isotherms (°C) for the month of July. Based on World Meteorological Organization (1979).

NORTH AMERICAN CORDILLERA

The mountain systems and the intervening plateaus and basins in western NA were grouped into the North American Cordillera Division by Lobeck (1948). The origins and history of this complex system are imperfectly understood. Much of the western third of NA consists of a collage of six major exotic terranes and their interacting margins, which were accreted to the old continental margin throughout Mesozoic time (Dott and Prothero, 1994). These terranes were carried northward by a subduction zone initiated in the Devonian, and their locations reflect the complex and changing geometry generated as the North American Plate overrode plate fragments

along the eastern margin of the Pacific Plate; tectonic influences extended as far inland as the Rocky Mountains.

The major structural components of the west coast of NA were in place by the Cretaceous. Presumably, the landscape was reminiscent of the modern Andes Mountains in South America. Subsequent erosion exposed the granitic batholiths underlying the initial volcanic belt in a chain extending from northern British Columbia, Canada, southward through the Sierra Nevada to the peninsular ranges of northern Baja California, Mexico; each of these units was uplifted again in the late Tertiary and Quaternary. Along the present west coast, the central

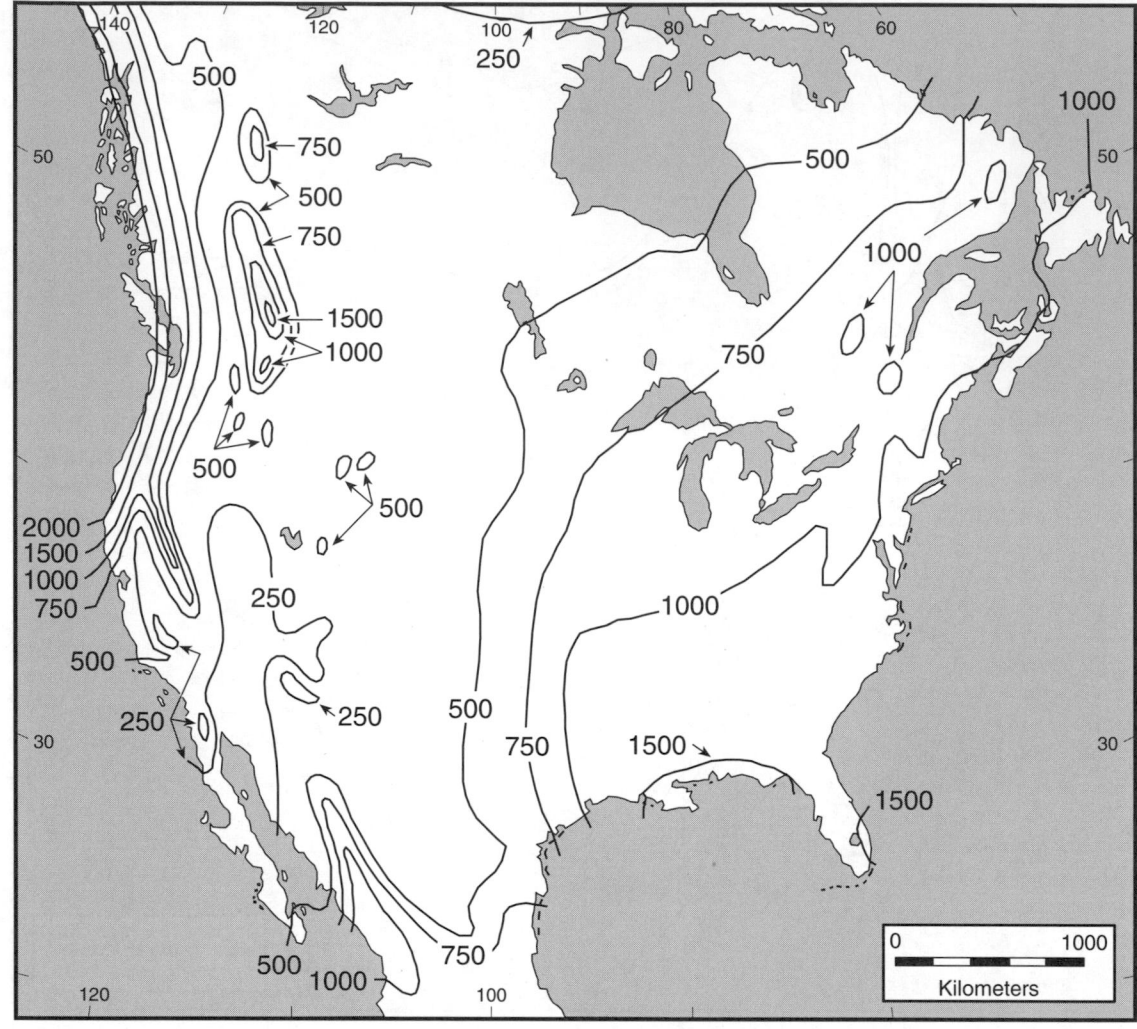

Fig. 2:6.   Mean annual precipitation (mm). Based on World Meteorological Organization (1979).

Pacific Ranges continue to grow as wedges, forming the linear Puget Sound Trough and Central Valley of California as deep basins paralleling the coast.

From the late Cretaceous to the early Eocene, the Sevier and Laramide orogenies developed east of the accretionary margin; they initially drove overthrust sheets of Paleozoic rocks eastward and then warped the underlying Archean basement into the long anticlines representing the modern Rocky Mountains. Beginning with the Miocene, contact with the Pacific Plate partially reoriented the subduction zone into a transform fault (San Andreas Fault), the subsequent northward motion of which has profoundly affected the region (Dickinson and Snyder, 1979; Severinghaus and Atwater, 1990). Anchored in the north, the Cascade Ranges and the

Sierra Nevada have been rotated westward, thereby opening the Basin and Range Province like a fan and producing its characteristic horst and graben topography. This rotation also initiated the uplift of the Colorado Plateau and perhaps the eruption of the basalts that make up much of the Columbia Plateau. This transform faulting also separated the peninsula of Baja California from the Mexican mainland. Within this physiographically complex division, we recognize three provinces.

Pacific–Cascade–Sierra Nevada Mountain System

This mountain system closely approximates the Pacific coast from southwestern Alaska to the southern tip of the Peninsula of Baja California; the complex system, with an area of about 1,689,000 km², has three major regions, which are discussed separately from west to east. The Pacific and Cascade-Sierra ranges arose from a Triassic volcanic arc erupting through previously accreted terranes (Dott and Prothero, 1994) to produce an Andes-like coastal cordillera. This complex was eroded down to its granitic batholith core during the Cretaceous and was separated from the coast by the emerging Pacific Ranges, which were composed of a prism of sea-floor deposits scraped from subducting plates. The Cascade–Sierra Nevada ranges were uplifted again in the Miocene and Pliocene, whereas the Pacific Ranges were uplifted primarily in the Pliocene. The intervening troughs are structural and have existed throughout the Cenozoic.

Pacific Ranges: This long, slender series of mountains includes the ranges closely bordering the Pacific Coast as well as offshore ranges that form the Alexander Archipelago in Alaska and the Queen Charlotte Islands and Vancouver Island in Canada. Tectonically, the mountainous peninsula of Baja California now is known to be associated with the southern part of the Pacific Ranges, but Lobeck (1948) considered that region to be a part of the Cascade–Sierra Nevada Ranges. These mountains drop precipitously to the coast throughout most of their length; thus the coastal region is included in the ranges. Herein, five natural regions of the Pacific Ranges are recognized:

1. Pacific Ranges in Alaska and northwestern Canada include the Aleutian Range, Alaskan Range, Canadian Coast Range, and the offshore islands southward to include Vancouver Island. The highest mountains in the Pacific Range are in this section; Mt. McKinley at 6194 m is the highest mountain in NA. Mean temperatures in July range from about 10°C in Alaska to 15°C on Vancouver Island, whereas mean temperatures in January range from about –15°C in Alaska to 0°C on Vancouver Island. Most of the coastal region and windward slopes of the mountains receive 1500–2000 mm of precipitation annually. This cool, moist region supports a coniferous forest dominated by hemlock and Sitka spruce to elevations of no more than 600 m in Alaska and 1000 m in southern Canada; sequentially above these elevations are alpine meadows, barren rock, and permanent snow.

2. Pacific Ranges (North) includes the coastal ranges north of the latitude of the Klamath Mountains in Oregon and Washington. This region includes the mountains in the Olympic Peninsula; the highest mountain in the range is Mt. Olympus (2824 m). Climate is moderated by the Pacific Ocean, so annual temperature variation is not extreme, experiencing means of about 15°C in July and 5°C in January. At low elevations on the windward slopes, annual precipitation exceeds 2000 mm, whereas precipitation diminishes to no more than 1500 mm annually at higher elevations. The dominant vegetation is coniferous forest.

3. Pacific Ranges (Central) includes the coastal ranges in California north of the San Francisco Bay area. Most mountains in this part of the range are less than 2000 m in height. Climate is moderated by the Pacific Ocean; annual temperature variation is slight—means of about 15°C in July and 5°C in January. Annual precipitation declines from nearly 2000 mm in the north to less than 750 mm in the south. The vegetation in the northern part is much the same as that in the Pacific Ranges (North), but to the south, Douglas fir (*Pseudotsuga menziesii*) and redwood (*Sequoia sempervirens*) are dominant.

4. Pacific Ranges (South) includes the Coast and Diablo ranges south of San Francisco Bay area, the Transverse Ranges, the Peninsular Ranges in California, and the Sierra Juárez and Sierra San Pedro Mártir in northern Baja California del Norte, Mexico. Most mountains in the northern part of the range are less than 1500 m, but Junipero Serra Peak is 1787 m; elevations increase to the south (Mt. Pinos, 2692 m), especially in the San Gabriel, San Bernardino, and San Jacinto ranges, where several peaks exceed 3000 m. Mean temperatures in July vary from about 15°C in the north to about 22°C in the south, and mean temperatures in January vary from about 8°C in the north to 12°C in the south. Rainfall declines from about 700 mm annually in the north to less than 300 mm annually in the south, with an increasing seasonality (dry summers and winter rains) in the south. The northern part of this southern range has some conifers (*Pinus* and *Pseudotsuga*) and oaks (*Quercus*), but most of the southern range supports a dense sclerophyllous vegetation, chaparral, dominated by buckthorns (*Ceanothus*), sages (*Artemesia*), and evergreen oaks (*Quercus*). However, at higher elevations in the Sierra San Pedro Mártir, coniferous forest with pines (*Pinus*) and cedar (*Libocedrus*) exists (Welsh, 1988). Such coniferous forest also occurs at high elevations in the southernmost Coast Range and in the San Gabriel and San Jacinto mountains.

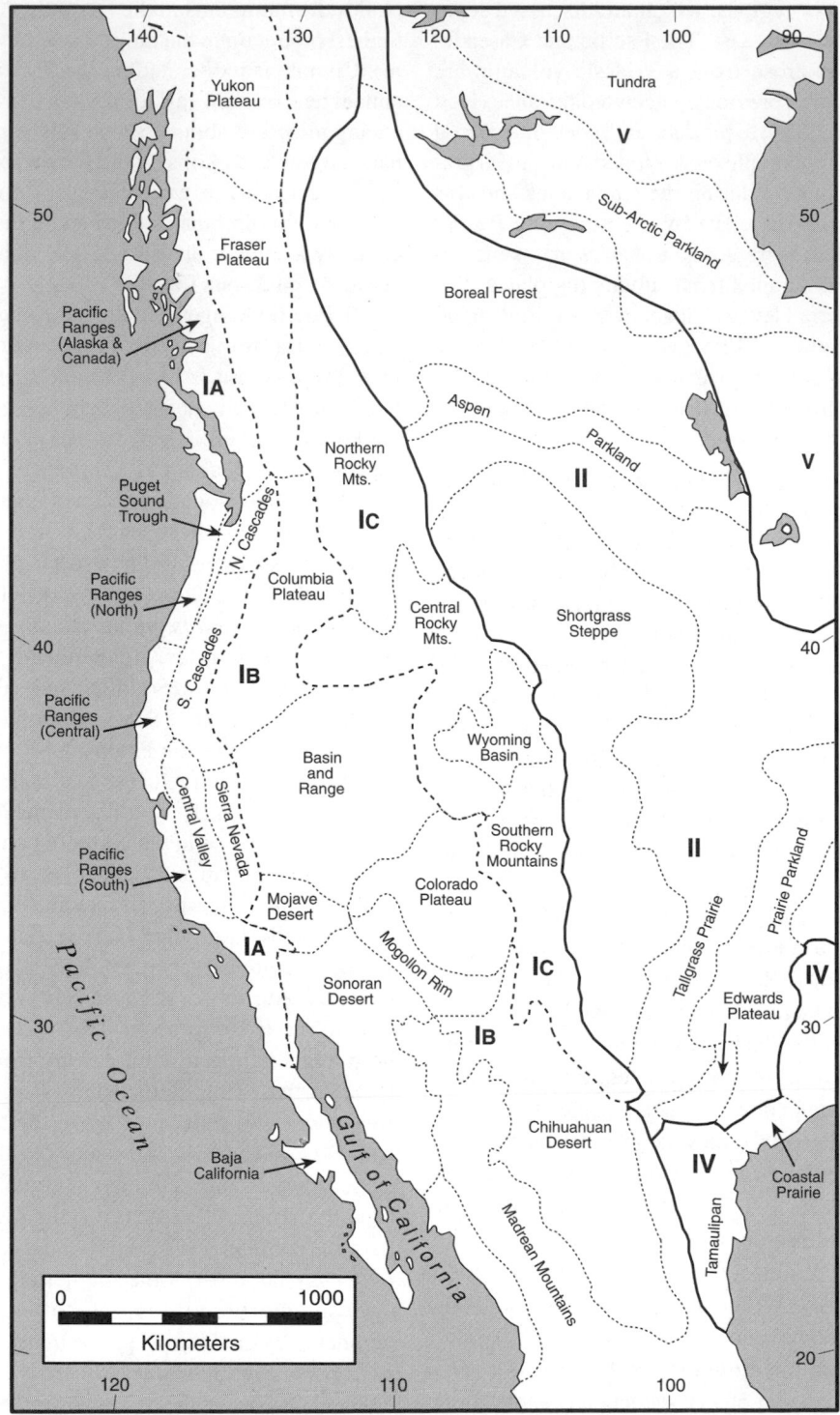

Fig. 2:7. Natural regions of western North America. Borders of regions are shown by dotted lines; borders of physiographic divisions are shown by heavy, solid lines, and borders of physiographic provinces within the North American Cordillera are shown by dashed lines. Physiographic divisions and provinces are identified as follows: IA = Pacific–Cascade–Sierra Nevada Mountain System; IB = Intermontane Plateaus; IC = Rocky Mountain System; II = Interior Plains; IV = Atlantic Coastal Plain; V = Laurentian Upland.

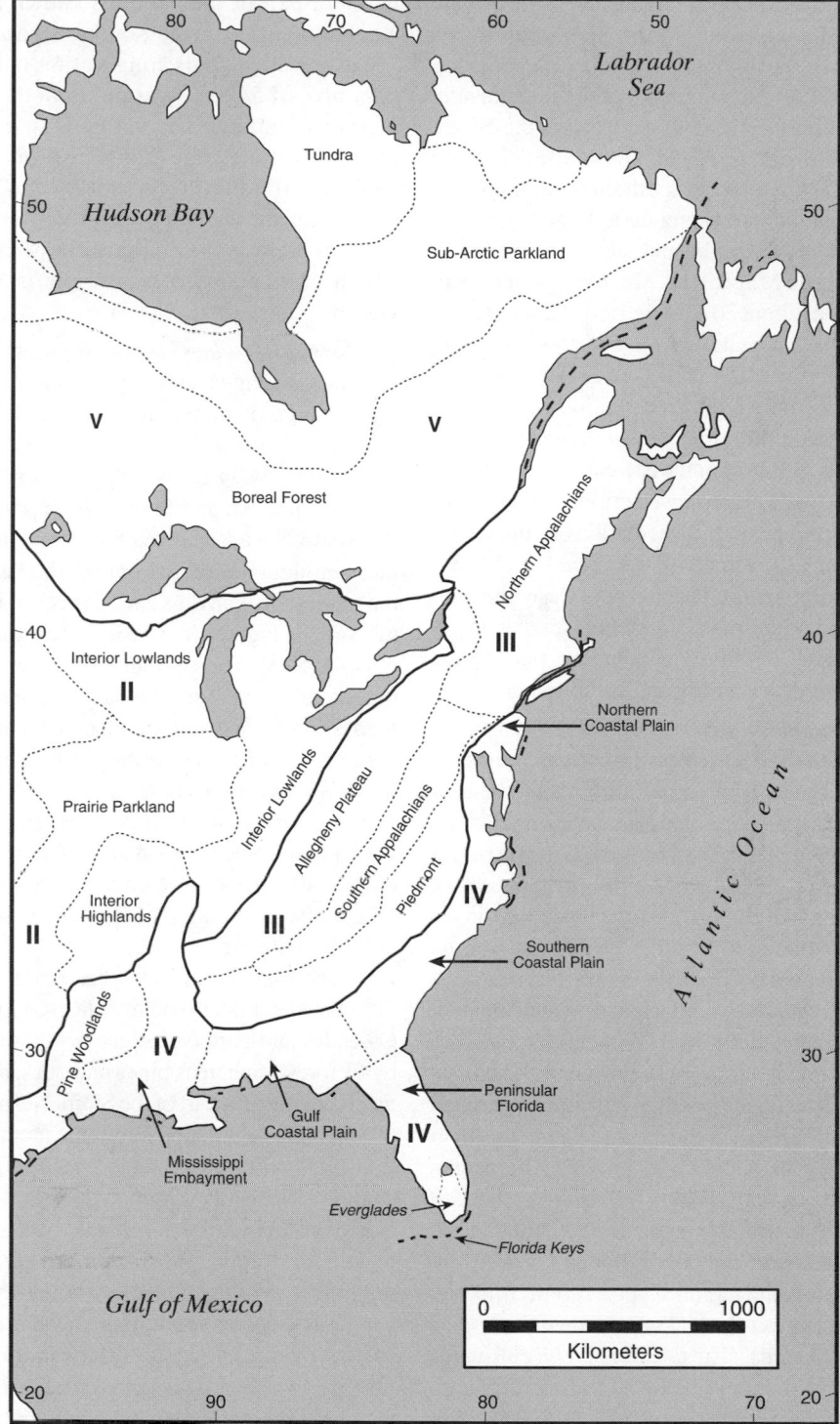

Fig. 2:8.   Natural regions of eastern North America. Borders of regions are shown by dotted lines, and borders of physiographic divisions are shown by heavy, solid lines. Physiographic regions are identified as follows: II = Interior Plains; III = Appalachian Highlands; IV = Atlantic Coastal Plain; and V = Laurentian Upland. The Everglades and Florida Keys are part of the Peninsular Florida Region.

5. Baja California includes the Peninsula of Baja California, Mexico, south of the Sierra San Pedro Mártir. Most of the mountains in Baja California are less than 1500 m. Physiographically, the peninsula consists of a plain in the north (west of the Sierra San Pedro Mártir); south of this is the Vizcaino Desert, south of which the southern mountains (Sierra de la Giganta and Sierra de la Laguna) extend to the tip of the peninsula, and are bordered on the west by Llano de Magdalena. Mean temperatures in July range from about 20°C in the northwest to nearly 30°C in the south, whereas mean temperatures in January range from about 10°C in the northwest to about 20°C in the south. Precipitation is less than 250 mm annually throughout most of the peninsula, but it exceeds 500 mm in the extreme southern part of the peninsula. The northwestern part of the peninsula supports a chaparral vegetation dominated by sages (*Artemesia*), scrub oak (*Quercus*), and Manzanita (*Arctostaphylos*). The Vizcaino Desert is dominated by a peculiar, mostly endemic desertic vegetation (Wiggins, 1960). The southern part of the peninsula supports arid tropical scrub forest.

Intermontane Troughs: The Pacific Ranges are connected with the Sierra Nevada–Cascade ranges in southern Oregon and northern California by the Klamath Mountains with elevations in excess of 2000 m. The southern end of the Sierra Nevada–Cascade ranges and the Transverse Ranges of the Pacific Ranges have a moderately high (1200–1600 m) connection via the Tehachapi Mountains. In other places, the Pacific Ranges are separated from the Sierra Nevada–Cascade ranges by lowlands, principal of which are:

1. The Puget Sound (or Tacoma) Trough lies between the Pacific Ranges (North) and the Northern Cascade Range; it extends southward from Puget Sound and terminates in the upper Willamette River Valley north of the Klamath Mountains. This region has temperatures only slightly warmer than those of the bordering mountains and receives 1000–1500 mm of precipitation annually. Although now extensively cultivated, the region supports coniferous forest dominated by hemlock (*Tsuga*), grand fir (*Abies*), and spruce (*Picea*). These extensively cultivated lowlands are in the rain shadow of the Pacific Ranges and consequently receive considerably less rainfall than do the Pacific slopes of the Pacific Ranges.

2. The Central Valley of California extends for about 700 km from the Klamath Mountains in the north to the Tehachapi Mountains in the south; the northern part of the valley is known as the Sacramento Valley, and the southern part, as the San Joaquin Valley. This long valley has mean temperatures of 5–10°C warmer than those of the adjacent mountains. The valley is in a rain shadow of the Pacific Ranges; annual rainfall usually is less than 500 mm in the north to as little as 250 mm in the south. The Central Valley was once an extensive, seasonal marshland bordered by bunch-grass plains; it is now almost entirely converted to agricultural uses.

Sierra Nevada–Cascade Ranges: Lying east of the Pacific Ranges and the intermontane troughs, a major range of mountains extends from the Canada-USA border to southern California. The northern part of this igneous range is known as the Cascade Mountains, and the southern part is known as the Sierra Nevada. Throughout most of their length, these mountains are higher than the Pacific Ranges; the highest peak in the Cascades is Mt. Rainier (4392 m), and the highest in the Sierra Nevada is Mt. Whitney (4418 m). Throughout their entire length, these mountains have mean January temperatures of less than 0°C, whereas in July the mean temperatures are about 15°C in the north to more than 20°C in the south. The western slopes of the Cascade Range and northern part of the Sierra Nevada receive up to about 1500 mm of precipitation annually, but in the southern part of the Sierra Nevada, precipitation declines to less than 500 mm annually. The eastern slopes of the mountains are much drier than the western slopes, for these high mountains cast a rain shadow over the basins and intermontane plateaus to the east. The Cascades and northern Sierra Nevada are dominated by fir forests, whereas pines and oaks are characteristic of the southern part of the Sierra Nevada. This chain of mountains is divided into three natural regions:

1. Northern Cascades extend from extreme southern British Columbia, Canada, to the Columbia River. Many of the volcanic peaks exceed 2000 m, and Mt. Rainier (4392 m) is the highest. Many of the higher peaks retain snow cover throughout the year, but at lower elevations, mean temperatures in July are about 15°C and in January about 0°C. The western (windward) slopes of the northern Cascades receive 1000–1500 mm of precipitation annually, but the eastern (leeward) slopes receive only 750–1000 mm annually. Coniferous forest is characteristic of the moist western slopes; the drier eastern slopes support pine-oak and juniper associations.

2. Southern Cascades extend from the Columbia River southward to Mt. Lassen. These mountains have many volcanic peaks in excess of 2000 m; the highest is Mt. Shasta (4317 m). Temperatures are slightly warmer than those in the Northern Cascades, and precipitation, especially on the eastern slopes, is slightly less. On the western slopes, coniferous forest is prevalent. The drier eastern slopes are like those of the Northern Cascades.

3. Sierra Nevada extends from the vicinity of Mt. Lassen in northern California southward to the Tehachapi Mountains in southern California. Many peaks are over 3500 m; Mt. Whitney (4276 m) is the highest. Mean temperatures in July approximate 20°C, but fall below 0°C in the winter. In the north, annual precipitation is 750–1000 mm on the western slopes and 500–750 mm on the eastern slopes, but in the south, only 500–700 mm of precipitation falls on the western slopes. Coniferous forests dominate the western slopes in the north, whereas these are mixed with oaks and manzanita (*Arctostaphylos*) in the south; pine-oak and juniper associations dominate the eastern slopes.

Intermontane Plateaus

Separating the Rocky Mountain System and the Cascade–Sierra Nevada Mountain System is an extensive series of basins and plateaus with a total surface area of about 2,732,000 km². This category also includes the Central Alaskan Uplands and Plains that descend to sea level. Otherwise, most of the plateaus are at moderate elevations (1500–2000 m), but in some places in the Colorado Plateau and Mogollon Rim elevations exceed 3000 m. Throughout the intermontane region, except in Alaska, there are many small north-south mountain ranges, some of which have elevations of more than 3500 m. The southern part of the intermontane region is the Sonoran Desert lying west of the Sierra Madre Occidental and the Chihuahuan Desert between Sierra Madre Occidental and the Sierra Madre Oriental in Mexico; the Chihuahuan Desert is continuous with the Mexican Plateau. The northern part (Fraser and Columbia plateaus) is made up of Cenozoic lava flows, whereas southward (e.g., Basin and Range Province) faulting that began in the Miocene reached its climax in the Pliocene, and probably is ongoing. On the other hand, the extensive Colorado Plateau, made up of Mesozoic rocks and overlain nearly completely by Eocene deposits, was uplifted in the Miocene; the cutting of deep canyons, such as the Grand Canyon,

occurred in the Pliocene and Pleistocene. Glaciation was present only on some of the higher ranges during the Pleistocene. Ten natural regions are recognized within the Intermontane Plateaus; in an approximate north-to-south order, these are:

Yukon Plateau: This northern plateau is mostly at elevations of 350–700 m and drains into the Pacific Ocean by the Yukon River. Numerous northwest-southeast mountain ranges rise above the plateau to elevations of 2024 m in the St. Cyr Range and 2027 m in the Dawson Range. Average temperatures in July are about 10°C and in January about −20°C; annual precipitation is about 500 mm. The plateau supports coniferous forest with spruce (*Picea*) and pine (*Pinus*), and deciduous trees, mainly aspen (*Populus tremuloides*) and paper birch (*Betula glandulosa*).

Fraser Plateau: This narrow plateau between the northern Rocky Mountains and the Pacific Ranges is mostly at elevations of 200–400 m and is drained principally by the Fraser River. Average temperatures in July are about 10°C and in January about −15°C; annual precipitation is about 500 mm. The plateau mainly supports coniferous forest, with Douglas fir (*Pseudotsuga menziesii*) dominant in the north and ponderosa pine (*Pinus ponderosa*) dominant in the south.

Columbia Plateau: This broad basin at elevations of 150–650 m is drained by the Columbia River and its major tributary, the Snake River. Some small mountain ranges and basaltic buttes are present, but the major upland area is the Blue Mountains in the southeastern part of the Plateau; these mountains, which attain elevations of 2900 m, are nearly confluent with the Rocky Mountains. Temperatures on the plateau are extreme—means of 20°C in July and −5°C in January; precipitation is less than 250 mm annually. Temperatures in the Blue Mountains are lower, and precipitation there exceeds 500 mm annually. Vegetation is principally semidesert scrub.

Basin and Range: This large inland basin is mostly at elevations of 500–1000 m, but in places ascends to nearly 2000 m. Many lake basins are present; some contain fresh water, and others, such as Great Salt Lake, are saline. The basin is dominated by laval plains, basaltic buttes, and horst-and-graben topography with north-south ranges of mountains. Some of the dominant ones are the Steens

Mountains (2967 m), Shoshone Mountains (3143 m), Ruby Mountains (3471 m), Monitor Range (3309 m), and Schell Creek Range (3622 m). Throughout most of the basin, mean temperatures in July are 20–25°C and drop to 0–5°C in January. Rainfall, except in the mountain ranges, is about 200–250 mm annually. The vegetation in the basin consists of more or less spinescent, microphyll shrubs, principally *Atriplex* and sagebrush (*Artemesia*), whereas greasewood (*Sarcobatus*) is characteristic of saline soils. Lower montane slopes characteristically support pinyon pine and juniper (*Juniperus*) that grade into ponderosa pine and then into aspen (*Populus tremuloides*) with fir (*Abies* and *Pseudotsuga*) on higher slopes, and spruce (*Picea*) at the highest elevations.

Colorado Plateau: This extensive plateau is mostly at elevations of 1500–2300 m, but the western part (Kaibab Plateau) is over 2500 m. The plateau extends from the Rocky Mountains westward to the Mojave Desert and is bordered on the south by the Mogollon Rim. The region is characterized by many buttes, mesas, and deep canyons (e.g., the Grand Canyon). Mean monthly temperatures in July are 20–25°C, and in January –5–0°C; precipitation varies from about 250 to 500 mm annually. Vegetation varies from shrublands dominated by sagebrush (*Artemesia*) with greasewood (*Sarcobatus*) and saltbush (*Atriplex*) on saline soils, through juniper and pinyon pine to ponderosa pine (*Pinus ponderosa*) at higher elevations.

Mogollon Rim: This east-west mountain range reaches elevations of more than 3000 m on several peaks—Whitewater Baldy (3321 m) and Baldy Peak (3476 m) in the east and Humphreys Peak (3851 m) near the western end of the system. The latter, one of the San Francisco Peaks, was the high-elevation part of the transect studied by C. Hart Merriam and his associates in 1889; that study formed the basis for the life-zone system (Merriam, 1890). Over most of the rim, mean temperatures in July are no more than 20°C, whereas they dip to –5°C or lower in January; the higher peaks retain snow throughout most of the year. Precipitation is mostly about 500 mm annually. Principal vegetation consists of pine and spruce at lower elevations and grading into firs (*Abies* and *Pseudotsuga*) on higher slopes, and spruce (*Picea* or *Abies*) above 3000 m.

Madrean Mountains: This category includes the northern part of the Sierra Madre Occidental in Mexico and the northern ranges (Chiricahua, Huachuca, Peloncillo, and Santa Rita mountains) of this mountain system in southern Arizona, southwestern New Mexico, and adjacent Mexico. The northern part of the Sierra Madre Occidental, separating the Sonoran Desert on the west from the Chihuahuan Desert to the east, mostly is at elevations less than 2500 m, but there are peaks in excess of 2800 m in the Chiricahua, Huachuca, and Santa Rita mountains. Mean monthly temperatures in July are about 25°C and in January about 5–10°C, except that cooler temperatures prevail at the higher elevations. The lower slopes of these mountains support a juniper-pine association, which grades into pine-oak woodland at higher elevations and ponderosa pine (*Pinus ponderosa*) at still higher elevations.

Mojave Desert: This relatively small region consists of series of closed basins separated by small ranges of mountains. Elevations range from 3368 m in the Panamint Range to –86 m in Death Valley. Mean monthly temperatures in the basins are 25–30°C in July to 10–20°C in January; precipitation is extremely low (80–150 mm annually) and highly variable. Vegetation consists of xerophyllous plants.

Sonoran Desert: This large desert extending from the Gulf of Mexico west of the Sierra Madre Oriental into southern Arizona and southeastern California ranges in elevation from sea level to about 1500 m. Mean monthly temperature in July commonly exceeds 30°C, whereas in January the mean temperatures are 10–20°C. Precipitation does not exceed 250 mm annually. According to Lowe (1964), two major vegetation associations exist. The paloverde-sahuaro association is characteristic on rocky slopes, whereas the creosote bush–bur sage association is characteristic of less rocky soils in areas of low relief. On alkaline soils, saltbush (*Atriplex*) is dominant; mesquite (*Prosopis*) grows along rivers.

Chihuahuan Desert: This large desert plateau extends between the Sierra Madre Occidental and the Sierra Madre Oriental in Mexico, where it is continuous with the Mexican Plateau at higher elevations to the south of 22° N lat. The northern terminus is in the Rio Grande drainage in New Mexico. Most of the desert is at elevations of 1000–1500 m, but some basins are 500 m lower. Numerous small mountain ranges rise to higher elevations—up to 2500 m. Mean monthly temperatures are 25–30°C for July and 10–15°C for January. Mean annual precipitation is 75–300 mm. Morafka (1977) recognized three major vegetation formations: (1) sclerophyllous scrub forest characterized by creosote bush (*Larrea*) and tar bush (*Flourensia*); (2) saxicolous scrub dominated by succulents (*Agave* and *Euphorbia*); and (3) desert grassland.

Rocky Mountain System

This massive mountain system, with an area of about 2,650,000 km², forms the continental divide, with the Great Plains on the east and the Intermontane Plateaus to the west. This major mountain range was first uplifted in the late Devonian, but the major uplift (Sevier Orogeny) occurred in the early Cretaceous. The Laramide Orogeny in the late Cretaceous–mid-Eocene resulted in uplift by buckling of the Archean basement, mostly to the east of the Sevier Orogeny. The Laramide basins were filled by the late Eocene; the Rocky Mountains were uplifted again in the late Miocene–Pliocene. Many peaks in these rugged mountains exceed 4000 m in elevation. Montane glaciers were present throughout most of the system in the Pleistocene. Many of the higher peaks in the northern Rocky Mountains have permanent snow, and snow is present throughout most of the year on higher peaks in the Southern Rocky Mountains. Four natural areas are recognized in the Rocky Mountains.

Northern Rocky Mountains: This unit includes the Brooks Range in Alaska, the ranges in Canada and the mountains in Idaho and Montana north of the headwaters of the Snake and Missouri rivers. In Alaska and northern Canada, only a few peaks exceed 2000 m in height, whereas in southern Canada and extreme northern USA, the mountains are much higher with peaks as high as 3950 m. The Northern Rocky Mountains have been sculpted by montane glaciers, and glaciers persist on many of the higher mountains in the south, but not on the lower mountains to the north. In the Rocky Mountains of southern Canada and extreme northern USA, mean temperatures on the low and moderate slopes of the mountains in July are about 15°C, but in January the temperature is –10°C to –20°C. Annual precipitation ranges from about 500 mm on the lower slopes to more than 1500 mm on some of the high ridges. The northernmost mountains, the Brooks Range in Alaska, are treeless, whereas the Rocky Mountains in southern Canada and extreme northern USA support coniferous forest, with fir

(*Abies*) and spruce (*Picea*) on high and humid slopes.

Central Rocky Mountains: This part of the Rocky Mountain System forms a broad C around the Wyoming Basin and includes the Bighorn Mountains to the northeast and the Wasatch and Uinta mountains to the west. Many peaks are in excess of 4000 m, and most show evidence of glaciation. Mean annual temperature in July is 15–20°C and in January –5°C to –10°C, with colder temperatures prevailing at higher elevations. Annual precipitation exceeds 500 mm only on the highest peaks. Throughout most of this region, lodgepole pine (*Pinus contorta*) is dominant; at lower elevations, aspen (*Populus tremuloides*) is characteristic, whereas on dry lower slopes, junipers are dominant.

Wyoming Basin: Nearly surrounded by the Central Rocky Mountains, the Wyoming Basin lies at elevations of 2300–2500 m. The continental divide passes through the basin; the northeastern part of the basin is drained by the Sweetwater River, which eventually flows into the Mississippi River and the Gulf of Mexico, whereas the southwestern part of the basin is drained by the Green River, a tributary of the Colorado River, which flows into the Gulf of Mexico. The midcontinental climate is much the same as the surrounding mountains, except that it is slightly warmer and drier. Vegetation consists principally of grassland with scattered junipers.

Southern Rocky Mountains: This region extends from the northern end of the Sangre de Cristo Mountains in Colorado southward to include the Sacramento Mountains, and Guadalupe Mountains to terminate in the Chisos Mountains in the Big Bend Region of Texas, where the Southern Rocky Mountains are replaced south of the narrow Rio Grande Valley by the Sierra Madre Oriental of Mexico. The mountains are much higher (more than 4000 m) in the north than in the south, where Emory Peak (2285 m) is the highest. The northern ranges (e.g., Sangre de Cristo Range) were glaciated in the Pleistocene; glaciers were present as far south as the Sierra Blanca in New Mexico. Mean monthly temperatures in July are 15–25°C, and in January, 0–10°C; precipitation is less than 500 mm annually. Below the alpine meadows, coniferous forest is dominated by spruce (*Picea*) and fir (*Abies*) to about 3000 m, below which firs and pines are mixed with aspen (*Populus tremuloides*). At about 2100 m, the coniferous forest gives way to pinyon pine (*Pinus edulis*) and junipers. The pinyon-juniper association occurs at elevations above 1450 m in the Chisos Mountains, where it is mixed with oaks (*Quercus*).

INTERIOR PLAINS

This vast interior lowland region extends from the Laurentian Upland and Appalachian Plateau on the east to the Rocky Mountains on the west, and from the Arctic Sea nearly to the Gulf of Mexico, where it meets the Atlantic Coastal Plain. Three physiographic provinces are recognized.

Great Plains

This long region, with a surface area of about 1,489,000 km², extends from about 55° N lat. southward to the Atlantic Coastal Plain. It rises gradually from about 300 m in the east, where it interdigitates with the Interior Lowlands, to an average of about 1600 m in the west, where it meets the Rocky Mountains. Underlain mostly by Cretaceous limestones (deposited in the Midcontinental Sea), Tertiary deposits (thickest in the west) provide little relief, except for an eroded mountain dome, the Black Hills. The southern end of the Great Plains is the Edwards Plateau, which, unlike the rest of the Great Plains, lacks Tertiary deposits overlying the Cretaceous limestone. The northern three fourths of the plains were glaciated in the Pleistocene. The central and southern Great Plains are crossed by several major rivers (e.g., Missouri, Arkansas) that are tributaries of the Mississippi River, which flows into the Gulf of Mexico. The northern part of the plains is drained by the Peace River, a tributary of the Mackenzie River, which flows into the Arctic Ocean. A strong latitudinal gradient in temperature results in the mean temperature in January of –25°C in the north to 15°C in the south, whereas summer temperature differences are less—15°C in the north to more than 25°C in the south. Generally, there is a decrease in precipitation from about 750 mm in the east to about 300 mm in the west. Five natural regions are recognized within the Great Plains.

Shortgrass Steppe: Also called the Shortgrass Prairie, this region is one of high plains on the alluvium from the Rocky Mountains. The region is characterized by low rainfall, high winds, and extremes in temperature. It is bordered entirely on the east and narrowly on the north by the more humid Tallgrass Prairie. The Shortgrass Steppe is dominated by grasses, but sagebrush (*Artemesia*) also is common. In the southern half of the region, a yucca (*Yucca glauca*) and low cacti of the genus *Opuntia* are present.

Tallgrass Prairie: Lower in elevation than the Shortgrass Steppe to the west, the Tallgrass Prairie has less extreme temperatures and higher precipitation. It is bordered on the east by the Prairie Parkland and on the north by the Aspen Parkland. Vegetation consists of a variety of tall grasses.

Prairie Parkland: This is a broad ecotone between the eastern deciduous hardwood forest in the Interior Lowlands and the Tallgrass Prairie. Even though most of the area is cultivated, there remain stands of forest mixed with grasses and sedges of the Tallgrass Prairie (especially species of *Andropogon, Carex,* and *Panicum*); in some places, called glades, pure grassland exists. The forest is deciduous and consists almost entirely of trees characteristic of the eastern deciduous hardwood forest.

Aspen Parkland: This is a narrow transition zone between the Tallgrass Prairie and the Boreal Forest. Mean temperatures in July are about 15–20°C and in January –15°C to –20°C; annual precipitation is less than 500 mm. The dominant tree is the quaking aspen (*Populus tremuloides*), and the ground cover is mainly sedges of the genus *Carex;* moist areas support willows (*Salix interior*).

Edwards Plateau: This biogeographically distinct plateau lies between the southern end of the Tallgrass Prairie, the southwestern border of the Prairie Parkland, and the Atlantic Coastal Plain. It is characterized by subsurface drainage into deeply incised canyons, with an abundance of caves and springs. The Edwards Plateau lies at elevations of 600–750 m and terminates at the Balcones Escarpment, which drops to the lowlands at elevations of about 160 m. Rather open forest consists of oak, juniper, and pinyon pine, with an understory of grasses.

Interior Lowlands

This large region, with a surface area of about 2,646,000 km², embraces the Great Lakes (except Lake Superior) and forms a broad transition zone between the Appalachian Highlands on the east and the Great Plains on the west. At its northern extreme, it includes the Arctic Plains of Alaska and the Mackenzie Lowlands of northwestern Canada; eastward, it borders the Laurentian Shield southward to the Great Lakes. Topographically, the region consists of slightly rolling hills to extensive flat areas mostly at elevations of 200–400 m. The bedrocks are mostly

Paleozoic limestones and shales. Nearly all of the region was glaciated in the Pleistocene. Latitudinal differences in temperature are extreme, especially in the winter. The mean temperature in July is about 10°C in the north to more than 25°C in the south, whereas the mean temperature in January is less than −25°C in the north to about 10°C in the south. Throughout the southern part of the Interior Lowlands, annual precipitation declines from about 1000 mm in the east to about 500 mm in the west. In the north, the decline is from 750 mm to less than 500 mm. From north to south, the vegetation changes from coniferous forest dominated by fir and spruce to pine and hemlock in the eastern part of the region. South of the coniferous forest is deciduous hardwood forest—beech and maple in the east with basswood mixed with these in the west. South of the beech-maple-basswood forests are oak-hickory forests.

Interior Highlands

This small (119,000 km^2), uplifted region in south-central USA is surrounded by lowlands. The northern part of the region, the Ozark Plateau, is characterized by a plateau topography developed on nearly horizontal strata. The Ozark Plateau is bordered on the south by the narrow Arkansas River Valley, which separates the northern area from the Ouachita Mountains to the south. The Ouachita Mountains were formed by folded Paleozoic strata by an orogeny in the Permian; subsequently, the mountains have been eroded extensively. The highest mountains (up to 839 m) are in the Ouachita Mountain section of the region. The region has many spring-fed streams; some springs are thermal. Mean monthly temperatures vary from about 5°C in January to about 25°C in July; annual precipitation is about 750 mm. The Interior Highlands supports a mixture of deciduous hardwood forests—primarily oaks (*Quercus*) and hickory (*Carya*); at higher elevations, pine forest prevails.

Appalachian Highlands

This long, northeast-southwest region encompassing about 1,355,000 km^2 extends from the Island of Newfoundland to the Mississippi Embayment. These highlands separate the Atlantic Coastal Plain from the Interior Lowlands. Although previous orogenies (Acadian, Caledonian, Taconian) helped shape the region, the present Appalachian Mountains resulted from a major orogeny, the Ap-

palachian Revolution, extending from the late Pennsylvanian through the late Triassic. Although Lobeck (1948) distinguished five geological divisions, four natural regions can be recognized in the Appalachian Highlands.

Northern Appalachians

The northern part of the Appalachian Highlands (sometimes referred to as the Maritime or Arcadian Province) includes the region from the Island of Newfoundland southward through the Gaspé Peninsula of Canada and extreme northeastern USA (all of the New England states and eastern part of New York). This region consists of uplifted Paleozoic strata that were extensively glaciated in the Pleistocene. However, rugged peaks are present; the highest is Mt. Washington (1917 m). Topographically, this northern region is separated from the Southern Appalachians (the Newer or Folded Appalachian Province of Lobeck [1948]) by a series of basins generated by continental separation in the late Triassic. Mean temperatures in July are 15–20°C and in January, 0°C to −15°C; precipitation throughout most of the region is less than 1000 mm annually, but the coastal region in the southeastern part of the region receives precipitation in excess of 1000 mm annually. At high elevations, coniferous forests are dominated by fir and spruce, whereas at lower elevations, hemlock and pine gradually are replaced by deciduous trees, principally beech and maples.

Southern Appalachians

These prominent mountains extend southwestward from the Northern Appalachians and essentially parallel the Atlantic coast. The axis of the orogenic belt curves westward parallel to the Gulf coast and underlies both the Ouachita Region of the Interior Highlands and the Edwards Plateau before extending southward into Mexico under the Sierra Madre Oriental. The Southern Appalachians were formed from folded and overthrust-faulted Paleozoic rocks during the Permian closure of the Iapetus Ocean. Major ridges are aligned in northeast-southwest direction. These mountains have been eroded extensively, but they were not glaciated in the Pleistocene. Several peaks in the southern Appalachians exceed 2000 m; Mt. Mitchell (2063 m) is the highest. Temperatures in July are 20–25°C and in January 0–5°C; annual precipitation exceeds 1000 mm. At elevations above about 1300 m, the forest is dominated by conifers, principally fir, with spruce at lower and drier sites; lower on the slopes, the coniferous forest interdigitates with deciduous hardwoods.

Piedmont

This region, lying to the east and south of the Southern Appalachians, consists of eroded older Paleozoic rocks overlain by rich residual soils. This peneplained region has moderate relief to elevations of about 300 m at the edge of the Appalachian Mountains; the Piedmont descends gradually to the eastern and southern edges, where the steeper gradient to the coastal plain is known as the Fall Line. Throughout most of the Piedmont, temperatures in July are above 25°C; in the north in January, temperatures are about 0°C, but they increase to about 10°C in the south. Annual precipitation exceeds 1000 mm. The forest is composed primarily of deciduous hardwoods, but loblolly pine is present in drier areas.

Allegheny Plateau

West of the Appalachian Mountains is a broad region (considerably narrower in the south) also known as the Appalachian Plateau. This plateau, which is formed from thrust-faulted Paleozoic rocks overlain by rich soils, descends in elevation through a series of escarpments from about 800 m to the Interior Lowlands. Temperatures in July are 20–25°C and in January 0–5°C; annual precipitation varies from about 750 mm at lower elevations to 1000 mm at higher elevations. At elevations above about 600 m, pine, hemlock, and spruce are mixed with deciduous trees. At lower elevations, the forest is dominated by oaks mixed with various species of maples.

ATLANTIC COASTAL PLAIN

This region, which has a surface area of about 913,000 km^2 in the Nearctic, consists of the lowlands along the Atlantic Ocean and Gulf of Mexico. In the north, the plain includes Cape Cod and Long Island; southward, the plain gradually widens to include all of peninsular Florida and the Gulf Coast of the USA and Mexico to encompass the Yucatan Peninsula. Most of the coastal plain consists of loose or only partly consolidated formations of sand, gravel, clay, or marl, underlain by Cretaceous rocks. Topographically, there is an almost imperceptible increase in elevation inland from the coast; nearly the entire region is at elevations of less than 250 m. The submergence of the northern part of the plain in the Cretaceous resulted in many large bays and drowned river mouths. A major northward intrusion of about

1000 km is the Mississippi alluvial plain; formerly, much of this area was an embayment. Only the extreme northeastern part of the Coastal Plain was glaciated in the Pleistocene, when fluctuating sea levels alternately expanded and constricted the area. Temperatures gradually increase from north to south. Mean temperature in January ranges from 0°C in the north to more than 15°C in peninsular Florida and in the southwest, whereas mean temperature in July ranges from 20°C in the north to more than 25°C in the south. With the exception of the southwestern part of the coastal plain, mean annual precipitation exceeds 1000 mm, and in places along the eastern part of the Gulf Coast and in southeastern peninsular Florida, rainfall exceeds 1500 mm annually. Throughout most of the Atlantic Coastal Plain, native vegetation consists of deciduous forest or (depending on edaphic conditions) pine forests. South of Lake Okeechobee in southern Florida, the Everglades (= Tropical Savanna) form a broad, grassy marsh. In the drier southwestern part of the plain, the deciduous forest gives way to prairie parkland and then to scrub forest. Eight natural regions are recognized.

Northern Coastal Plain

The Northern Coastal Plain is arbitrarily separated from the Southern Coastal Plain at the northern edge of Chesapeake Bay. From that point, this narrow region extends northward to include Long Island and Cape Cod. The northern part was glaciated in the Pleistocene. Temperatures in July are 20–25°C and in January 0–5°C; annual precipitation exceeds 1000 mm. The dominant vegetation is deciduous hardwood forest, but some sandy soils support pines.

Southern Coastal Plain

This part of the Atlantic Coastal Plain is narrow in the north and wider in the south, where it terminates at the base of the Florida Peninsula. Much of the seaward margin is fringed with strips of offshore bars, behind which are lagoons and marshes. Temperatures in July exceed 25°C, whereas the monthly mean for January is 5°C in the north to more than 10°C in the south; annual precipitation is more than 1000 mm. In the northernmost part of the Southern Coastal Plain, the dominant vegetation is deciduous hardwood forest dominated by oaks and hickories. To the south on sandy soils, pines dominate; in some areas they are mixed with oak. Elsewhere, forests consist mainly of mixed evergreen hardwoods. Cypress (*Taxodium*) are the dominant trees in permanent swamps; the largest of these are the Dismal Swamp (1942 km²) and the Okefenokee Swamp (1550 km²).

Peninsular Florida

This low (highest point 105 m) peninsula about 600 km in length separates the Atlantic Ocean and the Gulf of Mexico. During the Sangamon Interglacial of the Pleistocene, peninsular Florida was reduced to a series of small islands. Extending in an arc southwest from the southern part of the peninsula is a chain of islands, the Florida Keys. With the exception of the Everglades in the southern part of the peninsula, where marl and muck are present, the soils in the peninsula are shallow sands overlaying Miocene and Pliocene limestones. Temperature in July is in excess of 25°C, whereas in January, the mean temperature is above 15°C throughout most of the peninsula, but frosts are not uncommon in the northern part of the peninsula. The native vegetation throughout much of the peninsula consists of pine-oak scrub. The understory in these areas consists of saw palmetto (*Serenoa repens*) and wire grass (*Aristida stricta*). Throughout most of southern Florida, the northern pines are replaced by *Pinus caribaea.* As in the Southern Coastal Plain, there are hammocks of live oaks, cypress swamps dominated by *Taxodium,* and the ubiquitous aerial Spanish moss (*Tillandsia usneoides*). A dominant feature in southern Florida is the flooded prairie (the Everglades) extending from Lake Okeechobee south to the sea; the sawgrass prairie is interrupted by islands of cabbage palms (*Sabal palmetto*). Also, in the extreme southeastern part of the peninsula and on the upper keys, hammocks of tropical trees with Bahamian and Cuban affinities develop in pockets of rich soils (Duellman and Schwartz, 1958).

Gulf Coastal Plain

This region is defined as that part of the coastal plain west of peninsular Florida and east of the Mississippi River Valley. Mean temperatures in July exceed 25°C; in January, mean temperatures generally are above 10°C. Annual precipitation throughout most of the region exceeds 1500 mm. Most of the forest consists of pines, but in some areas, there are pure stands of post oak. Hammocks of evergreen hardwoods are common, as are cabbage palms (*Sabal palmetto*) and cypress swamps, dominated by *Taxodium.*

Mississippi Embayment

The alluvial plains of the central and southern Mississippi River form a broad plain extending nearly 1000 km inland from the Gulf of Mexico to the confluence of the Ohio River at an elevation of 90 m. The southern part of the region, which was submerged during various interglacials in the Pleistocene, is a lowland barrier between the Appalachian Highlands and the Interior Highlands. Mean temperatures in July are at least 25°C, but in January they vary from less than 5°C in the north to more than 10°C in the south; annual precipitation increases from about 1000 mm in the north to 1500 mm in the south. The northern part of the embayment has deciduous forest. In some areas, there are large stands of oaks. The vegetation in the southern part of the embayment is much like that on the Gulf Coastal Plain.

Pine Woodlands

A rather extensive area of flatland in western Louisiana and eastern Texas forms a transition zone between the more humid and biologically diverse Gulf Coastal Plain and Mississippi Embayment to the east, the Prairie Parkland to the northwest, and the Coastal Prairie to the southwest. Mean temperatures exceed 25°C in July and 10°C in January; annual precipitation varies from about 750 mm in the west to nearly 1000 mm in the east. The sandy soil supports nearly pure stands of longleaf pine (*Pinus palustris*) and wire grass (*Aristida stricta*).

Coastal Prairie

The coastal plain of southern Texas lies between the more humid and cooler Pine Woodland and the drier and hotter Tamaulipan Region. The mean temperature exceeds 25°C in July and 10°C in January; annual precipitation varies from about 500 mm in the west to nearly 750 mm in the east. Grasses in the Coastal Prairie primarily are a subset of species in the Tallgrass Prairie.

Tamaulipan

This subtropical region includes southwestern Texas, the Rio Grande Embayment, and the coastal plain in northeastern Mexico, principally in the state of Tamaulipas. The region is characterized by sandy flatland. The mean temperature exceeds 25°C in July and 15°C in January; annual precipitation is in excess of 750 mm. The native vegetation in the northeastern part of the region consists of acacia-grassland. South of the Rio Grande River, grasses are less numerous and thorn forest is prevalent.

LAURENTIAN UPLAND

In the broadest sense, the Laurentian Shield includes Greenland and the Arctic Archipelago, but herein we are concerned only with the continental portion, termed the Laurentian Upland Province by Lobeck (1948). This huge region has a surface area of about 6,123,000 km^2 and encompasses the eastern two thirds of Canada and some small areas of the USA in the vicinity of the Great Lakes. The topography is characterized by a peneplain developed on ancient granitic rocks (mostly Proterozoic) by subsequent erosion and most recently by glaciation that has resulted in many lakes in basins in the rock. The entire region was glaciated in the Pleistocene. Most of the region is at elevations of less than 500 m, but scattered hills in the eastern part of the region exceed 1000 m. The southern part of the Laurentian Shield is characterized by a continental climate with cool summers; the mean temperatures in July do not exceed 20°C, whereas mean temperature in January is no more than –10°C. Precipitation is about 1000 mm annually in the southeastern part and decreases westward and northward to less than 250 mm annually at the Arctic Circle. The Laurentian Upland supports three kinds of vegetation—Boreal forest, Sub-Arctic parkland, and Tundra. Boreal forest is dominated by balsam fir (*Abies balsamea*) mixed with spruces (primarily *Picea glauca*); in the southeastern part of the Laurentian Upland, the fir is mixed with scattered trees, such as aspen (*Populus tremuloides*), yellow birch (*Betula alleghaniensis*), and red maple (*Acer rubrum*). The Sub-Arctic parkland has scattered dwarf trees, principally spruce (*Picea*) and tamarack (*Larix laricina*). Various kinds of low plants make up the tundra vegetation in areas of permafrost to the north of the Sub-Arctic parkland.

REGIONAL PATTERNS OF DISTRIBUTION

The most general analysis of amphibian distributions is at the level of physiographic divisions; the distribution of species in these five divisions is given in the Appendix. Comparison of the numbers of species and endemics in each division reveals that the numbers of species in each division, except the Laurentian Upland, are nearly the same (Table 2:2). As expected, the lowest number of species (14) is at high latitudes in the Laurentian Upland, which has no endemic species.

Within the context of the Nearctic, five genera of anurans occur in only one physiographic division, but all of these have most of their distributions in the neotropics. These include *Pternohyla* (1 species) in the North American Cordillera, *Smilisca* (1), *Leptodactylus* (1), *Hypopachus* (1), and *Rhinophrynus* (1) in the Atlantic Coastal Plain.

The number of species shared by various physiographic regions is 3–27 ($\bar{x} = 10.6$) among anurans and 0–24 ($\bar{x} = 8.4$) among salamanders; comparable coefficients of biogeographic resemblance (CBR) are 11.8–63.5 ($\bar{x} \doteq 34.3$) for anurans and 00.0–41.0 ($\bar{x} = 18.2$) for salamanders (Table 2:3). Almost without exception, the lowest CBRs are between the North American Cordillera and other divisions. The highest for all amphibians are between the Interior Plains and the Atlantic Coastal Plain, followed by that between the Atlantic Coastal Plain and the Appalachian Highlands; the latter is only slightly greater than the CBR between the Interior Plains and the Appalachian Highlands.

Comparison of surface areas of the physiographic regions with the numbers of species in the regions reveals a nearly perfect negative correlation between size of the division and the number of species per area (Table 2:4). However, rather than being a biological phenomenon, this negative correlation reflects the vagaries of geological processes and cli-

Table 2:2. Species of amphibians in five physiographic divisions in North America (excluding introduced taxa). Numbers in each column are anurans + salamanders = total. These numbers reflect only the occurrence within these physiographic divisions in the Nearctic; thus, a species that has a wide distribution in the Neotropics and occurs only in the Atlantic Coastal Plain in the Nearctic is scored as an endemic to that region. Likewise, some species are shared with other physiographic regions in Mexico.

Division	Total species			Shared species			Endemic species			Percent endemic species		
North American Cordillera	44	+ 40	= 84	15	+ 1	= 16	29	+ 39	= 68	65.9	+ 97.5	= 81.0
Interior Plains	38	+ 47	= 85	34	+ 29	= 63	4	+ 18	= 22	10.5	+ 38.3	= 25.9
Atlantic Coastal Plain	47	+ 46	= 93	29	+ 29	= 58	18	+ 17	= 35	38.3	+ 37.0	= 37.6
Appalachian Highlands	19	+ 71	= 90	18	+ 33	= 51	1	+ 38	= 39	00.5	+ 53.5	= 43.3
Laurentian Upland	7	+ 7	= 14	7	+ 7	= 14	0	+ 0	= 0	00.0	+ 00.0	= 00.0

Table 2:3. Distribution of native species of amphibians in the five physiographic regions of North America. Abbreviations in headings to columns correspond to regions in first column. In each cell in the matrix, the top number pertains to anurans, the middle number pertains to salamanders, and the bottom number is for all amphibians. The number of species in each region is shown in boldface in the common cell; the number of species that are common to two regions is in the upper right, and the Coefficient of Biogeographic Resemblance is in italics in the lower left.

Physiographic Division	NAC	INP	ACP	APH	LAU
North American Cordillera (NAC)	**44**	14	8	5	3
	40	1	1	1	0
	84	15	9	6	3
Interior Plains (INP)	*34.1*	**38**	27	16	6
	02.3	**47**	19	20	6
	17.8	**85**	46	36	12
Atlantic Coastal Plain (ACP)	*17.5*	*63.5*	**47**	15	5
	02.3	*40.9*	**46**	24	6
	10.2	*51.7*	**93**	39	11
Appalachian Highlands (APH)	*15.9*	*56.1*	*45.5*	**19**	7
	01.7	*33.9*	*41.0*	**71**	6
	06.9	*41.1*	*42.6*	**90**	13
Laurentian Upland (LAU)	*11.8*	*26.7*	*18.5*	*53.8*	**7**
	00.0	*22.2*	*22.6*	*15.4*	**7**
	06.1	*24.2*	*20.6*	*25.0*	**14**

mates, as well as the somewhat arbitrary divisions that we recognize. The density of species in a given division is skewed in favor of Atlantic Coastal Plain, all of which (except brackish coastal marshes) is habitable by amphibians. In contrast, the other divisions extend northward into high latitudes, where the numbers of species of amphibians decline appreciably.

Herein, for each natural region, the amphibian fauna is assessed with respect to numbers of taxa and their patterns of distribution within the region and adjacent regions.

NORTH AMERICAN CORDILLERA

The massive mountain ranges and the intervening plateaus and troughs in western NA are home to 44 species of anurans and 40 species of salamanders; 29 of the anurans and all but one of the salamanders are endemic. Fourteen of the anurans are shared with the Interior Plains, whereas no more than eight species of anurans are shared with any other region. Only one species of salamander, *Ambystoma tigrinum,* occurs in other regions. Within the North American Cordillera, the Pacific–Cascade–Sierra Nevada Mountain System shares four species of anurans (*Bufo boreas, Pseudacris regilla, Rana catesbeiana,* and *R. luteiventris*) and two species of salamanders (*Ambystoma macrodactylum* and *A. tigrinum*) with both the Intermontane Plateaus and the Rocky Mountain System, whereas these western ranges also share two anurans (*Ascaphus truei* and *Rana sylvatica*) with the Rocky Mountain System. Two other anurans (*Bufo punctatus* and *Scaphiopus couchii*) that occur only in the Baja California section are shared with the Intermontane Plateaus. Two anurans (*Hyla*

Table 2:4. Area:species relationships in different physiographic divisions in North America.

Region	Area (10^6 km^2)	Percent total area	No. of species	Percent species	Species/ (10^6 km^2)	Rank
North American Cordillera	7.071	27.86	84	34.9	11.8	4
Interior Plains	4.254	21.58	85	35.3	20.0	3
Atlantic Coastal Plain	0.913	4.63	93	38.6	101.9	1
Appalachian Highlands	1.355	6.87	90	37.3	66.4	2
Laurentian Shield	6.123	31.06	14	5.7	2.2	5

Table 2:5. Distribution of species af amphibians in the Pacific–Cascade–Sierra Nevada ranges and major intervening lowlands. + = present; – = absent; I = introduced. Abbreviations of regions: BCA = Baja California, CVC = Central Valley of California, NCR = Cascade Range north of Columbia River, PRA = Pacific Ranges in Alaska and northwestern Canada, PRC = Pacific Ranges in California north of San Francisco Bay, PRN = Pacific Ranges north of California (Oregon and Washington), PRS = Pacific Ranges south of San Francisco Bay, PST = Puget Sound Trough, SCR = Cascade Range south of Columbia River, SN = Sierra Nevada. Species designated by asterisk (*) also occur in other regions. Under Elevation, 0 = near sea level.

Species	Region										Elevation (m)
	PRA	PRN	PRC	PRS	BCA	CVC	PST	NCR	SCR	SN	
ANURA: ASCAPHIDAE:											
Ascaphus truei*	–	+	+	–	–	–	+	+	+	–	0–1980
ANURA: BUFONIDAE:											
Bufo boreas*	+	+	+	+	–	+	+	+	+	+	0–3600
Bufo californicus	–	–	–	+	–	–	–	–	–	–	0–1830
Bufo canorus	–	–	–	–	–	–	–	–	–	+	1460–3630
Bufo punctatus*	–	–	–	–	+	–	–	–	–	–	0–1980
ANURA: HYLIDAE:											
Pseudacris cadaverina	–	–	–	+	+	–	–	–	–	–	0–2290
Pseudacris regilla*	–	+	+	+	+	+	+	+	+	+	0–2900
ANURA: PELOBATIDAE:											
Scaphiopus couchii*	–	–	–	–	+	–	–	–	–	–	0–1710
Spea hammondii	–	–	–	+	+	+	–	–	–	–	0–910
ANURA: PIPIDAE:											
Xenopus laevis*	–	–	–	I	–	–	–	–	–	–	0–200
ANURA: RANIDAE:											
Rana aurora	–	+	+	+	–	+	+	+	+	–	0–2440
Rana boylii	–	+	+	+	–	–	–	+	+	+	0–2130
Rana cascadae	–	+	–	–	–	–	–	+	+	–	800–2740
Rana catesbeiana*	–	+	+	+	+	+	+	+	+	+	0–1800
Rana clamitans*	–	I	–	–	–	–	I	–	–	–	0–200
Rana luteiventris*	+	–	–	–	–	–	–	–	–	–	0-600
Rana muscosa	–	–	–	+	–	–	–	–	+	+	370–3650
Rana pretiosa	–	+	–	–	–	–	–	+	+	–	1200–3050
Rana sylvatica*	+	–	–	–	–	–	–	–	–	–	0–600
CAUDATA: AMBYSTOMATIDAE:											
Ambystoma californiense	–	–	–	+	–	+	–	–	–	–	0–800
Ambystoma gracile	+	+	+	–	–	–	+	+	+	–	0–3110
Ambystoma macrodactylum*	+	+	–	–	–	–	+	+	+	–	0–3000
Ambystoma tigrinum*	–	–	–	–	–	–	–	+	+	–	90–1400
CAUDATA: DICAMPTODONTIDAE:											
Dicamptodon copei	–	+	–	–	–	–	–	+	–	–	0–1350
Dicamptodon ensatus	–	–	+	–	–	–	–	–	–	–	0–1500
Dicamptodon tenebrosus	–	+	+	–	–	–	+	+	+	–	0–2160
CAUDATA: PLETHODONTIDAE:											
Aneides ferreus	–	+	+	–	–	–	–	–	–	–	0–1700
Aneides flavipunctatus	–	–	+	–	–	–	–	–	+	–	0–1650
Aneides lugubris	–	–	+	+	–	–	–	–	–	+	0–1520
Batrachoseps aridus	–	–	–	+	–	–	–	–	–	–	850
Batrachoseps attenuatus	–	–	+	+	–	+	–	–	+	+	0–1120
Batrachoseps gabrieli	–	–	–	+	–	–	–	–	–	–	1150–1550
Batrachoseps nigriventris	–	–	–	+	–	+	–	–	–	+	0–2260
Batrachoseps pacificus	–	–	–	+	–	+	–	–	–	+	0–2440
Batrachoseps simatus	–	–	–	–	–	–	–	–	–	+	430–1920
Batrachoseps stebbinsi	–	–	–	–	–	–	–	–	–	+	600–1400
Batrachoseps wrighti	–	–	–	–	–	–	–	–	+	–	15–1340
Ensatina eschscholtzii	–	+	+	+	–	–	+	+	+	+	0–2440
Hydromantes brunus	–	–	–	–	–	–	–	–	–	+	370–760
Hydromantes platycephalus	–	–	–	–	–	–	–	–	–	+	1220–3660
Hydromantes shastae	–	–	–	–	–	–	–	–	+	–	300–910
Plethodon dunni	–	+	–	–	–	–	–	–	+	–	0–1000

Table 2:5 Continued

Species	Region PRA	PRN	PRC	PRS	BCA	CVC	PST	NCR	SCR	SN	Elevation (m)
Plethodon elongatus	–	+	–	–	–	–	–	–	–	–	0–1200
Plethodon larselli	–	–	–	–	–	–	–	+	+	–	12–300
Plethodon stormi	–	+	–	–	–	–	–	–	–	–	488–1078
Plethodon vandykei	–	+	–	–	–	–	–	+	–	–	0–1550
Plethodon vehiculum	–	+	–	–	–	–	+	+	+	–	0–1250
CAUDATA: RHYACOTRITONIDAE:											
Rhyacotriton cascadae	–	–	–	–	–	–	–	–	+	–	200–1200
Rhyacotriton kezeri	–	+	–	–	–	–	–	–	–	–	0–1200
Rhyacotriton olympicus	–	+	–	–	–	–	–	–	–	–	0–1000
Rhyacotriton variegatus	–	–	+	–	–	–	–	–	+	–	0–1200
CAUDATA: SALAMANDRIDAE:											
Taricha granulosa	+	+	+	–	–	–	+	+	+	–	0–2800
Taricha rivularis	–	–	+	–	–	–	–	–	–	–	0–1200
Taricha torosa	–	–	+	+	–	–	–	–	+	+	0–2000

eximia and *Pseudacris triseriata*) that occur in the Rocky Mountain System also inhabit the Intermontane Plateaus. The vastly different provinces making up the North American Cordillera contain different assemblages of amphibians; consequently, the Pacific–Cascade–Sierra Nevada Mountain System, the Intermontane Plateaus, and the Rocky Mountain System are treated separately.

Pacific–Cascade–Sierra Nevada Mountain System

These western mountain ranges harbor 17 native species of anurans and 35 species of salamanders; of these 52 species, nine anurans and 34 salamanders are endemic. Three anurans (*Xenopus laevis, Rana catesbeiana, R. clamitans*) have been introduced. The percentage of endemic species is relatively high in each of three mountain ranges composing this mountain system—25.0% in the Pacific Ranges, 14.8% in the Cascade Range, and 31.3% in the Sierra Nevada. Endemism is higher in salamanders than in anurans, and these mountains are the only places where the salamander genera *Ensatina, Rhyacotriton, Taricha,* and *Hydromantes* are known, except the latter also occurs in southern Europe. No Neotropical species extend northward into this system.

The occurrence of amphibians in the 10 natural regions recognized in this mountain system is summarized, as follows.

1. Pacific Range (Alaska and northwestern Canada).—Three anurans and three salamanders; no endemics.

2. Pacific Range (North).—Eight anurans and 14 salamanders; *Plethodon elongatus, P. stormi, Rhyacotriton kezeri,* and *Rhyacotriton olympicus* endemic.

3. Pacific Range (Central).—Six anurans and 12 salamanders; *Dicamptodon ensatus* and *Taricha rivularis* endemic.

4. Pacific Range (South).—Nine anurans and eight salamanders; *Bufo californicus, Batrachoseps aridus,* and *B. gabrieli* endemic.

5. Baja California.—Six anurans and no salamanders; no endemics.

6. Puget Sound Trough.—Five anurans and six salamanders; no endemics.

7. Central Valley of California.—Five anurans and four salamanders; no endemics.

8. Northern Cascades.—Eight anurans and 10 salamanders; *Dicamptodon copei* endemic.

9. Southern Cascades.—Nine anurans and 16 salamanders; *Batrachoseps wrighti, Hydromantes shastae,* and *Rhyacotriton cascadae* endemic.

10. Sierra Nevada.—Six anurans and 10 salamanders; *Bufo canorus, Batrachoseps simatus, B. stebbinsi, Hydromantes brunus,* and *H. platycephalus* endemic.

The distributions of all of the species are shown in Table 2:5. Although *Plethodon larselli* is scored as inhabiting the northern and southern Cascades, this species actually is known only from elevations of 12–300 m in the Columbia River Gorge that separates the two ranges (Nussbaum et al., 1983). With the exception of the northernmost part of the Pacific

Range in northwestern Canada and Alaska, the northern ranges have more species, especially of salamanders, than the southern regions, especially the Pacific Range (South) and Baja California. This latitudinal difference also is evident in the lowland troughs; although the same number of species of anurans (5) are present in both the Puget Sound Trough and the Central Valley of California, the latter has two fewer salamanders than the Puget Sound Trough.

Species richness seems to be best correlated with rainfall. The Pacific Ranges north of the San Francisco Bay area, the Cascade Ranges, and the northern part of the Sierra Nevada receive more precipitation than the southern ranges. In contrast, within these northern areas, decreased temperature with increased elevation seems to play an insignificant role; most species have elevational ranges from near sea level to more than 1500 m. The upper distributional limits of four species of anurans (*Bufo boreas, B. canorus, Rana muscosa,* and *R. pretiosa*) and three species of salamanders (*Ambystoma gracile, A. macrodactylum,* and *Hydromantes platycephalus*) exceed 3000 m (Table 2:5); the highest elevations are reached in the Sierra Nevada by *Bufo canorus* (3630 m), *Rana muscosa* (3650 m), and *Hydromantes platycephalus* (3660 m), whereas the greatest range in elevation exists in *Bufo boreas* (0–3600 m).

With a few exceptions, anurans have broader distributions than salamanders. For example, *Bufo boreas* occurs in nine regions, and *Pseudacris regilla* and *Rana catesbeiana* occur in all regions except the northernmost part of the Pacific Ranges. However, *Bufo canorus* is restricted to the Sierra Nevada, and *Pseudacris cadaverina* exits only in the Pacific Range south of San Francisco Bay and Baja California. The number of species shared between regions is 4–16. In the northern part of these mountains, the numbers of species shared by the regions of the Pacific Ranges with the parts of the Cascade Range and the Sierra Nevada are equal to, or greater than, the numbers of species shared by the regions of the Pacific Ranges or between the northern and southern Cascade ranges and between the southern Cascade Range and the Sierra Nevada (Fig. 2:9).

The peninsula of Baja California has relatively few amphibians. Of the 12 species known to occur in the peninsula, eight (*Bufo boreas, B. californicus, Spea hammondii, Rana aurora, R. catesbeiana, Aneides lugubris, Batrachoseps pacificus,* and *Ensatina eschscholtzii*) inhabit only the northwestern

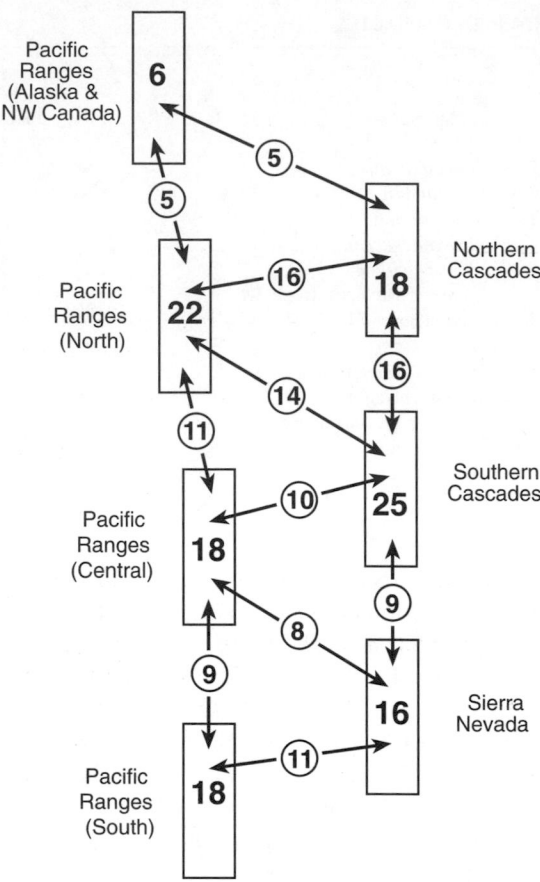

Fig. 2:9. Diagrammatic representation of seven mountainous regions in the Pacific–Cascade–Sierra Nevada ranges. Numbers in boxes are the number of species in each mountain range; numbers in circles indicate species shared by ranges at ends of arrows.

part of the peninsula, which Savage (1960) recognized as the Californian Herpetofaunal Area.

All of these species are shared with the adjacent southern part of the Pacific Ranges. *Pseudacris cadaverina,* also widespread in the southern Pacific Range, extends well into the peninsular desert in the peninsula (Fig. 2:10). *Pseudacris regilla* has a disjunct distribution with a gap between populations in the northwestern part of the peninsula and one in the southern part of the peninsula. The other two anurans (*Bufo punctatus* and *Scaphiopus couchii*) arc widespread in the peninsula; both have extensive distributions in the Intermontane Plateau Region.

The islands off the coast of western Canada have populations of several species of amphibians. According to Cook (1984), three anurans (*Bufo boreas,*

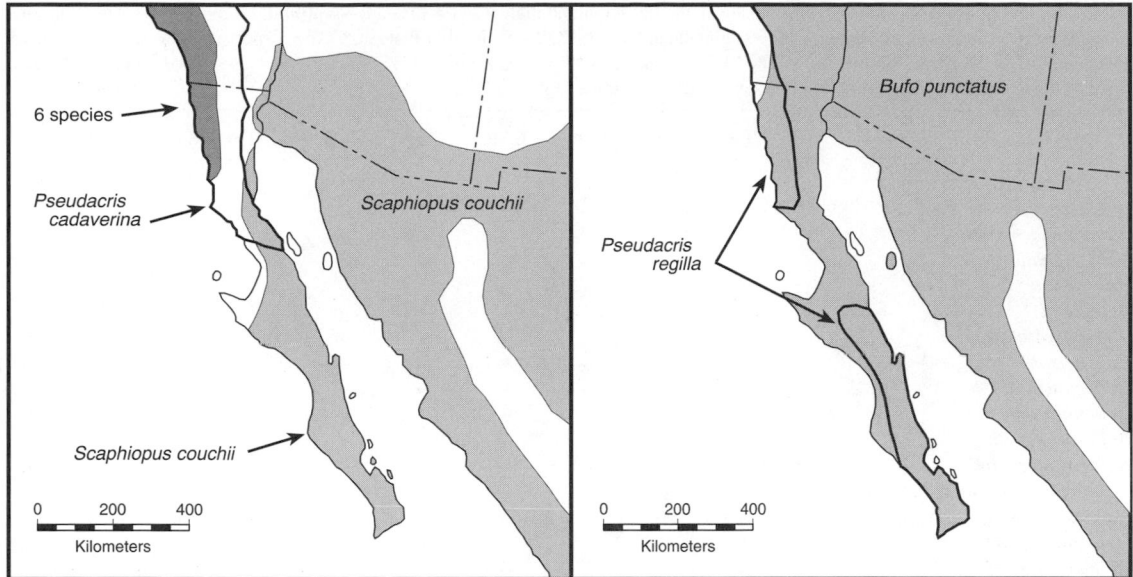

Fig. 2:10.   Patterns of distribution of amphibians in the peninsula of Baja California. The eight species that occur only in the northwestern part of the peninsula are listed in the text.

Pseudacris regilla, and *Rana aurora*) and six salamanders (*Ambystoma gracile, A. macrodactylum, Aneides ferreus, Ensatina eschscholtzii, Plethodon vehiculum,* and *Taricha granulosa*) occur on Vancouver Island. *Bufo boreas* occurs on Queen Charlotte Island, and *Rana luteiventris* and *Taricha granulosa* occur on the islands in the Alexander Archipelago. All of these species inhabit the Pacific Ranges on the mainland. The Channel Islands off the coast of southern California are inhabited by *Pseudacris regilla* and *Batrachoseps pacificus,* and *B. nigriventris* occurs only on Santa Cruz Island of the Channel Islands. The other two species also occur on the Coronado Islands off the coast of northern Baja California. *Aneides lugubris* occurs on the Coronado Islands and the Fallaron Islands off the coast of central California. Two of the desert-inhabiting species of anurans (*Bufo punctatus* and *Scaphiopus couchii*) in Baja California occur on Isla Espíritu Santo-Partida (and the former also on Isla Tiburón) in the Gulf of California (Soulé and Sloan, 1966).

Intermontane Plateaus

This series of plateaus spanning a latitudinal distance of more than 30° is inhabited by 38 native species of amphibians and the introduced *Xenopus laevis* and *Rana catesbeiana.* Only three of these are salamanders. Eight anurans and one salamander are endemic to the region; the salamander (*Batra-*

choseps campi) and three of the anurans (*Bufo exsul, B. nelsoni,* and *Rana fischeri*) are endemic to the Basin and Range Region. *Bufo microscaphus, Spea intermontana,* and *Rana yavapaiensis* are endemic to the Intermontane Plateaus but occur in several regions therein, whereas *Rana onca* is endemic to the Mojave Desert, and *Rana subaquavocalis* is endemic to the Chiricahua Mountains (Table 2:6). The plateaus share seven species of anurans and two of salamanders with the Pacific–Cascade–Sierra Nevada ranges, and the plateaus share 12 anurans and two salamanders with the Rocky Mountain System. Eight of these anurans (*Bufo punctatus, B. woodhousii, Pseudacris triseriata, Spea multiplicata, Scaphiopus couchii, Rana catesbeiana, R. pipiens,* and *R. sylvatica*) and the one salamander (*Ambystoma tigrinum*) also occur in the Great Plains. Seven other species of anurans (*Bufo cognatus, B. debilis, B. speciosus, Acris crepitans, Eleutherodactylus augusti, Gastrophryne olivacea,* and *Spea bombifrons*) are shared with the Great Plains but not the Rocky Mountain System.

Within the Intermontane Plateaus, the numbers of species shared by the 11 included regions are highly variable (Table 2:6; Fig. 2:11). The Yukon and Fraser plateaus in Canada are inhabited by *Bufo boreas, Rana luteiventris,* and *R. sylvatica,* species that also inhabit the northern parts of the Pacific–

Table 2:6. Distribution of species of amphibians in the Intermontane Plateaus. + = present; – = absent; I = introduced. Abbreviations of regions: BAR = Basin and Range, CHI = Chihuahuan Desert, COL = Columbia Plateau, COP = Colorado Plateau, FRA = Fraser Plateau, MAD = Madrean Mountains, MOJ = Mojave Desert, MOR = Mogollon Rim, SON = Sonoran Desert, YUK = Yukon Plateau. Species designated by an asterisk (*) also occur in other regions.

Species	YUK	FRA	COL	BAR	COP	MOR	MAD	MOJ	SON	CHI
ANURA: BUFONIDAE:										
Bufo alvarius	–	–	–	–	–	–	–	–	+	+
Bufo boreas	+	+	+	+	–	–	–	–	–	–
Bufo cognatus*	–	–	–	+	–	–	–	+	+	+
Bufo debilis*	–	–	–	–	–	–	+	–	–	+
Bufo exsul	–	–	–	+	–	–	–	–	–	–
Bufo microscaphus	–	–	–	+	+	+	+	+	–	–
Bufo nelsoni	–	–	–	+	–	–	–	–	–	–
Bufo punctatus*	–	–	–	+	–	+	+	+	+	+
Bufo retiformis	–	–	–	–	–	–	+	–	+	–
Bufo speciosus*	–	–	–	–	–	–	+	–	–	+
Bufo woodhousii*	–	–	+	+	+	–	+	–	+	+
ANURA: HYLIDAE:										
Acris crepitans*	–	–	–	–	–	–	–	–	–	+
Hyla arenicolor*	–	–	–	+	+	+	+	–	+	+
Hyla eximia*	–	–	–	–	–	+	+	–	–	–
Pseudacris regilla*	–	–	+	+	–	–	–	–	–	–
Pseudacris triseriata*	–	–	–	+	+	+	–	–	–	–
Pternohyla fodiens*	–	–	–	–	–	–	–	–	+	–
ANURA: LEPTODACTYLIDAE:										
Eleutherodactylus augusti*	–	–	–	–	–	+	+	–	–	+
Eleutherodactylus guttilatus*	–	–	–	–	–	–	–	–	–	+
ANURA: MICROHYLIDAE:										
Gastrophryne olivacea*	–	–	–	–	–	–	–	–	–	+
Gastrophryne usta*	–	–	–	–	–	–	–	–	+	–
ANURA: PELOBATIDAE:										
Scaphiopus couchii*	–	–	–	–	–	+	–	–	+	+
Spea bombifrons*	–	–	–	–	+	–	+	–	–	+
Spea intermontana	–	–	+	+	+	–	–	–	–	–
Spea multiplicata*	–	–	–	–	+	–	+	–	+	+
ANURA: PIPIDAE:										
Xenopus laevis*	–	–	–	–	–	–	–	–	I	–
ANURA: RANIDAE:										
Rana berlandieri*	–	–	–	–	–	–	–	–	–	+
Rana catesbeiana*	–	–	+	+	+	–	–	+	+	+
Rana chiricahuensis	–	–	–	–	–	+	+	–	–	–
Rana fischeri	–	–	–	+	–	–	–	–	–	–
Rana luteiventris*	+	+	+	+	–	–	–	–	–	–
Rana onca	–	–	–	–	–	–	–	+	–	–
Rana pipiens*	–	–	+	+	+	–	–	–	–	–
Rana subaquavocalis	–	–	–	–	–	–	+	–	–	–
Rana sylvatica*	+	+	–	–	–	–	–	–	–	–
Rana tarahumarae	–	–	–	–	–	–	+	–	–	–
Rana yavapaiensis	–	–	–	–	–	–	+	+	+	+
CAUDATA: AMBYSTOMATIDAE:										
Ambystoma macrodactylum*	–	–	+	–	–	–	–	–	–	–
Ambystoma tigrinum*	–	–	+	+	+	–	–	–	+	+
CAUDATA: PLETHODONTIDAE:										
Batrachoseps campi	–	–	–	+	–	–	–	–	–	–

Cascade–Sierra Nevada ranges and the Rocky Mountains. Of these three northern species, only *Rana sylvatica* is absent from the Columbia Plateau and northern part of the Basin and Range Region. Other primarily northern species inhabiting the Co-lumbia and Colorado plateaus and the northern part of the Basin and Range Region are *Spea intermontana* and *Rana pipiens*. In the southern part of the Intermontane Plateaus, *Bufo punctatus, B. wood-housii,* and *Ambystoma tigrinum* are nearly ubiqui-

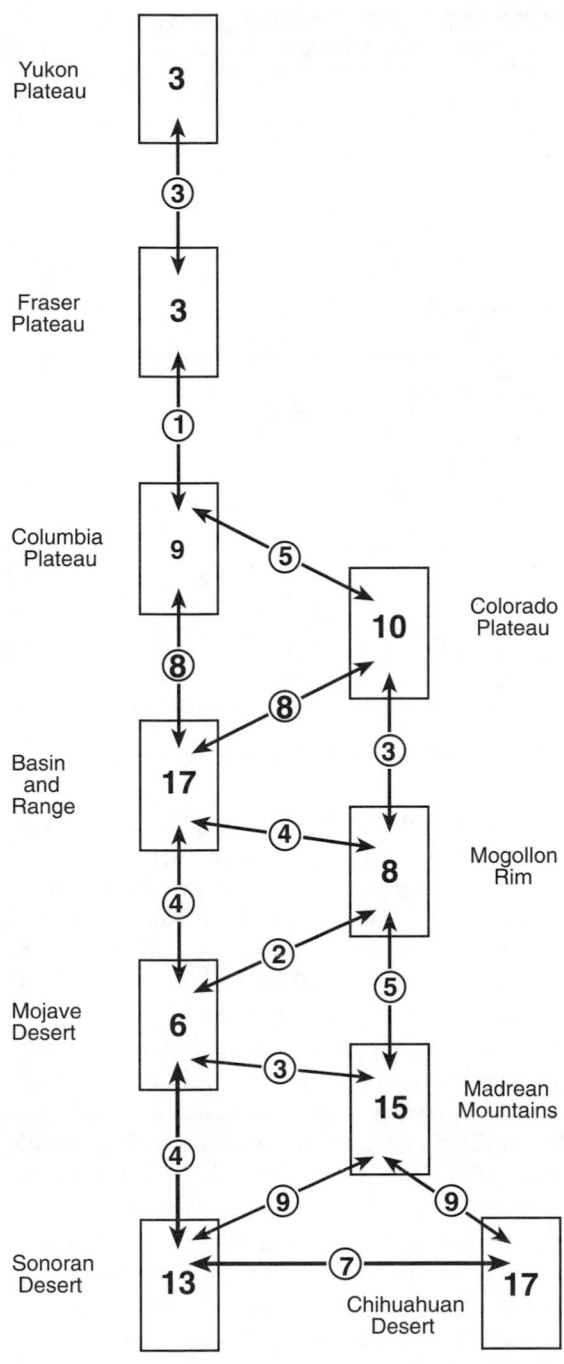

Fig. 2:11. Diagrammatic representation of nine regions of the Intermontane Plateaus. Numbers in boxes are the number of species in each region; numbers in circles indicate species shared by regions at ends of arrows.

tous, and all but *B. punctatus* extend as far north as the Columbia Plateau.

Numerous north-south mountain ranges in the Basin and Range Region, the Colorado Plateau, and/or the east-west Mogollon Rim are places where some southern species reach the northern limits of their distributions; these include *Bufo microscaphus, Hyla arenicolor, H. eximia, Eleutherodactylus augusti, Spea multiplicata, Scaphiopus couchii,* and *Rana chiricahuensis.* Among the 15 species of anurans known from the Chiricahua, Huachuca, and Santa Rita mountains, *Rana subaquavocalis* is endemic to the Chiricahua Mountains, and nine species are shared with the Sonoran Desert to the south and west and the Chihuahuan Desert to the east, but only four species (*Bufo punctatus, B. woodhousii, Spea multiplicata,* and *Rana yavapaiensis*) are shared by the three regions.

The Sonoran and Chihuahuan deserts share the widespread *Ambystoma tigrinum* and have only 13 and 17 species of anurans, respectively. Of these, *Pternohyla fodiens* and *Gastrophryne usta* occur only in the Sonora Desert in the Nearctic; both species range southward well into Mexico. Likewise, in the Intermontane Plateaus, *Acris crepitans, Eleutherodactylus guttilatus,* and *Rana berlandieri* occur only in the Chihuahuan Desert, but they range eastward into other physiographic regions.

Thus, parts of the Intermontane Plateaus and mountain ranges included therein are inhabited by species characteristic of: (1) northern alpine habitats, (2) southwestern arid and semiarid habitats, and (3) primarily neotropical taxa that reach the northern limits of their distributions in the region (Fig. 2:12).

Rocky Mountain System

The entire Rocky Mountain amphibian fauna consists of 15 anurans and six salamanders. One anuran (*Bufo baxteri*) and four salamanders (*Dicamptodon aterrimus, Aneides hardii, Plethodon idahoensis,* and *P. neomexicanus*) are endemic to the Rocky Mountains. Six species (*Ascaphus truei, Bufo boreas, Pseudacris regilla, Rana luteiventris, Ambystoma macrodactylum,* and *A. tigrinum*) in the northern Rocky Mountains are shared with the mountains in the northern part of the Pacific–Cascade–Sierra Nevada ranges. Six species (*Bufo woodhousii, Pseudacris triseriata, Spea bombifrons, Rana catesbeiana, R. pipiens,* and *Ambystoma tigrinum*) that inhabit the Wyoming Basin and/or the Central Rocky Mountains also occur on the Colorado Plateau in the Intermontane Plateaus, but only two wide-

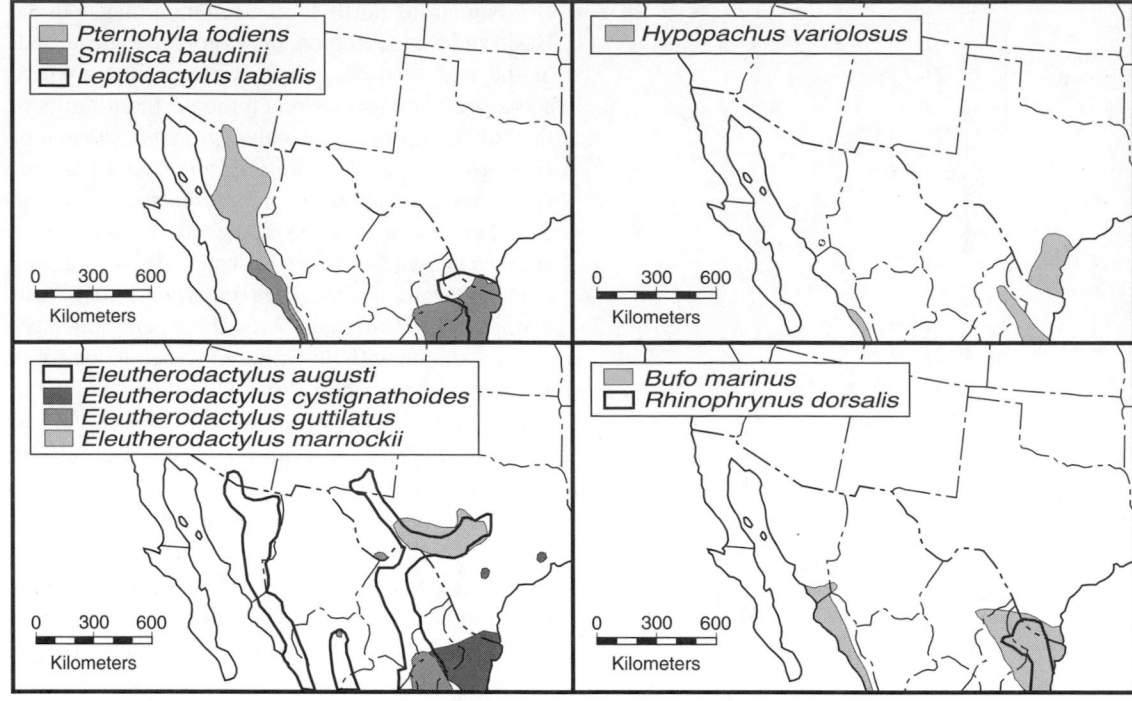

Fig. 2:12.   Northern limits of distribution of Neotropical amphibians in southwestern North America.

spread anurans (*Bufo punctatus* and *Pseudacris tris-eriata*) are among the eight species of amphibians inhabiting the Mogollon Rim in the Intermontane Plateaus. Four species (*Bufo debilis, B. punctatus, B. woodhousii,* and *Eleutherodactylus gutillatus*) that occur in the Southern Rocky Mountains are shared with the Madrean Mountains. Six of the anurans in the Southern Rocky Mountains (*Bufo cognatus, B. debilis, B. punctatus, B. woodhousii, Pseudacris tris-eriata,* and *Rana catesbeiana*) also inhabit the Intermontane Plateaus and the Great Plains, lowland regions to the west and east, respectively, of the Rocky Mountains.

Rana pipiens and *Ambystoma tigrinum* are the only species that occur in all four regions of the Rocky Mountain System, but the southern terminus of the former is in the northern part of the Southern Rocky Mountains. *Bufo boreas* has an equal latitudinal range, but it is absent in the Wyoming Basin, where it apparently is replaced by the endemic *Bufo baxteri.* The Wyoming Basin also is the only place where *Spea bombifrons* occurs in the Rocky Mountains. All of the other species have more restricted latitudinal ranges in the Rocky Mountains (Fig. 2:13).

Because of the latitudinal gradient in temperature and associated vegetation zones, the upper elevational limits reached by species changes from north to south. For example, in the Northern Rocky Mountains in Alberta, Canada, *Ambystoma tigrinum* reaches elevations of 2800 m, and *Bufo boreas* 2300 m (Russell and Bauer, 1993). In the Central Rocky Mountains in Wyoming, the upper limits of these two species are at 2800 and 2865 m (Koch and Peterson, 1995), whereas in New Mexico the upper limits are 3355 and 3200 m, respectively (Degenhardt et al., 1996). In Colorado, *Pseudacris triseriata* has been recorded at 3670 m, *Ambystoma tigrinum* at 3658 m, and *Rana pipiens* 3553 m (Hammerson, 1986).

The two endemic plethodontid salamanders in the Southern Rocky Mountains are enigmatic. *Aneides hardii* is known from elevations of 2400–3570 m in spruce-fir forest and above treeline on three peaks in the Sacramento Mountains; *Plethodon neomexicanus* is restricted to pine-fir-spruce forest at elevations of 2200–2900 m in the Jemez Mountains (Degenhardt et al., 1996). *Eleutherodactylus guttilatus,* with its major distribution in the Sierra Madre Oriental in Mexico, has a disjunct population in the Chisos Mountains.

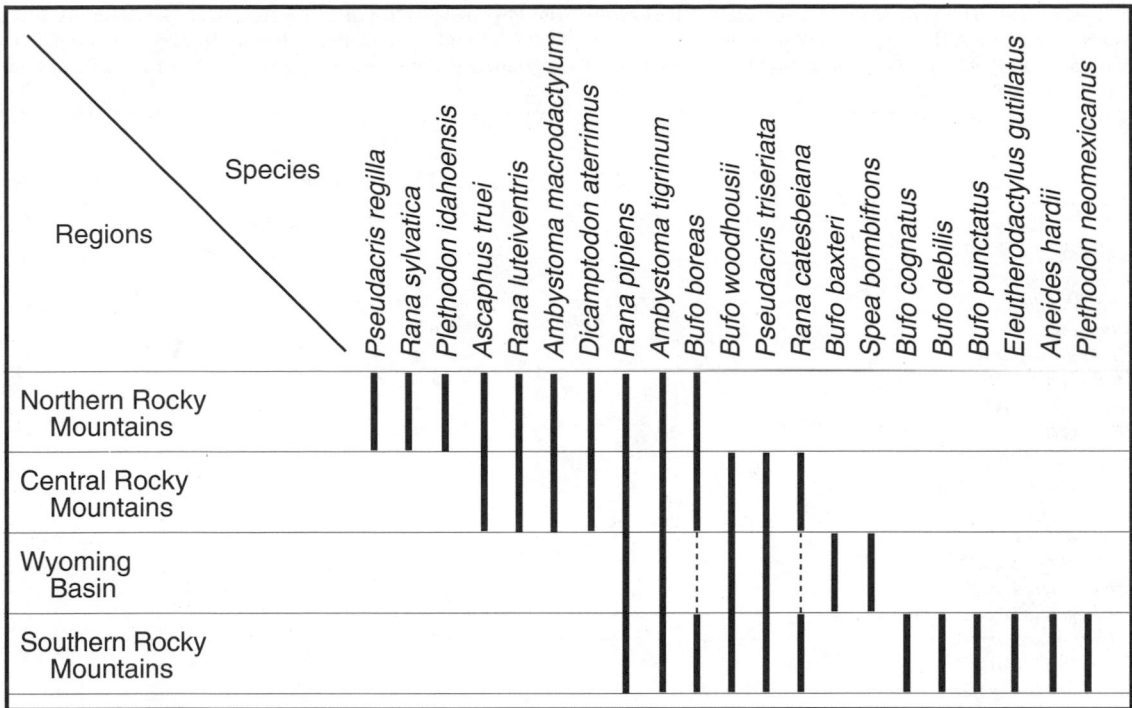

Fig. 2:13.   Latitudinal distribution of species of amphibians in the Rocky Mountains. The vertical bars represent the distributions of the species.

INTERIOR PLAINS

The Interior Plains have 38 species of anurans and 47 of salamanders, of which four anurans and 18 salamanders are endemic (Table 2:7). Patterns of distribution are discussed independently under the three provinces.

Great Plains

Although the Prairie Parkland is listed as a subdivision of the Great Plains, it is treated as a broad zone of merger of eastern and western species in the Interior Plains. Thus, for purposes of analysis of the amphibian fauna of the Great Plains, only species that occur in the Prairie Parkland and other subdivisions of the Great Plains are included. This definition results in the recognition of 29 species of anurans, eight of which are *Bufo,* and eight salamanders in the Great Plains. With a few exceptions, the amphibian fauna of the Great Plains consists of species having widespread distributions. Twenty species are shared with the Atlantic Coastal Plain; 11 of these are among the 13 species shared with the Interior Lowlands. Among the 17 species shared with the Intermontane Plateaus, 10 also occur in the Rocky Mountains. Eight species extend northward into the Boreal Forest.

Six surface-dwelling or cave-dwelling aquatic salamanders (4 *Eurycea* and 2 *Typhlomolge*) are endemic to the Edwards Plateau, where there is an extensive system of mutually isolated subterranean streams (Sweet, 1982). One other salamander (*Plethodon albagula*) and two anurans (*Eleutherodactylus marnockii* and *E. cystignathoides*) occur only on the Edwards Plateau in the Great Plains, but the *Plethodon* and *E. cystignathoides* also inhabit the Atlantic Coastal Plain, and *E. augusti* occurs in the Intermontane Plateaus. *Eleutherodactylus cystignathoides* has been introduced into Fort Worth and Tyler, Texas (J. A. Campbell, pers. comm.).

The other 28 species in the Great Plains are widespread. Of the 20 species inhabiting the Shortgrass Steppe, 18 occur in the Tallgrass Prairie, and 14 in the Prairie Parkland (Fig. 2:14). Twelve species that inhabit both the Shortgrass Steppe and Tallgrass Prairie also occur on the Edwards Plateau, a region that shares 15 species with the Prairie Parkland and 17 with the Atlantic Coastal Plain.

Latitudinally within the USA, six species (*Bufo cognatus, B. woodhousii, Acris crepitans, Hyla chrysoscelis, H. versicolor,* and *Spea bombifrons*) range throughout the Great Plains. Five species (*Bufo he-*

Table 2:7. Distribution of species of amphibians in the Interior Plains Physiographic Division. + = present, and − = absent. Abbreviations of regions: ASP = Aspen Parkland, BOR = Boreal Forest, EDP = Edwards Plateau, INH = Interior Highlands, INL = Interior Lowlands, PRP = Prairie Parkland, SGS = Shortgrass Steppe, TGP = Tallgrass Prairie. Species designated by an asterisk (*) also occur in other regions.

Species	Region							
	BOR	ASP	SGS	TGP	EDP	PRP	INL	INH
ANURA: BUFONIDAE:								
Bufo americanus*	+	−	−	+	−	+	+	+
Bufo cognatus*	−	−	+	+	−	+	−	−
Bufo debilis*	−	−	+	−	+	−	−	−
Bufo fowleri*	−	−	−	−	−	+	+	+
Bufo hemiophrys	+	+	+	+	−	−	+	−
Bufo punctatus*	−	−	+	+	+	−	−	−
Bufo speciosus*	−	−	+	+	+	+	−	−
Bufo valliceps*	−	−	+	+	+	+	−	−
Bufo woodhousii*	−	−	+	+	+	+	−	−
ANURA: HYLIDAE:								
Acris crepitans*	−	−	+	+	+	+	+	+
Hyla avivoca*	−	−	−	−	−	−	+	−
Hyla chrysoscelis*	+	+	−	−	+	+	+	+
Hyla cinerea*	−	−	−	−	−	−	+	−
Hyla gratiosa*	−	−	−	−	−	−	+	−
Hyla versicolor*	+	+	−	−	+	+	+	+
Pseudacris clarkii*	−	−	+	+	+	−	−	−
Pseudacris crucifer*	−	−	−	−	−	+	+	+
Pseudacris streckeri*	−	−	−	−	+	+	+	−
Pseudacris triseriata*	+	+	+	+	−	+	+	+
ANURA: LEPTODACTYLIDAE:								
Eleutherodactylus augusti*	−	−	−	−	+	−	−	−
Eleutherodactylus cystignathoides*	−	−	−	−	+	−	−	−
Eleutherodactylus marnockii	−	−	+	−	+	−	−	−
ANURA: MICROHYLIDAE:								
Gastrophryne carolinensis*	−	−	−	−	+	+	+	+
Gastrophryne olivacea*	−	−	+	+	+	+	−	−
ANURA: PELOBATIDAE:								
Scaphiopus couchii*	−	−	+	+	+	+	−	−
Scaphiopus holbrookii*	−	−	−	−	−	+	+	−
Scaphiopus hurterii*	−	−	−	−	+	+	−	+
Spea bombifrons*	−	−	+	+	−	+	−	−
Spea multiplicata*	−	−	+	−	−	−	−	−
ANURA: RANIDAE:								
Rana areolata*	−	−	−	−	−	+	+	−
Rana berlandieri*	−	−	+	+	+	+	−	−
Rana blairi	−	−	+	+	−	+	+	−
Rana catesbeiana*	−	−	+	+	+	+	+	+
Rana clamitans*	+	−	−	−	−	−	+	+
Rana palustris*	−	−	−	−	−	−	+	+
Rana pipiens*	+	+	+	+	−	+	+	−
Rana sphenocephala*	−	−	−	−	−	−	+	+
Rana sylvatica*	+	+	−	+	−	−	+	+
CAUDATA: AMBYSTOMATIDAE:								
Ambystoma annulatum	−	−	−	−	−	−	−	+
Ambystoma barbouri*	−	−	−	−	−	−	+	−
Ambystoma jeffersonianum*	−	−	−	−	−	−	+	−
Ambystoma laterale*	+	−	−	−	−	−	+	−
Ambystoma maculatum*	−	−	−	−	−	−	+	+
Ambystoma nothagenes	−	−	−	−	−	−	+	−
Ambystoma opacum*	−	−	−	−	−	−	+	+
Ambystoma platineum*	−	−	−	−	−	−	+	−

Table 2:7 Continued

Species	Region							
	BOR	ASP	SGS	TGP	EDP	PRP	INL	INH
Ambystoma talpoideum*	–	–	–	–	–	–	+	–
Ambystoma texanum*	–	–	–	–	–	–	+	–
Ambystoma tigrinum*	+	–	+	+	+	+	+	+
Ambystoma tremblayi*	–	–	–	–	–	–	+	–
CAUDATA: CRYPTOBRANCHIDAE:								
Cryptobranchus alleganiensis*	–	–	–	–	–	–	+	+
CAUDATA: PLETHODONTIDAE:								
Desmognathus brimleyorum	–	–	–	–	–	–	–	+
Desmognathus conanti*	–	–	–	–	–	–	+	–
Desmognathus fuscus*	–	–	–	–	–	–	l	–
Eurycea cirrigera*	–	–	–	–	–	–	+	–
Eurycea longicauda*	–	–	–	–	–	–	+	+
Eurycea lucifuga*	–	–	–	–	–	–	+	+
Eurycea multiplicata	–	–	–	–	–	–	–	+
Eurycea nana	–	–	–	–	+	–	–	–
Eurycea neotenes	–	–	–	–	+	–	–	–
Eurycea sosorum	–	–	–	–	+	–	–	–
Eurycea tridentifera	–	–	–	–	+	–	–	–
Eurycea tynerensis	–	–	–	–	–	–	–	+
Hemidactylium scutatum*	–	–	–	–	–	–	+	+
Plethodon albagula	–	–	–	–	+	–	–	+
Plethodon angusticlavius	–	–	–	–	–	–	–	+
Plethodon caddoensis	–	–	–	–	–	–	–	+
Plethodon cinereus*	+	–	–	–	–	–	+	–
Plethodon dorsalis*	–	–	–	–	–	–	+	–
Plethodon fourchensis	–	–	–	–	–	–	–	+
Plethodon glutinosus*	–	–	–	–	–	–	+	–
Plethodon kiamichi	–	–	–	–	–	–	–	+
Plethodon mississippi*	–	–	–	–	–	–	+	–
Plethodon ouachitae	–	–	–	–	–	–	–	+
Plethodon richmondi*	–	–	–	–	–	–	+	–
Plethodon sequoyah	–	–	–	–	–	–	–	+
Plethodon serratus*	–	–	–	–	–	–	–	+
Pseudotriton montanus*	–	–	–	–	–	–	+	–
Pseudotriton ruber*	–	–	–	–	–	–	+	–
Typhlomolge rathbuni	–	–	–	–	+	–	–	–
Typhlomolge robusta	–	–	–	–	+	–	–	–
Typhlotriton spelaeus	–	–	–	–	–	–	–	+
CAUDATA: PROTEIDAE:								
Necturus louisianensis*	–	–	–	–	–	–	–	+
Necturus maculosus*	+	–	–	–	–	–	+	–
CAUDATA: SALAMANDRIDAE:								
Notophthalmus viridescens*	+	–	–	–	–	–	+	+
CAUDATA: SIRENIDAE:								
Siren intermedia*	–	–	–	–	–	–	+	–

miophrys, Pseudacris triseriata, Rana pipiens, R. sylvatica, and Ambystoma tigrinum) occur only in the northern part of the Great Plains, whereas all of the others are in the central (latitudinally) and southern parts of the plains.

Interior Lowlands

This region has an amphibian fauna of 22 anurans and 27 salamanders, of which only Ambysto-

ma nothagenes is endemic. The fauna is dominated by Ambystoma (11 species) and Rana (8 species). In contrast to the Great Plains which has eight species of Bufo, this genus is represented by only two species in the Interior Lowlands.

With the exception of a few species of salamanders (e.g., Ambystoma barbouri, A. nothagenes, and Plethodon richmondi), most of the species in

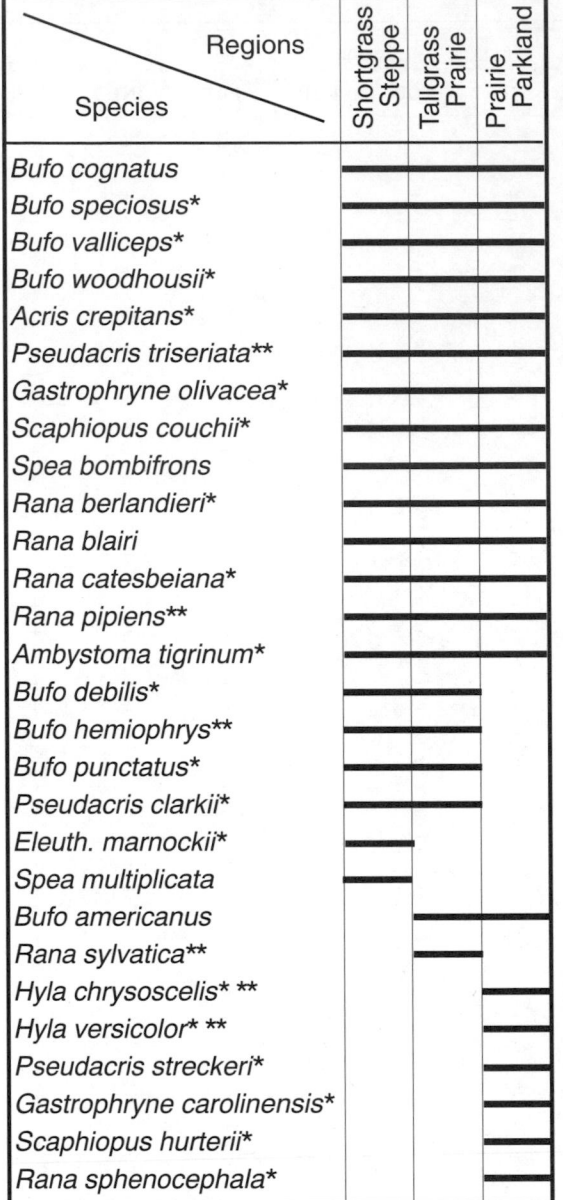

Species / Regions	Shortgrass Steppe	Tallgrass Prairie	Prairie Parkland
Bufo cognatus	▬		
Bufo speciosus*	▬		
Bufo valliceps*	▬		
Bufo woodhousii*	▬		
Acris crepitans*	▬		
Pseudacris triseriata**	▬		
Gastrophryne olivacea*	▬		
Scaphiopus couchii*	▬		
Spea bombifrons	▬		
Rana berlandieri*	▬		
Rana blairi	▬		
Rana catesbeiana*	▬		
Rana pipiens**	▬		
Ambystoma tigrinum*	▬		
Bufo debilis*	▬		
Bufo hemiophrys**	▬		
Bufo punctatus*	▬		
Pseudacris clarkii*	▬		
Eleuth. marnockii*	▬		
Spea multiplicata	▬		
Bufo americanus		▬	
Rana sylvatica**		▬	
Hyla chrysoscelis* **			▬
Hyla versicolor* **			▬
Pseudacris streckeri*			▬
Gastrophryne carolinensis*			▬
Scaphiopus hurterii*			▬
Rana sphenocephala*			▬

Fig. 2:14. Longitudinal distributions of species of amphibians in the Great Plains. The horizontal bars represent the distributions of the species. Specific names followed by an asterisk (*) also occur on the Edwards Plateau; those followed by two asterisks (**) occur in the northern Aspen Parkland, as well as the regions indicated by bars.

in this region have widespread distributions. For example, all of the species shared with the Boreal Forest occur on the Allegheny Plateau, and eight of those species also occur on the Atlantic Coastal Plain. The near absence of endemism is indicative of the degree of sharing of species with adjacent re-

gions; more species are shared with the Allegheny Plateau (part of the Applachian Highlands) than with any other region (Fig. 2:15). The Mississippi Embayment extends northward into the Interior Lowlands; several species (e.g., *Hyla avivoca, Rana sphenocephala, Plethodon mississippi,* and *Siren intermedia*) enter the Interior Lowlands via the Mississippi Embayment. *Pseudacris streckeri,* an inhabitant of the southern part of the Great Plains, has disjunct populations in the Interior Lowlands. Likewise, *Rana areolata,* an inhabitant of the Atlantic Coastal Plain, has disjunct populations in the Interior Lowlands.

Interior Highlands

The relatively small region of the Interior Highlands is home to 14 species of anurans and 22 of salamanders, 12 of which are endemic. Of the 14 species of anurans, all also occur in the Appalachian Highlands, and all but *Rana palustris* and *R. sylvatica* occur on the Atlantic Coastal Plain; all but *Scaphiopus hurterii* occur in the Interior Lowlands, and nine species also occur in the Great Plains. The population of *Rana sylvatica* in the Interior Highlands is disjunct from the populations in the Interior Lowlands and Applachian Highlands.

Among the salamanders, three species of *Ambystoma* (*A. maculatum, opacum,* and *tigrinum*), *Eurycea longicauda, Hemidactylium scutatum, Plethodon serratus,* and *Notophthalmus viridescens* are widespread in the Interior Lowlands, Appalachian Highlands, and the Atlantic Coastal Plain, whereas *Eurycea lucifuga* has an extensive distribution in the Interior Lowlands, as does *Cryptobranchus alleganiensis,* but the range of the latter is disjunct from that in the Interior Highlands. *Plethodon albagula* has disjunct populations in the Prairie Parkland and on the Edwards Plateau in the Great Plains, and *Necturus louisianensis* also occurs on the Atlantic Coastal Plain. Thus, the amphibian fauna of the Interior Highlands has about equal affinities with the Interior Lowlands, Appalachian Highlands, and Atlantic Coastal Plain, but has notably fewer affinities with the Great Plains (Fig. 2:16).

Of the 12 endemic species of salamanders, only *Ambystoma annulatum, Desmognathus brimleyorum, Eurycea multiplicata, E. tynerensis,* and *Typhlotriton speleaeus* are not members of the genus *Plethodon.* Three of the endemic *Plethodon* (*P. caddoensis, fourchensis,* and *ouachitae*) form the

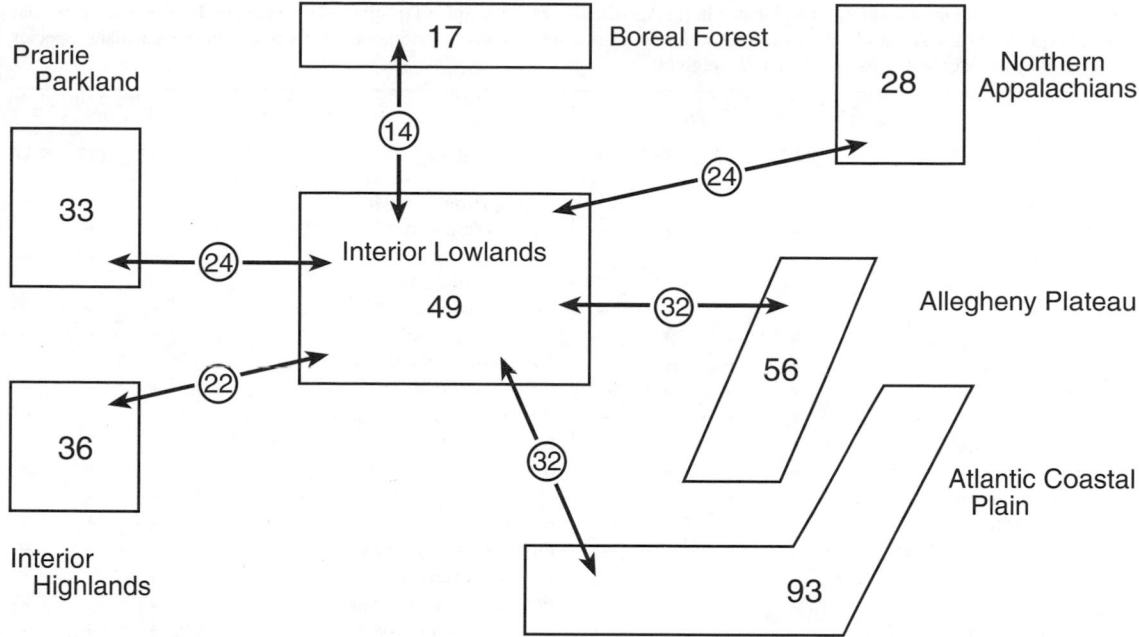

Fig. 2:15. Diagrammatic representation of the Mixed Forest in the Interior Lowlands and adjacent regions. Numbers in boxes are the number of species in each region; numbers in circles indicate species shared with the Mixed Forest.

Plethodon ouachitae complex, whereas *P. kiamichi* and *P. sequoyah,* together with *P. albagula,* are members of the speciose *Plethodon glutinosus* complex, most species of which are in the Appalachian Highlands (Highton, 1995). The other endemic *Plethodon* is *P. angusticlavius,* a member of the *Plethodon welleri* group centered in the Appalachian Highlands.

APPALACHIAN HIGHLANDS

These ancient highlands in eastern NA are home to only 19 species of anurans, of which *Pseudacris brachyphona* is the sole endemic. However, these highlands contain the richest salamander fauna in the world; of the 74 species that occur there, 57 are plethodontids, and 38 of these are endemic to the highlands (Table 2:8). Of the 19 species of anurans in the Appalachian Highlands, seven (36.8%) occur in all four of the recognized regions, as well as the Interior Lowlands and the Atlantic Coastal Plain, but only seven (9.5%) of the 74 salamanders have such wide distributions. The four natural regions are treated separately.

Northern Appalachians

This region harbors 28 species of amphibians (12 anurans and 16 salamanders). No species is endemic to the region, but this is the only place in the Appalachian Highlands where *Rana septentrionalis,* a species widespread in the eastern Boreal Forests, occurs. The northern Appalachian Region shares 26 species each with the Allegheny Plateau and the Southern Appalachians, and 19 species with the Piedmont. With the exception of *Rana septentrionalis,* all of the anurans are shared with the Interior Lowlands, the Allegheny Plateau, and the Southern Appalachians, and all but *R. pipiens, septentrionalis,* and *sylvatica* are shared with the Piedmont. Twelve of the salamanders are widespread species that inhabit the Interior Lowlands, Allegheny Plateau, and

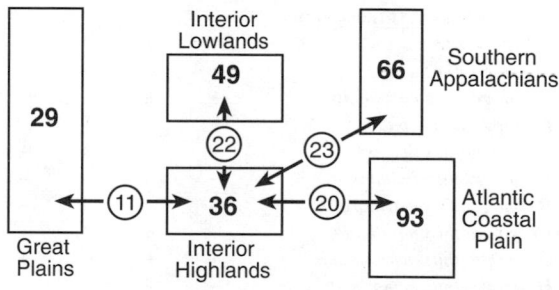

Fig. 2:16. Diagrammatic representation of the Interior Highlands and adjacent regions. Numbers in boxes are the number of species in each region; numbers in circles indicate species shared with the Interior Highlands.

Table 2:8. Distribution of species of amphibians in the Appalachian Physiographic Division. + = present, and – = absent. Abbreviations of regions: ALL = Allegheny Plateau, NAP = Northern Appalachians, PIE = Piedmont, SAP = Southern Appalachians. Species designated by an asterisk (*) also occur in other regions.

Species	Region				Species	Region			
	ALL	SAP	PIE	NAP		ALL	SAP	PIE	NAP
ANURA:BUFONIDAE:					Desmognathus wrighti	–	+	–	–
Bufo americanus*	+	+	+	+	Eurycea bislineata*	+	+	+	+
Bufo fowleri*	+	+	+	+	Eurycea cirrigera*	+	+	+	–
ANURA: HYLIDAE:					Eurycea junaluska	–	+	–	–
Acris crepitans*	+	–	+	–	Eurycea longicauda*	+	+	+	+
Acris gryllus*	–	–	+	–	Eurycea lucifuga*	+	+	+	–
Hyla chrysoscelis*	+	+	+	+	Eurycea quadridigittata*	–	–	+	–
Hyla cinerea*	+	–	+	–	Eurycea wilderae	–	+	–	–
Hyla versicolor*	+	+	+	+	Gyrinophilus gulolineatus	+	–	–	–
Pseudacris brachyphona	+	+	+	–	Gyrinophilus palleucus	+	–	–	–
Pseudacris crucifer*	+	+	+	+	Gyrinophilus porphyriticus*	+	+	+	–
Pseudacris triseriata*	+	+	+	+	Gyrinophilus subterraneus	+	–	–	–
ANURA: MICROHYLIDAE:					Hemidactylium scutatum*	+	+	+	+
Gastrophryne carolinensis*	+	+	+	–	Phaeognathus hubrichti	–	–	–	–
ANURA: PELOBATIDAE:					Plethodon aureolus	–	+	–	–
Scaphiopus holbrookii*	+	+	+	–	Plethodon chattahoochee	–	+	–	–
ANURA: RANIDAE:					Plethodon chlorobryonis*	–	+	+	–
Rana catesbeiana*	+	+	+	+	Plethodon cinereus*	+	+	+	+
Rana clamitans*	+	+	+	+	Plethodon cylindraceus*	–	+	+	–
Rana palustris*	+	+	+	+	Plethodon dorsalis	+	–	–	–
Rana pipiens*	+	+	–	+	Plethodon glutinosus*	+	+	+	+
Rana septentrionalis*	–	–	–	+	Plethodon hoffmani	+	+	–	–
Rana sphenocephala*	–	+	+	–	Plethodon hubrichti	–	+	–	–
Rana sylvatica*	+	+	–	+	Plethodon jordani	–	+	–	–
CAUDATA: AMBYSTOMATIDAE:					Plethodon kentucki	+	–	–	–
Ambystoma barbouri*	+	–	–	–	Plethodon mississippi*	–	–	+	–
Ambystoma jeffersonianum*	+	+	–	+	Plethodon nettingi	+	–	–	–
Ambystoma laterale*	–	–	–	+	Plethodon ocmulgee*	–	–	+	–
Ambystoma maculatum*	+	+	+	+	Plethodon oconaluftee	–	+	–	–
Ambystoma opacum*	+	+	+	–	Plethodon petraeus	+	–	–	–
Ambystoma platineum*	+	–	–	–	Plethodon punctatus	–	+	–	–
Ambystoma talpoideum*	+	+	+	–	Plethodon richmondi*	+	+	–	–
Ambystoma texanum*	+	–	+	–	Plethodon savannah	–	–	+	–
Ambystoma tigrinum*	+	+	+	+	Plethodon serratus*	–	+	–	–
Ambystoma tremblayi*	+	–	–	–	Plethodon shenandoah	–	+	–	–
CAUDATA: CRYPTOBRANCHIDAE:					Plethodon ventralis	–	–	+	–
Cryptobranchus alleganiensis*	+	+	–	–	Plethodon websteri	–	–	+	–
CAUDATA: PLETHODONTIDAE:					Plethodon wehrlei	+	+	–	+
Aneides aeneus	+	+	–	–	Plethodon welleri	–	+	–	–
Desmognathus aeneus	–	+	+	–	Plethodon yonahlossee	–	+	–	–
Desmognathus carolinensis	+	+	+	–	Pseudotriton diasticus	+	–	–	–
Desmognathus conanti*	+	+	+	–	Pseudotriton montanus*	+	–	+	–
Desmognathus fuscus*	+	+	+	+	Pseudotriton ruber*	+	+	+	+
Desmognathus imitator	–	+	–	–	CAUDATA: PROTEIDAE:				
Desmognathus marmoratus	–	+	–	–	Necturus alabamensis*	–	–	+	–
Desmognathus monticola	+	+	+	–	Necturus beyeri*	–	–	+	–
Desmognathus ochrophaeus	+	+	–	+	Necturus lewisi*	–	–	+	–
Desmognathus ocoee	–	+	–	–	Necturus maculosus*	+	+	–	+
Desmognathus orestes	–	+	–	–	Necturus punctatus*	–	–	+	–
Desmognathus quadramaculatus	–	+	–	–	CAUDATA: SALAMANDRIDAE:				
Desmognathus santeetlah	–	+	–	–	Notophthalmus viridescens*	+	+	+	+
Desmognathus welteri	+	+	–	–					

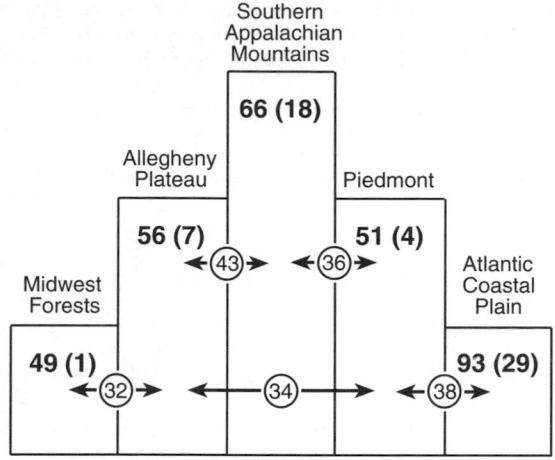

Fig. 2:17. Diagrammatic representation of the Appalachian Highlands and adjacent regions. Numbers in boxes are the total number of species (number of endemics in parentheses) in each region; numbers in circles indicate species shared with the different regions.

Southern Appalachians, and all but three of these also occur in the Piedmont. Seven of the species of anurans and seven of the salamanders in the Northern Appalachians also occur to the north in the Boreal Forest.

Southern Appalachians

Of the 66 species of amphibians in the Southern Appalachians, only 15 are anurans, and of the 51 species of salamanders, 43 are plethodontids, of which 18 are endemic. In the Southern Appalachians, the greatest diversity of *Plethodon* exists—26 species, of which nine are endemic. The highest diversity in two other genera of plethodontids also occurs there; of the 14 species of *Desmognathus,* seven are endemic, and two of the seven species of *Eurycea* are endemic. Many species are shared with the Allegheny Plateau (43 species) and the Piedmont (36), whereas fewer species are shared with the lowlands to the east and west (Fig. 2:17).

Allegheny Plateau

The Allegheny Plateau is inhabited by 16 species of anurans and 40 species of salamanders; among the latter are seven endemics—*Gyrinophilus gulolineatus, G. palleucus, G. subterraneus, Plethodon kentucki, P. nettingi, P. petraeus,* and *Pseudotriton distichus.* Among the 43 species shared with the Southern Appalachians are four plethodontid salamanders (*Aneides aeneus, Desmognathus welteri, Plethodon dorsalis, P. hoffmani*) that occur

only in those two regions. Likewise, many species (32) are shared with the Interior Lowlands to the west; two salamanders (*Ambystoma barbouri* and *A. platineum*) occur only in those two regions.

Piedmont

The amphibian fauna of the Piedmont consists of few endemics—four plethodontid salamanders (*Phaeognathus hubrichti, Plethodon savannah, P. ventralis,* and *P. websteri*)—and a mixture of species shared with the Southern Appalachians, Atlantic Coastal Plain, or both of those regions. Ten species of anurans and 14 species of salamanders occur in all three regions. Of the 36 species shared by the Southern Appalachians and the Piedmont, 10 anurans and 17 salamanders also occur on the Atlantic Coastal Plain; two of the salamanders (*Plethodon chlorobryonis* and *P. cylindraceus*) are endemic to the three regions. One frog (*Acris gryllus*) and six salamanders (*Eurycea quadridigittata, Plethodon ocmulgee, Necturus alabamensis, N. beyeri, N. lewisi,* and *N. punctatus*) occur only in the Atlantic Coastal Plain and the Piedmont.

ATLANTIC COASTAL PLAIN

The Atlantic Coastal Plain contains the largest number of species of amphibians in the Nearctic (Table 2:9). Of the 47 native species of anurans, 18 are endemic, and 17 of the 46 species of salamanders are endemic. Among the anurans are three species (*Osteopilus septentrionalis, Eleutherodactylus coqui,* and *E. planirostris*) that are native to the West Indies but have been introduced into southeastern USA; also, *Bufo marinus,* a native to the southwestern part of the coastal plain, also has been introduced into southern Florida.

Endemism is highest in frogs of the genera *Pseudacris* (4 species) and *Rana* (4), followed by *Bufo* (3) and *Hyla* (3). Among salamanders, the monotypic *Haideotriton* and *Stereochilus,* the three species of *Amphiuma,* and the two species of *Pseudobranchus* are endemic to the Atlantic Coastal Plain, as are eight other salamanders of the genera *Ambystoma* (2 species), *Plethodon* (3), *Notophthalmus* (2), and *Siren* (1).

One of the most distinctive features of the amphibian fauna of the Atlantic Coastal Plain is the presence of, and diversity in, families of obligate neotenic salamanders. Of the four families of obligate neotenes, only *Cryptobranchus* does not occur on the Atlantic Coastal Plain (Fig. 2:18B). The three

Table 2:9. Distribution of species of amphibians on the Atlantic Coastal Plain. + = present; – = absent; I = introduced. Abbreviations of regions: CPR = Coastal Prairie, GCP = Gulf Coastal Plain, MEM = Mississippi Embayment; NCP = Northern Coastal Plain, PFL = Peninsular Florida, PWO = Pine Woodlands, SCP = Southern Coastal Plain, TAM = Tamaulipan. Species designated by an asterisk (*) also occur in other regions.

Species	Region							
	NCP	SCP	PFL	GCP	MEM	PWO	CPR	TAM
ANURA: BUFONIDAE:								
Bufo americanus*	+	–	–	–	+	–	–	–
Bufo fowleri*	+	+	–	+	+	+	–	–
Bufo houstonensis	–	–	–	–	–	+	–	–
Bufo marinus	–	–	I	–	–	–	–	+
Bufo quercicus	–	+	+	+	–	–	–	–
Bufo speciosus*	–	–	–	–	–	–	+	+
Bufo terrestris	–	+	+	+	–	–	–	–
Bufo valliceps*	–	–	–	+	+	+	+	+
Bufo woodhousii*	–	–	–	–	+	+	+	+
ANURA: HYLIDAE:								
Acris crepitans*	+	–	–	+	+	+	+	+
Acris gryllus*	–	+	+	+	+	–	–	–
Hyla andersonii	+	+	–	+	–	–	–	–
Hyla avivoca*	–	–	–	+	+	–	–	–
Hyla chrysoscelis*	+	+	–	+	+	+	–	–
Hyla cinerea*	+	+	+	+	+	+	+	+
Hyla femoralis	–	+	+	+	–	–	–	–
Hyla gratiosa*	+	+	+	+	–	–	–	–
Hyla squirella	–	+	+	+	+	+	+	–
Hyla versicolor*	+	+	+	+	+	+	–	–
Osteopilus septentrionalis	–	–	I	–	–	–	–	–
Pseudacris brimleyi	–	+	–	–	–	–	–	–
Pseudacris clarkii*	–	–	–	–	–	+	+	+
Pseudacris crucifer*	+	+	+	+	+	+	–	–
Pseudacris nigrita	–	+	+	+	–	–	–	–
Pseudacris ocularis	–	+	+	+	–	–	–	–
Pseudacris ornata	–	+	+	+	–	–	–	–
Pseudacris streckeri*	–	–	–	–	–	+	+	+
Pseudacris triseriata*	+	+	–	+	+	+	+	–
Smilisca baudinii	–	–	–	–	–	–	–	+
ANURA: LEPTODACTYLIDAE:								
Eleutherodactylus coqui	–	–	I	I	–	–	–	–
Eleutherodactylus cystignathoides*	–	–	–	–	–	+	–	–
Eleutherodactylus planirostris	–	–	I	I	–	–	–	–
Leptodactylus labialis	–	–	–	–	–	–	–	+
ANURA: MICROHYLIDAE:								
Gastrophryne carolinensis*	+	+	+	+	+	+	+	–
Gastrophryne olivacea*	–	–	–	–	–	+	+	+
Hypopachus variolosus	–	–	–	–	–	–	+	+
ANURA: PELOBATIDAE:								
Scaphiopus couchii*	–	–	–	–	–	–	+	+
Scaphiopus holbrookii*	+	+	+	+	+	–	–	–
Scaphiopus hurterii*	–	–	–	–	–	+	+	+
Spea bombifrons*	–	–	–	–	–	–	+	+
ANURA: RANIDAE:								
Rana areolata*	–	+	+	+	+	+	–	–
Rana berlandieri*	–	–	–	–	–	–	+	+
Rana catesbeiana*	+	+	+	+	+	+	+	+
Rana clamitans*	+	+	+	+	+	+	–	–
Rana grylio	–	+	+	+	+	–	–	–
Rana heckscheri	–	+	+	+	–	–	–	–
Rana okaloosae	–	–	–	+	–	–	–	–
Rana palustris*	+	+	–	+	+	+	–	–

Table 2:9. Continued

Species	Region							
	NCP	SCP	PFL	GCP	MEM	PWO	CPR	TAM
Rana sphenocephala*	+	+	+	+	+	+	+	−
Rana sylvatica*	+	−	−	−	−	−	−	−
Rana virgatipes	+	+	−	−	−	−	−	−
ANURA: RHINOPHRYNIDAE:								
Rhinophrynus dorsalis	−	−	−	−	−	−	−	+
CAUDATA: AMBYSTOMATIDAE:								
Ambystoma cingulatum	−	+	+	+	−	−	−	−
Ambystoma laterale*	+	−	−	−	−	−	−	−
Ambystoma mabeei	−	+	−	−	−	−	−	−
Ambystoma maculatum*	+	+	−	+	+	+	−	−
Ambystoma opacum*	+	+	−	+	+	+	−	−
Ambystoma talpoideum*	−	+	+	+	+	+	−	−
Ambystoma texanum*	−	−	−	+	+	+	+	−
Ambystoma tigrinum*	+	+	+	+	−	+	+	+
CAUDATA: AMPHIUMIDAE:								
Amphiuma means	−	+	+	+	−	−	−	−
Amphiuma pholeter	−	−	+	+	−	−	−	−
Amphiuma tridactylum	−	−	−	+	+	+	−	−
CAUDATA: PLETHODONTIDAE:								
Desmognathus apalachicolae	−	−	−	+	−	−	−	−
Desmognathus auriculatus	−	+	+	+	+	+	−	−
Desmognathus conanti*	−	−	−	+	+	+	−	−
Desmognathus fuscus*	+	−	−	−	−	−	−	−
Eurycea bislineata*	+	−	−	−	−	−	−	−
Eurycea cirrigera*	−	+	−	+	+	−	−	−
Eurycea longicauda*	−	+	−	+	+	−	−	−
Eurycea quadridigittata*	−	+	+	+	+	+	−	−
Haideotriton wallacei	−	−	−	+	−	−	−	−
Hemidactylium scutatum*	+	+	−	+	−	−	−	−
Plethodon chlorobryonis*	−	+	−	−	−	−	−	−
Plethodon cinereus*	+	+	−	−	−	−	−	−
Plethodon cylindraceus*	−	+	−	−	−	−	−	−
Plethodon grobmani	−	+	+	−	−	−	−	−
Plethodon kisatchie	−	−	−	−	−	+	−	−
Plethodon mississippi*	−	−	−	+	+	−	−	−
Plethodon ocmulgee*	−	+	−	−	−	−	−	−
Plethodon serratus*	−	−	−	−	−	+	−	−
Plethodon variolatus	−	+	−	−	−	−	−	−
Pseudotriton montanus*	+	+	+	+	−	−	−	−
Pseudotriton ruber*	+	−	−	+	−	−	−	−
Stereochilus marginatus	−	+	−	−	−	−	−	−
CAUDATA: PROTEIDAE:								
Necturus alabamensis*	−	−	−	+	−	−	−	−
Necturus beyeri*	−	−	−	+	−	−	−	−
Necturus lewisi*	−	+	−	−	−	−	−	−
Necturus louisianensis*	−	−	−	−	+	+	−	−
Necturus punctatus*	−	+	−	−	−	−	−	−
CAUDATA: SALAMANDRIDAE:								
Notophthalmus meridionalis	−	−	−	−	−	−	+	+
Notophthalmus perstriatus	−	+	+	−	−	−	−	−
Notophthalmus viridescens*	+	+	+	+	+	+	+	−
CAUDATA: SIRENIDAE:								
Pseudobranchus axanthus	−	−	+	−	−	−	−	−
Pseudobranchus striatus	−	+	+	+	−	−	−	−
Siren intermedia*	−	+	+	+	+	+	+	+
Siren lacertina	+	+	+	+	−	−	−	−

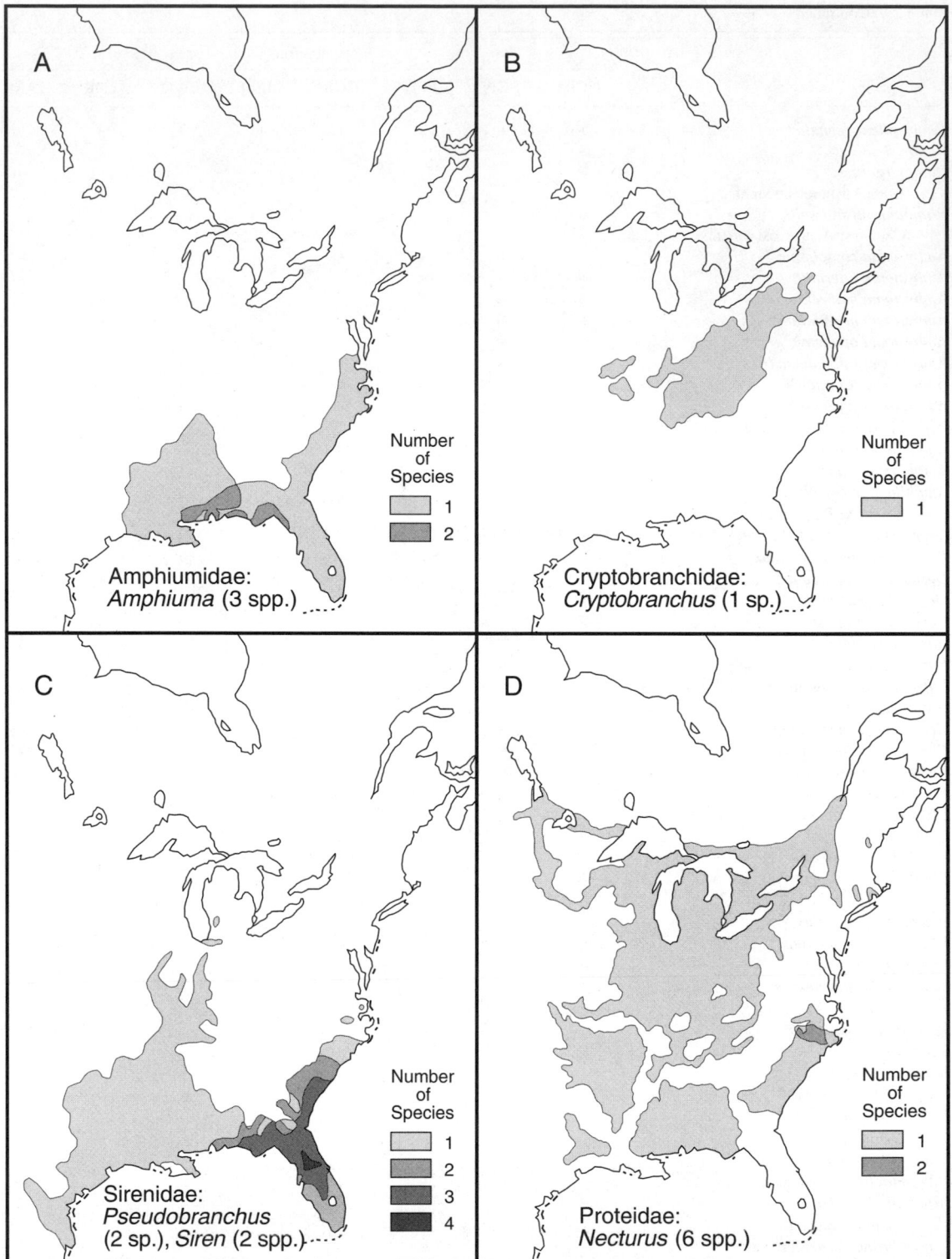

Fig. 2:18.   Distributions of families of obligately neotenic salamanders. Different degrees of shading indicate numbers of sympatric species.

species of *Amphiuma* (Amphiumidae) and the two species of *Pseudobranchus* (Sirenidae) are endemic to the Atlantic Coastal Plain; *Amphiuma means* extends northward in the Mississippi Embayment (Fig. 2:18A). *Siren lacertina* (Sirenidae) is endemic to the coastal plain, whereas *S. intermedia* extends northward in the Mississippi Embayment well into the Interior Lowlands (Fig. 2:18C). Five of the six species of *Necturus* (Proteidae) occur on the coastal plain; four of these are shared with the Piedmont Region of the Appalachian Highlands, and one is shared with the Interior Highlands (Fig. 2:18D). The only other species, *Necturus maculosus,* is widespread west and north of the Atlantic Coastal Plain. The Proteidae is represented by only one other species, *Proteus anguinus,* in southeastern Europe, and the only other genus of Cryptobranchidae (*Andrias*) is represented in eastern Asia by two species. (See Borkin, this volume.)

Only two species of anurans (*Hyla cinerea* and *Rana catesbeiana*) occur in all eight natural regions in the Atlantic Coastal Plain; *Gastrophryne carolinensis* and *Rana sphenocephala* occur in all except the arid Tamaulipan Region in the southwest, and *Siren intermedia* occurs in all except the northern coastal plain. Although *Ambystoma tigrinum* is distributed throughout most of NA and is present in seven regions of the Atlantic Coastal Plain, it is absent in the Mississippi Embayment.

The Northern Coastal Plain has 30 species of amphibians, none of which is endemic. On the Atlantic Coastal Plain, three species (*Acris crepitans, Rana sylvatica,* and *Eurycea bislineata*) occur only in the Northern Coastal Plain; except for a small, disjunct population south of Chesapeake Bay, *Bufo americanus* occurs nowhere else on the Atlantic Coastal Plain except in the Mississippi Embayment. Twenty-three species are shared with the Southern Coastal Plain. Among these are three species of anurans (*Hyla andersonii, H. gratiosa,* and *Rana virgatipes*) that have rather extensive distributions in the Southern Coastal Plain but have disjunct populations in the Northern Coastal Plain. Of the 30 species in the Northern Coastal Plain, 13 anurans and 10 salamanders are shared with the Piedmont, and 11 anurans and eight salamanders are shared with the Northern Appalachians.

The Northern Coastal Plain is arbitrarily separated from the Southern Coastal Plain at the southern edge of Chesapeake Bay, which forms a limit to the distribution of 10 species (e.g., *Amphiuma means, Eurycea cirrigera, Bufo terrestris, Acris gryllus,* and *Hyla squirella*). The area of the Chesapeake Bay also is the southern limit on the coastal plain of *Acris crepitans, Rana sylvatica,* and *Eurycea bislineata; Bufo americanus,* which has an extensive distribution to the north of Chesapeake Bay has a disjunct population to the south. No species is endemic to the northern coastal plain.

Of the 54 species of amphibians known from the Southern Coastal Plain, four (*Pseudacris brimleyi, Ambystoma mabeei, Plethodon variolatus,* and *Stereochilus marginatus*) are endemic. Forty-one of the species in the Southern Coastal Plain are shared with the Gulf Coastal Plain, where five species (*Rana okaloosae, Amphiuma tridactylum, Desmognathus apalachicolae,* and *Haideotriton wallacei*) are endemic. Of the 35 native species in peninsular Florida, only one (*Pseudobranchus axanthus*) is endemic, and 33 and 32 species, respectively, are shared with the Southern Coastal Plain and the Gulf Coastal Plain (Fig. 2:19). Twelve anurans and 15 salamanders are shared by the Southern Coastal Plain and the Piedmont of the Appalachian Highlands. The Fall Line (border between the Piedmont and the coastal plain) is a boundary for several pairs of species (e.g., *Desmognathus fuscus* and *D. auriculatus, Bufo americanus* and *B. terrestris,* and *Pseudacris triseriata* and *P. nigrita* in the Piedmont and coastal plain, respectively.

With the exception of the endemic salamander (*Pseudobranchus axanthus*), the native amphibian fauna of peninsular Florida is a subset of the species occurring on the Southern Coastal Plain and the Gulf Coastal Plain. Only *Plethodon grobmani* and *Notophthalmus perstriatus* are shared by the peninsula and the Southern Coastal Plain but not the Gulf Coastal Plain, whereas *Amphiuma pholeter* is the only species shared by the peninsula and the Gulf Coastal Plain but absent on the Southern Coastal Plain. A noticeable peninsula effect is evident in the amphibian fauna in peninsular Florida; the number of species diminishes from north to south (Fig. 2:20). Also, there is a diminished number of species inhabiting the Everglades (= Tropical Savanna); of the 19 native species in southern Florida, *Eurycea quadridigittata* barely enters the northern part of the Everglades, and *Hyla femoralis, H. gratiosa, Rana areolata,* and *Scaphiopus holbrookii* do not enter the Everglades.

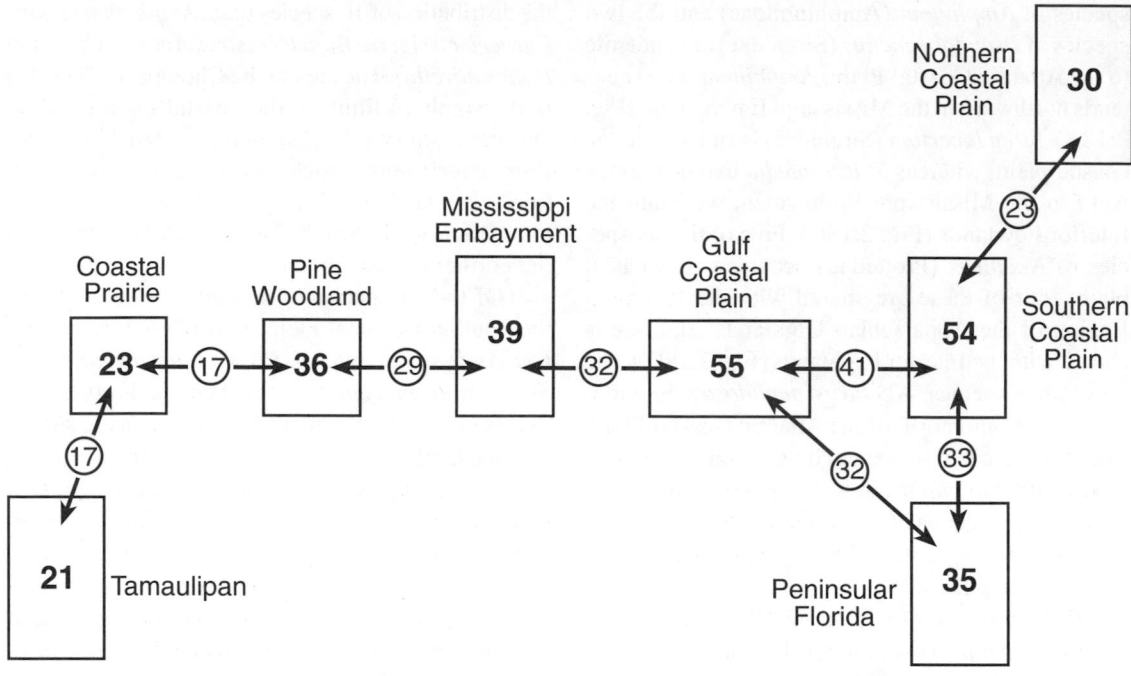

Fig. 2:19.   Diagrammatic representation of the Atlantic Coastal Plain. Numbers in boxes are the number of native species in each region; numbers in circles indicate species shared with different regions.

According to Wilson and Porras (1983), *Osteopilus septentrionalis* was first reported in southern Florida in 1931, *Bufo marinus* in 1957, and *Eleutherodactylus coqui* in 1975. *Osteopilus* evidently was introduced from Cuba or the Bahama Islands, and *E. coqui* from Puerto Rico, whereas the provenance of released *Bufo marinus* is unknown. The treatment of *Eleutherodactylus planirostris* as an introduction is questionable. This species has a broad distribution in Cuba, the Bahama Islands, and the Caicos Islands in the Caribbean Sea. It was first reported from "south Florida" in 1875, when there was little human travel between southern Florida and the islands in the Caribbean, although Spanish galleons anchored at St. Augustine in northern Florida in the 1500s; thus, the frog may have arrived in Florida by rafting from the islands. No matter how it arrived, *E. planirostris* is not a native Nearctic species.

Of the 55 species in the Gulf Coastal Plain, 22 do not extend west of the Mississippi River. Fourteen anurans and 14 salamanders are shared with the Piedmont of the Appalachian Highlands. The lower Mississippi River is the western border for seven species of anurans (*Bufo quercicus, B. terrestris, Acris gryllus, Hyla avivoca, H. gratiosa, H. femor-*

alis, and *Scaphiopus holbrookii*) and six species of salamanders (*Amphiuma means, Eurycea cirrigera, E. longicauda, Plethodon mississippi, Pseudotriton ruber,* and *P. montanus*). To the north, four of these species (*Hyla avivoca, Scaphiopus holbrookii, Eurycea longicauda,* and *Plethodon mississippi*) have disjunct distributions in the Interior Plains west of the Mississippi River. *Rana grylio,* which is widely distributed east of the river, extends only short distance west of the river. In contrast, no species widely distributed to the west of the Mississippi River reaches the eastern limits of its distribution at the river, but *Bufo woodhousii,* a western species, has a limited distribution to the east of the river. Five species (*Hyla avivoca, H. cinerea, Amphiuma tridactylum, Ambystoma talpoideum,* and *Siren intermedia*) that have broad distributions on the coastal plain extend northward in the Mississippi Embayment and thereby invade the Interior Plains.

To the west of the Mississippi Embayment is the region termed the Pine Woodlands (Piney Woods of Bailey, 1976). This region is inhabited by 21 species of anurans and 15 species of salamanders; *Bufo houstonensis* and *Plethodon kisatchie* are endemic. Most (29) of the species in the pine forests are shared

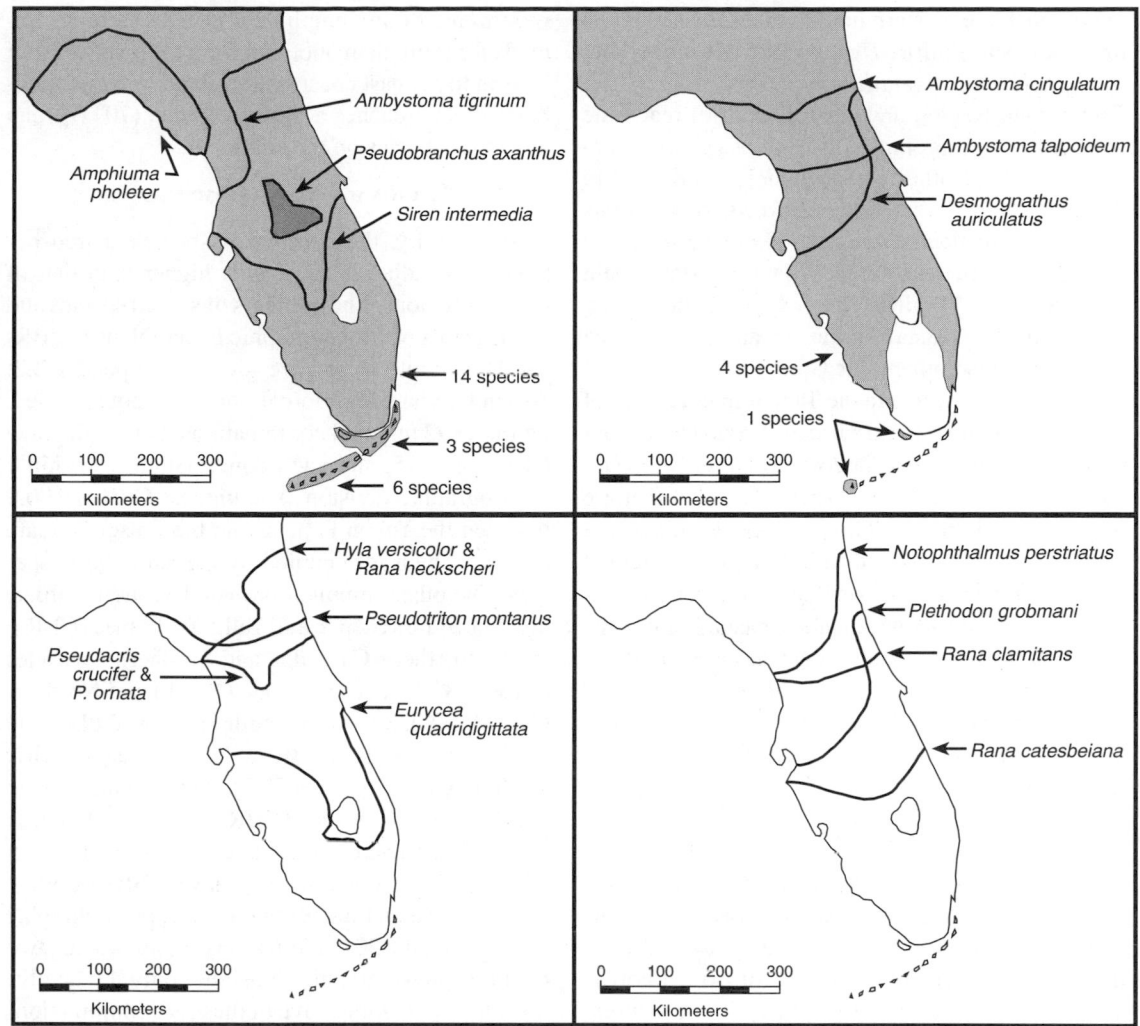

Fig. 2:20. Patterns of distribution of native amphibians in peninsular Florida; lines indicate southern limits of ranges. The 14 species that occur throughout the peninsula are: the salamanders *Amphiuma means, Notophthalmus viridescens, Pseudobranchus striatus,* and *Siren lacertina;* and the anurans *Acris gryllus, Bufo quercicus, B. terrestris, Gastrophryne carolinensis, Hyla cinerea, H. squirella, Pseudacris nigrita, P. ocularis, Rana grylio,* and *R. sphenocephala.* Of the anurans, *Bufo quercicus, Gastrophryne carolinensis,* and *Hyla squirella* occur on the northern and southern Florida Keys. The four species that occur throughout the peninsula except for the Tropical Savanna (the Everglades) are *Hyla femoralis, H. gratiosa, Rana areolata,* and *Scaphiopus holbrookii.* Of these, the latter also occurs on the southern Florida Keys, and *Hyla femoralis* also occurs on Marco Island off the southwestern coast of the peninsula.

with the Mississippi Embayment; six anurans and eight salamanders that are widely distributed to the east have the western limits of their distributions in the pine forests, whereas four anurans (*Pseudacris clarkii, P. streckeri, Gastrophryne olivacea,* and *Scaphiopus hurterii*) have the eastern limits of their distributions there. The Pine Woodlands is the only place of occurrence on the Atlantic Coastal Plain of two species; the leptodactylid frog *Eleutherodacty-*

lus cystignathoides has its major distribution in the Chihuahuan Desert in the Intermontane Plateaus, and *Plethodon serratus* occurs in the Interior Highlands and the Appalachian Highlands.

Ecologically, the Coastal Prairie of southern Texas is part of the Great Plains, but physiographically it is part of the Atlantic Coastal Plain. None of the 23 species of amphibians is endemic. *Hyla squirella, Rana sphenocephala,* and *Ambystoma tex-*

anum reach the western limits of their distributions in the Coastal Prairie. *Hypopachus variolosus* and *Notophthalmus meridionalis* (endemic to the Tamaulipan Region and Coastal Prairie) reach the northeastern limits of their distributions in the Coastal Prairie. Four other species widely distributed to the south and west (*Bufo speciosus, Scaphiopus couchii, Spea bombifrons,* and *Rana berlandieri*) extend no farther east on the Atlantic Coastal Plain than the Coastal Prairie. Thus, of the 12 species occurring in the Coastal Prairie, 15 anurans and one salamander also inhabit the Great Plains.

Of the 21 species in the Tamaulipan Region of northeastern Mexico and adjacent Texas (USA), four anurans (*Bufo marinus, Smilisca baudinii, Leptodactylus labialis,* and *Rhinophrynus dorsalis*) that are widely distributed in Middle America reach the northern limits of their distributions in this region (Fig. 2:10). The Tamaulipan Region also is the southwestern terminus in the Atlantic Coastal Plain of the distributions of five anurans (*Hyla cinerea, Pseudacris clarkii, P. streckeri, Scaphiopus couchii,* and *Spea bombifrons*) and three salamanders (*Ambystoma tigrinum, Notophthalmus meridionalis,* and *Siren intermedia*).

LAURENTIAN UPLAND

Although the Laurentian Upland Province of the Laurentian Shield constitutes nearly one-third of the continental Nearctic, it contains only 14 (5.7%) of the species, which are inhabitants of Boreal Forest. There are no endemics. All species in the Laurentian Shield Region also occur in the Northern Appalachians, and all of these but two (*Rana septentrionalis* and *Eurycea bislineata*) also occur in the Interior Lowlands. Three species (*Bufo americanus, Pseudacris triseriata,* and *Rana sylvatica*) also occur in the northern Great Plains; all except *Bufo americanus* are shared with the northern Pacific Ranges. *Rana sylvatica* is shared with the Rocky Mountains, and *Pseudacris triseriata* is shared with the Intermontane Plateaus and the Rocky Mountains. Seven species (*Bufo americanus, Pseudacris crucifer, P. triseriata, Rana pipiens, R. septentrionalis, R. sylvatica,* and *Ambystoma laterale*) extend northward from the Boreal Forest into the Sub-Arctic Parkland. Three species (*Pseudacris triseriata, Rana pipiens,* and *R. sylvatica*) enter the southern margin of the Tundra. *Rana sylvatica* has the greatest east-

west range of any amphibian in NA; its range extends for more than 6000 km from the west coast of Alaska to the east coast of the Island of Newfoundland; it also reaches a higher latitude (70° N) than any other species on the continent.

REGIONAL FAUNAL COMPARISONS

Generally, the number of species shared between two adjacent regions is higher than that of disjunct regions. The numbers of shared species and Coefficients of Biogeographic Resemblance (CBR) for 44 natural regions are given in Appendix 2:2; Aspen Parkland and Boreal Forest are not included. Of the 1892 regional combinations, only eight have CBRs of ≥ 75, and each combination is within a physiographic division. The highest CBR is 100.0 between the Yukon Plateau and the Fraser Plateau, each of which is occupied by the same three species. The other combinations (in descending order) are: Shortgrass Steppe and Tallgrass Prairie (CBR = 90.0), Northern Cascades and Southern Cascades (CBR = 82.1), Pacific Range (North) and Northern Cascades (CBR = 80.0), Southern Coastal Plain and Gulf Coastal Plain (CBR = 77.3), Coastal Prairie and Tamaulipan (CBR = 77.3), Puget Sound Trough and Northern Cascades (CBR = 75.9), and Pacific Range South and Central Valley California (CBR = 75.0). Only 71 combinations have CBRs of 50.0–74.9; of these, 41 are within a given physiographic division. Of the 30 combinations encompassing two or more physiographic divisions, only 11 involve nonadjacent regions. Five of these are combinations with the Interior Lowlands and involve the Southern Appalachians, Piedmont, Northern Appalachians, Northern Coastal Plain, and Gulf Coastal Plain. The latter is bridged by the Mississippi Embayment, whereas the others are bridged by the Allegheny Plateau. Low CBRs (<10.0) are numerous between some widely disjunct regions, because of the broad distributions of three species—*Ambystoma tigrinum, Pseudacris triseriata,* and *Rana catesbeiana.*

In some combinations of natural regions, definite geographic trends are evident. Fifteen of the 16 species of anurans inhabiting the Allegheny Plateau also occur in the Interior Lowlands. Of those 15 species, 13 continue westward into the Prairie Parkland; five continue westward into the Tallgrass Prairie, and four enter the Shortgrass Steppe. Thus, the Prairie Parkland is intermediate between the forested regions to the east and the grasslands to the west. Likewise, the Piedmont shares 36 species with the Southern Appalachians and 23 species with the Southern Coastal Plain; the Piedmont is a physiographically and biogeographically intermediate region between the Appalachians and the Coastal Plain.

There is a continuous gradient in diminishing numbers of eastern species from east to west along the Gulf Coast. Of the 55 species in the Gulf Coastal Plain, 31 extend westward into the Mississippi Embayment, and 25 of those continue into the Pine Woodlands. Of those 25 species, 12 continue westward into the Coastal Prairie and six into the Tamaulipan Region.

Although the Interior Highlands has a disjunct population of the frog *Rana sylvatica*, the anuran fauna of these highlands is common to the surrounding regions. Quite the opposite is true among the salamanders inhabiting the Interior Highlands; 12 of the 22 species of salamanders are endemic to the region, thereby giving the region one of the highest degrees of endemism in NA.

Within the topographically complex North American Cordillera, some of the regions within the Intermontane Plateaus Province have biogeographic affinities with the Rocky Mountains to the east, and one region has moderate affinities with both the mountains to the east and those to the west. The CBRs between the Columbia Plateau and the Northern Rocky Mountains and Central Rocky Mountains are 63.2 and 73.2, respectively, whereas the CBRs with the Northern Cascades and the Southern Cascades are 29.6 and 26.6, respectively. The Colorado Plateau has CBRs of 50.0 and 62.5 with the Central Rocky Mountains and the Wyoming Basin, respectively.

AREAS OF HIGH DIVERSITY AND ENDEMISM

Comparison of the amphibian faunas in the natural regions reveals that the greatest numbers of species occur in regions in the southeastern part of the continent—Southern Appalachians (66 species), Alleghany Plateau (56), Gulf Coastal Plain (55), and Southern Coastal Plain (54) (Table 2:10). Twenty of the 45 natural regions have no endemics. The number of endemic species is highest (18, 27.3%) in the Southern Appalachians, followed by 11 (30.6%) in the Interior Highlands, seven (12.5%) in the Alleghany Plateau, six (21.4%) in the Edwards Plateau, five (31.3%) in the Sierra Nevada, four in the Pacific Ranges (North) (18.2%), four (7.8%) in the Piedmont, and four (7.4%) in the Southern Coastal Plain.

Endemism in individual natural regions is notably higher among salamanders than among anurans; endemic anurans are most notable on the Atlantic Coastal Plain and in the Pacific–Cascade–Sierra Nevada ranges (Table 2:11). The highest degree of endemism is in the North American Cordillera. This is especially evident among salamanders and includes six endemic genera—*Dicamptodon* (4 species), *Batrachoseps* (9), *Ensatina* (1), *Hydromantes* (3), *Rhyacotriton* (4), and *Taricha* (3); the only endemic genus of anurans is the monotypic *Ascaphus*. High endemism also exists among salamanders in the Appalachian Highlands, but only two genera—*Gyrinophilus* (4 species) and *Phaeognathus*

(1)—are endemic; the other endemics are members of the plethodontid genera *Aneides, Desmognathus, Eurycea,* and *Plethodon,* all of which occur in other physiographic divisions. The only genera endemic to other divisions are the salamanders *Amphiuma* (3 species), *Haideotriton* (1), *Stereochilus* (1) and *Pseudobranchus* (2) in the Atlantic Coastal Plain and *Typhlomolge* (2) and *Typhlotriton* (1) in the Interior Lowlands. However, it should be noted that *Haideotriton, Typhlomolge,* and *Typhlotriton* apparently are nested within *Eurycea* (D. M. Hillis, pers. comm.).

The various vegetation zones that have been noted lack endemic species; their amphibian faunas are composed of some of the species from adjacent regions. Three of the zones are northern. The Aspen Parkland to the north of the Tallgrass Prairie is a narrow intermediate zone between the prairie and the Boreal Forest; the latter also shares species with the adjacent Interior Lowlands and Northern Appalachians, as well as some species with more distant regions. The seven species inhabiting the Sub-Arctic Parkland are a subset of those in the Boreal Forest, and the three species that enter the Tundra are a subset of those in the Sub-Arctic Parkland. The 15 species inhabiting the Everglades in southern Florida are a subset of those in peninsular Florida.

Table 2:10. Natural regions of North America and numbers of native and endemic taxa of amphibians.

PHYSIOGRAPHIC DIVISION, PROVINCE, Natural Region, or Vegetation Zone.	Species		
	Total no.	Endemics No.	%
NORTH AMERICAN CORDILLERA	84	68	80.9
PACIFIC–CASCADE–SIERRA NEVADA	50	41	82.0
Pacific Range (AL & CA)	6	0	0.0
Pacific Range (North)	22	4	18.2
Pacific Range (Central)	18	2	11.1
Pacific Range (South)	15	0	0.0
Baja California	12	0	0.0
Puget Sound Trough	11	0	0.0
Central Valley of California	9	0	0.0
Northern Cascades	18	1	5.5
Southern Cascades	22	2	9.1
Northern Sierras	13	1	7.7
Southern Sierras	16	5	31.3
INTERMONTANE PLATEAUS	41	11	26.8
Yukon Plateau	3	0	0.0
Fraser Plateau	3	0	0.0
Columbia Plateau	9	0	0.0
Basin & Range	19	6	31.5
Colorado Plateau	10	0	0.0
Mogollon Rim	8	0	0.0
Madrean Mountains	15	2	13.3
Mojave Desert	6	1	16.7
Sonoran Desert	13	0	0.0
Chihuahuan Desert	17	0	0.0
ROCKY MOUNTAIN SYSTEM	15	4	26.6
Northern Rocky Mts.	10	0	0.0
Central Rocky Mts.	10	1	10.0
Wyoming Basin	6	1	16.7
Southern Rocky Mts.	12	2	16.7

Table 2:10. Continued

PHYSIOGRAPHIC DIVISION, PROVINCE, Natural Region, or Vegetation Zone.	Species		
	Total no.	Endemics No.	%
INTERIOR PLAINS	85	22	38.6
GREAT PLAINS	29	6	20.7
Aspen Parkland	6	0	0.0
Shortgrass Steppe	20	0	0.0
Tallgrass Prairie	20	0	0.0
Prairie Parkland[1]	33	0	0.0
Edwards Plateau	28	6	21.4
INTERIOR LOWLANDS	49	1	2.0
INTERIOR HIGHLANDS	36	11	30.6
APPALACHIAN HIGHLANDS	90	39	43.3
Alleghany Plateau	56	7	12.5
Southern Appalachians	66	18	27.3
Piedmont	51	4	7.8
Northern Appalachians	28	0	0.0
ATLANTIC COASTAL PLAIN	93	35	37.6
Northern Coastal Plain	30	0	0.0
Southern Coastal Plain	54	4	7.4
Peninsular Florida	35	1	2.9
Everglades	15	0	0.0
Gulf Coastal Plain	55	3	5.5
Mississippi Embayment	39	0	0.0
Pine Woodlands	36	2	5.6
Coastal Prairie	23	0	0.0
Tamaulipan	21	0	0.0
LAURENTIAN UPLAND	14	0	0.0
Boreal Forest	14	0	0.0
Sub-Arctic Parkland	7	0	0.0
Tundra	3	0	0.0

[1]Includes species listed under Great Plains and Interior Lowlands.

TAXONOMIC PATTERNS OF DISTRIBUTION

The distributions of anurans tend to incorporate more physiographic divisions than do those of salamanders. Two anurans (2.2%, *Pseudacris triseriata* and *Rana sylvatica*) occur in all five physiographic divisions, whereas no salamander occurs in all five divisions. Eight (8.9%) anurans occur in four divisions, 11 (12.2%) in three divisions, and 16 (17.8%) in two divisions; the other 53 (58.9%) species are endemic to one division. In contrast, five (3.3%) salamanders occur in four divisions, 13 (8.6%) in three divisions, and 21 (13.9%) in two divisions; the other 112 (74.2%) species are endemic to one division. The mean number of physiographic divisions inhabited by anurans is 1.78 and by salamanders, 1.41. The differences between the mean extent of distributions of anurans and salamanders is largely influenced by the restricted distributions of many species of plethodontid salamanders in the North American Cordillera and the Appalachian Highlands. Only one salamander, *Ambystoma tigrinum,* which occurs in all physiographic divisions except the Laurentian Upland, has a distribution of the

extent of several anurans (e.g., *Acris crepitans, Pseudacris triseriata, Rana catesbeiana, R. pipiens,* and *R. sylvatica*).

Comparison of Figures 2:21 and 2:22 clearly shows different patterns of species diversity (= richness or density). The greatest number of species of both anurans and salamanders is in southeastern NA, but anurans have their greatest diversity in the lowlands, whereas salamanders are most diverse in the highlands. In both groups there is drastic diminution of species westward; however, centers of high diversity in both groups exist on the Edwards Plateau, and anuran diversity is relatively high in places in southwestern USA and adjacent Mexico. Diversity in both anurans and especially salamanders increases in the humid mountains in western NA. These generalized patterns are better illustrated by examining the distributional patterns of selected taxa.

Of the four most speciose genera of North American anurans, *Bufo* (21 species) and *Rana* (26 species) have the widest distributions. However, their patterns

Table 2:11. Regions of high endemism. Endemic genera in boldface; number of endemic species in parentheses.

Pacific–Cascades–Sierra Nevada	Edwards Plateau	Interior Highlands	Appalachian Highlands	Atlantic Coastal Plain
ANURA:				
Bufo (2)				*Bufo* (3)
Pseudacris (1)				*Hyla* (3)
Rana (5)				*Pseudacris* (4)
Spea (1)				*Rana* (4)
CAUDATA:				
Ambystoma (2)	*Eurycea* (4)	*Ambystoma* (1)	*Desmognathus* (12)	**Amphiuma** (3)
Aneides (3)	**Typhlomolge** (2)	*Desmognathus* (1)	*Eurycea* (2)	*Ambystoma* (2)
Batrachoseps (6)		*Eurycea* (2)	*Gyrinophilus* (3)	*Desmognathus* (2)
Dicamptodon (3)		*Plethodon* (6)	**Phaeognathus** (1)	**Haideotriton** (1)
Ensatina (1)		**Typhlotriton** (1)	*Plethodon* (17)	*Notophthalmus* (2)
Hydromantes (3)			*Pseudotriton* (1)	*Plethodon* (2)
Plethodon (6)				**Pseudobranchus** (2)
Rhyacotriton (4)				*Siren* (1)
Taricha (3)				**Stereochilus** (1)

of diversity are quite different. The greatest number of species of *Bufo* is in the southwestern part of the USA and adjacent Mexico; some areas have as many as seven species (Fig. 2:23A). In contrast, by far the greatest diversity of *Rana* is in the southeastern lowlands, where as many as eight species occur together in the Southern Coastal Plain (Fig. 2:23D), but where only two species of *Bufo* occur.

Eight of the 10 species of *Hyla* inhabiting the Nearctic Region occur in eastern NA, where there is a gradual diminishment of species from the Southern Coastal Plain and Gulf Coastal Plain westward and northward (Fig. 2:23B). The two species in western NA have affinities with species occurring farther south in Mexico. Likewise, *Pseudacris* has its greatest number of species in the southeastern part of the continent (Fig. 2:23C). However, there are additional species in the south-central part of NA and two species in western NA.

Among the salamanders, *Ambystoma* has a pattern similar to that of *Rana* with the largest number of species in the Southern Coastal Plain, several in the Ohio Valley in the Interior Lowlands, and additional species in the humid northwest (Fig. 2:24A); one species, *Ambystoma tigrinum,* occurs throughout the middle of the continent. The other genera of North American salamanders are more restricted in their distributions. Six genera (*Batrachoseps, Dicamptodon, Ensatina, Hydromantes, Rhyacotriton,* and *Taricha*) are essentially restricted to the Pacific–Cascade–Sierra Nevada ranges in western NA (Fig. 2:24B, C). Among the other plethodontids, *Desmognathus* and *Eurycea* are most diverse in the Appalachian Highlands, especially the southern Appalachians; both have

species in the Interior Highlands, and *Eurycea* has four species in the Edwards Plateau (Fig. 2:24C). By far, the greatest diversity of *Plethodon* is in the Appalachian Highlands, where 26 of the 44 species occur (Fig. 2:24D); however, an additional six species occur in the Interior Highlands, and six occur only in the humid forests in the Pacific-Cascade ranges. Additionally, there are two isolated species in the Rocky Mountains. The opposite pattern exists in *Aneides,* with three species in the Pacific–Cascade–Sierra Nevada ranges, one isolated species in the Southern Rocky Mountains, and one in the Appalachian Highlands (Fig. 2:24C).

DISJUNCT DISTRIBUTIONS

The distributions of amphibians in the Nearctic are sufficiently well known that gaps in their geographical ranges apparently are real and not an artifact of inadequate collecting. Some species have many isolated populations scattered over a large geographic area. For example, *Ambystoma talpoideum* has a continuous distribution in the Southern and Gulf Coastal plains, with a continuation northward in the Mississippi Embayment, but the species is present in many isolated areas in the Appalachian Highlands (Conant and Collins, 1991: Map 217). On the other hand, *Hemidactylium scutatum* has a broad range across the eastern half of southern Canada and northern USA and southward in the Appalachian Highlands, but with many isolated populations as far south as the Gulf Coastal Plain and westward into the Interior Highlands (Conant and Collins, 1991: Map 258).

Major disjunctions are apparent in several genera of plethodontid salamanders. As noted previously,

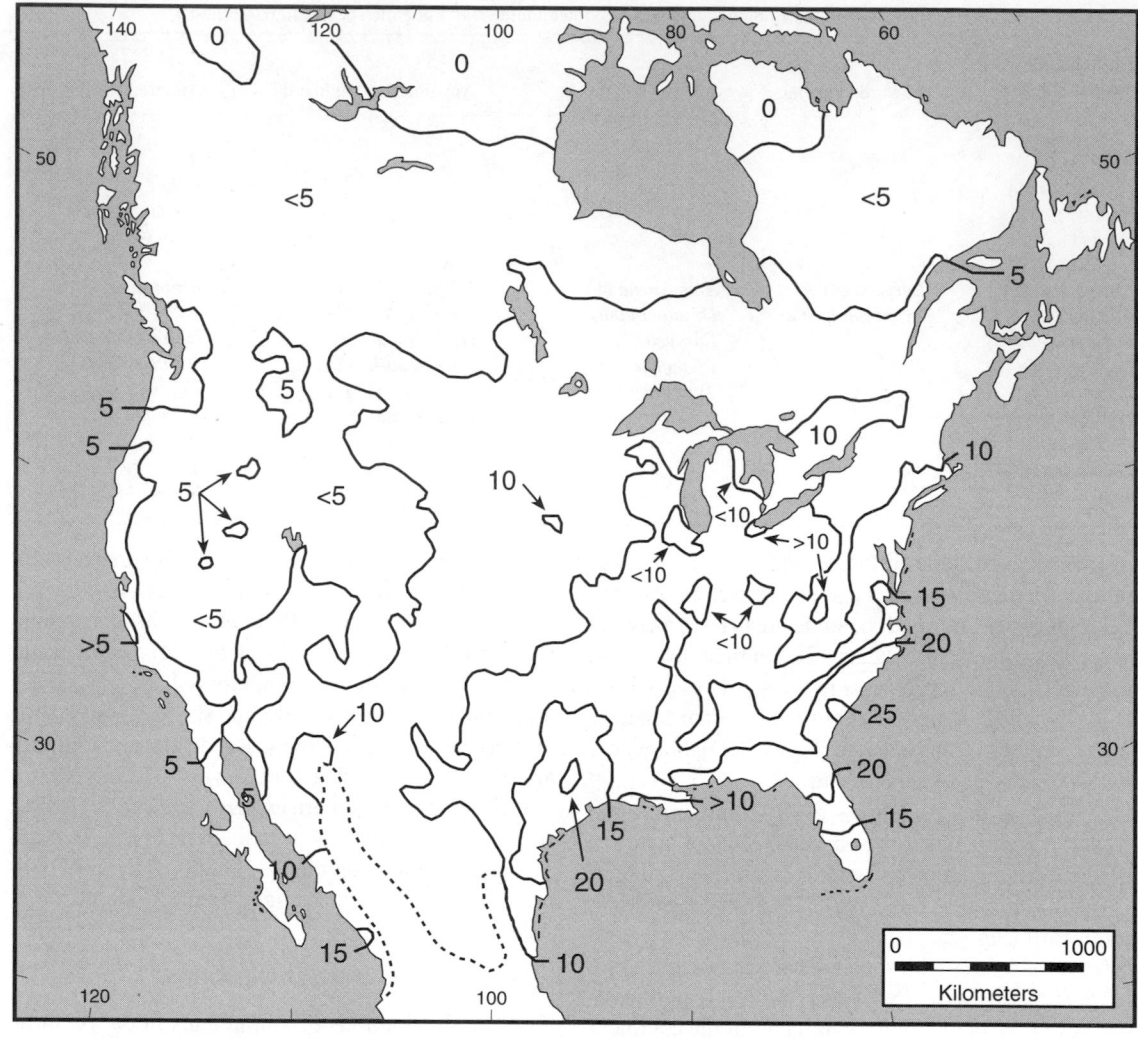

Fig. 2:21.   Species densities of anurans.

three species of *Aneides* occur in the humid mountains in the Pacific–Cascade ranges; *A. aeneus* is
widely distributed in the Appalachian Highlands, and
A. hardii exists as a presumed relict species in the
Southern Rocky Mountains (Fig. 2:24C). More than
half of the species of *Plethodon* occur in the Appalachian Highlands, but many isolated species occur
elsewhere—Interior Highlands, Edwards Plateau,
Northern Rocky Mountains, Southern Rocky Mountains, and Pacific-Cascade ranges (Fig. 2:24D). Within the *Eurycea* generic complex (*Eurycea, Haideotriton, Typhlomolge,* and *Typhlotriton*), many species
occur in the eastern USA (Appalachian Highlands,
Atlantic Coastal Plain, and Interior Lowlands), but
two of these species extend westward into the Inte-

rior Highlands, where three other species are endemic; six isolated endemics occur on the Edwards Plateau (Fig. 2:24C). *Batrachoseps* has seven species
in the Pacific–Sierra Nevada ranges, one isolated
species in the Southern Cascades, and one isolated
species in the Basin and Range (Fig. 2:24B).

A repeated pattern is found in species inhabiting the northern Pacific–Cascade ranges and the
Northern Rocky Mountains. For example, a broad
disjunction exists in the range of *Ascaphus truei* (Fig.
2:25A); three species of *Dicamptodon* inhabit the
Pacific–Cascade ranges, and one occurs in the Northern Rocky Mountains (Fig. 2:25D). Some species,
such as *Bufo boreas,* have a continuous distribution
from the Pacific-Cascade ranges across the Fraser

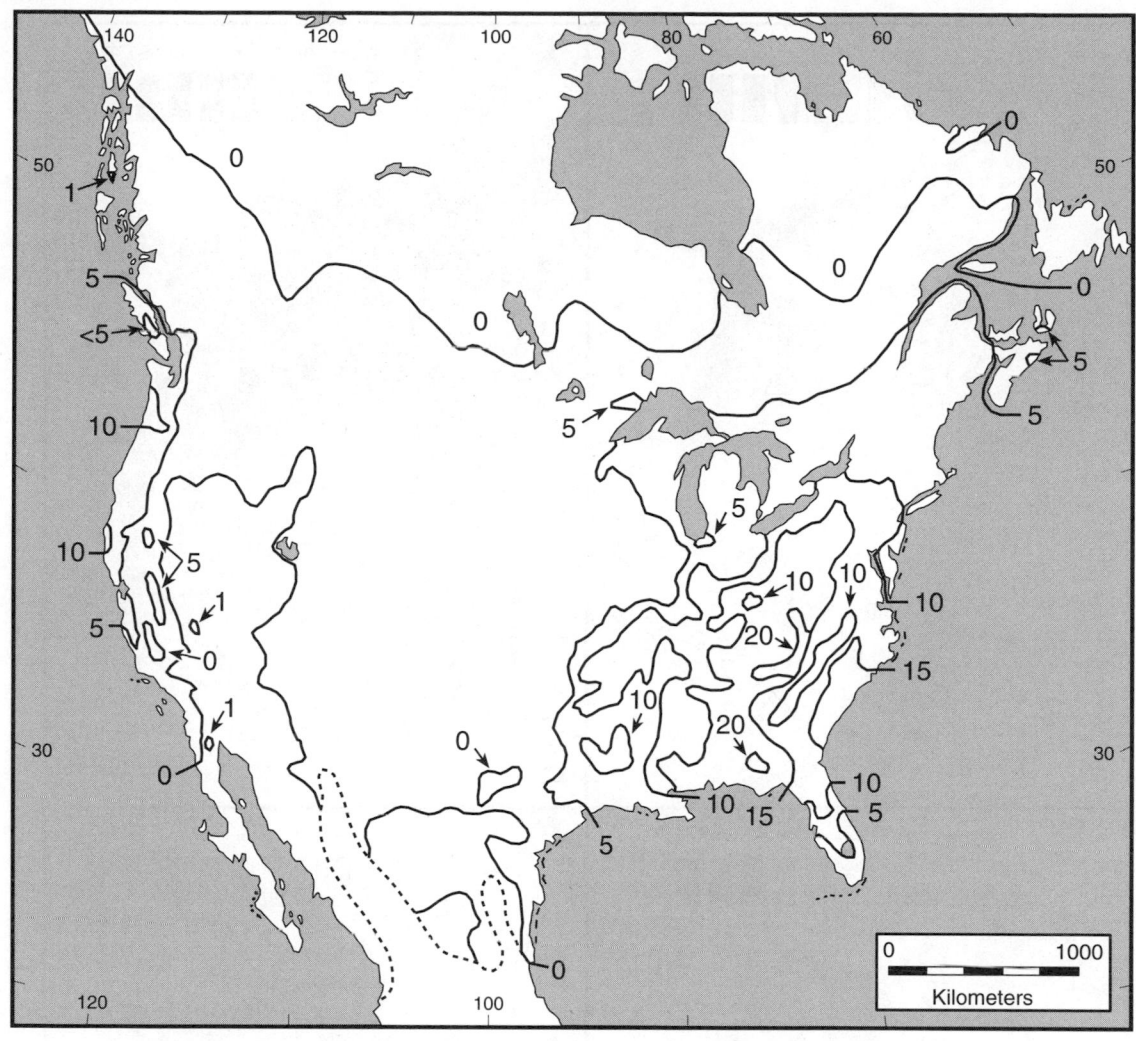

Fig. 2:22. Species densities of salamanders.

and Columbian plateaus, but have isolated populations in the Central and Southern Rocky Mountains (Fig. 2:25B). Another widespread, primarily boreal species, *Rana sylvatica,* also has disjunct populations in the Rocky Mountains, as well as in the Interior Highlands (Fig. 2:25C).

Some species that primarily inhabit the mountains of Mexico have disjunct populations in mountain ranges in southwestern USA. These include *Bufo microscaphus, Hyla eximia, Eleutherodactylus augusti,* and *E. gutillatus* (Figs. 2:12 and 2:25A). Amphibians inhabiting the lowlands tend to have more continuous distributions than do those that occur principally in highlands. Yet, there are some noteworthy discontinuous distributions. *Rana palustris* and *Cryptobranchus alleganiensis* are widely distributed to the east and west of the Mississippi River, but they are absent in the Mississippi River Valley proper (Figs. 2:25B and 2:25D). In the Atlantic Coastal Plain, *Hyla gratiosa* and *Rana virgatipes* have extensive distributions in the Southern Coastal Plain and disjunct populations in the Northern Coastal Plain, whereas *Hyla andersonii* is restricted to three populations, one each in the Northern, Southern, and Gulf Coastal plains (Fig. 2:25A). *Pseudacris streckeri* has an extensive distribution in the Great Plains but has isolated populations east of the Mississippi River (Fig. 2:25A).

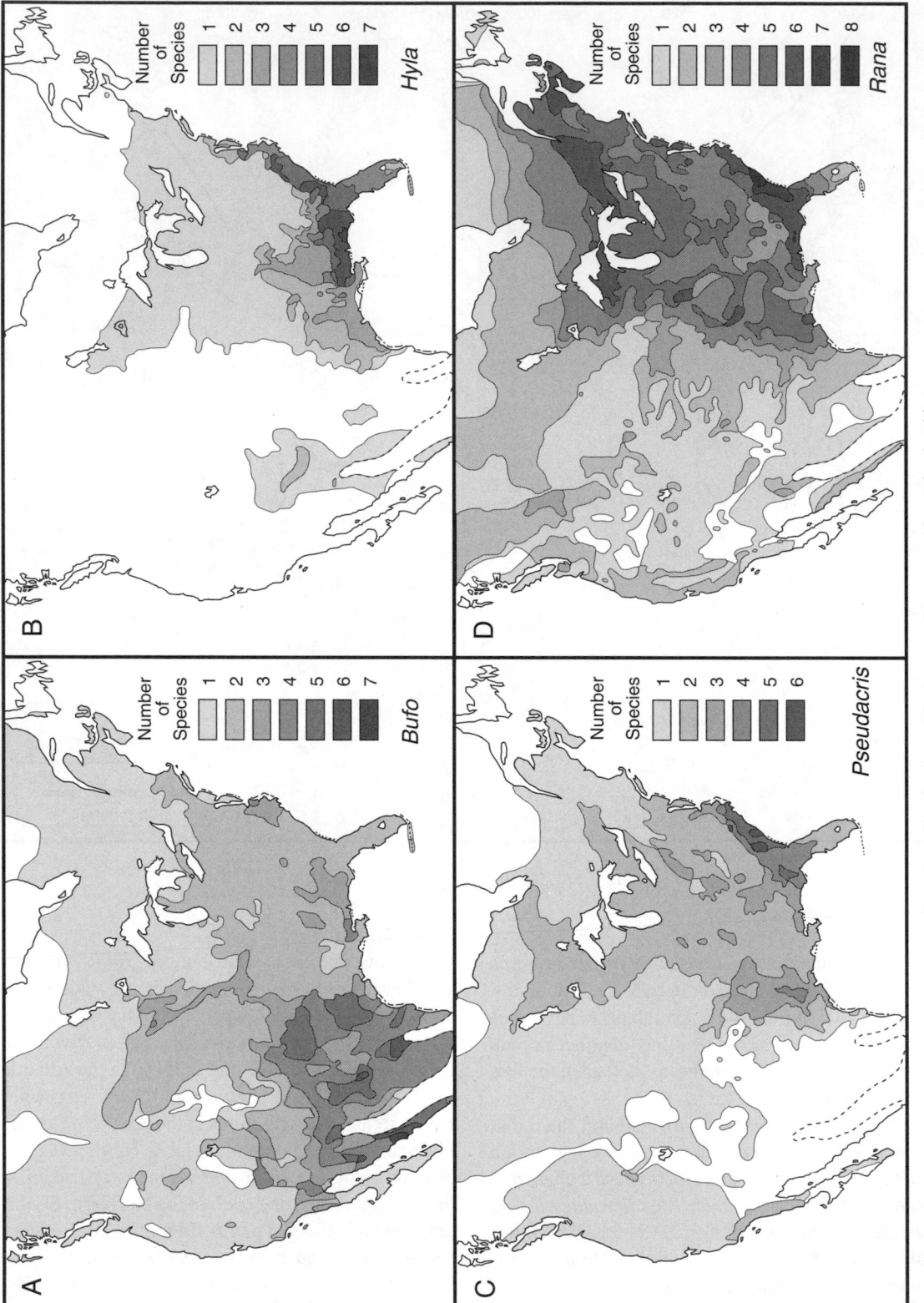

Fig. 2:23. Distributions of four genera of anurans with species densities.

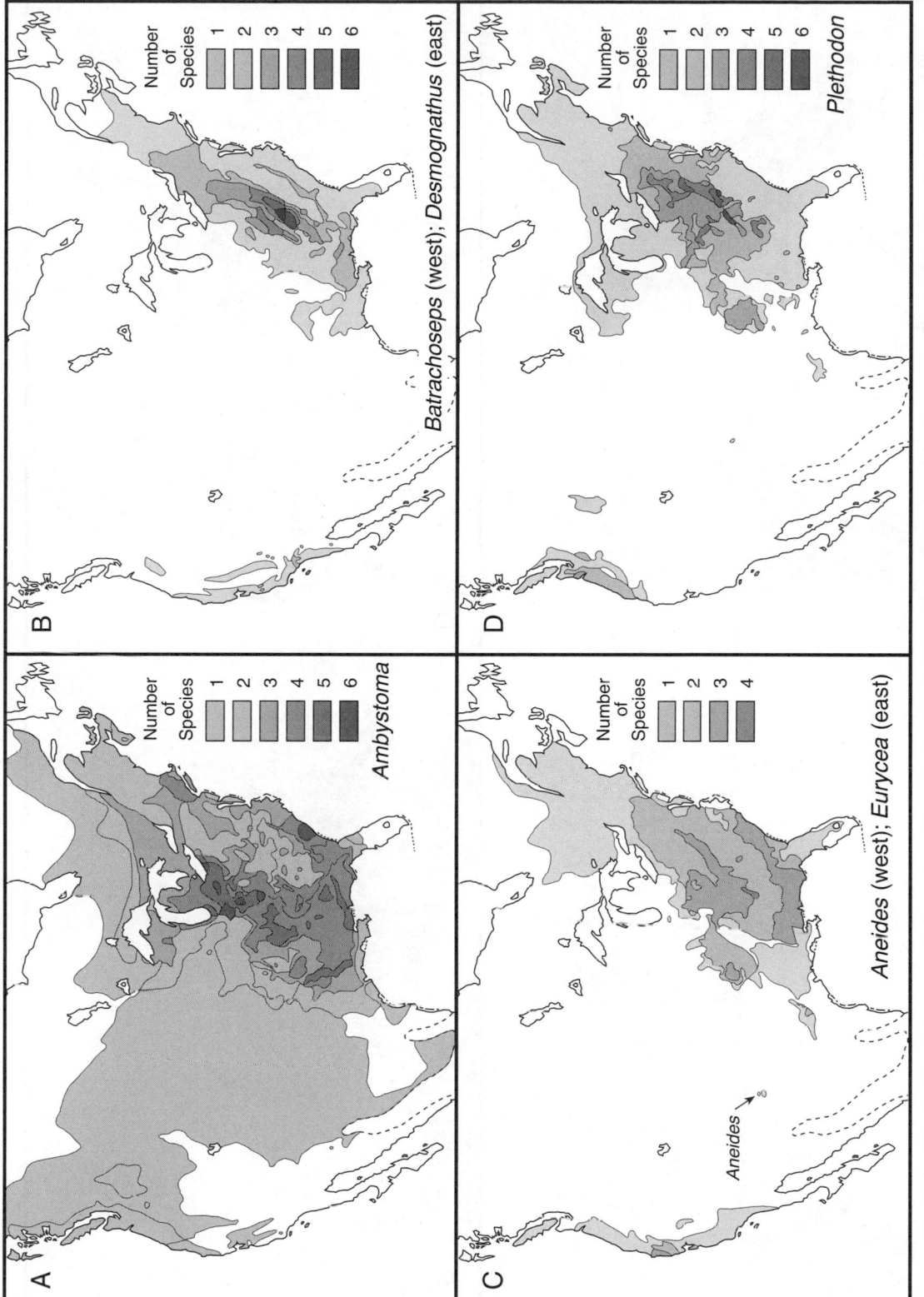

Fig. 2.24. Distributions of six genera of salamanders with species densities. In Figure C, the distribution of *Aneides aeneus* in eastern North America is not shown.

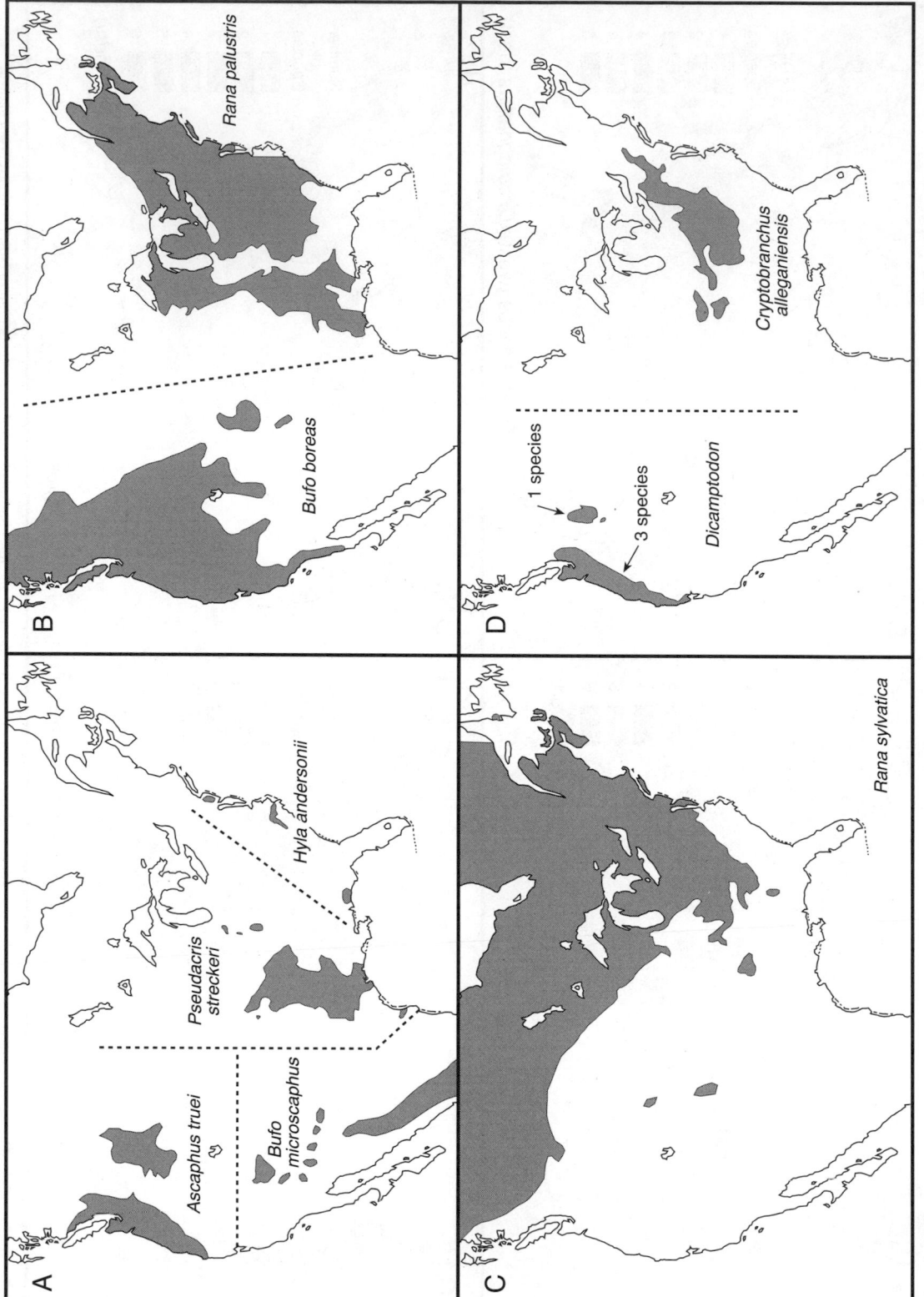

Fig. 2:25.   Disjunct distributions of some Nearctic amphibians.

CORRELATIONS OF DISTRIBUTION PATTERNS

TOPOGRAPHY

Upon examining the entire Nearctic amphibian fauna, it is evident that there is high diversity in the Pacific–Cascade–Sierra Nevada ranges and the Appalachian Highlands, and somewhat lower diversity in the Interior Highlands and Rocky Mountains. Even though throughout most of the length of the Pacific Ranges, the mountains (and their amphibian inhabitants) descend nearly to sea level, all species inhabiting those ranges are considered to be montane. (See Table 2:5 for elevational ranges.) However, nine of 18 anurans and 34 of 35 salamanders there do not invade lowlands elsewhere and can be considered strictly montane species. Thus, 50% of the anurans and 97% of the salamanders are strictly montane species in the Pacific–Cascade–Sierra Nevada ranges. In the Appalachian Highlands, only one of 19 species of anurans (*Pseudacris brachyphona*) is restricted to the highlands; all other species of anurans there have extensive distributions in the lowlands. However, 38 of the 74 species of salamanders in the Appalachian Highlands are strictly montane; thus, only 5% of the anurans but 51% of the salamanders are strictly montane. All of the anurans in the Interior Highlands have widespread distributions in the lowlands, whereas 12 (54%) of the 22 species of salamanders there are strictly montane. Of the 15 species of anurans in the Rocky Mountains, six (40%) are strictly montane, as are three (75%) of four species of salamanders. With the realization that some species (*Ascaphus truei, Bufo boreas, Rana luteiventris,* and *Ambystoma macrodactylum*) are common to both the Rocky Mountains and the Pacific–Cascade–Sierra Nevada ranges, the number of strictly montane species of amphibians in the Nearctic is widely disparate between anurans and salamanders. About 58% of the salamanders in the Nearctic are restricted to mountains, whereas only 20% of the anurans are strictly montane.

Perhaps one of the most fascinating distributional phenomena exhibited by Nearctic amphibians is the great elevational distributions of many species. Six species range from sea level to at least 3000 m—*Bufo boreas* (3600 m), *Pseudacris triseriata* (3670 m), *Rana pipiens* (3553 m), *Rana sylvatica* (3050 m), *Ambystoma macrodactylum* (3000 m), and *Ambystoma tigrinum* (3658 m). All but *Bufo boreas* and *Ambystoma macrodactylum* have extensive distributions in the lowlands to the east of the Rocky Mountains.

CLIMATE

The greatest diversity of both anurans and salamanders is in regions having an excess of 1000 mm of precipitation annually—Atlantic Coastal Plain, Appalachian Highlands, and Pacific-Cascade ranges. A distinct east-west moisture gradient along the Gulf Coastal Plain correlates with a decrease in the number of species of amphibians—55 species in the east where annual precipitation is 1000–1500 mm to 21 species in the west where there is 500–750 mm of precipitation. A similar gradient is evident from the forested Interior Lowlands in the east to the Shortgrass Steppe in the west; along this gradient, rainfall decreases from 750–1000 mm to less than 500 mm, and the number of amphibians decreases from 49 to 20. Likewise, there is a gradient in annual precipitation from more than 1500 mm in the northern part of the Pacific Ranges to less than 750 mm in the southern part of the Pacific Ranges; this diminishes further to less than 250 mm in Baja California. The corresponding decrease in the number of species of amphibians is 22 to 17 to 6. By far the lowest diversity is in dry regions, such as Baja California (6 species), the Mojave Desert (6 species), and the Sonoran Desert (13 species).

The amphibian faunas of dry areas are dominated by *Bufo, Spea,* and *Scaphiopus,* taxa that are remarkably tolerant of arid conditions; most of these taxa are capable of burrowing to levels in the earth where they can acquire moisture during long dry periods. In contrast, the highest diversity of *Hyla, Pseudacris,* and *Rana* is in wet areas, principally the Southern and Gulf Coastal plains (Fig. 2:23). With the exception of *Ambystoma tigrinum,* salamanders seldom occur in regions receiving less than 500 mm of precipitation annually, and their greatest diversity is in regions receiving more than 1000 mm of precipitation annually; facultative neoteny may have played a role in the survival of *A. tigrinum* in dry regions.

There are no clear correlations of distributional limits with temperatures. However, obligate neotenes

Table 2:12. Northern limits of distributions of amphibians that have wide distributions at latitudes higher than 50° N lat. in North America. Arranged in descending order of highest latitudes reached. Precipitation and temperatures are for the regions of the northernmost parts of the ranges.

Species	Degrees N lat.	Natural region(s)	Annual regional precipitation (mm)	Mean annual temperature (°C)	
				January	July
Rana sylvatica	70	Cordillera/tundra	250	−25	+05
Pseudacris triseriata	63	Boreal forest	250	−25	+15
Bufo boreas	62	Pacific Range/Cordillera	1500	−10	+15
Rana pipiens	62	Boreal forest	250	−25	+15
Bufo hemiophrys	61	Boreal forest	250	−30	+15
Taricha granulosa	61	Pacific ranges	1500	−10	+10
Rana luteiventris	60	Cordillera	1500	−20	+10
Ambystoma macrodactylum	58	Cordillera	1500	−20	+10
Ambystoma gracile	55	Pacific Range/Cordillera	1500	−15	+10
Ambystoma laterale	55	Boreal forest	750	−25	+15
Rana septentrionalis	55	Boreal forest	500	−20	+15
Bufo americanus	55	Boreal forest	500	−25	+10
Pseudacris regilla	54	Pacific ranges	2000	0	+10
Ambystoma tigrinum	54	Great Plains	250	−15	+15
Pseudacris crucifer	53	Boreal forest	500	−20	+10

usually live in waters that do not freeze, either because they inhabit waters that are not subjected to freezing temperatures or lentic conditions prevent freezing (e.g., streams inhabited by *Cryptobranchus*). Sirenids and amphiumids are in the former category, but in the northern part of its range in the Interior Lowlands, *Siren lacertina* is subjected to freezing temperatures and must hibernate. Of the six species of *Necturus,* only *N. maculosus* inhabits regions where lakes freeze over in the winter, but the salamanders remain active below the ice (Bishop, 1941), as do facultatively neotenic populations of *Ambystoma tigrinum.*

Fifteen species (10 anurans and 5 salamanders) have extensive ranges north of 50° N Lat. (Table 2:12). With the exception of species along the Pacific coast, where winter temperatures are moderated by the ocean, these species survive winters with mean temperatures of less than −20°C. Some of these species survive the long, harsh winters by hibernating below the freezing level, but at least four species (*Hyla versicolor, Pseudacris crucifer, P. triseriata,* and *Rana sylvatica*) synthesize cryoprotectants (glucose or glycerol) that act as antifreeze in the blood; thus, the frogs can tolerate long periods of below-freezing temperatures (Pinder et al., 1992). Of the other species that occur at northern latitudes, *Bufo americanus, B. boreas, Pseudacris regilla, Rana pipiens,* and *R. septentrionalis* cannot tolerate

long-term freezing; the physiological abilities of the other species are not known.

Six of these northern species (*Bufo boreas, Pseudacris regilla, Rana luteiventris, Ambystoma gracile, A. macrodactylum,* and *Taricha granulosa*) extend into high latitudes (54–62° N. lat.) in western NA; four others (*Bufo hemiophrys, Pseudacris triseriata, Rana pipiens,* and *Ambystoma tigrinum*) reach their northern limits (54–63° N. lat.) in midcontinental regions (Fig. 2:26). Four species (*Bufo americanus, Pseudacris crucifer, Rana septentrionalis,* and *Ambystoma laterale*) reach their northern limits at 53–55° N lat. in eastern NA. The northernmost frog on the continent, *Rana sylvatica,* has a range extending north of 50° N lat. across the continent; it is the only species that extends north of the Arctic Circle and reaches the shores of the Arctic Ocean.

Only two of these species with extensive northern distributions (*Bufo hemiophrys* and *Rana septentrionalis*) are confined to northern regions; the southern limits of their ranges are at about 44° N lat. Many of the other species have broad latitudinal distributions. For example, the ranges of *Pseudacris crucifer* and *P. triseriata* encompass the Boreal Forest to the nearly subtropical conditions of the Gulf Coastal Plain; the ranges of *Bufo boreas* and *Ambystoma tigrinum* extend into Mexico, and the ranges of *Rana pipiens* and *R. sylvatica* extend far southward in mountainous regions.

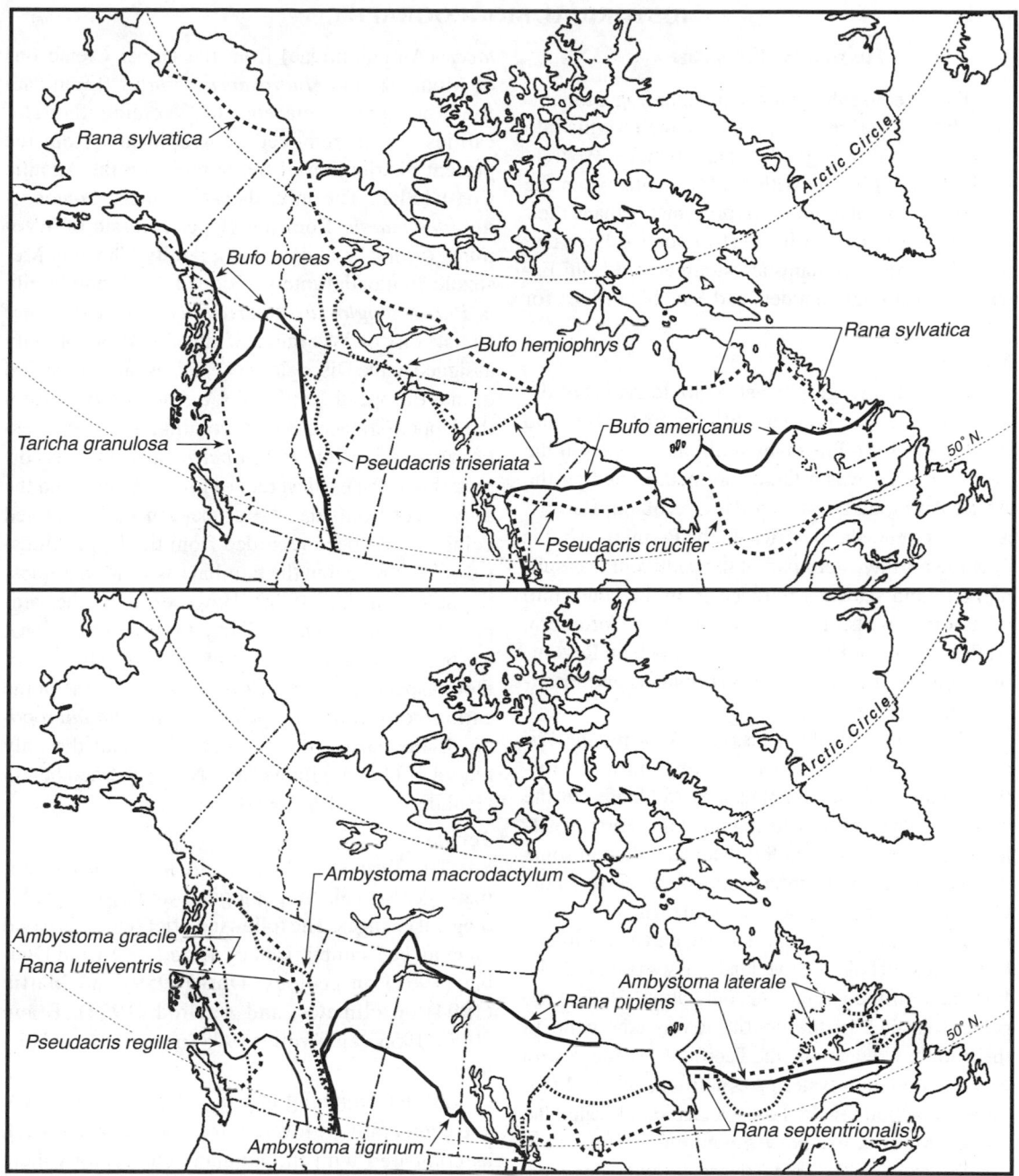

Fig. 2:26.   Northern limits of distribution of 15 species of amphibians that have extensive ranges north of 50° N lat. Distributions in the USA are not shown.

VEGETATION

The association of various species of amphibians with vegetation formations has been treated in a general way in the sections on Natural Regions of NA and Regional Patterns of Distribution. Generally, amphibians are not closely associated with particular species of plants. Further detail is beyond the scope of this treatment.

HISTORICAL BIOGEOGRAPHY

HISTORICAL PHENOMENA

Paleogeography and paleoclimatology, together with the fossil record, provide some insights into the historical biogeography of amphibians. We treat the history beginning with the Mesozoic, when the modern groups of amphibians presumably made their first appearance. Unless referenced otherwise, records of fossil remains are summarized from Estes (1981) for salamanders and Sanchiz (1998) for anurans.

Mesozoic

The paleogeography and tectonic events of the Mesozoic are summarized briefly from Dott and Prothero (1994). For many millions of years in the Paleozoic, NA was a large continent including the Arctic Archipelago of Canada, Greenland, Europe, and parts of northern Africa. In the Permian, what is now the northwestern part of the continent was submerged, and seaways intruded from the south into what is now parts of the North American Cordillera. At the end of the Permian, the Appalachian Revolution uplifted the Appalachian Highlands and Interior Highlands.

The major events in eastern NA during the Triassic and Jurassic involved erosion, uplift, and further erosion of the Appalachian Highlands. In the west, seaways invaded the region of the North American Cordillera from the north and south, and some uplift occurred in the intervening lands. Early in the Cretaceous, submergence occurred on a grand scale; continental margins were flooded, and a vast interior sea divided NA into two land masses (Fig. 2:27A). For much of the Cretaceous only about 50% of the present surface of the continent was emergent. In the middle Cretaceous, the Pacific–Cascade–Sierra Nevada mountain system was uplifted, and the Laramide Revolution in the Late Cretaceous elevated the Rocky Mountains; both mountain regions were greatly eroded in the Late Cretaceous and Paleogene.

Atmospheric circulation was very different from the present. Partly this was the result of a lower temperature differential between the poles and the equator; the differential has been estimated to have been as little as 17°C in the mid-Cretaceous, whereas the present differential is 41°C (Kerr, 1984).

The fossil record in NA provides little information about modern families of amphibians. The only members of living families are *Proamphiuma cre-*

tacea (Amphiumidae) from the Upper Cretaceous of Montana, and *Habrosaurus dilatus* (Sirenidae) from the Upper Cretaceous of Wyoming; these localities are far removed geographically from the present distribution of the families on the Atlantic Coastal Plain. The only other salamander, *Comonecturoides marshi* from the Upper Jurassic of Wyoming, cannot be assigned to a family. The only Mesozoic anuran definitely assigned to a living family is *Paradiscoglossus americanus* from the Upper Cretaceous of Wyoming; if this fossil is correctly assigned to the Discoglossidae, this is the only record of that European family in NA. Another fossil from the Upper Cretaceous of Wyoming, has been questionably assigned to *Palaeobatrachus;* if that is correct, this is the earliest record for the family and the only record outside of Europe. An undetermined pelobatid has been recorded from the Upper Jurassic of Wyoming, but the familial association is questionable (Sanchiz, 1998). However, two other amphibian fossils from the Early Jurassic of Arizona, are highly significant to the evolutionary history of their respective groups. *Prosalirus vitis* is one of the earliest known true frogs, and *Eocaecilia micropodia* is the earliest known caecilian, and the only record of Gymnophiona from NA as defined herein (Jenkins and Walsh, 1993).

Tertiary

The Tertiary history of NA reveals some dramatic geological, climatological, and associated biological changes. The following brief review is based on evidence summarized by King (1958) and Dunbar (1961) on geology, Dorf (1959) and Martin (1994) on climates, and Axelrod (1953), Braun (1950, 1955), and Wolfe (1978) on vegetational history.

At the beginning of the Tertiary, the continent had comparatively low relief, especially in the west, as compared with the mountainous regions there today (Fig. 2:27B). The climate was much more moderate than today, with far less marked changes in seasons. Humid conditions prevailed in the midcontinental region, as evidenced by the presence in the Eocene of trees such as *Sequoia* and *Metasequoia,* now found in humid forests. Tropical and subtropical conditions occurred far north of their present positions (Fig. 2:27C); the taxonomic composition of the northern tropical forest in NA has its

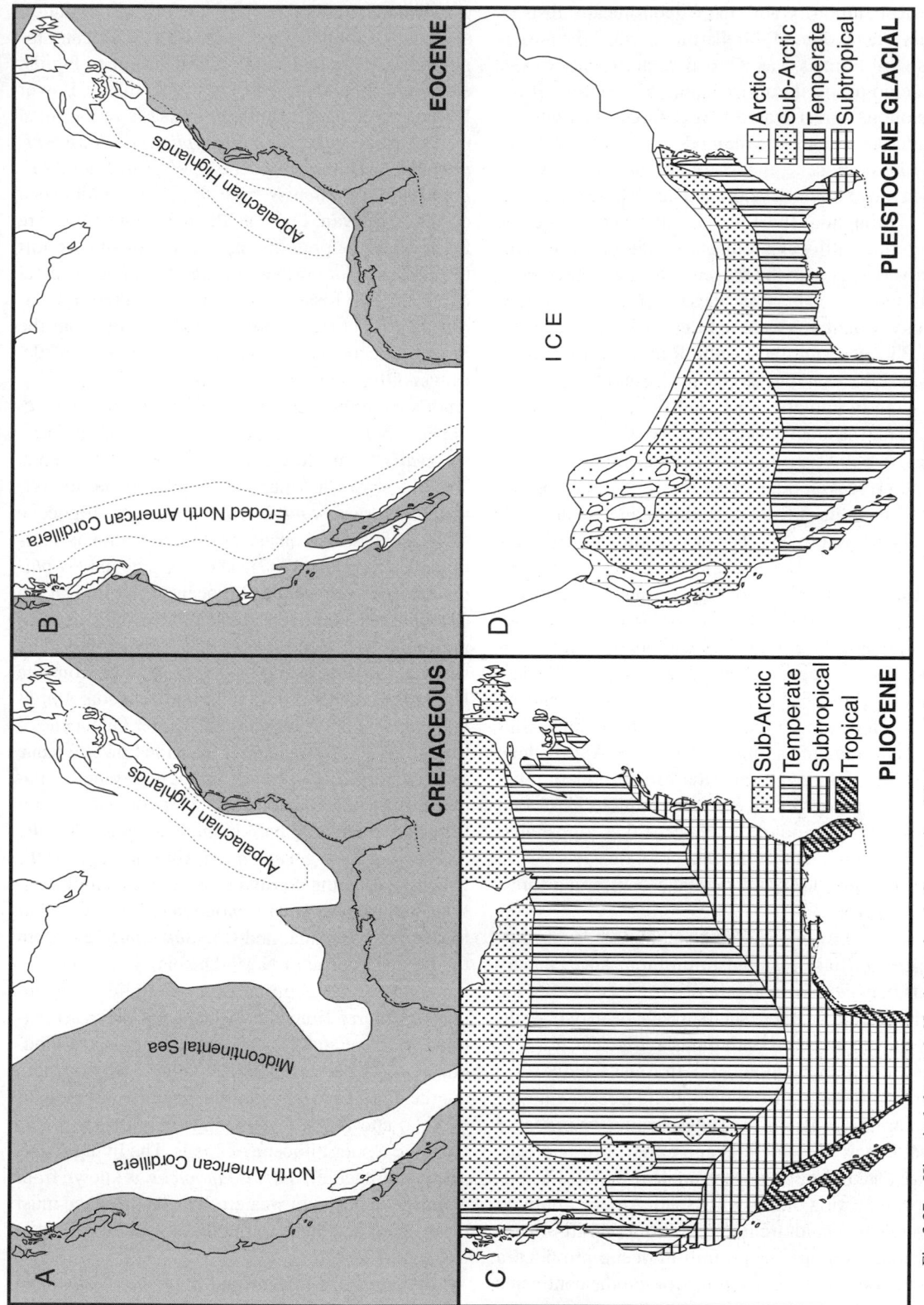

Fig. 2.27. Historical changes in North America. A. Composite of Cretaceous showing maximum submergence. B. Eocene. C. Pliocene climatic zones (probably also approximates Pleistocene interglacials). D. Composite of Pleistocene glacials showing climatic zones. A and B adapted from Dunbar (1961); C and D adapted from Dorf (1959).

greatest similarity with the Paleotropical forests of Malaysia (Wolfe, 1980). Farther north and continuous with Eurasia was a broad Arcto-tertiary forest, made up primarily of deciduous trees. North of the deciduous forest was a boreal forest far richer in coniferous species than the modern boreal forest. Probably in the Eocene, the Madro-tertiary mixed woodland type of forest developed on the Mexican Plateau.

Throughout the Tertiary, but especially beginning in the Miocene, climate steadily deteriorated, partly as a result of the orogeny of the western mountain systems. The Cascades, Sierra Nevada, and Rocky Mountains were uplifted in the Miocene and the Pliocene, and the Pacific Ranges were uplifted in the Pliocene. These changes resulted in the depression of climatic zones and floras southward. A broad rain shadow developed east of the rising mountains, thereby forcing the deciduous forest eastward of newly developing grasslands and isolating this forest from its Asiatic component and depauperate relicts in western NA. The increasingly rising, cooler, and drier western part of the continent was invaded at lower elevations by various elements derived from the Madro-tertiary forest and simultaneously at higher elevations by boreal Arcto-tertiary elements. Probably in the Late Miocene or Early Pliocene, a sclerophyll woodland-pine association of Madro-tertiary origin invaded the Gulf Coastal Plain and southern part of the Appalachian Highlands; this was ancestral to the modern pine-oak associations in the eastern deciduous hardwood forests. By the end of the Pliocene there apparently was no equivalent to the modern extensive, nearly uniform, northern and montane coniferous forest associations.

Plate movements and changing sea levels during the Tertiary resulted in different land connections between NA and other continents. During the early Paleogene, there seem to have been two northern land connections with Eurasia (McKenna, 1975). The Bering connection between Alaska and Siberia existed off and on from the middle Eocene into the Quaternary, whereas the DeGeer route via probably what is now Ellesmere Island to northwestern Eurasia existed in the Paleocene until the middle Eocene. The connection of NA with South America initially was via an island-arc in the Miocene; uplift of Panamanian Isthmus in the mid-Pliocene provided a continuous land connection between the continents (Pitman et al., 1993).

Paleogene fossils contribute little to our understanding of biogeography of North American amphibians. An anuran fossil of undetermined family, *Eorubeta nevadensis,* is known from the Lower Eocene of Nevada. Rhinophrynidae is represented by two fossil genera (*Chelomophrynus* and *Eorhinophrynus*) from the Eocene of Wyoming, and *Rhinophrynus canadensis* from the Lower Oligocene of Saskatchewan, Canada; all of these localities are far north of the present range of family and support the evidence for northern occurrence of subtropical climates. Two fossils assigned to the otherwise European extinct pelobatid *Eopelobates* are from the Lower Eocene of Wyoming (*E. guthrei*), and the Lower Oligocene of South Dakota (*E. grandis*). Another family presently restricted to the Palearctic, Pelodytidae, is represented by *Tephrodytes brassicarvalis* from the Upper Oligocene of Montana. The oldest North American fossil assigned to Hylidae, *Hyla swanstoni,* is from the Lower Oligocene of Saskatchewan, Canada, but this allocation is questionable (Sanchíz, 1997). The living genus *Scaphiopus* is known from the Oligocene of North Dakota (*Scaphiopus skinneri*). Two living genera of salamanders are represented by *Cryptobranchus saskatchewanensis* and *Necturus krausei* from the Upper Paleocene of Saskatchewan, Canada. *Amphiuma jepseni* is known from the Upper Paleocene of Wyoming, and *Siren dunni* is from the Lower Eocene of Wyoming; the presence of these neotenic salamanders is indicative of warm temperate, if not subtropical, conditions there in the Paleogene. *Taricha lindoei* and *T. oligocenica* are from the Upper Oligocene of Oregon. Fossils assigned to Dicamptodontidae are *Ambystomichus montanensis* from the Paleocene of Montana, and *Chrysotriton tiheni* from the Lower Eocene of North Dakota.

With one exception, (the pelodytid *Miopelodytes gilmorei* from the Miocene of Nevada), extinct Miocene and Pliocene genera belong to modern North American families. Other extinct genera are the hylid *Proacris mintoni* from the Miocene of Florida, and the ambystomatid *Amphitriton brevis* from the Upper Pliocene of Texas. The living Asiatic cryptobranchid, *Andrias japonicus,* is known from Miocene deposits in western NA; its presence must be associated with the Beringia connection with Asia in the mid-Tertiary.

Otherwise, Miocene and Pliocene fossils have been assigned to living North American genera, the earliest records for which in NA are *Scaphiopus, Spea, Bufo, Acris, Hyla, Rana, Ambystoma, Aneides, Plethodon,* and *Notophthalmus* in the Miocene, and *Dicamptodon, Batrachoseps,* and *Pseudobranchus* in the Pliocene. Living genera (*Amphiuma, Taricha,* and *Siren*) that appeared in the Paleogene also are represented in Miocene and/or Pliocene deposits. Fossils from the Miocene and Pliocene have been assigned to some living species—*Spea bombifrons* from the Miocene and Pliocene, *Rana "pipiens"* from the Miocene, and *Bufo alvarius, B. cognatus, B. compactilis, B. woodhousii, Acris crepitans, Rana areolata, R. catesbeiana, R. sylvatica, Ambystoma maculatum,* and *A. tigrinum* from the Pliocene. With the exception of *Aneides* and *Plethodon* from the Lower Miocene of Montana, all of the Miocene and Pliocene fossils of living genera are within, or close to, the ranges of the genera today. The presence of *Plethodon* and *Aneides* in the Lower Miocene of Montana, is indicative of temperate forests in the region at that time.

Quaternary

By the end of the Pliocene, the long period of climatic deterioration with increasing seasonality was culminated by formation of continental and mountain glaciers. Classically, four major glacial periods—Nebraskan, Kansan, Illinoian, and Wisconsin—and three interglacial periods—Aftonian, Yarmouth, and Sangamon—have been recognized (Flint, 1957). More recent interpretations (Dott and Prothero, 1994) suggest that most of the earlier half of the Pleistocene may have no record in the glaciated parts of NA; moreover, oxygen-isotope curves indicate a larger number of glacial advances in the early Pleistocene (1–2 million years ago) on an approximate 100,000-yr cycle.

Glacial periods resulted in dramatic southward shifts in climatic zones and climatic depression in mountainous regions (Fig. 2:27D), as well as lowering of sea level, thereby greatly expanding the coastal regions, especially along the Atlantic and Gulf coasts. During interglacial periods climatic zones probably were much the same as during the Pliocene (Fig. 2:27C), and sea level was higher than at present. Sea level was as much as 100 m higher in some interglacial periods and, thus, inundated much of the coastal plain in eastern NA and created a major intrusion in the Mississippi Embayment. Despite these great periodic shifts, climates were more equable than at present; the drastic seasonal shifts in climates seem to have developed in the Holocene.

Although the recurrent glacial ebb and flow resulted in latitudinal and elevational shifts in the biota (especially well documented by fossil pollens), fossil evidence strongly supports the notion that the biotas were not simply compressed latitudinally or elevationally during glacial periods; nor were the reestablished biotas during the interglacials necessarily the same as those at the same place during preceding interglacials. Consequently, many Pleistocene deposits contain the remains of species that are widely allopatric today.

Data from the time of the Greatlakean advance (= Wisconsin glaciation in earlier terminology) (8,000–11,000 yr ago) and the Holocene show differential development of climates and associated vegetation. For example, woodland communities were present in the Chihuahuan, Sonoran, and Mojave deserts in the late Wisconsin; warm desert species were common in woodlands at low elevations, and mixed conifer and subalpine forests existed at high elevations in southwestern USA (Van Devender and Spaulding, 1979). Montane plant communities acquired their modern aspects and the more mesophytic species disappeared at lower elevations in southwestern USA about 11,000 yr ago; early Holocene xeric woodlands and an inferred winter precipitation regime persisted until about 8,000 yr ago. According to Van Devender and Spaulding (1979), the present circulation patterns, rainfall regimes, and biotic distributions probably formed as a result of the melting of the continental ice sheets; communities in southwestern NA seemed to have responded quickly to these climatic changes in contrast to the more gradual response in central and eastern NA.

In other cases, Wisconsin (= Greatlakean) plant communities resemble present communities at higher latitudes or elevations. For example, Wells and Stewart (1987) noted that the taiga-like community 18,000–14000 yr ago in western Kansas and Nebraska was made up of a mixture of coniferous trees and aspen parkland resembling the modern subalpine forest in the Rocky Mountains, whereas today the region is semiarid shortgrass steppe. Extensive studies of Holocene changes in plant communities in the Great Basin by Wells (1983) document a northward shift of 500–640 km during the last 8,000–10,000 yr and likewise show during the same time the isolation of species of conifers on the mountains in the region.

The Pleistocene herpetofaunas of NA were summarized by Holman (1995), and the details are not repeated here. When the paleofaunistics are examined, several points are noteworthy. First, many late Pleistocene and Holocene amphibian faunas contain assemblages of species that occur at, or near, the same place today. For example, all seven of the species in the Williston 3A fauna in Florida, are present there today, as are the seven species in Groesbeck Creek fauna in Texas, and the nine species in the Clark Cave fauna in Virginia.

Second, some middle- and late- Pleistocene faunas include not only species that occur there today but also extralimital species. For example, the Arredondo fauna in Florida, contains nine species present in the area today, plus *Bufo woodhousii*, which occurs to the north and west. The Albert Ahrens fauna in Nebraska, contains five species that occur there today, plus four (*Bufo americanus, Hyla versicolor, Rana clamitans,* and *R. sylvatica*) that occur primarily to the north and east and one (*Pseudacris clarkii*) that only occurs farther south. The Lubbock Lake fauna in Texas, contains seven species that occur in the vicinity today, plus *Rana palustris* that only exists today far to the east. The coexistence of "ecologically incompatible" taxa was the basis for an equability model proposed by Hibbard (1960). This model predicts that mild winters would allow northern expansion of southern species, and that cool summers would allow northern species to survive south of their present range. Climatic equability was used to explain disharmonious Pleistocene herpetofaunas in southern USA and the Great Plains by Holman (1976, 1980).

Third, the amphibian fauna of NA was quite stable during the Pleistocene, as compared with the avian and mammalian faunas (Holman, 1991), but there are virtually no data on amphibians west of the cordillera. In contrast to endotherms, there is no evidence of familial or generic extinctions in the North American amphibian fauna during the Quaternary, and there were relatively few range adjustments in the herpetofauna, compared with those of mammals.

PHYLOGENY AND PATTERNS OF DISTRIBUTION

Phylogenetic analyses have been generated for several monophyletic groups of amphibians inhabiting NA. The results of these analyses can be applied to biogeography by associating branching patterns of cladograms of taxa with their occurrence in different regions. Prior to a synthesis of the patterns, we provide the independent data sets.

Anurans of the genus *Bufo* were included in an analysis of many taxa in Bufonidae by Graybeal (1997), whose phylogenetic arrangement was based on data derived from DNA sequences. Her analysis included 16 of the 20 Nearctic species and 25 extralimital species. The Nearctic *Bufo baxteri, californicus, hemiophrys,* and *woodhousii* were not included; her sample of "*B. woodhousii*" was from North Carolina, and therefore referable to *B. fowleri.* According to Graybeal's (1997) strict consensus tree, Nearctic *Bufo* are members of four major clades (Fig. 2:28). The first major clade includes only *B. alvarius* and *B. valliceps;* the former is only in the Sonoran and Chihuahuan deserts, and the latter is widely distributed in Mexico and Central America, but enters the Nearctic on the Gulf Coastal Plain as far east as the Mississippi Embayment and into the Interior Plains in the southern part of the Shortgrass Steppe and Tallgrass Prairie, as well as the Edwards Plateau. The other Nearctic *Bufo* form a trichotomy: (1) *Bufo boreas* group in the North American Cordillera and Intermontane Plateaus; (2) *Bufo bocourti* in the Guatemalan highlands in Central America; and (3) the remaining Nearctic *Bufo.* Within the latter, *B. punctatus* is the sister clade to the other taxa, among which *B. debilis* and *B. retiformis* form a clade that is the sister group to all of the others. These three species occur in southwestern USA and on the Mexican Plateau. The distributions of the remaining seven species are entirely east of the Sierra Nevada and show pairs of eastern and western species—*B. cognatus* (western) and *B. americanus* (eastern); *B. microscaphus* (western) and *B. terrestris* (eastern).

Da Silva (1997) added new morphological data to Cocroft's (1994) phylogenetic analysis of North American hylid frogs based on morphological, biochemical, and behavioral data sets, and produced a cladogram showing three major clades of Nearctic hylids (Fig. 2:29). *Acris,* with one species widely distributed in eastern NA and another in the Atlantic Coastal Plain and Piedmont of the Appalachian Highlands, is the sister group to all other Nearctic *Hyla,* within which are two major clades. The first clade consists of two primarily Mexican species (*H.*

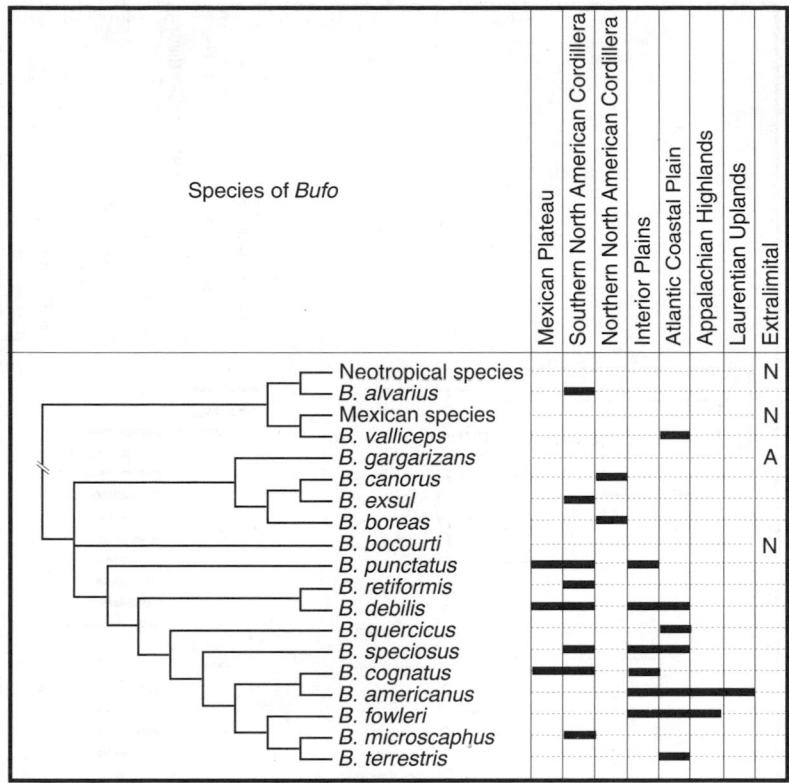

Fig. 2:28. Phylogenetic arrangement of North American *Bufo* with species occurrence in physiographic divisions. Adapted from Graybeal (1997); her *Bufo woodhousii* = *B. fowleri*. In the column Extralimital, A = eastern Asia, and N = Neotropical.

arenicolor and *H. eximia*) and is the sister clade to all other hylids in NA. The second major clade includes the two species of *Acris* as the sister group to all other NA *Hyla*. Within *Hyla*, the cryptic *H. chrysoscelis* and *H. versicolor*, widely distributed in eastern NA, are sister taxa to two species on the Atlantic Coastal Plain (*H. andersonii* and *H. avivoca*). The other clade within *Hyla* contains four species (*H. cinerea, femoralis, gratiosa*, and *squirella*) that are distributed primarily on the Atlantic Coastal Plain. The third major clade identifies *Pseudacris*, within which the two western species (*P. cadaverina* and *P. regilla*) form a sister group that seems to be basal to all other *Pseudacris*, of which the widespread eastern *P. crucifer* is the sister species to the remaining members of the genus. The distributions of these include the Atlantic Coastal Plain, except for *P. brachyphona* in the Appalachian Highlands. *Pseudacris brachyphona* is the sister species of *P. triseriata*, which is widely distributed east of the Sierra Nevada. However, in both Cocroft's (1994) and da Sil-

va's (1997), analyses, *P. brachyphona* was nested with subspecies (*P. t. kalmi* and *P. t. feriarum*) of *P. triseriata*.

A phylogenetic analysis of New World *Rana* based on morphology and rDNA (Hillis and Davis, 1986; Hillis, 1988) reveals a basal trichotomy of: (1) the Palearctic *Rana temporaria* and *R. sylvatica*, which is widespread in the northern Nearctic; (2) the *Rana boylii* group restricted to the North American Cordillera, but mainly distributed in the cool montane regions of that division; and (3) all other New World *Rana* (Fig. 2:30). Within the latter, the *Rana catesbeiana* group is basal to the other clades; this group includes the sister species *R. catesbeiana* and *R. clamitans*, both widely distributed in eastern NA, nested within three other species (*R. grylio, heckscheri*, and *virgatipes*) restricted to the Atlantic Coastal Plain. Among the remaining clades, one major clade includes the *Rana palmipes* group (at least six species in Middle and South America) as the sister group to a clade of three species, two of

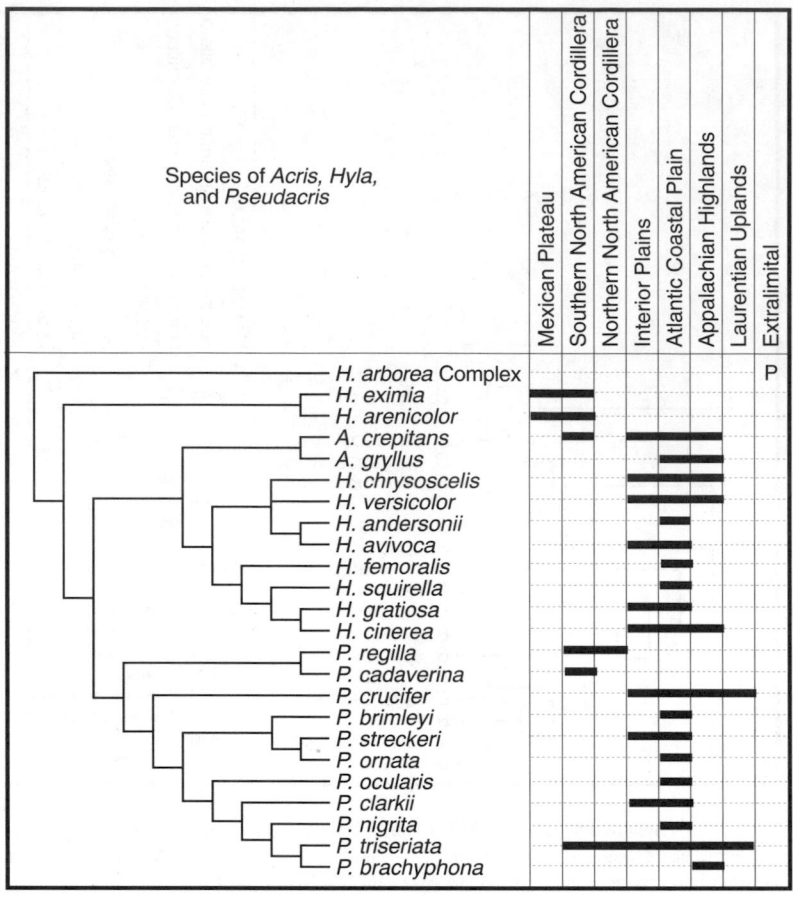

Fig. 2:29. Phylogenetic arrangement of North American hylid frogs with species occurrence in physiographic divisions. Adapted from da Silva (1997). In the column Extralimital, P = Palearctic.

which (*R. pustulosa* and *R. zweifeli*) are on the Mexican Plateau, and the other (*R. tarahumarae*) occurs in the Madrean Mountains in southwestern NA. All of the other New World *Rana* are members of the *Rana pipiens* complex, as defined by Hillis (1988), who noted that each of the two divisions of the complex contained two clades—one North American and one Middle American. The first division, the *Rana palustris* group, has a clade of two species (*Rana areolata* and *R. palustris*) in eastern NA and a clade of four species (*R. chiricahuensis, dunni, montezumae,* and *trilobata*) on the Mexican Plateau, one of which (*R. chiricahuensis,* the sister species to the other species in the clade) extends northward into the mountains in the southern part of Intermontane Plateaus. Within the second division, the *Rana pipiens* group, a clade of three species is present in eastern NA; of these, *R. sphenocephala* principally on

the Atlantic Coastal Plain, is basal to *R. pipiens,* which is widely distributed in northeastern NA, and its sister species *R. blairi,* which is principally in the Great Plains. A second clade in this division contains species that are restricted to Mexico, except for *R. yavapaiensis,* which occurs in the mountains in the southern part of the Intermontane Plateaus, and *R. berlandieri,* which has an extensive distribution in eastern Mexico and enters the Nearctic in the Chihuahuan Desert, the western part of the Atlantic Coastal Plain, and the southernmost part of the Great Plains.

Using data from allozymes and morphology, Shaffer et al. (1991) generated a maximum-parsimony cladogram of *Ambystoma,* exclusive of the unisexual species *A. nothagenes, platineum,* and *tremblayi* (Fig. 2:31). Their analysis reveals three major clades. In a basal clade, *A. talpoideum* in south-

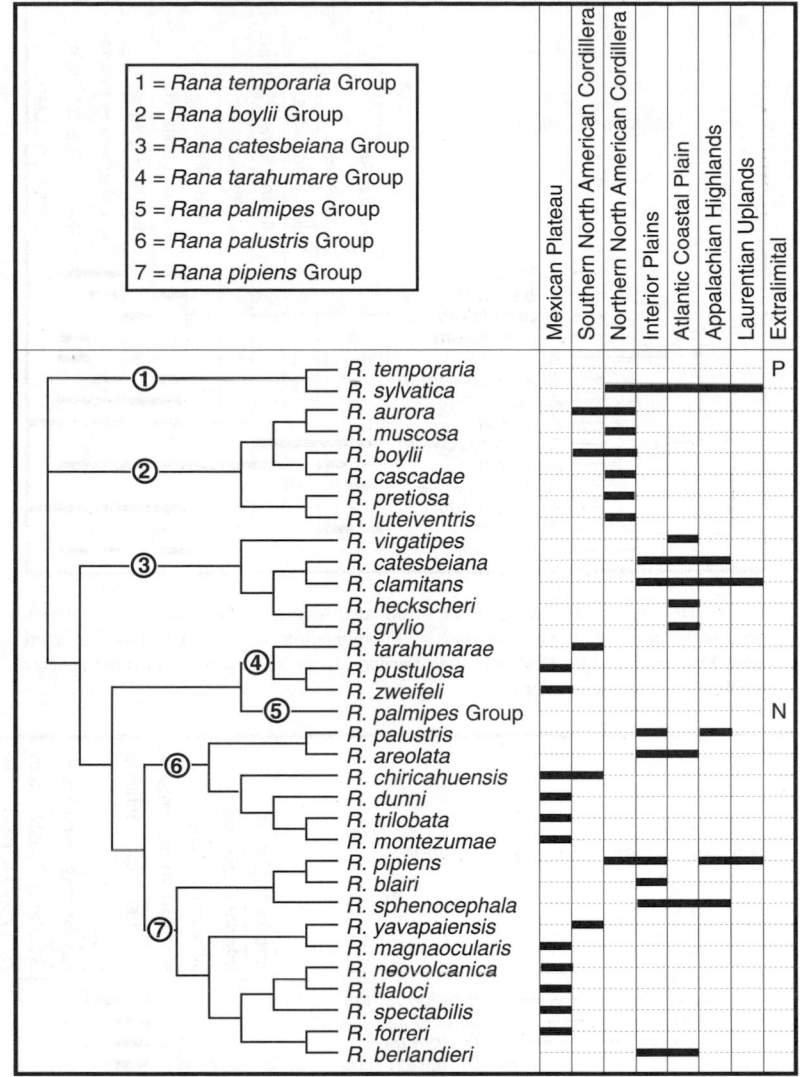

Fig. 2:30. Phylogenetic arrangement of North American *Rana* with taxonomic occurrence in physiographic divisions. Arrangement based on Hillis (1988) and Hillis and Davis (1986). In the column Extralimital, N = Neotropical, and P = Palearctic. *Rana trilobata* = *R. megapoda* of Hillis (1988) and Hillis and Davis (1986).

eastern NA is the sister species to *A. maculatum* in eastern NA and *A. gracile* in northwestern NA. The second major clade includes the *Ambystoma texanum* group, the basal member of which is *A. mabeei* on the Atlantic Coastal Plain; two pairs of species in this clade are *A. texanum*, which is widespread in eastern NA, and *A. barbouri*, which has a small distribution in the Interior Lowlands. The other pair consists of *A. cingulatum* on the Southern and Gulf Coastal plains and *A. annulatum* in the Interior Highlands. The third clade consists of three pairs of species—*A. jeffersonianum* and *A. opacum* in eastern NA, *A. laterale* in the northeast and *A. macrodactylum* in the northwest, and *A. californiense* in the Pacific Ranges and *A. tigrinum*, which occurs throughout most of the southern half of the continent east of the Sierra Nevada, as well as in the Cascade Range in the northwest. *Ambystoma tigrinum* is part of a complex of species centered on the Mexican Plateau (Shaffer, 1983, 1984).

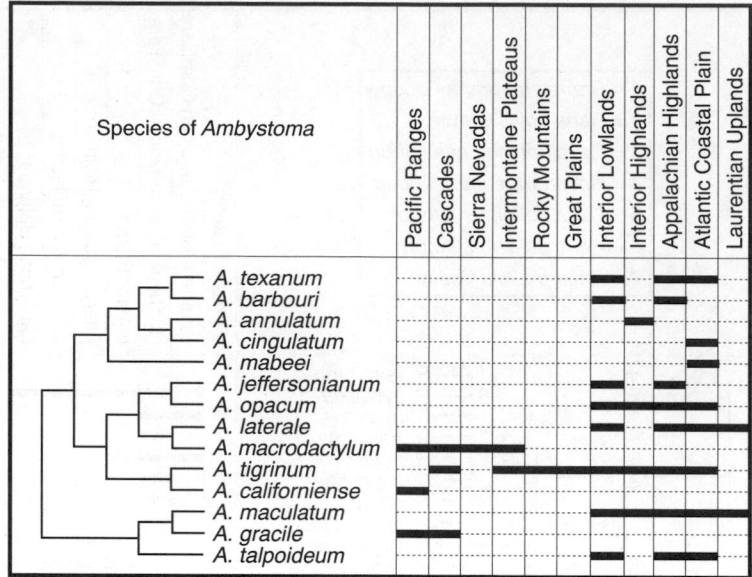

Fig. 2:31.   Phylogenetic arrangement of bisexual species of *Ambystoma* with species occurrence in physiographic divisions and combinations of natural regions. Adapted from Shaffer et al. (1991). *Ambystoma tigrinum* is the sister species to other species inhabiting the Mexican plateau.

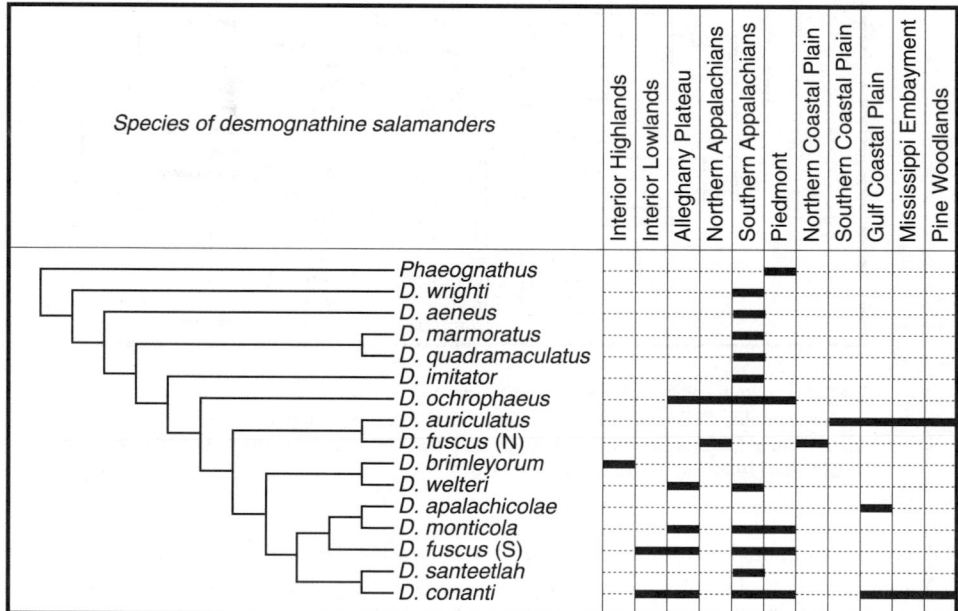

Fig. 2:32.   Phylogenetic arrangement of desmognathine salamanders of the family Plethodontidae with taxonomic occurrence in natural regions. Adapted from Titus and Larson (1996). The following species recognized herein are not included: *D. carolinensis, ocoee,* and *orestes;* two populations of *D. "fuscus"* are designated north (N) and south (S).

Titus and Larson's (1996) analysis of mDNA of desmognathine salamanders was combined with morphological characters and life-history traits to hypothesize the evolutionary history of the group (Fig. 2:32). Their analysis showed that the basal clade consists only of *Phaeognathus hubrichti,* which has a restricted distribution in the southern Piedmont. Species in the next four succeeding clades are restricted to the Southern Appalachians, whereas the following clade, consisting only of *Desmognathus ochrophaeus* occurs throughout the Appalachian Highlands. A final major clade contains four pairs of species, three of which have one member restricted to one or more regions in the Appalachian Highlands and the other in another region—*D. fuscus* (highlands) and *D. auriculatus* (coastal plain), *D. monticola* (highlands) and *D. apalachicolae* (coastal plain), and *D. welteri* (highlands) and *D. brimleyorum* (Interior Highlands). The fourth pair consists of *D. santeetlah* in the Southern Appalachians and *D. conanti* in the Appalachian Highlands, Interior Lowlands, and coastal plains.

Regrettably, no phylogenetic analysis is available for *Plethodon,* but it is possible to assemble a dendrogram of phenetic relationships based on allozyme data (Fig. 2:33). These data show the distinctiveness of the eastern *Plethodon* (*P. cinereus, glutinosus, wehrlei,* and *welleri* groups) from the western *Plethodon* (*P. elongatus, neomexicanus, vandykei,* and *vehiculum* groups). With two exceptions, all western *Plethodon* arc restricted to the northern parts of the Pacific and Cascade ranges. *Plethodon idahoensis* in the Northern Rocky Mountains is shown to form a species pair with *P. vandykei* in the Pacific northwest, and *P. neomexicanus* in the Southern Rocky Mountains forms a pair with *P. larselli* in the Cascade Range. All but three species in the *Plethodon cinereus* group are restricted to the Appalachian Highlands; *P. cinereus* and *P. richmondi* also occur in the Interior Lowlands, and *P. cinereus* and *P. serratus* also occur on the coastal plain and the latter also is in the Interior Highlands. Four of the species in the *Plethodon welleri* group are restricted to the Appalachian Highlands, and one species (*P. angusticlavus*) occurs only in the Interior Highlands. Both species in the *Plethodon wehrlei* group are in the Appalachian Highlands.

Nine of the 21 species in the *Plethodon glutinosus* group occur only in the Appalachian Highlands, Only two species (*P. glutinosus* and *P. missis-* *sippi*) occur in the Appalachian Highlands and in the Interior Lowlands, and the latter also occurs in the Gulf Coastal Plain and Mississippi Embayment. Two other Appalachian species (*P. chlorobryonis* and *P. ocmulgee*) also occur on the coastal lowlands, to which three species (*P. grobmani, kisatchie,* and *variolatus*) are restricted. Six species are endemic (or nearly so) to the Interior Highlands. One group of three such species (*P. caddoensis, fourchensis,* and *ouachitae*) are distinct allozymically from other members of the *Plethodon glutinosus* group, whereas the other three species are nested within the rest of the group; *P. albagula* (also on the Edwards Plateau) and *P. kiamichi* are shown to be nearest relatives, whereas the third species, *P. sequoyah,* is considered to be most closely related to *P. kisatchie* in the adjacent Pine Woodlands of the Atlantic Coastal Plain.

The analyses of the anurans and *Ambystoma* emphasize the significance of the southern part of the North American Cordillera (Sierra Madre Occidental and Mexican Plateau in Mexico) to the biogeography of Nearctic amphibians. At least four lineages of *Bufo* (*B. alvarius, valliceps, punctatus,* and *debilis + retiformis*) presumably evolved on the Mexican Plateau and dispersed northward (and in some cases also eastward) into NA, as was the case with *Hyla eximia* group (*H. arenicolor + H. eximia*) and *Rana tarahumarae.* In two other groups of *Rana* (*R. palustris* and *R. pipiens* groups), the basal species are in eastern NA, whereas the more derived species are in southwestern USA and Mexico; also the *Ambystoma tigrinum* complex, which is speciose on the Mexican Plateau, is nested with a lineage of *Ambystoma* in eastern NA. At least in the cases of *Ambystoma* and the two groups of *Rana,* divergence between the eastern North American taxa and lineages composed of the numerous taxa on the Mexican Plateau cannot be interpreted as resulting from Pleistocene relictualism on the Mexican Plateau, inasmuch as fossils assigned to the "*Rana pipiens* complex" and to an "*Ambystoma tigrinum*-like" salamander are known from the Miocene of NA (Estes, 1981; Sanchiz, 1998).

Whereas the Appalachian Highlands and Interior Highlands have anurans faunas made up of a subset of species in adjacent lowlands (except *Pseudacris brachyphona* endemic to the Appalachian Highlands), this pattern is reversed in plethodontid salamanders. Within *Plethodon,* the greatest diver-

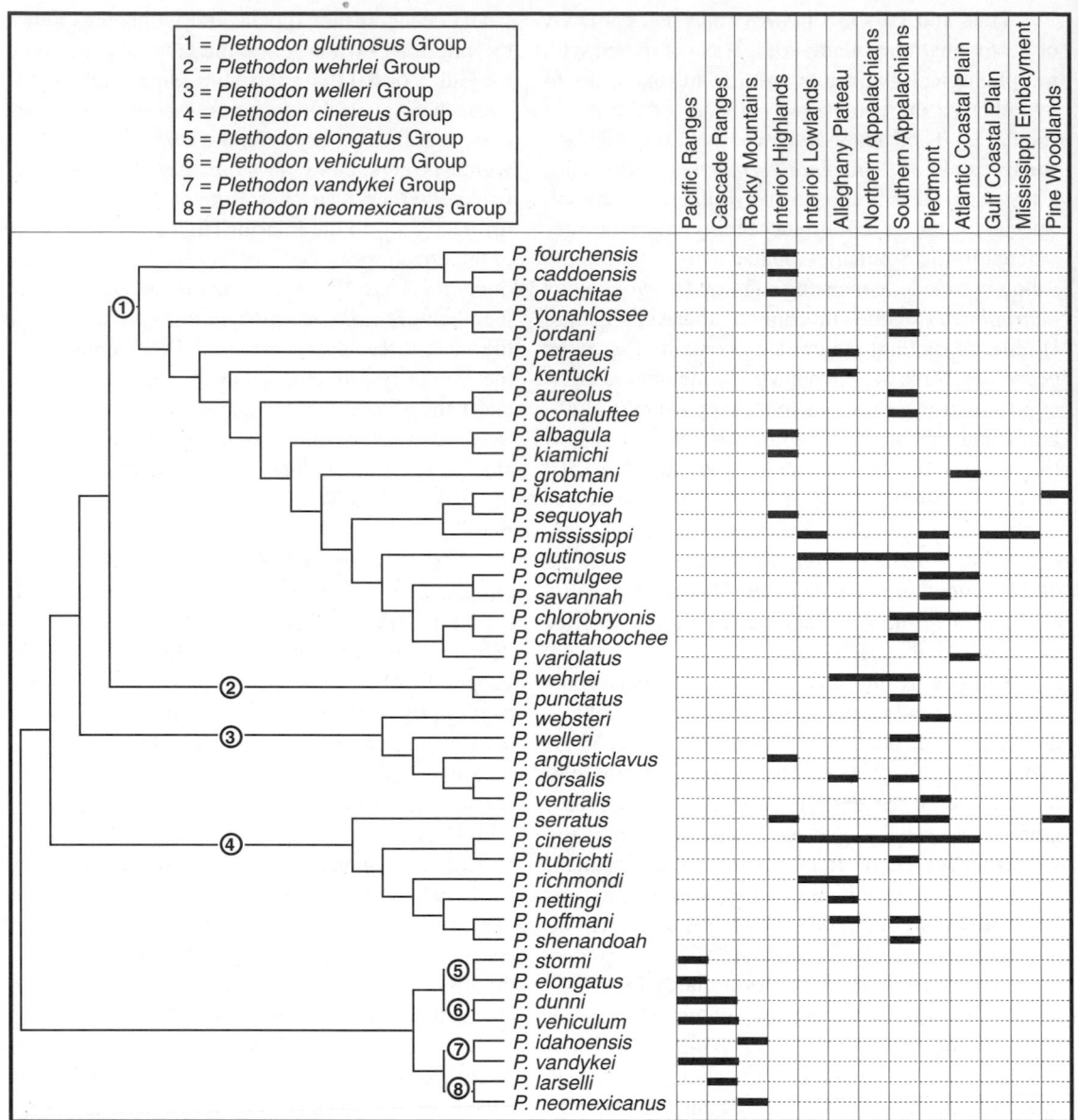

Fig. 2:33. Phenetic arrangement of *Plethodon* with species occurrence in natural regions. Reconstructed from allozyme-based "trees" and data in Highton and Larson (1979), Wynn et al. (1988), Highton et al. (1989), and Highton (1995, 1997).

sity exists in the Appalachian Highlands, especially the Southern Appalachians; 18 of the 24 species of *Plethodon* in the Appalachian Highlands are endemic to the physiographic division, and these species are representatives of all four species groups of eastern *Plethodon*. Eight species of *Plethodon* are endemic to the Interior Highlands; the species are members of three species groups, all of which inhabit the Appalachian Highlands. Based on the proposed inter-

specific relations among members of this genus, five separate lineages of *Plethodon* are shared by the Interior Highlands and the Appalachian Highlands (Fig. 2:33), from which lineages presumably dispersed to the Interior Highlands. Conceivably, some of these lineages may have entered the Interior Highlands during times of Pleistocene glacial depression, but such a late date does not seem to be reasonable for the lineage consisting of three species (*P. cad-*

doensis, P. fourchensis, and *P. ouachitae*) endemic to the Interior Highlands. All species of *Plethodon* that occur in the lowlands either also inhabit the Appalachian Highlands and presumably dispersed into the lowlands (*P. chlorobryonis, cinereus, glutinosus, mississippi, ocmulgee, richmondi,* and *serratus*) or are nested within groups that are restricted to highlands (*P. grobmani, kisatchi,* and *variolatus*).

The distinct and closely related groups of *Plethodon* in western NA apparently are relicts of a lineage that was across the continent prior to the end of the Miocene, as evidenced by the presence of Miocene *Plethodon* in Montana (Estes, 1981). Nested within the six species in the Pacific northwest are two relictual species in the Rocky Mountains; each of these has a separate closest relative in the Pacific northwest.

With the exception of taxa in the Pacific northwest, the historical pattern in *Desmognathus* parallels that of *Plethodon.* Most species of the genus, and the sister taxon, *Phaeognathus hubrichti,* are endemic to the Appalachian Highlands. One species (*D. brimleyorum*) is endemic to the Interior Highlands, and its sister species (*D. welteri*) is restricted to the Appalachian Highlands. Of the four species that occur in the lowlands, two (*D. conanti* and *D. fuscus*) also occur in the Appalachian Highlands, whereas two species (*D. apalachicolae* and *D. auriculatus*), each with sister species restricted to the Appalachian Highlands, occur only in the lowlands.

Although no phylogeny of plethodontine salamanders of the Tribe Hemidactyliini has been proposed, the pattern of distribution of these salamanders is much the same as *Plethodon* and *Desmognathus. Hemidactylium,* with a broad, overall distribution consisting of many disjunct populations in eastern NA, may be separate from other taxa in the tribe, which do seem to form a monophyletic group (Wake, 1993). The four species of *Gyrinophilus* are endemic to the Appalachian Highlands; two species (one endemic) of *Pseudotriton* and seven species (2 endemic) of *Eurycea* also occur there. Those *Eurycea* that are not endemic to the Appalachian Highlands occur in the adjacent lowlands, except for two species, plus the monotypic *Typhlotriton,* in the Interior Highlands, and four species on the Edwards Plateau, where the two species of *Typhlomolge* are endemic. The monotypic genera *Haideotriton* and *Stereochilus* are endemic to the Atlantic Coastal Plain.

Thus, Nearctic plethodontid salamanders, with the exception of four western genera, seem to have radiated from the Appalachian Highlands. Of the western genera, *Aneides* is the only genus not restricted to the far west; it has three species (*A. ferreus, flavipunctatus,* and *lugubris*) in the Pacific–Cascade–Sierra Nevada ranges, one (*A. hardii*) in the Southern Rocky Mountains, and one (*A. aeneus*) in the Appalachian Highlands. According to Larson et al. (1981), *Aneides* is most closely related to western *Plethodon* (thereby rendering *Plethodon* paraphyletic), and according to their immunological data, the two genera diverged in the Eocene-Oligocene. Likewise, the differentiation of species of *Aneides* took place in the Miocene; that age of the genus is supported by the existence of Miocene fossils from Montana (Estes, 1981). The western monotypic genus *Ensatina* is considered to be the sister group to *Aneides* + *Plethodon* and to have diverged from that lineage in the Paleocene (Larson et al., 1981).

The other two western genera of plethodontine salamanders (*Batrachoseps* and *Hydromantes*) belong to the Bolitoglossini, a tribe that is highly diverse in the neotropics; *Hydromantes* has congeners in Mediterranean Europe. (See Borkin, this volume.) It is thought that a lineage ancestral to the Bolitoglossini entered Mexico from North America no later than the early Tertiary (Wake, 1987). Hendrickson (1986) proposed that the North American bolitoglossines arrived there via northward movements of continental fragments associated with the fault systems along the west coast of Mexico and NA; he argued that the congruence of distributions of various species of *Batrachoseps* with historical geological units in California strongly substantiated the theory of paleotransportation. Wake (1996) recognized two major clades in *Batrachoseps*—one containing only *B. wrighti* from the Cascade Range and *B. campi* from the Inyo Mountains, California; these widely disjunct species are essentially at the northern and eastern extremes of the generic range. The genetic distance between *B. campi* and congeners is great, but closest to *B. wrighti;* genetic divergence between isolated populations in the comparatively small Inyo Range is comparable to that between species in the major mountain ranges (Yanev, 1980; Yanev and Wake, 1981).

The salamandrid genera *Notophthalmus* and *Taricha* are sister groups (Titus and Larson, 1995). According to biochemical data the minimum age of divergence between the eastern *Notophthalmus* and the western *Taricha* is approximately 20 mya in the Miocene, whereas the minimum age of divergence between the northern *N. viridescens* and a southern lineage (*N. meridionalis + perstriatus*) was approximately 15 mya, and the separation of the southern species was 7–8 mya (Reilly, 1990). Fossils assigned to *Taricha* from the Oligocene of Oregon and Miocene of Montana extend these minimal times of divergences. Reilly (1990) postulated that *Notophthalmus* differentiated into a northern population (*N. viridescens)* in the temperate forests of northeastern NA and a southern population (*N. meridionalis + perstriatus*) in the subtropical forests in southern NA in the Miocene, and that subsequent climatic deterioration resulted in the isolation of populations of the latter in the southeastern coastal plain (*N. perstriatus*) and in the southwestern coastal plain (*N. meridionalis*).

A combination of latitudinal zonation of climates and formation of grasslands in the central part of the continent southward to the Gulf of Mexico beginning in the Miocene, together with intermittent inundations of the Mississippi Embayment in the Miocene-Pleistocene, is reflected in patterns of speciation and distribution in the Gulf Coastal Plain from peninsular Florida to northeastern Mexico. Several sister species have distributions separated by the Coastal Prairies and/or Mississippi Embayment; examples are *Scaphiopus holbrookii* and *S. hurterii, Gastrophryne carolinensis* and *G. olivacea, Pseudacris ornata* and *P. streckeri,* and *Notophthalmus meridionalis* and *N. perstriatus.*

The dating of divergences among lineages of amphibians by biochemical means (e.g., Larson et al., 1981; Maxson and Wilson, 1975; Wake et al., 1978; Wallace et al., 1973) generally is congruent with historical events in earth history and the limited data from the fossil record. The molecular clock

interpretation of protein differences strongly supports the differentiation of many intrafamilial lineages among North American amphibians as having occurred in the early to mid-Tertiary, whereas specific differentiation occurred as early as the Miocene.

However, apparently not all species are so old. Highton et al. (1989) estimated Pleistocene divergence among some species in the *Plethodon glutinosus* group, and Good and Wake (1992) noted that although Pleistocene glacial events were instrumental in the formation of patterns of distribution in *Rhyacotriton,* biochemical data support an earlier differentiation of the species. Lowered climatic zones during glacial times have been discounted for the invasion of mountain ranges in the southwestern part of the Basin and Range Region from the Sierra Nevada by *Batrachoseps;* instead, these salamanders are thought to have inhabited fissures in rocks since the Tertiary (Marlow et al., 1979).

Nonetheless, Pleistocene events and Holocene climatic changes were important in shaping present distributions. With few exceptions (*Bufo hemiophrys, Rana septentrionalis,* and the unisexual *Ambystoma nothagenys, platineum,* and *tremblayi),* no species of amphibians in eastern NA are restricted to glaciated regions. However, many species do occupy these formerly glaciated regions. The low genetic variability in populations of *Plethodon cinereus* and in species of the *P. glutinosus* complex inhabiting glaciated regions compared to that in several fragmented populations of the same species inhabiting unglaciated regions has been attributed to rapid invasion of the glaciated region by a northern population of each species (Highton and Webster, 1976; Highton et al., 1989). Species now inhabiting high mountains that were glaciated in western NA (e.g., *Bufo canorus, Rana muscosa,* and *Hydromantes platycephalus*) must have ascended to high elevations after the Wisconsin glacial period.

CONSERVATION OF THE AMPHIBIAN FAUNA

The conservation status of North American amphibians received little attention until the late 1980s (Blaustein and Wake, 1990; Wake, 1991), when a few earlier reports (e.g., Moyle, 1973; Baxter et al., 1982; Sweet, 1983; Corn and Fogelman, 1984; Hayes and Jennings, 1986; Bradford, 1989) and a large number of unpublished observations (reviewed in Stebbins and Cohen, 1995) indicated widespread declines and regional extirpations, principal-

ly among anurans. Prior to 1990, a few species of salamanders had been afforded state and federal protection in the USA; in each instance, protection was based on greatly restricted distributions or threats of habitat loss or degradation. However, the prospect of a more general effect attracted both public and scientific attention. The latter interest rendered negative survey data publishable; this resulted in a flood of reports of population declines and extirpations (summarized by Vial and Saylor, 1993; Blaustein et al., 1994; Carey, 1993; Corn, 1994; Phillips, 1994; Blaustein and Wake, 1995; Drost and Fellers, 1996; Fisher and Shaffer, 1996) and a continuing debate on causes and significance (e.g., Pechman et al., 1991; Blaustein, 1994; Pechman and Wilbur, 1994; Stebbins and Cohen, 1995; Green, 1997). This is an active field of inquiry at present; for current information, the reader should consult the publication *Froglog* (IUCB/SSC, 1993 et seq.). Only a general outline of the conservation status of North American amphibians is presented here.

By virtue of their physiology, amphibians commonly are restricted to stable, mesic microenvironments, persistence of which depends on complex interactions of geological, climatic, and floristic factors on both local and regional scales. As such, many species appear to be sensitive indicators of environmental change; amphibians tend to disappear early when areas are converted from natural habitat by agriculture, urbanization, etc. At the same time, many species can persist in local, favorable microsites, with the result that distributional relics commonly occur either as outliers of the former range or, in the longer term, as regional endemics or relict taxa.

The loss of amphibian diversity to habitat alienation should surprise no one; what seems to be unusual is the widespread occurrence of declines and local extinctions in areas where little or no tangible environmental degradation can be documented in the period encompassing the effect. Declines of this nature in NA seem to correspond to an emerging global pattern in being more pronounced at high elevations, and on the western margins of continents (Wake, 1991). These declines are consistent with the expected effects of major climatic changes, especially those affecting patterns of rainfall. In western NA, species of *Rana* and *Bufo* seem to be particularly susceptible; effects are less evident or absent among other anurans and in most salamanders. It can be argued that many of the amphibians in decline in NA are distributional or ecological relicts in a broad sense; these species are dependent on relatively extensive (or regionally influenced) mesic habitats that have been progressively restricted during the present interglacial interval. In the Cordillera, the Pacific-Sierra Nevada-Cascade ranges and the areas between their southern extensions, the extent and nature of mesic microhabitats presumably have been undergoing rapid alteration during the past

10,000–12,000 years; the rate of change has been accelerated considerably by human activities in recent decades. Stebbins and Cohen (1995) presented a useful review of specific cases involving direct and indirect human influences, as did several authors in the volume edited by Green (1997) focusing on Canada.

Western North American species of *Bufo* (including *B. boreas* in the Cordillera, *baxteri, californicus, canorus, hemiophrys,* and *microscaphus*) and *Rana* (including *R. aurora, boylii, cascadae, chiricahuensis, fisheri, luteiventris, muscosa, onca, pretiosa, subaquavocalis, sylvatica* in the central Rocky Mountains, *tarahumarae,* and *yavapaiensis*) have declined in parts or in all of their ranges; in some cases, species are locally extinct. Each of these species requires either relatively extensive aquatic systems or geographically restricted habitats, such as springs or streams; the integrity of these systems and habitats depends on regional influences. Both are subject to short-term climatic fluctuations, and many are susceptible to water-management activities, such as groundwater pumping, drainage for agriculture, or impoundment for water-supply or flood-control purposes, which singly or in combination have drastically altered preexisting seasonal patterns. Logging, intensive livestock grazing, pesticide applications, introduction of predatory fishes (such as trout at high elevations and various centrarchids elsewhere) and bullfrogs (*Rana catesbeiana*) contribute further to environmental degradation and to increased mortality of larval and juvenile frogs.

Some combination of these anthropogenic effects characterizes nearly every instance of decline in amphibian populations in western North America, but unresolved issues remain in a number of cases. For example, it is not clear why some taxa are affected more than others (often in sympatry), nor why an apparent threshold phenomenon has occurred in the absence of any clear evidence of augmentation in potentially causal factors. Also, precipitous declines often occur over short intervals and in manners that suggest action by some pathogen, rather than as a dwindling owing to progressive environmental degradation reflected in reduced recruitment. Fungal and bacterial pathogens identified in various cases seem to be secondary infections, as opposed to primary causes.

The current focus on declining amphibians is energized by these relatively dramatic examples, but beyond this focus a broader pattern is emphasized less often. This is simply that anthropogenic effects are degrading or alienating a rapidly increasing fraction of the continent, resulting in the further fragmentation of suitable habitat across most of the centers of diversity and endemism documented previously. There is general agreement that habitat protection is the most effective conservation measure, but the terms of such protection often are misconstrued to incorporate uses incompatible with maintenance of amphibian diversity (e.g., grazing, flood control, fish restocking, or intensive recreational use).

No species of amphibian in NA has reached extinction as a consequence of European settlement, but a number of western *Rana* are perilously close; they have been extirpated from significant fractions of their former ranges and exist now as tiny and isolated relict populations. Other species, mainly salamanders, have not declined in historic times, but many are restricted to single localities (e.g., *Batrachoseps aridus, Eurycea nana, E. sosorum, Gyrinophilus subterraneus, Typhlomolge rathbuni,* and *T. robusta*). Any of these taxa could become extinct through accidental or intentional single events or in response to increasing climatic fluctuations. Additionally, several species of *Batrachoseps* in the west and of *Plethodon* in the east have quite restricted ranges in areas that are highly susceptible to disturbance. Other taxa (e.g., species of *Ambystoma, Spea,* and *Scaphiopus*) retain wider distributions but are experiencing fragmentation of ranges through widespread destruction or degradation of breeding sites.

Formal legislative protection of threatened and endangered species of North American amphibians has lagged considerably behind the documented need for such actions. At the present time, only 13 species have been listed under the United States Endangered Species Act (Table 2:13); there are no comparable designations in Canada or Mexico. Many states and provinces have listings of protected amphibians (Levelle, 1997); these vary widely in their inclusiveness and effect. In addition to designating species that are threatened throughout their ranges, many state and provincial regulations cover widely distributed species having ranges that barely enter the jurisdictions involved. Relatively few states and provinces have proactive legislation (i.e., regulations precluding habitat destruction) and rely instead on restrictions on the taking of individuals. The gener-

Table 2:13. Amphibians listed under the United States Endangered Species Act. E = endangered; T = threatened.

Species	Status
Bufo baxteri	E
Bufo californicus	E
Bufo houstonensis	E
Rana aurora draytoni	T
Ambystoma macrodactylum croceum	E
Ambystoma tigrinum stebbinsi	E
Batrachoseps aridus	E
Eurycea nana	T
Eurycea sosorum	E
Phaeognathus hubrichti	T
Plethodon nettingi	T
Plethodon shenandoah	T
Typhlomolge rathbuni	E

al lack of effective state and provincial laws, and the incomplete and cumbersome federal listing process under the United States Endangered Species Act seem to reflect both a (diminishing) perception that amphibians have a low priority, and more importantly the experience that effective conservation measures are potentially disruptive of entrenched economic interests. Water-supply issues tend to be the conflicts most clearly involved, although land-use practices, such as logging, mining, grazing, and urban conversion, also are affected frequently.

In general, we have a mixed outlook for amphibian conservation in NA. The level of awareness of potential problems is relatively high in the USA and Canada; most threatened and declining species are being monitored to a degree matched in few other parts of the world. However, the range and scope of events threatening the survival of many species (e.g., western *Rana*) is large and includes climatic and biotic factors that are difficult or impossible to counteract effectively. Probably some of these species will become extinct in the forseeable future, despite the measures applied. On the other hand, many amphibians are widely distributed and relatively secure under present patterns of land use; moreover, many regional endemics occur in areas with some measure of protection against habitat loss owing both to restrictions on land use and to the intrinsic resilience of their local environments. A few species have profited from land-use practices (e.g., irrigation or livestock-watering impoundments in arid regions) and as a result have increased their distributions.

Acknowledgments: As students of North American amphibians, we and our colleagues owe a great debt of gratitude to Roger Conant and Robert C. Stebbins for their painstaking efforts in providing outstanding field guides containing the excellent distribution maps that were the basis for our analyses. We are grateful to Jonathan A. Campbell and Linda Trueb for their critical review of the manuscript. Duellman is indebted to Larry D. Martin and Jeffrey R. Parmelee for providing some pertinent literature.

LITERATURE CITED

AXELROD, D. I. 1953. Evolution of the Madro-tertiary geoflora. Bot. Rev. 24:433–509.

BAILEY, R. G. 1976. *Ecoregions of the United States.* (Map only). Ogden, Utah: U. S. Forest Service.

BAILEY, R. G. 1978. *Ecoregions of the United States.* Ogden, Utah: U. S. Forest Service.

BAXTER, G. T., M. R. STROMBERG, AND C. K. DODD, JR. 1982. The status of the Wyoming toad (*Bufo hemiophrys baxteri*). Environ. Conserv. 9:348, 388.

BISHOP, S. C. 1941. The salamanders of New York. New York State Mus. Bull. 324:1–365.

BLAUSTEIN, A. R. 1994. Chicken Little or Nero's fiddle? A perspective on declining amphibian populations. Herpetologica 50:85–97.

BLAUSTEIN, A. R., AND D. B. WAKE. 1990. Declining amphibian populations. A global phenomenon? Trends Ecol. Evol. 5:203–204.

BLAUSTEIN, A. R., AND D. B. WAKE. 1995. The puzzle of declining amphibian populations. Sci. Amer. 272:56–61.

BLAUSTEIN, A. R., D. B. WAKE, AND W. P. SOUSA. 1994. Amphibian declines: judging stability, persistence, and susceptibility of populations to local and global extinctions. Conserv. Biol. 8:60–71.

BRADFORD, D. F. 1989. Allotopic distribution of native frogs and introduced fishes in the high Sierra Nevada lakes of California: implications of the negative effect of fish introductions. Copeia 1989:775–778.

BRAUN, E. L. 1950. *Deciduous Forests of Eastern North America.* Philadelphia: Blakiston Co.

BRAUN, E. L. 1955. The phytogeography of unglaciated eastern United States and its interpretation. Bot. Rev. 21:297–375.

CAREY, C. 1993. Hypotheses concerning the causes of the disappearance of boreal toads from the mountains of Colorado. Conserv. Biol. 7:355–362.

CHIPPINDALE, P. T., A. H. PRICE, AND D. M. HILLIS. 1993. A new species of perennibranchiate salamander (*Eurycea:* Plethodontidae) from Austin, Texas. Herpetologica 49:248–259.

CLEMENTS, F. E., AND V. E. SHELFORD. 1939. *Bioecology.* New York: John Wiley & Sons.

COCKROFT, R. B. 1994. A cladistic analysis of chorus frog phylogeny (Hylidae: *Pseudacris*). Herpetologica 50:420–437.

COLLINS, J. T. 1991. Viewpoint: a new taxonomic arrangement for some North American amphibians and reptiles. Herpetol. Rev. 22:42–43.

COLLINS, J. T. 1992. The evolutionary species concept: a reply to Van Devender et al. and Montanucci. Herpetol. Rev. 23:43–46.

COLLINS, J. T. 1997. Standard common and current scientific names for North American amphibians and reptiles, 4th ed. Herpetol. Circular 25:1–37.

CONANT, R., AND J. T. COLLINS. 1991. *A Field Guide to Reptiles and Amphibians. Eastern and Central North America,* 3rd Ed. Boston: Houghton Mifflin Co.

COOK, F. R. 1984. *Introduction to Canadian Amphibians and Reptiles.* Ottawa, Canada: National Museum of Natural Sciences.

CORN, P. S. 1994. What we know and don't know about amphibian declines in the west. Pp. 59–67 *in* W. W. Covington and L. F. DeBano (eds.), *Sustainable Ecological Systems: Implementing an Ecological Approach to Land Management.* Ft. Collins, Colorado: USDA Forest Service, Rocky Mountain Forest and Range Experiment Station, Gen. Tech. Rept. RM-247.

CORN, P. S., AND J. C. FOGELMAN. 1984. Extinction of montane populations of the northern leopard frog (*Rana pipiens*) in Colorado. J. Herpetol. 18:147–152.

CROWLEY, J. M. 1967. Biogeography. Canadian Geographer 11:312–326.

DA SILVA, H. R. 1997. Two character states new for hylines and the taxonomy of the genus *Pseudacris.* J. Herpetol. 31:609–613.

DEGENHARDT, W. G., C. W. PAINTER, AND A. H. PRICE. 1996. *Amphibians and Reptiles of New Mexico.* Albuquerque: Univ. New Mexico Press.

DICE, L. R. 1943. *The Biotic Provinces of North America.* Ann Arbor: Univ. Michigan Press.

DICKINSON, W. R., AND W. S. SNYDER. 1979. Geometry of subducted slabs related to the San Andreas transform. J. Geol. 887:609–627.

DORF, E. 1959. Climatic changes of the past and present. Contr. Mus. Paleont. Univ. Michigan 13:181–210.

DOTT, R. H., Jr., and D. R. PROTHERO. 1994. *Evolution of the Earth,* Ed. 5. New York: McGraw-Hill Book Co.

DOWLING, H. G. 1993. Viewpoint: a reply to Collins (1991, 1992). Herpetol. Rev. 24:11–13.

DROST, C. A., AND G. M. FELLERS. 1996. Collapse of a regional frog fauna in the Yosemite area of the California Sierra Nevada. Conserv. Biol. 10:414–425.

DUELLMAN, W. E. 1990. Herpetofaunas in neotropical rainforests: comparative composition, history, and resource utilization. Pp. 455–505 *in* A. H. Gentry (ed.), *Four Neotropical Rainforests.* New Haven: Yale Univ. Press.

DUELLMAN, W. E. 1993. Amphibian species of the world: additions and corrections. Spec. Publ. Nat. Hist. Mus. Univ. Kansas 21:1–372.

DUELLMAN, W. E., AND A. SCHWARTZ. 1958. Amphibians and reptiles of southern Florida. Bull. Florida St. Mus. 3:181–324.

DUNBAR, C. O. 1961. *Historical Geology,* 2nd Ed. New York: John Wiley & Sons, Inc.

EISENMANN, E. 1955. The species of Middle American birds. Trans. Linn. Soc. New York 7:1–128.

ESTES, R. 1981. *Encyclopedia of Paleoherpetology. Part 2. Gymnophiona, Caudata.* New York: Gustav Fischer Verlag.

FENNEMAN, N. M. 1917. Physiographic divisions of the United States. An. Assoc. Am. Geogr. 6:19–98.

FISHER, R. N., AND H. B. SHAFFER. 1996. The decline of amphibians in California's Great Central Valley. Conserv. Biol. 10:1387–1397.

FLINT, R. F. 1957. *Glacial and Pleistocene Geology.* New York: John Wiley and Sons.

FROST, D. R. 1985. Amphibian species of the world. A taxonomic and geographical reference. Lawrence, Kansas: Allen Press, Inc. and Assoc. Syst. Coll.

FROST, D. R., A. G. KLUGE, AND D. M. HILLIS. 1992. Species in contemporary herpetology: Comments on phylogenetic inference and taxonomy. Herpetol. Rev. 23:46–54.

GOOD, D. A. 1989. Hybridization and cryptic species on *Dicamptodon* (Caudata: Dicamptodontidae). Evolution 43:728–744.

GOOD, D. A., AND D. B. WAKE. 1992. Geographic variation and speciation in the torrent salamanders of the genus *Rhyacotriton* (Caudata: Rhyacotritonidae). Univ. California Publ. Zool. 126:i–xii, 1–91.

GRAYBEAL, A. 1997. Phylogenetic relationships of bufonid frogs and tests of alternate macroevolutionary hypotheses characterizing their radiation. Zool. J. Linnean Soc. 119:297–338.

GREEN, D. M. (ed.). 1997. Amphibians in decline: Canadian studies of a global problem. Herpetol. Conserv. 1:1–338.

GREEN, D. M., H. KAISER, T. F. SHARBEL, J. KEARSLEY, AND K. R. MCALLISTER. 1997. Cryptic species of spotted frogs, *Rana pretiosa* complex, in western North America. Copeia 1997:1–8.

HAIRSTON, N. G., SR. 1993. On the validity of the name *teyahalee* as applied to a member of the *Plethodon glutinosus* complex (Caudata: Plethodontidae): a new name. Brimleyana 18:65–69.

HAMMERSON, G. A. 1986. *Amphibians and Reptiles in Colorado.* Denver: Colorado Division of Wildlife.

HAYES, M. P., AND M. R. JENNINGS. 1986. Decline of ranid frog species in western North America: Are bullfrogs (*Rana catesbeiana*) responsible? J. Herpetol. 20:490–509.

HEDGES, S. B. 1986. An electrophoretic analysis of holarctic hylid frog evolution. Syst. Zool. 35:1–21.

HENDRICKSON, D. A. 1986. Congruence of bolitoglossine biogeography and phylogeny with geological history: paleotransport on displaced suspect terranes? Cladistics 2:113–129.

HERSKOVITZ, P. 1958. A geographical classification of neotropical mammals. Fieldiana:Zool. 36:581–629.

HIBBARD, C. W. 1960. Pliocene and Pleistocene climates in North America. An. Rept. Michigan Acad. Arts. Sci. Letters 62:5–30.

HIGHTON, R. 1995. Speciation in eastern North American salamanders of the genus *Plethodon.* An. Rev. Ecol. Syst. 26:579–600.

HIGHTON, R. 1997. Geographic protein variation and speciation in the *Plethodon dorsalis* complex. Herpetologica 53:345–356.

HIGHTON, R., AND A. LARSON. 1979. The genetic relationships of salamanders of the genus *Plethodon.* Syst. Zool. 28:579–599.

HIGHTON, R., G. C. MAHA, AND E. R. MAXSON. 1989. Biochemical evolution in the slimy salamanders of the *Plethodon glutinosus* complex in the eastern United States. Illinois Biol. Monogr. 57:1–153.

HIGHTON, R., AND T. P. WEBSTER. 1976. Geographic protein variation and divergence in populations of the salamander *Plethodon cinereus.* Evolution 30:33–45.

HILLIS, D. M. 1988. Systematics of the *Rana pipiens* complex: puzzle and paradigm. An. Rev. Ecol. Syst. 19:39–63.

HILLIS, D. M., AND S. K. DAVIS. 1986. Evolution of ribosomal DNA: 50 million years of recorded history in the frog genus *Rana.* Evolution 40:1275–1288.

HOLMAN, J. A. 1976. Paleoclimatic implications of "ecologically incompatible" herpetological species (late Pleistocene, southeastern United States). Herpetologica 32:290–295.

HOLMAN, J. A. 1980. Paleoclimatic implications of Pleistocene herpetofaunas of eastern and central North America. Trans. Nebraska Acad. Sci. 8:131–140.

HOLMAN, J. A. 1991. North American Pleistocene herpetofaunal stability and its impact on the interpretation of modern herpetofaunas: an overview. Pp. 227–235 *in* J. R. Purdue, W. E. Klippel, and B. W. Styles (eds.), *Beamers, Bobwhites, and Blue-Points: Tributes to the Career of Paul W. Parmalee.* Springfield, Illinois: Illinois State Museum.

HOLMAN, J. A. 1995. *Pleistocene Amphibians and Reptiles in North America.* New York: Oxford Univ. Press.

JENKINS, F. A., JR., AND D. M. WALSH. 1993. An Early Jurassic caecilian with limbs. Nature 365:246–249.

KENDEIGH, S. C. 1954. History and evaluation of various concepts of plant and animal communities in North America. Ecology 35:152–171.

KENDEIGH, S. C. 1961. *Animal Ecology.* Englewood Cliffs, New Jersey: Prentice-Hall.

KERR, R. A. 1984. How to make a warm Cretaceous climate. Science 223:677–678.

KIESTER, A. R. 1971. Species densities of North American amphibians and reptiles. Syst. Zool. 20:127–137.

KING, P. B. 1958. Evolution of modern surface features of western North America. Pp. 3–60 *in* C. L. Hubbs (ed.), *Zoogeography.* Washington, D.C.: Am. Assoc. Advan. Sci.

KOCH, E. D., AND C. W. PETERSON. 1995. *Amphibians and Reptiles of Yellowstone and Grand Teton National Parks.* Salt Lake City: Univ. Utah Press.

LARSON, A., D. B. WAKE, L. R. MAXSON, AND R. HIGHTON. 1981. A molecular phylogenetic perspective on the origins of morphological novelties in the salamanders of the tribe Plethodontini (Amphibia: Plethodontidae). Evolution 35:405–422.

LEVELLE, J. P. 1997. *A Field Guide to Reptiles and the Law.* Malabar, Florida: Krieger Publishing Co.

LOBECK, A. K. 1948. *The Physiographic Provinces of North America* [Map]. New York: The Geographical Press, Columbia Univ.

LOWE, C. H. 1964. Arizona landscapes and habitats. Pp. 1–132 *in* C. H. Lowe (ed.), *The Vertebrates of Arizona.* Tucson: Univ. Arizona Press.

MARLOW, R. W., J. M. BRODE, AND D. B. WAKE. 1979. A new salamander, genus *Batrachoseps,* from the Inyo Mountains of California, with a discussion of relationships within the genus. Contr. Sci. Nat. Hist. Mus. Los Angeles Co. 308:1–17.

MARTIN, L. D. 1994. Cenozoic climatic history from a biological perspective. Inst. Tertiary-Quaternary Stud., TER-QUA Symp. Ser. 2:39–56.

MAXSON, L. R., AND A. C. WILSON. 1975. Albumin evolution and organismal evolution in tree frogs (Hylidae). Syst. Zool. 24:1–15.

MCKENNA, M. C. 1975. Fossil mammals and Early Eocene North Atlantic land continuity. Ann. Missouri Bot. Gard. 62:335–353.

MERRIAM, C. H. 1890. Results of a biological survey of the San Francisco Mountains region and desert of the Little Colorado in Arizona. U. S. Dept. Agr., N. Am. Fauna 3:1–136.

MERRIAM, C. H. 1894. Laws of temperature control of the geographic distribution of terrestrial animals and plants. Natl. Geogr. Mag. 6:229–238.

MOLE, P. E., AND J. KEZER. 1993. Karyology and systematics of the salamander genus *Pseudobranchus* (Sirenidae). Copeia 1993:39–47.

MONTANUCCI, R. M. 1992. Commentary on a proposed taxonomic arrangement for some North American amphibians and reptiles. Herpetol. Rev. 23:9–10.

MORAFKA, D. J. 1977. *A Biogeographic Analysis of the Chihuahuan Desert through its Herpetofauna.* The Hague: W. Junk.

MOYLE, P. M. 1973. Effects of introduced bullfrogs, *Rana catesbeiana,* on native frogs in the San Joaquin Valley, California. Copeia 1973:18–22.

NUSSBAUM, R. A., E. D. BRODIE, JR., AND R. M. STORM. 1983. *Amphibians and Reptiles of the Pacific Northwest.* Moscow, Idaho: Univ. Idaho Press.

PECHMANN, J. H. K., D. E. SCOTT, R. D. SEMLITSCH, J. P. CALDWELL, L. J. VITT, AND J. W. GIBBONS. 1991. Declining amphibian populations: the problem of separating human impacts from natural fluctuations. Science 253:892–895.

PECHMANN, J. H. K., AND J. M. WILBUR. 1991. Putting declining amphibian populations in perspective: natural fluctuations and human impacts. Herpetologica 50:65–84.

PETERS, J. A. 1955. Use and misuse of the biotic province concept. Am. Nat. 89:21–28.

PHILLIPS, K. 1994. *Tracking the Vanishing Frogs.* New York: St. Martin's Press.

PINDER, A. W., K. B. STOREY, AND G. R. ULTSCH. 1992. Estivation and hibernation. Pp. 250–274 in M. E., Feder and W. W. Burggren (eds.). *Environmental Physiology of the Amphibians.* Chicago: Univ. Chicago Press.

PITMAN, W. C., III, S. CANDE, J. LABRECQUE, AND J. PINDELL, 1993. Fragmentation of Gondwana: the separation of Africa from South America. Pp. 15–34 in P. Goldblatt (ed.), *Biological Relationships between Africa and South America.* New Haven: Yale Univ. Press.

PLATZ, J. E. 1993. *Rana subaquavocalis,* a remarkable new species of leopard frog (*Rana pipiens* complex) from southeastern Arizona that calls under water. J. Herpetol. 27:154–162.

REILLY, S. M. 1990. Biochemical systematics and evolution of the eastern North American newts, genus *Notophthalmus* (Caudata: Salamandridae). Herpetologica 46:51–59.

RUSSELL, A. P., AND A. M. BAUER. 1993. *The Amphibians and Reptiles of Alberta.* Calgary: Univ. Alberta Press.

SANCHIZ, B. 1998. *Encyclopedia of Paleoherpetology. Part 4. Salientia.* New York: Gustav Fischer Verlag.

SAVAGE, J. M. 1960. Evolution of a peninsular Herpetofauna. Syst. Zool. 0:184–212.

SAVAGE, J. M. 1966. The origins and history of the Central American herpetofauna. Copeia 1966:719–766.

SEVERINGHAUS, J., AND T. ATWATER. 1990. Cenozoic geometry and thermal state of the subducting slabs beneath western North America. Mem. Geol. Soc. Am. 176:1–22.

SHAFFER, H. B. 1983. Biosystematics of *Ambystoma rosaceum* and *A. tigrinum* in northwestern Mexico. Copeia 1983:67–78.

SHAFFER, H. B. 1984. Evolution in a paedomorphic lineage. I. An electrophoretic analysis of the Mexican ambystomatid salamanders. Evolution 38:1194–1206.

SHAFFER, H. B., J. M. CLARK, AND F. KRAUS. 1991. When molecules and morphology clash: a phylogenetic analysis of the North American ambystomatid salamanders (Caudata: Ambystomatidae). Syst. Zool. 40:284–303.

SHAFFER, H. B., AND M. L. McKNIGHT. 1996. The polytypic species revisited: genetic differentiation and molecular phylogenetics of the tiger salamander *Ambystoma tigrinum* (Amphibia: Caudata) complex. Evolution 50:417–433.

SOULÉ, M., AND A. J. SLOAN. 1966. Biogeography and distribution of the reptiles and amphibians on islands in the Gulf of California, Mexico. Trans. San Diego Soc. Nat. Hist. 14:137–156.

STEBBINS, R. C. 1951. *Amphibians of Western North America.* Berkeley, California: Univ. California Press.

STEBBINS, R. C. 1985. *A Field Guide to Western Reptiles and Amphibians,* 2nd Ed. Boston: Houghton Mifflin Co.

STEBBINS, R. C., AND N. W. COHEN. 1995. *A Natural History of Amphibians.* Princeton, New Jersey: Princeton University Press.

SWEET, S. S. 1982. A distributional analysis of epigean populations of *Eurycea neotenes* in central Texas, with comments on the origin of troglobitic populations. Herpetologica 38:430–444.

SWEET, S. S. 1983. Mechanics of a natural extinction event: *Rana boylii* in southern California. Salt Lake City, Utah: Program Society Study Amphibians Reptiles, Herpetologists' League (abstract).

TILLEY, S. G., AND M. J. MAHONEY. 1996. Patterns of genetic differentiation in salamanders of the *Desmognathus ochrophaeus* complex (Amphibia: Plethodontidae). Herpetol. Monogr. 10:1–42.

TITUS, T. A., AND A. LARSON. 1995. A molecular phylogenetic perspective on the evolutionary radiation of the salamander family Salamandridae. Syst. Biol. 44:125–151.

TITUS, T. A., AND A. LARSON. 1996. Molecular phylogenetics of desmognathine salamanders (Caudata: Plethodontidae): a reevaluation of evolution in ecology, life history, and morphology. Syst. Biol. 45:451–472.

VAN DEVENDER, T. R., C. H. LOWE, H. K. McCRYSTAL, AND H. E. LAWLER. 1992. Viewpoint: reconsider suggested systematic arrangements for some North American amphibians and reptiles. Herpetol. Rev. 22:10–14.

VAN DEVENDER, T. R., AND W. G. SPAULDING. 1979. Development of vegetation and climate in the southwestern United States. Science 204:701–710.

VESTAL, A. G. 1914. Internal relations of terrestrial associations. Am. Nat. 48:413–445.

VIAL, J. L., AND L. SAYLOR. 1993. The status of amphibian populations. Corvallis, Oregon: Declining Amphibian Populations Task Force, Working Document 1:1–98.

WAKE, D. B. 1987. Adaptive radiation of salamanders in Middle American cloud forests. Ann. Missouri Bot. Gard. 74:242–264.

WAKE, D. B. 1991. Declining amphibian populations. Science 253:860.

WAKE, D. B. 1993. Phylogenetic and taxonomic issues relating to salamanders of the family Plethodontidae. Herpetologica 49:229–237.

WAKE, D. B. 1996. A new species of *Batrachoseps* (Amphibia: Plethodontidae) from the San Gabriel Mountains, southern California. Contr. Sci. Los Angeles Co. Mus. 463:1–12.

WAKE, D. B., L. R. MAXSON, AND G. Z. WURST. 1978. Genetic differentiation, albumin evolution, and their biogeographic implications in plethodontid salamanders of California and southern Europe. Herpetologica 32:529–539.

WALLACE, D. G., M.-C. KING, AND A. C. WILSON. 1973. Albumin differences among ranid frogs: taxonomic and phylogenetic implications. Syst. Zool. 22:1–13.

WELLS, P. V. 1983. Paleobiogeography of mountain islands in the Great Basin since the last glaciopluvial. Ecol. Monogr. 53:341–382.

WELLS, P. V., AND J. D. STEWART. 1987. Cordilleran-boreal taiga and fauna on the central Great Plains of North America, 14,000–18,000 years ago. Am. Midl. Nat. 118:94–106.

WELSH, H. H., JR. 1988. An ecogeographic analysis of the herpetofauna of the Sierra San Pedro Mártir region, Baja California, with a contribution to the biogeography of the Baja California herpetofauna. Proc. California Acad. Sci., 46:1–72.

WIGGINS, I. L. 1960. The origin and relationships of the land flora. Syst. Zool. 9:148–165.

WILSON, L. D., AND L. PORRAS. 1983. The ecological impact of man on the South Florida herpetofauna. Spec. Publ. Mus. Nat. Hist. Univ. Kansas 9:1–89.

WOLFE, J. A. 1978. A paleobotanical interpretation of Tertiary climates in the Northern Hemisphere. Am. Sci. 66:694–703.

WOLFE, J. A. 1980. Tertiary climates and floristic relationships at high latitudes in the northern hemisphere. Paleogeogr. Paleoclimatol. Paleoecol. 30:313–323.

WORLD METEOROLOGICAL ASSOCIATION. 1979. Climatic Atlas of North and Central America, Vol. 1. Geneva: UNESCO.

WYNN, A. H., R. HIGHTON, AND J. J. JACOBS. 1988. A new species of rock-crevice dwelling Plethodon from Pigeon Mountain, Georgia. Herpetologica 44:135–143.

YANEV, K. P. 1980. Biogeography and distribution of three parapatric salamander species in coastal and borderland California. Pp. 531–550 in D. M. Power (ed.), The California Islands: Proceedings of a Multidiciplinary Symposium. Santa Barbara, California: Santa Barbara Museum of Natural History.

YANEV, K. P., AND D. W. WAKE. 1981. Genic differentiation in a relict desert salamander. Herpetologica 37:16–28.

APPENDIX 2:1

DISTRIBUTION OF SPECIES OF AMPHIBIANS IN PHYSIOGRAPHIC DIVISIONS IN NORTH AMERICA

Symbols in columns: + = present; – = absent; I = introduced. Names followed by an asterisk (*) are taxa that usually are recognized as subspecies, but probably should be accorded specific status. Abbreviations for physiographic divisions are: ACP = Atlantic Coastal Plain, APH = Appalachian Highlands, LAU = Laurentian Upland, INP = Interior Plains, NAC = North American Cordillera.

Taxon and date named	Physiographic divisions					Taxon and date named	Physiographic divisions				
	NAC	INP	ACP	APH	LAU		NAC	INP	ACP	APH	LAU
ANURA: ASCAPHIDAE:						Smilisca baudinii 1841	–	–	+	–	–
Ascaphus truei 1899	+	–	–	–	–	ANURA: LEPTODACTYLIDAE:					
ANURA: BUFONIDAE:						Eleutherodactylus augusti 1879	+	+	–	–	–
Bufo alvarius 1859	+	–	–	–	–	Eleutherodactylus coqui 1966	–	–	I	–	–
Bufo americanus 1836	–	+	+	+	+	Eleuth. cystignathoides 1877	–	+	+	–	–
Bufo baxteri 1968	+	–	–	–	–	Eleutherodactylus guttilatus 1879	+	–	–	–	–
Bufo boreas 1852	+	–	–	–	–	Eleutherodactylus marnockii 1878	–	+	–	–	–
Bufo californicus 1915	+	–	–	–	–	Eleutherodactylus planirostris 1862	–	–	I	–	–
Bufo canorus 1916	+	–	–	–	–	Leptodactylus labialis 1877	–	–	+	–	–
Bufo cognatus 1823	+	+	–	–	–	ANURA: MICROHYLIDAE:					
Bufo debilis 1854	+	–	–	–	–	Gastrophryne carolinensis 1836	–	+	+	+	–
Bufo exsul 1942	+	–	–	–	–	Gastrophryne olivacea 1857	–	+	+	–	–
Bufo fowleri 1882	–	+	+	+	–	Gastrophryne usta 1866	+	–	–	–	–
Bufo hemiophrys 1886	–	+	–	–	–	Hypopachus variolosus 1866	–	–	+	–	–
Bufo houstonensis 1953	–	–	+	–	–	ANURA: PELOBATIDAE:					
Bufo marinus 1758	–	–	+	–	–	Scaphiopus couchii 1854	+	+	+	–	–
Bufo microscaphus 1867	+	–	–	–	–	Scaphiopus holbrookii 1835	–	+	+	+	–
Bufo punctatus 1852	+	+	–	–	–	Scaphiopus hurterii 1910*	–	+	+	–	–
Bufo quercicus 1840	–	–	+	–	–	Spea bombifrons 1863	+	+	+	–	–
Bufo retiformis 1951	+	–	–	–	–	Spea hammondii 1859	+	–	–	–	–
Bufo speciosus 1854	+	+	+	–	–	Spea intermontana 1883	+	–	–	–	–
Bufo terrestris 1789	–	–	+	–	–	Spea multiplicata 1863	+	+	–	–	–
Bufo valliceps 1833	–	+	+	–	–	ANURA: PIPIDAE:					
Bufo woodhousii 1854	+	+	–	–	–	Xenopus laevis 1802	I	–	–	–	–
ANURA: HYLIDAE:						ANURA: RANIDAE:					
Acris crepitans 1854	+	+	+	+	–	Rana areolata 1852	–	+	+	–	–
Acris gryllus 1825	–	–	+	+	–	Rana aurora 1852	+	–	–	–	–
Hyla andersonii 1854	–	–	+	–	–	Rana berlandieri 1854	+	+	+	–	–
Hyla arenicolor 1886	+	–	–	–	–	Rana blairi 1973	–	+	–	–	–
Hyla avivoca 1928	–	+	+	–	–	Rana boylii 1854	+	–	–	–	–
Hyla chrysoscelis 1880	–	+	+	+	–	Rana cascadae 1939	+	–	–	–	–
Hyla cinerea 1792	–	+	+	+	–	Rana catesbeiana 1802	+	+	+	+	–
Hyla eximia 1854	+	–	–	–	–	Rana chiricahuensis 1979	+	–	–	–	–
Hyla femoralis 1800	–	–	+	–	–	Rana clamitans 1801	I	+	+	+	+
Hyla gratiosa 1856	–	+	+	–	–	Rana fischeri 1893*	–	+	–	–	–
Hyla squirella 1800	–	–	+	–	–	Rana grylio 1901	–	–	+	–	–
Hyla versicolor 1825	–	+	+	+	–	Rana heckscheri 1924	–	–	+	–	–
Osteopilus septentrionalis 1841	–	–	I	–	–	Rana luteiventris 1913	+	–	–	–	–
Pseudacris brachyphona 1889	–	–	–	+	–	Rana muscosa 1917	+	–	–	–	–
Pseudacris brimleyi 1933	–	–	+	–	–	Rana okaloosae 1985	–	–	+	–	–
Pseudacris cadaverina 1866	+	–	–	–	–	Rana onca 1875	+	–	–	–	–
Pseudacris clarkii 1854	–	+	+	–	–	Rana palustris 1825	–	+	–	+	–
Pseudacris crucifer 1838	–	+	+	+	+	Rana pipiens 1782	+	+	–	+	+
Pseudacris nigrita 1825	–	–	+	–	–	Rana pretiosa 1853	+	–	–	–	–
Pseudacris ocularis 1801	–	–	+	–	–	Rana septentrionalis 1854	–	–	–	+	+
Pseudacris ornata 1836	–	–	+	–	–	Rana sphenocephala 1889	–	+	+	+	–
Pseudacris regilla 1852	+	–	–	–	–	Rana subaquavocalis 1993	+	–	–	–	–
Pseudacris streckeri 1923	–	+	+	–	–	Rana sylvatica 1825	+	+	+	+	+
Pseudacris triseriata 1838	+	+	+	+	+	Rana tarahumarae 1917	+	–	–	–	–
Pternohyla fodiens 1882	+	–	–	+	–	Rana virgatipes 1891	–	–	+	–	–

Appendix 2:1 Continued

Taxon and date named	Physiographic divisions					Taxon and date named	Physiographic divisions				
	NAC	INP	ACP	APH	LAU		NAC	INP	ACP	APH	LAU
Rana yavapaiensis 1984	+	−	−	−	−	*Desmognathus ochrophaeus* 1859	−	−	−	+	−
ANURA: RHINOPHRYNIDAE:						*Desmognathus ocoee* 1949	−	−	−	+	−
Rhinophrynus dorsalis 1841	−	−	+	−	−	*Desmognathus orestes* 1996	−	−	−	+	−
CAUDATA: AMBYSTOMATIDAE:						*Desmog. quadramaculatus* 1840	−	−	−	+	−
Ambystoma annulatum 1886	−	+	−	−	−	*Desmognathus santeetlah* 1981	−	−	−	+	−
Ambystoma barbouri 1989	−	+	−	+	−	*Desmognathus welteri* 1950	−	−	−	+	−
Ambystoma californiense 1853	+	−	−	−	−	*Desmognathus wrighti* 1936	−	−	−	+	−
Ambystoma cingulatum 1868	−	−	+	−	−	*Ensatina eschscholtzii* 1850	+	−	−	−	−
Ambystoma gracile 1859	+	−	−	−	−	*Eurycea bislineata* 1818	−	−	+	+	+
Ambystoma jeffersonianum 1827	−	+	−	+	−	*Eurycea cirrigera* 1830	−	+	+	+	−
Ambystoma laterale 1856	−	+	+	−	+	*Eurycea junaluska* 1976	−	−	−	+	−
Ambystoma mabeei 1928	−	−	+	−	−	*Eurycea longicauda* 1818	−	+	+	+	−
Ambystoma macrodactylum 1849	+	−	−	−	−	*Eurycea lucifuga* 1822	−	+	−	+	−
Ambystoma maculatum 1802	−	+	+	+	+	*Eurycea multiplicata* 1869	−	+	−	−	−
Ambystoma nothagenes 1985	−	+	−	−	−	*Eurycea nana* 1941	−	+	−	−	−
Ambystoma opacum 1807	−	+	+	+	−	*Eurycea neotenes* 1937	−	+	−	−	−
Ambystoma platineum 1868	−	+	−	+	−	*Eurycea quadridigittata* 1842	−	−	+	+	−
Ambystoma talpoideum 1838	−	+	+	+	−	*Eurycea sosorum* 1993	−	+	−	−	−
Ambystoma texanum 1855	−	+	+	−	−	*Eurycea tridentifera* 1965	−	+	−	−	−
Ambystoma tigrinum 1825	+	+	+	+	−	*Eurycea tynerensis* 1939	−	+	−	−	−
Ambystoma tremblayi 1943	−	+	−	+	−	*Eurycea wilderae* 1920	−	−	−	+	−
CAUDATA: AMPHIUMIDAE:						*Gyrinophilus gulolineatus* 1965	−	−	−	+	−
Amphiuma means 1821	−	−	+	−	−	*Gyrinophilus palleucus* 1954	−	−	−	+	−
Amphiuma pholeter 1964	−	−	+	−	−	*Gyrinophilus porphyriticus* 1827	−	−	−	+	−
Amphiuma tridactylum 1827	−	−	+	−	−	*Gyrinophilus subterraneus* 1977*	−	−	−	+	−
CAUDATA: CRYPTOBRANCHIDAE:						*Haideotriton wallacei* 1939	−	−	+	−	−
Cryptobranchus alleganiensis 1803	−	+	−	+	−	*Hemidactylium scutatum* 1838	−	+	+	+	+
CAUDATA: DICAMPTODONTIDAE:						*Hydromantes brunus* 1954	+	−	−	−	−
Dicamptodon aterrimus 1868	+	−	−	−	−	*Hydromantes platycephalus* 1916	+	−	−	−	−
Dicamptodon copei 1970	+	−	−	−	−	*Hydromantes shastae* 1953	+	−	−	−	−
Dicamptodon ensatus 1833	+	−	−	−	−	*Phaeognathus hubrichti* 1961	−	−	−	+	−
Dicamptodon tenebrosus 1852	+	−	−	−	−	*Plethodon albagula* 1944	−	+	−	−	−
CAUDATA: PLETHODONTIDAE:						*Plethodon angusticlavus* 1944	−	+	−	−	−
Aneides aeneus 1881	−	−	−	+	−	*Plethodon aureolus* 1984	−	−	−	+	−
Aneides ferreus 1869	+	−	−	−	−	*Plethodon caddoensis* 1951	−	+	−	−	−
Aneides flavopunctatus 1870	+	−	−	−	−	*Plethodon chattahoochee* 1989	−	−	−	+	−
Aneides hardii 1941	+	−	−	−	−	*Plethodon chlorobryonis* 1951	−	−	+	+	−
Aneides lugubris 1849	+	−	−	−	−	*Plethodon cinereus* 1818	−	+	+	+	+
Batrachoseps aridus 1970	+	−	−	−	−	*Plethodon cylindraceus* 1825	−	−	+	+	−
Batrachoseps attenuatus 1833	+	−	−	−	−	*Plethodon dorsalis* 1889	−	−	−	+	−
Batrachoseps campi 1979	+	−	−	−	−	*Plethodon dunni* 1934	+	−	−	−	−
Batrachoseps gabrieli 1996	+	−	−	−	−	*Plethodon elongatus* 1916	+	−	−	−	−
Batrachoseps nigriventris 1869	+	−	−	−	−	*Plethodon fourchensis* 1979	−	+	−	−	−
Batrachoseps pacificus 1865	+	−	−	−	−	*Plethodon glutinosus* 1818	−	+	−	APH	−
Batrachoseps simatus 1968	+	−	−	−	−	*Plethodon grobmani* 1949	−	−	+	−	−
Batrachoseps stebbinsi 1968	+	−	−	−	−	*Plethodon hoffmani* 1972	−	−	−	+	−
Batrachoseps wrighti 1937	+	−	−	−	−	*Plethodon hubrichti* 1957	−	−	−	+	−
Desmognathus aeneus 1947	−	−	−	+	−	*Plethodon idahoensis* 1940	+	−	−	−	−
Desmognathus apalachicolae 1989	−	−	+	−	−	*Plethodon jordani* 1901	−	−	−	+	−
Desmognathus auriculatus 1838	−	−	+	−	−	*Plethodon kentucki* 1951	−	−	−	+	−
Desmognathus brimleyorum 1895	−	+	−	−	−	*Plethodon kiamichi* 1989	−	+	−	−	−
Desmognathus carolinensis 1916	−	−	−	+	−	*Plethodon kisatchie* 1989	−	−	+	−	−
Desmognathus conanti 1958	−	+	+	+	−	*Plethodon larselli* 1953	+	−	−	−	−
Desmognathus fuscus 1818	−	+	+	+	−	*Plethodon mississippi* 1989	−	+	−	−	−
Desmognathus imitator 1927	−	−	−	+	−	*Plethodon neomexicanus* 1950	+	−	−	−	−
Desmognathus marmoratus 1899	−	−	−	+	−	*Plethodon nettingi* 1938	−	−	−	+	−
Desmognathus monticola 1916	−	−	+	+	−	*Plethodon ocmulgee* 1989	−	−	+	+	−

Appendix 2:1 Continued

Taxon and date named	Physiographic divisions					Taxon and date named	Physiographic divisions				
	NAC	INP	ACP	APH	LAU		NAC	INP	ACP	APH	LAU
Plethodon oconaluftee 1950, 1993	–	–	–	+	–	CAUDATA: PROTEIIDAE:					
Plethodon ouachitae 1933	–	+	–	–	–	*Necturus alabamensis* 1937	–	–	+	+	–
Plethodon petraeus 1988	–	–	–	+	–	*Necturus beyeri* 1937	–	–	+	+	–
Plethodon punctatus 1972	–	–	–	+	–	*Necturus lewisi* 1924	–	–	+	+	–
Plethodon richmondi 1938	–	+	–	+	–	*Necturus louisianensis* 1938*	–	+	+	–	–
Plethodon savannah 1989	–	–	–	+	–	*Necturus maculosus* 1818	–	+	–	+	+
Plethodon sequoyah 1989	–	+	–	–	–	*Necturus punctatus* 1850	–	–	+	+	–
Plethodon serratus 1944	–	+	+	+	–	CAUDATA: RHYACOTRITONIDAE:					
Plethodon shenandoah 1967	–	–	–	+	–	*Rhyacotriton cascadae* 1992	+	–	–	–	–
Plethodon stormi 1965	+	–	–	–	–	*Rhyacotriton kezeri* 1992	+	–	–	–	–
Plethodon vandykei 1906	+	–	–	–	–	*Rhyacotriton olympicus* 1917	+	–	–	–	–
Plethodon variolatus 1818	–	–	+	–	–	*Rhyacotriton variegatus* 1951	+	–	–	–	–
Plethodon vehiculum 1860	+	–	–	–	–	CAUDATA: SALAMANDRIDAE:					
Plethodon ventralis 1997	–	–	–	+	–	*Notophthalmus meridionalis* 1880	–	–	+	–	–
Plethodon websteri 1979	–	–	–	+	–	*Notophthalmus perstriatus* 1941	–	–	+	–	–
Plethodon wehrlei 1917	–	–	–	+	–	*Notophthalmus viridescens* 1820	–	+	+	+	+
Plethodon welleri 1931	–	–	–	+	–	*Taricha granulosa* 1849	+	–	–	–	–
Plethodon yonahlossee 1917	–	–	–	+	–	*Taricha rivularis* 1935	+	–	–	–	–
Pseudotriton diasticus 1941*	–	–	–	+	–	*Taricha torosa* 1833	+	–	–	–	–
Pseudotriton montanus 1849	–	+	+	+	–	CAUDATA: SIRENIDAE:					
Pseudotriton ruber 1801	–	+	+	+	–	*Pseudobranchus axanthus* 1942	–	–	+	–	–
Stereochilus marginatus 1856	–	–	+	–	–	*Pseudobranchus striatus* 1824	–	–	+	–	–
Typhlomolge rathbuni 1897	–	+	–	–	–	*Siren intermedia* 1826	–	+	+	–	–
Typhlomolge robusta 1978	–	+	–	–	–	*Siren lacertina* 1766	–	–	+	–	–
Typhlotriton spelaeus 1893	–	+	–	–	–						

APPENDIX 2:2

DISTRIBUTION OF AMPHIBIANS IN 44 NATURAL REGIONS IN NORTH AMERICA

The number of species in each region is shown in boldface type in the common cell; the number of species that are shared by two regions is in the upper right, and the Coefficient of Biogeographic Resemblance is in italics in the lower left. Abbreviations of regions: ALL = Alleghany Plateau, ASP = Aspen Parkland, BAR = Basin and Range, BCA = Baja California, CHI = Chihuahuan Desert, COL = Columbia Plateau, COP = Colorado Plateau, CPR = Coastal Prairie, CVC = Central Valley of California, CRM = Central Rocky Mountains, EDP = Edwards Plateau, FRA = Fraser Plateau, GCP = Gulf Coastal Plain, INH = Interior

Region	PRA	PRN	PRC	PRS	BCA	PST	CVC	NCA	SCA	SNE	YUK	FRA	COL	BAR	COP	MOR	MAD	MOJ	SON	CHI	NRM	CRM
PRA	**6**	5	3	1	0	4	1	5	4	1	2	2	1	1	0	0	0	0	0	0	4	3
PRN	*35.7*	**22**	11	6	2	11	4	16	14	5	1	1	3	3	1	0	0	1	1	1	4	4
PRC	*15.0*	*55.0*	**18**	9	2	9	5	10	14	5	1	1	3	3	1	0	0	1	1	1	3	3
PRS	*08.3*	*30.0*	*50.0*	**18**	4	5	9	6	9	8	1	1	3	3	1	0	0	1	1	1	2	2
BCA	*00.0*	*14.3*	*16.7*	*33.3*	**6**	2	3	2	2	2	0	0	2	3	1	2	1	2	3	3	0	1
PST	*47.1*	*66.7*	*62.0*	*34.5*	*23.5*	**11**	4	11	11	4	1	1	3	3	1	0	0	1	1	1	4	4
CVC	*13.3*	*25.8*	*37.0*	*66.7*	*40.0*	*40.0*	**9**	4	5	5	1	1	3	3	1	0	0	1	1	1	2	2
NCA	*25.8*	*59.6*	*55.6*	*33.3*	*16.7*	*75.9*	*29.6*	**18**	16	5	1	1	4	4	2	0	0	1	2	2	4	5
SCA	*25.8*	*59.6*	*65.1*	*41.9*	*12.9*	*61.1*	*29.4*	*74.4*	**25**	7	1	1	4	4	2	0	0	1	2	2	4	5
SNE	*09.1*	*26.3*	*35.3*	*47.1*	*18.2*	*29.6*	*40.0*	*29.4*	*39.0*	**16**	1	1	3	3	1	0	0	1	1	1	2	2
YUK	*44.4*	*08.0*	*09.5*	*9.5*	*00.0*	*14.3*	*16.6*	*09.5*	*07.1*	*10.5*	**3**	3	2	2	1	0	0	0	0	0	3	2
FRA	*44.4*	*08.0*	*09.5*	*9.5*	*00.0*	*14.3*	*16.6*	*09.5*	*07.1*	*10.5*	*100.0*	**3**	2	2	1	0	0	0	0	0	3	2
COL	*13.3*	*19.4*	*22.2*	*22.2*	*26.6*	*30.0*	*33.3*	*29.6*	*23.5*	*24.0*	*36.3*	*18.2*	**9**	7	5	0	0	1	3	3	6	7
BAR	*08.7*	*15.4*	*17.1*	*17.1*	*26.1*	*21.4*	*23.1*	*22.9*	*19.0*	*18.2*	*20.0*	*20.0*	*53.8*	**17**	8	4	4	4	6	6	5	7
COP	*00.0*	*06.3*	*07.1*	*07.7*	*12.5*	*09.5*	*10.5*	*14.3*	*11.4*	*07.7*	*00.0*	*00.0*	*52.6*	*59.3*	**10**	3	5	4	5	6	2	5
MOR	*00.0*	*00.0*	*00.0*	*00.0*	*28.5*	*00.0*	*00.0*	*00.0*	*00.0*	*00.0*	*00.0*	*00.0*	*00.0*	*32.0*	*33.3*	**8**	6	2	3	3	0	1
MAD	*00.0*	*00.0*	*00.0*	*00.0*	*09.5*	*00.0*	*00.0*	*00.0*	*00.0*	*00.0*	*00.0*	*00.0*	*00.0*	*25.0*	*40.0*	*52.2*	**15**	3	9	9	0	1
MOJ	*00.0*	*07.1*	*08.3*	*08.3*	*13.3*	*07.1*	*07.7*	*11.4*	*09.5*	*09.1*	*00.0*	*00.0*	*13.3*	*34.8*	*50.0*	*28.6*	*28.6*	**6**	4	4	0	1
SON	*00.0*	*05.7*	*06.5*	*06.5*	*31.6*	*08.3*	*09.1*	*12.9*	*10.5*	*06.9*	*00.0*	*00.0*	*28.6*	*40.0*	*43.5*	*28.6*	*64.3*	*42.1*	**13**	7	1	3
CHI	*00.0*	*05.1*	*05.7*	*05.7*	*26.1*	*07.1*	*07.7*	*11.4*	*09.5*	*06.1*	*00.0*	*00.0*	*23.1*	*35.3*	*44.4*	*24.1*	*56.3*	*34.8*	*46.6*	**17**	1	2
NRM	*50.0*	*25.0*	*21.4*	*14.3*	*00.0*	*38.1*	*21.1*	*28.6*	*22.9*	*15.4*	*46.2*	*46.2*	*63.2*	*37.0*	*20.0*	*00.0*	*00.0*	*00.0*	*08.7*	*07.4*	**10**	7
CRM	*37.5*	*25.0*	*21.4*	*14.3*	*12.5*	*38.1*	*21.1*	*35.7*	*28.6*	*15.4*	*30.8*	*30.8*	*73.7*	*51.9*	*50.0*	*11.1*	*08.0*	*12.5*	*26.1*	*14.8*	*70.0*	**10**
WYB	*00.0*	*00.0*	*00.0*	*00.0*	*00.0*	*00.0*	*00.0*	*00.83*	*06.5*	*00.0*	*00.0*	*00.0*	*40.0*	*34.8*	*62.5*	*14.3*	*19.0*	*00.0*	*21.1*	*26.1*	*25.0*	*50.0*
SRM	*11.1*	*11.8*	*13.3*	*13.3*	*22.2*	*17.4*	*19.0*	*20.0*	*16.2*	*14.3*	*13.3*	*13.3*	*47.6*	*55.2*	*45.5*	*20.0*	*22.2*	*33.3*	*40.0*	*34.5*	*27.3*	*45.5*
ASP	*16.7*	*00.0*	*00.0*	*00.0*	*00.0*	*00.0*	*00.0*	*00.0*	*00.0*	*00.0*	*22.2*	*22.2*	*13.3*	*17.4*	*25.0*	*14.3*	*00.0*	*00.0*	*00.0*	*00.0*	*25.0*	*25.0*
SGS	*00.0*	*04.8*	*05.3*	*05.3*	*23.1*	*06.5*	*06.9*	*10.5*	*08.9*	*05.6*	*08.7*	*08.7*	*27.6*	*37.8*	*46.7*	*14.3*	*34.3*	*23.1*	*42.4*	*64.9*	*13.3*	*33.3*
TGP	*07.7*	*04.8*	*05.3*	*05.3*	*23.1*	*06.5*	*06.9*	*10.5*	*08.9*	*05.6*	*00.0*	*00.0*	*27.6*	*37.8*	*40.0*	*14.3*	*28.6*	*23.1*	*36.4*	*59.5*	*20.0*	*33.3*
EDP	*00.0*	*04.0*	*04.3*	*04.3*	*17.6*	*05.1*	*05.4*	*08.7*	*07.5*	*04.5*	*00.0*	*00.0*	*16.2*	*17.8*	*21.0*	*11.1*	*18.6*	*11.8*	*24.4*	*35.6*	*05.3*	*15.8*
PRP	*00.0*	*03.6*	*03.9*	*03.9*	*10.3*	*04.5*	*04.8*	*07.7*	*06.9*	*04.8*	*00.0*	*00.0*	*19.0*	*24.0*	*27.9*	*04.9*	*16.7*	*10.3*	*21.7*	*40.0*	*09.3*	*18.6*
INL	*03.6*	*02.8*	*03.0*	*03.0*	*03.6*	*03.3*	*03.4*	*06.0*	*05.4*	*03.1*	*03.8*	*03.8*	*10.3*	*12.1*	*13.6*	*03.5*	*00.0*	*03.6*	*06.5*	*09.1*	*10.2*	*13.6*
INH	*04.8*	*03.4*	*03.7*	*03.7*	*04.8*	*04.2*	*04.4*	*07.4*	*06.5*	*03.8*	*05.1*	*05.1*	*08.9*	*11.3*	*13.0*	*04.5*	*00.0*	*04.8*	*08.2*	*11.3*	*08.7*	*13.0*
ALL	*03.2*	*02.6*	*02.7*	*02.7*	*03.2*	*02.9*	*03.1*	*05.4*	*04.9*	*02.8*	*03.3*	*03.3*	*09.2*	*11.0*	*12.1*	*03.1*	*00.0*	*03.2*	*05.8*	*05.5*	*09.1*	*12.1*
SAP	*02.8*	*02.3*	*02.3*	*02.4*	*02.8*	*02.6*	*02.7*	*04.8*	*04.4*	*02.4*	*02.9*	*02.9*	*08.0*	*09.6*	*10.5*	*02.7*	*00.0*	*02.8*	*0.51*	*0.48*	*07.9*	*10.5*
PIE	*00.0*	*02.7*	*02.9*	*02.9*	*03.5*	*03.2*	*03.3*	*05.8*	*05.3*	*03.0*	*00.0*	*00.0*	*06.7*	*08.8*	*09.8*	*03.4*	*00.0*	*03.5*	*06.3*	*05.9*	*06.6*	*09/8*
NAP	*05.9*	*04.0*	*04.3*	*04.3*	*05.9*	*05.1*	*05.4*	*08.7*	*07.5*	*04.5*	*06.5*	*06.5*	*16.2*	*17.8*	*21.1*	*05.6*	*00.0*	*05.9*	*09.8*	*08.9*	*15.8*	*21.1*
NCP	*05.6*	*03.8*	*04.2*	*04.2*	*05.6*	*04.9*	*05.1*	*08.3*	*07.3*	*04.3*	*06.1*	*06.1*	*10.2*	*12.8*	*15.0*	*05.3*	*00.0*	*05.6*	*09.3*	*12.8*	*10.0*	*15.0*
SCP	*00.0*	*02.6*	*02.8*	*02.8*	*03.3*	*03.1*	*03.2*	*05.6*	*05.1*	*02.9*	*00.0*	*00.0*	*06.3*	*08.6*	*09.4*	*03.2*	*00.0*	*03.3*	*06.0*	*05.6*	*03.1*	*09.4*
PFL	*00.0*	*03.5*	*03.8*	*03.8*	*04.9*	*04.3*	*04.5*	*07.5*	*06.7*	*03.9*	*00.0*	*00.0*	*09.1*	*07.7*	*08.9*	*00.0*	*00.0*	*04.9*	*08.3*	*07.7*	*04.4*	*08.9*
GCP	*00.0*	*02.6*	*02.7*	*02.8*	*03.3*	*03.0*	*03.1*	*05.5*	*05.0*	*02.8*	*00.0*	*00.0*	*06.3*	*08.3*	*09.2*	*03.2*	*00.0*	*03.3*	*05.9*	*08.3*	*03.1*	*09.2*
MEM	*00.0*	*03.3*	*03.5*	*03.5*	*04.4*	*04.0*	*04.2*	*03.5*	*03.1*	*03.6*	*00.0*	*00.0*	*08.3*	*10.7*	*12.2*	*04.2*	*03.7*	*04.4*	*07.7*	*10.7*	*00.0*	*12.2*
PWO	*00.0*	*03.4*	*03.7*	*03.7*	*04.8*	*04.2*	*04.4*	*07.0*	*06.6*	*03.8*	*00.0*	*00.0*	*13.3*	*15.1*	*17.4*	*04.5*	*03.9*	*04.8*	*12.2*	*18.9*	*04.3*	*17.4*
CPR	*00.0*	*04.4*	*04.8*	*04.9*	*13.8*	*05.9*	*06.3*	*09.8*	*08.3*	*05.1*	*00.0*	*00.0*	*18.8*	*20.0*	*24.2*	*12.9*	*15.8*	*06.9*	*22.2*	*45.0*	*06.1*	*24.2*
TAM	*00.0*	*04.2*	*05.1*	*05.1*	*14.8*	*06.3*	*06.7*	*10.3*	*09.9*	*05.4*	*00.0*	*00.0*	*20.0*	*15.8*	*25.8*	*13.8*	*16.7*	*07.4*	*23.5*	*47.4*	*06.5*	*19.4*
LAU	*10.0*	*00.0*	*00.0*	*00.0*	*00.0*	*00.0*	*00.0*	*00.0*	*00.0*	*00.0*	*11.8*	*11.8*	*08.7*	*12.9*	*16.7*	*09.1*	*00.0*	*00.0*	*00.0*	*00.0*	*16.6*	*16.7*

Highlands, INL = Interior Lowlands, LAU = Laurentian Uplands (= Boreal Forest), MAD = Madrean Mountains, MEM = Mississippi Embayment, MOJ = Mojave Desert, MOR = Mogollon Rim, NAP = Northern Appalachians, NCA = Northern Cascades, NCP = Northern Coastal Plain, NRM = Northern Rocky Mountains, PFL = Peninsular Florida, PIE = Piedmont, PRA = Pacific Range (Alaska and Canada), PRC = Pacific Range (Central), PRN = Pacific Range (North), PRP = Prairie Parkland, PRS = Pacific Range (South), PST = Puget Sound Trough, PWO = Pine Woodland, SAP = Southern Appalachians, SCA = Southern Cascades, SCP = Southern Coastal Plain, SGS = Shortgrass Steppe, SNE = Sierra Nevada, SON = Sonoran Desert, SRM = Southern Rocky Mountains, TAM = Tamaulipan, TGP = Tallgrass Prairie, WYB = Wyoming Basin, YUK = Yukon Plateau. Matrix continuous across two pages.

Region	WYB	SRM	ASP	SGS	TGP	EDP	PRP	INL	INH	ALL	SAP	PIE	NAP	NCP	SCP	PFL	GCP	MEM	PWO	CPR	TAM	LAU
PRA	0	1	1	0	1	0	0	1	1	1	1	0	1	1	0	0	0	0	0	0	0	1
PRN	0	2	0	1	1	1	1	1	1	1	1	1	1	1	1	1	1	1	1	1	1	0
PRC	0	2	0	1	1	1	1	1	1	1	1	1	1	1	1	1	1	1	1	1	1	0
PRS	0	2	0	1	1	1	1	1	1	1	1	1	1	1	1	1	1	1	1	1	1	0
BCA	0	2	0	3	3	3	2	1	1	1	1	1	1	1	1	1	1	1	1	2	2	0
PST	0	2	0	1	1	1	1	1	1	1	1	1	1	1	1	1	1	1	1	1	1	0
CVC	0	2	0	1	1	1	1	1	1	1	1	1	1	1	1	1	1	1	1	1	1	0
NCA	1	3	0	2	2	2	2	2	2	2	2	2	2	2	2	2	2	1	2	2	2	0
SCA	1	3	0	2	2	2	2	2	2	2	2	2	2	2	2	2	2	1	2	2	2	0
SNE	0	2	0	1	1	1	1	1	1	1	1	1	1	1	1	1	1	1	1	1	1	0
YUK	0	1	1	1	0	0	0	1	1	1	1	0	1	1	0	0	0	0	0	0	0	1
FRA	0	1	1	1	0	0	0	1	1	1	1	0	1	1	0	0	0	0	0	0	0	1
COL	3	5	1	4	3	4	4	3	2	3	3	2	3	2	2	2	2	2	3	3	3	1
BAR	4	8	2	7	7	4	6	4	3	4	4	3	4	3	3	2	3	3	4	4	3	2
COP	5	5	2	7	6	4	6	4	3	4	4	3	4	3	3	2	3	3	4	4	4	2
MOR	1	2	1	2	2	2	1	1	1	1	1	1	1	1	1	0	1	1	1	2	2	1
MAD	2	3	0	6	5	4	4	0	0	0	0	0	0	0	0	0	0	1	1	3	3	0
MOJ	0	3	0	3	3	2	2	1	1	1	1	1	1	1	1	1	1	1	1	1	1	0
SON	2	5	0	7	6	5	5	2	2	2	2	2	2	2	2	2	2	2	3	4	4	0
CHI	3	5	0	12	11	8	10	3	3	2	2	2	2	3	2	2	3	3	5	9	9	0
NRM	2	3	2	2	3	1	2	3	2	3	3	2	3	2	1	1	1	0	1	1	1	2
CRM	4	5	2	5	5	3	4	4	3	4	4	3	4	3	3	2	3	3	4	4	3	2
WYB	**6**	4	2	5	5	2	5	3	2	3	3	2	3	2	2	1	3	2	3	3	3	2
SRM	*44.4*	**12**	2	8	7	4	7	5	3	4	4	3	4	3	3	2	4	4	5	5	4	2
ASP	*33.3*	*23.2*	**6**	3	4	2	4	6	4	5	5	3	5	4	3	1	3	3	3	1	0	3
SGS	*38.5*	*50.0*	*23.1*	**20**	18	13	14	7	4	5	4	4	4	4	4	3	5	5	8	12	11	2
TGP	*38.5*	*43.8*	*30.1*	*90.0*	**20**	12	15	15	6	7	6	5	6	7	4	2	5	7	6	12	11	4
EDP	*11.8*	*26.7*	*11.8*	*54.2*	*50.0*	**28**	15	14	8	6	5	6	4	6	5	4	7	7	11	12	11	0
PRP	*25.6*	*31.1*	*20.5*	*52.8*	*54.5*	*49.2*	**33**	24	11	12	10	11	9	11	9	7	11	11	10	11	9	3
INL	*10.9*	*16.4*	*21.8*	*20.3*	*43.5*	*36.4*	*58.5*	**49**	22	32	34	30	24	25	25	15	30	26	21	11	5	11
INH	*09.5*	*12.5*	*19.0*	*14.3*	*21.4*	*25.0*	*31.9*	*51.8*	**36**	21	23	19	15	18	16	8	17	19	18	8	4	8
ALL	*09.7*	*11.8*	*16.1*	*13.2*	*18.4*	*14.3*	*27.0*	*61.0*	*45.7*	**56**	43	34	25	24	21	11	20	21	19	9	4	12
SAP	*09.3*	*10.3*	*09.3*	*09.3*	*13.9*	*10.6*	*20.2*	*59.1*	*45.1*	*70.6*	**66**	36	25	22	23	11	22	19	17	6	2	12
PIE	*07.0*	*09.5*	*10.5*	*11.3*	*14.1*	*15.2*	*26.2*	*60.0*	*43.7*	*53.5*	*61.5*	**51**	18	23	27	14	27	27	21	9	4	9
NAP	*17.6*	*20.0*	*29.4*	*16.7*	*25.0*	*14.3*	*29.5*	*62.3*	*46.9*	*59.5*	*53.2*	*45.6*	**28**	18	14	6	14	13	11	4	2	12
NCP	*11.1*	*14.3*	*22.2*	*16.0*	*28.0*	*20.7*	*34.9*	*63.3*	*54.5*	*55.8*	*45.8*	*56.8*	*62.1*	**30**	23	13	23	17	16	7	4	11
SCP	*06.7*	*09.1*	*10.0*	*10.8*	*10.8*	*12.2*	*20.7*	*48.5*	*35.6*	*38.2*	*35.8*	*51.4*	*34.1*	*54.8*	**54**	33	41	25	21	9	5	7
PFL	*04.9*	*08.5*	*04.9*	*10.9*	*07.3*	*12.7*	*20.6*	*35.7*	*22.5*	*24.2*	*21.8*	*32.6*	*19.0*	*34.7*	*74.2*	**35**	32	17	15	8	4	2
GCP	*09.8*	*11.9*	*09.8*	*13.6*	*13.3*	*16.9*	*25.0*	*57.7*	*37.4*	*36.0*	*36.4*	*50.9*	*33.7*	*54.1*	*82.8*	*35.6*	**55**	32	26	12	6	6
MEM	*08.9*	*15.7*	*13.3*	*16.9*	*23.7*	*20.9*	*30.6*	*59.1*	*50.1*	*44.2*	*36.2*	*60.0*	*38.8*	*49.3*	*53.8*	*45.9*	*68.1*	**39**	29	12	6	6
PWO	*14.3*	*20.8*	*14.3*	*28.6*	*21.4*	*34.4*	*29.0*	*49.4*	*50.0*	*41.3*	*33.3*	*38.3*	*34.4*	*48.5*	*46.7*	*42.3*	*57.1*	*77.3*	**36**	17	11	5
CPR	*20.7*	*28.6*	*06.9*	*55.8*	*55.8*	*47.1*	*39.3*	*30.6*	*27.1*	*22.8*	*13.5*	*24.3*	*15.7*	*26.4*	*23.4*	*27.6*	*30.8*	*38.7*	*57.6*	**23**	17	2
TAM	*22.2*	*24.2*	*00.0*	*53.7*	*53.7*	*44.9*	*33.3*	*14.3*	*14.0*	*10.4*	*04.6*	*11.1*	*08.2*	*15.7*	*13.3*	*14.3*	*15.8*	*20.0*	*38.6*	*77.3*	**21**	0
LAU	*20.0*	*15.4*	*30.0*	*11.8*	*23.5*	*00.0*	*12.8*	*34.9*	*32.0*	*34.3*	*30.0*	*27.7*	*57.1*	*50.0*	*20.6*	*08.2*	*17.4*	*22.6*	*20.0*	*10.8*	*00.0*	**14**

3. Distribution Patterns of Amphibians in Middle America

JONATHAN A. CAMPBELL

Department of Biology
The University of Texas at Arlington
Arlington, Texas 76019, USA

ABSTRACT   Middle America is characterized by an exceedingly diverse amphibian fauna. This region contains a total of 3 orders, 15 families, 63 genera, and 598 species of amphibians. The distributions of these species are analyzed by topographic areas, vegetation type, elevation, and country. The highlands of Costa Rica–western Panama, western Nuclear Central America, and the Sierra Madre Oriental in Mexico contain the greatest number of species with 133, 126, and 118, respectively. In the lowlands, those of lower Central America contain the greatest diversity of species with the Pacific of Costa Rica–Panama having 109 species and the Atlantic of Nicaragua–Panama harboring 104 species. Mesic forests up to 2700 m contain more species than any other type of forest. Subtropical Rainforest plus Subtropical Wet Forest provide habitats for the greatest number of species (263) of any forest in the region, but Lower Montane Wet Forest and Tropical Wet plus Moist Forest also possess high numbers of 202 and 183 species, respectively. The number of anuran species increases from 144–156 in the lowlands (0–600 m) to 164–183 at moderate elevations (600–1200 m); the number varies from 146 to 162 species at slightly higher elevations (1330–1600 m), and at higher elevations (1600–3000 m) the number of species declines from 141 to nine species. Only one species of anuran occurs above 3500 m. Salamanders show somewhat the same pattern, but with a more dramatic increase of species in the highlands; they reach their greatest diversity at a much higher elevation, and begin their decline at yet higher elevations. The number of salamander species in the lowlands (0–600 m) is only 25–29, but this number steadily increases to 50–55 species at elevations of 2400–2700 m; from a high of 55 species at between 2400–2500 m, the number declines to 12 species at 3500 m. Several species of salamanders are found at elevations above 4000 m and one species occurs above 4300 m. Caecilians are a tropical group restricted to elevations below 1500 m; the greatest number of species occurs between sea level and 300 m (9 species), but a moderate number of species (3–5) occurs from 300 m up to 1400 m. Mexico has the greatest number of amphibian species (318); the numbers of species in other Middle American countries in descending order are 173 (Panama), 160 (Costa Rica), 124 (Guatemala), 85 (Honduras), 59 (Nicaragua), 38 (Belize), and 30 (El Salvador).

Overall, species densities appear to be greatest in the south. However, within localized areas of mesic montane forests, amphibian species richnesses are comparable throughout Middle America, and, in some instances, are actually higher in the north. Approximately 78% of the anuran species, 97% of the salamander species, and 75% of the caecilian species are endemic to the region. A total of 84% of amphibian species and 35% of amphibian genera are endemic. It is difficult to overemphasize the importance of the role played by the wet montane Middle American forests in speciation of amphibians. These forests contain the greatest adaptive radiation of salamanders and anurans, and conservation efforts are most urgently needed for the inhabitants of these forests.

The herpetofauna of Middle America is composed of various groups that have entered the region at different times. Unlike the situation for certain groups of freshwater fishes that have similar patterns of distribution and relationships, it should not be anticipated that very many groups of terrestrial organisms in this region, even those with ostensibly similar habitats, are likely to have highly congruent phylogenies. This would imply that these groups have shared common distribution patterns and ecologies during much of their histories.

Key words: Amphibia; Mexico; Central America; Patterns of distribution; Biogeography; Endemicity; Species richness.

INTRODUCTION

That region extending from Mexico through Central America to Panama, usually referred to as Middle America, can best be described and characterized by its bountiful diversity. This "land of contrasts," as it is sometimes called, boasts an exceedingly intricate terrain and biota. The geological history of the region is one of the most complex for any part of the earth, with forces related to tectonic plates and volcanics forming a rugged landscape. Many broken mountain ranges and series of volcanic peaks arch through the region. Climate in Middle America is affected not only by latitude, which ranges from about 7° to 30° N, but also, in part, from the influences and interactions of the major oceanic currents of the Atlantic and Pacific, elevation and resulting adiabatic lapse rates, and the profound seasonality of the Northeastern Trade Winds (Vivó Escoto, 1964).

Many attempts have been made to map or classify the vegetation of the region (Beard, 1944, 1953, 1955; Holdridge, 1947, 1959a, 1959b, 1962a, 1962b, 1967; Holdridge and Budowski, 1959; Leopold, 1950; Lundell, 1937; Rzedowski, 1986; Tosi, 1969; Wagner, 1964). Owing to the complexity of edaphic conditions, climate, and relief, a correspondingly complex interdigitation of vegetational associations may occur over very short distances.

The diversity of vertebrates in Middle America is astounding, especially when due consideration is given of its relatively small land area. Middle America, including Mexico and the Central American republics through Panama, encompasses a mainland total of only about 2,495,710 km^2. Nonetheless, this region contains a wealth of species rivaled by few other parts of the world, and many of these species are endemic to the area. Middle America has representatives of 35 of the approximately 50 families of nonmarine mammals occurring in the Western Hemisphere (Hershkovitz, 1958). Birds are represented in Middle America by about 1500 species (about 17% of the world total), which are placed in 94 families (about 56% of the world total; numbers taken from Eisenmann, 1955, and Stuart, 1964). No complete list of reptiles exists for Middle America. Villa et al. (1988) reported only 455 species of reptiles from "Middle America," but these authors did not include the area west and north of the Isthmus of Tehuantepec. When this area is included, Stuart's (1964) estimate of 1200 species of reptiles for Middle America may be reasonable.

This wealth of species no doubt has been brought about by the geographical location of Middle America, a tropical region of high relief forming a bridge between the extensive temperate lands to the north and the continent of South America to the south. The biogeography of the fauna of Middle America is a study of many vicariance and dispersal events that have shaped present-day ranges. A complex geological history and still controversial association of Middle America with Caribbean islands has formed many complex, and sometimes seemingly conflicting patterns of distribution.

Pliocene and Pleistocene volcanics undoubtedly caused numerous local extinctions, fragmented previously continuous distributions, and at the same time created new highlands. The effects of glaciation during the Pleistocene had dramatic effects on the climate of the region, at times restricting tropical habitats, while increasing temperate forests. Today, several corridors are evident in the region, including a more-or-less continuous rainforest on the Atlantic lowlands from southern Veracruz, Mexico, southward into South America, and a belt of seasonal, subhumid forest extending along the west coast of Mexico, throughout much of Pacific Central America, into Panama, with only relatively small hiates of wet forest in Chiapas–southwestern Guatemala, southeastern Costa Rica, and parts of Panama.

I have attempted to analyze amphibian distributions in Middle America by topographic region, vegetation type, and elevation. Although our knowledge for many species is still woefully incomplete and these analyses are admittedly crude, this endeavor hopefully will serve to point out important trends and point to certain areas that perhaps are in need of special attention in this age of environmental degradation and declining populations of amphibians.

MATERIALS AND METHODS

The primary database, consisting of a list of species of amphibians occurring in Middle America and their distributions, was derived mostly from two sources: (1) material that I have examined from the American Museum of Natural History (AMNH), California Academy of Science (CAS), the Field Museum of Natural History (FMNH), The University of Kansas (KU), the Museum of Vertebrate Zoology (MVZ), The University of Michigan Museum of Zoology (UMMZ), the United States National Museum (USNM), and The University of Texas at Arlington (UTA), and (2) the literature, including major summaries of the amphibian fauna of Middle America (Blair, 1972; Brame, 1968; Brocchi, 1881–1883; Cope, 1887; Duellman, 1970, 1993; Frost, 1985, 1994; Günther, 1885–1902; Smith and Smith, 1976; Villa et al., 1988; Wake and Lynch, 1976), country or regional treatises (Campbell and Vannini, 1989a; Cope, 1876; Duellman, 1961b, 1963a, 1965a; Dunn 1931b, 1931d, 1933; Flores-Villela, 1993; Hardy and McDiarmid, 1969; Hartweg and Oliver, 1940; Hayes et al., 1989; Henderson and Hoevers, 1975; Johnson, 1989; Kellogg, 1932; Lee, 1980; Martin, 1958; Martínez-Cortés, 1984; Meerman, 1993; Mertens, 1952; Meyer and Wilson, 1971a; Myers and Rand, 1969; Neill and Allen, 1959a; Noble, 1918; Pérez-Higareda et al., 1987; Rand and Myers, 1990; Savage and Villa, 1986; Schmidt, 1941; Scott et al., 1983; Smith and Smith, 1976; Smith and Taylor, 1948; Stuart, 1963; Taylor, 1952a, 1952b; Taylor and Smith, 1945; Timmerman and Hayes, 1981; Villa, 1972). Additionally, many more specific taxonomic and biogeographic papers were consulted. (See references in Appendix 3:1.) This work is limited to the mainland of Middle America. There are few, if any, species of amphibians endemic to nearby islands, *Rana miadis* from Little Corn Island off Nicaragua perhaps being an exception (Barbour and Loveridge, 1929).

For the sake of reasonable completeness, I have listed a number of undescribed species that have been alluded to in the published literature, and also have included a few undescribed species with which I am currently involved. However, I must admit that I am aware of many other undescribed species and, therefore, the current list of Middle American amphibians should be viewed only as an ephemeral effort that will soon be outdated. For example, a number of populations of *Chiropterotriton* have not been allocated and may represent new species (Darda, 1994). Collins et al. (1980) suggested that *"Ambystoma tigrinum"* in Mexico is a multispecies conglomerate, a view supported by Shaffer (1993). Groups that remain poorly known include members of the *"pipiens"* group of *Rana*, the *rugulosus* group of *Eleutherodactylus* (sensu lato), and bolitoglossine salamanders. Had it not been for the contributions of Hillis and colleagues, Lynch and Savage, and Wake and colleagues, respectively, these groups would be in near taxonomic chaos. I estimate that 5–10% of the amphibian fauna of Middle America remains to be described, and this estimate may be conservative.

I realize that country lists of amphibians have little biological significance (Appendix 3:2); however, they may be useful to various conservation and governmental agencies. They no doubt will be important to persons having a nationalistic bent. I have attempted to define amphibian distributions using the vegetational associations of Holdridge (1967), described below. The few aquatic species (e.g., *Ambystoma dumerilii*) are associated with the habitat surrounding the bodies of water that they inhabit. Elevational distributions were gleaned from the literature and data associated with museum specimens. In some instances, published records must be viewed with some suspicion. For example, Shannon and Werler (1955a) and Hanken (1983) gave elevations for *Thorius dubitus* and *T. troglodytes* as 3050 m and

3280 m, respectively. I have visited the localities in question and put them considerably lower, at around 2290 m. The elevation given by Hanken (1983) is a typographical error for 2380 (D. B. Wake, pers. comm). Thus, I restrict the upper elevational range for these two species at 2500 m.

Some parts of Middle America have been surveyed in great detail and are as well known as many parts of the United States, but other regions I consider almost *terra incognita*. Perhaps the regions in the most need of study are the Sierra de los Cuchumatanes of Guatemala, the mountain ranges in eastern Honduras and western Nicaragua, and the highlands of eastern Panama.

The geographical region of Middle America covered in this analysis is highly arbitrary and was determined in part by political boundaries and in part by natural features. The northern limits of this work were determined in consultation with the authors of the section for North America. It was decided that they would include in their section the amphibians of the Peninsula of Baja California, the Sonoran and Chihuahuan Deserts, and the dry forest of the northeastern Gulf Coastal Plain of northeastern Mexico, often considered the "Tamaulipan Province." The amphibian fauna of these areas is composed mostly of distinctly temperate groups.

The northern limits for Middle America in this analysis include the Pacific lowlands of northwestern Mexico from southern Sonora, across the Sierra Madre Occidental at the level of the Río Papagochic Valley, southward along the eastern flank of the Sierra Madre Occidental to the Mesa Central, eastward across the northern boundary of the Mesa Central to the Sierra Madre Oriental, northward to include the Sierra Madre Oriental, and then eastward across the Gulf lowlands in southern Tamaulipas (Fig. 3:1). This northern boundary for Middle America, although somewhat arbitrary, does coincide with major physiographic and/or ecological features. The southern limit of this work is the Panama–Colombia border.

THE NATURAL LANDSCAPE

One of the factors contributing to the great biodiversity in Middle America is that it contains such a complex and varied physiography and a correspondingly complex and varied climate and vegetation. Add to this a geographic position mostly in the tropics, but not so far removed as to escape from faunal invasions of subtropical and temperature lands, and a complex geological history that has frequently fragmented and modified portions of ancestral habitats. Elevations range from sea level to over 5700 m. Climates range from hot and tropical to regions of perpetual snow and ice. Among the various kinds of vegetation are starkly beautiful deserts receiving less than 500 mm of rainfall per year to the enchanting, almost mystical, cloud forests that on most days are bathed in mist and in some places may recieve up to about 7000 mm of precipitation per year. Seasonal fluctuations may involve the effects of winter temperatures in the north; there may be distinctly wet and dry periods owing to the effects of the Northeast Trade Winds, or temperature and rainfall may be relatively equitable year round. Against this environmental matrix, Middle American amphibians and reptiles have evolved numerous adaptations and have invaded practically every type of habitat.

Topographic Areas

Highlands

The highlands of Middle America can be divided into three major regions, the highlands north and west of the Isthmus of Tehuantepec, the highlands between the Isthmus of Tehuantepec and the Nicaraguan lowlands at about the latitude of Lago de Nicaragua, and the mountains of Costa Rica and western Panama. Additionally, several ranges of lesser extent occur in eastern Panama.

Sierra Madre Oriental: This major mountain system extends in the north from the state of Coahuila southward to form the northern highlands in Oaxaca, Mexico. Cerro Zempoaltepec, at 3390 m the highest point in Oaxaca, is near the southern terminus of the range. Over most of the Sierra Madre Oriental, sharp ridges are aligned roughly north–south. The crest is mostly between 2000 m and 3000 m, but may exceed 4000 m in some areas. Rocks forming this range are mostly limestone, but a volcanic influence is evident where this region merges with the Mexican Plateau.

Sierra Madre Occidental: This range extends from near the United States–Mexico border to the Río Grande de Santiago that forms the border be-

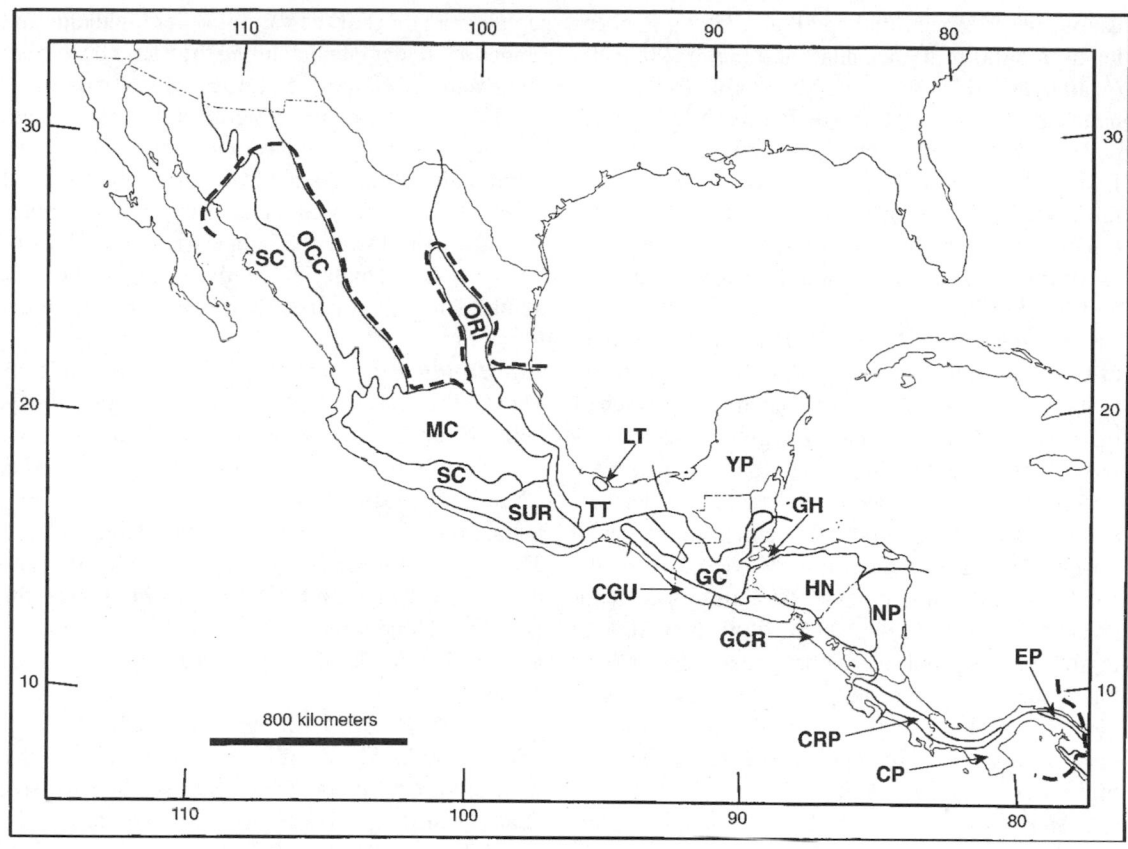

Fig. 3:1. Topographic regions of Middle America. Abbreviations are as follows: ORI, Sierra Madre Oriental; OCC, Sierra Madre Occidental; MC, Mesa Central; SUR, Sierra Madre del Sur and associated highlands; LT, Sierra de Los Tuxtlas; GC, western Nuclear Central American highlands; HN, eastern Nuclear Central American highlands; CRP, Isthmian Central American highlands; EP, highlands of eastern Panama; TT, Gulf lowlands from Tamaulipas to Tabasco; YP, Yucatan Platform; GH, Caribbean lowlands of eastern Guatemala and northern Honduras; NP, Caribbean lowlands from Nicaragua to Panama; SC, Pacific lowlands from Sinaloa to western Chiapas; CGU, Pacific lowlands from eastern Chiapas to south-central Guatemala; GCR, Pacific lowlands from southeastern Guatemala to southwestern Costa Rica; and CP, Pacific lowlands from central Costa Rica through Panama. Heavy dashed lines represent geographic limits of this work.

tween Nayarit and Jalisco. The Pacific side of the Sierra Madre Occidental is noted for its rugged character and has many deep canyons (*barrancas*). The main crest of this range reaches between 2500 m and 3000 m in many places.

Mexican Plateau: This is the largest physiographic province, if taken in its totality, in Middle America. It extends from the border region of the United States southward to the Río Balsas Basin. In the north, the Mexican Plateau is sometimes referred to as the Altiplanicie Septentrional and is relatively low, usually at elevations less than 1000 m. The Altiplanicie Septentrional extends from the states of Aguascalientes and San Luis Potosí northward and is mostly dry country and includes the southern

portion of the Chihuahuan Desert. The southern portion of the Mexican Plateau, also called the Altiplanicie Meridional, or more commonly, the Mesa Central, is higher; elevations gradually increase to over 2000 m over most of the terrain. The Mexican Plateau is bounded to the east and west by major mountain systems, the Sierra Madre Oriental and the Sierra Madre Occidental, respectively. Along the southern edge of the Mesa Central is a region of spectacular volcanos extending in the west from the Nevado and Volcán de Colima across central Mexico to the Cofre de Perote and Volcán de Orizaba in the east. This dramatic series of volcanos is known as the Cordillera Neovolcánica or the Transverse Volcanic Axis. The highest points along these volcanos are Orizaba (Citlaltepetl) at 5750 m, Popocatepetl at 5400 m, Ixtaccihuatl at 5286 m, Malinche (Matlalcueyetl) at 4461 m, Nevado de Toluca (Xinantecatl) at 4392 m, Cofre de Perote (Naucampatepetl) at 4280 m, and Nevado de Colima at 4265 m. Just to the north of the Cordillera Neovolcánica, there are a number of lakes located on the southern portion of the Mesa Central. Some of the largest of these lakes are Chapala, Pátzcuaro, Cuitzeo, and Xochimilco.

Mesa del Sur: These highlands lie to the south of the Mesa Central and Río Balsas depression and to the north of the Isthmus of Tehuantepec. Much of these rugged highlands have a highly dissected surface. The southern highlands of the Mesa del Sur are generally referred to as the Sierra Madre del Sur; this range extends from the Tehuantepec region well up into the state of Guerrero, Mexico, just to the southeast of the mouth of the Río Balsas. The major crest of the Sierra Madre del Sur is at about 2000 m in some places and occasionally exceeds 3000 m, but there are several low passes near the Oaxaca–Guerrero border. The highest point of the Sierra Madre del Sur is Cerro Teotepec, which at 3550 m is the highest point in Guerrero.

Sierra de Los Tuxtlas: These isolated volcanic highlands rise out of the Gulf Coastal Plain in southern Veracruz. Although not of major extent, this region is treated separately owing to its geographic isolation from other highland regions and because of its apparent faunal affinities with both the Chiapan highlands to the south and with the Oaxacan highlands to the west (Campbell, 1984). The highest point in the range is the summit of Volcán San Martín Tuxtla at about 1600 m; two other volcanos, Santa Marta and Martín Pajapan exceed 1200 m in elevation.

Chiapan–Guatemalan Highlands: The dominant topographical features of northern Central America include several more-or-less east–west trending mountain systems (e.g., Sierra de las Minas, Sierra de Chuacús, Sierra de los Cuchumatanes) and drainages (e.g., Río Motagua, Río Polochic, upper Río Grijalva) that are associated with fault zones (Malfait and Dinkelman, 1972; Plafker, 1976). The western Nuclear Central American highlands include the Sierra Madre de Chiapas that extends parallel to the Pacific Ocean from southeastern Oaxaca, Mexico, to about the Chiapas-Guatemala border with peaks ranging from 2000 m to 3100 m; the Meseta Central of Chiapas; the southern volcanic cordillera of Guatemala; and most of the various other highlands and ranges of Guatemala, including the Sierra de los Cuchumatanes, Montaña de Cuilco, Guatemalan Plateau, Sierra de Chuacús, Sierra de las Minas, and the Verapaz highlands. I have also included the Maya Mountains of Belize and Guatemala with this area; much of the amphibian fauna of these mountains is typical of the lower piedmont areas of eastern or northern Guatemala, and only a single endemic amphibian species currently is recognized from the Maya Mountains.

Honduran–northern Nicaraguan Highlands: The eastern Nuclear Central American highlands occur south of the lower Río Motagua Valley and east of a line in Guatemala connecting Zacapa, Chiquimula, Concepción Las Minas, and the Guatemalan–El Salvador border at the Pacific Coast. These highlands include the Sierra de Merendón and the eastern Chiquimulan highlands in Guatemala, all of the Honduran and El Salvadoran highlands, and the highlands of Nicaragua north of Lago de Nicaragua.

Costa Rican–western Panamanian Highlands: The highlands of Isthmian Central America include the Cordillera of Guanacaste, which has five peaks exceeding 1500 m; the Cordillera de Tilarán, which has a crest exceeding 1500 m over much of its length; the Cordillera Central, which contains several high volcanos that attain elevations between 2705 m and 3432 m; the Cordillera de Talamanca, which has ten peaks exceeding 3000 m, including Chirripó Grande, which at 3820 is the highest point in lower Central America; and the Serranía de Tabasará. which is mostly above 1200 m.

Eastern Panamanian Highlands: These highlands include the Serranía del Darién near the Caribbean Coast, which does not reach impressive elevations (mostly below 800 m, one peak reaching 1000 m); the Serranía de Pirre, Serranía de Jungurudó, and Cerro Tacarcuna near the Colombian border that all exceed 1500 m; and the small, somewhat isolated Serranía de Maje and Serranía del Sapo near the Pacific Coast, which are mostly below 1500 m. The ranges of eastern Panama presumably are associated with the Andean orogeny.

Lowlands

The lowlands of Middle America include the coastal plains and associated foothills of the Gulf of Mexico, Caribbean Sea, and Pacific Ocean.

Atlantic Lowlands: These lowlands extend from the Río Grande on the United States–Mexico border southward to the Isthmus of Tehuantepec, include all of the Yucatán Peninsula, and then extend southeastward through all of Central America. In Tamaulipas and Nuevo León in northern Mexico, these lowlands are relatively broad, but they narrow to practically little more than a strip of beach in central Veracruz where the Cordillera Volcánica nearly reaches the Gulf of Mexico. The crest of the Continental Divide in the Isthmus of Tehuantepec region reaches only about 250 m, and is the only area in Mexico where there is considerable faunal interchange between both Atlantic and Pacific lowland faunas. The lowlands of the Yucatán Peninsula are part of the Atlantic lowlands. Relief is slight with maximum elevations of only about 350 m above sea level in the southern part of the region and about 50 m in the north. The southern part of the peninsula is drained by several large rivers flowing to either the Gulf of Mexico or the Caribbean Sea. However, in the north, surface streams become conspicuously absent. The Atlantic lowlands of eastern Guatemala and most of northwestern and north-central Honduras are relatively narrow but characterized by narrow incursions inland owing to many major river valleys. The coastal plain widens considerably in northeastern Honduras and eastern Nicaragua, but narrows once again in Costa Rica and Panama.

For the purposes of geographical analysis, I have divided the Atlantic lowlands of Middle America into four units: (1) the Gulf Coastal lowlands from southern Tamaulipas to Chiapas and Tabasco, Mexico, a region covered mostly by Tropical Dry and Wet Forest; (2) the Yucatán Peninsula of Mexico, including northern Guatemala and northern Belize; (3) the Caribbean lowlands of southern Belize, eastern Guatemala, and northern Honduras; and (4) the Caribbean lowlands from northern Nicaragua through Panama.

Pacific Lowlands: These lowlands extend from Sonora in northern Mexico, where they are the widest, southeastward throughout Mexico and Central America. In general, the Pacific lowlands are narrower than those of the Atlantic. The major inland intrusion of the Pacific lowlands is the Río Balsas Basin of west-central Mexico, extending eastward to the state of Morelos. The hot and dry Río Balsas Basin separates the western part of the Mesa Central from the Mesa del Sur. The Pacific Coastal Plain

is especially narrow west of the Cordillera Neovolcánica, the Sierra Madre del Sur, and south of the highlands of western El Salvador and parts of Costa Rica.

For purposes of the geographical analysis, I have divided the Pacific lowlands into four units: (1) The lowlands from southern Sonora to southwestern Chiapas, Mexico, that are characterized by xeric to subhumid conditions. I have adopted the somewhat unorthodox position of including the Central Depression of Chiapas with the Pacific lowlands (Fig. 3:1), even though I am aware that the Río Grijalva most assuredly flows into the Gulf of Mexico. I have done this because only a low ridge of the Continental Divide (> 800 m) in the vicinity of Rizo de Oro, Chiapas, separates the Pacific lowlands from the Central Depression, and the amphibian fauna of the semiarid Central Depression of Chiapas has been derived mostly from the Pacific lowlands, a relationship substantiated by previous comparisons of the herpetofaunas of the two regions (Johnson, 1989). (2) A rather abrupt transition from subhumid to humid vegetational types occurs at about the level of Tonalá in Chiapas and this humid forest extends along the piedmont and adjacent portion of the coastal plain to about the level of Escuintla in Guatemala. (3) From southeastern Guatemala to northwestern Costa Rica the Pacific lowlands are covered by a subhumid vegetation. (4) From about west-central Costa Rica through Panama several distinctive habitats occur, ranging from very humid to subhumid, but not as dry as are those habitats to the north.

GEOLOGICAL HISTORY

Much of the geologic history of southern Mexico and Central America remains poorly known and controversial. Inferences about the evolution of species and the waxing and waning of ancestral distributions are open to diverse speculation. Nevertheless, it is the practice of biogeographers to propose theories that may be scrutinized subsequently, and I herein attempt to summarize briefly some of what is known about the historical geology and paleoclimates of Middle America.

The entire region has had a history of being at or near plate boundaries since the late Jurassic, and the tectonic effects are dramatic. Dominant in the plate tectonics of the region has been transform faulting (strike-slip faulting), which has given rise to displaced terranes (regions of distinctive geology, bounded by faults). Displaced terranes are regions of lithosphere moved horizontally along strike-slip faults. One of the two most important displaced terranes is the entire Chortis block, thought to have originally been a part of the south coast of Mexico (Donnelly, 1989; Burkart, 1983). Thus, Acapulco in Guerrero and Puerto Angel in Oaxaca were adjacent to Guastatoya in Guatemala and San Pedro Sula in Honduras prior to the Eocene (early Tertiary). Another displaced terrane is the Maya (Yucatán) block, a region bounded by the Motagua and Polochic faults of Guatemala, by the Orizaba fault zone of southern Mexico, and by faults located mainly offshore, along the northern and eastern margins of the Yucatán Peninsula. The Maya block is now moving to the southeast with respect to the North American Plate (Burkart, 1983).

The tectonics of southern Mexico and northern Guatemala have operated so as to elevate repeatedly the same regions and to depress repeatedly other regions during the Cretaceous and Cenozoic. The Mexican Plateau, Sierra Madre Oriental, Sierra Madre del Sur, the Sierra Madre de Chiapas, the Sierra de Chuacús, and the Sierra de las Minas were repeatedly elevated or made topographically positive. These regions were connected before the mid-Miocene (10 million years ago), after which they were displaced 130 km left-laterally across the Polochic fault (Burkart, 1983).

Many changes in the configuration, relief, climate, and biota occurred during the late Cretaceous when Middle America entered a phase of intense mountain building with the appearance of new mountain ranges and plateaus in Mexico and Nuclear Central America. These mountains, resulting primarily from folding, greatly modified Middle America during the Laramide Revolution (late Cretaceous to early Tertiary; Maldonado-Koerdell, 1964), during which time the Sierra Madre Oriental and Rocky Mountains to the north also were elevated.

Coincident with the Laramide Revolution, the Mexican Plateau, the Sierra Madre Oriental, and the Sierra Madre del Sur were uplifted, and these landmasses have remained emergent up to the present. During the Cretaceous most of Middle America was covered by shallow seas which deposited limestones over wide areas and evaporites in restricted basins, predominantly on the Isthmus of Tehuantepec, Yucatán Peninsula, and Petén of Guatemala (Walper, 1960). Parts of Nuclear Central America, the Si-

erra Madre de Chiapas, the Sierra Madre Oriental, and the Sierra Madre del Sur were exposed periodically. These exposed lands have been visualized as being of relatively low topographic relief with mesic, tropical conditions and having high temperatures and precipitation (Dorf, 1959; Savage, 1966). The Central American paleopeninsula persisted in isolation through most of the Cenozoic with an island archipelago, the Guanarivas Ridge, situated to the south of a line connecting the peninsula with South America. This peninsula, acting as a cul-de-sac for species dispersing into the region, has been suggested to have increased faunal diversity (Savage, 1966; Schmidt, 1943; Smith, 1949).

Although some modern genera may have evolved by Cretaceous times (Estes, 1985; Savage, 1966; Tihen, 1964), it is likely that many montane species in Middle America did not make an appearance until the late Miocene at the earliest, the latter portion of which saw the region uplifted to respectable elevations for the first time (Childs and Beebe, 1963; Dengo, 1968; Williams, 1960). Prior to the Miocene, Middle America probably possessed highlands that were not extensive, of little relief, and of relatively low elevation (Dengo, 1968; Maldonado-Koerdell, 1964). The mountain system that includes the Sierra Madre de Chiapas, Sierra de Chuacús, and Sierra de las Minas contains the oldest exposed rocks, mostly Paleozoic metamorphics and sediments, in Nuclear Central America. It is likely that the distributions of extant lineages, if already adapted to montane elevations at this time, were relatively restricted, and their ranges were not greatly fragmented.

The highlands of Nuclear Central America (here defined as the uplands between the Isthmus of Tehuantepec and the Nicaraguan depression), including Chiapas and southeastern Oaxaca, Mexico, developed during the Miocene with increasing elevations into the Pliocene (Schuchert, 1935; Stuart, 1950; Savage, 1982). The Sierra de los Cuchumatanes and Meseta Central of Chiapas are composed mostly of Mesozoic sediments and are thought to have been uplifted during the late Cretaceous or early Tertiary (Dengo, 1968; Anderson et al., 1973). To what extent the region was elevated in the early Tertiary is also controversial, but it appears that after the period of mountain building ended in the early Tertiary the region underwent a long period of erosion (McBirney, 1963).

The time of greatest change in climate and vegetation during the Tertiary began in the late Miocene and continued into the Pliocene. It was during this time that the first of two periods of volcanism began to rock Nuclear Central America. These volcanoes erupted from fissures along a broad belt, some 50–70 km wide paralleling the Pacific Coast and laid down volcanic rocks on a broad surface of rugged relief (Williams, 1960; Williams et al., 1964). These eruptions occurred to the south of the ridge of the Chuacús-Minas mountain system that was already present and produced a broad plateau in western Guatemala. The surface configuration prior to this volcanic activity had been eroded to one of strong relief and a ridge close to the present Continental Divide dominated the landscape along the axis formed by the Sierra de Chuacús and the Sierra de las Minas (Williams, 1960). Middle Tertiary volcanism, coupled with a reduction in the temperature owing to increased elevation, created areas subjected to cold temperatures and the distinct zonation of climate and vegetation on mountain slopes that has persisted until the present (Maldonado-Koerdell, 1964; Stuart, 1957). Similar Miocene events also were occurring in Mexico; great igneous activity was building the Sierra Madre Occidental, and the Mexican Plateau was raised to its present elevation (Duellman, 1965b; Maldonado-Koerdell, 1964). The Neovolcanic Axis, extending roughly along latitude 19°N, seems also to have had its beginning during this time with the great cones, such as Orizaba, being formed later in the Pliocene (Dengo, 1968; Maldonado-Koerdell, 1964).

The isolated Sierra de los Tuxtlas on the Gulf Coast of southern Veracruz is of late Pliocene age with extensive modification owing to Quaternary volcanism (West, 1964). This volcanic region often has been thought to represent the eastern outlier of the Neovolcanic axis (Ordóñez, 1936), but more likely it is associated with an ancient syncline that is unrelated to the Neovolcanic Axis (Murray, 1961).

The question of whether a late Tertiary marine barrier existed in the Tehuantepec region remains unsettled, but certainly one of the most dramatic vicariance events that fragmented the ancestral range of many amphibians occurred in the Isthmus of Tehuantepec region. This low region has been an important area of faunal interchange between the Atlantic and Pacific lowlands, and has served as an effective barrier between the Oaxacan and Nuclear Central American highlands (Duellman, 1960a). Durham et al. (1955) concluded on the basis of sedimentary evidence that Nuclear Central America remained connected with southern Mexico throughout this period, but that the Isthmus was perhaps of lower relief and reduced to almost half of its present width; lacustrine deposits suggest greater local relief and heavier rainfall in the region during the Pleistocene than today. However, as pointed out by Stuart (1966), a shallow channel washed by a scouring sea might not have left any trace of itself. Certainly, the presence of fossil camels and horses of early Pliocene age in the Mejocote Valley of southwestern Honduras supports the contention that no marine barrier was present in the Isthmus of Tehuantepec at this time (Olson and McGrew, 1941). It has been suggested that the Isthmus region might have been inundated by a marine portal during the early to mid-Tertiary (Donnelly, 1989; Maldonado-Koerdell, 1964; Sykes et al., 1982).

The high elevations characteristic of Nuclear Central America and the Mexican Plateau were attained in the Pliocene. The occurrence of marine sediments of Miocene age at about 2300 m in Chiapas indicates a tremendous amount of uplift in some areas (Schuchert, 1935; Stuart, 1950). It is probable that the marine embayment of Amatique that extended through northern Guatemala, and today is indicated by Lago de Izabal, continued to be a barrier and to influence differentiation to the north and south (Schuchert, 1935; Campbell and Smith, 1992). The barrier presented by the embayment across the Nicaraguan Depression persisted until late Pliocene, dissecting Central America from about the Río San Juan almost to the Gulf of Fonseca on the Pacific (Lloyd, 1963). The rugged landscape that had been carved into the older rocks was modified by inundations of Tertiary volcanism, and the region became one of more moderate relief (Williams et al., 1964). The middle Pliocene was marked by an extended period of volcanic quiescence and severe erosion, creating features in the landscape still much in evidence today in central Mexico and the highlands of Central America (Williams et al., 1964). Remnants of a deeply weathered erosion surface in the western portion of the Sierra de las Minas at about 2000 m are indications of the broad uplift and subsequent erosion that have occurred since the Pliocene (McBirney, 1963).

The lands uplifted during early Pliocene underwent a period of erosion during mid- to late

Pliocene during which time several Atlantic drainage river systems became deeply entrenched. These rivers have their headwaters in the Minas–Chuacús–Madre de Chiapas axis that are composed of resistant igneous and metamorphic rocks and make up the present-day Continental Divide. Formerly, some of these rivers appear to have flowed roughly parallel to these ranges, but, owing to differential erosional properties of the regions, were subsequently captured by other stream systems. Thus, the headwaters of the Río Polochic were captured by the Río Negro and have formed the Salamá Valley and the deep Río Negro Gorge that isolate the highland faunas of the Cuchumatanes from those of the Alta and Baja Verapaz highlands. The Río Cuilco and Río Selegua probably became deeply entrenched at this time and partially isolated highland species in the Cuilco massif (Wake and Lynch, 1982). The Río Grijalva also may have changed its course in the region presently called the Cañon del Sumidero. From extant distributional patterns it is tempting to speculate that the deep gorge of El Sumidero was formed at about the same time as many of the other streams flowing through the Cretaceous limestone of northern Central America, thereby dissecting the ridge that may have provided the bridge between faunas of the Mesa Central of Chiapas and the highlands of southeastern Oaxaca.

Several recent studies have demonstrated the close relationships of species inhabiting the Atlantic and Pacific versants of northern Nuclear Central America; genera with closely related species inhabiting both versants include *Duellmanohyla* (Campbell and Smith, 1992), *Plectrohyla* (Duellman and Campbell, 1992), *Ptychohyla* (Duellman, 1970; Campbell and Smith, 1992), *Eleutherodactylus* (Savage, 1987), *Bolitoglossa* (Wake and Lynch, 1976), and *Dendrotriton* (Wake and Lynch, 1976; Lynch and Wake, 1978; Collins-Rainboth and Buth, 1990).

By the Pliocene, in addition to the development of elevational climatic zones, drastic changes in the climatic patterns of the lowlands had developed, including increasingly arid conditions and greater seasonal temperature ranges initiated in the Pliocene and extending to Recent times. Desert vegetation developed in the northern portion of the Mexican Plateau (Dorf, 1959). Subhumid vegetation types advanced southward along the Pacific Coast and in the rainshadow valleys that extended across Nuclear Central America. This no doubt fragmented many

highland mesic forests, and may have eliminated others altogether. The effects of this drying trend on the highland forests was compounded by the effects of late Tertiary volcanism in Central America which must have had a profound influence on its biota, fragmenting the distributions of many species and eliminating others. Late Tertiary volcanic ejecta created thick deposits all along the southern Guatemalan highlands from the Mexican border east-southeastward into El Salvador, and over half of the Nuclear Central American highlands are covered by extrusives from this time.

Several contributing factors of the Pleistocene drastically modified existing distributions and molded the ranges of today. Foremost among these were the renewal of intense and widespread volcanic activity, actually re-initiated in the late Pliocene, and fluctuations in climate brought on by advances and recessions of glaciers. Pleistocene and Recent volcanism in Nuclear Central America has been mostly restricted to a narrow belt along the southern margin of the Tertiary belt and has produced the spectacular strato-volcanos along the Pacific slope of the western portion of the Guatemalan Plateau (McBirney, 1963; Williams, 1960). There is evidence that these Quaternary volcanos were produced in a progressively southeastern succession from about the Chiapas border (Tacaná, Tajumulco) to south of Guatemala City (Fuego, Agua, Pacaya) (Wake and Lynch, 1982). The physiography of the region was greatly modified by the heavy showers of pumice emitted from these eruptions that covered intermontane basins, especially those formed by the parallel belt of eroded, late Tertiary volcanic and sedimentary rocks lying to the north (McBirney, 1963; Williams, 1960). The formation of the more recent Quaternary volcanos did not greatly increase the extent of the Central American highlands, but did increase elevations along the southern portion of the Guatemalan Plateau and produce numerous scattered highland "islands." These volcanoes lie for the most part on a Tertiary pedestal. For example, Volcán de Agua, which attains 3766 m, lies on a Tertiary basement that ranges from about 1100 m on the south to about 1900 m on the northeast side (Williams, 1960). Whereas the volcanoes of the southwestern Guatemalan highlands are confined to a narrow belt, those in southeastern Guatemala are more widely scattered and do not attain comparable elevations to those of the west.

No doubt the formation of these volcanoes, especially those in southwestern Guatemala, not only had a severe effect on local faunas through extinction, but also greatly influenced local weather patterns, producing cloud forest conditions on their windward (south) slopes and creating rain shadow conditions in some of the interior valleys of Guatemala, thereby further isolating mesic habitats on opposite versants in the process (Campbell and Vannini, 1988).

Pleistocene climatic fluctuation caused vegetational shifts that brought about extensions, fragmentations, coalescences, and extirpations of various montane forests and portions of their herpetofaunas. The development of high volcanos along the southern Nuclear Central American highlands also must have altered considerably local wind currents and rainfall patterns of the region, especially on the Guatemalan Plateau.

The relationship between temperatures and precipitation during the various stages of glaciation of the Pleistocene are complex. Although it has been traditional to correlate the alternating climatic fluctuations of cold, moist (glacial) and warm, dry (interglacial) conditions that are supposed to have existed at northern latitudes with Middle American paleoclimates, the reverse may be true; cool, dry periods may have alternated with warm, moist ones (Martin and Harrell, 1957). Interpretations of data from paleobotanical studies suggest that the glaciopluvial periods of the more northerly latitudes in North America may have coincided with periods of aridity in the tropics (Raven and Axelrod, 1974, 1975). Duellman (1965b) suggested that in Mexico and northern Central America there may have been changes in the general patterns of high and low pressure systems that modified the alternating patterns of cool-moist versus warm-dry periods that prevailed in North America during the Pleistocene. In the generalized paleotemperature curve presented by Emiliani and Rosa (1969), low temperatures appear to coincide with periods of aridity documented by the palynological studies of van der Hammen (1974) in northern South America. Some of the faunal evidence for Pleistocene events in Middle America was summarized by Duellman (1960a, 1966a), Martin (1955a), Martin and Harrell (1957), Savage (1966), and Stuart (1950, 1966).

During the height of the glacial advances, there is evidence for small glaciers throughout Middle America, and their existence has been documented for some of the highest peaks in Mexico and Costa Rica (Anderson et al., 1973; Maldonado-Koerdell, 1964; West, 1964; Weyl, 1955; White, 1962). That large vertical shifts in the environment occurred during Pleistocene fluctuations seems to be indisputable, but the extent of these shifts remains controversial. An estimate of a downward vertical displacement of vegetation in the New World tropics for as much as 1000 m was suggested by Graham (1973), Martin (1958), and Simpson (1975, 1979). Such a drastic shift has been disputed by Stuart (1951) and Savage (1966), who contended that such a depression would eliminate all tropical habitats from the region—an event not supported by present tropical faunal distributions. Even the ameliorating influences of oceanic currents and more extensive coast exposed from a lowering of sea level as proposed by Duellman (1960a, 1965b) would not seem to surmount the effects of such a drastic depression of habitats. Maximum depression of mean annual temperature was suggested to have been no more than about 5° C (Stuart, 1957) or 6° C (Savage, 1966). Because the adiabatic lapse rate is from 6 to 10° C/km, depending on the amount of moisture in the air, it is possible that depression of vegetational belts might have approached 1000 m below those of the present day under certain conditions in some regions.

It is probable that most of the cloud forests of the major mountain systems were connected at various times in the past via narrow belts of continuous forest, especially because most were connected via topographic ridge systems exceeding 1000 m. Duellman's (1960a) contention that a cloud forest–like corridor may have existed across the Isthmus of Tehuantepec during periods of Pleistocene history may be justified. Although the topographic ridges connecting the westernmost extension of the Sierra Madre de Chiapas with the southern Mexican highlands descends to about 250 m, the hiatus separating the nearest 1000 m contours on either side is scarcely 60 km. Certainly if a continuous cloud forest corridor did not exist, a narrow strip of mesic forest acting as a filter barrier allowed the dispersal of some cloud forest species, particularly those that are not restricted to cloud forest (sensu stricto).

Cloud forest formation is not necessarily dependent on large amounts of precipitation, but rather a low evapotranspiration rate. Some Middle American cloud forests receive less than 2000 mm of rain annually. The depression of temperatures by 5–6° C may have caused cloud or fog formations along the lower slopes and foothills of ridge systems that resulted in the development of cloud forest conditions. The elevational distribution of Middle American cloud forests is dependent on latitude, exposure to prevailing winds, and local rainfall patterns. For example, cloud forest occurs below 1000 m at some places on the Caribbean versant in Central America and also on the Pacific versant in Costa Rica and Panama. In contrast, cloud forest is found at much higher elevations on many of the Pacific highlands, such as the Sierra Madre del Sur of Guerrero and the volcanoes of southern Guatemala.

Old vicariates across the Isthmus, strongly differentiated, include the genus *Plectrohyla* from the *Hyla bistincta* group (Duellman and Campbell, 1992) and members of the genus *Ptychohyla* (Campbell and Smith, 1992). It appears that the isolation across this barrier is relatively old. Certain salamander genera provide evidence of an even older separation than the anuran examples noted above (D. B. Wake, pers. comm.). *Chiropterotriton, Thorius,* and *Lineatriton* are isolated in the mountains to the north of the Isthmus of Tehuantepec, whereas *Dendrotriton, Bradytriton,* and *Nyctanolis* occur only to the south.

There is abundant evidence that periods in the past were severely affected by aridity. In northern South America, van der Hammen (1974) and his associates have documented several periods of aridity. Perhaps the most convincing evidence comes from consideration of the distributions of closely related species presently confined to subhumid habitats in Middle America. It is assumed that the common ancestors of these groups inhabited a comparable environment to that of their descendants and that these ancestors possessed a wider distribution in the past. An example of subhumid distributional relicts includes frogs of the genus *Triprion* (Trueb, 1970).

CLIMATE

Most of the region under consideration in Middle America is tropical, although especially in the north and at higher elevations, temperate influences prevail. Regardless of what has sometimes been reported in the literature, precipitation over the entire area is seasonal, but the degree of seasonality varies widely. For example, over 80% of annual precipitation is received from May through October over por-

tions of the Mexican Plateau, the mountains of southern Mexico and Nuclear Central America, and the Pacific lowands of Middle America from Sonora, Mexico, to Guanacaste, Costa Rica.

The central and southern portions of the Gulf Coast in Veracruz, Campeche, Tabasco, and Chiapas in Mexico are wet and relatively aseasonal, usually with some rain falling every month. The lowlands of Belize have a rainy season extending from June to October, with May being either wet or dry. The dry period is from February to April, and from November to January usually little rain falls owing to the northers that bring cool, dry air. A distinct gradient in rainfall pattern is evident in Belize with 4570 mm falling in Barranco in the south and 1320 mm falling at Corozal in the north. Parts of the northern Yucatán Peninsula receive less than 1000 mm of precipitation per year. The rainy season for much of Honduras extends from May to late November, but is less seasonal in Atlantic lowlands, which usually receive at least 100 mm of rain every month. (The driest months are from March to June.) The wettest lowlands in Middle America occur along the Caribbean in Nicaragua and Costa Rica. A rainfall gradient is present from north to south in Nicaragua, with 2540 mm per annum at the Río Coco in the north and 6500 mm near San Juan del Norte in the south.

The Pacific lowlands of Guatemala receive 1500–2500 mm of rain annually; this rain falls from May through October; precipitation is infrequent from November to early May. The Pacific lowlands of El Salvador receive 1700–1800 mm of precipitation per year, mostly during a rainy season that lasts from May through October. The Pacific lowlands of Nicaragua receive 1000–1500 mm from May through November. The Pacific slopes and Pacific lowlands of Nicaragua have a well-defined dry season, extending from December through April. The Peninsula de Nicoya in Costa Rica gets about 1300–2300 mm of precipitation per year, depending on locality. The other Pacific lowlands of Costa Rica receive 2500-4000 mm of rain per year. Most low and moderate elevations in Panama have a dry period from late December to April.

The wettest areas in Middle America are the Atlantic-facing cloud forests such as the Sierra Juárez of Oaxaca, the northwestern Sierra de los Cuchumatanes, and northeast-facing slopes of the Cordillera de Talamanca, which receive between

6000–7000 mm of rain annually. The cloud forests along the Pacific also are wet, but not as wet as those facing the Atlantic. The Pacific-facing cloud forests of Guerrero, Oaxaca, southeastern Chiapas, and southwestern Guatemala receive about 4000 mm of rain per year. Caribbean slopes of Guatemala receive 4000–5000 mm of rain annually; it may rain every month and there is no well-defined dry season. The windward (Pacific) slopes of the higher mountains and volcanos of El Salvador receive 2000–3000 mm of rain per year. The piedmont and eastern slopes of the central highlands of Nicaragua also receive abundant rainfall, between 1500–2200 mm per year.

Some of the interior valleys may be quite dry owing to rainshadow effects and may receive less than 1000 mm of rain annually. A series of these rainshadow valleys extends across northern Central America forming a subhumid corridor (Stuart, 1954a).

Rain falls throughout most of the year over most of Costa Rica with most areas receiving 3000–6000 mm (rainfall in general higher in the north), but a few areas experience a brief dry season from about January to May, and in Guanacaste the dry season is long and harsh, extending from about October to May. Some areas such as the western Atlantic coast of Panama and the Portobelo area have a wet climate throughout the year and are considered tropical wet climates, but most areas in lowland Panama, even though they may receive enough rain to be classified as tropical wet climates, receive their rain seasonally and are considered tropical monsoon climates. In Panama, some areas on the Pacific coast have relatively harsh dry seasons.

Temperature in Middle America is largely a function of elevation. The various informal climatic regimes recognized by Guatemalans serve to illustrate the elevational influence on temperature. The lowlands at elevations from sea level to about 1200 m are regarded as *tierra caliente,* with average daytime highs of 29–32° C and average annual temperatures of 24–27° C. Elevations of 1200–2000 m are regarded as *tierra templada* and experience average daytime highs of 16–28° C and average annual tempratures of 15–22° C. At higher elevations, *tierra fria* is subject to cold nighttime temperatures, but temperatures during the day may be surprisingly high. For example, at elevations above 3000 m the mean annual temperature may be less than 10° C, but temperatures during the day may be as high as 24° C.

In Honduras, at elevations of 600–1500 m mean annual temperatures of 18–24° C are experienced; above this elevation mean annual temperatures are less than 18° C. In Costa Rica, on the Atlantic side at elevations of 600–1600 m temperatures are 16–22° C; on the Pacific side at comparable elevations average temperatures are several degrees higher. At elevations above 1600, mean annual temperatures range from 10 to 16° C, or even lower.

Temperatures on the Gulf and Caribbean lowlands are warm and vary relatively little throughout the year. Monthly means of temperature vary at Belize City from 23° C in December to 28° C in July. The north coastal lowlands of Honduras experiences a mean annual temperature of 26–28° C, but is subject to the cooling effects of northers from October to April. The mean annual temperature on the Caribbean coast in Nicaragua is 25–28° C. The mean annual temperature of the Caribbean lowlands of Costa Rica is 22–27° C. Mean annual temperatures in the Pacific lowlands (0–600 m) of El Salvador vary from 22 to 29° C; in Nicaragua these temperatures vary from 26 to 30° C, and in Guanacaste they are from 22 to 28° C.

NATURAL VEGETATION

In addition to its geographical position and complex geological history, the diverse physiography, climate, and edaphic conditions in Middle America contribute to the extremely complex flora. This diversity ranges in the lowlands from desert to tropical rainforest and in the highlands from mesquite grassland to cloud forest or fir forest.

With few exceptions (but see Martin and Harrell, 1957), associations of particular species of amphibians are usually made with aggregations (sensu Wagner, 1964), associations (sensu Dice, 1952), formation series (sensu Beard, 1944), formations (sensu Leopold, 1950), or life zones (sensu Holdridge, 1967) of plant species, rather than individual species of plants. Such a physiognomic approach may not be necessarily preferrable to a floristic one, depending on the questions a particular investigator seeks to answer. However, systematic considerations aside, attempts to correlate animal species with vegetational classifications are useful in trying to explain biogeographic patterns and have a practical aspect when dealing with conservation issues.

Several attempts have been made to classify the vegetation of the region, either in its entirety or in

Table 3:1. Approximate elevations and general distributions of Holdridge Life Zones in Middle America. See text for more complete explanation of distributions.

Life Zone	Elevation	General Distribution of Major Tracts
Tropical Wet Forest and Tropical Moist Forest	0–600 m	Atlantic lowlands from southern Veracruz southward through Panama; more restricted on Pacific but occurring in Chiapas-Guatemala and lower Central America
Tropical Dry Forest	0–600 m	Northern Yucatán Peninsula, west coast of Mexico and Central America to northwestern Costa Rica
Tropical Very Dry Forest	0–600 m	Río Balsas Valley, middle Río Motagua Valley, northwestern Guatemala
Subtropical Rainforest and Subtropical Wet Forest	600–1500 m	Windward slopes along Atlantic and Pacific slopes from Tamaulipas and Guerrero in Mexico southward through Panama
Subtropical Moist Forest	600–1500 m	Piedmont area of mountains of western Mexico and interior of Central America from southeastern Guatemala to northwestern Nicaragua
Subtropical Dry Forest	600–1500 m	Foothills of western Mexico and many of the rainshadow valleys of southern Mexico and northern Central America
Lower Montane Wet Forest	1500–2700 m	Windward slopes along Atlantic and Pacific slopes of major mountains and ranges from Tamaulipas and Guerrero in Mexico southward through Panama
Lower Montane Moist Forest	1500–2700 m	Much of the Sierra Madre Oriental and Occidental, Mexican Plateau, highlands of Oaxaca and Guerrero, highlands of Chiapas and Guatemala
Lower Montane Dry Forest	1550–2700 m	Much of the Mexican Plateau and interior slopes of the Sierra Madre Oriental and Occidental
Montane Wet Forest	>2700 m	Along ridges and on peaks of higher mountains and ranges from Oaxaca, Mexico, to Panama
Montane Moist Forest	>2700 m	Higher portions of the Sierra Madre Oriental and Occidental, Cordillera Neovolcánica, highlands of Oaxaca and Guerrero, Mexico

part (Breedlove, 1973; Harshberger, 1911; Holdridge, 1967; Myers, 1969; Wagner, 1964). The vegetation of Mexico was treated by Leopold (1950) and Rzedowski (1986), and that of Central America was treated by Lauer (1959) and Holdridge (1959a, 1959b, 1962a, 1962b), who provided a series of detailed maps. Besides Holdridge, the region has been chorographed by Leopold (1950), Rzedowski (1986), Tosi (1969), and Wagner (1964).

Features used to classify vegetation are somewhat subjective and vary between habitats and authors. Vegetation may be classified morphologically, genetically, or environmentally, and may include such features as presence and/or abundance of particular species, canopy height, number of stories, deciduous nature of components, life cycles, shape and texture of leaves, soil, and the supposed climate (Wagner, 1964).

Beard (1944, 1955) recognized six plant formation series in Middle America; these were subsequently mapped by Wagner (1964:223, his Fig. 1). Wagner's (1964) outline of plant formations is not particularly useful in that it is too general for an accurate assessment of most amphibian distributions. For example, the montane formation series includes such diverse vegetation cover as pine-oak forest, fir forest, mesquite-grassland, alpine meadows, and cloud forest. Furthermore, his map is not delineated with sufficient accuracy as to allow association of most amphibian species to only the appropriate habitat(s).

For assessing the distributions of amphibians, I have found that the system advanced by Holdridge (1947) to be the most useful, and all of the Central American countries have been mapped according to the Holdridge system (Fig. 3:2). Unfortunately, Mexico has never been mapped using the Holdridge classifications, and I have had to rely on my own field experience in this country, which includes every state and mountain range, to associate species with particular habitats. For a summary of the approximate elevational and geographic distributions of the Holdridge Life Zones in Middle America, see Table 3:1.

Tropical forests

These forests occur from sea level to about 600 m and are the most extensive type of forest in Middle America. Tropical forests may be classified as Wet, Moist, Dry, or Very Dry, according to amount and seasonality of rainfall.

Tropical Wet and Tropical Moist Forests (TWM): These forests are widespread on the Atlantic lowlands from about the Los Tuxtlas region of southern Veracruz, Mexico, southward through the southern portion of the Yucatán Peninsula and much of the Caribbean slopes and lowlands of eastern Guatemala, northern Honduras, northern and eastern Nicaragua, northern Costa Rica, and eastern Panama. These mesic forests are less extensive on the Pacific and occur from the Pacific piedmont of Chiapas, Mexico, to about south-central Guatemala, and along some portions of the Pacific lowlands of Costa Rica and Panama. Most of the area delineated above is Tropical Moist Forest. Tropical Wet Forest is quite limited in extent in Middle America and occurs only in several localized areas of southeastern Nicaragua and northwestern and southeastern Costa Rica. Because of its limited extent and the similarity of its amphibian fauna with Tropical Moist Forest, this type of forest was combined with Tropical Moist Forest for the biogeographic analysis. Much of these forests are often referred to as lowland rainforest. Some of the more mesic portions of Tropical Evergreen Forest may be considered Tropical Moist Forest. Some of the dominant trees include species in the genera *Achras, Swietenia, Ficus, Brosimum, Bursera, Lucuma, Dipholis,* and *Ceiba.*

Tropical Dry Forest (TD): On the Atlantic versant this type of forest covers much of the northern portion of the Yucatán Peninsula and the coastal and foothill region of eastern Mexico; it also is present in portions of many of the "rainshadow" valleys that form the "subhumid corridor" of Stuart (1954a), including the Grijalva, Motagua, Chixoy, Chamelecón–Ulúa, Aguán, Negro (of Honduras), and Patuca valleys. A corridor of Tropical Dry Forest extends along the west coast of Mexico from Sonora to Chiapas, and then on the Pacific Coast from Guatemala, El Salvador, Honduras, and Nicaragua to northwestern Costa Rica, including all of the Península de Nicoya. Several isolated areas of Tropical Dry Forest occur along the Pacific lowlands of southern Panama. This type of forest is sometimes referred to, in part, as dry Tropical Evergreen, Tropical Deciduous or Tropical Thorn Forest.

Following two pages:

Fig. 3.2. The Natural Life Zones of Middle America based on the Holdridge (1947, 1959a, 1959b, 1962a, 1962b, 1967) and Holdridge and Budowski (1959) system of classification. From Campbell and Lamar (1989) with permission of Cornell University Press.

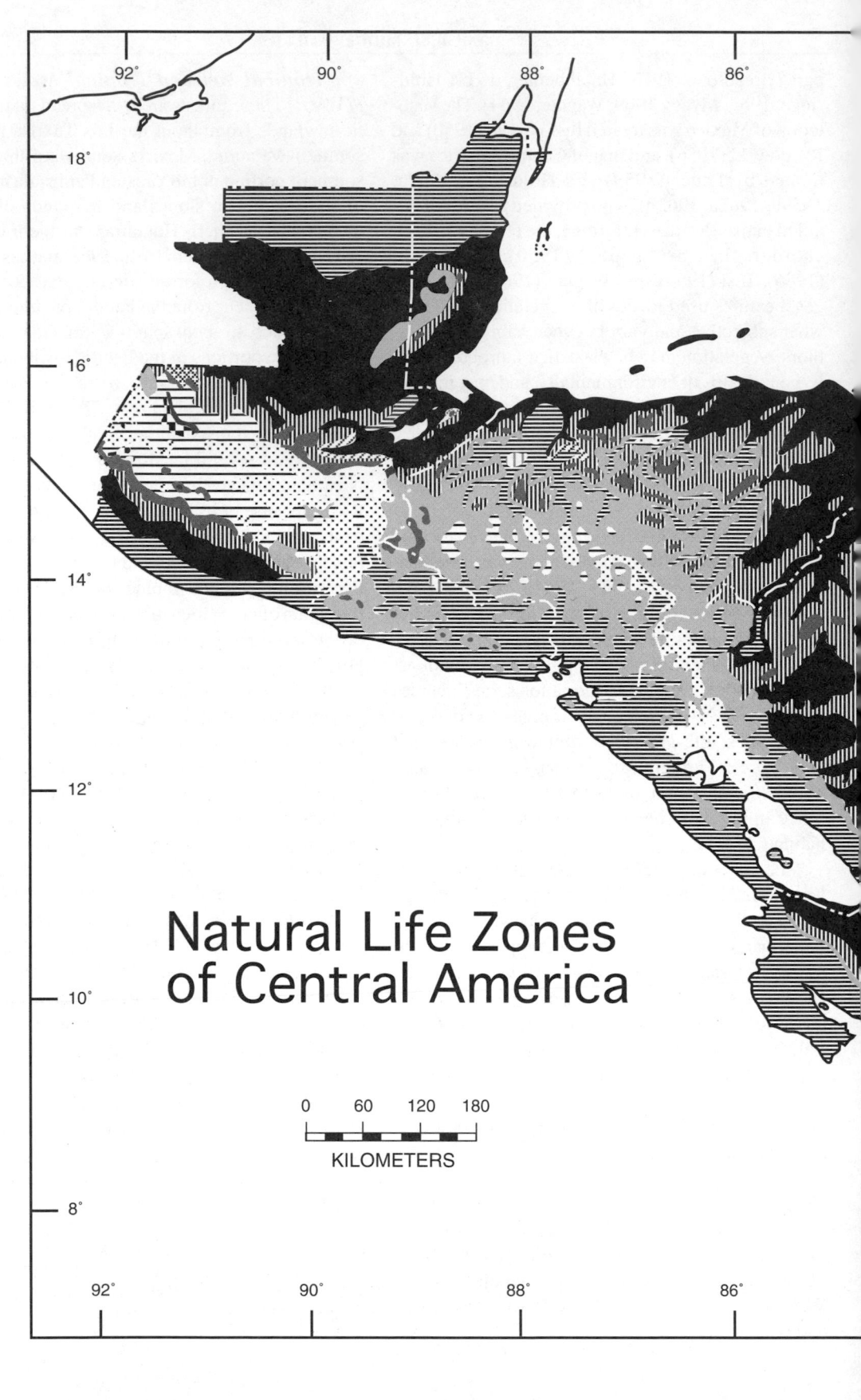

Natural Life Zones
of Central America

0   60   120   180
KILOMETERS

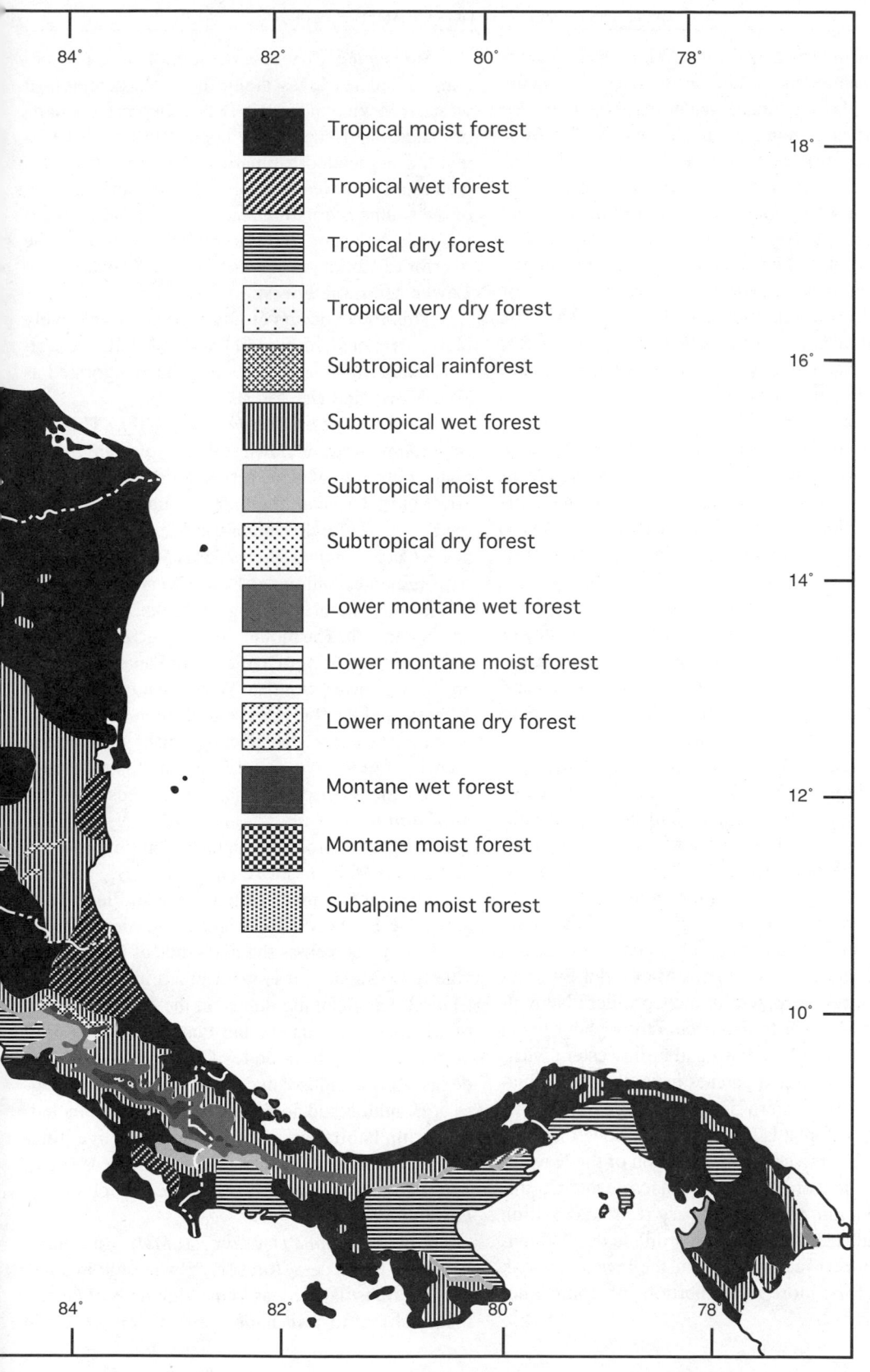

Tropical Very Dry Forest (TVD): This forest is restricted in extent in Middle America and occurs only in a relatively small area north of the Lagos de Nicaragua and Managua, in the middle Río Motagua Valley, and much of the Río Balsas Valley. Localized areas covered by vegetation approaching Tropical Very Dry Forest occur in a few places along the west coast of Mexico, the Zapotitlán Salinas Basin of southern Puebla, the northern end of the Yucatán Peninsula, and the rainshadow valleys of northern Central America (see TD above). This type of vegetation is sometimes called arid tropical scrub, and is characterized by various species of cacti, acacias, and leguminous trees.

Subtropical Forest

In Middle America this type of forest occurs roughly from about 600 to 1500 m, although local conditions often greatly alter these somewhat arbitrary elevational ranges. Four kinds of subtropical forest occur in the region, including Rainforest, Wet, Moist, and Dry.

Subtropical Rainforest and Subtropical Wet Forest (SRW): These forests occur along the lower and intermediate windward slopes of most of the major ranges in eastern and southern Mexico and Central America on both the Atlantic and Pacific versants. Subtropical Rainforest is localized, occurring only in northern Oaxaca and Chiapas, Mexico, the northwestern Sierra de los Cuchumatanes of Guatemala, perhaps several small ranges in southern Nicaragua, and the northern slopes of the Cordillera de Tilarán in Costa Rica. This forest type includes the lower portion of the cloud forest.

Subtropical Moist Forest (SM): This forest occurs along the piedmont of much of the Sierra Madre Occidental, the Sierra Madre del Sur, and the southern escarpment of the Cordillera Neovolcánica of Mexico. It also occurs throughout much of the interior of Honduras and northwestern Nicaragua, with restricted patches in southeastern Guatemala and the western side of the Maya Mountains. This type of forest is not common in lower Central America, but covers a small portion of the leeward slopes of the Cordillera de Guanacaste and Cordillera de Tilarán, as well as a very restricted portion of the southern slopes of the Cordillera de Talamanca. This forest includes part of the lower pine-oak forest and the more mesic portions of tropical deciduous forest.

Subtropical Dry Forest (SD): This forest occurs in isolated areas along the foothill region of western Mexico in the Río Balsas Depression, parts of Oaxaca and Puebla, the upper Río Grijalva Valley and associated tributaries, the upper Río Motagua and Río Negro Valleys of Guatemala, much of the southeastern highlands of Guatemala and adjacent El Salvador, and various valleys through the interior of Honduras and northwestern Nicaragua.

Lower Montane Forest

This forest occurs at intermediate to relatively high elevations (1500–2700 m). In Middle America, Lower Montane Forests may be categorized as Wet, Moist, and Dry forests.

Lower Montane Wet Forest (LMW): This forest occurs on the windward slopes of many of the major ranges in Middle America, including the Sierra Madre Oriental, the Sierra Madre del Sur, and the Sierra Madre de Chiapas in Mexico; the Sierra de los Cuchumatanes, Sierra de las Minas, Alta Verapaz highlands, and upper Pacific versant of Guatemala; and a number of higher peaks in Honduras and Nicaragua. The mountain systems running from about central Costa Rica to western Panama are covered with Lower Montane Wet Forest on both their Atlantic and Pacific sides at appropriate elevations. This forest includes the upper portion of what is commonly referred to as cloud forest, and is characterized by many species of hardwood trees (e.g., *Liquidambar, Cornus, Fagus, Tilia, Nyssa,* and *Quercus),* tree ferns, and abundant epiphytes.

Lower Montane Moist Forest (LMM): This forest occurs across much of the Mexican Plateau and along the Sierra Madre Oriental and Sierra Madre Occidental, as well as the highlands of Oaxaca and Guerrero, Mexico. It is widespread in the Nuclear Central American highlands on the Meseta Central of Chiapas, the Guatemalan Plateau, and the drier portions of the Sierra de los Cuchumatanes, Sierra de las Minas, and isolated highland peaks and ridges of Honduras and northwestern Nicaragua. Included in this habitat are most of the extensive pine-oak forests of Mexico and northern Central America. Dominant genera of trees include *Pinus* and *Quercus.*

Lower Montane Dry Forest (LMD): In Central America, this type of forest is present only in a few very small, isolated areas in the highlands of Guatemala subject to rainshadow effects. However, in

Mexico this type of vegetation is much more widespread and includes the mesquite-grasslands of the Mexican Plateau and interior slopes of the Sierra Madre Oriental and Sierra Madre Occidental. This type of forest includes *Prosopis,* grama grasses, various acacias, yuccas, and often various species of cacti.

Montane forests

These forests occur mainly above 2700 m in Middle America. Only two kinds of montane forests, Wet and Moist, occur in the region.

Montane Wet Forest (MW): This type of forest is restricted to small areas in Middle America where it occurs only along the ridges and peaks of the higher mountains. It exists in the Sierra Juárez and a few of the highest peaks in Oaxaca, the Sierra Madre of Chiapas, the Sierra de los Cuchumatanes, the Sierra de las Minas, Cerro Santa Bárbara in Honduras, and the Cordilleras Central and de Talamanca in Costa Rica and Panama.

Montane Moist Forest (MM): This forest occurs in the Sierra Madre Oriental and Occidental, the Cordillera Neovolcánica, and the highlands of Oaxaca and Guerrero. In Central America, it covers only a few peaks and ridges. This type of forest includes what is sometimes referred to as boreal forest, upper pine-oak forest, open pine and bunchgrass associations, fir forest, and pine-alder-fir forest. Some of the dominant trees occurring in parts of this forest include *Pinus, Juniperus, Abies,* and *Alnus.*

COMPOSITION OF THE AMPHIBIAN FAUNA

Taxonomic Composition

There are at least 598 species of amphibians in Middle America, contained in 3 orders, 15 families and 63 genera. These numbers represent 100% of the orders, 37% of the families, about 15% of the genera, and about 13% of the species of amphibians in the world (calculations based on world totals provided by Duellman, 1993, and Frost, 1985, 1994, and updated by recent descriptions). The number of Middle America families, genera, and species of amphibians contained in the various orders are given in Table 3:2.

Endemics

Middle America harbours no endemic families, but one, Rhinophrynidae, is essentially restricted to the region; it extends northward only to extreme southern Texas. The number of endemic genera (Table 3:3) includes 23% of the Middle American anuran genera, 67% of the salamander genera, and 50% of the caecilian genera. To arrive at these statistics, I have considered the genera *Syrrhophus, Rhinophry-*

nus, and *Dermophis* as Middle American endemics because their ranges lie mostly within Middle America or because most of the species contained in these taxa are restricted to Middle America. The number of amphibian species endemic to Middle America includes 78% of the anuran species, 97% of the salamander species, and 75% of the caecilian species of the region.

Species Richness

A list of species occurring in each Middle American country is given in Appendix 3:2, and the number of species in these countries by family is given in Table 3:4. Fifteen families of amphibians occur in Middle America; of these the Salamandridae and

Table 3:2. Composition of the Middle American amphibian fauna.

Order	Families	Genera	Species
Anura	10	44	389
Caudata	4	15	197
Gymnophiona	1	4	12
Total	15	63	598

Table 3:3. Genera of amphibians endemic to Middle America. (See Figs 3:6–3:10 for distributions.)

Anura	Caudata	Gymnophiona
Atelophryniscus	*Bradytriton*	*Dermophis*[1]
Crepidophryne	*Chiropterotriton*	*Gymnopis*
Anotheca	*Dendrotriton*	
Duellmanohyla	*Ixalotriton*	
Pachymedusa	*Lineatriton*	
Plectrohyla	*Nototriton*	
Ptychohyla	*Nyctanolis*	
Triprion	*Parvimolge*	
Syrrhophus[2]	*Pseudoeurycea*	
Rhinophrynus[1]	*Thorius*	

[1]Range extending slightly outside of Middle America.
[2]Includes species formerly placed in *Tomodactylus.*

Table 3:4. Number of species by family in Middle American countries. Parenthetical percentages = number of species/total number of species in family occurring in Middle America.

Family	Country							
	Mexico	Belize	Guatemala	El Salvador	Honduras	Nicaragua	Costa Rica	Panama
Bufonidae (41 sp.)	20 (49%)	3 (7%)	9 (22%)	5 (12%)	8 (20%)	6 (15%)	14 (34%)	15 (35%)
Centrolenidae (13 sp.)	1 (8%)	1 (8%)	1 (8%)	1 (8%)	2 (15%)	5 (38%)	13 (100%)	12 (92%)
Dendrobatidae (1 sp.)	0	0	0	0	0	2 (12%)	7 (41%)	15 (88%)
Hylidae (145 sp.)	79 (54%)	13 (9%)	38 (26%)	10 (7%)	27 (19%)	17 (12%)	40 (28%)	46 (32%)
Leptodactylidae (128 sp.)	57 (45%)	10 (8%)	28 (22%)	5 (4%)	21 (16%)	13 (10%)	38 (30%)	46 (36%)
Microhylidae (10 sp.)	5 (50%)	2 (20%)	4 (40%)	3 (30%)	3 (30%)	3 (30%)	4 (40%)	4 (40%)
Pelobatidae (2 sp.)	2 (100%)	0	0	0	0	0	0	0
Pipidae (1 sp.)	0	0	0	0	0	0	0	1 (100%)
Ranidae (31 sp.)	25 (81%)	3 (10%)	5 (16%)	2 (6%)	4 (13%)	6 (19%)	6 (19%)	5 (16%)
Rhinophrynidae (1 sp.)	1 (100%)	1 (100%)	1 (100%)	1 (100%)	1 (100%)	1 (100%)	1 (100%)	0
Ambystomatidae (18 sp.)	18 (100%)	0	0	0	0	0	0	0
Plethodontidae (178 sp.)	106 (60%)	4 (2%)	36 (20%)	2 (1%)	18 (10%)	4 (2%)	33 (19%)	19 (11%)
Salamandridae (1 sp.)	1 (100%)	0	0	0	0	0	0	0
Sirenidae (1 sp.)	1 (100%)	0	0	0	0	0	0	0
Caeciliidae (12 sp.)	2 (17%)	1 (8%)	2 (17%)	1 (8%)	2 (17%)	2 (17%)	4 (33%)	9 (75%)
Total	318	38	124	30	86	59	160	172

Table 3:5. Comparison of numbers of species of amphibians and surface area in Middle America by country and with other geographic regions in the New World.

Region	Area (km^2)	Families	Genera	Species	Species/ Genera	Area/ Species
Middle America[1]	1,650,000	15	63	598	9.5	2,759
Mexico[1]	1,126,837	13	36	318	8.8	3,544
Belize	22,963	9	19	38	2.0	604
Guatemala	108,889	9	28	124	4.4	878
El Salvador	21,041	9	19	30	1.6	701
Honduras	112,088	9	25	85	3.4	1,319
Nicaragua	130,000	10	22	59	2.7	2,203
Costa Rica	51,100	10	32	160	5.0	319
Panama	77,082	10	39	173	4.4	446
United States[2]	7,834,000	15	26	225	8.7	34,817
West Indies[3]	218,633	4	8	167	20.9	1,309
South America[4]	17,793,000	16	140	1661	11.9	10,712

[1]As defined herein; for northern boundary, see Fig. 3:2.
[2]Includes only lower contiguous 48 states; number of species based on Conant and Collins (1991), Stebbins (1985), and more recent taxonomic papers.
[3]Includes only native species based on Hedges (this volume).
[4]Includes all of continental South America; number of species based on Duellman (this volume).

Sirenidae barely enter the region in northeastern Mexico, and the Pipidae ranges only as far north as eastern Panama. The Pelobatidae is widespread across the Mexican Plateau, but is represented by only two species. The other families are widespread in the region with the Hylidae, Leptodactylidae, and Plethodontidae collectively representing about 75% of the amphibian diversity. Many taxa are particularly well represented in, or restricted to, certain regions: *Bufo, Hyla, Syrrhopus, Rana, Ambystoma, Chiropterotriton, Pseudoeurycea,* and *Thorius* in Mexico; *Plectrohyla* and *Bolitoglossa* in Chiapas and Guatemala; and *Atelopus,* centrolenids, dendrobatids, *Oedipina,* and caecilians in Costa Rica and Panama. Mexico has the greatest number of amphibian species (318); the numbers of species in other Middle American countries in descending order are 173 (Panama), 160 (Costa Rica), 124 (Guatemala), 85 (Honduras), 59 (Nicaragua), 38 (Belize), and 30 (El Salvador).

It is perhaps no coincidence that several groups of amphibians which have experienced dramatic radiations in Middle America have independently evolved a life cycle that includes eggs that undergo direct development. Over 178 species of bolitoglossine salamanders occur in the region and have adapt-

ed to habitats from sea level to above timberline. All but one species, which is undescribed, are terrestrial or arboreal and exist in a variety of habitats. Eleutherodactyline frogs also lay terrestrial eggs and reach their greatest diversity in the wet forests of lowland and montane habitats. Another speciose group, hylid frogs, also are adapted to a variety of habitats but tend to be more abundant in relatively mesic forests; they have evolved a plethora of reproductive strategies but not terrestrial eggs.

A comparison of the number of species, genera, and families of Middle American amphibians reveals that this region is equal to the United States for total number of families, and possesses about two and a half times the number of genera and species, even though Middle America is only about one fifth the size of the United States (Table 3:5). South America, with almost 1600 species of amphibians, is indisputably the "frog continent." South America harbors over two and a half times the number of amphibian species and over twice the number of genera than does Middle America; however, Middle America is less than one-tenth of the size of South America. The number of amphibian families between these two regions is close, with 16 in South America and 15 in Middle America (Table 3:5).

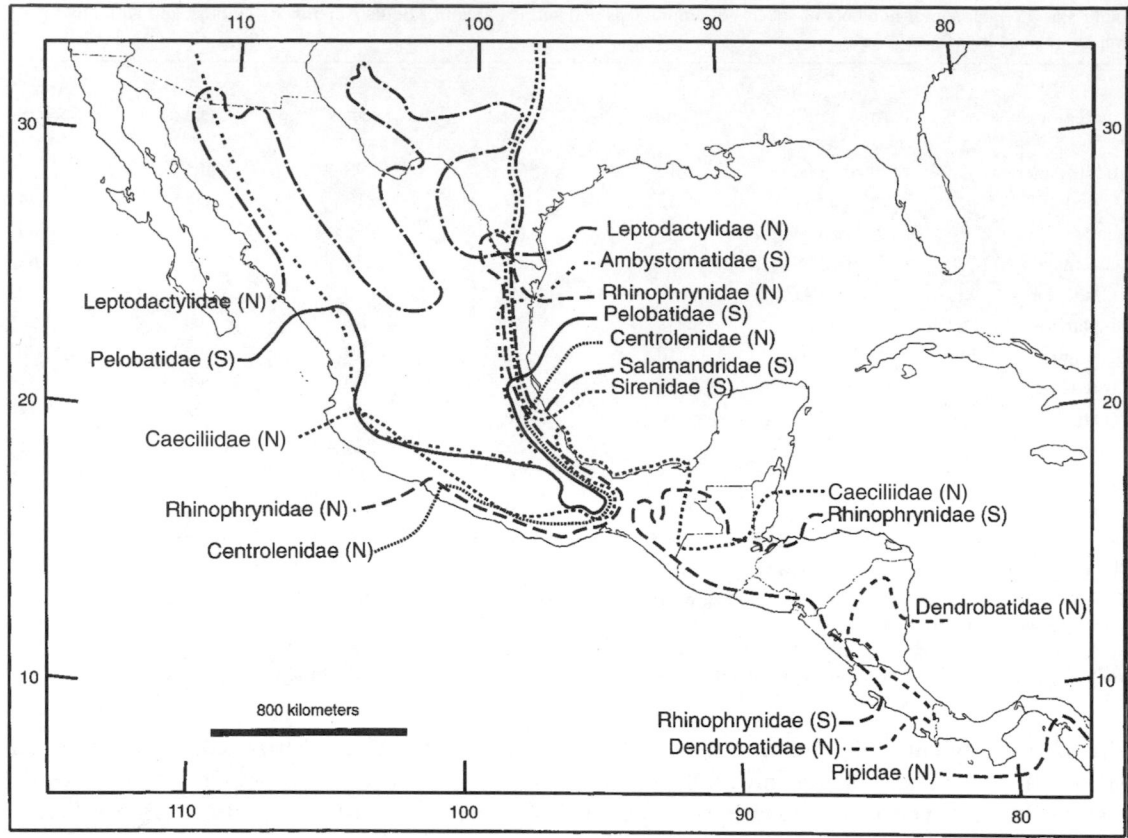

Fig. 3:3. Distributional limits for Middle American families of amphibians. (N) and (S) refer to northern and southern familial limits, respectively.

PATTERNS OF DISTRIBUTION

A number of amphibian families reach their southern limits of distribution in Middle America, including the Salamandridae, Sirenidae, Ambystomatidae, and Pelobatidae. All of these families are mostly temperate in distribution in North America and reach the southern limits of their distribution on the Mexican Plateau (ambystomatids), the Oaxacan highlands (pelobatids), or along the eastern Gulf Coast (salamandrids, sirenids). The northern limits of the families Pipidae, Dendrobatidae, Centrolenidae, and Caeciliidae all fall within Middle America (Fig. 3:3). The monotypic family Rhinophrynidae is essentially restricted to Middle America with both the northern and southern limits of distribution within the region. A few species of the large Neotropical family Leptodactylidae range northward out of Middle America into northern Mexico and the southern United States, but the family, by and large, is restricted to Middle America, the

Caribbean area, and South America. In comparison with reptiles, most species of amphibians have much more restricted distributions (Campbell and Vannini, 1989a).

DISTRIBUTION BY TOPOGRAPHIC REGION

For purposes of analysis, Middle America was subdivided into 17 geographic areas, nine in the highlands and eight in the lowlands (Fig. 3:1). The distributions of Middle American species of amphibians by topographic areas are given in Appendix 3:3, and the number of species inhabiting these areas by family is provided in Table 3:6. The highlands of Costa Rica–western Panama, western Nuclear Central America, and the Sierra Madre Oriental contain the greatest number of species with 133, 126, and 118, respectively. In the lowlands, those of lower Central America contain the greatest diversity of species with the Pacific of Costa Rica–Panama having 109 spe-

Table 3:6. Number of species by family inhabiting different topographic areas in Middle America. Abbreviations of areas are: CG = western nuclear Central American highlands; CGU = Pacific lowlands from eastern Chiapas to south-central Guatemala; CP = Pacific lowlands from ventral Costa Rica through Panama; CRP = Isthmian Central American highlands; EP = highlands of eastern Panama; GCR = Pacific lowlands from southeastern Guatemala to southwestern Costa Rica; GH = Caribbean lowlands of eastern Guatemala and northern Honduras; HN = eastern nuclear Central American highlands; LT = Sierra de Los Tuxtlas; MC = Meseta Central; NP = Caribbean lowlands from Nicaragua to Panama; OCC = Sierra Madre Occidental; ORI = Sierra Madre Oriental; SC = Pacific lowlands from Sinaloa to western Chiapas; SUR = Sierra Madre del Sur; TT = Gulf lowlands from Tamaulipas to Tabasco; YP = Yucatan Platform.

Family	Highlands									Atlantic Lowlands				Pacific Lowlands			
	ORI	OCC	MC	SUR	LT	CG	HN	CRP	EP	TT	YP	GH	NP	SC	CGU	GCR	CP
Bufonidae	4	9	9	5	3	9	5	14	7	3	3	4	8	9	3	4	6
Centrolenidae	1	0	0	1	1	1	1	11	5	1	1	1	10	1	1	1	11
Dendrobatidae	0	0	0	0	0	0	0	7	8	0	0	0	10	0	0	0	11
Hylidae	37	8	10	28	15	38	27	29	9	9	10	10	25	12	4	10	24
Leptodactylidae	14	10	12	13	10	29	16	33	9	10	9	16	25	11	4	5	33
Microhylidae	1	2	2	1	3	2	2	1	1	4	2	1	2	2	1	1	3
Pelobatidae	1	2	2	1	0	0	0	0	0	1	0	0	0	1	0	0	0
Pipidae	0	0	0	0	0	0	0	0	0	0	0	0	0	0	0	0	1
Ranidae	3	6	11	5	1	4	2	3	1	5	3	3	5	4	2	2	2
Rhinophrynidae	0	0	0	0	1	0	0	0	0	1	1	1	0	1	1	1	0
Ambystomatidae	1	2	17	0	0	0	0	0	0	1	0	0	0	0	0	0	0
Plethodontidae	56	1	9	19	7	40	19	33	4	4	5	8	14	0	2	3	9
Salamandridae	0	0	0	0	0	0	0	0	0	1	0	0	0	0	0	0	0
Sirenidae	0	0	0	0	0	0	0	0	0	1	0	0	0	0	0	0	0
Caeciliidae	0	0	0	1	1	3	1	2	0	1	1	2	5	2	1	1	9
Total	118	40	72	74	42	126	73	133	44	42	35	46	104	43	19	28	109

Table 3:7. Numbers of species of amphibians by family inhabiting different vegetational associations in Middle America. Abbreviations are: LMD = Lower Montane Dry Forest; LMM = Lower Montane Moist Forest; LMW = Lower Montane Wet Forest; MM = Montane Moist Forest; MW = Montane Wet Forest; SD = Subtropical Dry Forest; SM = Subtropical Moist Forest; SRW = Subtropical Rainforest and Subtropical Wet Forest; TD = Tropical Dry Forest; TVD = Tropical Very Dry Forest; TWM = Tropical Wet Forest and Tropical Moist Forest.

Family	Vegetational Association										
	TWM	TD	TVD	SRW	SM	SD	LMW	LMM	LMD	MW	MM
Bufonidae	10	11	7	22	6	10	12	6	8	0	2
Centrolenidae	11	0	0	11	0	0	3	0	0	0	0
Dendrobatidae	15	0	0	12	0	0	2	0	0	0	0
Hylidae	45	28	11	76	29	10	61	34	3	8	11
Leptodactylidae	47	25	8	67	22	15	43	24	5	1	0
Microhylidae	7	6	1	2	4	2	1	1	1	0	0
Pelobatidae	0	1	1	0	2	2	0	1	2	0	0
Pipidae	1	0	0	0	0	0	0	0	0	0	0
Ranidae	9	8	2	5	15	4	7	14	5	0	1
Rhinophrynidae	1	1	0	0	0	0	0	0	0	0	0
Ambystomatidae	0	1	1	0	1	1	0	14	4	0	9
Plethodontidae	26	10	1	62	14	5	73	39	0	23	38
Salamandridae	0	1	1	0	0	0	0	0	0	0	0
Sirenidae	0	1	0	0	0	0	0	0	0	0	0
Caeciliidae	11	2	0	6	3	1	0	0	0	0	0
Total	183	95	33	263	96	50	202	133	28	32	61

cies and the Atlantic of Nicaragua-Panama harboring 104 species.

Distribution by Vegetation Type

The distribution of Middle American amphibian species by vegetation type is given in Appendix 3:4, and the number of species inhabiting these different kinds of vegetation by family is provided in Table 3:7. Mesic forests up to elevations of 2700 m contain more species than any other type of forest. Subtropical Rainforest plus Subtropical Wet Forest provide habitats for the greatest number of species (263) of any forests in the region, but Lower Montane Wet Forest and Tropical Wet plus Moist Forest also possess high numbers of 202 and 183, respectively.

Elevational Distribution

The number of species of anurans and salamanders adapted to specific elevations in Middle America follows a pattern previously noted in Guatemala and Belize (Campbell and Vannini, 1989a), with a moderate number in the lowlands that increases at moderate or intermediate elevations and then declines precipitously at high elevations (Fig. 3:4). The number of anuran species increases from 144 to 156 in the lowlands (0–600 m) to 164 to 183 at moderate elevations (600–1200 m); the number varies from 146 to 162 species at slightly higher elevations (1330–1600 m), and at higher elevations (1600–3000 m), the number declines from 141 to nine species. Only one species of anuran occurs above 3500 m.

Salamanders show somewhat the same pattern but with a more dramatic increase of species in the highlands; they reach their greatest diversity at a much higher elevation, and begin their decline at yet higher elevations. The number of salamander species in the lowlands (0–600 m) is only 25–29, but this number steadily increases to 50–55 species at elevations of 2400–2700 m; from a high of 55 species at between 2400–2500 m, the number declines to 12 species at 3500 m. Several species of salamanders are found at elevations above 4000 m and one species occurs above 4300 m.

Caecilians are a tropical group restricted to elevations below 1500 m; the greatest number of species occurs between sea level and 300 m (9 spe-

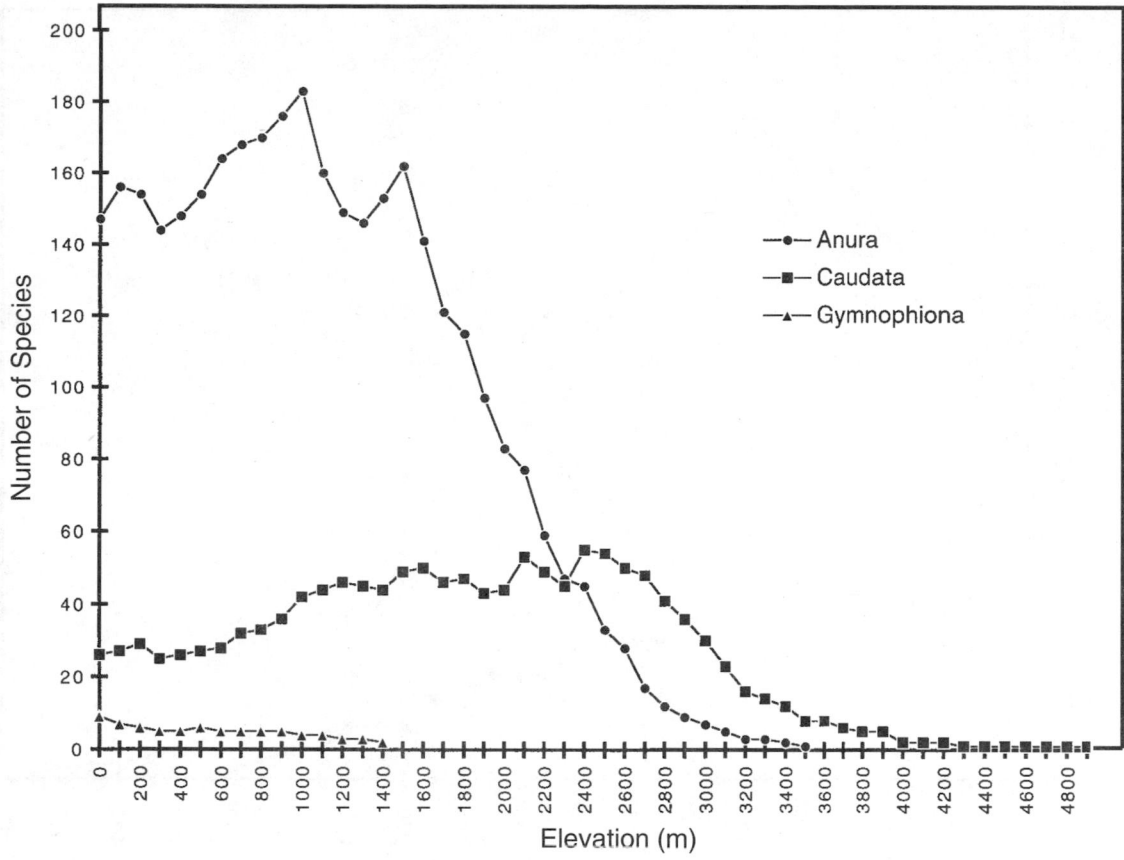

Fig.3:4.   Variation in overall species richness of Middle American amphibians with elevation.

cies), but a moderate number of species (3–5) occurs from 300 m up to 1400 m.

A comparison of the elevational distributions of anuran species inhabiting wet and dry forests reveals that at most elevations there are 3–5 times as many species in wet forests (Fig. 3:5). Also, in dry habitats the number of species is relatively constant between sea level and 1300 m (34–40 species) and then diminishes to one species at 3000 m. The number of species in wet forests increases from 108–119 at low elevations (0–600 m) to a high of 144 species between 1000–1100 m and another spike is apparent at the 1500–1600 m interval with 133 species.

AREAS OF HIGH SPECIES DIVERSITY AND ENDEMICITY

Most of the species of Middle American amphibians are endemic to the region. Therefore the areas of high species diversity also tend to be the areas of high endemicity. An analysis of the distributions of endemic amphibian genera (Figs. 3:6–

3:10) shows that those for anurans and salamanders are mostly restricted to the highlands, especially in southern Mexico and Nuclear Central America. There are only two genera of caecilians that may be considered endemic to Middle America, *Gymnopis* and *Dermophis* (Fig. 3:6); the latter barely extends into South America. Members of these caecilian genera are mostly lowland or foothill in distribution, but may occur up to about 1500 m in some places.

The only lowland endemic anuran genera occurring in the region are found in the north and include *Syrrhophus* (in part) and *Triprion* (Figs. 3:7 and 3:8, respectively). These genera have fragmented distributions and occur both east and west of the Isthmus of Tehuantepec. The lowland depression of the Isthmus has not served to isolate any of the highland endemic genera of southern Mexico; *Anotheca*, *Duellmanohyla*, and *Ptychohyla* are widespread in Middle America and, besides southern Mexico, also occur in the Nuclear Central American and lower

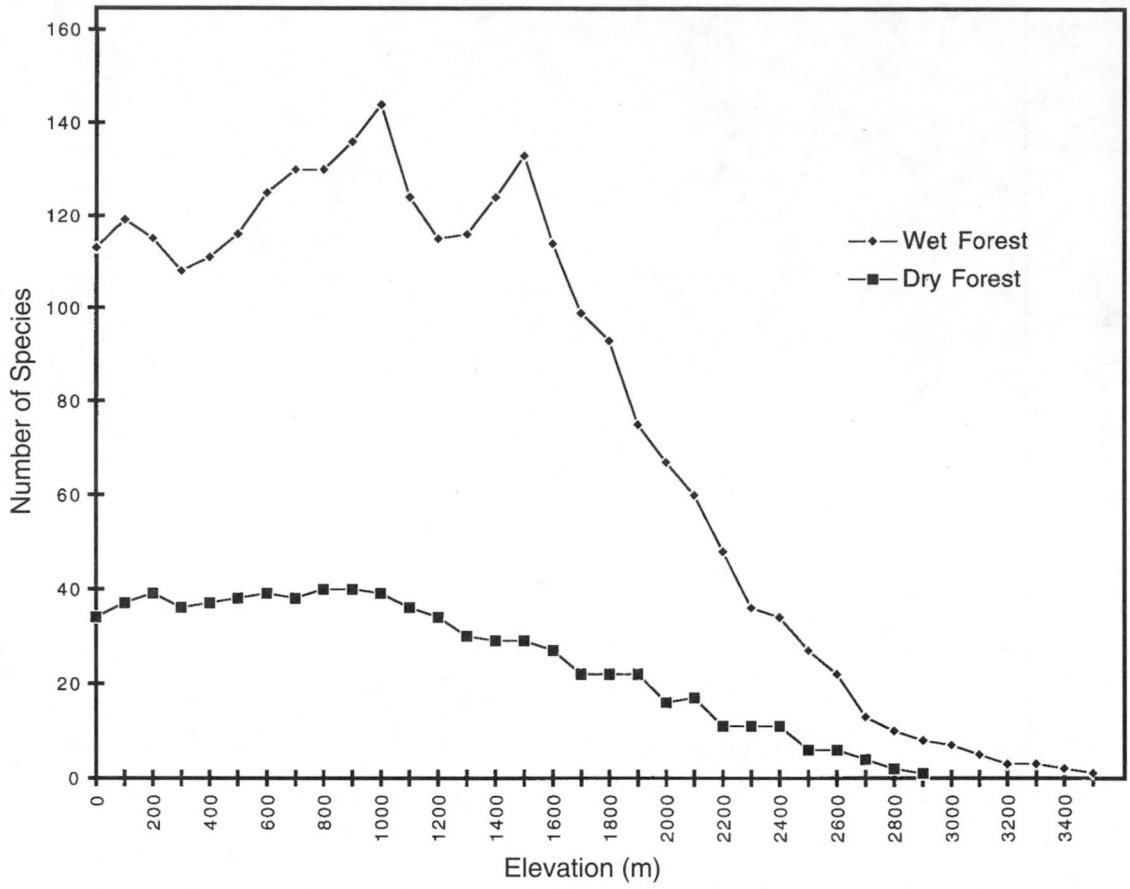

Fig. 3:5. Comparison of variation in species richness of Middle American anurans with elevation in wet and dry habitats.

Central American highlands. Several anuran genera, *Plectrohyla* and *Atelophryniscus,* are restricted to the Nuclear Central American highlands, and *Crepidophryne* is restricted to the Talamancan highlands of lower Central America.

All endemic genera of Middle American salamanders are essentially upland in distribution (Figs. 3:9 and 3:10), although several genera may occur locally at only several hundred meters elevation. *Chiropterotriton, Lineatriton, Parvimolge,* and *Thorius* are all restricted to the highlands of Mexico north and west of the Isthmus of Tehuantepec, whereas *Bradytriton, Dendrotriton, Ixalotriton,* and *Nyctanolis* occur only in the highlands of Nuclear Central America east and south of the Isthmus. *Nototriton* and *Pseudoeurycea* inhabit the highlands on either side of the Isthmus of Tehuantepec, and *Nototriton* inhabits both the Nuclear Central American and Lower Central American highlands.

The greatest species diversity and highest level of endemicity for amphibians in Middle America occurs along the windward, mesic slopes of major mountain ranges between elevations of 800 and 2800 m. Species of anurans tend to reach their greatest diversity at the lower portion of this range, between elevations of 800–1800 m, whereas species of salamanders are most abundant at elevations of 2400–2800 m. I have identified (Table 3:6) three highland areas with especially diverse amphibian faunas: the Sierra Madre Oriental of Mexico (118 species), the highlands of western Nuclear Central America (126 species), and the highlands of Costa Rica–Panama (133 species). Two lowland areas also are worthy of mention—the Atlantic lowlands of lower Central America from Nicaragua through Panama (104 species) and the Pacific lowlands of Costa Rica–Panama (109 species). Lowland areas in general tend to be rich in anurans and caecilians, whereas the high-

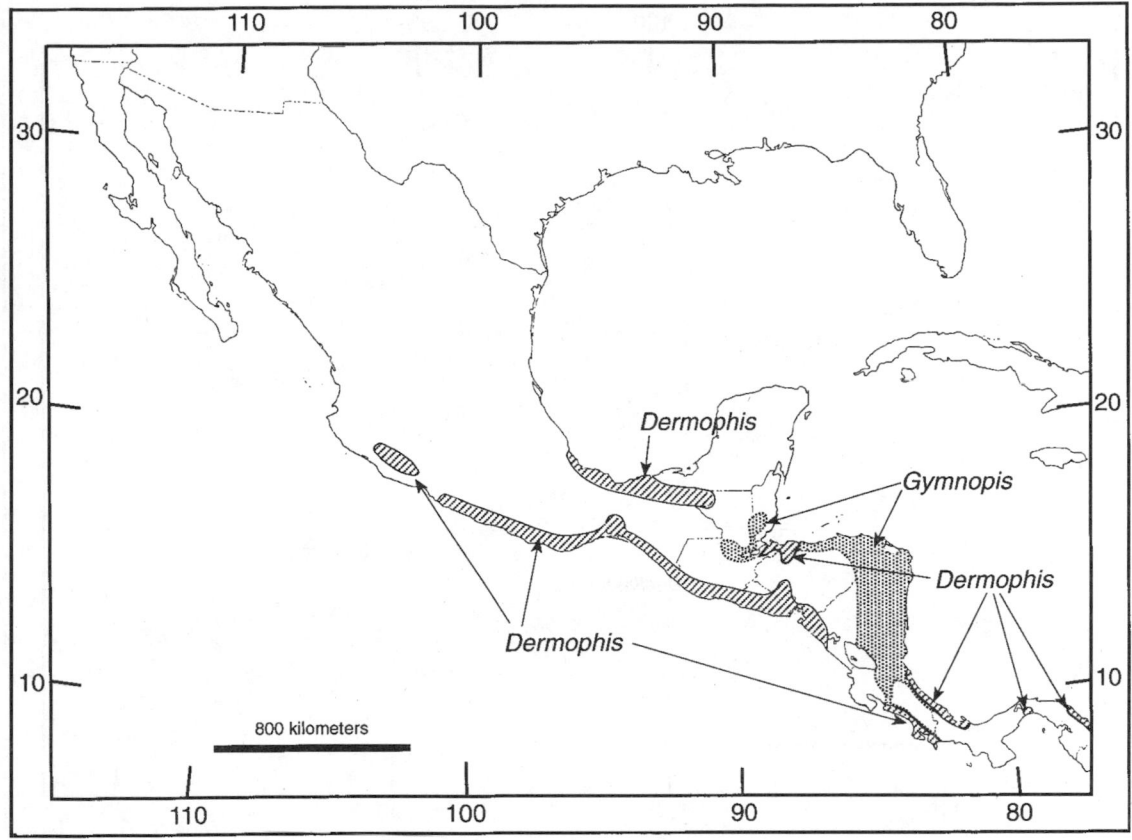

Fig. 3:6.   Distribution of endemic genera of Middle American caecilians.

lands tend to harbor a relatively high number of sala-manders. Not surprisingly, the major ecogeograph-ic features that fragment the highland distribution of Middle American amphibians are the lowlands of the Isthmus of Tehuantepec, the Nicaraguan De-pression, and the interior valleys of the larger river systems.

From geographical, elevational, and vegetation-al data (Appendix 3:4, Tables 3:6 and 3:7), it is ob-vious that the areas in Middle America of greatest amphibian diversity and endemism are (1) the east-ern slopes of the southern portion of the Sierra Ma-dre Oriental (17–20°N latitude); (2) the northern es-carpments of the major mountain ranges of the northern Nuclear Central American highlands from the Meseta Central of Chiapas to the Sierra Nombre de Dios in Honduras; and (3) the highlands of Costa Rica and western Panama, which include the cordi-lleras de Guanacaste, Tilarán, Central, and Talaman-ca. These latter ranges are arranged more-or-less in a linear series. At moderate to high elevations, mesic forests prevail on both Caribbean and Pacific slopes, although the Caribbean side tends to be wetter.

The wet, highland habitats of Middle American mountain ranges for the most part are isolated from one another, and many microendemic species are commonly restricted to a small portion of a single range. The lowland, mesic forests (Tropical Wet and Tropical Moist Forests) throughout Middle Ameri-ca, but especially in the south, also contain a large number of amphibian species. However, these spe-cies tend to be relatively widespread and few, if any, are microendemic.

Patterns of amphibian species diversity may be determined by comparing assemblages at specific localities (Table 3:8). Caecilians and usually sala-manders are absent from the Tropical Dry Forests of Middle America, in which the number of amphib-ians is relatively low and shows no obvious increase at more southerly latitudes. The fewest amphibians recorded for a Tropical Dry Forest site is Sarteneja in northern Belize with 15 species, and the largest

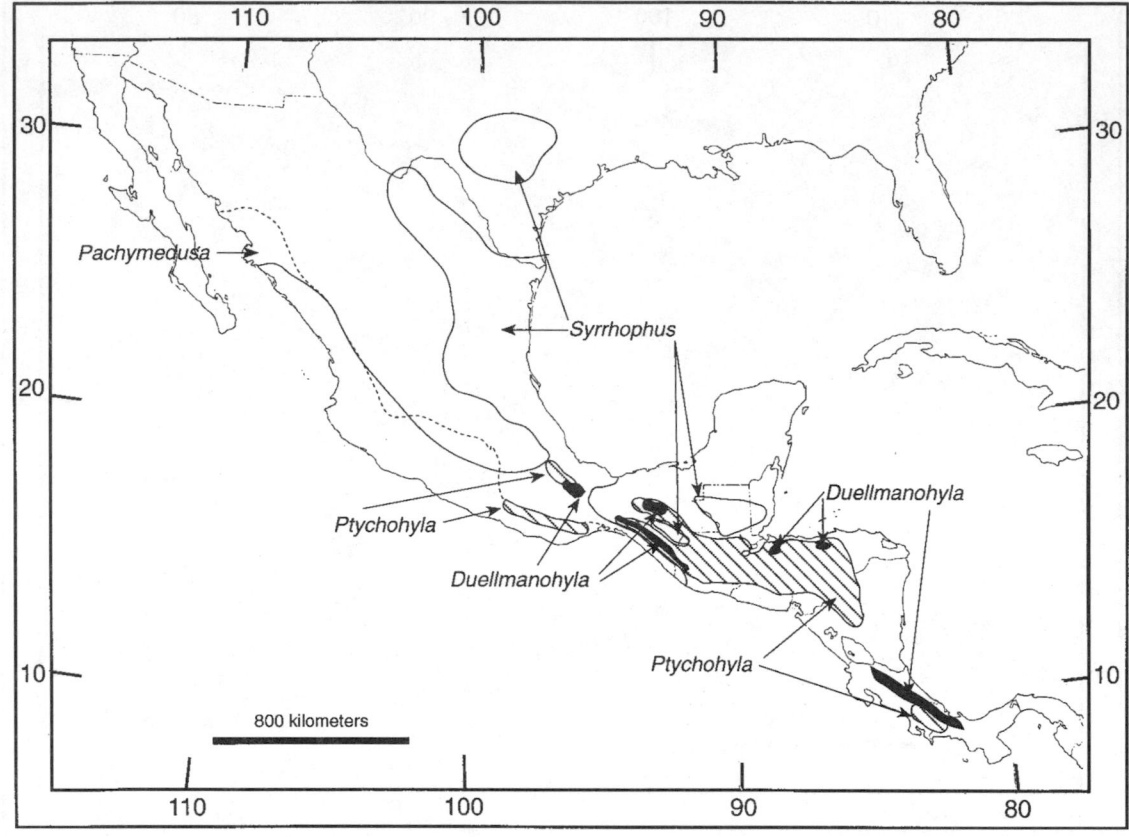

Fig. 3:7.   Distribution of certain endemic genera of Middle American anurans.

number is 23 at Finca Taboga in Costa Rica, which is rivaled by 21 species at Mazatlán in Mexico. The number of species in the Tropical Moist Forest around Tikal, Guatemala, is comparable to Tropical Dry Forest sites; the forest surrounding Tikal is sometimes considered Tropical Dry Forest but, unlike the Tropical Dry Forest sites listed in Table 3:8, the forest at Tikal is evergreen and most of the trees are not highly deciduous. The greatest species diversity of amphibians at a particular site is in humid forests where caecilians and salamanders occur and there is a substantial radiation of anurans. The Tropical Wet Forest near Chinajá, Guatemala, harbors 23 species of amphibians, including one caecilian and two salamanders. The number of species more than doubles in Tropical Wet Forest sites of Costa Rica and Panama. Subtropical and Lower Montane Wet Forest sites also have a corresponding increase in numbers of species at lower latitudes.

AMPHIBIAN CONSERVATION IN MIDDLE AMERICA

Amphibians have undergone a remarkable adaptive radiation and exhibit a greater diversity of life histories than any other group of vertebrates (Duellman, 1988). They are the only tetrapods that have a biphasic live cycle and, although they have successfully colonized land, they are nevertheless rather delicate creatures, dependent on water or high humidity for reproduction. It is little wonder that even relatively small perturbations of the environment might cause serious consequences for the survival of some amphibians. A number of possible reasons for amphibian decline have been suggested—chemical pollutants; habitat destruction; thinning of the ozone layer which has led to increased ultraviolet radiation; increased atmospheric CO_2 that has led to global warming; introduction of competitors, predators, or diseases; climatic stress which has led certain microparasites to reach epidemic proportions;

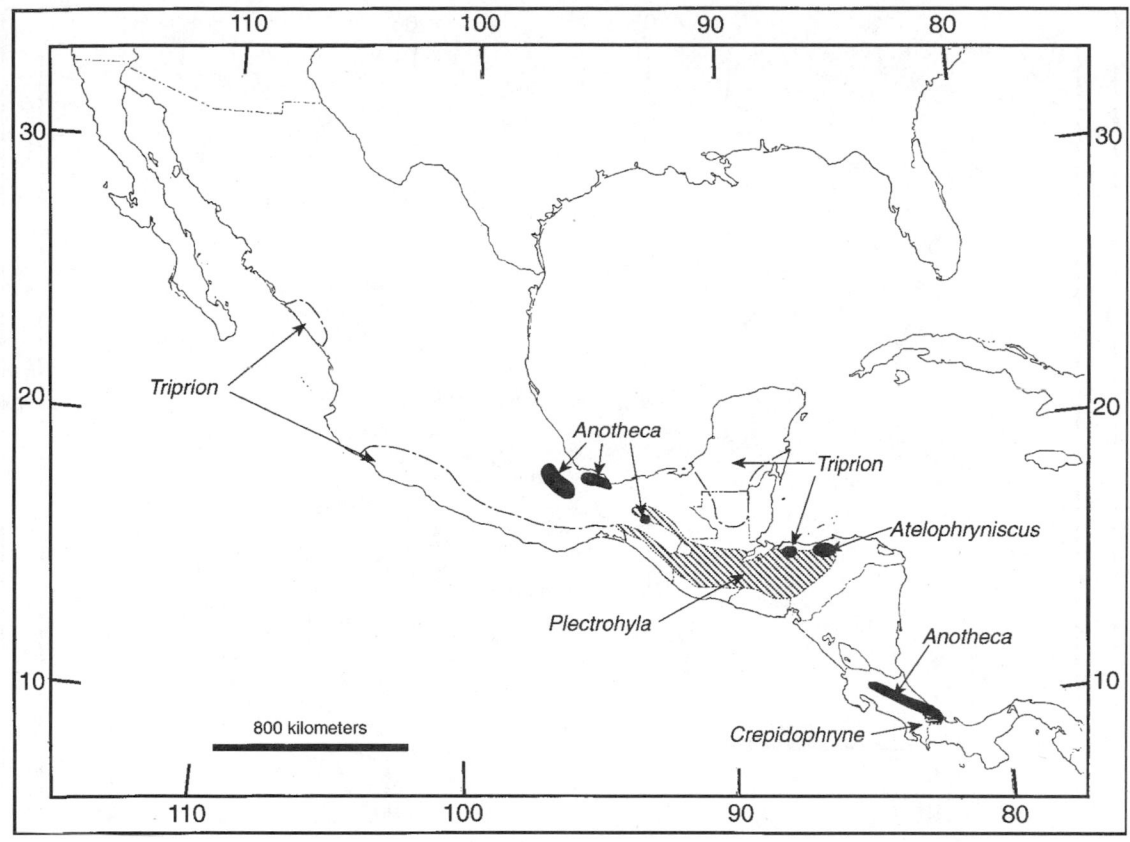

Fig. 3:8.   Distribution of certain endemic genera of Middle American anurans.

and overexploitation for food, scientific use, or the pet trade (Stebbins and Cohen, 1995).

Information concerning amphibian declines in Middle America is scarce, but undoubtedly various populations in the region have suffered from this apparently worldwide phenomenon. The best documented case is *Bufo periglenes* of Costa Rica. This toad was described comparatively recently (Savage, 1967a) and large breeding aggregations were observed regularly in the 1970s and early 1980s. Between 1987 and 1989, the species suffered a population crash, and only one individual was seen in 1989 (Crump et al., 1992; Pounds and Crump, 1994). Since then, the species has not been observed and is almost certainly extinct. Furthermore, in the Monteverde area of Costa Rica where *Bufo periglenes* occurred, field studies have revealed that 19 of the 49 species of anurans that were present a decade ago have now mysteriously disappeared (Pounds and Fogden, 1996).

Other examples of declining frog populations in Middle America, for which I have first-hand ex-

perience, are some of the streamfrogs of the genus *Eleutherodactylus (rugulosus* group) in eastern Guatemala. In the early 1980s, two species in this group were abundant along the streams flowing out of the Montañas del Mico, and I regularly observed during the course of an evening dozens or even hundreds of these frogs sitting on boulders or banks of streams. In about 1985 and 1986, several visits to the area revealed that these frogs were becoming quite scarce, and this scarcity has continued through my last visit in June 1996 when a single individual was seen. Another frog that seems to have declined is *Hyla bromeliacia* in the vicinity of Purulhá, Baja Verapaz, Guatemala. These frogs and their larvae were in practically every large bromeliad in the region up until about the mid-1980s. Since then, I have not seen a single individual near Purulhá, although they remain relatively common in other mountain ranges in Guatemala. Most of the habitat for *Eleutherodactylus* in eastern Guatemala and *Hyla bromeliacia* in the Sierra de las Minas has remained rela-

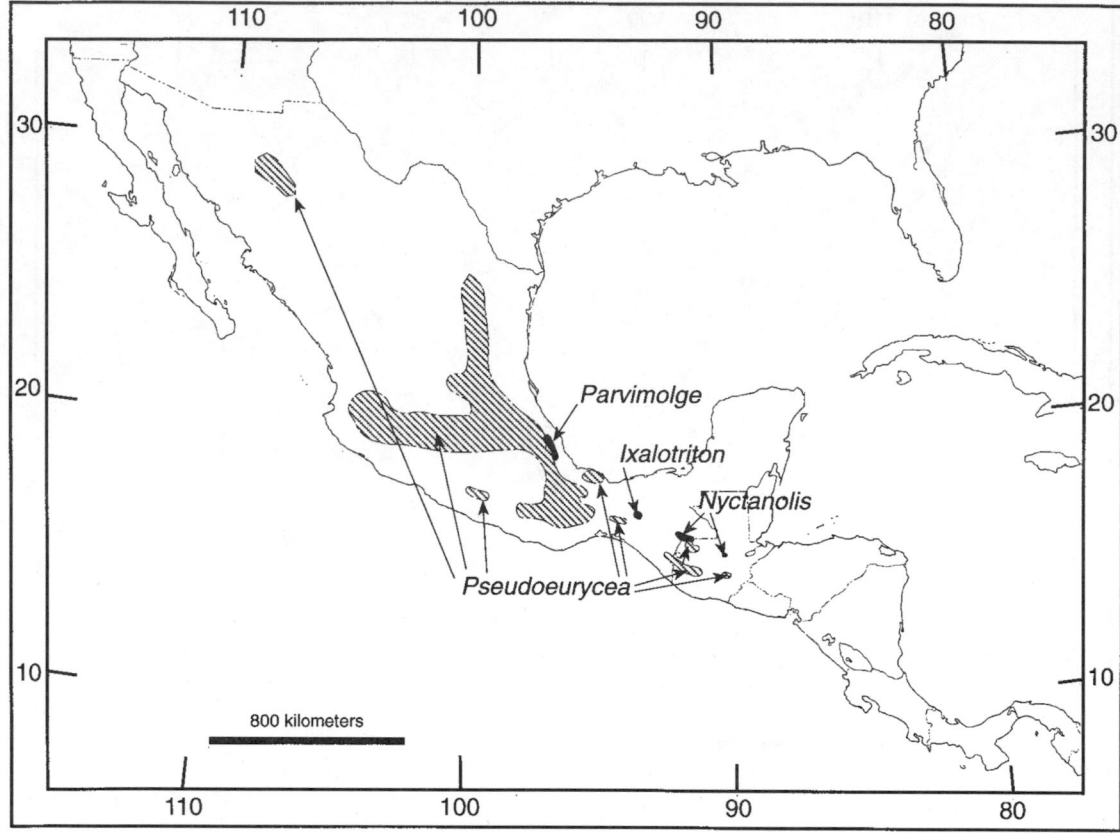

Fig. 3:9.   Distribution of certain endemic genera of Middle American salamanders.

tively pristine, so the common explanation often invoked of habitat destruction is not the answer.

What can be done to help protect amphibian populations in Middle America? Probably very little. The burgeoning human population of the region and the economic realities that drive agrarian societies make most direct conservation efforts unfeasible. Fortunately, many species of amphibians do rather well at surviving in human-degraded environments. In general, those species occurring in the drier, more open habits appear to be surviving if not thriving. It is some of the species that live in the wet environments and that are adapted to "deep forest" conditions that are in the most danger. It is also these habitats that contain the greatest amphibian diversity in Middle America. Many species of hylids, centrolenids, *Eleutherodactylus,* some bufonids, and arboreal salamanders disappear almost immediately when primary forest habitats are cleared. A series of park systems are in place in most countries, although many of these parks in actuality receive no protec-

tion and are in the process of being cleared. Nevertheless, a continuing strengthening of national park systems seems the best hope for many of these species. If, in addition to these parks, more emphasis could be put on protecting the watersheds that provide potable water for many cities and towns, it would be to the mutual benefit of humans and amphibians. Coffee fincas, until recently, provided an important habitat for certain amphibians, although herbicides and insecticides undoubtedly took their toll. The recent practice of many farmers of switching to sun-tolerant varieties of coffee and of clearing the old shade trees has been detrimental to amphibians and other shade-loving species.

Several ill-conceived programs have been instigated in Middle American countries with U.S. government support including one to eradicate the Mediterranean fruit fly and one to help control the Africanized honey bee. Both of these have involved the widespread and indiscriminate use of insecticides and the results probably have been detrimental to

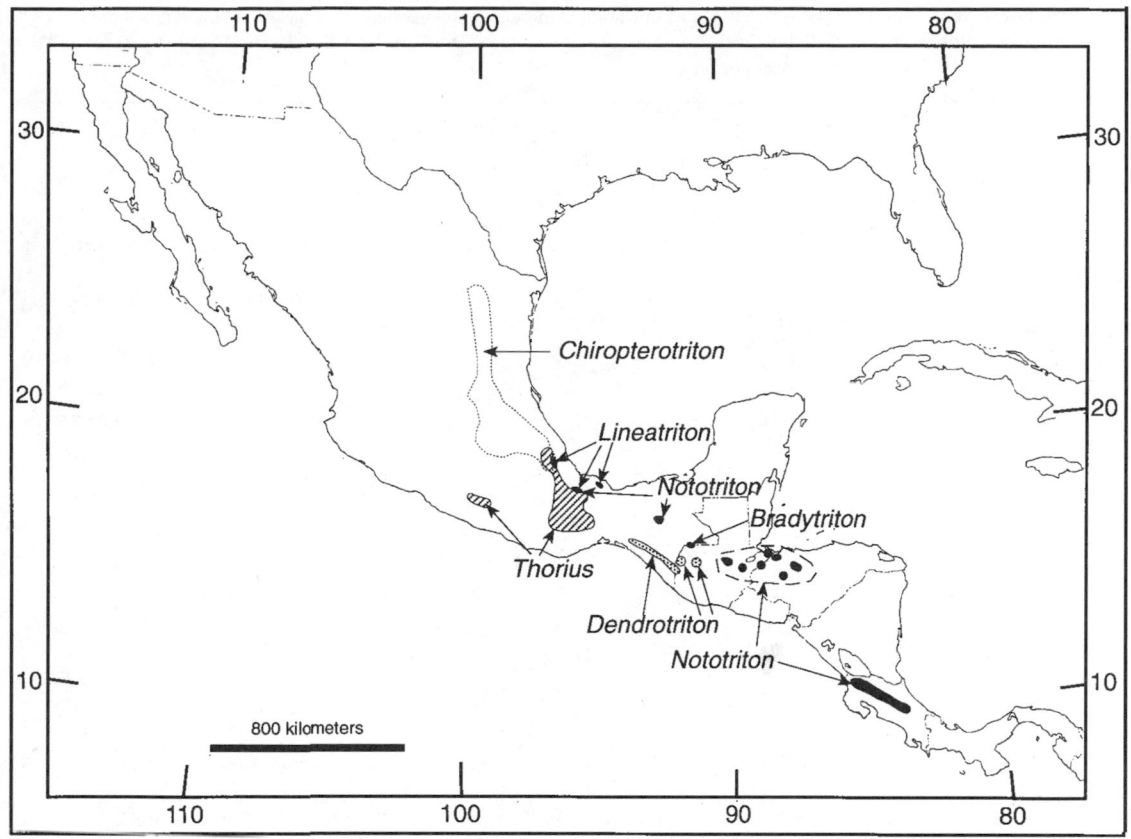

Fig. 3:10.   Distribution of certain endemic genera of Middle American salamanders.

amphibian populations and inconsequential to their intended targets. These kinds of programs should be discouraged whenever possible. On the other hand, careful and responsible scientific collecting and research needs to be encouraged. All too often, various bureaucrats, who are ignorant of any aspect of biology either view themselves as erudite defenders of wildlife or just don't care, and they are the ones responsible for issuing scientific permits to conduct research or to collect, export, or import specimens. There are some governmental agencies, both in Middle American countries and the USA, that have an enlightened attitude, but this change has been slow in coming and is not widespread. Other problems include "sapismo" which is still rampant throughout the region and hinders both national and international research (Mares, 1991). Finally, the creation of "protected" or "red" lists are an exercise in futility in all Middle American countries. Of what possible consequence is it to protect species of tropical amphibians, of which we know very little of their ecology or population characteristics, from a handful of scientists that might want to study them, when they receive no protection from the lay human population within the country?

DISCUSSION

The data presented here are intended to serve only as a general reference for those interested in distributional patterns and composition of the Middle American amphibian fauna. The data at hand permit only a broad picture of the region as a whole.

The Middle American amphibian fauna is an assemblage containing groups with diverse historical backgrounds (Stuart, 1964; Savage, 1966, 1982). If the criteria for the recognition of a particular faunal region is the presence of endemic taxa or the

Table 3:8. Local amphibian diversity at specific sites in Middle America. Abbreviations for vegetation associations are: LMW = Lower Montane Wet Forest; SWF = Subtropical Wet Forest; TDF = Tropical Dry Forest; TVD = Tropical Very Dry Forest; TMF = Tropical Moist Forest; TWF = Tropical Wet Forest.

Locality	Associ– iation	Lat. north	Elev. (m)	Taxonomic Group			Total	Source
				Gymno- phiona	Cau- data	Anura		
Pisté, Yucatán, Mexico	TVD	20°42'	10	0	1	16	17	Duellman, 1965a; Lee, 1996
Apatzingán, Michoacán, Mexico	TVD	19°06'	335	0	0	17	17	Duellman, 1961b, 1965b
Mazatlán, Sinaloa, Mexico	TDF	23°13'	10	0	0	21	21	Hardy & McDiarmid, 1969
Chamela, Jalisco, Mexico	TDF	19°31'	0–200	0	0	19	19	Ramírez-Bautista, 1994
Sartenejal, Corozal, Belize	TDF	18°20'	10	0	1	14	15	Meerman, 1993
Tehuantepec, Oaxaca, Mexico	TDF	16°20'	35	0	0	18	18	Duellman, 1960a
Finca Taboga, Guana- caste, Costa Rica	TDF	10°20'	40	1	0	22	23	Scott et al., 1983
Tikal, El Petén, Guatemala	TMF	17°20	200– 300	0	1	18	19	Campbell, 1998
Chinajá, Alta Verapaz, Guatemala	TWF	16°02'	140	1	2	20	23	Duellman, 1963a
La Selva, Heredia, Costa Rica	TWF	10°25'	90	1	3	44	48	Scott et al., 1983
Rincón de Osa, Punta- renas, Costa Rica	TWF	08°42'	10	1	4	42	47	Scott et al., 1983
Barro Colorado Island, Panama, Panama	TWF	09°10'	150	1	2	49	52	Rand & Myers, 1990
Rancho del Cielo, Tamaulipas, Mexico	SWF	23°06'	1000– 1500	0	5	8	13	Campbell, 1982; Martin, 1958
Los Tuxtlas, Veracruz, Mexico	SWF	18°30'	600– 1500	1	6	30	37	Campbell, 1982; Pérez- Higareda et al., 1987
Vista Hermosa, Oaxaca, Mexico	LMW	17°51'	1500– 1800	0	5	16	21	Campbell, 1982; Wake, 1987
Purulhá, Baja Verapaz, Guatemala	LMW	15°26'	1500– 1800	1	5	17	23	Campbell, 1982
Monteverde, Puntarenas- Alajuela, Costa Rica	LMW	10°17'	1470– 1859	1	5	31	37	Hayes et al., 1989
Las Cruces, Puntarenas, Costa Rica	LMW	08°48'	1500	1	1	30	32	Scott et al., 1983

radiation of many groups within the region, then Middle America is a good candidate based on its amphibian fauna. In Middle America, 84% of amphibian species and 35% of amphibian genera are endemic. A number of major radiations of amphibians has occurred in Middle America. Perhaps the most notable radiation is that of the bolitoglossine salamanders (Wake, 1970, 1987), of which there are 178 species in Middle America allocated to, in part, 10 genera endemic to the region. Other notable ra-

diations of Middle American amphibians are hylid frogs, especially those groups adapted to breed in cascading mountain streams, and leptodactylid frogs of the genera *Syrrhophus* and *Eleutherodactylus,* both of which lay terrestrial eggs that undergo direct development.

Other workers have arrived at somewhat different conclusions on how to treat the Middle American herpetofauna. Schmidt (1954) assigned Central America to the Holarctic Region based on the fact that its geological relationships are to the north. He considered Central America as being "transitional" between the Arctogaean and Neogaean Realms. Although during the Tertiary Middle America formed a paleopeninsula into which many northern groups dispersed, Stuart (1964) pointed out that zoogeographic regions must be recognized on the basis of existing faunal distributions and not on past geological connections or ancient faunal assemblages, and he pointed out that the Central American herpetofauna was mostly of Neotropical origin. Stuart (1964:338) did not consider Middle America to be a separate faunal region, but viewed it as containing fauna from two regions, the Nearctic and the Neotropic. However, Savage (1966) regarded the Middle American herpetofauna as a distinctive unit, equivalent in rank to the Nearctic and Neotropical units. The extent of Middle America as defined by Savage (1966, 1982) is roughly the same as in this paper.

That Middle America has served as an important link between North and South America is especially apparent by the number of vertebrate families that reach their southern or northern limits of distribution in this region. Ten of the fifteen families of amphibians occurring in Middle America reach either their northern or southern limits within the region (Fig. 3:3).

Kiester (1971) analyzed species density of North American (i.e., United States) amphibians and found that, in general, amphibian species densities in this region tend to increase toward the equator; are higher in the east, which is a more mesic area; are positively correlated with areas of high topographic relief; and increased moving inland from coastal shores. Many of the same trends are apparent in amphibian species densities in Middle America. Overall, species densities seem to be greatest in the south, with Panama and Costa Rica harboring 170 and 160 species, respectively; to the north, Guatemala and Honduras possess 124 and 82 species, respectively, despite the fact that these countries are larger than either Costa Rica or Panama. However, within localized areas of mesic montane forests, amphibian species richnesses are comparable throughout Middle America, and in some instances actually higher in the north. Amphibian species densities in Middle America are positively correlated with precipitation and therefore highest in Tropical Wet and Moist Forests, Subtropical Rainforests, Subtropical Wet Forests, and Lower Montane Wet Forests. These wet forests are more amply distributed along the coastal plain and mountain escarpments of the Atlantic, although mesic forests occur on major mountain masses facing the Pacific and locally in some places on the Pacific coastal plain. Areas of high topographic relief have a greater amphibian species diversity in Middle America than the lowlands as indicated in Fig. 3:4. Finally, the number of amphibian species increases as one goes from coastal areas toward inland areas (Fig. 3:4). The reason for this apparent increase was suggested by Kiester (1971) to be simply an artifact caused by the reduced area within coastal quadrats. Analysis of elevational distributions in this study was performed by tallying known elevational ranges for each species (Appendix 3:3) and not by quadrat sampling. Campbell and Vannini (1989a) suggested that a slightly lower number of species at elevations near sea level might be the result of the effects of tropical storms or a tolerance of more species to better-drained terrains.

Subhumid environments, which are not particularly rich in amphibian species, were estimated to cover at least two-thirds of Middle America (Stuart, 1964). Despite the restricted nature of mesic habitats, the amphibian fauna of Middle America is rich and varied. The coastal lowlands of both the Atlantic and Pacific have been important corridors for amphibian dispersal in Middle America, with humid habitats prevailing on the Atlantic and subhumid forests covering much of the Pacific. Both Dunn (1940b) and Griscom (1932) noted that the mesic forest of Caribbean Panama continued southward along the Pacific Coast of Colombia and Ecuador; and that the relatively dry Pacific forest of Panama extends along the Caribbean Coast of Colombia and Venezuela.

It is difficult to overemphasize the importance of the role played by the wet montane Middle American forests in speciation of amphibians. These forests contain the greatest adaptive radiation of salamanders (Wake, 1987) and anurans (Campbell and Vannini, 1989a; Duellman, 1970; Savage, 1987; Smith and Taylor, 1948) in the region. The isolation of biotas inhabiting disjunct mesic uplands in Middle America is relatively great, and as pointed out by Janzen (1967), as barriers to dispersal, mountain passes tend to be "higher" in the tropics than in temperate regions. That is to say, that organisms inhabiting specific elevational zones in the tropics tend to be adapted to relatively narrow temperature regimes, whereas montane species of temperate regions are exposed to a wide range of temperatures.

Cloud forests were probably more restricted in extent and more fragmented at various times in the past than today. Many of the smaller patches of cloud forest, especially those on mountains of relatively low elevation, were probably eliminated altogether. This, in part, may explain the depauperate nature of the southeastern Guatemalan highlands and isolated highland areas in El Salvador and Honduras. While it is tempting to speculate that current distributional patterns of many groups were formed by Pleistocene events, Gadow's (1913) statement that "the key to the distribution of any group lies in the geologic configuration of that epoch in which it made its first appearance" is worth emphasizing. It is relevant that the herpetofauna of Middle America is composed of various groups which have entered the region at different times (Dunn, 1931a; Schmidt, 1943; Stuart,

1950, 1951; Savage, 1966, 1982). Unlike the situation for certain groups of freshwater fishes that have similar patterns of distribution and relationships, it should not be anticipated that very many groups of terrestrial organisms in this region, even those with ostensibly similar habitats, are likely to have highly congruent phylogenies. This would imply that these groups have shared common distribution patterns and ecologies during much of their histories.

Acknowledgments: I am grateful to W. E. Duellman for inviting me to write this chapter and for prodding me along for well over a year. The painful process of reading this manuscript in its entirety was undertaken by W. E. Duellman, J. R. Mendelson, and D. B. Wake, all of whom made many helpful comments, but of course cannot be held accountable for any errors remaining. This material is based in part upon work supported by the Texas Advanced Research Program under Grant No. 003656-001.

LITERATURE CITED

ADLER, K. 1965. Three new frogs of the genus *Hyla* from the Sierra Madre del Sur of México. Occas. Pap. Mus. Zool. Univ. Michigan 642:1–18.

ADLER, K., AND D. M. DENNIS. 1972. New tree frogs of the genus *Hyla* from the cloud forests of western Guerrero, México. Occas. Pap. Mus. Nat. Hist. Univ. Kansas 7:1–19.

ANDERSON, J. D. 1961. The life history and systematics of *Ambystoma rosaceum*. Copeia 1961:371–377.

ANDERSON, J. D. 1978. *Ambystoma rosaceum* Taylor. Tarahumara salamander. Cat. Am. Amph. Rept. 206.1–2.

ANDERSON, J. D., AND W. Z. LIDICKER, JR. 1963. A contribution of our knowledge of the herpetofauna of the Mexican state of Aguascalientes. Herpetologica 19:40–51.

ANDERSON, T. H., B. BURKART, R. E. CLEMONS, O. H. BOHNENBERGER, AND D. N. BLOUNT. 1973. Geology of the western Altos Cuchumatanes, northwestern Guatemala. Bull. Geol. Soc. Am. 84:805–826.

ANDERSON, J. D., AND R. G. WEBB. 1978. Life history aspects of the Mexican salamander, *Ambystoma rosaceum* (Amphibia, Urodela, Ambystomatidae). J. Herpetol. 12:89–93.

AROSEMENA, F. A., AND R. IBANEZ D. 1991. Geographic distribution: *Hyla angustilineata*. Herpetol. Rev. 22:133.

BAKER, R. H., R. G. WEBB, AND E. STERN. 1971. Amphibians, reptiles, and mammals from north-central Chiapas. An. Inst. Biol. Univ. Nac. Auton., Mexico 42, Ser. Zool. 1:77–86.

BARBOUR, T. 1923. Notes on reptiles and amphibians from Panama. Occas. Pap. Mus. Zool. Univ. Michigan 129:1–16.

BARBOUR, T. 1925. A new frog and a new snake from Panama. Occas. Pap. Boston Soc. Nat. Hist. 5:155–156.

BARBOUR, T. 1928. New Central American frogs. Proc. New England Zool. Club. 10:25–31.

BARBOUR, T., AND E. R. DUNN. 1921. Herpetological novelties. Proc. Biol. Soc. Washington 34:157–162.

BARBOUR, T., AND A. LOVERIDGE. 1929. Vertebrates from the Corn Islands. Bull. Mus. Comp. Zool. 69:138–146.

BAYLOR, E. R., AND L. C. STUART. 1961. A new race of *Bufo valliceps* from Guatemala. Proc. Biol. Soc. Washington 74:195–202.

BEARD, J. S. 1944. Climax vegetation in tropical America. Ecology 25:127–158.

BEARD, J. S. 1953. The savanna vegetation of northern tropical America. Ecol. Monogr. 23:149–215.

BEARD, J. S. 1955. The classification of tropical American vegetation-types. Ecology 36:89–100.

BLAIR, W. F. 1965. *Rana johni,* substitute name for the frog *Rana moorei* Blair. Copeia 1965:517.

BLAIR, W. F. 1972. *Bufo* of North and Central America. Pp. 93–101 *in* W. F. Blair (ed.), *Evolution in the Genus* Bufo. Austin, Texas: Univ. Texas Press.

BOGERT, C. M. 1962. Isolation mechanisms in toads of the *Bufo debilis* group in Arizona and western Mexico. Am. Mus. Novit. 2100:1–37.

BOGERT, C. M. 1967. New salamanders of the plethodontid genus *Pseudoeurycea* from the Sierra Madre del Sur of Mexico. Am. Mus. Novit. 2314:1–27.

BOGERT, C. M. 1968. A new genus and species of dwarf boa from southern Mexico. Am. Mus. Novit. 2354:1–38.

BOGERT, C. M. 1969. The eggs and hatchlings of the Mexican leptodactylid frog *Eleutherodactylus decoratus* Taylor. Am. Mus. Novit. 2376:1–9.

BOLAÑOS, F., D. C. ROBINSON, AND D. B. WAKE. 1987. A new species of salamander (genus *Bolitoglossa*) from Costa Rica. Rev. Biol. Trop. 35:87–92.

BRAME, A. H., JR. 1960. The salamander genus, *Oedipina*, in northern Central America. Bull. S. California Acad. Sci. 59:153–162.

BRAME, A. H., JR. 1963. A new Costa Rican salamander (genus *Oedipina*) with a re-examination of *O. collaris* and *O. serpens*. Contrib. Sci. Nat. Hist. Mus. Los Angeles Co. 65:1–12.

BRAME, A. H., JR. 1965. Redescription of the Costa Rican salamander, *Magnadigita cerroensis*, with remarks on *Magnadigita marmorea*. Abh. Ber. Naturkd. Vorgesch. Magdeburg 11:105–118.

BRAME, A. H., JR. 1968. Systematics and evolution of the Mesoamerican salamander genus *Oedipina*. J. Herpetol. 2:1–64.

BRAME, A. H., JR., AND W. E. DUELLMAN. 1970. A new salamander (genus *Oedipina*) of the *uniformis* group from western Panama. Contrib. Sci. Nat. Hist. Mus. Los Angeles Co. 201:1–8.

BRAME, A. H., JR., AND D. B. WAKE. 1963a. The salamanders of South America. Contrib. Sci. Nat. Hist. Mus. Los Angeles Co. 69:1–72.

BRAME, A. H., JR., AND D. B. WAKE. 1963b. Redescription of the plethodontid salamander *Bolitoglossa lignicolor* (Peters), with remarks on the status of *B. palustris* Taylor. Proc. Biol. Soc. Washington 76:289–296.

BRAME, A. H., JR., AND D. B. WAKE. 1972. New species of salamanders (genus *Bolitoglossa*) from Colombia, Ecuador and Panama. Contrib. Sci. Nat. Hist. Mus. Los Angeles Co. 219:1–34.

BRANDON, R. A. 1970a. Size range, size at maturity, and reproduction of *Ambystoma (Bathysiredon) dumerilii* (Dugès), a paedogenetic Mexican salamander endemic to Lake Patzcuaro, Michoacán. Copeia 1970:385–388.

BRANDON, R. A. 1970b. Courtship, spermatophores and eggs of the Mexican achoque, *Ambystoma (Bathysiredon) dumerili* (Dugès). Zool. J. Linn. Soc. 49:247–254.

BRANDON, R. A., AND R. G. ALTIG. 1973. Eggs and small larvae of two species of *Rhyacosiredon*. Herpetologica 29:349–351.

BRANDON, R. A., E. J. MARUSKA, AND W. T. RUMPH. 1981. A new species of neotenic Ambystoma (Amphibia, Caudata) endemic to Laguna Alchichica, Puebla, Mexico. Bull. S. California Acad. Sci. 80:112–125.

BREDER, C. M., JR. 1946. Amphibians and reptiles of the Rio Chucunaque drainage, Darien, Panama, with notes on their life histories and habitats. Bull. Am. Mus. Nat. Hist. 86:375–436.

BREEDLOVE, D. E. 1973. The phytogeography and vegetation of Chiapas, (Mexico). Pp. 149–165 *in* A. Graham (ed.), *Vegetation and Vegetational History of Northern Latin America*. Amsterdam: Elsevier Sci. Publ. Co.

BROCCHI, P. 1881–1883. Mission Scientifique au Mexique et dans l'Amérique Centrale. Batraciens, pt. 3, Sec. 2. 122 pp.

BRODIE, E. D., JR., AND J. A. CAMPBELL. 1993. A new salamander of the genus *Oedipina* (Caudata: Plethodontidae) from the Pacific versant of Guatemala. Herpetologica 49:259–265.

BUMZAHEM, C. B. 1955. A new species of frog in the genus *Eleutherodactylus* from Guatemala. Copeia 1955:118–119.

BUMZAHEM, C. B., AND H. M. SMITH. 1954. Additional records and descriptions of Mexican frogs of the genus *Plectrohyla*. Herpetologica 10:61–66.

BUMZAHEM, C. B., AND H. M. SMITH. 1955. Additional notes and descriptions of plethodontid salamanders from Mexico. Herpetologica 11:73–75.

BURKART, B. 1983. Neogene North American–Caribbean Plate boundary across northern Central America: offset along the Polochic fault. Tectonophysics 99:251–270.

BUSACK, S. D. 1966. Notes on a herpetological collection from the Azuero Peninsula, Panama. Copeia 1966:371.

CALDWELL, J. 1974. A re-evaluation of the *Hyla bistincta* species group, with descriptions of three new species (Anura: Hylidae). Occas. Pap. Mus. Nat. Hist. Univ. Kansas 28:1–37.

CAMPBELL, H. W., AND R. S. SIMMONS. 1962. Notes on the eggs and larvae of *Rhyacosiredon altamirani* (Dugès). Herpetologica 18:131–133.

CAMPBELL, H. W., AND S. R. TELFORD. 1971. Observations on two species of the *Hyla rubra* group in Panama. J. Herpetol. 5:52–55.

CAMPBELL, J. A. 1982. The Biogeography of the Cloud Forest Herpetofauna of Middle America, with Special Reference to the Sierra de las Minas of Guatemala. Ph.D. Dissertation, University of Kansas, Lawrence, Kansas.

CAMPBELL, J. A. 1984. A new species of *Abronia* (Sauria: Anguidae) with comments on the herpetogeography of southern Mexico. Herpetologica 40:373–381.

CAMPBELL, J. A. 1994a. A new species of *Eleutherodactylus* (Anura: Leptodactylidae) of the *biporcatus* group from eastern Guatemala. Herpetologica 50:296–302.

CAMPBELL, J. A. 1994b. New species of *Eleutherodactylus* of the *milesi* (Anura: Leptodactylidae) group from Guatemala. Herpetologica 50:398–411.

CAMPBELL, J. A. 1998. *The Amphibians and Reptiles of the Tikal-Flores and Adjacent Regions of Guatemala, Belize, and Mexico*. Norman, Oklahoma: Univ. Oklahoma Press.

CAMPBELL, J. A., AND E. D. BRODIE, JR. 1992. A new species of treefrog (Hylidae) from the Sierra de los Cuchumatanes of Guatemala. J. Herpetol. 26:187–190.

CAMPBELL, J. A., AND T. M. KUBIN. 1990. A key to the larvae of *Plectrohyla* (Hylidae), with a description of the tadpole presumed to be *Plectrohyla avia*. Southwest. Nat. 35:91–94.

CAMPBELL, J. A., AND W. W. LAMAR. 1989. *The Venomous Reptiles of Latin America*. Ithaca, New York: Cornell Univ. Press.

CAMPBELL, J. A., W. W. LAMAR, AND D. M. HILLIS. 1989. A new species of diminutive *Eleutherodactylus* (Leptodactylidae) from Oaxaca, Mexico. Proc. Biol. Soc. Washington 102:491–499.

CAMPBELL, J. A., AND D. P. LAWSON. 1992. *Hyla bocourti* (Mocquard, 1899), a valid species of frog (Anura: Hylidae) from Guatemala. Proc. Biol. Soc. Washington 105:393–399.

CAMPBELL, J. A., J. M. SAVAGE, AND J. R. MEYER. 1994. A new species of *Eleutherodactylus* (Anura: Leptodactylidae) of the *rugulosus* group from Guatemala and Belize. Herpetologica 50:412–419.

CAMPBELL, J. A., AND E. N. SMITH. 1992. A new frog of the genus *Ptychohyla* (Hylidae) from the Sierra de Santa Cruz, Guatemala, and description of a new genus of Middle American stream-breeding treefrogs. Herpetologica 48:153–167.

CAMPBELL, J. A., AND J. P. VANNINI. 1988. A new subspecies of Beaded Lizard, *Heloderma horridum* (Sauria: Helodermatidae), from the Motagua Valley of Guatemala. J. Herpetol. 22(4):457–468.

CAMPBELL, J. A., AND J. P. VANNINI. 1989a. Distribution of amphibians and reptiles in Guatemala and Belize. Proc. West. Found. Vert. Zool. 4:1–21.

CAMPBELL, J. A., AND J. P. VANNINI. 1989b. Listado preliminar de la herpetofauna de Finca El Palmar, Quezaltenango, Guatemala. Fund. Interam. Invest. Trop. Publ. Ocas. 1:1–10.

CARR, A. F., JR. 1963. Outline for a classification of animal habitats in Honduras. Bull. Am. Mus. Nat. Hist. 94:567–594.

CHILDS, O. E., AND B. W. BEEBE (eds.). 1963. *Backbone of the Americas*. Tulsa, Oklahoma: Am. Assoc. Petrol. Geol. Mem. 2.

COCHRAN, D. M., AND C. J. GOIN. 1970. Frogs in Colombia. Bull. U.S. Natl. Mus. 288:1–655.

COLLINS, J. P., J. B. MITTON, AND B. A. PIERCE. 1980. *Ambystoma tigrinum*: a multispecies conglomerate? Copeia 1980:938–941.

COLLINS-RAINBOTH, A., AND D. G. BUTH. 1990. A reevaluation of the systematic relationships among species of the genus *Dendrotriton* (Caudata: Plethodontidae). Copeia 1990:995–960.

CONANT, R., AND J. T. COLLINS. 1991. *A Field Guide to Reptiles and Amphibians of Eastern and Central North America*. Boston: Houghton Mifflin Co.

COPE, E. D. 1866. Fourth contribution to the herpetology of tropical America. Proc. Acad. Nat. Sci. Philadelphia 18:123–132.

COPE, E. D. 1876 "1875." On the batrachia and reptilia of Costa Rica. Proc. Acad. Nat. Sci. Philadelpha (2)8:93–154.

COPE, E. D. 1887. Catalogue of batrachians and reptiles of Central America and Mexico. Bull. U.S. Natl. Mus. 32:1–98.

CRUMP, M. L. 1983. *Dendrobates granuliferus* and *D. Pumilio*. Pp. 396–398 *in* D. H. Janzen (ed.), *Costa Rican Natural History*. Chicago: Univ. Chicago Press.

CRUMP, M. L., F. R. HENSLEY, AND K. L. CLARK. 1992. Apparent decline of the golden toad: underground or extinct? Copeia 1992:413–420.

CRUZ DIAZ, G. A., AND L. D. WILSON. 1983. *Bufo haematiticus* Cope: an addition to the anuran fauna of Honduras. Herpetol. Rev. 14:31.

DARDA, D. M. 1994. Allozyme variation and morphological evolution among Mexican salamanders of the genus *Chiropterotriton* (Caudata: Plethodontidae). Herpetologica 50:164–187.

DARLING, D. M., AND H. M. SMITH. 1954. A collection of reptiles and amphibians from eastern Mexico. Trans. Kansas Acad. Sci. 57:180–195.

DAVIS, W. B., AND J. R. DIXON. 1955. Notes on Mexican toads of the genus *Tomodactylus* with descriptions of two new species. Herpetologica 11:154–160.

DAVIS, W. B., AND J. R. DIXON. 1957. Notes on Mexican amphibians, with description of a new *Microbatrachylus*. Herpetologica 13:145–147.

DAVIS, W. B., AND J. R. DIXON. 1964. Amphibians of the Chilpancingo Region, México. Herpetologica 20:225–233.

DAVIS, W. B., AND H. M. SMITH. 1953. Amphibians of the Mexican state of Morelos. Herpetologica 8:144–149.

DENGO, G. 1968. Estructura geológica, historia tectónica, y morfología de América Central. Centro Regional de Ayuda Técnica Publ., Mexico, 50 pp.

DICE, L. R. 1952. Natural Communities. Univ. Michigan Press, Ann Arbor, Michigan.

DIXON, J. R. 1957. Geographic variation and distribution of the genus *Tomodactylus* in Mexico. Texas J. Sci. 9:379–409.

DIXON, J. R. 1963. A new species of salamander of the genus *Ambystoma* from Jalisco, Mexico. Copeia 1963:99–101.

DIXON, J. R., AND R. G. WEBB. 1966. A new *Syrrhophus* from Mexico (Amphibia: Leptodactylidae). Los Angeles Co. Mus. Contrib. Sci. 102:1–5.

DONNELLY, T. W. 1989. Geologic history of the Caribbean and Central America. Pp. 299–321 *in* A. W. Bally and A. R. Palmer (eds.), *The Geology of North America—An Overview*. Boulder, Colorado: Geological Society of America.

DORF, E. 1959. Climatic changes of the past and present. Contrib. Mus. Paleontol. Univ. Michigan 13:181–210.

DUBOIS, A., AND W. R. HEYER. 1992. *Leptodactylus labialis*, the valid name for the American white-lipped frog (Amphibia: Leptodactylidae). Copeia 1992:584–585.

DUELLMAN, W. E. 1960a. A distributional study of the amphibians of the Isthmus of Tehuantepec, Mexico. Univ. Kansas Publ. Mus. Nat. Hist. 13:19–72.

DUELLMAN, W. E. 1960b. Synonymy, variation, and distribution of *Ptychohyla leonhard-schultzei* Ahl. Studies of American hylid frogs IV. Herpetologica 16:191–197.

DUELLMAN, W. E. 1960c. Redescription of *Hyla valancifer*. Studies of American hylid frogs, III. Herpetologica 16:55–57.

DUELLMAN, W. E. 1961a. Descriptions of two new species of frogs, genus *Ptychohyla*. Studies of American hylid frogs, V. Univ. Kansas Publ. Mus. Nat. Hist. 13:349–357.

DUELLMAN, W. E. 1961b. The amphibians and reptiles of Michoacán, Mexico. Univ. Kansas Publ. Mus. Nat. Hist. 15:1–148.

DUELLMAN, W. E. 1961c. Descriptions of new species of tree frog from Mexico. Studies of American hylid frogs, VI. Herpetologica 17:1–5.

DUELLMAN, W. E. 1962 [dated 1961]. A new species of fringe-limbed tree frog from México. Studies of American hylid frogs VIII. Trans. Kansas Acad. Sci. 64:349–352.

DUELLMAN, W. E. 1963a. Amphibians and reptiles of the rainforest of southern El Petén, Guatemala. Univ. Kansas Publ. Mus. Nat. Hist. 15:205–249.

DUELLMAN, W. E. 1963b. A review of the Middle American tree frogs of the genus *Ptychohyla*. Univ. Kansas Publ. Mus. Nat. Hist. 15:297–349.

DUELLMAN, W. E. 1964a. Description of a new species of tree frog from Veracruz, Mexico. Herpetologica 19:225–228.

DUELLMAN, W. E. 1964b. Status and identities of the tree frogs *Hyla godmani* (*H. rickardsi*) and *Hyla glandulosa* (*Plectrohyla cotzicensis*). Copeia 1964:455–456.

DUELLMAN, W. E. 1964c. A review of the frogs of the *Hyla bistincta* group. Univ. Kansas Publ. Mus. Nat. Hist. 15:469–491.

DUELLMAN, W. E. 1965a. Amphibians and reptiles from the Yucatan Peninsula, Mexico. Univ. Kansas Publ. Mus. Nat. Hist. 15:577–614.

DUELLMAN, W. E. 1965b. A biogeographic account of the herpetofauna of Michoacán, México. Univ. Kansas Publ. Mus. Nat. Hist. 15:627–709.

DUELLMAN, W. E. 1965c. A new species of tree frog from Oaxaca, Mexico. Herpetologica 21:32–34.

DUELLMAN, W. E. 1965d. Frogs of the *Hyla taeniopus* group. Copeia 1965:159–168.

DUELLMAN, W. E. 1966a. The Central American herpetofauna: an ecological perspective. Copeia 1966:700–719.

DUELLMAN, W. E. 1966b. Taxonomic notes on some Mexican and Central American hylid frogs. Univ. Kansas Publ. Mus. Nat. Hist. 17:263–279.

DUELLMAN, W. E. 1966c. A new species of fringe-limbed tree frog, genus *Hyla*, from Darién, Panamá. Univ. Kansas Publ. Mus. Nat. Hist. 15:257–262.

DUELLMAN, W. E. 1968a. The taxonomic status of some American hylid frogs. Herpetologica 24:194–207.

DUELLMAN, W. E. 1968b. Descriptions of new hylid frogs from Mexico and Central America. Univ. Kansas Publ. Mus. Nat. Hist. 17:559–578.

DUELLMAN, W. E. 1970. The hylid frogs of Middle America. Monogr. Mus. Nat. Hist. Univ. Kansas 1:1–753.

DUELLMAN, W. E. 1971. The burrowing toad, *Rhinophrynus dorsalis,* on the Caribbean lowlands of Central America. Herpetologica 27:55–56.

DUELLMAN, W. E. 1972. A review of the Neotropical frogs of the *Hyla bogotensis* group. Occas. Pap. Mus. Nat. Hist. Univ. Kansas 11:1–31.

DUELLMAN, W. E. 1978. The fringe–limbed tree frog, *Hyla valancifer* (Amphibia, Anura, Hylidae), in Guatemala. J. Herpetol. 12:407.

DUELLMAN, W. E. 1983. A new species of marsupial frog (Hylidae: *Gastrotheca*) from Colombia and Ecuador. Copeia 1983:868–874.

DUELLMAN, W. E. 1988. Patterns of species diversity in anuran amphibians in the American tropics. Ann. Missouri Bot. Gard. 75:79–104.

DUELLMAN, W. E. 1993. Amphibian species of the world: additions and corrections. Spec. Publ. Mus. Nat. Hist. Univ. Kansas 21:1–372.

DUELLMAN, W. E., AND J. A. CAMPBELL. 1982. A new frog of the genus *Ptychohyla* (Hylidae) from the Sierra de las Minas, Guatemala. Herpetologica 38:374–380.

DUELLMAN, W. E., AND J. A. CAMPBELL. 1984. Two new species of *Plectrohyla* from Guatemala (Anura: Hylidae). Copeia 1984:390–397.

DUELLMAN, W. E., AND J. A. CAMPBELL. 1992. Hylid frogs of the genus *Plectrohyla*: systematics and phylogenetic relationships. Misc. Publ. Mus. Zool. Univ. Michigan 181:1–32.

DUELLMAN, W. E., AND J. R. DIXON. 1959. A new frog of the genus *Tomodactylus* from Michoacan, Mexico. Texas J. Sci. 11:78–82.

DUELLMAN, W. E., AND M. J. FOUQUETTE, JR. 1968. Middle American hylid frogs of the *Hyla microcephala* group. Univ. Kansas Publ. Mus. Nat. Hist. 17:517–557.

DUELLMAN, W. E., AND D. L. HOYT. 1961. Description of a new species of *Hyla* from Chiapas, Mexico. Copeia 1961:414–417.

DUELLMAN, W. E., AND L. KLAAS. 1964. The biology of the hylid frog *Triprion petasatus*. Copeia 1964:308–321.

DUELLMAN, W. E., AND L. TRUEB. 1966. Neotropical hylid frogs, genus *Smilisca*. Univ. Kansas Publ. Mus. Nat. Hist. 17:281–375.

DUELLMAN, W. E., AND L. TRUEB. 1967. Two new species of tree frogs (genus *Phyllomedusa*) from Panamá. Copeia 1967:125–131.

DUELLMAN, W. E., AND J. B. TULECKE. 1960. The distribution, variation and life history of the frog *Cochranella viridissima* in Mexico. Am. Midl. Nat. 63:392–397.

DUELLMAN, W. E., AND A. VELOSO. 1977. Phylogeny of *Pleurodema* (Anura: Leptodactylidae): a biogeographic model. Occas. Pap. Mus. Nat. Hist. Univ. Kansas 64:1–46.

DUNDEE, H. A., D. A. WHITE, AND V. RICO-GRAY. 1986. Observations on the distribution and biology of some Yucatán Peninsula amphibians and reptiles. Bull. Maryland Herp. Soc. 22:37–150.

DUNN, E. R. 1921. Two new Central American salamanders. Proc. Biol. Soc. Washington 34:143–146.

DUNN, E. R. 1922a. A new salamander from Mexico. Proc. Biol. Soc. Washington 35:5–6.

DUNN, E. R. 1922b. Notes on some tropical ranae. Proc. Biol. Soc. Washington 35:221–222.

DUNN, E. R. 1924a. Some Panamanian frogs. Occas. Pap. Mus. Zool. Univ. Michigan 151:1–17.

DUNN, E. R. 1924b. New salamanders of the genus *Oedipus* with a synoptical key. Field Mus. Nat. Hist. Zool. Ser. 12:93–100.

DUNN, E. R. 1924c. New amphibians from Panama. Occas. Pap. Boston Soc. Nat. Hist. 5:93–95.

DUNN, E. R. 1926. The salamanders of the family Plethodontidae. Smith College 50th Anniv. Publ. 7:1–441.

DUNN, E. R. 1928. A new genus of salamanders from Mexico. Proc. New England Zool. Club 10:85–86.

DUNN, E. R. 1931a. The herpetological fauna of the Americas. Copeia 1931:106–119.

DUNN, E. R. 1931b. Preliminary list of the reptiles and amphibians of the Canal Zone and the provinces of Panama and Colon, R. P. Pp. 15–18 *in* T. Barbour (ed.), Seventh Annual Report of Barro Colorado Island Biological Laboratory, Panama Canal Zone Nat. Research Council. 24 pp. (mimeographed).

DUNN, E. R. 1931c. New frogs from Panama and Costa Rica. Occas. Pap. Boston Soc. Nat. Hist. 5:385–401.

DUNN, E. R. 1931d. The amphibians of Barro Colorado Island. Occas. Pap. Boston Soc. Nat. Hist. 5:403–421.

DUNN, E. R. 1933. Amphibians and reptiles from El Valle de Anton, Panama. Occas. Pap. Boston Soc. Nat. Hist. 8:65–79.

DUNN, E. R. 1934. Two new frogs from Darien. Am. Mus. Novit. 747:1–2.

DUNN, E. R. 1936. The amphibians and reptiles of the Mexican expedition of 1934. Proc. Acad. Nat. Sci. Philadelphia 88:471–477.

DUNN, E. R. 1937. The amphibian and reptile fauna of bromeliads in Costa Rica and Panama. Copeia 1937:163–167.

DUNN, E. R. 1940a. New and noteworthy herpetological material from Panama. Proc. Acad. Nat. Sci. Philadelphia 92:105–122.

DUNN, E. R. 1940b. Some aspects of herpetology in lower Central America. Trans. New York Acad. Sci. (2)2:156–158.

DUNN, E. R. 1942. A new species of frog (Eleutherodactylus) from Costa Rica. Notul. Natur. 104:1–2.

DUNN, E. R., AND J. T. EMLEN, JR. 1932. Reptiles and amphibians from Honduras. Proc. Acad. Nat. Sci. Philadelphia 84:21–32.

DUNN, E. R., H. TRAPIDO, AND H. EVANS. 1948. A new species of the microhylid frog genus *Chiasmocleis* from Panama. Am. Mus. Novit. 1367:1–8.

DURHAM, J. W., A. R. V. ARELLANO, AND J. H. PECK, JR. 1955. Evidence for no Cenozoic Isthmus of Tehuantepec seaways. Bull. Geol. Soc. Am. 66:977–992.

EISENMANN, E. 1955. The species of Middle American birds. Trans. Linn. Soc. New York 7:1–128.

ELIAS, P. 1984. Salamanders of the northwestern highlands of Guatemala. Contrib. Sci. Nat. Hist. Mus. Los Angeles Co. 348:1–20.

ELIAS, P., AND D. B. WAKE. 1983. *Nyctanolis pernix*, a new genus and species of plethodontic salamander from northwestern Guatemala and Chiapas, Mexico. Pp. 1–12 *in* A. G. J. Rhodin and K. Miyata (eds.), *Advances in Herpetology and Evolutionary Biology: Essays in Honor of Ernest E. Williams*. Cambridge, Massachusetts: Museum of Comparative Zoology.

EMILIANI, C., AND E. ROSA. 1969. Caribbean cores P6304–8 and P6304–9: new analysis of absolute chronology. A reply. Science 166:1551–1552.

ESTES, R. 1985. Herpetofaunas of North and South America during the Late Cretaceous and Cenozoic: evidence for interchange? Pp. 139–197 in F. G. Stehli and S. D. Webb (eds.), *The Great American Biotic Interchange.* New York: Plenum Press.

EVANS, H. E. 1947. Notes on Panamanian reptiles and amphians. Copeia 1947:166–170.

FIRSCHEIN, I. L. 1950. A new toad from Mexico with a redefinition of the *cristatus* group. Copeia 1950:81–87.

FIRSCHEIN, I. L. 1951. Rediscovery of the broad-headed frog *Eleutherodactylus laticeps* (Dumeril) of Mexico. Copeia 1951:268–274.

FIRSCHEIN, I. L., AND H. M. SMITH. 1956. A new fringe-limbed *Hyla* (Amphibia: Anura) from a new faunal district of Mexico. Herpetologica 12:17–21.

FLORES-VILLELA, O. 1993. Herpetofauna Mexicana: Annotated list of the species of amphibians and reptiles of Mexico, recent taxonomic changes, and new species. Carnegie Mus. Nat. Hist. Spec. Publ. 17:1–73.

FORD, L. S., AND J. M. SAVAGE. 1984. A new frog of the genus *Eleutherodactylus* (Leptodactylidae) from Guatemala. Occas. Pap. Mus. Nat. Hist. Univ. Kansas 110:1–9.

FOSTER, M. S., AND R. W. MCDIARMID. 1983. *Rhinophrynus dorsalis.* Pp. 419–421 in D. H. Janzen (ed.), *Costa Rican Natural History.* Chicago: Univ. Chicago Press.

FOUQUETTE, M. J., JR. 1961. Status of the frog *Hyla albomarginata* in Central America. Fieldiana Zool. 39:595–601.

FOUQUETTE, M. J., JR. 1969. Rhinophrynidae, *Rhinophrynus dorsalis.* Cat. Am. Amph. Rept. 78.1–2.

FROST, D. R. (ed.) 1985. *Amphibian Species of the World.* Lawrence, Kansas: Allen Press and Assoc. Syst. Coll.

FROST, D. R. 1994. Amphibian Species of the World. Vers. 1994/6. Electronic manuscript compiled under the auspices of the Herpetologists' League. Privately distributed.

FROST, J. S. 1982. Functional genetic similarity between geographically separated populations of Mexican leopard frogs (*Rana pipiens* complex). Syst. Zool. 31:57–67.

FROST, J. S., AND J. T. BAGNARA. 1976. A new species of leopard frog (*Rana pipiens* complex) from northwestern Mexico. Copeia 1976:332–338.

FUNKHOUSER, A. 1957. A review of the Neotropical tree-frogs of the genus *Phyllomedusa.* Occas. Pap. Nat. Hist. Mus. Stanford Univ. 5:1–90.

GADOW, H. F. 1913. *The Wanderings of Animals.* Cambridge: Cambridge Univ. Press.

GAIGE, H. T. 1929. Three new tree-frogs from Panama and Bolivia. Occas. Pap. Mus. Zool. Univ. Michigan 207:1–6.

GAIGE, H. T., N. HARTWEG, AND L. C. STUART. 1937. Notes on collections of amphibians and reptiles from eastern Nicaragua. Occas. Pap. Mus. Zool. Univ. Michigan 14:1–18.

GAIGE, H. T., AND L. C. STUART. 1934. A new *Hyla* from Guatemala. Occas. Pap. Mus. Zool. Univ. Michigan 281:1–3.

GALLARDO, J. M. 1965. The species *Bufo granulosus* Spix (Salientia: Bufonidae) and its geographic variations. Bull. Mus. Comp. Zool. 134:107–138.

GEHLBACH, F. R. 1959. New plethodontid salamanders of the genus *Thorius* from Puebla, Mexico. Copeia 1959:203–206.

GOOD, D. A., AND D. B. WAKE. 1993. Systematic studies of the Costa Rican moss salamander, genus *Nototriton,* with descriptions of three new species. Herpetol. Monogr. 7:131–159.

GOSE, W. A., AND D. K. SWARTZ. 1977. Paleomagnetic results from Cretaceous sediments in Honduras—tectonic implications. Geology 5:505–508.

GRAHAM, A. 1973. History of the arborescent temperate element in the northern Latin American biota. Pp. 301–314 in A. Graham (ed.), *Vegetation and Vegetational History in Northern Latin America.* New York: Elsevier Scientific Publishing Co.

GRISCOM, L. 1932. The distribution of bird-life in Guatemala. A contribution to a study of the origin of Central American bird-life. Bull. Am. Mus. Nat. Hist. 64:1–439.

GÜNTHER, A. C. L. G. 1885–1902. *Reptilia and Batrachia. Biologia Centrali-Americana.* London: Taylor and Francis.

HANKEN, J. 1983. Genetic variation in a dwarfed lineage, the Mexican salamander genus *Thorius* (Amphibia: Plethodontidae): taxonomic, ecologic and evolutionary implications. Copeia 1983:1051–1073.

HANKEN, J., AND D. B. WAKE. 1982. Genetic differentiation among plethodontid salamanders (genus *Bolitoglossa*) in Central and South America: implications for the South American invasion. Herpetologica 38:272–287.

HANKEN, J., AND D. B. WAKE. 1994. Five new species of minute salamanders, genus *Thorius* (Caudata: Plethodontidae), from northern Oaxaca, Mexico. Copeia 1994:573–590.

HARDY, L. M., AND R. W. MCDIARMID. 1969. The amphibians and reptiles of Sinaloa, Mexico. Univ. Kansas Publ. Mus. Nat. Hist. 18:39–252.

HARSHBERGER, J. W. 1911. *A Phytogeographic Survey of North America.* Vol. 3 in A. Engier and O. Drude (eds.), *Die Vegetation der Erde.* Leipzig: W. Engelmann.

HARTWEG, N. 1941. Notes on the genus *Plectrohyla,* with descriptions of new species. Occas. Pap. Mus. Zool. Univ. Michigan 437:1–10.

HARTWEG, N., AND J. A. OLIVER. 1940. A contribution to the herpetology of the Isthmus of Tehuantepec, IV. Misc. Publ. Mus. Zool. Univ. Michigan 47:1–31.

HARTWEG, N., AND G. ORTON. 1941. Notes on the tadpoles of the genus *Plectrohyla.* Occas. Pap. Mus. Zool. Univ. Michigan 438:1–6.

HAYES, M. P., J. A. POUNDS, AND D. C. ROBINSON. 1986. The fringe-limbed frog *Hyla fimbrimembra* (Anura: Hylidae): new records from Costa Rica. Florida Sci. 49:193–198.

HAYES, M. P., J. A. POUNDS, AND W. W. TIMMERMAN. 1989. An annotated list and guide to the amphibians and reptiles of Monteverde, Costa Rica. Soc. Study Amph. Rept. Herpetol. Circ. 17:1–67.

HAYES, M. P., AND P. H. STARRETT. 1980. Notes on a collection of centrolenid frogs from the Colombian Chocó. Bull. S. California Acad. Sci. 79:89–96.

HEATWOLE, H. 1963. The frog genus *Pipa* in Panama. Copeia 1963:436–438.

HEDGES, S. B. 1989. Evolution and biogeography of West Indian frogs of the genus *Eleutherodactylus:* slow-evolving loci and the major groups. Pp. 305–370 in C. A. Woods (ed.), *Biogeography of the West Indies: Past, Present, and Future.* Gainesville, Florida: Sandhill Crane Press.

HENDERSON, R. W., AND L. G. HOEVERS. 1975. A checklist and key to the amphibians and reptiles of Belize, Central America. Milwaukee Pub. Mus. Contrib. Biol. Geol. 5:1–63.

HERSHKOVITZ, P. 1958. A geographical classification of Neotropical mammals. Fieldiana: Zool. 36:581–620.

HEYER, W. R. 1967. A herpetofaunal study of an ecological transect through the Cordillera de Tilarán, Costa Rica. Copeia 1967:259–271.

HEYER, W. R. 1970a. Studies on the genus *Leptodactylus* (Amphibia: Leptodactylidae). II. Diagnosis and distribution of the *Leptodactylus* of Costa Rica. Rev. Biol. Trop. 16:171–205.

HEYER, W. R. 1970b. Studies on the frogs of the genus *Leptodactylus* (Amphibia, Leptodactylidae). VI. Biosystematics of the *melanonotus* group. Contrib. Sci. Nat. Hist. Mus. Los Angeles Co. 191:1–48.

HEYER, W. R. 1971. *Leptodactylus labialis*. Cat. Am. Amph. Rept. 104.1-104.3.

HEYER, W. R. 1978. Systematics of the *fuscus* group of the frog genus *Leptodactylus* (Amphibia, Leptodactylidae). Sci. Bull. Nat. Hist. Mus. Los Angeles Co. 29:1–85.

HEYER, W. R. 1979. Systematics of the *pentadactylus* species group of the frog genus *Leptodactylus* (Amphibia, Leptodactylidae). Smithsonian Contrib. Zool. 301:1–43.

HIDALGO, H. 1982. *Centrolenella fleischmanni* (Boettger): new to the anuran fauna of El Salvador. Herpetol. Rev. 13:54–55.

HILLIS, D. M. 1981. Premating isolating mechanisms among three species of the *Rana pipiens* complex in Texas and southern Oklahoma. Copeia 1981:312–319.

HILLIS, D. M. 1985. Evolutionary genetics and systematics of New World frogs of the genus *Rana:* An analysis of ribosomal DNA, allozymes, and morphology. Ph.D. Dissertation, Univ. Kansas.

HILLIS, D. M. 1988. Systematics of the *Rana pipiens* complex: puzzle and paradigm. Ann. Rev. Ecol. Syst. 19:39–63.

HILLIS, D. M., AND S. K. DAVIS. 1986. Evolution of ribosomal DNA: fifty million years of recorded history in the frog genus *Rana.* Evolution 40:1275–1288.

HILLIS, D. M., AND R. DE SÁ. 1988. Phylogeny and taxonomy of the *Rana palmipes* group (Salientia: Ranidae). Herp. Monogr. 2:1–26.

HILLIS, D. M., AND J. S. FROST. 1985. Three new species of leopard frogs (*Rana pipiens* complex) from the Mexican Plateau. Occas. Pap. Mus. Nat. Hist. Univ. Kansas 117:1–14.

HILLIS, D. M., D. R. FROST, AND J. S. FROST. 1983. Allocation and distribution of *Rana trilobata* Mocquard. J. Herpetol. 17:73–75.

HILLIS, D. M., J. S. FROST, AND R. G. WEBB. 1984. A new species of frog of the *Rana tarahumarae* group from southwestern Mexico. Copeia 1984:398–403.

HILLIS, D. M., J. S. FROST, AND D. A. WRIGHT. 1983. Phylogeny and biogeography of the *Rana pipiens* complex: a biochemical evaluation. Syst. Zool. 32:132–143.

HOEVERS, L. G., AND R. W. HENDERSON. 1974. Additions to the herpetofauna of Belize (British Honduras). Milwaukee Pub. Mus. Contrib. Biol. Geol. 2:1–6.

HOLDRIDGE, L. R. 1947. Determination of world plant formations from simple climatic data. Science 105:367–368.

HOLDRIDGE, L. R. 1959a. Mapa ecológico de Guatemala, C. A. 1:1,000,000. Instituto Interamericano de Ciencias Agrícolas de la Organización de Estados Americanos, Proyecto 39, Programa de Cooperación Técnica, San José, Costa Rica (2 sheets).

HOLDRIDGE, L. R. 1959b. Mapa ecológico de El Salvador, C. A. 1:1,000,000. Instituto Interamericano de Ciencias Agrícolas de la Organización de Estados Americanos, Proyecto 39, Programa de Cooperación Técnica, San José, Costa Rica (1 sheet).

HOLDRIDGE, L. R. 1962a. Mapa ecológico de Honduras, C.A. 1:1,000,000. Org. Estad. Am. Baltimore: A. Hoen and Co.

HOLDRIDGE, L. R. 1962b. Mapa ecológico de Nicaragua, C.A. 1:1,000,000. Agencia Desarrollo Internac. Gob. Estados Unidos de América (2 sheets).

HOLDRIDGE, L. R. 1967. *Life Zone Ecology*. San José, Costa Rica: Trop. Sci. Center.

HOLDRIDGE, L. R., AND G. BUDOWSKI. 1959. Mapa ecológico de Panamá, C.A. 1:1,000,000. Instituto Interamericano de Ciencias Agrícolas de la Organización de Estados Americanos, Proyecto 39, Programa de Cooperación Técnica, San José, Costa Rica (2 sheets).

HOLMAN, J. A. 1964. New and interesting amphibians and reptiles from Guerrero and Oaxaca, Mexico. Herpetologica 20:48–54.

HOYT, D. L. 1965. A new frog of the genus *Tomodactylus* from Oaxaca, Mexico. J. Ohio Herpetol. Soc. 5:19–22.

IBAÑEZ D., R. 1988. Geographic distribution: *Centrolenella pulverata.* Herp. Rev. 19:59.

IBAÑEZ D., R., C. A. JARAMILLO, AND F. E. AROSEMENA. 1994. A new species of *Eleutherodactylus* (Anura Leptodactylidae) from Panamá. Amphibia-Reptilia 15:337–341.

IBAÑEZ D., R., C. A. JARAMILLO, F. A. SOLIS. 1995. Una especie Nueva de *Atelopus* (Amphibia: Bufonidae) de Panamá. Caribbean J. Sci. 31:57–64.

IBAÑEZ D., R., C. A. JARAMILLO, F. A. SOLIS, AND F. E. JARAMILLO. 1991. Geographic distribution: *Hyla fimbrimembra*. Herpetol. Rev. 22:123–124.

IBAÑEZ D., R. AND A. S. RAND. 1990. Geographic distribution: *Eleutherodactylus johnstonei.* Herpetol. Rev. 21:37.

JANZEN, D. H. 1967. Why mountain passes are higher in the tropics. Am. Nat. 101:233–249.

JARAMILLO, F. E., C. A. JARAMILLO, AND R. IBAÑEZ D. 1988a. Geographic distribution: *Centrolenella colybiphyllum.* Herpetol. Rev. 19:59.

JARAMILLO, F. E., C. A. JARAMILLO, AND R. IBAÑEZ D. 1988b. Geographic distribution: *Centrolenella vireovittata.* Herpetol. Rev. 19:59.

JOHNSON, J. D. 1984 [dated 1983]. New records of reptiles and amphibians from Chiapas, México. Trans. Kansas Acad. Sci. 76:223–225.

JOHNSON, J. D. 1989. A biogeographic analysis of the herpetofauna of northwestern nuclear Central America. Milwaukee Pub. Mus. Contrib. Biol. Geol. 76:1–66.

JOHNSON, J. D., A. ELY, AND R. G. WEBB. 1977. Biogeographical and taxonomic notes on some herpetozoa from the northern highlands of Chiapas, Mexico. Trans. Kansas Acad. Sci. 79:131–139.

JOHNSON, J. D., AND J. M. SAVAGE. 1995. A new species of the *Eleutherodactylus rugulosus* group (Leptodactylidae) from Chiapas, Mexico. J. Herpetol. 29:501–506.

KELLOGG, R. 1932. Mexican tailless amphibians in the United States National Museum. Bull. U.S. Natl. Mus. 160:1–224.

KIESTER, A. R. 1971. Species density of North American amphibians and reptiles. Syst. Zool. 20:127–137.

KLUGE, A. G. 1979. The gladiator frogs of Middle America and Colombia—a reevaluation of their systematics. Occas. Pap. Mus. Zool. Univ. Michigan 688:1–24.

KLUGE, A. G. 1981. The life history, social organization and parental behavior of *Hyla rosenbergi* Boulenger, a nest-building gladiator frog. Misc. Publ. Mus. Zool. Univ. Michigan 160:1–170.

KREBS, S. L., AND R. A. BRANDON. 1984. A new species of salamander (family Ambystomatidae) from Michoacan, Mexico. Herpetologica 40:238–245.

LAFRENTZ, K. 1930. Ein neuer Plethodont-Salamander aus Mexiko. Abh. Ber. Mus. Nat. Heimatk. Magdeburg 6:150–152.

LAHANAS, P. N., AND J. M. SAVAGE. 1992. A new species of caecilian from the Península de Osa of Costa Rica. Copeia 1992:703–708.

LANGEBARTEL, D. A., AND F. A. SHANNON. 1956. A new frog (Syrrhophus) from the Sinaloan lowlands of Mexico. Herpetologica 12:161–165.

LARSON, A. 1983a. A molecular phylogenetic perspective on the origines of a lowland tropical salamander fauna I. Phylogenetic inferences from protein comparisons. Herpetologica 39:85–99.

LARSON, A. 1983b. A molecular phylogenetic perspective on the origines of a lowland tropical salamander fauna II. Patterns of morphological evolution. Evolution 37:1141–1153.

LAUER, W. 1959. Klimatische und pflanzengeographische Grundzüge Zentralamerikas. Erdkunde 13:344–354.

LAZCANO-BARRERO, M. A. 1992. First record of Bolitoglossa mulleri (Caudata: Plethodontidae) from Mexico. Southwest. Nat. 37:315–316.

LEE, J. C. 1976. Rana maculata Brocchi: an addition to the herpetofauna of Belize. Herpetologica 32:211–214.

LEE, J. C. 1980. An ecogeographic analysis of the herpetofauna of the Yucatan Peninsula. Misc. Publ. Mus. Nat. Hist. Univ. Kansas 67:1–75.

Lee, J. C. 1996. The Amphibians and Reptiles of the Yucatán Peninsula. Ithaca, New York: Cornell Univ. Press.

LEÓN, J. R. 1969. The systematics of the frogs of the Hyla rubra group in Middle America. Univ. Kansas Publ. Mus. Nat. Hist. 18:505–545.

LEOPOLD, A. S. 1950. Vegetation zones of Mexico. Ecology 31:507–518.

LLOYD, J. J. 1963. Tectonic history of the south Central-American orogen. Pp. 88–100 in O. E. Childs and B. W. Beebe (eds.), Backbone of the Americas. Tulsa, Oklahoma: Am. Assoc. Petrol. Geol. Mem. 2..

LUNDELL, C. L. 1937. The vegetation of Petén. Carnegie Inst. Washington Publ. 478:1–244.

LYNCH, J. D. 1964. A small collection of anuran amphibians from Panama, with the description of two new species of Eleutherodactylus (Leptodactylidae). J. Ohio Herpetol. Soc. 4:65–68.

LYNCH, J. D. 1965a. A review of the rugulosus group of Eleutherodactylus in northern Central America. Herpetologica 21:102–113.

LYNCH, J. D. 1965b. Two new species of Eleutherodactylus from Mexico (Amphibia: Leptodactylidae). Herpetologica 20:246–252.

LYNCH, J. D. 1965c. The races of the microhylid frog, Gastrophryne usta, in Mexico. Trans. Kansas Acad. Sci. 68:396–400.

LYNCH, J. D. 1965d. A new species of Eleutherodactylus (Leptodactylidae: Anura) from southeastern Chiapas, Mexico. Nat. Hist. Misc. 181:1–6.

LYNCH, J. D. 1965e. A review of the eleutherodactylid frog genus Microbatrachylus (Leptodactylidae). Nat. Hist. Misc. 182:1–12.

LYNCH, J. D. 1966. A new species of Eleutherodactylus from Chiapas, Mexico (Amphibia: Leptodactylidae). Trans. Kansas Acad. Sci. 69:76–78.

LYNCH, J. D. 1967a. Two new species of Eleutherodactylus from Guatemala and Mexico (Amphibia: Leptodactylidae). Trans. Kansas Acad. Sci. 70:177–183.

LYNCH, J. D. 1967b. Two new Eleutherodactylus from western Mexico (Amphibia: Leptodactylidae). Proc. Biol. Soc. Washington 80:211–218.

LYNCH, J. D. 1967c. Synonymy, distribution and variation in Eleutherodactylus decoratus of Mexico (Amphibia: Leptodactylidae). Trans. Illinois Acad. Sci. 60:299–304.

LYNCH, J. D. 1970a. A taxonomic revision of the leptodactylid frog genus Syrrhophus Cope. Univ. Kansas Publ. Mus. Nat. Hist. 20:1–45.

LYNCH, J. D. 1970b. Taxonomic notes on some Mexican frogs (Eleutherodactylus: Leptodactylidae). Herpetologica 26:172–180.

LYNCH, J. D. 1975. A review of the broad-headed eleutherodactyline frogs of South America (Leptodactylidae). Occas. Pap. Mus. Nat. Hist. Univ. Kansas 38:1–46.

LYNCH, J. D. 1980a. A new frog of the genus Eleutherodactylus from western Panama. Trans. Kansas Acad. Sci. 83:101–105.

LYNCH, J. D.1980b. Systematic status and distribution of some poorly known frogs of the genus Eleutherodactylus from the Chocoan lowlands of South America. Herpetologica 36:175–189.

LYNCH, J. D. 1985. A new species of Eleutherodactylus from western Panama. Herpetologica 41:443–447.

LYNCH, J. D. 1986. The definition of the Middle American clade of Eleutherodactylus based on jaw musculature (Amphibia: Leptodactylidae). Herpetologica 42:248–258.

LYNCH, J. D., AND T. H. FRITTS. 1965. A new species of Eleutherodactylus from eastern Mexico. Trans. Illinois Acad. Sci. 58:46–49.

LYNCH, J. D., AND C. W. MYERS. 1983. Frogs of the fitzingeri group of Eleutherodactylus in eastern Panama and Chocoan South America (Leptodactylidae). Bull. Am. Mus. Nat. Hist. 175:481–572.

LYNCH, J. D., AND H. M. SMITH. 1966a. New or unusual amphibians and reptiles from Oaxaca, Mexico, II. Trans. Kansas Acad. Sci. 69:58–75.

LYNCH, J. D., AND H. M. SMITH. 1966b. A new toad from western Mexico. Southwest. Nat. 11:19–23.

LYNCH, J. F., AND D. B. WAKE. 1975. Systematics of the Chiropterotriton bromeliacia group (Amphibia: Caudata), with descriptions of two new species from Guatemala. Contrib. Sci. Nat. Hist. Mus. Los Angeles Co. 265:1–45.

LYNCH, J. F., AND D. B. WAKE. 1978. A new species of Chiropterotriton (Amphibia: Cuadata) from Baja Verapaz, Guatemala, with comments on relationships among Central American members of the genus. Contrib. Sci. Nat. Hist. Mus. Los Angeles Co. 294:1–22.

LYNCH, J. F., AND D. B. WAKE. 1989. Two new species of Pseudoeurycea (Amphibia: Caudata) from Oaxaca, Mexico. Contrib. Sci. Nat. Hist. Mus. Los Angeles Co. 411:11–12.

LYNCH, J. F., D. B. WAKE, AND S. Y. YANG. 1983. Genic and morphological differentiation in Mexican Pseudoeurycea (Caudata: Plethodontidae), with a description of a new species. Copeia 1983:884–894.

LYNCH, J. F., S. Y. YANG, AND T. J. PAPENFUSS. 1977. Studies of Neotropical salamanders of the genus Pseudoeurycea, I: Systematic status of Pseudoeurycea unguidentis. Herpetologica 33:46–52.

MALDONADO-KOERDELL, M. 1964. Geohistory and paleogeography of Middle America. Pp. 3–32 in R. Wauchope and R. C. West (eds.), Handbook of Middle American Indians, Vol. 1,

Natural Environment and Early Cultures. Austin, Texas: Univ. Texas Press.

MALFAIT, B. T., AND M. G. DINKELMAN. 1972. Circum-Caribbean tectonic and igneous activity and the evolution of the Caribbean plate. Bull. Geol. Soc. Am. 83:251–272.

MARES, M. A. 1991. How scientists can impede the development of their discipline: egocentrism, small pool size, and the evolution of "sapismo." Pp 57–75 *in* M. A. Mares and D. J. Schmidly (eds.), *Latin American Mammalogy.* Norman, Oklahoma: Univ. Oklahoma Press.

MARTIN, P. S. 1955a. Zonal distribution of vertebrates in a Mexican cloud forest. Am. Nat. 89:347–361.

MARTIN, P. S. 1955b. Herpetological records for the Gomez Farias region of southwestern Tamaulipas, Mexico. Copeia 1955:173–180.

MARTIN, P. S. 1958. A biogeography of reptiles and amphibians in the Gomez Farias region, Tamaulipas, Mexico. Misc. Publ. Mus. Zool. Univ. Michigan 101:1–102.

MARTIN, P. S., AND B. E. HARRELL. 1957. The Pleistocene history of temperate biotas in Mexico and eastern United States. Ecology 38:468–480.

MARTÍNEZ-CORTÉS. V. 1984. Investigación preliminar de los anfibios (Salientia) de Quebrada de Arena y área adyacentes: listado anotado. Natura (Panama) 4:30–33.

MAXSON, L. R., AND C. W. MYERS. 1985. Albumin evolution in tropical poison frogs (Dendrobatidae). Biotropica 17:50–56.

McBIRNEY, A. R. 1963. Geology of a part of the central Guatemalan Cordillera. Univ. California Publ. Geol. Sci. 38:177–242.

McCOY, C. J. 1966. Additions to the herpetofauna of southern El Petén, Guatemala. Herpetologica 22:306–308.

McCOY, C. J. 1991 "1990." Addition to the herpetofauna of Belize, Central America. Caribbean J. Sci. 26:164–166.

McCOY, C. J., E. J. CENSKY, AND R. R. VAN DEVENDER. 1986. Distribution records for amphibians and reptiles in Belize, Central America. Herpetol. Rev. 17:28–29.

McCOY, C. J., AND D. H. VAN HORN. 1962. Herpetozoa from Oaxaca and Chiapas. Herpetologica 18:180–186.

McCOY, C. J., AND C. F. WALKER. 1966. A new salamander of the genus *Bolitoglossa* from Chiapas. Occas. Pap. Mus. Zool. Univ. Michigan 649:1–11.

McCRANIE, J. R., J. M. SAVAGE, AND L. D. WILSON. 1989. Description of two new species of the *Eleutherodactylus milesi* group (Amphibia: Anura: Leptodactylidae) from northern Honduras. Proc. Biol. Soc. Washington 102:483–490.

McCRANIE, J. R., AND L. D. WILSON. 1981. A new hylid frog of the genus *Plectrohyla* from a cloud forest in Honduras. Occas. Pap. Mus. Nat. Hist. Univ. Kansas 92:1–7.

McCRANIE, J. R., AND L. D. WILSON. 1985. *Plectrohyla matudai* Hartweg and *Norops petersi* (Bocourt): additions to the herpetofauna of Honduras. Herpetol. Rev. 16:107–108.

McCRANIE, J. R., AND L. D. WILSON. 1986. A new species of redeyed treefrog of the *Hyla uranochroa* group (Anura: Hylidae) from northern Honduras. Proc. Biol. Soc. Washington 99:51–55.

McCRANIE, J. R., AND L. D. WILSON. 1987. The biology of the herpetofauna of the pine-oak woodlands of the Sierra Madre Occidental of México. Milwaukee Pub. Mus. Contrib. Biol. Geol. 72:1–30.

McCRANIE, J. R., AND L. D. WILSON. 1993a. A review of the *Bolitoglossa dunni* group (Amphibia: Caudata) from Honduras

with the description of three new species. Herpetologica 49:1–15.

McCRANIE, J. R., AND L. D. WILSON. 1993b. Taxonomic changes associated with the names *Hyla spinipollex* Schmidt and *Ptychohyla merazi* Wilson and McCranie (Anura: Hylidae). Southwest. Nat. 38:100–104.

McCRANIE, J. R., AND L. D. WILSON. 1995a. A new species of salamander of the *Bolitoglossa dunni* group (Caudata: Plethodontidae) from northern Honduras. Herpetologica 51:131–140.

McCRANIE, J. R., AND L. D. WILSON. 1995b. A new species of salamander of the genus *Bolitoglossa* (Caudata: Plethodontidae) from Parque Nacional El Cusuco, Honduras. J. Herpetol. 29:447–452.

McCRANIE, J. R., L. D. WILSON, AND L. PORRAS. 1980. A new species of *Leptodactylus* from the cloud forests of Honduras. J. Herpetol. 14:361–367.

McCRANIE, J. R., L. D. WILSON, AND K. L. WILLIAMS. 1989. A new genus and species of toad (Anura: Bufonidae) with an extraordinary stream adapted tadpole from northern Honduras. Occas. Pap. Mus. Nat. Hist. Univ. Kansas 129:1–18.

McCRANIE, J. R., L. D. WILSON, AND K. L. WILLIAMS. 1993a. New species of tree frog of the genus *Hyla* (Anura: Hylidae) from northern Honduras. Copeia 1993:1057–1062.

McCRANIE, J. R., L. D. WILSON, AND K. L. WILLIAMS. 1993b. A new species of *Oedipina* (Amphibia: Caudata: Plethodontidae) from northern Honduras. Proc. Biol. Soc. Washington 106:385–389.

McDIARMID, R. W. 1983. *Centrolenella fleischmanni.* Pp. 389–390 *in* D. H. Janzen (ed.), *Costa Rican Natural History.* Chicao: Univ. Chicago Press.

MEERMAN, J. 1993. Checklist of the reptiles and amphibians of the Shipstern Nature Reserve and Sarteneja, Corozal District, Belize. Occas. Pap. Belize Nat. Hist. Soc. 2:65–69.

MENDELSON, J. R., III. 1990. Notas sobre una coleción de anfibios y reptiles de Pueblo Viejo, Alta Verapaz, Guatemala. Fund. Interam. Invest. Trop. Publ. Ocas. 3:1–18.

MENDELSON, J. R., III. 1994. A new species of toad (Anura: Bufonidae) from the lowlands of eastern Guatemala. Occas. Pap. Mus. Nat. Hist. Univ. Kansas 166:1–21.

MENDELSON, J. R., III, AND J. A. CAMPBELL. 1994. Two new species of the *Hyla sumichrasti* group (Amphibia: Anura) from Mexico. Proc. Biol. Soc. Washington 107:398–409.

MERTENS, R. 1930. Bemerkungen über die von Herrn Dr. K. Lafrentz in Mexico gesammelten Amphibien und Reptilien. Abhand. Ber. Mus. Nat. Heimatk. Magdeburg 6:153–161.

MERTENS, R. 1952. Die Amphibien und Reptilien von El Salvador, auf Grund der Reisen von R. Mertens und A. Zilch. Abhand. Senckenb. Naturf. Ges. 487:1–83.

MEYER, J. R. 1966. Records and observations on some amphibians and reptiles from Honduras. Herpetologica 22:172–181.

MEYER, J. R., AND L. D. WILSON. 1971a. A distributional checklist of the amphibians of Honduras. Contrib. Sci. Nat. Hist. Mus. Los Angeles Co. 218:1–47.

MEYER, J. R., AND L. D. WILSON. 1971b. Taxonomic studies and notes on some Honduran amphibians and reptiles. Bull. So. California Acad. Sci. 70:1–47.

MIYAMOTO, M. M. 1983a. Biochemical variation in the frog *Eleutherodactylus bransfordii:* geographic patterns and cryptic species. Syst. Zool. 32:43–51.

MIYAMOTO, M. M. 1983b. Frogs of the *Eleutherodactylus rugu-*

losus group: a cladistic study of allozyme, morphological and karyological data. Syst. Zool. 32:109–124.

MURRAY, G. E. 1961. *Atlantic and Gulf Coast Geology of North America.* New York: Harper and Co.

MYERS, C. W. 1966. The distribution and behavior of a tropical horned frog, *Ceratohyla panamensis* Stejneger. Herpetologica 22:68–71.

MYERS, C. W. 1969. The ecological geography of cloud forest in Panama. Am. Mus. Novit. 2396:1–52.

MYERS, C. W. 1982. Spotted Poison frogs: Descriptions of three new *Dendrobates* from western Amazonia, and resurrection of a lost species from "Chiriqui." Am. Mus. Novit. 2721:1–23.

MYERS, C. W. 1987. New generic names for some neotropical poison frogs (Dendrobatidae). Pap. Avul. Zool., São Paulo 36:301–306.

MYERS, C. W. 1991. Distribution of the dendrobatid frog *Colostethus chocoensis* and description of a related species occurring macrosympatrically. Am. Mus. Novita. 3010:1–15.

MYERS, C. W., AND J. W. DALY. 1976. Preliminary evaluation of skin toxins and vocalizations in taxonomic and evolutionary studies of poison-dart frogs (Dendrobatidae). Bull. Am. Mus. Nat. Hist. 157:1–262.

MYERS, C. W., AND J. W. DALY. 1983. Dart-poison frogs. Sci. Am. 248:120–133.

MYERS, C. W., J. W. DALY, and V. MARTÍNEZ. 1984. An arboreal poison frog (*Dendrobates*) from western Panama. Am. Mus. Novit. 2783:1–20.

MYERS, C. W., AND W. E. DUELLMAN. 1982. A new species of *Hyla* from Cerro Colorado, and other tree frog records and geographical notes from western Panama. Am. Mus. Novit. 2752:1–32.

MYERS, C. W., AND A. S. RAND. 1969. Checklist of amphibians and reptiles of Barro Colorado Islands, Panama, with comments on faunal change and sampling. Smithson. Contrib. Zool. 10:1–11.

NEILL, W. T. 1965. New and noteworthy amphibians and reptiles from British Honduras. Bull. Florida St. Mus. 9:77–130.

NEILL, W. T., AND E. R. ALLEN. 1959a. Studies on the amphibians and reptiles of British Honduras. Publ. Res. Div. Ross Allen's Rept. Inst. 2:1–76.

NEILL, W. T., AND E. R. ALLEN. 1959b. Additions to the British Honduras herpetofaunal list. Herpetologica 15:235–240.

NEILL, W. T., AND E. R. ALLEN. 1961. Further studies on the herpetology of British Honduras. Herpetologica 17:37–52.

NELSON, C. E. 1972a. *Gastrophryne elegans.* Cat. Am. Amph. Rept. 121:1–2.

NELSON, C. E. 1972b. *Gastrophryne olivacea.* Cat. Am. Amph. Rept. 122.1–4

NELSON, C. E. 1972c. *Gastrophryne usta.* Cat. Am. Amph. Rept. 123:1–2.

NELSON, C. E. 1972d. Systematic studies on the North American microhylid genus *Gastrophryne.* J. Herpetol. 6:111–137.

NELSON, C. E. 1972e. Distribution and biology of *Chiasmocleis panamensis* (Amphibia: Microhylidae). Copeia 1972:895–898.

NELSON, C. E. 1973a. *Gastrophryne pictiventris.* Cat. Am. Amph. Rept. 135:1–2.

NELSON, C. E. 1973b. Systematics of Middle American upland populations of *Hypopachus* (Anura: Microhylidae). Herpetologica 29:6–17.

NELSON, C. E. 1974. Further studies on the systematics of *Hypopachus* (Anura: Microhylidae). Herpetologica 30:250–274.

NELSON, C. E., AND D. L. HOYT. 1961. New Central American records for *Rhinophrynus dorsalis.* Herpetologica 17:216.

NELSON, C. E., AND M. A. NICKERSON. 1966. Notes on some Mexican and Central American amphibians and reptiles. Southwest. Nat. 11:128–131.

NOBLE, G. K. 1918. The amphibians collected by the American Museum expedition to Nicaragua in 1916. Bull. Am. Mus. Nat. Hist. 38:311–347.

NOVAK, R. M., AND D. C. ROBINSON. 1975. Observations on the reproduction and ecology of the tropical montane toad, *Bufo holdridgei* Taylor in Costa Rica. Rev. Biol. Trop. 23:213–237.

NUSSBAUM, R. A. 1988. On the status of *Copeotyphlinus syntremus, Gymnopis oligozona,* and *Minascaecilia sartoria*: a comedy of errors. Copeia 1988:921–928.

OLSON, E. C., AND P. O. McGREW. 1941. Mammalian fauna from the Pliocene of Honduras. Bull. Geol. Soc. Am. 52:1219–1244.

OLSON, R. E. 1984. Geographic distribution: *Centrolenella fleischmanni* (glass frog). Herpetol. Rev. 15:76.

ORDÓÑEZ, E. 1936. Principal physiographic provinces of Mexico. Bull. Am. Assoc. Petrol. Geol. 20:1277–1307.

O'SHEA, M. T. 1986. Geographical distribution: *Centrolenella albomaculata.* Herpetol. Rev. 17:49.

O'SHEA, M. T. 1989. New departmental records for northeastern Honduran herpetofauna. Herpetol. Rev. 20:16.

PAPENFUSS, T. J., AND D. B. WAKE. 1987. Two new species of plethodontid salamanders (genus *Nototriton*) from Mexico. Acta Zool. Mexicana 21:1–16.

PAPENFUSS, T. J., D. B. WAKE, AND K. ADLER. 1983. Salamanders of the genus *Bolitoglossa* from the Sierra Madre del Sur of southern Mexico. J. Herpetol. 17:295–307.

PÉREZ-HIGAREDA, G., R. C. VOGT, AND O. FLORES-VILLELA. 1987 [no date given]. Lista anotada de los anfibios y reptiles de la región de Los Tuxtlas, Veracruz. Inst. Biol. Univ. Autón. Nac. México, 23 pp.

PETERS, W. 1874 "1873." Über eine neue Schildkrotenart, *Cinosternon Effeldtii* und einige andere neue oder weniger bekannte Amphibien. Monatsber. K. Preuss. Akad. Wiss. Berlin 1873:603–618.

PETERSON, H. W., AND H. M. SMITH. 1983. Geographic distribution: *Syrrhophus pallidus.* Herpetol. Rev. 14:27.

PLAFKER, G. 1976. Tectonic aspects of the Guatemala earthquake of 4 February 1976. Science 193:1201–1208.

PLATZ, J. E., AND J. S. FROST. 1984. *Rana yavapaiensis,* a new species of leopard frog (*Rana pipiens* complex). Copeia 1984:940–948.

PLATZ, J. E., AND J. S. MECHAM. 1979. *Rana chiricahuensis,* a new species of leopard frog (*Rana pipiens* complex) from Arizona. Copeia 1979:383–390.

PORRAS, L., AND L. D. WILSON. 1987. A new species of *Hyla* from the highlands of Honduras and El Salvador. Copeia 1987:478–482.

PORTER, K. R. 1963. Distribution and taxonomic status of seven species of Mexican *Bufo.* Herpetologica 19:229–247.

PORTER, K. R. 1966. Mating calls of six Mexican and Central American toads (genus *Bufo*). Herpetologica 22:60–67.

PORTER, K. R. 1967. *Bufo cycladen* (Bufonidae). A case of nomen dubium. Southwest. Nat. 12:200–201.

PORTER, K. R. 1970. *Bufo valliceps.* Cat. Am. Amph. Rept. 94:1–4.

POUNDS, J. A., AND M. L. CRUMP. 1994. Amphibian declines and climatic disturbance: the case of the golden toad and the harlequin frog. Conserv. Biol. 8:72–85.

POUNDS, J. A., AND M. P. FOGDEN. 1996. Conservation of the golden toad: a brief history. Bull. British Herpetol. Soc. 55:5–7.

RABB, G. B. 1955. A new salamander of the genus *Parvimolge* from Mexico. Breviora 42:1–9.

RABB, G. B. 1956. A new plethodontid salamander from Nuevo León, Mexico. Fieldiana: Zool. 39:11–20.

RABB, G. B. 1958. On certain Mexican salamanders of the plethodontid genus *Chiropterotriton*. Occas. Pap. Mus. Zool. Univ. Michigan 587:1–37.

RABB, G. B. 1959. A new frog of the genus *Plectrohyla* from the Sierra de los Tuxtlas, Mexico. Herpetologica 15:45–47.

RABB, G. B. 1960. A new salamander of the genus *Chiropterotriton* from Chiapas, Mexico, with notes on related species. Copeia 1960:304–311.

RABB, G. B. 1965. A new salamander of the genus *Chiropterotriton* (Caudata: Plethodontidae) from Mexico. Breviora 235:1–7.

RAMÍREZ-BAUTISTA. 1994. Manual y claves ilustradas de los anfibios y reptiles de la región de Chamela, Jalisco México. Univ. Nac. Auton. México Inst. Biol. Cuad. 23:1–127.

RAND, A. S. 1957. Notes on amphibians and reptiles from El Salvador. Fieldiana Zool. 34:505–534.

RAND, A. S. 1983. *Physalaemus pustulosus*. Pp. 412–414 *in* D. H. Janzen (ed.), *Costa Rican Natural History*. Chicago: Univ. Chicago Press.

RAND, A. S., AND C. W. MYERS. 1990. The herpetofauna of Barro Colorado Island, Panama: An ecological summary. Pp. 386–409 *in* A. H. Gentry (ed.), *Four Neotropical Rainforests*. New Haven, Connecticut: Yale University Press.

RAVEN, P. H., AND D. I. AXELROD. 1974. Angiosperm biogeography and past continental movements. Ann. Missouri Bot. Gard. 61:539–673.

RAVEN, P. H., AND D. I. AXELROD. 1975. History of the flora and fauna of Latin America. Am. Sci. 63:420–429.

REGAL, P. J. 1966. A new plethodontid salamander from Oaxaca, Mexico. Am. Mus. Novit. 2266:1–8.

REILLY, S. M., AND R. A. BRANDON. 1994. Partial paedomorphosis in the Mexican stream ambystomatids and taxonomic status of the genus *Rhyacosiredon* Dunn. Copeia 1994:656–662.

ROBINSON, D. C. 1976. A new dwarf salamander of the genus *Bolitoglossa* (Plethodontidae) from Costa Rica. Proc. Biol. Soc. Washington 89:289–294.

ROBINSON, D. C. 1983. *Rana palmipes*. Pp. 415–416 *in* D. H. Janzen (ed.), *Costa Rican Natural History*. Chicago: Univ. Chicago Press.

Rosen, D. E. 1978. Vicariant patterns and historical explanation in biogeography. Syst. Zool. 27:159–188.

RUTHVEN, A. G. 1914. Description of a new engystomatid frog of the genus *Hypopachus*. Proc. Biol. Soc. Washington 27:77–80.

RYAN, M. J. 1985. *The Túngara Frog*. Chicago: Univ. Chicago Press.

RZEDOWSKI, J. 1986. *Vegetación de México*. Mexico City: Editorial Limusa.

SANDERS, O. 1973. A new leopard frog (*Rana berlandieri brownorum*) from southern Mexico. J. Herpetol. 7:87–92.

SANDERS, O., AND H. M. SMITH. 1951. Geographic variation in toads of the *debilis* group of *Bufo*. Field and Laboratory 19:141–160.

SAVAGE, J. M. 1965. A new bromeliad frog of the genus *Eleutherodactylus* from Costa Rica. Bull. S. California Acad. Sci. 64:106–110.

SAVAGE, J. M. 1966. The origins and history of the Central American herpetofauna. Copeia 1966:719–766.

SAVAGE, J. M. 1967a "1966." An extraordinary new toad (Bufo) from Costa Rica. Rev. Biol. Trop. 14:153–167.

SAVAGE, J. M. 1967b. A new tree-frog (Centrolenidae) from Costa Rica. Copeia 1967:325–331.

SAVAGE, J. M. 1968a. A new red-eyed tree frog (family Hylidae) from Costa Rica, with a review of the *Hyla uranochroa* group. Bull. S. California Acad. Sci. 67:1–20.

SAVAGE, J. M. 1968b. The dendrobatid frogs of Central America. Copeia 1968:745–776.

Savage, J. M. 1968c. The distribution and synonymy of the Neotropical frog, *Eleutherodactylus moro*. Copeia 1968:878–879.

SAVAGE, J. M. 1972a. The systematic status of *Bufo simus* O. Schmidt with description of a new toad from western Panama. J. Herpetol. 6:25–33.

SAVAGE, J. M. 1972b. The harlequin frogs, genus *Atelopus*, of Costa Rica and western Panama. Herpetologica 28:77–94.

SAVAGE, J. M. 1974a. On the leptodactylid frogs called *Eleutherodactylus palmatus* (Boulenger) and the status of *Hylodes fitzingeri* O. Schmidt. Herpetologica 30:289–299.

SAVAGE, J. M. 1974b. Type localities for species of amphibians and reptiles described from Costa Rica. Rev. Biol. Trop. 22:71–122.

SAVAGE, J. M. 1975. Systematics and distribution of the Mexican and Central American stream frogs related to *Eleutherodactylus rugulosus*. Copeia 1975:254–306.

SAVAGE, J. M. 1980. A new frog of the genus *Eleutherodactylus* (Leptodactylidae) from the Monteverde forest preserve, Costa Rica. Bull. S. California Acad. Sci. 79:13–19.

SAVAGE, J. M. 1981a. The systematic status of Central American frogs confused with *Eleutherodactylus cruentus*. Proc. Biol. Soc. Washington 94:413–420.

SAVAGE, J. M. 1981b. *Eleutherodactylus bransfordii* (Cope): an addition to the frog fauna of Honduras. Herpetol. Rev. 12:14.

SAVAGE, J. M. 1982. The enigma of the Central American herpetofauna: a dispersal or vicariance? Ann. Missouri Bot. Gard. 69:464–547.

SAVAGE, J. M. 1984. A new species of montane rain frog, genus *Eleutherodactylus* (Leptodactylus [sic]), from Guerrero, Mexico. Amphibia-Reptilia 5:253–260.

SAVAGE, J. M. 1987. Systematics and distribution of the Mexican and Central American rainfrogs of the *Eleutherodactylus gollmeri* group (Amphibia: Leptodactylidae). Fieldiana: Zool., new series 33:1–57.

SAVAGE, J. M., AND J. E. DEWEESE. 1979. A new species of leptodactylid frog, genus *Eleutherodactylus*, from the Cordillera de Talamanca, Costa Rica. Bull. S. California Acad. Sci. 78:107–115.

SAVAGE, J. M., AND J. E. DEWEESE. 1981 "1980." The status of the Central American leptodactylid frogs *Eleutherodactylus melanostictus* (Cope) and *Eleutherodactylus platyrhynchus* (Günther). Proc. Biol. Soc. Washington 93:928–942.

SAVAGE, J. M., AND S. B. EMERSON. 1970. Central American frogs allied to *Eleutherodactylus bransfordii* (Cope): a problem of polymorphism. Copeia 1970:623–644.

SAVAGE, J. M., AND W. R. HEYER. 1967. Variation and distribution in the tree-frog genus *Phyllomedusa* in Costa Rica, Central America. Beitr. Neotrop. Fauna 5:111–131.

SAVAGE, J. M., AND W. R. HEYER. 1969. The tree-frogs of Costa Rica: diagnosis and distribution. Rev. Biol. Trop. 16:1–127.

SAVAGE, J. M., AND A. G. KLUGE. 1961. Rediscovery of the strange Costa Rica toad *Crepidius epioticus* Cope. Rev. Biol. Trop. 9:39–51.

SAVAGE, J. M., J. R. MCCRANIE, AND L. D. WILSON. 1988. New upland stream frogs of the *Eleutherodactylus rugulosus* group (Amphibia: Anura: Leptodactylidae) from Honduras. Bull. S. California Acad. Sci. 87:50–56.

SAVAGE, J. M., AND P. H. STARRETT. 1967. A new fringe-limbed tree-frog (family Centrolenidae) from lower Central America. Copeia 1967:604–609.

SAVAGE, J. M., AND J. VILLA. 1986. An Introduction to the Herpetofauna of Costa Rica. Soc. Stud. Amph. Rept. Contrib. Herp. 3:1–207.

SAVAGE, J. M., AND M. H. WAKE. 1972. Geographic variation and systematics of the Middle American caecilians, genera *Dermophis* and *Gymnopis*. Copeia 1972:680–695.

SCHMIDT, K. P. 1933a. New reptiles and amphibians from Honduras. Field Mus. Nat. Hist., Zool. Ser. 20:15–22.

SCHMIDT, K. P. 1933b. Amphibians and reptiles collected by the Smithsonian Biological Survey of the Panama Canal Zone. Smithsonian Misc. Collect. 89:1–20.

SCHMIDT, K. P. 1936a. Guatemalan salamanders of the genus *Oedipus*. Field Mus., Zool. Ser. 20:135–166.

SCHMIDT, K. P. 1936b. New amphibians and reptiles from Honduras in the Museum of Comparative Zoology. Proc. Biol. Soc. Washington 49:43–50.

SCHMIDT, K. P. 1939. New Central American frogs of the genus *Hypopachus*. Field Mus. Nat. Hist. Zool. Ser. 24:1–5.

SCHMIDT, K. P. 1941. The amphibians and reptiles of British Honduras. Field Mus. Nat. Hist., Zool. Ser. 22:475–510.

SCHMIDT, K. P. 1943. Corollary and commentary for "Climate and Evolution." Am. Midl. Nat. 30:241–253.

SCHMIDT, K. P. 1954. Faunal realms, regions, and provinces. Quart. Rev. Biol. 29:322–331.

SCHMIDT, K. P., AND L. C. STUART. 1941. The herpetological fauna of the Salama Basin, Baja Verapaz, Guatemala. Field Mus. Nat. Hist., Zool. Ser. 24:233–247.

SCHUCHERT, C. 1935. *Historical Geology of the Antillean-Caribbean Region.* New York: John Wiley and Sons.

SCOTT, N. J. 1983a. *Agalychnis callidryas.* Pp. 374–375 *in* D. H. Janzen (ed.), *Costa Rican Natural History.* Chicago: Univ. Chicago Press.

SCOTT, N. J. 1983b. *Bolitoglossa subpalmata.* Pp. 382–383 *in* D. H. Janzen (ed.), *Costa Rican Natural History.* Chicago: Univ. Chicago Press.

SCOTT, N. J. 1983c. *Bufo haematiticus.* Pp. 385 *in* D. H. Janzen (ed.), *Costa Rican Natural History.* Chicago: Univ. Chicago Press.

SCOTT, N. J. 1983d. *Eleutherodactylus bransfordii.* Pp. 399 *in* D. H. Janzen (ed.), *Costa Rican Natural History.* Chicago: Univ. Chicago Press.

SCOTT, N. J. 1983e. *Eleutherodactylus diastema.* Pp. 399 *in* D. H. Janzen (ed.), *Costa Rican Natural History.* Chicago: Univ. Chicago Press.

SCOTT, N. J. 1983f. *Hyla boulengeri.* Pp. 401 *in* D. H. Janzen (ed.), *Costa Rican Natural History.* Chicago: Univ. Chicago Press.

SCOTT, N. J. 1983g. *Leptodactylus pentadactylus.* Pp. 405–406 *in* D. H. Janzen (ed.), *Costa Rican Natural History.* Chicago: Univ. Chicago Press.

SCOTT, N. J., J. M. SAVAGE, AND D. C. ROBINSON. 1983. Checklist of reptiles and amphibians. Pp. 367–374 *in* D. H. Janzen (ed.), *Costa Rican Natural History.* Chicago: Univ. Chicago Press.

SCOTT, N. J., AND A. STARRETT. 1974. An unusual breeding aggregation of frogs with notes on the ecology of *Agalychnis spurrelli* (Anura: Hylidae). Bull S. California Acad. Sci. 73:86–94.

SHAFFER, H. B. 1983. Biosystematics of *Ambystoma rosaceum* and *A. tigrinum* in northwestern Mexico. Copeia 1983:67–78.

SHAFFER, H. B. 1993. Phylogenetics of model organisms: the laboratory axolotl, *Ambystoma mexicanum.* Syst. Zool. 42:508–522.

SHANNON, F. A. 1951. Notes on a herpetological collection from Oaxaca and other localities in Mexico. Proc. U. S. Natl. Mus. 101:465–484.

SHANNON, F. A., AND J. E. WERLER. 1955a. Report on a small collection of amphibians from Veracruz, with a description of a new species of *Pseudoeurycea.* Herpetologica 11:81–85.

SHANNON, F. A., AND J. E. WERLER. 1955b. Notes on amphibians of the Los Tuxtlas range of Veracruz, Mexico. Trans. Kansas Acad. Sci. 58:360–386.

SHREVE, B. 1936. A new *Atelopus* from Panama and a new *Hemidactylus* from Colombia. Occas. Pap. Boston Soc. Nat. Hist. 8:269–271.

SILVERSTONE, P. A. 1975. A revision of the poison-arrow frogs of the genus *Dendrobates* Wagler. Sci. Bull. Nat. Hist. Mus. Los Angeles Co. 21:1–55.

SILVERSTONE, P. A. 1976. A revision of the poison-arrow frogs of the genus *Phyllobates* Bibron *in* Sagra (family Dendrobatidae). Sci. Bull. Nat. Hist. Mus. Los Angeles Co. 27:1–53.

SIMPSON, B. B. 1975. Pleistocene changes in the flora of the high tropical Andes. Paleobiology 1:273–294.

SIMPSON, B. B. 1979. Quaternary biogeography of the high mountain regions of South America. Pp. 157–188 *in* W. E. Duellman (ed.), The South American herpetofauna: its origin, evolution, and dispersal. Monogr. Univ. Kansas Mus. Nat. Hist. 7:1–485.

SMITH, H. M. 1938. Notes on reptiles and amphibians from Yucatan and Campeche, Mexico. Occas. Pap. Mus. Zool. Univ. Michigan 388:1–22.

SMITH, H. M. 1939. Notes on Mexican reptiles and amphibians. Field Mus. Nat. Hist., Zool. Ser. 24:15–35.

SMITH, H. M. 1949. Herpetogeny in Mexico and Guatemala. Anal. Assoc. Am. Geograph. 39:219–238.

SMITH, H. M. 1957. A new casque-headed frog (*Pternohyla*) from Mexico. Herpetologica 13:1–4.

SMITH, H. M. 1959. Herpetozoa from Guatemala, I. Herpetologica 15:210–216.

SMITH, H. M., AND R. A. BRANDON. 1968. Data nova herpetologica Mexicana. Trans. Kansas Acad. Sci. 71:49–61.

SMITH, H. M., AND W. L. BURGER. 1955. Range extensions of certain amphibians and reptiles of southern Mexico. Herpetologica 11:75–77.

SMITH, H. M., AND L. E. LAUFE. 1940. Mexican amphibians and reptiles in the Texas Cooperative Wildlife Collections. Trans. Kansas Acad. Sci. 48:325–354.

SMITH, H. M., AND L. E. LAUFE. 1945. Notes on a herpetological collection from Oaxaca. Herpetologica 3:1–13.

SMITH, H. M., J. D. LYNCH, AND R. ALTIG. 1965. New and noteworthy herpetozoa from southern Mexico. Nat. Hist. Misc. 180:1–4.

SMITH, H. M., AND W. L. NECKER. 1943. Alfredo Dugès' types of Mexican reptiles and amphibians. Anal. Esc. Nac. Cienc. Biol. 3:179–233.

SMITH, H. M., AND R. B. SMITH. 1976. *Synopsis of the Herpetofauna of Mexico, Vol. IV: Source Analysis and Index for Mexican Amphibians.* North Benington, Vermont: John Johnson.

SMITH, H. M., AND E. H. TAYLOR. 1948. An annotated checklist and key to the Amphibia of Mexico. Bull. U.S. Natl. Mus. 194:1–118.

SMITH, P. W. 1952. A new toad from the highlands of Guatemala and Chiapas. Copeia 1952:175–177.

SNYDER, D. H. 1972. *Hyla juanitae,* a new treefrog from southern México, and its relationship to *H. pinorum.* J. Herpetol. 6:5–15.

STAFFORD, P. J. 1994. Amphibians and reptiles of the upper Raspaculo River Basin, Maya Mountains, Belize. Bull. British Herp. Soc. 47:23–29.

STARRETT, P. H. 1966. Rediscovery of *Hyla pictipes* Cope, with description of a new montane stream *Hyla* from Costa Rica. Bull. S. California Acad. Sci. 65:17–28.

STARRETT, P. H., AND J. M. SAVAGE. 1973. The systematic status and distribution of Costa Rican glass-frogs, genus *Centrolenella* (family Centrolenidae), with description of a new species. Bull. S. California Acad. Sci. 72:57–78.

STEBBINS, R. C. 1985. *A Field Guide to Western Reptiles and Amphibians.* Boston: Houghton Mifflin Co.

STEBBINS, R. C., AND N. W. COHEN. 1995. *A Natural History of Amphibians.* Princeton, New Jersey: Princeton Univ. Press.

STEJNEGER, L. 1911. Descriptions of three new batrachians from Costa Rica and Panama. Proc. U.S. Natl. Mus. 41:285–288.

STRAUGHAN, I. R., AND J. W. WRIGHT. 1969. A new stream breeding frog from Oaxaca, Mexico (Anura, Hylidae). Contrib. Sci. Nat. Hist. Mus. Los Angeles Co. 169:1–12.

STUART, L. C. 1934. A contribution to a knowledge of the herpetological fauna of El Peten, Guatemala. Occas. Pap. Mus. Zool. Univ. Michigan 292;1–18.

STUART, L. C. 1935. A contribution to a knowledge of the herpetology of a portion of the savanna region of central Petén, Guatemala. Misc. Publ. Mus. Zool. Univ. Michigan 29:1–56.

STUART, L. C. 1937. Some further notes on the amphibians and reptiles of the Peten forest of northern Guatemala. Copeia 1937:67–70.

STUART, L. C. 1941. Two new species of *Eleutherodactylus* from Guatemala. Proc. Biol. Soc. Washington 54:197–200.

STUART, L. C. 1942. Descriptions of two new species of *Plectrohyla* Brocchi with comments on several forms of tadpoles. Occas. Pap. Mus. Zool. Univ. Michigan 455:1–14.

STUART, L. C. 1943a. Taxonomic and geographic comments on Guatemalan salamanders of the genus *Oedipus.* Misc. Publ. Mus. Zool. Univ. Michigan 56:1–33.

STUART, L. C. 1943b. Comments on the herpetofauna of the Sierra de las Cuchumatanes of Guatemala. Occas. Pap. Mus. Zool. Univ. Michigan 471:1–28.

STUART, L. C. 1948a. The amphibians and reptiles of Alta Verapaz, Guatemala. Misc. Publ. Mus. Zool. Univ. Michigan 69:5–109.

STUART, L. C. 1948b. Another new *Plectrohyla* from Guatemala. Proc. Biol. Soc. Washington 61:17–18.

STUART, L. C. 1950. A geographic study of the herpetofauna of Alta Verapaz, Guatemala. Contrib. Lab. Vert. Biol. Univ. Michigan 45:1–77.

STUART, L. C. 1951. The herpetofauna of the Guatemalan Plateau, with special reference to its distribution on the southwestern highlands. Contrib. Lab. Vert. Biol. Univ. Michigan 49:1–71.

STUART, L. C. 1952. Some new amphibians from Guatemala. Proc. Biol. Soc. Washington 65:1–12.

STUART, L. C. 1954a. A description of a subhumid corridor across northern Central America, with comments on its herpetofaunal indicators. Contrib. Lab. Vert. Biol. Univ. Michigan 65:1–26.

STUART, L. C. 1954b. Herpetofauna of the southeastern highlands of Guatemala. Contrib. Vert. Biol. Univ. Michigan 68:1–65.

STUART, L. C. 1954c. Descriptions of some new amphibians and reptiles from Guatemala. Proc. Biol. Soc. Washington 67:159–178.

STUART, L. C. 1957. Herpetofaunal dispersal routes through northern Central America. Copeia 1957:89–94.

STUART, L. C. 1961. Some observations on the natural history of tadpoles of *Rhinophrynus dorsalis* Duméril and Bibron. Herpetologica 17:73–79.

STUART, L. C. 1963. A checklist of the herpetofauna of Guatemala. Misc. Publ. Mus. Zool. Univ. Michigan 122:1–150.

STUART, L. C. 1964. Fauna of Middle America. Pp. 316–362 *in* R. Wauchope and R. C. West (eds.), *Handbook of Middle American Indians, Vol. 1, Natural Environment and Early Cultures.* Austin, Texas: Univ.Texas Press.

STUART, L. C. 1966. The environment of the Central American cold-blooded vertebrate fauna. Copeia 1966:684–699.

SYKES, L. R., W. R. McCANN, AND A. C. KAFKA. 1982. Motion of the Caribbean plates during the last 7 million years and implications for earlier Cenozoic movements. J. Geophys. Res. 87:10656–10676.

TANNER, W. W. 1950. A new genus of plethodontid salamander from Mexico. Great Basin Nat. 10:37–44.

TANNER, W. W. 1957. Notes on a collection of amphibians and reptiles from southern Mexico, with a description of a new *Hyla.* Great Basin Nat. 30:41–45.

TANNER, W. W. 1962. A new *Bolitoglossa* (salamander) from southern Panama. Herpetologica 18:18–20.

TANNER, W. W. 1989. Amphibians of western Chihuahua. Great Basin Nat. 49:38–70.

TANNER, W. W., AND A. H. BRAME. 1961. Descripton of a new salamander from Panama. Great Basin Nat. 21:23–26.

TANNER, W. W., AND W. G. ROBISON, JR. 1960. A collection of herptiles from Urique, Chihuahua. Great Basin Nat. 19:75–82.

TAYLOR, E. H. 1936. New species of amphibia from Mexico. Trans. Kansas Acad. Sci. 29:349–363.

TAYLOR, E. H. 1937. New species of hylid frogs from Mexico with comments on the rare *Hyla bistincta* Cope. Proc. Biol. Soc. Washington 50:43–54.

TAYLOR, E. H. 1938 "1936." Notes on the herpetological fauna of the Mexican state of Sinaloa. Univ. Kansas Sci. Bull. 22:505–537.

TAYLOR, E. H. 1939a "1938." Concerning Mexican salamanders. Univ. Kansas Sci. Bull. 25:259–313.

TAYLOR, E. H. 1939b "1938." New species of Mexican tailless amphibians. Univ. Kansas Sci. Bull. 25:385–405.

TAYLOR, E. H. 1939c "1938." Frogs of the *Hyla eximia* group in Mexico, with descriptions of two new species. Univ. Kansas Sci. Bull. 25:421–445.

TAYLOR, E. H. 1940a "1939." New salamanders from Mexico,

with a discussion of certain known forms. Univ. Kansas Sci. Bull. 26:407–439.

TAYLOR, E. H. 1940b "1939." New species of Mexican anura. Univ. Kansas Sci. Bull. 26:385–405.

TAYLOR, E. H. 1940c. A new *Rhyacosiredon* (Caudata) from western Mexico. Herpetologica 1:171–176.

TAYLOR, E. H. 1940d. Herpetological miscellany, No. I. Univ. Kansas Sci. Bull. 26:489–571.

TAYLOR, E. H. 1940e. Two new anuran amphibians from Mexico. Proc. U.S. Natl. Mus. 89:43–123.

TAYLOR, E. H. 1941a. Herpetological miscellany, No. II. Univ. Kansas Sci. Bull. 27:105–139.

TAYLOR, E. H. 1941b. New amphibians from the Hobart M. Smith collections. Univ. Kansas Sci. Bull. 27:141–167.

TAYLOR, E. H. 1941c. New plethodont salamanders from Mexico. Herpetologica 2:57–68.

TAYLOR, E. H. 1941d. Two new species of Mexican plethodontid salamanders. Proc. Biol. Soc. Washington 54:81–86.

TAYLOR, E. H. 1941e. Two new ambystomid salamanders from Chihuahua. Copeia 1941:143–146.

TAYLOR, E. H. 1942a. New tailless Amphibia from Mexico. Univ. Kansas Sci. Bull. 28:67–89.

TAYLOR, E. H. 1942b. New Caudata and Salientia from Mexico. Univ. Kansas Sci. Bull. 28:295–323.

TAYLOR, E. H. 1942c. The frog genus *Diaglena* with a description of a new species. Univ. Kansas Sci. Bull. 28:57–65.

TAYLOR, E. H. 1943a. Herpetological novelties from Mexico. Univ. Kansas Sci. Bull. 29:343–361.

TAYLOR, E. H. 1943b. A new ambystomatid salamander adapted to brackish water. Copeia 1943:151–156.

TAYLOR, E. H. 1944. A new ambystomatid salamander from the Plateau of Mexico. Univ. Kansas Sci. Bull. 30:57–61.

TAYLOR, E. H. 1948a. New Costa Rican salamanders. Proc. Biol. Soc. Washington 61:177–180.

TAYLOR, E. H. 1948b. Two new hylid frogs from Costa Rica. Copeia 1948:233–238.

TAYLOR, E. H. 1949a. Costa Rican frogs of the genera *Centrolene* and *Centrolenella.* Univ. Kansas Sci. Bull. 33:257–270.

TAYLOR, E. H. 1949b. New salamanders from Costa Rica. Univ. Kansas Sci. Bull. 33:279–288.

TAYLOR, E. H. 1949c. New or unusual Mexican amphibians. Am. Mus. Novit 1437:1–21.

TAYLOR, E. H. 1950. A new bromeiliad frog from the Mexican state of Veracruz. Copeia 1950:274–276.

TAYLOR, E. H. 1951. A new Veracrucian salamander. Univ. Kansas Sci. Bull. 34:189–193.

TAYLOR, E. H. 1952a. The salamanders and caecilians of Costa Rica. Univ. Kansas Sci. Bull. 34:695–791.

TAYLOR, E. H. 1952b. The frogs and toads of Costa Rica. Univ. Kansas Sci. Bull. 35:577–942.

TAYLOR, E. H. 1954. Additions to the known herpetological fauna of Costa Rica with comments on other species. No. I. Univ. Kansas Sci. Bull. 36:597–639.

TAYLOR, E. H. 1955. Additions to the known herpetological fauna of Costa Rica with comments on other species. No. II. Univ. Kansas Sci. Bull. 37:499–575.

TAYLOR, E. H. 1958a. Notes on Costa Rican Centrolenidae with descriptions of new forms. Univ. Kansas Sci. Bull. 39:41–68.

TAYLOR, E. H. 1958b. Additions to the known herpetological fauna of Costa Rica with comments on other species. No. III. Univ. Kansas Sci. Bull. 39:3–40.

TAYLOR, E. H. 1968. *The Caecilians of the World: A Taxonomic Review.* Lawrence, Kansas: Univ. Kansas Press.

TAYLOR, E. H. 1969. A new Panamanian caecilian. Univ. Kansas Sci. Bull. 48:315–323.

TAYLOR, E. H., AND H. M. SMITH. 1945. Summary of the collections of amphibians made in Mexico under the Walter Rathbone Traveling Scholarship. Proc. U.S. Natl. Mus. 95:521–613.

TIHEN, J. A. 1964. Tertiary changes in the herpetofauna of temperate North America. Senck. Biol. 45:265–279.

TIMMERMAN, W. W., AND M. P. HAYES. 1981. *The Reptiles and Amphibians of Monteverde: An Annotated List to the Herpetofauna of Monteverde, Costa Rica.* Monteverde and San José, Costa Rica: Pensión Quetzal (Monteverde) and Tropical Sci. Cent. (San José).

TOAL, K. R., III. 1994. A new species of *Hyla* (Anura: Hylidae) from the Sierra de Juárez, Oaxaca, Mexico. Herpetologica 50:187–193.

TOSI, J. A., JR. 1969. Mapa ecológico de Costa Rica. 1:750,000. Inst. Geogr. Nac., San José, Costa Rica.

TRUEB, L. 1968. Variation in the tree frog *Hyla lancasteri*. Copeia 1968:285–299.

TRUEB, L. 1970. The evolutionary relationships of the casque-headed treefrogs with co-ossified skulls (family Hylidae). Univ. Kansas Publ. Mus. Nat. Hist. 18:547–716.

TRUEB, L. 1971. Phylogenetic relationships of certain neotropical toads with the description of a new genus (Anura: Bufonidae). Contrib. Sci. Mus. Nat. Hist. Los Angeles Co. 216:1–40.

TRUEB, L. 1974. Systematic relationships of Neotropical horned frogs, genus *Hemiphractus* (Anura: Hyliade). Occas. Pap. Mus. Nat. Hist. Univ. Kansas 29:1–60.

TRUEB, L. 1984. Description of a new species of *Pipa* (Anura: Pipidae) from Panama. Herpetologica 40:225–234.

TRUEB, L. AND D. C. CANNATELLA. 1986. Systematics, morphology, and phylogeny of genus *Pipa* (Anura: Pipidae). Herpetologica 42:412–449.

VAN DER HAMMEN, T. 1974. The pleistocene changes of vegetation and climate in tropical South America. J. Biogeogr. 1:3–26.

VAN DEVENDER, T. R., P. A. HOLM, AND C. H. LOWE. 1989. Life history notes: *Pseudoeurycea belli sierraoccidentalis*. Herpetol. Rev. 20:48–49.

VIAL, J. L. 1963. A new plethodontid salamander (*Bolitoglossa sooyorum*) from Costa Rica. Rev. Biol. Trop. 11:89–97.

VIAL, J. L. 1966. Variation in altitudinal populations of the salamander, *Bolitoglossa subpalmata*, on the Cerra de la Muerte, Costa Rica. Rev. Biol. Trop. 14:111–121.

VIAL, J. L. 1968. The ecology of the tropical salamander, *Bolitoglossa subpalmata*, in Costa Rica. Rev. Biol. Trop. 15:13–115.

VILLA, J. 1972. *Anfibios de Nicaragua*. Managua, Nicaragua: Inst. Geogr. Nac. Banco Centr. Nicaragua.

VILLA, J. 1979. Synopsis of the biology of the Middle American highland frog *Rana maculata* Brocchi. Milwaukee Pub. Mus. Contrib. Biol. Geol. 21:1–17.

VILLA, J. 1984a. Geographic distribution: *Rhinophrynus dorsalis*. Herpetol. Rev. 15:52.

VILLA, J. 1984b. Biology of a Neotropical glass frog, *Centrolenella fleischmanni* (Boettger), with special reference to its frogfly associates. Milwaukee Pub. Mus. Contrib. Biol. Geol. 55:1–60.

VILLA, J., L. D. WILSON, AND J. D. JOHNSON. 1988. *Middle American Herpetology: A Bibliographic Checklist.* Columbia, Missouri: Univ. Missouri Press.

VIVÓ ESCOTO, J. A. 1964. Weather and climate of Mexico and Central America. Pp. 187–215 *in* R. Wauchope and R. C. West (eds.), *Handbook of Middle American Indians, Vol. 1, Natural Environment and Early Cultures.* Austin, Texas: Univ. Texas Press.

WAGNER, P. L. 1964. Natural vegetation of Middle America. Pp. 216–264 *in* R. Wauchope and R. C. West (eds.), *Handbook of Middle American Indians, Vol. 1, Natural Environment and Early Cultures.* Austin, Texas: Univ. Texas Press.

WAKE, D. B. 1970. The abundance and diversity of tropical salamanders. Am. Nat. 104:211–213.

WAKE, D. B. 1987. Adaptive radiation of salamanders in Middle American cloud forests. Ann. Missouri Bot. Gard. 74:242–264.

WAKE, D. B., AND A. H. BRAME, JR. 1962. A new species of salamander from Colombia and the status of *Geotriton andicola* Posada Arango. Los Angeles Co. Mus. Contr. Sci. 49:1–8.

WAKE, D. B., AND A. H. BRAME, JR. 1963. A new species of Costa Rican salamander, genus *Bolitoglossa.* Rev. Biol. Trop. 11:63–73.

WAKE, D. B., AND A. H. BRAME, JR. 1966. A new species of lungless salamander (genus *Bolitoglossa*) from Panama. Fieldiana Zool. 51:1–10.

WAKE, D. B., AND A. H. BRAME, JR. 1969. Systematics and evolution of Neotropical salamanders of the *Bolitoglossa helmrichi* group. Contrib. Sci. Nat. Hist. Mus. Los Angeles Co. 175:1–40.

WAKE, D. B., A. H. BRAME, JR., AND W. E. DUELLMAN. 1973. New species of salamanders, genus *Bolitoglossa,* from Panama. Contrib. Sci. Nat. Hist. Mus. Los Angeles Co. 248:1–19.

WAKE, D. B., A. H. BRAME, JR., AND C. W. MYERS. 1970. *Bolitoglossa taylori,* a new salamander from cloud forest of the Serranía de Pirre, eastern Panama. Am. Mus. Novit. 2430:1–18.

WAKE, D. B., AND P. ELIAS. 1983. New genera and a new species of Central American salamanders, with a review of the tropical genera (Amphibia, Caudata, Plethodontidae). Contrib. Sci. Nat. Hist. Mus. Los Angeles Co. 345:1–19.

WAKE, D. B., AND J. D. JOHNSON. 1989. A new genus and species of plethodontid salamander from Chiapas, Mexico. Contrib. Sci. Nat. Hist. Mus. Los Angeles Co. 411:1–10.

WAKE, D. B., AND J. F. LYNCH. 1976. The distribution, ecology, and evolutionary history of plethodontid salamanders in tropical America. Contrib. Sci. Nat. Hist. Mus. Los Angeles Co. 25:1–65.

WAKE, D. B., AND J. F. LYNCH. 1982. Evolutionary relationships among Central American salamanders of the *Bolitoglossa franklini* group, with a description of a new species from Guatemala. Herpetologica 38:257–272.

WAKE, D. B., AND J. F. LYNCH. 1988. The taxonomic status of *Bolitoglossa resplendens* (Amphibia: Caudata). Herpetologica 44:105–108.

WAKE, D. B., T. J. PAPENFUSS, AND J. F. LYNCH. 1992. Distribution of salamanders along elevational transects in Mexico and Guatemala. Tulane Stud. Zool. Bot. Suppl. Pub. 1:303–319.

WAKE, D. B., S. Y. YANG, AND T. J. PAPENFUSS. 1980. Natural hybridization and its evolutionary implications in Guatemalan plethodontid salamanders of the genus *Bolitoglossa.* Herpetologica 36:335–345.

WAKE, M. H. 1983. *Gymnopis multiplicata, Dermophis mexicanus,* and *Dermophis parviceps.* Pp. 400–401 *in* D. H. Janzen (ed.), *Costa Rican Natural History.* Chicago: Univ. Chicago Press.

WAKE, M. H., AND J. A. CAMPBELL. 1983. A new genus and species of caecilian from the Sierra de las Minas of Guatemala. Copeia 1983:857–863.

WALKER, C. F. 1955. A new salamander of the genus *Pseudoeurycea* from Tamaulipas. Occas. Pap. Mus. Zool. Univ. Michigan 567:1–8.

WALPER, J. 1960. Geology of the Coban–Purulha area, Alta Verapaz, Guatemala. Am. Assoc. Petrol. Geol. Bull. 44:1273–1315.

WEBB, R. G. 1960. Notes on some amphibians and reptiles from northern Mexico. Trans. Kansas Acad. Sci. 63:289–298.

WEBB, R. G. 1962. A new species of frog (genus *Tomodactylus*) from western México. Univ. Kansas Publ. Mus. Nat. Hist. 15:175–181.

WEBB, R. G. 1972. Resurrection of *Bufo mexicanus* for a highland toad in western Mexico. Herpetologica 28:1–6.

WEBB, R. G. 1978. A systematic review of the Mexican frog *Rana sierramadrensis* Taylor. Contrib. Sci. Nat. Hist. Mus. Los Angeles Co. 300:1–13.

WEBB, R. G. 1984. Herpetogeography in the Mazatlán-Durango region of the Sierra Madre Occidental, Mexico. Pp. 217–241 *in* R. A. Seigel et al. (eds.), Vertebrate ecology and systematics. Spec. Publ. Mus. Nat. Hist. Univ. Kansas 10:1–278.

WEBB, R. G. 1988. Frogs of the *Rana tarahumarae* group in eastern Mexico. Occas. Pap. Mus. Texas Tech Univ. 212:1–15.

WEBB, R. G., AND R. H. BAKER. 1962. Terrestrial vertebrates of the Pueblo Nuevo area of southwestern Durango, Mexico. Am. Midl. Nat. 68:325–333.

WEBB, R. G., AND R. H. BAKER. 1969. Vertebrados terrestres del suroeste de Oaxaca. An. Inst. Biol. Univ. Nac. Autón. México 40, Ser. Zool. 1:139–152.

WERLER, J. E., AND H. M. SMITH. 1952. Notes on a collection of reptiles and amphibians from Mexico, 1951–1952. Texas J. Sci. 14:551–573.

WERNER, F. 1896. Beiträge zur Kenntniss der Reptilien und Batrachier von Centralamerika und Chile, sowie einiger seltenerer Schlangenarten. Verh. Zool. Bot. Ges. Wien 46:344–365.

WEST, R. C. 1964. Surface configuration and associated geology of Middle America. Pp. 33–83 *in* R. Wauchope and R. C. West (eds.), *Handbook of Middle American Indians, Vol. 1, Natural Environment and Early Cultures.* Austin, Texas: Univ. Texas Press.

WEYL, R. 1955. Vestigos de una glaciación del Pleistoceno en la Cordillera de Talamanca, Costa Rica, C.A. Informe Semestral del Inst. Geogr. Costa Rica, San José:9–32.

WHITE, S. E. 1962. Late Pleistocene glacial sequence for the west side of Iztaccihuatl, Mexico. Bull. Geol. Soc. Am. 73:935–958.

WILLIAMS, H. 1960. Volcanic history of the Guatemalan highlands. Univ. California Publ. Geol. Sci. 38:1–36.

WILLIAMS, H., A. R. MCBIRNEY, AND G. DENGO. 1964. Geologic reconnaissance of southeastern Guatemala. Univ. California Publ. Geol. Sci. 50:1–56.

WILSON, L. D., AND G. A. CRUZ DIAZ. 1986. Two additions to the herpetofauna of Honduras: *Smilisca sordida* and *Drymobius melanotropis.* Southwest. Nat. 31:249–250.

WILSON, L. D., AND J. R. MCCRANIE. 1979. Notes on the herpetofauna of two mountain ranges in México (Sierra Fria, Aguascalientes, and Sierra Morones, Zacatecas). J. Herpetol. 13:271–278.

WILSON, L. D., AND J. R. MCCRANIE. 1985. A new species of red-eyed *Hyla* of the *uranochroa* group (Anura: Hylidae) from the Sierra de Omoa of Honduras. Herpetologica 41:133–140.

WILSON, L. D., AND J. R. MCCRANIE. 1989. A new species of *Ptychohyla* of the *euthysanota* group from Honduras, with comments on the status of the genus *Ptychohyla* Taylor (Anura: Hylidae). Herpetologica 45:10–17.

WILSON, L. D., AND J. R. MCCRANIE. 1991. Additional departmental records for the herpetofauna of Honduras. Herp. Rev. 22:69–71.

WILSON, L. D., AND J. R. MCCRANIE. 1994. Comments on the occurrence of a salamander and three lizard species in Honduras. Amph. Rept. 15:416–421.

WILSON, L. D., J. R. MCCRANIE, AND G. A. CRUZ. 1994. A new species of *Plectrohyla* (Anura: Hylidae) from a premontane rainforest in northern Honduras. Proc. Biol. Soc. Washington 107:67–78.

WILSON, L. D., J. R. MCCRANIE, AND K. L. WILLIAMS. 1985. Two new species of fringe-limbed frogs from Nuclear Middle America. Herpetologica 41:141–150.

WILSON, L. D., J. R. MCCRANIE, AND K. L. WILLIAMS. 1986. *Plectohyla glandulosa* (Boulenger): an addition to the anuran fauna of Honduras. Herpetol. Rev. 17:8–9.

WILSON, L. D., L. PORRAS, AND J. R. MCCRANIE. 1986. Distributional and taxonomic comments on some members of the Honduran herpetofauna. Milwaukee Pub. Mus. Contr. Biol. Geol. 66:1–18.

WOODALL, H. T. 1941. A new Mexican salamander of the genus *Oedipus*. Occas. Pap. Mus. Zool. Univ. Michigan 444:1–4.

ZUG, G. R. 1983. Bufo marinus. Pp. 386–387 *in* D. H. Janzen (ed.), *Costa Rican Natural History.* Chicago: Univ. Chicago Press.

ZUG, R. G., AND P. B. ZUG. 1979. The marine toad, *Bufo marinus:* a natural history resumé of native populations. Smithsonian Contrib. Zool. 284:1–58.

ZWEIFEL, R. G. 1954. A new frog of the genus *Rana* from western Mexico with a key to the Mexican species of the genus. Bull. S. California Acad. Sci. 53:131–141.

ZWEIFEL, R. G. 1955. Ecology, distribution and systematics of frogs of the *Rana boylei* group. Univ. California Publ. Zool. 54:207–292.

ZWEIFEL, R. G. 1957. A new frog of the genus *Rana* from Michoacán, Mexico. Copeia 1957:78–83.

ZWEIFEL, R. G. 1964. Distribution and life history of a Central American frog, *Rana vibicaria.* Copeia 1964:300–308.

ZWEIFEL, R. G. 1967. *Eleutherodactylus augusti.* Cat. Am. Amphib. Rept. 1967:411–414.

APPENDIX 3:1

TAXONOMIC GUIDE TO REFERENCES

The following pertinent references on the systematics and distribution of species of amphibians occurring in Middle America is arranged alphabetically by order, family, genus, and species. Only currently recognized names are given; some references cite names that are synonyms.

ANURA: BUFONIDAE
Atelophryniscus chrysophorus
McCranie et al., 1989
Atelopus certus
Barbour, 1923; Savage, 1972b
Atelopus chiriquiensis
Savage, 1972b; Shreve, 1936
Atelopus glyphus
Breder, 1946; Cochran and Goin, 1970; Dunn, 1931d; Savage, 1972b; Schmidt, 1933b
Atelopus limosus
Ibañez, Jaramillo, and Solís, 1995
Atelopus senex
Savage, 1972b; Savage, 1974b; Taylor, 1952b, 1955
Atelopus varius
Dunn, 1931a, 1931b, 1931d, 1933, 1940a; Martínez-Cortés, 1984; Savage, 1972b, 1974b; Taylor, 1952b, 1955
Bufo bocourti
Blair, 1972; Campbell and Vannini, 1989a, 1989b; Günther, 1885–1902; Johnson, 1989, 1989; Nelson and Nickerson, 1966; Smith and Burger, 1955, Stuart, 1943b, 1951, 1963
Bufo campbelli
Mendelson, 1994
Bufo canaliferus
Blair, 1972; Campbell and Vannini, 1989a, 1989b; Duellman, 1960a; Johnson, 1989; Kellogg, 1932; Mertens, 1952; Porter, 1963; Rand, 1957; Stuart, 1963
Bufo cavifrons
Firschein, 1950; Porter, 1963; Shannon and Werler, 1955b
Bufo coccifer
Blair, 1972; Davis and Dixon, 1964; Duellman, 1960a, 1961b, 1965b; Dunn and Emlen, 1932; Frost, 1985; Hartweg and Oliver, 1940; Johnson, 1989; Lynch and Smith, 1966b; Mertens, 1952; Meyer and Wilson, 1971a; Porter, 1963, 1967; Rand, 1957; Stuart, 1954b, 1963; Taylor, 1952b; Villa, 1972
Bufo cognatus
Kellogg, 1932; Smith and Taylor, 1948; Stebbins, 1985; Tanner, 1989; Webb, 1984
Bufo compactilis
Duellman, 1961b, 1965b; Kellogg, 1932; McCranie and Wilson, 1987; Smith and Laufe, 1940; Smith and Taylor, 1948; Webb, 1984
Bufo coniferus
Breder, 1946; Cochran and Goin, 1970; Dunn, 1931a, 1931b, 1931d; Martínez-Cortés, 1984; Noble, 1918; Porter, 1966; Schmidt, 1933db; Taylor, 1952b; Villa, 1972
Bufo cristatus
Blair, 1972; Cope, 1866; Kellogg, 1932; Porter, 1963; Smith and Taylor, 1948
Bufo debilis
Blair, 1972; Bogert, 1962; Kellogg, 1932; Sanders and Smith, 1951; Smith and Taylor, 1948; Stebbins, 1985; Webb, 1984
Bufo fastidiosus
Cope, 1876; Günther, 1885–1902; Savage, 1972a; Taylor, 1952b
Bufo gemmifer
Davis and Dixon, 1964; Porter, 1963; Smith and Taylor, 1948; Taylor, 1940d
Bufo granulosus
Busack, 1966; Cochran and Goin, 1970; Gallardo, 1965; Rand and Myers, 1990; Schmidt, 1933b
Bufo haematiticus
Barbour, 1923; Blair, 1972; Breder, 1946; Brocchi, 1881–1883; Cruz and Wilson, 1983; Dunn, 1931d; Evans, 1947; Martínez-Cortés, 1984; Noble, 1918; Rand and Myers, 1990; Schmidt, 1933b; Scott, 1983c; Taylor, 1952b; Villa, 1972

Bufo holdridgei
Blair, 1972; Novak and Robinson, 1975; Savage, 1972a, 1974b; Taylor, 1952b
Bufo ibarrai
Blair, 1972; Porter, 1966; Stuart, 1954b, 1954c, 1963
Bufo kelloggi
Bogert, 1962; Davis and Dixon, 1957; Hardy and McDiarmid, 1969; Sanders and Smith, 1951; Smith and Taylor, 1948; Taylor, 1938; Webb, 1984
Bufo luetkenii
Blair, 1972; Johnson, 1989, 1989; Mertens, 1952; Meyer, 1966; Meyer and Wilson, 1971a; Nelson and Nickerson, 1966; Porter, 1966; Stuart, 1954b, 1963; Taylor, 1952b; Villa, 1972; Wilson and McCranie, 1991
Bufo marinus
Breder, 1946; Campbell and Vannini, 1989a, 1989b; Davis and Smith, 1953; Duellman, 1960a, 1961b, 1965b; Dunn and Emlen, 1932; Frost, 1985; Hardy and McDiarmid, 1969; Hartweg and Oliver, 1940; Henderson and Hoevers, 1975; Johnson, 1989; Lee, 1980; Martin, 1958; Rand, 1957; Rand and Myers, 1990; Schmidt, 1933b, 1941; Schmidt and Stuart, 1941; Smith, 1938; Smith and Laufe, 1940; Smith and Taylor, 1948; Stuart, 1934, 1935, 1948a, 1954b, 1963; Tanner, 1989; Tanner and Robison, 1960; Taylor, 1938, 1952b; Webb, 1984; Webb and Baker, 1969; Zug, 1983; Zug and Zug, 1979
Bufo marmoreus
Blair, 1972; Davis and Dixon, 1964; Duellman, 1960a, 1961b, 1965b; Frost, 1985; Hardy and McDiarmid, 1969; Hartweg and Oliver, 1940; Johnson, 1989; Kellogg, 1932; Porter, 1966; Smith and Burger, 1955; Smith and Taylor, 1948; Webb, 1984; Webb and Baker, 1969
Bufo mazatlanensis
Hardy and McDiarmid, 1969; Porter, 1963; Smith and Taylor, 1948; Tanner, 1989; Tanner and Robison, 1960; Taylor, 1940d; Webb, 1984; Webb and Baker, 1962
Bufo melanochloris
Blair, 1972; Frost, 1985; Taylor, 1952b, 1958b
Bufo microscaphus
Stebbins, 1985; Tanner, 1989; Webb, 1972, 1984; Webb and Baker, 1962
Bufo occidentalis
Davis and Dixon, 1964; Davis and Smith, 1953; Duellman, 1961b, 1965b; Hardy and McDiarmid, 1969; Kellogg, 1932; McCranie and Wilson, 1987; Porter, 1966; Smith and Laufe, 1940; Smith and Taylor, 1948; Tanner, 1989; Webb, 1984; Webb and Baker, 1962; Wilson and McCranie, 1979, 1987
Bufo periglenes
Blair, 1972; Frost, 1985; Savage, 1967a, 1972a
Bufo peripatetes
Auth, 1994; Frost, 1985; Savage, 1972a
Bufo perplexus
Davis and Dixon, 1964; Davis and Smith, 1953; Duellman, 1961b, 1965b; Porter, 1966; Smith and Burger, 1955; Smith and Taylor, 1948; Taylor, 1943a
Bufo punctatus
Anderson and Lidicker, 1963; Dunn, 1936; Hardy and McDiarmid, 1969; Kellogg, 1932; McCranie and Wilson, 1987; Smith and Laufe, 1940; Smith and Taylor, 1948; Tanner, 1989; Tanner and Robison, 1960; Stebbins, 1985; Webb, 1984
Bufo tacanensis
Blair, 1972; Frost, 1985; Johnson, 1989; P. W. Smith, 1952; Stuart, 1963
Bufo typhonius
Breder, 1946; Cochran and Goin, 1970; Dunn, 1931a, 1931b, 1931d; Evans, 1947; Myers and Rand, 1969; Rand and Myers, 1990; Schmidt, 1933b

Bufo valliceps
Baylor and Stuart, 1961; Blair, 1972; Duellman, 1960a, 1965a; Dundee et al., 1986; Frost, 1985; Gaige et al., 1937; Johnson, 1989; Martin, 1955b; Lee, 1980; Porter, 1963, 1970; Rand, 1957; Schmidt, 1941; Smith, 1938; Smith and Laufe, 1940; Stuart, 1934, 1935, 1943b, 1948a, 1954b, 1963; Tanner, 1957; Villa, 1972; Werler and Smith, 1952
Bufo woodhousii
Blair, 1972; Kellogg, 1932; McCranie and Wilson, 1987; Smith and Taylor, 1948; Stebbins, 1985; Tanner, 1989
Crepidophryne epioticus
Cope, 1876; Günther, 1885–1902; Savage, 1972a, 1974b; Savage and Kluge, 1961; Taylor, 1952b
Rhamphophryne acrolopha
Frost, 1985; Trueb, 1971

ANURA: CENTROLENIDAE

Centrolene ilex
Hayes and Starrett, 1980; Savage, 1967b, 1974b; Savage and Starrett, 1967; Starrett and Savage, 1973
Centrolene prosoblepon
Dunn 1931a, 1931b; Martínez-Cortés, 1984; McCranie and Wilson, 1987; Rand and Myers, 1990; Savage, 1974b; Starrett and Savage, 1973; Taylor, 1949a, 1952b, 1958a; Villa, 1972
Cochranella albomaculata
O'Shea, 1986; Savage, 1974b; Starrett and Savage, 1973; Taylor, 1949a,1952b, 1958a
Cochranella euknemos
Martínez-Cortés, 1984; Savage, 1974b; Savage and Starrett, 1967; Starrett and Savage, 1973
Cochranella granulosa
Frost, 1985; Rand and Myers, 1990; Savage, 1974b; Starrett and Savage, 1973; Taylor, 1949a, 1952b, 1958a; Villa, 1972
Cochranella spinosa
Frost, 1985; Myers and Rand, 1969; Rand and Myers, 1990; Savage, 1974b; Starrett and Savage, 1973; Taylor, 1949a, 1952b, 1958a
Hyalinobatrachium chirripoi
Auth, 1994; Savage, 1974b; Starrett and Savage, 1973; Taylor, 1958a
Hyalinobatrachium colymbiphyllum
Jaramillo et al., 1988a; Rand and Myers, 1990; Savage, 1974b; Starrett and Savage, 1973; Taylor, 1949a, 1952b, 1958
Hyalinobatrachium fleischmanni
Duellman and Tulecke, 1960; Henderson and Hoevers, 1975; Hidalgo, 1982; Lee, 1980; McCoy et al., 1986; McDiarmid, 1983; Meyer and Wilson, 1971a; Olson, 1984; Rand and Myers, 1990; Savage, 1974b; Starrett and Savage, 1973; Stuart, 1937, 1948a, 1950, 1963; Taylor, 1942a, 1949a, 1952b; Villa, 1972, 1984b; Webb and Baker, 1969
Hyalinobatrachium pulveratum
Dunn, 1931a, 1931b; Ibañez, 1988; Starrett and Savage, 1973; Taylor, 1952b, 1958a; Villa, 1972
Hyalinobatrachium talamancae
Savage, 1974b; Starrett and Savage, 1973; Taylor, 1952b, 1958a
Hyalinobatrachium valerioi
Dunn, 1931c; Frost, 1985; Savage, 1974b; Starrett and Savage, 1973; Taylor, 1949b, 1952b, 1958a
Hyalinobatrachium vireovittata
Frost, 1985; Jaramillo et al., 1988b; Savage, 1974b; Starrett and Savage, 1973

ANURA: DENDROBATIDAE

Colostethus chocoensis
Cochran and Goin, 1970; Myers, 1991
Colostethus flotator
Dunn, 1931c, 1931d; Evans, 1947; Myers, 1991; Myers and Rand, 1969; Rand and Myers, 1990; Schmidt, 1933b; Taylor, 1952b
Colostethus inguinalis
Breder, 1946; Cochran and Goin, 1970; Dunn, 1933; Evans, 1947; Frost, 1985; Myers, 1991; Rand and Myers, 1990; Savage, 1968b

Colostethus latinasus
Cochran and Goin, 1970; Dunn, 1931c, 1931d; Frost, 1985; Savage, 1968b; Schmidt, 1933b
Colostethus nubicola
Dunn, 1924a, 1931c; Martínez-Cortés, 1984; Myers, 1991; Myers and Rand, 1969; Savage, 1968b
Colostethus pratti
Breder, 1946; Cochran and Goin, 1970; Martínez-Cortés, 1984; Myers, 1991; Myers and Duellman, 1982; Savage, 1968; Taylor, 1952b
Colostethus talamancae
Barbour and Dunn, 1921; Breder, 1946; Dunn, 1931d; Evans, 1947; Martínez-Cortés, 1984; Myers, 1991; Rand and Myers, 1990; Savage, 1968b; Taylor, 1952b
Dendrobates arboreus
Maxson and Myers, 1985; Myers et al., 1984
Dendrobates auratus
Breder, 1946; Crump, 1983; Dunn, 1931d; Evans, 1947; Maxson and Myers, 1985; Myers, 1991; Myers and Daly, 1976; Myers and Rand, 1969; Noble, 1918; Rand and Myers, 1990; Savage, 1968b; Schmidt, 1933b; Silverstone, 1975; Taylor, 1952b; Villa, 1972
Dendrobates granuliferus
Crump, 1983; Myers and Daly, 1976; Savage, 1968b, 1974b; Silverstone, 1975
Dendrobates pumilio
Crump, 1983; Maxson and Myers, 1985; Myers and Daly, 1976; Noble, 1918; Savage, 1968b; Schmidt, 1933b; Silverstone, 1975; Taylor, 1952b; Villa, 1972
Dendrobates speciosus
Martínez-Cortés, 1984; Maxson and Myers, 1985; Myers and Duellman, 1982; Myers et al., 1984; Savage, 1968; Silverstone, 1975
Epipedobates maculatus
Myers, 1982, 1987; Peters, 1874; Silverstone, 1975
Minyobates fulguritus
Myers, 1987, 1991; Silverstone, 1975
Minyobates minutus
Cochran and Goin, 1970; Myers, 1987, 1991; Myers and Daly, 1976; Savage, 1968b; Silverstone, 1975
Phyllobates lugubris
Cochran and Goin, 1970; Dunn, 1931c; Maxson and Myers, 1985; Myers and Daly, 1983; Savage, 1968b; Silverstone, 1976; Taylor, 1952b
Phyllobates vittatus
Maxson and Myers, 1985; Myers and Daly, 1983; Silverstone, 1976

ANURA: HYLIDAE

Agalychnis annae
Duellman, 1970; Savage, 1974b; Savage and Heyer, 1967, 1969
Agalychnis calcarifer
Duellman, 1970; Dunn, 1931d; Funkhouser, 1957; Myers and Duellman, 1982; Myers and Rand, 1969; Savage and Heyer, 1967, 1969
Agalychnis callidryas
Duellman, 1960a, 1963a, 1965a, 1970; Funkhouser, 1957; Henderson and Hoevers, 1975; Johnson, 1989; Lee, 1980; Meyer and Wilson, 1971a; Myers and Rand, 1969; Savage and Heyer, 1967; Schmidt, 1941; Scott, 1983a; Stuart, 1948a, 1963; Villa, 1972
Agalychnis litodryas
Duellman, 1970; Duellman and Trueb, 1967
Agalychnis moreletii
Campbell and Vannini, 1989a; Duellman, 1970; Funkhouser, 1957; Henderson and Hoevers, 1975; Johnson, 1989; Johnson et al., 1977; Lee, 1980; Mertens, 1952; Meyer and Wilson, 1971a; Schmidt, 1941; Stuart, 1943b, 1948a, 1963
Agalychnis saltator
Duellman, 1970; Funkhouser, 1957; Savage, 1974b; Savage and Heyer, 1967, 1969; Taylor, 1955; Villa, 1972
Agalychnis spurrelli
Dunn, 1931a, 1931b; Duellman, 1970; Funkhouser, 1957;

Savage and Heyer, 1967; Scott and Starrett, 1974
Anotheca spinosa
  Duellman, 1968a, 1970; Johnson, 1989; Johnson et al., 1977; Kellogg, 1932; Martínez-Cortés, 1984; Perez-Higareda et al., 1987; Savage, 1974b; Savage and Heyer, 1969; Shannon and Werler, 1955b; Stejneger, 1911; Taylor, 1952b
Duellmanohyla chamulae
  Campbell and Smith, 1992; Duellman, 1961a, 1963b, 1970
Duellmanohyla ignicolor
  Campbell and Smith, 1992; Duellman, 1961a, 1963b, 1970
Duellmanohyla lythrodes
  Duellman, 1970; Myers and Duellman, 1982; Savage, 1968a, 1974b
Duellmanohyla rufioculis
  Duellman, 1970; Savage, 1968a, 1974b; Savage and Heyer, 1969; Taylor, 1952b
Duellmanohyla salvavida
  McCranie and Wilson, 1986
Duellmanohyla schmidtorum
  Campbell and Vannini, 1989a, 1989b; Duellman, 1961a, 1963b, 1970; Johnson, 1989; Stuart, 1954c, 1963
Duellmanohyla soralia
  Wilson and McCranie, 1985
Duellmanohyla uranochroa
  Dunn, 1924a; Duellman, 1966b, 1970; Martínez-Cortés, 1984; Myers and Duellman, 1982; Savage, 1974b, 1968a; Savage and Heyer, 1969; Taylor, 1952b
Gastrotheca cornuta
  Duellman, 1966b, 1970, 1983; Myers and Duellman, 1982
Gastrotheca nicefori
  Duellman, 1970; Myers and Duellman, 1982
Hemiphractus fasciatus
  Duellman, 1970; Martínez-Cortés, 1984; Myers, 1966; Myers and Duellman, 1982; Trueb, 1974
Hyla altipotens
  Duellman, 1968b, 1970
Hyla angustilineata
  Arosemena and Ibanez, 1991; Duellman, 1970; Savage and Heyer, 1969; Taylor, 1952b
Hyla arborescandens
  Caldwell, 1974; Duellman, 1970; Smith and Taylor, 1948; Taylor, 1939b
Hyla arenicolor
  Anderson and Lidicker, 1963; Caldwell, 1974; Davis and Dixon, 1964; Davis and Smith, 1953; Duellman, 1961b, 1965b, 1970; Kellogg, 1932; McCranie and Wilson, 1987; Smith and Taylor, 1948; Stebbins, 1985; Tanner, 1989; Webb, 1984; Wilson and McCranie, 1979
Hyla bistincta
  Adler, 1965; Duellman, 1961b, 1965b, 1964c, 1970; Kellogg, 1932; Shannon and Werler, 1955a; Smith and Taylor, 1948; Straughan and Wright, 1969; Taylor, 1937; Webb, 1984
Hyla boans
  Duellman, 1970
Hyla bocourti
  Campbell and Lawson, 1992; Frost, 1985; Stuart, 1948a, 1950, 1963
Hyla bogertae
  Caldwell, 1974; Duellman, 1970; Straughan and Wright, 1969
Hyla bromeliacia
  Duellman, 1970; Meyer and Wilson, 1971a; Schmidt, 1933a; Stuart, 1943b, 1948a, 1963
Hyla calvicollina
  Toal, 1994
Hyla catracha
  Porras and Wilson, 1987
Hyla cembra
  Caldwell, 1974
Hyla chaneque
  Duellman, 1961c, 1965d, 1970; Johnson, 1989; Lynch and Smith, 1966a; Perez-Higareda et al., 1987
Hyla charadricola
  Caldwell, 1974; Duellman, 1964c,1970; Straughan and Wright, 1969

Hyla chimalapa
  Mendelson and Campbell, 1994
Hyla chryses
  Adler, 1965; Caldwell, 1974; Duellman 1970; Straughan and Wright, 1969
Hyla colymba
  Duellman, 1966b, 1970, 1972; Dunn, 1931c; Martínez-Cortés, 1984; Myers and Duellman, 1982; Savage, 1974b; Savage and Heyer, 1969
Hyla crassa
  Caldwell, 1974; Duellman, 1964c, 1970; Smith and Taylor, 1948; Straughan and Wright, 1969; Taylor, 1940b
Hyla crepitans
  Duellman, 1970; Rand and Myers, 1990
Hyla cyanomma
  Caldwell, 1974
Hyla debilis
  Duellman, 1970; Martínez-Cortés, 1984; Myers and Duellman, 1982; Savage, 1968a, 1974b; Savage and Heyer, 1969; Taylor, 1952b
Hyla dendroscarta
  Duellman, 1970; Pérez-Higareda et al., 1987; Smith and Taylor, 1948; Taylor, 1940e
Hyla ebraccata
  Duellman, 1966b, 1970; Henderson and Hoevers, 1975; Johnson, 1989; Lee, 1980; Schmidt, 1941; Stuart, 1963; Savage and Heyer, 1969; Taylor, 1942a; Villa, 1972
Hyla echinata
  Duellman, 1962, 1970
Hyla euphorbiacea
  Duellman, 1970; Smith and Taylor, 1948; Taylor, 1939c
Hyla eximia
  Duellman, 1961b, 1965b, 1970; Kellogg, 1932; McCranie and Wilson, 1987; Smith and Taylor, 1948; Stebbins, 1985; Tanner, 1989; Taylor, 1939c; Webb, 1984; Wilson and McCranie, 1979
Hyla fimbrimembra
  Duellman, 1970; Hayes et al., 1986; Ibañez et al., 1991; Savage, 1974b; Savage and Heyer, 1969; Taylor, 1948b, 1952
Hyla godmani
  Duellman, 1964b, 1970
Hyla graceae
  Myers and Duellman, 1982
Hyla hazelae
  Duellman, 1970; Smith and Taylor, 1948; Taylor, 1940b; Webb and Baker, 1969
Hyla insolita
  McCranie et al., 1993a
Hyla juanitae
  Snyder, 1972
Hyla labeculata
  Shannon, 1951
Hyla lancasteri
  Barbour, 1928; Duellman, 1966b, 1970; Martínez-Cortés, 1984; Myers and Duellman, 1982; Savage, 1974b; Savage and Heyer, 1969; Trueb, 1968
Hyla loquax
  Duellman, 1960a, 1965a, 1966b, 1970; Gaige and Stuart, 1934; Henderson and Hoevers, 1975; Johnson, 1989; Johnson et al., 1977; Lee, 1980; Meyer and Wilson, 1971a; Savage and Heyer, 1969; Schmidt, 1936b, 1941; Stuart, 1934, 1935, 1948a, 1963; Taylor, 1949c, 1952b; Villa, 1972
Hyla melanomma
  Campbell and Brodie, 1992; Davis and Dixon, 1964; Duellman, 1966b, 1970; Duellman and Hoyt, 1961; Johnson, 1989; Smith and Brandon, 1968
Hyla microcephala
  Busack, 1966; Duellman, 1960a, 1965a, 1970; Duellman and Fouquette, 1968; Frost, 1985; Henderson and Hoevers, 1975; Johnson, 1989; Lee, 1980; Meyer and Wilson, 1971a; Rand and Myers, 1990; Savage, 1974b; Savage and Heyer, 1969; Smith and Brandon, 1968; Stuart, 1935, 1948a, 1963; Taylor, 1940d; Villa, 1972
Hyla miliaria
  Duellman, 1970; Dunn, 1936; Myers and Duellman, 1982; Savage, 1974b; Savage and Heyer, 1969; Smith and Brandon,

1968; Taylor, 1952b; Villa, 1972
Hyla minera
Duellman, 1978; Wilson et al., 1985
Hyla miotympanum
Duellman, 1970; Dunn, 1936; Johnson, 1989; Kellogg, 1932; Martin 1955a, 1955b, 1958; Pérez-Higareda et al., 1987; Smith and Brandon, 1968
Hyla mixe
Duellman, 1965c, 1970
Hyla mixomaculata
Duellman, 1970; Taylor, 1950
Hyla mykter
Adler and Dennis, 1972; Caldwell, 1974
Hyla nubicola
Duellman, 1964a, 1970; Smith and Brandon, 1968
Hyla pachyderma
Caldwell, 1974; Duellman, 1964c, 1970; Smith and Taylor, 1948; Straughan and Wright, 1969; Taylor, 1942b
Hyla palmeri
Myers and Duellman, 1982
Hyla pellita
Duellman, 1968b,1970
Hyla pentheter
Adler, 1965; Caldwell, 1974; Duellman, 1970; Straughan and Wright, 1969; Webb and Baker, 1969
Hyla perkinsi
Campbell and Brodie, 1992
Hyla phlebodes
Duellman, 1970; Duellman and Fouquette, 1968; Myers and Rand, 1969; Savage, 1974b; Savage and Heyer, 1969; Taylor, 1952b; Villa, 1972
Hyla picadoi
Duellman, 1970; Dunn, 1937; Savage and Heyer, 1969; Savage, 1974b; Taylor, 1952b
Hyla picta
Duellman, 1960a, 1965a, 1970; Henderson and Hoevers, 1975; Johnson, 1989; Lee, 1980; Meyer and Wilson, 1971a; Smith and Taylor, 1948; Stuart, 1935, 1948a, 1963
Hyla pictipes
Auth, 1994; Duellman, 1966b, 1970; Savage, 1968a; Savage and Heyer, 1969; Starrett, 1966
Hyla pinorum
Duellman, 1970; Smith and Taylor, 1948; Snyder, 1972; Taylor 1937
Hyla plicata
Duellman, 1961b, 1965b, 1968a, 1970; Kellogg, 1932; Smith and Taylor, 1948
Hyla pseudopuma
Duellman, 1968b, 1970; Günther, 1885–1902; Savage, 1974b; Savage and Heyer, 1969; Taylor, 1952b
Hyla pugnax
Kluge, 1979
Hyla rivularis
Duellman, 1970; Myers and Duellman, 1982; Savage, 1968a, 1974b; Savage and Heyer, 1969; Taylor, 1952b
Hyla robertmertensi
Duellman, 1960a, 1970; Duellman and Fouquette, 1968; Johnson, 1989; Mertens, 1952; Stuart, 1954b, 1963; Taylor, 1937
Hyla robertsorum
Caldwell, 1974; Duellman, 1964c, 1970; Smith and Taylor, 1948; Straughan and Wright, 1969; Taylor, 1940b
Hyla rosenbergi
Breder, 1946; Duellman, 1970; Kluge, 1979, 1981; Savage and Heyer, 1969; Taylor, 1954
Hyla rufitela
Dunn, 1924a; Duellman, 1970; Fouquette, 1961; Myers and Rand, 1969; Savage and Heyer, 1969; Villa, 1972
Hyla sabrina
Caldwell, 1974
Hyla salvaje
Wilson et al., 1985
Hyla sartori
Duellman, 1970; Duellman and Fouquette, 1968
Hyla siopela
Caldwell, 1974; Duellman, 1968b, 1970

Hyla smaragdina
Duellman, 1961b, 1965b, 1970; Hardy and McDiarmid, 1969; Webb, 1984
Hyla smithii
Davis and Dixon, 1964; Davis and Smith, 1953; Duellman, 1961b, 1965b, 1970; Hardy and McDiarmid, 1969; Smith and Taylor, 1948; Taylor, 1938; Webb, 1984
Hyla subocularis
Duellman, 1970; Dunn, 1934
Hyla sumichrasti
Duellman, 1960a, 1970; Johnson, 1984, 1989; Mendelson and Campbell, 1994
Hyla taeniopus
Duellman, 1965d, 1970; Günther, 1885–1902; Kellogg, 1932; Smith and Brandon, 1968; Smith and Taylor, 1948
Hyla thorectes
Adler, 1965; Duellman, 1970
Hyla thysanota
Duellman, 1966c, 1970
Hyla tica
Duellman, 1970; Martínez-Cortés, 1984; Myers and Duellman, 1982; Savage, 1968a; Savage and Heyer, 1969; Savage, 1974b; Starrett, 1966
Hyla trux
Adler and Dennis, 1972
Hyla valancifer
Duellman, 1960c, 1970; Firschein and Smith, 1956; Pérez-Higareda et al., 1987; Wilson et al., 1985
Hyla walkeri
Campbell and Lawson, 1992; Duellman, 1970; Johnson, 1989; Smith and Brandon, 1968; Stuart, 1954b, 1954c, 1963
Hyla xanthosticta
Duellman, 1968b, 1970; Savage, 1974b, Savage and Heyer, 1969
Hyla xera
Mendelson and Campbell, 1994
Hyla zeteki
Duellman, 1970; Gaige, 1929; Martínez-Cortés, 1984; Myers and Duellman, 1982; Savage and Heyer, 1969
Hyla sp.A
Campbell and Camarillo, ms. [northern Oaxaca]
Pachymedusa dacnicolor
Davis and Dixon, 1957; Davis and Smith, 1953; Duellman, 1961b, 1965b, 1970; Funkhouser, 1957; Hardy and McDiarmid, 1969; Taylor, 1938; Webb, 1984; Webb and Baker, 1969
Phrynohyas venulosa
Duellman, 1961b, 1965a, 1965, 1970; Dundee et al., 1986; Frost, 1985; Hardy and McDiarmid, 1969; Henderson and Hoevers, 1975; Johnson, 1989; Lee, 1980; Mertens, 1952; Myers and Rand, 1969; Neill, 1965; Neill and Allen, 1959a, 1959b; Savage and Heyer, 1969; Shannon and Werler, 1955b; Stuart, 1935, 1963
Phyllomedusa lemur
Duellman, 1970; Funkhouser, 1957; Martínez-Cortés, 1984; Savage and Heyer, 1967, 1969
Phyllomedusa venusta
Duellman, 1970; Duellman and Trueb, 1967
Plectrohyla acanthodes
Duellman and Campbell, 1992
Plectrohyla avia
Bumzahem and Smith, 1954; Campbell and Kubin, 1990; Campbell and Vannini, 1989a, 1989b; Duellman, 1970; Duellman and Campbell, 1992; Johnson, 1989; Stuart, 1952, 1963
Plectrohyla chrysopleura
Wilson et al., 1994
Plectrohyla dasypus
Duellman and Campbell, 1992; McCranie and Wilson, 1981
Plectrohyla glandulosa
Duellman, 1964b; Duellman and Campbell, 1992; Stuart, 1948b, 1951, 1954b, 1963; Wilson et al., 1986a
Plectrohyla guatemalensis
Brocchi, 1881–1883; Bumzahem and Smith, 1954; Duellman, 1970; Duellman and Campbell, 1992; Johnson, 1989; Mertens, 1952; Meyer and Wilson, 1971a; Rand, 1957; Stuart, 1951, 1954b, 1963; Wilson and McCranie, 1991

Plectrohyla hartwegi
Duellman, 1968b, 1970; Duellman and Campbell, 1992
Plectrohyla ixil
Duellman, 1970; Duellman and Campbell, 1992; Johnson, 1989; Stuart, 1942, 1943b, 1963
Plectrohyla lacertosa
Bumzahem and Smith, 1954; Duellman, 1970; Duellman and Campbell, 1992; Johnson, 1989
Plectrohyla matudai
Bumzahem and Smith, 1954; Campbell and Vannini, 1989a, 1989b; Duellman, 1970; Duellman and Campbell, 1992; Hartweg, 1941; Hartweg and Orton, 1941; Johnson, 1989; Lynch and Smith, 1966a; McCranie and Wilson, 1985; Smith and Brandon, 1968; Stuart, 1954b, 1963; Taylor, 1949c
Plectrohyla pokomchi
Duellman and Campbell, 1984, 1992
Plectrohyla pycnochila
Duellman, 1970; Duellman and Campbell, 1992; Johnson, 1989; Rabb, 1959
Plectrohyla quecchi
Duellman, 1970; Duellman and Campbell, 1992; Stuart, 1942, 1948a, 1963
Plectrohyla sagorum
Duellman, 1970; Duellman and Campbell, 1992; Hartweg, 1941; Johnson, 1989; Stuart, 1963
Plectrohyla tecunumani
Duellman and Campbell, 1984, 1992
Plectrohyla teuchestes
Duellman and Campbell, 1992
Pternohyla dentata
Duellman, 1970; Smith, 1957; Trueb, 1970
Pternohyla fodiens
Davis and Dixon, 1957; Duellman, 1961b, 1965b, 1970; Hardy and McDiarmid, 1969; Kellogg, 1932; Smith and Taylor, 1948; Stebbins, 1985; Taylor, 1938; Trueb, 1970; Webb, 1984
Ptychohyla erythromma
Campbell and Smith, 1992; Davis and Dixon, 1964; Duellman, 1970; Rand, 1957; Smith and Taylor, 1948; Taylor, 1937
Ptychohyla euthysanota
Campbell and Vannini, 1989a, 1989b; Duellman, 1961a, 1963b, 1970; Johnson, 1989; Lynch and Smith, 1966a; Mertens, 1952; Stuart, 1963; Taylor, 1942a, 1949c
Ptychohyla hypomykter
McCranie and Wilson, 1993b; Schmidt, 1936b; Stuart, 1963
Ptychohyla legleri
Duellman, 1970; Savage, 1968a, 1974b; Savage and Heyer, 1969
Ptychohyla leonhardschultzei
Duellman, 1960b, 1961a, 1963b, 1970
Ptychohyla macrotympanum
Campbell and Smith, 1992; Duellman, 1961a, 1963b, 1970; Tanner, 1957
Ptychohyla panchoi
Campbell and Smith, 1992; Duellman and Campbell, 1982
Ptychohyla salvadorensis
Campbell and Smith, 1992; Duellman, 1970; Mertens, 1952; Meyers and Wilson, 1971a
Ptychohyla sanctaecrucis
Campbell and Smith, 1992
Ptychohyla spinipollex
McCranie and Wilson, 1993b; Schmidt, 1936b; Wilson and McCranie, 1989
Scinax boulengeri
Campbell and Telford, 1971; Duellman, 1970; Leon, 1969; Myers and Rand, 1969; Noble, 1918; Savage and Heyer, 1969; Scott, 1983f; Taylor, 1952b; Villa, 1972
Scinax elaeochroa
Duellman, 1966b, 1970; Leon, 1969; Savage, 1974b; Savage and Heyer, 1969; Taylor, 1952b, 1958b; Villa, 1972
Scinax rostrata
Duellman, 1970; Leon, 1969
Scinax rubra
Campbell and Telford, 1971; Duellman, 1970; Leon, 1969; Rand and Myers, 1990

Scinax staufferi
Campbell and Vannini, 1989a; Duellman, 1960, 1963, 1970; Frost, 1985; Henderson and Hoevers, 1975; Johnson, 1989; Johnson et al., 1977; Lee, 1980; Leon, 1969; Mertens, 1952; Neill and Allen, 1959a; Rand, 1957; Schmidt, 1941; Stuart, 1935, 1948a, 1948; 1963; Villa, 1972
Smilisca baudinii
Campbell and Vannini, 1989a, 1989b; Davis and Smith, 1953; Duellman, 1961b, 1963a, 1965b, 1970; Duellman and Trueb, 1966; Frost, 1985; Gaige et al., 1937; Hardy and McDiarmid, 1969; Henderson and Hoevers, 1975; Johnson, 1989; Lee, 1980; Martin 1958; Mertens, 1952; Meyer and Wilson, 1971a; Rand, 1957; Savage, 1974b; Savage and Heyer, 1969; Schmidt, 1941; Schmidt and Stuart, 1941; Smith, 1938; Smith and Taylor, 1948; Stuart, 1934, 1948, 1963; Taylor, 1938; Villa, 1972; Webb, 1984; Webb and Baker, 1969
Smilisca cyanosticta
Duellman, 1968a, 1970; Duellman and Trueb, 1966; Johnson, 1989; Lee, 1980; Perez-Higareda et al., 1987; Shannon and Werler, 1955; Stuart, 1963
Smilisca phaeota
Duellman, 1968a, 1970; Duellman and Trueb, 1966; Gaige et al., 1937; Martínez-Cortés, 1984; Meyer and Wilson, 1971a; Myers and Rand, 1969; Savage and Heyer, 1969; Villa, 1972
Smilisca puma
Duellman, 1968a, 1970; Duellman and Trueb, 1966; Savage, 1974b; Savage and Heyer, 1969; Villa, 1972
Smilisca sila
Duellman, 1968a, 1970; Duellman and Trueb, 1966; Myers and Rand, 1969; Savage and Heyer, 1969
Smilisca sordida
Duellman, 1968a, 1970; Duellman and Trueb, 1966; Myers and Duellman, 1982; Savage and Heyer, 1969; Savage, 1974b; Wilson and Cruz, 1986
Triprion petasatus
Cope, 1866; Duellman, 1965a, 1970; Duellman and Klaas, 1964; Henderson and Hoevers, 1975; Hoevers and Henderson, 1974; Kellogg, 1932; Lee, 1980; McCoy et al., 1986; Smith, 1938; Stuart, 1935, 1963; Taylor, 1942c; Trueb, 1970; Wilson et al., 1986b
Triprion spatulatus
Duellman, 1960a, 1961b, 1965b, 1970; Hardy and McDiarmid, 1969; Taylor, 1938, 1942c; Trueb, 1970; Webb, 1984.

ANURA: LEPTODACTYLIDAE
Eleutherodactylus achatinus
Breder, 1946; Dunn, 1934; Lynch and Myers, 1983
Eleutherodactylus adamastus
Campbell, 1994b
Eleutherodactylus alfredi
Campbell and Vannini, 1989a; Campbell et al., 1989; Duellman, 1960a, 1965a; Johnson, 1989; Lee, 1980
Eleutherodactylus altae
Dunn, 1942; Hayes et al., 1989; Savage, 1974b; Taylor, 1952b
Eleutherodactylus ancianao
Savage et al., 1988
Eleutherodactylus andi
Hayes et al., 1989; Savage, 1974a, 1974b, 1980
Eleutherodactylus angelicus
Savage, 1975

Eleutherodactylus antillensis
    Auth, 1994; Frost, 1985
Eleutherodactylus aphanus
    Campbell, 1994a
Eleutherodactylus augusti
    Anderson and Lidicker, 1963; Davis and Dixon, 1957, 1964;
Duellman, 1961b, 1965b; Hardy and McDiarmid, 1969; Hartweg
and Oliver, 1940; Lynch, 1986; Stebbins, 1985;Wilson and Mc-
Cranie, 1979; Webb, 1984; Webb and Baker, 1962; Zweifel, 1967
Eleutherodactylus aurilegulus
    Savage et al., 1988
Eleutherodactylus azueroensis
    Miyamoto, 1983b; Savage, 1975
Eleutherodactylus batrachylus
    Flores-Villela, 1993; Taylor, 1940d
Eleutherodactylus berkenbuschi
    Brocchi, 1881–1883; Duellman, 1960a; Miyamoto, 1983b;
Savage, 1975; Savage and DeWeese, 1979; Taylor, 1952b
Eleutherodactylus biporcatus
    Brocchi, 1881–1883; Dunn, 1931a, 1931b; Hayes et al.,
1989; Lynch, 1975; Martínez-Cortés, 1984; Meyer and Wilson,
1971a; Myers and Rand, 1969; Rand and Myers, 1990; Savage,
1974b; Villa, 1972
Eleutherodactylus bocourti
    Brocchi, 1881–1883; Campbell and Vannini, 1989a; Stu-
art, 1948a, 1963
Eleutherodactylus bransfordii
    Hayes et al., 1989; Martínez-Cortés, 1984; Miyamoto,
1983a; Noble, 1918; Rand and Myers, 1990; Savage, 1974b,
1981b; Savage and Emerson, 1970; Scott, 1983d; Taylor, 1952b;
Villa, 1972
Eleutherodactylus brocchi
    Brocchi, 1881–1883; Campbell and Vannini, 1989a; Lynch,
1965a; Miyamoto, 1983b; Savage, 1975; Stuart, 1948a, 1963
Eleutherodactylus bufoniformis
    Breder, 1946; Cochran and Goin, 1970; Dunn, 1931a,
1931b; Lynch, 1975; Myers and Rand, 1969; Rand and Myers,
1990; Taylor, 1952b
Eleutherodactylus caryophyllaceus
    Barbour, 1928; Hayes et al., 1989; Martínez-Cortés, 1984;
Savage and DeWeese, 1981; Taylor, 1952b
Eleutherodactylus cerasinus
    Dunn, 1931d; Lynch, 1964; Martínez-Cortés, 1984; Rand
and Myers, 1990; Savage, 1981a, 1982; Taylor, 1952b; Villa, 1972
Eleutherodactylus chac
    Henderson and Hoevers, 1975; Savage, 1987; Schmidt, 1941
Eleutherodactylus chrysozetetes
    McCranie et al., 1989
Eleutherodactylus crassidigitus
    Hayes et al., 1989; Lynch and Myers, 1983; Martínez-
Cortés, 1984; Rand and Myers, 1990; Taylor, 1952b
Eleutherodactylus cruentus
    Brocchi, 1881–1883; Cochran and Goin, 1970; Frost, 1985;
Hayes et al., 1989; Lynch, 1964; Martínez-Cortés, 1984; Myers
and Rand, 1969; Savage, 1974b, 1981a; Taylor, 1952b
Eleutherodactylus cruzi
    McCranie et al., 1989
Eleutherodactylus cuaquero
    Hayes et al., 1989; Savage, 1980
Eleutherodactylus daryi
    Campbell and Vannini, 1989a; Ford and Savage, 1984
Eleutherodactylus decoratus
    Campbell et al., 1989; Lynch, 1967c; Taylor, 1942a
Eleutherodactylus diastema
    Breder, 1946; Dunn, 1931a, 1931b; Hayes et al., 1989; Mar-
tínez-Cortés, 1984; Myers and Rand, 1969; Rand and Myers,
1990; Scott, 1983e; Taylor, 1952b; Villa, 1972
Eleutherodactylus emcelae
    Lynch, 1985
Eleutherodactylus escoces
    Miyamoto, 1983b; Savage, 1974b, 1975
Eleutherodactylus fitzingeri
    Breder, 1946; Dunn, 1931b, 1931d; Hayes et al., 1989;
Lynch and Myers, 1983; Myers and Rand, 1969; Rand and My-
ers, 1990; Savage, 1974a, 1980; Taylor, 1952b; Villa, 1972

Eleutherodactylus fleischmanni
    Dunn, 1940a; Miyamoto, 1983b; Savage, 1975; Taylor,
1952b, 1958b
Eleutherodactylus gaigei
    Breder, 1946; Dunn, 1931c; Frost, 1985; Lynch, 1980b,
1986; Myers and Rand, 1969; Rand and Myers, 1990; Taylor,
1952b
Eleutherodactylus glaucus
    Campbell et al., 1989a; Johnson, 1984, 1989; Lynch, 1967a
Eleutherodactylus gollmeri
    Martínez-Cortés, 1984; Savage, 1974b, 1987; Taylor, 1952b
Eleutherodactylus greggi
    Bumzahem, 1955; Ford and Savage, 1984; Johnson, 1989;
Lynch, 1965d; Savage, 1975; Stuart, 1963
Eleutherodactylus guerreroensis
    Campbell et al., 1989; Lynch, 1967b
Eleutherodactylus hobartsmithi
    Duellman, 1961b, 1965b; Flores-Villela, 1993; Hardy and
McDiarmid, 1969; Lynch, 1965e; Taylor, 1936, 1940d; Webb,
1984
Eleutherodactylus hylaeformis
    Hayes et al., 1989; Savage, 1974b; Savage and Villa, 1986;
Taylor, 1952b
Eleutherodactylus johnstonei
    Auth, 1994; Ibañez and Rand, 1990.
Eleutherodactylus jota
    Lynch, 1980a
Eleutherodactylus laticeps
    Firschein, 1951; Frost, 1985; Henderson and Hoevers,
1975; Johnson, 1989; Lynch and Fritts, 1965; Meyer and Wil-
son, 1971a; Neill, 1965; Savage, 1987; Schmidt, 1941
Eleutherodactylus latidiscus
    Auth, 1994
Eleutherodactylus lineatus
    Brocchi, 1881–1883; Campbell and Vannini, 1989a, 1989b;
Frost, 1985; Johnson, 1989; Savage, 1987; Stuart, 1941, 1943b,
1948a, 1963
Eleutherodactylus longirostris
    Dunn, 1931a, 1931b; Lynch and Myers, 1983; Myers and
Rand, 1969
Eleutherodactylus matudai
    Johnson, 1989; Lynch, 1965a; Savage, 1975; Stuart, 1963;
Taylor, 1941b
Eleutherodactylus megalotympanum
    Campbell et al., 1989; Shannon and Werler, 1955b
Eleutherodactylus melanostictus
    Brocchi, 1881–1883; Cope, 1887; Hayes et al., 1989; Mar-
tínez-Cortés, 1984; Savage, 1974b; Savage and DeWeese, 1981;
Taylor, 1952b
Eleutherodactylus merendonensis
    Meyer and Wilson, 1971a, 1971b; Miyamoto, 1983b; Sav-
age, 1975; Schmidt, 1933a
Eleutherodactylus mexicanus
    Baker et al., 1971; Bogert, 1968; Davis and Dixon, 1957;
Johnson, 1989; Lynch, 1970b; Savage, 1987; Smith and Laufe,
1945; Taylor, 1942
Eleutherodactylus milesi
    Campbell, 1994b; Lynch, 1965a; McCranie et al, 1989;
Meyer and Wilson, 1971a; O'Shea, 1989; Savage, 1975; Schmidt,
1933a
Eleutherodactylus mimus
    Savage, 1974b, 1975, 1987; Meyer and Wilson, 1971a;
Taylor, 1955; Villa, 1972
Eleutherodactylus monnichorum
    Dunn, 1940a; Taylor, 1958b
Eleutherodactylus moro
    Savage, 1965, 1968c, 1974b
Eleutherodactylus museosus
    Ibañez et al., 1994
Eleutherodactylus noblei
    Barbour and Dunn, 1921; Savage, 1974b, 1975, 1987; Tay-
lor, 1952b; Villa, 1972; Wilson and McCranie, 1991
Eleutherodactylus occidentalis
    Hardy and McDiarmid, 1969; Lynch, 1970b, 1986; Mc-
Cranie and Wilson, 1987; Taylor, 1941a; Webb, 1984; Webb and

Baker, 1962
Eleutherodactylus omiltemanus
Davis and Dixon, 1964; Ford and Savage, 1984; Lynch, 1970b; Martínez-Cortés, 1984
Eleutherodactylus pardalis
Barbour, 1928; Savage and Villa, 1986; Taylor, 1952b
Eleutherodactylus planirostris
Flores-Villela, 1993
Eleutherodactylus podiciferus
Hayes et al., 1989; Martínez-Cortés, 1984; Savage, 1974b; Savage and Villa, 1986; Taylor, 1952b
Eleutherodactylus polymniae
Campbell et al., 1989
Eleutherodactylus pozo
Johnson and Savage, 1995; Johnson et al., 1977 [*"brocchi,"* Chiapas]
Eleutherodactylus psephosypharus
Campbell et al., 1994
Eleutherodactylus punctariolus
Dunn, 1940a; Martínez-Cortés, 1984; Miyamoto, 1983b; Savage, 1975; Savage and Villa, 1986
Eleutherodactylus pygmaeus
Davis and Dixon, 1964; Duellman, 1960a, 1961b, 1965b; Flores-Villela, 1993; Johnson, 1989; Lynch, 1965e; Stuart, 1963; Taylor, 1936, 1940d; Webb and Baker, 1969
Eleutherodactylus raniformis
Lynch and Myers, 1983
Eleutherodactylus rayo
Martínez-Cortés, 1984; Savage and De Weese, 1979
Eleutherodactylus rhodopis
Campbell and Vannini, 1989a, 1989b; Duellman, 1960a; Flores-Villela, 1993; Johnson, 1989; Johnson et al., 1977; Lynch and Fritts, 1965; Mertens, 1952; Rand, 1957; Savage, 1987; Shannon and Werler, 1955b; Smith, 1959; Stuart, 1948a, 1963; Villa et al., 1988
Eleutherodactylus ridens
Barbour, 1928; Hayes et al., 1989; Lynch, 1980b; Meyer and Wilson, 1971a; Noble, 1918; Rand and Myers, 1990; Savage, 1981a; Taylor, 1952b; Villa, 1972; Wilson and McCranie, 1991
Eleutherodactylus rostralis
Duellman, 1963a; Savage, 1987; Werner, 1896
Eleutherodactylus rugulosus
Adler and Dennis, 1972; Campbell and Savage, ms.; Davis and Dixon, 1964; Duellman, 1960a; Johnson, 1989; Lynch, 1965a; Savage, 1975; Webb and Baker, 1969
Eleutherodactylus saltator
Adler, 1965; Adler and Dennis, 1972; Lynch, 1970b; Taylor, 1941b
Eleutherodactylus sandersoni
Campbell and Savage, in prep.; Neill and Allen, 1961; Schmidt, 1941
Eleutherodactylus sartori
Lynch, 1965e; Johnson, 1989; Taylor, 1942b
Eleutherodactylus silvicola
Campbell et al., 1989; Lynch 1967b
Eleutherodactylus spatulatus
Bogert, 1969; Campbell et al., 1989; Lynch, 1965b, 1965c, 1967c, 1970b
Eleutherodactylus stadelmani
Campbell, 1994b; Schmidt, 1936b
Eleutherodactylus stejnegerianus
Frost, 1985; Miyamoto, 1983a; Taylor, 1952b
Eleutherodactylus stuarti
Campbell and Vannini, 1989a, 1989b; Campbell et al., 1989; Johnson, 1984, 1989; Lynch, 1967a, 1970b
Eleutherodactylus taeniatus
Breder, 1946; Evans, 1947; Lynch, 1980b; Myers and Rand, 1969; Rand and Myers, 1990
Eleutherodactylus talamancae
Dunn, 1931c; Lynch and Myers, 1983; Martínez-Cortés, 1984; Taylor, 1952b; Villa, 1972
Eleutherodactylus tarahumaraensis
Lynch, 1986; Tanner, 1989; Taylor, 1940d; Webb, 1984
Eleutherodactylus taurus
Miyamoto, 1983b; Savage, 1974b, 1975; Taylor, 1958b

Eleutherodactylus taylori
Campbell et al., 1989; Johnson, 1989; Lynch, 1966
Eleutherodactylus trachydermus
Campbell, 1994
Eleutherodactylus uno
Savage, 1984
Eleutherodactylus vocalis
Duellman, 1961b; Hardy and McDiarmid, 1969; Lynch, 1965a; Miyamoto, 1983b; Savage, 1975; Taylor, 1940b; Webb, 1960, 1984; Webb and Baker, 1962
Eleutherodactylus vocator
Rand and Myers, 1990; Savage, 1974b; Taylor, 1955; Villa et al., 1988
Eleutherodactylus xucanebi
Campbell et al., 1989; Campbell and Vannini, 1989a; Stuart, 1941, 1948a, 1963
Eleutherodactylus yucatanensis
Campbell et al., 1989; Lee, 1980; Lynch, 1965b
Eleutherodactylus sp. A
Savage and Villa, 1986 ["sp. 1 Savage," p. 10]; Scott et al., 1983 ["sp. 1," p. 368]
Eleutherodactylus sp. B
Campbell and Savage, in prep. [Atlantic lowlands, *rugulosus* group]
Eleutherodactylus sp. C
Campbell and Savage, in prep. [Zacapa, *rugulosus* group]
Eleutherodactylus sp. D
Campbell and Savage, in prep. [Higher Pacific versant, *rugulosus* group]
Eleutherodactylus sp. E
Campbell and Savage, in prep.; Campbell and Vannini, 1989b [Lower Pacific Versant, *rugulosus* group]
Leptodactylus bolivianus
Barbour, 1923; Breder, 1946; Busack, 1966; Dunn 1931a, 1931b, 1931d; Evans, 1947; Heyer, 1970a; Myers and Rand, 1969; Schmidt, 1933b; Taylor, 1952b, 1954
Leptodactylus fuscus
Heyer, 1978
Leptodactylus labialis
Breder, 1946; Brocchi, 1881–1883; Busack, 1966; Davis and Dixon, 1964; Dubois and Heyer, 1992; Duellman, 1960a, 1961b, 1965a, 1965b; Dundee et al., 1986; Hartweg and Oliver, 1940; Henderson and Hoevers, 1975; Heyer, 1970a, 1971, 1978, 1979; Johnson, 1989; Lee, 1980; Martínez-Cortés, 1984; Mertens, 1952; Neill and Allen, 1961; Noble, 1918; Rand, 1957; Stuart, 1934, 1935, 1948a, 1954b, 1963; Werler and Smith, 1952
Leptodactylus melanonotus
Breder, 1946; Brocchi, 1881–1883; Campbell and Vannini, 1989a; Davis and Dixon, 1964; Duellman, 1960a, 1961b, 1965a; Gaige et al., 1937; Hardy and McDiarmid, 1969; Heyer, 1970a, 1970b; Johnson, 1989; Lee, 1980; Mertens, 1952; Neill and Allen, 1959a, 1961; Noble, 1918; Rand, 1957; Schmidt, 1941; Stuart, 1935, 1937, 1948a, 1954b, 1963; Taylor, 1938, 1952b; Villa, 1972; Webb, 1984; Webb and Baker, 1969; Werler and Smith, 1952
Leptodactylus pentadactylus
Barbour, 1923; Breder, 1946; Gaige et al., 1937; Günther, 1885–1902; Heyer, 1970a, 1979; Meyer and Wilson, 1971a; Myers and Rand, 1969; Noble, 1918; Schmidt, 1933a; Scott, 1983g; Taylor, 1952b; Villa, 1972
Leptodactylus poecilochilus
Breder, 1946; Busack, 1966; Dunn, 1940a; Heyer, 1970a, 1978
Leptodactylus silvanimbus
McCranie et al., 1980; Wilson et al., 1986b
Physalaemus pustulosus
Breder, 1946; Busack, 1966; Cochran and Goin, 1970; Duellman, 1960a, 1965a; Hartweg and Oliver, 1940; Henderson and Hoevers, 1975; Johnson, 1989; Lee, 1980; Mertens, 1952; Rand, 1957, 1983; Ryan, 1985; Schmidt, 1933b; Smith and Taylor, 1948; Stuart, 1954b, 1963; Villa, 1972; Werler and Smith, 1952
Pleurodema brachyops
Busack, 1966; Cochran and Goin, 1970; Duellman and Veloso, 1977

Syrrhophus albolabris
Davis and Dixon, 1955, 1964; Dixon, 1957; Flores-Villela, 1993; Taylor, 1943a

Syrrhophus angustidigitorum
Dixon, 1957; Duellman, 1961b, 1965b; Flores-Villela, 1993; Taylor, 1940d

Syrrhophus cystignathoides
Flores-Villela, 1993; Lynch, 1970a; Martin 1955a, 1955b, 1958

Syrrhophus dennisi
Lynch, 1970a

Syrrhophus dilatus
Davis and Dixon, 1955, 1964; Dixon, 1957

Syrrhophus guttilatus
Lynch, 1970a

Syrrhophus interorbitalis
Hardy and McDiarmid, 1969; Langebartel and Shannon, 1956

Syrrhophus leprus
Duellman, 1960a, 1963a; Henderson and Hoevers, 1975; Johnson, 1989; Lee, 1980; Lynch, 1970a; Mendelson, 1990; Neill, 1965; Neill and Allen, 1961; Shannon and Werler, 1955b; Stuart, 1963; Werler and Smith, 1952

Syrrhophus longipes
Lynch, 1970a; Martin 1955a, 1955b, 1958

Syrrhophus maurus
Davis and Dixon, 1955; Dixon, 1957; Duellman, 1961b, 1965b; Flores-Villela, 1993; Hedges, 1989

Syrrhophus modestus
Davis and Dixon, 1957; Hardy and McDiarmid, 1969; Lynch, 1970a; Taylor, 1942b

Syrrhophus nitidus
Davis and Dixon, 1955, 1964; Davis and Smith, 1953; Dixon, 1957; Duellman, 1961b, 1965b; Hardy and McDiarmid, 1969; McCranie and Wilson, 1987; Webb, 1984; Webb and Baker, 1962, 1969

Syrrhophus nivicolimae
Dixon and Webb, 1966; Lynch, 1970a

Syrrhophus pallidus
Lynch, 1970a; Peterson and Smith, 1983; Webb and Baker, 1962

Syrrhophus pipilans
Davis and Dixon, 1964; Duellman, 1960a; Johnson, 1989; Lynch, 1970a; Stuart, 1963; Taylor, 1940e; Webb and Baker, 1969

Syrrhophus rubrimaculatus
Campbell and Vannini, 1989; Johnson, 1989; Lynch, 1970a; Taylor and Smith, 1945

Syrrhophus rufescens
Duellman, 1961b, 1965b; Duellman and Dixon, 1959

Syrrhophus saxatilis
Flores-Villela, 1993; Hardy and McDiarmid, 1969; Webb, 1962, 1984; Webb and Baker, 1962

Syrrhophus syristes
Hoyt, 1965

Syrrhophus teretistes
Lynch, 1970a; Webb, 1984

Syrrhophus verrucipes
Lynch, 1970a

ANURA: MICROHYLIDAE

Chiasmocleis panamensis
Dunn et al., 1948; Nelson, 1972e; Rand and Myers, 1990

Elachistocleis ovalis
Cochran and Goin, 1970; Dunn, 1931a, 1931b; Frost, 1985; Nelson, 1972e

Gastrophryne elegans
Baker et al., 1971; Duellman, 1965a; Henderson and Hoevers, 1975; Johnson, 1989; Lee, 1980; McCoy et al., 1986; Meyer and Wilson, 1971a; Nelson, 1972a, 1972d; Stuart, 1934, 1963

Gastrophryne olivacea
Hardy and McDiarmid, 1969; Nelson, 1972b, 1972d; Stebbins, 1985; Taylor, 1938; Webb, 1960, 1984

Gastrophryne pictiventris
Nelson, 1972d, 1973a; Taylor, 1952b; Villa, 1972

Gastrophryne usta
Davis and Dixon, 1964; Duellman, 1960a; Johnson, 1989; Hardy and McDiarmid, 1969; Hartweg and Oliver, 1940; Lynch, 1965c; Nelson, 1972c, 1972d; Taylor and Smith, 1945; Stuart, 1963; Webb, 1984

Hypopachus barberi
Johnson, 1989; Nelson, 1973b, 1974; Schmidt, 1939; Stuart, 1948a, 1951, 1954b, 1963; Wilson et al., 1986b

Hypopachus variolosus
Davis and Dixon, 1964; Davis and Smith, 1953; Duellman, 1961b, 1963a, 1965a, 1965b; Dundee et al., 1986; Hardy and McDiarmid, 1969; Henderson and Hoevers, 1975; Johnson, 1989; Lee, 1980; Meerman, 1993; Mertens, 1952; Meyer and Wilson, 1971a; Nelson, 1974; Schmidt, 1939; Smith, 1938; Stuart, 1935, 1954a, 1963; Taylor, 1952b; Villa, 1972; Webb, 1984

Nelsonophryne aterrima
Cochran and Goin, 1970; Frost, 1985; Hayes et al., 1989; Martínez-Cortés, 1984; Taylor, 1952b

Relictivomer pearsei
Busack, 1966; Cochran and Goin, 1970; Frost, 1985; Nelson and Nickerson, 1966; Ruthven, 1914

ANURA: PELOBATIDAE

Scaphiopus couchii
Davis and Dixon, 1957; Hardy and McDiarmid, 1969; Kellogg, 1932; Smith and Taylor, 1948; Stebbins, 1985; Tanner, 1989; Taylor, 1938; Webb, 1984

Spea multiplicata
Anderson and Lidicker, 1963; Brocchi, 1881–1883; Davis and Dixon, 1957, 1964; Davis and Smith, 1953; Duellman, 1961b, 1965b; Kellogg, 1932; McCranie and Wilson, 1987; Smith and Taylor, 1948; Stebbins, 1985; Tanner, 1989; Webb, 1984; Wilson and McCranie, 1979

ANURA: PIPIDAE

Pipa myersi
Heatwole, 1963; Trueb, 1984; Trueb and Cannatella, 1986

ANURA: RANIDAE

Rana berlandieri
Anderson and Lidicker, 1963; Dundee et al., 1986; D. Frost, 1985; J. Frost, 1982; Henderson and Hoevers, 1975; Hillis, 1985, 1988; Hillis and Davis, 1986; Hillis et al., 1983; Lee, 1980; Martin, 1955b, 1958; Meerman, 1993; Schmidt, 1941; Stuart, 1934, 1935, 1948a, 1963; Wilson and McCranie, 1979; Villa, 1972

Rana brownorum
Frost, 1985; Hillis, 1981; Sanders, 1973

Rana catesbeiana
Frost, 1985

Rana chiricahuensis
Anderson and Lidicker, 1963; Hillis, 1988; Hillis et al., 1983; McCranie and Wilson, 1987; Platz and Mecham, 1979; Stebbins, 1985

Rana dunni
Duellman, 1961b, 1965b; Hillis, 1988; Hillis et al., 1983; Zweifel, 1957

Rana forreri
J. Frost, 1982; Günther, 1885–1902; Hillis, 1981, 1985, 1988; Hillis et al., 1983; Mertens, 1952; Stuart, 1963; Taylor, 1952b; Villa, 1972

Rana johni
Blair, 1965; Hillis et al., 1984; Webb, 1988

Rana juliani
Henderson and Hoevers, 1975; Hillis and de Sá, 1988; Lee, 1976, 1980

Rana maculata
Campbell and Vannini, 1989a; Günther, 1885–1902; Johnson, 1989; Hillis, 1985; Hillis and Davis, 1986; Hillis and de Sá, 1988; Mertens, 1952; Meyer and Wilson, 1971a; Rand, 1957; Schmidt and Stuart, 1941; Smith, 1959; Smith and Brandon, 1968; Stuart, 1948a, 1951, 1954b, 1963; Villa, 1972, 1979; Zweifel, 1954

Rana magnaocularis
Frost and Bagnara, 1976; Hillis, 1988; Hillis and Davis,

1986; Hillis et al., 1883; McCranie and Wilson, 1987

Rana montezumae
Anderson and Lidicker, 1963; Duellman, 1961b, 1965b; Günther, 1885–1902; Hillis and Davis, 1986; Hillis et al., 1983; Wilson and McCranie, 1979; Zweifel, 1957

Rana neovolcanica
Hillis, 1988; Hillis and Frost, 1985; Hillis et al., 1983 ["Chapala form"]

Rana omiltemana
Günther, 1885–1902; Davis and Dixon, 1964; Hillis, 1988

Rana pueblae
Hillis et al., 1984; Webb, 1988; Zweifel, 1955

Rana pustulosa
Duellman, 1961b, 1965b; Hillis and Davis, 1986; Hillis et al., 1983, 1984; McCranie and Wilson, 1987; Webb, 1984; Zweifel, 1954

Rana sierramadrensis
Davis and Dixon, 1964; Hillis and de Sá, 1988; Taylor, 1939b; Webb, 1978; Webb and Baker, 1969; Zweifel, 1954

Rana spectabilis
Hillis, 1988; Hillis and Davis, 1986; Hillis et al., 1983 ["Hidalgo form"]; Hillis and Frost, 1985

Rana tarahumarae
Hillis and Davis, 1986; Hillis et al., 1984; McCranie and Wilson, 1987; Stebbins, 1985; Tanner, 1989

Rana taylori
Hillis, 1985, 1988; Noble, 1918; Savage, 1974b; Smith, 1959; Taylor, 1952b; Villa et al., 1988

Rana tlaloci
Hillis, 1988; Hillis and Frost, 1985; Hillis et al., 1983 ["Xochimilco form"]

Rana trilobata
Duellman, 1961b, 1965b; Hillis, 1988; Hillis et al., 1983 ["*megapoda*"]; Taylor, 1942b; Zweifel, 1957

Rana vaillanti
Barbour, 1923; Brocchi, 1881–1883; Campbell and Vannini, 1989a; Duellman, 1960a, 1963a, 1965a; Dunn, 1922b, 1931a, 1931b; Gaige et al., 1937; Günther, 1885–1902; Henderson and Hoevers, 1975; Hillis and de Sá, 1988; Hillis and Davis, 1986; Johnson, 1989; Lee, 1980; Meyer and Wilson, 1971a; Myers and Rand, 1969; Neill and Allen, 1959a; Noble, 1918; Robinson, 1983; Schmidt, 1933b, 1941; Stuart, 1935, 1937, 1948a, 1954b, 1963; Taylor, 1952b, 1954; Villa, 1972; Zweifel, 1954

Rana vibicaria
Dunn, 1922b; Hillis and Davis, 1986; Hillis and De Sá, 1988; Martínez-Cortés, 1984; Noble, 1918; Savage, 1974b; Taylor, 1952b; Zweifel, 1964

Rana warschewitschii
Barbour, 1925; Dunn, 1931d; Gaige, et al., 1937; Hillis and De Sá, 1988; Martínez-Cortés, 1984; Meyer and Wilson, 1971a; Taylor, 1952b

Rana yavapaiensis
Platz and Frost, 1984; Stebbins, 1985

Rana zweifeli
Davis and Dixon, 1964; Davis and Smith, 1953; Hillis et al., 1984

Rana sp. A
Hillis, 1988; Hillis et al., 1983 [Beta division; "Atenquique long form"]

Rana sp. B
Hillis, 1988; Hillis et al., 1983 [Beta division; "Atenquique short form"]

Rana sp. C
Hillis, 1988 [Nuclear CA highlands]

Rana sp. D
Hillis, 1988 [Isthmian CA highlands]

Rana sp. E
Hillis, 1988 [Lowland Caribbean Panama]

ANURA: RHINOPHRYNIDAE

Rhinophrynus dorsalis
Campbell and Vannini, 1989a; Duellman, 1960a, 1961b, 1965a, 1971; Foster and McDiarmid, 1983; Fouquette, 1969; Hartweg and Oliver, 1940; Henderson and Hoevers, 1975;

Johnson, 1989; Kellogg, 1932; Lee, 1980; McCoy, 1966; Mertens, 1952; Meyer and Wilson, 1971a; Neill, 1965; Nelson and Hoyt, 1961; Nelson and Nickerson, 1966; Smith and Taylor, 1948; Stuart, 1934, 1935, 1961, 1963; Villa, 1972, 1984a; Webb and Baker, 1969

CAUDATA: AMBYSTOMATIDAE

Ambystoma altamirani
Brandon and Altig, 1973; Campbell and Simmons, 1962; Dunn, 1928; Reilly and Brandon, 1994; Shaffer, 1993; Smith and Necker, 1943; Smith and Taylor, 1948; Taylor, 1939a

Ambystoma amblycephalum
Duellman, 1961b, 1965b; Shaffer, 1993; Taylor, 1940a

Ambystoma andersoni
Frost, 1985; Krebs and Brandon, 1984; Shaffer, 1993

Ambystoma bombypellum
Frost, 1985; Taylor, 1940a

Ambystoma dumerilii
Brandon, 1970a, 1970b; Duellman, 1961b, 1965b; Frost, 1985; Shaffer, 1993; Smith, 1939; Smith and Necker, 1943; Smith and Taylor, 1948; Taylor, 1940a

Ambystoma flavipiperatum
Campbell, unpubl.; Dixon, 1963; Frost, 1985; Shaffer, 1993

Ambystoma granulosum
Frost, 1985; Shaffer, 1993; Taylor, 1944a

Ambystoma leorae
Reilly and Brandon, 1994; Smith and Taylor, 1948; Taylor, 1943a

Ambystoma lermaense
Frost, 1985; Shaffer, 1993; Smith and Taylor, 1948; Taylor, 1940a

Ambystoma mexicanum
Frost, 1985; Smith, 1939; Smith and Taylor, 1948; Taylor, 1939a, 1940a

Ambystoma ordinarium
Duellman, 1961b, 1965b; Frost, 1985; Shaffer, 1993; Taylor, 1940a

Ambystoma rivularis
Brandon and Altig, 1973; Reilly and Brandon, 1994; Shaffer, 1993; Smith and Taylor, 1948; Taylor, 1940c

Ambystoma rosaceum
Anderson, 1961, 1978; Anderson and Webb, 1978; Frost, 1985; Shaffer, 1983, 1993; Tanner, 1989; Taylor, 1941e; Webb, 1984; Webb and Baker, 1962

Ambystoma subsalsum
Brandon et al., 1981; Smith and Laufe, 1940; Smith and Taylor, 1948; Taylor, 1943b

Ambystoma taylori
Brandon et al., 1981; Shaffer, 1993

Ambystoma tigrinum
Brandon et al., 1981; Duellman, 1961b, 1965b; McCranie and Wilson, 1987; Shaffer, 1983, 1993; Stebbins, 1985; Webb, 1984

Ambystoma velasci
Frost, 1985; Smith and Necker, 1943

Ambystoma zempoalensis
Reilly and Brandon, 1994; Smith and Taylor, 1948; Taylor and Smith, 1945

CAUDATA: PLETHODONTIDAE

Bolitoglossa alvaradoi
Savage, 1974b; Taylor, 1954; Wake, 1987; Wake and Lynch, 1976

Bolitoglossa arborescandens
Savage, 1974b; Taylor, 1954; Wake and Lynch, 1976

Bolitoglossa biseriata
Brame and Wake, 1963a; Tanner, 1962; Wake and Lynch, 1976

Bolitoglossa carri
McCranie and Wilson, 1993

Bolitoglossa celaque
McCranie and Wilson, 1993

Bolitoglossa cerroensis
Brame, 1965; Savage, 1974b; Taylor, 1952a; Wake, 1987;

Wake and Lynch, 1976
Bolitoglossa colonnea
    Breder, 1946; Dunn, 1924b, 1926; Taylor, 1952a; Wake, 1987; Wake and Brame, 1963
Bolitoglossa compacta
    Wake et al., 1973
Bolitoglossa conanti
    McCranie and Wilson, 1993
Bolitoglossa cuchumatana
    Elias, 1984; Larson, 1983a; Wake and Brame, 1969; Wake and Lynch, 1976; Stuart, 1943a, 1943b, 1963
Bolitoglossa cuna
    Wake and Lynch, 1976; Wake et al., 1973
Bolitoglossa diaphora
    McCranie and Wilson, 1995b
Bolitoglossa diminuta
    Bolaños et al., 1987; Frost, 1985; Papenfuss and Wake, 1987; Robinson, 1976; Wake, 1987; Wake and Elias, 1983
Bolitoglossa dofleini
    Duellman, 1963a; Larson, 1983a, 1983b; Lee, 1980; McCoy, 1991; Meyer and Wilson, 1971a; Stuart, 1943a, 1943b, 1948b, 1963; Wake et al., 1992; Wake and Lynch, 1976
Bolitoglossa dunni
    Campbell, unpub. data; Larson, 1983a, 1983b; McCranie and Wilson, 1993; Schmidt, 1933a; Wake and Lynch, 1976
Bolitoglossa engelhardti
    Campbell and Vannini, 1989a; Johnson, 1989; Larson, 1983a, 1983b; Mertens, 1952; Rand, 1957; Schmidt, 1936a; Stuart, 1963; Wake and Lynch, 1976; Wake et al., 1992
Bolitoglossa epimela
    Savage, 1974b; Wake, 1987; Wake and Brame, 1963; Wake and Lynch, 1976
Bolitoglossa flavimembris
    Johnson, 1989; Larson, 1983a; Schmidt, 1936a; Stuart, 1963; Wake and Brame, 1969; Wake and Lynch, 1976; Wake et al., 1992
Bolitoglossa flaviventris
    Johnson, 1989; Schmidt, 1936a; Smith and Taylor, 1948; Stuart, 1963; Wake and Lynch, 1976; Wake et al, 1992
Bolitoglossa franklini
    Bumzahem and Smith, 1955; Schmidt, 1936a; Wake and Lynch, 1976, 1982; Wake et al., 1980, 1992
Bolitoglossa gracilis
    Bolaños et al., 1987
Bolitoglossa hartwegi
    Elias, 1984; Johnson, 1989; Larson, 1983a; Wake and Brame, 1969; Wake and Lynch, 1976; Wake et al., 1992
Bolitoglossa helmrichi
    Campbell and Vannini, 1989a; Elias, 1984; Larson, 1983a; Schmidt, 1936a; Stuart, 1943a, 1948a, 1963; Wake and Brame, 1969; Wake and Lynch, 1976
Bolitoglossa hermosa
    Papenfuss et al., 1983
Bolitoglossa jacksoni
    Elias, 1984; Wake et al., 1992
Bolitoglossa lignicolor
    Brame and Wake, 1963b; Dunn, 1931d, 1940a; Taylor, 1952a, 1954; Wake and Lynch, 1976
Bolitoglossa lincolni
    Campbell and Vannini, 1989a; Elias, 1984; Johnson, 1989; Larson, 1983a, 1983b; McCoy and Walker, 1966; Stuart, 1943a, 1943b, 1963; Wake and Lynch, 1982, 1988; Wake et al., 1980, 1992
Bolitoglossa macrinii
    Lafrentz, 1930; Papenfuss et al., 1983; Taylor, 1939a, 1949c; Wake et al., 1992
Bolitoglossa marmorea
    Brame, 1965; Tanner and Brame, 1961; Wake and Lynch, 1976
Bolitoglossa medemi
    Brame and Wake, 1972; Wake and Lynch, 1976
Bolitoglossa meliana
    Wake and Lynch, 1982
Bolitoglossa mexicana
    Baker et al., 1971; Campbell and Vannini, 1989a; Duell-

man, 1965a; Elias, 1984; Henderson and Hoevers, 1975; Johnson, 1989; Johnson et al., 1977; Lee, 1980; McCoy and Van Horn, 1962; Meyer and Wilson, 1971a; Neill and Allen, 1959a; Schmidt, 1941; Shannon and Werler, 1955b; Smith and Taylor, 1948; Stuart, 1935, 1943, 1963; Wake and Lynch, 1976; Wake et al., 1992
Bolitoglossa minutula
    Wake et al., 1973
Bolitoglossa morio
    Dunn, 1926; Elias, 1984; Frost, 1985; Larson, 1983a, 1983b; Schmidt, 1936a; Stuart, 1951, 1952, 1963; Wake and Lynch, 1976; Wake et al., 1992
Bolitoglossa mulleri
    Brocchi, 1881–1883; Duellman, 1963a; Elias, 1984; Lazcano-Barrero, 1992; Stuart, 1943a, 1948a, 1963; Wake and Lynch, 1976; Wake et al., 1992
Bolitoglossa nigrescens
    Hanken and Wake, 1982; Savage, 1974b; Taylor, 1949b, 1952a; Wake, 1987
Bolitoglossa occidentalis
    Campbell and Vannini, 1989a; Duellman, 1960a; Johnson, 1989; Larson, 1983a, 1983b; Meyer and Wilson, 1971a; Smith and Taylor, 1948; Stuart, 1963; Taylor, 1941b; Wake and Elias, 1983; Wake and Lynch, 1976; Wake et al., 1992
Bolitoglossa odonnelli
    Stuart, 1943a, 1948a, 1963; Wake and Elias, 1983
Bolitoglossa phalarosoma
    Wake and Brame, 1962; Wake and Lynch, 1976; Wake et al., 1970
Bolitoglossa platydactyla
    Duellman, 1960a; Pérez-Higareda et al., 1987; Smith and Taylor, 1948; Wake and Lynch, 1976; Wake et al., 1992; Werler and Smith, 1952
Bolitoglossa porrasorum
    McCranie and Wilson, 1995a
Bolitoglossa riletti
    Holman, 1964; Papenfuss et al., 1983; Webb and Baker, 1969
Bolitoglossa robusta
    Dunn, 1926; Smith et al., 1965; Taylor, 1952a; Wake, 1987; Wake and Lynch, 1976
Bolitoglossa rostrata
    Campbell and Vannini, 1989a; Dunn, 1926; Elias, 1984; Johnson, 1989; Larson, 1983a, 1983b; Schmidt, 1936a; Stuart, 1943a, 1943b, 1951, 1963; Wake and Lynch, 1976; Wake et al., 1992
Bolitoglossa rufescens
    Campbell and Vannini, 1989a; Elias, 1984; Johnson, 1989; Lee, 1980; Larson, 1983a, 1983b; McCoy, 1966, 1991; Meerman, 1993; Meyer and Wilson, 1971a; O'Shea, 1989; Schmidt, 1936a; Shannon and Werler, 1955b; Smith and Taylor, 1948; Stuart, 1943a, 1948a, 1963; Wake and Lynch, 1976; Wake et al., 1992
Bolitoglossa salvinii
    Brocchi, 1881–1883; Campbell and Vannini, 1989a, 1989b; Schmidt, 1936a; Stuart, 1963; Wake and Lynch, 1976; Wake et al., 1992
Bolitoglossa schizodactyla
    Wake and Brame, 1966; Wake and Lynch, 1976; Wake et al., 1973
Bolitoglossa schmidti
    Dunn and Emlen, 1932; Dunn, 1924b, 1926; Meyer and Wilson, 1971a; Wake and Lynch, 1976
Bolitoglossa sooyorum
    Savage, 1974b; Vial, 1963; Wake, 1987; Wake and Lynch, 1976
Bolitoglossa striatula
    Meyer and Wilson, 1971a; Noble, 1918; Taylor, 1952a; Villa, 1972; Wake, 1987; Wake and Lynch, 1976; Wake et al., 1973
Bolitoglossa stuarti
    Johnson, 1989; Wake and Brame, 1969; Wake and Lynch, 1976; Wake et al., 1992
Bolitoglossa subpalmata
    Dunn, 1926; Frost, 1985; Hanken and Wake, 1982; Savage, 1974b; Scott, 1983b; Taylor, 1952a; Vial, 1966, 1968; Wake, 1987

Bolitoglossa taylori
Wake and Lynch, 1976; Wake et al., 1970
Bolitoglossa veracrucis
Duellman, 1960a; Taylor, 1951; Wake and Lynch, 1976
Bolitoglossa yucatana
Dundee et al., 1986; Lee, 1980; Smith and Taylor, 1948; Wake and Lynch, 1976
Bolitoglossa sp. A
Savage and Villa, 1986["sp. 1," p. 9]; Scott et al., 1983 ["sp. 1," p. 367]
Bolitoglossa sp. B
Savage and Villa, 1986 ["sp. 2," p. 9]; Scott et al., 1983 ["sp. 2," p. 367]
Bolitoglossa sp. C
Savage and Villa, 1986 ["sp. 3," p. 9]
Bolitoglossa sp. D
Wake and Lynch, 1982 [Tres Picos; pp. 262–263]
Bolitoglossa sp. E
Johnson, 1989 ["sp.," p. 60]
Bradytriton silus
Elias, 1984; Wake and Elias, 1983; Wake et al., 1992
Chiropterotriton arboreus
Darda, 1994; Rabb, 1958; Smith and Taylor, 1948; Taylor, 1941c
Chiropterotriton chiropterus
Darda, 1994; Davis and Smith, 1953; Dunn, 1926; Smith and Taylor, 1948; Taylor, 1939a; Wake et al., 1992
Chiropterotriton chondrostega
Darda, 1994; Rabb, 1958; Smith and Taylor, 1948; Taylor, 1941a
Chiropterotriton cracens
Darda, 1994; Martin, 1955a, 1955b, 1958 [all as *chondrostega*]; Rabb, 1958
Chiropterotriton dimidiatus
Darda, 1994; Rabb, 1958; Smith and Taylor, 1948; Taylor, 1940a
Chiropterotriton lavae
Darda, 1994; Smith and Taylor, 1948; Taylor, 1942b; Wake et al., 1992
Chiropterotriton magnipes
Darda, 1994; Rabb, 1965
Chiropterotriton mosaueri
Darda, 1994; Rabb, 1958; Smith and Taylor, 1948; Woodall, 1941
Chiropterotriton multidentatus
Darda, 1994; Martin, 1955a, 1955b, 1958; Rabb, 1958; Smith and Taylor, 1948; Taylor, 1939a
Chiropterotriton orculus
Darda, 1994
Chiropterotriton priscus
Darda, 1994; Rabb, 1956
Chiropterotriton terrestris
Darda, 1994; Rabb, 1958; Smith and Taylor, 1948; Taylor, 1941a
Chiropterotriton sp. A
Darda, 1994 [Distrito Federal, Mexico]
Chiropterotriton sp. B
Darda, 1994 [Mexico]
Chiropterotriton sp. C
Darda, 1944 [Veracruz]
Chiropterotriton sp. D
Darda, 1994 [Veracruz]
Chiropterotriton sp. E
Darda, 1994 [Puebla]
Chiropterotriton sp. F
Darda, 1994 [Veracruz]
Chiropterotriton sp. G
Darda, 1994 [Puebla]
Chiropterotriton sp. H
Darda, 1994 [Oaxaca]; Papenfuss and Wake, 1987; Wake et al., 1992 ["sp. nov.," fig. 6]
Chiropterotriton sp. I
Darda, 1994 [Oaxaca]
Chiropterotriton sp. J
Darda, 1994 [Hidalgo]

Dendrotriton bromeliacia
Lynch and Wake, 1975; Rabb, 1960; Stuart, 1963; Schmidt, 1936a; Wake and Lynch, 1976; Wake et al., 1992
Dendrotriton cuchumatanus
Elias, 1984; Lynch and Wake, 1975; Wake and Elias, 1983; Wake and Lynch, 1976; Wake et al., 1992
Dendrotriton megarhinus
Johnson, 1989; Lynch and Wake, 1975; Rabb, 1960; Wake and Elias, 1983; Wake and Lynch, 1976
Dendrotriton rabbi
Elias, 1984; Lynch and Wake, 1975; Wake and Elias, 1983; Wake and Lynch, 1976; Wake et al., 1992
Dendrotriton xolocalcae
Johnson, 1989; Lynch and Wake, 1975; Rabb, 1960; Smith and Taylor, 1948; Taylor, 1941b; Wake and Elias, 1983; Wake and Lynch, 1976; Wake et al., 1992
Ixalotriton niger
Wake and Johnson, 1989; Wake et al., 1992
Lineatriton lineola
Campbell, 1982; Cope, 1866; Dunn, 1926; Pérez-Higareda et al., 1987; Shannon and Werler, 1955; Smith and Taylor, 1948; Tanner, 1950; Wake and Elias, 1983; Wake et al., 1992
Nototriton abscondens
Good and Wake, 1993; Savage, 1974b; Taylor, 1948a, 1954
Nototriton adelos
Papenfuss and Wake, 1987; Wake et al., 1992
Nototriton alvarezdeltoroi
Papenfuss and Wake, 1987; Wake et al., 1992
Nototriton barbouri
Lynch and Wake, 1975; Meyer and Wilson, 1971a; Schmidt, 1936b; Wake and Lynch, 1976
Nototriton guanacaste
Good and Wake, 1993
Nototriton major
Good and Wake, 1993
Nototriton nasalis
Dunn, 1924b, 1926; Lynch and Wake, 1975; Meyer and Wilson, 1971a; Wake and Lynch, 1976
Nototriton picadoi
Dunn, 1926; Good and Wake, 1993; Papenfuss and Wake, 1987; Stejneger, 1911; Taylor, 1958b; Wake, 1987
Nototriton richardi
Good and Wake, 1993; Papenfuss and Wake, 1987; Savage, 1974b; Taylor, 1949b, 1952a, 1958b; Wake, 1987
Nototriton tapanti
Good and Wake, 1993
Nototriton veraepacis
Lynch and Wake, 1978; Papenfuss and Wake, 1987
Nyctanolis pernix
Elias, 1984; Elias and Wake, 1983; Johnson, 1989; Lynch and Wake, 1989; Wake and Elias, 1983; Wake et al., 1992
Oedipina alfaroi
Brame, 1968; Dunn, 1921; Frost, 1985; Savage, 1974b; Taylor, 1958; Wake, 1987; Wake and Lynch, 1976
Oedipina altura
Brame, 1968; Savage, 1974b; Wake, 1987; Wake and Lynch, 1976
Oedipina carablanca
Brame, 1968; Savage, 1974b; Wake, 1987
Oedipina collaris
Brame, 1963, 1968; Savage, 1974b; Taylor, 1949b, 1958b; Villa, 1972; Wake, 1987
Oedipina complex
Brame, 1968; Brame and Wake, 1963a; Breder, 1946; Dunn, 1924c, 1926, 1931d, 1940a; Myers and Rand, 1969; Rand and Myers, 1990; Taylor, 1958b
Oedipina cyclocauda
Brame, 1968; Meyer and Wilson, 1971a; Savage, 1974b; Taylor, 1952a, 1958b; Villa, 1972; Wake, 1987
Oedipina elongata
Brame, 1960, 1968; Brocchi, 1881–1883; Campbell and Vannini, 1989a; Henderson and Hoevers, 1975; Johnson, 1989; Lee, 1980; Schmidt, 1936a, 1941; Stuart, 1943a, 1948a, 1950, 1963; Wake and Lynch, 1976

Oedipina gephyra
   McCranie et al., 1993b
Oedipina grandis
   Brame and Duellman, 1970; Wake and Lynch, 1976
Oedipina ignea
   Brame, 1960, 1968; Brodie and Campbell, 1993; Stuart, 1952; Wake and Lynch, 1976; Wilson and McCranie, 1994
Oedipina parvipes
   Brame, 1968; Brame and Wake, 1987; Dunn, 1931d; Myers and Rand, 1969; Rand and Myers, 1990; Savage, 1974b; Wake, 1987
Oedipina paucidentata
   Brame, 1968; Savage, 1974b; Wake, 1987
Oedipina poelzi
   Brame, 1963, 1968; Savage, 1974b; Wake, 1987; Wake and Lynch, 1976
Oedipina pseudouniformis
   Brame, 1968; Savage, 1974b; Villa, 1972; Wake, 1987
Oedipina stenopodia
   Brame, 1960; Brodie and Campbell, 1993; Campbell and Vannini, 1989b; Wake and Lynch, 1976; Wake et al., 1992
Oedipina stuarti
   Brame, 1968; Meyer and Wilson, 1971a; Wake and Lynch, 1976
Oedipina taylori
   Brame, 1960, 1968; Brodie and Campbell, 1993; Mertens, 1952; Rand, 1957; Stuart, 1952, 1954b, 1963; Wake and Lynch, 1976
Oedipina uniformis
   Brame, 1968; Savage, 1974b; Taylor, 1948a, 1952a; Wake, 1987; Wake and Lynch, 1976
Parvimolge townsendi
   Dunn, 1922b, 1926; Taylor, 1939a; Wake and Elias, 1983; Wake et al., 1992
Pseudoeurycea altamontana
   Davis and Smith, 1953; Smith and Taylor, 1948; Taylor, 1939a; Wake et al., 1992
Pseudoeurycea anitae
   Bogert, 1967; Wake et al., 1992
Pseudoeurycea bellii
   Davis and Dixon, 1964; Davis and Smith, 1953; Dunn, 1926; Campbell, 1982; Duellman, 1961b, 1965b; Martin, 1955a, 1955b, 1958; McCranie and Wilson, 1987; Smith and Taylor, 1948; Smith et al., 1965; Tanner, 1989; Taylor, 1939a; Van Devender et al., 1989; Wake et al., 1992; Wilson and McCranie, 1979
Pseudoeurycea brunnata
   Bumzahem and Smith, 1955; Johnson, 1989; Wake and Elias, 1983; Wake and Lynch, 1976; Wake et al., 1992
Pseudoeurycea cephalica
   Davis and Smith, 1953; Dunn, 1936; Martin, 1955a, 1955b, 1958; Smith and Taylor, 1948; Taylor, 1939a; Wake et al., 1992
Pseudoeurycea cochranae
   Smith and Taylor, 1948; Taylor, 1943a, 1949c; Wake, 1987; Wake et al., 1992
Pseudoeurycea conanti
   Bogert, 1967; Wake et al., 1992
Pseudoeurycea exspectata
   Stuart, 1954b, 1954c, 1963; Wake and Lynch, 1976
Pseudoeurycea firscheini
   Shannon and Werler, 1955a
Pseudoeurycea gadovii
   Dunn, 1926; Smith and Taylor, 1948; Taylor, 1939a; Wake et al., 1992
Pseudoeurycea galeanae
   Taylor, 1941d; Smith and Taylor, 1948
Pseudoeurycea goebeli
   Bumzahem and Smith, 1955; Schmidt, 1936a; Stuart, 1954c, 1963; Wake and Elias, 1983; Wake and Lynch, 1976; Wake et al., 1992
Pseudoeurycea juarezi
   Campbell, 1982; Regal, 1966; Wake and Lynch, 1976; Wake et al., 1992
Pseudoeurycea leprosa
   Davis and Smith, 1953; Günther, 1885–1902; Lynch et al.,

1983; Smith and Laufe, 1940; Smith and Taylor, 1948; Taylor, 1939a; Wake et al., 1992
Pseudoeurycea longicauda
   Lynch et al., 1983
Pseudoeurycea melanomolga
   Smith and Taylor, 1948; Taylor, 1941d; Wake et al., 1992
Pseudoeurycea mystax
   Bogert, 1967
Pseudoeurycea nigromaculata
   Shannon and Werler, 1955; Smith and Taylor, 1948; Taylor, 1941b; Werler and Smith, 1952
Pseudoeurycea parva
   Lynch and Wake, 1989
Pseudoeurycea praecellens
   Rabb, 1955; Wake and Elias, 1983; Wake et al., 1992
Pseudoeurycea rex
   Elias, 1984; Dunn, 1921, 1926; Johnson, 1989; Schmidt, 1936a; Stuart, 1943a, 1943b, 1951, 1954c, 1963; Wake and Lynch, 1976; Wake et al., 1992
Pseudoeurycea robertsi
   Duellman, 1961b, 1965b; Lynch et al., 1983; Smith and Taylor, 1948; Taylor, 1939a
Pseudoeurycea saltator
   Lynch and Wake, 1989; Wake et al., 1992
Pseudoeurycea scandens
   Martin, 1955b, 1958; Walker, 1955
Pseudoeurycea smithi
   Lynch et al., 1977; Smith and Taylor, 1948; Taylor, 1939a; Wake, 1987; Wake et al., 1992
Pseudoeurycea unguidentis
   Lynch et al., 1977; Smith and Taylor, 1948; Taylor, 1941c; Wake, 1987; Wake et al., 1992
Pseudoeurycea werleri
   Darling and Smith, 1954; Pérez-Higareda et al., 1987; Shannon and Werler, 1955b; Wake et al., 1992
Pseudoeurycea sp. A
   Wake and Campbell, in prep. [Oaxaca, Mexico]
Pseudoeurycea sp. B
   Wake and Lynch, 1976 [San Marcos]; Wake et al., 1992 ["sp.," Fig. 12]
Pseudoeurycea sp. C
   Wake et al., 1992 ["sp. nov. 1," Fig. 2]
Pseudoeurycea sp. D
   Wake et al., 1992 ["sp. nov. 2," Fig. 2]
Pseudoeurycea sp. E
   Wake et al., 1992 ["sp. nov. 1," Fig. 6]
Pseudoeurycea sp. F
   Wake et al., 1992 ["sp. nov. 2," Fig. 6]
Pseudoeurycea sp. G
   Adler, ms.
Pseudoeurycea sp. H
   Adler, ms.
Pseudoeurycea sp. I
   Adler, ms.
Pseudoeurycea sp. J
   Adler, ms.
Pseudoeurycea sp. K
   Adler, ms.
Thorius arboreus
   Hanken, 1983; Hanken and Wake, 1994
Thorius aureus
   Hanken, 1983; Hanken and Wake, 1994
Thorius boreas
   Hanken, 1983; Hanken and Wake, 1994
Thorius dubitus
   Hanken, 1983; Shannon and Werler, 1955a; Taylor, 1941a
Thorius insperatus
   Hanken and Wake, 1994
Thorius macdougalli
   Campbell, 1982; Hanken, 1983; Taylor, 1949c; Wake and Lynch, 1976; Wake et al., 1992
Thorius minutissimus
   Hanken, 1983; Taylor, 1949c; Wake et al., 1992
Thorius narismagnus
   Shannon and Werler, 1955b; Hanken and Wake, 1994

Thorius narisovalis
Hanken, 1983; Taylor, 1940a; Wake, 1987; Wake et al., 1992
Thorius pennatulus
Hanken, 1983; Pérez-Higareda et al., 1987; Taylor, 1940a; Wake et al., 1992
Thorius pulmonaris
Campbell, 1982; Hanken, 1983; Taylor, 1940a; Wake, 1987; Wake and Lynch, 1976; Wake et al., 1992
Thorius schmidti
Gehlbach, 1959; Hanken, 1983
Thorius smithi
Hanken and Wake, 1994
Thorius troglodytes
Hanken, 1983; Shannon and Werler, 1955a; Taylor, 1941a
Thorius sp. A
Hanken, 1983 [Veracruz: Puerto del Aire]
Thorius sp. B
Hanken, 1983 [Veracruz: Volcan Orizaba and Las Vigas]
Thorius sp. C
Adler, 1965; Hanken, 1983 [Guerrero]
Thorius sp. D
Hanken, 1983 [Oaxaca]; Wake et al., 1992
Thorius sp. E
Wake et al., 1992 [Veracruz; "sp. nov., " Fig. 2]

CAUDATA: SALAMANDRIDAE
Notophthalmus meridionalis
Conant and Collins, 1991; Smith and Taylor, 1948; Taylor, 1939a

CAUDATA: SIRENIDAE
Siren intermedia
Flores-Villela, 1993; Smith and Taylor, 1948

GYMNOPHIONA: CAECILIIDAE
Caecilia leucocephala
Taylor, 1968
Caecilia nigricans
Taylor, 1968; Savage and Wake, 1972
Caecilia tentaculata
Taylor, 1968; Savage and Wake, 1972
Caecilia volcani
Taylor, 1969; Savage and Wake, 1972
Dermophis mexicanus
Campbell and Vannini, 1989a, 1989b; Mertens, 1952; Meyer, 1966; Meyer and Wilson, 1971a; Nelson and Nickerson, 1966; Savage, 1974b; Savage and Wake, 1972; Smith and Taylor, 1948; Stuart, 1963; Taylor, 1968; Wake, 1983
Dermophis oaxacae
Johnson, 1984, 1989; Mertens, 1930; Savage and Wake, 1972; Smith and Taylor, 1948; Taylor, 1968
Dermophis parviceps
Dunn, 1924c; Savage, 1974b; Savage and Wake, 1972; Taylor, 1955, 1968; Wake, 1983
Gymnopis multiplicata
Meyer and Wilson, 1971a; Savage and Wake, 1972; Wake, 1983
Gymnopis syntrema
Campbell and Vannini, 1989a; Cope, 1866; Nussbaum, 1988; Savage and Wake, 1972; Stafford, 1994; Stuart, 1948a, 1950, 1963; Taylor, 1968; Wake and Campbell, 1982
Oscaecilia elongata
Savage and Wake, 1972; Taylor, 1968
Oscaecilia ochrocephala
Dunn 1931a, 1931b; Evans, 1947; Myers and Rand, 1969; Rand and Myers, 1990; Savage and Wake, 1972; Schmidt, 1933b; Taylor, 1968
Oscaecilia osae
Lahanas and Savage, 1992

APPENDIX 3:2

Distribution by country of species of amphibians in Middle America. Abbreviations are: BEL = Belize, COS = Costa Rica, GUA = Guatemala, HON = Honduras, MEX = Mexico, NIC = Nicaragua, PAN = Panama, SAL = El Salvador. + = confirmed presence; – = absent; ? = probable occurrence.

Taxon	MEX	BEL	GUA	SAL	HON	NIC	COS	PAN
ANURA: BUFONIDAE:								
Atelophryniscus chrysophorus	–	–	–	–	+	–	–	–
Atelopus certus	–	–	–	–	–	–	–	+
Atelopus chiriquiensis	–	–	–	–	–	–	+	+
Atelopus glyphus	–	–	–	–	–	–	–	+
Atelopus limosus	–	–	–	–	–	–	–	+
Atelopus senex	–	–	–	–	–	–	+	–
Atelopus varius	–	–	–	–	–	–	+	+
Bufo bocourti	+	–	+	–	–	–	–	–
Bufo campbelli	?	+	+	–	+	–	–	–
Bufo canaliferus	+	–	+	+	–	–	–	–
Bufo cavifrons	+	–	–	–	–	–	–	–
Bufo coccifer	+	–	+	+	+	+	+	+
Bufo cognatus	+	–	–	–	–	–	–	–
Bufo compactilis	+	–	–	–	–	–	–	–
Bufo coniferus	–	–	–	–	–	+	+	+
Bufo cristatus	+	–	–	–	–	–	–	–
Bufo debilis	+	–	–	–	–	–	–	–
Bufo fastidiosus	–	–	–	–	–	–	+	+
Bufo gemmifer	+	–	–	–	–	–	–	–
Bufo granulosus	–	–	–	–	–	–	–	+
Bufo haematiticus	–	–	–	–	+	+	+	+
Bufo holdridgei	–	–	–	–	–	–	+	–
Bufo ibarrai	–	–	+	–	+	–	–	–
Bufo kelloggi	+	–	–	–	–	–	–	–
Bufo luetkenii	–	–	+	+	+	+	+	–
Bufo marinus	+	+	+	+	+	+	+	+
Bufo marmoreus	+	–	–	–	–	–	–	–
Bufo mazatlanensis	+	–	–	–	–	–	–	–
Bufo melanochloris	–	–	–	–	–	–	+	–
Bufo microscaphus	+	–	–	–	–	–	–	–
Bufo occidentalis	+	–	–	–	–	–	–	–
Bufo periglenes	–	–	–	–	–	–	+	–
Bufo peripatetes	–	–	–	–	–	–	–	+
Bufo perplexus	+	–	–	–	–	–	–	–
Bufo punctatus	+	–	–	–	–	–	–	–
Bufo tacanensis	+	–	+	–	–	–	–	–
Bufo typhonius	–	–	–	–	–	–	–	+
Bufo valliceps	+	+	+	+	+	+	+	–
Bufo woodhousii	+	–	–	–	–	–	–	–
Crepidophryne epioticus	–	–	–	–	–	–	+	+
Rhamphophryne acrolopha	–	–	–	–	–	–	–	+
ANURA: CENTROLENIDAE:								
Centrolene ilex	–	–	–	–	–	+	+	+
Centrolene prosoblepon	–	–	–	–	–	+	+	+
Cochranella albomaculata	–	–	–	–	+	?	+	+
Cochranella euknemos	–	–	–	–	–	–	+	+
Cochranella granulosa	–	–	–	–	–	+	+	+
Cochranella spinosa	–	–	–	–	–	–	+	+
Hyalinobatrachium chirripoi	–	–	–	–	–	–	+	+
Hyalinobatrachium colymbiphyllum	–	–	–	–	–	–	+	+
Hyalinobatrachium fleischmanni	+	+	+	+	+	+	+	+
Hyalinobatrachium pulveratum	–	–	–	–	–	+	+	+

Appendix 3:2 continued

Taxon	MEX	BEL	GUA	SAL	HON	NIC	COS	PAN
Hyalinobatrachium talamancae	−	−	−	−	−	−	+	−
Hyalinobatrachium valerioi	−	−	−	−	−	−	+	+
Hyalinobatrachium vireovittata	−	−	−	−	−	−	+	+
ANURA: DENDROBATIDAE:								
Colostethus chocoensis	−	−	−	−	−	−	−	+
Colostethus flotator	−	−	−	−	−	−	−	+
Colostethus inguinalis	−	−	−	−	−	−	−	+
Colostethus latinasus	−	−	−	−	−	−	−	+
Colostethus nubicola	−	−	−	−	−	−	+	+
Colostethus pratti	−	−	−	−	−	−	−	+
Colostethus talamancae	−	−	−	−	−	−	+	+
Dendrobates arboreus	−	−	−	−	−	−	−	+
Dendrobates auratus	−	−	−	−	−	+	+	+
Dendrobates granuliferus	−	−	−	−	−	−	+	−
Dendrobates pumilio	−	−	−	−	−	+	+	+
Dendrobates speciosus	−	−	−	−	−	−	−	+
Epipedobates maculatus	−	−	−	−	−	−	−	+
Minyobates fulguritus	−	−	−	−	−	−	−	+
Minyobates minutus	−	−	−	−	−	−	−	+
Phyllobates lugubris	−	−	−	−	−	−	+	+
Phyllobates vittatus	−	−	−	−	−	−	+	−
ANURA: HYLIDAE:								
Agalychnis annae	−	−	−	−	−	−	+	?
Agalychnis calcarifer	−	−	−	−	−	−	+	+
Agalychnis callidryas	+	+	+	−	+	+	+	+
Agalychnis litodryas	−	−	−	−	−	−	−	+
Agalychnis moreletii	+	+	+	+	+	−	−	−
Agalychnis saltator	−	−	−	−	−	+	+	−
Agalychnis spurrelli	−	−	−	−	−	−	+	+
Anotheca spinosa	+	−	−	−	−	−	+	+
Duellmanohyla chamulae	+	−	−	−	−	−	−	−
Duellmanohyla ignicolor	+	−	−	−	−	−	−	−
Duellmanohyla lythrodes	−	−	−	−	−	−	+	+
Duellmanohyla rufioculis	−	−	−	−	−	−	+	−
Duellmanohyla salvavida	−	−	−	−	+	−	−	−
Duellmanohyla schmidtorum	+	−	+	−	−	−	−	−
Duellmanohyla soralia	−	−	+	−	+	−	−	−
Duellmanohyla uranochroa	−	−	−	−	−	−	+	+
Gastrotheca cornuta	−	−	−	−	−	−	+	+
Gastrotheca nicefori	−	−	−	−	−	−	−	+
Hemiphractus fasciatus	−	−	−	−	−	−	?	+
Hyla altipotens	+	−	−	−	−	−	−	−
Hyla angustilineata	−	−	−	−	−	−	+	+
Hyla arborescandens	+	−	−	−	−	−	−	−
Hyla arenicolor	+	−	−	−	−	−	−	−
Hyla bistincta	+	−	−	−	−	−	−	−
Hyla boans	−	−	−	−	−	−	−	+
Hyla bocourti	−	−	+	−	−	−	−	−
Hyla bogertae	+	−	−	−	−	−	−	−
Hyla bromeliacia	−	+	+	−	+	−	−	−
Hyla calvicollina	+	−	−	−	−	−	−	−
Hyla catracha	−	−	−	−	+	−	−	−
Hyla cembra	+	−	−	−	−	−	−	−
Hyla chaneque	+	−	−	−	−	−	−	−
Hyla charadricola	+	−	−	−	−	−	−	−
Hyla chimalapa	+	−	−	−	−	−	−	−
Hyla chryses	+	−	−	−	−	−	−	−
Hyla colymba	−	−	−	−	−	−	+	+

Appendix 3:2 continued

Taxon	MEX	BEL	GUA	SAL	HON	NIC	COS	PAN
Hyla crassa	+	–	–	–	–	–	–	–
Hyla crepitans	–	–	–	–	–	–	–	+
Hyla cyanomma	+	–	–	–	–	–	–	–
Hyla debilis	–	–	–	–	–	–	+	+
Hyla dendroscarta	+	–	–	–	–	–	–	–
Hyla ebraccata	+	+	+	–	–	+	+	+
Hyla echinata	+	–	–	–	–	–	–	–
Hyla euphorbiacea	+	–	–	–	–	–	–	–
Hyla eximia	+	–	–	–	–	–	–	–
Hyla fimbrimembra	–	–	–	–	–	–	+	+
Hyla godmani	+	–	–	–	–	–	–	–
Hyla graceae	–	–	–	–	–	–	–	+
Hyla hazelae	+	–	–	–	–	–	–	–
Hyla insolita	–	–	–	–	+	–	–	–
Hyla juanitae	+	–	–	–	–	–	–	–
Hyla labeculata	+	–	–	–	–	–	–	–
Hyla lancasteri	–	–	–	–	–	–	+	+
Hyla loquax	+	+	+	–	+	+	+	–
Hyla melanomma	+	–	–	–	–	–	–	–
Hyla microcephala	+	+	+	–	+	+	+	+
Hyla miliaria	–	–	–	–	–	+	+	+
Hyla minera	–	+	+	–	–	–	–	–
Hyla miotympanum	+	–	–	–	–	–	–	–
Hyla mixe	+	–	–	–	–	–	–	–
Hyla mixomaculata	+	–	–	–	–	–	–	–
Hyla mykter	+	–	–	–	–	–	–	–
Hyla nubicola	+	–	–	–	–	–	–	–
Hyla pachyderma	+	–	–	–	–	–	–	–
Hyla palmeri	–	–	–	–	–	–	–	+
Hyla pellita	+	–	–	–	–	–	–	–
Hyla pentheter	+	–	–	–	–	–	–	–
Hyla perkinsi	–	–	+	–	–	–	–	–
Hyla phlebodes	–	–	–	–	–	+	+	+
Hyla picadoi	–	–	–	–	–	–	+	+
Hyla picta	+	+	+	–	+	–	–	–
Hyla pictipes	–	–	–	–	–	–	+	+
Hyla pinorum	+	–	–	–	–	–	–	–
Hyla plicata	+	–	–	–	–	–	–	–
Hyla pseudopuma	–	–	–	–	–	–	+	+
Hyla pugnax	–	–	–	–	–	–	–	+
Hyla rivularis	–	–	–	–	–	–	+	+
Hyla robertmertensi	+	–	+	+	–	–	–	–
Hyla robertsorum	+	–	–	–	–	–	–	–
Hyla rosenbergi	–	–	–	–	–	–	+	+
Hyla rufitela	–	–	–	–	–	+	+	+
Hyla sabrina	+	–	–	–	–	–	–	–
Hyla salvaje	–	–	+	–	+	–	–	–
Hyla sartori	+	–	–	–	–	–	–	–
Hyla siopela	+	–	–	–	–	–	–	–
Hyla smaragdina	+	–	–	–	–	–	–	–
Hyla smithii	+	–	–	–	–	–	–	–
Hyla subocularis	–	–	–	–	–	–	–	+
Hyla sumichrasti	+	–	–	–	–	–	–	–
Hyla taeniopus	+	–	–	–	–	–	–	–
Hyla thorectes	+	–	–	–	–	–	–	–
Hyla thysanota	–	–	–	–	–	–	–	+
Hyla tica	–	–	–	–	–	–	+	+
Hyla trux	+	–	–	–	–	–	–	–

Appendix 3:2 continued

Taxon	MEX	BEL	GUA	SAL	HON	NIC	COS	PAN
Hyla valancifer	+	−	−	−	−	−	−	−
Hyla walkeri	+	−	+	−	−	−	−	−
Hyla xanthosticta	−	−	−	−	−	−	+	−
Hyla xera	+	−	−	−	−	−	−	−
Hyla zeteki	−	−	−	−	−	−	+	+
Hyla sp. A	+	−	−	−	−	−	−	−
Pachymedusa dacnicolor	+	−	−	−	−	−	−	−
Phrynohyas venulosa	+	+	+	+	+	+	+	+
Phyllomedusa lemur	−	−	−	−	−	−	+	+
Phyllomedusa venusta	−	−	−	−	−	−	−	+
Plectrohyla acanthodes	+	−	+	−	−	−	−	−
Plectrohyla avia	+	−	+	−	−	−	−	−
Plectrohyla chrysopleura	−	−	−	−	+	−	−	−
Plectrohyla dasypus	−	−	−	−	+	−	−	−
Plectrohyla glandulosa	−	−	+	+	+	−	−	−
Plectrohyla guatemalensis	+	−	+	+	+	−	−	−
Plectrohyla hartwegi	+	−	+	−	+	−	−	−
Plectrohyla ixil	+	−	+	−	−	−	−	−
Plectrohyla lacertosa	+	−	−	−	−	−	−	−
Plectrohyla matudai	+	−	+	−	+	−	−	−
Plectrohyla pokomchi	−	−	+	−	−	−	−	−
Plectrohyla pycnochila	+	−	−	−	−	−	−	−
Plectrohyla quecchi	−	−	+	−	−	−	−	−
Plectrohyla sagorum	+	−	+	+	?	−	−	−
Plectrohyla tecunumani	−	−	+	−	−	−	−	−
Plectrohyla teuchestes	−	−	+	−	+	−	−	−
Pternohyla dentata	+	−	−	−	−-	−	−	−
Pternohyla fodiens	+	−	−	−	−	−	−	−
Ptychohyla erythromma	+	−	−	−	−	−	−	−
Ptychohyla euthysanota	+	−	+	+	?	−	−	−
Ptychohyla hypomykter	−	−	+	−	+	+	−	−
Ptychohyla legleri	−	−	−	−	−	−	+	+
Ptychohyla leonhardschultzei	+	−	−	−	−	−	−	−
Ptychohyla macrotympanum	+	−	+	−	−	−	−	−
Ptychohyla panchoi	−	−	+	−	−	−	−	−
Ptychohyla salvadorensis	−	−	+	+	+	−	−	−
Ptychohyla sanctaecrucis	−	−	+	−	−	−	−	−
Ptychohyla spinipollex	−	−	−	−	+	−	−	−
Scinax boulengeri	−	−	−	−	−	+	+	+
Scinax elaeochroa	−	−	−	−	−	+	+	+
Scinax rostrata	−	−	−	−	−	−	−	+
Scinax rubra	−	−	−	−	−	−	−	+
Scinax staufferi	+	+	+	+	+	+	+	+
Smilisca baudinii	+	+	+	+	+	+	+	−
Smilisca cyanosticta	+	+	+	−	−	−	−	−
Smilisca phaeota	−	−	−	−	+	+	+	+
Smilisca puma	−	−	−	−	−	+	+	−
Smilisca sila	−	−	−	−	−	−	+	+
Smilisca sordida	−	−	−	−	+	+	+	+
Triprion petasatus	+	+	+	−	+	−	−	−
Triprion spatulatus	+	−	−	−	−	−	−	−
ANURA: LEPTODACTYLIDAE:								
Eleutherodactylus achatinus	−	−	−	−	−	−	−	+
Eleutherodactylus adamastus	−	−	+	−	−	−	−	−
Eleutherodactylus alfredi	+	−	+	−	−	−	−	−
Eleutherodactylus altae	−	−	−	−	−	−	+	+
Eleutherodactylus ancianao	−	−	−	−	+	−	−	−
Eleutherodactylus andi	−	−	−	−	−	−	+	+

Appendix 3:2 continued

Taxon	MEX	BEL	GUA	SAL	HON	NIC	COS	PAN
Eleutherodactylus angelicus	−	−	−	−	−	−	+	−
Eleutherodactylus antillensis	−	−	−	−	−	−	−	+
Eleutherodactylus aphanus	−	−	+	−	−	−	−	−
Eleutherodactylus augusti	+	−	−	−	−	−	−	−
Eleutherodactylus aurilegulus	−	−	−	−	+	−	−	−
Eleutherodactylus azueroensis	−	−	−	−	−	−	−	+
Eleutherodactylus batrachylus	+	−	−	−	−	−	−	−
Eleutherodactylus berkenbuschi	+	−	−	−	−	−	−	−
Eleutherodactylus biporcatus	−	−	−	−	+	+	+	+
Eleutherodactylus bocourti	−	−	+	−	−	−	−	−
Eleutherodactylus bransfordii	−	−	−	−	+	+	+	+
Eleutherodactylus brocchi	−	−	+	−	−	−	−	−
Eleutherodactylus bufoniformis	−	−	−	−	−	−	+	+
Eleutherodactylus caryophyllaceus	−	−	−	−	−	−	+	+
Eleutherodactylus cerasinus	−	−	−	−	−	+	+	+
Eleutherodactylus chac	−	+	+	−	+	−	−	−
Eleutherodactylus chrysozetetes	−	−	−	−	+	−	−	−
Eleutherodactylus crassidigitus	−	−	−	−	−	−	+	+
Eleutherodactylus cruentus	−	−	−	−	−	−	+	+
Eleutherodactylus cruzi	−	−	−	−	+	−	−	−
Eleutherodactylus cuaquero	−	−	−	−	−	−	+	−
Eleutherodactylus daryi	−	−	+	−	−	−	−	−
Eleutherodactylus decoratus	+	−	−	−	−	−	−	−
Eleutherodactylus diastema	−	−	−	−	−	+	+	+
Eleutherodactylus emcelae	−	−	−	−	−	−	−	+
Eleutherodactylus escoces	−	−	−	−	−	−	+	−
Eleutherodactylus fitzingeri	−	−	−	−	−	+	+	+
Eleutherodactylus fleischmanni	−	−	−	−	−	−	+	+
Eleutherodactylus gaigei	−	−	−	−	−	−	+	+
Eleutherodactylus glaucus	+	−	−	−	−	−	−	−
Eleutherodactylus gollmeri	−	−	−	−	−	−	+	+
Eleutherodactylus greggi	+	−	+	−	−	−	−	−
Eleutherodactylus guerreroensis	+	−	−	−	−	−	−	−
Eleutherodactylus hobartsmithi	+	−	−	−	−	−	−	−
Eleutherodactylus hylaeformis	−	−	−	−	−	−	+	−
Eleutherodactylus johnstonei	−	−	−	−	−	−	−	+
Eleutherodactylus jota	−	−	−	−	−	−	−	+
Eleutherodactylus laticeps	+	+	+	−	+	−	−	−
Eleutherodactylus latidiscus	−	−	−	−	−	−	−	+
Eleutherodactylus lineatus	+	−	+	−	−	−	−	−
Eleutherodactylus longirostris	−	−	−	−	−	−	−	+
Eleutherodactylus matudai	+	−	−	−	−	−	−	−
Eleutherodactylus megalotympanum	+	−	−	−	−	−	−	−
Eleutherodactylus melanostictus	−	−	−	−	−	−	+	+
Eleutherodactylus merendonensis	−	−	−	−	+	−	−	−
Eleutherodactylus mexicanus	+	−	−	−	−	−	−	−
Eleutherodactylus milesi	−	−	−	−	+	−	−	−
Eleutherodactylus mimus	−	−	−	−	+	+	+	?
Eleutherodactylus monnichorum	−	−	−	−	−	−	?	+
Eleutherodactylus moro	−	−	−	−	−	−	+	+
Eleutherodactylus museosus	−	−	−	−	−	−	−	+
Eleutherodactylus noblei	−	−	−	−	+	+	+	+
Eleutherodactylus occidentalis	+	−	−	−	−	−	−	−
Eleutherodactylus omiltemanus	+	−	−	−	−	−	−	−
Eleutherodactylus pardalis	−	−	−	−	−	−	+	+
Eleutherodactylus planirostris	+	−	−	−	−	−	−	−
Eleutherodactylus podiciferus	−	−	−	−	−	−	+	+
Eleutherodactylus polymniae	+	−	−	−	−	−	−	−

Appendix 3:2 continued

Taxon	MEX	BEL	GUA	SAL	HON	NIC	COS	PAN
Eleutherodactylus pozo	+	−	−	−	−	−	−	−
Eleutherodactylus psephosypharus	−	+	+	−	?	−	−	−
Eleutherodactylus punctariolus	−	−	−	−	−	−	+	+
Eleutherodactylus pygmaeus	+	−	+	−	−	−	−	−
Eleutherodactylus raniformis	−	−	−	−	−	−	−	+
Eleutherodactylus rayo	−	−	−	−	−	−	+	+
Eleutherodactylus rhodopis	+	+	+	+	+	−	−	−
Eleutherodactylus ridens	−	−	−	−	+	+	+	+
Eleutherodactylus rostralis	−	−	+	−	+	−	−	−
Eleutherodactylus rugulosus	+	−	−	−	−	−	−	−
Eleutherodactylus saltator	+	−	−	−	−	−	−	−
Eleutherodactylus sandersoni	−	+	+	−	−	−	−	−
Eleutherodactylus sartori	+	−	−	−	−	−	−	−
Eleutherodactylus silvicola	+	−	−	−	−	−	−	−
Eleutherodactylus spatulatus	+	−	−	−	−	−	−	−
Eleutherodactylus stadelmani	−	−	−	−	+	−	−	−
Eleutherodactylus stejnegerianus	−	−	−	−	−	−	+	+
Eleutherodactylus stuarti	+	−	+	−	−	−	−	−
Eleutherodactylus taeniatus	−	−	−	−	−	−	−	+
Eleutherodactylus talamancae	−	−	−	−	−	+	+	+
Eleutherodactylus tarahumaraensis	+	−	−	−	−	−	−	−
Eleutherodactylus taurus	−	−	−	−	−	−	+	+
Eleutherodactylus taylori	+	−	−	−	−	−	−	−
Eleutherodactylus trachydermus	−	−	+	−	−	−	−	−
Eleutherodactylus uno	+	−	−	−	−	−	−	−
Eleutherodactylus vocalis	+	−	−	−	−	−	−	−
Eleutherodactylus vocator	−	−	−	−	−	−	+	+
Eleutherodactylus xucanebi	−	−	+	−	−	−	−	−
Eleutherodactylus yucatanensis	+	−	−	−	−	−	−	−
Eleutherodactylus sp. A	−	−	−	−	−	−	+	−
Eleutherodactylus sp. B	−	+	+	−	−	−	−	−
Eleutherodactylus sp. C	−	−	+	−	−	−	−	−
Eleutherodactylus sp. D	−	−	+	−	−	−	−	−
Eleutherodactylus sp. E	+	−	+	+	−	−	−	−
Leptodactylus bolivianus	−	−	−	−	−	−	+	+
Leptodactylus fuscus	−	−	−	−	−	−	−	+
Leptodactylus labialis	+	+	+	+	+	+	+	+
Leptodactylus melanonotus	+	+	+	+	+	+	+	+
Leptodactylus pentadactylus	−	−	−	−	+	+	+	+
Leptodactylus poecilochilus	−	−	−	−	−	−	+	+
Leptodactylus silvanimbus	−	−	−	−	+	−	−	−
Physalaemus pustulosus	+	+	+	+	+	+	+	+
Pleurodema brachyops	−	−	−	−	−	−	−	+
Syrrhophus albolabris	+	−	−	−	−	−	−	−
Syrrhophus angustidigitorum	+	−	−	−	−	−	−	−
Syrrhophus cystignathoides	+	−	−	−	−	−	−	−
Syrrhophus dennisi	+	−	−	−	−	−	−	−
Syrrhophus dilatus	+	−	−	−	−	−	−	−
Syrrhophus guttilatus	+	−	−	−	−	−	−	−
Syrrhophus interorbitalis	+	−	−	−	−	−	−	−
Syrrhophus leprus	+	+	+	−	−	−	−	−
Syrrhophus longipes	+	−	−	−	−	−	−	−
Syrrhophus maurus	+	−	−	−	−	−	−	−
Syrrhophus modestus	+	−	−	−	−	−	−	−
Syrrhophus nitidus	+	−	−	−	−	−	−	−
Syrrhophus nivicolimae	+	−	−	−	−	−	−	−
Syrrhophus pallidus	+	−	−	−	−	−	−	−
Syrrhophus pipilans	+	−	+	−	−	−	−	−

Appendix 3:2 continued

Taxon	MEX	BEL	GUA	SAL	HON	NIC	COS	PAN
Syrrhophus rubrimaculatus	+	−	+	−	−	−	−	−
Syrrhophus rufescens	+	−	−	−	−	−	−	−
Syrrhophus saxatilis	+	−	−	−	−	−	−	−
Syrrhophus syristes	+	−	−	−	−	−	−	−
Syrrhophus teretistes	+	−	−	−	−	−	−	−
Syrrhophus verrucipes	+	−	−	−	−	−	−	−
ANURA: MICROHYLIDAE:								
Chiasmocleis panamensis	−	−	−	−	−	−	−	+
Elachistocleis ovalis	−	−	−	−	−	−	−	+
Gastrophryne elegans	+	+	+	−	+	+	+	−
Gastrophryne olivacea	+	−	−	−	−	−	−	−
Gastrophryne pictiventris	−	−	−	−	−	+	+	−
Gastrophryne usta	+	−	+	+	−	−	−	−
Hypopachus barberi	+	−	+	+	+	−	−	−
Hypopachus variolosus	+	+	+	+	+	+	+	−
Nelsonophryne aterrima	−	−	−	−	−	−	+	+
Relictivomer pearsei	−	−	−	−	−	−	−	+
ANURA: PELOBATIDAE:								
Scaphiopus couchii	+	−	−	−	−	−	−	−
Spea multiplicata	+	−	−	−	−	−	−	−
ANURA: PIPIDAE:								
Pipa myersi	−	−	−	−	−	−	−	+
ANURA: RANIDAE:								
Rana berlandieri	+	+	+	−	+	+	−	−
Rana brownorum	+	−	−	−	−	−	−	−
Rana catesbeiana	+	−	−	−	−	−	−	−
Rana chiricahuensis	+	−	−	−	−	−	−	−
Rana dunni	+	−	−	−	−	−	−	−
Rana forreri	+	−	+	+	?	+	+	−
Rana johni	+	−	−	−	−	−	−	−
Rana juliani	−	+	?	−	−	−	−	−
Rana maculata	+	−	+	+	+	+	−	−
Rana magnaocularis	+	−	−	−	−	−	−	−
Rana montezumae	+	−	−	−	−	−	−	−
Rana neovolcanica	+	−	−	−	−	−	−	−
Rana omiltemana	+	−	−	−	−	−	−	−
Rana pueblae	+	−	−	−	−	−	−	−
Rana pustulosa	+	−	−	−	−	−	−	−
Rana sierramadrensis	+	−	−	−	−	−	−	−
Rana spectabilis	+	−	−	−	−	−	−	−
Rana tarahumarae	+	−	−	−	−	−	−	−
Rana taylori	−	−	−	−	−	+	+	−
Rana tlaloci	+	−	−	−	−	−	−	−
Rana trilobata	+	−	−	−	−	−	−	−
Rana vaillanti	+	+	+	?	+	+	+	+
Rana vibicaria	−	−	−	−	−	−	+	+
Rana warschewitschii	−	−	−	−	+	+	+	+
Rana yavapaiensis	+	−	−	−	−	−	−	−
Rana zweifeli	+	−	−	−	−	−	−	−
Rana sp. A	+	−	−	−	−	−	−	−
Rana sp. B	+	−	−	−	−	−	−	−
Rana sp. C	+	−	+	−	−	−	−	−
Rana sp. D	−	−	−	−	−	−	+	+
Rana sp. E	−	−	−	−	−	−	−	+
ANURA: RHINOPHRYNIDAE:								
Rhinophrynus dorsalis	+	+	+	+	+	+	+	−
CAUDATA: AMBYSTOMATIDAE:								
Ambystoma altamirani	+	−	−	−	−	−	−	−

Appendix 3:2 continued

Taxon	MEX	BEL	GUA	SAL	HON	NIC	COS	PAN
Ambystoma amblycephalum	+	–	–	–	–	–	–	–
Ambystoma andersoni	+	–	–	–	–	–	–	–
Ambystoma bombypellum	+	–	–	–	–	–	–	–
Ambystoma dumerilii	+	–	–	–	–	–	–	–
Ambystoma flavipiperatum	+	–	–	–	–	–	–	–
Ambystoma granulosum	+	–	–	–	–	–	–	–
Ambystoma leorae	+	–	–	–	–	–	–	–
Ambystoma lermaense	+	–	–	–	–	–	–	–
Ambystoma mexicanum	+	–	–	–	–	–	–	–
Ambystoma ordinarium	+	–	–	–	–	–	–	–
Ambystoma rivularis	+	–	–	–	–	–	–	–
Ambystoma rosaceum	+	–	–	–	–	–	–	–
Ambystoma subsalsum	+	–	–	–	–	–	–	–
Ambystoma taylori	+	–	–	–	–	–	–	–
Ambystoma tigrinum	+	–	–	–	–	–	–	–
Ambystoma velasci	+	–	–	–	–	–	–	–
Ambystoma zempoalensis	+	–	–	–	–	–	–	–
CAUDATA: PLETHODONTIDAE:								
Bolitoglossa alvaradoi	–	–	–	–	–	–	+	–
Bolitoglossa arborescandens	–	–	–	–	–	–	+	–
Bolitoglossa biseriata	–	–	–	–	–	–	?	+
Bolitoglossa carri	–	–	–	–	+	–	–	–
Bolitoglossa celaque	–	–	–	?	+	–	–	–
Bolitoglossa cerroensis	–	–	–	–	–	–	+	–
Bolitoglossa colonnea	–	–	–	–	–	–	+	+
Bolitoglossa compacta	–	–	–	–	–	–	?	+
Bolitoglossa conanti	–	–	?	?	+	–	–	–
Bolitoglossa cuchumatana	–	–	+	–	–	–	–	–
Bolitoglossa cuna	–	–	–	–	–	–	–	+
Bolitoglossa diaphora	–	–	–	–	+	–	–	–
Bolitoglossa diminuta	–	–	–	–	–	–	+	–
Bolitoglossa dofleini	–	+	+	–	+	–	–	–
Bolitoglossa dunni	–	–	+	–	+	–	–	–
Bolitoglossa engelhardti	+	–	+	–	–	–	–	–
Bolitoglossa epimela	–	–	–	–	–	–	+	–
Bolitoglossa flavimembris	+	–	+	–	–	–	–	–
Bolitoglossa flaviventris	+	–	+	–	–	–	–	–
Bolitoglossa franklini	+	–	+	–	–	–	–	–
Bolitoglossa gracilis	–	–	–	–	–	–	+	–
Bolitoglossa hartwegi	+	–	+	–	–	–	–	–
Bolitoglossa helmrichi	–	–	+	–	–	–	–	–
Bolitoglossa hermosa	+	–	–	–	–	–	–	–
Bolitoglossa jacksoni	–	–	+	–	–	–	–	–
Bolitoglossa lignicolor	–	–	–	–	–	–	+	+
Bolitoglossa lincolni	+	–	+	–	–	–	–	–
Bolitoglossa macrinii	+	–	–	–	–	–	–	–
Bolitoglossa marmorea	–	–	–	–	–	–	?	+
Bolitoglossa medemi	–	–	–	–	–	–	–	+
Bolitoglossa meliana	–	–	+	–	–	–	–	–
Bolitoglossa mexicana	+	+	+	–	+	–	–	–
Bolitoglossa minutula	–	–	–	–	–	–	?	+
Bolitoglossa morio	–	–	+	–	–	–	–	–
Bolitoglossa mulleri	+	–	+	–	–	–	–	–
Bolitoglossa nigrescens	–	–	–	–	–	–	+	+
Bolitoglossa occidentalis	+	–	+	–	+	–	–	–
Bolitoglossa odonnelli	–	–	+	–	–	–	–	–
Bolitoglossa phalarosoma	–	–	–	–	–	–	–	+
Bolitoglossa platydactyla	+	–	–	–	–	–	–	–

Appendix 3:2 continued

Taxon	MEX	BEL	GUA	SAL	HON	NIC	COS	PAN
Bolitoglossa porrasorum	–	–	–	–	+	–	–	–
Bolitoglossa riletti	+	–	–	–	–	–	–	–
Bolitoglossa robusta	–	–	–	–	–	–	+	+
Bolitoglossa rostrata	+	–	+	–	–	–	–	–
Bolitoglossa rufescens	+	+	+	–	+	–	–	–
Bolitoglossa salvinii	–	–	+	+	–	–	–	–
Bolitoglossa schizodactyla	–	–	–	–	–	–	?	+
Bolitoglossa schmidti	–	–	–	–	+	–	–	–
Bolitoglossa sooyorum	–	–	–	–	–	–	+	–
Bolitoglossa striatula	–	–	–	–	+	+	+	–
Bolitoglossa stuarti	+	–	+	–	–	–	–	–
Bolitoglossa subpalmata	–	–	–	–	–	–	+	–
Bolitoglossa taylori	–	–	–	–	–	–	–	+
Bolitoglossa veracrucis	+	–	–	–	–	–	–	–
Bolitoglossa yucatana	+	–	–	–	–	–	–	–
Bolitoglossa sp. A	–	–	–	–	–	–	+	–
Bolitoglossa sp. B	–	–	–	–	–	–	+	–
Bolitoglossa sp. C	–	–	–	–	–	–	+	–
Bolitoglossa sp. D	+	–	–	–	–	–	–	–
Bolitoglossa sp. E	+	–	–	–	–	–	–	–
Bradytriton silus	?	–	+	–	–	–	–	–
Chiropterotriton arboreus	+	–	–	–	–	–	–	–
Chiropterotriton chiropterus	+	–	–	–	–	–	–	–
Chiropterotriton chondrostega	+	–	–	–	–	–	–	–
Chiropterotriton cracens	+	–	–	–	–	–	–	–
Chiropterotriton dimidiatus	+	–	–	–	–	–	–	–
Chiropterotriton lavae	+	–	–	–	–	–	–	–
Chiropterotriton magnipes	+	–	–	–	–	–	–	–
Chiropterotriton mosaueri	+	–	–	–	–	–	–	–
Chiropterotriton multidentatus	+	–	–	–	–	–	–	–
Chiropterotriton orculus	+	–	–	–	–	–	–	–
Chiropterotriton priscus	+	–	–	–	–	–	–	–
Chiropterotriton terrestris	+	–	–	–	–	–	–	–
Chiropterotriton sp. A	+	–	–	–	–	–	–	–
Chiropterotriton sp. B	+	–	–	–	–	–	–	–
Chiropterotriton sp. C	+	–	–	–	–	–	–	–
Chiropterotriton sp. D	+	–	–	–	–	–	–	–
Chiropterotriton sp. E	+	–	–	–	–	–	–	–
Chiropterotriton sp. F	+	–	–	–	–	–	–	–
Chiropterotriton sp. G	+	–	–	–	–	–	–	–
Chiropterotriton sp. H	+	–	–	–	–	–	–	–
Chiropterotriton sp. I	+	–	–	–	–	–	–	–
Chiropterotriton sp. J	+	–	–	–	–	–	–	–
Dendrotriton bromeliacia	–	–	+	–	–	–	–	–
Dendrotriton cuchumatanus	–	–	+	–	–	–	–	–
Dendrotriton megarhinus	+	–	–	–	–	–	–	–
Dendrotriton rabbi	–	–	+	–	–	–	–	–
Dendrotriton xolocalcae	+	–	–	–	–	–	–	–
Ixalotriton niger	+	–	–	–	–	–	–	–
Lineatriton lineola	+	–	–	–	–	–	–	–
Nototriton abscondens	–	–	–	–	–	–	+	–
Nototriton adelos	+	–	–	–	–	–	–	–
Nototriton alvarezdeltoroi	+	–	–	–	–	–	–	–
Nototriton barbouri	–	–	–	–	+	–	–	–
Nototriton guanacaste	–	–	–	–	–	–	+	–
Nototriton major	–	–	–	–	–	–	+	–
Nototriton nasalis	–	–	–	–	+	–	–	–
Nototriton picadoi	–	–	–	–	–	–	+	–

Appendix 3:2 continued

Taxon	MEX	BEL	GUA	SAL	HON	NIC	COS	PAN
Nototriton richardi	–	–	–	–	–	–	+	–
Nototriton tapanti	–	–	–	–	–	–	+	–
Nototriton veraepacis	–	–	+	–	–	–	–	–
Nyctanolis pernix	+	–	+	–	–	–	–	–
Oedipina alfaroi	–	–	–	–	–	–	+	+
Oedipina altura	–	–	–	–	–	–	+	–
Oedipina carablanca	–	–	–	–	–	–	+	–
Oedipina collaris	–	–	–	–	–	+	+	+
Oedipina complex	–	–	–	–	–	–	?	+
Oedipina cyclocauda	–	–	–	–	+	+	+	+
Oedipina elongata	+	+	+	–	–	–	–	–
Oedipina gephyra	–	–	–	–	+	–	–	–
Oedipina grandis	–	–	–	–	–	–	+	+
Oedipina ignea	–	–	+	?	+	–	–	–
Oedipina parvipes	–	–	–	–	–	–	+	+
Oedipina paucidentata	–	–	–	–	–	–	+	–
Oedipina poelzi	–	–	–	–	–	–	+	–
Oedipina pseudouniformis	–	–	–	–	–	+	+	–
Oedipina stenopodia	–	–	+	–	–	–	–	–
Oedipina stuarti	–	–	–	?	+	–	–	–
Oedipina taylori	–	–	+	+	–	–	–	–
Oedipina uniformis	–	–	–	–	–	–	+	–
Parvimolge townsendi	+	–	–	–	–	–	–	–
Pseudoeurycea altamontana	+	–	–	–	–	–	–	–
Pseudoeurycea anitae	+	–	–	–	–	–	–	–
Pseudoeurycea bellii	+	–	–	–	–	–	–	–
Pseudoeurycea brunnata	+	–	+	–	–	–	–	–
Pseudoeurycea cephalica	+	–	–	–	–	–	–	–
Pseudoeurycea cochranae	+	–	–	–	–	–	–	–
Pseudoeurycea conanti	+	–	–	–	–	–	–	–
Pseudoeurycea exspectata	–	–	+	–	–	–	–	–
Pseudoeurycea firscheini	+	–	–	–	–	–	–	–
Pseudoeurycea gadovii	+	–	–	–	–	–	–	–
Pseudoeurycea galeanae	+	–	–	–	–	–	–	–
Pseudoeurycea goebeli	+	–	+	–	–	–	–	–
Pseudoeurycea juarezi	+	–	–	–	–	–	–	–
Pseudoeurycea leprosa	+	–	–	–	–	–	–	–
Pseudoeurycea longicauda	+	–	–	–	–	–	–	–
Pseudoeurycea melanomolga	+	–	–	–	–	–	–	–
Pseudoeurycea mystax	+	–	–	–	–	–	–	–
Pseudoeurycea nigromaculata	+	–	–	–	–	–	–	–
Pseudoeurycea parva	+	–	–	–	–	–	–	–
Pseudoeurycea praecellens	+	–	–	–	–	–	–	–
Pseudoeurycea rex	+	–	+	–	–	–	–	–
Pseudoeurycea robertsi	+	–	–	–	–	–	–	–
Pseudoeurycea saltator	+	–	–	–	–	–	–	–
Pseudoeurycea scandens	+	–	–	–	–	–	–	–
Pseudoeurycea smithi	+	–	–	–	–	–	–	–
Pseudoeurycea unguidentis	+	–	–	–	–	–	–	–
Pseudoeurycea werleri	+	–	–	–	–	–	–	–
Pseudoeurycea sp. A	+	–	–	–	–	–	–	–
Pseudoeurycea sp. B	–	–	+	–	–	–	–	–
Pseudoeurycea sp. C	+	–	–	–	–	–	–	–
Pseudoeurycea sp. D	+	–	–	–	–	–	–	–
Pseudoeurycea sp. E	+	–	–	–	–	–	–	–
Pseudoeurycea sp. F	+	–	–	–	–	–	–	–
Pseudoeurycea sp. G	+	–	–	–	–	–	–	–
Pseudoeurycea sp. H	+	–	–	–	–	–	–	–

Appendix 3:2 continued

Taxon	MEX	BEL	GUA	SAL	HON	NIC	COS	PAN
Pseudoeurycea sp. I	+	−	−	−	−	−	−	−
Pseudoeurycea sp. J	+	−	−	−	−	−	−	−
Pseudoeurycea sp. K	+	−	−	−	−	−	−	−
Thorius arboreus	+	−	−	−	−	−	−	−
Thorius aureus	+	−	−	−	−	−	−	−
Thorius boreas	+	−	−	−	−	−	−	−
Thorius dubitus	+	−	−	−	−	−	−	−
Thorius insperatus	+	−	−	−	−	−	−	−
Thorius macdougalli	+	−	−	−	−	−	−	−
Thorius minutissimus	+	−	−	−	−	−	−	−
Thorius narismagnus	+	−	−	−	−	−	−	−
Thorius narisovalis	+	−	−	−	−	−	−	−
Thorius pennatulus	+	−	−	−	−	−	−	−
Thorius pulmonaris	+	−	−	−	−	−	−	−
Thorius schmidti	+	−	−	−	−	−	−	−
Thorius smithi	+	−	−	−	−	−	−	−
Thorius troglodytes	+	−	−	−	−	−	−	−
Thorius sp. A	+	−	−	−	−	−	−	−
Thorius sp. B	+	−	−	−	−	−	−	−
Thorius sp. C	+	−	−	−	−	−	−	−
Thorius sp. D	+	−	−	−	−	−	−	−
Thorius sp. E	+	−	−	−	−	−	−	−
CAUDATA: SALAMANDRIDAE:								
Notophthalmus meridionalis	+	−	−	−	−	−	−	−
CAUDATA: SIRENIDAE:								
Siren intermedia	+	−	−	−	−	−	−	−
GYMNOPHIONA: CAECILIIDAE:								
Caecilia leucocephala	−	−	−	−	−	−	−	+
Caecilia nigricans	−	−	−	−	−	−	−	+
Caecilia tentaculata	−	−	−	−	−	−	−	+
Caecilia volcani	−	−	−	−	−	−	−	+
Dermophis mexicanus	+	−	+	+	+	+	+	+
Dermophis oaxacae	+	−	−	−	−	−	−	−
Dermophis parviceps	−	−	−	−	−	−	+	+
Gymnopis multiplicata	−	−	−	−	+	+	+	+
Gymnopis syntrema	−	+	+	−	?	−	−	−
Oscaecilia elongata	−	−	−	−	−	−	−	+
Oscaecilia ochrocephala	−	−	−	−	−	−	−	+
Oscaecilia osae	−	−	−	−	−	−	+	−

APPENDIX 3:3

Distribution of Middle American amphibians in topographic areas. Abbreviations of areas are: CG = western nuclear Central American highlands; CGU = Pacific lowlands from eastern Chiapas to south-central Guatemala; CP = Pacific lowlands from ventral Costa Rica through Panama; CRP = Isthmian Central American highlands; EP = highlands of eastern Panama; GCR = Pacific lowlands from southeastern Guatemala to southwestern Costa Rica; GH = Caribbean lowlands of eastern Guatemala and northern Honduras; HN = eastern nuclear Central American highlands; LT = Sierra de Los Tuxtlas; MC = Meseta Central; NP = Caribbean lowlands from Nicaragua to Panama; OCC = Sierra Madre Occidental; ORI = Sierra Madre Oriental; SC = Pacific lowlands from Sinaloa to western Chiapas; SUR = Sierra Madre del Sur; TT = Gulf lowlands from Tamaulipas to Tabasco; YP = Yucatan Platform.

Taxon	Middle American Highlands									Atlantic Coast Lowlands				Pacific Coast Lowlands				Elevation
	ORI	OCC	MC	SUR	LT	CG	HN	CRP	EP	TT	YP	GH	NP	SC	CGU	GCR	CP	Meters
ANURA: BUFONIDAE:																		
Atelophryniscus chrysophorus	—	—	—	—	—	—	+	—	—	—	—	—	—	—	—	—	—	785–1760
Atelopus certus	—	—	—	—	—	—	—	—	+	—	—	—	—	—	—	—	+	1000
Atelopus chiriquiensis	—	—	—	—	—	—	—	+	+	—	—	—	—	—	—	+	—	1400–2100
Atelopus glyphus	—	—	—	—	—	—	—	—	+	—	—	—	—	—	—	—	+	1280–1500
Atelopus limosus	—	—	—	—	—	—	—	—	—	—	—	—	+	—	—	—	+	10–270
Atelopus senex	—	—	—	—	—	—	—	+	—	—	—	—	+	—	—	—	—	1100–2150
Atelopus varius	—	—	—	—	—	—	—	+	—	—	—	—	+	—	—	—	+	0–2150
Bufo bocourti	—	—	—	—	—	+	—	—	—	—	—	—	—	—	—	—	—	1859–3200
Bufo campbelli	—	—	—	—	—	+	+	—	—	—	+	+	—	—	+	—	—	120–1200
Bufo canaliferus	—	—	—	—	—	+	—	—	—	—	—	—	—	—	+	+	—	0–1175
Bufo cavifrons	+	—	—	—	—	+	+	—	—	—	—	—	—	+	—	—	—	900–1600
Bufo coccifer	—	—	—	+	—	+	+	+	+	—	—	—	—	+	+	+	—	0–2000
Bufo cognatus	—	+	+	—	—	—	—	—	—	—	—	—	—	—	—	—	—	1000–2440
Bufo compactilis	—	+	+	—	—	—	—	—	—	—	—	—	—	—	—	—	—	1900–2500
Bufo coniferus	+	—	—	—	—	—	—	+	+	—	—	—	+	—	—	—	+	100–1400
Bufo cristatus	+	—	—	—	—	—	—	+	—	—	—	—	—	—	—	—	—	1500–2000
Bufo debilis	—	+	+	—	—	+	—	—	—	—	—	—	—	—	—	—	—	1000–2134
Bufo fastidiosus	—	—	—	—	—	—	—	+	—	—	—	—	—	—	—	—	—	762–1810
Bufo gemmifer	—	—	—	—	—	—	—	—	—	—	—	—	—	+	—	—	+	0–823
Bufo granulosus	—	—	—	—	—	—	—	—	—	—	—	—	+	—	—	—	+	0–200
Bufo haematiticus	—	—	—	—	—	—	—	+	—	—	—	+	+	—	—	—	+	0–1050
Bufo holdridgei	—	—	—	—	—	—	—	+	—	—	—	—	—	—	—	—	—	1920–2286
Bufo ibarrai	—	—	—	—	—	+	+	—	—	—	—	—	—	—	—	—	—	1366–1730
Bufo kelloggi	—	—	—	—	—	—	—	—	—	—	—	—	—	+	—	—	—	0–579
Bufo luetkenii	—	—	—	—	—	+	+	+	—	—	+	+	—	—	—	+	+	0–1300
Bufo marinus	+	+	—	+	+	+	+	+	+	+	+	+	+	+	+	+	+	0–2000
Bufo marmoreus	—	—	+	+	—	—	—	—	—	—	—	—	—	+	—	—	—	0–1006

Appendix 3:3 continued

Taxon	Middle American Highlands									Atlantic Coast Lowlands				Pacific Coast Lowlands				Elevation
	ORI	OCC	MC	SUR	LT	CG	HN	CRP	EP	TT	YP	GH	NP	SC	CGU	GCR	CP	Meters
Bufo mazatlanensis	–	+	–	–	–	–	–	–	–	–	–	–	–	+	–	–	–	0–1100
Bufo melanochloris	–	–	–	–	–	–	–	+	–	–	–	–	–	–	–	–	–	600–1000
Bufo microscaphus	–	+	–	–	–	–	–	–	–	–	–	–	–	–	–	–	–	1500–2590
Bufo occidentalis	–	+	+	+	–	–	–	–	–	–	–	–	–	–	–	–	–	900–2670
Bufo periglenes	–	–	–	–	–	–	–	+	–	–	–	–	–	–	–	–	–	1590
Bufo peripatetes	–	–	–	–	–	–	–	+	–	–	–	–	–	–	–	–	–	1500
Bufo perplexus	–	+	+	–	–	–	–	–	–	–	–	–	–	+	–	–	–	180–1768
Bufo punctatus	–	+	+	–	–	–	–	–	–	–	–	–	–	+	–	–	–	0–2896
Bufo tacanensis	–	–	–	–	–	+	–	–	–	–	–	–	–	–	–	–	–	1500–1700
Bufo typhonius	–	–	–	–	–	–	–	–	+	–	–	–	+	+	–	–	+	0–1000
Bufo valliceps	–	–	–	–	+	+	+	+	–	+	+	+	–	+	+	+	–	0–1200
Bufo woodhousii	–	+	–	–	–	–	–	–	–	–	–	–	–	–	–	–	–	1500–2000
Crepidophryne epioticus	–	–	–	–	–	–	–	+	–	–	–	–	–	–	–	–	–	915–1524
Rhamphophryne acrolopha	–	–	–	–	–	–	–	–	+	–	–	–	–	–	–	–	–	1265–1481
ANURA: CENTROLENIDAE:																		
Centrolene ilex	–	–	–	–	–	–	–	+	–	–	–	–	+	–	–	–	+	250–775
Centrolene prosoblepon	–	–	–	–	–	–	–	+	+	–	–	–	+	+	–	–	+	0–1900
Cochranella albomaculata	–	–	–	–	–	–	–	+	+	–	–	–	+	+	–	–	+	0–1180
Cochranella euknemos	–	–	–	–	–	–	+	+	+	–	–	–	+	–	–	–	+	90–1500
Cochranella granulosa	–	–	–	–	–	–	–	+	+	–	–	–	+	+	–	–	+	0–1116
Cochranella spinosa	–	–	–	–	–	–	–	–	–	–	–	–	+	+	–	–	+	0–260
Hyalinobatrachium chirripoi	–	–	–	–	–	–	–	–	–	–	–	–	+	+	–	–	+	0–300
Hyalinobatrachium colymbiphyllum	–	–	–	–	–	–	+	+	–	–	–	–	+	+	–	–	+	0–1710
Hyalinobatrachium fleischmanni	+	–	–	+	+	–	+	+	–	+	+	–	+	+	+	+	+	0–1686
Hyalinobatrachium pulveratum	–	–	–	–	–	–	–	+	–	–	–	–	+	+	–	–	+	0–800
Hyalinobatrachium talamancae	–	–	–	–	–	–	–	+	–	–	–	–	+	–	–	–	+	1116
Hyalinobatrachium valerioi	–	–	–	–	–	–	–	+	+	–	–	–	+	+	–	–	+	0–1500
Hyalinobatrachium vireovittata	–	–	–	–	–	–	–	+	–	–	–	–	–	–	–	–	–	630–880
ANURA: DENDROBATIDAE:																		
Colostethus chocoensis	–	–	–	–	–	–	–	–	–	–	–	–	+	+	–	–	+	0–370
Colostethus flotator	–	–	–	–	–	–	–	–	–	–	–	–	+	+	–	–	+	0–370
Colostethus inguinalis	–	–	–	–	–	–	–	–	+	–	–	–	–	–	–	–	–	0–853
Colostethus latinasus	–	–	–	–	–	–	–	+	+	–	–	–	+	–	–	–	+	0–1400
Colostethus nubicola	–	–	–	–	–	–	–	+	+	–	–	–	+	+	–	–	+	0–1830
Colostethus pratti	–	–	–	–	–	–	–	+	–	–	–	–	+	+	–	–	+	0–1160
Colostethus talamancae	–	–	–	–	–	–	–	+	–	–	–	–	+	+	–	–	+	0–1050
Dendrobates arboreus	–	–	–	–	–	–	–	+	–	–	–	–	–	–	–	–	–	20–1300
Dendrobates auratus	–	–	–	–	–	–	+	+	–	–	–	–	+	+	–	–	+	0–800
Dendrobates granuliferus	–	–	–	–	–	–	–	–	–	–	–	–	–	–	–	–	+	0–700

Appendix 3:3 continued

Taxon	Middle American Highlands									Atlantic Coast Lowlands				Pacific Coast Lowlands				Elevation
	ORI	OCC	MC	SUR	LT	CG	HN	CRP	EP	TT	YP	GH	NP	SC	CGU	GCR	CP	Meters
Dendrobates pumilio	–	–	–	–	–	–	–	+	–	–	–	–	+	–	–	–	–	0–960
Dendrobates speciosus	–	–	–	–	–	–	–	+	–	–	–	–	+	–	–	–	–	1050–2150
Epipedobates maculatus	–	–	–	–	–	–	–	–	–	–	–	–	?	–	–	–	?	?
Minyobates fulgaritus	–	–	–	–	–	–	–	–	+	–	–	–	+	+	–	–	+	0–800
Minyobates minutus	–	–	–	–	–	–	–	–	+	–	–	–	+	–	–	–	+	0–1100
Phyllobates lugubris	–	–	–	–	–	–	–	–	–	–	–	–	+	–	–	–	–	0–650
Phyllobates vittatus	–	–	–	–	–	–	–	–	–	–	–	–	–	–	–	–	+	0–200
ANURA: HYLIDAE:																		
Agalychnis annae	–	–	–	–	–	–	–	–	–	–	–	–	+	–	–	–	–	50–1600
Agalychnis calcarifer	–	+	–	–	–	–	–	+	–	–	–	–	+	–	–	–	+	50–820
Agalychnis callidryas	–	–	–	–	+	+	+	+	–	+	+	+	+	–	–	+	+	0–960
Agalychnis litodryas	–	–	–	–	–	–	–	–	–	–	–	–	+	–	–	–	+	130
Agalychnis moreletii	+	+	–	+	+	+	+	+	–	–	–	–	+	–	–	–	–	240–2130
Agalychnis saltator	–	–	–	–	–	–	–	–	–	–	–	–	+	–	–	–	–	0–780
Agalychnis spurrelli	+	–	–	–	+	+	–	+	–	–	–	–	+	–	–	–	+	0–885
Anotheca spinosa	+	–	–	–	+	+	–	+	–	–	–	–	–	–	–	–	–	300–1800
Duellmanohyla chamulae	+	–	–	–	–	+	–	–	–	–	–	–	–	–	–	–	–	350–1700
Duellmanohyla ignicolor	+	–	–	–	–	+	–	–	–	–	–	–	–	–	–	–	–	1500–1865
Duellmanohyla lythrodes	–	–	–	–	–	–	–	+	–	–	–	–	+	–	–	–	–	170–800
Duellmanohyla rufioculis	–	–	–	–	–	–	–	+	–	–	–	–	–	–	–	–	–	650–1600
Duellmanohyla salvavida	–	–	–	–	–	–	+	–	–	–	–	–	–	–	–	–	–	880
Duellmanohyla schmidtorum	–	–	–	–	–	+	–	–	–	–	–	–	–	–	–	–	–	500–2200
Duellmanohyla soralia	–	–	–	–	–	–	+	–	–	–	+	–	–	–	–	–	–	420–1570
Duellmanohyla uranochroa	–	–	–	–	–	–	–	+	–	+	–	–	+	–	–	–	–	70–1750
Gastrotheca cornuta	–	–	–	–	–	–	–	+	+	–	–	–	–	–	–	–	+	300–1500
Gastrotheca nicefori	–	–	–	–	–	–	–	+	+	–	–	–	–	–	–	–	–	800–1660
Hemiphractus fasciatus	–	–	–	–	–	–	–	+	+	–	–	–	+	+	–	–	+	300–1600
Hyla altipotens	–	–	–	+	–	–	–	+	–	–	–	–	–	–	–	–	–	1100–1900
Hyla angustilineata	–	–	–	–	–	–	–	+	–	–	–	–	–	–	–	–	–	1500–2300
Hyla arborescandens	+	+	+	+	–	–	–	–	–	–	–	–	–	–	–	–	–	1600–3100
Hyla arenicolor	+	+	+	+	–	–	–	–	–	–	–	–	–	+	–	–	–	300–3000
Hyla bistincta	+	+	–	–	–	–	–	–	–	–	–	–	–	–	–	–	–	1400–2800
Hyla boans	–	–	–	–	–	+	–	–	–	–	–	–	–	–	–	–	+	0–300
Hyla bocourti	–	–	–	–	–	–	–	–	–	–	–	+	–	–	–	–	+	1325–1540
Hyla bogertae	–	–	–	+	–	+	–	–	–	–	–	–	–	–	–	–	–	2652
Hyla bromeliacia	–	–	–	–	–	+	+	–	–	–	–	–	–	–	–	–	–	900–1650
Hyla calvicollina	+	–	–	–	–	–	–	–	–	–	–	–	–	–	–	–	–	2518–2712
Hyla catracha	–	–	–	–	–	–	+	–	–	–	–	–	–	–	–	–	–	2010–2125
Hyla cembra	–	–	–	+	–	–	–	–	–	–	–	–	–	–	–	–	–	2160

Appendix 3:3 continued

Taxon	Middle American Highlands									Atlantic Coast Lowlands				Pacific Coast Lowlands				Elevation
	ORI	OCC	MC	SUR	LT	CG	HN	CRP	EP	TT	YP	GH	NP	SC	CGU	GCR	CP	Meters
Hyla chaneque	+	—	—	—	—	—	—	—	—	—	—	—	—	—	—	—	—	800–2200
Hyla charadricola	+	—	—	—	—	—	—	—	—	—	—	—	—	—	—	—	—	2000–2300
Hyla chimalapa	—	—	—	—	—	+	—	—	—	—	—	—	—	—	—	—	—	1000–1542
Hyla chryses	—	—	—	+	—	—	—	—	—	—	—	—	—	—	—	—	—	2350–2600
Hyla colymba	—	—	—	—	—	—	—	+	+	—	—	—	—	—	—	—	—	457–1410
Hyla crassa	+	—	—	+	—	—	—	—	—	—	—	—	—	—	—	—	—	1500–2300
Hyla crepitans	—	—	—	—	—	—	—	—	—	—	—	—	—	—	—	—	+	0–300
Hyla cyanomma	+	—	—	—	—	—	—	—	—	—	—	—	—	—	—	—	—	2650–2670
Hyla debilis	—	—	—	—	—	—	—	+	—	—	—	—	—	—	—	—	—	910–1700
Hyla dendroscarta	+	—	—	—	+	—	—	—	—	—	—	—	—	—	—	—	—	450–1900
Hyla ebraccata	—	—	—	—	+	—	—	—	—	+	+	+	+	—	—	—	+	0–1200
Hyla echinata	+	—	—	—	+	—	—	—	—	—	—	—	—	—	—	—	—	1500
Hyla euphorbiacea	+	+	—	+	—	—	—	—	—	—	—	—	—	—	—	—	—	1600–3150
Hyla eximia	+	+	+	+	—	—	—	—	—	—	—	—	—	—	—	—	—	900–2900
Hyla fimbrimembra	—	—	—	—	—	—	—	+	—	—	—	—	—	—	—	—	—	1200–2040
Hyla godmani	+	—	—	—	—	—	+	—	—	—	—	—	—	—	—	—	—	0–900
Hyla graceae	—	—	—	—	—	—	—	+	—	—	—	—	—	—	—	—	—	1100–1650
Hyla hazelae	+	—	—	+	—	—	—	—	—	—	—	—	—	—	—	—	—	2300–3000
Hyla insolita	—	—	—	—	—	—	+	—	—	—	—	—	—	—	—	—	—	1550
Hyla juanitae	+	—	—	+	—	—	—	—	—	—	—	—	—	—	—	—	—	750–1070
Hyla labeculata	+	—	—	+	—	—	—	—	—	—	—	—	—	—	—	—	—	1800
Hyla lancasteri	—	—	—	—	—	—	+	+	—	—	—	—	—	—	—	—	—	90–1920
Hyla loquax	—	—	—	—	+	—	+	—	—	+	+	+	+	—	—	—	—	0–1100
Hyla melanomma	+	—	—	+	—	+	—	—	—	+	+	+	—	—	—	—	—	853–2000
Hyla microcephala	—	—	—	—	+	+	+	—	—	+	+	+	—	—	—	+	+	0–1200
Hyla miliaria	—	—	—	—	—	—	+	+	+	—	—	—	—	—	—	—	—	600–1200
Hyla minera	—	—	—	—	—	+	+	—	—	—	—	—	—	—	—	—	—	700–1830
Hyla miotympanum	+	—	—	—	+	—	—	—	—	+	—	—	—	—	—	—	—	100–2280
Hyla mixe	+	—	—	—	—	—	—	—	—	—	—	—	—	—	—	—	—	1800
Hyla mixomaculata	+	—	—	—	—	—	—	—	—	—	—	—	—	—	—	—	—	900–1524
Hyla mykter	—	—	—	+	—	—	—	—	—	—	—	—	—	—	—	—	—	1985–2520
Hyla nubicola	+	—	—	—	—	—	—	—	—	—	—	—	—	—	—	—	—	900–1400
Hyla pachyderma	+	—	—	—	—	—	—	—	—	—	—	—	—	—	—	—	—	1600
Hyla palmeri	—	—	—	—	—	—	—	+	?	—	—	—	—	—	—	—	?	600
Hyla pellita	—	—	—	+	—	—	—	—	—	—	—	—	—	—	—	—	—	1500–1700
Hyla pentheter	—	—	—	+	—	—	—	—	—	—	—	—	—	—	—	—	—	1320–2000
Hyla perkinsi	—	—	—	—	—	+	—	—	—	—	—	—	—	—	—	—	—	1050–1080
Hyla phlebodes	—	—	—	—	—	—	—	—	—	—	—	—	+	—	—	+	+	0–700
Hyla picadoi	—	—	—	—	—	—	—	+	—	—	—	—	—	—	—	—	—	1900–2510

Appendix 3:3 continued

| Taxon | Middle American Highlands | | | | | | | | | Atlantic Coast Lowlands | | | | Pacific Coast Lowlands | | | | Elevation |
	ORI	OCC	MC	SUR	LT	CG	HN	CRP	EP	TT	YP	GH	NP	SC	CGU	GCR	CP	Meters
Hyla picta	+	–	–	–	+	–	–	–	–	+	+	+	–	–	–	–	–	0–1300
Hyla pictipes	–	–	–	–	–	–	–	+	–	+	+	–	–	–	–	–	–	1900–2500
Hyla pinorum	–	–	+	+	–	–	–	–	–	–	–	–	–	–	–	–	–	700–1070
Hyla plicata	+	–	+	+	–	–	–	–	–	–	–	–	–	–	–	–	–	1400–3600
Hyla pseudopuma	–	–	–	–	–	–	–	+	–	–	–	–	–	–	–	–	–	830–2400
Hyla pugnax	–	–	–	–	–	–	–	–	–	–	–	–	+	–	–	–	+	0–300
Hyla rivularis	–	–	–	–	–	–	–	+	–	–	–	–	–	–	–	–	–	1200–2840
Hyla robertmertensi	–	–	–	–	–	–	–	–	–	–	–	–	–	+	+	+	–	0–700
Hyla robertsorum	+	–	–	–	–	–	–	–	–	–	–	–	–	–	–	–	–	2250–3050
Hyla rosenbergi	–	–	–	–	–	–	–	–	–	–	–	–	–	–	–	–	+	0–300
Hyla rufitela	+	–	–	–	–	–	–	–	–	–	–	–	+	–	–	–	+	0–300
Hyla sabrina	+	–	–	–	–	–	–	–	–	–	–	–	–	–	–	–	–	1650–2020
Hyla salvaje	–	–	–	–	–	–	+	–	–	–	–	–	–	–	–	–	–	1370
Hyla sartori	–	–	–	–	–	–	–	–	–	–	–	–	–	+	–	–	–	0–300
Hyla siopela	+	–	–	–	–	–	–	–	–	–	–	–	–	–	–	–	–	2500–2550
Hyla smaragdina	–	+	+	–	–	–	–	–	–	–	–	–	–	+	–	–	–	100–1500
Hyla smithii	–	+	+	–	–	–	–	–	–	–	–	–	–	+	–	–	–	0–1000
Hyla subocularis	–	–	–	–	–	–	–	–	–	–	–	–	–	–	–	–	+	0–800
Hyla sumichrasti	–	–	–	+	–	+	–	–	–	–	–	–	–	+	–	–	–	200–1675
Hyla taeniopus	+	–	–	–	–	–	–	–	–	–	–	–	–	–	–	–	–	1200–2100
Hyla thorectes	–	–	–	+	–	–	–	–	–	–	–	–	–	–	–	–	–	1600–1900
Hyla thysanota	–	–	–	–	–	–	–	–	+	–	–	–	–	–	–	–	–	1265
Hyla tica	–	–	–	–	–	–	–	+	–	–	–	–	–	–	–	–	–	835–1920
Hyla trux	–	–	–	+	–	–	–	–	–	–	–	–	–	–	–	–	–	1760–2120
Hyla valancifer	–	–	–	–	+	–	–	–	–	–	–	–	–	–	–	–	–	1180–1372
Hyla walkeri	–	–	–	–	–	+	–	–	–	–	–	–	–	–	–	–	–	1450–2340
Hyla xanthosticta	–	–	–	–	–	–	–	+	+	–	–	–	–	–	–	–	–	2100
Hyla xera	–	–	–	+	–	–	–	–	–	–	–	–	–	–	–	–	–	1490
Hyla zeteki	–	–	–	–	–	–	–	+	–	–	–	–	–	–	–	–	–	1000–1800
Hyla sp. A	+	–	–	–	–	–	–	–	–	–	–	–	–	–	–	–	–	1219
Pachymedusa dacnicolor	–	+	–	+	–	–	–	–	–	–	–	–	–	–	–	–	–	0–1000
Phrynohyas venulosa	+	+	–	+	–	+	+	+	–	+	+	+	+	+	+	+	+	0–1500
Phyllomedusa lemur	–	–	–	–	–	–	–	–	–	–	–	–	–	–	–	+	–	650–1600
Phyllomedusa venusta	–	–	–	–	–	–	–	–	–	–	–	–	+	–	–	–	–	130
Plectrohyla acanthodes	–	–	–	–	–	+	–	–	–	–	–	–	–	–	–	–	–	1540–2250
Plectrohyla avia	–	–	–	–	–	+	–	–	–	–	–	–	–	–	–	–	–	1650–2200
Plectrohyla chrysopleura	–	–	–	–	–	–	+	–	–	–	–	–	–	–	–	–	–	930–990
Plectrohyla dasypus	–	–	–	–	–	–	+	–	–	–	–	–	–	–	–	–	–	1530–1690
Plectrohyla glandulosa	–	–	–	–	–	+	+	–	–	–	–	–	–	–	–	–	–	2400–3500

Appendix 3:3 continued

Taxon	Middle American Highlands									Atlantic Coast Lowlands				Pacific Coast Lowlands				Elevation
	ORI	OCC	MC	SUR	LT	CG	HN	CRP	EP	TT	YP	GH	NP	SC	CGU	GCR	CP	Meters
Plectrohyla guatemalensis	–	–	–	–	–	+	+	–	–	–	–	–	–	–	–	–	–	990–2800
Plectrohyla hartwegi	–	–	–	–	–	+	+	–	–	–	–	–	–	–	–	–	–	925–2700
Plectrohyla ixil	–	–	–	–	–	+	–	–	–	–	–	–	–	–	–	–	–	1100–1700
Plectrohyla lacertosa	–	–	–	–	–	+	–	–	–	–	–	–	–	–	–	–	–	2134
Plectrohyla matudai	–	–	–	–	–	+	+	–	–	–	–	–	–	–	–	–	–	700–2300
Plectrohyla pokomchi	–	–	–	–	–	+	–	–	–	–	–	–	–	–	–	–	–	1400–1900
Plectrohyla pycnochila	–	–	–	–	–	+	–	–	–	–	–	–	–	–	–	–	–	2400
Plectrohyla quecchi	–	–	–	–	–	+	–	–	–	–	–	–	–	–	–	–	–	615–1850
Plectrohyla sagorum	–	–	–	–	–	+	+	–	–	–	–	–	–	–	–	–	–	1450–2050
Plectrohyla tecunumani	–	–	–	–	–	+	–	–	–	–	–	–	–	–	–	–	–	3200–3400
Plectrohyla teuchestes	–	–	–	–	–	+	+	–	–	–	–	–	–	–	–	–	–	1000–1570
Pternohyla dentata	–	–	+	–	–	–	–	–	–	–	–	–	–	–	–	–	–	1800–1900
Pternohyla fodiens	–	+	+	–	–	–	–	–	–	–	–	–	–	+	–	–	–	0–1585
Ptychohyla erythromma	+	–	–	+	–	–	–	–	–	–	–	–	–	–	–	–	–	600–950
Ptychohyla euthysanota	–	–	–	–	–	+	+	–	–	–	–	–	–	–	–	–	–	660–2200
Ptychohyla hypomykter	–	–	–	–	–	+	+	–	–	–	–	+	–	–	–	–	–	340–2070
Ptychohyla legleri	–	–	–	–	–	–	–	+	–	–	–	–	–	–	–	–	–	700–1600
Ptychohyla leonhardschultzei	+	–	–	+	–	+	–	–	–	–	–	–	–	–	–	–	–	700–2000
Ptychohyla macrotympanum	–	–	–	–	–	+	–	–	–	–	–	–	–	–	–	–	–	700–1700
Ptychohyla panchoi	–	–	–	–	–	+	+	–	–	–	–	–	–	–	–	–	–	300–700
Ptychohyla salvadorensis	–	–	–	–	–	–	–	+	–	–	–	–	–	–	–	–	–	700–1870
Ptychohyla sanctaecrucis	–	–	–	–	–	+	–	–	–	–	–	–	–	–	–	–	–	366–1150
Ptychohyla spinipollex	–	–	–	–	–	–	+	–	–	–	–	–	–	–	–	–	–	810–1090
Scinax boulengeri	–	–	–	–	–	–	–	–	–	–	–	–	+	–	–	+	+	0–700
Scinax elaeochroa	–	–	–	–	–	–	–	–	–	–	–	–	+	–	–	+	+	0–1100
Scinax rostrata	–	–	–	–	–	–	–	–	–	–	–	–	+	–	–	+	+	0–300
Scinax rubra	–	–	–	–	–	–	–	–	–	–	–	–	–	–	–	+	+	0–300
Scinax staufferi	+	–	–	–	+	–	–	–	–	–	+	+	+	+	+	+	+	0–1200
Smilisca baudinii	+	+	–	+	+	–	–	–	–	+	+	+	+	–	+	+	–	0–1925
Smilisca cyanosticta	+	–	–	–	+	+	–	–	–	–	+	–	+	–	–	–	–	200–1200
Smilisca phaeota	–	–	–	–	–	–	–	–	–	–	–	+	+	–	–	–	+	0–1050
Smilisca puma	–	–	–	–	–	–	–	+	–	–	–	–	+	–	–	–	–	0–285
Smilisca sila	–	–	–	–	–	–	–	+	+	–	–	–	+	–	–	–	+	0–1300
Smilisca sordida	–	–	–	–	–	–	–	–	–	–	–	+	+	–	–	–	+	0–1435
Triprion petasatus	–	–	–	–	–	–	–	–	–	–	–	–	–	–	–	–	–	0–300
Triprion spatulatus	–	–	–	–	–	–	–	–	–	–	–	–	–	+	–	–	–	0–350
ANURA: LEPTODACTYLIDAE:																		
Eleutherodactylus achatinus	–	–	–	–	–	–	–	–	–	–	–	–	+	–	–	–	+	50–900
Eleutherodactylus adamastus	–	–	–	–	–	+	–	–	+	–	–	–	–	–	–	–	–	600–650

Appendix 3:3 continued

Taxon	Middle American Highlands									Atlantic Coast Lowlands				Pacific Coast Lowlands				Elevation
	ORI	OCC	MC	SUR	LT	CG	HN	CRP	EP	TT	YP	GH	NP	SC	CGU	GCR	CP	Meters
Eleutherodactylus alfredi	+	–	–	–	+	–	–	–	–	+	+	–	–	–	–	–	–	0–650
Eleutherodactylus altae	–	–	–	–	–	–	–	+	–	+	+	–	–	–	–	–	–	1200–1524
Eleutherodactylus ancianao	–	–	–	–	–	–	+	+	–	–	–	–	–	–	–	–	–	1770–1840
Eleutherodactylus andi	–	–	–	–	–	–	–	+	–	–	–	–	–	–	–	–	–	1150–1600
Eleutherodactylus angelicus	–	–	–	–	–	–	–	+	–	–	–	–	–	–	–	–	–	600–1700
Eleutherodactylus antillensis	–	–	–	–	–	–	–	–	–	–	–	–	+	–	–	–	–	0–300
Eleutherodactylus aphanus	–	–	–	–	–	+	–	–	–	–	–	–	–	–	–	–	–	591–786
Eleutherodactylus augusti	+	+	+	–	–	–	–	–	–	–	–	–	–	–	–	–	–	300–2715
Eleutherodactylus aurilegulus	–	–	–	–	–	–	+	–	–	–	–	–	–	–	–	–	–	780–1110
Eleutherodactylus azueroensis	+	–	–	–	–	–	–	–	–	–	–	–	–	–	–	–	+	0–940
Eleutherodactylus batrachylus	+	–	–	–	+	–	–	–	–	+	–	–	–	–	–	–	–	1524
Eleutherodactylus berkenbuschi	+	–	–	+	+	–	–	+	–	–	–	–	+	–	–	–	+	250–1990
Eleutherodactylus biporcatus	–	–	–	–	–	+	–	+	+	–	–	–	–	–	–	–	–	0–1150
Eleutherodactylus bocourti	–	–	–	–	–	+	–	+	–	–	–	–	+	–	–	–	–	1300–1700
Eleutherodactylus bransfordii	–	–	–	–	–	–	–	+	–	–	–	–	+	–	–	–	+	0–1600
Eleutherodactylus brocchi	–	–	–	–	–	+	–	+	–	–	–	–	+	–	–	–	–	1300–2000
Eleutherodactylus bufoniformis	–	–	–	–	–	+	–	–	–	–	–	–	+	–	–	–	+	0–300
Eleutherodactylus caryophyllaceus	–	–	–	–	–	+	–	+	–	–	–	–	+	–	–	–	+	1000–1700
Eleutherodactylus cerasinus	–	–	–	–	–	+	–	+	–	–	–	–	+	–	–	–	+	40–2000
Eleutherodactylus chac	–	–	–	–	–	+	+	–	–	–	–	+	–	–	–	–	–	0–780
Eleutherodactylus chrysozetetes	–	–	–	–	–	–	+	–	–	–	–	–	–	–	–	–	–	880–1110
Eleutherodactylus crassidigitus	–	–	–	–	–	–	–	+	+	–	–	–	+	–	–	–	+	0–2000
Eleutherodactylus cruentus	–	–	–	–	–	–	–	+	–	–	–	–	+	–	–	–	+	40–1676
Eleutherodactylus cruzi	–	–	–	–	–	–	+	–	–	–	–	–	–	–	–	–	–	1520
Eleutherodactylus cuaquero	–	–	–	–	–	–	–	+	–	–	–	–	–	–	–	–	–	1520–1859
Eleutherodactylus daryi	–	–	–	–	–	+	–	–	–	–	–	–	–	–	–	–	–	1500–2340
Eleutherodactylus decoratus	+	–	–	–	–	–	–	+	–	–	–	–	–	–	–	–	–	420–1830
Eleutherodactylus diastema	–	–	–	–	–	–	–	+	+	–	–	–	+	–	–	–	+	0–1680
Eleutherodactylus emcelae	–	–	–	–	–	–	–	+	–	–	–	–	–	–	–	–	–	910–1450
Eleutherodactylus escoces	–	–	–	–	–	–	–	+	+	–	–	–	–	–	–	–	–	1100–2100
Eleutherodactylus fitzingeri	–	–	–	–	–	–	–	+	–	–	–	–	+	–	–	–	+	0–1600
Eleutherodactylus fleischmanni	–	–	–	–	–	–	–	+	–	–	–	–	–	–	–	–	–	1050–2300
Eleutherodactylus gaigei	–	–	–	–	–	–	–	–	–	–	–	–	+	–	–	–	+	0–820
Eleutherodactylus glaucus	–	–	–	–	–	+	–	–	–	–	–	–	–	–	–	–	–	2100
Eleutherodactylus gollmeri	–	–	–	–	–	+	–	+	–	–	–	–	+	–	–	–	+	0–1600
Eleutherodactylus greggi	–	–	–	–	–	+	–	–	–	–	–	–	–	–	–	–	–	1829–2700
Eleutherodactylus guerreroensis	–	–	–	+	–	–	–	–	–	–	–	–	–	–	–	–	–	980
Eleutherodactylus hobartsmithi	–	+	+	–	–	–	–	–	–	–	–	–	–	–	–	–	–	1450–2750
Eleutherodactylus hylaeformis	–	–	–	–	–	–	–	+	–	–	–	–	–	–	–	–	–	1470–2134

Appendix 3:3 continued

Taxon	Middle American Highlands									Atlantic Coast Lowlands				Pacific Coast Lowlands				Elevation
	ORI	OCC	MC	SUR	LT	CG	HN	CRP	EP	TT	YP	GH	NP	SC	CGU	GCR	CP	Meters
Eleutherodactylus johnstonei	–	–	–	–	–	–	–	–	–	–	–	–	–	–	–	–	+	0–100
Eleutherodactylus jota	–	–	–	–	–	–	–	–	–	–	–	–	–	–	–	–	–	830
Eleutherodactylus laticeps	–	–	–	–	+	+	+	+	–	–	+	+	–	–	–	–	–	0–1800
Eleutherodactylus latidiscus	–	–	–	–	–	–	–	–	–	–	–	–	–	–	–	–	+	0–600
Eleutherodactylus lineatus	–	–	–	+	–	+	–	–	–	–	–	–	–	–	–	–	–	900–2250
Eleutherodactylus longirostris	–	–	–	–	–	–	–	–	+	–	–	–	–	–	–	–	+	320–1100
Eleutherodactylus matudai	–	–	–	–	–	+	–	–	–	–	–	–	–	–	–	–	–	1300–2290
Eleutherodactylus megalotympanum	–	–	–	–	–	–	–	–	–	–	–	–	–	–	–	–	–	900–1220
Eleutherodactylus melanostictus	–	–	–	–	–	–	–	+	–	–	–	–	–	–	–	–	–	1050–2483
Eleutherodactylus merendonensis	–	–	–	–	–	–	–	–	–	–	–	+	–	–	–	–	–	150–200
Eleutherodactylus mexicanus	+	–	–	–	–	–	–	–	–	–	–	–	–	–	–	–	–	1500–2520
Eleutherodactylus milesi	–	–	–	–	–	–	+	–	–	–	–	–	–	–	–	–	–	1050–1720
Eleutherodactylus mimus	–	–	–	–	–	–	–	+	–	–	–	+	+	–	–	–	–	0–940
Eleutherodactylus monnichorum	–	–	–	–	–	–	–	+	?	–	–	–	–	–	–	–	+	610–1829
Eleutherodactylus moro	–	–	–	–	–	–	–	+	–	–	–	–	–	–	–	–	–	550–1245
Eleutherodactylus museosus	–	–	–	–	–	–	–	+	–	–	–	–	–	–	–	–	+	700–1000
Eleutherodactylus noblei	–	–	–	–	–	–	+	–	–	–	–	+	+	–	–	–	+	0–1200
Eleutherodactylus occidentalis	+	+	+	–	–	–	–	–	–	–	–	–	–	+	–	–	–	0–2500
Eleutherodactylus omiltemanus	–	–	–	+	–	–	–	–	–	–	–	–	–	–	–	–	–	1829–2500
Eleutherodactylus pardalis	–	–	–	–	–	–	–	+	–	–	–	–	–	–	–	–	–	1050–1400
Eleutherodactylus planirostris	–	–	–	–	–	–	–	–	–	+	+	–	–	–	–	–	–	0–200
Eleutherodactylus podiciferus	+	–	–	–	–	–	–	+	–	–	–	–	–	–	–	–	–	1050–2134
Eleutherodactylus polymniae	+	–	–	–	–	–	–	–	–	–	–	–	–	–	–	–	–	1420
Eleutherodactylus pozo	–	–	–	–	–	+	–	–	–	–	–	–	–	–	–	–	–	700–1200
Eleutherodactylus psephosypharus	–	–	–	–	–	+	–	–	–	–	–	+	–	–	–	–	–	150–1170
Eleutherodactylus punctariolus	–	+	–	–	–	–	+	+	–	–	–	+	+	–	–	–	+	400–1800
Eleutherodactylus pygmaeus	+	+	+	+	+	–	–	–	–	–	–	+	–	–	–	–	–	80–2000
Eleutherodactylus raniformis	–	–	–	+	–	–	–	+	+	+	+	+	+	–	–	–	+	0–1400
Eleutherodactylus rayo	+	–	–	–	+	+	+	+	–	+	+	+	+	–	–	–	–	1050–1840
Eleutherodactylus rhodopis	–	–	–	–	–	+	+	–	–	–	–	+	+	–	–	–	–	0–1700
Eleutherodactylus ridens	–	–	–	–	–	+	+	+	–	–	–	+	+	–	–	–	+	0–1800
Eleutherodactylus rostralis	–	–	–	–	–	–	+	–	–	–	–	–	–	–	–	–	–	850–1300
Eleutherodactylus rugulosus	–	–	–	–	–	–	–	–	–	–	–	–	–	+	–	–	–	50–2120
Eleutherodactylus saltator	–	–	–	–	–	+	–	–	–	–	–	–	–	–	–	–	–	1760–2600
Eleutherodactylus sandersoni	–	–	–	–	+	+	–	–	–	–	–	+	–	–	–	–	–	0–900
Eleutherodactylus sartori	–	–	–	–	–	+	–	–	–	–	–	–	–	–	–	–	–	1829
Eleutherodactylus silvicola	+	–	–	–	–	–	–	–	–	–	–	–	–	–	–	–	–	1494
Eleutherodactylus spatulatus	+	–	–	–	–	–	–	–	–	–	–	–	–	–	–	–	–	1620–2235
Eleutherodactylus stadelmani	–	–	–	–	–	–	+	–	–	–	–	–	–	–	–	–	–	500–1460

Appendix 3:3 continued

Taxon	Middle American Highlands									Atlantic Coast Lowlands				Pacific Coast Lowlands				Elevation
	ORI	OCC	MC	SUR	LT	CG	HN	CRP	EP	TT	YP	GH	NP	SC	CGU	GCR	CP	Meters
Eleutherodactylus stejnegerianus	—	—	—	—	—	—	—	+	—	—	—	—	—	—	—	—	+	0–1170
Eleutherodactylus stuarti	—	—	—	—	—	+	—	—	—	—	—	—	—	—	—	—	+	1300–2250
Eleutherodactylus taeniatus	—	—	—	—	—	—	—	—	—	—	—	—	+	—	—	—	+	0–650
Eleutherodactylus talamancae	—	—	—	—	—	—	—	+	—	—	—	—	+	—	—	—	+	0–1050
Eleutherodactylus tarahumaraensis	—	+	—	—	—	—	—	—	—	—	—	—	—	—	—	—	—	2300–2500
Eleutherodactylus taurus	—	—	—	—	—	—	—	—	—	—	—	—	—	—	—	—	+	0–525
Eleutherodactylus taylori	—	—	—	—	—	+	—	—	—	—	—	—	—	—	—	—	—	1690
Eleutherodactylus trachydermus	—	—	—	—	—	+	—	—	—	—	—	+	—	—	—	—	—	900
Eleutherodactylus uno	—	—	—	+	—	—	—	—	—	—	—	—	—	—	—	—	—	2034
Eleutherodactylus vocalis	—	+	+	—	—	—	—	—	—	—	—	—	—	+	—	—	—	130–1650
Eleutherodactylus vocator	—	—	—	—	—	+	—	+	—	—	—	—	+	—	—	—	—	100–1106
Eleutherodactylus xucanebi	—	—	—	—	—	+	—	—	—	—	+	—	—	—	—	—	—	740–1700
Eleutherodactylus yucatanensis	—	—	—	—	—	—	—	—	—	—	+	—	+	—	—	—	—	0–200
Eleutherodactylus sp. A	—	—	—	—	—	—	—	—	—	—	—	—	—	—	—	—	—	?
Eleutherodactylus sp. B	—	—	—	—	—	+	—	—	—	—	—	+	—	—	—	—	—	0–800
Eleutherodactylus sp. C	—	—	—	—	—	—	—	—	—	—	—	+	—	—	—	—	—	300
Eleutherodactylus sp. D	—	—	—	—	—	+	—	—	—	—	—	—	—	—	—	—	—	1500–1800
Eleutherodactylus sp. E	—	—	—	—	—	+	—	—	—	—	—	—	—	—	—	—	—	500–1000
Leptodactylus bolivianus	—	—	—	—	—	—	—	—	—	—	—	—	—	—	—	—	+	0–800
Leptodactylus fuscus	—	—	—	—	—	—	—	—	—	—	—	—	—	—	—	—	+	0–600
Leptodactylus labialis	—	—	—	—	—	+	—	+	—	+	+	+	+	—	+	+	+	0–1300
Leptodactylus melanonotus	—	+	—	—	—	+	—	—	—	—	+	+	+	—	+	+	+	0–1440
Leptodactylus pentadactylus	—	—	—	—	—	—	—	+	+	—	+	+	+	—	+	+	+	0–1200
Leptodactylus poecilochilus	—	—	—	—	—	—	—	—	—	—	—	—	+	—	+	+	+	0–1150
Leptodactylus silvanimbus	—	—	—	—	+	—	+	—	—	+	—	—	—	—	+	+	+	1470–1870
Physalaemus pustulosus	—	—	—	—	+	+	—	—	—	+	—	—	—	—	+	+	+	0–1400
Pleurodema brachyops	—	—	—	—	—	—	—	—	—	—	—	—	—	—	—	—	+	0–200
Syrrhophus albolabris	—	—	—	+	—	—	—	—	—	—	—	—	—	—	—	—	—	820–1220
Syrrhophus angustidigitorum	—	—	+	—	—	—	—	—	—	—	—	—	—	—	—	—	—	1500–2682
Syrrhophus cystignathoides	—	—	—	—	—	—	—	—	—	—	—	—	—	—	—	—	—	100–1200
Syrrhophus dennisi	—	—	—	+	—	—	—	—	—	+	—	—	—	—	—	—	—	250
Syrrhophus dilatus	—	—	—	+	—	—	—	—	—	—	—	—	—	—	—	—	—	2134–2591
Syrrhophus guttilatus	+	—	+	—	—	—	—	—	—	—	—	—	—	—	—	—	—	600–2000
Syrrhophus interorbitalis	—	—	—	—	—	—	—	—	—	—	—	—	—	—	—	—	—	0–300
Syrrhophus leprus	—	—	—	—	+	—	—	—	—	+	+	+	—	+	—	—	—	0–715
Syrrhophus longipes	+	—	—	—	—	—	—	—	—	—	—	—	—	—	—	—	—	420–2000
Syrrhophus maurus	—	—	+	—	—	—	—	—	—	—	—	—	—	—	—	—	—	2100–2682
Syrrhophus modestus	—	—	+	+	—	—	—	—	—	—	—	—	—	+	—	—	—	0–914
Syrrhophus nitidus	+	+	+	+	—	—	—	—	—	—	—	—	—	+	—	—	—	20–2500

Appendix 3:3 continued

Taxon	Middle American Highlands									Atlantic Coast Lowlands				Pacific Coast Lowlands				Elevation
	ORI	OCC	MC	SUR	LT	CG	HN	CRP	EP	TT	YP	GH	NP	SC	CGU	GCR	CP	Meters
Syrrhophus nivicolimae	–	–	+	–	–	–	–	–	–	–	–	–	–	–	–	–	–	600–2400
Syrrhophus pallidus	–	–	–	–	–	–	–	–	–	–	–	–	–	+	–	–	–	0–300
Syrrhophus pipilans	–	–	–	+	–	–	–	–	–	–	–	–	–	+	–	–	–	0–1800
Syrrhophus rubrimaculatus	–	–	–	–	–	+	–	–	–	–	–	–	–	–	–	+	–	300–880
Syrrhophus rufescens	–	–	+	–	–	–	–	–	–	–	–	–	–	–	–	–	–	1800–2103
Syrrhophus saxatilis	–	+	–	+	–	–	–	–	–	–	–	–	–	–	–	–	–	1800–2100
Syrrhophus syristes	–	–	–	–	–	–	–	–	–	–	–	–	–	–	–	–	–	850–1620
Syrrhophus teretistes	–	+	–	–	–	–	–	–	–	–	–	–	–	–	–	–	–	840–1646
Syrrhophus verrucipes	+	–	+	–	–	–	–	–	–	–	–	–	–	–	–	–	–	1500–1850
ANURA: MICROHYLIDAE:																		
Chiasmocleis panamensis	–	–	–	–	–	–	–	–	–	–	–	–	–	–	–	–	+	0–200
Elachistocleis ovalis	–	–	–	–	–	–	–	–	–	–	–	–	+	–	–	–	+	0–300
Gastrophryne elegans	–	+	–	–	+	–	–	–	–	+	+	+	–	+	–	–	–	0–924
Gastrophryne olivacea	–	+	–	–	–	–	–	–	–	+	–	–	–	+	–	–	–	0–1300
Gastrophryne pictiventris	–	–	–	–	–	–	–	–	–	–	–	–	+	–	–	–	–	0–100
Gastrophryne usta	–	–	+	–	–	+	–	–	–	+	–	–	–	–	+	+	–	0–1006
Hypopachus barberi	+	+	–	+	–	+	+	–	–	+	+	–	–	–	–	–	–	1000–2300
Hypopachus variolosus	+	+	–	+	–	+	–	–	–	–	–	–	–	–	–	–	–	0–2200
Nelsonophryne aterrima	–	–	–	–	–	–	–	+	+	–	–	–	–	–	–	–	–	600–1450
Relictivomer pearsei	–	–	–	–	–	–	–	+	–	–	–	–	–	–	–	–	+	0–300
ANURA: PELOBATIDAE:																		
Scaphiopus couchii	+	+	+	–	–	–	–	–	–	+	–	–	–	+	–	–	–	0–1800
Spea multiplicata	–	+	+	+	–	–	–	–	–	–	–	–	–	–	–	–	–	600–2743
ANURA: PIPIDAE:																		
Pipa myersi	–	–	–	–	–	–	–	–	–	–	–	–	–	–	–	–	+	30
ANURA: RANIDAE:																		
Rana berlandieri	+	+	+	+	+	+	+	–	–	+	+	+	+	–	–	–	–	0–2500
Rana brownorum	–	–	–	–	–	–	–	–	–	+	+	–	–	–	–	–	–	0–300
Rana catesbeiana	–	–	–	–	–	–	–	–	–	+	–	–	–	–	–	–	–	0–300
Rana chiricahuensis	–	+	–	–	–	–	–	–	–	–	–	–	–	–	–	–	–	1200–2500
Rana dunni	–	–	+	–	–	–	–	–	–	–	–	–	–	–	–	–	–	2000–2165
Rana forreri	–	–	–	–	–	–	–	–	–	+	–	–	–	+	+	+	–	0–700
Rana johni	–	–	–	–	–	–	–	–	–	–	–	–	–	–	+	–	–	400
Rana juliani	–	–	–	–	–	+	–	–	–	–	–	–	–	–	+	–	–	450–920
Rana maculata	–	–	–	–	–	+	+	–	–	–	–	+	–	–	+	–	–	300–2700
Rana magnaocularis	–	–	–	–	–	–	–	–	–	–	–	–	–	+	–	–	–	0–1500
Rana montezumae	–	+	+	–	–	–	–	–	–	–	–	–	–	–	–	–	–	1500–2700
Rana neovolcanica	–	–	+	–	–	–	–	–	–	–	–	–	–	–	–	–	–	1500–2500
Rana omiltemana	–	–	–	+	–	–	–	–	–	–	–	–	–	–	–	–	–	2286–2438

Appendix 3:3 continued

Taxon	Middle American Highlands									Atlantic Coast Lowlands				Pacific Coast Lowlands				Elevation
	ORI	OCC	MC	SUR	LT	CG	HN	CRP	EP	TT	YP	GH	NP	SC	CGU	GCR	CP	Meters
Rana pueblae	+	–	–	–	–	–	–	–	–	–	–	–	–	–	–	–	–	1500
Rana pustulosa	–	+	+	–	–	–	–	–	–	–	–	–	–	+	–	–	–	500–2150
Rana sierramadrensis	–	–	–	+	–	–	–	–	–	–	–	–	–	–	–	–	–	823–1600
Rana spectabilis	+	+	+	+	–	–	–	–	–	–	–	–	–	–	–	–	–	1200–3200
Rana tarahumarae	–	+	–	–	–	–	–	–	–	–	–	–	–	–	–	–	–	460–1860
Rana taylori	–	–	–	–	–	–	–	–	–	–	–	–	+	–	–	–	–	0–368
Rana tlaloci	–	–	+	–	–	–	–	–	–	–	–	–	–	–	–	–	–	2225–2300
Rana trilobata	–	–	+	–	–	–	–	–	–	–	–	–	–	–	–	–	–	1500–1650
Rana vaillanti	–	–	–	–	–	+	+	–	–	+	+	+	+	–	+	+	–	0–990
Rana vibicaria	–	–	–	–	–	–	–	+	–	–	–	–	–	–	–	–	+	1200–2286
Rana warschewitschii	–	–	–	–	–	–	–	+	+	–	–	–	+	–	–	–	+	0–2500
Rana yavapaiensis	–	+	–	–	–	–	–	–	–	–	–	–	–	–	–	–	–	762–1524
Rana zweifeli	–	–	+	+	–	–	–	–	–	–	–	–	–	–	–	–	–	500–1700
Rana sp. A	–	–	+	–	–	–	–	–	–	–	–	–	–	–	–	–	–	1000
Rana sp. B	–	–	+	–	–	–	–	–	–	–	–	–	–	–	–	–	–	1000
Rana sp. C	–	–	–	–	–	+	–	–	–	–	–	–	–	–	–	–	–	1300–2000
Rana sp. D	–	–	–	–	–	–	–	+	–	–	–	–	–	–	–	–	–	1500–2000
Rana sp. E	–	–	–	–	–	–	–	–	–	–	–	–	+	–	–	–	–	0–600
ANURA: RHINOPHRYNIDAE:																		
Rhinophrynus dorsalis	–	–	–	–	+	–	–	–	–	+	+	+	–	+	+	+	–	0–670
CAUDATA: AMBYSTOMATIDAE:																		
Ambystoma altamirani	–	–	+	–	–	–	–	–	–	–	–	–	–	–	–	–	–	2000–3350
Ambystoma amblycephalum	–	–	+	–	–	–	–	–	–	–	–	–	–	–	–	–	–	2000–2800
Ambystoma andersoni	–	–	+	–	–	–	–	–	–	–	–	–	–	–	–	–	–	2000
Ambystoma bombypellum	–	–	+	–	–	–	–	–	–	–	–	–	–	–	–	–	–	2300–2600
Ambystoma dumerilii	–	–	+	–	–	–	–	–	–	–	–	–	–	–	–	–	–	2165
Ambystoma flavipiperatum	–	–	+	–	–	–	–	–	–	–	–	–	–	–	–	–	–	1493–2134
Ambystoma granulosum	–	–	+	–	–	–	–	–	–	–	–	–	–	–	–	–	–	2600–2750
Ambystoma leorae	–	–	+	–	–	–	–	–	–	–	–	–	–	–	–	–	–	2800–2900
Ambystoma lermaense	–	–	+	–	–	–	–	–	–	–	–	–	–	–	–	–	–	2500
Ambystoma mexicanum	–	–	+	–	–	–	–	–	–	–	–	–	–	–	–	–	–	2240–2450
Ambystoma ordinarium	–	–	+	–	–	–	–	–	–	–	–	–	–	–	–	–	–	2400–2750
Ambystoma rivulare	–	–	+	–	–	–	–	–	–	–	–	–	–	–	–	–	–	2600–3500
Ambystoma rosaceum	–	+	–	–	–	–	–	–	–	–	–	–	–	–	–	–	–	1675–3110
Ambystoma subsalsum	–	–	+	–	–	–	–	–	–	–	–	–	–	–	–	–	–	2500
Ambystoma taylori	–	–	+	–	–	–	–	–	–	–	–	–	–	–	–	–	–	2345
Ambystoma tigrinum	+	+	+	–	–	–	–	–	–	+	–	–	–	–	–	–	–	0–4000
Ambystoma velasci	–	–	+	–	–	–	–	–	–	–	–	–	–	–	–	–	–	2240–2500
Ambystoma zempoalensis	–	–	+	–	–	–	–	–	–	–	–	–	–	–	–	–	–	2750

Appendix 3:3 continued

Taxon	Middle American Highlands									Atlantic Coast Lowlands				Pacific Coast Lowlands				Elevation
	ORI	OCC	MC	SUR	LT	CG	HN	CRP	EP	TT	YP	GH	NP	SC	CGU	GCR	CP	Meters
CAUDATA: PLETHODONTIDAE:																		
Bolitoglossa alvaradoi	–	–	–	–	–	–	–	+	–	–	–	–	–	–	–	–	–	300–1116
Bolitoglossa arborescandens	–	–	–	–	–	–	–	+	–	–	–	–	–	–	–	–	–	1116
Bolitoglossa biseriata	–	–	–	–	–	–	–	–	–	–	–	–	–	–	–	–	+	50
Bolitoglossa carri	–	–	–	–	–	–	+	–	–	–	–	–	–	–	–	–	–	1840–2070
Bolitoglossa celaque	–	–	–	–	–	–	+	–	–	–	–	–	–	–	–	–	–	1930–2620
Bolitoglossa cerroensis	–	–	–	–	–	–	–	+	–	–	–	–	–	–	–	–	–	2134–2900
Bolitoglossa colonnea	–	–	–	–	–	–	–	+	+	–	–	–	+	–	–	–	+	183–1000
Bolitoglossa compacta	–	–	–	–	–	–	+	+	–	–	–	–	–	–	–	–	–	1810–2134
Bolitoglossa conanti	–	–	–	–	–	–	+	–	–	–	–	–	–	–	–	–	–	1370–1680
Bolitoglossa cuchumatana	–	–	–	–	–	+	–	–	–	–	–	–	–	–	–	–	–	1200–2500
Bolitoglossa cuna	–	–	–	–	–	–	–	–	–	–	–	–	+	–	–	–	–	0–50
Bolitoglossa diaphora	–	–	–	–	–	–	+	+	–	–	–	–	–	–	–	–	–	1540–2200
Bolitoglossa diminuta	–	–	–	–	–	–	–	+	–	–	–	–	–	–	–	–	–	1200–1300
Bolitoglossa dofleini	–	–	–	–	–	+	+	–	–	–	–	+	–	–	–	–	–	0–1300
Bolitoglossa dunni	–	–	–	–	–	–	+	–	–	–	–	–	–	–	–	–	–	1090–1900
Bolitoglossa engelhardti	–	–	–	–	–	+	–	+	–	–	–	–	–	–	–	–	–	1100–2600
Bolitoglossa epimela	–	–	–	–	–	–	–	+	–	–	–	–	–	–	–	–	–	775–1300
Bolitoglossa flavimembris	–	–	–	–	–	+	–	–	–	–	–	–	–	+	–	–	–	1800–2440
Bolitoglossa flaviventris	–	–	–	–	–	+	–	–	–	–	–	–	–	–	–	–	–	0–600
Bolitoglossa franklini	–	–	–	–	–	+	–	–	–	–	–	–	–	–	–	–	–	1500–3000
Bolitoglossa gracilis	–	–	–	–	–	–	–	+	–	–	–	–	–	–	–	–	–	1200–1800
Bolitoglossa hartwegi	–	–	–	–	–	+	–	–	–	–	–	–	–	–	–	–	–	1200–3000
Bolitoglossa helmrichi	–	–	–	–	–	+	–	–	–	–	–	–	–	–	–	–	–	1300–2290
Bolitoglossa hermosa	–	–	–	+	–	–	–	–	–	–	–	–	–	–	–	–	–	775–825
Bolitoglossa jacksoni	–	–	–	–	–	+	–	–	–	–	–	–	–	–	–	–	–	1400
Bolitoglossa lignicolor	–	–	–	–	–	–	–	+	–	–	–	–	+	–	–	–	+	0–1220
Bolitoglossa lincolni	–	–	–	–	–	+	–	–	–	–	–	–	–	–	–	–	–	2000–3000
Bolitoglossa macrinii	–	–	–	+	–	–	–	+	–	–	–	–	–	–	–	–	–	500–1860
Bolitoglossa marmorea	–	–	–	–	–	–	–	–	+	–	–	–	–	–	–	–	–	3200–3500
Bolitoglossa medemi	–	–	–	–	–	–	–	–	–	–	–	+	–	–	–	–	–	50–800
Bolitoglossa meliana	–	–	–	–	–	+	–	+	–	–	–	–	–	–	–	–	–	1550–2730
Bolitoglossa mexicana	–	–	–	–	+	+	+	–	–	+	+	–	–	–	–	–	–	0–1900
Bolitoglossa minutula	–	–	–	–	–	–	–	+	–	–	–	–	–	–	–	–	–	1810–2000
Bolitoglossa morio	–	–	–	–	–	+	–	–	–	–	–	–	–	–	–	–	–	1300–3000
Bolitoglossa mulleri	–	–	–	–	–	+	–	–	–	–	–	–	–	–	–	–	–	100–1500
Bolitoglossa nigrescens	–	–	–	–	–	–	+	+	–	–	–	–	–	–	–	–	–	1300–2600
Bolitoglossa occidentalis	–	–	–	–	–	+	–	–	–	–	–	–	–	–	–	+	–	0–1600
Bolitoglossa odonnelli	–	–	–	–	–	+	–	–	–	–	–	+	–	–	–	–	–	0–1200

Appendix 3:3 continued

Taxon	Middle American Highlands									Atlantic Coast Lowlands				Pacific Coast Lowlands				Elevation Meters
	ORI	OCC	MC	SUR	LT	CG	HN	CRP	EP	TT	YP	GH	NP	SC	CGU	GCR	CP	
Bolitoglossa phalarosoma	–	–	–	–	–	–	–	–	+	–	–	–	–	–	–	–	–	730–960
Bolitoglossa platydactyla	+	–	–	–	+	–	–	–	–	+	–	–	–	–	–	–	–	0–1300
Bolitoglossa porrasorum	–	–	–	–	–	–	+	–	–	–	–	–	–	–	–	–	–	1550–1920
Bolitoglossa riletti	–	–	–	–	+	–	–	–	–	–	–	–	–	–	–	–	–	700–1030
Bolitoglossa robusta	–	–	–	–	–	–	–	+	–	–	–	–	–	–	–	–	–	457–1830
Bolitoglossa rostrata	–	–	–	–	–	+	–	–	–	–	–	+	–	–	–	–	–	2500–3480
Bolitoglossa rufescens	+	–	–	–	+	+	+	–	–	+	+	+	–	–	–	–	–	0–1500
Bolitoglossa salvinii	–	–	–	–	–	–	+	–	–	–	–	–	–	–	–	–	–	600–1400
Bolitoglossa schizodactyla	–	–	–	–	–	–	–	+	–	–	–	–	+	–	–	–	+	0–850
Bolitoglossa schmidti	–	–	–	–	–	–	+	–	–	–	–	–	–	–	–	–	–	610–650
Bolitoglossa sooyorum	–	–	–	–	–	–	–	+	–	–	–	–	–	–	–	–	–	2400–2860
Bolitoglossa striatula	–	–	–	–	–	–	–	+	–	–	–	+	+	–	–	–	–	50–800
Bolitoglossa stuarti	–	–	–	–	–	+	–	–	–	–	–	–	–	–	–	–	–	950–1615
Bolitoglossa subpalmata	–	–	–	–	–	–	–	+	–	–	–	–	–	–	–	–	–	1375–3350
Bolitoglossa taylori	–	–	–	–	–	–	–	–	+	–	–	–	–	–	–	–	–	900–1440
Bolitoglossa veracrucis	–	–	–	–	–	–	–	–	–	+	–	–	–	–	–	–	–	107
Bolitoglossa yucatana	–	–	–	–	–	–	–	–	–	–	+	–	–	–	–	–	–	0–300
Bolitoglossa sp. A	–	–	–	–	–	–	–	–	–	–	–	–	+	–	–	–	–	?
Bolitoglossa sp. B	–	–	–	–	–	–	–	–	–	–	–	–	–	–	–	–	+	?
Bolitoglossa sp. C	–	–	–	–	–	–	–	–	–	–	–	–	–	–	–	–	–	?
Bolitoglossa sp. D	–	–	–	–	–	+	–	–	–	–	–	–	+	–	–	–	–	1500
Bolitoglossa sp. E	–	–	–	–	–	+	–	–	–	–	–	–	–	–	–	–	–	2000–2500
Bradytriton silus	–	–	–	–	–	+	–	–	–	–	–	–	–	–	–	–	–	1310–1370
Chiropterotriton arboreus	+	–	–	–	–	–	–	–	–	–	–	–	–	–	–	–	–	1950
Chiropterotriton chiropterus	+	–	–	–	–	–	–	–	–	–	–	–	–	–	–	–	–	2100–3658
Chiropterotriton chondrostega	+	–	–	–	–	–	–	–	–	–	–	–	–	–	–	–	–	1524–2042
Chiropterotriton cracens	+	–	–	–	–	–	–	–	–	–	–	–	–	–	–	–	–	914–1890
Chiropterotriton dimidiatus	+	–	–	–	–	–	–	–	–	–	–	–	–	–	–	–	–	2550–3300
Chiropterotriton lavae	+	–	–	–	–	–	–	–	–	–	–	–	–	–	–	–	–	2000–2100
Chiropterotriton magnipes	+	–	–	–	–	–	–	–	–	–	–	–	–	–	–	–	–	1170–1810
Chiropterotriton mosaueri	+	–	–	–	–	–	–	–	–	–	–	–	–	–	–	–	–	2195
Chiropterotriton multidentatus	+	–	–	–	–	–	–	–	–	–	–	–	–	–	–	–	–	1067–2743
Chiropterotriton orculus	+	–	–	–	–	–	–	–	–	–	–	–	–	–	–	–	–	2775
Chiropterotriton priscus	+	–	–	–	–	–	–	–	–	–	–	–	–	–	–	–	–	2438–3658
Chiropterotriton terrestris	+	–	–	–	–	–	–	–	–	–	–	–	–	–	–	–	–	2010–2700
Chiropterotriton sp. A	–	–	+	–	–	–	–	–	–	–	–	–	–	–	–	–	–	3000–3100
Chiropterotriton sp. B	–	–	+	–	–	–	–	–	–	–	–	–	–	–	–	–	–	3075
Chiropterotriton sp. C	+	–	–	–	–	–	–	–	–	–	–	–	–	–	–	–	–	2500
Chiropterotriton sp. D	+	–	–	–	–	–	–	–	–	–	–	–	–	–	–	–	–	3000

Appendix 3:3 continued

| Taxon | Middle American Highlands | | | | | | | | | Atlantic Coast Lowlands | | | | Pacific Coast Lowlands | | | | Elevation |
	ORI	OCC	MC	SUR	LT	CG	HN	CRP	EP	TT	YP	GH	NP	SC	CGU	GCR	CP	Meters
Chiropterotriton sp. E	+	–	–	–	–	–	–	–	–	–	–	–	–	–	–	–	–	1150
Chiropterotriton sp. F	+	–	–	–	–	–	–	–	–	–	–	–	–	–	–	–	–	3000
Chiropterotriton sp. G	+	–	–	–	–	–	–	–	–	–	–	–	–	–	–	–	–	2700
Chiropterotriton sp. H	+	–	–	–	–	–	–	–	–	–	–	–	–	–	–	–	–	1400–2100
Chiropterotriton sp. I	–	–	–	+	–	–	–	–	–	–	–	–	–	–	–	–	–	2850
Chiropterotriton sp. J	+	–	–	–	–	–	–	–	–	–	–	–	–	–	–	–	–	2800
Dendrotriton bromeliacia	–	–	–	–	–	+	–	–	–	–	–	–	–	–	–	–	–	1700–2750
Dendrotriton cuchumatanus	–	–	–	–	–	+	–	–	–	–	–	–	–	–	–	–	–	2860
Dendrotriton megarhinus	–	–	–	–	–	+	–	–	–	–	–	–	–	–	–	–	–	2100–2425
Dendrotriton rabbi	–	–	–	–	–	+	–	–	–	–	–	–	–	–	–	–	–	2100–2700
Dendrotriton xolocalcae	–	–	–	–	–	+	–	–	–	–	–	–	–	–	–	–	–	1635–2440
Ixalotriton niger	–	–	–	–	–	+	–	–	–	–	–	–	–	–	–	–	–	1068
Lineatriton lineola	+	–	–	–	+	–	–	–	–	–	–	–	–	–	–	–	–	200–1220
Nototriton abscondens	–	–	–	–	–	–	–	+	–	–	–	–	–	–	–	–	–	1200–2000
Nototriton adelos	+	–	–	–	–	–	–	–	–	–	–	–	–	–	–	–	–	1530–2050
Nototriton alvarezdeltoroi	–	–	–	–	–	+	–	–	–	–	–	–	–	–	–	–	–	1550–2000
Nototriton barbouri	–	–	–	–	–	–	+	–	–	–	–	–	–	–	–	–	–	750–1700
Nototriton guanacaste	–	–	–	–	–	+	–	+	–	–	–	–	–	–	–	–	–	1580
Nototriton major	–	–	–	–	–	+	–	+	–	–	–	–	–	–	–	–	–	1200
Nototriton nasalis	–	–	–	–	–	–	+	–	–	–	–	–	–	–	–	–	–	1372–2900
Nototriton picadoi	–	–	–	–	–	–	–	+	–	–	–	–	–	–	–	–	–	1300–2200
Nototriton richardi	–	–	–	–	–	–	–	+	–	–	–	–	–	–	–	–	–	1200–1980
Nototriton tapanti	–	–	–	–	–	–	–	+	–	–	–	–	–	–	–	–	–	1300
Nototriton veraepacis	–	–	–	–	–	+	–	–	–	–	–	–	–	–	–	–	–	1610–2290
Nyctanolis pernix	–	–	–	–	–	+	+	–	–	–	–	–	–	–	–	–	–	1290–1650
Oedipina alfaroi	–	–	–	–	–	–	–	–	–	–	–	–	+	–	–	–	–	19–50
Oedipina altura	–	–	–	–	–	–	–	+	–	–	–	–	–	–	–	–	–	2100–2400
Oedipina carablanca	–	–	–	–	–	–	–	–	–	–	–	–	+	–	–	–	–	0–300
Oedipina collaris	–	–	–	–	–	–	–	–	–	–	–	–	+	–	–	–	+	120–610
Oedipina complex	–	–	–	–	–	–	–	+	–	–	–	–	+	–	–	–	+	0–1250
Oedipina cyclocauda	–	–	–	–	–	–	+	+	–	–	–	+	+	–	–	–	–	0–1400
Oedipina elongata	–	–	–	–	–	+	+	–	–	–	+	–	–	–	–	–	–	0–770
Oedipina gephyra	–	–	–	–	–	–	+	–	–	–	–	–	–	–	–	–	–	1690–1810
Oedipina grandis	–	–	–	–	–	–	+	–	–	–	–	–	–	–	–	–	–	1810–1950
Oedipina ignea	–	–	–	–	–	–	+	–	–	–	–	–	–	–	–	–	–	1000–1750
Oedipina parvipes	–	–	–	–	–	–	–	+	–	–	–	–	–	–	–	–	+	0–865
Oedipina paucidentata	–	–	–	–	–	–	–	+	–	–	–	–	–	–	–	–	–	2100–2300
Oedipina poelzi	–	–	–	–	–	–	–	+	–	–	–	–	–	–	–	–	–	700–1790
Oedipina pseudouniformis	–	–	–	–	–	–	–	+	–	–	–	–	–	–	–	+	–	215–1220

Appendix 3:3 continued

| Taxon | Middle American Highlands | | | | | | | | | Atlantic Coast Lowlands | | | | Pacific Coast Lowlands | | | | Elevation |
	ORI	OCC	MC	SUR	LT	CG	HN	CRP	EP	TT	YP	GH	NP	SC	CGU	GCR	CP	Meters
Oedipina stenopodia	–	–	–	–	–	+	–	–	–	–	–	–	–	–	–	–	–	900–1500
Oedipina stuarti	–	–	–	–	–	–	+	–	–	–	–	–	–	–	–	+	–	0–975
Oedipina taylori	–	–	–	–	–	–	+	–	–	–	–	+	–	–	–	+	–	100–1050
Oedipina uniformis	–	–	–	–	–	–	–	+	–	–	–	–	+	–	–	–	+	0–2300
Parvimolge townsendi	+	–	–	–	–	–	–	–	–	–	–	–	–	–	–	–	–	1000–2440
Pseudoeurycea altamontana	–	+	–	–	–	–	–	–	–	–	–	–	–	–	–	–	–	3050–3200
Pseudoeurycea anitae	–	–	+	+	–	–	–	–	–	–	–	–	–	–	–	–	–	2100–2300
Pseudoeurycea bellii	+	+	+	+	–	–	–	–	–	–	–	–	–	–	–	–	–	1050–4267
Pseudoeurycea brunnata	+	–	–	–	–	+	–	–	–	–	–	–	–	–	–	–	–	2400–3000
Pseudoeurycea cephalica	+	+	–	–	–	–	–	–	–	–	–	–	–	–	–	–	–	1000–3960
Pseudoeurycea cochranae	–	–	–	+	–	–	–	–	–	–	–	–	–	–	–	–	–	2000–2900
Pseudoeurycea conanti	–	–	–	+	–	–	–	–	–	–	–	–	–	–	–	–	–	900–2500
Pseudoeurycea exspectata	–	–	–	–	–	+	–	–	–	–	–	–	–	–	–	–	–	2500–2600
Pseudoeurycea firscheini	+	–	–	–	–	–	–	–	–	–	–	–	–	–	–	–	–	2130–2290
Pseudoeurycea gadovii	+	+	–	–	–	–	–	–	–	–	–	–	–	–	–	–	–	2590–5000
Pseudoeurycea galeanae	+	–	–	–	–	–	–	–	–	–	–	–	–	–	–	–	–	1585–2134
Pseudoeurycea goebeli	–	–	–	–	–	+	–	–	–	–	–	–	–	–	–	–	–	2400–3200
Pseudoeurycea juarezi	+	–	+	–	–	–	–	–	–	–	–	–	–	–	–	–	–	2520–3160
Pseudoeurycea leprosa	+	+	+	–	–	–	–	–	–	–	–	–	–	–	–	–	–	2450–3200
Pseudoeurycea longicauda	–	–	+	–	–	–	–	–	–	–	–	–	–	–	–	–	–	2800–2970
Pseudoeurycea melanomolga	+	–	–	–	–	–	–	–	–	–	–	–	–	–	–	–	–	2134–3800
Pseudoeurycea mystax	–	–	–	+	–	–	–	–	–	–	–	–	–	–	–	–	–	1951
Pseudoeurycea nigromaculata	+	–	–	–	–	–	–	–	–	–	–	–	–	–	–	–	–	1200–1500
Pseudoeurycea parva	–	–	–	–	–	+	–	–	–	–	–	–	–	–	–	–	–	1600–1860
Pseudoeurycea praecellens	+	–	–	–	–	–	–	–	–	–	–	–	–	–	–	–	–	1000–1200
Pseudoeurycea rex	–	–	–	–	–	+	–	–	–	–	–	–	–	–	–	–	–	2450–4000
Pseudoeurycea robertsi	+	+	+	–	–	–	–	–	–	–	–	–	–	–	–	–	–	2900–3353
Pseudoeurycea saltator	+	–	–	–	–	–	–	–	–	–	–	–	–	–	–	–	–	1580–2050
Pseudoeurycea scandens	+	–	–	–	–	–	–	–	–	–	–	–	–	–	–	–	–	1000–1800
Pseudoeurycea smithi	+	–	–	+	–	–	–	–	–	–	–	–	–	–	–	–	–	2750–3050
Pseudoeurycea unguidentis	–	–	–	+	–	–	–	–	–	–	–	–	–	–	–	–	–	2200–3100
Pseudoeurycea werleri	–	–	–	–	+	–	–	–	–	–	–	–	–	–	–	–	–	900–1650
Pseudoeurycea sp. A (Oaxaca)	+	–	–	–	–	–	–	–	–	–	–	–	–	–	–	–	–	2158
Pseudoeurycea sp. B (San Marcos)	–	–	–	–	–	+	–	–	–	–	–	–	–	–	–	–	–	2500–2800
Pseudoeurycea sp. C (Veracruz)	+	–	–	–	–	–	–	–	–	–	–	–	–	–	–	–	–	2900–3100
Pseudoeurycea sp. D (Veracruz)	+	–	–	–	–	–	–	–	–	–	–	–	–	–	–	–	–	2000–2200
Pseudoeurycea sp. E (Oaxaca)	+	–	–	–	–	–	–	–	–	–	–	–	–	–	–	–	–	2800–2900
Pseudoeurycea sp. F (Oaxaca)	+	–	–	–	–	–	–	–	–	–	–	–	–	–	–	–	–	2900–3000
Pseudoeurycea sp. G (Guerrero)	–	–	–	+	–	–	–	–	–	–	–	–	–	–	–	–	–	3296

Taxon	Middle American Highlands									Atlantic Coast Lowlands				Pacific Coast Lowlands				Elevation
	ORI	OCC	MC	SUR	LT	CG	HN	CRP	EP	TT	YP	GH	NP	SC	CGU	GCR	CP	Meters
Pseudoeurycea sp. H (Guerrero)	—	—	—	—	—	—	—	—	—	—	—	—	—	—	—	—	—	3425
Pseudoeurycea sp. I (Guerrero)	—	—	—	+	—	—	—	—	—	—	—	—	—	—	—	—	—	2650–2966
Pseudoeurycea sp. J (Guerrero)	—	—	—	+	—	—	—	—	—	—	—	—	—	—	—	—	—	2560
Pseudoeurycea sp. K (Guerrero)	—	—	—	+	—	—	—	—	—	—	—	—	—	—	—	—	—	2200–2600
Thorius arboreus	+	—	—	—	—	—	—	—	—	—	—	—	—	—	—	—	—	2170–2755
Thorius aureus	+	—	—	—	—	—	—	—	—	—	—	—	—	—	—	—	—	2475–2930
Thorius boreas	+	—	—	—	—	—	—	—	—	—	—	—	—	—	—	—	—	2885–2950
Thorius dubius	+	—	—	—	—	—	—	—	—	—	—	—	—	—	—	—	—	2130–2500
Thorius insperatus	+	—	—	—	—	—	—	—	—	—	—	—	—	—	—	—	—	1500
Thorius macdougalli	+	—	—	—	—	—	—	—	—	—	—	—	—	—	—	—	—	2200–3160
Thorius minutissimus	—	—	—	+	—	—	—	—	—	—	—	—	—	—	—	—	—	2150–2800
Thorius narismagnus	—	—	—	—	+	—	—	—	—	—	—	—	—	—	—	—	—	1000–1220
Thorius narisovalis	+	—	—	+	—	—	—	—	—	—	—	—	—	—	—	—	—	2600–3185
Thorius pennatulus	+	—	—	—	—	—	—	—	—	—	—	—	—	—	—	—	—	800–1200
Thorius pulmonaris	+	—	—	+	—	—	—	—	—	—	—	—	—	—	—	—	—	2000–3160
Thorius schmidti	+	—	—	—	—	—	—	—	—	—	—	—	—	—	—	—	—	2470–2760
Thorius smithi	+	—	—	—	—	—	—	—	—	—	—	—	—	—	—	—	—	800–1550
Thorius troglodytes	+	—	—	—	—	—	—	—	—	—	—	—	—	—	—	—	—	2130–2500
Thorius sp. A (Veracruz)	+	—	—	—	—	—	—	—	—	—	—	—	—	—	—	—	—	2500
Thorius sp. B (Veracruz)	+	—	—	—	—	—	—	—	—	—	—	—	—	—	—	—	—	2500–2640
Thorius sp. C (Guerrero)	—	—	—	+	—	—	—	—	—	—	—	—	—	—	—	—	—	2500–3110
Thorius sp. D (Veracruz)	+	—	—	—	—	—	—	—	—	—	—	—	—	—	—	—	—	2500–2700
Thorius sp. E (Veracruz)	+	—	—	—	—	—	—	—	—	—	—	—	—	—	—	—	—	2450–2550
CAUDATA: SALAMANDRIDAE:																		
Notophthalmus meridionalis	—	—	—	—	—	—	—	—	—	+	—	—	—	—	—	—	—	0–300
CAUDATA: SIRENIDAE:																		
Siren intermedia	—	—	—	—	—	—	—	—	—	+	—	—	—	—	—	—	—	0–300
GYMNOPHIONA: CAECILIIDAE:																		
Caecilia leucocephala	—	—	—	—	—	—	—	—	—	—	—	—	—	—	—	—	+	0–300
Caecilia nigricans	—	—	—	—	—	—	—	—	—	—	—	—	+	—	—	—	—	914
Caecilia tentaculata	—	—	—	—	—	—	—	—	—	—	—	—	—	—	—	—	+	0–200
Caecilia volcani	—	—	—	—	—	—	—	—	—	—	—	—	—	—	—	—	+	550
Dermophis mexicanus	—	—	—	—	+	—	—	—	—	+	+	+	+	+	+	—	+	0–1500
Dermophis oaxacae	—	—	+	—	+	+	—	—	—	—	—	—	—	+	—	—	—	45–1500
Dermophis parviceps	—	—	—	—	—	—	—	+	—	—	—	—	+	—	—	—	+	50–1200
Gymnopis multiplicata	—	—	—	—	—	—	+	+	—	—	—	+	+	—	—	—	+	0–1400
Gymnopis syntrema	—	—	—	—	—	+	+	—	—	—	—	—	—	—	—	—	—	300–900
Oscaecilia elongata	—	—	—	—	—	—	—	—	—	—	—	—	—	—	—	—	—	50
Oscaecilia ochrocephala	—	—	—	—	—	—	—	—	—	—	—	—	+	—	—	—	+	0–300
Oscaecilia osae	—	—	—	—	—	—	—	—	—	—	—	—	—	—	—	—	+	0–10

APPENDIX 3:4

Distribution of species of amphibians in Middle America by vegetation type. Abbreviations are: LMD = Lower Montane Dry Forest; LMM = Lower Montane Moist Forest; LMW = Lower Montane Wet Forest; MM = Montane Moist Forest; MW = Montane Wet Forest; SD = Subtropical Dry Forest; SM = Subtropical Moist Forest; SRW = Subtropical Rainforest and Subtropical Wet Forest; TD = Tropical Dry Forest; TVD = Tropical Very Dry Forest; TWM = Tropical Wet Forest and Tropical Moist Forest.

| | Elevational Range and Vegetation Type | | | | | | | | | | |
| | 0–600 m | | | 600–1500 m | | | 1500–2700 m | | | >2700 m | |
Taxon	TWM	TD	TVD	SRW	SM	SD	LMW	LMM	LMD	MW	MM
ANURA: BUFONIDAE:											
Atelophryniscus chrysophorus	−	−	−	+	−	−	+	−	−	−	−
Atelopus certus	−	−	−	+	−	−	−	−	−	−	−
Atelopus chiriquiensis	−	−	−	+	−	−	+	−	−	−	−
Atelopus glyphus	−	−	−	+	−	−	−	−	−	−	−
Atelopus limosus	+	−	−	−	−	−	−	−	−	−	−
Atelopus senex	−	−	−	+	−	−	+	−	−	−	−
Atelopus varius	+	−	−	+	−	−	+	−	−	−	−
Bufo bocourti	−	−	−	+	−	−	+	+	−	−	+
Bufo campbelli	+	−	−	+	−	−	−	−	−	−	−
Bufo canaliferus	+	+	−	+	−	−	−	−	−	−	−
Bufo cavifrons	−	−	−	+	−	−	+	−	−	−	−
Bufo coccifer	−	+	+	−	+	+	−	−	+	−	−
Bufo cognatus	−	−	−	−	−	+	−	−	+	−	−
Bufo compactilis	−	−	−	−	−	−	−	−	+	−	−
Bufo coniferus	+	−	−	+	−	−	−	−	−	−	−
Bufo cristatus	−	−	−	+	−	−	−	−	−	−	−
Bufo debilis	−	−	−	−	−	+	−	−	+	−	−
Bufo fastidiosus	−	−	−	+	−	−	+	−	−	−	−
Bufo gemmifer	−	+	−	−	−	−	−	−	−	−	−
Bufo granulosus	+	−	−	−	−	−	−	−	−	−	−
Bufo haematiticus	+	−	−	+	−	−	−	−	−	−	−
Bufo holdridgei	−	−	−	−	−	−	+	−	−	−	−
Bufo ibarrai	−	−	−	+	+	−	+	+	−	−	−
Bufo kelloggi	−	+	−	−	−	−	−	−	−	−	−
Bufo luetkenii	−	+	+	−	−	+	−	−	−	−	−
Bufo marinus	+	+	+	+	+	+	−	+	+	−	−
Bufo marmoreus	−	+	+	−	−	+	−	−	−	−	−
Bufo mazatlanensis	−	+	−	−	−	+	−	−	−	−	−
Bufo melanochloris	−	−	−	+	−	−	−	−	−	−	−
Bufo microscaphus	−	−	−	−	−	−	−	+	−	−	−
Bufo occidentalis	−	−	−	−	+	+	−	+	+	−	−
Bufo periglenes	−	−	−	−	−	−	+	−	−	−	−
Bufo peripatetes	−	−	−	−	−	−	+	−	−	−	−
Bufo perplexus	−	+	+	−	−	+	−	−	−	−	−
Bufo punctatus	−	+	+	−	+	+	−	−	+	−	+
Bufo tacanensis	−	−	−	+	−	−	+	−	−	−	−
Bufo typhonius	+	−	−	+	−	−	−	−	−	−	−
Bufo valliceps	+	+	+	+	+	−	−	−	−	−	−
Bufo woodhousii	−	−	−	−	−	−	−	+	+	−	−
Crepidophryne epioticus	−	−	−	+	−	−	−	−	−	−	−
Rhamphophryne acrolopha	−	−	−	+	−	−	−	−	−	−	−
ANURA: CENTROLENIDAE:											
Centrolene ilex	+	−	−	+	−	−	−	−	−	−	−
Centrolene prosoblepon	+	−	−	+	−	−	+	−	−	−	−
Cochranella albomaculata	+	−	−	+	−	−	−	−	−	−	−
Cochranella euknemos	+	−	−	+	−	−	−	−	−	−	−
Cochranella granulosa	+	−	−	+	−	−	−	−	−	−	−

Appendix 3:4 continued

Taxon	Elevational Range and Vegetation Type										
	0–600 m			600–1500 m			1500–2700 m			>2700 m	
	TWM	TD	TVD	SRW	SM	SD	LMW	LMM	LMD	MW	MM
Cochranella spinosa	+	–	–	–	–	–	–	–	–	–	–
Hyalinobatrachium chirripoi	+	–	–	–	–	–	–	–	–	–	–
Hyalinobat. colymbiphyllum	+	–	–	+	–	–	+	–	–	–	–
Hyalinobatrachium fleischmanni	+	–	–	+	–	–	+	–	–	–	–
Hyalinobatrachium pulveratum	+	–	–	+	–	–	–	–	–	–	–
Hyalinobatrachium talamancae	–	–	–	+	–	–	–	–	–	–	–
Hyalinobatrachium valerioi	+	–	–	+	–	–	–	–	–	–	–
Hyalinobatrachium vireovittata	–	–	–	+	–	–	–	–	–	–	–
ANURA: DENDROBATIDAE:											
Colostethus chocoensis	+	–	–	–	–	–	–	–	–	–	–
Colostethus flotator	+	–	–	–	–	–	–	–	–	–	–
Colostethus inguinalis	+	–	–	+	–	–	–	–	–	–	–
Colostethus latinasus	+	–	–	+	–	–	–	–	–	–	–
Colostethus nubicola	+	–	–	+	–	–	+	–	–	–	–
Colostethus pratti	+	–	–	+	–	–	–	–	–	–	–
Colostethus talamancae	+	–	–	+	–	–	–	–	–	–	–
Dendrobates arboreus	+	–	–	+	–	–	–	–	–	–	–
Dendrobates auratus	+	–	–	+	–	–	–	–	–	–	–
Dendrobates granuliferus	+	–	–	+	–	–	–	–	–	–	–
Dendrobates pumilio	+	–	–	+	–	–	–	–	–	–	–
Dendrobates speciosus	–	–	–	+	–	–	+	–	–	–	–
Epipedobates maculatus	?	–	–	?	–	–	–	–	–	–	–
Minyobates fulguritus	+	–	–	+	–	–	–	–	–	–	–
Minyobates minutus	+	–	–	+	–	–	–	–	–	–	–
Phyllobates lugubris	+	–	–	–	–	–	–	–	–	–	–
Phyllobates vittatus	+	–	–	–	–	–	–	–	–	–	–
ANURA: HYLIDAE:											
Agalychnis annae	+	–	–	+	–	–	–	–	–	–	–
Agalychnis calcarifer	+	–	–	+	–	–	–	–	–	–	–
Agalychnis callidryas	+	+	–	+	+	–	–	–	–	–	–
Agalychnis litodryas	+	–	–	–	–	–	–	–	–	–	–
Agalychnis moreletii	+	–	–	+	+	–	+	–	–	–	–
Agalychnis saltator	+	–	–	+	–	–	–	–	–	–	–
Agalychnis spurrelli	+	–	–	+	–	–	–	–	–	–	–
Anotheca spinosa	+	–	–	+	–	–	+	–	–	–	–
Duellmanohyla chamulae	–	–	–	+	–	–	+	–	–	–	–
Duellmanohyla ignicolor	–	–	–	–	–	–	+	–	–	–	–
Duellmanohyla lythrodes	+	–	–	+	–	–	–	–	–	–	–
Duellmanohyla rufioculis	–	–	–	+	–	–	+	–	–	–	–
Duellmanohyla salvavida	–	–	–	+	–	–	–	–	–	–	–
Duellmanohyla schmidtorum	–	–	–	+	–	–	+	–	–	–	–
Duellmanohyla soralia	+	–	–	+	–	–	–	–	–	–	–
Duellmanohyla uranochroa	+	–	–	+	–	–	+	–	–	–	–
Gastrotheca cornuta	+	–	–	+	–	–	–	–	–	–	–
Gastrotheca nicefori	–	–	–	+	–	–	+	–	–	–	–
Hemiphractus fasciatus	+	–	–	+	–	–	–	–	–	–	–
Hyla altipotens	–	–	–	+	–	–	+	+	–	–	–
Hyla angustilineata	–	–	–	–	–	–	+	–	–	–	–
Hyla arborescandens	–	–	–	–	–	–	+	+	–	+	+
Hyla arenicolor	–	+	+	–	+	–	–	+	–	–	+
Hyla bistincta	–	–	–	–	–	–	+	+	–	+	+
Hyla boans	+	–	–	–	–	–	–	–	–	–	–
Hyla bocourti	–	–	–	+	+	–	–	–	–	–	–
Hyla bogertae	–	–	–	–	–	–	–	+	–	–	+
Hyla bromeliacia	–	–	–	+	–	–	–	–	–	–	–
Hyla calvicollina	–	–	–	–	–	–	+	–	–	+	–

Appendix 3:4 continued

Taxon	0–600 m			600–1500 m			1500–2700 m			>2700 m	
	TWM	TD	TVD	SRW	SM	SD	LMW	LMM	LMD	MW	MM
Hyla catracha	−	−	−	−	−	−	−	+	−	−	−
Hyla cembra	−	−	−	−	−	−	−	−	+	−	−
Hyla chaneque	−	−	−	+	−	−	+	−	−	−	−
Hyla charadricola	−	−	−	−	−	−	−	+	−	−	−
Hyla chimalapa	−	−	−	+	+	−	−	−	−	−	−
Hyla chryses	−	−	−	−	−	−	+	+	−	−	−
Hyla colymba	+	−	−	+	−	−	−	−	−	−	−
Hyla crassa	−	−	−	−	−	−	−	+	−	−	−
Hyla crepitans	+	−	−	−	−	−	−	−	−	−	−
Hyla cyanomma	−	−	−	−	−	−	+	−	−	−	−
Hyla debilis	−	−	−	+	−	−	+	−	−	−	−
Hyla dendroscarta	−	−	−	+	−	−	+	−	−	−	−
Hyla ebraccata	+	+	−	+	−	−	−	−	−	−	−
Hyla echinata	−	−	−	+	−	−	−	−	−	−	−
Hyla euphorbiacea	−	−	−	−	−	−	−	+	−	−	+
Hyla eximia	−	−	−	+	+	−	+	+	−	−	+
Hyla fimbrimembra	−	−	−	+	−	−	+	−	−	−	−
Hyla godmani	−	+	−	−	+	−	−	−	−	−	−
Hyla graceae	−	−	−	+	−	−	+	−	−	−	−
Hyla hazelae	−	−	−	−	−	−	+	−	−	+	+
Hyla insolita	−	−	−	−	−	−	−	+	−	+	−
Hyla juanitae	−	−	−	−	+	−	−	−	−	−	−
Hyla labeculata	−	−	−	−	−	−	+	−	−	−	−
Hyla lancasteri	+	−	−	+	−	−	+	−	−	−	−
Hyla loquax	+	+	+	−	+	−	−	−	−	−	−
Hyla melanomma	−	−	−	+	+	−	+	+	−	−	−
Hyla microcephala	+	+	+	+	+	+	−	−	−	−	−
Hyla miliaria	−	−	−	+	−	−	−	−	−	−	−
Hyla minera	−	−	−	+	−	−	+	−	−	−	−
Hyla miotympanum	+	+	−	+	+	−	+	+	−	−	−
Hyla mixe	−	−	−	−	−	−	+	−	−	−	−
Hyla mixomaculata	−	−	−	+	−	−	−	−	−	−	−
Hyla mykter	−	−	−	−	−	−	+	+	−	−	−
Hyla nubicola	−	−	−	+	−	−	−	−	−	−	−
Hyla pachyderma	−	−	−	−	−	−	+	−	−	−	−
Hyla palmeri	−	−	−	+	−	−	−	−	−	−	−
Hyla pellita	−	−	−	+	−	−	+	−	−	−	−
Hyla pentheter	−	−	−	−	+	−	−	+	−	−	−
Hyla perkinsi	−	−	−	+	−	−	−	−	−	−	−
Hyla phlebodes	+	+	−	−	−	−	−	−	−	−	−
Hyla picadoi	−	−	−	−	−	−	+	−	−	−	−
Hyla picta	+	+	−	+	+	−	−	−	−	−	−
Hyla pictipes	−	−	−	−	−	−	+	−	−	−	−
Hyla pinorum	−	−	−	+	+	−	−	−	−	−	−
Hyla plicata	−	−	−	−	−	−	−	+	−	−	+
Hyla pseudopuma	−	−	−	+	+	−	+	−	−	−	−
Hyla pugnax	+	−	−	−	−	−	−	−	−	−	−
Hyla rivularis	−	−	+	−	−	+	−	−	+	−	−
Hyla robertmertensi	+	+	−	−	−	−	−	−	−	−	−
Hyla robertsorum	−	−	−	−	−	−	−	+	−	−	+
Hyla rosenbergi	+	+	−	−	−	−	−	−	−	−	−
Hyla rufitela	+	−	−	−	−	−	−	−	−	−	−
Hyla sabrina	−	−	−	−	−	−	+	−	−	−	−
Hyla salvaje	−	−	−	+	−	−	−	−	−	−	−
Hyla sartori	−	+	−	−	−	−	−	−	−	−	−
Hyla siopela	−	−	−	−	−	−	−	+	−	−	−

Appendix 3:4 continued

| | Elevational Range and Vegetation Type | | | | | | | | | | |
| | 0–600 m | | | 600–1500 m | | | 1500–2700 m | | | >2700 m | |
Taxon	TWM	TD	TVD	SRW	SM	SD	LMW	LMM	LMD	MW	MM
Hyla smaragdina	–	+	+	–	–	+	–	–	–	–	–
Hyla smithii	–	+	+	–	+	+	–	–	–	–	–
Hyla subocularis	+	–	–	–	–	–	–	–	–	–	–
Hyla sumichrasti	–	+	–	–	–	+	–	–	–	–	–
Hyla taeniopus	–	–	–	+	–	–	+	–	–	–	–
Hyla thorectes	–	–	–	–	–	–	+	+	–	–	–
Hyla thysanota	–	–	–	+	–	–	–	–	–	–	–
Hyla tica	–	–	–	+	–	–	+	–	–	–	–
Hyla trux	–	–	–	–	–	–	+	+	–	–	–
Hyla valancifer	–	–	–	+	–	–	–	+	–	–	–
Hyla walkeri	–	–	–	–	–	–	–	+	–	–	–
Hyla xanthosticta	–	–	–	–	–	–	+	–	–	–	–
Hyla xera	–	–	–	–	–	+	–	–	–	–	–
Hyla zeteki	–	–	–	+	–	–	+	–	–	–	–
Hyla sp.A	–	–	–	–	+	–	–	–	–	–	–
Pachymedusa dacnicolor	–	+	+	–	+	+	–	–	–	–	–
Phrynohyas venulosa	+	+	+	–	+	+	–	–	–	–	–
Phyllomedusa lemur	–	–	–	+	–	–	+	–	–	–	–
Phyllomedusa venusta	+	–	–	–	–	–	–	–	–	–	–
Plectrohyla acanthodes	–	–	–	–	–	–	+	+	–	–	–
Plectrohyla avia	–	–	–	–	–	–	+	–	–	–	–
Plectrohyla chrysopleura	–	–	–	+	–	–	–	–	–	–	–
Plectrohyla dasypus	–	–	–	–	–	–	+	–	–	–	–
Plectrohyla glandulosa	–	–	–	–	–	–	+	+	–	+	+
Plectrohyla guatemalensis	–	–	–	+	+	–	+	+	–	+	–
Plectrohyla hartwegi	–	–	–	+	–	–	+	–	–	–	–
Plectrohyla ixil	–	–	–	+	–	–	+	–	–	–	–
Plectrohyla lacertosa	–	–	–	–	–	–	+	–	–	–	–
Plectrohyla matudai	–	–	–	+	–	–	+	+	–	–	–
Plectrohyla pokomchi	–	–	–	+	–	–	+	–	–	–	–
Plectrohyla pycnochila	–	–	–	–	–	–	–	+	–	–	–
Plectrohyla quecchi	–	–	–	+	–	–	+	–	–	–	–
Plectrohyla sagorum	–	–	–	–	–	–	+	–	–	–	–
Plectrohyla tecunumani	–	–	–	–	–	–	–	–	–	–	+
Plectrohyla teuchestes	–	–	–	+	–	–	–	–	–	–	–
Pternohyla dentata	–	–	–	–	–	–	–	–	+	–	–
Pternohyla fodiens	–	+	+	–	–	+	–	–	–	–	–
Ptychohyla erythromma	–	–	–	+	+	–	–	–	–	–	–
Ptychohyla euthysanota	–	–	–	+	–	–	+	+	–	–	–
Ptychohyla hypomykter	+	–	–	+	+	–	+	+	–	–	–
Ptychohyla legleri	–	–	–	+	–	–	+	–	–	–	–
Ptychohyla leonhardschultzei	–	–	–	+	–	–	+	+	–	–	–
Ptychohyla macrotympanum	–	–	–	+	–	–	+	+	–	–	–
Ptychohyla panchoi	+	–	–	+	–	–	–	–	–	–	–
Ptychohyla salvadorensis	–	–	–	+	+	–	–	+	–	–	–
Ptychohyla sanctaecrucis	+	–	–	+	–	–	–	–	–	–	–
Ptychohyla spinipollex	–	–	–	+	–	–	–	–	–	–	–
Scinax boulengeri	+	+	–	–	–	–	–	–	–	–	–
Scinax elaeochroa	+	+	–	+	+	–	–	–	–	–	–
Scinax rostrata	+	+	–	–	–	–	–	–	–	–	–
Scinax rubra	+	+	–	–	–	–	–	–	–	–	–
Scinax staufferi	+	+	+	–	+	+	–	–	–	–	–
Smilisca baudinii	+	+	+	+	+	+	+	+	+	–	–
Smilisca cyanosticta	+	–	–	+	–	–	–	–	–	–	–
Smilisca phaeota	+	–	–	+	–	–	–	–	–	–	–
Smilisca puma	+	–	–	–	–	–	–	–	–	–	–

Appendix 3:4 continued

| | Elevational Range and Vegetation Type | | | | | | | | | | |
| | 0–600 m | | | 600–1500 m | | | 1500–2700 m | | | >2700 m | |
Taxon	TWM	TD	TVD	SRW	SM	SD	LMW	LMM	LMD	MW	MM
Smilisca sila	+	+	−	+	+	−	−	−	−	−	−
Smilisca sordida	+	+	−	+	+	−	−	−	−	−	−
Triprion petasatus	+	+	−	−	−	−	−	−	−	−	−
Triprion spatulatus	−	+	+	−	−	−	−	−	−	±	−
ANURA: LEPTODACTYLIDAE:											
Eleutherodactylus achatinus	+	−	−	+	−	−	−	−	−	−	−
Eleutherodactylus adamastus	−	−	−	+	−	−	−	−	−	−	−
Eleutherodactylus alfredi	+	+	−	−	−	−	−	−	−	−	−
Eleutherodactylus altae	−	−	−	+	−	−	−	−	−	−	−
Eleutherodactylus ancianao	−	−	−	−	−	−	−	+	−	−	−
Eleutherodactylus andi	−	−	−	+	−	−	−	−	−	−	−
Eleutherodactylus angelicus	−	−	−	+	−	−	+	−	−	−	−
Eleutherodactylus antillensis	+	−	−	−	−	−	−	−	−	−	−
Eleutherodactylus aphanus	−	−	−	+	−	−	−	−	−	−	−
Eleutherodactylus augusti	−	+	+	−	+	+	−	+	+	−	−
Eleutherodactylus aurilegulus	−	−	−	+	−	−	−	−	−	−	−
Eleutherodactylus azueroensis	+	−	−	+	−	−	−	−	−	−	−
Eleutherodactylus batrachylus	−	−	−	−	+	−	−	−	−	−	−
Eleutherodactylus berkenbuschi	+	−	−	+	+	−	+	−	−	−	−
Eleutherodactylus biporcatus	+	−	−	+	−	−	−	−	−	−	−
Eleutherodactylus bocourti	−	−	−	+	−	−	+	−	−	−	−
Eleutherodactylus bransfordii	+	+	−	+	+	−	+	−	−	−	−
Eleutherodactylus brocchi	−	−	−	+	−	−	+	−	−	−	−
Eleutherodactylus bufoniformis	+	−	−	−	−	−	−	−	−	−	−
Eleuth. caryophyllaceus	−	−	−	+	−	−	+	−	−	−	−
Eleutherodactylus cerasinus	+	−	−	+	−	−	+	−	−	−	−
Eleutherodactylus chuc	+	−	−	+	−	−	−	−	−	−	−
Eleutherodactylus chrysozetetes	−	−	−	+	−	−	−	−	−	−	−
Eleutherodactylus crassidigitus	+	−	−	+	−	−	+	−	−	−	−
Eleutherodactylus cruentus	+	−	−	+	−	−	+	−	−	−	−
Eleutherodactylus cruzi	−	−	−	−	−	−	+	−	−	−	−
Eleutherodactylus cuaquero	−	−	−	−	−	−	+	−	−	−	−
Eleutherodactylus daryi	−	−	−	−	−	−	+	−	−	−	−
Eleutherodactylus decoratus	+	−	−	+	−	−	+	−	−	−	−
Eleutherodactylus diastema	+	−	−	+	−	−	+	−	−	−	−
Eleutherodactylus emcelae	−	−	−	+	−	−	−	−	−	−	−
Eleutherodactylus escoces	−	−	−	+	−	−	+	−	−	−	−
Eleutherodactylus fitzingeri	+	−	−	+	−	−	−	−	−	−	−
Eleutherodactylus fleischmanni	−	−	−	+	−	−	+	−	−	−	−
Eleutherodactylus gaigei	+	−	−	+	−	−	−	−	−	−	−
Eleutherodactylus glaucus	−	−	−	−	−	−	−	+	−	−	−
Eleutherodactylus gollmeri	+	−	−	+	−	−	−	−	−	−	−
Eleutherodactylus greggi	−	−	−	−	−	−	+	−	−	+	−
Eleutherodactylus guerreroensis	−	−	−	−	+	−	−	−	−	−	−
Eleutherodactylus hobartsmithi	−	−	−	−	−	+	−	+	−	−	−
Eleutherodactylus hylaeformis	−	−	−	−	−	−	+	−	−	−	−
Eleutherodactylus johnstonei	−	+	−	−	−	−	−	−	−	−	−
Eleutherodactylus jota	−	−	−	+	−	−	−	−	−	−	−
Eleutherodactylus laticeps	+	−	−	+	−	−	+	−	−	−	−
Eleutherodactylus latidiscus	+	−	−	−	−	−	−	−	−	−	−
Eleutherodactylus lineatus	−	−	−	+	−	−	+	−	−	−	−
Eleutherodactylus longirostris	+	−	−	+	−	−	−	−	−	−	−
Eleutherodactylus matudai	−	−	−	+	−	−	+	−	−	−	−
Eleuth. megalotympanum	−	−	−	+	−	−	−	−	−	−	−
Eleutherodactylus melanostictus	−	−	−	+	−	−	+	−	−	−	−
Eleutherodactylus merendonensis	+	−	−	−	−	−	−	−	−	−	−

Appendix 3:4 continued

| | Elevational Range and Vegetation Type | | | | | | | | | | |
| | 0–600 m | | | 600–1500 m | | | 1500–2700 m | | | >2700 m | |
Taxon	TWM	TD	TVD	SRW	SM	SD	LMW	LMM	LMD	MW	MM
Eleutherodactylus mexicanus	–	–	–	–	–	–	+	+	–	–	–
Eleutherodactylus milesi	–	–	–	+	–	–	+	–	–	–	–
Eleutherodactylus mimus	+	–	–	+	–	–	–	–	–	–	–
Eleutherodactylus monnichorum	–	–	–	+	–	–	+	–	–	–	–
Eleutherodactylus moro	–	–	–	+	–	–	–	–	–	–	–
Eleutherodactylus museosus	–	–	–	+	–	–	–	–	–	–	–
Eleutherodactylus noblei	+	–	–	+	–	–	–	–	–	–	–
Eleutherodactylus occidentalis	–	+	–	–	+	+	–	+	+	–	–
Eleutherodactylus omiltemanus	–	–	–	–	–	–	+	+	–	–	–
Eleutherodactylus pardalis	–	–	–	+	–	–	–	–	–	–	–
Eleutherodactylus planirostris	+	–	–	–	–	–	–	–	–	–	–
Eleutherodactylus podiciferus	–	–	–	+	–	–	+	–	–	–	–
Eleutherodactylus polymniae	–	–	–	+	–	–	–	–	–	–	–
Eleutherodactylus pozo	–	–	–	+	–	–	–	–	–	–	–
Eleutherodactylus psephosypharus	+	–	–	+	–	–	–	–	–	–	–
Eleutherodactylus punctariolus	+	–	–	+	–	–	+	–	–	–	–
Eleutherodactylus pygmaeus	+	+	–	+	+	–	+	+	–	–	–
Eleutherodactylus raniformis	+	–	–	+	–	–	–	–	–	–	–
Eleutherodactylus rayo	–	–	–	+	–	–	+	–	–	–	–
Eleutherodactylus rhodopis	+	+	–	+	–	–	+	–	–	–	–
Eleutherodactylus ridens	+	–	–	+	–	–	+	–	–	–	–
Eleutherodactylus rostralis	–	–	–	+	–	–	–	–	–	–	–
Eleutherodactylus rugulosus	–	+	+	–	+	+	+	+	–	–	–
Eleutherodactylus saltator	–	–	–	–	–	–	+	+	–	–	–
Eleutherodactylus sandersoni	+	–	–	+	–	–	–	–	–	–	–
Eleutherodactylus sartori	–	–	–	–	–	–	+	–	–	–	–
Eleutherodactylus silvicola	–	–	–	+	–	–	–	–	–	–	–
Eleutherodactylus spatulatus	–	–	–	–	–	–	+	+	–	–	–
Eleutherodactylus stadelmani	–	–	–	+	–	–	–	–	–	–	–
Eleutherodactylus stejnegerianus	+	–	–	+	–	–	–	–	–	–	–
Eleutherodactylus stuarti	–	–	–	+	–	–	+	–	–	–	–
Eleutherodactylus taeniatus	+	–	–	–	–	–	–	–	–	–	–
Eleutherodactylus talamancae	+	–	–	+	–	–	–	–	–	–	–
Eleuth. tarahumaraensis	–	–	–	–	–	–	–	+	–	–	–
Eleutherodactylus taurus	+	–	–	–	–	–	–	–	–	–	–
Eleutherodactylus taylori	–	–	–	–	–	–	+	–	–	–	–
Eleutherodactylus trachydermus	–	–	–	+	–	–	–	–	–	–	–
Eleutherodactylus uno	–	–	–	–	–	–	+	–	–	–	–
Eleutherodactylus vocalis	–	+	+	–	+	+	–	+	–	–	–
Eleutherodactylus vocator	+	–	–	+	–	–	–	–	–	–	–
Eleutherodactylus xucanebi	–	–	–	+	–	–	+	–	–	–	–
Eleutherodactylus yucatanensis	–	+	+	–	–	–	–	–	–	–	–
Eleutherodactylus sp. A	–	–	–	?	–	–	–	–	–	–	–
Eleutherodactylus sp. B	+	–	–	+	–	–	–	–	–	–	–
Eleutherodactylus sp. C	–	–	+	–	–	–	–	–	–	–	–
Eleutherodactylus sp. D	–	–	–	–	–	–	+	–	–	–	–
Eleutherodactylus sp. E	+	–	–	+	–	–	–	–	–	–	–
Leptodactylus bolivianus	+	–	–	–	–	–	–	–	–	–	–
Leptodactylus fuscus	+	+	–	–	–	–	–	–	–	–	–
Leptodactylus labialis	+	+	+	+	+	+	–	–	–	–	–
Leptodactylus melanonotus	+	+	+	–	+	+	–	–	–	–	–
Leptodactylus pentadactylus	+	+	–	+	+	–	–	–	–	–	–
Leptodactylus poecilochilus	+	+	–	+	–	–	–	–	–	–	–
Leptodactylus silvanimbus	–	–	–	–	–	–	+	–	–	–	–
Physalaemus pustulosus	+	+	–	–	+	+	–	–	–	–	–
Pleurodema brachyops	–	+	–	–	–	–	–	–	–	–	–

Appendix 3:4 continued

Taxon	Elevational Range and Vegetation Type										
	0–600 m			600–1500 m			1500–2700 m			>2700 m	
	TWM	TD	TVD	SRW	SM	SD	LMW	LMM	LMD	MW	MM
Syrrhophus albolabris	–	–	–	–	+	+	–	–	–	–	–
Syrrhophus angustidigitorum	–	–	–	–	–	–	–	+	–	–	–
Syrrhophus cystignathoides	–	+	–	–	+	+	–	–	–	–	–
Syrrhophus dennisi	–	+	–	–	–	–	–	–	–	–	–
Syrrhophus dilatus	–	–	–	–	–	–	–	+	–	–	–
Syrrhophus guttilatus	–	–	–	–	+	+	–	+	+	–	–
Syrrhophus interorbitalis	–	+	–	–	–	–	–	–	–	–	–
Syrrhophus leprus	+	–	–	–	+	–	–	–	–	–	–
Syrrhophus longipes	–	+	–	–	+	–	–	+	–	–	–
Syrrhophus maurus	–	–	–	–	–	–	–	+	–	–	–
Syrrhophus modestus	–	+	–	–	–	+	–	–	–	–	–
Syrrhophus nitidus	–	+	+	–	+	+	–	+	+	–	–
Syrrhophus nivicolimae	–	–	–	–	+	–	–	+	–	–	–
Syrrhophus pallidus	–	+	–	–	–	–	–	–	–	–	–
Syrrhophus pipilans	–	+	–	–	+	+	–	+	–	–	–
Syrrhophus rubrimaculatus	+	–	–	+	–	–	–	–	–	–	–
Syrrhophus rufescens	–	–	–	–	–	–	–	+	+	–	–
Syrrhophus saxatilis	–	–	–	–	–	–	–	+	–	–	–
Syrrhophus syristes	–	–	–	–	+	–	–	+	–	–	–
Syrrhophus teretistes	–	–	–	–	–	+	–	–	–	–	–
Syrrhophus verrucipes	–	–	–	–	–	–	+	–	–	–	–
ANURA: MICROHYLIDAE:											
Chiasmocleis panamensis	+	+	–	–	–	–	–	–	–	–	–
Elachistocleis ovalis	+	–	–	–	–	–	–	–	–	–	–
Gastrophryne elegans	+	+	–	–	+	–	–	–	–	–	–
Gastrophryne olivacea	–	+	–	–	–	+	–	–	–	–	–
Gastrophryne pictiventris	+	–	–	–	–	–	–	–	–	–	–
Gastrophryne usta	+	+	–	–	+	–	–	–	–	–	–
Hypopachus barberi	–	–	–	+	+	–	+	+	–	–	–
Hypopachus variolosus	+	+	+	–	+	+	–	–	+	–	–
Nelsonophryne aterrima	–	–	–	+	–	–	–	–	–	–	–
Relictivomer pearsei	+	+	–	–	–	–	–	–	–	–	–
ANURA: PELOBATIDAE:											
Scaphiopus couchii	–	+	+	–	+	+	–	–	+	–	–
Spea multiplicata	–	–	–	–	+	+	–	+	+	–	–
ANURA: PIPIDAE:											
Pipa myersi	+	–	–	–	–	–	–	–	–	–	–
ANURA: RANIDAE:											
Rana berlandieri	+	+	–	+	+	–	+	–	–	–	–
Rana brownorum	+	+	–	–	–	–	–	–	–	–	–
Rana catesbeiana	–	+	–	–	–	–	–	–	–	–	–
Rana chiricahuensis	–	–	–	–	+	–	–	+	+	–	–
Rana dunni	–	–	–	–	–	–	–	+	–	–	–
Rana forreri	+	+	+	–	–	–	–	–	–	–	–
Rana johni	–	+	–	–	–	–	–	–	–	–	–
Rana juliani	+	–	–	+	+	–	–	–	–	–	–
Rana maculata	+	–	–	+	+	–	+	+	–	–	–
Rana magnaocularis	–	+	+	–	–	+	–	–	–	–	–
Rana montezumae	–	–	–	–	–	–	–	+	+	–	–
Rana neovolcanica	–	–	–	–	–	–	–	+	+	–	–
Rana omiltemana	–	–	–	–	–	–	+	+	–	–	–
Rana pueblae	–	–	–	–	–	–	–	+	–	–	–
Rana pustulosa	–	–	–	–	+	+	–	+	+	–	–
Rana sierramadrensis	–	–	–	–	+	–	–	+	–	–	–
Rana spectabilis	–	–	–	–	+	–	–	+	–	–	+
Rana tarahumarae	–	–	–	–	+	–	–	+	–	–	–

Appendix 3:4 continued

| Taxon | Elevational Range and Vegetation Type | | | | | | | | | | |
| | 0–600 m | | | 600–1500 m | | | 1500–2700 m | | | >2700 m | |
	TWM	TD	TVD	SRW	SM	SD	LMW	LMM	LMD	MW	MM
Rana taylori	+	–	–	–	–	–	–	–	–	–	–
Rana tlaloci	–	–	–	–	–	–	–	+	–	–	–
Rana trilobata	–	–	–	–	–	–	–	–	+	–	–
Rana vaillanti	+	+	–	–	+	–	–	–	–	–	–
Rana vibicaria	–	–	–	+	+	–	+	–	–	–	–
Rana warschewitschii	+	+	–	+	+	–	+	–	–	–	–
Rana yavapaiensis	–	–	–	–	–	+	–	–	–	–	–
Rana zweifeli	–	–	–	–	+	+	+	+	–	–	–
Rana sp. A	–	–	–	–	+	–	–	–	–	–	–
Rana sp. B	–	–	–	–	+	–	–	–	–	–	–
Rana sp. C	–	–	–	–	+	–	–	+	–	–	–
Rana sp. D	–	–	–	–	–	–	+	–	–	–	–
Rana sp. E	+	–	–	–	–	–	–	–	–	–	–
ANURA: RHINOPHRYNIDAE:											
Rhinophrynus dorsalis	+	+	–	–	–	–	–	–	–	–	–
CAUDATA: AMBYSTOMATIDAE:											
Ambystoma altamirani	–	–	–	–	–	–	–	+	–	–	+
Ambystoma amblycephalum	–	–	–	–	–	–	–	+	+	–	+
Ambystoma andersoni	–	–	–	–	–	–	–	+	–	–	–
Ambystoma bombypellum	–	–	–	–	–	–	–	+	–	–	–
Ambystoma dumerilii	–	–	–	–	–	–	–	+	–	–	–
Ambystoma flavipiperatum	–	–	–	–	–	–	–	+	–	–	–
Ambystoma granulosum	–	–	–	–	–	–	–	+	–	–	+
Ambystoma leorae	–	–	–	–	–	–	–	–	–	–	+
Ambystoma lermaense	–	–	–	–	–	–	–	+	–	–	–
Ambystoma mexicanum	–	–	–	–	–	–	–	+	–	–	–
Ambystoma ordinarium	–	–	–	–	–	–	–	+	–	–	+
Ambystoma rivularis	–	–	–	–	–	–	–	+	–	–	+
Ambystoma rosaceum	–	–	–	–	–	–	–	+	–	–	+
Ambystoma subsalsum	–	–	–	–	–	–	–	–	+	–	–
Ambystoma taylori	–	–	–	–	–	–	–	–	+	–	–
Ambystoma tigrinum	–	+	+	–	+	+	–	+	+	–	+
Ambystoma velasci	–	–	–	–	–	–	–	+	–	–	–
Ambystoma zempoalensis	–	–	–	–	–	–	–	–	–	–	+
CAUDATA: PLETHODONTIDAE:											
Bolitoglossa alvaradoi	+	–	–	+	–	–	–	–	–	–	–
Bolitoglossa arborescandens	–	–	–	+	–	–	–	–	–	–	–
Bolitoglossa biseriata	+	–	–	–	–	–	–	–	–	–	–
Bolitoglossa carri	–	–	–	–	–	–	+	–	–	–	–
Bolitoglossa celaque	–	–	–	–	–	–	+	–	–	–	–
Bolitoglossa cerroensis	–	–	–	–	–	–	+	–	–	+	–
Bolitoglossa colonnea	+	–	–	+	–	–	–	–	–	–	–
Bolitoglossa compacta	–	–	–	–	–	–	+	–	–	–	–
Bolitoglossa conanti	–	–	–	–	+	–	–	+	–	–	–
Bolitoglossa cuchumatana	–	–	–	+	+	–	+	+	–	–	–
Bolitoglossa cuna	+	–	–	–	–	–	–	–	–	–	–
Bolitoglossa diaphora	–	–	–	–	–	–	+	–	–	–	–
Bolitoglossa diminuta	–	–	–	+	–	–	–	–	–	–	–
Bolitoglossa dofleini	+	–	–	+	–	–	–	–	–	–	–
Bolitoglossa dunni	–	–	–	+	–	–	+	–	–	–	–
Bolitoglossa engelhardti	–	–	–	+	–	–	+	–	–	–	–
Bolitoglossa epimela	–	–	–	+	–	–	–	–	–	–	–
Bolitoglossa flavimembris	–	–	–	–	–	–	+	–	–	–	–
Bolitoglossa flaviventris	+	+	–	–	–	–	–	–	–	–	–
Bolitoglossa franklini	–	–	–	–	–	–	+	–	–	+	–
Bolitoglossa gracilis	–	–	–	+	–	–	+	–	–	–	–

Appendix 3:4 continued

| | Elevational Range and Vegetation Type | | | | | | | | | | |
| | 0–600 m | | | 600–1500 m | | | 1500–2700 m | | | >2700 m | |
Taxon	TWM	TD	TVD	SRW	SM	SD	LMW	LMM	LMD	MW	MM
Bolitoglossa hartwegi	–	–	–	+	–	–	+	+	–	+	+
Bolitoglossa helmrichi	–	–	–	+	–	–	+	–	–	–	–
Bolitoglossa hermosa	–	–	–	–	+	–	–	–	–	–	–
Bolitoglossa jacksoni	–	–	–	+	–	–	–	–	–	–	–
Bolitoglossa lignicolor	+	–	–	+	–	–	–	–	–	–	–
Bolitoglossa lincolni	–	–	–	–	–	–	+	+	–	+	+
Bolitoglossa macrinii	–	+	–	–	+	–	–	+	–	–	–
Bolitoglossa marmorea	–	–	–	–	–	–	–	–	–	+	–
Bolitoglossa medemi	+	–	+	–	–	–	–	–	–	–	–
Bolitoglossa meliana	–	–	–	–	–	–	+	+	–	–	–
Bolitoglossa mexicana	+	–	–	+	–	–	+	–	–	–	–
Bolitoglossa minutula	–	–	–	–	–	–	+	–	–	–	–
Bolitoglossa morio	–	–	–	+	–	–	+	+	–	+	+
Bolitoglossa mulleri	+	–	–	+	–	–	–	–	–	–	–
Bolitoglossa nigrescens	–	–	–	+	–	–	+	–	–	–	–
Bolitoglossa occidentalis	+	+	–	+	–	–	–	–	–	–	–
Bolitoglossa odonnelli	+	–	–	+	–	–	–	–	–	–	–
Bolitoglossa phalarosoma	–	–	–	+	–	–	–	–	–	–	–
Bolitoglossa platydactyla	+	+	–	+	–	–	–	–	–	–	–
Bolitoglossa porrasorum	–	–	–	–	–	–	+	–	–	–	–
Bolitoglossa riletti	–	–	–	–	+	+	–	–	–	–	–
Bolitoglossa robusta	–	–	–	+	–	–	+	–	–	–	–
Bolitoglossa rostrata	–	–	–	–	–	–	+	–	–	+	+
Bolitoglossa rufescens	+	+	–	+	–	–	–	–	–	–	–
Bolitoglossa salvinii	–	–	–	+	–	–	–	–	–	–	–
Bolitoglossa schizodactyla	+	+	–	+	–	–	–	–	–	–	–
Bolitoglossa schmidti	–	–	–	+	–	–	–	–	–	–	–
Bolitoglossa sooyorum	–	–	–	–	–	–	+	–	–	+	–
Bolitoglossa striatula	+	–	–	+	–	–	–	–	–	–	–
Bolitoglossa stuarti	–	–	–	–	–	+	–	+	–	–	–
Bolitoglossa subpalmata	–	–	–	–	–	–	+	–	–	+	–
Bolitoglossa taylori	–	–	–	+	–	–	–	–	–	–	–
Bolitoglossa veracrucis	+	–	–	–	–	–	–	–	–	–	–
Bolitoglossa yucatana	–	+	+	–	–	–	–	–	–	–	–
Bolitoglossa sp. A	–	–	–	?	–	–	–	–	–	–	–
Bolitoglossa sp. B	–	–	–	?	–	–	–	–	–	–	–
Bolitoglossa sp. C	–	–	–	?	–	–	–	–	–	–	–
Bolitoglossa sp. D	–	–	–	+	–	–	–	–	–	–	–
Bolitoglossa sp. E	–	–	–	–	–	–	+	–	–	–	–
Bradytriton silus	–	–	–	+	–	–	–	–	–	–	–
Chiropterotriton arboreus	–	–	–	–	–	–	+	–	–	–	–
Chiropterotriton chiropterus	–	–	–	–	–	–	–	+	–	–	+
Chiropterotriton chondrostega	–	–	–	–	–	–	+	–	–	–	–
Chiropterotriton cracens	–	–	–	+	–	–	+	–	–	–	–
Chiropterotriton dimidiatus	–	–	–	–	–	–	–	+	–	–	+
Chiropterotriton lavae	–	–	–	–	–	–	+	–	–	–	–
Chiropterotriton magnipes	–	–	–	+	+	–	–	+	–	–	–
Chiropterotriton mosaueri	–	–	–	–	–	–	–	+	–	–	–
Chiropterotriton multidentatus	–	–	–	+	–	–	+	–	–	–	–
Chiropterotriton orculus	–	–	–	–	–	–	–	–	–	+	–
Chiropterotriton priscus	–	–	–	–	–	–	–	+	–	–	+
Chiropterotriton terrestris	–	–	–	–	–	–	+	–	–	–	–
Chiropterotriton sp. A	–	–	–	–	–	–	–	–	–	–	+
Chiropterotriton sp. B	–	–	–	–	–	–	–	–	–	–	+
Chiropterotriton sp. C	–	–	–	–	–	–	–	+	–	–	–
Chiropterotriton sp. D	–	–	–	–	–	–	–	–	–	–	+

Appendix 3:4 continued

Taxon	Elevational Range and Vegetation Type										
	0–600 m			600–1500 m			1500–2700 m			>2700 m	
	TWM	TD	TVD	SRW	SM	SD	LMW	LMM	LMD	MW	MM
Chiropterotriton sp. E	–	–	–	+	–	–	–	–	–	–	–
Chiropterotriton sp. F	–	–	–	–	–	–	–	–	–	–	+
Chiropterotriton sp. G	–	–	–	–	–	–	–	–	–	–	+
Chiropterotriton sp. H	–	–	–	–	–	–	+	–	–	–	–
Chiropterotriton sp. I	–	–	–	–	–	–	–	–	–	–	+
Chiropterotriton sp. J	–	–	–	–	–	–	–	–	–	–	+
Dendrotriton bromeliacia	–	–	–	–	–	–	+	–	–	–	–
Dendrotriton cuchumatanus	–	–	–	–	–	–	–	–	–	+	–
Dendrotriton megarhinus	–	–	–	–	–	–	+	–	–	–	–
Dendrotriton rabbi	–	–	–	–	–	–	+	–	–	–	–
Dendrotriton xolocalcae	–	–	–	–	–	–	+	–	–	–	–
Ixalotriton niger	–	–	–	–	+	–	–	–	–	–	–
Lineatriton lineola	–	–	–	+	+	–	–	–	–	–	–
Nototriton abscondens	–	–	–	+	–	–	+	–	–	–	–
Nototriton adelos	–	–	–	–	–	–	+	–	–	–	–
Nototriton alvarezdeltoroi	–	–	–	–	–	–	+	–	–	–	–
Nototriton barbouri	–	–	–	+	–	–	+	–	–	–	–
Nototriton guanacaste	–	–	–	–	–	–	+	–	–	–	–
Nototriton major	–	–	–	+	–	–	–	–	–	–	–
Nototriton nasalis	–	–	–	+	–	–	+	–	–	+	–
Nototriton picadoi	–	–	–	–	–	–	+	–	–	–	–
Nototriton richardi	–	–	–	+	–	–	+	–	–	–	–
Nototriton tapanti	–	–	–	+	–	–	–	–	–	–	–
Nototriton veraepacis	–	–	–	–	–	–	+	–	–	–	–
Nyctanolis pernix	–	–	–	+	–	–	+	–	–	–	–
Oedipina alfaroi	+	–	–	–	–	–	–	–	–	–	–
Oedipina altura	–	–	–	–	–	–	+	–	–	–	–
Oedipina carablanca	+	–	–	–	–	–	–	–	–	–	–
Oedipina collaris	+	–	–	–	–	–	–	–	–	–	–
Oedipina complex	+	+	–	–	+	–	–	–	–	–	–
Oedipina cyclocauda	+	–	–	+	–	–	–	–	–	–	–
Oedipina elongata	+	–	–	+	–	–	–	–	–	–	–
Oedipina gephyra	–	–	–	–	–	–	+	–	–	–	–
Oedipina grandis	–	–	–	–	–	–	+	–	–	–	–
Oedipina ignea	–	–	–	+	+	–	–	+	–	–	–
Oedipina parvipes	+	–	–	+	–	–	–	–	–	–	–
Oedipina paucidentata	–	–	–	–	–	–	+	–	–	–	–
Oedipina poelzi	–	–	–	+	–	–	+	–	–	–	–
Oedipina pseudouniformis	+	–	–	+	–	–	–	–	–	–	–
Oedipina stenopodia	–	–	–	+	–	–	–	–	–	–	–
Oedipina stuarti	–	+	–	–	+	+	–	–	–	–	–
Oedipina taylori	–	+	–	–	–	+	–	–	–	–	–
Oedipina uniformis	+	–	–	+	–	–	+	–	–	–	–
Parvimolge townsendi	–	–	–	+	–	–	+	–	–	–	–
Pseudoeurycea altamontana	–	–	–	–	–	–	–	–	–	–	+
Pseudoeurycea anitae	–	–	–	–	–	–	–	+	–	–	–
Pseudoeurycea bellii	–	–	–	+	+	+	+	+	–	–	+
Pseudoeurycea brunnata	–	–	–	–	–	–	+	–	–	+	–
Pseudoeurycea cephalica	–	–	–	–	+	–	–	+	–	–	+
Pseudoeurycea cochranae	–	–	–	–	–	–	–	+	–	–	+
Pseudoeurycea conanti	–	–	–	+	–	–	–	+	–	–	–
Peudoeurycea exspectata	–	–	–	–	–	–	+	–	–	–	–
Pseudoeurycea firscheini	–	–	–	–	–	–	+	–	–	–	–
Pseudoeurycea gadovii	–	–	–	–	–	–	+	+	–	+	+
Pseudoeurycea galeanae	–	–	–	–	–	–	–	+	–	–	–
Pseudoeurycea goebeli	–	–	–	–	–	–	+	–	–	+	–

Appendix 3:4 continued

	Elevational Range and Vegetation Type										
	0–600 m			600–1500 m			1500–2700 m			>2700 m	
Taxon	TWM	TD	TVD	SRW	SM	SD	LMW	LMM	LMD	MW	MM
Pseudoeurycea juarezi	–	–	–	–	–	–	+	–	–	+	–
Pseudoeurycea leprosa	–	–	–	–	–	–	–	+	–	–	+
Pseudoeurycea longicauda	–	–	–	–	–	–	–	–	–	–	+
Pseudoeurycea melanomolga	–	–	–	–	–	–	–	+	–	–	+
Pseudoeurycea mystax	–	–	–	–	–	–	–	+	–	–	–
Pseudoeurycea nigromaculata	–	–	–	+	–	–	–	–	–	–	–
Pseudoeurycea parva	–	–	–	–	–	–	+	–	–	–	–
Pseudoeurycea praecellens	–	–	–	+	–	–	–	–	–	–	–
Pseudoeurycea rex	–	–	–	–	–	–	+	+	–	+	+
Pseudoeurycea robertsi	–	–	–	–	–	–	–	–	–	–	+
Pseudoeurycea saltator	–	–	–	–	–	–	+	–	–	–	–
Pseudoeurycea scandens	–	–	–	–	+	–	–	+	–	–	–
Pseudoeurycea smithi	–	–	–	–	–	–	–	–	–	+	+
Pseudoeurycea unguidentis	–	–	–	–	–	–	–	+	–	–	+
Pseudoeurycea werleri	–	–	–	+	–	–	+	–	–	–	–
Pseudoeurycea sp. A	–	–	–	–	–	–	+	–	–	–	–
Pseudoeurycea sp. B	–	–	–	–	–	–	+	–	–	+	–
Pseudoeurycea sp. C	–	–	–	–	–	–	–	–	–	–	+
Pseudoeurycea sp. D	–	–	–	–	–	–	–	+	–	–	–
Pseudoeurycea sp. E	–	–	–	–	–	–	–	–	–	–	+
Pseudoeurycea sp. F	–	–	–	–	–	–	–	–	–	+	–
Pseudoeurycea sp. G	–	–	–	–	–	–	–	–	–	–	+
Pseudoeurycea sp. H	–	–	–	–	–	–	–	–	–	–	+
Pseudoeurycea sp. I	–	–	–	–	–	–	–	–	–	–	+
Pseudoeurycea sp. J	–	–	–	–	–	–	–	+	–	–	–
Pseudoeurycea sp. K	–	–	–	–	–	–	+	–	–	–	–
Thorius arboreus	–	–	–	–	–	–	+	–	–	–	–
Thorius aureus	–	–	–	–	–	–	+	–	–	+	–
Thorius boreas	–	–	–	–	–	–	–	–	–	–	+
Thorius dubitus	–	–	–	–	–	–	+	+	–	–	–
Thorius insperatus	–	–	–	+	–	–	–	–	–	–	–
Thorius macdougalli	–	–	–	–	–	–	–	+	–	+	+
Thorius minutissimus	–	–	–	–	–	–	–	+	–	+	–
Thorius narismagnus	–	–	–	+	–	–	–	–	–	–	–
Thorius narisovalis	–	–	–	–	–	–	–	+	–	–	+
Thorius pennatulus	–	–	–	+	–	–	–	–	–	–	–
Thorius pulmonaris	–	–	–	–	–	–	–	+	–	–	+
Thorius schmidti	–	–	–	–	–	–	+	–	–	–	–
Thorius smithi	–	–	–	+	–	–	–	–	–	–	–
Thorius troglodytes	–	–	–	–	–	–	+	+	–	–	–
Thorius sp. A	–	–	–	–	–	–	–	–	–	–	+
Thorius sp. B	–	–	–	–	–	–	–	+	–	–	–
Thorius sp. C	–	–	–	–	–	–	+	+	–	–	+
Thorius sp. D	–	–	–	–	–	–	–	+	–	–	+
Thorius sp. E	–	–	–	–	–	–	–	+	–	–	–
CAUDATA: SALAMANDRIDAE:											
Notophthalmus meridionalis	–	+	+	–	–	–	–	–	–	–	–
CAUDATA: SIRENIDAE:											
Siren intermedia	–	+	–	–	–	–	–	–	–	–	–
GYMNOPHIONA: CAECILIIDAE:											
Caecilia leucocephala	+	–	–	–	–	–	–	–	–	–	–
Caecilia nigricans	–	–	–	+	–	–	–	–	–	–	–
Caecilia tentaculata	+	–	–	–	–	–	–	–	–	–	–
Caecilia volcani	+	–	–	–	–	–	–	–	–	–	–
Dermophis mexicanus	+	+	–	+	+	+	–	–	–	–	–
Dermophis oaxacae	+	–	–	+	–	–	–	–	–	–	–

Appendix 3:4 continued

| | Elevational Range and Vegetation Type | | | | | | | | | | |
| | 0–600 m | | | 600–1500 m | | | 1500–2700 m | | | >2700 m | |
Taxon	TWM	TD	TVD	SRW	SM	SD	LMW	LMM	LMD	MW	MM
Dermophis parviceps	+	–	–	+	+	–	–	–	–	–	–
Gymnopis multiplicata	+	+	–	+	+	–	–	–	–	–	–
Gymnopis syntrema	+	–	–	+	–	–	–	–	–	–	–
Oscaecilia elongata	+	–	–	–	–	–	–	–	–	–	–
Oscaecilia ochrocephala	+	–	–	–	–	–	–	–	–	–	–
Oscaecilia osae	+	–	–	–	–	–	–	–	–	–	–

ADDENDUM

After the completion of this paper more than three years ago, several important publications have appeared or were discovered that described new species or provided new distributional records for Middle American amphibians. These new taxa and records are listed in the same order as the taxa appear in the appendices.

ANURA: BUFONIDAE

Bufo cavifrons was restricted to the Sierra de Los Tuxtlas in southern Veracruz, Mexico, by Mendelson (1997a), who referred the populations in the southern part of Sierra Madre Oriental, Mexico, to a new species, *B. spiculatus,* which occurs at elevations of 800–1689 m in subtropical wet forest. *Bufo macrocristatus* was resurrected from the synonymy of *B. valliceps* by Mendelson (1997b); *B. macrocristatus* occurs at elevations of 300–1767 m in tropical and subtropical wet forests on the Atlantic slopes of Chiapas, Mexico, and Guatemala. *Bufo tutelarius* was described from the highlands of southeastern Oaxaca and southern Chiapas in Mexico and from adjacent western Guatemala (Mendelson, 1997b); the species is known from elevations of 1050–2000 in lower montane wet and moist forests.

ANURA: CENTROLENIDAE

Centrolene prosoblepon was reported from eastern Honduras (McCranie and Wilson, 1997a), who described *Hyalinobatrachium cardiacalyptum* and *H. crybetes* from elevations of 200–225 m and 500–680 m, respectively, in tropical moist forest in eastern Honduras. *Hyalinobatrachium pulveratum* was reported from low elevations (95–100 m) in tropical moist forest in northeastern Honduras (McCranie, 1993).

ANURA: HYLIDAE

Agalychnis calcarifer was reported from Honduras (Wilson et al., 1998), and *Agalychnis saltator* and *Anotheca coronata* were reported from low elevations (95–100 m) in tropical moist forest in northeastern Honduras (McCranie, 1993). *Hyla ebraccata* was reported from tropical moist forest in the northern lowlands of Honduras by Cruz and Wilson (1990). *Hyla calypsa* was described from elevations of 1500–2100 m in lower montane wet forest in the Costa Rica–Panama highland (Lips, 1996). *Hyla celata* was described from elevations of 2460–2670 m in lower montane wet forest in the Sierra Madre del Sur of Oaxaca, Mexico (Toal and Mendelson, 1995). *Hyla labedactyla* was described from a locality at 2346 m in lower montane moist forest in the Sierra Madre del Sur of Oaxaca, Mexico (Mendelson and Toal, 1996). Honduran specimens formerly referred to *Plectrohyla teuchestes* were named *P. exquisita* by McCranie and Wilson (1998); the species is known from elevations of 1490–1680 m in lower montane wet forest in the highlands of western Honduras.

ANURA: LEPTODACTYLIDAE

Eleutherodactylus fitzingeri was reported from low elevations (95–100 m) in tropical moist forest in northeastern Honduras (McCranie, 1993). *Eleutherodactylus epochthidius, fecundus, omoaensis,* and *saltuarius* were described from elevations of 460–1800 m in tropical moist forest, subtropical wet forest, and lower montane wet forest in the mountains of Honduras (McCranie and Wilson, 1997b). Frogs formerly recognized as *E. bransfordii* in Honduras were recognized as a new species, *E. lauraster,* by

Savage et al. (1996); the species is known from elevations of 85–1200 m in tropical moist forest and subtropical wet forest in northeastern Honduras.

CAUDATA: PLETHODONTIDAE

Bolitoglossa decora was described from elevations of 1430–1780 m in lower montane forest in Honduras by McCranie and Wilson (1997c), and *B. longissima* was described from elevations of 1900–2200 in lower montane wet forest (peña wind scrub of Carr, 1950) in east-central Honduras by McCranie and Cruz (1996). *Bolitoglossa schmidti* was placed in the synonymy of *B. dofleini* by McCranie et al. (1996).

Three new species of *Nototriton* (*N. brodiei, munzoni,* and *wakei*) were described from elevations of 875–1570 m in subtropical and lower montane wet forests in eastern Guatemala by Campbell and Smith (1998). *Nototriton sanctibarbarus* was described from elevations of 1829–2744 m in lower montane wet forest in the highlands of Honduras (McCranie and Wilson, 1996), and *N. lignicola* was described from elevations of 1430–1780 m in lower montane forest in Honduras (McCranie and Wilson, 1997c).

Oedipina elongata was reported from Honduras (Wilson et al., 1998).

Five new species of *Pseudoeurycea* were named from the Sierra Madre del Sur in Guerrero, Mexico (Adler, 1996). The species and their letter designations that are used in the appendices are: *P. ahuitzotl* (species G), *P. mixcoatl* (Species K), *P. tenchalli* (Species J), *P. teotepec* (Species H), and *P. tlahcuiloh* (Species I).

Five new species of *Thorius* were described from Veracruz and Puebla, Mexico, by Hanken and Wake (1998); *T. lunaris, magnipes, minydemus, munificus,* and *spilogaster;* three of these species are designated in the appendices as Species A, B, and C.

DISCUSSION

The total number of described amphibians in Middle America, as delimited herein, now stands at 620 species, nearly a 4% increase over the 598 species I estimated only three years ago. At the current rate of discovery, my estimate of 5–10% of amphibian species remaining to be discovered in the region indeed may be conservative. The number of amphibians known to occur in certain countries has changed—Mexico with 324 species, Guatemala with 127, Honduras with 102 (this number will increase to at least 106; L. D. Wilson, pers. comm.), Costa Rica with 161, and Panama with 174. Although recent discoveries have altered some of the statistics in certain tables, the general patterns of distribution remain pertinent. I thank J. Hanken, J. R. McCranie, J. R. Mendelson, and L. D. Wilson for their assistance in preparing this addendum.

LITERATURE CITED

ADLER, K. 1996. The salamanders of Guerrero, Mexico, with descriptions of five new species of *Pseudoeurycea* (Caudata: Plethodontidae). Occas. Pap. Mus. Nat. Hist. Univ. Kansas 177:1–28.

CAMPBELL, J. A., AND E. N. SMITH. 1998. New species of *Nototriton* (Caudata: Plethodontidae) from eastern Guatemala. Sci. Pap. Nat. Hist. Mus. Univ. Kansas 6:1–8.

CARR, A. F. 1950. Outline for a classification of animal habitats in Honduras. Bull. Am. Mus. Nat. Hist. 94:563–594.

CRUZ, G. A., AND L. D. WILSON. 1990. *Hyla ebraccata* Cope: an addition to the anuran fauna of Honduras. Bull. Chicago Herpetol. Soc. 25:144.

HANKEN, J., AND D. B. WAKE. 1998. The biology of tiny animals: systematics of the minute salamanders (*Thorius:* Plethodontidae) from Veracruz and Puebla, Mexico, with descriptions of five new species. Copeia 1998:312–345.

LIPS, K. R. 1996. New treefrog from the Cordillera de Talamanca of Central America with a discussion of systematic relationships in the *Hyla lancasteri* group. Copeia 1996:615–626.

MCCRANIE, J. R. 1993. Additions to the herpetofauna of Honduras. Caribbean J. Sci. 29:254–255.

MCCRANIE, J. R., AND G. A. CRUZ. 1996. A new species of salamander of the *Bolitoglossa dunni* group (Caudata: Plethodontidae) from the Sierra de Agalta, Honduras. Caribbean J. Sci. 32:195–200.

MCCRANIE, J. R., D. B. WAKE, AND L. D. WILSON. 1996. The taxonomic status of *Bolitoglossa schmidti,* with comments on the biology of the Mesoamerican salamander *Bolitoglossa dofleini* (Caudata: Plethodontidae). Caribbean J. Sci. 32:395–398.

MCCRANIE, J. R., AND L. D. WILSON. 1996. A new species of salamander of the genus *Nototriton* (Caudata: Plethodontidae) from Montaña de Santa Bárbara, Honduras. Southwest. Nat. 41:111–115.

MCCRANIE, J. R., AND L. D. WILSON. 1997a. Two new species of centrolenid frogs on the genus *Hyalinobatrachium* from eastern Honduras. J. Herpetol. 31:10–16.

MCCRANIE, J. R., AND L. D. WILSON. 1997b. A review of the *Eleutherodactylus milesi*-like frogs (Anura: Leptodactylidae) from Honduras with the descriptions of four new species. Alytes 14:147–174.

MCCRANIE, J. R., AND L. D. WILSON. 1997c. Two new species of salamanders (Caudata: Plethodontidae) of the genera *Bolitoglossa* and *Nototriton* from Parque Nacional La Muralla, Honduras. Proc. Biol. Soc. Washington 110:366–372.

MCCRANIE, J. R., AND L. D. WILSON. 1998. Specific status of the Honduran frogs formerly referred to *Plectrohyla teuchestes* (Anura: Hylidae). J. Herpetol. 32:96–101.

MENDELSON, J. R., III. 1997a. A new species of toad (Anura: Bufonidae) from Oaxaca, Mexico, with comments on the status of *Bufo cavifrons* and *Bufo cristatus*. Herpetologica 53:268–286.

MENDELSON, J. R., III. 1997b. A new species of *Bufo* (Anura: Bufonidae) from the Pacific highlands of Guatemala and southern Mexico, with comments on the status of *Bufo valliceps macrocristatus*. Herpetologica 53:14–30.

MENDELSON, J. R., III, AND K. R. TOAL, III. 1996. A new species of *Hyla* (Anura: Hylidae) from the Sierra Madre del Sur of Oaxaca, Mexico, with comments on *Hyla chryses* and *Hyla mykter*. J. Herpetol. 30:326–333.

SAVAGE, J. M., J. R. MCCRANIE, AND M. R. ESPINAL. 1996. A new species of *Eleutherodactylus* from Honduras relat-

ed to *Eleutherodatylus bransfordii* (Anura: Leptodactyliae). Proc. Biol. Soc. Washington 109:366–372.

TOAL, K. R., III, AND J. R. MENDELSON, III. 1995. A new species of *Hyla* (Anura: Hylidae) from cloud forest in Oaxaca, Mexico, with comments on the status of the *Hyla bistincta* group. Occas. Pap. Mus. Nat. Hist. Univ. Kansas 174:1–120.

WILSON, L. D., J. R. MCCRANIE, AND M. R. ESPINAL. 1998. The ecogeography of the Honduran herpetofauna and the design of biotic reserves. *In* J. D. Johnson et al. (eds.), *Middle American Herpetology: Systematics, Natural History, and Conservation*. El Paso, Texas: Texas Western Press (in press).

4. Distribution Patterns of Amphibians in the West Indies

S. Blair Hedges

Department of Biology
208 Mueller Laboratory,
Pennsylvania State University
University Park, Pennsylvania 16802, USA

ABSTRACT   There are 174 species of amphibians known from the West Indies, 167 of these are native and eight introduced; 164 (98%) of the native species are endemic. The native fauna, all anurans, belongs to four families: Bufonidae (1 genus, 11 species), Dendrobatidae (1, 1), Hylidae (4, 11), and Leptodactylidae (2, 144). Most species (84%) of West Indian amphibians belong to the large leptodactylid genus *Eleutherodactylus*. The greatest diversity of bufonids (8 species) occurs in Cuba, and of hylids (5 species) in Jamaica. Except for two Cuban species occurring elsewhere, single-island endemism is 100% in the Greater Antilles, and most species are restricted to small areas (< 100 km^2) within an island, and 11 species (7%) are known from only type-localities. There are 50 native species (96% endemic) in Cuba, 22 species (100% endemic) in Jamaica, 63 species (100% endemic) in Hispaniola, 20 species (100% endemic) in the Puerto Rican Bank, and 10 species (90% endemic) in the Lesser Antilles. Only two species are native to the Bahamas Bank, and one species is native to the Cayman Islands; none is endemic. Ten percent of the amphibian fauna, including a new family for the West Indies, has been discovered in the last four years; this rate of discovery suggests that our knowledge of species diversity is far from complete.

Distributional data for West Indian amphibians are summarized and subjected to several analyses. Biogeographic regions (74) are defined for the Greater and Lesser Antilles and used in faunal similarity analyses to better understand regional patterns of distribution and endemism. Species density is mapped for the Greater Antilles in order to determine "hot spots" of species diversity. Altitudinal distributions are analyzed to search for possible trends in both species density and in body size, and distributional areas are examined to determine the typical extent of species distributions and possible taxonomic differences. Climatic, vegetational, topographic, and historical factors affecting distributional patterns are reviewed. Amphibian declines and extinctions are discussed, current conservation measures are reviewed, and recommendations are made to help protect the existing amphibian fauna.

Key words: Caribbean, Biogeography, Amphibia, Anura, Tropical, Conservation, Biodiversity, Islands.

INTRODUCTION

The West Indies are located between North and South America (12–27° N) and comprise a total land area of 223,846 km², which is similar to that of Great Britain. Historically, this region includes the Greater Antilles (Cuba, Jamaica, Hispaniola, and Puerto Rico), the Lesser Antilles, the Bahamas Bank, the Cayman Islands, and San Andrés and Providencia in the southwestern Caribbean (Fig. 4:1). These are the areas treated in this paper. Occasionally, some satellite islands of South America also are included in definitions of the West Indies, but these areas, such as Trinidad, Tobago, and the Netherlands Antilles, are zoogeographically more associated with the adjacent mainland. Because there are no amphibians endemic to the Bahamas or the Cayman Islands, the focus of this paper is on the amphibian fauna of the Greater and Lesser Antilles.

Climate in the West Indies is strongly influenced by the prevailing winds from the northeast, which typically bring moisture to the northern and eastern areas of each island and result in dry southern areas. The distribution and composition of the vegetation closely follows this rainfall pattern, often resulting in well-developed moist forests on northern and eastern slopes and dry, xerophytic vegetation in the south. Before human alteration, most islands were almost completely forested. Lowland rainforest with tall buttressed trees graded into montane rainforest on the lower slopes of mountains, with cloud forest at cloud level and elfin woodland on the summits. Wet limestone forest covered the karst regions, and dry scrub forest occupied the drier southern areas. Essentially no lowland rainforest remains anywhere in the West Indies, except for perhaps a few isolated buttressed trees (e.g., Cabezada, Guantánamo Province, Cuba); those forests were the first to disappear after discovery and colonization of these islands. Most other forest types are disappearing but still can be found in patches and isolated tracts throughout the Antilles. The destruction of wet limestone forest (with its sharp limestone rock substrate) and dry scrub forest (with its abundant spiny plants) has occurred at a slower pace partly because of the difficulty these pose for human access. For the same reason, some of the last remaining patches of montane rainforest (perhaps the most endangered forest type) exist only on the steepest mountain slopes.

The amphibian fauna of the West Indies (all anurans) is characterized by a relatively high number of native species (167 species) for its small land

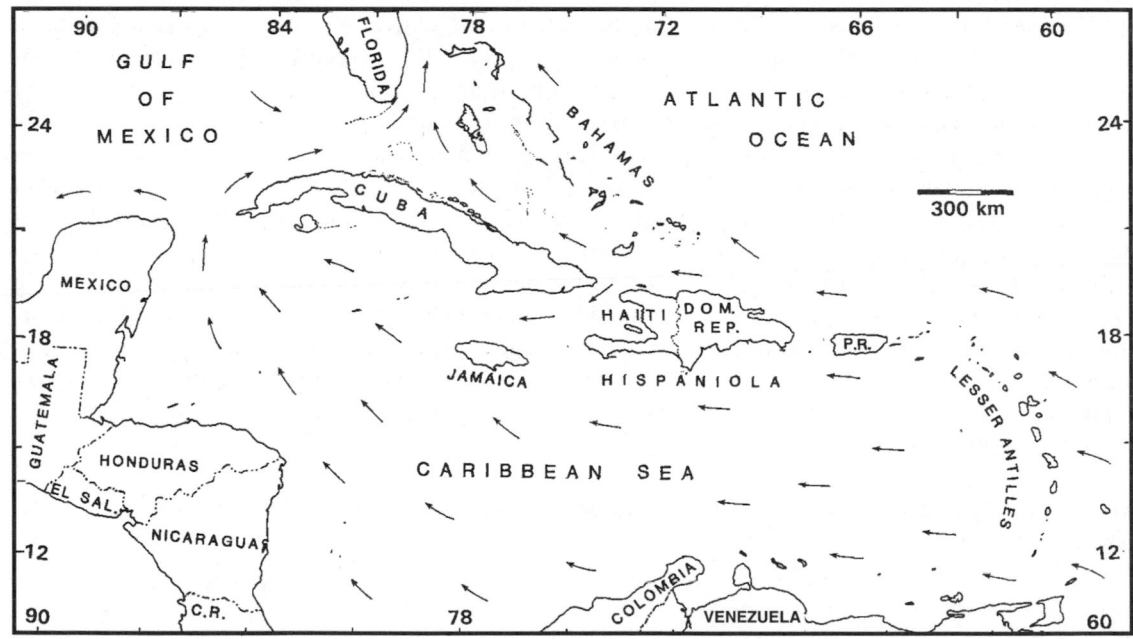

Fig. 4:1. The West Indies, showing major water current patterns.

area, and a low number of genera (7) and families (4). The majority of species are in the single lepto-dactylid genus *Eleutherodactylus*. As terrestrial breeders with direct development (at least one species is ovoviviparous), the members of this genus have invaded a diversity of ecological niches including forest floor, leaf litter, rock outcrops, caves, burrows, streams, tree trunks, tree holes, bromeliads, root holes in elfin forest, and even coastal mangrove swamp (Hedges, 1989a; Hedges and Thomas, 1992). Nearly every species is endemic to a single island and usually to a small area on an island.

Previously, distributional data for the species have been summarized in the form of a checklist (Schwartz and Henderson, 1988), a supplement to the checklist (Hedges and Thomas, 1989), and distributional maps of individual species (Schwartz and Henderson, 1991). The amphibians of Hispaniola have been discussed in relation to the abundance of South Island species (Schwartz, 1973) and the recognition of two paleoislands (Schwartz, 1980). Some general taxonomic and distributional information also has been reviewed, in comparison with West Indian reptiles (Schwartz, 1978). Biogeographic studies on herpetofaunas of some selected areas in the West Indies include Cuba (Rodríguez-Schettino, 1993; reptiles only), Cuban satellite islands (Estrada, 1986, 1993a, b; Garrido et al., 1986), and the Puerto Rican Bank (Heatwole et al., 1981).

MATERIALS AND METHODS

The definition of the West Indies follows Bond (1979). Distributional data on West Indian amphibians were gleaned from the literature and from personal field notes. The distribution maps in Schwartz and Henderson (1991) summarize most locality records up to about 1989, including my specimen and locality data from 1981 through 1988. This has been supplemented by my unpublished field records gathered since that time (1988–1994), and by new published distributional information (e.g., Henderson et al., 1992; Kaiser, 1992; Powell et al., 1992; Thomas and Joglar, 1995). Elevational data for Cuban species were omitted from Schwartz and Henderson (1991) so those data reported here (and many new distributional records) are mostly from my field records. In cases where I believe published locality records are doubtful as to correct identification (taxon or locality), I have modified the distributions accordingly. An effort was made to include not only all described species, but also those currently being described (designated by first letter or letters of proposed name) for the benefit of completeness. After maps of locality records were gathered, species distributions were defined by circumscribing those records with a line or lines and adjusting for unsuitable habitat (e.g., a wide valley separating two montane populations). Small-scale changes in elevation creating unsuitable habitat were not taken into account.

Biogeographic regions were defined by considering their topographic, floristic, and faunistic distinctiveness. For reptiles or other groups, some of these regions may be combined, or require further subdivision. In the descriptions of these regions, their map location, relation to other regions, rainfall data, and areas (in the case of islands) have been omitted because this information is presented elsewhere in the paper.

Faunal similarity, or Coefficient of Biogeographic Resemblance (CBR; Duellman, 1990) was measured using the formula CBR = $2C \div (N_1+N_2)$, where C = number of species in common between two regions and N_1 and N_2 are the number of species present in each region; faunal distance (D) is 1−CBR. For comparison with other studies, cluster analysis of CBR values was done with the UPGMA algorithm (Sneath and Sokal, 1973). However, in addition, D values (no corrections were made) were used with the neighbor joining algorithm (Saitou and Nei, 1987). The latter method often performs better than UPGMA in computer simulations of phylogenetic data (Nei, 1991) so it was used here to explore its usefulness with faunal similarity analysis. If species are shared disproportionately among regions, a faunal similarity/distance tree might have unequal branch lengths as is common with phylogenetic data. In those cases, a clustering algorithm that does not equalize branch lengths (such as neighbor joining) should be at an advantage. Because it is a new application for neighbor joining, both trees are presented here for each comparison. The program MEGA (Kumar et al., 1994) was used for tree construction. To avoid spurious results, only regions containing more than one species were used in the analyses.

Native species are defined here as those that occur naturally, endemics are those native species that occur nowhere else other than the region under consideration (except as introductions), and percent endemism is number of endemic species divided by the number of native species. These terms are applied in this paper at three different levels: the West Indies, major divisions within Antilles (e.g., Cuba, Lesser Antilles), or biogeographic regions. In some cases (e.g., regions within Lesser Antilles) the biogeographic region is an island or island group. The classification of West Indian amphibians used here follows Duellman (1993) and Hedges (1996).

Species density is defined as the number of overlapping species distributions, and is approximately equal to the number of sympatric species (except in some areas of rapid elevational change) but often is greater than the number of syntopic species. By reference to species distribution maps, species density was determined for specific locations at a relatively fine scale throughout Cuba, Jamaica, Hispaniola, and Puerto Rico, and contours (interval = 2 species) were fitted. Species distributional areas were determined by weighing distributions that had been carefully traced and cut from a uniform medium (mylar) and applying a conversion factor calculated by application to regions of known area. Where body size is used, values are the maximum snout-vent length in mm, as reported in Schwartz and Henderson (1991); for nearly all species, the maximum value occurs in females.

Geographic, geologic, climatic, and vegetational data were taken from a variety of sources, including Asprey and Robbins (1953), Box and Cameron (1989), Caribbean and Latin American Action (1993), Instituto Cubano de Geodesia y Cartografía (1978), Johnson (1988), MacPherson (1980), Marrero (1946), Maurrasse (1982), National Geographic Society (1981), Núñez Jiménez (1972), Picó (1974), Rickards (1980), SEA/DVS (1990), Showker, (1989), Vickers (1979), Woodring et al. (1924), World Resources Institute (1992, 1994), and 1:1,000,000 scale topographic maps of the American Geographical Society, and various other topographic and road maps of the islands. Published values for island areas were found to vary slightly and sometimes the average or modal value was used. In the maps of elevation and rainfall for the Greater Antilles presented here, different contour intervals had to be used for different islands in order to highlight variation (e.g., rainfall in Jamaica is higher than on other islands).

There are many different forest types described for the West Indies, partly reflecting the diversity of underlying rock types. However, much of the land area of the islands was covered with shallow marine waters at times during the Mesozoic and Cenozoic and therefore the predominate surface rock type is limestone. Forests overlying limestone are lower, have a more open canopy, and a greater number of vines and epiphytes than forests on other rock types. The forest types used here are from Asprey and Robbins (1953), Instituto Cubano de Geodesia y Cartografía (1978), Thompson et al. (1986a), and SEA/DVS (1990), with some modification. **Dry scrub forest** has a low canopy (2–10 m), often with thorny shrubs and cacti, and grows on bare rock or thin soil cover in areas receiving < 1 m rainfall annually; usually this is a coastal formation (low in elevation). **Wet limestone forest** essentially is rainforest growing on limestone rock, sometimes with a thin soil layer; the canopy is about 15–25 m and there are abundant vines and epiphytes; it usually occurs in areas of > 2 m rainfall annually. **Lowland rainforest** with high canopy and large, buttressed trees no longer exists in the West Indies (except in a few small patches); it was cleared in the early days of European colonization. **Montane rainforest** is characterized by a high (25–40 m), dense canopy and occurs on mountain slopes (typically 500–1500 m elevation) in areas receiving > 2 m rainfall annually. **Cloud forest** occurs in the upper elevations (usually > 1000 m) at the cloud level and thus it receives less light and greater humidity than other forest types. The canopy is low (10–15 m) and ferns (also tree ferns), mosses, and epiphytes are abundant. **Elfin woodland** also occurs at the cloud level on some ridges and mountain tops from as low as 750 m (Jamaica) to > 2400 m (Hispaniola). The vegetation is a low, windblown, tangle of mossy trees and shrubs about 5 m in height; ferns and epiphytes are abundant. **Pine forest** occurs naturally and in cultivation throughout the West Indies, but most often in the higher elevations (> 1000 m). **Mangrove woodland** (usually *Rhizophora mangle*) is a common forest type in coastal areas where trees average about 5 m in height.

The Cayman Islands have only a single, nonendemic, native amphibian species (*Osteopilus septentrionalis*) and the Bahamas Bank (including the Turks and Caicos Islands) also has a low diversity of amphibians (two native species, neither endemic); both regions will be mentioned only briefly. The single native (but nonendemic) amphibian species on Isla de San Andrés and Isla de Providencia (off the east coast of Nicaragua), *Leptodactylus insularum*, is included in counts of species but is not discussed further.

DESCRIPTION OF REGION ·

Each of the major islands, or groups of islands, is defined below, with a definition of each of the biogeographic regions contained therein. Highest elevations are noted; unless otherwise indicated, the lowest elevation is approximately sea level. Relevant features of geology, physiography, and vegetation, including current extent of forest cover (if known), are described.

DEFINITION OF REGIONS

Cuba

This is the largest island (105,007 km²) and makes up nearly one-half of the total area of the West Indies (Fig. 4:2). It is 1250 km long by 191 km at its widest point and 31 km at its narrowest point and has a maximum elevation of 1972 m. There are three major upland areas: the Cordillera de Guaniguanico in the west, Macizo del Escambray in the center, and the Sierra Maestra/Macizo de Sagua-Baracoa in the east. Most forest habitats are limited to those three areas; agriculture (especially sugar cane) predominates in the intervening lowland areas.

Península de Guanahacabibes: This low peninsula at the extreme western end of the island attains elevations of only 22 m. It is characterized by exposed dogtooth limestone rock (diente de perro), caves, and dry scrub forest.

Llanura Occidental: The western coastal plain at elevations of no more than 50 m is characterized by the near-absence of trees and the presence of extensive areas of sugar cane (*Saccharum officinarum*) and rice cultivation.

Cordillera de Guaniguanico: This interconnected upland region reaches 699 m in elevation and comprises two mountain chains with different soil, rock, and vegetation types. The western chain, the Sierra de los Organos (617 m), with its northern (Alturas Pizarrosas del Norte) and southern (Alturas Pizarrosas del Sur) adjoining ranges, primarily is an eroded limestone block with underlying metamorphic rocks of Jurassic age. It is dominated by pine (primarily *Pinus tropicalis*, but also *P. caribaea*) with dry scrub forest on the mogotes (limestone hillocks). The resulting "haystack" karst physiography, although similar to some areas in NW Puerto Rico, is unique in the West Indies and resembles some karst regions of southern China. The eastern chain, the Sierra del Rosario (699 m), is characterized by gently rolling hills, with outcrops of limestone, sedimentary, igneous, and metamorphic rocks.

Isla de Juventud and Archipiélago de los Canarreos: The largest satellite island, Isla de Juventud (formerly Isla de Pinos; 2,200 km²) is composed of Jurassic metamorphic rocks similar to those underlying the western end of Cuba (Sierra de los Organos) and has a southern limestone plain. Most of the island is low in elevation, but a central upland area rises to 303 m. A long chain of low-lying islands, the Archipiélago de los Canarreos, extends to the east.

Alturas de la Habana-Matanzas: This upland area includes the Alturas de Bujucal-Madruga-Coliseo. These are mostly low limestone ridges and hills (no higher than 381 m) with some outcrops of serpentine.

Llanura de Zapata: This region includes an extensive swampland area (Ciénaga de Zapata) at or near sea level, and a slightly higher (10 m) central area of limestone supporting dry scrub forest. Dogtooth limestone is a common substrate.

Alturas Centrales: Included here are the Alturas de Santa Clara, Alturas del Nordeste, and Alturas del Noroeste. This upland region, mostly above 50 m but no higher than 487 m, is underlain by igneous and extrusive volcanic rock, and has limestone outcrops. The dominant vegetation is sugar cane and grassland with some tree savannas of palm (*Roystonea*) and ceiba (*Ceiba*).

Archipiélago Sabana-Camaguey: These northern satellite islands form a long (450 km) archipelago of low-lying islands composed of primarily Quaternary limestone and sediments. Mangrove woodland is the predominate vegetation on the smaller, western keys (Archipiélago de Sabana), whereas the larger keys in the eastern portion (Archipiélago de Camagüey) are dominated by dry scrub forest on limestone and attain an elevation of no more than 62 m.

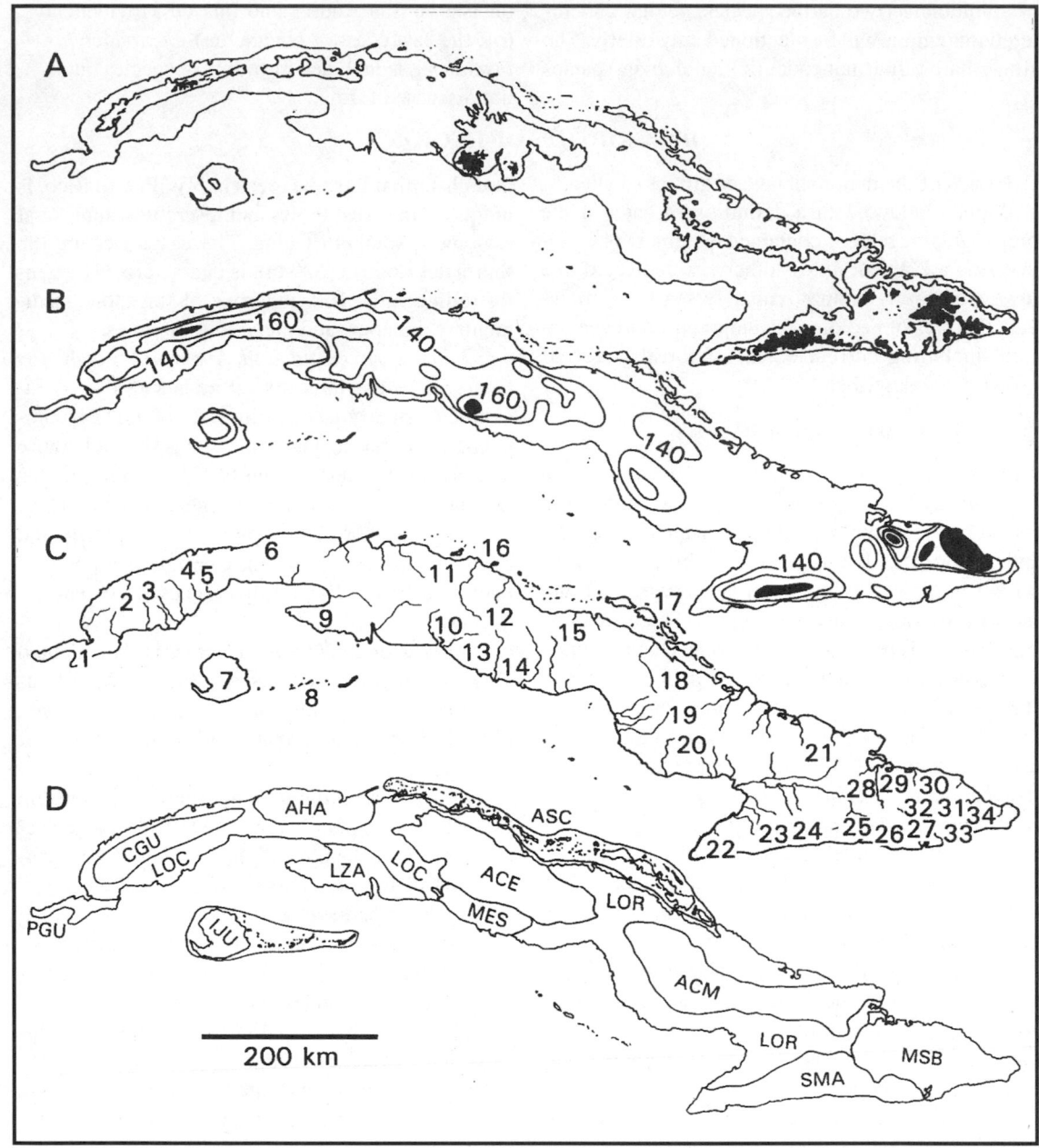

Fig. 4:2.   Cuba. (A) Topographic map, showing the 100 m contour interval and areas above 500 m (black). (B) Rainfall map, showing contour intervals of 140 cm, 160 cm, and areas over 200 cm (black) annual precipitation. (C) Drainage map, showing major rivers. Numbers are locations of peaks, mountain ranges, major cities, and other geographic areas mentioned in text: Península de Guanahacabibes (1), Sierra de los Organos (2), Viñales (3), Sierra del Rosario (4), Soroa (5), La Habana (6), Isla de Juventud (7), Archipiélago de los Canarreos (8), Ciénaga de Zapata (9), Cienfuegos (10), Alturas del Noroeste (11), Alturas de Santa Clara (12), Alturas de Trinidad (13), Alturas de Sancti Spíritus (14), Alturas del Nordeste (15), Archipiélago de Sabana (16), Archipiélago de Camagüey (17), Sierra de Cubitas (18), Peniplano de Florida-Camagüey-Tunas (19), Sierra de Najasa (20), Grupo de Maniabón (21), Meseta de Cabo Cruz (22), Pico Turquino (23), Cordillera del Turquino (24), Santiago de Cuba (25), Cordillera de la Gran Piedra (26), Guantánamo (27), Altiplanicie de Nipe (28), Sierra del Cristal (29), Cuchillas de Moa (30), Cuchillas de Toa (31), Meseta del Guaso (32), Sierra de Mariana (33), and Sierra del Purial (34). (D) Biogeographic regions. Abbreviations: ACE Alturas de Central (Alturas de Santa Clara, Alturas del Nordeste, Alturas del Noroeste), ACM Alturas de Camaguey-Maniabón (Sierra de Cubitas, Peniplano de Florida-Camaguey-Tunas, Sierra de Najasa, Grupo de Maniabón), AHA Alturas de la Habana-Matanzas (including

Macizo del Escambray: This major central highland region includes a larger western area, the Alturas de Trinidad (1140 m), and a smaller eastern area, the Alturas de Sancti Spíritus (842 m), separated by the valley of the Río Agabama. Jurassic igneous rocks underlie both regions, but limestone outcrops occur throughout the region. Upper elevations are characterized by montane rainforest (now largely replaced by coffee, *Coffea arabica*).

Llanura Oriental: This low eastern coastal plain attains elevations of only 50 m. It is similar to the Llanura Occidental in being characterized by extensive areas of sugar cane cultivation and the near-absence of trees.

Alturas de Camagüey-Maniabón: Included in this region of moderate relief are the Sierra de Cubitas (330 m), Peniplano de Florida-Camagüey-Tunas (297 m), Sierra de Najasa (301 m), and Grupo de Maniabón (275 m). All primarily are underlain by igneous (serpentine and granite) rocks and extrusive volcanics. Vegetation is a mixture of mostly tree savanna and dry scrub forest, with sugar cane cultivation in the lower elevations.

Sierra Maestra: With a large area above 1000 m elevation, this is the major eastern highland region and it includes a large limestone platform (Meseta de Cabo Cruz, 401 m), a major central massif (Cordillera del Turquino, 1972 m), and a partially isolated eastern extension (Cordillera de la Gran Piedra, 1214 m). Pico Turquino (1972 m) is the highest point in Cuba. The underlying rocks primarily are Tertiary volcanic and sedimentary formations, including limestone. The dominant natural vegetation is montane rainforest (now mostly coffee cultivation on the northern slopes), and there is cloud forest and elfin woodland at the highest elevations.

Macizo de Sagua-Baracoa: Separated from the Sierra Maestra by the Río Guantánamo and Valle Central, this major upland region is a complex of seven smaller upland areas: the Altaplanicie de Nipe (995 m), the Sierra del Cristal (1231 m), Meseta del Guaso (862 m), Sierra de Mariana (747), Cuchillas de Moa (1175 m), Cuchillas de Toa (921 m), and the Sierra del Purial (1176 m). The underlying rocks primarily are a mixture of Mesozoic igneous and metamorphic rocks overlain by montane rainforest and

pine forest (*Pinus cubensis*) and Cenozoic limestone overlain by pine forest and wet limestone forest. Cloud forest occurs in some of the higher elevations in the wettest areas.

Jamaica

With an area of 10,992 km^2, this island lies to the south of eastern Cuba; it is 230 km long by 80 km wide, and has a maximum elevation of 2256 m (Fig. 4:3). Most of the island is a dissected Tertiary limestone platform and therefore hilly and mountainous; a high mountain chain (Blue Mountains), involving other rock types, exists in the east. It is the wettest island in the Greater Antilles; some areas have more than 5 m of precipitation annually. Although Jamaica appears lush and green, this is due in large part to an introduced flora and high rainfall; few natural forests remain, and these are concentrated in the Cockpit Country of west-central Jamaica and in the Blue Mountains.

Western Lowlands: As in most of Jamaica, this region is largely underlain by mid-Tertiary limestone. There are a few patches of dry scrub forest and some coastal mangrove woodland, and a large (250 km^2) fresh-water sedge marsh (Black River Morass), but most areas are under cultivation (primarily sugar cane). Elevations do not exceed 300 m.

Western Uplands: Although partly isolated from each other, these upland areas all lie on the western side (Hanover Block) of north-south trending fault zone (Montpelier-Newmarket Graben). Although the southern areas (801 m) are higher in elevation, Dolphin Head (545 m) has the better-developed wet limestone forest.

Santa Cruz Mountains: This is an isolated upland area in south-central Jamaica with little or no original wet limestone forest remaining. The highest elevation is 725 m.

Cockpit Country: This rugged karst plateau, with typical elevations of 600-700 m (946 m elevation at highest point), is the largest remaining area of wet limestone forest in Jamaica. It is bordered on the west by the Montpelier-Newmarket Graben, and on the north by the drier north coast, but the southern and eastern boundaries are less well defined.

North Coast: Coconut cultivation dominates the relatively dry north coast, but there are some occa-

Facing page:

Alturas de Bujucal-Madruga-Coliseo), ASC Archipiélago Sabana-Camaguey, CGU Cordillera de Guaniguanico, IJU Isla de Juventud and Archipiélago de los Canarreos, LOC Llanura Occidental, LOR Llanura Oriental, LZA Llanura de Zapata (including Cienagas de Zapata), MES Macizo del Escambray, MSB Macizo de Sagua-Baracoa, PGU Península de Guanahacabibes, SMA Sierra Maestra.

sional patches of dry scrub forest. Elevations do not exceed 300 m.

Central Uplands: This region is similar in karst topography and elevation to the adjoining Cockpit Country, but with greater human population density and less wet limestone forest; the highest point is 838 m elevation. The eastern border is a major fault, the Wagwater Trough.

Manchester Plateau: This region includes the Don Figuero Mountains and is characterized by gently rolling hills (up to 995 m elevation) with less exposed rock and greater soil depth than the adjoining Central Uplands to the north. Some small patches of wet limestone forest remain; human population density is high, and most areas are cultivated.

Southern Lowlands: Because of the rainshadow effect of the Blue Mountains and Central Uplands, some of the driest areas in Jamaica are in this region, which includes areas below 300 m elevation. Cultivation and cattle grazing is extensive, although some small areas of dry scrub forest remain.

Portland Ridge Peninsula: This relatively low limestone ridge (up to 160 m elevation) with many caves is overlain by dry scrub forest on a dogtooth limestone substrate. It was an island during times of elevated sea level in the Pleistocene.

Hellshire Hills: Similar in some respects to the Portland Ridge Peninsula, this region largely is covered by dry scrub forest on limestone rock; the highest elevation is 258 m.

Blue Mountains: With several peaks above 2000 m, this is the major mountain chain of Jamaica; highest peak = Blue Mountain Peak (2256 m). The underlying rocks are igneous, metamorphic, and clastic sedimentaries; these are overlain by remnant montane rainforest (< 1350 m) grading to cloud forest (1350–1500 m) and elfin woodland (> 1500 m). Cultivation of pine and coffee is common at higher elevations. The northern slopes are considerably wetter and more forested than the southern slopes.

John Crow Mountains: The wettest areas in Jamaica are in this region, which is a northwest-southeast-oriented limestone block, tilted up to the southwest and separated (mostly) from the Blue Mountains by the Rio Grande Valley. Possibly because of the high rainfall in this region, the expected wet limestone forest is replaced by montane rainforest, and vegetational zones typical of the Blue Mountains (including elfin woodland) occur at low-

er elevations (Asprey and Robbins, 1953). The highest elevation is 1202 m.

Hispaniola

Lying between Cuba and Puerto Rico, Hispaniola has an area of 76,470 km^2; it is 650 km long by 255 km wide with a maximum elevation of 3175 m (Fig. 4:4). The western one-third of the island is occupied by Haiti and the remainder by the Dominican Republic. Numerous mountain ranges and valleys dissect this island, which is divided into two faunistically defined paleoislands ("North Island" and "South Island"), separated by a below sea-level trough, the Cul de Sac and Valle de Neiba. Natural forests are essentially absent from Haiti, and this has caused widespread silting and destruction of the already overfished coral reefs. Most remaining forests in the Dominican Republic are confined to the Cordillera Central, but some relatively small tracts remain in other upland areas.

Navassa Island: This is a small island with a steep, rock shoreline located about 60 km west of the Tiburon Peninsula of Haiti.

Massif de la Hotte: The region defined here forms the core (Montagnes de la Hotte) of the more inclusive Massif de la Hotte, and includes the highest elevations (up to 2347 m). Lower elevations largely are barren, and remnants of montane rainforest and wet limestone forest exist on the slopes and foothills; some pine forest and cloud forest still may be present on the highest peaks.

Presqu'île de la Tiburon: The Tiburon Peninsula includes some limestone hills (up to 1340 m), representing the eastern extension of the Massif de la Hotte; at one time they may have been covered with wet limestone forest.

Massif de la Selle–Sierra de Baoruco: Two names have applied to this single mountain range that straddles the international border and is formed largely by an uplifted limestone platform that has been tilted up toward the north. This has resulted in a gradual southern incline and a steeper and more rugged northern slope; both have dry scrub forest at lower elevations. The mostly pine-clad ridge, rising up to 2690 m elevation, has occasional patches of cloud forest, especially in the extreme eastern end.

Península de Barahona: The dry southern peninsula of Hispaniola is a terraced limestone platform supporting dry scrub forest; the maximum elevation is 331 m.

Plaine du Cul de Sac–Valle de Neiba: Separating

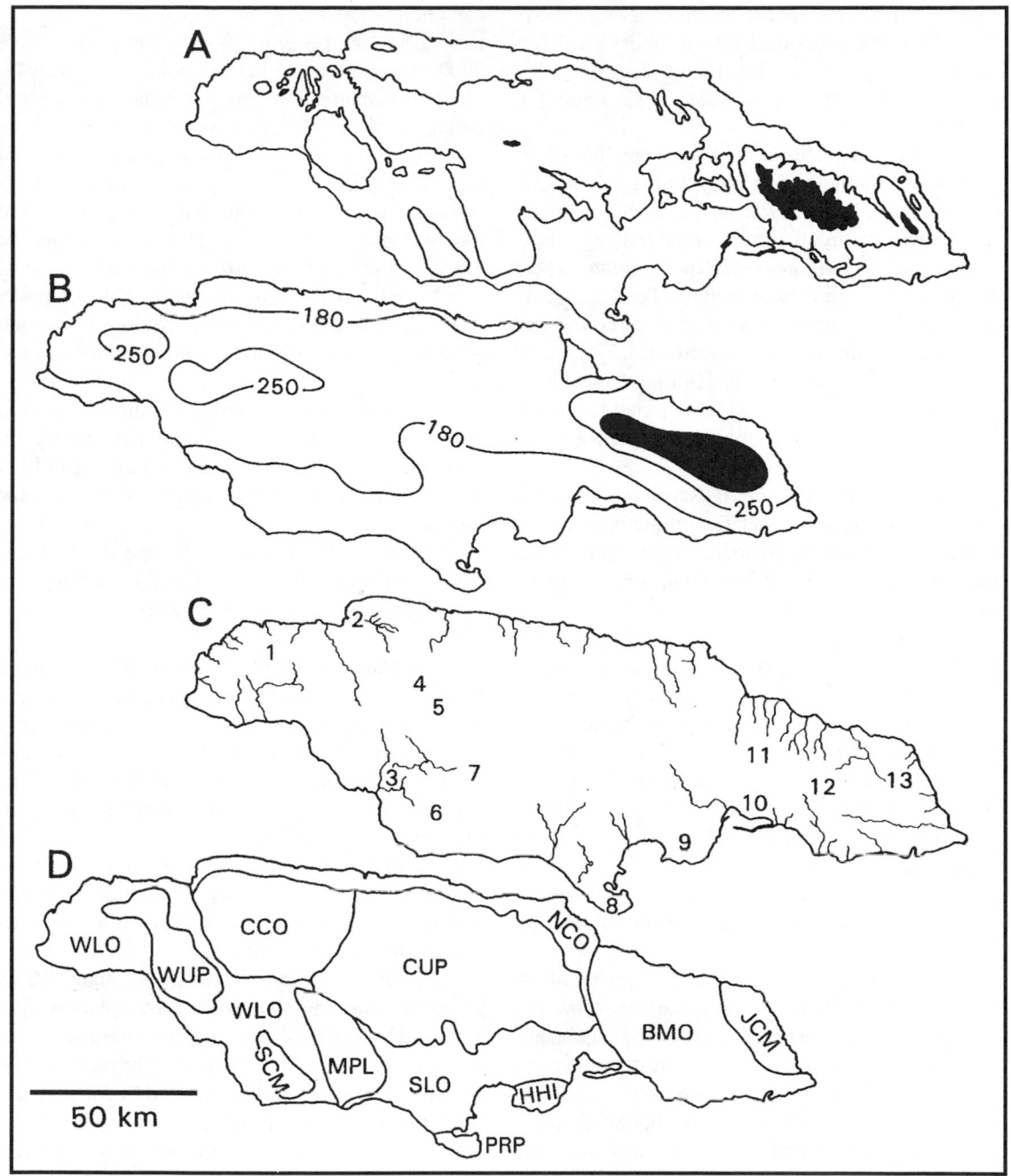

Fig. 4:3. Jamaica. (A) Topographic map, showing 300 m contour interval and areas above 1,000 m (black). (B) Rainfall map, showing contour intervals of 180 cm, 250 cm, and areas over 380 cm (black) annual precipitation. (C) Drainage map, showing major rivers. Numbers are locations of peaks, mountain ranges, major cities, and other geographic areas mentioned in text: Dolphin Head (1), Montego Bay (2), Black River Morass (3), Cockpit Country (4), Quick Step (5), Santa Cruz Mountains (6), Don Figuerero Mountains (7), Portland Ridge Peninsula (8), Hellshire Hills (9), Kingston (10), Hardwar Gap (11), Blue Mountain Peak (12), and John Crow Mountains (13). (D) Biogeographic regions. Abbreviations: BMO Blue Mountains, CCO Cockpit Country, CUP Central Uplands, HHI Hellshire Hills, JCM John Crow Mountains, MPL Manchester Plateau, NCO North Coast, PRP Portland Ridge Peninsula, SCM Santa Cruz Mountains, SLO Southern Lowlands, WLO Western Lowlands, WUP Western Uplands.

the two paleoislands of Hispaniola along a major fault, this long, mostly barren valley (below 100 m elevation) with occasional patches of dry scrub forest lies below sea level in many places and has several large hypersaline lakes (e.g., Lago Enriquillo, −42 m).

Île Gonâve: This is the largest satellite island of Hispaniola and has an upland area (Montagnes de la Gonâve, rising to 702 m); the island is densely populated and largely devoid of natural vegetation.

Chaîne des Matheux–Sierra de Neiba: These ranges, and the Haitian Montagnes Trou D'eau, are oriented in a northwest-southeast direction and attain moderate to high elevations (1575 m and 2279 m, respectively). The Haitian ranges have been largely deforested, but some cloud forest and wet limestone forest are present in the Sierra de Neiba.

Sierra de Martín García: Surrounded by xeric lowlands in a region of relatively low rainfall, this isolated range with a maximum elevation of 1368 m supports some pine and hardwood forest at upper elevations.

Valle de San Juan: This relatively dry valley (below 400 m elevation) is the rain shadow of the Cordillera Central and presently an agricultural region with little remaining natural vegetation.

Plateau Central: At an elevation of about 300–400 m, this plateau is an extension of the Valle de San Juan and separates the Massif du Nord from the Montagnes Noires; there is little or no remaining natural vegetation.

Massif des Montagnes Noires: These are mostly limestone ridges rising to 1793 m elevation, now deforested.

Plaine de l'Artibonite: This floodplain of the large Rivière Artibonite, with elevations below 100 m, supports some agriculture but no natural vegetation.

Presqu'île du Nord Ouest: The dry northwest peninsula of Haiti resembles a barren lunar landscape when viewed from the air, but some small patches of secondary forest remain on steep slopes of the interior ranges; the maximum elevation is 840 m.

Île de la Tortue: Once thickly forested with large hardwood trees but now essentially barren, this large satellite island is mostly limestone; the maximum elevation is 378 m.

Massif du Nord: This is the major mountain range of northern Haiti and represents the northwest extension of the Cordillera Central; the maximum elevation is 1210 m.

Cordillera Central: With many peaks above 2000 m, including the highest point in the West Indies (Pico Duarte, 3087 m), this is the major central mountain range of Hispaniola. Much of it has been deforested, but the core of this range and the highest elevations typically support pine (*Pinus occidentalis*) on a limestone substrate. Some cloud forest and montane rainforest exist in patches throughout the range, primarily in the northwest sector.

Plaine du Nord–Valle de Cibao: This is a mesic, fertile agricultural valley with elevations below 300 m and with essentially no remaining natural vegetation.

Cordillera Septentrional: The northern mountain range of the Dominican Republic is a nearly linear northwest-southeast-oriented ridge rising to 1249 m elevation and with scattered patches of montane rainforest.

Península de Samana: This rugged, wet, limestone peninsula with several mountains has only secondary forest remaining; the maximum elevation is 606 m.

Los Haitises: Wedged between the Cordillera Central and the Cordillera Oriental, this highly dissected karst area is similar to the Cockpit Country of Jamaica but lower in elevation. Typical elevations are 150–250 m and wet limestone forest still exists, primarily on the tops of some mogotes; the maximum elevation is 467 m.

Cordillera Oriental: This eastern mountain chain of Hispaniola, attaining a maximum elevation of 736 m, is continuous with Los Haitises but has little remaining natural forest.

Llanura Costera del Caribe: In many places along the coast, this Caribbean coastal plain is characterized by low (0–20 m) limestone terraces with a substrate of dogtooth limestone. The depauperate flora of the terraces is characterized by *Bucida buceras,* whereas agriculture (primarily sugar cane) predominates in the more fertile inland areas; the maximum elevation is 200 m.

Puerto Rican Bank

This archipelago extends from the Isla Mona in the west to Anegada in the east. Except for Isla Mona and St. Croix, virtually all are separated by relatively shallow water. The main island of Puerto Rico is the largest in the chain but is the smallest of the Greater Antilles (8,768 km²); it is 179 km long by 58 km

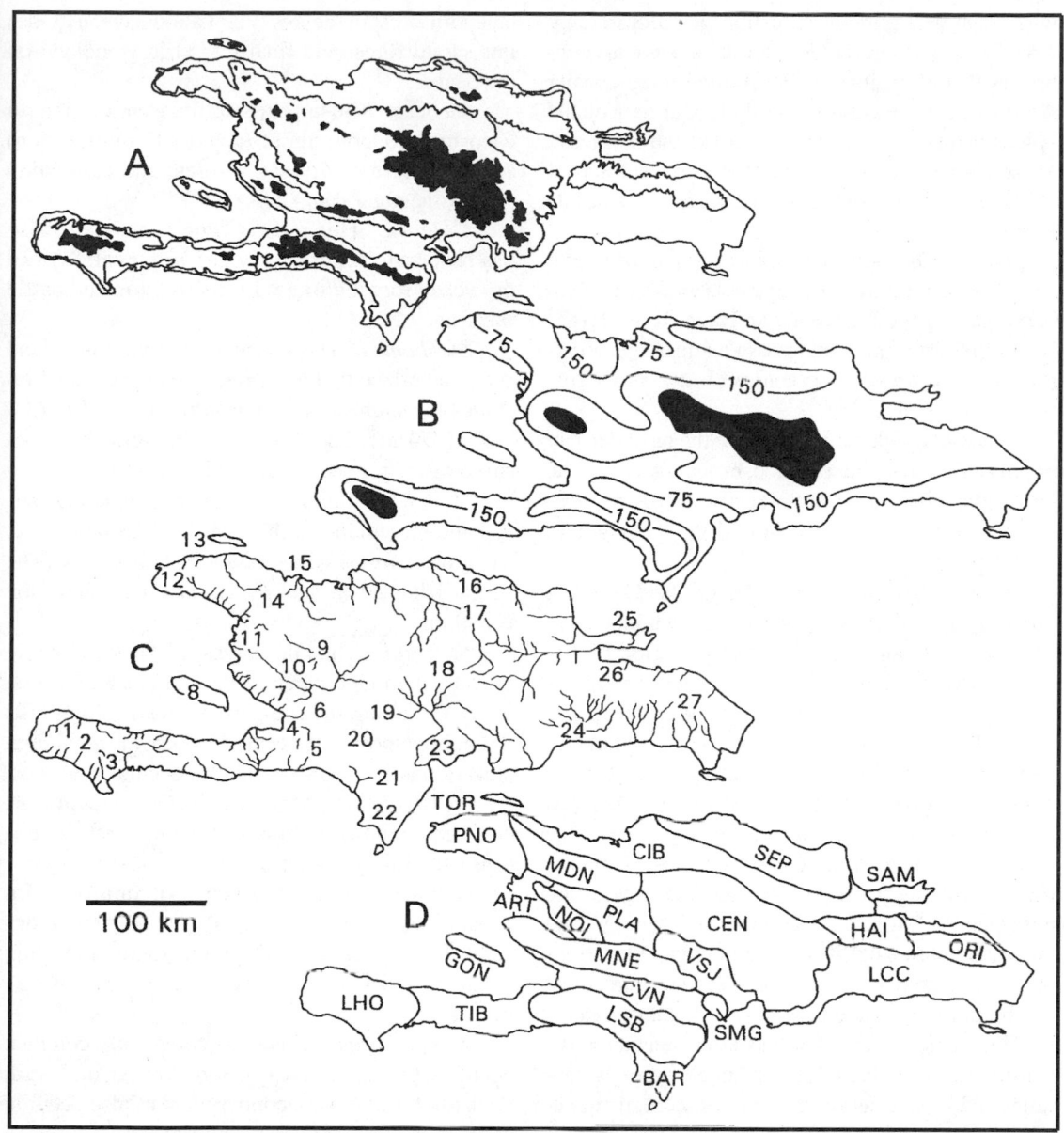

Fig. 4:4. Hispaniola. (Λ) Topographic map, showing 200 m contour interval and areas above 1,000 m (black). (B) Rainfall map, showing contour intervals of 75 cm, 150 cm, and areas over 225 cm (black) annual precipitation. (C) Drainage map, showing major rivers. Numbers are locations of peaks, mountain ranges, major cities, and other geographic areas mentioned in text: Castillon (1), Montagnes de la Hotte (2), Les Cayes (3), Port-au-Prince (4), Massif de la Selle (5), Montagnes Trou D'eau (6), Chaîne des Matheux (7), Île Gonâve (8), Plateau Central (9), Massif des Montagnes Noires (10), Plaine de l'Artibonite (11), Presqu'île du Nord Ouest (12), Île de la Tortue (13), Massif du Nord (14), Cap-Haïtien (15), Cordillera Septentrional (16), Santiago (17), Pico Duarte, Cordillera Central (18), Sierra de Neiba (19), Lago Enriquillo (20), Sierra de Baoruco (21), Península de Barahona (22), Sierra de Martín García (23), Santo Domingo (24), Península de Samana (25), Los Haitises (26), and Cordillera Oriental (27). (D) Biogeographic regions. Abbreviations: ART Plaine de l'Artibonite, BAR Península de Barahona, CEN Cordillera Central, CIB Plaine du Nord-Valle de Cibao, CVN Plaine du Cul de Sac–Valle de Neiba, GON Île Gonâve, HAI Los Haitises, LCC Llanura Costera del Caribe, LHO Massif de la Hotte, LSB Massif de la Selle–Sierra de Baoruco, MDN Massif du Nord, MNE Chaîne des Matheux–Sierra de Neiba, NOI Massif des Montagnes Noires, ORI Cordillera Oriental, PLA Plateau Central, PNO Presqu'île du Nord Ouest, SAM Península de Samana, SEP Cordillera Septentrional, SMG Sierra de Martín García, TIB Presqu'île de la Tiburon, TOR Île de la Tortue, VSJ Valle de San Juan.

wide with a maximum elevation of 1338 m (Fig. 4:5). The physiography of this island is not as complex as that of the three larger islands of the Greater Antilles; it has a central cordillera and an isolated upland region (Sierra de Luquillo) in the northeast. Most natural forests, and one of the largest tracts of elfin woodland in the Caribbean, are in the Sierra de Luquillo.

Isla Mona: Although politically part of Puerto Rico, this island between Hispaniola and Puerto Rico is not part of the Puerto Rican Bank. It is a largely uninhabited limestone plateau attaining a maximum elevation of 85 m and with dry scrub forest.

Coastal Lowlands: These are the most densely populated areas in Puerto Rico, especially along the north coast; the maximum elevation is 100 m. Sugar cane cultivation is common along the southern coastal plain.

Central Uplands: The highland "backbone" of Puerto Rico, or Cordillera Central, extends in an east-west direction and includes a major region of serpentine rocks (Maricao forest) in the west; the maximum elevation is 1338 m. The northern, and especially northwestern, extension of the upland region is a dissected limestone plateau with karst topography similar to the Cockpit Country of Jamaica and Los Haitises of Hispaniola; in some areas, the wide separation of the mogotes resembles the Viñales region of western Cuba (although the mogotes are not as large). Montane rainforest has been replaced mostly by coffee cultivation in the upper elevations of the main cordillera, whereas wet limestone forest can be found on some mogotes in the karst areas.

Sierra de Cayey: This eastern extension of the Cordillera Central rising to 903 m elevation, sometimes referred to as the humid east-central mountains, is more densely populated and therefore less forested.

Cuchilla de Pandura: This relatively low range (up to 525 m elevation) consists mostly of weathered igneous rocks in the form of giant bolders (some > 10 m in diameter) which in turn form a network of bolder caves. Most natural forest has been removed, and human population density is high.

El Yunque: The largest area of natural forest in Puerto Rico is in this region, which takes the name of one of the three major peaks (El Yunque, 1065 m; highest peak = El Toro, 1074 m) in the Sierra de Luquillo. The mostly igneous rock supports montane rainforest in the lower elevations which grades into cloud forest and finally to elfin woodland on the peaks.

Vieques: The largest satellite island of Puerto Rico has moderate relief (up to 301 m elevation) and a mixture of dry scrub forest and cultivation (primarily sugar cane).

Culebra: This mostly limestone island also has moderate relief (up to 195 m elevation) and dry scrub forest along with cultivation and cattle farming.

St. Thomas: This volcanic and limestone island is characterized by highly dissected terrain, and includes the highest elevation in the U.S. Virgin Islands (474 m). There is very little agriculture because of the rugged terrain and poor soil.

St. John: Similar to St. Thomas in being rugged and mountainous, this island is less populated, has greater forest cover, and a National Park (The Virgin Islands National Park); maximum elevation is 389 m.

St. Croix: This, the largest of the U.S. Virgin Islands, is mountainous in the northeast and relatively flat along the south coast, with rolling hills throughout most of the central region; the maximum elevation is 355 m. Most rainfall occurs in the west leaving the eastern portion relatively arid. Some forests remain in the northwest, but most of the interior is used for agriculture.

Tortola: There are patches of rainforest on Mount Sage in the north (a "Protected Area"); dry scrub forest occurs mostly in the southern portion of this hilly island; the maximum elevation is 521 m.

Virgin Gorda: This is a geologically complex island with a small area of forest cover on Virgin Peak (414 m). Some secondary forests also occur in other high elevation areas and bordering drainage systems. The southwestern portion of the island is relatively flat and used for agriculture.

Anegada: This is a nearly flat limestone island attaining a maximum elevation of only 8 m.

Lesser Antilles

This chain of islands (5840 km^2 total) begins just east of the Puerto Rican Bank and extends in an arc southward to the continental shelf of South America at Trinidad and Tobago (Fig. 4:6). The largest island, Guadeloupe (1,510 km^2), also is the highest (1467 m). Most islands have at least some small patches of forest remaining; Dominica is the most

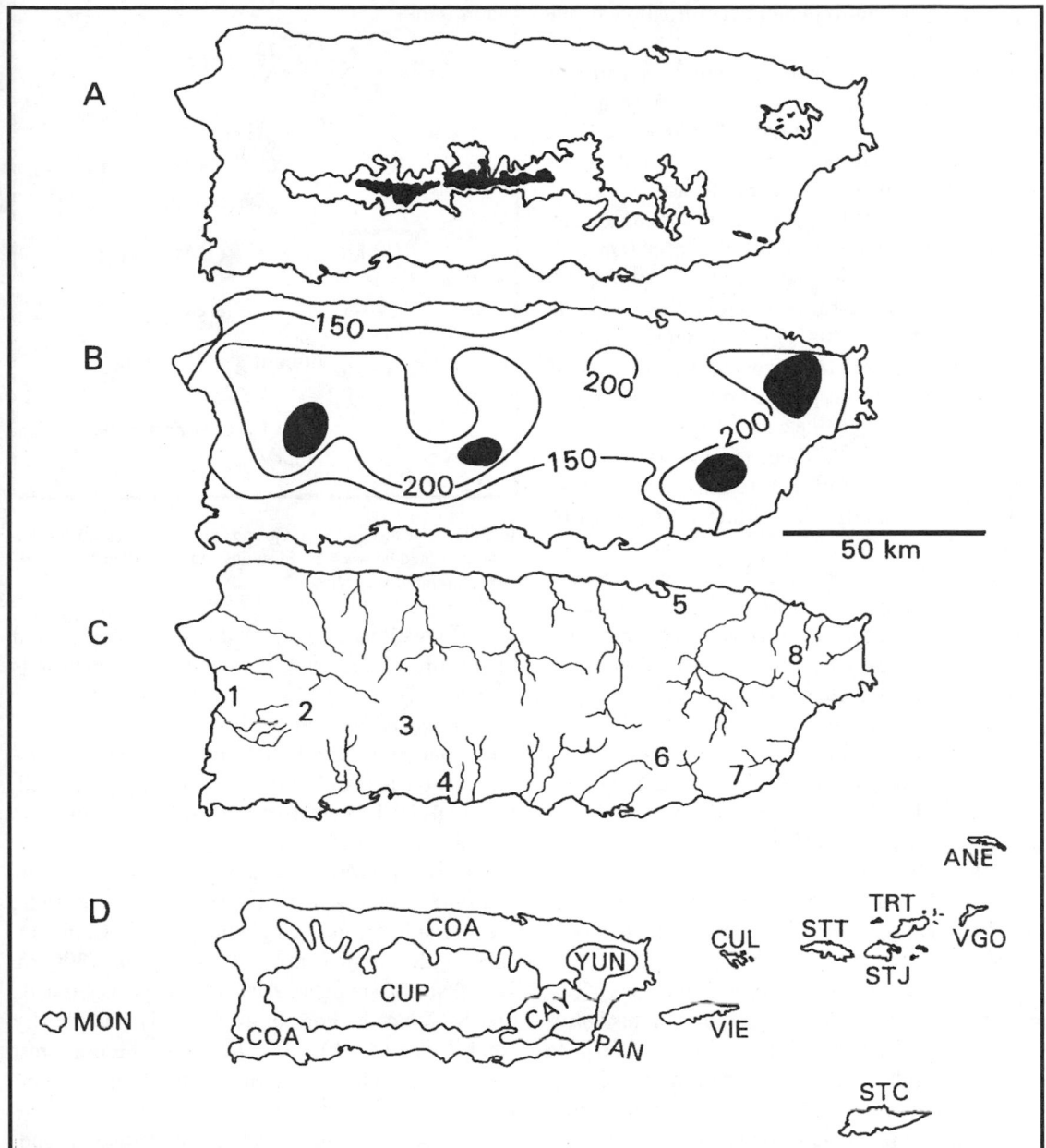

Fig. 4:5. Puerto Rican Bank. (A) Topographic map of Puerto Rico, showing 500 m contour interval and areas above 1,000 m (black). (B) Rainfall map of Puerto Rico, showing contour intervals of 150 cm, 200 cm, and areas over 250 cm (black) annual precipitation. (C) Drainage map of Puerto Rico, showing major rivers. Numbers are locations of peaks, mountain ranges, major cities, and other geographic areas mentioned in text: Mayaguez (1), Maricao forest (2), Cordillera Central (3), Ponce (4), San Juan (5), Sierra de Cayey (6), Cuchilla de Pandura (7), and El Yunque (8). (D) Biogeographic regions of the Puerto Rican Bank, including Mona Island and St. Croix. Abbreviations: ANE Anegada, CAY Sierra de Cayey, COA Coastal Lowlands, CUL Culebra, CUP Central Uplands, MON Isla Mona, PAN Cuchilla de Pandura, STC St. Croix, STJ St. John, STT St. Thomas, TRT Tortola, VGO Virgin Gorda, VIE Vieques, YUN El Yunque.

forested, Barbados the least. Hurricanes, and on a less-frequent time scale, volcanic eruptions, are natural causes of environmental perturbation in the Lesser Antilles.

Anguilla Bank: Included in this region are Anguilla (59 m maximum elevation), St. Martin (392 m), and St.-Barthélémy (424 m). These are three rocky islands of only moderate elevation. Anguilla is covered mostly with dry scrub forest.

Saba Bank: This is a small, rocky, volcanic island attaining a maximum elevation of 870 m.

St. Eustatius Bank: Included here are the islands of St. Eustatius (600 m maximum elevation), St. Christopher or "St. Kitts" (1156 m), and Nevis (985 m). All three are relatively rocky, volcanic islands with montane rainforest in the upper elevations and dry scrub forest along the coasts.

Barbuda Bank: Barbuda (62 m maximum elevation) and Antigua (402 m) are low, mostly denuded limestone islands with rolling hills and cattle farms.

Montserrat: Although primary forest has been removed from this volcanic island, there is some dry scrub forest (low elevations) secondary rainforest (mid-elevations of north slopes) and elfin woodland (summits); the maximum elevation is 914 m.

Guadeloupe Bank: The largest island, Basse-Terre (1467 m maximum elevation), is volcanic and has some montane rainforest; Grande-Terre is a low and essentially denuded limestone island; La Desirade (273 m) is a sharp ridge of limestone and exposed Jurassic rock.

Marie Galante: This is a low, deforested limestone island attaining a maximum elevation of only 204 m.

Dominica: The largest remaining tracts of primary forest in the Lesser Antilles are on this volcanic island, where 60% of the land is forested with either dry scrub forest (west coast), montane rainforest (south and central), or elfin woodland (summits); the maximum elevation is 1447 m.

Martinique: Moderately large tracts of forest, totalling 12% of land area, remain on this volcanic island which rises to 1397 m, but all have been affected by human activities. These tracts include dry scrub forest (south), montane rainforest (intermediate elevations), and cloud forest (Mt. Pelee and west-central).

St. Lucia: This volcanic island supports some forest habitats, including dry scrub forest (north, and

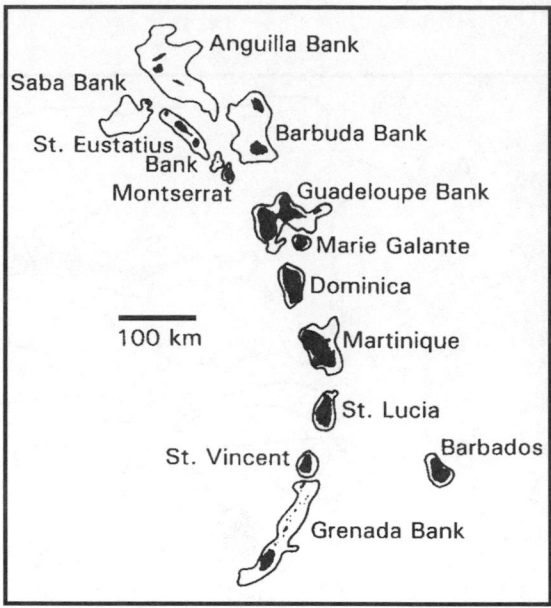

Fig. 4:6.   Islands and island groups of the Lesser Antilles. Redonda, located between St. Eustatius and Montserrat, has no reported amphibian species.

lower elevations near coast), rainforest (slopes), and cloud forest (summits); the maximum elevation is 950 m.

St. Vincent: There are some secondary forests, but apparently there is only one remaining undisturbed montane rainforest area which is located along the ridge in the center of this volcanic island; the maximum elevation is 1234 m.

Grenada Bank: Grenada (840 m maximum elevation) is the largest of these volcanic islands; the Grenadines, including among others Carriacou (294 m), Ronde (164 m), and Union Island (305 m), is a chain of small, low islands between Grenada and St. Vincent. Forests on Grenada include dry scrub forest (south), secondary montane rainforest (intermediate elevations), and elfin woodland (summits).

Barbados: This densely populated low island, formed by uplifted ocean floor, is devoid of forest; the maximum elevation is 340 m.

Bahamas Bank

This cluster of islands lies to the north of Cuba and Hispaniola and to the east of the southern tip of Florida. It includes two political units, the Bahamas and the Turks and Caicos, which have a total area of 11,296 km². All of the thousands of islands on this bank are low (0–40 m elevation) and with a calcareous substrate. Some well-developed lowland rainforest existed at the time of discovery, but only secondary forests are present today.

Cayman Islands

These are three small islands (Grand Cayman, Cayman Brac, and Little Cayman) located about 250 km south of central Cuba and about 250 km northwest of Jamaica; the total area is about 260 km². They represent the emergent portion of the east-west-oriented Cayman Ridge (the southern edge of the North American tectonic plate) and lie immediately to the north of the Cayman Trough. The substrate is mostly coral and limestone and the highest point is only 45 m elevation (Cayman Brac). Dry scrub forest is the predominate vegetation, although a few pockets of primary hardwood forest (including mahogany) remain.

HISTORICAL GEOLOGY

The complex geologic history of the Caribbean still is the subject of intense study but some general features are evident. The Greater Antilles are old islands that originated in the early Cretaceous through island arc accretion in a zone between North and South America (Perfit and Williams, 1989; Pindell and Barrett, 1990). During the mid-Cretaceous, submarine basaltic extrusions in the eastern Pacific created an unusually thick and buoyant oceanic plate that later formed the Caribbean Plate after moving to the north and east behind the Greater Antillean arc (Burke et al., 1978). In the late Cretaceous (70 mya), the island arc connection between North and South America began to break apart as the northeastward movement progressed and a trailing subduction zone formed in the Pacific which would later become the isthmus of Central America. By the early Tertiary (60 mya), the Greater Antilles arc had reached the Bahamas Platform and started a collision which plugged the subduction zone and eventually (50 mya) sutured Cuba, northern Hispaniola, and Puerto Rico to the North American Plate. As a result, a fault zone and small spreading center (Cayman Trough) south of the arc now became the northern boundary of the Caribbean Plate and the direction of plate movement shifted from northeast to east. Jamaica and the southern portion of Hispaniola were carried eastward with the Caribbean Plate to their present locations; the two major portions of Hispaniola became sutured in the Miocene. Although volcanic rock is present, mountain building in the Great-

er Antilles largely has been caused by uplift as a result of compressional forces; there are no active volcanoes.

Subduction of the Atlantic Plate beneath the eastern edge of the Caribbean Plate initiated the Lesser Antillean arc system in the mid- or late Cretaceous. Back-arc spreading shortly after the Cretaceous-Tertiary boundary split the arc into the Aves Ridge (now inactive and almost completely submerged) and the present Lesser Antilles (Bouysse, 1988; Maury et al., 1990). Since then, there has been further subdivision within the Lesser Antilles. An older outer arc (Marie-Galante, Grande Terre of Guadeloupe, La Désirade, Antigua, Barbuda, St. Bartholemew, St. Martin, Dog, and Sombrero) is characterized by low limestone islands that are relatively flat, whereas most other islands (inner arc) in the Lesser Antilles are volcanic and some have active volcanos; there is one active submarine volcano north of Grenada (Maury et al., 1990). The two arcs diverge north of Martinique. This double arc pattern is the result of an initial period of volcanic activity in the Eocene (outer arc), a quiescent period for about 10 my in the late Oligocene and early Miocene, and then renewed volcanic activity since the mid-Miocene (Maury et al., 1990). Barbados differs from the other islands in being the emergent portion of the Lesser Antilles Accretionary Ridge and is, therefore, not volcanic. The presence of Mesozoic basement rocks at several locations in the Lesser Antilles supports a Mesozoic origin for the arc (Bouysse, 1988; Maury et al., 1990; Pindell and Barrett, 1990).

It was suggested recently that the northern (north of Martinique) and southern Lesser Antilles have had separate geologic histories that have influenced the evolution of anoline lizards in the West Indies (Roughgarden, 1995). The southern Lesser Antilles were proposed to be a separate plate "or piece of the South American Plate that has become sutured to the present-day eastward-moving Caribbean Plate." This was based in part on a suggestion in the geologic literature that the northern and southern portions of the Lesser Antilles were separated by a major east-west transform fault (Bouysse, 1984). Motion along the fault was believed to have caused the separation between the Aves arc and Lesser Antilles arc. However, the location of that proposed fault was north of Dominica (not Martinique) and Bouysse since has abandoned the hypothesis

Table 4:1. Numbers of species of anurans in the West Indies.

Taxon	Total	Native[1]	Endemic[2]	Introduced	%Endemic[3]
Bufonidae:					
Bufo	12	11	11	1	92
Dendrobatidae:					
Colostethus	1	1	1	0	100
Hylidae:					
Calyptahya	1	1	1	0	100
Hyla	7	5	5	2	100
Osteopilus	4	4	4	0	100
Pseudacris	1	0	0	1	0
Scinax	1	0	0	1	0
Leptodactylidae:					
Eleutherodactylus	139	139	139	0	100
Leptodactylus	5	5	3	0	60
Microhylidae:					
Gastrophryne	1	0	0	1	0
Ranidae:					
Rana	3	0	0	3	0
Total	174	167	164	8	98

[1]Occurs naturally within the West Indies.
[2]Occurs naturally within the West Indies and nowhere else (except as introduced).
[3]Number of endemic species divided by number of native species.

based on more recent geologic evidence that favors back-arc spreading of a single continuous island arc (Bouysse, 1988; Holcombe et al., 1990; Maury et al., 1990). A geologic model by Speed (1985) also was cited by Roughgarden (1995) as supporting evidence, but Speed's model indicates a different origin only for the extreme southern islands (e.g., Grenada) in the Lesser Antilles.

The location and extent of exposed land through time is important in understanding the historical biogeography of the West Indies. Unfortunately, very little of this information is known. In the mid- to late Cretaceous (90-70 mya), paleocoastline data suggest that areas in North and South America adjacent to the proto-Antilles were below sea level (Smith et al., 1994) suggesting that the proto-Antilles also were submerged, or at least did not form a land connection with the mainland. Jamaica is believed to have been completely (or nearly so) submerged during the Oligocene based on the extensive limestone sediments of that period virtually covering the island (reviewed in Hedges, 1989a). The southern portion of Hispaniola also may have been submerged during the Oligocene, although a large amount of subsequent mountain building since has erased much of that evidence (Maurrasse, 1982). Most of Cuba, northern Hispaniola, and Puerto Rico also were submerged during that time, although some areas appear to have been emergent throughout the Cenozoic (Bowin, 1975; Maurrasse, 1982; Ituralde-Vinent, 1988; Lewis and Draper, 1990). The Bahamas Bank, presently a low limestone platform, probably was completely submerged during the interglacial periods of the Pleistocene.

THE AMPHIBIAN FAUNA

The number of native amphibian species inhabiting the West Indies (167 species) is remarkably high considering the relatively small total area. Even more surprising is the small number of native families (4) and genera (7) represented; all are anurans (Table 4:1). Only two native West Indian species (Leptodactylus insularum and L. validus; Heyer, 1994) occur naturally outside the area. The majority of species (139 species or 84%) are from a single leptodactylid genus Eleutherodactylus, which also is the largest genus of vertebrates with over 500 species (Duellman, 1993). Other components of the amphibian fauna include: a relatively small radiation of bufonids (11 species: Bufo peltocephalus group), centered primarily in Cuba, but with representatives in Hispaniola and in the Puerto Rican Bank, some hylid frogs placed in three genera (10 species: Calyptahyla, Hyla, and Osteopilus), other leptodactylids (5

Table 4:2. Numbers of species of anurans on islands and island groups in the West Indies. Introduced taxa are noted by an asterisk.

Region	Total	Native	Endemic	Introduced	% Endemic
Cuba:	52	50	48	2	96
Bufo	8	8	8	0	100
Osteopilus	1	1	0	0	0
Pseudacris*	1	0	0	1	0
Eleutherodactylus	41	41	40	0	98
Rana*	1	0	0	1	0
Jamaica:	26	22	22	4	100
Bufo*	1	0	0	1	0
Calyptahyla	1	1	1	0	100
Hyla	2	2	2	0	100
Osteopilus	2	2	2	0	100
Eleutherodactylus	19	17	17	2	100
Rana*	1	0	0	1	0
Hispaniola:	65	63	63	2	100
Bufo	3	2	2	1	100
Hyla	3	3	3	0	100
Osteopilus	1	1	1	0	100
Eleutherodactylus	56	56	56	0	100
Leptodactylus	1	1	1	0	100
Rana*	1	0	0	1	0
Puerto Rican Bank:	25	20	20	5	100
Bufo	2	1	1	1	100
Hyla*	1	0	0	1	0
Osteopilus*	1	0	0	1	0
Scinax*	1	0	0	1	0
Eleutherodactylus	18	18	18	0	100
Leptodactylus	1	1	1	0	100
Rana*	1	0	0	1	0
Lesser Antilles:	14	10	9	4	90
Bufo*	1	0	0	1	0
Colostethus	1	1	1	0	100
Osteopilus*	1	0	0	1	0
Scinax*	1	0	0	1	0
Eleutherodactylus	8	7	7	1	100
Leptodactylus	2	2	1	0	50
Bahamas Bank:	6	2	0	4	0
Hyla*	1	0	0	1	0
Osteopilus	1	1	0	0	0
Eleutherodactylus	1	1	0	0	0
Gastrophryne*	1	0	0	1	0
Rana*	2	0	0	2	0

species: *Leptodactylus*), and a dendrobatid (*Colostethus*) on Martinique in the Lesser Antilles.

Introductions of plants and animals are common in the West Indies, and there are eight introduced amphibian species (Table 4:1). Unfortunately, two large species, *Bufo marinus* and *Rana catesbeiana*, have been introduced successfully on most of the major and many of the smaller islands.

The high level of endemism seen in the West Indian amphibian fauna also extends to the level of individual islands (Table 4:2), where completely endemic faunas are found in Jamaica (22 species), His-

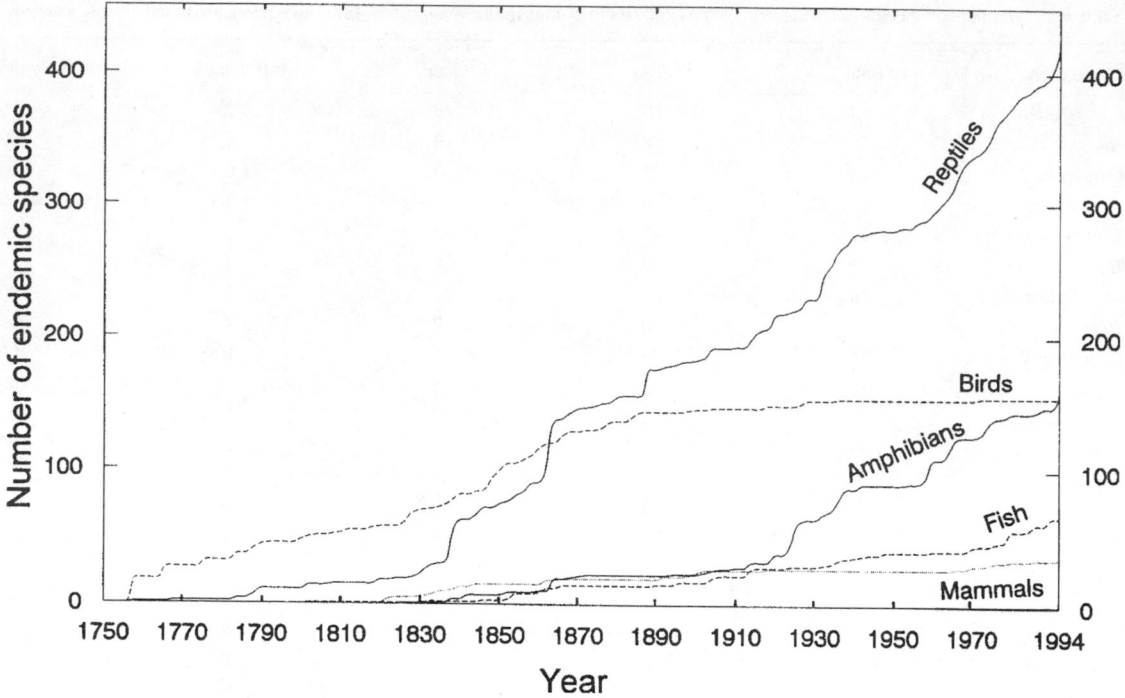

Fig. 4:7.   Discovery curve of West Indian amphibian species in relation to curves of other West Indian vertebrate groups. Plotted is the number of species known at each time interval.

paniola (63 species), and the Puerto Rican Bank (20 species). High levels of endemism also are found in Cuba (96% of 50 species) and the Lesser Antilles (90% of 10 species). Only two species (*Eleuthero-dactylus planirostris* and *Osteopilus septentriona-lis*) are native to the Bahamas Bank and only one (*Osteopilus septentrionalis*) is native to the Cayman Islands; none is endemic. As with the amphibian fauna in general, the genus *Eleutherodactylus* is the major component of the amphibian fauna in each island, although there are some subtle differences. In Cuba, *Bufo* (8 species, 16% of total) is a significant component, whereas in Jamaica, it is the hylid frogs (5 species, 23% of total).

Trend curves of the rate of species description (Steyskal, 1965) can provide a gauge of our current understanding (in terms of species diversity) of a taxonomic group. Such discovery curves for West Indian vertebrates (Fig. 4:7) show that some groups, such as mammals and birds, have reached a plateau suggesting that nearly all of the extant species have been discovered. However, the curves for amphibians and reptiles rise steeply indicating that our knowledge of these groups is far from complete. For example, at least 16 new species of frogs have been discovered in the last four years; these represent 10% of the known amphibian fauna and include a family (Dendrobatidae) new to the West Indies.

PATTERNS OF DISTRIBUTION

Cuba

There are 50 native amphibian species in Cuba (Appendix 4:1), 48 of which are endemic (*Eleuth-erodactylus planirostris* and *Osteopilus septentrion-alis* also occur elsewhere). The total number of species inhabiting each biogeographic region of Cuba varies from five in the low Llanura de Zapata to highs of 26 and 27 in the two eastern highland regions (Sierra Maestra and Macizo de Sagua-Baracoa, re-

spectively). Three other regions with high numbers of species are the Cordillera de Guaniguanico (17 species), the Macizo del Escambray (17 species), and the Alturas de Camaguey-Maniabón (16 species). These five peaks also are seen in the species density map (Fig. 4:8). Steep gradients in species density (6–20 species) are evident along the northern edges of the two eastern regions and correspond closely to rainfall and elevation. The depression in species den-

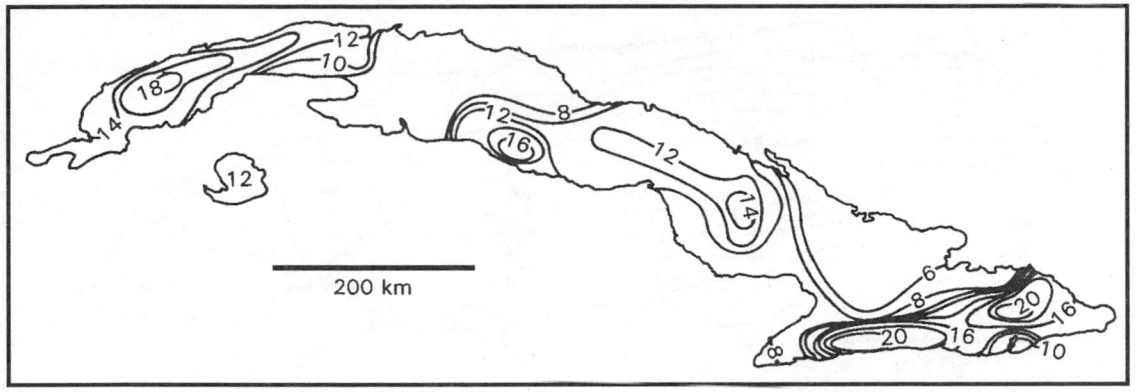

Fig. 4:8.   Amphibian species density in Cuba (contour interval = two species).

sity between the two eastern peaks (Fig. 4:8) corresponds to the Valle Central, which separates the Sierra Maestra from the Macizo de Sagua-Baracoa. Noteworthy is the relatively high number of species (up to 15) occurring at low elevations in several parts of Cuba; this reflects the radiation of bufonids (*Bufo*) that mostly inhabit low areas on the island and possibly also reflects the high rainfall in some coastal regions of eastern Cuba.

Three of the five peaks in species density are not reflected in regional endemism (Appendix 4:1). The two eastern regions each have seven endemics; these account for 28% of the species on the island, whereas three other regions (Península de Guanahacabibes, Llanura de Zapata, and Macizo del Escambray) each have only one endemic species, and one region (Cordillera de Guaniguanico) has two endemics. The highest numbers of sympatric species at single localities are 16 at Soroa, Pinar del Río (200 m), and 13 at Gran Piedra, Santiago de Cuba (1100 m).

Analysis of amphibian faunal similarity identifies three major clusters corresponding to the western, central, and eastern regions (Fig. 4:9A). Amphibian faunal similarity among these clusters is low (CBR = 0.29–0.49). Even within these clusters, some adjacent regions have large differences in faunal composition. For example, the two eastern highland regions, separated by only 45 km, have only 68% (CBR = 0.68) of their species in common. The amphibian fauna in the Macizo del Escambray is more like that in the Alturas de Camaguey-Maniabón (CBR = 0.91) than in the adjacent Alturas Centrales (CBR = 0.66). The regions of western Cuba associate into two groups: the upland regions (Cordillera

de Guaniguanico, Alturas de la Habana-Matanzas) and the lowland regions (Península de Guanahacabibes, Llanura Occidental, Isla de Juventud). The position of the Llanura de Zapata differs between the two analyses (Fig. 4:9A, B), probably because of the small number of species (5) in that region. The relationships of the three major clusters (western, central, and eastern) also differs between the two trees.

JAMAICA

All 22 native amphibian species in Jamaica are endemic, and four other species are introduced (Ap-

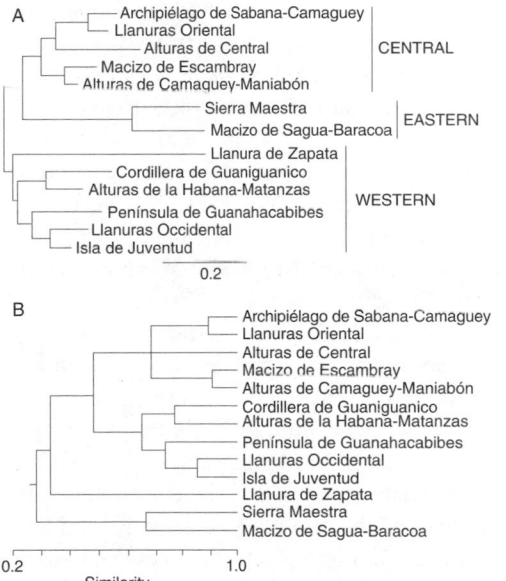

Fig. 4:9.   Relationships of the biogeographic regions of Cuba based on amphibian faunal similarity. (A) Neighbor-joining analysis (scale in distance units). (B) UPGMA analysis (scale in similarity units).

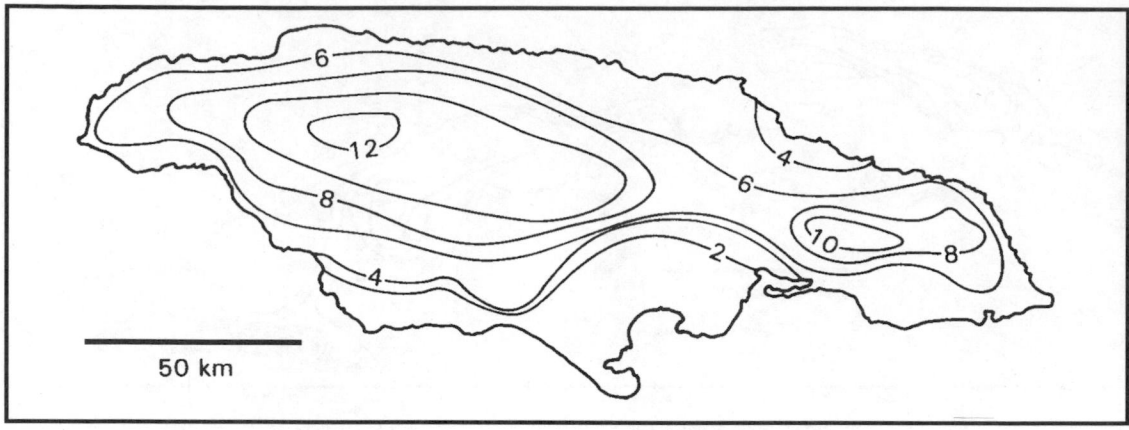

Fig. 4:10. Amphibian species density in Jamaica (contour interval = two species).

pendix 4:2). The total number of species inhabiting each biogeographic region of Jamaica ranges from one each in the three dry southern regions (Southern Lowlands, Portland Ridge Peninsula, and Hellshire Hills) to a high of 15 in the Cockpit Country of west-central Jamaica. The highland regions have considerably more species (5–15, $\bar{X}$ = 9.9) than the lowland regions (1–6, $\bar{X}$ = 2.4). This pattern also is seen in the species density map of Jamaica (Fig. 4:10), where overlapping ranges reach peaks of 12 species in the Cockpit Country and 10 species in the Blue Mountains. Three regions have endemic species: Cockpit Country (2), Portland Ridge Peninsula (1), and the Blue Mountains (2). The highest numbers of sympatric species at single localities are: 12 species at Quick Step, Trelawny (400 m) and 11 species at Hardwar Gap, St. Andrew/Portland (1200 m).

Analysis of faunal similarity defines two major clusters in Jamaica (Fig. 4:11): the western and central regions (CBR = 0.59) and the eastern regions (Blue Mountains and John Crow Mountains) (CBR = 0.70). The average similarity between these two clusters is only 0.35. This illustrates the strong regional endemism of amphibian species within Jamaica. Within the cluster of western and central regions, the neighbor-joining tree separates most of the upland regions (Western Uplands, Cockpit Country, Manchester Plateau, and Central Uplands) into one group and the two lowland regions (and Santa Cruz Mountains) into a separate cluster. Also, that analysis joins the Cockpit Country and Central Uplands together as "sister regions." These relationships, although concordant with geography and some

species distributions, were not seen in the UPGMA analysis (Fig. 4:11B).

HISPANIOLA

There are 63 endemic native amphibian species, and two introduced species in Hispaniola (Appendix 4:3). The number of species inhabiting each biogeographic region of Hispaniola ranges from only one in three Haitian regions (Île Gonâve, Plaine de l'Artibonite, Massif des Montagnes Noires) to the highest (32) in any region in the West Indies, in the

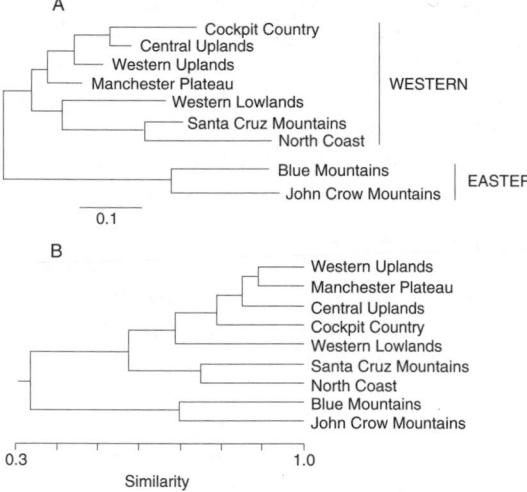

Fig. 4:11. Relationships of the biogeographic regions of Jamaica based on amphibian faunal similarity. (A) Neighbor-joining analysis (scale in distance units). (B) UPGMA analysis (scale in similarity units). Three regions with only one species in each were omitted.

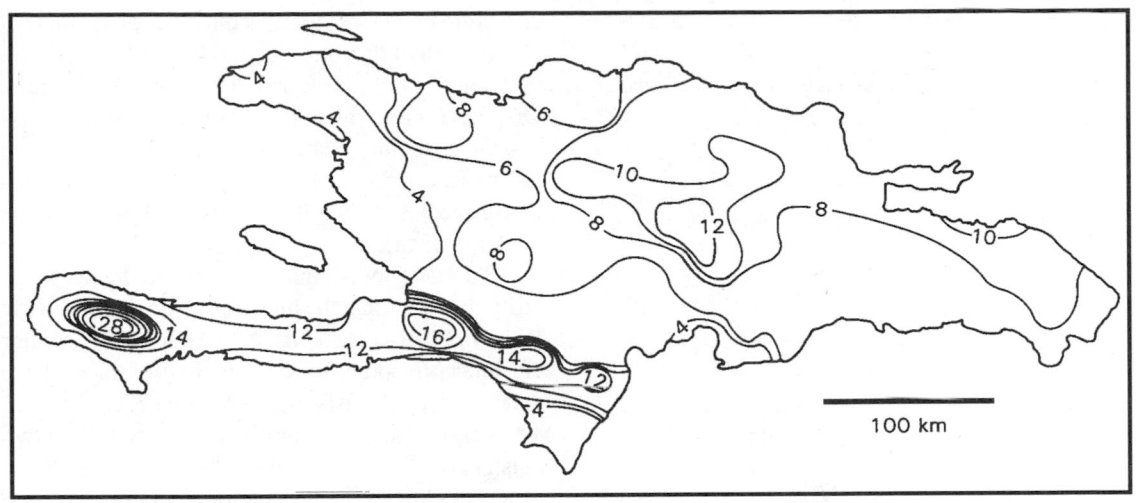

Fig. 4:12.   Amphibian species density in Hispaniola (contour interval = two species).

Massif de la Hotte (also in Haiti). Two other mountain ranges have at least 20 species: the Massif de la Selle–Sierra de Baoruco (26) and the Cordillera Central (20). Although these three regions represent the peaks of species diversity within Hispaniola, the species density map (Fig. 4:12) reveals that the high density of the Cordillera Central primarily is because of the large area of the defined region (Fig. 4:4), which encompasses species with nonoverlapping ranges. The actual species density of overlapping ranges does not exceed 12 (Fig. 4:12) and represents only a modest rise above the surrounding regions with 8–10 species. For similar reasons, species density reaches only 16 in the La Selle–Baoruco range. However, species density in the Massif de la Hotte drops only slightly, from 32 (regional) to 28 (species density); this rises steeply from the surrounding areas of the Tiburon Peninsula that have only 12 species.

Regional endemism within Hispaniola shows patterns similar to species density. Approximately one-fourth (16 species) of Hispaniolan amphibian species occur only in the Massif de la Hotte region, where they represent 50% of the anuran fauna. About 16% (10 species) and 10% (6 species), respectively, of Hispaniolan species occur in the Massif de la Selle–Sierra de Baoruco and Cordillera Central regions. In the poorly known northwest peninsula of Haiti, the Presqu'île du Nord Ouest, three of the five anurans are endemic species. Other regions with endemic species are the Chaîne des Matheux–Sier-

ra de Neiba (2), the Llanura Costera del Caribe (2), Île de la Tortue (1), the Plaine du Nord–Valle de Cibao (1), and the Massif du Nord (1). The highest numbers of sympatric species at single localities are 24 at Castillon, Dept. de la Grande'Anse, Haiti (950 m) and 11 at Furcy, Dept. de l'Ouest, Haiti (1500 m).

The neighbor-joining analysis of faunal relationships of the biogeographic regions of Hispaniola (Fig. 4:13A) defines two major clusters (CBR approximately 0.50) which corresponds to the classical "North Island" and "South Island" of previous workers (Mertens, 1939; Williams, 1961; Schwartz, 1980) and is supported by molecular phylogenetic studies of amphibians (Hedges, 1989b; Hass and Hedges, 1991). The other regions not joining these two clusters have relatively small numbers of species (3–6) and primarily are low, dry regions. Strong regional differentiation is illustrated by the two major South Island regions which are separated by only 110 km. Although clustering as sister regions, they have a CBR of only 0.52. Relationships within the large cluster of North Island regions agrees closely with geography. For example, the four regions of eastern Hispaniola (Península de Samana, Haitises, Cordillera Oriental, and Llanura Costera del Caribe) form a cluster, as do two adjacent regions in the northern part of the island (Cordillera Septentrional and Plaine du Nord–Valle de Cibao). However, the UPGMA analysis (Fig. 4:13B) separates the two South Island upland regions and does not agree as well with geography as does the neighbor-joining analysis.

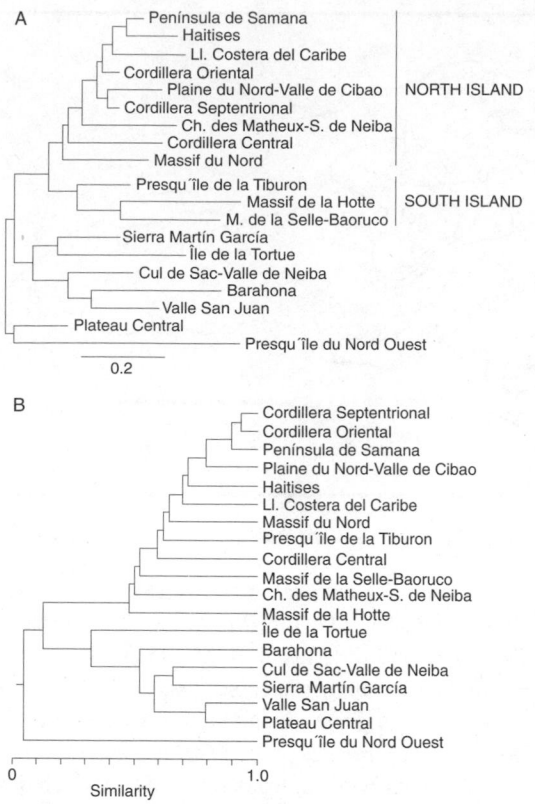

Fig. 4:13. Relationships of the biogeographic regions of Hispaniola based on amphibian faunal similarity. (A) Neighbor-joining analysis (scale in distance units). (B) UPGMA analysis (scale in similarity units). Three regions with only one species in each were omitted.

PUERTO RICAN BANK

There are 20 endemic native amphibian species and five introduced species in the Puerto Rican Bank (Appendix 4:4). The number of species inhabiting each biogeographic region of the Puerto Rican Bank ranges from only one species on Mona and Anegada to 13 species in the Central Uplands and 14 species at El Yunque. The species density map of Puerto Rico (Fig. 4:14) also identifies those latter two regions as having high species density but with greater resolution; the high densities are shown to be associated with higher elevations on El Yunque, the Cordillera Central, and the Sierra de Cayey. The steepest contours of species density occur around El Yunque, where density ranges from 6 to 14 species in a distance of less than 10 km.

Regional endemism within the Puerto Rican Bank is relatively low compared to the other Great-

er Antillean islands; there is only one endemic species in each of the four regions (Island Mona, Sierra de Cayey, Cuchilla de Panduras, and El Yunque). Disjunct ranges, rather than regional endemism, explain the distinct peaks in species density seen in Puerto Rico (Fig. 4:14). The highest number of sympatric species at a single locality is 11 at El Yunque Peak (1065 m).

The neighbor-joining analysis of faunal similarity indicates that the biogeographic regions of the Puerto Rican Bank (Fig. 4:15A) form two major clusters corresponding to the regions of Puerto Rico and those of the islands to the east (Fig. 4:15A). The amphibian faunal similarity between these two clusters is low (CBR = 0.44–0.56). Within the Puerto Rican cluster, the three highland regions form a group, with a strong association between El Yunque and the Central Uplands (CBR = 0.96). The Sierra de Cayey, with its single endemic (*Eleutherodactylus jasperi*), is slightly more differentiated (CBR = 0.82). With the cluster of eastern islands (Fig. 4:15A), the two satellite islands of Puerto Rico (Vieques and Culebra) are identical in faunal similarity, as are St. Thomas and St. Croix. The three U.S. Virgin Islands form a group (CBR = 0.89 among islands) and the two British Virgin Islands form a group (CBR = 0.57 between islands). Some of these relationships, which are largely concordant with geography, are not evident in the UPGMA analysis (Fig. 4:15B).

LESSER ANTILLES

Ten of the 14 species inhabiting the Lesser Antilles are native (including nine endemics) and four are introduced (Appendix 4:5). The number of species inhabiting each biogeographic region of the Leser Antilles ranges from one species on five of the island banks to four species in Guadeloupe. Five island banks have endemic species: Guadeloupe (2), Dominica (1), Martinique (1), St. Vincent (1), and Grenada (1).

The relationships among island banks in the Lesser Antilles, based on faunal similarity, differ between the two analyses (Fig. 4:16). Both define a southern cluster containing St. Vincent and the Grenada Bank, and a northern-central cluster containing the Barbuda Bank, Marie Galante, Martinique, and the Guadeloupe Bank. However, the neighbor-joining analysis (Fig. 4:16A) joins Montserrat and Dominica and places them as the most diver-

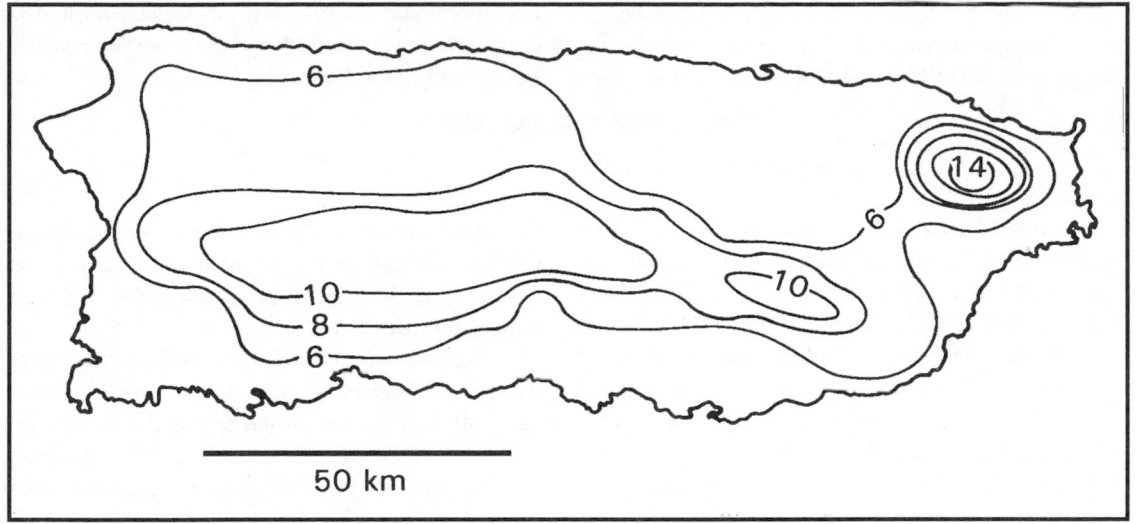

Fig. 4:14. Amphibian species density in Puerto Rico (contour interval = two species).

gent cluster, whereas the UPGMA analysis (Fig. 4:16B) places Montserrat as the sister region to the northern cluster, with Dominica joining that more inclusive group. The southern regions (St. Vincent and the Grenada Bank) have a CBR of only 0.32 with the other regions in the Lesser Antilles.

BAHAMAS BANK

Only two species of amphibians are native to the Bahamas Bank; neither is endemic. *Osteopilus septentrionalis* is distributed throughout the area, with the exception of Mayaguana Island and the Turks and Caicos Islands. *Eleutherodactylus planirostris* also is widely distributed, but apparent-

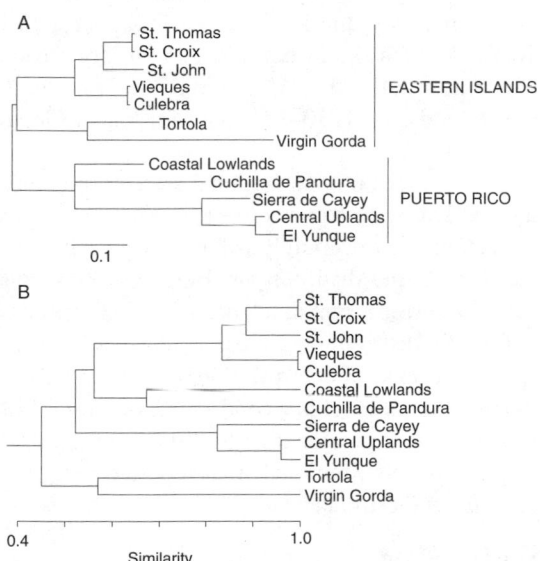

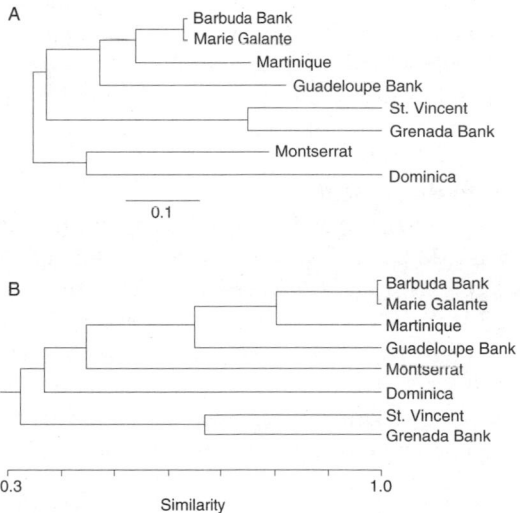

Fig. 4:15. Relationships of the biogeographic regions of the Puerto Rican Bank based on amphibian faunal similarity. (A) Neighbor-joining analysis (scale in distance units). (B) UPGMA analysis (scale in similarity units). Two regions with only one species in each were omitted.

Fig. 4:16. Relationships of the biogeographic regions (islands and island banks) of Lesser Antilles based on amphibian faunal similarity. (A) Neighbor-joining analysis (scale in distance units). UPGMA analysis (scale in similarity units). Five regions with only one species in each were omitted.

ly is absent from Crooked Island, Acklins Island, and Mayaguana Island. Both are common and widespread species in Cuba. One hylid (*Hyla squirella*), one microhylid (*Gastrophryne carolinensis*), and two ranids (*Rana grylio* and *R. sphenocephala*) have been introduced to the Bahamas.

DISTRIBUTIONAL AREA

GEOGRAPHIC PATTERNS

Cuba

The total area of Cuba, including Isla de Juventud and other satellite islands, is 110,922 km^2. The average area of species distributions is 21,130 km^2, but one-half of the species (25) have distributions smaller than 6000 km^2 (5% of island area), and about one-quarter (13) have distributions occupying less than 1100 km^2 (1% of island area). Five species (8% of the amphibian fauna) are known only from their type localities (Appendix 4:1, Fig. 4:17; the single locality for *Eleutherodactylus pezopetrus* encompasses a range in elevation).

Jamaica

The total area of Jamaica is 10,990 km^2. The average area of species distributions is 2630 km^2, but one-half (11) of the species have distributions of less than 1000 km^2 (< 9% of the area of the island), and eight of those species inhabit areas less than 550 km^2 (Appendix 4:2, Fig. 4:17).

Hispaniola

The total area of Hispaniola (including satellite islands) is 76,470 km^2. The average area of species distributions is 9597 km^2, but one-half of the species have distributions smaller than 1330 km^2 (2% of the total island area), and one-third of the species occupy less than 1% of the total area of the island. Six species (10% of fauna) are known only from their type localities (Appendix 4:3, Fig. 4:17).

Puerto Rican Bank

The total area of the Puerto Rican Bank (including Isla Mona and St. Croix, which are not on the bank) is 9511 km^2 (Puerto Rico = 8768 km^2, Mona = 57 km^2, Vieques = 138 km^2, Culebra = 27 km^2, U.S. Virgin Islands = 347 km^2, British Virgin Islands = 174 km^2). The average area of species distributions is 2567 km^2, although one half of the species have distributions smaller than 1050 km^2 (11% of the total bank area). The distributions of six species (30% of fauna) are no larger than 160 km^2 or less than 1.7% of the total area (Appendix 4:4, Fig. 4:17).

Lesser Antilles

The total area of the Lesser Antilles is 5840 km^2. The average area of species distributions is 1029 km^2, this high figure is owing largely to the wide ranges of *Eleutherodactylus johnstonei* and *E. martinicensis*. Five species (50% of the fauna) have distributions no greater than 170 km^2 or 3% of the total area of the Lesser Antilles (Appendix 4:5).

GENERAL PATTERNS

Patterns of distributional area of individual islands are similar in that many species have small distributions, and only a few species are widespread. Considering all West Indian species, the same highly skewed pattern is evident; the median distributional area is 1545 km^2, and the mode is less than 100 km^2 (Fig. 4:17). If anuran families are examined separately, the median distributional area is 10,425 km^2 for the 11 bufonids, 25 km^2 for the single dendrobatid, 10,990–39,500 km^2 for the 10 hylids, and only 1250–1315 km^2 for the 143 leptodactylids.

The values of distributional area reported here have at least two sources of error. First, additional collecting almost certainly will reveal that some species have larger distributions. However, the second source of error is that much of the habitat contained within the distributional areas is unsuitable for the species because of deforestation or fine-scale elevational differences. This second source of error is likely to be much greater than the first; and therefore many of the distributional areas reported here probably are overestimates.

ALTITUDINAL DISTRIBUTION

GEOGRAPHIC PATTERNS

Cuba

Considering only altitude, amphibian species density in Cuba (*n* = 50 species) is highest (31) near sea level and then decreases with elevation to a low of three species at the highest point in Cuba (Pico Turquino), near 2000 m (Fig. 4:18A). Concordant with this pattern is a decrease in mean maximum body size (SVL), from a high of 60.5 mm (*n* = 27) at sea level to 31.3 mm (*n* = 3) at 2000 m (Fig. 4:18A).

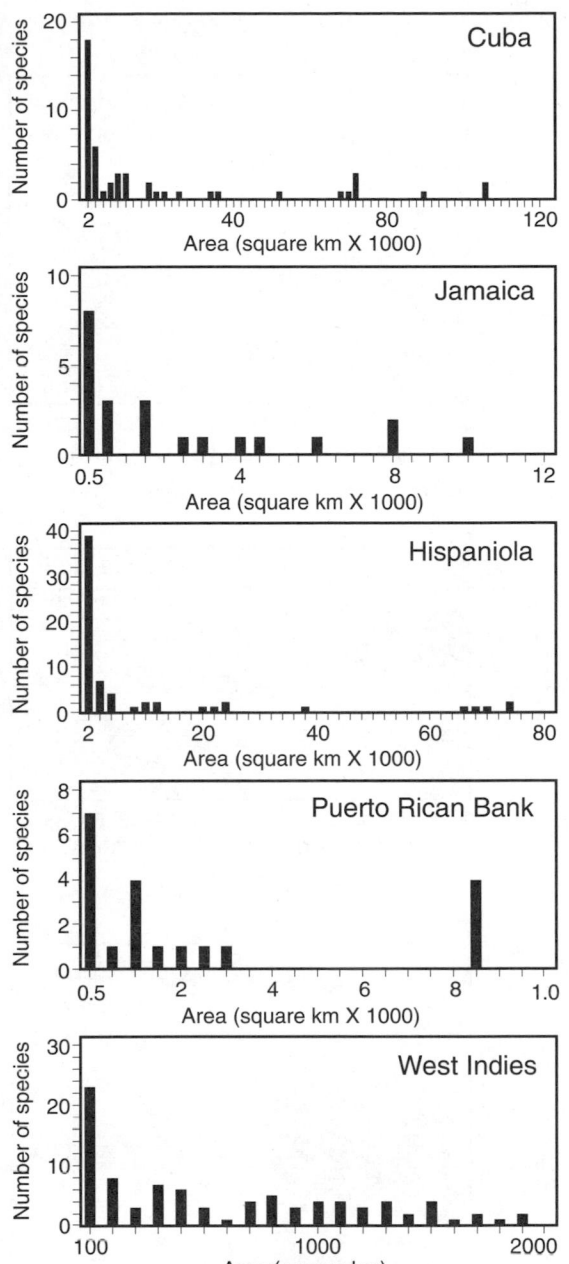

Fig. 4:17.   Histogram showing the number of species of West Indian amphibians at intervals of distributional area; intervals are 2,000 km² for Cuba and Hispaniola, and 500 km² for Jamaica and Puerto Rico. Arrows indicate the total area of each island. Combined analysis (West Indies) includes species with distributions less than 2,000 km² in area.

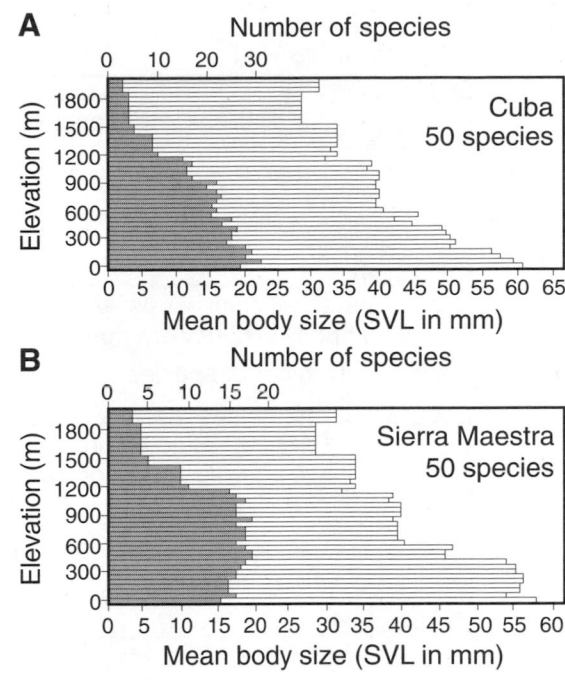

Fig. 4:18.   Species density (shaded bars) and mean maximum body size (snout-vent length, SVL; white bars) at different elevational intervals in Cuba. Elevation is in 50 m intervals (e.g., 0-49, 50-99, etc.). (A) Islandwide. (B) Sierra Maestra.

By restricting the analysis to the 26 species in the Sierra Maestra, a somewhat different pattern is seen (Fig. 4:18B); there species density increases slightly from 14 at sea level to 18 at 450 m, where it re-

mains relatively constant until about 1050 m, after which point it drops sharply to three species at 2000 m. The difference in species density pattern between the two analyses probably is because of the inclusion of the vast lowland areas of Cuba (and associated species) in the first analysis. However, body size shows the same trend (decreasing with increasing elevation) in the Sierra Maestra as for Cuba in general (Fig. 4:18B).

Jamaica

Amphibian species density in Jamaica ($n = 22$ species) increases from 10 at sea level to 17 species at 450-600 m and then decreases to one species at the highest point (Blue Mountain Peak), at about 2200 m (Fig. 4:19A). The peak at 450–600 m corresponds to the central limestone platform of Jamaica, and thus includes the endemic species of the Western Uplands, Cockpit Country, and Central Uplands. In general, body size decreases from 53.6 mm at sea level to 26.0 mm at 2200 m. However, there is a "depression" in the slope at about 450–600 m apparently caused by the inclusion of the small Cockpit endemics (especially *Eleutherodactylus griphus* and *E. sisyphodemus*). When the Blue Mountains

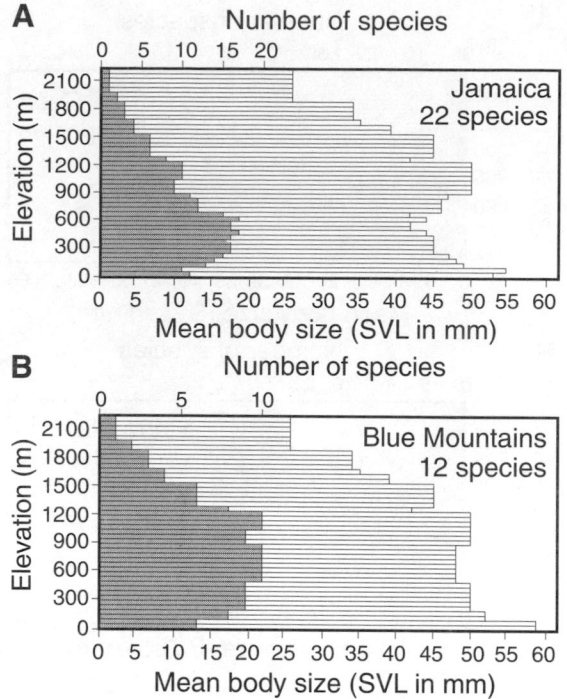

Fig. 4:19.   Species density and mean maximum body size (snout-vent length, SVL) at different elevations in Jamaica. (A) Island-wide. (B) Blue Mountains.

are examined separately (Fig. 4:19B), the mid-elevational bulge in species density (10) now extends from 500 m to 1200 m, and body size shows a more continuous decline with increasing elevation.

Hispaniola

Species density in Hispaniola (*n* = 63 species) shows a strong mid-elevational bulge between 600 m and 1600 m, with a peak of 32 species at 1050–1100 m (Fig. 4:20A). Body size shows a relatively continuous decline with increasing elevation, from 51.0 mm at sea-level to 28.0 mm at 2450 m. The slightly higher value of 35.0 mm between 2450 m and 3050 m relates to the presence of a single species (*Eleutherodactylus patricae*) at the highest elevations of Pico Duarte.

The effect of elevation on species density and body size is perhaps best seen in the fauna of the Massif de la Hotte (Fig. 4:20B), because it is an isolated and well-defined massif that contains the highest density of amphibian species in the West Indies. Twelve species occur at sea-level; density increases to 24 species at 1050–1100 m before decreasing to six species at 2300 m. In contrast to the sharp mid-

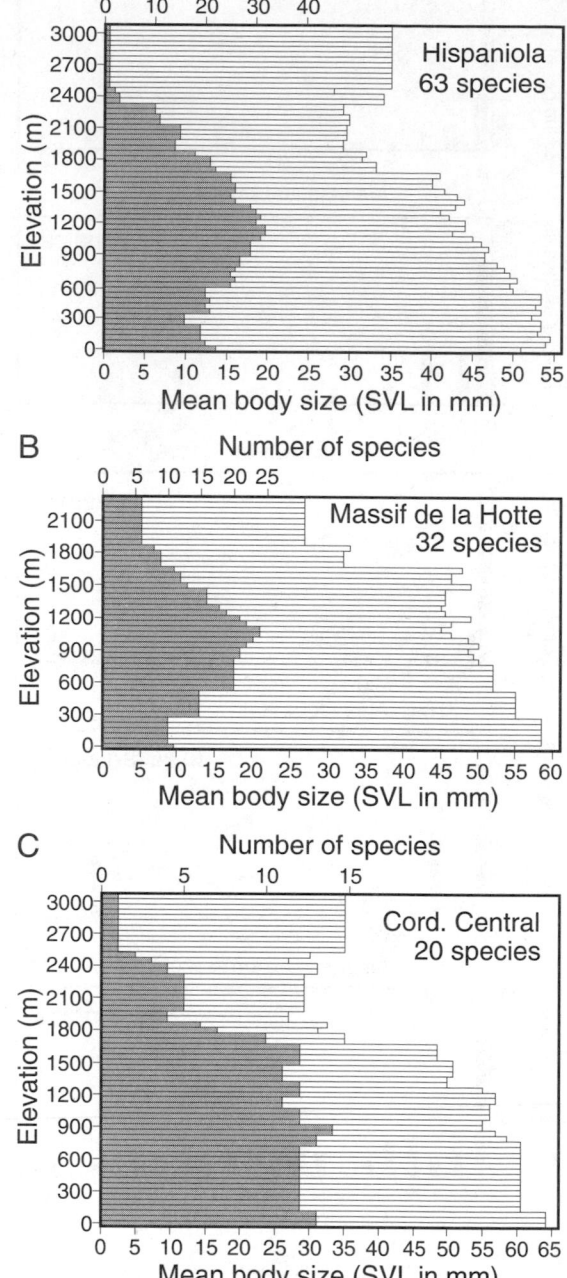

Fig. 4:20.   Species density and mean maximum body size (SVL) at different elevations in Hispaniola. (A) Islandwide. (B) Massif de la Hotte. (C) Cordillera Central.

elevation peak in species density, body size decreases more or less continuously with elevation, from 55.1 mm at sea-level to 26.5 mm at 2300 m.

A

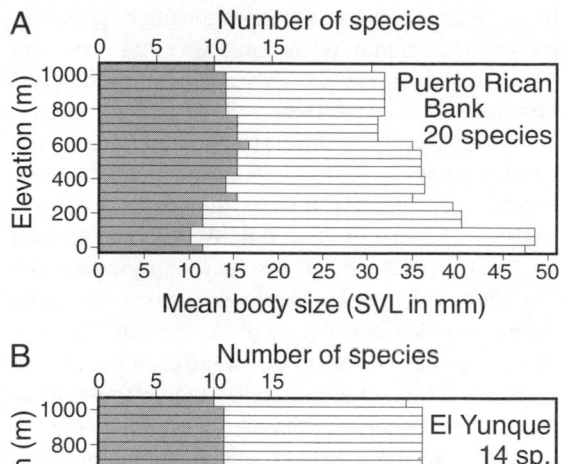

B

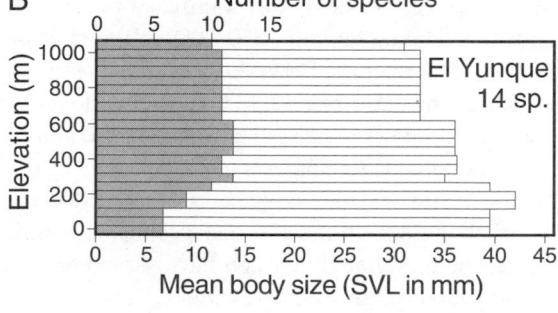

Fig. 4:21. Species density and mean maximum body size (SVL) at different elevations in the Puerto Rican Bank. (A) Puerto Rican Bank. (B) El Yunque.

The Cordillera Central exhibits a different pattern of species density change with elevation. Species density remains fairly constant (12–14) between sea level and 1650 m before decreasing to one species at 2550–3050 m. However, body size shows the typical decreasing trend (Fig. 4:20C).

Puerto Rican Bank

Analysis of the Puerto Rican Bank (n = 20 species) and a more restricted analysis of El Yunque (n = 14) both show species density increasing to a high of about 12–13 species at 500–600 m and decreasing only slightly (10 species) at the highest elevations (1050 m). Body size in both analyses decreases with increasing elevation, from 47.4 mm to 30.7 mm (Fig. 4:21A) and from 39.4 mm to 30.7 mm (Fig. 4:21B), respectively. This somewhat truncated pattern, with little decrease in the upper elevations, probably is because the highest elevations in Puerto Rico are lower than those on the other three large islands and that El Yunque Peak (1065 m) corresponds approximately to a general peak in species density with increasing elevation (see below).

Lesser Antilles

Although the total number of species (10 species) is relatively small, the same patterns that are

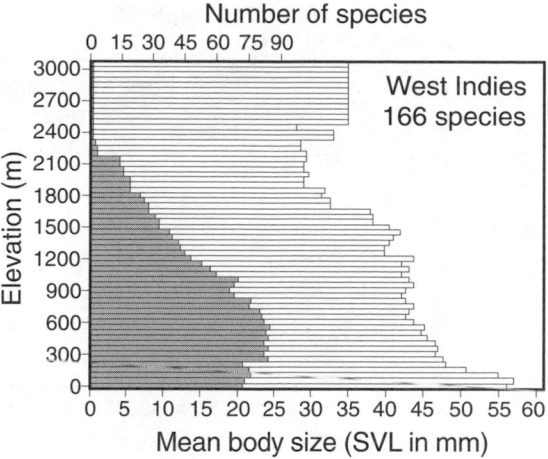

Fig. 4:22. Species density and mean maximum body size (SVL) at different elevations throughout the West Indies.

seen in the larger island faunas are evident in the Lesser Antilles. There are more species (9) at middle elevations than at sea level (5) or at the highest elevations (1); mean body size shows a general decrease with increasing elevation (73.2 mm to 20.0 mm).

GENERAL PATTERNS

When all West Indian species are examined, a mid-elevational bulge in species density is seen: from 72 species (sea level) to 82–83 species (300-600 m) and then to 1 species at 3050 m (Fig. 4:22). However, this pattern is misrepresentative of the elevational pattern observed in a typical mountain range in the West Indies, where the mid-elevational increase in density occurs at higher elevations. A restricted but more representative analysis of 80 species in four upland areas ($\geq$ 2000 m), including the Sierra Maestra, Blue Mountains, Massif de la Hotte, and Cordillera Central, shows that the mid-elevational bulge extends from 550 m to 1150 m, with a slight peak at 1050 m (Fig. 4:23). The reason for this difference between the complete analysis (Fig. 4:22) and the montane analysis (Fig. 4:23) is probably because the former includes lower elevation areas of locally high species density (e.g., as in the case of the Cockpit Country noted above) that cause an elevational "cap" in species density.

The relationship between mean maximum body size and elevation is virtually the same in all analyses: body size decreases with increasing elevation. When all West Indian species are considered (Fig. 4:22), body size decreases from 56.0 mm SVL (sea

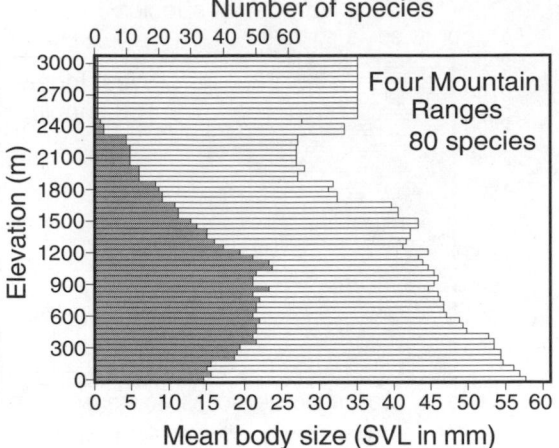

Fig. 4:23.   Species density and mean maximum body size (SVL) at different elevations in four montane areas in the West Indies: Sierra Maestra (Cuba), Blue Mountains (Jamaica), Massif de la Hotte and Cordillera Central (Hispaniola). Duplicate species within Hispaniola were excluded.

level) to 28.0 mm (2450 m), not considering the single species (35 mm) occurring above that elevation. The rate of decrease in body size is approximately 1 mm per 100 m of increasing elevation.

If the elevational limits (upper and/or lower) of species coincide, then it is possible that some ecological or topographical factor may be involved. An analysis of range limits for all West Indian species (n = 164 species) reveals one major and several minor peaks (Fig. 4:24). The major peak is the common lower elevational limit of sea level in 72 (44%) of the species. The next most significant peak is the upper elevational limit of 1050-1099 m found in 14 (8.5%) of the species. However, that peak is largely an expression of the fauna at El Yunque, Puerto Rico, the upper limit of which is restricted by the height of the mountain. An analysis of range limits of the 80 species in the four montane areas ($\geq$ 2000 m) noted above does not show that peak. In general, range limits, although not completely random, do not seem to be significantly clumped.

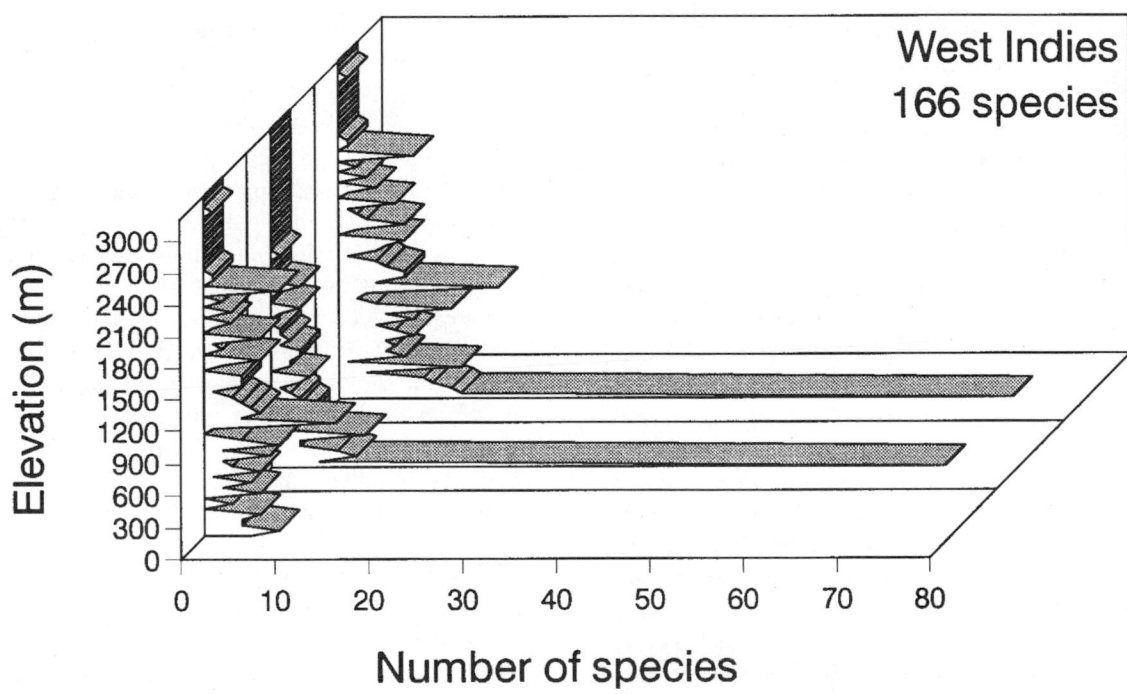

Fig. 4:24.   Elevational limits of amphibian species in the West Indies; front = upper limits, middle = lower limits, and back = combined range limits.

CORRELATIONS OF DISTRIBUTION PATTERNS

TOPOGRAPHY AND CLIMATE

Amphibian distribution in the West Indies is closely associated with both topography and rainfall. This is evident in comparing the species density maps (Figs. 4:8, 4:10, 4:12, and 4:14) with corresponding maps of topography and rainfall (Figs. 4:2–5), and in the altitudinal analysis. However, because of the relationship between topography and rainfall, it is difficult to separate these two variables to determine the exact relationship of each to species density and distribution.

The fact that high species density occurs near sea level only in regions of high rainfall (e.g., northern slopes of Cuchillas de Moa and Cuchillas de Toa in Cuba; northern slopes of Blue Mountains and John Crow Mountains in Jamaica) suggests that rainfall is an important factor in the distribution of these amphibians, and that high elevation is not required. Supporting that contention is the near-absence of amphibian species in the driest areas (e.g., Hellshire Hills of Jamaica; Barahona Peninsula of Hispaniola). However, this relationship between rainfall and amphibian distribution is not tied to a requirement for reproduction in water, because the majority (84%) of West Indian amphibians have direct development and can reproduce far away from standing water.

VEGETATION

Most of the West Indies were at one time covered with forests, and virtually all extant West Indian taxa can be considered forest-dwelling species. Because 80–99% of the original forest has been removed from the Greater Antilles (see below), most amphibians presently are found in forest remnants and are not distributed evenly throughout their geographic ranges. Next to nothing is known about the ecological requirements of most West Indian amphibians; therefore, presently it is not possible to ascertain the association of species with vegetation types except in a general or anecdotal way. However, my impression, gained through field work, is that most species are not closely tied to specific species of trees or vegetational zones. Examples of some possible exceptions are *Hyla marianae* (*Tillandsia* bromeliads), *Eleutherodactylus unicolor* (elfin woodland of Puerto Rico) and *E. caribe* (mangrove forest of southwestern Haiti).

HISTORICAL PHENOMENA

Mesozoic

The origin of the West Indian biota has been a topic of considerable interest over the years (literature reviewed in Williams, 1989; Hedges, 1996). The recognition that the geologic history of the Antilles extends back into the Mesozoic, and that there has been considerable plate movement within the Caribbean, led Rosen (1975) to propose a vicariance theory for the origin of the biota. A review of the extensive literature on the debate over the vicariance versus dispersal theories for the origin of the Antillean biota is beyond the scope of this paper. Several studies of West Indian amphibians have addressed phylogeny with morphological data (e.g., Joglar, 1989; Pregill, 1981). However, pertinent to this discussion are the studies that provide estimates of time of divergence from molecular data (there are no Mesozoic fossils of West Indian amphibians).

Hedges (1989b) examined variation in slow-evolving protein loci in 82 species of West Indian *Eleutherodactylus* (and two mainland species). Phylogenetic analysis defined three major groups within the West Indies, and each was assigned the rank of subgenus. The subgenus *Euhyas* (83 species) primarily is a western Caribbean group that includes all Jamaican, most Cuban and most South Island (Hispaniola) species that are mostly terrestrial. The subgenus *Eleutherodactylus* includes mostly arboreal species (50) and is distributed throughout the West Indies; it includes all taxa in the Puerto Rican Bank and Lesser Antilles. *Pelorius* is a subgenus of six large species on Hispaniola. Indications from DNA sequence data are that these three West Indian subgenera (and possibly *Syrrhophus* of Central America) form a single monophyletic group within the genus (Hedges and Youngblood, in prep.).

It was proposed that the large amount of molecular divergence between the subgenera *Euhyas* and *Eleutherodactylus* represents a phylogenetic divergence event in the late Mesozoic, possibly related to the breakup of the proto-Antilles (Hedges, 1989b). Albumin immunological data on West Indian and mainland *Eleutherodactylus* indicated that this divergence event probably occurred in the late Cretaceous, ca. 70 mya (Hass and Hedges, 1991). A review of the origin of all West Indian amphibians and reptiles indicates that the genus *Eleutherodactylus* is one of the few groups, if not the only group, to show a Mesozoic origin (Hedges, 1996).

Tertiary

The only known Tertiary fossil amphibians from the West Indies are those specimens of *Eleutherodactylus* embedded in Hispaniolan amber, Oligocene to mid-Miocene in age (Poinar and Cannatella, 1987; Poinar, 1992). These fossils establish the presence of amphibians in the West Indies by the mid-Tertiary.

Albumin immunological data suggested that the hylid, bufonid, and leptodactyline frogs all arrived to the Antilles by overwater dispersal (on flotsam) from South America in the Tertiary (Hedges et al., 1992b; Hedges, 1996). Most reptile lineages in the West Indies also show an origin in South America, and this pattern probably is due to the direction of water currents (Hedges, 1996). It was suggested that this predominant pattern of dispersal, rather than vicariance, may have been related to extinctions caused by the bolide impact that occurred in the Caribbean region at 65 mya (Hedges et al., 1992b).

In *Eleutherodactylus,* dispersal from the Antilles (subgenus *Euhyas*) to Central America (subgenus *Syrrhophus*) during the Tertiary has been suggested (Hass and Hedges, 1991; Hedges, 1996). No albumin immunological data are available for the recently described dendrobatid species, although indirect evidence from albumin IDs in the genus *Colostethus,* relationships, and distributions, suggest that it originated by dispersal from South America in the Tertiary (Hedges, 1996).

Hedges (1989b) proposed a biogeographic scenario to explain the distribution and relationships of *Eleutherodactylus* in the western Caribbean; this was based on allozyme and immunological data (Hedges, 1989a,b; Hass and Hedges, 1991) and is supported by chromosome data (Bogart and Hedges, 1995). In the early Miocene (20 mya), the South Island of Hispaniola had not yet collided with the remainder of Hispaniola, and dispersal of the subgenus *Euhyas* to Jamaica and the South Island probably occurred. Speciation during the late Miocene and Pliocene led to the large radiation of *Euhyas* existing today in these two areas. Collision of the South Island with the North Island of Hispaniola (containing frogs of the subgenus *Eleutherodactylus*) in the late Miocene (7 mya) led to limited dispersal between the North and South Islands. Jamaican *Eleuth-*

erodactylus thus represent a monophyletic group within the subgenus *Euhyas,* and most Cuban and South Island species belong to that subgenus. Therefore the major differences seen between the North and South Island amphibian faunas probably are not the result of sea-level changes in the Quaternary (see below), but rather are the consequence of completely isolated islands colliding in the Tertiary with minimal subsequent exchange of faunas.

Quaternary

Pregill and Olson (1981) suggested that arid savanna, grassland, and xeric scrub forest habitats predominated in the West Indies during the Pleistocene glaciations. This conclusion primarily was based on evidence from palynological and climatic studies of areas adjacent to the West Indies, and from some West Indian deposits indicating frequent fires during that period. Because West Indian amphibians predominantly are forest-dwelling animals, these vegetational changes would have resulted in a significant reduction in the area occupied by species, and probably would have led to extinctions. At the same time, this reduction in area would have increased isolation of populations and may have led to geographic differentiation and possibly to speciation.

Besides the climatic and vegetation changes, sea-level fluctuation both joined and isolated land areas during the Pleistocene. Estimates of sea-level change suggest dramatic changes in exposed land area. Pregill and Olson (1981) speculated that the diversity of species within *Eleutherodactylus* in the South Island of Hispaniola largely is the result of Pleistocene climatic cycles and fragmentation of ranges leading to speciation. Although it is likely that climatic cycles were important in the speciation and radiation of West Indian amphibians, immunological distances > 3 (1.8 mya) between species suggest that most divergences predated the Pleistocene (Hedges, 1989a, b; Hass and Hedges, 1991; Hedges et al., 1992b); this agrees with molecular data from most South American amphibians examined (Duellman et al., 1988; Maxson and Heyer, 1988; Hass et al., 1995), although speciation may have been more recent in the high Andes (Duellman et al., 1988). It is more likely that intraspecific variation seen in some West Indian amphibians (e.g., *Eleutherodactylus audanti, E. ruthae, E. schmidti, Bufo longinasus*) is the result of differentiation associated with Pleistocene climatic cycles.

The absence of endemic amphibians in the Bahamas may provide additional support for the importance of pre-Quaternary speciation in West Indian amphibians. These islands were submerged, either mostly or entirely, during interglacial periods and therefore their present terrestrial flora and fauna probably arrived since that time and could provide a "gauge" of the time required for speciation. Some geographic differentiation in one of the two native species (*Eleutherodactylus planirostris*) has taken place to the extent that an endemic subspecies is recognized, but there does not seem to have been enough time for speciation to have occurred. The presence of endemic species of reptiles in the Bahamas suggests that speciation may occur more rapidly in reptiles than in amphibians, but also it could simply mean that reptiles were able to colonize the Bahamas more quickly than amphibians. Nonetheless, this concept of finding an area that was "wiped clean" at a known time period and examining rates of colonization and speciation in different groups might be useful in other regions of the world.

Disjunct Distributions

Typically, West Indian amphibians have well-defined distributions restricted to a particular area on an island. The few distributions that can be unambiguously characterized as disjunct tend to be upland species isolated in different mountain ranges within an island. For example, in Cuba, this includes *Eleutherodactylus limbatus, Bufo longinasus,* and *B. taladai.* Some Cuban species that appeared to have disjunct distributions (Schwartz and Henderson, 1991) now are known to have wider distributions based on recent collections (e.g., *E. varians;* Hedges, unpubl.).

In Hispaniola, several species (*Eleutherodactylus nortoni, E. oxyrhynchus,* and *E. semipalmatus*) are found only in the Massif de la Hotte and Massif de la Selle, but not in the intervening area of the Tiburon Peninsula. This may be due to ecological requirements, or more likely, due to the fact that little more than barren rock and dirt now exists in the intervening areas of suitable elevation because of deforestation. The distribution of the burrowing species *E. ruthae* appears to be truly disjunct, and four subspecies have been described to recognize geographic variation in the species. It is a species of low to moderate elevations, and it is possible that upland areas have posed barriers to gene flow. This is almost certainly the explanation for the disjunct distributions of the Hispaniolan toad *Bufo guentheri* and the Puerto Rican toad, *B. lemur,* both of which are restricted to low elevations. One high elevation species of Hispaniola, *E. audanti,* is distributed in disjunct populations on the North and South Islands and has differentiated into three recognized subspecies.

COLLECTING BIAS

For reasons of economics, politics, and accessibility, collecting activities in the West Indies have not been evenly distributed throughout the region; this fact may introduce a bias in the analysis of distributions. Puerto Rico, Jamaica, and the Lesser Antilles have received greater attention than Hispaniola, whereas Cuba has received by far the least amount of attention by collectors. Within Hispaniola, the much more difficult logistics, greater expense, and greater personal risk of an expedition to Haiti has resulted in a far greater amount of attention given to the Dominican Republic. The result of this type of bias is an underestimation of distributions, including area and elevational range, in more poorly known countries such as Cuba and Haiti.

ISLAND BIOGEOGRAPHY

The number of species on an island is a product of many factors, but usually it is positively correlated, to some degree, with island area. Based on empirical plots of this relationship, Darlington (1957) discovered that a tenfold increase in the area typically leads to a doubling of the number of species. In an effort to explain this general relationship, an equilibrium model of biogeography was proposed (MacArthur and Wilson, 1963) which suggests that rates of colonization and extinction are the primary factors determining the number of species present. Other factors include distance from source area and the size of the island. More recently, environmental heterogeneity has been suggested as the primary factor responsible for the species-area effect (Williamson, 1981, 1988).

Darlington's (1957) example of the species-area effect was the amphibians and reptiles of the West Indies, and this data set, unrevised, has been reproduced repeatedly (e.g., MacArthur and Wilson, 1967:8; Cox et al., 1976:103; Wilson, 1992:222). Because the original data on species abundance on

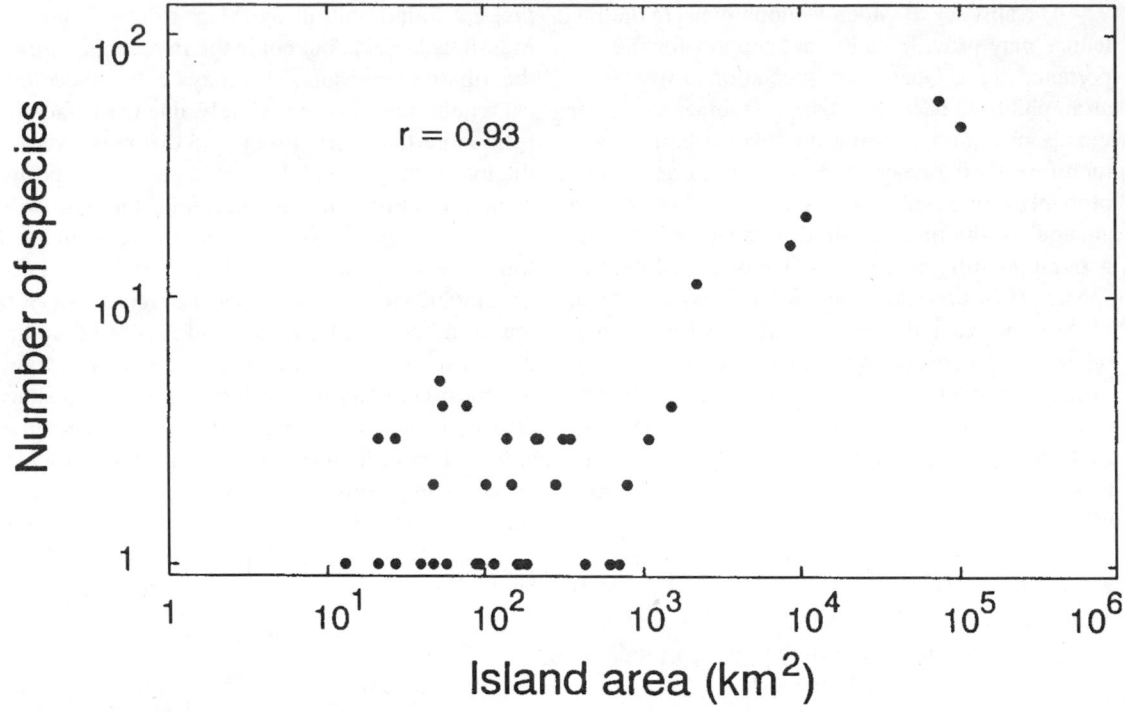

Fig. 4: 25.   Relationship between island area and number of native amphibian species in the West Indies (areas as in Table 4:3).

each island are nearly four decades old, a revised plot of this relationship in West Indian amphibians is shown here (Fig. 4:25; Table 4:3). Although the number of species of West Indian amphibians is correlated with island area, this relationship is unlikely to be explained by MacArthur and Wilson's (1967) equilibrium model of immigration and extinction or by extinction alone (Jones, 1980). For example, all 17 native Jamaican species of *Eleutherodactylus* are the result of a single colonization event and subsequent adaptive radiation (Hedges, 1989a, b). Although not yet confirmed, preliminary data suggest that the amphibian faunas of the other Greater Antillean islands represent one or a few large adaptive radiations (Hedges, 1989b; Hass and Hedges, 1991). As noted earlier, Hispaniola is a composite of two

islands; it is likely that considerable speciation took place prior to the collision of the North and South Islands (the postcollision exchange of species poses difficulties in treating these two areas as separate islands in a species-area analysis). The West Indian bufonids apparently represent a single colonization from the mainland (Pregill, 1981); there is no phylogenetic evidence of any dispersal of *Eleutherodactylus* from the mainland to the Greater Antilles during the entire Cenozoic. In fact, dispersal in the reverse direction has been proposed to explain the origin of the subgenus *Syrrhophus* in Central America (Hedges, 1989b). The increased topographic variability and greater availability of niches for adaptive radiation on larger islands is the most likely explanation for the species-area relationship in West Indian amphibians.

CONSERVATION OF THE AMPHIBIAN FAUNA

CURRENT STATUS

In response to current concerns over the possible disappearance of some amphibian species in several regions of the world, the conservation status of West Indian amphibians was reviewed by Hedges (1993). Each species was listed, along with the year that it was most recently encountered. No general

decline at the species level was found, although eight species were listed that had not been seen in recent years: *Eleutherodactylus orcutti* (Jamaica); *E. lucioi, E. neodreptus, E. warreni,* and *Bufo fluviatica* (Hispaniola), and *E. jasperi, E. karlschmidti,* and *E. lentus* (Puerto Rican Bank). However, sufficient information was available for only one of those species

Table 4:3. Numbers of species of native amphibians on islands in the West Indies (> 10 km²).

Island	Island Area (km²)	No. Species
Cuba	105,007	50
Hispaniola	76,470	63
Jamaica	10,992	22
Puerto Rico	8,768	17
Isla de Juventud	2,200	12
Guadeloupe	1,510	4
Martinique	1,079	3
Dominica	790	3
Île Gonâve	702	1
St. Lucia	616	1
Barbados	430	1
St. Vincent	345	3
Grenada	311	3
Antigua	280	2
St. Croix	218	3
Île de la Tortue	209	3
Grand Cayman	184	1
St. Kitts	166	1
Barbuda	161	1
Marie Galante	149	2
Isla Vieques	138	3
Isla Saona	114	1
Montserrat	102	2
Nevis	93	1
Anguilla	91	1
St. Martin	88	1
St. Thomas	77	4
Isla Mona	57	1
Tortola	54	4
St. John	52	5
Ile à Vache	47	2
Ile Grand Cayemite	47	1
Isla Beata	47	1
Anegada	39	1
Isla Culebra	27	3
La Desirade	27	1
Virgin Gorda	21	3
St. Eustatius	21	1
St. Barthélémy	21	1
Saba	13	1

(*E. karlschmidti*) to conclude that it has probably disappeared. At the population level, virtually all West Indian amphibians presumably have declined substantially as a consequence of deforestation and the resulting loss of habitat. Introduced predators such as the mongoose (*Herpestes auropunctatus*),

black rat (*Rattus rattus*), and feral cat (*Felis domesticus*) have been linked to the extinction of species of West Indian reptiles (Henderson, 1992) and also may be adversely affecting native amphibians, even in undisturbed forest (Hedges and Thomas, 1991; Hedges, 1993). The large introduced amphibians, *Rana catesbeiana* and *Bufo marinus* known to prey on smaller amphibians, are well-established on most islands and probably have adversely affected the native species. Population level surveys have been conducted for some Puerto Rican species, although there is no agreement as to the reasons for the declines (Burrowes and Joglar, 1991; Joglar and Burrowes, 1991; Moreno, 1991; Rivero, 1991).

The list of last reported records (Hedges, 1993: Table 1) requires some revision following recent field work in Hispaniola (1993) and Cuba (1994). In northern Hispaniola, nine species were encountered (*Eleutherodactylus abbotti, E. audanti, E. flavescens, E. inoptatus, E. montanus, E. patricae, E. pictissimus, E. weinlandi,* and *Osteopilus dominicensis*). In Cuba, 26 described species were encountered: *Bufo empusa, B. gundlachi, B. peltocephalus, B. taladai, Eleutherodactylus acmonis, E. albipes, E. atkinsi, E. auriculatus, E. cubanus, E. cuneatus, E. dimidiatus, E. guantanamera, E. gundlachi, E. intermedius, E. ionthus, E. leberi, E. limbatus, E. melacara, E. planirostris, E. ricordii, E. ronaldi, E. toa, E. turquinensis, E. varians, E. varleyi,* and *Osteopilus septentrionalis*. No species was conspicuously absent in areas where it was expected to occur. Noteworthy was the rediscovery in 1994 (Hedges et al., 1996) of two Cuban species, *E. cubanus* and *E. turquinensis,* previously known only from the type series collected in the mid-1930s (although these two species were cited in Hedges, 1993, as having been collected recently, that material has been reidentified as a new species and *E. cuneatus,* respectively).

Estimates of forest cover for islands of the West Indies are among the lowest in the world, and this is tied to relatively high human population densities in the region (Hedges and Woods, 1993; Fig. 4:26). Of the ten tropical countries with the highest rates of net annual deforestation during 1981–1990, three are in the West Indies: Dominican Republic (2.5%), Haiti (3.9%), and Jamaica (5.3%); Jamaica's is the highest in the world (World Resources Institute, 1994). Because of the fact that a species will "exist" until its last population disappears, the decline in forest cover has not been accompanied by a similar

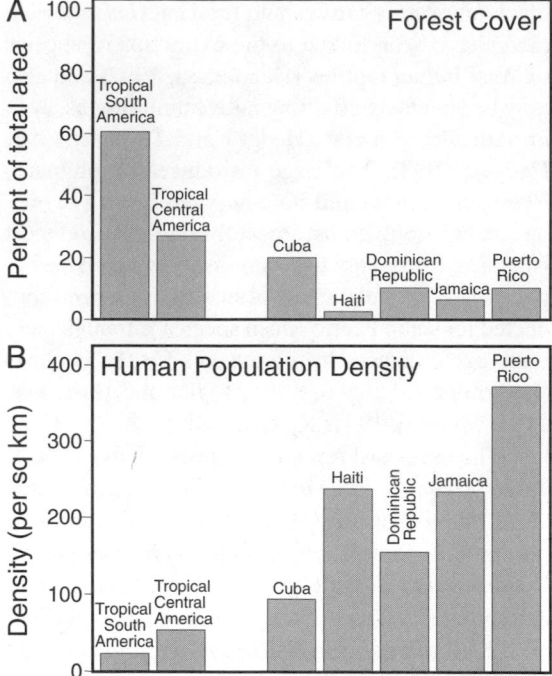

Fig. 4:26.   Forest cover and human population size in the West Indies, Central America, and South America. After Hedges and Woods (1993); original sources are Caribbean/Latin American Action (1993), Cohen (1984), Johnson (1988), Paryski et al. (1989), and World Resources Institute (1992). The figure of 5% for Jamaica was disputed (Goreau, 1993) in favor of a much higher value, 46% published elsewhere (Rampair, 1986). However, that higher value was not included in the reports by the Tropical Forests Resource Assessment Project of the Food and Agriculture Organization of the United Nations, and the United Nations Environment Programme, summarized in World Resources Institute (1992) as 6% closed forest for Jamaica. Another study of Jamaican forest cover (Thompson et al., 1986b) reported a value of 7% for "fully stocked Broadleaf Forest," which is in close agreement with the value accepted by the United Nations. More recent information on forest cover is available (World Resources Institute, 1994), but closed forest ("trees cover a high proportion of the ground") is not distinguished from open forest ("at least 10% tree cover and a continuous grass layer on the forest floor"), so the earlier values are shown here.

decline in species (Hedges, 1993). However, the critically low levels of forest cover remaining in some areas, such as in Jamaica (5%) and Haiti (1%), indicate that species extinctions will occur in the near future unless the trend is reversed. It was suggested that Haiti may soon become the Earth's "first major biodiversity disaster" (Hedges, 1993).

Nearly all countries in the West Indies currently have some protected areas that typically encom-pass the largest or last remaining stands of natural vegetation. For amphibian species diversity and en-demism, some of the most important of these pro-tected areas are: Parque Nacional Sierra Maestra (Cuba), Blue Mountains National Park (Jamaica), Parc National Pic Macaya and Parc National Morne la Visite (Haiti), Parque Nacional Sierra de Baoru-co, Parque Nacional Armando Bermudez, and Parque Nacional José del Carmen Ramírez (Domin-ican Republic), and Caribbean National Forest (Pu-erto Rico).

Unfortunately, protection of the natural resourc-es is not complete in any of these areas, and it es-sentially doesn't exist in some areas such as the Haitian national parks. In Cuba, the severe econom-ic difficulties of the last several years has taken a toll on the forests. Felling of trees for firewood, char-coal, and building materials is taking place in the Sierra Maestra National Park and other protected areas in Cuba. Jamaica has only recently established a system of protected areas, and it remains to be seen whether or not they will be effective. Haiti is the poorest country in the Western Hemisphere ($400 per capita annual income), has a burgeoning human population of 7 million, and has exhausted virtually all of its forests. On an expedition in 1991 to the center of Pic Macaya National Park, Richard Tho-mas and I saw trees being cut and removed from one of the last remaining stands of natural forest in the country. Although it is illegal to fell trees in the Dominican Republic, clearing and burning is wide-spread and common, and trees are being cut within the National Parks. Protection of the forests in Puer-to Rico has been more effective than elsewhere, al-though the Caribbean National Forest has "lost its virginity" because of the abundance of introduced predators.

RECOMMENDATIONS

General

It cannot be overstated that the loss of biodiver-sity in the West Indies, as in other regions, is closely tied to the loss of forest cover, and therefore it is of prime importance to halt the deforestation. Greater efforts should be made to enforce park boundaries and prevent illegal felling of trees. For the longterm, some form of human population control must be implemented or else the economic pressures and de-mands of a growing population will lead to contin-ued deforestation.

Specific

Additional protected areas: Many "hot spots" of West Indian amphibian diversity and endemism already are contained within protected areas, and therefore emphasis should be placed on enforcement. Important areas for amphibian diversity not yet protected (including proposed areas) are: in Cuba, (1) Sierra de los Organos (including region of Viñales), (2) Sierra del Rosario (including region of Soroa), (3) Macizo del Escambray, (4) Macizo de Sagua-Baracoa (including the Meseta del Guaso, Cuchillas de Toa, Cuchillas de Moa, and Sierra del Purial); in Jamaica, (5) Dolphin Head National Park, and (6) Cockpit Country National Park; in Haiti, (7) enlargement of Parc National Pic Macaya and Parc National Morne La Visite; in the Dominican Republic, (8) Sierra de Neiba National Park, (9) eastward extension of Sierra de Baoruco National Park (including Loma Remigio and Loma Trocha de Pey); and in Puerto Rico, (11) Cuchilla de Panduras.

Removal of introduced predators: In view of the disappearance of *Eleutherodactylus karlschmidti* and apparent decline of several amphibian species in undisturbed forest on El Yunque, the U.S. Forest Service should make feral animal removal/extirpation a major concern for the Caribbean National Forest. Removal of introduced predators also should be a concern in other areas of the West Indies.

Promotion of scientific collecting: All species of West Indian amphibians were described from specimens that had been collected and preserved, and in most cases, those specimens were exported to another country for study. Despite the obvious importance of this activity, it is discouraged by the governments of both the country of origin of the specimens and the country of origin of the researcher through unrealistic restrictions and permit procedures (Hedges and Thomas, 1991). Often, collection of only a few specimens of each species is permitted even though additional specimens would cause no harm to populations or species and would increase the quality of the information gathered (Hedges and Thomas, 1991). It is not uncommon to spend one week out of a four-week expedition simply to obtain a collecting or export permit for a small number of specimens while, at the same time, thousands or millions of these organisms have perished in that country through deforestation.

The majority of researchers surveying tropical biodiversity are not required to do so as part of their employment. Many are university professors whose primary obligations are to teaching, and are free to choose any research topic or location. For this reason, countries that have excessively complicated and stringent permitting procedures suffer because scientists simply avoid them and focus their attention on other countries.

Scientific collecting should be promoted, rather than discouraged. The immediate benefit for the country is a much greater knowledge of its biodiversity, which will aid in planning and implementing conservation policies. Regulation is important, but permit procedures should be streamlined and have greater flexibility so that legitimate scientists can obtain permits quickly and collect and export a sufficient number of specimens for systematic study. Importation procedures also should be improved so that scientists with proper documentation can enter any port without encountering long, unnecessary delays.

Acknowledgments: I thank the many individuals, especially Richard Thomas, who have accompanied and assisted me on expeditions throughout the West Indies over the years. The staff of the Division of Amphibians and Reptiles, United States National Museum of Natural History (Ronald Crombie, Ronald Heyer, Addison Wynn, and George Zug), have been especially helpful in the accessioning of specimens. I also thank my Cuban colleagues, especially Alberto Estrada (Ministerio de Agricultura), Orlando Garrido and Gilberto Silva (Museo Nacional de Historia Natural), and Laredo Gonzalez (Academia de Ciencias de Cuba) for support of my research in Cuba; my Dominican colleagues Sixto and Yvonne Inchaustequi (Grupo Jaragua) for support of my research in the Dominican Republic; Hinrich Kaiser for use of his unpublished data on Lesser Antillean anurans; Carla Hass for assistance with many of the figures; Richard Highton and Linda Maxson for providing facilities during earlier stages of the research; and Carla Hass for comments on the manuscript. I also thank Bill Duellman for the invitation to participate in this volume. This research was supported by grants from the National Science Foundation (BSR 8307115 to Richard Highton, and BSR 8906325 and BSR 9123556 to S.B.H.).

LITERATURE CITED

ARDEN, D. D. 1975. The geology of Jamaica and the Nicaraguan Rise. Pp. 617–661 *in* Nairn, A. E. M. and F. G. Stehli (eds.), *Ocean Basins and Margins, Vol. 3, Gulf Coast, Mexico, and the Caribbean.* Plenum Press, New York.

ASPREY, G. F., AND R. G. ROBBINS. 1953. The vegetation of Jamaica. Ecol. Monogr. 23:359–413.

BOGART, J. P., AND S. B. HEDGES. 1995. Rapid chromosome evolution in Jamaican frogs of the genus *Eleutherodactylus* (Leptodactylidae). J. Zool. (London) 235:9–31.

BOND, J. 1979. *Birds of the West Indies.* London, England: Collins Publishers.

BOUYSSE, PH. 1984. The Lesser Antilles arc: structure and geodynamic evolution. Pp. 83–103 *in* B. Biju-Duval, J. C. Moore, et al. (eds.), *Initial Reports of the Deep Sea Drilling Project, Vol. 78A.* Washington, D.C.: U. S. Government Printing Office.

BOUYSSE, PH. 1988. Opening of the Grenada back-arc basin and evolution of the Caribbean during the Mesozoic and early Paleogene. Tectonophysics 149:121–143.

BOWIN, C. 1975. The geology of Hispaniola. Pp. 501–522 *in* Nairn, A.E.M. and F.G. Stehli (eds.), *Ocean Basins and Margins, Vol. 3, Gulf Coast, Mexico, and the Caribbean.* New York, NY: Plenum Press.

BOX, B., AND S. CAMERON (eds.). 1989. *Caribbean Islands Handbook.* New York, NY: Prentice-Hall.

BURKE, K., P. J. FOX, AND A. M. C. SENGOR. 1978. Buoyant ocean floor and the evolution of the Caribbean. J. Geophys. Res. 83:3949–3954.

BURROWES, P. AND R. JOGLAR. 1991. A survey of the population status and an ecological evaluation of three Puerto Rican frogs. Pp. 42–46 *in* J. A. Moreno (ed.), Status y Distribución de los Reptiles y Anfibios de la Región de Puerto Rico. Depart. Recur. Nat. Puerto Rico, Pub. Cien. Misc. 1:1–67.

CARIBBEAN AND LATIN AMERICAN ACTION. 1993. *Caribbean Basin Databook.* Washington, D.C.: Caribbean and Latin American Action.

COHEN, W. B. 1984. *Environmental Degradation in Haiti: An Analysis of Aerial Photography.* Port-au-Prince, Haiti: Report, USAID/Haiti.

COX, C. B., I. N. HEALEY, AND P. D. MOORE. 1976. *Biogeography.* New York: John Wiley and Sons.

DARLINGTON, P. J. 1957. *Zoogeography: The Geographical Distribution of Animals.* New York, NY: Wiley.

DUELLMAN, W. E. 1990. Herpetofaunas in Neotropical rainforests: comparative composition, history, and resource use. Pp. 455–505 *in* A. H. Gentry (ed.), *Four Neotropical Rainforests.* New Haven, CT: Yale University Press.

DUELLMAN, W. E. 1993. *Amphibian Species of the World: Additions and Corrections.* Spec. Publ. Univ. Kansas, Mus. Nat. Hist. 21:1–372.

DUELLMAN, W. E., L. R. MAXSON, AND C. JESIOLOWSKI. 1988. Evolution of marsupial frogs (Hylidae: Hemiphractinae): immunological evidence. Copeia 1988:527–543.

ESTRADA, A. R. 1986. Anfibios, reptiles, y aves de Cayo Guajaba, Archipiélago Sabana-Camaguey, Cuba. Poeyana 328:1–34.

ESTRADA, A. R. 1993a. Herpetofauna del Archipiélago de los Canarreos, Cuba. Poeyana 431:1–19.

ESTRADA, A. R.. 1993b. Anfibios y reptiles de Cayo Coco, Archipiélago de Sabana-Camaguey, Cuba. Poeyana 432:1–21.

GARRIDO, O. H., A. R. ESTRADA, AND A. LLANES. 1986. Anfibios, reptiles y aves de Cayo Guajaba, Archipiélago de Sabana-Camagüey, Cuba. Poeyana 328:1–34.

GOREAU, T. J. 1993. Forest Cover. Nature 365:688.

HASS, C. A., J. F. DUNSKI, L. R. MAXSON, AND M. S. HOOGMOED. 1995. Divergent lineages within the *Bufo margaritifera* complex (Amphibia: Anura; Bufonidae) revealed by albumin immunology. Biotropica 27:238–249.

HASS, C. A., AND S. B. HEDGES. 1991. Albumin evolution in West Indian frogs of the genus *Eleutherodactylus* (Leptodactylidae): Caribbean biogeography and a calibration of the albumin immunological clock. J. Zool. (London) 225:413–426.

HEATWOLE, H., R. LEVINS, AND M. D. BYER. 1981. Biogeography of the Puerto Rican Bank. Atoll Research Bull. 251:1–55.

HEDGES, S. B. 1989a. An island radiation: allozyme evolution in Jamaican frogs of the genus *Eleutherodactylus* (Anura, Leptodactylidae). Carib. J. Sci. 25:123–147.

HEDGES, S. B. 1989b. Evolution and biogeography of West Indian frogs of the genus *Eleutherodactylus:* slow-evolving loci and the major groups. Pp. 305–370 *in* C. A. Woods (ed.), *Biogeography of the West Indies: Past, Present, and Future.* Gainesville, FL: Sandhill Crane Press.

HEDGES, S. B. 1993. Global amphibian declines: a perspective from the Caribbean. Biodiversity and Conservation 2:290–303.

HEDGES, S. B. 1996. The origin of West Indian amphibians and reptiles. Pp. 95–128 *in* R. Powell and R. W. Henderson (eds), *Contributions to West Indian Herpetology: A Tribute to Albert Schwartz.* Ithaca, NY: Society for the Study of Amphibians and Reptiles.

HEDGES, S. B., A. R. ESTRADA, AND R. THOMAS. 1992a. Three new species of *Eleutherodactylus* from eastern Cuba, with notes on vocalizations of other species (Anura, Leptodactylidae). Herpetol. Monogr. 6:68–83.

HEDGES, S. B., L. GONZALEZ, AND A. R. ESTRADA. 1996. Rediscovery of the Cuban frogs *Eleutherodactylus cubanus* and *E. turquinensis* (Anura: Leptodactylidae). Carib. J. Sci. 31:327–322.

HEDGES, S. B., C. A. HASS, AND L. R. MAXSON. 1992b. Caribbean biogeography: molecular evidence for dispersal in West Indian terrestrial vertebrates. Proc. Natl. Acad. Sci. (U.S.A.) 89:1909–1913.

HEDGES, S. B., AND R. THOMAS. 1989. Supplement to West Indian Amphibians and Reptiles: a checklist. Milwaukee Publ. Mus. Contr. Biol. Geol. 77:1–11.

HEDGES, S. B., AND R. THOMAS. 1991. The importance of systematic research in the conservation of amphibian and reptile populations. Pp. 56-61 *in* J. A. Moreno (ed.), Status y Distribución de los Reptiles y Anfibios de la Región de Puerto Rico. Depart. Recur. Nat. Puerto Rico, Pub. Cien. Misc. 1:1–67.

HEDGES, S. B., AND R. THOMAS. 1992. A new marsh-dwelling species of *Eleutherodactylus* from Haiti (Anura: Leptodactylidae). J. Herpetol. 26:191–195.

HEDGES, S. B., AND C. A. WOODS. 1993. Caribbean hot spot. Nature 364:375.

HENDERSON, R. W. 1992. Consequenses of predator introductions and habitat destruction on amphibians and reptiles in the post-Columbus West Indies. Carib. J. Sci. 28:1–10.

HENDERSON, R. W., J. DAUDIN, G. T. HAAS, AND T. J. McCARTHY. 1992. Significant distribution records for some amphibians and reptiles in the Lesser Antilles. Carib. J. Sci. 28:101–103.

HEYER, W. R. 1994. Variation within the *Leptodactylus podicipinus-wagneri* complex of frogs (Amphibia: Leptodactylidae). Smithsonian Contr. Zool. 546:1–124.

HOLCOMBE, T. L., J. W. LADD, G. WESTBROOK, N. T. EDGAR, AND C. L. BOWLAND. 1990. Caribbean marine geology; ridges and basins of the plate interior. Pp. 231–260 *in* G. Dengo and J. E. Case (eds.), *The Caribbean Region.* Boulder, CO: Geological Society of America.

INSTITUTO CUBANO DE GEODESIA Y CARTOGRAFÍA. 1978. *Atlas de Cuba.* Havana, Cuba: Instituto Cubano de Geodesia y Cartografía.

ITURALDE-VINENT, M. A. 1988. *Naturaleza Geológica de Cuba.* Havana, Cuba: Editorial Científico-Técnica.

JOGLAR, R. L. 1989. Phylogenetic relationships of the West Indian frogs of the genus *Eleutherodactylus:* a morphological analysis. Pp. 371–408 *in* C. A. Woods (ed.), *Biogeography of the West Indies: Past, Present, and Future.* Gainesville, FL: Sandhill Crane Press.

JOGLAR, R. L., AND P. BURROWES. 1991. El efecto del Huracan Hugo sobre una communidad de anfibios en El Yunque, Puerto Rico y algunas recomendaciones para la proteccion de las especies del genero Eleutherodactylus. Pp. 47–53 *in* J. A. Moreno (ed.), Status y distribución de los reptiles y anfibios de la región de Puerto Rico. Depart. Recur. Nat. Puerto Rico, Pub. Cien. Misc. 1:1–67.

JOHNSON, T. H. 1988. Biodiversity and conservation in the Caribbean: profiles of selected islands. Intern. Coun. for Bird Preser., Monogr. 1:1–144.

JONES, K. L. 1980. Frog diversity in the Lesser Antilles. Carib. J. Sci. 16:19–22.

KAISER, H. 1992. The trade-mediated introduction of *Eleutherodactylus martinicensis* (Anura: Leptodactylidae) on St. Barthélémy, French Antilles and its implications for Lesser Antillean biogeography. J. Herpetol. 26:264–273.

KAISER, H., L. A. COLOMA, AND H. M. GRAY. 1994. A new species of *Colostethus* (Anura: Dendrobatidae) from Martinique, French Antilles. Herpetologica 50:23–32.

KAISER, H., D. M. GREEN, AND M. SCHMID. 1994. Systematics and biogeography of eastern Caribbean frogs (*Eleutherodactylus:* Leptodactylidae), with a description of a new species from Dominica. Canadian J. Zool. 72:2217–2237.

KUMAR, S., K. TAMURA, AND M. NEI. 1994. MEGA: Molecular Evolutionary Genetics Analysis software for microcomputers. CABIOS 10:189–191.

LEWIS, J. F., AND G. DRAPER. 1990. Geology and tectonic evolution of the northern Caribbean margin. Pp. 77–140 *in* G. Dengo and J. E. Case (eds.), *The Caribbean Region.* Boulder, CO: Geological Society of America.

MACARTHUR, R. H., AND E. O. WILSON. 1963. An equilibrium theory of insular zoogeography. Evolution 17:373–387.

MACARTHUR, R. H., AND E. O. WILSON. 1967. *The Theory of Island Biogeography.* Princeton, NJ: Princeton University Press.

MACPHERSON, J. 1980. *Caribbean Lands.* Essex, England: Longman.

MARRERO, L. 1946. *Elementos de Geografía de Cuba.* Havana, Cuba: Editorial Minerva.

MAURRASSE, F. J-M. R. 1982. *Survey of the Geology of Haiti.* Miami, FL: Miami Geological Society.

MAURY, R. C., G. K. WESTBROOK, P. E. BAKER, PH. BOUYSSE, AND D. WESTERCAMP. 1990. Geology of the Lesser Antilles. Pp. 141–166 *in* G. Dengo and J. E. Case (eds.), *The Caribbean Region.* Boulder, CO: Geological Society of America.

MAXSON, L. R., AND W. R. HEYER. 1988. Molecular systematics of the frog genus *Leptodactylus* (Amphibia: Leptodactylidae). Fieldiana: Zool. 41:1–13.

MERTENS, R. 1939. Herpetologische Ergebnisse einer Reise nach der Insel Hispaniola Westindien. Abh. Senckenberg. Naturf. Ges. 449:1–84.

MORENO, J. 1991. Status survey of the Golden Coqui *Eleutherodactylus jasperi.* Pp. 37–41 *in* J. A. Moreno (ed.), Status y Distribución de los Reptiles y Anfibios de la Región de Puerto Rico. Depart. Recur. Nat. Puerto Rico, Pub. Cien. Misc. 1:1–67.

NATIONAL GEOGRAPHIC SOCIETY. 1981. *National Geographic Atlas of the World.* Washington, D. C.: National Geographic Society.

NEI, M. 1991. Relative efficiencies of different tree-making methods for molecular data. Pp. 90–128 *in* M. M. Miyamoto and J. Cracraft (eds.), *Phylogenetic Analysis of DNA Sequences.* Oxford, England: Oxford University Press.

NÚÑEZ JIMÉNEZ, A. 1972. *Geografía de Cuba.* Havana, Cuba: Pueblo y Educación.

PARYSKI, C. A. WOODS, AND F. SERGILE. 1989. Conservation strategies and the preservation of biological diversity in Haiti. Pp. 855–78 *in* C. A. Woods (ed.), *Biogeography of the West Indies: Past, Present, and Future.* Gainesville, FL: Sandhill Crane Press.

PERFIT, M. R., AND E. E. WILLIAMS. 1989. Geological constraints and biological retrodictions in the evolution of the Caribbean Sea and its islands. Pp. 47–102 *in* C. A. Woods (ed.), *Biogeography of the West Indies: Past, Present, and Future.* Gainesville, FL: Sandhill Crane Press.

PICÓ, R. 1974. *The Geography of Puerto Rico.* Chicago, IL: Aldine Publ. Co.

PINDELL, J. L., AND S. F. BARRETT. 1990. Geological evolution of the Caribbean region: A plate-tectonic perspective. Pp. 405–432 *in* G. Dengo and J. E. Case (eds.), *The Geology of North America. Vol. H. The Caribbean Region.* Boulder, CO: Geological Society of America.

POINAR, G. O., JR. 1992. *Life in Amber.* Stanford, CA: Stanford University Press.

POINAR, G. O., JR., AND D. C. CANNATELLA. 1987. An upper Eocene frog from the Dominican Republic and its implication for Caribbean biogeography. Science 237:1215–1216.

POWELL, R., R. J. PASSARO, AND R. W. HENDERSON. 1992. Noteworthy herpetological records from Saint Maarten, Netherlands Antilles. Carib. J. Sci. 28:234–235.

PREGILL, G. K. 1981. Cranial morphology and the evolution of West Indian toads (Salientia: Bufonidae): resurrection of the genus *Peltophryne* Fitzinger. Copeia 1981:273–285.

PREGILL, G. K. AND S. L. OLSON. 1981. Zoogeography of the West Indian vertebrates in relation to Pleistocene climatic cycles. Ann. Rev. Ecol. Syst. 12:75–98.

RAMPAIR, S. 1986. Jamaica Resource Assessment-1980. Pp. 77–80 *in* D. A. Thompson, P. K. Bretting, and M. Humphreys (eds.), *Forests of Jamaica.* Kingston, Jamaica: Jamaican Soc. Scientists and Technol.

RICKARDS, C. (ed.). 1980. *Caribbean Year Book.* Toronto, Canada: Caribook Limited.

RIVERO, J. A. 1991. Divagaciones sobre las especies de coquies en peligro de extincion y las causas posibles de esa situa-cion. Pp 54–55 *in* J. A. Moreno (ed.), Status y distribución de los reptiles y anfibios de la región de Puerto Rico. De-part. Recur. Nat. Puerto Rico, Pub. Cien. Misc. 1:1–67.

RODRÍGUEZ-SCHETTINO, L. 1993. Areas faunísticas de Cuba según la distribución ecogeográfica actual y el endemismo de los reptiles. Poeyana 436:1–17.

ROSEN, D. E. 1975. A vicariance model of Caribbean biogeogra-phy. Syst. Zool. 24:431–464.

ROUGHGARDEN, J. 1995. *Anolis lizards of the Caribbean.* New York, NY: Oxford Univ. Press.

SAITOU, N. AND M. NEI. 1987. The neighbor-joining method: a new method for reconstructing phylogenetic trees. Mol. Biol. Evol. 4:406–425.

SCHWARTZ, A. 1973. Six new species of *Eleutherodactylus* (Anu-ra, Leptodactylidae) from Hispaniola. J. Herpetology 7:249–273.

SCHWARTZ, A. 1978. Some aspects of the herpetogeography of the West Indies. Acad. Nat. Sci. Phila. Spec. Publ. 13:31–51.

SCHWARTZ, A. 1980. The herpetogeography of Hispaniola, West Indies. Stud. Faun. Curaçao Other Carib. Islands 189:86–127.

SCHWARTZ, A., AND R. W. HENDERSON. 1988. West Indian amphib-ians and reptiles: a checklist. Milwaukee Publ. Mus. Contr. Biol. Geol. 74:1–264.

SCHWARTZ, A., AND R. W. HENDERSON. 1991. *West Indian Amphib-ians and Reptiles: Distributions, Descriptions, and Natural History.* Gainesville, FL: University of Florida Press.

SEA/DVS. 1990. *La Biodiversidad Biológica en la República Dominicana.* Santo Domingo, Dominican Republic: Sec-retaría de Estado de Agricultura, SURENA/DVS.

SHOWKER, K. 1989. *Caribbean.* New York, NY: Stewart, Tabori, and Chang.

SMITH, A. G., D. G. SMITH, AND B. M. FUNNEL. 1994. *Atlas of Mesozoic and Cenozoic Coastlines.* Cambridge, England: Cambridge Univ. Press.

SNEATH, P. H. A., AND R. R. SOKAL. 1973. *Numerical Taxonomy.* San Francisco, CA: W. H. Freeman.

SPEED, R. C. 1985. Cenozoic collision of the Lesser Antilles Arc and continental South America and the origin of the El Pi-lar Fault. Tectonophysics 4:41–69.

STEYSKAL, G. C. 1965. Trend curves of the rate of species de-scription in zoology. Science 149:880–882.

THOMAS, R., AND R. JOGLAR. 1995. *The Herpetology of Puerto Rico: Past, Present, and Future.* Publ. New York Acad. Sci., in press.

THOMPSON, D. A., P. K. BRETTING, AND M. HUMPHREYS. (eds.), 1986a. *Forests of Jamaica.* Kingston, Jamaica: Jamaican Soc. Scientists Technol.

THOMPSON, D. A., D. L. WRIGHT, AND O. EVELYN. 1986b. Forest resources in Jamaica. Pp. 81–90 *in* D. A. Thompson, P. K. Bretting, and M. Humphreys (eds.), *Forests of Jamaica.* Kingston, Jamaica: Jamaican Soc. Scientists Technol.

VICKERS, D. O. 1979. The rainfall of Jamaica. J. Geol. Soc. Jamaica 18:5–26.

WILLIAMS, E. E. 1961. Notes on Hispaniolan herpetology. 3. The evolution and relationships of the *Anolis semilineatus* group. Breviora 136:1–8.

WILLIAMS, E. E. 1989. Old problems and new opportunities in West Indian biogeography. Pp. 1–46 *in* C. A. Woods (ed.), *Biogeography of the West Indies: Past, Present, and Fu-ture.* Gainesville, FL: Sandhill Crane Press.

WILLIAMSON, M. 1981. *Island Populations.* Oxford, England: Oxford Univ. Press.

WILLIAMSON, M. 1988. Relationships of species number to area, distance and other variables. Pp. 91–115 *in* A. A. Myers and P. S. Giller (eds.), *Analytical Biogeography.* New York, NY: Chapman and Hall.

WILSON, E. O. 1992. *The Diversity of Life.* Cambridge, MA: Harvard Univ. Press.

WOODRING, W. P., J. S. BROWN, AND W. S. BURBANK. 1924. *Geol-ogy of the Republic of Haiti.* Port-au-Prince, Haiti: Dept. Public Works.

WORLD RESOURCES INSTITUTE. 1992. *World Resources 1992–93.* New York, NY: Oxford Univ. Press.

WORLD RESOURCES INSTITUTE. 1994. *World Resources 1994–95.* New York, NY: Oxford Univ. Press.

APPENDIX 4:1

DISTRIBUTION OF SPECIES OF AMPHIBIANS IN CUBA

Symbols in columns: – absent, + present (native), I introduced. Abbreviations for regions: AC Alturas de Central (Alturas de Santa Clara, Alturas del Nordeste, Alturas del Noroeste), AH Alturas de la Habana-Matanzas (including Alturas de Bujucal-Madruga-Coliseo), AS Archipiélago Sabana-Camaguey, CG Cordillera de Guaniguanico, CM Alturas de Camaguey-Maniabón (Sierra de Cubitas, Peniplano de Florida-Camaguey-Tunas, Sierra de Nejasa, Grupo de Maniabón), IJ Isla de Juventud and Archipiélago de los Canarreos, LC Llanura Occidental, LR Llanura Oriental, LZ Llanura de Zapata (including Cienagas de Zapata), ME Macizo del Escambray, MS Macizo de Sagua-Baracoa, PG Península de Guanahacabibes, SM Sierra Maestra.

Taxon	Elevation (meters)	Distribution Area (km²)	PG	LC	CG	IJ	AH	LZ	AC	AS	ME	LR	CM	SM	MS
BUFONIDAE:															
Bufo cataulaciceps	0–50	6,950	+	+	–	+	–	–	–	–	–	–	–	–	–
Bufo empusa	0–70	73,360	–	+	–	+	+	+	–	+	–	+	+	–	+
Bufo fustiger	0–155	16,600	+	+	+	–	+	–	–	–	–	–	–	–	–
Bufo gundlachi	0–70	75,290	+	+	–	+	+	–	–	+	+	+	+	–	–
Bufo longinasus	100–820	1,545	–	–	+	–	–	–	–	+	–	–	–	–	+
Bufo peltocephalus	0–410	70,270	–	–	–	+	–	–	+	+	+	+	+	+	+
Bufo taladai	0–560	10,425	–	–	–	–	–	–	–	–	+	–	+	+	+
Bufo species "Z"	0–50	1	–	–	–	–	–	+	–	–	–	–	–	–	–
HYLIDAE:															
Osteopilus septentrionalis	0–1110	110,922	+	+	+	+	+	+	+	+	+	+	+	+	+
Pseudacris crucifer (I)	0	—	–	–	–	–	I	–	–	–	–	–	–	–	–
LEPTODACTYLIDAE:															
Eleutherodactylus acmonis	30–1150	10,810	–	–	–	–	–	–	–	–	–	–	–	+	+
Eleutherodactylus albipes	1300–1974	385	–	–	–	–	–	–	–	–	–	–	–	+	–
Eleutherodactylus atkinsi	0–1212	93,365	+	+	+	+	–	–	–	+	+	+	+	+	+
Eleutherodactylus auriculatus	0–1150	33,205	+	+	+	+	+	–	–	–	+	–	+	+	+
Eleutherodactylus bartonsmithi	30–212	2,315	–	–	–	–	–	–	–	–	–	–	–	–	+
Eleutherodactylus bresslerae	30–212	1,160	–	–	–	–	–	–	–	–	–	–	–	–	+
Eleutherodactylus cubanus	1060–1400	385	–	–	–	–	–	–	–	–	–	–	–	+	–
Eleutherodactylus cuneatus	0–1515	16,215	–	–	–	–	–	–	–	–	–	–	–	+	+
Eleutherodactylus dimidiatus	0–1375	75,290	+	–	+	–	–	–	–	+	+	+	+	+	+
Eleutherodactylus eileenae	0–700	36,680	+	–	+	–	+	–	+	+	–	+	–	+	–
Eleutherodactylus emiliae	350–400	385	–	–	–	–	–	–	–	+	–	–	–	–	–
Eleutherodactylus etheridgei	0–151	1,160	–	–	–	–	–	–	–	–	–	–	–	+	+
Eleutherodactylus glamyrus	800–1972	1,160	–	–	–	–	–	–	–	–	–	–	–	+	–
Eleutherodactylus greyi	0–820	18,530	–	–	–	–	–	–	+	+	+	–	–	–	–
Eleutherodactylus guanahacabibes	0–20	1,350	+	–	–	–	–	–	–	–	–	–	–	–	–
Eleutherodactylus guantanamera	60–1150	7,335	–	–	–	–	–	–	–	–	–	–	–	+	+
Eleutherodactylus gundlachi	650–1375	3,860	–	–	–	–	–	–	–	–	–	–	–	+	+
Eleutherodactylus iberia	600	1	–	–	–	–	–	–	–	–	–	–	–	–	+
Eleutherodactylus intermedius	454–1818	1,930	–	–	–	–	–	–	–	–	–	–	–	+	+
Eleutherodactylus ionthus	0–1230	3,090	–	–	–	–	–	–	–	–	–	–	–	+	–
Eleutherodactylus klinikowskii	75–182	3,090	–	–	+	–	–	–	–	–	–	–	–	–	–
Eleutherodactylus leberi	394–465	2,700	–	–	–	–	–	–	–	–	–	–	–	+	+
Eleutherodactylus limbatus	50–1150	24,710	–	–	+	–	+	–	–	–	+	–	+	+	+
Eleutherodactylus mariposa	720	1	–	–	–	–	–	–	–	–	–	–	–	–	+
Eleutherodactylus melacara	845–1974	1,160	–	–	–	–	–	–	–	–	–	–	–	+	–
Eleutherodactylus pezopetrus	100–270	1	–	–	–	–	–	–	–	–	–	–	–	–	+
Eleutherodactylus pinarensis	0–381	20,850	+	+	+	+	+	+	+	–	–	–	–	–	–
Eleutherodactylus planirostris	0–727	110,922	+	+	+	+	+	–	+	+	+	+	+	+	–
Eleutherodactylus ricordii	290–1150	8,495	–	–	–	–	–	–	–	–	–	–	–	+	+
Eleutherodactylus species "R"	0–830	75,290	+	+	+	+	+	–	+	+	+	+	+	+	–
Eleutherodactylus ronaldi	212–1060	9,650	–	–	–	–	–	–	–	–	–	–	–	+	+
Eleutherodactylus symingtoni	70–155	8,880	–	–	+	–	–	–	–	–	–	–	–	–	–
Eleutherodactylus tetajulia	600	1	–	–	–	–	–	–	–	–	–	–	–	–	+
Eleutherodactylus thomasi	0–390	10,425	–	–	–	–	–	–	+	+	+	+	+	–	–
Eleutherodactylus toa	195–900	5,405	–	–	–	–	–	–	–	–	–	–	–	–	+
Eleutherodactylus tonyi	10–50	1,350	–	–	–	–	–	–	–	–	–	–	–	+	–
Eleutherodactylus turquinensis	455–1400	770	–	–	–	–	–	–	–	–	–	–	–	+	–
Eleutherodactylus varians	0–845	38,220	+	–	+	+	+	+	+	–	+	–	+	–	–
Eleutherodactylus varleyi	0–845	55,985	–	+	+	+	–	–	+	+	+	+	+	+	+
Eleutherodactylus zeus	75–182	3,090	–	–	+	–	–	–	–	–	–	–	–	–	–
Eleutherodactylus zugi	155–390	1,545	–	–	+	–	+	–	–	–	–	–	–	–	–
RANIDAE:															
Rana catesbeiana (I)	0	—	–	–	–	I	I	–	–	–	–	–	–	–	–
Total native species in region (A)			13	11	17	12	14	5	9	10	17	10	16	26	27
Endemic species in region (B)			1	0	2	0	0	1	0	0	1	0	0	7	7
Percent regional endemism (B/A)			8	0	12	0	0	20	0	0	6	0	0	27	26
Percent total endemism (B/50)			2	0	4	0	0	2	0	0	2	0	0	14	14

APPENDIX 4:2

DISTRIBUTION OF SPECIES OF AMPHIBIANS IN JAMAICA

Symbols in columns: – absent, + present (native), I introduced. Abbreviations for regions: BM Blue Mountains, CC Cockpit Country, CU Central Uplands, HH Hellshire Hills, JC John Crow Mountains, MP Manchester Plateau, NC North Coast, PR Portland Ridge Peninsula, SC Santa Cruz Mountains, SL Southern Lowlands, WL Western Lowlands, WU Western Uplands.

Taxon	Elevation (meters)	Distribution Area (km²)	Biogeographic Region											
			WL	WU	SC	CC	NC	CU	MP	SL	PR	HH	BM	JC
BUFONIDAE:														
Bufo marinus (I)	0–?	—	I	–	–	–	I	–	–	–	–	–	–	–
HYLIDAE:														
Calyptahyla crucialis	0–1200	10,990	–	+	+	+	–	+	+	–	–	–	+	–
Hyla marianae	121–879	1595	–	–	–	+	–	+	–	–	–	–	–	–
Hyla wilderi	121–879	4540	–	+	+	+	–	+	+	–	–	–	+	+
Osteopilus brunneus	0–1515	8430	+	+	+	+	+	+	+	–	–	–	+	+
Osteopilus new species	0–450	240?	–	–	–	+	–	–	–	–	–	+	–	+
LEPTODACTYLIDAE:														
Eleutherodactylus alticola	1650–2248	80	–	–	–	–	–	–	–	–	–	–	+	–
Eleutherodactylus andrewsi	545–1970	485	–	–	–	–	–	–	–	–	–	–	+	+
Eleutherodactylus cavernicola	10–15	55	–	–	–	–	–	–	–	–	+	–	–	–
Eleutherodactylus cundalli	0–636	6325	+	+	+	+	+	+	+	–	–	–	–	–
Eleutherodactylus fuscus	121–682	700	–	+	–	+	–	–	–	–	–	–	–	–
Eleutherodactylus glaucoreius	0–1652	1515	–	–	–	–	–	–	–	–	–	–	+	+
Eleutherodactylus gossei	0–1515	8350	+	+	+	+	+	+	+	+	–	–	+	+
Eleutherodactylus grabhami	152–667	2675	+	+	–	+	–	+	+	–	–	–	–	–
Eleutherodactylus griphus	250–636	160	–	–	–	+	–	–	–	–	–	–	–	–
Eleutherodactylus jamaicensis	121–1288	3405	–	+	–	+	–	+	–	–	–	–	+	+
Eleutherodactylus johnstonei (I)	0–1200	—	I	–	–	I	I	I	I	I	–	–	I	I
Eleutherodactylus junori	606–833	945	–	–	–	+	–	+	–	–	–	–	–	–
Eleutherodactylus luteolus	0–682	1810	+	+	–	+	–	+	–	–	–	–	–	–
Eleutherodactylus nubicola	1061–1879	160	–	–	–	–	–	–	–	–	–	–	+	–
Eleutherodactylus orcutti	227–1212	485	–	–	–	–	–	–	–	–	–	–	+	+
Eleutherodactylus pantoni	0–1636	4135	+	+	–	+	–	+	+	–	–	–	+	–
Eleutherodactylus pentasyringos	0–1273	730	–	–	–	–	–	–	–	–	–	–	+	+
Eleutherodactylus planirostris (I)	0–100?	—	I	I	I	I	I	I	I	I	–	–	I	I
Eleutherodactylus sisyphodemus	450	40	–	–	–	+	–	–	–	–	–	–	–	–
RANIDAE:														
Rana catesbeiana (I)	0–?	—	I	–	–	–	–	–	–	–	–	–	–	–
Total species native to region (A)			6	10	5	15	3	11	8	1	1	1	12	9
Endemic species in region (B)			0	0	0	2	0	0	0	0	1	0	2	0
Percent regional endemism (B/A)			0	0	0	13	0	0	0	0	100	0	17	0
Percent total endemism (B/22)			0	0	0	9	0	0	0	0	5	0	9	0

APPENDIX 4:3

DISTRIBUTION OF SPECIES OF AMPHIBIANS IN HISPANIOLA

Symbols in columns: – absent, + present (native), I introduced. Abbreviations for regions: AR Plaine de l'Artibonite, BA Península de Barahona, CE Cordillera Central, CI Plaine du Nord–Vallée de Cibao, CV Plaine du Cul de Sac–Valle de Neiba, GO Île Gonâve, HA Los Haitíses, LC Llanura Costera del Caribe, LH Massif de la Hotte, LS Massif de la Selle–Sierra de Baoruco, MD Massif du Nord, MN Chaîne des Matheux–Sierra de Neiba, NO Massif des Montagnes Noires, OR Cordillera Oriental, PL Plateau Central, PN Presqu'île du Nord Ouest, PS Península de Samaná, SE Cordillera Septentrional, SM Sierra de Martín García, TI Presqu'île de la Tiburon, TO Île de la Tortue, VS Valle de San Juan. No species of amphibians are known from Navassa Island.

Taxon	Elevation (meters)	Distribution Area (km²)	LH	TI	LS	BA	CV	GO	MN	SM	VS	PL	NO	AR	PN	TO	MD	CE	CI	SE	PS	HA	OR	LC
BUFONIDAE:																								
Bufo fluviatica	150–175	625	–	–	–	–	+	–	–	–	–	–	–	–	–	–	–	–	+	–	–	–	–	–
Bufo guentheri	0–107	24,135	–	+	–	+	+	–	–	–	+	+	–	+	+	+	+	+	+	–	–	+	+	+
Bufo marinus (I)	0	—	–	–	–	–	I	–	–	–	–	–	–	–	–	–	–	–	–	–	–	–	–	–
HYLIDAE:																								
Hyla heilprini	0–1856	74,275	+	+	+	–	–	+	+	–	–	+	–	–	+	–	+	+	+	+	–	+	–	–
Hyla pulchrilineata	0–1091	39,500	+	+	+	–	–	–	–	–	+	+	–	–	–	–	+	+	+	+	+	+	+	+
Hyla vasta	0–1697	74,590	+	+	+	+	–	–	–	–	+	+	–	+	+	+	+	+	+	+	–	+	+	+
Osteopilus dominicensis	0–1212	70,690	+	+	+	+	+	+	+	+	+	+	+	–	+	+	+	+	+	+	+	+	+	+
LEPTODACTYLIDAE:																								
Eleutherodactylus abbotti	0–1818	67,555	+	+	+	–	+	–	+	+	+	+	–	–	–	–	+	+	+	+	+	+	+	+
Eleuth. alcoae	0–606	2,300	–	–	+	+	+	–	–	+	–	–	–	–	–	–	–	–	–	–	–	–	–	–
Eleuth. amadeus	1000–2340	390	+	–	–	–	–	–	–	–	–	–	–	–	–	–	–	–	–	–	–	–	–	–
Eleuth. apostates	333–1640	815	+	–	–	–	–	–	–	–	–	–	–	–	–	–	–	–	–	–	–	–	–	–
Eleuth. armstrongi	152–1697	2,190	+	–	+	–	–	–	–	–	–	–	–	–	–	–	+	–	–	–	–	–	–	–
Eleuth. audanti	800–2500	4,700	+	+	+	–	–	+	+	+	+	+	–	–	–	–	+	+	–	–	–	–	–	–
Eleuth. auriculatoides	788–1879	3,760	–	–	–	–	–	–	–	–	–	–	–	–	–	–	+	+	+	–	–	–	–	–
Eleuth. bakeri	890–2325	815	+	–	–	–	–	–	–	–	–	–	–	–	–	–	–	–	–	–	–	–	–	–
Eleuth. brevirostris	575–2375	920	+	–	–	–	–	–	–	–	–	–	–	–	–	–	–	–	–	–	–	–	–	–
Eleuth. caribe	0	1	+	–	–	–	–	–	–	–	–	–	–	–	–	–	–	–	–	–	–	–	–	–
Eleuth. chlorophenax	990–1290	910	+	–	–	–	–	–	–	–	–	–	–	–	–	–	–	–	–	–	–	–	–	–
Eleuth. corona	1120	1	+	–	–	–	–	–	–	–	–	–	–	–	–	–	–	–	–	–	–	–	–	–
Eleuth. counouspeus	303–760	4,075	+	–	–	–	–	–	–	–	–	–	–	–	–	–	–	–	–	–	–	–	–	–
Eleuth. darlingtoni	1720–2200	235	–	–	+	–	–	–	–	–	–	–	–	–	–	–	–	–	–	–	–	–	–	–
Eleuth. dolomedes	1120–1120	1	+	–	–	–	–	–	–	–	–	–	–	–	–	–	–	–	–	–	–	–	–	–
Eleuth. eunaster	575–1300	890	+	–	–	–	–	–	–	–	–	–	–	–	–	–	–	–	–	–	–	–	–	–
Eleuth. flavescens	0–909	22,730	–	–	+	–	–	–	–	–	–	–	–	–	–	–	+	+	+	+	+	+	+	+
Eleuth. fowleri	1045–1303	815	–	–	+	–	–	–	–	–	–	–	–	–	–	–	–	–	–	–	–	–	–	–
Eleuth. furcyensis	803–2100	1,960	+	–	+	–	–	–	–	–	–	–	–	–	–	–	–	–	–	–	–	–	–	–
Eleuth. glandulifer	303–1886	1,330	+	–	–	–	–	–	–	–	–	–	–	–	–	–	–	–	–	–	–	–	–	–
Eleuth. glanduliferoides	1515–2121	280	–	–	+	–	–	–	–	–	–	–	–	–	–	–	–	–	–	–	–	–	–	–
Eleuth. glaphycompus	576–1480	5,640	+	+	–	–	–	–	–	–	–	–	–	–	–	–	+	–	–	–	–	–	–	–

Taxon	Elevation	Distribution	LH	TI	LS	BA	CV	GO	MN	SM	VS	PL	NO	AR	PN	TO	MD	CE	CI	SE	PS	HA	OR	LC
Eleuth. grahami	20–330	78	–	–	–	–	–	–	–	–	–	–	–	–	+	–	–	–	–	–	–	–	–	–
Eleuth. haitianus	1545–2455	1.72	–	+	+	–	–	–	–	–	–	–	–	–	–	–	+	+	–	–	–	–	–	–
Eleuth. heminota	0–1697	10,660	+	+	+	–	–	–	–	–	–	–	–	–	–	–	–	–	–	–	–	–	–	–
Eleuth. hypostenor	667–1061	2,040	–	+	+	–	–	–	–	–	–	–	–	–	–	–	–	–	–	–	–	–	–	–
Eleuth. inoptatus	0–1697	69,580	+	+	+	–	–	–	+	+	–	+	–	–	+	+	+	+	+	+	+	+	+	+
Eleuth. jugans	1242–2146	1,220	+	–	+	–	–	–	–	–	–	–	–	–	–	–	–	–	–	–	–	–	–	–
Eleuth. lamprotes	818–1455	750	–	–	+	–	–	–	–	–	–	–	–	–	–	–	–	–	–	–	–	–	–	–
Eleuth. leoncei	1182–2303	1,315	–	–	+	–	–	–	–	–	–	–	–	–	–	–	–	–	–	–	–	–	–	–
Eleuth. lucioi	100	1	–	–	–	–	–	–	–	–	–	–	–	+	–	–	–	–	–	–	–	–	–	–
Eleuth. minutus	879–2300	2,115	–	–	–	–	–	–	–	–	–	–	–	–	–	–	+	+	–	–	–	–	–	–
Eleuth. montanus	1270–2424	2,540	–	–	–	–	–	–	–	–	–	–	–	–	–	–	+	+	–	–	–	–	–	–
Eleuth. sp "N"	1500–1800	440	–	–	–	–	–	–	+	–	–	–	–	–	–	–	–	–	–	–	–	–	–	–
Eleuth. neodreptus	1121	1	–	+	+	–	–	–	–	–	–	–	–	–	–	–	–	–	–	–	–	–	–	–
Eleuth. nortoni	576–1515	2,350	+	+	+	–	–	–	–	–	–	–	–	–	–	–	–	–	–	–	–	–	–	–
Eleuth. oxyrhynchus	333–1212	1,680	+	+	+	–	–	–	–	–	–	–	–	–	–	–	–	–	–	–	–	–	–	–
Eleuth. parabates	1455–1870	440	–	–	–	–	–	–	+	–	–	–	–	–	–	–	–	–	–	–	–	–	–	–
Eleuth. parapelates	950–1050	470	+	–	–	–	–	–	–	–	–	–	–	–	–	–	–	–	–	–	–	–	–	–
Eleuth. patricae	2000–3050	4,700	–	–	–	–	–	–	–	–	–	–	–	–	–	–	–	+	–	–	–	–	–	–
Eleuth. paulsoni	0–750	8,780	+	+	+	+	–	–	–	–	+	–	–	–	–	–	–	+	–	–	–	–	–	+
Eleuth. pictissimus	0–1758	20,380	+	+	+	+	+	–	+	+	+	+	–	–	–	–	–	+	+	–	–	–	–	+
Eleuth. pituinus	1212–1770	1,065	–	–	–	–	–	–	–	–	–	–	–	–	–	–	–	+	–	–	–	–	–	–
Eleuth. poolei	550–650	310	–	–	–	–	–	–	–	–	–	–	–	–	–	+	–	–	–	–	–	–	–	–
Eleuth. probolaeus	0–60	1,490	–	–	–	–	–	–	–	–	–	–	–	–	+	–	–	–	–	–	–	–	–	+
Eleuth. rhodesi	30	45	–	–	–	–	–	–	–	–	–	–	–	–	–	–	–	–	–	–	–	–	–	–
Eleuth. ruffemoralis	727–1370	550	+	+	+	–	–	–	+	–	–	–	–	–	–	–	–	–	–	–	–	–	–	–
Eleuth. ruthae	0–900	12,85	–	–	–	–	–	–	–	–	–	–	–	–	–	–	–	+	+	+	+	–	+	+
Eleuth. schmidti	0–1758	11,600	+	–	–	–	–	–	–	–	–	–	–	–	–	+	+	+	+	+	+	–	–	–
Eleuth. sciagraphus	1060–1181	80	+	–	+	–	–	–	–	–	–	–	–	–	–	–	–	–	–	–	–	–	–	–
Eleuth. semipalmatus	303–1697	1,250	+	–	+	–	–	–	–	–	–	–	–	–	–	–	–	–	–	–	–	–	–	–
Eleuth. thorectes	1700–2340	4	+	–	–	–	–	–	–	–	–	–	–	–	–	–	–	–	–	–	–	–	–	–
Eleuth. ventriilineatus	1700–2340	45	+	–	–	–	–	–	–	–	–	–	–	–	–	–	–	–	–	–	–	–	–	–
Eleuth. warreni	400	1	–	–	–	–	–	–	–	–	–	–	+	–	+	–	–	–	–	–	–	–	–	–
Eleuth. weinlandi	0–788	24,610	–	–	+	–	–	–	+	–	–	–	–	–	–	–	–	+	+	+	+	–	+	+
Eleuth. wetmorei	0–1324	12,070	+	+	+	–	–	–	+	–	–	–	–	–	–	+	+	+	–	–	–	–	–	–
Leptodactylus dominicensis	0–60	1,220	–	–	–	–	–	–	–	–	–	–	–	–	+	–	–	–	–	–	–	–	–	+
RANIDAE:																								
Rana catesbeiana (I)	0	—	–	–	–	–	–	–	–	–	–	–	–	–	–	–	–	–	–	I	–	I	–	I
Total native species in region (A)			32	13	26	3	5	1	10	4	4	6	1	1	5	3	9	20	11	10	8	6	9	12
Endemic species in region (B)			16	0	10	0	0	0	2	0	0	0	1	0	3	1	1	6	1	0	0	0	0	2
Percent regional endemism (B/A)			50	0	39	0	0	0	20	0	0	0	0	0	60	33	11	30	9	0	0	0	0	17
Percent total endemism (B/63)			25	0	16	0	0	0	3	0	0	0	0	0	5	2	2	10	2	0	0	0	0	3

APPENDIX 4:4

DISTRIBUTION OF SPECIES OF AMPHIBIANS IN THE PUERTO RICAN BANK

Symbols in columns: – absent, + present (native), I introduced. Abbreviations for regions: ANE Anegada, CAY Sierra de Cayey, COA Coastal Lowlands, CUL Culebra, CUP Central Uplands, MON Isla Mona, PAN Cuchilla de Pandura, STC St. Croix, STJ St. John, STT St. Thomas, TRT Tortola, VGO Virgin Gorda, VIE Vieques, YUN El Yunque.

Taxon	Elevation (meters)	Distribution Area (km²)	MON	COA	CUP	CAY	PAN	YUN	VIE	CUL	STT	STJ	STC	TRT	VGO	ANE
BUFONIDAE:																
Bufo lemur	0–50	1000	–	–	–	–	–	–	–	–	–	–	–	–	+	–
Bufo marinus (I)	0	—	–	I	–	–	–	–	–	–	I	I	I	–	–	–
HYLIDAE:																
Hyla cinerea (I)	0	—	–	I	–	–	–	–	–	–	–	–	–	–	–	–
Osteopilus septentrionalis (I)	0	—	–	I	–	–	–	–	–	–	I	–	–	–	–	–
Scinax rubra (I)	0	—	–	I	–	–	–	–	–	–	–	–	–	–	–	–
LEPTODACTYLIDAE:																
Eleutherodactylus antillensis	0–1212	9355	–	+	+	+	+	+	+	+	+	+	+	+	+	–
Eleutherodactylus brittoni	100–636	2500	–	+	+	+	–	+	–	–	–	–	–	–	–	–
Eleutherodactylus cochranae	0–333	9059	–	+	+	+	–	+	+	+	+	+	–	+	–	–
Eleutherodactylus cooki	91–242	135	–	–	–	–	+	–	–	–	–	–	–	–	–	–
Eleutherodactylus coqui	0–1128	8768	–	+	+	+	+	+	–	–	–	–	–	–	–	–
Eleutherodactylus eneidae	303–1151	1725	–	–	+	+	–	+	–	–	–	–	–	–	–	–
Eleutherodactylus gryllus	303–1182	1050	–	–	+	–	–	+	–	–	–	–	–	–	–	–
Eleutherodactylus hedricki	455–1152	160	–	–	+	+	–	+	–	–	–	–	–	–	–	–
Eleutherodactylus jasperi	647–785	110	–	–	–	+	–	–	–	–	–	–	–	–	–	–
Eleutherodactylus karlschmidti	182–630	1050	–	–	+	+	–	+	–	–	–	–	–	–	–	–
Eleutherodactylus lentus	0–10	347	–	–	–	–	–	–	–	–	–	+	+	–	–	–
Eleutherodactylus locustus	273–1050	565	–	–	+	–	–	+	–	–	–	–	–	–	–	–
Eleutherodactylus monensis	0–10	25	+	–	–	–	–	–	–	–	–	–	–	–	–	–
Eleutherodactylus portoricensis	273–1182	1480	–	–	+	+	+	+	–	–	–	–	–	–	–	–
Eleutherodactylus richmondi	40–1152	3395	–	+	+	+	+	+	–	–	–	–	–	–	–	–
Eleutherodactylus schwartzi	0–227	130	–	–	–	–	–	–	–	–	+	+	–	+	–	–
Eleutherodactylus unicolor	670–1039	40	–	–	–	–	–	+	–	–	–	–	–	–	–	–
Eleutherodactylus wightmanae	303–1182	2020	–	+	+	+	–	+	–	–	–	–	–	–	–	–
Leptodactylus albilabris	0–1030	9433	–	+	+	+	+	+	+	+	+	+	+	+	+	+
RANIDAE:																
Rana catesbeiana (I)	0	—	–	I	–	–	–	–	–	–	–	–	–	–	–	–
Total native species in region (A)			1	7	13	11	5	14	3	3	4	5	3	4	3	1
Endemic species in region (B)			1	0	0	1	1	1	0	0	0	0	0	0	0	0
Percent regional endemism (B/A)			100	0	0	9	20	7	0	0	0	0	0	0	0	0
Percent total endemism (B/20)			5	0	0	5	5	5	0	0	0	0	0	0	0	0

APPENDIX 4:5

DISTRIBUTION OF SPECIES OF AMPHIBIANS IN THE LESSER ANTILLES

Symbols in columns: – absent, + present (native), E extirpated, I introduced. Abbreviations for regions: AN Anguilla Bank, BB Barbuda Bank, BA Barbados, DO Dominica, GR Grenada Bank, GU Guadeloupe Bank, MG Marie Galante, MA Martinique, MO Montserrat, SA Saba Bank, SE St. Eustatius Bank, SL St. Lucia, SV St. Vincent.

Taxon	Elevation (meters)	Distribution Area (km²)	AN	SA	SE	BB	MO	GU	MG	DO	MA	SL	SV	GR	BA
BUFONIDAE:															
Bufo marinus (I)	0	—	–	–	I	I	I	I	–	–	I	I	I	I	I
DENDROBATIDAE:															
Colostethus chalcopis	500	25	–	–	–	–	–	–	–	–	+	–	–	–	–
HYLIDAE:															
Osteopilus septentrionalis (I)	0	—	I	–	–	–	–	–	–	–	–	–	–	–	–
Scinax rubra (I)	0	—	–	–	–	–	–	–	–	–	–	I	–	–	–
LEPTODACTYLIDAE:															
Eleutherodactylus amplinympha	300–1200	150	–	–	–	–	–	–	–	+	–	–	–	–	–
Eleutherodactylus barlagnei	121–1200	530	–	–	–	–	–	+	–	–	–	–	–	–	–
Eleutherodactylus euphronides	300–840	16	–	–	–	–	–	–	–	–	–	–	–	+	–
Eleutherodactylus johnstonei	0–922	5050	+	+	+	+	+	+	+	–	+	+	+	+	+
Eleutherodactylus martinicensis	0–1200	2860	I	–	–	+	–	+	+	+	+	–	–	–	–
Eleutherodactylus pinchoni	182–1467	855	–	–	–	–	–	+	–	–	–	–	–	–	–
Eleutherodactylus planirostris (I)	0	—	–	–	–	–	–	–	–	–	–	–	–	I	–
Eleutherodactylus shrevei	275–922	90	–	–	–	–	–	–	–	–	–	–	+	–	–
Leptodactylus fallax	0–300	17	–	–	E	–	+	–	–	+	E	–	–	–	–
Leptodactylus validus	0–735	395	–	–	–	–	–	–	–	–	–	–	+	+	–
Total native species in region (A)			1	1	1	2	2	4	2	3	3	1	3	3	1
Endemic species in region (B)			0	0	0	0	0	2	0	1	1	0	1	1	0
Percent regional endemism (B/A)			0	0	0	0	0	50	0	33	33	0	33	33	0
Percent total endemism (B/10)			0	0	0	0	0	20	0	10	10	0	10	10	0

ADDENDUM

Subsequent to my reading proof, two additional species of *Eleutherodactylus* have been named from Cuba. Their elevational range, distribution area, and biogeographic region (see Appendix 4:1) are, respectively: *E. jaumei*—150–200 m, 1 km², SM; *E. principalis*—300–1000 m, 468 km², MS.

5. Distribution Patterns of Amphibians in South America

WILLIAM E. DUELLMAN

Natural History Museum and
Department of Systematics and Ecology
The University of Kansas
Lawrence, Kansas 66045-2454, USA

ABSTRACT   More than 1700 species of amphibians are known from South America, and this number of species is growing rapidly with the discovery of many new species, especially in the highlands. For purposes of analysis, 12 regions are identified on the continent; one of these, the Andes, is a composite of six subregions. Only 61 species are shared with Central America and two with the Lesser Antilles. The largest number of species (754) is in the Andes followed by the Amazonia-Guiana Region (335) and the Atlantic Forest Domain (334). More than 90% of the species are endemic in five regions—the Andes, Guiana Highlands, Atlantic Forest Domain, Austral Temperate Forest, and Atacama Desert. Species diversity and endemism is relatively low in subhumid regions, such as the Llanos, Cerrado-Caatinga-Chaco, and Pampean-Monte regions. Endemism is low in the Chocó, because 60 species are shared with Central America. Geographic ranges tend to be much greater among species inhabiting lowlands than among those in the highlands. Thus, preservation of relatively few, large areas in the lowlands will preserve a large percentage of the amphibian fauna, many smaller, strategically placed reserves are needed to preserve the highland species.

Key words: Amphibia; South America; Biogeography; Patterns of distribution; Conservation.

INTRODUCTION

As viewed from outer space, South America may be perceived as a large teardrop dangling from the southern tip of the North American continent. However, geologically and biologically this perception is entirely unreal. Geological evidence reveals that South America was a large island for about 50 million years after its separation from Africa until its tenuous connection with what is now lower Central America (Stehli and Webb, 1985; Pittman et al., 1993). This proposed long period of isolation is strongly supported by biological evidence: mammals (Simpson, 1980; Marshall and Sempere, 1993), birds (Vuilleumier, 1985; Vuilleumier and Andors, 1993), reptiles and amphibians (Duellman, 1979a, 1993b, Vanzolini and Heyer, 1985).

No synthesis of South American amphibians exists. Taxonomic lists and varying degrees of patterns of distribution were presented for some regions—tropical lowland forests (Lynch, 1979), Guianan Region (Hoogmoed, 1979), dry lowlands of the northern part of the continent (Rivero-Blanco and Dixon, 1979), the Chaco (Gallardo, 1979), Patagonia (Cei, 1979), the austral temperate forests (Formas, 1979), and the Andes (Duellman, 1979b). Heyer (1988) reviewed the distribution of anurans in the cis-Andean tropics.

During the past two decades intensive work at selected sites has resulted in thorough studies of anuran faunas—Santa Cecilia, Ecuador (Duellman, 1978); El Manteco, Venezuela (Hoogmoed and Gor-

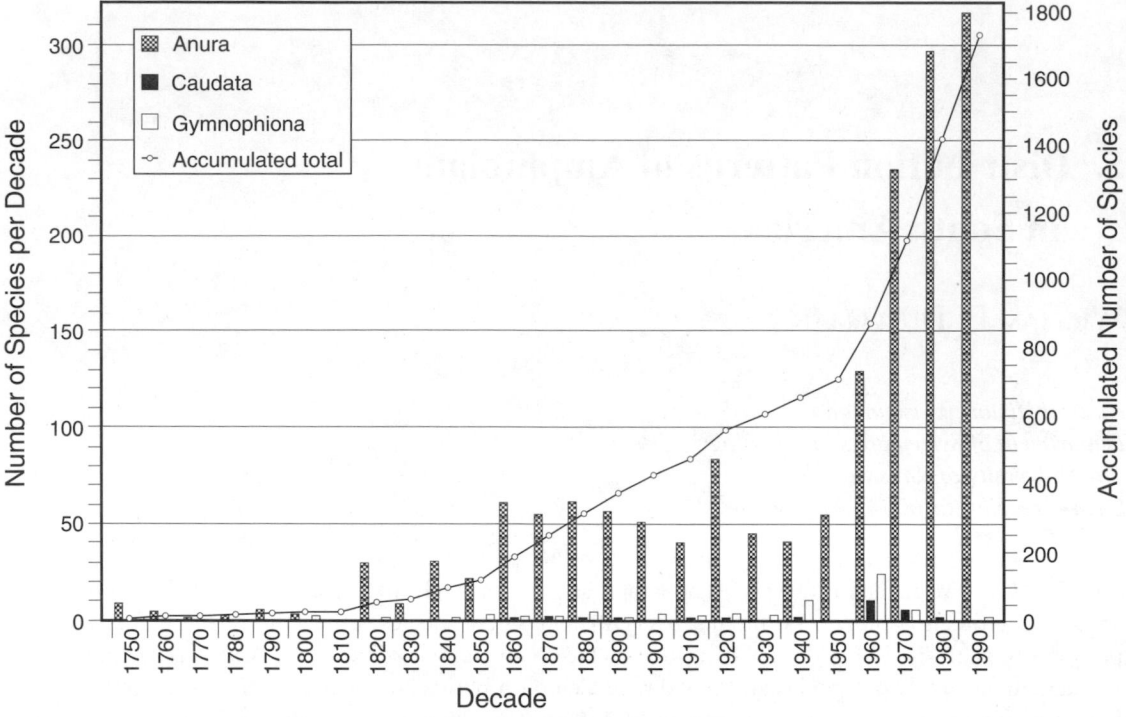

Fig. 5:1.   Discovery rate of amphibians known in South America by decade in which the species was named. The decade 1990 includes only 1990 through part of 1997.

zula, 1979); Panguana, Peru (Schlüter, 1984); Trois Souts, French Guiana (Lescure, 1986); Pajas Blancas, Uruguay (Gudynas and Rudolf, 1987); Boracéia, Brazil (Heyer et al, 1990); Manaus, Brazil (Zimmerman and Rodrigues, 1990); Punta Lara, Argentina (Basso, 1990); Cuzco Amazónico, Peru (Duellman and Salas, 1991); Cocha Cashu, Peru (Rodríguez, 1992); Puerto Almacén, Bolivia (De la Riva, 1993a); Iquitos region, Peru (Rodríguez and Duellman, 1994); Parque Nacional Henri Pittier, Venezuela (Puppo et al., 1995), northern Departamento Loreto, Peru (Duellman and Mendelson, 1995); Balta, Peru (Duellman and Thomas, 1996); Pakitza, Peru (Morales and McDiarmid, 1996); and La Escalera region in southeastern Venezuela (Duellman, 1997).

During the past decade, lists of species with distributional data have been published for several countries—Bolivia (De la Riva, 1990), Colombia (Ruíz-Carranza et al., 1996), Ecuador (Coloma, 1991), Peru (Rodríguez et al., 1993; Morales, 1995), Uruguay (Klappenbach and Langone, 1992), and Venezuela (La Marca, 1997).

Despite the large number of amphibians known from South America, our knowledge of the fauna is mediocre at best. Large areas are essentially unknown; many small areas of possible high diversity have never been collected. That the taxonomic diversity may be woefully underestimated is attested to by the fact that 33.2% of the presently recognized species have been described since 1979 (Fig. 5:1). With these limitations firmly in mind, I have attempted to summarize our knowledge of distributions of amphibians within the complex physiographic and climatic regions of South America.

MATERIALS AND METHODS

The number of taxa and their major distributions were compiled from Frost (1985) and Duellman (1993a), updated by all pertinent publications that have come to my attention. These additional references are: Ardila-Robalo et al. (1993), Ayarzagüena and Señaris (1993, 1996), Ayarzagüena et al.

(1993), Basso (1990), Bastos and Pombal (1995, 1996), Caramaschi (1996), Caramaschi and Cruz (1997), Caramaschi and Velosa (1996), Cascon and Lima Verde (1994), Coloma (1995), Cruz et al. (1997), De la Riva (1993b, 1994a, 1994b, 1996), De la Riva and Lynch (1997), Duellman and Coloma (1993), Duellman and Mendelson (1995), Duellman and Schulte (1993), Duellman and Wiens (1993), Duellman and Wild (1993), Duellman and Yoshpa (1996), Duellman at al. (1997), Dwyer (1995), Flores (1993), Flores and Rodríguez (1997), Flores and Vigle (1994), Formas (1997), Giaretta and Sazima (1993), Gluesenkamp (1995), Gomes and Peixoto (1996), Grant et al. (1997), Graybeal and Cannatella (1995), Haddad and Martins (1994), Haddad and Pombal (1995), Haddad et al. (1996), Harvey (1996), Harvey and Keck (1995), Harvey and Smith (1993, 1994), Heyer (1994), Heyer et al. (1996), Hoogmoed et al. (1994), Izecksohn (1993a, 1993b), Jungfer and Schiesari (1995), Kaplan (1994, 1997a, 1997b), Kaplan and Ruíz (1997), Köhler and Jungfer (1995), La Marca (1994a, 1994b, 1996a, 1996b), Lavilla and Ergueta (1995), Lobo (1994), Lötters (1992, 1996), Lynch (1993a, 1993b, 1994a, 1994b, 1995, 1996a, 1996b, 1996c, 1997), Lynch and Duellman (1995, 1997), Lynch and La Marca (1993), Lynch and Rueda-Almonacid (1997), Lynch and Ruíz-Carranza (1996a, 1996b), Lynch et al. (1994, 1996), Meinhardt and Parmelee (1996), Mercadal de Barrio and Barrio (1993), Morales (1994), Morales and Schulte (1993), Myers and Donnelly (1996, 1997), Ortiz and Vidal (1993), Pombal (1993), Pombal and Caramaschi (1995), Pombal and Haddad (1993), Pombal et al. (1995), Pyburn (1993), Reichle and Köhler (1997), Rivero and Morales (1992), Rivero and Serna (1991), Rodríguez (1994), Rodríguez and Myers (1993), Rueda-Almonacid (1994), Ruíz-Carranza and Osorno-Muñoz (1994), Ruíz-Carranza et al. (1994), Ruíz-Carranza and Lynch (1995a–d, 1996), Ruíz-Carranza et al. (1996, 1997), Salas and Sinsch (1996), Señaris (1993), Señaris and Ayarzagüena (1993), Señaris et al. (1994, 1996), Wiens (1993), Wild (1994, 1995) and Wilkinson (1996). Thus, this compilation is based on published information only. Nominal species with no known localities have not been included. All families, genera, and species are listed in Appendix 5:1, together with their distributions in the regions of South America defined herein.

The definition of regions was achieved by a process of reduction. Initially, I defined 26 regions based primarily on the morphoclimatic domains defined by Ab'Saber (1977) with appropriate modifications (e.g., distinguishing between northern and southern tropical Andes). Taxa were scored for occurrence and endemism in each of these regions. Subsequent analysis revealed that many regions lacked reality in terms of amphibian distributions. For example, based on distribution of anurans, the Caatinga in northeastern Brazil, Cerrados of central and southern Brazil, and the Chaco of Paraguay and northern Argentina form a continuum of overlapping ranges of species. Likewise, the amphibian fauna in the Guianan lowlands in northeastern South America primarily is a subset of the fauna in the Amazon Basin, and the species on the continental islands of Trinidad and Tobago mostly are a subset of the fauna of the Guiana lowlands.

Two of the 12 resulting regions present special problems. Originally, I attempted to separate the Atlantic Coastal Forest from the Brazilian Highlands, which in the form of coastal ranges (e.g., Serra do Mar, Serra da Mantiquiera) descend to the coast. Many species of frogs inhabiting this region are associated with cascading streams which originate in the highlands and bisect the coastal forest. My inability to assign species with confidence to one or the other of these regions resulted in their combination into one region, herein referred to as the Atlantic Forest Domain. Originally, I subdivided the Andes into regions approximating those recognized by Duellman (1979b), but because of the greater similarities among the amphibians of the various parts of the Andes to one another in contrast to other regions, I recognize the Andes as one region with subregions based on topography. Furthermore, for purposes of analysis, a distinction was made between the forested slopes and the supratreeline habitats.

Theoretically, a more objective method would be the determination of occurrence of species in a system of grids. This was done in 2°-grids for 10 groups of common frogs east of the Andes by Heyer (1988), who found that 14% of the 283 grids lacked any records of the species in the 10 groups, and only 5% of the grids had species in all 10 groups represented. The most meaningful of Heyer's conclusions was that distributional data, even for the most common species, are woefully inadequate for fine-scaled analyses. The application of a grid system to areas of high relief, such as the Andes, would have to be much smaller than 2° in order not to include lowlands to elevations above treeline in a single grid.

The areas of regions were determined from a vegetation map (Hueck and Seibert, 1972) by use of a Micro-Plan II™ image analysis system (Laboratory Computer Systems, Inc., Cambridge, Massachusetts). All taxa were scored with respect to distributions within the 12 regions (also subregions and habitats in the Andes). In many cases, especially in the lowlands, the boundaries of these regions are not distinct; one region gradually merges with another. Consequently, the regions as they are mapped are generalized, and the reader must keep in mind that transition zones exist. For purposes of analysis of distribution of amphibians among vegetation types, each species was assigned to one vegetation type; if the species occurs in more than one vegetation type, it was assigned to the type in which it is most widely distributed. Faunal similarities among regions were calculated by using the Coefficient of Biogeographic Resemblance (Duellman, 1990): $2C/(N_1 + N_2)$, where N_1 is the number of species in Region 1, N_2 is the number of species in Region 2, and C is the number of species in common to the two regions.

THE AMPHIBIAN FAUNA

The amphibian fauna of South America is the richest in the world; it consists of 1742 recognized species placed in 140 genera of 16 families. By far, the largest number of species (94.4%) are anurans (Table 5:1). The total surface area of South America is 17,793,000 km^2; the species density of amphibians is 97.9 per 10^6 km^2, and that of anurans is 92.4 per 10^6 km^2. Although salamanders make up a small percentage (1.4%) of the fauna, more than one-third of the species of anurans, and slightly less than one half of the species of caecilians known worldwide occur in South America. The only extra–South American regions with which species are shared naturally are Central America (61 species) and the Lesser Antilles (2 species).

Three globally widespread genera occur in South America; of these, *Hyla* is represented worldwide by 301 species, of which 195 (64.8%) occur in South America. *Bufo* contains 212 species, of which 47 (22.2%) occur in South America. The diversity in these two genera contrasts sharply with *Rana*, represented worldwide by 224 species, of which only three (1.3%) occur in South America. The most speciose family in South America is the Leptodactylidae, represented by 49 genera containing 712 species; the genus *Eleutherodactylus* accounts for 352 of these species. Other large genera in South America are *Leptodactylus* (59 species), *Telmatobius* (45), *Physalaemus* (37), and *Cycloramphus* (25). The second largest family is the Hylidae with 25 genera containing 438 species, of which *Hyla* accounts for 195 species and *Scinax* for 80 species. The third family is the Dendrobatidae with eight genera and 170 species, of which 99 are *Colostethus*. The fourth family, Bufonidae, is represented by 10 genera and 137 species, of which 55 are *Atelopus* and 47 are *Bufo*. The Centrolenidae is represented by three genera with 121 species, of which 61 are *Cochranella* and 37 are *Centrolene*. The Microhylidae is represented by 18 genera with only 38 species, of which 14 are *Chiasmocleis*. The other five families of anurans (Allophrynidae, Brachycephalidae, Pipidae, Pseudidae, and Rhinodermatidae) each contain six or fewer species. Of the 123 genera of anurans, the four most speciose genera (*Eleutherodactylus, Hyla, Colostethus,* and *Scinax*) contain 44.1% of the species of South American frogs.

Salamanders in South America are represented by two genera of bolitoglossine plethodontids; both genera, *Bolitoglossa* with 22 species in South America, and *Oedipina* with two species in South America, are more speciose in Central America. Among the Gymnophiona, the Caeciliidae contains nine genera with 50 species, 28 of which are *Caecilia*. Of the nine species of Rhinatrematidae, eight are *Epicrionops*. The aquatic caecilians of the family Typhlonectidae are represented by four genera containing 15 species, of which six are *Chthonerpeton* and five are *Typhlonectes*.

DEFINITION OF REGIONS

Twelve natural, biogeographic regions are defined for continental South America (Figs. 5:2–3). The continental islands of Trinidad and Tobago are included with the Amazonia-Guiana Region; the Caribbean coastal islands (e.g., Isla de Margarita), islands immediately off the coast of southeastern Brazil (e.g., Ilha Grande and Ilha de São Sebastião), and the Chilean archipelago, including Tierra del Fuego, are combined with the adjacent mainland. No amphibians occur on the Islas Galápagos, so those islands are excluded. Although these regions encompass dispersal centers as used by Müller (1973), they are not equivalent geographically, biologically, or conceptually.

Table 5:1. Taxonomic composition of the amphibian fauna of South America. Numbers in each column are total, endemics, and percent endemic.

Order	Families			Genera			Species		
Anura	12	4	33.3	123	97	78.9	1644	1573	95.7
Caudata	1	0	0.0	2	0	0.0	24	20	83.3
Gymnophiona	3	2	66.7	15	12	80.0	74	72	97.3
Total	16	6	37.5	140	109	77.9	1742	11665	95.6

TROPICAL LOWLANDS

Chocó

This narrow coastal area extends from the Golfo de Urabá in northwestern Colombia to southern Ecuador and extreme northern Peru, exclusive of the southern part of the Guayas Peninsula in southwestern Ecuador. This region receives high rainfall—in excess of 8,000 mm annually with no dry season in parts of Colombia. Except for the southern part, which is a progressively dry forest transition to the Atacama Desert, the Chocoan Region originally supported continuous humid to very humid lowland tropical rainforest that is continuous with that in Panama. The lowland forest gradually changes to humid montane rainforest on the Pacific slopes of the Cordillera Occidental in Colombia and Ecuador; species primarily with distributions below 600 m are considered to be Chocoan.

Caribbean Coastal Forest

The subhumid and semiarid coastal lowlands to elevations of about 400 m in most places (up to 1000 m locally) extend from the vicinity of Cartagena, Colombia, to the Golfo de Paria in northeastern Venezuela; this region also includes the lower part of the Río Magdalena Valley in Colombia. Inland the region is bordered by the Andes and the Cordillera de la Costa. Maximum rainfall ranges from 1000 to 1600 mm annually with a prolonged dry season, usually from November until April. The vegetation varies from tropical dry forest to tropical thorn woodland and desert scrub with tropical evergreen and semideciduous gallery forest (Sarmiento, 1976).

Amazonia-Guiana

The lowland area (> 600 m) of the drainage of the Rio Amazonas and its many tributaries make up the Amazon Basin, which is continuous with the Guianan lowlands in northeastern South America. The continental islands of Trinidad and Tobago lying off the north coast of Venezuela are included in this region, which is the largest region in South America

with an area of 5,960,000 km². The region generally is characterized by low relief and many large, meandering rivers, which during the rainy season overflow their banks and inundate vast areas of land. The basin has an equatorial climate with high temperatures and humidity. Rainfall is highest (< 4000 mm annually) and least seasonal near the equator in the western part of the basin; eastward and at higher latitudes rainfall becomes progressively lower (2000 mm annually) and heaviest seasonally, usually October–May. With the exception of islands of savannas extending southeastward from southern Venezuela and the Guianas, the region naturally was covered by humid tropical rainforest. However, this rainforest is not uniform. Tree diversity is lower on soils that are predominantly sandy and associated with the Guianan Shield than on alluvial soils eroded from the Andes. Moreover, the fluctuation in water levels influences forest types. Terra firme forests are those that are not normally inundated either because of accumulated rainfall or overflow of rivers. Varzea forest is inundated during the rainy season and at times of high water levels in rivers. Along the lower Rio Amazonas, tidal effects result in twice-daily flooding of some varzea forests. Other forests, igapó, are permanently flooded. Associated with meandering rivers are many oxbow lakes, the shores of many of which have a monospecies forest of *Mauritia* palms, which may be widely distributed in adjacent inundated forest. Within the northeastern part of this region, table mountains (tepuis) rise abruptly from the lowlands; these areas make up the Guiana Highlands, which are treated separately.

Atlantic Forest Domain

This region consists of the narrow zone of the Atlantic coastal forest in eastern Brazil and the adjacent uplands, usually referred to as the Brazilian Highlands; the whole region encompasses an area of about 1,150,000 km². The highlands consist of numerous ranges of mountains, most of which do

Fig. 5:2.   South America showing political units and major rivers.

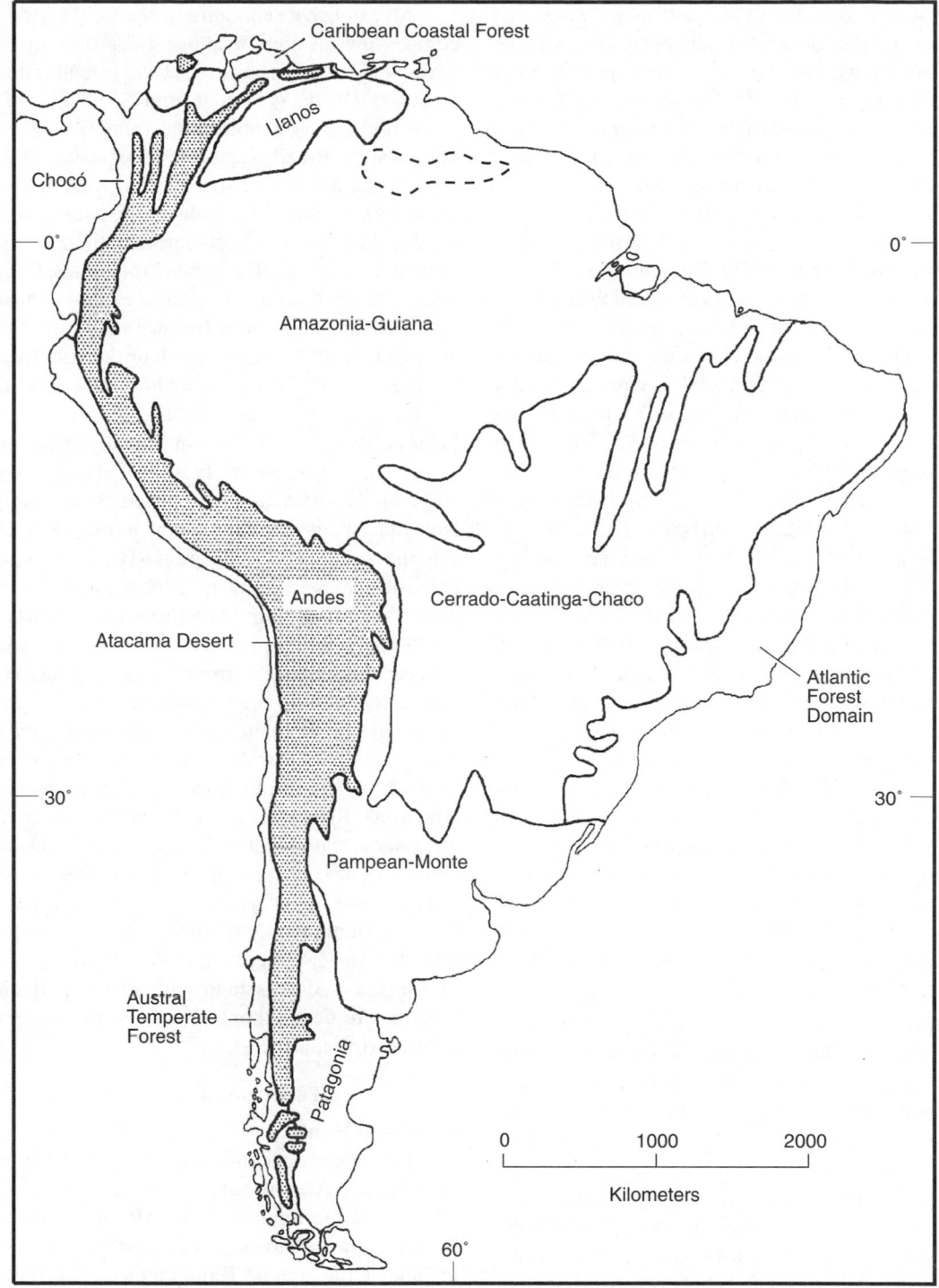

Fig. 5:3.   Natural regions of South America based primarily on the morphoclimatic domains of Ab'Saber (1977). The Guiana Highlands lie within the area of the dashed line.

not attain elevations of more than 2000 m. These ranges tend to parallel the coast; inland ranges are separated from the coastal ranges by river valleys. The highlands extend from southern Bahia to Santa Catarina and northeastern Rio Grande do Sul. The coastal forest extends from Rio Grande do Norte (with isolated patches northwestward in Ceara and Piauí) southward through Santa Catarina. The average annual temperature declines from 25–26° C in the north to 20–21° C in the south. Rainfall is highest in southern Bahia (1850–2000 mm annually) and in the south (1500–1900 mm) with lower rainfall centrally from Santos to Rio de Janeiro (860–1450 mm). There is a pronounced dry season of 1–4 months in the north, but rainfall is more evenly distributed in the south. The coastal ranges receive 2000–2400 mm of rain, with about 75% of the rain falling in November–April; inland the rainfall is less, and the majority falls in September–March. During the winter, temperatures drop below 0° C at the highest elevations. The coastal region supports a unique type of vegetation, the maritime scrub formation (*restinga*). This formation consists of abundant rosette-like and creeping plants on the beaches and xerophytic scrub with many bromeliads and cacti on the dunes; commonly brackish lagoons are present inland from the dunes. Elsewhere, the narrow coastal plain supported humid tropical rainforest (now mostly destroyed); this forest was continuous with the humid evergreen forest at elevations of 200–1800 m on the coastal slopes of the maritime ranges of the highlands. Inland there is a more open woodland with grassland predominating at elevations above 1800 m. The western edge of the region blends into the cerrados and southwestward into the Mesopotamian region.

Llanos

The open formations of the Orinoco drainage extending from near the mouth of the Río Orinoco westward nearly to the Andes in Colombia are collectively referred to as the Llanos, which vary from open grasslands (a fire climax) to savanna, and gallery forest (Beard, 1953; Ramia, 1967). Most of the region, which encompasses an area of about 575,000 km², receives less than 1500 mm of rain annually, and nearly all of that falls between April and November. However, up to 4000 mm of rain falls annually in parts of the western Llanos in Colombia, but the sandy soil prevents the accumulation of water.

Cerrado-Caatinga-Chaco

An extensive subhumid to semiarid region encompassing an area of about 4,655,000 km² and mostly below 500 m above sea level extends through the interior of the continent from northeastern Brazil to northern Argentina. The semiarid region in northeastern Brazil, Caatinga, is characterized by high temperatures and low rainfall (300–800 mm annually), mostly falling during a period of 1–3 months. Most trees are deciduous, and there are many cacti and locally small arboreal bromeliads. Continuous with the Caatinga is a more extensive area of mixed savanna and open forest formations, Cerrado, that extends from just south of the Rio Amazonas in eastern Brazil southward to south-central Brazil. Rainfall varies from about 1,600 mm annually in the north to about 1,400 mm annually in the south, where two dry seasons (January–February, June–August) are characteristic, in contrast to August–November being the dry months to the north. During the rainy season, especially in the south, extensive areas are inundated; the most extensive of these, the pantanal is in Mato Grosso and Mato Grosso do Sul in southern Brazil. Fires are common in the cerrados, and many fire-resistant trees are in the grasslands. Most of the nonpalm species of trees are deciduous. South of the Cerrado and west of the Río Paraguay, the seasonally arid Chaco extends from southern Bolivia through Paraguay to north-central Argentina. Rainfall is low (750–900 mm annually) and concentrated in a short rainy season (December–February). Although temperatures are high during most of the year, in the southern part of the Chaco, nighttime temperatures may drop below 0° C in May–September. Vegetation consists of a mixture of grassland with scattered deciduous trees to dense thorn forest and primarily evergreen riparian forest.

Temperate Lowlands

Pampean-Monte

This region encompasses the loosely defined Oriental and Mesopotamian subprovinces of Lutz (1972) and the regions of the Mesopotamian and Monte batrachofaunas of Cei (1980). The region includes Uruguay and Rio Grande do Sul (Brazil) and extends westward to the Andes and southward to the Río Colorado in Argentina. Most of the area of 1,506,000 km² is at elevations below 500 m, but peaks in some extra-Andean cordilleras, such as the

Sierra de San Luis and Sierra Grande de Córdoba in Argentina, exceed 2000 m. In the northeast, the climate is subtropical with up to 30 days of frost in the winter; rainfall varies from 1250 to 1750 mm annually with no month receiving less than 30 mm. South of the Río de la Plata the climate is temperate and characterized by moderately low rainfall, most of which falls between November and March; freezing temperatures during June–August are common, especially in the south. The western part of the region receives less rainfall; nearly all of it falls in October–April. In the northeast, the mesophytic vegetation consists of semideciduous forests, wet savanna, and grassland. The latter is characteristic of eastern Argentina from the Río de la Plata to the Río Colorado, where deciduous trees form narrow gallery forests along rivers. In the western part of the region, the vegetation tends to be deciduous thorn forest.

Patagonia

Extending from southern Mendoza on the west and from the Río Colorado on the east to the Straits of Magellan, this large cis-Andean region (998,000 km^2) varies in elevation from about 2000 m at the base of the Andes to sea level to the east and south. Much of the region consists of eroded plains dissected by rivers draining the Andean slopes, but there are many basaltic tablelands and numerous lakes, especially in the western part. Patagonia is characterized by severe seasonal drought with five cold winter months and cool dry summers. Vegetation consists of grasses, a variety of low bushes, and some creeping plants; to the south small trees are present along rivers.

Austral Temperate Forest

This narrow region between the Pacific Ocean and elevations in excess of 1000 m in the Andes extends from 36° S to the tip of the continent. In a few areas the forests penetrate low passes in the Andes and enter the eastern slopes of the Andes in Argentina. The region is characterized by many cascading rivers and, especially in the south, numerous lakes. Climatically, the region is cool and humid. There is a gradual decline in mean annual temperature from about 13° C in the north to 6° C in the south. Rainfall may exceed 4000 mm annually in some areas. Throughout the region, most of the forests are dominated by various species of *Nothofagus*, whereas at higher elevations the coniferous

Araucaria and *Fitzroya* are common. To the north there is a transition zone with the Atacama Desert; this zone is characterized by grasses and legumes, especially *Acacia caven*.

Atacama

One of the driest regions of the world, the Atacama Desert stretches from central Chile (30° S) to northern Peru and includes the narrow Pacific coastal region and the western slopes of the Andes to elevations of about 2000 m. Temperature is moderated by the effects of the cold Humboldt Current that flows northward along the coast, so that average monthly temperatures vary between 3° C and 5° C. Although coastal fog may provide some moisture, there are areas in the Atacama Desert where rainfall has never been recorded. Extensive areas of this desert are completely devoid of plant life, whereas in some places on the Andean slopes where fog banks provide moisture low bushes and herbs exist. The desert is traversed by several small rivers, along which there are narrow borders of riparian vegetation—grasses, bushes, and small trees, mostly willows (*Alnus*). In northern Peru and extreme southern Ecuador there is a gradual transition to the humid tropical rainforests of the Chocoan Region; this transition zone consists of scattered scrubby trees in the south to tropical dry forest and tropical semideciduous forest northward.

HIGHLANDS

Andes

By far the dominant physiographic feature of the continent is the vast mountain system that extends for nearly 8000 km along the northern and western edges of the continent and encompasses an area of about 1,751,000 km^2. The Andes are divided into six major subregions, which are discussed from northeast to south; the numbered paragraphs correspond to numbered regions in Figure 5:4.

1. The Cordillera de la Costa parallels the Caribbean coast of Venezuela. The main coastal range rises abruptly from the narrow coastal lowlands to elevations in excess of 2000 m (highest point 2765 m); the broader interior highlands to the south have few peaks over 1000 m. The isolated easternmost range in the Paria Peninsula reaches heights of only slightly more than 800 m, whereas the disjunct Cerro Turumiquire rises to 2630 m. Rainfall is highly variable and lowest in November–April; more than 2000 mm falls on the coastal slopes. Cloud forest

dominates on the slopes at elevations of more than 800 m on the coastal slopes of the coastal range, 600 m in the Serranía de Paria, and 1500 m Cerro Turumiquire; the inland slopes of the interior highlands support tropical dry forest.

2. The Venezuelan Andes have a length of about 400 km and breadth of about 100 km. Pico Bolívar at 5002 m is the highest peak. The Venezuelan Andes are separated from the Cordillera Oriental of the northern Andes by the Depresión de Cúcuta with elevations of less than 600 m. The tropical habitat, cloud forest, begins at elevations of 600–1000 m on the northern slopes and 1200–1700 m on the southern slopes and extends to elevations of 2400–2900 m. The alpine Venezuelan Andean region is composed of a series of islands of subparamo and páramo at lower elevations of 2400–2900 and extending to snow line at 4600–4900 m.

3. In northern Colombia, the Sierra Nevada de Santa Marta rises abruptly from the Caribbean coast to elevations in excess of 5000 m. The lower slopes are incorporated in the dry Caribbean coastal forest, whereas cloud forest occurs at elevations of 1300–2700 m. Above treeline, páramo vegetation extends to snowline at about 4900 m.

4. The northern Andes have a latitudinal extent of about 1800 km from the Caribbean lowlands at 10°50' N to the Huancabamba Depression at 4°30' S. As described by Duellman (1979b), the northern Andes are made up of five major north-south ranges diverging from the Nudo de Pasto in southern Colombia and adjacent Ecuador. North of the Nudo de Pasto are three ranges. The western range, the Cordillera Occidental, extends for about 650 km; this narrow range (< 50 km) between the Pacific lowlands and the valley of the Río Cauca lacks continuous high ridges, and only three peaks exceed 4000 m. The Cordillera Central extending north for about 750 km from the Nudo de Pasto has a width of about 100 km; it is bordered on the west by the valley of the Río Cauca and the east by the valley of the Río Magdalena. Extensive areas are above 3000 m, and four peaks with permanent snow exceed 5000 m. The Cordillera Oriental extends north-northeast from the Nudo de Pasto for about 1200 km and reaches widths of 200 km. In the southern three-fourths of the cordillera there are large areas above 3000 m. South of the Nudo de Pasto, the Cordillera Occidental of Ecuador extends southward for about 500

km. The elevation of the southern two-thirds of the cordillera exceeds 3000 m, and many volcanoes in the cordillera exceed 4500 m. The Cordillera Oriental of Ecuador extends south from the Nudo de Pasto for about 620 km to the Huancabamba Depression. This cordillera consists of many snow-covered volcanoes separated by deep river valleys. Between the eastern and western cordilleras in Ecuador are 10 intermontane basins at elevations of 2000–3100 m that are completely or partially separated by transverse ridges connected or not with the eastern and western cordilleras. The tropical northern Andes range from elevations of 800–1000 m to 2500–2900 m on the Pacific slopes and on the eastern slopes of Cordillera Oriental, where luxuriant cloud forests are present. These humid montane forests also occur locally in the Cordillera Central. At higher elevations (usually above 3000 m) to snow line, the alpine habitat is páramo with annual precipitation of 1000–2000 mm, little seasonal fluctuation in temperature, but daily variation of 10° C or more.

5. The Huancabamba Depression is a complex system of relatively low mountain ranges and basins; in the depression the major cordilleras either terminate or are fragmented into isolated ranges usually less than 3500 m high and separated by valleys mostly between 1000 and 2000 m above sea level.

6. The southern Andes extend from the northern part of the Huancabamba Depression in northern Peru southward to the northern limits of the austral forests in the Andes of Chile and Argentina. The Cordillera Occidental forms the backbone of the Andes south of the Huancabamba Depression; it extends southward for about 2800 km with only one pass below 4000 m. The eastern Cordilleras in Peru and Bolivia are made up of many high ranges separated by long north-south valleys. Large areas in the southern Andes are at elevations above 3000 m, which usually is about treeline on the eastern slopes; most of the western slopes are treeless or have only scrubby xerophilous trees. The alpine southern Andes are above treeline. The high elevations in the Cordillera Occidental are dry and support low xerophytic vegetation, whereas the supra-treeline habitats in the eastern cordilleras are wet páramo in the north and usually dry grassland (puna) to the south. The tropical southern Andes are characterized by cloud forest primarily on the Amazonian slopes at elevations of 1000–2800 m from the Huancabamba

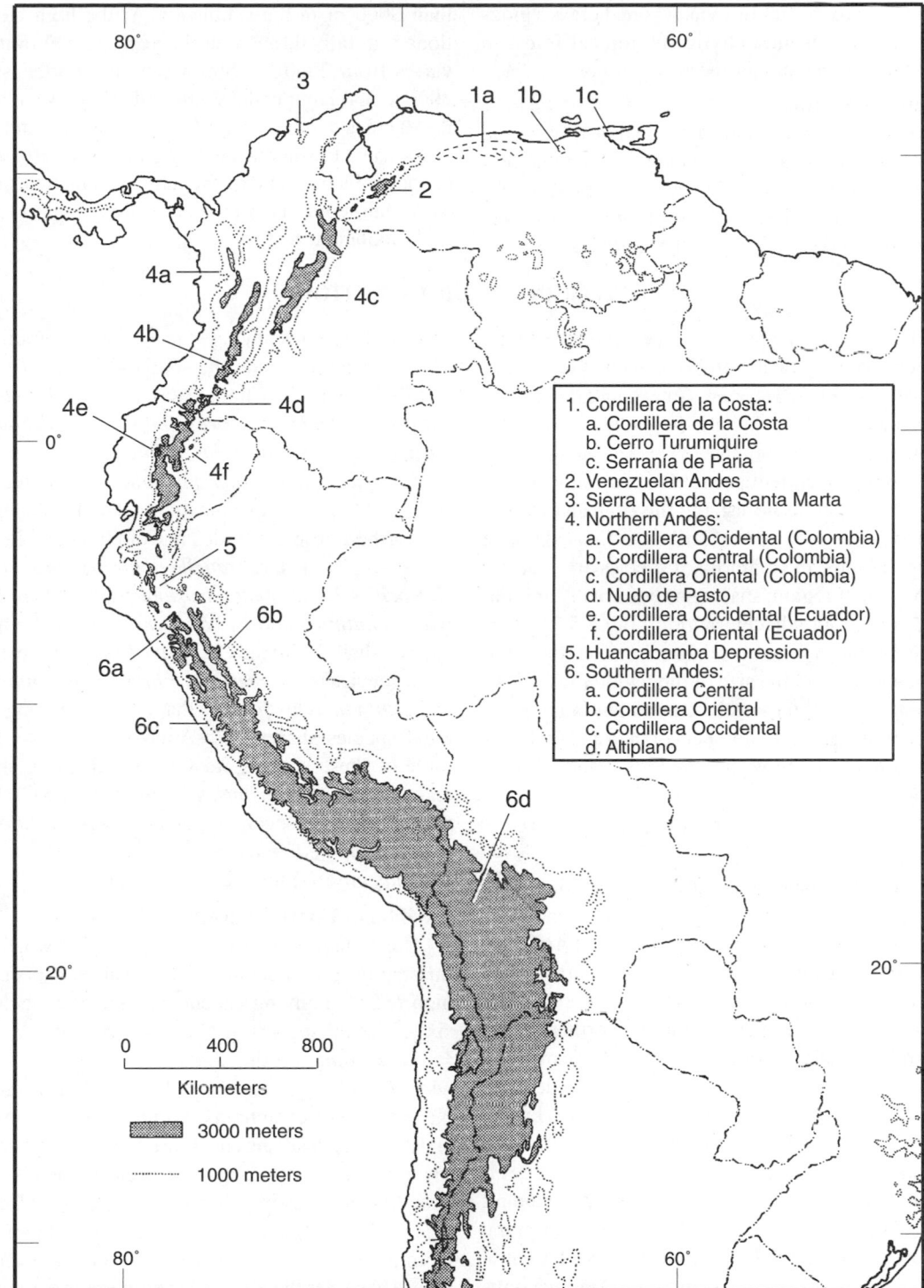

80°   60°

3   1a 1b 1c

2

4a

4c

4b

4e   4d

4f

1. Cordillera de la Costa:
   a. Cordillera de la Costa
   b. Cerro Turumiquire
   c. Serranía de Paria
2. Venezuelan Andes
3. Sierra Nevada de Santa Marta
4. Northern Andes:
   a. Cordillera Occidental (Colombia)
   b. Cordillera Central (Colombia)
   c. Cordillera Oriental (Colombia)
   d. Nudo de Pasto
   e. Cordillera Occidental (Ecuador)
   f. Cordillera Oriental (Ecuador)
5. Huancabamba Depression
6. Southern Andes:
   a. Cordillera Central
   b. Cordillera Oriental
   c. Cordillera Occidental
   d. Altiplano

5

6b

6a

6c

6d

0°

20°   20°

0   400   800
Kilometers

▓▓▓ 3000 meters
······· 1000 meters

80°   60°

Fig. 5:4. Subregions of the Andes.

Depression to central Bolivia, where there is a gradual transition to mesophytic subtropical forest in southern Bolivia and northern Argentina.

Guiana Highlands

Scattered from central Suriname to extreme southern Venezuela are tablelands (tepuis) that presumably are remnants of a Mesozoic plateau, but some are granitic. The tops of these mountains range from only about 1000 m in the northeast to more than 3000 m in the southwest. At the high elevations rain falls throughout the year; annual rainfall varies from 2000 to 2500 mm in the northeast to 3000 to 3300 mm in the southwest. Shallow soils on the flat tops of many of these mountains support only grasses and low herbaceous vegetation, whereas the deeper soils on some of the larger mountains, such as Auyán-tepui and Cerro de Neblina support humid montane forest.

PATTERNS OF DISTRIBUTION

The number of species and percentage of endemism in each region are highly variable (Table 5:2). The Andes with 753 species, Amazonia-Guiana with 335 species, and the Atlantic Forest Domain with 334 have the highest number of species, but the greatest amount of endemism (100%) is found in the Austral Temperate Forest and the Atacama Desert. With the exception of the Caribbean Coastal Forest and the Llanos, the majority of species are endemic to a particular region; shared species usually are continuous with an adjacent region (Table 5:3). Furthermore, there is no correlation between the number of species and the size of the region (Table 5:4). Herein, for each region the amphibian fauna is assessed with respect to numbers of taxa and their patterns of distribution within the region and adjacent regions.

TROPICAL LOWLANDS

Chocó

The amphibian fauna of the Chocó consists of 126 species, of which only 42.1% are endemic. Among the endemics are many species of *Atelopus, Colostethus, Dendrobates, Eleutherodactylus,* and centrolenids; many of these seem to be related to species in Central America or in the northern tropical Andes. Other endemics are trans-Andean vicariants of groups otherwise known in cis-Andean regions; examples are *Ceratophrys stolzmanni* (Lynch, 1982), *Physalaemus coloradorum* and *P. pustulatus* (Cannatella and Duellman, 1984), and *Trachycephalus jordani* (Trueb, 1970). Sixty species (53 anurans, 4 salamanders, 3 caecilians) are shared with Central America. Some of these are members of Central American groups that exist in South America only, or mostly, in the Chocoan Region; examples are *Oedipina* (Wake and Lynch, 1976), *Agalychnis* (Duellman, 1970), and the *Craugastor* clade of *Eleutherodactylus* (Lynch, 1986a).

Many species having extensive distributions in Central America reach the southern limits of their distributions in the Chocó; examples are *Bufo haematiticus, Centrolene prosoblepon, Dendrobates auratus, Eleutherodactylus fitzingeri, Hyla ebraccata, Nelsonophryne aterrima,* and *Rana vaillanti.* At least 14 species also inhabit the Pacific slopes of the northern tropical Andes in Colombia and Ecuador; examples are *Cochranella ramirezi, Colostethus chocoensis, Dendrobates histrionicus, Eleutherodactylus achatinus,* and *E. caprifer.* Among the nine species shared with the Caribbean Coastal Forest are three species (*Bufo marinus, Phrynohyas venulosa,* and *Leptodactylus fuscus*) that also are among the eight species shared with Amazonia-Guiana. All three of these species are shared with the Llanos, and all but *Bufo marinus,* with the Cerrado-Caatinga-Chaco. Four species (*Atelopus elegans, Bufo typhonius, Epipedobates boulengeri,* and *Eleutherodactylus gularis*) inhabit Isla Gorgona.

Caribbean Coastal Forest

Less than half of the 47 species known from this region are endemic. Endemism is especially high (81.8%) among caecilians; endemics include six terrestrial species of *Caecilia, Oscaecilia,* and *Parvicaecilia,* and three aquatic species of *Nectocaecilia* and *Typhlonectes.* Salamanders are absent. No one group of anurans is represented by many endemics. Instead, the endemic anurans mostly seem to be vicariants of groups that occur in the cis-Andean lowlands; examples are *Ceratophrys calcarata* (Lynch, 1982), *Hyla luteoocellata* (Duellman and Crump, 1974), and *Pipa parva* (Trueb and Cannatella, 1986). Of the 27 species of anurans occurring in other regions, most are widespread species (e.g., *Bufo marinus, Hyla crepitans, Phrynohyas venulosa,* and *Leptodactylus fuscus*). Of these species, 16 are shared with dry regions of Central America, 17 are

Table 5:2. Species of amphibians in 12 regions in South America. Numbers in each column are anurans + salamanders + caecilians = total.

Region	Total species	Shared species	Endemic species	Percent endemic
Chocó	109 + 7 + 10 = 126	65 + 4 + 6 = 73	46 + 3 + 4 = 53	42.2 + 42.9 + 40.0 = 42.1
Caribbean Coastal Forest	36 + 0 + 11 = 47	27 + 0 + 2 = 29	9 + 0 + 9 = 18	25.0 + 00.0 + 81.8 = 38.3
Amazonia-Guiana	305 + 2 + 28 = 335	54 + 1 + 3 = 58	249 + 1 + 25 = 275	81.6 + 50.0 + 89.3 = 82.6
Atlantic Forest Domain	322 + 0 + 12 = 334	23 + 0 + 1 = 24	300 + 0 + 10 = 310	92.9 + 00.0 + 90.9 = 92.8
Llanos	30 + 0 + 1 = 31	23 + 0 + 1 = 24	7 + 0 + 0 = 7	23.3 + 00.0 + 00.0 = 22.5
Cerrado-Caatinga-Chaco	98 + 0 + 1 = 99	46 + 0 + 0 = 46	52 + 0 + 1 = 53	53.1 + 00.0 + 100.0 = 53.5
Pampean-Monte	83 + 0 + 2 = 85	37 + 0 + 1 = 38	46 + 0 + 1 = 47	55.4 + 00.0 + 50.0 = 55.3
Patagonia	10 + 0 + 0 = 10	2 + 0 + 0 = 2	8 + 0 + 0 = 8	80.0 + 00.0 + 00.0 = 80.0
Austral Temperate Forests	32 + 0 + 0 = 32	0 + 0 + 0 = 0	32 + 0 + 0 = 32	100.0 + 00.0 + 00.0 = 100.0
Atacama	5 + 0 + 0 = 5	0 + 0 + 0 = 0	5 + 0 + 0 = 5	100.0 + 00.0 + 00.0 = 100.0
Andes	720 + 16 + 17 = 753	37 + 1 + 0 = 38	683 + 15 + 17 = 715	94.9 + 93.8 + 100.0 = 95.0
Guiana Highlands	76 + 0 + 0 = 76	5 + 0 + 0 = 5	71 + 0 + 0 = 71	93.7 + 00.0 + 00.0 = 93.7

Table 5:3. Distribution of species of amphibians in 12 regions in South America. Abbreviations in headings to columns correspond to regions in first column. The number of species in each region is shown in boldface in common cell; the numbers of species that are in common to two regions are in the upper right, and the Coefficient of Biogeographic Resemblance is in italics in the lower left.

Region	CHO	CCF	A-G	AFD	LLA	CCC	PAM	PAT	ATF	ATA	AND	GUH
Chocó (CHO)	**126**	9	8	0	3	2	0	0	0	0	13	0
Caribbean Coastal Forest (CCF)	0.10	**47**	17	4	14	8	1	0	0	0	2	1
Amazon-Guiana (A-G)	0.03	0.09	**335**	9	21	19	5	0	0	0	22	5
Atlantic Forest Domain (AFD)	0.00	0.02	0.03	**334**	2	13	17	0	0	0	0	1
Llanos (LLA)	0.04	0.36	0.11	0.01	**31**	9	1	0	0	0	1	1
Cerrado-Caatinga-Chaco (CCC)	0.02	0.12	0.09	0.06	0.14	**99**	28	1	0	0	1	1
Pampean-Monte (PAM)	0.00	0.02	0.02	0.08	0.02	0.30	**85**	1	0	0	1	0
Patagonia (PAT)	0.00	0.00	0.00	0.00	0.00	0.02	0.02	**10**	0	0	1	0
Austral Temperate Forest (ATF)	0.00	0.09	0.00	0.00	0.00	0.00	0.00	0.00	**32**	0	0	0
Atacama (ATA)	0.00	0.00	0.00	0.00	0.00	0.00	0.00	0.00	0.00	**5**	0	0
Andes (AND)	0.03	0.01	0.04	0.00	0.01	0.01	0.01	0.01	0.00	0.00	**753**	2
Guiana Highlands (GUH)	0.00	0.02	0.02	0.01	0.02	0.01	0.00	0.00	0.00	0.00	0.01	**76**

Table 5:4. Area:species relationships in different natural regions in South America.

Region	Area 10^6 km^2	Percent total Area	No. of species	Percent species	Species/ 10^6 km^2	Rank
Chocó	0.194	1.092	126	7.23	649.5	2
Caribbean Coastal Forest	0.176	0.098	47	2.70	267.0	5
Amazonia-Guiana	5.960	33.497	335	19.23	56.2	8
Atlantic Forest Domain	1.150	6.46	334	19.17	290.4	4
Llanos	0.575	3.232	31	1.80	53.9	9
Cerrado-Caatinga-Chaco	4.655	26.161	99	5.68	21.3	10
Pampean-Monte	1.506	8.463	85	4.90	56.4	7
Patagonia	0.998	0.989	10	0.06	10.0	12
Austral Temperate Forest	0.349	1.961	32	1.84	91.7	6
Atacama	0.297	1.700	5	0.03	16.8	11
Andes	1.751	9.839	753	43.21	430.0	3
Guiana Highlands	0.017	0.009	76	4.36	4470.1	1

shared with Amazonia-Guiana, and 14 with the Llanos. *Hyla pugnax* is shared with humid forests in Central America (La Marca, 1996a).

Amazonia-Guiana

This immense region mostly supporting humid tropical rainforest contains 335 species (305 anurans, 2 salamanders, and 28 caecilians); one of the salamanders, 25 of the caecilians, and 249 of the anurans are endemic to the region. Thirteen genera of anurans and two genera of caecilians are endemic to the region; these are (number of species in parentheses): *Allophryne* (1), *Nyctimantis* (1), *Scarthyla* (1), *Edalorhina* (2), *Hydrolaetare* (1), *Lithodytes* (1), *Phyzelaphryne* (1), *Vanzolinius* (1), *Adelastes* (1), *Altigius* (1), *Hamptophryne* (1), *Synapturanus* (3), *Syncope*(2), *Brasilotyphlus* (1), and *Rhinatrema* (1). The 305 species of anurans are distributed among 50 genera, but 191 species (62.6%) belong to six genera—*Hyla* (57 species), *Eleutherodactylus* (53), *Leptodactylus* (29), *Scinax* (20), *Epipedobates* (17), and *Dendrobates* (15)

More than one-half (58.1%) of the species in the northeastern (Guianan) part of the region also occur in the Amazon Basin; these include many widespread species, such as *Bufo marinus, Hyla boans, H. geographica, Phrynohyas venulosa, Leptodactylus pentadactylus,* and *Pseudis paradoxa.* But many of the Guianan species (e.g., *Hyla leucophyllata, Phyllomedusa bicolor, Eleutherodactylus zeuctotylus,* and *Lithodytes lineatus*) are shared only with the Amazon Basin.

Although some of the endemic species, such as *Hyla parviceps, Ceratophrys cornuta, Eleutherodactylus fenestratus,* and *Pipa snethlageae,* are widely distributed in the region, most endemics are more localized, as are some species not endemic to the region. By far the largest number of endemics are in the western part of the Amazon Basin, especially in Ecuador and Peru. One salamander (*Bolitoglossa equatoriana*), ten caecilians (e.g., *Caecilia dissossea, Microcaecilia albiceps, Oscaecilia bassleri,* and *Nectocaecilia petersii*), and 108 anurans are endemic to the western Amazon Basin; endemic anurans include many species of dendrobatids and hylids, and several species of *Eleutherodactylus,* other leptodactylids, and microhylids.

In the northeastern (Guianan) part of the region, 48 species (39 anurans, 9 caecilians) are endemic. The endemics are forest-inhabitants and include *Atelopus franciscus,* three species of centrolenids, three dendrobatids, seven hylids, *Adelophryne gutturosa, Adenomera lutzi,* four species of *Eleutherodactylus,* three of *Leptodactylus,* two species of microhylids, and nine species of caecilians, including three of the five species of *Microcaecilia.* Five anurans (*Bufo anderssoni, Colostethus stepheni, Hyla imitator, H. minima,* and *H. pauiniensis*) and one caecilian (*Typhlonectes cunhai*) are endemic to the central part of the basin. Five anurans (*Bufo castaneoticus, Dendrobates castaneoticus, D. galactonotus, Hyla anataliasiasi,* and *H. inframaculata*) and two caecilians (*Nectocaecilia ladigesi* and *Typhlonectes obesus*) are endemic to the eastern part of the Amazon Basin, and five anurans (*Colostethus sanmartini, Aparasphenodon venezolanus, Hyla tintinnabulum, Scinax baumgardneri,* and *Leptodactylus lithonae-tes*) are endemic to the northern part of the basin.

In the western part of the Amazon Basin, 21 species ascend the slopes of the Andes and inhabit montane rainforest at elevations in excess of 1000 m.

Fourteen of these species are shared with the northern tropical Andes (north of the Huancabamba Depression); examples are *Hemiphractus scutatus, Eleutherodactylus lanthanites, E. nigrovittatus, E. sulcatus, Ischnocnema quixensis,* and *Bolitoglossa altamazonica.* Only seven species have ranges including the upper Amazon Basin and the Andean slopes south of the Huancabamba Depression—the Southern Tropical Andes; examples are: *Eleutherodactylus cruralis, E. peruvianus, E. rhabdolaemus,* and *Bolitoglossa altamazonica.*

Of the species inhabiting the nonforest environments in the northeastern (Guianan) part of the region, 21 are shared with the Llanos, 17 with the Caribbean Coastal Forest, and 17 with the Cerrado-Caatinga-Chaco. Eleven species (*Bufo granulosus, B. marinus, Hyla crepitans, H. microcephala, Phrynohyas venulosa, Scinax rostrata, S. rubra, Leptodactylus bolivianus, L. pentadactylus, Physalaemus pustulosus,* and *Pleurodema brachyops*), most of which inhabit nonrainforest environments, are widely distributed and occur in Central America. All of these, except *Leptodactylus pentadactylus,* also occur in the Llanos and in the Caribbean Coastal Forest, and four of these species (*Bufo granulosus, Hyla crepitans, Phrynohyas venulosa,* and *Leptodactylus bolivianus*) also occur in the Cerrado-Caatinga-Chaco.

Ten species occur both in Amazonia-Guiana and the Atlantic Forest Domain; these are *Hyla albomarginata, H. crepitans, H. geographica, H. microcephala, H. minuta, Scinax x-signata, S. rubra, Adenomera hylaedactyla, Leptodactylus mystaceus,* and *Siphonops annulatus.* Five species (*Scinax fuscomarginata, Leptodactylus ocellatus, Lysapsus limellus, Pseudis paradoxa,* and *Siphonops annulatus*) also occur in the Pampean-Monte region. Two hylid frogs (*Hyla crepitans* and *H. ornatissima*), one centrolenid (*Cochranella oyampiensis*), and one microhylid (*Otophryne robusta*) are shared with the Guiana Highlands.

The continental islands of Trinidad and Tobago have 31 species of anurans, 10 of which are present on both islands; one of these (*Flectonotus fitzgeraldi*) is shared with the northeasternmost part of the Cordillera de la Costa of the Andes in northern Venezuela. Trinidad has an additional 17 species of anurans, including three endemics—*Mannophryne trinitatis, Phyllodytes auratus,* and *Leptodactylus nesiotus; Phyllomedusa trinitatis* also occurs in the Caribbean Coastal Forest. Tobago has an additional four species; *Mannophryne olmonae* is endemic and *Hyalinobatrachium orientalis, Eleutherodactylus rozei,* and *E. terraebolivaris,* which also occur in the Cordillera de la Costa of the Andes in northern Venezuela. Twenty-one species are shared with Amazonia-Guiana Lowlands; 14 of these also are shared with Caribbean Coastal Forest. Most of these are widespread species, some of which (e.g., *Bufo granulosus* and *Pseudis paradoxa*) extend southward into the Chaco. Two species on Trinidad and Tobago, *Leptodactylus validus* and the widespread *Scinax rubra,* are the only South American species that also occur in the Lesser Antilles.

Atlantic Forest Domain

Of the 334 species (322 anurans, 12 caecilians), 310 species (300 anurans, 10 caecilians) are endemic. Of the endemic species, 80 (25.8%) constitute 22 endemic genera (number of species in parentheses)—*Brachycephalus* (2), *Psyllophryne* (1), *Frostius* (1), *Hylomantis* (2), *Phasmahyla* (4), *Phrynomedusa* (3), *Crossodactylodes* (3), *Cycloramphus* (25), *Hylodes* (18), *Macrogenioglottus* (1), *Megaelosia* (2), *Paratelmatobius* (3), *Scythrophrys* (1), *Thoropa* (5), *Arcovomer* (1), *Dasypops* (1), *Hyophryne* (1), *Myersiella* (1), *Stereocyclops* (1), *Luetkenotyphlus* (1), and *Mimosiphonops* (1). Most species in two other genera are endemic to the region; six (5 endemic) of the seven species of *Crossodactylus* occur there, as do all six species of *Siphonops,* but two species also occur in the Amazonia-Guiana and Pampean-Monte. Speciose genera with many endemics include *Hyla* (50 endemic species), *Scinax* (41), *Eleutherodactylus* (26), *Physalaemus* (16), and *Leptodactylus* (11). In contrast to Amazonia-Guiana, there arc few species of bufonids (8 species), centrolenids (3 endemic *Hyalinobatrachium*), and dendrobatids (4 endemic *Colostethus*).

The narrow coastal region of eastern Brazil is inhabited by 168 species of anurans and 10 of caecilians; about 90% of the species are endemic to the region. Local endemism is common, and many species are restricted to the southern part of the region (Rio de Janeiro to Santa Catarina) and others in the northern part of the region (principally Espírito Santo and Bahia). For example, 74 anurans and seven caecilians are restricted to the southern part, which is characterized by many *Hyla* and *Scinax,* plus *Aplastodiscus, Gastrotheca microdisca, Hyalinobatrach-*

ium, brachycephalids and most coastal species of *Siphonops.* Two caecilians and 68 anurans are restricted to the northern part of the coastal region, which is home to endemic genera, such as *Frostius, Dasypops, Hylomantis,* and *Stereocyclops,* and three of the six species of *Adelophryne,* plus many species of leptodactylids and hylids, including most coastal species of *Phyllodytes* and *Proceratophrys.* However, some species (e.g., *Macrogenioglottus alipioi, Aparasphenodon brunoi, Hyla berthalutzae,* and *Osteocephalus langsdorffii*) extend throughout most of the coastal region. A few species (e.g., *Elachistocleis piauiensis*) are restricted to isolated patches of mesic forest to the northwest of the coastal forest.

The amphibian fauna of highlands (coastal ranges and interior highlands) consists of 154 species of anurans, of which 142 are endemic. The largest number of endemics are leptodactylids (80 species), of which the following genera are endemic (or nearly so): *Crossodactylodes, Cycloramphus, Hylodes, Megaelosia, Paratelmatobius, Scythrophrys,* and *Thoropa.*

The few species that are shared include wide-ranging taxa, such as *Hyla albomarginata, H. geographica, H. microcephala, H. minuta, Adenomera hylaedactyla,* and *Siphonops annulatus,* which also occur at least in Amazonia-Guiana. The greatest number of species (17) are shared with the Pampean-Monte Region; these include *Bufo crucifer, B. paracnemis, Aplastodiscus perviridis, Hyla albopunctata, H. claresignata, H. faber, H. pardalis, H. semiguttata, Osteocephalus langsdorffii, Crossodactylus dispar, Leptodactylus geminus, L. mystaceus, L. mystacinus, Physalaemus albifrons, P. cuvieri, P. centralis,* and *Siphonops annulatus.* Thirteen species are shared with the Cerrado-Caatinga-Chaco; these include *Bufo paracnemis, Hyla albopunctata, H. crepitans, H. melanargyrea, H. pardalis, Phyllomedusa burmeisteri, Scinax x-signata, Trachycephalus nigromaculatus, Adenomera hylaedactyla, Leptodactylus geminus, L. labyrinthicus, L. mystaceus,* and *Physalaemus albifrons.* Four species (*Hyla crepitans, H. microcephala, Scinax rubra,* and *S. x-signata*) are shared with the Llanos.

Llanos

This extensive region of mostly grassland and savanna has a depauperate amphibian fauna consisting of only 31 species (30 anurans, 1 aquatic caecil-

ian). The seven endemic anurans are *Colostethus ranoides, Hyla mathiassoni, Scinax blairi, S. kennedyi, S. wandae, Physalaemus enesefae,* and *Pseudopaludicola llanera* (Rivero-Blanco and Dixon, 1979; Lynch, 1989). Most of these are restricted to the western part of the Llanos in Colombia. Most of the 23 nonendemic anurans are shared with the nearby tropical lowlands—Amazonia-Guiana (21) and Caribbean Coastal Forest (14), but some others are widespread in tropical dry forests and savannas. Nine species are shared with the Cerrado-Caatinga-Chaco, and four of these (*Bufo granulosus, Phrynohyas venulosa, Phyllomedusa hypocondrialis,* and *Pseudis paradoxa*) extend to the southern part of that region.

Cerrado-Caatinga-Chaco

Salamanders are absent, and caecilians are represented by one aquatic species, *Chthonerpeton perisodus.* Of the 98 species of anurans, 52 are endemic. Among these are four endemic genera—*Corythomantis* (*C. greeningi* in the northeast), *Dermatonotus* (*D. muelleri* in the central and south), *Chacophrys pierotti* and the three species of *Lepidobatrachus* in the southern Chacoan region. Otherwise, most endemic anurans are members of widespread genera; examples are *Bufo ocellatus, Hyla alvarengai, H. oliveirai, H. varelae, Phyllodytes tuberculosus, Phyllomedusa centralis, P. sauvagii, Scinax pachychrus, S. trachythorax, Trachycephalus atlas, Ceratophrys cranwelli, C. joazeirensis, Leptodactylus bufonius, L. laticeps, Odontophrynus carvalhoi, Physalaemus albonotatus, Pleurodema diplolistre, Proceratophrys cristiceps,* and *Chiasmocleis centralis.* The enigmatic leptodactylid genus *Barycholos* is represented by one endemic species (*B. savagei*); the other member of the genus (*B. pulcher*) is endemic to the Chocó. Centrolenids are absent, and dendrobatids are represented by two species, *Colostethus goianus* (endemic) and *Epipedobates flavopictus.* Likewise, only a single species of *Eleutherodactylus* (the endemic *E. heterodactylus*) is present.

Nineteen species are shared with Amazonia-Guiana; some of these are widespread species (e.g., *Bufo granulosus, Leptodactylus bolivianus,* and *Pseudis paradoxa*) that are among the eight species shared with the Caribbean Coastal Forest and Llanos, whereas others (e.g., *Hyla crepitans, Scinax x-signata,* and *Leptodactylus mystaceus*) are among the 13 species shared with the Atlantic Forest Domain. The largest number of species (28) is shared with the Pampean-Monte Region. Most of these are species that occur in the southern (Chacoan) part of the region and include *Bufo arenarum, Hyla nana, Scinax fuscomarginata, Leptodactylus chaquensis, Odontophrynus americanus, Physalaemus biligoniger, Pseudopaludicola falcipes,* and *Elachistocleis bicolor.*

Because of the long dry season and short, intense rainy season, anurans in this region are explosive breeders. As detailed by Gallardo (1979), most of the anurans have adaptations to xeric environments—foam nests (e.g., *Leptodactylus, Physalaemus,* and *Pleurodema*), burrowing (e.g., *Ceratophrys, Chacophrys, Lepidobatrachus, Odontophrynus,* and *Pleurodema*), aestivation in rodent burrows (e.g., *Bufo arenarum, Leptodactylus bufonius, L. chaquensis,* and *L. laticeps*), in termitaria (*Dermatonotus* and *Elachistocleis*), and in self-made cocoons (*Ceratophrys, Lepidobatrachus,* and *Phyllomedusa*). *Corythomantis, Trachycephalus,* and some of the other hylids pass the dry season in arboreal bromeliads.

TEMPERATE LOWLANDS

Pampean-Monte

Of the 83 species of anurans and two species of caecilians, 46 anurans and one caecilian (*Chthonerpeton indistinctum*) are endemic. The monotypic genera *Argenteohyla* and *Limnomedusa* are endemic to the region. Endemism is especially noteworthy in three genera; seven of the eight species of *Melanophryniscus,* seven of the 12 species of *Physalaemus ,* and all five species of *Pleurodema* in the region are endemic. Other endemics include four *Bufo,* eight *Hyla,* one *Phrynohyas,* one *Phyllomedusa,* four *Scinax,* one *Ceratophrys,* three *Odontophrynus,* three *Proceratophrys,* and *Pseudis minuta.* Centrolenids, dendrobatids, and *Eleutherodactylus* are absent.

The greatest number of shared species (28) is with Cerrados-Caatinga-Chaco, as discussed in the preceding paragraphs. The only other region with a high number of shared species is the Atlantic Forest Domain (16). The northeastern part of the Pampean-Monte Region is the southern terminus of distribution of several species characteristic of the southern part of the Atlantic Coastal Forest; these include *Aplastodiscus perviridis, Hyla claresignata, H. semiguttata, Osteocephalus langsdorffii, Crossodactylus dispar,* and *Physalaemus centralis.* The Pampean-Monte Region also is the southern limit of distribution of many tropical genera—*Hyla, Scinax, Physalaemus, Lysapsus,* and *Pseudis.* The *Hyla pulchella* group is shared with the Andes, and *Bufo arenarum* is shared with Patagonia.

Patagonia

Of the 10 species of anurans inhabiting this region, only two are shared with other regions; *Bufo spinulosus* is shared with the southern Andes, and *Bufo arenarum* is shared with the Pampean-Monte Region. Except for the widespread *Pleurodema bufoninum* and the enigmatic *Zachaenus roseus,* the endemic species are highly localized in aquatic environments and include five species of *Atelognathus* and *Somuncuria somuncurensis.*

Austral Temperate Forest

All 32 species of anurans are endemic, as are six of the telmatobiine leptodactylid genera (*Batrachyla, Caudiverbera, Eupsophus, Hylorina, Insuetophrynus,* and *Telmatobufo*) and the family Rhinodermatidae. The only widespread genera having representatives in this region are *Bufo* (2 species) and *Pleurodema* (1 species). *Alsodes* is shared with the Andean region, and *Atelognathus,* with Patagonia.

Atacaman

Only five species, all endemic, occur in isolated riparian situations in this desert. Four of the species are members of the genus *Bufo,* and the fifth is *Colostethus littoralis.*

HIGHLANDS

Andes

The entire range of the Andes Mountains and associated cordilleras are home to the largest number of species (753) and endemics (715) of any region on the continent. Of the 720 species of anurans, 683 (94.9%) are endemic, as are 93.8% of the 16 species of salamanders and all of the caecilians. Eleven genera containing 86 species are endemic to the Andes; these are: *Nephelobates* (5), *Andinophryne* (3), *Osornophryne* (5), *Truebella* (2), *Aromobates* (1), *Cryptobatrachus* (3), *Atopophrynus* (1), *Batrachophrynus* (2), *Geobatrachus* (1), *Phrynopus* (20), and *Telmatobius* (45). Many other genera of anurans are highly speciose in the Andes. Of the 55 South American species of *Atelopus,* 43 are endemic to the Andes. Likewise, 36 of 37 South American species of *Centrolene,* 46 of 61 *Cochranella,* 62 of 99

Colostethus, and 36 of 42 *Gastrotheca* are endemic. The most speciose genus in the Andes is *Eleutherodactylus* with 352 species, of which 241 are endemic.

The distribution of amphibians in the Andes is discussed with respect to five of the six subregions defined in the previous section; the Huancabamba Depression is included in the southern Andes. Moreover, distributions are discussed with respect to two major habitats—the tropical forests on the slopes (tropical Andes) and the supra-treeline habitats (alpine Andes) at high elevations.

Cordillera de la Costa: Of the 40 species of amphibians in this subregion (all tropical), 23 anurans and 1 salamander (*Bolitoglossa borburata*) are endemic. Most of the endemics are in the principal Cordillera de la Costa, but *Colostethus mandelorum* and *Eleutherodactylus turumiriquirensis* are restricted to Cerro Turumiquiri, and *Cochranella castroviejoi, C. vozmedianoi,* and *Mannophryne riveroi* is restricted to the mountains in the Península de Paria. The endemics include two species of *Cochranella,* three of *Hyalinobatrachium,* three of *Colostethus,* three of *Mannophryne,* five of *Eleutherodactylus,* two of *Gastrotheca,* and one each of *Atelopus, Bufo, Hyla, Phyllomedusa,* and *Bolitoglossa.* Three species (*Flectonotus fitzgeraldi, Eleutherodactylus rozei,* and *E. terraebolivaris*) are shared only with Trinidad and Tobago; *Hyla crepitans* also is shared with these islands, as well as the Guiana Highlands, Caribbean Coastal Forest, Llanos, and other nonforest regions in northeastern South America. *Hyalinobatrachium orientalis* and *Flectonotus pygmaeus* are shared with the tropical Venezuelan Andes; the former also is shared with the Guiana Highlands and the latter with the tropical northern Andes.

Venezuelan Andes: In the tropical Venezuelan Andes, only anurans constitute the 29 species in the region; 27 species are endemic. The endemics include three species of *Atelopus,* eight of centrolenids, the monotypic genus *Aromobates,* three species of *Colostethus,* one of *Mannophryne,* three of *Nephelobates,* two of *Hyla,* and six of *Eleutherodactylus. Flectonotus pygmaeus* and *Gastrotheca nicefori* are shared with the northern tropical Andes, and the former is shared with the Cordillera de la Costa.

Of the 17 species of anurans and one of salamander in the alpine Venezuelan Andean region, all except one anuran are endemic. The endemics include three species of *Atelopus,* four of *Colostethus,* two of *Nephelobates,* five of *Eleutherodactylus,* and one each of *Mannophryne, Hyla,* and *Bolitoglossa.* The only nonendemic, *Centrolene buckleyi,* also occurs in the Northern Andes.

Sierra Nevada de Santa Marta: All 19 species are endemic. These include eight species of *Eleutherodactylus,* six of *Atelopus,* one each of *Centrolene, Colostethus,Cryptobatrachus,* and *Bolitoglossa,* and the monotypic genus *Geobatrachus.*

Northern Andes: The amphibian fauna of the tropical Northern Andes consists of 405 species, of which 347 are endemic. Most species (379) are anurans, but there are nine species of salamanders and 17 of caecilians. Seven genera are highly speciose in the region; of these, *Eleutherodactylus* (178 species) is the most conspicuous, followed by *Colostethus* (38), *Cochranella* (31), *Hyla* (29), *Centrolene* (23), *Atelopus* (19), and *Gastrotheca* (14). *Andinophryne* with three species is endemic. All salamanders are members of the genus *Bolitoglossa.* Most of the caecilians (11) are members of the genus *Caecilia,* but there are five species of *Epicrionops.*

Of the 37 shared species, many inhabit the adjacent lowlands. Thirteen species (e.g., *Cochranella ramirezi, Colostethus chocoensis, Eleutherodactylus anomalus, E. chalceus,* and *E. walkeri*) are shared with the Chocó. Nine species (e.g., *Hemiphractus bubalus, Eleutherodactylus nigrovittatus, E. quaquaversus,* and *E. trachyblepharis*) are shared with the Amazon Basin. *Bufo typhonius, Eleutherodactylus peruvianus,* and *Bolitoglossa altamazonica* are shared among the Amazon Basin and the tropical Southern Andes, which has 22 species in common with the tropical Northern Andes. These include 10 species (e.g., *Cochranella siren, Colostethus nexipus, Gastrotheca testudinea, Hyla phyllognatha, Osteocephalus verruciger,* and *Caecilia attenuata*) that inhabit the eastern slopes of the Andes to the north and south of the Huancabamba Depression and seven species (e.g., *Colostethus elachyhistus, Epipedobates tricolor, Eleutherodactylus colodactylus, E. lymani,* and *E. phoxocephalus*) that occur in the mountains of southern Ecuador and northern Peru. Three hylids (*Gastrotheca nicefori, Hyla miliaria,* and *H. palmeri*) are shared with the mountains of lower Central America, and the former and *Flectonotus pygmaeus* are shared with the tropical Venezuelan Andes. Only five species are shared with the alpine Northern Andes.

The large array of taxa in the tropical Northern Andes is a reflection of the complex topography. If the region is broken down into the components recognized by Duellman (1979b), the largest number of species is in the Cordillera Occidental in Colombia (130, 92 endemic), followed by the Cordillera Oriental of Ecuador (102, 93 endemic), Cordillera Occidental of Ecuador (97, 59 endemic), the Cordillera Oriental of Colombia (72, 54 endemic), and the Cordillera Central of Colombia (63, 48 endemic). Of the 130 species in Cordillera Occidental in Colombia, 10 are shared with the Cordillera Central, 27 with the Cordillera Occidental in Ecuador, and one each with the other cordilleras. Of the 72 species in the Cordillera Oriental of Colombia, 15 are shared with the Cordillera Oriental in Ecuador, five with the Cordillera Central in Colombia; one of these also is shared with the Cordillera Occidental in Colombia. Of the 63 species in the Cordillera Central in Colombia, two are shared with both Ecuadorian cordilleras and one in each of the Ecuadorian cordilleras. Only two of the 102 species in the Cordillera Oriental in Ecuador are shared with the Cordillera Occidental.

Three endemic species of salamanders (*Bolitoglossa*) and 75 species of anurans inhabit the supra-treeline regions of the Northern Andes.

The monotypic genus *Atopophrynus* and four of the six species of *Osornophryne* are among the 73 species restricted to supra-treeline habitats. The most speciose genera are *Eleutherodactylus* (39 species), *Gastrotheca* (9), *Atelopus* (8), and *Colostethus* (7). Five species (*Centrolene buckleyi, Gastrotheca monticola, Eleutherodactylus buckleyi, E. chloronotus,* and *E. w-nigrum*) are shared with the tropical Northern Andes; the first two species also occur in the Southern Andes, and *Centrolene buckleyi* occurs in the Venezuelan Andes. As in the tropical Northern Andes, the alpine Northern Andes are fragmented into isolated high elevations in five cordilleras plus the Nudo de Pasto in south-central Colombia and adjacent Ecuador. Only seven species, all endemic, occur in the Cordillera Occidental in Colombia, whereas in the Cordillera Central 18 of 22 species are endemic. The Cordillera Oriental in Colombia has 17 species, of which 14 are endemic. The Cordillera Occidental in Ecuador has 15 spe-

cies, of which only four are endemic; the Cordillera Oriental in Ecuador has 28 species, of which 18 are endemic.

Southern Andes: The amphibian fauna of the tropical and subtropical portion of the Southern Andes consists of 128 species of anurans, two of salamanders, and two of caecilians; 99 of the anurans are endemic, as are *Bolitoglossa digitigrada* and *Epicrionops peruvianus.* The most speciose genera are *Eleutherodactylus* (31 species, 21 endemic), *Cochranella* (13, 12 endemic), *Bufo* (10, 9 endemic), *Hyla* (10, 9 endemic), *Gastrotheca* (8, 4 endemic), *Colostethus* (8, 5 endemic), and *Centrolene* (7, 6 endemic). The bufonid genus *Truebella* with two species is endemic. Most shared species are with the tropical Northern Andes (see above). Seven species (e.g., *Epipedobates bassleri, E. petersi,* and *E. tricolor*) are shared with the upper Amazon Basin. Four species (*Gastrotheca excubitor, G. monticola, Phyllonastes heyeri,* and *Telmatobius truebae*) are shared with the alpine Southern Andes.

Of the 70 species of anurans making up the amphibian fauna of the alpine Southern Andes, 65 species are endemic. With the exception of *Atelopus* (1 species), *Bufo* (4 species), *Gastrotheca* (6 species), and *Pleurodema* (2 species), the fauna consists of telmatobiine leptodactylids, including *Telmatobius* represented by 35 species, *Phrynopus* by 11 species, *Alsodes* by six species, and the endemic *Batrachophrynus* with two species. In addition to the four species shared with the tropical Southern Andes (see above), *Bufo spinulosus* is shared with Patagonia.

Guiana Highlands

Only anurans are known from the Guiana Highlands; 71of the 76 species are endemic. Five genera are endemic, or nearly so; these include *Metaphryniscus* (1 species), *Oreophrynella* (4 species), *Dischidodactylus* (2 species), *Tepuihyla* (7 species), and *Stefania* (13 species); 1 species of *Stefania* occurs in the Guiana lowlands. Other endemics include seven centrolenids, four dendrobatids, 12 hylids, seven *Eleutherodactylus,* and one *Leptodactylus.* *Hyalinobatrachium orientalis* is shared with the Andes (Cordillera de la Costa and the tropical Venezuelan Andes); *Hyla crepitans* is shared with cis-Andean tropical lowlands and the Cordillera de la Costa of the Andes, and *Cochranella oyampiensis, Hyla ornatissima,* and *Otophryne robusta* are shared with the Guiana lowlands.

Table 5:5. Numbers, percentages, and densities of species of amphibians in different vegetation types in South America.

Vegetation type	Area (10⁶ km²)	Percent total area	No. of Species	Percent Species	Species/ (10⁶ km²)
Lowland tropical rainforest	6.365	35.8	608	34.90	95.5
Montane tropical rainforest	0.760	4.3	745	42.77	980.0
Tropical dry forest	4.595	25.8	116	6.66	25.2
Temperate forest	0.446	2.5	32	1.84	71.7
Tropical savanna-grassland	3.100	17.3	32	1.84	10.3
Temperate grassland	0.610	4.0	23	1.32	37.7
Montane grassland	1.540	8.6	180	10.33	116.9
Desert	0.297	1.7	5	0.28	16.8

AREAS OF HIGH SPECIES DIVERSITY AND ENDEMISM

Herein I use the term "species diversity," or simply diversity, solely as the number of species. Two kinds of diversity are applied here:

• Alpha diversity is the number of species at a single site. Thus, the diversity of anurans at Santa Cecilia in Amazonian Ecuador was reported as 81 (according to present taxonomy 84 species) by Duellman (1978) and at Boracéia in the Atlantic Coastal Forest of Brazil, 65 (Heyer et al., 1990).

• Beta diversity is the number of species in a given region (in two or more sites in a given region). These are the numbers that have been given for each of the regions defined herein (Table 5:2).

REGIONAL DIVERSITY AND ENDEMISM

Beta diversity is highest in the Andes (753 species) and lowest in the Atacama Desert (5 species) (Table 5:2). In general, humid tropical forested regions have greater diversity than tropical dry forests or open formations and temperate habitats (Table 5:5).

In five of the 12 regions in South America, 90% or more of the species of amphibians are endemic (Table 5:2). All species in the Austral Temperate Forest and Atacama Desert are endemic, but species diversity is low, for the regions contain only 32 and 5 species, respectively. The genera *Batrachyla, Caudiverbera, Eupsophus, Hylorina, Insuetophrynus, Telmatobufo* (all leptodactylids) and the sole rhinodermatid genus *Rhinoderma* with two species are endemic to the Austral Temperate Forest.

The Atlantic Forest Domain with 92.8% endemism has 299 endemic species of anurans and 10 of caecilians. Twenty genera of anurans and two of caecilians are endemic to this region.

The other two regions having more than 90% endemism are highland regions. Seventy-one of the 76 species (93.4%) of frogs known from the Guiana Highlands are endemic. This region still is poorly known, and probably many more species will be discovered when collections are made on previously uncollected tepuis. Four genera (*Metaphryniscus, Oreophrynella, Tepuihyla,* and *Dischidodactylus*) are restricted to the Guiana Highlands, and all but one of the 14 species of *Stefania* are endemic.

With the exception of the Austral Temperate Forest and Atacama Desert, the highest degree of endemism within any of the 12 regions is in the Andes, where 715 of the 753 (95.0%) species are endemic. This high diversity and endemism can be attributed to the complex topography in combination with high humidity and low evapotranspiration rates, which sustain humid montane forests or cloud forests and provide conditions highly suitable for amphibians. Species of *Colostethus, Eleutherodactylus,* and centrolenids are especially numerous. Although the alpine regions in the Venezuelan, Northern, and Southern Andes have fewer species than their tropical counterparts (18, 78, and 70 species, respectively), endemism is higher (94.4%, 93.6%, and 92.9%, respectively). The alpine environments exist as habitat islands above treeline, and many species of *Atelopus, Colostethus, Eleutherodactylus, Phrynopus,* and *Telmatobius* are restricted to one or adjacent groups of habitat islands. Practically all nonendemics are shared only with the tropical habitats at lower elevations.

Among the strictly tropical lowland regions, the Amazonia-Guiana Region has the highest species diversity, but only 275 of 335 (82.6%) of the species are endemic. As noted by many authors (e.g., Lynch, 1979), the highest species diversity and greatest endemism is in the upper Amazon Basin in Ecuador and Peru.

Although the Chocó has a large number of species (126), only 46 (42.1%) are endemic to the region. This low degree of endemism is a reflection of the fact that this region is historically and biogeographically part of Central America, with which 60 species of amphibians are shared.

The Caribbean Coastal Forest and the Llanos have comparatively few species and endemics—Caribbean Coastal Forest with 47 species (38.3% endemic) and Llanos 31 (22.5% endemic). For the large area encompassed, the arid and semiarid Cerrados-Caatinga-Chaco has relatively few species (99) with only 53.5% endemic. The temperate Pampean-Monte and Patagonian regions contain 85 and 10 species, respectively. In the former, 47 species (55.3%) are endemic, whereas eight species (80.0%) are endemic to the Patagonian Region.

LOCAL DIVERSITY AND ENDEMISM

Data on alpha diversity are sufficient to make only broad generalizations. Data on anurans provided by Duellman (1988, 1993b) show that alpha diversity is highest at sites in the humid tropical lowlands: 29–81 ($\bar{x}$ = 52.9) at 13 sites in Amazonia-Guiana, 35–49 ($\bar{x}$ = 42.6) at three sites in the Chocó, and 65 at one site in the Atlantic Forest Domain. In contrast, sites in the nonrainforest lowlands have lower diversity: 16–26 ($\bar{x}$ = 21) at two sites in the Llanos, and 22–29 ($\bar{x}$ = 25.3) at three sites in the Cerrado-Caatinga-Chaco. Likewise, diversity is lower in montane habitats: 20–39 ($\bar{x}$ = 26.8) at four sites in montane cloud forest in the Andes and 5–15 ($\bar{x}$ = 8.1) at seven sites above treeline in the Andes.

Within Amazonia-Guiana there is a gradient from higher rainfall distributed more evenly throughout the year in the western part of the Amazon Basin to lower rainfall less evenly distributed in the central and eastern part of the basin and still less rainfall in the Guianas. A similar pattern is reflected in alpha diversity of anurans: 55–81 ($\bar{x}$ = 62.3) species at six sites in the western part of the basin, 32–59 ($\bar{x}$ = 46.2) species at four sites in the central and eastern part of the basin, and 29–65 ($\bar{x}$ = 43.0) species at three sites in the Guianas.

Likewise, altitudinal gradients are evident in alpha diversity. In an equatorial transect across the Andes, the alpha diversity of anurans is 49–81 ($\bar{x}$ = 65.0) at two sites below 1000 m, 20–25 ($\bar{x}$ = 23.0) at three sites between 1000 and 3000 m, and 4–7 ($\bar{x}$ = 5.5) at two sites above 3000 m (Duellman, 1988). In more localized situations, the pattern is similar. For example, in a transect of the Cordillera de Huancabamba in the Andes of northern Peru, seven species of anurans occur in cloud forest at an elevation of 1735 m, and one of these is among the nine species known from an elevation of 2770 m; three of these are among the four species known at elevations above 3000 m (Duellman and Wild, 1993).

Patterns of distribution of *Eleutherodactylus* on the Amazonian and western slopes of the Andes in Ecuador reveal that species at higher elevations generally have narrower latitudinal ranges than those at but a few species at high elevations (e.g., *E. w-nigrum*) have relatively broad latitudinal ranges. Similar patterns are evident among other genera that inhabit the Andes and the Amazon Basin. These are evident in centrolenids, *Colostethus,* and *Hyla.*

Few species in the Andes have broad distributions; the two having the greatest distributions are *Centrolene buckleyi* and *Gastrotheca nicefori.* Many species are endemic to relative small areas—islands of páramo (e.g., *Phrynopus* [Cannatella, 1984]), lakes within otherwise dry environments (e.g., *Atelognathus* [Cei, 1980]), or drainage systems (e.g., *Telmatobius* [Cei, 1986]). Isolated mountain ranges include many endemics. A few species (e.g., centrolenids, *Colostethus, Bufo, Gastrotheca, Phyllomedusa, and Eleutherodactylus* [Duellman and Toft, 1979; Duellman and Lynch, 1988; Duellman and Wild, 1993]) are endemic to ranges adjacent to the Andes, such as the cordilleras de Colán, Cóndor, Cutucú, and Huancabamba, and the Serranía de Sira that are at least partially isolated from the principal Andean cordilleras. However, all 16 species restricted to elevations of more than 1000 m in the completely isolated Sierra Nevada de Santa Marta are endemic to that range (Ruthven, 1922; Lynch and Ruíz-Carranza, 1985; Rueda-Almonacid, 1994; Kaplan, 1997a). The most striking endemism is in the Guiana Highlands, where 71 species are endemic; most are isolated on one of the many table mountains (tepuis) that rise from the lowlands. These endemics include bufonids (Señaris et al., 1994), centrolenids (Ayarzagüena, 1992; Myers and Donnelly, 1997), *Colostethus* (La Marca, 1996b), hylids (Ayarzagüena et al., 1992; Ayarzagüena and Señaris, 1993; Duellman and Hoogmoed, 1984, 1992; Senãris et al., 1996; Myers and Donnelly, 1997), and *Eleutherodactylus* (Myers and Donnelly, 1996, 1997). (Also see Duellman, 1997.)

CORRELATIONS OF DISTRIBUTION PATTERNS

It is evident that distinctly different patterns of distribution exist among the large and diverse amphibian fauna. These patterns can be correlated with topographic features, climate, and vegetation; the latter is a reflection of the first two, plus soils. Also historical phenomena have played an important role in the development of these patterns.

TOPOGRAPHY

The major topographic features of South America are three highland areas. Of these, the most imposing are the Andes, which have a length of about 8000 km and span 66 degrees of latitude. The combined area of the six regions of the Andes recognized herein plus their associated, but isolated, ranges, the Sierra Nevada de Santa Marta and Cordillera de la Costa, is 1,750,600 km². The total number of amphibians in the Andes and associated ranges is 753, of which 715 (95.0%) are endemic; the other species are shared with the adjacent lowlands, except three, which also are in the highlands of Central America. The highest density of species is in the northern Andes (tropical and alpine Andes in Colombia and Ecuador), where 486 species are known. This area encompassing 445,000 km² (2.5% of the area of the continent) is home to 28% of the species of amphibians known from the continent.

The numerous table mountains (tepuis) making up the Guiana Highlands have a total area of only 16,500 km², and they are home to 76 species of anurans, of which 71 (93.7%) are endemic. The larger highland part of the Atlantic Forest Domain (961,800 km²) has 156 species, of which 144 (92.3%) are endemic.

In combination, the three highland areas have a total area of 2,728,900 km² (15.3% of the area of the continent). Yet they are home to 985 species, of which 930 are endemic. Thus, 53.4% of the species of amphibians in South America live in highland regions, where there is a density of 361 species per 10⁶ km².

The vast majority of the continent consists of lowlands, which for purposes of comparison can be divided into three major regions: (1) trans-Andean lowlands consisting of the Chocó, Caribbean Coastal Forest, Atacama Desert, and Austral Temperate Forest; (2) cis-Andean lowlands consisting of Llanos, Amazonia-Guiana, Cerrados-Caatinga-Chaco, Pampean-Monte, and Patagonian regions; and (3) Atlantic Coastal Forest. The Trans-Andean lowlands consisting of 1,182,300 km² have 197 species of amphibians, of which 171 occur nowhere else on the continent; 73 species are shared with Central America. The cis-Andean lowlands having an area of 13,693,600 km² contain 456 species, of which 413 are endemic; the Atlantic Coastal Forest has an area of 188,200 km² and has 181 species, of which 164 are endemic. Thus, the entire lowland region of South America has 834 species of amphibians (47.9% of the total for the continent), where there is a density of 55.4 species per 10⁶ km².

These rather crude calculations reveal that topography plays a major role in determining the diversity and degree of endemism among amphibians in South America. More than one-half of the species are restricted to highland regions that constitute only 15.3% of the land area of the continent.

As expected, there is a linear correlation between numbers of species and numbers of endemics in any given region. However, there is no correlation between numbers of species or numbers of endemic and the sizes of the regions (Table 5:4).

CLIMATE

The climates of South America range from wet tropical with rainfall throughout the year to cold and dry; these climates are classified into 17 types in the Köppen system (Eidt, 1968; Fig. 5:5). As expected, because of their ecological tolerances and need for moisture, the diversity of amphibians is highly correlated with moisture. On a broad scale this has been demonstrated for amphibians in North America (Kiester, 1971) and for a moisture gradient in the Yucatan Peninsula in Mexico and Guatemala (Lee, 1980). In the lowlands, the greatest numbers of species occur in the humid lowland tropical forests (Chocó, Amazonia-Guiana, and Atlantic Forest Domain), where temperatures are high and rainfall is heavy. But even within the Amazonia-Guiana Region there is a gradient from high species richness in the wetter, aseasonal western part of the Amazon

Basin to fewer species in the somewhat drier eastern part of the basin and Guiana Lowlands with less rainfall that is seasonal (Lynch, 1979; Duellman, 1988). Duellman and Thomas (1996) showed a positive correlation between anuran species richness and mean annual rainfall at nine sites in the upper Amazon Basin; they also noted (as did Duellman, 1988) that anurans having reproductive modes dependent on high atmospheric humidity (e.g., dendrobatids, centrolenids, and *Eleutherodactylus*) are most numerous at sites having heavy rainfall throughout the year. This correlation of decreased amphibian richness with lower rainfall is reflected in lower numbers of species in the Caribbean Coastal Forest, Llanos, and Cerrado-Caatinga-Chaco, where annual rainfall is less and the dry season is longer.

In contrast to the tropical lowlands, both moisture and temperature influence patterns of distribution in the temperate lowlands. For example, there is a gradual decrease in species of amphibians in the cis-Andean lowlands from the Pampean-Monte Region in southern Brazil and northern Argentina through the Pampas and southward through Patagonia. Likewise, in the Austral Temperate Forest there is a gradual decrease from 15 species of frogs from 37° S to one species at 53° S; throughout this region relative humidity generally is 82–86%, but the mean annual temperature declines from 13.3° C to 5.9° C (Formas, 1979). On the other hand, at 37° S in Patagonia, which is dry, only two species of amphibians occur.

Altitudinal decreases in temperature are reflected in a decrease in the number of amphibian species at higher elevations. Among the 32 species of anurans in the lowlands of the Austral Temperate Forest, six extend up to an elevation of 1000 m, and only two reach 2000 m. In an equatorial transect from the Amazon Basin to the crest of the Cordillera Oriental of the Andes, the number of anuran species diminishes from 84 at 340 m to 23 at 1500 m, six at 3000 m, and four at 4000 m (Duellman and Trueb, 1986: Fig. 5-22-4). Along this transect, not only does mean annual temperature decrease from 25.1° C at 340 to 9.1° C at 3160 m, but the precipitation decreases from 2190 mm to 1310 mm annually. However, with decreased temperature and insolation there is a decrease in the evaporation, so that at 3000 m on the Amazon slopes of the Andes the climate is cool and humid. At high elevations in the Andes there exists a gradual diminution of species of frogs from equatorial regions to southern latitudes (Péfaur and Duellman, 1980). In the alpine Northern Andes, anuran communities consist of 5–15 species, whereas in the alpine Southern Andes, communities consist of 3–8 species.

Lynch's (1986b) analysis of Duellman's (1979b) data on altitudinal distributions of Andean species shows that in the northern Andes (Colombia and Ecuador) 77.3% of the species of frogs have altitudinal ranges of < 1000 m, whereas in Peru and Bolivia 72.7% and in Argentina and Chile only 62.5% are in that category. Janzen's (1967) thesis was used to explain this phenomenon: In temperate lowlands, the broad annual climatic variation necessitates that organisms have broad physiological tolerances that serve as a preadaptation for invasion of a wide range of altitudes. Conversely, in tropical lowlands, the narrow climatic variability selects for narrow physiological tolerances. Thus, tropical lowland species are able to select for a narrower range of altitudes.

Vegetation

For purposes of this analysis, the vegetation types mapped by Hueck and Seibert (1972) were reduced to eight major categories, which may be viewed as biomes (Fig. 5:6). These are the same categories defined by Duellman (1993b), in which, for example, all habitats above treeline are grouped as montane grassland, and the tropical dry forest includes the xeric and semiarid habitats in the Cerrado-Caatinga-Chaco Region. According to this scheme, the greatest numbers of species are in the montane tropical rainforest followed by the lowland tropical forest (Table 5:6), but the density of species is nearly 10 times greater in the montane tropical rainforest than in the lowland tropical rainforest. Montane grassland has a greater number of species and higher density than temperate grassland, and tropical dry forest has a greater number of species and higher density than tropical savanna-grassland. Although the temperate forest has a low number of species (32), the density of species (71.7/10⁶ km²) is nearly as great as that in the lowland tropical rainforest (95.5/10⁶ km²). Desert with only five species has a density that exceeds only that of the much more extensive tropical savanna-grassland.

The comparative densities should not be misinterpreted. The total number of species in two vegetation types may be similar (e.g., 608 in lowland trop-

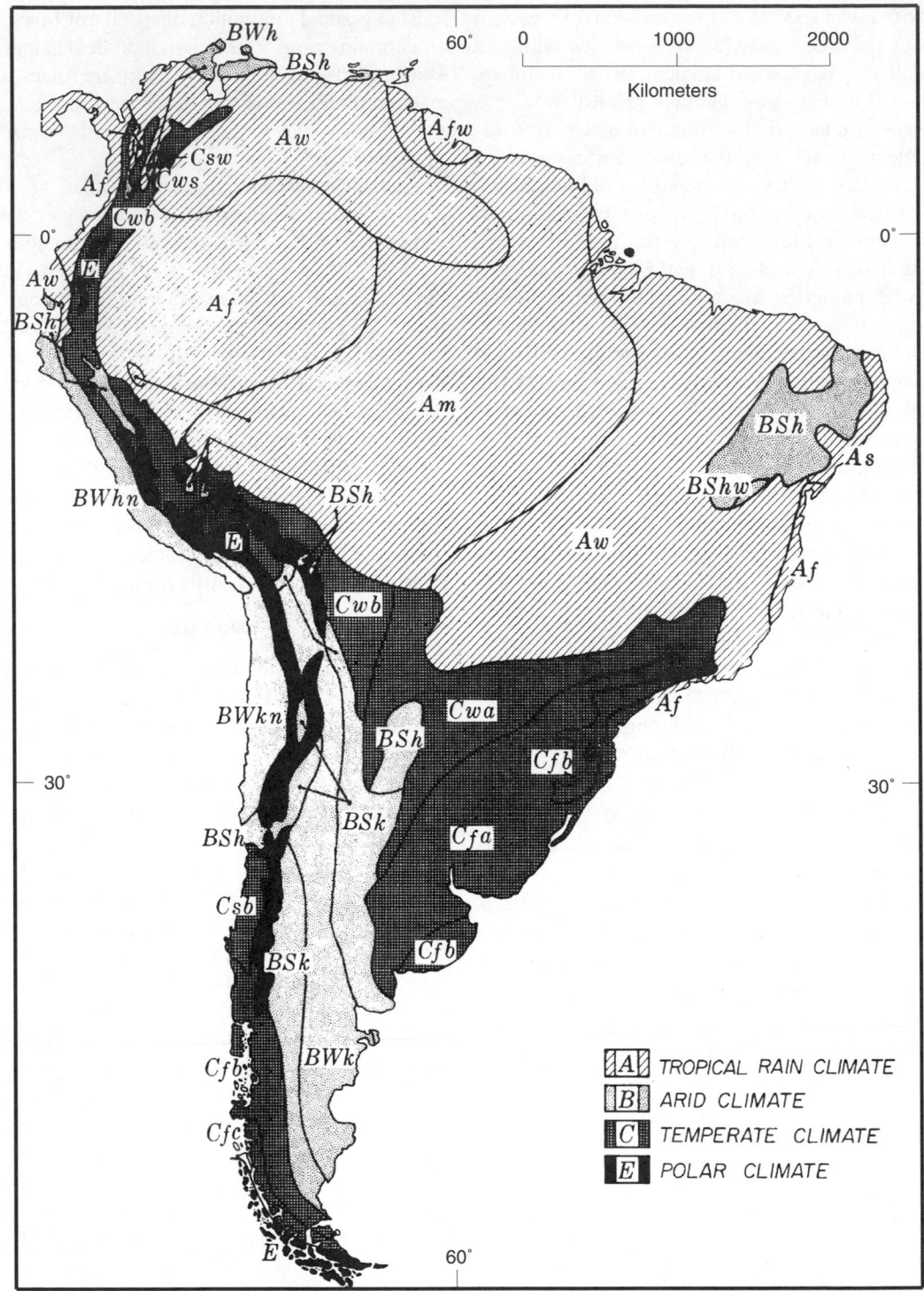

Fig. 5:5. Climates of South America (adapted from Eidt, 1968).

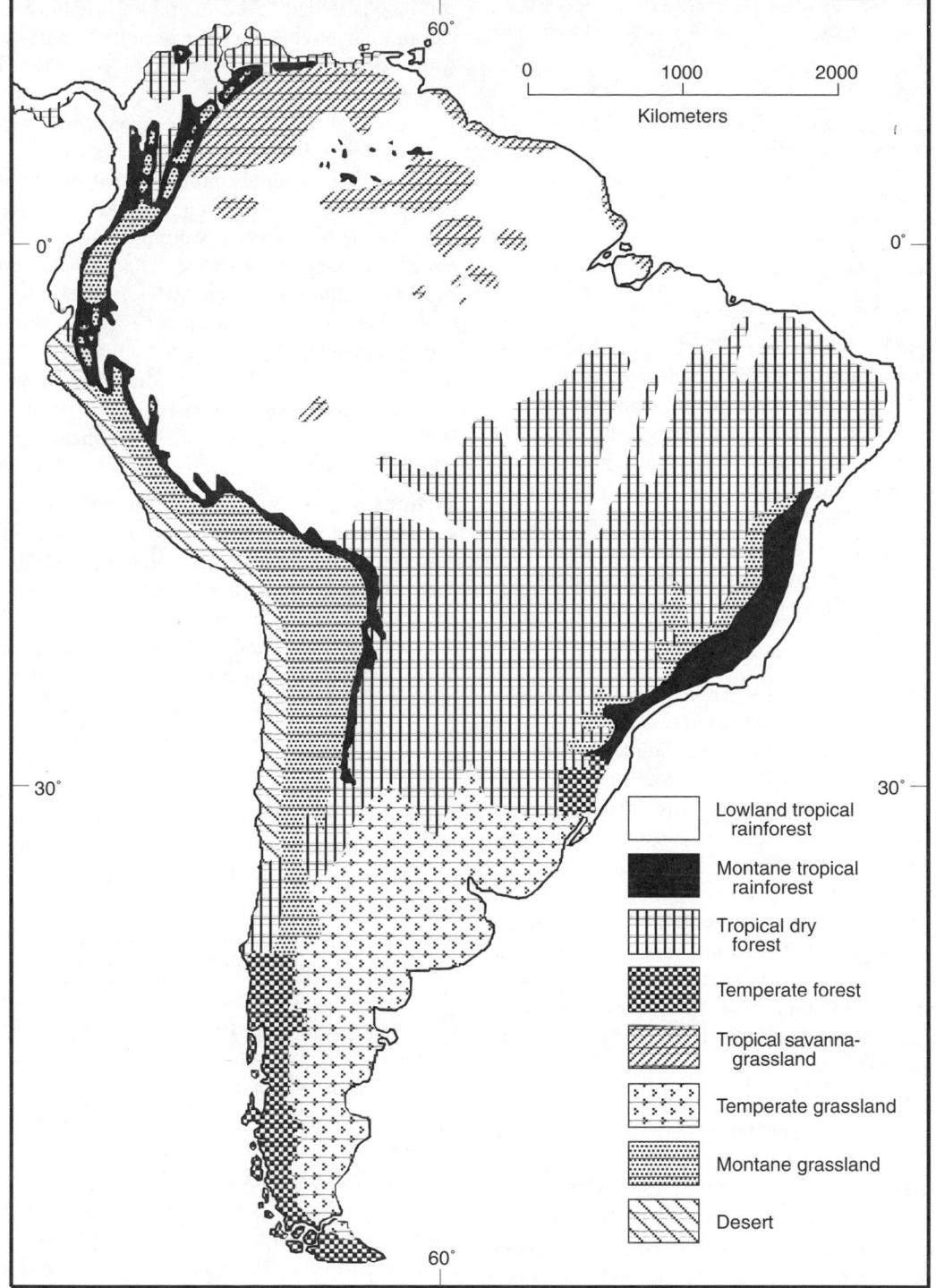

Fig. 5:6. Eight major vegetation types of South America (adapted from Hueck and Seibert, 1972).

Table 5:6. Numbers of species in anuran communities in different vegetation types in South America (expanded from Duellman, 1993b).

Vegetation type	N	Range	Mean
Lowland tropical rainforest	15	35–81	55.1
Montane tropical rainforest	5	20–39	25.4
Tropical dry forest	3	22–39	25.3
Temperate forest	4	5–8	6.5
Tropical savanna-grassland	2	16–26	21.0
Temperate grassland	3	3–17	7.7
Montane grassland	6	5–9	7.0
Desert	3	1	1.0

ical rainforest and 745 in montane tropical rainforest), but because of differences in area encompassed by the vegetation types, species densities (number of species per 10^6 km²) can be vastly different. The high density of species in the tropical montane rainforest and montane grassland is the reflection of small geographic ranges of most of the species. Sympatric occurrence of more species is typical of lowland vegetation types, whereas fewer species occur sympatrically in montane environments (Table 5:6).

HISTORICAL PHENOMENA

Subsequent to its separation from Africa in the Cretaceous and from the Antarctica-Australian landmass in the middle Oligocene, South America was an isolated landmass, except for a brief connection with Middle America in the Late Cretaceous, until its connection with North America via the Isthmus of Panama in the Pliocene (Pittman et al., 1993). Even prior to its separation from other continents, the South American biota was differentiating, and during the period of isolation, much of the biota of South America differentiated, and distributions were influenced by major climatic and physiographic changes. These events can be grouped into three sequential periods:

Mesozoic

Climatic and vegetation patterns changed dramatically during the Cretaceous. Evidence strongly suggests that the interior of the large African-American continent was arid prior to the birth of the South Atlantic Ocean, which brought maritime and mesic climates to western Africa and eastern South America for the first time (Axelrod, 1972); moreover, the southern part of the continent enjoyed subtropical, mesic climates. Prior to a major uplift in the Cretaceous and a coincidental subsidence of the Amazon Basin, the Guianan and Brazilian highlands were

continuous so that the basin drained into the Pacific Ocean; this was closed when an initial uplift of the Andes occurred in the Late Cretaceous (Beurlen, 1970).

Although the meager Mesozoic fossil record contains only pipids and leptodactylids among living groups, it is highly probable that most of the living families were extant at the end of the Mesozoic. Bufonids and hylids were present in the Paleocene, so conceivably ancestors to bufonid genera such as *Melanophryniscus, Metaphryniscus,* and *Oreophrynella* existed in the areas of the Guianan and Brazilian highlands prior to their uplifts. A similar chronology based on distribution patterns and postulated ages of divergence based on immunological distances was suggested for hemiphractine hylids (Duellman et al., 1988).

Tertiary

Subsequent to the Eocene, temperate South America gradually became cooler and drier. In the early Tertiary, southern South America was more equable than at present; the austral forests extended to at least 30° S latitude in Chile and across Patagonia in the Oligocene (Jeannel, 1967). In the Miocene there was progressive climatic deterioration in Patagonia, and an Atlantic transgression, the Paranense Sea, inundated most of the Pampean-Monte Region and southern part of the Cerrado-Caatinga-Chaco Region (Báez and Scillato Yané, 1979). Major uplift of the southern Andes took place in the Miocene, whereas the northern Andean oro-geny occurred primarily in the Pliocene; also, the Brazilian Highlands were uplifted farther in the Tertiary (Simpson, 1979). These events and the final separation of Antarctica from South America resulting in the cold Humboldt Current sweeping up the Pacific coast of South America created the major physiographic and climatic patterns evident today.

Probably most of the genera and many of the species of South American amphibians differentiated in the Tertiary, but some were in existence in the Cretaceous. Based on immunological distances among leptodactylids in the Brazilian Shield, estimates for times of divergence are no more recent than the Eocene (Maxson and Heyer, 1982), whereas times of divergence of species of *Telmatobius* and *Gastrotheca* in the southern Andes and of *Gastrotheca* in the northern Andes are Miocene and Pliocene, respectively, times of major uplifts of those portions of the Andes (Scanlan et al., 1980; Duellman et al., 1988). Phylogenetic analyses of *Ceratophrys* (Lynch, 1982) strongly suggest the speciation in that group occurred during the Tertiary; immunological distance data support hypotheses of differentiation of species of *Leptodactylus* (Heyer and Maxson, 1982) and *Cycloramphus* (Heyer and Maxson, 1983) in the Tertiary. The differentiation seems to be correlated with the expansion and fragmentation of sub-humid environments for *Ceratophrys* and *Leptodactylus,* and the orogenies of the coastal ranges in southeastern Brazil for *Cycloramphus.*

The uplift of the Andes effectively isolated lowland amphibians in the cis-Andean lowlands from populations in the trans-Andean lowlands This is evident from the number of vicariant pairs or groups of taxa in the lowlands on either side of the Andes. *Ceratophrys* (Lynch, 1982), *Physalaemus* (Cannatella and Duellman, 1984), *Gastrotheca* (Duellman et al., 1988), and *Trachycephalus* (Trueb, 1970) are examples.

One geological factor that greatly influenced the amphibian fauna of South America (especially the northwestern part) was the connection of that continent with Central America via the Isthmus of Panama in the Pliocene. According to Pittman et al. (1993), the island-arc system that was to become the Isthmus of Panama probably made contact with northwestern South America in the Miocene, thereby establishing an Aleutian-type connection between South America and Central America; by the mid-Pliocene the Isthmus of Panama had been uplifted, and the Panamanian corridor was established. Although the island-arc may have provided a passage for some species, the major interchange of amphibians between Central America and South America most likely did not occur until after the connection of the continents, but at least *Bolitoglossa* may have entered South America in the Miocene (Hanken and Wake, 1982). The exchange of amphibians has been reviewed by Savage (1982) and Vanzolini and Heyer (1985). Only a few Central American genera (*Agalychnis, Smilisca, Rana, Oedipina,* and the *Craugastor* clade of *Eleutherodactylus*) entered South America at that time. Of these, only *Agalychnis craspedopus, Rana palmipes,* and several species of the *Craugastor* clade of *Eleutherodactylus* are not confined to the Chocó.

Quaternary

Although dramatic climatological changes and concomitant distributions of the biota in the Pleistocene have been well documented in Eurasia and North America, only recently have data revealed that the tropical regions, especially the lowland rainforest, have not persisted unchanged for millions of years. Changing climates during the Quaternary have been modeled by Fairbridge (1972), and these models have been applied to South America by Haffer (1974) and Simpson (1979), and Ab'Saber (1977) mapped the presumed natural regions of the continent at the time of the most recent extensive arid climatic phase (Fig. 5:7).

Substantial geological and biogeographic evidence exists for dramatic climatic changes in the highlands during glacial periods during the Pleistocene, but the evidence for changes in the lowlands is debatable. Two or three glaciations occurred in various parts of the Andes; during times of maximum glaciation in the northern Andes, temperatures were depressed 6–7°C and environments were shifted downward by 1000–2000 m (Vuilleumier, 1971; van der Hammen, 1974). Glaciations depressed snow lines by as much as 1500 m in the Peruvian Andes (Hastenrath, 1967; Simpson, 1979). Because the Andes were not elevated to their present elevations until the Pliocene and Pleistocene, the supra-treeline habitats (páramo and puna) did not form until the Pliocene (Simpson, 1975); during the Pleistocene the treeline was lower during glacial times and higher during interglacial intervals. Thus, amphibian inhabitants of the páramo and puna could not have differentiated in such habitats before that time, and during the glacial periods of lowered treeline, the supra-treeline habitats were more extensive and contiguous than they are now. Conversely, with the elevation of treeline during interglacial phases, the supra-treeline habitats were more constricted and isolated. This scenario predicts a classic allopatric speciation model, which is substantiated by present patterns of distribution of *Osornophryne, Eleutherodactylus,* and *Phrynopus,* and immunological distances of *Gastrotheca* (Duellman et al., 1988).

Cladistic analyses of some monophyletic groups inhabiting the Andes reveal apparently different historical phenomena, as noted by Lynch (1986b), who reviewed hypotheses for the origin of the Andean herpetofauna. Some groups have all members of their respective lineages in the Andes; examples are *Phrynopus* (Cannatella, 1984), the *devillei* assem-

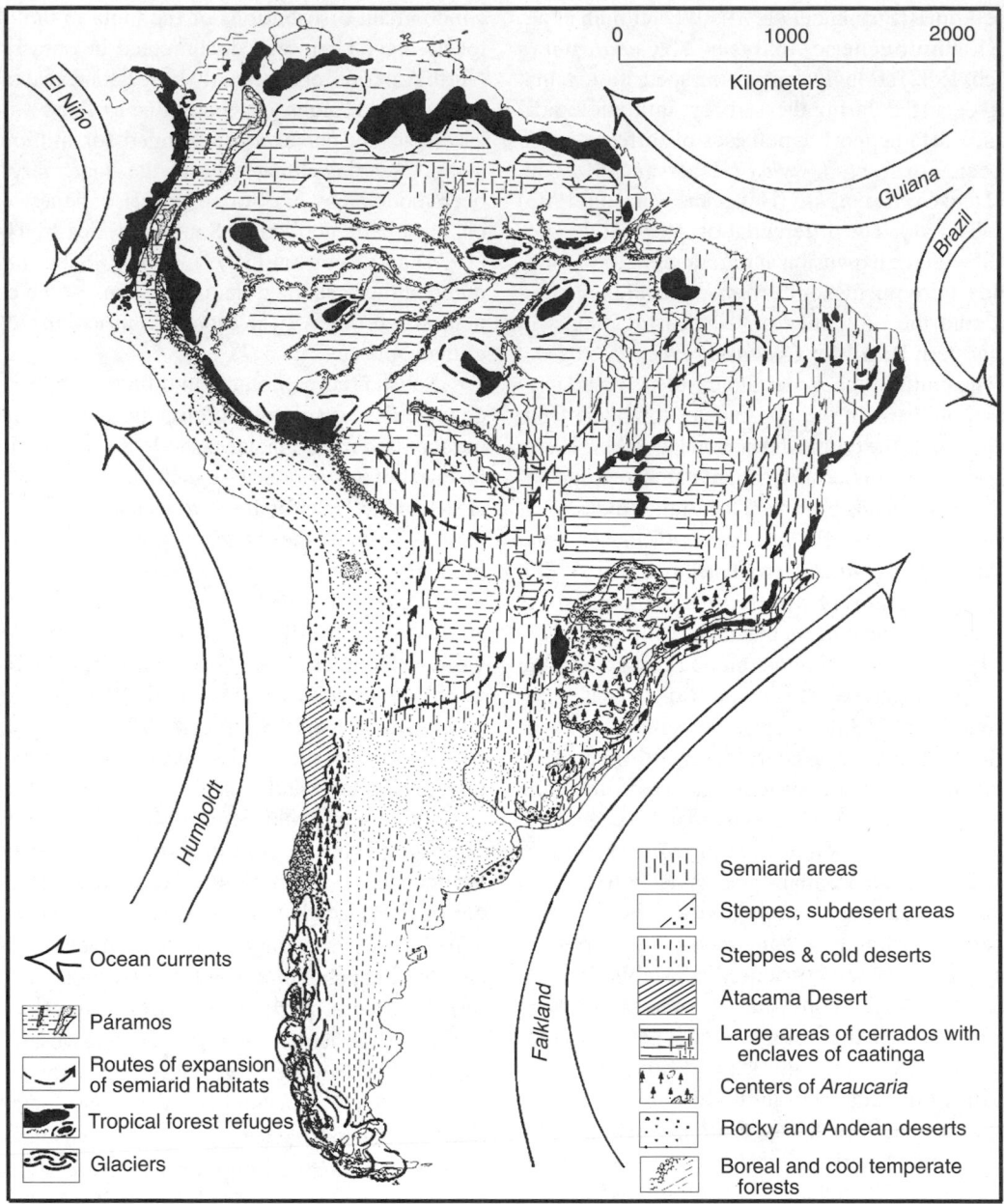

Fig. 5:7. Natural regions of South America 13,000–18,000 years ago, as postulated by Ab'Saber (1977).

bly of the *Eleutherodactylus unistrigatus* group (Lynch, 1983), the *myersi* assembly of the *Eleutherodactylus unistrigatus* group (Wiens and Coloma, 1992), the *Gastrotheca marsupiata* and *plumbea* groups (Duellman and Hillis, 1987), the *Hyla larinopygion* group (Duellman and Hillis, 1990), and various clades of *Eleutherodactylus* (Lynch and Duellman, 1997; Lynch et al., 1997). Other groups are principally Andean but have a few extra-Andean members; examples are the *Eleutherodactylus nigrovittatus* clade (Lynch et al., 1997), the *Hyla bogotensis* group (Duellman, 1972), and the *adspersa* subgroup of the *Bolitoglossa adspersa* group (Brame and Wake, 1963). Some extra-Andean groups have one or more Andean members; examples are *Flectonotus* (Duellman and Gray, 1983), the *Eleutherodactylus sulcatus* group (Lynch, 1981) and *Pleurodema* (Duellman and Veloso, 1977).

The major problem with interpretation of diverse kinds of evidence in the lowlands is centered on the Amazon Basin. The Pleistocene forest refuge hypothesis of Haffer (1969), supported by Vuilleumier (1971), Haffer (1979) and many authors in Prance (1982), is based on geological, palynological, and biogeographic evidence; this hypothesis, which applies to all of tropical lowland America and which has been applied to tropical Africa and Asia (Flenley, 1979; Livingston, 1993), states that during glacial periods the tropical lowlands were drier than during interglacial (pluvial) periods, that tropical rainforest and its inhabitants were restricted to regions of high rainfall, and that these isolated areas of rainforest were separated by nonforest environments (scrub forest or savanna). This climatic oscillation resulting in sequences of fragmentation and reunification of forests and nonforests has been used to explain the highly diverse Amazonian biota, distribution patterns, and zones of parapatry and hybridization (Prance, 1982). A speciation model applied to several groups of frogs (Duellman, 1982) supports the refuge hypothesis, but its validity is controversial because the ages of the species are not known.

Two other hypotheses regarding patterns of speciation and distribution in the Amazon Basin are based on geological evidence. It is highly likely that the river systems in the upper Amazon Basin were dramatically modified in the Quaternary (Räsänen et al., 1987); these perturbations were addressed as possible explanations of patterns of distribution of some groups of frogs by Duellman and Wiens (1993). Frailey et al. (1988) provided evidence for a large Pleistocene/Holocene lake extending from central Amazonia westward nearly to the Andes; they speculated that the biota in the upper Amazon Basin is in a state of disequilibrium resulting from dispersal into the region by species from all sides of the lake.

Colinvaux (1993 and references cited therein) disputed the refuge hypothesis by pointing out that pollen profiles in the Amazon Basin do not reveal alternating pluvial and arid periods, but he provided fossil evidence for cool montane forests to have been depressed about 1500 m in the equatorial Andes. These observations are not necessarily contradictory to other hypotheses. Most pollen profiles are from sites near major rivers, where riparian vegetation would have persisted during arid phases, and Kronberg et al. (1991) provided geochemical evidence for aridity during the last glacial phase in the upper

Rio Purús and lower Rio Acre in southwestern Brazil. Duellman (1983) used data from climatic lapse rates to hypothesize climatic compression on the slopes of the Andes and provided examples of various groups of lowland anurans that have vicariants in the cooler Andean forests. This same idea was stated as the disturbance-vicariance hypothesis by Colinvaux (1993), who emphasized that climatic cycles force endless successions of invasion and dispersal on all sides of the basin and that vicariance ensures that the resulting swarms of local species will persist as endemics to localized areas of the Amazon Basin.

There is an appalling lack of testable hypotheses of phylogenetic relationships for lowland groups of amphibians. In an early analysis of the *Hyla parviceps* group, Duellman and Crump (1974) showed three Amazonian species (four more recognized subsequently) with one vicariant each in the Atlantic Forest Domain (one additional species recognized subsequently), the Caribbean Coastal Forest, and the Chocó. Timing of the vicariance event was suggested to be during the Quaternary. Data on frogs of the genus *Ceratophrys* was analyzed cladistically by Lynch (1982), who showed that one clade contained *C. cranwelli* (Chaco) as the sister species of *C. ornata* (Pampean) + *C. aurita* (Atlantic Forest Domain). The second clade contained *C. stolmanni* (Chocó) as the sister species of *C. calcarata* (Caribbean Coastal Forest) + *C. cornuta* (Amazonia-Guiana). Tertiary divergence of these taxa was implied. Aquatic frogs of the genus *Pipa* (Trueb and Cannatella, 1986) presumably differentiated into two stocks. One of these was in trans-Andean northwestern South America and subsequently differentiated into two species, whereas the other stock was in the Orinoco-Amazon region and subsequently differentiated into a species in eastern Brazil and four species in the Amazonia-Guiana Region. No time of divergence was suggested.

Thus, although evidence is meager regarding speciation of amphibians associated with Quaternary perturbations, these events probably greatly influenced the patterns of distributions witnessed today. Elevational climatic changes in the highlands resulted in montane habitat islands harboring species of amphibians, especially in the Andes, whereas changes in rainfall in the lowlands resulted in disjunct distributions of species and/or their relatives in isolated forested and nonforested areas.

DISJUNCT DISTRIBUTIONS

Several examples of notable disjunct patterns of distribution exist among South American amphibians. Some of these may be artifacts of classification; examples are *Barycholos* with one species in the Chocó and one species in eastern Amazonia, *Ischnocnema* with one species in the Atlantic Forest Domain and three species on the Amazonian slopes of the Andes and in the upper Amazon Basin, and *Zachaenus* with two species in the Atlantic Forest Domain and one in extreme southern Patagonia. On the other hand, the distinctive frogs of the genus *Phyllodytes* are represented by one species on Trinidad and six species in eastern Brazil; all inhabit bromeliads. One species, *P. tuberculosus,* in the arid Caatinga may be a relict of a former distribution of the genus that encompassed eastern South America from southeastern Brazil to the Guianas and Trinidad.

Most of the Central America groups that entered South America at the end of the Pliocene are restricted to northwestern South America, principally the Chocó. However, two groups have vicariants in the lowlands on either side of the Andes. *Agalychnis craspedopus* is the only member of the genus in the upper Amazon Basin; it shares derived features of the tadpoles with the Chocoan-Central America *A. calcarifer* (Hoogmoed and Cadle, 1991). Likewise, *Rana palmipes* is the only member of that northern genus in the Amazon Basin; it is the sister taxon to *R. vaillanti* and *R. bwana* on the Pacific lowlands (Hillis and de Sá, 1988).

Other major disjunctions seem to have resulted from older geological-climatic events. The subgenus *Stombus* of *Ceratophrys* contains three species— *C. cornuta* in Amazonia-Guiana, *C. calcarata* in the Caribbean Coastal Forest, and *C. stolzmanni* on the Pacific lowlands of southern Ecuador and northern Peru. Other members of the genus are in the Atlantic Forest Domain, Cerrado-Caatinga-Chaco, and Pampean-Monte regions (Lynch, 1982). Because the two species inhabiting rainforest (*C. cornuta* in Amazonia-Guiana and *C. aurita* in the Atlantic Coastal Forest) are far removed topographically in his phylogenetic tree, Lynch (1982) argued that distributions did not support the Pleistocene refugia hypothesis and that speciation took place prior to the Quaternary.

Frogs of the genus *Flectonotus* occur in the southeastern part of the Atlantic Forest Domain (3 species), mountains of the Península de Paria in Venezuela and on Trinidad and Tobago (1 species), and in the Cordillera de la Costa, Venezuelan Andes, and adjacent part of the Cordillera Oriental of the Andes in northern Colombia (1 species). A phylogenetic analysis by Duellman and Gray (1983) revealed that *F. goeldii* was the sister species to *F. ohausi* and *F. fissilis,* and that these Brazilian species were the sister group to the two Venezuelan species, *F. fitzgeraldi* and *F. pygmaeus.* Using geological data on the timing of the uplifts of the mountain ranges and the subsidence of the intervening lower Amazon Basin, Duellman and Gray (1983) postulated that the Brazilian and Venezuelan lineages diverged in the early Tertiary (57–60 million years ago).

Of 43 species of *Gastrotheca,* 41 are associated with the Andes (2 of these also in the Chocó and 1 also in the upper Amazon Basin). An immense geographic hiatus exists between those species and two (*G. fissipes* and *G. microdisca*) in extreme eastern and southeastern Brazil. These two species apparently are not one another's closest relatives (Duellman, 1974). A comparable situation exists in *Trachycephalus* with *T. jordani* (Chocó) as the sister species of *T. atlas* (Caatinga) + *T. nigromaculatus* (Atlantic Forest Domain and Cerrado) (Trueb, 1970). No times have been postulated for the vicariance of the *Gastrotheca* and *Trachycephalus.*

A common disjunct pattern among taxa in the tropical lowlands is one of numerous members of a group in Amazonia-Guiana with one or two vicariants in the Atlantic Coastal Forest; these are inhabitants of humid tropical rainforests, and the distributions of the vicariants are separated from Amazonia-Guiana by the dry Cerrado and/or Caatinga. This pattern is evident in *Adelophryne* (Hoogmoed et al, 1994), *Adenomera* (Heyer, 1973), *Osteocephalus* (Trueb and Duellman, 1971; Duellman, 1974), and in the *Hyla leucophyllata, marmorata,* and *parviceps* groups (Duellman, 1974; Bokermann, 1964; Duellman and Crump, 1974). In the cases of the *Hyla leucophyllata* and *parviceps* groups, vicariants also are present in the Chocó. The pattern of Chocoan vicariants also is present in *Hemiphractus* (Trueb, 1974) and *Physalaemus* (Cannatella and Duellman, 1984). On the other hand, *Sphaenorhynchus* has only three species in Amazonia-Guiana but seven in the Atlantic Forest Domain.

A similar pattern exists in some taxa that are not restricted to rainforests. Of 12 species in the

Leptodactylus podicipinus-wagneri complex (Heyer, 1994), 11 species occur on the Amazonian slopes of the Andes and in Amazonia-Guiana (including Trinidad), the Llanos, and Cerrado-Caatinga-Chaco, whereas one species (*L. natalensis*) is restricted to the Atlantic coast of Brazil. Of 11 species in the *Leptodactylus pentadactylus* complex (Heyer, 1979), one species (*L. flavopictus*) is restricted to the Atlantic Coastal Forest, whereas the others are distributed primarily in Amazonia-Guiana; one of these (*L. pentadactylus*) also occurs in the Chocó and Central America, and another (*L. labyrinthicus*) also occurs in the Atlantic Forest Domain.

Frogs of the genus *Hyalinobatrachium* are most speciose in northwestern South America, where they inhabit the Andes, Chocó, and western part of the Amazon Basin. However, three species united by one synapomorphy (Ruíz-Carranza and Lynch, 1991) are restricted to the Atlantic Forest Domain;

no biogeographic scenario was proposed. Similarly, *Eleutherodactylus* has 259 species in the Andes and 53 species in Amazonia-Guiana, with a gradual diminution of species from west to east (Lynch, 1979), but there are 26 endemic species of *Eleutherodactylus* in the Atlantic Forest Domain. Likewise there is a great decrease in the number of dendrobatids eastward from the Amazon slopes of the Andes and upper Amazon Basin; only *Colostethus* (4 endemic species) is present in the Atlantic Forest Domain. The centrolenids, dendrobatids, and *Eleutherodactylus* all deposit eggs out of water and thus are dependent on high humidity provided by high rainfall and the low evaporation rates characteristic of lower levels in closed-canopy forests. Lower rainfall, greater seasonality, and more open forests in eastern Amazonia presumably restrict the occurrence of many of these species in that region (Lynch, 1979; Duellman, 1988).

CONSERVATION OF THE AMPHIBIAN FAUNA

The increasing rate of destruction of natural environments in South America is typical of that in tropical regions worldwide where burgeoning human populations seek food, fuel, and land for agricultural use. Add to these activities perturbations by commercial logging, mining, and petroleum extraction (and resulting pollution) plus massive devastation wrought by hydroelectric dams and the amount of habitat destruction or modification is catastrophic. Moreover, in recent years populations of some species of amphibians have declined dramatically, in some cases for no determined environmental cause (Blaustein and Wake, 1990; see also Blaustein, 1994; McCoy, 1994; and Pechmann and Wilbur, 1994).

A summary of the status of amphibian populations (Vial and Saylor, 1993) reveals that in South America 57 species of anurans and two of salamanders are classified as threatened or endangered, whereas eight anurans are classified as critical, and 11 additional species are vulnerable. These amount to 78 species or about 4.5% of the species on the continent. Twenty-eight species of anurans are identified as exhibiting declines in abundance, and 16 others are noted as being locally absent. These amount to 44 species or about 2.5% of the South American species. Twelve of the species reported as locally absent and six reported as declining are in southeastern Brazil and Uruguay, whereas all other

species in these categories are at high elevations in the Andes. In contrast, studies of amphibian abundance over a period of 6 yr in undisturbed lowland rainforest in the upper Amazon Basin revealed no compelling evidence for declines (Duellman, 1995).

The survival of many species of amphibians (and other organisms) depends on the establishment of protected areas, whether these be biosphere reserves, ecological stations, or national parks, where natural environmental conditions are maintained. As of 1992 there were 667 protected areas in South America covering 98,906 km^2 or about 5.5% of the continent (Barzetti, 1993). As is evident from the map (Fig. 5:8), the largest protected areas are in the Amazon Basin, followed by the tropical Andes and the austral temperate forests, and there are numerous small areas in the Atlantic Forest Domain.

The available data do not allow a determination of the percentages of natural regions or vegetation types that are in protected areas. Likewise, it is not possible to deduce how many species of amphibians are in the protected areas, but it is unlikely that more than 50% of the species inhabit these areas. In order to increase that percentage, it is necessary to establish protected areas in regions of high species diversity and endemism. Within vegetation types, amphibian species density is greatest per unit area in the austral temperate forest and cloud forest. Nu-

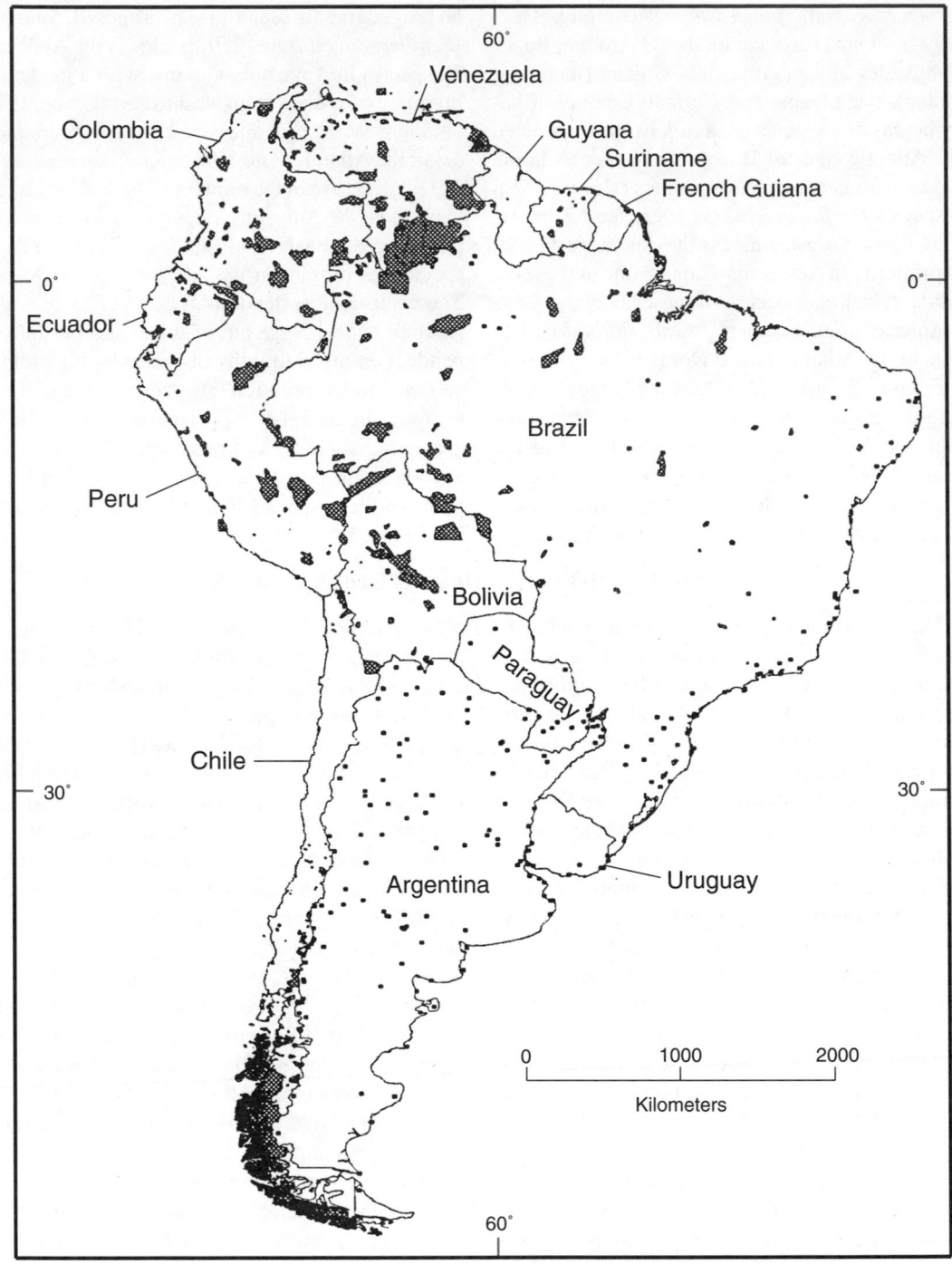

Fig. 5:8. Protected areas in South America (modified from Barzetti, 1993).

merous large parks and reserves exist in the austral temperate forest both in Argentina and Chile, but protection is not effective in Chile where harvesting of lumber is rampant. As of 1986, 16 protected areas existed in the tropical Andes, and 15 new areas were proposed (Saavedra and Freese, 1986). Some of the Andean areas, such as Manu National Park in Peru and the Cayambe-Coca National Park in Ecuador extend from supra-treeline habitats to Amazonian rainforest and thus encompass a wide altitudinal range of montane environments.

Among frogs of the genus *Eleutherodactylus,* the most speciose group in Andean cloud forests, species at higher elevations in the Andes in Ecuador have smaller latitudinal ranges than those at lower elevations (Lynch and Duellman, 1980, 1997). Experience with Andean anurans suggests that this phenomenon is the general rule, but *Centrolene buckleyi* is a notable exception. If this is true, the few large protected zones that exist now or have been proposed will exclude the ranges of many species. Therefore, for preservation of the greatest number of species of amphibians in montane regions (cloud forest and supra-treeline habitats), more small protected areas are better than a few large ones.

Island biogeography theory has played an important role in the conceptual approach to establishment of protected areas (see Boecklen and Gotelli, 1984, for review). Theoretically, larger areas will harbor more species over a longer period of time than will smaller areas, but different kinds of organisms require different-sized areas. Amphibians generally have home ranges that are minuscule contrasted to those of jaguars, spider monkeys, or Amazon parrots, all groups that receive much attention from conservationists. The only empirical data on sizes of protected areas for amphibians was presented by Zimmerman and Bierregaard (1986), who noted that in different-sized forest reserves near Manaus, Brazil, 90% of the species in the region were found in woodlots of only 350 ha; furthermore, they emphasized the need to include breeding sites in protected areas. Thus, it would seem that many small protected areas of only 1000 ha that include ponds and streams (unaffected by pollution) in areas of known species richness might effectively preserve a greater percentage of the amphibian fauna than is protected now. This is especially desirable in the Andes, where so many species have restricted distributions.

DISCUSSION

The validity of the data is questionable with respect to the number of taxa in many regions. Whereas more than two decades have passed since new species have been found in the Caribbean Coastal Forest or in the Llanos and only about 5%of the species have been discovered in the past two decades in the Cerrado-Caatinga-Chaco and Pampean-Monte regions, the rate of species discovery is high in many regions, especially the highlands and the lowland rainforests. Species discovery rates for the latter reveal that 14.3% of the species in the Chocó have been discovered since 1979; the values for Amazonia-Guiana and the Atlantic Forest Domain are 19.4% and 29.3%, respectively (Fig. 5:9). The highlands had a much higher rate of species discovery in the 1980s and 1990s, when 49.9% of the species in the Andes and 67.1% of the species in the Guiana Highlands were described.

These discrepancies are a reflection on the history of sampling of different regions. For, example, many collections were made along the northeastern coast of South America prior to 1900, by which time 57% of the species in the Caribbean Coastal Forest

and 58% in the Guianan lowlands had been described; also by that time 75% of the species on Trinidad and Tobago were known to science. These numbers are much higher than those for the lowland rainforests, for by 1900, only 37% of the species in the Chocó and 36% of the species in the Amazon Basin had been described. Surprisingly, by 1900 only 19% of the species in the Atlantic Forest Domain were known. Even these numbers contrast markedly with the highlands, where by 1900 only 12% of the species in the Andes and 2% of those in the Guiana Highlands had been described. In 1900, none of the presently known 19 species endemic to the Sierra Nevada de Santa Marta was known.

As collecting effort (= sampling time) increases in a given area, the number of new species added to the fauna asymptotically approaches some ceiling; for example, see Duellman's (1978) species accumulation curve for the herpetofauna at Santa Cecilia, Ecuador. It is tempting to apply equations to describe the cumulative species-effort relationship and thereby predict the size of a fauna, as reviewed by Soberón and Llorente (1993). Although such

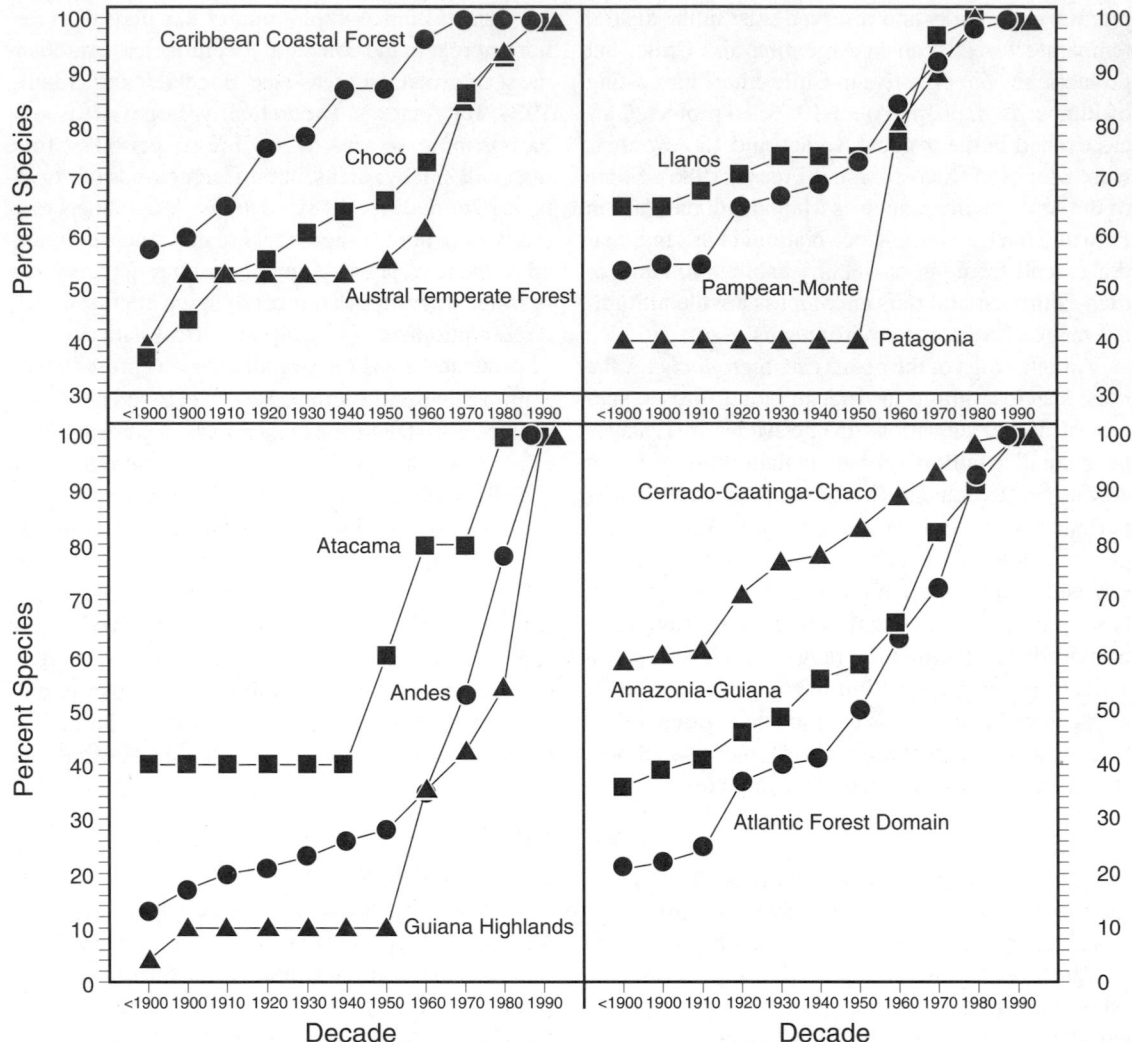

Fig. 5:9.   Percentage accumulation of amphibian species (1900–1996) in twelve regions in South America; the decade 1990 includes 1990 through part of 1996.

equations have predictive value with a fauna for which the sampling effort is known, they cannot be applied appropriately to the data on accumulated number of species in the different regions discussed here, because collecting effort is not known. In some cases decades have passed without any collecting in some regions; for example, the number of amphibian species in Patagonia remained static at four species for six decades, until it was doubled in the 1960s.

During the past two decades collecting efforts have intensified, especially by South American biologists. Previously collected areas have been studied more intensively, and collections have been made in many previously unknown areas. The result has

been an astonishing rate of discovery of new species, so that nearly one-third of the South American amphibians have been described in the 1980s and 1990s. Although there is no mathematical equation applicable to the existing data, I predict that by the second millennium the number of species of amphibians known in South America will approach 2000. Unfortunately, by the turn of the century many of these species probably will be extinct.

As Heyer's (1988) analysis of amphibians of South America east of the Andes shows, many areas of South America have not been collected. Patterns of distribution are biased accordingly. Geographical patterns of collecting have definite effects on biogeographic interpretation. For example, there is a positive correlation between plots of collecting density of plants and proposed Pleistocene refugia (Nelson et al., 1990). Thus, patterns of distribution, even in the lowlands, are subject to change as more extensive sampling is accomplished. For example, three species of anurans first discovered in, and described from, southern Amazonian Peru in 1988 and 1989 (*Phyllomedusa atelopoides*, Duellman et al., 1988; *Hyla allenorum* and *H. koechlini*, Duellman and Trueb, 1989) subsequently have been found in the Iquitos region, about 1000 km to the north-northwest.

These data suggest that we can anticipate the discovery of many more species, especially in the highlands, and the extension of ranges of many known species, especially in the lowlands. These discoveries will result from much more extensive inventories of the amphibian fauna. However, to gain a real understanding of the evolution of the richest amphibian fauna in the world and its component communities, systematists must determine the phylogenetic relationships of the taxa, and ecologists and biogeographers must incorporate these evolutionary histories into their analyses. Although standard alcohol-preserved museum specimens document geographic occurrence and provide material for morphological study, other kinds of data are obtainable only from nontraditional material (e.g., chromosome preparations and frozen tissues for molecular study). These and important life history data usually must be obtained from specific field studies incorporating the collection of such data. Thus, systematic biologists have a strong vested interest in the preservation of the fauna for future study.

Acknowledgments: First and foremost, a great debt of gratitude is owed to the hundreds of biologists who for more than 200 yr have been accumulating data on the systematics and distribution of South American amphibians; only a small fraction of the published work has been cited. Secondly, gratitude is expressed to the compilers of *Amphibian Species of the World* (Frost, 1985, as updated by Duellman, 1993a); these works have made the accumulation of data a much simpler process. Thirdly, I am indebted to many colleagues with whom I share a love of Neotropical amphibians and with whom I have discussed over the years many ideas relating to the systematics and biogeography of these organisms. I am no longer aware of which ideas are whose and take full responsibility for statements made herein, but I feel especially indebted to David C. Cannatella, José M. Cei, W. Ronald Heyer, Marinus S. Hoogmoed, Raymond F. Laurent, John D. Lynch, Roy W. McDiarmid, and Linda Trueb. Lastly, I thank Linda Trueb for her patience and assistance with the production of the figures herein. For critical review of the manuscript, I thank Jonathan A. Campbell, Hélio da Silva, Ignacio de la Riva, W. Ronald Heyer, John D. Lynch, and Linda Trueb. Many of the data incorporated herein resulted from extensive field work in South America supported at various times by the National Geographic Society, National Science Foundation, and the Natural History Museum of The University of Kansas.

LITERATURE CITED

Ab'Saber, A. N. 1977. Os domínios morfoclimáticos na América do sul. Primeira aproximação. Geomorfologia 53:1–23.

Ardila-Robayo, M. C., P. M. Ruíz-Carranza, and S. H. Roa-Trujillo. 1993. Una nueva especie de *Hyla* del grupo *larinopygion* (Amphibia: Anura: Hylidae) del sur de la Cordillera Central de Colombia. Rev. Acad. Colombiana Cienc. 18:559–566.

Axelrod, D. I. 1972. Edaphic aridity as a factor in angiosperm evolution. Amer. Nat. 106:311–320.

Ayarzagüena, J. 1992. Los centrolenidos de la Guayana Venezolana. Publ. Asoc. Amigos Doñana 1:1–48.

Ayarzagüena, J., and C. Señaris. 1993. Dos nuevas especies de *Hyla* (Anura: Hylidae) para las cumbres tepuyanas del Estado Amazonas, Venezuela. Mem. Soc.Cien. Nat. La Salle 53:127–146.

Ayarzagüena, J., and C. Señaris. 1996. Dos nuevas especies de *Cochranella* (Anura; Centrolenidae) para Venezuela. Publ. Asoc. Amigos Doñana 8:1–16.

Ayarzagüena, J., J. C. Señaris, and S. Gorzula. 1992. El grupo *Osteocephalus rodriguezi* de las tierras altas de la Guayana Venezolana: descripción de cinco nuevas especies. Mem. Soc. Cien. Nat. La Salle 52:113–142.

Ayarzagüena, J., J. C. Señaris, and S. Gorzula. 1993 "1992." Un nuevo género para las especies del "grupo *Osteocephalus rodriguezi*" (Anura: Hylidae). Mem. Soc. Cienc. Nat. La Salle 52:213–221.

Báez, A. M., and G. J. Scillato Yané. 1979. Late Cenozoic environmental changes in temperate Argentina. Pp. 141–156 *in* W. E. Duellman (ed.), The South American herpetofauna: its origin, evolution, and dispersal. Monogr. Mus. Nat. Hist. Univ. Kansas 7:1–485.

Barzetti, V. 1993. *Parks and Progress*. Washington, D.C.: IUCN-The World Conservation Union.

Basso, N. G. 1990. Estrategias adaptivas en una comunidad subtropical de anuros. Cuad. Herpetol. Ser. Monogr. 1:1–70.

Bastos, R. P., and J. P. Pombal, Jr. 1995. New species of *Crossodactylus* (Anura: Leptodactylidae) from the Atlantic rainforest of southeastern Brazil. Copeia 1995:436–439.

Bastos, R. P., and J. P. Pombal, Jr. 1996. A new species of *Hyla* (Anura: Hylidae) from eastern Brazil. Amphibia-Reptilia 17:325–331.

BEURLEN, K. 1970. *Geologie von Brasilien.* Berlin: Borntraeger.

BEARD, J. S. 1953. The savanna vegetation of northern tropical America. Ecol. Monogr. 23:149–215.

BLAUSTEIN, A. R. 1994. Chicken Little or Nero's fiddle? A perspective on declining amphibian populations. Herpetologica 50:85–97.

BLAUSTEIN, A. R., AND D. B. WAKE. 1990. Declining amphibian populations: a global phenomenon? Trends Ecol. Evol. 5:203–204.

BOECKLEN, W. J., AND N. J. GOTELLI. 1984. Island biogeographic theory and conservation practice: species-area or specious-area relationship? Biol. Conserv. 29:63–80.

BOKERMANN, W. C. A. 1964. Notes on frogs of the *Hyla marmorata* group with description of a new species (Amphibia, Hylidae). Senckenberg. Biol. 45:243–254.

BRAME, A. H., JR., AND D. B. WAKE. 1963. The salamanders of South America. Contrib. Sci. Los Angeles Co. Mus. 69:1–72.

CANNATELLA, D. C. 1984. Two new species of the leptodactylid frog genus *Phrynopus,* with comments on the phylogeny of the genus. Occas. Pap. Mus. Nat. Hist. Univ. Kansas 131:1–16.

CANNATELLA, D. C., AND W. E. DUELLMAN. 1984. Leptodactylid frogs of the *Physalaemus pustulosus* group. Copeia 1984:902–921.

CARAMASCHI, U. 1996. Nova espécie de *Odontophrynus* Reinhardt & Lütken, 1862 do Brasil central (Amphibia, Anura, Leptodactylidae). Bol. Mus. Nac. Rio de Janeiro 367:1–7.

CARAMASCHI, U., AND C. A. G. DA CRUZ. 1997. Redescription of *Chiasmocleis albopunctata* (Boettger) and description of a new species of *Chiasmocleis* (Anura: Microhylidae). Herpetologica 53:259–268.

CARAMASCHI, U., AND A. VELOSA. 1996. Nova espécie de *Hyla* Laurenti, 1768 do leste brasileiro (Amphibia, Anura, Hylidae). Bol. Mus. Nac. Rio de Janeiro 365:1–7.

CASCON, P., AND J. S. LIMA VERDE. 1994. Uma nova espécie de *Chthonerpeton* do nordeste brasileiro (Amphibia, Gymnophiona, Typhlonectidae). Rev. Brasileiro Biol. 54:549–553.

CEI, J. M. 1979. The Patagonian herpetofauna. Pp. 309–339 *in* W. E. Duellman (ed.), The South American herpetofauna: its origin, evolution, and dispersal. Monogr. Mus. Nat. Hist. Univ. Kansas 7:1–485.

CEI, J. M. 1980. Amphibians of Argentina. Monitor. Zool. Italiano (N.S.), Monogr. 2:1–609.

CEI, J. M. 1986. Speciation and adaptive radiation in Andean *Telmatobius* frogs. Pp. 374–386 *in* F. Vuilleumier and M. Monasterio (eds.), *High Altitude Tropical Biogeography.* New York: Oxford Univ. Press.

COLINVAUX, P. 1993. Pleistocene biogeography and diversity in tropical forests of South America. Pp. 373–499 *in* P. Goldblatt (ed.), *Biological Relationships between Africa and South America.* New Haven: Yale Univ. Press.

COLOMA, L. A. 1991. *Anfibios del Ecuador: Lista de Especies, Ubicación Altitudinal y Referencias Bibliográficas.* Quito:Rept. Técn. EcoCiencia.

COLOMA, L. A. 1995. Ecuadorian frogs of the genus *Colostethus* (Anura: Dendrobatidae). Misc. Publ. Nat. Hist. Mus. Univ. Kansas 87:1–72.

CRUZ, C. A. G. DA., U. CARAMASCHI, AND E. IZECKSOHN. 1997. The genus *Chiasmocleis* Méhely 1904 (Anura, Microhylidae), with description of three new species. Alytes 15:49–71.

DE LA RIVA, I. 1990. Lista preliminar comentada de los anfibios

de Bolivia con datos sobre su distribución. An. Mus. Reg. Sci. Nat. Turino 8:261–319.

DE LA RIVA, I. 1993a. Ecología de una comunidad neotropical de anfibios durante la estación lluviosa. Doctoral dissertation. Madrid: Univ. Complutense de Madrid.

DE LA RIVA, I. 1993b. A new species of *Scinax* (Anura:Hylidae) from Argentina and Bolivia. J. Herpetol. 27:41–46.

DE LA RIVA, I. 1994a. A new aquatic frog of the genus *Telmatobius* (Anura: Leptodactylidae) from Bolivian cloud forests. Herpetologica 50:38–45.

DE LA RIVA, I. 1994b. Description of a new species of *Telmatobius* from Bolivia (Amphibia: Anura: Leptodactylidae). Graellsia 50:161–164.

DE LA RIVA, I. 1996. The specific name of *Adenomera* (Anura: Leptodactylidae) in the Paraguay River Basin. J. Herpetol. 30:556–558.

DE LA RIVA, I., AND J. D. LYNCH. 1997. New species of *Eleutherodactylus* from Bolivia (Amphibia: Leptodactylidae). Copeia 1997:151–157.

DUELLMAN, W. E. 1970. The hylid frogs of Middle America. Monogr. Mus. Nat. Hist. Univ. Kansas 1:1–753.

DUELLMAN, W. E. 1972. A review of neotropical frogs of the *Hyla bogotensis* group. Occas. Pap. Mus. Nat. Hist. Univ. Kansas 22:1–31.

DUELLMAN, W. E. 1974. A reassessment of the taxonomic status of some neotropical hylid frogs. Occas. Pap. Mus. Nat. Hist. Univ. Kansas 27:1–27.

DUELLMAN, W. E. 1978. The biology of an equatorial herpetofauna in Amazonian Ecuador. Misc. Publ. Mus. Nat. Hist. Univ. Kansas 65:1–352.

DUELLMAN, W. E. 1979a. The South American herpetofauna: a panoramic view. Pp. 1–28 *in* W. E. Duellman (ed.), The South American herpetofauna: its origin, evolution, and dispersal. Monogr. Mus. Nat. Hist. Univ. Kansas 7:1–485.

DUELLMAN, W. E. 1979b. The herpetofauna of the Andes: patterns of distribution, origin, differentiation, and present communities. Pp. 371–460 *in* W. E. Duellman (ed.), The South American herpetofauna: its origin, evolution, and dispersal. Monogr. Mus. Nat. Hist. Univ. Kansas 7:1–485.

DUELLMAN, W. E. 1982. Quaternary climatic-ecological fluctuations in the lowland tropics: frogs and forests. Pp. 389–402 *in* G. T. Prance (ed.), *Biological Diversification in the Tropics.* New York: Columbia Univ. Press.

DUELLMAN, W. E. 1983. Compresión climática cuaternaria en los Andes: efectos sobre la especiación. Pp. 177–201 + 269, 271, 278 *in* P. J. Salinas (ed.), *Zoologico Neotropical, Actas VIII Congreso Latinoamericano de Zoologia.* Mérida, Venezuela: Univ. de los Andes.

DUELLMAN, W. E. 1984. Taxonomy of Brazilian hylid frogs of the genus *Gastrotheca.* J. Herpetol. 18:302–312.

DUELLMAN, W. E. 1988. Patterns of species diversity in anuran amphibians in the American tropics. Ann. Missouri Bot. Gard. 75:79–104.

DUELLMAN, W. E. 1990. Herpetofaunas in neotropical rainforests: comparative composition, history, and resource utilization. Pp. 455–505 *in* A. H. Gentry (ed.), *Four Neotropical Rainforests.* New Haven: Yale Univ. Press.

DUELLMAN, W. E. 1993a. Amphibian species of the world: additions and corrections. Spec. Publ. Mus. Nat. Hist. Univ. Kansas 21:1–372.

DUELLMAN, W. E. 1993b. Amphibians in Africa and South America: evolutionary history and ecological comparisons.

Pp. 200–243 *in* P. Goldblatt (ed.), *Biological Relationships between Africa and South America*. New Haven: Yale Univ. Press.

DUELLMAN, W. E. 1995. Temporal fluctuation in abundances of anuran amphibians in a seasonal Amazonian rainforest. J. Herpetol. 29:13–21.

DUELLMAN, W. E. 1997. Amphibians of La Escalera region, southeastern Venezuela: taxonomy, ecology, and biogeography. Sci. Pap. Nat. Hist. Mus.Univ. Kansas 2:1–52.

DUELLMAN, W. E., J. E. CADLE, AND D. C. CANNATELLA. 1989. A new species of terrestrial *Phyllomedusa* (Anura: Hylidae) from southern Peru. Herpetologica 44:91–95.

DUELLMAN, W. E., AND L. A. COLOMA. 1993. *Hyla staufferorum,* a new species of treefrog in the *Hyla larinopygion* group from the cloud forests of Ecuador. Occas. Pap. Mus. Nat. Hist. Univ. Kansas 161:1–11.

DUELLMAN, W. E., AND M. L. CRUMP. 1974. Speciation in frogs of the *Hyla parviceps* group in the upper Amazon Basin. Occas. Pap. Mus. Nat. Hist. Univ. Kansas 23:1–40.

DUELLMAN, W. E., I. DE LA RIVA, AND E. R. WILD. 1997. Frogs of the *Hyla armata* and *Hyla pulchella* groups in the Andes of South America, with definitions and analyses of phylogenetic relationships. Sci. Pap. Nat. Hist. Mus. Univ. Kansas 3:1–41.

DUELLMAN, W. E., AND P. GRAY. 1983. Developmental biology and systematics of the egg-brooding hylid frogs, genera *Flectonotus* and *Fritziana*. Herpetologica 39:333–359.

DUELLMAN, W. E., AND D. M. HILLIS. 1987. Marsupial frogs (Anura: Hylidae: *Gastrotheca*) of the Ecuadorian Andes: resolution of taxonomic problems and phylogenetic relationships. Herpetologica 43:141–173.

DUELLMAN, W. E., AND D. M. HILLIS. 1990. Systematics of the *Hyla larinopygion* group. Occas. Pap. Mus. Nat. Hist. Univ. Kansas 134:1–23.

DUELLMAN, W. E., AND M. S. HOOGMOED. 1984. The taxonomy and phylogenetic relationships of the hylid frog genus *Stefania*. Misc. Publ. Mus. Nat. Hist. Univ. Kansas 75:1–39.

DUELLMAN, W. E., AND M. S. HOOGMOED. 1992. Some hylid frogs from the Guiana Highlands, northeastern South America: new species, distributional records, and a generic reallocation. Occas. Pap. Mus. Nat. Hist. Univ. Kansas 147:1–21.

DUELLMAN, W. E., AND J. D. LYNCH. 1988. Anuran amphibians from the Cordillera de Cutucú, Ecuador. Proc. Acad. Nat. Sci. Philadelphia 140:125–142.

DUELLMAN, W. E., L. R. MAXSON, AND C. A. JESIOLOWSKI. 1988. Evolution of marsupial frogs (Hylidae: Hemi-phractinae): Immunological evidence. Copeia 1988:527–543.

DUELLMAN, E. E., AND J. R. MENDELSON III. 1995. Amphibians and reptiles from northern Departamento Loreto, Peru: taxonomy and biogeography. Univ. Kansas Sci. Bull. 55:329–376.

DUELLMAN, W. E., AND A. W. SALAS. 1991. Annotated checklist of the amphibians and reptiles of Cuzco Amazónico, Peru. Occas. Pap. Mus. Nat. Hist. Univ. Kansas 143:1–13.

DUELLMAN, W. E., AND R. SCHULTE. 1993. New species of centrolenid frogs from northern Peru. Occas. Pap. Mus. Nat. Hist. Univ. Kansas 155:1–33.

DUELLMAN, W. E., AND R. THOMAS. 1996. Anuran amphibians from a seasonally dry forest in southeastern Peru and comparisons of the anurans among sites in the upper Amazon Basin. Occas. Pap. Nat. Hist. Mus. Univ. Kansas 180:1–34.

DUELLMAN, W. E., AND C. A. TOFT. 1979. Anurans from the Ser-

ranía de Sira, Amazonian Perú: taxonomy and biogeography. Herpetologica 35:60–70.

DUELLMAN, W. E., AND L. TRUEB. 1986. *Biology of Amphibians.* New York: McGraw-Hill Book Co.

DUELLMAN, W. E., AND L. TRUEB. 1989. Two new treefrogs of the *Hyla parviceps* group from the Amazon Basin in southern Peru. Herpetologica 45:1–10.

DUELLMAN, W. E., AND A. VELOSO M. 1977. Phylogeny of *Pleurodema* (Anura: Leptodactylidae): a biogeographic model. Occas. Pap. Mus. Nat. Hist. Univ. Kansas 64:1–46.

DUELLMAN, W. E., AND J. J. WIENS. 1993. Hylid frogs of the genus *Scinax* Wagler, 1830, in Amazonian Ecuador and Peru. Occas. Pap. Mus. Nat. Hist. Univ. Kansas 153:1–57.

DUELLMAN, W. E., AND E. R. WILD. 1993. Anuran amphibians from the Cordillera de Huancabamba, northern Peru: systematics, ecology, and biogeography. Occas. Pap. Mus. Nat. Hist. Univ. Kansas 157:1–53.

DUELLMAN, W. E., AND M. YOSHPA. 1996. A new species of *Tepuihyla* (Anura: Hylidae) from Guyana. Herpetologica 52:275–281.

DWYER, C. M. 1995. A new species of *Eleutherodactylus* from Perú (Anura: Leptodactylidae). Amphib.-Rept. 16:245–256.

EIDT, R. C. 1968. The climatology of South America. Pp. 54–81 *in* E. J. Fittkau, J. Illies, H. Klinge, G. H. Schwabe, and H. Sioli (eds.), *Biogeography and Ecology in South America. Vol. 1.* The Hague: Dr. W. Junk N. V.

FAIRBRIDGE, R. W. 1972. Climatology of a glacial cycle. Quatern. Res. 2:283–302.

FLENLEY, J. R. 1979. *A Geological History of Tropical Rainforest.* London: Butterworths.

FLORES, G. 1993. A new species of earless *Eleutherodactylus* (Anura: Leptodactylidae) from the Pacific slopes of the Ecuadorian Andes, with comments on the *Eleutherodactylus surdus* assembly. Herpetologica 49:427–434.

FLORES, G., AND L. O. RODRÍGUEZ. 1997. Two new species of the *Eleutherodactylus conspicillatus* group (Anura: Leptodactylidae) from Peru. Copeia 1997:388–394.

FLORES, G., AND G. O. VIGLE. 1994. A new species of *Eleutherodactylus* (Anura: Leptodactylidae) from the lowland rainforests of Amazonian Ecuador, with notes on the *Eleutherodactylus frater* assembly. J. Herpetol. 28:416–424.

FORMAS, J. R. 1979. La herpetofauna de los bosques temperados de Sudamérica. Pp. 341–369 *in* W. E. Duellman (ed.), The South American herpetofauna: its origin, evolution, and dispersal. Monogr. Mus. Nat. Hist. Univ. Kansas 7:1–485.

FORMAS, J. R. 1997. A new species of *Batrachyla* from southern Chile. Herpetologica 53:6–13.

FRAILEY, C. D., E. L. LAVINA, A. RANCY, AND J. PEREIRA DE SOUZA FILHO. 1988. A proposed Pleistocene/Holocene lake in the Amazon Basin and its significance to Amazonian geology and biogeography. Acta Amazonica 18:119–143.

FROST, D. R. 1985. *Amphibian Species of the World.* Lawrence, Kansas: Association of Systematics Collections and Allen Press, Inc.

GALLARDO, J. M. 1979. Composición, distribución y origen de la herpetofauna chaqueña. Pp. 299–307 *in* W. E. Duellman (ed.), The South American herpetofauna: its origin, evolution, and dispersal. Monogr. Mus. Nat. Hist. Univ. Kansas 7:1–485.

GIARETTA, A. A., AND I. SAZIMA. 1993. Nova espécie de *Proceratophrys* Mir. Rib. do sul de Minas Gerais, Brasil (Amphibia, Anura, Leptodactylidae). Rev. Brasileira Biol. 53:13–19.

GLUESENKAMP, A. C. 1995. A new species of *Osornophryne*

(Anura: Bufonidae) from Volcán Sumaco, Ecuador with notes on other members of the genus. Herpetologica 51:268–279.

GOMES, M . R., AND O. L. PEIXOTO. 1996. Nova espécie de *Hyla* do grupo *marmorata* de Sergipe, nordeste do Brasil (Amphibia, Anura, Hylidae). Iheringia (Zool.) 80:33–38.

GRANT, T., E. C. HUMPHREY., AND C. W. MYERS. 1997. The median lingual process of frogs: a bizarre character of Old World ranoids discovered in South American dendrobatids. Am. Mus. Novit. 3212:1–40.

GRAYBEAL, A., AND D. C. CANNATELLA. 1995. A new taxon of Bufonidae from Peru, with descriptions of two new species and a review of the phylogenetic status of supraspecific bufonid taxa. Herpetologica 51:105–131.

GUDYNAS, E., AND J. C. RUDOLF. 1987. La herpetofauna de la localidad costera de "Pajas Blancas" (Uruguay): lista sistemática comentada y estructura ecológica de la comunidad. Comun. Mus. Cienc. Pontif. Univ. Rio Grande do Sul 46:173–194.

HADDAD, C. F. B., AND M. MARTINS. 1994. Four species of Brazilian poison frogs related tp *Epipedobates pictus* (Dendrobatidae): taxonomy and natural history observations. Herpetologica 50:282–295.

HADDAD, C. F. B. , AND J. P. POMBAL, JR. 1995. A new species of *Hylodes* from southeastern Brazil (Amphibia: Leptodactylidae). Herpetologica 51:279–286.

HADDAD, C. F. B., J. P. POMBAL, JR., AND R. P. BASTOS. 1996. New species of *Hylodes* from the Atlantic Forest of Brazil (Amphibia: Leptodactylidae). Copeia 1996:965–969.

HAFFER, J. 1969. Speciation in Amazon forest birds. Science 161:131–137.

HAFFER, J. 1974. Avian speciation in tropical South America. Publ. Nuttall Ornithol. Club, Cambridge 14:1–390.

HAFFER, J. 1979. Quaternary biogeography of tropical lowland South America. Pp. 107–140 *in* W. E. Duellman (ed.), The South American herpetofauna: its origin, evolution, and dispersal. Monogr. Mus. Nat. Hist. Univ. Kansas 7:1–485.

HANKEN, J., AND D. B. WAKE. 1982. Genetic differentiation among plethodontid salamanders (genus *Bolitoglossa*) in Central and South America: implications for the South American invasion. Herpetologica 38:272–287.

HARVEY, M. B. 1996. A new species of glass frog (Anura: Centrolenidae: *Cochranella*) from Bolivia, and the taxonomic status of *Cochranella flavidigitata*. Herpetologica 52:427–435.

HARVEY, M. B., AND M. B. KECK. 1995. A new species of *Ischnocnema* (Anura: Leptodactylidae) from high elevations in the Andes of central Bolivia. Herpetologica 51:56–66.

HARVEY, M. B., AND E. N. SMITH. 1993. A new species of aquatic *Bufo* (Anura: Bufonidae) from cloud forests in the Serranía de Siberia, Bolivia. Proc. Biol. Soc. Washington 106:442–449.

HARVEY, M. B., AND E. N. SMITH. 1994. A new species of *Bufo* (Anura: Bufonidae) from cloud forests in Bolivia. Herpetologica 50:32–38.

HASTENRATH, S. I. 1967. Observations on the snow line in the Peruvian Andes. J. Glaciol. 6:541–550.

HEYER, W. R. 1973. Systematics of the marmoratus group of the frog genus *Leptodactylus* (Amphibia, Leptodactylidae). Contrib. Sci. Nat. Hist. Mus. Los Angeles Co. 251:1–50.

HEYER, W. R. 1979. Systematics of the *pentadactylus* species group of the frog genus *Leptodactylus* (Amphibia: Leptodactylidae). Smithsonian Contrib. Zool. 301:1–43.

HEYER, W. R. 1988. On frog distribution patterns east of the Andes. Pp. 225–273 *in* P. E. Vanzolini and W. R. Heyer (eds.). *Proceedings of a Workshop on Neotropical Distribution Patterns.* Rio de Janeiro: Acad. Brasil. Cien.

HEYER, W. R. 1994. Variation within the *Leptodactylus podicipinus-wagneri* complex of frogs (Amphibia: Leptodactylidae). Smithsonian Contr. Zool. 546:1–124.

HEYER, W. R., J. M. GARCÍA-LOPEZ, AND A. J. CARDOSO. 1996. Advertisement call variation in the *Leptodactylus mystaceus* species complex (Amphibia; Leptodactylidae) with a description of a new species. Amphibia-Reptilia 17:7–31.

HEYER, W. R., AND L. R. MAXSON. 1982. Distributions, relationships, and zoogeography of lowland frogs. The *Leptodactylus* complex in South America, with special reference to Amazonia. Pp. 375–388 *in* G. T. Prance. (ed.), Biological Diversification in the Tropics. New York: Columbia Univ. Press.

HEYER, W. R., AND L. R. MAXSON. 1983. Relationships, zoogeography, and speciation mechanisms of frogs of the genus *Cycloramphus* (Amphibia, Leptodactylidae). Arq. Zool., Mus. Zool. Univ. São Paulo 30:341–373.

HEYER, W. R., A. S. RAND, C. A. G. CRUZ, O. L. PEIXOTO, AND C. E. NELSON. 1990. Frogs of Boracéia. Arq. Zool. Mus. Zool Univ. São Paulo 31:231–410.

HILLIS, D. M., AND R. DE SÁ. 1988. Phylogeny and taxonomy of the *Rana palmipes* group (Salientia: Ranidae). Herpetol. Monogr. 2:1–26.

HOOGMOED, M. S. 1979. The herpetofauna of the Guianan Region. Pp. 241–279 *in* W. E. Duellman (ed.), The South American herpetofauna: its origin, evolution, and dispersal. Monogr. Mus. Nat. Hist. Univ. Kansas 7:1–485.

HOOGMOED, M. S., D. M. BORGES, AND P. CASCON. 1994. Three new species of the genus *Adelophryne* (Amphibia: Anura: Leptodactylidae) from northeastern Brazil, with remarks on the other species in the genus. Zool. Meded. 68:271–300.

HOOGMOED, M. S., AND J. E. CADLE. 1991. Natural history and distribution of *Agalychnis craspedopus* (Funkhouser, 1957) (Amphibia: Anura: Hylidae). Zool. Meded. Leiden 65:129–142.

HOOGMOED, M. S., AND S. J. GORZULA. 1979. Checklist of the savanna inhabiting frogs of the El Manteco region with notes on the ecology and the description of a new species of treefrog (Hylidae, Anura). Zool. Meded. Leiden 54:183–216.

HUECK, K., AND P. SEIBERT. 1972. *Vegetationskarte von Südamerika.* Stuttgart: Gustav Fischer Verlag.

IZECKSOHN E. 1993a. Nova espécie de *Dendrophryniscus* da região Amazônica (Amphibia, Anura, Bufonidae). Rev. Brasilera Zool. 10:407–412.

IZECKSOHN, E. 1993b. Três novas espécies de *Dendrophryniscus* Jiménez de la Espada das regiões sudeste e sul do Brazil (Amphibia, Anura, Bufonidae). Rev. Brasileira Zool.10:473–488.

JANZEN, D. H. 1967. Why mountain passes are higher in the tropics. Amer. Nat. 101:233–249.

JEANNEL, R. 1967. Biogeographie de l'Amerique australe. Pp. 401–460 *in* C. Delamere Deboutteville and E. Rapoport (eds.). *Biologie de l'Amerique Australe, Vol. 3.* Paris: C. N. R. S. Groupe Francais Argiles C. R. Reun. Etud.

JUNGFER, K.-H., AND L. C. SCHIESARI. 1995. Description of a central Amazonian and Guianan tree frog, genus *Osteocephalus* (Anura: Hylidae), with oophagus tadpoles. Alytes 13:1–13.

KAPLAN, M. 1994. A new species of frog of the genus *Hyla* from the Cordillera Oriental in Northern Colombia with comments on the taxonomy of *Hyla minuta*. J. Herpetol. 28:79–87.

KAPLAN, M. 1997a. A new species of *Colostethus* from the Sierra Nevada de Santa Marta (Colombia) with comments on intergeneric relationships within the dendrobatidae. J. Herpetol. 31:369–375.

KAPLAN, M. 1997b. On the status of *Hyla bogerti* Cochran and Goin. J. Herpetol. 31:536–541.

KAPLAN, M., AND P. M. RUÍZ. 1997. Two new species of *Hyla* from the Andes of central Colombia and their relationships to other small Andean *Hyla*. J. Herpetol. 31:230–244.

KIESTER, A. R. 1971. Species density of North American amphibians and reptiles. Syst. Zool., 20:127–137.

KLAPPENBACH, M. A., AND J. A. LANGONE. 1992. Lista sistemática y sinonímica de los anfibios de Uruguay. Con comentarios y notas sobre su distribución. An. Mus. Nac. Hist. Nat. Montevideo 8:163–222.

KÖHLER, J., AND K.-H. JUNGFER. 1995. Eine neue Art und ein Erstnachweis von Fröschen der Gattung *Eleutherodactylus* aus Bolivien. Salamandra 31:149–156.

KRONBERG, B. I., R. E. BENCHIMOL, AND M. I. BIRD. 1991. Geochemistry of the Acre subbasin sediments: window on ice-age Amazonia. Interciencia 16:138–141.

LA MARCA, E. 1994a. Descripción de un género nuevo de ranas (Amphibia: Dendrobatidae) de la Cordillera de Mérida, Venezuela. Anuario Invest. Inst. Geogr. Conserv. Recur. Nat. Univ. Los Andes 1991:39–41.

LA MARCA, E. 1994b. Descripción de una nueva especie de *Atelopus* (Amphibia: Anura: Bufonidae) de selva nublada andina de Venezuela. Mem. Soc. Cien. Nat. La Salle 54 (142):101–107.

LA MARCA, E. 1996a. First record of *Hyla pugnax* (Amphibia: Anura: Hylidae) from Venezuela. Bull. Maryland Herpetol. Soc. 32:35–37.

LA MARCA, E. 1996b. Ranas del género *Colostethus* (Amphibia: Anura: Dendrobatidae) de la Guyana venezolana, con la descripción de siete especies nuevas. Publ. Asoc. Amigos Doñana 9:1–64.

LA MARCA, E. 1997. Lista actualizada de los anfibios de Venezuela. Pp. 103–120 *in* E. La Marca (ed.), *Catálogo Zoológico de Venezuela,* Vol. 1. Mérida, Venezuela: Museo de Ciencia y Tecnología.

LAVILLA, E. O., AND P. ERGUETA S. 1995. Una nueva especies de *Telmatobius* (Anura: Leptodactylidae) de la ceja de montaña de La Paz (Bolivia). Alytes 13:45–51.

LEE, J. C. 1980. An ecogeographic analysis of the herpetofauna of the Yucatan Peninsula. Misc. Publ. Mus. Nat. Hist. Univ. Kansas 67:1–75.

LESCURE, J.-L. 1986. Les amphibiens anoures de la forêt guyanaise (Région de Trois Souts, Guyane Française). Mem. Mus. Natl. Hist. Nat. Paris, NS, Ser. A, Zool. 132:43–52.

LIVINGSTON, D. A. 1993. Evolution of African climate. Pp. 455–472 *in* P. Goldblatt (ed.), *Biological Relationships between Africa and South America.* New Haven: Yale Univ. Press.

LOBO, F. 1994. Descripción de una nueva especie de *Pseudopaludicola* (Anura: Leptodactylidae), redescripción de *P. falcipes* (Hensel, 1867) y *P. saltica* (Cope, 1887) y osteología de las tres especies. Cuad. Herpetol. 8:177–199.

LÖTTERS, S. 1992. Ein neuer Harlekin-Frosch (Anura: Bufonidae: *Atelopus*) aus dem Chocó, West-Kolumbien. Sauria 14:27–30.

LÖTTERS, S. 1996. *The Neotropical Toad Genus* Atelopus. Kóln, Germany: M. Vences & F. Glaw Verlags GbR.

LUTZ, B. 1972. Geographical and ecological notes on Cisandine to Platine frogs. J. Herpetol. 6:83–100.

LYNCH, J. D. 1976. The species groups of South American frogs of the genus *Eleutherodactylus* (Leptodactylidae). Occas. Pap. Mus. Nat. Hist. Univ. Kansas 61:1–24.

LYNCH, J. D. 1979. The amphibians of the lowland tropical forests. Pp. 189–215 *in* W. E. Duellman (ed.), The South American herpetofauna: its origin, evolution, and dispersal. Monogr. Mus. Nat. Hist. Univ. Kansas 7:1–485.

LYNCH, J. D. 1981. The systematic status of *Amblyphrynus ingeri* (Amphibia: Leptodactylidae) with the description of an allied species in western Colombia. Caldasia 13:313–332.

LYNCH, J. D. 1982. Relationships of the frogs of the genus *Ceratophrys* (Leptodactylidae) and their bearing on hypotheses of Pleistocene forest refugia in South America and punctuated equilibrium. Syst. Zool. 31:166–179.

LYNCH, J. D. 1983. A new leptodactylid frog from the Cordillera Oriental of Colombia. Pp. 52–57 *in* A. G. J. Rhodin and K. Miyata (eds.), *Advances in Herpetology and Evolutionary Biology.* Cambridge: Harvard Univ. Press.

LYNCH, J. D. 1986a. The definition of the Middle American clade of *Eleutherodactylus* based on jaw musculature (Amphibia: Leptodactylidae). Herpetologica 42:248–258.

LYNCH, J. D. 1986b. Origins of the high Andean herpetological fauna. Pp. 478–499 *in* F. Vuilleumier and M. Monasterio (eds.), *High Altitude Tropical Biogeography.* New York: Oxford Univ. Press.

LYNCH, J. D. 1989. A review of the leptodactylid frogs of the genus *Pseudopaludicola* in northern South America. Copeia 1989:577–588.

LYNCH, J. D. 1993a. A new centrolenid frog from the Andes of western Colombia. Rev. Acad. Colombiana Cien. 18:567–570.

LYNCH, J. D. 1993b. A new harlequin frog from the Cordillera Oriental of Colombia (Anura, Bufonidae, *Atelopus*). Alytes 11:77–87.

LYNCH, J. D. 1994a. Two new species of the *Eleutherodactylus conspicillatus* group (Amphibia: Leptodactylidae) from the Cordillera Oriental de Colombia. Rev. Acad. Colombiana Cienc. 19:187–193.

LYNCH, J. D. 1994b. A new species of frog (genus *Eleutherodactylus:* Leptodactylidae) from a cloud forest in Departamento de Santander, Colombia. Rev. Acad. Colombiana Cienc. 19:205–208.

LYNCH, J. D. 1995. Three new species of *Eleutherodactylus* (Amphibia: Leptodactylidae) from paramos of the Cordillera Occidental of Colombia. J. Herpetol. 29:513–521.

LYNCH, J. D. 1996a. New frog (*Eleutherodactylus:* Leptodactylidae) from the Andes of eastern Colombia, part of a remarkable pattern of distribution. Copeia 1996:103–108.

LYNCH, J. D. 1996b. Replacement names for three homonyms in the genus *Eleutherodactylus* (Anura: Leptodactylidae). J. Herpetol. 30:278–280.

LYNCH, J. D. 1996c. New frogs of the genus *Eleutherodactylus* (family Leptodactylidae) from the San Antonio region of the Colombian Cordillera Occidental. Rev. Acad. Colombiana Cien. 20:331–345.

LYNCH, J. D. 1997. Intrageneric relationships of mainland *Eleutherodactylus* II. A review of the *Eleutherodactylus sulcatus* group. Rec. Acad. Colombian Cienc. 21:353–372.

LYNCH, J. D., AND W. E. DUELLMAN. 1980. The *Eleutherodactylus* of the Amazonian slopes of the Ecuadorian Andes

(Anura: Leptodactylidae). Misc. Publ. Mus. Nat. Hist. Univ. Kansas 69:1–86.

LYNCH, J. D., AND W. E. DUELLMAN. 1995. A new fat little frog (Leptodactylidae: *Eleutherodactylus*) from the lofty Andean grasslands of southern Ecuador. Occas. Pap. Nat. Hist. Mus. Univ. Kansas 173:1–7.

LYNCH, J. D., AND W. E. DUELLMAN. 1997. Frogs of the genus *Eleutherodactylus* (Leptodactylidae) in western Ecuador: systematics, ecology, and biogeography. Misc. Publ. Nat. Hist. Mus. Univ. Kansas 23:1–236.

LYNCH, J. D., AND E. LA MARCA. 1993. Synonymy and variation in *Eleutherodactylus bicumulus* (Peters) from northern Venezuela, with a description of a new species (Amphibia: Leptodactylidae). Carib. J. Sci. 29:133–146.

LYNCH, J. D., AND J. V. RUEDA-ALMONACID. 1997. Three new frogs (*Eleutherodactylus:* Leptodactylidae) from cloud forests in eastern Departamento Caldas, Colombia. Rev. Acad. Colombiana Cien. 21:131–142.

LYNCH, J. D., AND P. M. RUÍZ-CARRANZA. 1985. A synopsis of the frogs of the genus *Eleutherodactylus* from the Sierra Nevada de Santa Marta, Colombia. Occas. Pap. Mus. Zool. Univ. Michigan 711:1–59.

LYNCH, J. D., AND P. M. RUÍZ-CARRANZA. 1996a. New sister-species of *Eleutherodactylus* from the Cordillera Occidental of southwestern Colombia (Amphibia: Salientia: Leptodactylidae). Rev. Acad. Colombiana Cienc. 20:347–363.

LYNCH, J. D., AND P. M. RUÍZ-CARRANZA. 1996b. A remarkable new centrolenid frog from Colombia with a review of nuptial excrescences in the family. Herpetologica 52:525–535.

LYNCH, J. D., P. M. RUÍZ-CARRANZA, AND M. C. ARDILA-ROBAYO. 1994. The identities of the Colombian frogs confused with *Eleutherodactylus latidiscus* (Boulenger)(Amphibia: Anura: Leptodactylidae). Occas. Pap. Nat. Hist. Mus. Univ. Kansas 170:1–42.

LYNCH, J. D., P. M. RUÍZ-CARRANZA, AND M. C. ARDILA-ROBAYO. 1996. Three new species of *Eleutherodactylus* (Amphibia: Leptodactylidae) from high elevations of the Cordillera Central of Colombia. Caldasia 18:329–342.

LYNCH, J. D., P. M. RUÍZ-CARRANZA, AND M. C. ARDILA-ROBAYO. 1997. Biogeographic patterns of Colombian frogs and toads. Rev. Acad. Colombiana Cien. 21:237–248

MARSHALL, L. G., AND T. SEMPERE. 1993. Evolution of the neotropical land mammal fauna in its geochronologic, stratigraphic, and tectonic context. Pp. 329–392 *in* P. Goldblatt (ed.), *Biological Relationships between Africa and South America.* New Haven: Yale Univ. Press.

MAXSON, L. R., AND W. R. HEYER. 1982. Leptodactylid frogs and the Brazilian Shield: an old and continuing adaptive relationship. Biotropica 14:10–15.

MCCOY, E. D. 1994. "Amphibian decline": a scientific dilemma in more ways than one. Herpetologica 50:98–103.

MEINHARDT, D. J., AND J. R. PARMELEE. 1996. A new species of *Colostethus* (Anura: Dendrobatidae) from Venezuela. Herpetologica 52:70–77.

MERCADAL DE BARRIO, I. T., AND A. BARRIO. 1993. Una nueva especie de *Proceratophrys* (Leptodactylidae) del nordeste de Argentina. Amph.-Rept. 14:13–18.

MORALES, V. R. 1994. Taxonomía sobre algunos *Colostethus* (Anura: Dendrobatidae) de Sudamérica, con descripción de dos especies nuevas. Rev. Española Herpetol. 8:95–103.

MORALES, V. R. 1995. Checklist and taxonomic bibliography of the amphibians from Perú. Smithsonian Herpetol. Info. Serv. 107–120.

MORALES, V. R., AND R. W. MCDIARMID. 1996. Annotated checklist of the amphibians and reptiles of Pakitza, Manu National Park Reserve Zone, with comments on the herpetofauna of Madre de Dios, Perú. Pp. 503–522 *in* D. E. Wilson and A. Sandoval (eds.). *Manu. The Biodiversity of Southeastern Peru.* Washington, D.C.: Smithsonian Institution.

MORALES, V. R., AND R. SCHULTE. 1993. Dos especies nuevas de *Colostethus* (Anura, Dendrobatidae) en las vertientes de la Cordillera Oriental del Perú y del Ecuador. Alytes 11:97–106.

MÜLLER, P. 1973. *The Dispersal Centres of Terrestrial Vertebrates in the Neotropical Realm.* The Hague: Dr. W. Junk B. V.

MYERS, C. W., AND M. H. DONNELLY. 1996. A new herpetofauna from Cerro Yaví: first results of the Robert G. Goulet American Museum-TERRAMAR Expedition to the northwestern tepuis. Am. Mus. Novit. 3172:1–56.

MYERS, C. W., AND M. H. DONNELLY. 1997. A tepui herpetofauna on a granitic mountain (Tamacuari) in the borderland between Venezuela and Brazil. Am. Mus. Novit. 3213:1–71.

NELSON, B. W., C. A. C. FERREIRA, M. F. DA SILVA, AND M. L. KAWASAKI. 1990. Endemism centres: refugia and botanical collection density in Brazilian Amazonia. Nature 345:714–716.

ORTIZ, J. C., AND H. I. VIDAL. 1993. Una nueva especie de Leptodactylidae (*Eupsophus*) de la Cordillera de Nahuelbuta (Chile). Acta Zool. Lilloana 41:75–79.

PECHMANN, J. H. K., AND H. W. WILBUR. 1994. Putting declining amphibian populations in perspective: natural fluctuations and human impacts. Herpetologica 50:65–84.

PÉFAUR, J. E., AND W. E. DUELLMAN. 1980. Community structure in high Andean herpetofaunas. Trans. Kansas Acad. Sci. 83:45–65.

PITTMAN, W. C., III, S. CANDE, J. LABRECQUE, AND J. PINDELL. 1993. Fragmentation of Gondwana: the separation of Africa from South America. Pp. 15–34 *in* P. Goldblatt (ed.), *Biological Relationships between Africa and South America.* New Haven: Yale Univ. Press.

POMBAL, J. P., JR. 1993. New species of *Aparasphenodon* (Anura: Hylidae) from southeastern Brazil. Copeia 1993:1088–1091.

POMBAL, J. P., JR., AND U. CARAMASCHI. 1995. Posição taxonômica de *Hyla pseudopseudis* Miranda-Ribeiro, 1937 e *Hyla saxicola* Bokermann, 1964 (Anura: Hylidae). Bol. Mus. Nac. Rio de Janeiro 363:1–8.

POMBAL, J. P., JR., AND C. F. B. HADDAD. 1993. *Hyla luctuosa,* a new treefrog from southeastern Brazil (Amphibia: Hylidae). Herpetologica, 49:16–21.

POMBAL, J. P., JR., C. F. B. HADDAD, AND S. KASAHARA. 1995. A new species of *Scinax* (Anura: Hylidae) from southeastern Brazil, with comments on the genus. J. Herpetol. 29:1–6.

PRANCE, G. T. (ed.) 1982. *Biological Diversification in the Tropics.* New York: Columbia Univ. Press.

PUPPO, J. M., A. FERNÁNDEZ-BADILLO, E. LA MARCA, AND R. V. GARCÍA. 1995. Fauna del Parque Nacional Henri Pittier, Venezuela: composición y distribución de los anfibios. Acta Cienc. Venezolana 46:294–302.

PYBURN, W. F. 1993. A new tree frog of the genus *Scinax* from the Vaupes River of northwestern Brazil. Texas J. Sci. 44:405–411.

RAMIA, M. 1967. Tipos de sabanas en los llanos de Venezuela. Bol. Soc. Venezolana Cienc. Nat. 112:264–287.

RÄSÄNEN, M. E., J. S. SALO, AND R. J. KALLIOLA. 1987. Fluvial perturbance in the western Amazon Basin: regulation by long-term sub-Andean tectonics. Science 238:1398–1400.

REICHLE, S., AND J. KÖHLER. 1997. A new species of *Eleutherodactylus* (Anura: Leptodactylidae) from the Andean slopes of Bolivia. Amphibia-Reptilia 18:333–337.

RIVERO, J. A., AND V. R. MORALES. 1992. Descripción de una especie nueva de *Atelopus* A M. C. Duméril & Bibron 1841 (Amphibia: Bufonidae). Brenesia 38:29–36.

RIVERO, J. A., AND M. A. SERNA. 1991. Tres nuevas especies de *Colostethus* (Anfibia [sic]: Dendrobatidae) de Colombia. Trianea 4:481–495.

RIVERO-BLANCO, C., AND J. R. DIXON. 1979. Origin and distribution of the herpetofauna of the dry lowland regions of northern South America. Pp. 281–298 *in* W. E. Duellman (ed.), The South American herpetofauna: its origin, evolution, and dispersal. Monogr. Mus. Nat. Hist. Univ. Kansas 7:1–485.

RODRÍGUEZ, L. O. 1992. Structure et organization du peuplement d'anoures de Cocha Cashu, Parc National Manu, Amazonie Péruvienne. Rev. Ecol. 47:151–197.

RODRÍGUEZ, L. O. 1994. A new species of the *Eleutherodactylus conspillatus* group (Leptodactylidae) from Peru, with comments on its call. Alytes 12:49–63.

RODRÍGUEZ, L. O., J. H. CÓRDOVA, AND J. ICOCHEA. 1993. Lista preliminar de los anfibios del Peru. Publ. Mus. Hist. Nat. Univ. Nac. Mayor San Marcos, A (Zool.) 45:1–22.

RODRÍGUEZ, L. O., AND W. E. DUELLMAN. 1994. Guide to the frogs of the Iquitos region, Amazonian Peru. Spec. Publ. Nat. Hist. Mus. Univ. Kansas 22:1–80, pls. 1–12.

RODRÍGUEZ, L. O., AND C. W. MYERS. 1993. A new poison frog from Manu National Park, southeastern Peru (Dendrobatidae, *Epipedobates*). Am. Mus. Novit. 3068:1–15.

RUEDA-ALMONACID, J. V. 1994. Una nueva especie de Atelopus Dumeril & Bibron, 1841 (Amphibia: Anura: Bufonidae) para la Sierra Nevada de Santa Marta, Colombia. Trianea 5:101–108.

RUÍZ-CARRANZA, P. M., M. C. ARDILA-ROBAYO, AND J. I. HERNÁNDEZ-CAMACHO. 1994. Tres nuevas especies de *Atelopus* A. M. C. Duméril & Bibron 1841 (Amphibia: Bufonidae) de la Sierra Nevada de Santa Marta, Colombia. Rev. Acad. Colombiana Cienc. 19:153–163.

RUÍZ-CARRANZA, P. M., M. C. ARDILA-ROBAYO, AND J. D. LYNCH. 1996. Lista actualizada de la fauna de Amphibia de Colombia. Rev. Acad. Colombiana Cienc. 20:365–415.

RUÍZ CARRANZA, P. M., M. C. ARDILA-ROBAYO, J. D. LYNCH, AND J. H. RESTREPO. 1997. Una nueva especie de *Gastrotheca* (Amphibia: Anura: Hylidae) de la Cordillera Occidental de Colombia. Rev. Acad. Colombiana Cien. 21:373–378.

RUÍZ-CARRANZA, P. M., AND J. D. LYNCH. 1991. Ranas Centrolenidae de Colombia 1. Propuesta de una nueva clasificación genérica. Lozania 57:1–30.

RUÍZ-CARRANZA, P, M, AND J. D. LYNCH. 1995a. Cuatro nuevas especies de *Cochranella* de la Cordillera Central. Lozania 62:1–23.

RUÍZ-CARRANZA, P, M, AND J. D. LYNCH. 1995b. Ranas Centrolenidae de Colombia VI. Cuatro nuevas especies de *Cochranella* de la Cordillera Occidental. Lozania 63:1–15.

RUÍZ-CARRANZA, P, M, AND J. D. LYNCH. 1995c. Ranas Centrolenidac de Colombia VII. Redescripción de *Centrolene andinum* (Rivero, 1968). Lozania 64:1–12.

RUÍZ-CARRANZA, P, M, AND J. D. LYNCH. 1995d. Ranas Centrolenidae de Colombia VIII. Cuatro nuevas especies de *Centrolene* de la Cordillera Central. Lozania 65:1–16.

RUÍZ-CARRANZA, P, M, AND J. D. LYNCH. 1996. Ranas Centrolenidae de Colombia IX. Dos nuevas especies del suroestes de Colombia. Lozania 68:1–11.

RUÍZ-CARRANZA, P, M, J. D. LYNCH, AND M. C. ARDILA-R. 1997. Seis nuevas especies de *Eleutherodactylus* Duméril & Bibron, 1841 (Amphibia: Leptodactylidae) del norte de la Cordillera Occidental de Colombia. Rev. Acad. Colombiana Cien. 21:155–174.

RUÍZ-CARRANZA, P. M., AND M. OSORNO-MUÑOZ. 1994. Tres nuevas especies de *Atelopus* A. M. C. Duméril & Bibron 1841 (Amphibia: Bufonidae) de la Cordillera Central de Colombia. Rev. Acad. Colombiana Cienc. 19:165–170.

RUÍZ-CARRANZA, P, M, C. M. VELEZ, AND M. C. ARDILA-R. 1995. Una nueva especie de *Atelopus* A. M. C. Duméril & Bibron 1841 (Amphibia: Bufonidae). Caldasia 18:113–118.

RUTHVEN, A. G. 1922. The amphibians and reptiles of the Sierra Nevada de Santa Marta, Colombia. Misc. Publ. Mus. Zool. Univ. Michigan 8:1–69.

SAAVEDRA, C., AND C. FREESE. 1986. *Prioridades Biológicas de Conservacón en los Andes Tropicales*. Gland, Switzerland: International Union for the Conservation of Nature.

SALAS, A. W., AND U. SINSCH. 1996. Two new *Telmatobius* species (Leptodactylidae: Telmatobiinae) of Ancash, Peru. Alytes 14:1–26.

SARMIENTO, G. 1976. Evolution of arid vegetation in tropical America. Pp. 65–99 *in* D. W. Goodall (ed.), *Evolution of Desert Biota*. Austin: Univ. Texas Press.

SAVAGE, J. M. 1982. The enigma of the Central American herpetofauna: dispersals or vicariance? Ann. Missouri Bot. Gard. 69:464-547.

SCANLAN, B. E., L. R. MAXSON, AND W. E. DUELLMAN. 1980. Albumin evolution in marsupial frogs (Hylidae: *Gastrotheca*). Evolution 34:222–229.

SCHLÜTER, A. 1984. ökologische Untersuchungen an einem Stillgewasser im tropischen Regenwald von Peru unter besonderer Berucksichtigung der Amphibien. Doctoral dissertation. Hamburg: Univ. Hamburg.

SEÑARIS, J. C. 1993. Una nueva especie de *Oreophrynella* (Anura: Bufonidae) de la cima del Auyan-Tepui, Estado Bolívar, Venezuela. Mem. Soc. Cien. Nat. La Salle 53 (140):177–183.

SEÑARIS, J. C., AND J. AYARZAGÜENA. 1993. Una nueva especie de *Centrolenella* (Anura: Centrolenidae) del Auyan-tepui, Edo. Bolívar, Venezuela. Mem. Soc. Cien. Nat. La Salle 53:121–126.

SEÑARIS, J. C., J. AYARZAGÜENA, AND S. GORZULA. 1994. Los sapos de la familia Bufonidae (Amphibia: Anura) de las tierras altas de la Guayana Venezolana: descripción de un nuevo género y tres especies. Publ. Asoc. Amigos Doñana 3:1–37.

SEÑARIS, J. C., J. AYARZAGÜENA, AND S. GORZULA. 1996. Revision taxonómica del género *Stefania* (Anura: Hylidae) en Venezuela con la descripción de cinco nuevas especies. Publ. Asoc. Amigos Doñana 7:1–57.

SIMPSON, B. 1975. Pleistocene changes in the flora of the high tropical Andes. Paleobiology 1:273–294.

SIMPSON, B. 1979. Quaternary biogeography of the high montane regions of South America. Pp. 157–188 *in* W. E. Duellman (ed.), The South American herpetofauna: its origin, evolution, and dispersal. Monogr. Mus. Nat. Hist. Univ. Kansas 7:1–485.

SIMPSON, G. G. 1980. *Splendid Isolation: The Curious History of South American Mammals*. New Haven: Yale Univ. Press.

SOBERÓN M., J., AND J. LLORENTE B. 1993. The use of species accumulation functions for the prediction of species richness. Conserv. Biol. 7:480–488.

STEHLI, F. G., AND S. D. WEBB (EDS.). 1985. *The Great American Biotic Interchange.* New York: Plenum Press.

TRUEB, L. 1970. Evolutionary relationships of casque-headed tree frogs with co-ossified skulls (family Hylidae). Univ. Kansas Publ. Mus. Nat. Hist. 18:547–716.

TRUEB, L. 1974. Systematic relationships of neotropical frogs, genus *Hemiphractus* (Anura: Hylidae). Occas. Pap. Mus. Nat. Hist. Univ. Kansas 29:1–60.

TRUEB, L., AND D. C. CANNATELLA. 1986. Systematics, morphology, and phylogeny of genus *Pipa* (Anura: Pipidae). Herpetologica 42:412–449.

TRUEB, L., AND W. E. DUELLMAN. 1971. A synopsis of neotropical hylid frogs, genus *Osteocephalus.* Occas. Pap. Mus. Nat. Hist. Univ. Kansas 1:1–47.

VAN DER HAMMEN, T. 1974. The Pleistocene changes in vegetation and climate in tropical South America. J. Biogeogr. 1:3–26.

VANZOLINI, P. E., AND W. R. HEYER. 1985. The American herpetofauna and the interchange. Pp. 475–487 *in* F. G. Stehli and S. D. Webb (eds.), *The Great American Biotic Interchange.* New York: Plenum Press.

VIAL, J. L., AND L. SAYLOR. 1993. The status of amphibian populations. Declining Amphibian Populations Task Force, Working Document 1:1–98.

VUILLEUMIER, B. S. 1971. Pleistocene changes in the fauna and flora of South America. Science, 173:771–780.

VUILLEUMIER, F. 1985. Fossil and Recent avifaunas and the interamerican interchange. Pp. 387–424 *in* F. G. Stehli and S. D. Webb (eds.), *The Great American Biotic Interchange.* New York: Plenum Press.

VUILLEUMIER, F., AND A. V. ANDORS. 1993. Avian biological relationships between Africa and South America. Pp. 289–328 *in* P. Goldblatt (ed.), *Biological Relationships between Africa and South America.* New Haven: Yale Univ. Press.

WAKE, D. B., AND J. F. LYNCH. 1976. The distribution, ecology, and evolutionary history of plethodontid salamanders in tropical America. Sci. Bull. Nat. Hist. Mus. Los Angeles County 25:1–65.

WIENS, J. J. 1993. Systematics of the leptodactylid frog genus *Telmatobius* in the Andes of northern Ecuador. Occas. Pap. Mus. Nat. Hist. Univ. Kansas 162:1–76.

WIENS, J. J., AND L. A. COLOMA. 1992. A new species of the *Eleutherodactylus myersi* (Anura: Leptodactylidae) assembly from Ecuador. J. Herpetol. 26:196–207.

WILD, E. R. 1994. Two new species of centrolenid frogs from the Amazonian slope of the Cordillera Oriental, Ecuador. J. Herpetol. 28:299–310.

WILD, E. R. 1995. New genus and species of Amazonian microhylid frog with a phylogenetic analysis of New World genera. Copeia 1995:837–849.

WILKINSON, M. 1996. Resolution of the taxonomic status of *Nectocaecilia haydee* (Roze) and a revised key to the genera of the Typhlonectidae (Amphibia: Gymnophiona). J. Herpetol. 30:413–415.

ZIMMERMAN, B. L., AND R. O. BIERREGAARD. 1986. Relevance of the equilibrium theory of island biogeography and species-area relations to conservation with a case from Amazonia. J. Biogeogr. 13:133–143.

ZIMMERMAN, B. L., AND M. T. RODRIGUES. 1990. Frogs, snakes, and lizards of the INPA-WWF reserves near Manaus, Brazil. Pp. 426–454 *in* A. H. Gentry (ed.), *Four Neotropical Rainforests.* New Haven: Yale Univ. Press.

APPENDIX 5:1

DISTRIBUTION OF SPECIES OF AMPHIBIANS IN NATURAL REGIONS IN SOUTH AMERICA

Symbols in columns: + present, – absent, CEN central, compass direction (e.g., NE = only in northeast), ? questionable. Symbols in boldface indicate endemism. Abbreviations for regions: AFD Atlantic forest domain, A-G Amazon Basin–Guiana lowlands, AND Andes, ATA Atacama Desert, ATF Austral temperate forest; CAR Caribbean lowlands, CCC Caatinga-Cerrado-Chaco, CHO Chocoan lowlands, GUH Guiana Highlands, LLA Llanos, PAM Pampean-Monte, PAT Patagonian. In column A-G, TO Tobago, TR Trinidad, TT Trinidad and Tobago = restricted to those islands.

Taxon and date named	CHO	CAR	A-G	AFD	LLA	CCC	PAM	PAT	ATF	ATA	AND	GUH
ANURA: ALLOPHRYNIDAE:												
Allophryne ruthveni 1926	–	–	+	–	–	–	–	–	–	–	–	–
ANURA: BRACHYCEPHALIDAE:												
Brachycephalus ephippium 1824	–	–	–	+	–	–	–	–	–	–	–	–
Brachycephalus nodoterga 1920	–	–	–	+	–	–	–	–	–	–	–	–
Psyllophryne didactyla 1971	–	–	–	+	–	–	–	–	–	–	–	–
ANURA: BUFONIDAE:												
Andinophryne atelopoides 1981	–	–	–	–	–	–	–	–	–	–	N	–
Andinophryne colomai 1985	–	–	–	–	–	–	–	–	–	–	N	–
Andinophryne olallai 1985	–	–	–	–	–	–	–	–	–	–	N	–
Atelopus arsyecue 1994	–	–	–	–	–	–	–	–	–	–	N	–
Atelopus arthuri 1973	–	–	–	–	–	–	–	–	–	–	N	–
Atelopus balios 1973	+	–	–	–	–	–	–	–	–	–	–	–
Atelopus bomolochos 1973	–	–	–	–	–	–	–	–	–	–	N	–
Atelopus boulengeri 1904	–	–	–	–	–	–	–	–	–	–	N	–
Atelopus carauta 1978	–	–	–	–	–	–	–	–	–	–	N	–
Atelopus carbonerensis 1972	–	–	–	–	–	–	–	–	–	–	NE	–
Atelopus carrikeri 1916	–	–	–	–	–	–	–	–	–	–	N	–
Atelopus chocoensis 1992	+	–	–	–	–	–	–	–	–	–	–	–
Atelopus chrysocollaris 1994	–	–	–	–	–	–	–	–	–	–	NE	–
Atelopus coynei 1980	–	–	–	–	–	–	–	–	–	–	N	–
Atelopus cruciger 1856	–	–	–	–	–	–	–	–	–	–	NE	–
Atelopus ebanoides 1963	–	–	–	–	–	–	–	–	–	–	N	–
Atelopus elegans 1882	+	–	–	–	–	–	–	–	–	–	–	–
Atelopus erythropus 1903	–	–	–	–	–	–	–	–	–	–	S	–
Atelopus farci 1993	–	–	–	–	–	–	–	–	–	–	N	–
Atelopus flavescens 1841	–	–	NE	–	–	–	–	–	–	–	–	–
Atelopus franciscus 1974	–	–	NE	–	–	–	–	–	–	–	–	–
Atelopus glyphus 1931	+	–	–	–	–	–	–	–	–	–	–	–
Atelopus halihelos 1973	–	–	–	–	–	–	–	–	–	–	N	–
Atelopus ignescens 1849	–	–	–	–	–	–	–	–	–	–	N	–
Atelopus laetissimus 1994	–	–	–	–	–	–	–	–	–	–	N	–
Atelopus leoperezii 1994	–	–	–	–	–	–	–	–	–	–	N	–
Atelopus longibrachius 1963	+	–	–	–	–	–	–	–	–	–	–	–
Atelopus longirostris 1868	–	–	–	–	–	–	–	–	–	–	N	–
Atelopus lynchi 1981	–	–	–	–	–	–	–	–	–	–	N	–
Atelopus mindoensis 1973	–	–	–	–	–	–	–	–	–	–	N	–
Atelopus minutulus 1988	–	–	–	–	–	–	–	–	–	–	N	–
Atelopus mucubajiensis 1972	–	–	–	–	–	–	–	–	–	–	NE	–
Atelopus muisca 1991	–	–	–	–	–	–	–	–	–	–	N	–
Atelopus nahumae 1994	–	–	–	–	–	–	–	–	–	–	N	–
Atelopus negreti 1995	–	–	–	–	–	–	–	–	–	–	N	–
Atelopus nepiozomus 1973	–	–	–	–	–	–	–	–	–	–	N	–
Atelopus oxyrhynchus 1903	–	–	–	–	–	–	–	–	–	–	NE	–
Atelopus pachydermus 1857	?	–	–	–	–	–	–	–	–	–	N	–
Atelopus palmatus 1945	–	–	–	–	–	–	–	–	–	–	N	–
Atelopus pedimarmoratus 1963	–	–	–	–	–	–	–	–	–	–	N	–
Atelopus peruensis 1985	–	–	–	–	–	–	–	–	–	–	S	–
Atelopus pictiventris 1986	–	–	–	–	–	–	–	–	–	–	N	–
Atelopus pinangoi 1982	–	–	–	–	–	–	–	–	–	–	NE	–
Atelopus planispinus 1875	–	–	–	–	–	–	–	–	–	–	N	–

APPENDIX 5:1 CONTINUED

Taxon and date named	CHO	CAR	A-G	AFD	LLA	CCC	PAM	PAT	ATF	ATA	AND	GUH
Atelopus quimbaya 1994	–	–	–	–	–	–	–	–	–	–	N	–
Atelopus sanjosei 1989	–	–	–	–	–	–	–	–	–	–	N	–
Atelopus seminifer 1874	–	–	W	–	–	–	–	–	–	–	–	–
Atelopus sernai 1994	–	–	–	–	–	–	–	–	–	–	N	–
Atelopus simulatus 1994	–	–	–	–	–	–	–	–	–	–	N	–
Atelopus sorianoi 1983	–	–	–	–	–	–	–	–	–	–	NE	–
Atelopus spumarius 1871	–	–	+	–	–	–	–	–	–	–	–	–
Atelopus spurrelli 1914	+	–	–	–	–	–	–	–	–	–	–	–
Atelopus subornatus 1899	–	–	–	–	–	–	–	–	–	–	N	–
Atelopus tamaense 1990	–	–	–	–	–	–	–	–	–	–	N	–
Atelopus tricolor 1902	–	–	–	–	–	–	–	–	–	–	S	–
Atelopus varius 1856	+	–	–	–	–	–	–	–	–	–	–	–
Atelopus walkeri 1963	–	–	–	–	–	–	–	–	–	–	N	–
Atelopus willimani 1969	–	–	SW	–	–	–	–	–	–	–	–	–
Bufo achalensis 1972	–	–	–	–	–	–	+	–	–	–	–	–
Bufo amboroensis 1993	–	–	–	–	–	–	–	–	–	–	S	–
Bufo anderssoni 1941	–	–	CEN	–	–	–	–	–	–	–	–	–
Bufo arborescandens 1992	–	–	–	–	–	–	–	–	–	–	S	–
Bufo arenarum 1867	–	–	–	–	–	S	+	N	–	–	–	–
Bufo arequipensis 1959	–	–	–	–	–	–	–	–	–	+	–	–
Bufo arunco 1782	–	–	–	–	–	–	–	–	–	+	–	–
Bufo atacamensis 1961	–	–	–	–	–	–	–	–	–	+	–	–
Bufo blombergi 1951	+	–	–	–	–	–	–	–	–	–	–	–
Bufo caeruleostictus 1859	–	–	–	–	–	–	–	–	–	–	N	–
Bufo castaneoticus 1991	–	–	E	–	–	–	–	–	–	–	–	–
Bufo ceratophrys 1882	–	–	W	–	–	–	–	–	–	–	–	–
Bufo coniferus 1862	+	–	–	–	–	–	–	–	–	–	–	–
Bufo cophotis 1900	–	–	–	–	–	–	–	–	–	–	S	–
Bufo corynetes 1991	–	–	–	–	–	–	–	–	–	–	S	–
Bufo crucifer 1821	–	–	–	+	–	–	+	–	–	–	–	–
Bufo dapsilis 1945	–	–	W	–	–	–	–	–	–	–	–	–
Bufo dorbignyi 1841	–	–	–	–	–	–	+	–	–	–	–	–
Bufo fernandezae 1957	–	–	–	–	–	–	+	–	–	–	–	–
Bufo fissipes 1903	–	–	–	–	–	–	–	–	–	–	S	–
Bufo gallardoi 1992	–	–	–	–	–	–	–	–	–	–	S	–
Bufo glaberrimus 1868	?	–	W	–	–	–	–	–	–	–	–	–
Bufo gnustae 1967	–	–	–	–	–	–	–	–	–	–	S	–
Bufo granulosus 1824	–	+	NE	–	+	+	–	–	–	–	–	–
Bufo guttatus 1799	–	–	NE	–	–	–	–	–	–	–	–	–
Bufo haematiticus 1862	+	–	–	–	–	–	–	–	–	–	–	–
Bufo hypomelas 1913	+	–	–	–	–	–	–	–	–	–	–	–
Bufo ictericus 1824	–	–	–	+	–	–	–	–	–	–	–	–
Bufo iserni 1875	–	–	–	–	–	–	–	–	–	–	S	–
Bufo justinianoi 1994	–	–	–	–	–	–	–	–	–	–	S	–
Bufo limensis 1901	–	–	–	–	–	–	–	–	–	+	–	–
Bufo marinus 1758	+	+	+	–	+	–	–	–	–	–	–	–
Bufo nasicus 1903	–	–	NE	–	–	–	–	–	–	–	–	–
Bufo nesiotes 1979	–	–	–	–	–	–	–	–	–	–	S	–
Bufo ocellatus 1859	–	–	–	–	–	+	–	–	–	–	–	–
Bufo paracnemis 1925	–	–	–	+	–	+	+	–	–	–	–	–
Bufo poeppigii 1845	–	–	–	–	–	–	–	–	–	–	S	–
Bufo pygmaeus 1952	–	–	–	+	–	–	–	–	–	–	–	–
Bufo quechua 1961	–	–	–	–	–	–	–	–	–	–	S	–
Bufo rubropunctatus 1843	–	–	–	–	–	–	–	–	+	–	–	–
Bufo rufus 1877	–	–	–	–	–	–	+	–	–	–	–	–
Bufo rumbolli 1992	–	–	–	–	–	–	–	–	–	–	S	–
Bufo spinulosus 1834	–	–	–	–	–	–	–	+	–	–	S	–
Bufo sternosignatus 1859	–	–	–	–	–	–	–	–	–	–	NE	–

APPENDIX 5:1 CONTINUED

Taxon and date named	CHO	CAR	A-G	AFD	LLA	CCC	PAM	PAT	ATF	ATA	AND	GUH
Bufo typhonius 1758	+	–	+	–	–	–	–	–	–	–	+	–
Bufo variegatus 1870	–	–	–	–	–	–	–	–	+	–	–	–
Bufo vellardi 1959 (1978)	–	–	–	–	–	–	–	–	–	–	S	–
Bufo veraguensis 1857	–	–	–	–	–	–	–	–	–	–	S	–
Dendrophryniscus berthalutzae 1993	–	–	–	+	–	–	–	–	–	–	–	–
Dendrophryniscus bokermanni 1993	–	–	+	–	–	–	–	–	–	–	–	–
Dendrophryniscus brevipollicatus 1871	–	–	–	+	–	–	–	–	–	–	–	–
Dendrophryniscus carvalhoi 1993	–	–	–	+	–	–	–	–	–	–	–	–
Dendrophryniscus leucomystax 1968	–	–	–	+	–	–	–	–	–	–	–	–
Dendrophryniscus minutus 1941	–	–	+	–	–	–	–	–	–	–	–	–
Dendrophryniscus stawiarskyi 1993	–	–	–	+	–	–	–	–	–	–	–	–
Frostius pernambucensis 1962	–	–	–	+	–	–	–	–	–	–	–	–
Melanophryniscus cambaraensis 1979	–	–	–	–	–	–	+	–	–	–	–	–
Melanophryniscus devincenzii 1968	–	–	–	–	–	–	+	–	–	–	–	–
Melanophryniscus macrogranulosus 1973	–	–	–	–	–	–	+	–	–	–	–	–
Melanophryniscus moreirae 1920	–	–	–	+	–	–	–	–	–	–	–	–
Melanophryniscus orejasmirandai 1986	–	–	–	–	–	–	+	–	–	–	–	–
Melanophryniscus rubriventris 1947	–	–	–	–	–	–	+	–	–	–	–	–
Melanophryniscus sanmartini 1968	–	–	–	–	–	–	+	–	–	–	–	–
Melanophryniscus stelzneri 1875	–	–	–	–	–	+	+	–	–	–	–	–
Melanophryniscus tumifrons 1905	–	–	–	–	–	–	+	–	–	–	–	–
Metaphryniscus sosai 1994	–	–	–	–	–	–	–	–	–	–	–	+
Oreophrynella cryptica 1993	–	–	–	–	–	–	–	–	–	–	–	+
Oreophrynella hubneri 1987	–	–	–	–	–	–	–	–	–	–	–	+
Oreophrynella macconnelli 1900	–	–	–	–	–	–	–	–	–	–	–	+
Oreophrynella nigra 1994	–	–	–	–	–	–	–	–	–	–	–	+
Oreophrynella quelchii 1895	–	–	–	–	–	–	–	–	–	–	–	+
Oreophrynella vasquezi 1994	–	–	–	–	–	–	–	–	–	–	–	+
Osornophryne antisana 1987	–	–	–	–	–	–	–	–	–	–	N	–
Osornophryne bufoniformis 1904	–	–	–	–	–	–	–	–	–	–	N	–
Osornophryne guacamayo 1987	–	–	–	–	–	–	–	–	–	–	N	–
Osornophryne percrassa 1976	–	–	–	–	–	–	–	–	–	–	N	–
Osornophryne sumacoensis 1995	–	–	–	–	–	–	–	–	–	–	N	–
Osornophryne talipes 1986	–	–	–	–	–	–	–	–	–	–	N	–
Rhamphophryne acrolopha 1971	+	–	–	–	–	–	–	–	–	–	–	–
Rhamphoprhyne festae 1904	–	–	W	–	–	–	–	–	–	–	–	–
Rhamphophryne lindae 1990	–	–	–	–	–	–	–	–	–	–	N	–
Rhamphophryne macrorhina 1971	–	–	–	–	–	–	–	–	–	–	N	–
Rhamphophryne nicefori 1970	–	–	–	–	–	–	–	–	–	–	N	–
Rhamphophryne proboscidea 1882	–	–	–	?	–	–	–	–	–	–	–	–
Rhamphophryne rostrata 1920	–	–	–	–	–	–	–	–	–	–	N	–
Rhamphophryne tenrec 1990	–	–	–	–	–	–	–	–	–	–	N	–
Rhamphophryne truebae 1990	–	–	–	–	–	–	–	–	–	–	?	–
Truebella skoptes 1995	–	–	–	–	–	–	–	–	–	–	N	–
Truebella tothastes 1995	–	–	–	–	–	–	–	–	–	–	N	–
ANURA: CENTROLENIDAE:												
Centrolene acanthidiocephalum 1989	–	–	–	–	–	–	–	–	–	–	N	–
Centrolene altitudinalis 1968	–	–	–	–	–	–	–	–	–	–	NE	–
Centrolene andinum 1968	–	–	–	–	–	–	–	–	–	–	NE	–
Centrolene antioquiensis 1920	–	–	–	–	–	–	–	–	–	–	N	–
Centrolene audax 1973	–	–	–	–	–	–	–	–	–	–	N	–
Centrolene azulae 1989	–	–	–	–	–	–	–	–	–	–	S	–
Centrolene bacatum 1994	–	–	–	–	–	–	–	–	–	–	N	–
Centrolene ballux 1989	–	–	–	–	–	–	–	–	–	–	N	–
Centrolene buckleyi 1882	–	–	–	–	–	–	–	–	–	–	+	–
Centrolene fernandoi 1993	–	–	–	–	–	–	–	–	–	–	S	–
Centrolene geckoideum 1872	–	–	–	–	–	–	–	–	–	–	N	–
Centrolene gemmatum 1985	–	–	–	–	–	–	–	–	–	–	N	–

Appendix 5:1 CONTINUED

Taxon and date named	CHO	CAR	A-G	AFD	LLA	CCC	PAM	PAT	ATF	ATA	AND	GUH
Centrolene gorzulai 1992	–	–	–	–	–	–	–	–	–	–	–	+
Centrolene grandisonae 1970	–	–	–	–	–	–	–	–	–	–	N	–
Centrolene guanacarum 1995	–	–	–	–	–	–	–	–	–	–	N	–
Centrolene helodermatum 1981	–	–	–	–	–	–	–	–	–	–	N	–
Centrolene hesparium 1990	–	–	–	–	–	–	–	–	–	–	S	–
Centrolene huilense 1995	–	–	–	–	–	–	–	–	–	–	N	–
Centrolene hybrida 1991	–	–	–	–	–	–	–	–	–	–	N	–
Centrolene ilex 1967	+	–	–	–	–	–	–	–	–	–	–	–
Centrolene lemniscatum 1993	–	–	–	–	–	–	–	–	–	–	S	–
Centrolene lynchi 1980	–	–	–	–	–	–	–	–	–	–	N	–
Centrolene mariae 1979	–	–	–	–	–	–	–	–	–	–	S	–
Centrolene medemi 1970	–	–	–	–	–	–	–	–	–	–	N	–
Centrolene muelleri 1993	–	–	–	–	–	–	–	–	–	–	S	–
Centrolene notostictum 1991	–	–	–	–	–	–	–	–	–	–	N	–
Centrolene paezorum 1986	–	–	–	–	–	–	–	–	–	–	N	–
Centrolene peristictum 1973	–	–	–	–	–	–	–	–	–	–	N	–
Centrolene petrophilum 1991	–	–	–	–	–	–	–	–	–	–	N	–
Centrolene pipilatum 1973	–	–	–	–	–	–	–	–	–	–	N	–
Centrolene prosoblepon 1892	+	–	–	–	–	–	–	–	–	–	N	–
Centrolene puyoensis 1989	–	–	–	–	–	–	–	–	–	–	N	–
Centrolene quindianum 1995	–	–	–	–	–	–	–	–	–	–	N	–
Centrolene robledoi 1995	–	–	–	–	–	–	–	–	–	–	N	–
Centrolene sanchezi 1991	–	–	–	–	–	–	–	–	–	–	N	–
Centrolene savagei 1991	–	–	–	–	–	–	–	–	–	–	N	–
Centrolene scirtetes 1989	–	–	–	–	–	–	–	–	–	–	N	–
Cochranella adiazeta 1991	–	–	–	–	–	–	–	–	–	–	N	–
Cochranella albomaculata 1949	+	–	–	–	–	–	–	–	–	–	–	–
Cochranella ametarsia 1987	–	–	W	–	–	–	–	–	–	–	–	–
Cochranella anomala 1973	–	–	–	–	–	–	–	–	–	–	N	–
Cochranella armata 1996	–	–	–	–	–	–	–	–	–	–	N	–
Cochranella auyantepuiana 1993	–	–	–	–	–	–	–	–	–	–	–	+
Cochranella balionota 1981	–	–	–	–	–	–	–	–	–	–	N	–
Cochranella bejaranoi 1980	–	–	–	–	–	–	–	–	–	–	S	–
Cochranella cariticommata 1994	–	–	–	–	–	–	–	–	–	–	N	–
Cochranella castroviejoi 1995	–	–	–	–	–	–	–	–	–	–	NE	–
Cochranella chami 1995	–	–	–	–	–	–	–	–	–	–	N	–
Cochranella chancas 1993	–	–	–	–	–	–	–	–	–	–	S	–
Cochranella christinae 1995	–	–	–	–	–	–	–	–	–	–	N	–
Cochranella cochranae 1961	–	–	–	–	–	–	–	–	–	–	N	–
Cochranella croceopodes 1993	–	–	–	–	–	–	–	–	–	–	S	–
Cochranella daidalea 1991	–	–	–	–	–	–	–	–	–	–	N	–
Cochranella duidaensis 1992	–	–	–	–	–	–	–	–	–	–	–	+
Cochranella euhystrix 1990	–	–	–	–	–	–	–	–	–	–	S	–
Cochranella euknemos 1967	+	–	–	–	–	–	–	–	–	–	–	–
Cochranella flavidigitata 1992	–	–	–	–	–	–	–	–	–	–	S	–
Cochranella flavopunctata 1973	–	–	–	–	–	–	–	–	–	–	N	–
Cochranella garciae 1995	–	–	–	–	–	–	–	–	–	–	N	–
Cochranella geijskei 1966	–	–	NE	–	–	–	–	–	–	–	–	–
Cochranella griffithsi 1961	–	–	–	–	–	–	–	–	–	–	N	–
Cochranella helenae 1992	–	–	–	–	–	–	–	–	–	–	–	+
Cochranella ignota 1990	–	–	–	–	–	–	–	–	–	–	N	–
Cochranella littoralis 1996	+	–	–	–	–	–	–	–	–	–	–	–
Cochranella luminosa 1995	–	–	–	–	–	–	–	–	–	–	N	–
Cochranella luteopunctata 1996	–	–	–	–	–	–	–	–	–	–	N	–
Cochranella megacheira 1973	–	–	–	–	–	–	–	–	–	–	N	–
Cochranella megista 1985	–	–	–	–	–	–	–	–	–	–	N	–
Cochranella midas 1973	–	–	W	–	–	–	–	–	–	–	–	–
Cochranella nephelophila 1991	–	–	–	–	–	–	–	–	–	–	N	–

Taxon and date named	CHO	CAR	A-G	AFD	LLA	CCC	PAM	PAT	ATF	ATA	AND	GUH
Cochranella nota 1996	–	–	–	–	–	–	–	–	–	–	S	–
Cochranella ocellata 1918	–	–	–	–	–	–	–	–	–	–	S	–
Cochranella ocellifera 1899	–	–	–	–	–	–	–	–	–	–	N	–
Cochranella orejuela 1989	–	–	–	–	–	–	–	–	–	–	N	–
Cochranella oreonympha 1991	–	–	–	–	–	–	–	–	–	–	N	–
Cochranella oyampiensis 1975	–	–	NE	–	–	–	–	–	–	–	–	+
Cochranella phenax 1982	–	–	–	–	–	–	–	–	–	–	S	–
Cochranella pluvialis 1982	–	–	–	–	–	–	–	–	–	–	S	–
Cochranella posadae 1995	–	–	–	–	–	–	–	–	–	–	N	–
Cochranella prasina 1981	–	–	–	–	–	–	–	–	–	–	N	–
Cochranella punctulata 1995	–	–	–	–	–	–	–	–	–	–	N	–
Cochranella ramirezi 1991	+	–	–	–	–	–	–	–	–	–	N	–
Cochranella resplendens 1973	–	–	**W**	–	–	–	–	–	–	–	–	–
Cochranella ritae 1952	–	–	**CEN**	–	–	–	–	–	–	–	–	–
Cochranella riveroi 1992	–	–	–	–	–	–	–	–	–	–	–	+
Cochranella ruizi 1993	–	–	–	–	–	–	–	–	–	–	N	–
Cochranella savagei 1991	–	–	–	–	–	–	–	–	–	–	N	–
Cochranella saxiscandens 1993	–	–	–	–	–	–	–	–	–	–	S	–
Cochranella siren 1973	–	–	–	–	–	–	–	–	–	–	+	–
Cochranella solitaria 1991	–	–	–	–	–	–	–	–	–	–	N	–
Cochranella spiculata 1976	–	–	–	–	–	–	–	–	–	–	S	–
Cochranella spinosa 1949	+	–	–	–	–	–	–	–	–	–	–	–
Cochranella susatamai 1995	–	–	–	–	–	–	–	–	–	–	N	–
Cochranella tangarana 1993	–	–	–	–	–	–	–	–	–	–	S	–
Cochranella truebae 1976	–	–	–	–	–	–	–	–	–	–	S	–
Cochranella vozmedianoi 1996	–	–	–	–	–	–	–	–	–	–	NE	–
Cochranella xanthocheridia 1995	–	–	–	–	–	–	–	–	–	–	N	–
Hyalinobatrachium antisthenesi 1963	–	–	–	–	–	–	–	–	–	–	NE	–
Hyalinobatrachium aureoguttatum 1989	+	–	–	–	–	–	–	–	–	–	N	–
Hyalinobatrachium bergeri 1980	–	–	–	–	–	–	–	–	–	–	S	–
Hyalinobatrachium chirripoi 1958	+	–	–	–	–	–	–	–	–	–	–	–
Hyalinobatrachium crurifasciatum 1997	–	–	–	–	–	–	–	–	–	–	–	+
Hyalinobatrachium duranti 1985	–	–	–	–	–	–	–	–	–	–	NE	–
Hyalinobatrachium eurygnathum 1925	–	–	–	+	–	–	–	–	–	–	–	–
Hyalinobatrachium fleischmanni 1893	+	–	NE	–	–	–	–	–	–	–	–	–
Hyalinobatrachium fragilis 1985	–	–	–	–	–	–	–	–	–	–	NE	–
Hyalinobatrachium iaspidiensis 1992	–	–	–	–	–	–	–	–	–	–	–	+
Hyalinobatrachium lemur 1993	–	–	–	–	–	–	–	–	–	–	S	–
Hyalinobatrachium loreocarinatum 1985	–	–	–	–	–	–	–	–	–	–	NE	–
Hyalinobatrachium munozorum 1973	–	–	**W**	–	–	–	–	–	–	–	–	–
Hyalinobatrachium orientalis 1968	–	–	–	–	–	–	–	–	–	–	+	+
Hyalinobatrachium ostracodermoides 1985	–	–	–	–	–	–	–	–	–	–	NE	–
Hyalinobatrachium pallidum 1985	–	–	–	–	–	–	–	–	–	–	NE	–
Hyalinobatrachium parvulum 1895	–	–	–	+	–	–	–	–	–	–	–	–
Hyalinobatrachium pellucidum 1973	–	–	–	–	–	–	–	–	–	–	N	–
Hyalinobatrachium pleurolineatum 1985	–	–	–	–	–	–	–	–	–	–	NE	–
Hyalinobatrachium revocatum 1985	–	–	–	–	–	–	–	–	–	–	NE	–
Hyalinobatrachium taylori 1968	–	–	NE	–	–	–	–	–	–	–	–	–
Hyalinobatrachium uranoscopum 1924	–	–	–	+	–	–	–	–	–	–	–	–
Hyalinobatrachium valerioi 1931	+	–	–	–	–	–	–	–	–	–	–	+
ANURA: DENDROBATIDAE:												
Aromobates nocturnus 1991	–	–	–	–	–	–	–	–	–	–	NE	–
Colostethus abditaurantius 1975	–	–	–	–	–	–	–	–	–	–	N	–
Colostethus agilis 1985	–	–	–	–	–	–	–	–	–	–	N	–
Colostethus alacris 1989	–	–	–	–	–	–	–	–	–	–	N	–
Colostethus alagoanus 1967	–	–	–	+	–	–	–	–	–	–	–	–
Colostethus alboguttatus 1903	–	–	–	–	–	–	–	–	–	–	NE	–
Colostethus anthracinus 1971	–	–	–	–	–	–	–	–	–	–	N	–

APPENDIX 5:1 CONTINUED

Taxon and date named	CHO	CAR	A-G	AFD	LLA	CCC	PAM	PAT	ATF	ATA	AND	GUH
Colostethus argyrogaster 1993	–	–	–	–	–	–	–	–	–	–	S	–
Colostethus atopoglossus 1997	–	–	–	–	–	–	–	–	–	–	N	–
Colostethus awa 1995	+	–	–	–	–	–	–	–	–	–	N	–
Colostethus ayarzaguenai 1996	–	–	–	–	–	–	–	–	–	–	–	+
Colostethus beebei 1923	–	–	NE	–	–	–	–	–	–	–	–	–
Colostethus betancuri 1991	–	–	–	–	–	–	–	–	–	–	N	–
Colostethus bocagei 1871	–	–	–	–	–	–	–	–	–	–	N	–
Colostethus brachystriatus 1986	–	–	–	–	–	–	–	–	–	–	N	–
Colostethus breviquartus 1986	–	–	–	–	–	–	–	–	–	–	N	–
Colostethus bromelicola 1956	–	–	–	–	–	–	–	–	–	–	NE	–
Colostethus brunneus 1887	–	–	+	–	–	–	–	–	–	–	–	+
Colostethus capixaba 1967	–	–	–	+	–	–	–	–	–	–	–	–
Colostethus capurinensis 1993	–	–	–	–	–	–	–	–	–	–	NE	–
Colostethus carioca 1967	–	–	–	+	–	–	–	–	–	–	–	–
Colostethus cevallosi 1991	–	–	–	–	–	–	–	–	–	–	N	–
Colostethus chocoensis 1912	+	–	–	–	–	–	–	–	–	–	N	–
Colostethus degranvillei 1975	–	–	+	–	–	–	–	–	–	–	–	–
Colostethus delatorreae 1995	–	–	–	–	–	–	–	–	–	–	N	–
Colostethus dunni 1961	–	–	–	–	–	–	–	–	–	–	NE	–
Colostethus duranti 1985	–	–	–	–	–	–	–	–	–	–	NE	–
Colostethus edwardsi 1982	–	–	–	–	–	–	–	–	–	–	N	–
Colostethus elachyhistus 1971	–	–	–	–	–	–	–	–	–	–	+	–
Colostethus exasperatus 1988	–	–	–	–	–	–	–	–	–	–	N	–
Colostethus faciopunctulatus 1991	–	–	W	–	–	–	–	–	–	–	–	–
Colostethus fallax 1991	–	–	–	–	–	–	–	–	–	–	+	–
Colostethus fraterdanieli 1971	–	–	–	–	–	–	–	–	–	–	N	–
Colostethus fujax 1993	–	–	–	–	–	–	–	–	–	–	N	–
Colostethus fuliginosus 1871	–	–	–	–	–	–	–	–	–	–	N	–
Colostethus furviventris 1991	–	–	–	–	–	–	–	–	–	–	N	–
Colostethus goianus 1975	–	–	–	–	–	+	–	–	–	–	–	–
Colostethus guanayensis 1996	–	–	–	–	–	–	–	–	–	–	–	+
Colostethus humilis 1978	–	–	–	–	–	–	–	–	–	–	NE	–
Colostethus idiomelas 1991	–	–	–	–	–	–	–	–	–	–	S	–
Colostethus imbricolus 1975	+	–	–	–	–	–	–	–	–	–	–	–
Colostethus infraguttatus 1898	–	–	–	–	–	–	–	–	–	–	N	–
Colostethus inguinalis 1868	+	–	–	–	–	–	–	–	–	–	–	–
Colostethus intermedius 1945	–	–	W	–	–	–	–	–	–	–	–	–
Colostethus jacobuspetersi 1991	–	–	–	–	–	–	–	–	–	–	N	–
Colostethus juanii 1994	–	–	NW	–	–	–	–	–	–	–	–	–
Colostethus kingsburyi 1918	–	–	–	–	–	–	–	–	–	–	N	–
Colostethus lacrimosus 1991	+	–	–	–	–	–	–	–	–	–	–	–
Colostethus latinasus 1863	+	–	–	–	–	–	–	–	–	–	–	–
Colostethus lehmanni 1971	–	–	–	–	–	–	–	–	–	–	N	–
Colostethus leopardalis 1976	–	–	–	–	–	–	–	–	–	–	NE	–
Colostethus littoralis 1984	–	–	–	–	–	–	–	–	–	+	–	–
Colostethus machalilla 1995	+	–	–	–	–	–	–	–	–	–	–	–
Colostethus mandelorum 1932	–	–	–	–	–	–	–	–	–	–	NE	–
Colostethus maquipucuna 1995	–	–	–	–	–	–	–	–	–	–	N	–
Colostethus marchesianus 1941	–	–	W	–	–	–	–	–	–	–	+	–
Colostethus marmoreoventris 1991	–	–	–	–	–	–	–	–	–	–	N	–
Colostethus mcdiarmidi 1992	–	–	–	–	–	–	–	–	–	–	S	–
Colostethus mertensi 1964	–	–	–	–	–	–	–	–	–	–	N	–
Colostethus mittermeiri 1991	–	–	–	–	–	–	–	–	–	–	S	–
Colostethus murisipanensis 1996	–	–	–	–	–	–	–	–	–	–	–	+
Colostethus mystax 1988	–	–	–	–	–	–	–	–	–	–	N	–
Colostethus nexipus 1986	–	–	–	–	–	–	–	–	–	–	+	–
Colostethus nubicola 1924	+	–	–	–	–	–	–	–	–	–	–	–
Colostethus olfersioides 1925	–	–	–	+	–	–	–	–	–	–	–	–

Taxon and date named	CHO	CAR	A-G	AFD	LLA	CCC	PAM	PAT	ATF	ATA	AND	GUH
Colostethus palmatus 1899	–	–	–	–	–	–	–	–	–	–	N	–
Colostethus parimae 1996	–	–	–	–	–	–	–	–	–	–	–	+
Colostethus parkerae 1996	–	–	–	–	–	–	–	–	–	–	–	+
Colostethus peculiaris 1991	–	–	–	–	–	–	–	–	–	–	N	–
Colostethus peruvianus 1941	–	–	W	–	–	–	–	–	–	–	S	–
Colostethus pinguis 1989	–	–	–	–	–	–	–	–	–	–	N	–
Colostethus poecilonotus 1991	–	–	–	–	–	–	–	–	–	–	S	–
Colostethus praderioi 1996	–	–	–	–	–	–	–	–	–	–	–	+
Colostethus pratti 1899	+	–	–	–	–	–	–	–	–	–	–	–
Colostethus pumilus 1991	–	–	–	–	–	–	–	–	–	–	N	–
Colostethus ramosi 1971	–	–	–	–	–	–	–	–	–	–	N	–
Colostethus ranoides 1918	–	–	–	–	+	–	–	–	–	–	–	–
Colostethus roraima 1996	–	–	–	–	–	–	–	–	–	–	–	+
Colostethus ruizi 1982	–	–	–	–	–	–	–	–	–	–	N	–
Colostethus ruthveni 1996	–	–	–	–	–	–	–	–	–	–	N	–
Colostethus saltuensis 1980	–	–	–	–	–	–	–	–	–	–	NE	–
Colostethus sanmartini 1986	–	–	N	–	–	–	–	–	–	–	–	–
Colostethus sauli 1974	–	–	W	–	–	–	–	–	–	–	–	–
Colostethus serranus 1985	–	–	–	–	–	–	–	–	–	–	NE	–
Colostethus shrevei 1961	–	–	–	–	–	–	–	–	–	–	–	+
Colostethus shuar 1988	–	–	–	–	–	–	–	–	–	–	N	–
Colostethus stepheni 1989	–	–	CEN	–	–	–	–	–	–	–	–	–
Colostethus subpunctatus 1899	–	–	–	–	–	–	–	–	–	–	N	–
Colostethus sylvaticus 1920	–	–	–	–	–	–	–	–	–	–	S	–
Colostethus talamancae 1875	+	–	–	–	–	–	–	–	–	–	–	–
Colostethus tamacuarensis 1997	–	–	–	–	–	–	–	–	–	–	–	+
Colostethus tepuyensis 1996	–	–	–	–	–	–	–	–	–	–	–	+
Colostethus thorntoni 1970	–	–	–	–	–	–	–	–	–	–	N	–
Colostethus toachi 1995	+	–	–	–	–	–	–	–	–	–	N	–
Colostethus trilineatus 1883	–	–	W	–	–	–	–	–	–	–	–	–
Colostethus utcubambensis 1994	–	–	–	–	–	–	–	–	–	–	S	–
Colostethus vergeli 1940	–	–	–	–	–	–	–	–	–	–	N	–
Colostethus vertebralis 1899	–	–	–	–	–	–	–	–	–	–	N	–
Colostethus whymperi 1882	–	–	–	–	–	–	–	–	–	–	N	–
Colostethus yaguara 1991	–	–	–	–	–	–	–	–	–	–	N	–
Dendrobates auratus 1855	+	–	–	–	–	–	–	–	–	–	–	–
Dendrobates azureus 1969	–	–	NE	–	–	–	–	–	–	–	–	–
Dendrobates biolat 1992	–	–	W	–	–	–	–	–	–	–	–	–
Dendrobates captivus 1982	–	–	W	–	–	–	–	–	–	–	–	–
Dendrobates castaneoticus 1990	–	–	E	–	–	–	–	–	–	–	–	–
Dendrobates fantasticus 1884	–	–	W	–	–	–	–	–	–	–	–	–
Dendrobates galactonotus 1864	–	–	E	–	–	–	–	–	–	–	–	–
Dendrobates histrionicus 1846	+	–	–	–	–	–	–	–	–	–	N	–
Dendrobates imitator 1986	–	–	W	–	–	–	–	–	–	–	–	–
Dendrobates lamasi 1992	–	–	W	–	–	–	–	–	–	–	–	–
Dendrobates lehmanni 1976	–	–	–	–	–	–	–	–	–	–	N	–
Dendrobates leucomelas 1864	–	–	NE	–	+	–	–	–	–	–	–	–
Dendrobates mysteriosus 1982	–	–	–	–	–	–	–	–	–	–	S	–
Dendrobates occulator 1976	+	–	–	–	–	–	–	–	–	–	–	–
Dendrobates quinquevittatus 1864	–	–	W	–	–	–	–	–	–	–	–	–
Dendrobates reticulatus 1884	–	–	W	–	–	–	–	–	–	–	–	–
Dendrobates rufulus 1990	–	–	–	–	–	–	–	–	–	–	–	+
Dendrobates sirensis 1991	–	–	–	–	–	–	–	–	–	–	S	–
Dendrobates tinctorius 1799	–	–	NE	–	–	–	–	–	–	–	–	–
Dendrobates truncatus 1861	–	+	–	–	–	–	–	–	–	–	–	–
Dendrobates vanzolinii 1982	–	–	W	–	–	–	–	–	–	–	–	–
Dendrobates variabilis 1988	–	–	W	–	–	–	–	–	–	–	–	–
Dendrobates ventrimaculatus 1935	–	–	+	–	–	–	–	–	–	–	–	–

Taxon and date named	CHO	CAR	A-G	AFD	LLA	CCC	PAM	PAT	ATF	ATA	AND	GUH
Epipedobates andinus 1987	–	–	–	–	–	–	–	–	–	–	N	–
Epipedobates azureiventris 1985	–	–	–	–	–	–	–	–	–	–	S	–
Epipedobates bassleri 1941	–	–	W	–	–	–	–	–	–	–	S	–
Epipedobates bilinguis 1989	–	–	W	–	–	–	–	–	–	–	–	–
Epipedobates bolivianus 1902	–	–	–	–	–	–	–	–	–	–	S	–
Epipedobates boulengeri 1909	+	–	–	–	–	–	–	–	–	–	–	–
Epipedobates braccatus 1864	–	–	S	–	–	–	–	–	–	–	–	–
Epipedobates cainarachi 1989	–	–	W	–	–	–	–	–	–	–	–	–
Epipedobates erythromos 1980	+	–	–	–	–	–	–	–	–	–	–	–
Epipedobates espinosai 1956	–	–	–	–	–	–	–	–	–	–	N	–
Epipedobates femoralis 1884	–	–	+	–	–	–	–	–	–	–	–	–
Epipedobates flavopictus 1925	–	–	S	–	–	+	–	–	–	–	–	–
Epipedobates hahneli 1883	–	–	W	–	–	–	–	–	–	–	–	–
Epipedobates ingeri 1970	–	–	W	–	–	–	–	–	–	–	–	–
Epipedobates macero 1993	–	–	W	–	–	–	–	–	–	–	–	–
Epipedobates myersi 1981	–	–	W	–	–	–	–	–	–	–	–	–
Epipedobates parvulus 1882	–	–	W	–	–	–	–	–	–	–	–	–
Epipedobates petersi 1976	–	–	W	–	–	–	–	–	–	–	S	–
Epipedobates pictus 1838	–	–	SW	–	–	–	–	–	–	–	–	–
Epipedobates pulchripectus 1976	–	–	NE	–	–	–	–	–	–	–	–	–
Epipedobates silverstonei 1979	–	–	–	–	–	–	–	–	–	–	S	–
Epipedobates smaragdinus 1976	–	–	W	–	–	–	–	–	–	–	–	–
Epipedobates tricolor 1899	–	–	–	–	–	–	–	–	–	–	+	–
Epipedobates trivittatus 1824	–	–	+	–	–	–	–	–	–	–	–	–
Epipedobates zaparo 1976	–	–	W	–	–	–	–	–	–	–	–	–
Mannophryne collaris 1912	–	–	–	–	–	–	–	–	–	–	NE	–
Mannophryne herminae 1893	–	–	–	–	–	–	–	–	–	–	NE	–
Mannophryne neblina 1956	–	–	–	–	–	–	–	–	–	–	NE	–
Mannophryne oblitteratus 1984	–	–	–	–	–	–	–	–	–	–	NE	–
Mannophryne olmonae 1983	–	–	TO	–	–	–	–	–	–	–	–	–
Mannophryne riveroi 1964	–	–	–	–	–	–	–	–	–	–	NE	–
Mannophryne trinitatis 1887	–	–	TR	–	–	–	–	–	–	–	–	–
Mannophryne yustizi 1989	–	–	–	–	–	–	–	–	–	–	NE	–
Minyobates abditus 1976	–	–	–	–	–	–	–	–	–	–	N	–
Minyobates altobueyensis 1975	+	–	–	–	–	–	–	–	–	–	–	–
Minyobates bombetes 1980	–	–	–	–	–	–	–	–	–	–	N	–
Minyobates fulguritus 1975	+	–	–	–	–	–	–	–	–	–	–	–
Minyobates minutus 1935	+	–	–	–	–	–	–	–	–	–	–	–
Minyobates opisthomelas 1899	–	–	–	–	–	–	–	–	–	–	N	–
Minyobates steyermarki 1971	–	–	–	–	–	–	–	–	–	–	–	+
Minyobates viridis 1976	–	–	–	–	–	–	–	–	–	–	N	–
Minyobates virolinensis 1992	–	–	–	–	–	–	–	–	–	–	N	–
Nephelobates haydeeae 1976	–	–	–	–	–	–	–	–	–	–	NE	–
Nephelobates mayorgai 1980	–	–	–	–	–	–	–	–	–	–	NE	–
Nephelobates meridensis 1972	–	–	–	–	–	–	–	–	–	–	NE	–
Nephelobates molinarii 1985	–	–	–	–	–	–	–	–	–	–	NE	–
Nephelobates orostoma 1980	–	–	–	–	–	–	–	–	–	–	NE	–
Phyllobates aurotaenia 1913	+	–	–	–	–	–	–	–	–	–	–	–
Phyllobates bicolor 1841	+	–	–	–	–	–	–	–	–	–	–	–
Phyllobates terribilis 1978	+	–	–	–	–	–	–	–	–	–	–	–
ANURA: HYLIDAE:												
Agalychnis calcarifer 1902	+	–	–	–	–	–	–	–	–	–	–	–
Agalychnis craspedopus 1957	–	–	W	–	–	–	–	–	–	–	–	–
Agalychnis litodryas 1967	+	–	–	–	–	–	–	–	–	–	–	–
Agalychnis spurrelli 1913	+	–	–	–	–	–	–	–	–	–	–	–
Aparasphenodon bokermanni 1993	–	–	–	+	–	–	–	–	–	–	–	–
Aparasphenodon brunoi 1920	–	–	–	+	–	–	–	–	–	–	–	–
Aparasphenodon venezolanus 1950	–	–	N	–	–	–	–	–	–	–	–	–

Taxon and date named	CHO	CAR	A-G	AFD	LLA	CCC	PAM	PAT	ATF	ATA	AND	GUH
Aplastodiscus perviridis 1950	–	–	–	+	–	–	+	–	–	–	–	–
Argenteohyla siemersi 1937	–	–	–	–	–	–	+	–	–	–	–	–
Corythomantis greeningi 1896	–	–	–	–	–	+	–	–	–	–	–	–
Cryptobatrachus boulengeri 1916	–	–	–	–	–	–	–	–	–	–	N	–
Cryptobatrachus fuhrmanni 1914	–	–	–	–	–	–	–	–	–	–	N	–
Cryptobatrachus nicefori 1970	–	–	–	–	–	–	–	–	–	–	N	–
Flectonotus fissilis 1920	–	–	–	+	–	–	–	–	–	–	–	–
Flectonotus fitzgeraldi 1933	–	–	TR	–	–	–	–	–	–	–	NE	–
Flectonotus goeldii 1895	–	–	–	+	–	–	–	–	–	–	–	–
Flectonotus ohausi 1907	–	–	–	+	–	–	–	–	–	–	–	–
Flectonotus pygmaeus 1893	–	–	–	–	–	–	–	–	–	–	NE	–
Gastrotheca abdita 1987	–	–	–	–	–	–	–	–	–	–	S	–
Gastrotheca anatomia 1997	–	–	–	–	–	–	–	–	–	–	N	–
Gastrotheca andaquiensis 1976	–	–	–	–	–	–	–	–	–	–	N	–
Gastrotheca angustifrons 1898	+	–	–	–	–	–	–	–	–	–	–	–
Gastrotheca argenteovirens 1892	–	–	–	–	–	–	–	–	–	–	N	–
Gastrotheca aureomaculata 1970	–	–	–	–	–	–	–	–	–	–	N	–
Gastrotheca bufona 1970	–	–	–	–	–	–	–	–	–	–	N	–
Gastrotheca christiani 1967	–	–	–	–	–	–	–	–	–	–	S	–
Gastrotheca chrysosticta 1976	–	–	–	–	–	–	–	–	–	–	S	–
Gastrotheca cornuta 1898	+	–	–	–	–	–	–	–	–	–	–	–
Gastrotheca dendronastes 1983	–	–	–	–	–	–	–	–	–	–	N	–
Gastrotheca dunni 1977	–	–	–	–	–	–	–	–	–	–	N	–
Gastrotheca espeletia 1987	–	–	–	–	–	–	–	–	–	–	N	–
Gastrotheca excubitor 1972	–	–	–	–	–	–	–	–	–	–	S	–
Gastrotheca fissipes 1888	–	–	–	+	–	–	–	–	–	–	–	–
Gastrotheca galeata 1978	–	–	–	–	–	–	–	–	–	–	S	–
Gastrotheca gracilis 1969	–	–	–	–	–	–	–	–	–	–	S	–
Gastrotheca griswoldi 1941	–	–	–	–	–	–	–	–	–	–	S	–
Gastrotheca guentheri 1882	–	–	–	–	–	–	–	–	–	–	N	–
Gastrotheca helenae 1944	–	–	–	–	–	–	–	–	–	–	N	–
Gastrotheca lateonota 1988	–	–	–	–	–	–	–	–	–	–	S	–
Gastrotheca litonedis 1987	–	–	–	–	–	–	–	–	–	–	N	–
Gastrotheca longipes 1882	–	–	W	–	–	–	–	–	–	–	–	–
Gastrotheca marsupiata 1841	–	–	–	–	–	–	–	–	–	–	S	–
Gastrotheca microdisca 1910	–	–	–	+	–	–	–	–	–	–	–	–
Gastrotheca monticola 1920	–	–	–	–	–	–	–	–	–	–	+	–
Gastrotheca nicefori 1933	–	–	–	–	–	–	–	–	–	–	+	–
Gastrotheca ochoai 1972	–	–	–	–	–	–	–	–	–	–	S	–
Gastrotheca orophylax 1980	–	–	–	–	–	–	–	–	–	–	N	–
Gastrotheca ovifera 1854	–	–	–	–	–	–	–	–	–	–	NE	–
Gastrotheca pacchamama 1987	–	–	–	–	–	–	–	–	–	–	S	–
Gastrotheca peruana 1900	–	–	–	–	–	–	–	–	–	–	S	–
Gastrotheca plumbea 1882	–	–	–	–	–	–	–	–	–	–	N	–
Gastrotheca pseustes 1987	–	–	–	–	–	–	–	–	–	–	N	–
Gastrotheca psychrophila 1975	–	–	–	–	–	–	–	–	–	–	N	–
Gastrotheca rebeccae 1988	–	–	–	–	–	–	–	–	–	–	S	–
Gastrotheca riobambae 1913	–	–	–	–	–	–	–	–	–	–	N	–
Gastrotheca ruizi 1986	–	–	–	–	–	–	–	–	–	–	N	–
Gastrotheca testudinea 1871	–	–	–	–	–	–	–	–	–	–	+	–
Gastrotheca trachyceps 1987	–	–	–	–	–	–	–	–	–	–	N	–
Gastrotheca walkeri 1980	–	–	–	–	–	–	–	–	–	–	NE	–
Gastrotheca weinlandii 1892	–	–	–	–	–	–	–	–	–	–	+	–
Gastrotheca williamsoni 1922	–	+	–	–	–	–	–	–	–	–	–	–
Hemiphractus bubalus 1871	–	–	+	–	–	–	–	–	–	–	N	–
Hemiphractus fasciatus 1862	+	–	–	–	–	–	–	–	–	–	N	–
Hemiphractus johnsoni 1917	–	–	–	–	–	–	–	–	–	–	+	–
Hemiphractus proboscideus 1871	–	–	W	–	–	–	–	–	–	–	–	–

APPENDIX 5:1 CONTINUED

Taxon and date named	CHO	CAR	A-G	AFD	LLA	CCC	PAM	PAT	ATF	ATA	AND	GUH
Hemiphractus scutatus 1824	–	–	W	–	–	–	–	–	–	–	N	–
Hyla acreana 1964	–	–	SW	–	–	–	–	–	–	–	–	–
Hyla albofrenata 1924	–	–	–	+	–	–	–	–	–	–	–	–
Hyla alboguttata 1882	–	–	W	–	–	–	–	–	–	–	–	–
Hyla albolineata 1939	–	–	–	+	–	–	–	–	–	–	–	–
Hyla albomarginata 1824	–	+	NE	+	–	–	–	–	–	–	–	–
Hyla albonigra 1841 (1923)	–	–	–	–	–	–	–	–	–	–	S	–
Hyla albopunctata 1824	–	–	–	+	–	+	+	–	–	–	–	–
Hyla albopunctulata 1882	–	–	W	–	–	–	–	–	–	–	–	–
Hyla albosignata 1838	–	–	–	+	–	–	–	–	–	–	–	–
Hyla alemani 1964	–	+	–	–	–	–	–	–	–	–	–	–
Hyla allenorum 1989	–	–	W	–	–	–	–	–	–	–	–	–
Hyla alvarengai 1956	–	–	–	–	–	+	–	–	–	–	–	–
Hyla alytolylax 1972	–	–	–	–	–	–	–	–	–	–	N	–
Hyla anataliasiasi 1973	–	–	SE	–	–	–	–	–	–	–	–	–
Hyla anceps 1929	–	–	–	+	–	–	–	–	–	–	–	–
Hyla andina 1924	–	–	–	–	–	–	–	–	–	–	S	–
Hyla aperomea 1982	–	–	–	–	–	–	–	–	–	–	S	–
Hyla arianae 1985	–	–	–	–	–	–	+	–	–	–	–	–
Hyla arildae 1985	–	–	–	+	–	–	–	–	–	–	–	–
Hyla armata 1902	–	–	–	–	–	–	–	–	–	–	S	–
Hyla aromatica 1993	–	–	–	–	–	–	–	–	–	–	–	+
Hyla astartea 1967	–	–	–	+	–	–	–	–	–	–	–	–
Hyla atlantica 1996	–	–	–	+	–	–	–	–	–	–	–	–
Hyla baileyi 1953	–	–	–	+	–	–	–	–	–	–	–	–
Hyla balzani 1898	–	–	–	–	–	–	–	–	–	–	S	–
Hyla battersbyi 1961	–	–	–	–	–	–	–	–	–	–	NE	–
Hyla benitezi 1961	–	–	–	–	–	–	–	–	–	–	–	+
Hyla berthalutzae 1962	–	–	–	+	–	–	–	–	–	–	–	–
Hyla bifurca 1945	–	–	W	–	–	–	–	–	–	–	–	–
Hyla biobeba 1974	–	–	–	+	–	–	–	–	–	–	–	–
Hyla bipunctata 1824	–	–	–	+	–	–	–	–	–	–	–	–
Hyla bischoffi 1887	–	–	–	+	–	–	–	–	–	–	–	–
Hyla boans 1758	+	–	+	–	–	–	–	–	–	–	–	–
Hyla bogerti 1970	–	–	–	–	–	–	–	–	–	–	N	–
Hyla bogotensis 1882	–	–	–	–	–	–	–	–	–	–	N	–
Hyla bokermanni 1960	–	–	+	–	–	–	–	–	–	–	–	–
Hyla branneri 1948	–	–	–	+	–	–	–	–	–	–	–	–
Hyla brevifrons 1974	–	–	+	–	–	–	–	–	–	–	–	–
Hyla caingua 1990	–	–	–	–	–	–	+	–	–	–	–	–
Hyla calcarata 1848	–	–	+	–	–	–	–	–	–	–	–	–
Hyla callipeza 1989	–	–	–	–	–	–	–	–	–	–	N	–
Hyla callipygia 1984	–	–	–	+	–	–	–	–	–	–	–	–
Hyla carnifex 1969	–	–	–	–	–	–	–	–	–	–	N	–
Hyla carvalhoi 1981	–	–	–	+	–	–	–	–	–	–	–	–
Hyla caucana 1993	–	–	–	–	–	–	–	–	–	–	N	–
Hyla cavicola 1984	–	–	–	+	–	–	–	–	–	–	–	–
Hyla charazani 1970	–	–	–	–	–	–	–	–	–	–	S	–
Hyla chlorostea 1992	–	–	–	–	–	–	–	–	–	–	S	–
Hyla circumdata 1867	–	–	–	+	–	–	–	–	–	–	–	–
Hyla claresignata 1939	–	–	–	+	–	–	+	–	–	–	–	–
Hyla clepsydra 1925	–	–	–	+	–	–	–	–	–	–	–	–
Hyla columbiana 1892	–	–	–	–	–	–	–	–	–	–	N	–
Hyla crepitans 1824	–	+	+	+	+	+	–	–	–	–	NE	+
Hyla cymbalum 1963	–	–	–	+	–	–	–	–	–	–	–	–
Hyla decipiens 1925	–	–	–	+	–	–	–	–	–	–	–	–
Hyla dentei 1967	–	–	NE	–	–	–	–	–	–	–	–	–
Hyla denticulenta 1972	–	–	–	–	–	–	–	–	–	–	N	–

Taxon and date named	CHO	CAR	A-G	AFD	LLA	CCC	PAM	PAT	ATF	ATA	AND	GUH
Hyla dutrai 1996	–	–	–	+	–	–	–	–	–	–	–	–
Hyla ebraccata 1874	+	–	–	–	–	–	–	–	–	–	–	–
Hyla elegans 1824	–	–	–	+	–	–	–	–	–	–	–	–
Hyla faber 1821	–	–	–	+	–	–	+	–	–	–	–	–
Hyla fasciata 1859	–	–	+	–	–	–	–	–	–	–	–	–
Hyla fluminea 1984	–	–	–	+	–	–	–	–	–	–	–	–
Hyla fuentei 1968	–	–	NE	–	–	–	–	–	–	–	–	–
Hyla garagoensis 1991	–	–	–	–	–	–	–	–	–	–	N	–
Hyla geographica 1824	–	–	+	+	–	–	–	–	–	–	–	–
Hyla giesleri 1950	–	–	–	+	–	–	–	–	–	–	–	–
Hyla gouveai 1992	–	–	–	+	–	–	–	–	–	–	–	–
Hyla grandisonae 1966	–	–	NE	–	–	–	–	–	–	–	–	–
Hyla granosa 1882	–	–	+	–	–	–	–	–	–	–	–	–
Hyla gryllata 1973	+	–	–	–	–	–	–	–	–	–	–	–
Hyla guentheri 1886	–	–	–	–	–	–	+	–	–	–	–	–
Hyla haddadi 1996	–	–	–	+	–	–	–	–	–	–	–	–
Hyla hadroceps 1992	–	–	NE	–	–	–	–	–	–	–	–	–
Hyla haraldschultzi 1962	–	–	W	–	–	–	–	–	–	–	–	–
Hyla helenae 1919	–	–	NE	–	–	–	–	–	–	–	–	–
Hyla hobbsi 1970	–	–	NW	–	–	–	–	–	–	–	–	–
Hyla hutchinsi 1984	–	–	NW	–	–	–	–	–	–	–	–	–
Hyla hylax 1985	–	–	–	+	–	–	–	–	–	–	–	–
Hyla ibitiguara 1983	–	–	–	+	–	–	–	–	–	–	–	–
Hyla ibitipoca 1990	–	–	–	+	–	–	–	–	–	–	–	–
Hyla imitator 1921	–	–	CEN	–	–	–	–	–	–	–	–	–
Hyla inframaculata 1882	–	–	E	–	–	–	–	–	–	–	–	–
Hyla inparquezi 1993	–	–	–	–	–	–	–	–	–	–	–	+
Hyla izecksohni 1979	–	–	–	+	–	–	–	–	–	–	–	–
Hyla jahni 1961	–	–	–	–	–	–	–	–	–	–	NE	–
Hyla kanaima 1965	–	–	–	–	–	–	–	–	–	–	–	+
Hyla karenanneae 1993	–	–	NW	–	–	–	–	–	–	–	–	–
Hyla koechlini 1989	–	–	W	–	–	–	–	–	–	–	–	–
Hyla labialis 1863	–	–	–	–	–	–	–	–	–	–	N	–
Hyla lanciformis 1870	–	+	+	–	–	–	–	–	–	–	–	–
Hyla langei 1965	–	–	–	–	–	–	+	–	–	–	–	–
Hyla larinopygion 1973	–	–	–	–	–	–	–	–	–	–	N	–
Hyla lascinia 1969	–	–	–	–	–	–	–	–	–	–	N	–
Hyla leali 1964	–	–	W	–	–	–	–	–	–	–	–	–
Hyla lemai 1971	–	–	–	–	–	–	–	–	–	–	–	+
Hyla leptolineata 1977	–	–	–	–	–	–	+	–	–	–	–	–
Hyla leucophyllata 1783	–	–	+	–	–	–	–	–	–	–	–	–
Hyla leucopygia 1984	–	–	–	+	–	–	–	–	–	–	–	–
Hyla limai 1962	–	–	–	+	–	–	–	–	–	–	–	–
Hyla lindae 1978	–	–	–	–	–	–	–	–	–	–	N	–
Hyla loveridgei 1961	–	–	–	–	–	–	–	–	–	–	–	+
Hyla luctuosa 1993	–	–	–	+	–	–	–	–	–	–	–	–
Hyla luteoocellata 1927	–	+	–	–	–	–	–	–	–	–	–	–
Hyla lynchi 1991	–	–	–	–	–	–	–	–	–	–	N	–
Hyla marginata 1887	–	–	–	–	–	–	+	–	–	–	–	–
Hyla marianitae 1992	–	–	–	–	–	–	–	–	–	–	S	–
Hyla marmorata 1768	–	–	+	–	–	–	–	–	–	–	–	–
Hyla martinsi 1964	–	–	–	+	–	–	–	–	–	–	–	–
Hyla mathiassoni 1970	–	–	–	–	+	–	–	–	–	–	–	–
Hyla melanargyrea 1887	–	–	NE	–	–	+	–	–	–	–	–	–
Hyla melanopleura 1912	–	–	–	–	–	–	–	–	–	–	S	–
Hyla meridensis 1961	–	–	–	–	–	–	–	–	–	–	NE	–
Hyla microcephala 1886	–	+	+	+	+	–	–	–	–	–	–	–
Hyla microderma 1977	–	–	W	–	–	–	–	–	–	–	–	–

Taxon and date named	CHO	CAR	A-G	AFD	LLA	CCC	PAM	PAT	ATF	ATA	AND	GUH
Hyla microps 1872	–	–	–	+	–	–	–	–	–	–	–	–
Hyla miliaria 1886	–	–	–	–	–	–	–	–	–	–	N	–
Hyla minima 1933	–	–	**CEN**	–	–	–	–	–	–	–	–	–
Hyla minuscula 1971	–	–	NE	–	+	–	–	–	–	–	–	–
Hyla minuta 1872	–	–	+	+	–	–	–	–	–	–	–	–
Hyla miyatai 1990	–	–	**W**	–	–	–	–	–	–	–	–	–
Hyla multifasciata 1859	–	–	NE	–	–	–	–	–	–	–	–	–
Hyla musica 1949	–	–	–	+	–	–	–	–	–	–	–	–
Hyla nahdereri 1963	–	–	–	+	–	–	–	–	–	–	–	–
Hyla nana 1889	–	–	–	–	–	+	+	–	–	–	–	–
Hyla nanuzae 1974	–	–	–	+	–	–	–	–	–	–	–	–
Hyla novaisi 1968	–	–	–	+	–	–	–	–	–	–	–	–
Hyla oliveirai 1963	–	–	–	–	–	+	–	–	–	–	–	–
Hyla ornatissima 1923	–	–	NE	–	–	–	–	–	–	–	–	+
Hyla pacha 1990	–	–	–	–	–	–	–	–	–	–	N	–
Hyla padreluna 1997	–	–	–	–	–	–	–	–	–	–	N	–
Hyla palaestes 1997	–	–	–	–	–	–	–	–	–	–	S	–
Hyla palmeri 1908	–	–	–	–	–	–	–	–	–	–	N	–
Hyla pantosticta 1982	–	–	–	–	–	–	–	–	–	–	N	–
Hyla pardalis 1824	–	–	–	+	–	+	+	–	–	–	–	–
Hyla parviceps 1882	–	–	+	–	–	–	–	–	–	–	–	–
Hyla pauiniensis 1977	–	–	**CEN**	–	–	–	–	–	–	–	–	–
Hyla pelidna 1989	–	–	–	–	–	–	–	–	–	–	N	–
Hyla pellucens 1901	+	–	–	–	–	–	–	–	–	–	–	–
Hyla phlebodes 1906	+	–	–	–	–	–	–	–	–	–	–	–
Hyla phyllognatha 1941	–	–	–	–	–	–	–	–	–	–	+	–
Hyla piceigularis 1982	–	–	–	–	–	–	–	–	–	–	N	–
Hyla picturata 1882	+	–	–	–	–	–	–	–	–	–	–	–
Hyla pinima 1974	–	–	–	+	–	–	–	–	–	–	–	–
Hyla platydactyla 1905	–	–	–	–	–	–	–	–	–	–	NE	–
Hyla polytaenia 1869	–	–	–	–	–	–	+	–	–	–	–	–
Hyla praestans 1983	–	–	–	–	–	–	–	–	–	–	N	–
Hyla prasina 1856	–	–	–	+	–	–	–	–	–	–	–	–
Hyla psarolaima 1990	–	–	–	–	–	–	–	–	–	–	N	–
Hyla pseudopseudis 1937	–	–	–	+	–	–	–	–	–	–	–	–
Hyla ptychodactyla 1990	–	–	–	–	–	–	–	–	–	–	N	–
Hyla pugnax 1857	–	+	–	–	–	–	–	–	–	–	–	–
Hyla pulchella 1841	–	–	–	–	–	–	+	–	–	–	S	–
Hyla pulidoi 1968	–	–	–	–	–	–	–	–	–	–	–	+
Hyla punctata 1799	–	+	+	–	–	+	–	–	–	–	–	–
Hyla raniceps 1862	–	–	NE	–	–	+	–	–	–	–	–	–
Hyla rhodopepla 1859	–	–	**W**	–	–	–	–	–	–	–	–	–
Hyla riveroi 1970	–	–	**W**	–	–	–	–	–	–	–	–	–
Hyla roeschmanni 1938	–	–	**SW**	–	–	–	–	–	–	–	–	–
Hyla roraima 1992	–	–	–	–	–	–	–	–	–	–	–	+
Hyla rosenbergi 1898	+	–	–	–	–	–	–	–	–	–	–	–
Hyla rossalleni 1959	–	–	+	–	–	–	–	–	–	–	–	–
Hyla rubicundula 1862	–	–	–	–	–	+	+	–	–	–	–	–
Hyla rubracyla 1970	+	–	–	–	–	–	–	–	–	–	–	–
Hyla ruschii 1987	–	–	–	+	–	–	–	–	–	–	–	–
Hyla sarampiona 1982	–	–	–	–	–	–	–	–	–	–	N	–
Hyla sarayacuensis 1935	–	–	**W**	–	–	–	–	–	–	–	–	–
Hyla saxicola 1964	–	–	–	+	–	–	–	–	–	–	–	–
Hyla sazimai 1983	–	–	–	+	–	–	–	–	–	–	–	–
Hyla schubarti 1963	–	–	**SW**	–	–	–	–	–	–	–	–	–
Hyla secedens 1963	–	–	–	+	–	–	–	–	–	–	–	–
Hyla semiguttata 1925	–	–	–	+	–	–	+	–	–	–	–	–
Hyla senicula 1868	–	–	–	+	–	–	–	–	–	–	–	–

APPENDIX 5:1 CONTINUED

Taxon and date named	CHO	CAR	A-G	AFD	LLA	CCC	PAM	PAT	ATF	ATA	AND	GUH
Hyla sibleszi 1971	–	–	–	–	–	–	–	–	–	–	–	+
Hyla simmonsi 1989	–	–	–	–	–	–	–	–	–	–	N	–
Hyla soaresi 1983	–	–	–	+	–	–	–	–	–	–	–	–
Hyla staufferorum 1993	–	–	–	–	–	–	–	–	–	–	N	–
Hyla stingi 1994	–	–	–	–	–	–	–	–	–	–	N	–
Hyla subocularis 1934	+	–	–	–	–	–	–	–	–	–	–	–
Hyla timbeba 1987	–	–	SW	–	–	–	–	–	–	–	–	–
Hyla tintinnabulum 1941	–	–	N	–	–	–	–	–	–	–	–	–
Hyla torrenticola 1978	–	–	–	–	–	–	–	–	–	–	N	–
Hyla triangulum 1869	–	–	W	–	–	–	–	–	–	–	–	–
Hyla tritaeniata 1965	–	–	–	–	–	+	–	–	–	–	–	–
Hyla truncata 1959	–	–	–	+	–	–	–	–	–	–	–	–
Hyla tuberculosa 1882	–	–	W	–	–	–	–	–	–	–	–	–
Hyla uruguaya 1944	–	–	–	–	–	+	–	–	–	–	–	–
Hyla varelae 1992	–	–	–	–	–	+	–	–	–	–	–	–
Hyla virolinensis 1997	–	–	–	–	–	–	–	–	–	–	N	–
Hyla walfordi 1962	–	–	SW	–	–	–	–	–	–	–	–	–
Hyla warreni 1992	–	–	–	–	–	–	–	–	–	–	–	+
Hyla wavrini 1936	–	–	NE	–	–	–	–	–	–	–	–	–
Hyla weygoldti 1985	–	–	–	+	–	–	–	–	–	–	–	–
Hyla xapuriensis 1987	–	–	SW	–	–	–	–	–	–	–	–	–
Hylomantis aspera 1872	–	–	–	+	–	–	–	–	–	–	–	–
Hylomantis granulosa 1988	–	–	–	+	–	–	–	–	–	–	–	–
Nyctimantis rugiceps 1882	–	–	W	–	–	–	–	–	–	–	–	–
Osteocephalus buckleyi 1882	–	–	+	–	–	–	–	–	–	–	–	–
Osteocephalus cabrerai 1970	–	–	W	–	–	–	–	–	–	–	–	–
Osteocephalus elkejungingerae 1981	–	–	+	–	–	–	–	–	–	–	S	–
Osteocephalus langsdorffii 1841	–	–	–	+	–	–	+	–	–	–	–	–
Osteocephalus leprieurii 1841	–	–	+	–	–	–	–	–	–	–	–	–
Osteocephalus oophagus 1995	–	–	+	–	–	–	–	–	–	–	–	–
Osteocephalus pearsoni 1929	–	–	–	–	–	–	–	–	–	–	S	–
Osteocephalus planiceps 1874	–	–	W	–	–	–	–	–	–	–	–	–
Osteocephalus subtilis 1987	–	–	SW	–	–	–	–	–	–	–	–	–
Osteocephalus taurinus 1862	–	–	+	–	–	–	–	–	–	–	–	–
Osteocephalus verruciger 1901	–	–	–	–	–	–	–	–	–	–	+	–
Phasmahyla cochranae 1966	–	–	–	+	–	–	–	–	–	–	–	–
Phasmahyla exilis 1980	–	–	–	+	–	–	–	–	–	–	–	–
Phasmahyla guttata 1925	–	–	–	+	–	–	–	–	–	–	–	–
Phasmahyla jandaia 1978	–	–	–	+	–	–	–	–	–	–	–	–
Phrynohyas coriacea 1867	–	–	+	–	–	–	–	–	–	–	–	–
Phrynohyas imitatrix 1926	–	–	–	+	–	–	–	–	–	–	–	–
Phrynohyas mesophaea 1867	–	–	–	+	–	–	–	–	–	–	–	–
Phrynohyas resinifictrix 1907	–	–	+	–	–	–	–	–	–	–	–	–
Phrynohyas venulosa 1768	+	+	+	–	+	+	–	–	–	–	–	–
Phrynomedusa appendiculata 1925	–	–	–	+	–	–	–	–	–	–	–	–
Phrynomedusa fimbriata 1923	–	–	–	+	–	–	–	–	–	–	–	–
Phrynomedusa marginata 1976	–	–	–	+	–	–	–	–	–	–	–	–
Phyllodytes acuminatus 1966	–	–	–	+	–	–	–	–	–	–	–	–
Phyllodytes auratus 1917	–	–	TR	–	–	–	–	–	–	–	–	–
Phyllodytes brevirostris 1988	–	–	–	+	–	–	–	–	–	–	–	–
Phyllodytes kautskyi 1988	–	–	–	+	–	–	–	–	–	–	–	–
Phyllodytes luteolus 1824	–	–	–	+	–	–	–	–	–	–	–	–
Phyllodytes melanomystax 1992	–	–	–	+	–	–	–	–	–	–	–	–
Phyllodytes tuberculosus 1966	–	–	–	–	–	+	–	–	–	–	–	–
Phyllomedusa atelopoides 1988	–	–	W	–	–	–	–	–	–	–	–	–
Phyllomedusa ayeaye 1966	–	–	–	+	–	–	–	–	–	–	–	–
Phyllomedusa baltea 1979	–	–	–	–	–	–	–	–	–	–	S	–
Phyllomedusa bicolor 1772	–	–	+	–	–	–	–	–	–	–	–	–

APPENDIX 5:1 CONTINUED

Taxon and date named	CHO	CAR	A-G	AFD	LLA	CCC	PAM	PAT	ATF	ATA	AND	GUH
Phyllomedusa boliviana 1902	–	–	SW	–	–	+	–	–	–	–	–	–
Phyllomedusa buckleyi 1882	–	–	–	–	–	–	–	–	–	–	N	–
Phyllomedusa burmeisteri 1882	–	–	–	+	–	+	–	–	–	–	–	–
Phyllomedusa centralis 1965	–	–	–	–	–	+	–	–	–	–	–	–
Phyllomedusa coelestis 1874	–	–	W	–	–	–	–	–	–	–	–	–
Phyllomedusa danieli 1988	–	–	–	–	–	–	–	–	–	–	N	–
Phyllomedusa distincta 1950	–	–	–	+	–	–	–	–	–	–	–	–
Phyllomedusa duellmani 1982	–	–	–	–	–	–	–	–	–	–	S	–
Phyllomedusa ecuatoriana 1982	–	–	–	–	–	–	–	–	–	–	N	–
Phyllomedusa hulli 1995	–	–	W	–	–	–	–	–	–	–	–	–
Phyllomedusa hypocondrialis 1802	–	–	+	–	+	+	–	–	–	–	–	–
Phyllomedusa iheringii 1885	–	–	–	–	–	–	+	–	–	–	–	–
Phyllomedusa medinai 1962	–	–	–	–	–	–	–	–	–	–	NE	–
Phyllomedusa palliata 1872	–	–	W	–	–	–	–	–	–	–	–	–
Phyllomedusa perinesos 1973	–	–	–	–	–	–	–	–	–	–	N	–
Phyllomedusa psilopygion 1980	+	–	–	–	–	–	–	–	–	–	–	–
Phyllomedusa rohdei 1926	–	–	–	+	–	–	–	–	–	–	–	–
Phyllomedusa sauvagii 1882	–	–	–	–	–	+	–	–	–	–	–	–
Phyllomedusa tarsius 1868	–	–	+	–	–	–	–	–	–	–	–	–
Phyllomedusa tetraploida 1992	–	–	–	+	–	–	–	–	–	–	–	–
Phyllomedusa tomopterna 1868	–	–	+	–	–	–	–	–	–	–	–	–
Phyllomedusa trinitatis 1926	–	+	TR	–	–	–	–	–	–	–	–	–
Phyllomedusa vaillanti 1882	–	–	+	–	–	–	–	–	–	–	–	–
Scarthyla ostinodactyla 1988	–	–	W	–	–	–	–	–	–	–	–	–
Scinax acuminata 1862	–	–	–	–	–	+	–	–	–	–	–	–
Scinax agilis 1983	–	–	–	+	–	–	–	–	–	–	–	–
Scinax albicans 1967	–	–	–	+	–	–	–	–	–	–	–	–
Scinax alcatraz 1973	–	–	–	+	–	–	–	–	–	–	–	–
Scinax alleni 1870	–	–	+	–	–	–	–	–	–	–	–	–
Scinax angrensis 1973	–	–	–	+	–	–	–	–	–	–	–	–
Scinax argyreornata 1926	–	–	–	+	–	–	–	–	–	–	–	–
Scinax ariadne 1967	–	–	–	+	–	–	–	–	–	–	–	–
Scinax atrata 1988	–	–	–	+	–	–	–	–	–	–	–	–
Scinax aurata 1821	–	–	–	+	–	–	–	–	–	–	–	–
Scinax baumgardneri 1961	–	–	N	–	–	–	–	–	–	–	–	–
Scinax berthae 1962	–	–	–	–	–	–	+	–	–	–	–	–
Scinax blairi 1972	–	–	–	–	+	–	–	–	–	–	–	–
Scinax boesemani 1966	–	–	NE	–	–	–	–	–	–	–	–	–
Scinax caldarum 1968	–	–	–	+	–	–	–	–	–	–	–	–
Scinax canastrensis 1982	–	–	–	+	–	–	–	–	–	–	–	–
Scinax cardosoi 1991	–	–	–	+	–	–	–	–	–	–	–	–
Scinax carnevalli 1989	–	–	–	+	–	–	–	–	–	–	–	–
Scinax castroviejoi 1993	–	–	–	–	–	–	–	–	–	–	S	–
Scinax catharinae 1888	–	–	–	+	–	–	–	–	–	–	–	–
Scinax chiquitana 1990	–	–	SW	–	–	–	–	–	–	–	–	–
Scinax crospedospila 1925	–	–	–	+	–	–	–	–	–	–	–	–
Scinax cruentomma 1972	–	–	W	–	–	–	–	–	–	–	–	–
Scinax cuspidata 1925	–	–	–	+	–	–	–	–	–	–	–	–
Scinax cynocephala 1841	–	–	NE	–	–	–	–	–	–	–	–	–
Scinax danae 1986	–	–	–	–	–	–	–	–	–	–	–	+
Scinax duartei 1951	–	–	–	+	–	–	–	–	–	–	–	–
Scinax ehrhardti 1924	–	–	–	–	–	–	+	–	–	–	–	–
Scinax eurydice 1968	–	–	–	+	–	–	–	–	–	–	–	–
Scinax exigua 1986	–	–	–	–	–	–	–	–	–	–	–	+
Scinax flavoguttata 1939	–	–	–	+	–	–	–	–	–	–	–	–
Scinax funerea 1974	–	–	W	–	–	–	–	–	–	–	–	–
Scinax fuscomarginata 1925	–	–	S	–	–	+	+	–	–	–	–	–
Scinax fuscovaria 1925	–	–	–	–	–	+	–	–	–	–	–	–

Taxon and date named	CHO	CAR	A-G	AFD	LLA	CCC	PAM	PAT	ATF	ATA	AND	GUH
Scinax garbei 1926	–	–	**W**	–	–	–	–	–	–	–	–	–
Scinax goinorum 1962	–	–	**SW**	–	–	–	–	–	–	–	–	–
Scinax hayii 1909	–	–	–	+	–	–	–	–	–	–	–	–
Scinax heyeri 1986	–	–	–	+	–	–	–	–	–	–	–	–
Scinax hiemalis 1987	–	–	–	+	–	–	–	–	–	–	–	–
Scinax humilis 1954	–	–	–	+	–	–	–	–	–	–	–	–
Scinax icterica 1993	–	–	**SW**	–	–	–	–	–	–	–	–	–
Scinax jureia 1991	–	–	–	+	–	–	–	–	–	–	–	–
Scinax kautskyi 1991	–	–	–	+	–	–	–	–	–	–	–	–
Scinax kennedyi 1973	–	–	–	–	+	–	–	–	–	–	–	–
Scinax lindsayi 1992	–	–	**NW**	–	–	–	–	–	–	–	–	–
Scinax littoralis 1991	–	–	–	+	–	–	–	–	–	–	–	–
Scinax littorea 1988	–	–	–	+	–	–	–	–	–	–	–	–
Scinax longilinea 1968	–	–	–	+	–	–	–	–	–	–	–	–
Scinax luizotavioi 1989	–	–	–	+	–	–	–	–	–	–	–	–
Scinax machadoi 1973	–	–	–	+	–	–	–	–	–	–	–	–
Scinax maracaya 1980	–	–	–	+	–	–	–	–	–	–	–	–
Scinax melloi 1988	–	–	–	+	–	–	–	–	–	–	–	–
Scinax nasica 1862	–	–	–	–	–	+	+	–	–	–	–	–
Scinax nebulosa 1824	–	–	**NE**	–	–	–	–	–	–	–	–	–
Scinax obtriangulata 1973	–	–	–	+	–	–	–	–	–	–	–	–
Scinax opalina 1968	–	–	–	+	–	–	–	–	–	–	–	–
Scinax oreites 1993	–	–	–	–	–	–	–	–	–	–	**S**	–
Scinax pachychrus 1937	–	–	–	–	–	+	–	–	–	–	–	–
Scinax parkeri 1929	–	–	**SW**	–	–	–	–	–	–	–	–	–
Scinax pedromedinai 1991	–	–	**SW**	–	–	–	–	–	–	–	–	–
Scinax perereca 1995	–	–	–	+	–	–	–	–	–	–	–	–
Scinax perpusilla 1939	–	–	–	+	–	–	–	–	–	–	–	–
Scinax proboscidea 1933	–	–	**NE**	–	–	–	–	–	–	–	–	–
Scinax quinquefasciata 1913	+	–	–	–	–	–	–	–	–	–	–	–
Scinax ranki 1987	–	–	–	+	–	–	–	–	–	–	–	–
Scinax rizibilis 1964	–	–	–	+	–	–	–	–	–	–	–	–
Scinax rostrata 1863	–	+	**NE**	–	+	–	–	–	–	–	–	–
Scinax rubra 1768	–	+	+	–	+	–	–	–	–	–	–	–
Scinax similis 1952	–	–	–	+	–	–	–	–	–	–	–	–
Scinax squalirostris 1925	–	–	–	–	–	–	+	–	–	–	–	–
Scinax strigilata 1824	–	–	–	+	–	–	–	–	–	–	–	–
Scinax sugillata 1973	+	–	–	–	–	–	–	–	–	–	–	–
Scinax trachythorax 1936	–	–	–	–	–	+	–	–	–	–	–	–
Scinax trapicheiroi 1954	–	–	–	+	–	–	–	–	–	–	–	–
Scinax trilineata 1979	–	–	**NE**	–	+	–	–	–	–	–	–	–
Scinax vauterii 1843	–	–	–	–	–	–	+	–	–	–	–	–
Scinax vigilans 1971	–	+	–	–	–	–	–	–	–	–	–	–
Scinax v–signata 1968	–	–	–	+	–	–	–	–	–	–	–	–
Scinax wandae 1971	–	–	–	–	+	–	–	–	–	–	–	–
Scinax x–signata 1824	–	–	**NE**	+	+	+	–	–	–	–	–	–
Smilisca phaeota 1862	+	–	–	–	–	–	–	–	–	–	–	–
Smilisca sila 1966	–	+	–	–	–	–	–	–	–	–	–	–
Sphaenorhynchus bromelicola 1966	–	–	–	+	–	–	–	–	–	–	–	–
Sphaenorhynchus carneus 1868	–	–	**W**	–	–	–	–	–	–	–	–	–
Sphaenorhynchus dorisae 1957	–	–	**W**	–	–	–	–	–	–	–	–	–
Sphaenorhynchus lacteus 1802	–	–	+	–	+	–	–	–	–	–	–	–
Sphaenorhynchus orophilus 1938	–	–	–	+	–	–	–	–	–	–	–	–
Sphaenorhynchus palustris 1966	–	–	–	+	–	–	–	–	–	–	–	–
Sphaenorhynchus pauloalvini 1973	–	–	–	+	–	–	–	–	–	–	–	–
Sphaenorhynchus planicola 1938	–	–	–	+	–	–	–	–	–	–	–	–
Sphaenorhynchus prasinus 1973	–	–	–	+	–	–	–	–	–	–	–	–
Sphaenorhynchus surdus 1953	–	–	–	+	–	–	–	–	–	–	–	–

APPENDIX 5:1 CONTINUED

Taxon and date named	CHO	CAR	A-G	AFD	LLA	CCC	PAM	PAT	ATF	ATA	AND	GUH
Stefania evansi 1904	–	–	NE	–	–	–	–	–	–	–	–	–
Stefania ginesi 1968	–	–	–	–	–	–	–	–	–	–	–	+
Stefania goini 1968	–	–	–	–	–	–	–	–	–	–	–	+
Stefania marahuaquensis 1961	–	–	–	–	–	–	–	–	–	–	–	+
Stefania oculosa 1996	–	–	–	–	–	–	–	–	–	–	–	+
Stefania percristata 1996	–	–	–	–	–	–	–	–	–	–	–	+
Stefania riae 1984	–	–	–	–	–	–	–	–	–	–	–	+
Stefania riveroi 1996	–	–	–	–	–	–	–	–	–	–	–	+
Stefania roraima 1984	–	–	–	–	–	–	–	–	–	–	–	+
Stefania satelles 1996	–	–	–	–	–	–	–	–	–	–	–	+
Stefania scalae 1970	–	–	–	–	–	–	–	–	–	–	–	+
Stefania schubarti 1996	–	–	–	–	–	–	–	–	–	–	–	+
Stefania tamacuarina 1997	–	–	–	–	–	–	–	–	–	–	–	+
Stefania woodleyi 1968	–	–	–	–	–	–	–	–	–	–	–	+
Tepuihyla aecii 1992	–	–	–	–	–	–	–	–	–	–	–	+
Tepuihyla edelcae 1992	–	–	–	–	–	–	–	–	–	–	–	+
Tepuihyla galani 1992	–	–	–	–	–	–	–	–	–	–	–	+
Tepuihyla luteolabris 1992	–	–	–	–	–	–	–	–	–	–	–	+
Tepuihyla rimarum 1992	–	–	–	–	–	–	–	–	–	–	–	+
Tepuihyla rodriguezi 1968	–	–	–	–	–	–	–	–	–	–	–	+
Tepuihyla talbergae 1996	–	–	–	–	–	–	–	–	–	–	–	+
Trachycephalus atlas 1966	–	–	–	–	–	+	–	–	–	–	–	–
Trachycephalus jordani 1891	+	–	–	–	–	–	–	–	–	–	–	–
Trachycephalus nigromaculatus 1838	–	–	–	+	–	+	–	–	–	–	–	–
ANURA: LEPTODACTYLIDAE:												
Adelophryne adiastola 1984	–	–	W	–	–	–	–	–	–	–	–	–
Adelophryne baturitensis 1994	–	–	–	+	–	–	–	–	–	–	–	–
Adelophryne gutturosa 1984	–	–	NE	–	–	–	–	–	–	–	–	–
Adelophryne maranguapensis 1994	–	–	–	+	–	–	–	–	–	–	–	–
Adelophryne pachydactyla 1994	–	–	–	+	–	–	–	–	–	–	–	–
Adelophryne tridactyla 1995	–	–	W	–	–	–	–	–	–	–	–	–
Adenomera andreae 1923	–	–	+	–	–	–	–	–	–	–	–	–
Adenomera bokermanni 1973	–	–	–	+	–	–	–	–	–	–	–	–
Adenomera diptyx 1885	–	–	–	–	–	+	–	–	–	–	–	–
Adenomera hylaedactyla 1868	–	–	+	+	–	+	–	–	–	–	–	–
Adenomera lutzi 1975	–	–	NE	–	–	–	–	–	–	–	–	–
Adenomera marmorata 1867	–	–	–	+	–	–	–	–	–	–	–	–
Adenomera martinezi 1956	–	–	–	–	–	+	–	–	–	–	–	–
Alsodes barrioi 1981	–	–	–	–	–	–	–	–	+	–	–	–
Alsodes gargola 1970	–	–	–	–	–	–	–	–	+	–	–	–
Alsodes illotus 1922	–	–	–	–	–	–	–	–	–	–	S	–
Alsodes laevis 1902	–	–	–	–	–	–	–	–	–	–	S	–
Alsodes montanus 1902	–	–	–	–	–	–	–	–	–	–	S	–
Alsodes monticola 1843	–	–	–	–	–	–	–	–	+	–	–	–
Alsodes nodosus 1841	–	–	–	–	–	–	–	–	–	–	S	–
Alsodes pehuenche 1976	–	–	–	–	–	–	–	–	–	–	S	–
Alsodes tumultuosus 1979	–	–	–	–	–	–	–	–	–	–	S	–
Alsodes vanzolinii 1974	–	–	–	–	–	–	–	–	+	–	–	–
Alsodes verrucosus 1902	–	–	–	–	–	–	–	–	+	–	–	–
Alsodes vittatus 1902	–	–	–	–	–	–	–	–	+	–	–	–
Atelognathus grandisonae 1975	–	–	–	–	–	–	–	–	+	–	–	–
Atelognathus nitoi 1973	–	–	–	–	–	–	–	–	+	–	–	–
Atelognathus patagonicus 1962	–	–	–	–	–	–	–	+	–	–	–	–
Atelognathus prebasalticus 1968	–	–	–	–	–	–	–	+	–	–	–	–
Atelognathus reverberii 1969	–	–	–	–	–	–	–	+	–	–	–	–
Atelognathus salai 1984	–	–	–	–	–	–	–	+	–	–	–	–
Atelognathus solitarius 1970	–	–	–	–	–	–	–	+	–	–	–	–
Atopophrynus syntomopus 1982	–	–	–	–	–	–	–	–	–	–	N	–

Taxon and date named	CHO	CAR	A-G	AFD	LLA	CCC	PAM	PAT	ATF	ATA	AND	GUH
Barycholos pulcher 1898	+	–	–	–	–	–	–	–	–	–	–	–
Barycholos savagei 1980	–	–	–	–	–	+	–	–	–	–	–	–
Batrachophrynus brachydactylus 1873	–	–	–	–	–	–	–	–	–	–	S	–
Batrachophrynus macrostomus 1873	–	–	–	–	–	–	–	–	–	–	S	–
Batrachyla antartandica 1967	–	–	–	–	–	–	–	–	+	–	–	–
Batrachyla fitzroya 1994	–	–	–	–	–	–	–	–	+	–	–	–
Batrachyla leptopus 1843	–	–	–	–	–	–	–	–	+	–	–	–
Batrachyla nibaldoi 1997	–	–	–	–	–	–	–	–	+	–	–	–
Batrachyla taeniata 1854	–	–	–	–	–	–	–	–	+	–	–	–
Caudiverbera caudiverbera 1758	–	–	–	–	–	–	–	–	+	–	–	–
Ceratophrys aurita 1823	–	–	–	+	–	–	–	–	–	–	–	–
Ceratophrys calcarata 1890	–	+	–	–	–	–	–	–	–	–	–	–
Ceratophrys cornuta 1758	–	–	+	–	–	–	–	–	–	–	–	–
Ceratophrys cranwelli 1980	–	–	–	–	–	+	–	–	–	–	–	–
Ceratophrys joazeirensis 1986	–	–	–	–	–	+	–	–	–	–	–	–
Ceratophrys ornata 1843	–	–	–	–	–	–	+	–	–	–	–	–
Ceratophrys stolzmanni 1882	+	–	–	–	–	–	–	–	–	–	–	–
Chacophrys pierotti 1948	–	–	–	–	–	+	–	–	–	–	–	–
Crossodactylodes bokermanni 1983	–	–	–	+	–	–	–	–	–	–	–	–
Crossodactylodes izecksohni 1983	–	–	–	+	–	–	–	–	–	–	–	–
Crossodactylodes pintoi 1938	–	–	–	+	–	–	–	–	–	–	–	–
Crossodactylus aeneus 1924	–	–	–	+	–	–	–	–	–	–	–	–
Crossodactylus bokermanni 1985	–	–	–	+	–	–	–	–	–	–	–	–
Crossodactylus caramaschii 1995	–	–	–	+	–	–	–	–	–	–	–	–
Crossodactylus dispar 1925	–	–	–	+	–	–	+	–	–	–	–	–
Crossodactylus gaudichaudii 1841	–	–	–	+	–	–	–	–	–	–	–	–
Crossodactylus schmidti 1961	–	–	–	–	–	–	+	–	–	–	–	–
Crossodactylus trachystomus 1862	–	–	–	+	–	–	–	–	–	–	–	–
Cycloramphus asper 1899	–	–	–	+	–	–	–	–	–	–	–	–
Cycloramphus bandeirensis 1983	–	–	–	+	–	–	–	–	–	–	–	–
Cycloramphus bolitoglossus 1897	–	–	–	+	–	–	–	–	–	–	–	–
Cycloramphus boraceiensis 1983	–	–	–	+	–	–	–	–	–	–	–	–
Cycloramphus brasiliensis 1864	–	–	–	+	–	–	–	–	–	–	–	–
Cycloramphus carvalhoi 1983	–	–	–	+	–	–	–	–	–	–	–	–
Cycloramphus catarinensis 1983	–	–	–	+	–	–	–	–	–	–	–	–
Cycloramphus cedrensis 1983	–	–	–	+	–	–	–	–	–	–	–	–
Cycloramphus diringshofeni 1957	–	–	–	+	–	–	–	–	–	–	–	–
Cycloramphus dubius 1920	–	–	–	+	–	–	–	–	–	–	–	–
Cycloramphus duseni 1914	–	–	–	+	–	–	–	–	–	–	–	–
Cycloramphus eleutherodactylus 1920	–	–	–	+	–	–	–	–	–	–	–	–
Cycloramphus fuliginosus 1838	–	–	–	+	–	–	–	–	–	–	–	–
Cycloramphus granulosus 1929	–	–	–	+	–	–	–	–	–	–	–	–
Cycloramphus izecksohni 1983	–	–	–	+	–	–	–	–	–	–	–	–
Cycloramphus jordanensis 1983	–	–	–	+	–	–	–	–	–	–	–	–
Cycloramphus juimirim 1989	–	–	–	+	–	–	–	–	–	–	–	–
Cycloramphus lutzorum 1983	–	–	–	+	–	–	–	–	–	–	–	–
Cycloramphus migueli 1988	–	–	–	+	–	–	–	–	–	–	–	–
Cycloramphus mirandaribeiroi 1983	–	–	–	+	–	–	–	–	–	–	–	–
Cycloramphus ohausi 1907	–	–	–	+	–	–	–	–	–	–	–	–
Cycloramphus rhyakonastes 1983	–	–	–	+	–	–	–	–	–	–	–	–
Cycloramphus semipalmatus 1920	–	–	–	+	–	–	–	–	–	–	–	–
Cycloramphus stejnegeri 1924	–	–	–	+	–	–	–	–	–	–	–	–
Cycloramphus valae 1983	–	–	–	+	–	–	–	–	–	–	–	–
Dischidodactylus colonelloi 1985	–	–	–	–	–	–	–	–	–	–	–	+
Dischidodactylus duidensis 1968	–	–	–	–	–	–	–	–	–	–	–	+
Edalorhina nasuta 1912	–	–	W	–	–	–	–	–	–	–	–	–
Edalorhina perezi 1870	–	–	W	–	–	–	–	–	–	–	–	–
Eleutherodactylus aaptus 1980	–	–	W	–	–	–	–	–	–	–	–	–

APPENDIX 5:1 CONTINUED

Taxon and date named	CHO	CAR	A-G	AFD	LLA	CCC	PAM	PAT	ATF	ATA	AND	GUH
Eleutherodactylus acatallelus 1983	–	–	–	–	–	–	–	–	–	–	N	–
Eleutherodactylus acerus 1980	–	–	–	–	–	–	–	–	–	–	N	–
Eleutherodactylus achatinus 1898	+	–	–	–	–	–	–	–	–	–	N	–
Eleutherodactylus actites 1979	–	–	–	–	–	–	–	–	–	–	N	–
Eleutherodactylus acuminatus 1935	–	–	W	–	–	–	–	–	–	–	–	–
Eleutherodactylus acutirostris 1984	–	–	–	–	–	–	–	–	–	–	N	–
Eleutherodactylus aemulatus 1997	–	–	–	–	–	–	–	–	–	–	N	–
Eleutherodactylus affinis 1899	–	–	–	–	–	–	–	–	–	–	N	–
Eleutherodactylus alalocophus 1991	–	–	–	–	–	–	–	–	–	–	N	–
Eleutherodactylus alberico 1996	–	–	–	–	–	–	–	–	–	–	N	–
Eleutherodactylus altamazonicus 1921	–	–	W	–	–	–	–	–	–	–	–	–
Eleutherodactylus anatipes 1983	–	–	–	–	–	–	–	–	–	–	N	–
Eleutherodactylus andicola 1891	–	–	–	–	–	–	–	–	–	–	S	–
Eleutherodactylus anolirex 1983	–	–	–	–	–	–	–	–	–	–	N	–
Eleutherodactylus anomalus 1898	+	–	–	–	–	–	–	–	–	–	N	–
Eleutherodactylus anonymus 1928	–	+	–	–	–	–	–	–	–	–	–	–
Eleutherodactylus anotis 1955	–	–	–	–	–	–	–	–	–	–	NE	–
Eleutherodactylus apiculatus 1990	–	–	–	–	–	–	–	–	–	–	N	–
Eleutherodactylus appendiculatus 1894	–	–	–	–	–	–	–	–	–	–	N	–
Eleutherodactylus atratus 1979	–	–	–	–	–	–	–	–	–	–	N	–
Eleutherodactylus aurantiguttatus 1997	–	–	–	–	–	–	–	–	–	–	N	–
Eleutherodactylus avius 1997	–	–	–	–	–	–	–	–	–	–	–	+
Eleutherodactylus babax 1989	–	–	–	–	–	–	–	–	–	–	N	–
Eleutherodactylus bacchus 1984	–	–	–	–	–	–	–	–	–	–	N	–
Eleutherodactylus balionotus 1979	–	–	–	–	–	–	–	–	–	–	N	–
Eleutherodactylus baryecuus 1979	–	–	–	–	–	–	–	–	–	–	N	–
Eleutherodactylus bearsei 1992	–	–	–	–	–	–	–	–	–	–	S	–
Eleutherodactylus bellona 1992	–	–	–	–	–	–	–	–	–	–	N	–
Eleutherodactylus bernali 1986	–	–	–	–	–	–	–	–	–	–	N	–
Eleutherodactylus bicolor 1983	–	–	–	–	–	–	–	–	–	–	N	–
Eleutherodactylus bicumulus 1863	–	–	–	–	–	–	–	–	–	–	NE	–
Eleutherodactylus bilineatus 1975	–	–	–	+	–	–	–	–	–	–	–	–
Eleutherodactylus binotatus 1824	–	–	–	+	–	–	–	–	–	–	–	–
Eleutherodactylus biporcatus 1863	+	–	–	–	–	–	–	–	–	–	–	+
Eleutherodactylus bockermanni 1970	–	–	SW	–	–	–	–	–	–	–	–	–
Eleutherodactylus boconoensis 1973	–	–	–	–	–	–	–	––	–	–	NE	–
Eleutherodactylus bogotensis 1863	–	–	–	–	–	–	–	–	–	–	N	–
Eleutherodactylus bolbodactylus 1925	–	–	–	+	–	–	–	–	–	–	–	–
Eleutherodactylus boulengeri 1981	–	–	–	–	–	–	–	–	–	–	N	–
Eleutherodactylus brevifrons 1981	–	–	–	–	–	–	–	–	–	–	N	–
Eleutherodactylus briceni 1903	–	–	–	+	–	–	–	–	–	–	–	–
Eleutherodactylus bromeliaceus 1979	–	–	–	–	–	–	–	–	–	–	+	–
Eleutherodactylus buccinator 1994	–	–	W	–	–	–	–	–	–	–	–	–
Eleutherodactylus buckleyi 1882	–	–	–	–	–	–	–	–	–	–	N	–
Eleutherodactylus bufoniformis 1896	+	–	–	–	–	–	–	–	–	–	–	–
Eleutherodactylus cabrerai 1970	–	–	–	–	–	–	–	–	–	–	N	–
Eleutherodactylus cacao 1992	–	–	–	–	–	–	–	–	–	–	N	–
Eleutherodactylus cadenai 1986	–	–	–	–	–	–	–	–	–	–	N	–
Eleutherodactylus cajamarcensis 1920	–	–	–	–	–	–	–	–	–	–	+	–
Eleutherodactylus calcaratus 1908	–	–	–	–	–	–	–	–	–	–	N	–
Eleutherodactylus calcarulatus 1976	–	–	–	–	–	–	–	–	–	–	N	–
Eleutherodactylus cantitans 1996	–	–	–	–	–	–	–	–	–	–	–	+
Eleutherodactylus caprifer 1977	+	–	–	–	–	–	–	–	–	–	N	–
Eleutherodactylus carmelitae 1922	–	–	–	–	–	–	–	–	–	–	N	–
Eleutherodactylus carranguerorum 1994	–	–	–	–	–	–	–	–	–	–	N	–
Eleutherodactylus carvalhoi 1952	–	–	+	–	–	–	–	–	–	–	–	–
Eleutherodactylus caryophyllaceus 1928	+	–	–	–	–	–	–	–	–	–	–	–
Eleutherodactylus cavernibardus 1997	–	–	–	–	–	–	–	–	–	–	–	+

Taxon and date named	CHO	CAR	A-G	AFD	LLA	CCC	PAM	PAT	ATF	ATA	AND	GUH
Eleutherodactylus celator 1976	–	–	–	–	–	–	–	–	–	–	N	–
Eleutherodactylus cerastes 1975	+	–	–	–	–	–	–	–	–	–	N	–
Eleutherodactylus ceuthophilus 1993	–	–	–	–	–	–	–	–	–	–	S	–
Eleutherodactylus chalceus 1873	+	–	–	–	–	–	–	–	–	–	N	–
Eleutherodactylus cheiroplethus 1990	–	–	–	–	–	–	–	–	–	–	N	–
Eleutherodactylus chiasonotus 1977	–	–	NE	–	–	–	–	–	–	–	–	–
Eleutherodactylus chloronotus 1970	–	–	–	–	–	–	–	–	–	–	N	–
Eleutherodactylus chlorosoma 1984	–	–	–	–	–	–	–	–	–	–	NE	–
Eleutherodactylus chrysops 1996	–	–	–	–	–	–	–	–	–	–	N	–
Eleutherodactylus citriogaster 1992	–	–	W	–	–	–	–	–	–	–	–	–
Eleutherodactylus colodactylus 1979	–	–	–	–	–	–	–	–	–	–	+	–
Eleutherodactylus colomai 1996	–	–	–	–	–	–	–	–	–	–	N	–
Eleutherodactylus colostichos 1982	–	–	–	–	–	–	–	–	–	–	NE	–
Eleutherodactylus condor 1980	–	–	–	–	–	–	–	–	–	–	N	–
Eleutherodactylus conspicillatus 1859	–	–	W	–	–	–	–	–	–	–	–	–
Eleutherodactylus cornutus 1871	–	–	–	–	–	–	–	–	–	–	N	–
Eleutherodactylus cosnipatae 1978	–	–	–	–	–	–	–	–	–	–	S	–
Eleutherodactylus cremnobates 1980	–	–	–	–	–	–	–	–	–	–	N	–
Eleutherodactylus crenunguis 1976	–	–	–	–	–	–	–	–	–	–	N	–
Eleutherodactylus cristinae 1985	–	–	–	–	–	–	–	–	–	–	N	–
Eleutherodactylus croceoinguinis 1968	–	–	W	–	–	–	–	–	–	–	N	–
Eleutherodactylus crucifer 1899	–	–	–	–	–	–	–	–	–	–	N	–
Eleutherodactylus cruralis 1902	–	–	SW	–	–	–	–	–	–	–	S	–
Eleutherodactylus cryophilus 1979	–	–	–	–	–	–	–	–	–	–	N	–
Eleutherodactylus cryptomelas 1979	–	–	–	–	–	–	–	–	–	–	+	–
Eleutherodactylus curtipes 1882	–	–	–	–	–	–	–	–	–	–	N	–
Eleutherodactylus danae 1978	–	–	–	–	–	–	–	–	–	–	S	–
Eleutherodactylus degener 1996	–	–	–	–	–	–	–	–	–	–	N	–
Eleutherodactylus deinops 1996	–	–	–	–	–	–	–	–	–	–	N	–
Eleutherodactylus delicatus 1917	–	–	–	–	–	–	–	–	–	–	N	–
Eleutherodactylus delius 1995	–	–	–	–	–	–	–	–	–	–	N	–
Eleutherodactylus devillei 1880	–	–	–	–	–	–	–	–	–	–	N	–
Eleutherodactylus diadematus 1875	–	–	W	–	–	–	–	–	–	–	–	–
Eleutherodactylus diaphonus 1986	–	–	–	–	–	–	–	–	–	–	N	–
Eleutherodactylus diastema 1876	+	–	–	–	–	–	–	–	–	–	–	–
Eleutherodactylus diogenes 1996	–	–	–	–	–	–	–	–	–	–	N	–
Eleutherodactylus discoidalis 1895	–	–	–	–	–	–	–	–	–	–	S	–
Eleutherodactylus dissimulatus 1996	–	–	–	–	–	–	–	–	–	–	N	–
Eleutherodactylus dolops 1980	–	–	–	–	–	–	–	–	–	–	N	–
Eleutherodactylus dorsopictus 1987	–	–	–	–	–	–	–	–	–	–	N	–
Eleutherodactylus douglasi 1996	–	–	–	–	–	–	–	–	–	–	N	–
Eleutherodactylus duellmani 1980	–	–	–	–	–	–	–	–	–	–	N	–
Eleutherodactylus elassodiscus 1973	–	–	–	–	–	–	–	–	–	–	N	–
Eleutherodactylus elegans 1863	–	–	–	–	–	–	–	–	–	–	N	–
Eleutherodactylus epipedus 1984	–	–	–	+	–	–	–	–	–	–	–	–
Eleutherodactylus eremitus 1980	–	–	–	–	–	–	–	–	–	–	N	–
Eleutherodactylus eriphus 1980	–	–	–	–	–	–	–	–	–	–	N	–
Eleutherodactylus ernesti 1987	–	–	–	–	–	–	–	–	–	–	N	–
Eleutherodactylus erythromerus 1984	–	–	–	+	–	–	–	–	–	–	–	–
Eleutherodactylus erythropleura 1896	–	–	–	–	–	–	–	–	–	–	N	–
Eleutherodactylus eugeniae 1996	–	–	–	–	–	–	–	–	–	–	N	–
Eleutherodactylus eurydactylus 1992	–	–	W	–	–	–	–	–	–	–	–	–
Eleutherodactylus fenestratus 1864	–	–	+	–	–	–	–	–	–	–	–	–
Eleutherodactylus fitzingeri 1858	+	+	–	–	–	–	–	–	–	–	–	–
Eleutherodactylus floridus 1996	–	–	–	–	–	–	–	–	–	–	N	–
Eleutherodactylus frater 1899	–	–	–	–	–	–	–	–	–	–	N	–
Eleutherodactylus fraudator 1987	–	–	–	–	–	–	–	–	–	–	S	–
Eleutherodactylus gaigei 1931	+	+	–	–	–	–	–	–	–	–	–	–

APPENDIX 5:1 CONTINUED

Taxon and date named	CHO	CAR	A-G	AFD	LLA	CCC	PAM	PAT	ATF	ATA	AND	GUH
Eleutherodactylus galdi 1870	–	–	–	–	–	–	–	–	–	–	N	–
Eleutherodactylus ganonotus 1988	–	–	–	–	–	–	–	–	–	–	N	–
Eleutherodactylus gentryi 1996	–	–	–	–	–	–	–	–	–	–	N	–
Eleutherodactylus ginesi 1964	–	–	–	–	–	–	–	–	–	–	NE	–
Eleutherodactylus gladiator 1976	–	–	–	–	–	–	–	–	–	–	N	–
Eleutherodactylus glandulosus 1880	–	–	–	–	–	–	–	–	–	–	N	–
Eleutherodactylus gracilis 1986	–	–	–	–	–	–	–	–	–	–	N	–
Eleutherodactylus grandiceps 1984	–	–	–	–	–	–	–	–	–	–	N	–
Eleutherodactylus grandoculis 1904	–	–	NE	–	–	–	–	–	–	–	–	–
Eleutherodactylus gualteri 1974	–	–	–	+	–	–	–	–	–	–	–	–
Eleutherodactylus guentheri 1864	–	–	–	+	–	–	–	–	–	–	–	–
Eleutherodactylus gularis 1898	+	–	–	–	–	–	–	–	–	–	–	–
Eleutherodactylus gutturalis 1977	–	–	NE	–	–	–	–	–	–	–	–	–
Eleutherodactylus hamiotae 1993	–	–	–	–	–	–	–	–	–	–	N	–
Eleutherodactylus hectus 1990	–	–	–	–	–	–	–	–	–	–	N	–
Eleutherodactylus helonotus 1975	–	–	–	–	–	–	–	–	–	–	N	–
Eleutherodactylus hernandezi 1983	–	–	–	–	–	–	–	–	–	–	N	–
Eleutherodactylus heterodactylus 1937	–	–	–	–	–	+	–	–	–	–	–	–
Eleutherodactylus hoehnei 1959	–	–	–	+	–	–	–	–	–	–	–	–
Eleutherodactylus holti 1948	–	–	–	+	–	–	–	–	–	–	–	–
Eleutherodactylus hybotragus 1992	+	–	–	–	–	–	–	–	–	–	–	–
Eleutherodactylus ignicolor 1980	–	–	–	–	–	–	–	–	–	–	N	–
Eleutherodactylus illotus 1996	–	–	–	–	–	–	–	–	–	–	N	–
Eleutherodactylus imitatrix 1978	–	–	W	–	–	–	–	–	–	–	–	–
Eleutherodactylus incanus 1980	–	–	–	–	–	–	–	–	–	–	N	–
Eleutherodactylus incomptus 1980	–	–	–	–	–	–	–	–	–	–	N	–
Eleutherodactylus ingeri 1961	–	–	–	–	–	–	–	–	–	–	N	–
Eleutherodactylus inguinalis 1940	–	–	NE	–	–	–	–	–	–	–	–	–
Eleutherodactylus insignitus 1917	–	–	–	–	–	–	–	–	–	–	N	–
Eleutherodactylus inustitatus 1980	–	–	–	–	–	–	–	–	–	–	N	–
Eleutherodactylus izecksohni 1988	–	–	–	+	–	–	–	–	–	–	–	–
Eleutherodactylus jaimei 1992	–	–	–	–	–	–	–	–	–	–	N	–
Eleutherodactylus johannesdei 1987	–	–	–	–	–	–	–	–	–	–	N	–
Eleutherodactylus johnwrighti 1980 (1996)	–	–	–	–	–	–	–	–	–	–	N	–
Eleutherodactylus jorgevelosai 1994	–	–	–	–	–	–	–	–	–	–	N	–
Eleutherodactylus juanchoi 1996	–	–	–	–	–	–	–	–	–	–	N	–
Eleutherodactylus juipoca 1978	–	–	–	+	–	–	–	–	–	–	–	–
Eleutherodactylus kaptotroides 1988	–	–	–	–	–	–	–	–	–	–	N	–
Eleutherodactylus karcharias 1997	–	–	–	–	–	–	–	–	–	–	S	–
Eleutherodactylus labiosus 1994	+	–	–	–	–	–	–	–	–	–	–	–
Eleutherodactylus lacrimosus 1875	–	–	W	–	–	–	–	–	–	–	–	–
Eleutherodactylus lacteus 1923	–	–	–	+	–	–	–	–	–	–	–	–
Eleutherodactylus lancinii 1968	–	–	–	–	–	–	–	–	–	–	NE	–
Eleutherodactylus lanthanites 1975	–	–	W	–	–	–	–	–	–	–	N	–
Eleutherodactylus lasalleorum 1995	–	–	–	–	–	–	–	–	–	–	N	–
Eleutherodactylus latens 1989	–	–	–	–	–	–	–	–	–	–	N	–
Eleutherodactylus laticlavius 1990	–	–	–	–	–	–	–	–	–	–	N	–
Eleutherodactylus latidiscus 1898	+	–	–	–	–	–	–	–	–	–	–	–
Eleutherodactylus lentiginosus 1984	–	–	–	–	–	–	–	–	–	–	NE	–
Eleutherodactylus leoni 1976	–	–	–	–	–	–	–	–	–	–	N	–
Eleutherodactylus leptolophus 1980	–	–	–	–	–	–	–	–	–	–	N	–
Eleutherodactylus leucopus 1976	–	–	–	–	–	–	–	–	–	–	N	–
Eleutherodactylus librarius 1994	–	–	W	–	–	–	–	–	–	–	–	–
Eleutherodactylus lichenoides 1997	–	–	–	–	–	–	–	–	–	–	N	–
Eleutherodactylus lindae 1978	–	–	–	–	–	–	–	–	–	–	S	–
Eleutherodactylus lirellus 1995	–	–	–	–	–	–	–	–	–	–	S	–
Eleutherodactylus lividus 1980	–	–	–	–	–	–	–	–	–	–	N	–
Eleutherodactylus longirostris 1898	+	–	–	–	–	–	–	–	–	–	N	–

Taxon and date named	CHO	CAR	A-G	AFD	LLA	CCC	PAM	PAT	ATF	ATA	AND	GUH
Eleutherodactylus loustes 1979	–	–	–	–	–	–	–	–	–	–	N	–
Eleutherodactylus luscombei 1995	–	–	W	–	–	–	–	–	–	–	–	–
Eleutherodactylus luteolateralis 1976	–	–	–	–	–	–	–	–	–	–	N	–
Eleutherodactylus lutitus 1984	–	–	–	–	–	–	–	–	–	–	N	
Eleutherodactylus lymani 1920	–	–	–	–	–	–	–	–	–	–	+	–
Eleutherodactylus lynchi 1977	–	–	–	–	–	–	–	–	–	–	N	–
Eleutherodactylus lythrodes 1980	–	–	W	–	–	–	–	–	–	–	–	–
Eleutherodactylus maculosus 1991	–	–	–	–	–	–	–	–	–	–	N	–
Eleutherodactylus malkini 1980	–	–	+	–	–	–	–	–	–	–	–	–
Eleutherodactylus mantipus 1908	–	–	–	–	–	–	–	–	–	–	N	–
Eleutherodactylus marmoratus 1900	–	–	NE	–	–	–	–	–	–	–	–	–
Eleutherodactylus mars 1996	–	–	–	–	–	–	–	–	–	–	N	–
Eleutherodactylus martiae 1974	–	–	W	–	–	–	–	–	–	–	N	–
Eleutherodactylus maussi 1893	–	–	–	–	–	–	–	–	–	–	NE	–
Eleutherodactylus medemi 1994	–	–	–	–	–	–	–	–	–	–	N	–
Eleutherodactylus megalops 1917	–	–	–	–	–	–	–	–	–	–	N	–
Eleutherodactylus melanoproctus 1984	–	–	–	–	–	–	–	–	–	–	N	–
Eleutherodactylus memorans 1997	–	–	–	–	–	–	–	–	–	–	–	+
Eleutherodactylus mendax 1978	–	–	–	–	–	–	–	–	–	–	S	–
Eleutherodactylus mercedesae 1987	–	–	–	–	–	–	–	–	–	–	S	–
Eleutherodactylus merostictus 1984	–	–	–	–	–	–	–	–	–	–	N	–
Eleutherodactylus miyatai 1984	–	–	–	–	–	–	–	–	–	–	N	–
Eleutherodactylus modipeplus 1981	–	–	–	–	–	–	–	–	–	–	N	–
Eleutherodactylus molybrignus 1986	–	–	–	–	–	–	–	–	–	–	N	–
Eleutherodactylus mondolfoi 1984	–	–	–	–	–	–	–	–	–	–	N	–
Eleutherodactylus moro 1965	+	–	–	–	–	–	–	–	–	–	–	–
Eleutherodactylus muricatus 1980	–	–	–	–	–	–	–	–	–	–	N	–
Eleutherodactylus myersi 1963	–	–	–	–	–	–	–	–	–	–	N	–
Eleutherodactylus nasutus 1925	–	–	–	+	–	–	–	–	–	–	–	–
Eleutherodactylus nebulosus 1992	–	–	–	–	–	–	–	–	–	–	S	–
Eleutherodactylus necerus 1975	–	–	–	–	–	–	–	–	–	–	N	–
Eleutherodactylus necopinus 1997	–	–	–	–	–	–	–	–	–	–	N	–
Eleutherodactylus nervicus 1994	–	–	–	–	–	–	–	–	–	–	N	–
Eleutherodactylus nicefori 1970	–	–	–	–	–	–	–	–	–	–	N	–
Eleutherodactylus nigriventris 1925	–	–	–	+	–	–	–	–	–	–	–	–
Eleutherodactylus nigrogriseus 1945	–	–	–	–	–	–	–	–	–	–	N	–
Eleutherodactylus nigrovittatus 1945	–	–	W	–	–	–	–	–	–	–	N	–
Eleutherodactylus nyctophylax 1976	–	–	–	–	–	–	–	–	–	–	N	–
Eleutherodactylus obmutescens 1980	–	–	–	–	–	–	–	–	–	–	N	–
Eleutherodactylus ocellatus 1990	–	–	–	–	–	–	–	–	–	–	N	–
Eleutherodactylus ockendeni 1912	–	–	+	–	–	–	–	–	–	–	N	–
Eleutherodactylus ocreatus 1981	–	–	–	–	–	–	–	–	–	–	N	–
Eleutherodactylus octavioi 1965	–	–	–	+	–	–	–	–	–	–	–	–
Eleutherodactylus oeus 1984	–	–	–	+	–	–	–	–	–	–	–	–
Eleutherodactylus oparcobates 1994	–	–	–	–	–	–	–	–	–	–	N	–
Eleutherodactylus orcesi 1972	–	–	–	–	–	–	–	–	–	–	N	–
Eleutherodactylus orestes 1979	–	–	–	–	–	–	–	–	–	–	N	–
Eleutherodactylus ornatissimus 1911	+	–	–	–	–	–	–	–	–	–	–	–
Eleutherodactylus orphnolaimus 1970	–	–	W	–	–	–	–	–	–	–	–	–
Eleutherodactylus palmeri 1912	–	–	–	–	–	–	–	–	–	–	N	–
Eleutherodactylus paramerus 1984	–	–	–	–	–	–	–	–	–	–	NE	–
Eleutherodactylus parvillus 1976	+	–	–	–	–	–	–	–	–	–	–	–
Eleutherodactylus parvus 1853	–	–	–	+	–	–	–	–	–	–	–	–
Eleutherodactylus pastazensis 1945	–	–	–	–	–	–	–	–	–	–	N	–
Eleutherodactylus paulodutrai 1975	–	–	–	+	–	–	–	–	–	–	–	–
Eleutherodactylus paululus 1974	–	–	W	–	–	–	–	–	–	–	–	–
Eleutherodactylus pecki 1988	–	–	–	–	–	–	–	–	–	–	N	–
Eleutherodactylus peraticus 1980	–	–	–	–	–	–	–	–	–	–	N	–

Taxon and date named	CHO	CAR	A-G	AFD	LLA	CCC	PAM	PAT	ATF	ATA	AND	GUH
Eleutherodactylus percultus 1979	–	–	–	–	–	–	–	–	–	–	N	–
Eleutherodactylus permixus 1994	–	–	–	–	–	–	–	–	–	–	N	–
Eleutherodactylus peruvianus 1941	–	–	W	–	–	–	–	–	–	–	+	–
Eleutherodactylus petrobardus 1991	–	–	–	–	–	–	–	–	–	–	S	–
Eleutherodactylus philipi 1995	–	–	–	–	–	–	–	–	–	–	N	–
Eleutherodactylus phoxocephalus 1979	–	–	–	–	–	–	–	–	–	–	+	–
Eleutherodactylus phragmipleuron 1987	–	–	–	–	–	–	–	–	–	–	N	–
Eleutherodactylus piceus 1996	–	–	–	–	–	–	–	–	–	–	N	–
Eleutherodactylus platydactylus 1903	–	–	–	–	–	–	–	–	–	–	S	–
Eleutherodactylus pleurostriatus 1984	–	–	–	–	–	–	–	–	–	–	NE	–
Eleutherodactylus plicifer 1888	–	–	–	+	–	–	–	–	–	–	–	–
Eleutherodactylus pluvicanorus 1997	–	–	–	–	–	–	–	–	–	–	S	–
Eleutherodactylus polychrus 1997	–	–	–	–	–	–	–	–	–	–	N	–
Eleutherodactylus prolatus 1980	–	–	–	–	–	–	–	–	–	–	N	–
Eleutherodactylus prolixodiscus 1978	–	–	–	–	–	–	–	–	–	–	N	–
Eleutherodactylus proserpens 1979	–	–	–	–	–	–	–	–	–	–	N	–
Eleutherodactylus pruinatus 1996	–	–	–	–	–	–	–	–	–	–	–	+
Eleuth. pseudoacuminatus 1935	–	–	W	–	–	–	–	–	–	–	–	–
Eleutherodactylus pteridophilus 1996	–	–	–	–	–	–	–	–	–	–	N	–
Eleutherodactylus pugnax 1973	–	–	–	–	–	–	–	–	–	–	N	–
Eleutherodactylus pulidoi 1984	–	–	–	–	–	–	–	–	–	–	N	–
Eleutherodactylus pulvinatus 1968	–	–	–	–	–	–	–	–	–	–	–	+
Eleutherodactylus pusillus 1967	–	–	–	+	–	–	–	–	–	–	–	–
Eleutherodactylus pycnodermis 1979	–	–	–	–	–	–	–	–	–	–	N	–
Eleutherodactylus pyrrhomerus 1976	–	–	–	–	–	–	–	–	–	–	N	–
Eleutherodactylus quaquaversus 1974	–	–	W	–	–	–	–	–	–	–	N	–
Eleutherodactylus quinquagesimus 1980	–	–	–	–	–	–	–	–	–	–	N	–
Eleutherodactylus racemus 1980	–	–	–	–	–	–	–	–	–	–	N	–
Eleutherodactylus ramagii 1888	–	–	–	+	–	–	–	–	–	–	–	–
Eleutherodactylus randorum 1985	–	–	–	+	–	–	–	–	–	–	–	–
Eleutherodactylus raniformis 1896	+	+	–	–	–	–	–	–	–	–	–	–
Eleutherodactylus repens 1984	–	–	–	–	–	–	–	–	–	–	N	–
Eleutherodactylus restrepoi 1996	–	–	–	–	–	–	–	–	–	–	N	–
Eleutherodactylus reticulatus 1955	–	–	–	–	–	–	–	–	–	–	NE	–
Eleutherodactylus rhabdolaemus 1978	–	–	SW	–	–	–	–	–	–	–	S	–
Eleutherodactylus rhodoplichus 1993	–	–	–	–	–	–	–	–	–	–	S	–
Eleutherodactylus ridens 1866	+	–	–	–	–	–	–	–	–	–	–	–
Eleutherodactylus riveroi 1993	–	–	–	–	–	–	–	–	–	–	NE	–
Eleutherodactylus riveti 1911	–	–	–	–	–	–	–	–	–	–	N	–
Eleutherodactylus rosadoi 1988	+	–	–	–	–	–	–	–	–	–	–	–
Eleutherodactylus roseus 1918	+	–	–	–	–	–	–	–	–	–	–	–
Eleutherodactylus rozei 1961	–	–	TO	–	–	–	–	–	–	–	NE	–
Eleutherodactylus rubicundus 1875	–	–	–	–	–	–	–	–	–	–	N	–
Eleutherodactylus ruedai 1997	–	–	–	–	–	–	–	–	–	–	N	–
Eleutherodactylus ruidus 1979	–	–	–	–	–	–	–	–	–	–	N	–
Eleutherodactylus ruizi 1981	–	–	–	–	–	–	–	–	–	–	N	–
Eleutherodactylus ruthveni 1985	–	–	–	–	–	–	–	–	–	–	N	–
Eleutherodactylus salaputium 1978	–	–	–	–	–	–	–	–	–	–	S	–
Eleutherodactylus samaipatae 1995	–	–	SW	–	–	–	–	–	–	–	–	–
Eleutherodactylus sanctamartae 1917	–	–	–	–	–	–	–	–	–	–	N	–
Eleutherodactylus satagius 1996	–	–	–	–	–	–	–	–	–	–	N	–
Eleutherodactylus savagei 1981	–	–	–	–	–	–	–	–	–	–	N	–
Eleutherodactylus schultei 1990	–	–	–	–	–	–	–	–	–	–	S	–
Eleutherodactylus scitulus 1978	–	–	–	–	–	–	–	–	–	–	S	–
Eleutherodactylus scoloblepharus 1991	–	–	–	–	–	–	–	–	–	–	N	–
Eleutherodactylus scolodiscus 1990	–	–	–	–	–	–	–	–	–	–	N	–
Eleutherodactylus scoparus 1996	–	–	–	–	–	–	–	–	–	–	N	–
Eleutherodactylus sernai 1984	–	–	–	–	–	–	–	–	–	–	N	–

Taxon and date named	CHO	CAR	A-G	AFD	LLA	CCC	PAM	PAT	ATF	ATA	AND	GUH
Eleutherodactylus signifer 1997	–	–	–	–	–	–	–	–	–	–	N	–
Eleutherodactylus silverstonei 1996	–	–	–	–	–	–	–	–	–	–	N	–
Eleutherodactylus simoteriscus 1996	–	–	–	–	–	–	–	–	–	–	N	–
Eleutherodactylus simoterus 1980	–	–	–	–	–	–	–	–	–	–	N	–
Eleutherodactylus siopelus 1990	–	–	–	–	–	–	–	–	–	–	N	–
Eleutherodactylus skymainos 1997	–	–	W	–	–	–	–	–	–	–	–	–
Eleutherodactylus sobetes 1980	–	–	–	–	–	–	–	–	–	–	N	–
Eleutherodactylus spanios 1985	–	–	–	+	–	–	–	–	–	–	–	–
Eleutherodactylus spilogaster 1984	–	–	–	–	–	–	–	–	–	–	N	–
Eleutherodactylus spinosus 1979	–	–	–	–	–	–	–	–	–	–	N	–
Eleutherodactylus stenodiscus 1955	–	–	–	–	–	–	–	–	–	–	NE	–
Eleutherodactylus sternohylax 1993	–	–	–	–	–	–	–	–	–	–	S	–
Eleutherodactylus subsigilatus 1902	+	–	–	–	–	–	–	–	–	–	–	–
Eleutherodactylus sulcatus 1874	–	–	W	–	–	–	–	–	–	–	N	–
Eleutherodactylus sulculus 1990	–	–	–	–	–	–	–	–	–	–	N	–
Eleutherodactylus supernatis 1980	–	–	–	–	–	–	–	–	–	–	N	–
Eleutherodactylus surdus 1882	–	–	–	–	–	–	–	–	–	–	N	–
Eleutherodactylus taeniatus 1912	+	–	–	–	–	–	–	–	–	–	–	–
Eleutherodactylus tamsitti 1970	–	–	–	–	–	–	–	–	–	–	N	–
Eleutherodactylus tayrona 1985	–	–	–	–	–	–	–	–	–	–	N	–
Eleutherodactylus tenebrionis 1980	+	–	–	–	–	–	–	–	–	–	–	–
Eleutherodactylus terraebolivaris 1961	–	–	TO	–	–	–	–	–	–	–	NE	–
Eleutherodactylus thectopternus 1975	–	–	–	–	–	–	–	–	–	–	N	–
Eleutherodactylus thymalopsoides 1976	–	–	–	–	–	–	–	–	–	–	N	–
Eleutherodactylus thymelensis 1972	–	–	–	–	–	–	–	–	–	–	N	–
Eleutherodactylus toftae 1978	–	–	SW	–	–	–	–	–	–	–	–	–
Eleutherodactylus trachyblepharis 1918	–	–	W	–	–	–	–	–	–	–	N	–
Eleutherodactylus trepidotus 1968	–	–	–	–	–	–	–	–	–	–	N	–
Eleutherodactylus tribulosus 1997	–	–	–	–	–	–	–	–	–	–	N	–
Eleutherodactylus truebae 1997	–	–	–	–	–	–	–	–	–	–	N	–
Eleutherodactylus tubernasus 1984	–	–	–	–	–	–	–	–	–	–	NE	–
Eleutherodactylus turumiriquirensis 1961	–	–	–	–	–	–	–	–	–	–	NE	–
Eleutherodactylus unistrigatus 1859	–	–	–	–	–	–	–	–	–	–	N	–
Eleutherodactylus uranobates 1991	–	–	–	–	–	–	–	–	–	–	N	–
Eleutherodactylus urichi 1894	–	–	TT	–	–	–	–	–	–	–	–	–
Eleutherodactylus vanadise 1984	–	–	–	–	–	–	–	–	–	–	NE	–
Eleutherodactylus variabilis 1968	–	–	W	–	–	–	–	–	–	–	–	–
Eleutherodactylus veletis 1997	–	–	–	–	–	–	–	–	–	–	N	–
Eleutherodactylus venancioi 1959	–	–	–	+	–	–	–	–	–	–	–	–
Eleutherodactylus ventrimarmoratus 1912	–	–	W	–	–	–	–	–	–	–	+	–
Eleutherodactylus verecundus 1990	–	–	–	–	–	–	–	–	–	–	N	–
Eleutherodactylus versicolor 1979	–	–	–	–	–	–	–	–	–	–	N	–
Eleutherodactylus vertebralis 1886	–	–	–	–	–	–	–	–	–	–	N	–
Eleutherodactylus vicarius 1983	–	–	–	–	–	–	–	–	–	–	N	–
Eleutherodactylus vidua 1979	–	–	–	–	–	–	–	–	–	–	N	–
Eleutherodactylus vilarsi 1941	–	–	W	–	–	–	–	–	–	–	–	–
Eleutherodactylus vinhai 1975	–	–	–	+	–	–	–	–	–	–	–	–
Eleutherodactylus viridicans 1977	–	–	–	–	–	–	–	–	–	–	N	–
Eleutherodactylus viridis 1997	–	–	–	–	–	–	–	–	–	–	N	–
Eleutherodactylus vocator 1955	+	–	–	–	–	–	–	–	–	–	–	–
Eleutherodactylus walkeri 1974	+	–	–	–	–	–	–	–	–	–	N	–
Eleutherodactylus wiensi 1993	–	–	–	–	–	–	–	–	–	–	S	–
Eleutherodactylus w–nigrum 1892	–	–	–	–	–	–	–	–	–	–	N	–
Eleutherodactylus xestus 1995	–	–	–	–	–	–	–	–	–	–	N	–
Eleutherodactylus xylochobates 1996	–	–	–	–	–	–	–	–	–	–	N	–
Eleutherodactylus yaviensis 1996	–	–	–	–	–	–	–	–	–	–	–	+
Eleutherodactylus zeuctotylus 1977	–	–	+	–	–	–	–	–	–	–	–	–
Eleutherodactylus zimmermanae 1991	–	–	+	–	–	–	–	–	–	–	–	–

APPENDIX 5:1 CONTINUED

Taxon and date named	CHO	CAR	A-G	AFD	LLA	CCC	PAM	PAT	ATF	ATA	AND	GUH
Eleutherodactylus zongoensis 1997	–	–	–	–	–	–	–	–	–	–	S	–
Eleutherodactylus zygodactylus 1983	+	–	–	–	–	–	–	–	–	–	–	–
Euparkerella brasiliensis 1926	–	–	–	+	–	–	–	–	–	–	–	–
Euparkerella cochranae 1988	–	–	–	+	–	–	–	–	–	–	–	–
Euparkerella robusta 1988	–	–	–	+	–	–	–	–	–	–	–	–
Euparkerella tridactyla 1988	–	–	–	+	–	–	–	–	–	–	–	–
Eupsophus calcaratus 1881	–	–	–	–	–	–	–	–	+	–	–	–
Eupsophus contulmoensis 1989	–	–	–	–	–	–	–	–	+	–	–	–
Eupsophus emiliopugini 1989	–	–	–	–	–	–	–	–	+	–	–	–
Eupsophus insularis 1902	–	–	–	–	–	–	–	–	+	–	–	–
Eupsophus migueli 1978	–	–	–	–	–	–	–	–	+	–	–	–
Eupsophus nahuelbutensis 1992	–	–	–	–	–	–	–	–	+	–	–	–
Eupsophus roseus 1841	–	–	–	–	–	–	–	–	+	–	–	–
Eupsophus vertebralis 1961	–	–	–	–	–	–	–	–	+	–	–	–
Geobatrachus walkeri 1915	–	–	–	–	–	–	–	–	–	–	N	–
Holoaden bradei 1958	–	–	–	+	–	–	–	–	–	–	–	–
Holoaden luederwaldti 1920	–	–	–	+	–	–	–	–	–	–	–	–
Hydrolaetare schmidti 1959	–	–	+	–	–	–	–	–	–	–	–	–
Hylodes asper 1924	–	–	–	+	–	–	–	–	–	–	–	–
Hylodes babax 1982	–	–	–	+	–	–	–	–	–	–	–	–
Hylodes charadranaetes 1986	–	–	–	+	–	–	–	–	–	–	–	–
Hylodes glabrus 1926	–	–	–	+	–	–	–	–	–	–	–	–
Hylodes heyeri 1996	–	–	–	+	–	–	–	–	–	–	–	–
Hylodes lateristrigatus 1912	–	–	–	+	–	–	–	–	–	–	–	–
Hylodes magalhaesi 1964	–	–	–	+	–	–	–	–	–	–	–	–
Hylodes meridionalis 1927	–	–	–	+	–	–	–	–	–	–	–	–
Hylodes mertensi 1956	–	–	–	+	–	–	–	–	–	–	–	–
Hylodes nasus 1823	–	–	–	+	–	–	–	–	–	–	–	–
Hylodes ornatus 1967	–	–	–	+	–	–	–	–	–	–	–	–
Hylodes otavioi 1983	–	–	–	+	–	–	–	–	–	–	–	–
Hylodes perplicatus 1926	–	–	–	+	–	–	–	–	–	–	–	–
Hylodes phyllodes 1986	–	–	–	+	–	–	–	–	–	–	–	–
Hylodes pulcher 1951	–	–	–	+	–	–	–	–	–	–	–	–
Hylodes regius 1979	–	–	–	+	–	–	–	–	–	–	–	–
Hylodes sazimai 1995	–	–	–	+	–	–	–	–	–	–	–	–
Hylodes vanzolinii 1982	–	–	–	+	–	–	–	–	–	–	–	–
Hylorina sylvatica 1843	–	–	–	–	–	–	–	–	+	–	–	–
Insuetophrynus acarpicus 1970	–	–	–	–	–	–	–	–	+	–	–	–
Ischnocnema quixensis 1872	–	–	W	–	–	–	–	–	–	–	N	–
Ischnocnema sanctaecrucis 1995	–	–	–	–	–	–	–	–	–	–	S	–
Ischnocnema saxatilis 1990	–	–	–	–	–	–	–	–	–	–	S	–
Ischnocnema simmonsi 1974	–	–	–	–	–	–	–	–	–	–	N	–
Ischnocnema verrucosa 1862	–	–	–	+	–	–	–	–	–	–	–	–
Lepidobatrachus asper 1899	–	–	–	–	–	+	–	–	–	–	–	–
Lepidobatrachus laevis 1899	–	–	–	–	–	+	–	–	–	–	–	–
Lepidobatrachus llanensis 1963	–	–	–	–	–	+	–	–	–	–	–	–
Leptodactylus bolivianus 1898	–	+	+	–	+	+	–	–	–	–	–	–
Leptodactylus bufonius 1894	–	–	–	–	–	+	–	–	–	–	–	–
Leptodactylus camaquara 1978	–	–	–	+	–	–	–	–	–	–	–	–
Leptodactylus chaquensis 1950	–	–	–	–	–	+	+	–	–	–	–	–
Leptodactylus colombiensis 1994	–	–	–	–	–	–	–	–	–	–	N	–
Leptodactylus cunicularius 1978	–	–	–	+	–	–	–	–	–	–	–	–
Leptodactylus dantasi 1959	–	–	SW	–	–	–	–	–	–	–	–	–
Leptodactylus didymus 1996	–	–	SW	–	–	–	–	–	–	–	–	–
Leptodactylus diedrus 1994	–	–	NW	–	–	–	–	–	–	–	–	–
Leptodactylus elenae 1978	–	–	SW	–	–	+	–	–	–	–	–	–
Leptodactylus flavopictus 1926	–	–	–	+	–	–	–	–	–	–	–	–
Leptodactylus furnarius 1978	–	–	–	+	–	–	–	–	–	–	–	–

Taxon and date named	CHO	CAR	A-G	AFD	LLA	CCC	PAM	PAT	ATF	ATA	AND	GUH
Leptodactylus fuscus 1799	+	+	+	−	+	+	−	−	−	−	−	−
Leptodactylus geminus 1973	−	−	−	+	−	+	+	−	−	−	−	−
Leptodactylus gracilis 1841	−	−	−	−	−	+	+	−	−	−	−	−
Leptodactylus griseigularis 1981	−	−	−	−	−	−	−	−	−	−	S	−
Leptodactylus jolyi 1978	−	−	−	+	−	−	−	−	−	−	−	−
Leptodactylus knudseni 1972	−	−	+	−	−	−	−	−	−	−	−	−
Leptodactylus labialis 1877	−	+	−	−	+	−	−	−	−	−	−	−
Leptodactylus labrosus 1875	+	−	−	−	−	−	−	−	−	−	−	−
Leptodactylus labyrinthicus 1824	−	+	−	+	−	+	−	−	−	−	−	−
Leptodactylus laticeps 1918	−	−	−	−	−	+	−	−	−	−	−	−
Leptodactylus latinasus 1875	−	−	−	−	−	+	+	−	−	−	−	−
Leptodactylus leptodactyloides 1945	−	−	+	−	−	−	−	−	−	−	−	−
Leptodactylus lithonaetes 1995	−	−	N	−	−	−	−	−	−	−	−	−
Leptodactylus longirostris 1882	−	−	NE	−	−	−	−	−	−	−	−	−
Leptodactylus macrosternum 1926	−	−	NE	−	+	+	−	−	−	−	−	−
Leptodactylus marambaiae 1976	−	−	−	+	−	−	−	−	−	−	−	−
Leptodactylus melanonotus 1861	+	−	−	−	−	−	−	−	−	−	−	−
Leptodactylus myersi 1995	−	−	NE	−	−	−	−	−	−	−	−	−
Leptodactylus mystaceus 1824	−	−	+	+	−	+	−	−	−	−	−	−
Leptodactylus mystacinus 1861	−	−	−	+	−	+	+	−	−	−	−	−
Leptodactylus natalensis 1930	−	−	−	+	−	−	−	−	−	−	−	−
Leptodactylus nesiotus 1994	−	−	TR	−	−	−	−	−	−	−	−	−
Leptodactylus notoaktites 1978	−	−	−	+	−	−	−	−	−	−	−	−
Leptodactylus ocellatus 1758	−	−	E	−	−	+	+	−	−	−	−	−
Leptodactylus pallidirostris 1930	−	−	NE	−	+	−	−	−	−	−	−	−
Leptodactylus pascoensis 1994	−	−	−	−	−	−	−	−	−	−	S	−
Leptodactylus pentadactylus 1768	+	−	+	−	−	−	−	−	−	−	−	−
Leptodactylus petersii 1864	−	−	+	−	−	+	−	−	−	−	−	−
Leptodactylus plaumanni 1936	−	−	−	+	−	−	−	−	−	−	−	−
Leptodactylus podicipinus 1862	−	−	+	−	−	+	−	−	−	−	−	−
Leptodactylus poecilochilus 1862	−	+	−	−	−	−	−	−	−	−	−	−
Leptodactylus pustulatus 1870	−	−	−	−	−	+	−	−	−	−	−	−
Leptodactylus rhodomystax 1883	−	−	+	−	−	−	−	−	−	−	−	−
Leptodactylus rhodonotus 1868	−	−	W	−	−	−	−	−	−	−	−	−
Leptodactylus riveroi 1983	−	−	+	−	−	−	−	−	−	−	−	−
Leptodactylus rugosus 1923	−	−	NE	−	−	−	−	−	−	−	−	−
Leptodactylus sabanensis 1994	−	−	−	−	−	−	−	−	−	−	−	+
Leptodactylus spixi 1983	−	−	−	+	−	−	−	−	−	−	−	−
Leptodactylus stenodema 1875	−	−	+	−	−	−	−	−	−	−	−	−
Leptodactylus syphax 1978	−	−	−	−	+	−	−	−	−	−	−	−
Leptodactylus tapiti 1978	−	−	−	+	−	−	−	−	−	−	−	−
Leptodactylus troglodytes 1926	−	−	−	−	−	+	−	−	−	−	−	−
Leptodactylus validus 1887	−	−	TT	−	−	−	−	−	−	−	−	−
Leptodactylus ventrimaculatus 1902	+	−	−	−	−	−	−	−	−	−	−	−
Leptodactylus viridis 1973	−	−	−	+	−	−	−	−	−	−	−	−
Leptodactylus wagneri 1862	−	−	W	−	−	−	−	−	−	−	−	−
Limnomedusa macroglossa 1841	−	−	−	−	−	−	+	−	−	−	−	−
Lithodytes lineatus 1841	−	−	+	−	−	−	−	−	−	−	−	−
Macrogenioglottus alipioi 1946	−	−	−	+	−	−	−	−	−	−	−	−
Megaelosia goeldii 1912	−	−	−	+	−	−	−	−	−	−	−	−
Megaelosia lutzae 1985	−	−	−	+	−	−	−	−	−	−	−	−
Odontophrynus achalensis 1984	−	−	−	−	−	−	+	−	−	−	−	−
Odontophrynus americanus 1841	−	−	−	−	−	+	+	−	−	−	−	−
Odontophrynus barrioi 1982	−	−	−	−	−	−	+	−	−	−	−	−
Odontophrynus carvalhoi 1965	−	−	−	−	−	+	−	−	−	−	−	−
Odontophrynus cultripes 1862	−	−	−	−	−	+	+	−	−	−	−	−
Odontophrynus lavillai 1985	−	−	−	−	−	+	−	−	−	−	−	−
Odontophrynus moratoi 1980	−	−	−	−	−	−	+	−	−	−	−	−

APPENDIX 5:1 CONTINUED

Taxon and date named	CHO	CAR	A-G	AFD	LLA	CCC	PAM	PAT	ATF	ATA	AND	GUH
Odontophrynus occidentalis 1896	–	–	–	–	–	+	–	–	–	–	–	–
Odontophrynus salvatori 1996	–	–	–	–	–	+	–	–	–	–	–	–
Paratelmatobius gaigeae 1938	–	–	–	+	–	–	–	–	–	–	–	–
Paratelmatobius lutzii 1958	–	–	–	+	–	–	–	–	–	–	–	–
Paratelmatobius poecilogaster 1990	–	–	–	+	–	–	–	–	–	–	–	–
Phrynopus bagrecitoi 1986	–	–	–	–	–	–	–	–	–	–	S	–
Phrynopus bracki 1990	–	–	–	–	–	–	–	–	–	–	S	–
Phrynopus brunneus 1975	–	–	–	–	–	–	–	–	–	–	N	–
Phrynopus columbianus 1899	–	–	–	–	–	–	–	–	–	–	N	–
Phrynopus cophites 1975	–	–	–	–	–	–	–	–	–	–	S	–
Phrynopus flavomaculatus 1938	–	–	–	–	–	–	–	–	–	–	N	–
Phrynopus juninensis 1938	–	–	–	–	–	–	–	–	–	–	S	–
Phrynopus kempffi 1992	–	–	–	–	–	–	–	–	–	–	S	–
Phrynopus laplacai 1968	–	–	–	–	–	–	–	–	–	–	S	–
Phrynopus lucida 1984	–	–	–	–	–	–	–	–	–	–	S	–
Phrynopus montium 1938	–	–	–	–	–	–	–	–	–	–	S	–
Phrynopus nanus 1963	–	–	–	–	–	–	–	–	–	–	N	–
Phrynopus nebulanastes 1984	–	–	–	–	–	–	–	–	–	–	S	–
Phrynopus parkeri 1975	–	–	–	–	–	–	–	–	–	–	S	–
Phrynopus peraccai 1975	–	–	–	–	–	–	–	–	–	–	N	–
Phrynopus pereger 1975	–	–	–	–	–	–	–	–	–	–	S	–
Phrynopus peruanus 1874	–	–	–	–	–	–	–	–	–	–	S	–
Phrynopus peruvianus 1921	–	–	–	–	–	–	–	–	–	–	S	–
Phrynopus simonsii 1900	–	–	–	–	–	–	–	–	–	–	S	–
Phrynopus wettsteini 1932	–	–	–	–	–	–	–	–	–	–	S	–
Phyllonastes heyeri 1986	–	–	–	–	–	–	–	–	–	–	+	–
Phyllonastes lochites 1976	–	–	–	–	–	–	–	–	–	–	N	–
Phyllonastes lynchi 1991	–	–	–	–	–	–	–	–	–	–	S	–
Phyllonastes myrmecoides 1976	–	–	W	–	–	–	–	–	–	–	–	–
Physalaemus aguirrei 1966	–	–	–	+	–	–	–	–	–	–	–	–
Physalaemus albifrons 1824	–	–	–	+	–	+	+	–	–	–	–	–
Physalaemus albonotatus 1864	–	–	–	–	–	+	–	–	–	–	–	–
Physalaemus barrioi 1967	–	–	–	+	–	–	–	–	–	–	–	–
Physalaemus biligoniger 1861	–	–	–	–	–	+	+	–	–	–	–	–
Physalaemus bokermanni 1985	–	–	–	+	–	–	–	–	–	–	–	–
Physalaemus centralis 1962	–	–	–	+	–	–	+	–	–	–	–	–
Physalaemus cicada 1966	–	–	–	+	–	–	–	–	–	–	–	–
Physalaemus coloradorum 1984	+	–	–	–	–	–	–	–	–	–	–	–
Physalaemus crombiei 1989	–	–	–	+	–	–	–	–	–	–	–	–
Physalaemus cuqui 1993	–	–	–	–	–	+	–	–	–	–	–	–
Physalaemus cuvieri 1826	–	–	–	+	–	+	+	–	–	–	–	–
Physalaemus deimaticus 1986	–	–	–	+	–	–	–	–	–	–	–	–
Physalaemus enesefae 1965	–	–	–	–	+	–	–	–	–	–	–	–
Physalaemus ephippifer 1864	–	–	NE	–	–	–	–	–	–	–	–	–
Physalaemus evangelistai 1967	–	–	–	+	–	–	–	–	–	–	–	–
Physalaemus fernandezi 1926	–	–	–	–	–	–	+	–	–	–	–	–
Physalaemus fuscomaculatus 1864	–	–	–	–	–	+	–	–	–	–	–	–
Physalaemus gracilis 1883	–	–	–	–	–	–	+	–	–	–	–	–
Physalaemus henselii 1872	–	–	–	–	–	–	+	–	–	–	–	–
Physalaemus jordanensis 1967	–	–	–	+	–	–	–	–	–	–	–	–
Physalaemus kroyeri 1862	–	–	–	+	–	–	–	–	–	–	–	–
Physalaemus lisei 1977	–	–	–	–	–	–	+	–	–	–	–	–
Physalaemus maculiventris 1925	–	–	–	+	–	–	–	–	–	–	–	–
Physalaemus moreirae 1937	–	–	–	+	–	–	–	–	–	–	–	–
Physalaemus nanus 1888	–	–	–	–	–	–	+	–	–	–	–	–
Physalaemus nattereri 1863	–	–	–	–	–	+	+	–	–	–	–	–
Physalaemus obtectus 1966	–	–	–	+	–	–	–	–	–	–	–	–
Physalaemus olfersii 1856	–	–	–	+	–	–	–	–	–	–	–	–

APPENDIX 5:1 CONTINUED

Taxon and date named	CHO	CAR	A-G	AFD	LLA	CCC	PAM	PAT	ATF	ATA	AND	GUH
Physalaemus petersi 1872	–	–	+	–	–	–	–	–	–	–	–	–
Physalaemus pustulatus 1941	+	–	–	–	–	–	–	–	–	–	–	–
Physalaemus pustulosus 1864	–	+	+	–	+	–	–	–	–	–	–	–
Physalaemus riograndensis 1960	–	–	–	–	–	–	+	–	–	–	–	–
Physalaemus rupestris 1991	–	–	–	+	–	–	–	–	–	–	–	–
Physalaemus santafecinus 1965	–	–	–	–	–	–	+	–	–	–	–	–
Physalaemus signifer 1853	–	–	–	+	–	–	–	–	–	–	–	–
Physalaemus soaresi 1965	–	–	–	+	–	–	–	–	–	–	–	–
Phyzelaphryne miriamae 1977	–	–	S	–	–	–	–	–	–	–	–	–
Pleurodema bibroni 1838	–	–	–	–	–	–	+	–	–	–	–	–
Pleurodema borellii 1895	–	–	–	–	–	+	–	–	–	–	–	–
Pleurodema brachyops 1869	–	+	NE	–	+	–	–	–	–	–	–	–
Pleurodema bufoninum 1843	–	–	–	–	–	–	+	–	–	–	–
Pleurodema cinereum 1877	–	–	–	–	–	–	–	–	–	–	S	–
Pleurodema diplolistre 1870	–	–	–	–	+	–	–	–	–	–	–
Pleurodema guayapae 1964	–	–	–	–	–	–	+	–	–	–	–	–
Pleurodema kriegi 1926	–	–	–	–	–	–	+	–	–	–	–	–
Pleurodema marmoratum 1841	–	–	–	–	–	–	–	–	–	–	S	–
Pleurodema nebulosum 1861	–	–	–	–	–	–	+	–	–	–	–	–
Pleurodema thaul 1826	–	–	–	–	–	–	–	–	+	–	–	–
Pleurodema tucumanum 1927	–	–	–	–	–	–	+	–	–	–	–	–
Proceratophrys appendiculata 1873	–	–	–	+	–	–	–	–	–	–	–	–
Proceratophrys avelinoi 1994	–	–	–	–	–	–	+	–	–	–	–	–
Proceratophrys bigibossa 1872	–	–	–	–	–	–	+	–	–	–	–	–
Proceratophrys boiei 1825	–	–	–	+	–	–	–	–	–	–	–	–
Proceratophrys cristiceps 1884	–	–	–	–	–	+	–	–	–	–	–	–
Proceratophrys cristinae 1973	–	–	–	–	–	–	+	–	–	–	–	–
Proceratophrys fryi 1873	–	–	–	+	–	–	–	–	–	–	–	–
Proceratophrys goyana 1937	–	–	–	–	–	+	–	–	–	–	–	–
Proceratophrys laticeps 1981	–	–	–	+	–	–	–	–	–	–	–	–
Proceratophrys moehringi 1985	–	–	–	+	–	–	–	–	–	–	–	–
Proceratophrys palustris 1993	–	–	–	+	–	–	–	–	–	–	–	–
Proceratophrys precrenulata 1937	–	–	–	+	–	–	–	–	–	–	–	–
Proceratophrys schirchi 1937	–	–	–	+	–	–	–	–	–	–	–	–
Pseudopaludicola boliviana 1927	–	–	I	–	–	+	–	–	–	–	–	–
Pseudopaludicola ceratophyes 1984	–	–	W	–	–	–	–	–	–	–	–	–
Pseudopaludicola falcipes 1867	–	–	–	–	+	+	–	–	–	–	–	–
Pseudopaludicola llanera 1989	–	–	–	+	–	–	–	–	–	–	–	–
Pseudopaludicola mineira 1996	–	–	–	–	+	–	–	–	–	–	–	–
Pseudopaludicola mystacalis 1887	–	–	–	–	+	+	–	–	–	–	–	–
Pseudopaludicola pusilla 1916	–	+	–	–	–	–	–	–	–	–	–	–
Pseudopaludicola saltica 1887	–	–	–	–	+	+	–	–	–	–	–	–
Pseudopaludicola ternetzi 1937	–	–	–	–	+	–	–	–	–	–	–	–
Scythrophrys sawayae 1953	–	–	–	+	–	–	–	–	–	–	–	–
Somuncuria somuncurensis 1969	–	–	–	–	–	–	–	+	–	–	–	–
Telmatobius albiventris 1940	–	–	–	–	–	–	–	–	–	–	S	–
Telmatobius arequipensis 1955	–	–	–	–	–	–	–	–	–	–	S	–
Telmatobius atacamensis 1962	–	–	–	–	–	–	–	–	–	–	S	–
Telmatobius atahualpai 1993	–	–	–	–	–	–	–	–	–	–	S	–
Telmatobius brevipes 1951	–	–	–	–	–	–	–	–	–	–	S	–
Telmatobius brevirostris 1955	–	–	–	–	–	–	–	–	–	–	S	–
Telmatobius carrillae 1988	–	–	–	–	–	–	–	–	–	–	S	–
Telmatobius ceiorum 1970	–	–	–	–	–	–	–	–	–	–	S	–
Telmatobius cirrhacelis 1979	–	–	–	–	–	–	–	–	–	–	N	–
Telmatobius colanensis 1993	–	–	–	–	–	–	–	–	–	–	S	–
Telmatobius contrerasi 1977	–	–	–	–	–	–	–	–	–	–	S	–
Telmatobius crawfordi 1940	–	–	–	–	–	–	–	–	–	–	S	–
Telmatobius culeus 1875	–	–	–	–	–	–	–	–	–	–	S	–

APPENDIX 5:1 CONTINUED

Taxon and date named	CHO	CAR	A-G	AFD	LLA	CCC	PAM	PAT	ATF	ATA	AND	GUH
Telmatobius degener 1993	–	–	–	–	–	–	–	–	–	–	S	–
Telmatobius edaphonastes 1994	–	–	–	–	–	–	–	–	–	–	S	–
Telmatobius halli 1938	–	–	–	–	–	–	–	–	–	–	S	–
Telmatobius hauthali 1895	–	–	–	–	–	–	–	–	–	–	S	–
Telmatobius hockingi 1996	–	–	–	–	–	–	–	–	–	–	S	–
Telmatobius hypselocephalus 1989	–	–	–	–	–	–	–	–	–	–	S	–
Telmatobius ignavus 1920	–	–	–	–	–	–	–	–	–	–	S	–
Telmatobius intermedius 1951	–	–	–	–	–	–	–	–	–	–	S	–
Telmatobius jahuira 1995	–	–	–	–	–	–	–	–	–	–	S	–
Telmatobius jelskii 1873	–	–	–	–	–	–	–	–	–	–	S	–
Telmatobius laticeps 1977	–	–	–	–	–	–	–	–	–	–	S	–
Telmatobius latirostris 1951	–	–	–	–	–	–	–	–	–	–	S	–
Telmatobius marmoratus 1841	–	–	–	–	–	–	–	–	–	–	S	–
Telmatobius mayoloi 1996	–	–	–	–	–	–	–	–	–	–	S	–
Telmatobius necopinus 1993	–	–	–	–	–	–	–	–	–	–	S	–
Telmatobius niger 1920	–	–	–	–	–	–	–	–	–	–	N	–
Telmatobius oxycephalus 1946	–	–	–	–	–	–	–	–	–	–	S	–
Telmatobius pefauri 1976	–	–	–	–	–	–	–	–	–	–	S	–
Telmatobius peruvianus 1835	–	–	–	–	–	–	–	–	–	–	S	–
Telmatobius pinguiculus 1989	–	–	–	–	–	–	–	–	–	–	S	–
Telmatobius platycephalus 1989	–	–	–	–	–	–	–	–	–	–	S	–
Telmatobius rimac 1954	–	–	–	–	–	–	–	–	–	–	S	–
Telmatobius schreiteri 1946	–	–	–	–	–	–	–	–	–	–	S	–
Telmatobius scrocchii 1986	–	–	–	–	–	–	–	–	–	–	S	–
Telmatobius simonsi 1940	–	–	–	–	–	–	–	–	–	–	S	–
Telmatobius stephani 1973	–	–	–	–	–	–	–	–	–	–	S	–
Telmatobius thompsoni 1993	–	–	–	–	–	–	–	–	–	–	S	–
Telmatobius truebae 1993	–	–	–	–	–	–	–	–	–	–	S	–
Telmatobius vellardi 1959	–	–	–	–	–	–	–	–	–	–	N	–
Telmatobius verrucosus 1899	–	–	–	–	–	–	–	–	–	–	S	–
Telmatobius yuracare 1994	–	–	–	–	–	–	–	–	–	–	S	–
Telmatobius zapahuirensis 1982	–	–	–	–	–	–	–	–	–	–	S	–
Telmatobufo australis 1972	–	–	–	–	–	–	–	–	+	–	–	–
Telmatobufo bullocki 1952	–	–	–	–	–	–	–	–	+	–	–	–
Telmatobufo venustus 1899	–	–	–	–	–	–	–	–	+	–	–	–
Thoropa lutzi 1938	–	–	–	+	–	–	–	–	–	–	–	–
Thoropa megalotympanum 1984	–	–	–	+	–	–	–	–	–	–	–	–
Thoropa miliaris 1824	–	–	–	+	–	–	–	–	–	–	–	–
Thoropa petropolitana 1907	–	–	–	+	–	–	–	–	–	–	–	–
Thoropa saxatilis 1988	–	–	–	+	–	–	–	–	–	–	–	–
Vanzolinius discodactylus 1883	–	–	W	–	–	–	–	–	–	–	–	–
Zachaenus carvalhoi 1983	–	–	–	+	–	–	–	–	–	–	–	–
Zachaenus parvulus 1853	–	–	–	+	–	–	–	–	–	–	–	–
Zachaenus roseus 1890	–	–	–	–	–	–	–	+	–	–	–	–
ANURA: MICROHYLIDAE:												
Adelastes hylonomus 1986	–	–	N	–	–	–	–	–	–	–	–	–
Altigius alios 1995	–	–	W	–	–	–	–	–	–	–	–	–
Arcovomer passarelli 1954	–	–	–	+	–	–	–	–	–	–	–	–
Chiasmocleis albopunctata 1885	–	–	–	–	–	+	+	–	–	–	–	–
Chiasmocleis anatipes 1974	–	–	W	–	–	–	–	–	–	–	–	–
Chiasmocleis atlantica 1997	–	–	–	+	–	–	–	–	–	–	–	–
Chiasmocleis bassleri 1949	–	–	W	–	–	–	–	–	–	–	–	–
Chiasmocleis capixaba 1997	–	–	–	+	–	–	–	–	–	–	–	–
Chiasmocleis carvalhoi 1997	–	–	–	+	–	–	–	–	–	–	–	–
Chiasmocleis centralis 1952	–	–	–	–	–	+	–	–	–	–	–	–
Chiasmocleis hudsoni 1940	–	–	NE	–	–	–	–	–	–	–	–	–
Chiasmocleis leucosticta 1888	–	–	–	+	–	–	–	–	–	–	–	–
Chiasmocleis mehelyi 1997	–	–	–	–	–	+	–	–	–	–	–	–

Taxon and date named	CHO	CAR	A-G	AFD	LLA	CCC	PAM	PAT	ATF	ATA	AND	GUH
Chiasmocleis panamensis 1948	+	+	–	–	–	–	–	–	–	–	–	–
Chiasmocleis schubarti 1952	–	–	–	+	–	–	–	–	–	–	–	–
Chiasmocleis shudikarensis 1949	–	–	NE	–	–	–	–	–	–	–	–	–
Chiasmocleis ventrimaculata 1945	–	–	W	–	–	–	–	–	–	–	–	–
Ctenophryne geayi 1904	–	–	+	–	–	–	–	–	–	–	–	–
Ctenophryne minor 1989	+	–	–	–	–	–	–	–	–	–	–	–
Dasypops schirchi 1924	–	–	–	+	–	–	–	–	–	–	–	–
Dermatonotus muelleri 1885	–	–	–	–	–	+	–	–	–	–	–	–
Elachistocleis bicolor 1838	–	–	+	–	–	+	+	–	–	–	–	–
Elachistocleis ovalis 1799	–	+	NE	–	+	–	–	–	–	–	–	–
Elachistocleis piauiensis 1983	–	–	–	+	–	–	–	–	–	–	–	–
Elachistocleis surinamensis 1802	–	–	NE	–	–	–	–	–	–	–	–	–
Hamptophryne boliviana 1927	–	–	+	–	–	–	–	–	–	–	–	–
Hyophryne histrio 1954	–	–	–	+	–	–	–	–	–	–	–	–
Myersiella microps 1841	–	–	–	+	–	–	–	–	–	–	–	–
Nelsonophryne aequatorialis 1904	–	–	–	–	–	–	–	–	–	–	N	–
Nelsonophryne aterrima 1900	+	–	–	–	–	–	–	–	–	–	–	–
Otophryne robusta 1900	–	–	NE	–	–	–	–	–	–	–	–	+
Relictivomer pearsei 1914	–	+	–	–	–	–	–	–	–	–	–	–
Stereocyclops incrassatus 1870	–	–	–	+	–	–	–	–	–	–	–	–
Synapturanus mirandaribeiroi 1975	–	–	NE	–	–	–	–	–	–	–	–	–
Synapturanus rabus 1976	–	–	NW	–	–	–	–	–	–	–	–	–
Synapturanus salseri 1975	–	–	NW	–	–	–	–	–	–	–	–	–
Syncope antenori 1973	–	–	W	–	–	–	–	–	–	–	–	–
Syncope carvalhoi 1975	–	–	W	–	–	–	–	–	–	–	–	–
ANURA: PIPIDAE:												
Pipa arrabali 1976	–	–	NE	–	–	–	–	–	–	–	–	–
Pipa aspera 1924	–	–	NE	–	–	–	–	–	–	–	–	–
Pipa carvalhoi 1937	–	–	–	–	–	+	–	–	–	–	–	–
Pipa parva 1923	–	+	–	–	–	–	–	–	–	–	–	–
Pipa pipa 1758	–	–	+	–	+	–	–	–	–	–	–	–
Pipa snethlageae 1914	–	–	+	–	–	–	–	–	–	–	–	–
ANURA: PSEUDIDAE:												
Lysapsus limellus 1862	–	–	+	–	–	+	+	–	–	–	–	–
Pseudis minuta 1859	–	–	–	–	–	–	+	–	–	–	–	–
Pseudis paradoxa 1758	–	+	+	–	+	+	+	–	–	–	–	–
ANURA: RANIDAE:												
Rana bwana 1988	+	–	–	–	–	–	–	–	–	–	–	–
Rana palmipes 1824	–	–	+	–	–	–	–	–	–	–	–	–
Rana vaillanti 1877	+	–	–	–	–	–	–	–	–	–	–	–
ANURA: RHINODERMATIDAE:												
Rhinoderma darwinii 1841	–	–	–	–	–	–	–	–	+	–	–	–
Rhinoderma rufum 1902	–	–	–	–	–	–	–	–	+	–	–	–
CAUDATA: PLETHODONTIDAE:												
Bolitoglossa adspersa 1863	–	–	–	–	–	–	–	–	–	–	N	–
Bolitoglossa altamazonica 1874	–	–	+	–	–	–	–	–	–	–	N	–
Bolitoglossa biseriata 1962	+	–	–	–	–	–	–	–	–	–	–	–
Bolitoglossa borburata 1942	–	–	–	–	–	–	–	–	–	–	NE	–
Bolitoglossa capitana 1963	–	–	–	–	–	–	–	–	–	–	N	–
Bolitoglossa chica 1963	+	–	–	–	–	–	–	–	–	–	–	–
Bolitoglossa digitigrada 1982	–	–	–	–	–	–	–	–	–	–	S	–
Bolitoglossa equatoriana 1972	–	–	W	–	–	–	–	–	–	–	–	–
Bolitoglossa hypacra 1962	–	–	–	–	–	–	–	–	–	–	N	–
Bolitoglossa medemi 1972	+	–	–	–	–	–	–	–	–	–	–	–
Bolitoglossa nicefori 1963	–	–	–	–	–	–	–	–	–	–	N	–
Bolitoglossa orestes 1962	–	–	–	–	–	–	–	–	–	–	NE	–
Bolitoglossa palmata 1897	–	–	–	–	–	–	–	–	–	–	N	–
Bolitoglossa pandi 1963	–	–	–	–	–	–	–	–	–	–	N	–

Taxon and date named	CHO	CAR	A-G	AFD	LLA	CCC	PAM	PAT	ATF	ATA	AND	GUH
Bolitoglossa peruviana 1883	–	–	–	–	–	–	–	–	–	–	+	–
Bolitoglossa phalarosoma 1962	–	–	–	–	–	–	–	–	–	–	N	–
Bolitoglossa ramosi 1972	–	–	–	–	–	–	–	–	–	–	N	
Bolitoglossa savagei 1963	–	–	–	–	–	–	–	–	–	–	N	–
Bolitoglossa silverstonei 1972	+	–	–	–	–	–	–	–	–	–	–	–
Bolitoglossa sima 1911	+	–	–	–	–	–	–	–	–	–	–	–
Bolitoglossa vallecula 1963	–	–	–	–	–	–	–	–	–	–	N	–
Bolitoglossa walkeri 1972	–	–	–	–	–	–	–	–	–	–	N	–
Oedipina complex 1924	+	–	–	–	–	–	–	–	–	–	–	–
Oedipina parvipes 1879	+	–	–	–	–	–	–	–	–	–	–	–
GYMNOPHIONA: CAECILIIDAE:												
Brasilotyphlus braziliensis 1945	–	–	NE	–	–	–	–	–	–	–	–	–
Caecilia abitaguae 1942	–	–	–	–	–	–	–	–	–	–	N	–
Caecilia albiventris 1803	–	–	NE	–	–	–	–	–	–	–	–	–
Caecilia antioquiensis 1968	–	–	–	–	–	–	–	–	–	–	N	–
Caecilia attenuata 1968	–	–	–	–	–	–	–	–	–	–	+	–
Caecilia bokermanni 1968	–	–	W	–	–	–	–	–	–	–	–	–
Caecilia caribea 1942	–	+	–	–	–	–	–	–	–	–	–	–
Caecilia corpulenta 1968	–	–	W	–	–	–	–	–	–	–	–	–
Caecilia crassisquama 1968	–	–	–	–	–	–	–	–	–	–	N	–
Caecilia degenerata 1942	–	–	–	–	–	–	–	–	–	–	N	–
Caecilia disossea 1968	–	–	W	–	–	–	–	–	–	–	–	–
Caecilia dunni 1938	+	–	W	–	–	–	–	–	–	–	–	–
Caecilia flavopunctata 1963	–	+	–	–	–	–	–	–	–	–	–	–
Caecilia gracilis 1802	–	–	NE	–	–	–	–	–	–	–	–	–
Caecilia guntheri 1942	+	–	–	–	–	–	–	–	–	–	–	–
Caecilia inca 1973	–	–	W	–	–	–	–	–	–	–	–	–
Caecilia leucocephala 1968	+	–	–	–	–	–	–	–	–	–	–	–
Caecilia marcusi 1984	–	–	SW	–	–	–	–	–	–	–	–	–
Caecilia nigricans 1902	+	+	–	–	–	–	–	–	–	–	–	–
Caecilia occidentalis 1968	–	–	–	–	–	–	–	–	–	–	N	–
Caecilia orientalis 1968	–	–	–	–	–	–	–	–	–	–	N	–
Caecilia pachycnema 1859	+	+	–	–	–	–	–	–	–	–	N	–
Caecilia perdita 1968	+	–	–	–	–	–	–	–	–	–	–	–
Caecilia pressula 1968	–	–	NE	–	–	–	–	–	–	–	–	–
Caecilia subdermalis 1968	–	–	–	–	–	–	–	–	–	–	N	–
Caecilia subnigricans 1942	–	+	–	–	–	–	–	–	–	–	–	–
Caecilia tentaculata 1758	–	–	+	–	–	–	–	–	–	–	–	–
Caecilia tenuissima 1973	+	–	–	–	–	–	–	–	–	–	–	–
Caecilia thompsoni 1902	–	–	–	–	–	–	–	–	–	–	N	–
Dermophis parviceps 1924	+	–	–	–	–	–	–	–	–	–	–	–
Luetkenotyphlus brasiliensis 1852	–	–	–	+	–	–	–	–	–	–	–	–
Microcaecilia albiceps 1882	–	–	W	–	–	–	–	–	–	–	–	–
Microcaecilia rabei 1963	–	–	NE	–	–	–	–	–	–	–	–	–
Microcaecilia supernumeraria 1969	–	–	–	+	–	–	–	–	–	–	–	–
Microcaecilia taylori 1979	–	–	NE	–	–	–	–	–	–	–	–	–
Microcaecilia unicolor 1864	–	–	NE	–	–	–	–	–	–	–	–	–
Mimosiphonops vermiculatus 1968	–	–	–	+	–	–	–	–	–	–	–	–
Oscaecilia bassleri 1942	–	–	W	–	–	–	–	–	–	–	–	–
Oscaecilia equatorialis 1973	+	–	–	–	–	–	–	–	–	–	–	–
Oscaecilia koepckeorum 1984	–	–	W	–	–	–	–	–	–	–	–	–
Oscaecilia polyzona 1879	–	+	–	–	–	–	–	–	–	–	–	–
Oscaecilia zweifeli 1968	–	–	NE	–	–	–	–	–	–	–	–	–
Parvicaecilia nicefori 1925	–	+	–	–	–	–	–	–	–	–	–	–
Parvicaecilia pricei 1944	–	+	–	–	–	–	–	–	–	–	–	–
Siphonops annulatus 1820	–	–	+	+	–	–	+	–	–	–	–	–
Siphonops confusionis 1968	–	–	–	+	–	–	–	–	–	–	–	–
Siphonops hardyi 1888	–	–	+	–	–	–	–	–	–	–	–	–

Taxon and date named	CHO	CAR	A-G	AFD	LLA	CCC	PAM	PAT	ATF	ATA	AND	GUH
Siphonops insulanus 1911	–	–	–	+	–	–	–	–	–	–	–	–
Siphonops leucoderus 1968	–	–	–	+	–	–	–	–	–	–	–	–
Siphonops paulensis 1892	–	–	–	+	–	–	–	–	–	–	–	–
GYMNOPHIONA: RHINATREMATIDAE:												
Epicrionops bicolor 1883	–	–	–	–	–	–	–	–	–	–	N	–
Epicrionops columbianus 1938	–	–	–	–	–	–	–	–	–	–	N	–
Epicrionops lativittatus 1968	–	–	+	–	–	–	–	–	–	–	–	–
Epicrionops marmoratus 1968	–	–	–	–	–	–	–	–	–	–	N	–
Epicrionops niger 1942	–	–	NE	–	–	–	–	–	–	–	–	–
Epicrionops parkeri 1942	–	–	–	–	–	–	–	–	–	–	N	–
Epicrionops peruvianus 1902	–	–	–	–	–	–	–	–	–	–	S	–
Epicrionops petersi 1968	–	–	–	–	–	–	–	–	–	–	N	–
Rhinatrema bivittatum 1829	–	–	+	–	–	–	–	–	–	–	–	–
GYMNOPHIONA: TYPHLONECTIDAE:												
Chthonerpeton arii 1994	–	–	–	+	–	–	–	–	–	–	–	–
Chthonerpeton braestrupi 1968	–	–	–	+	–	–	–	–	–	–	–	–
Chthonerpeton indistinctum 1861	–	–	–	–	–	–	+	–	–	–	–	–
Chthonerpeton onorei 1986	–	–	–	–	–	–	–	–	–	–	N	–
Chthonerpeton perisodus 1987	–	–	–	–	–	+	–	–	–	–	–	–
Chthonerpeton viviparum 1929	–	–	–	+	–	–	–	–	–	–	–	–
Nectocaecilia cooperi 1970	–	+	–	–	–	–	–	–	–	–	–	–
Nectocaecilia ladigesi 1968	–	–	E	–	–	–	–	–	–	–	–	–
Nectocaecilia petersii 1882	–	–	W	–	–	–	–	–	–	–	–	–
Potomotyphlus kaupii 1859	–	–	E	–	+	–	–	–	–	–	–	–
Typhlonectes compressicauda 1841	–	–	+	–	–	–	–	–	–	–	–	–
Typhlonectes cunhai 1991	–	–	CEN	–	–	–	–	–	–	–	–	–
Typhlonectes natans 1879	–	+	–	–	–	–	–	–	–	–	–	–
Typhlonectes obesus 1968	–	–	E	–	–	–	–	–	–	–	–	–
Typhlonectes venezuelensis 1914	–	+	–	–	–	–	–	–	–	–	–	–

ADDENDUM

Since this work was last revised in late 1997, several new taxa have been recognized in South America. Moreover, I discovered that I had omitted two species of *Megaelosia*. These taxa and their distributions are listed below. The new taxa reinforce the high species richness in the Andes and especially in the Atlantic Forest Domain.

BRACHYCEPHALIDAE

A third species of *Brachycephalus* (*B. pernix*) was discovered in the southern part of the Atlantic · Forest Domain in Paraná, Brazil (Pombal et al., 1998).

DENDROBATIDAE

Epipedobates rubriventris was described from cloud forest in the Cordillera Azul on theAmazonian slopes of the southern Andes by Lötters et al. (1997).

HYLIDAE

Hyla truncata in the Atlantic Forest Domain of eastern Brazil was placed in a new genus, *Xenohyla,* by Izecksohn (1998).

LEPTODACTYLIDAE

Eleutherodactylus manezinho was named from the southern part of the Atlantic Forest Domain in Santa Catarina, Brazil, by Garcia (1996). *Megaelo-sia bocainensis* and *M. massarti* were recognized by Giaretta et al. (1993), and *M. boticariana* was named by Giaretta and Aguiar (1998); all three species are from the southern part of the Atlantic Forest Domain in southeastern Brazil. *Physalaemus caeta* was described from the Atlantic Forest Domain in northeastern Brazil by Pombal and Madureira (1997).

LITERATURE CITED

GARCIA, P. C. A. 1996. Nova espécie de *Eleutherodactylus* Duméril & Bibron, 1891 (sic) do Estado de Santa Catarina, Brasil (Amphibia; Anura; Leptodactylidae). Biociências 4:57–56.

GIARETTA, A. A., AND O. AGUIAR, JR. 1998. A new species of *Megaelosia* from the Mantiquiera Range, southeastern Brazil. J. Herpetol. 32:80–83.

GIARETTA, A. A., W. C. A. BOKERMANN, AND C. F. B. HADDAD. 1993. A review of the genus *Megaelosia* (Anura: Leptodactylidae) with a description of a new species. J. Herpetol. 27:276–285.

IZECKSOHN, E. 1998 "1996." Novo gênero de Hylidae brasileiro (Amphibia, Anura). Rev. Univ. Rural Sér. Ciênc. Vida 18:47–52.

LÖTTERS, S., P. DEBOLD, K. HENLE, F. GLAW, AND M. KNELLER. 1997. Eine neuer Pfeilgiftfrosch aus der *Epipedobates pictis*-Gruppe vom Osthang der Cordillera Azul in Perú. Herpetofauna 19:25–34.

POMBAL, J. P., JR., AND C. A. MADUREIRA. 1997. A new species of *Physalaemus* (Anura: Leptodactylidae) from the Atlantic Rain Forest of northeastern Brazil. Alytes 15:105–112.

POMBAL, J. P., JR., E. M. WISTUBA, AND M. R. BORNSCHEIN. 1998. A new species of brachycephalid (Anura) from the Atlantic Rain Forest of Brazil. J. Herpetol. 32:70–74.

6. Distribution of Amphibians in North Africa, Europe, Western Asia, and the Former Soviet Union

Leo J. Borkin

Department of Herpetology
Zoological Institute
Russian Academy of Sciences
199034 St. Petersburg, Russia

ABSTRACT   The amphibians of the North Atlantic archipelagos (Macaronesia), North Africa, and West Asia eastward to Pakistan and the northwestern Himalayas, Europe, and the former Soviet Union are analyzed zoogeographically. Based on amphibian distributions, the southern limits of the Palearctic Realm are proposed. Accordingly, the southern Sahara (including the Sinai Peninsula), the Arabian Peninsula, southeastern Iran, southern and eastern Afghanistan, and Pakistan (except western peripheral mountains) are parts of the Afrotropical or Oriental realms. The taxonomic composition, endemism, and faunal components are outlined for the entire realm. Current taxonomy of the component families is reviewed. Major sources of the amphibian fauna and directions of dispersal are outlined. The reality of the Palearctic as a distinct realm is stressed, whereas the concept of the Holarctic is rejected. A complete list of the genera (34) and species (175) of amphibians of the Palearctic, as well as regional lists for Europe, the former Soviet Union, and the Near East are appended. Six regions are recognized, and provinces are proposed within the Mediterranean, European, and West Asian regions. The composition of the amphibian fauna is given for each region and province; faunal resemblance within and between regions varies significantly. Thirty distribution patterns are defined, as are 30 patterns of discontinuity. High species diversity, endemism, and/or peculiar zoogeographic importance are noted for 51 key areas.

Key words: Palearctic Realm; Composition, endemism, and distribution of amphibian fauna; Conservation.

INTRODUCTION

This review of the patterns of distribution of amphibians in the Palearctic Realm is mainly descriptive zoogeography. Except for some geological events, causal factors are not considered. Traditionally, most European (including Russian) zoogeographers interpreted historical regularities in distribution of amphibians (and reptiles), using classic ideology (centers of origin and dispersal, refu-

gia, etc.)—dispersal zoogeography that was influenced greatly by De Lattin (1967). However, recently, cladistic approaches have been used (e.g., Oosterbroek and Arntzen, 1992; Arntzen, 1995).

Within the framework of the descriptive zoogeography, two approaches were used in the past: (1) Based on a geographic classification of regions, faunal lists were compiled. With regard to amphibians and/or reptiles, this approach could be viewed as the herpetological description of the territory. (2) Distribution patterns of animals were defined; some such studies included analyses of peculiarities of the territory (either historical or ecological). This is herpetogeography sensu stricto. The first approach pertains primarily to a given region, and the second one focuses on distributions of taxa. These differences were clearly emphasized by a number of biogeographers (e.g., Stegmann, 1938; Takhtajan, 1978). In his classic text on Soviet zoogeography, Heptner (1936) pointed out that neither land shape and its geographic subdivisions, soils, or floristic regions could be used as the basis for zoogeography.

The preparation of this review proved to be markedly complicated. The main problems are:

1. The vast area (Fig. 6:1) lying between the Atlantic and Pacific oceans contains very different landscapes and elevations, ranging from hot deserts of North Africa to the severe tundra and permafrost of northeastern Asia, and from depressions below sea level to the "Roof of the World" (Pamirs-Himalaya). This region contains more than 70 countries. Permanent "stability of instability" in some southern countries, sometimes dangerous for foreign visitors, commonly results in poor sampling of interesting areas.

2. The territory covered herein is heterogeneous biogeographically, even though most of it is referred to the Palearctic Realm. Some regions considered below, such as southwestern Asia (the southern part of the Arabian Peninsula and Pakistan) could be included in the Oriental Realm, the amphibian fauna of which is treated by Inger (this volume) and in part by Zhao (this volume). Moreover, the southeastern part of the Palearctic (Mongolia, China, Korean Peninsula, and Japan) is the subject of Zhao's attention (this volume). Therefore, I tried to avoid any overlap with these chapters.

3. Significant growth of systematic studies and collecting activities have resulted in advances in taxonomy and regional faunistics. Taxonomic changes that have occurred in many genera (e.g., *Bufo, Discoglossus, Rana,* and *Hydromantes*) have greatly influenced our perception of faunal composition of some regions. For example, in Sclater's time, only 15 species of anurans were known from the Palearctic (Palacký, 1898), in contrast to 106 species recognized now (Appendix 6:1). The number of spe-

cies of amphibians recognized in Europe remained stable for nearly half a century—43 species in two comprehensive works by Schreiber (1912) and Mertens and Wermuth (1960), but increased 1.8 times in the past 36 years (Veith, 1996a; Appendix 6:2). A comparison of the works by Liu and Hu (1961) and Zhao and Adler (1993) reveals that the number of anurans known from China doubled in only three decades. Likewise, the number of Pakistani amphibians (Table 6:6) increased 1.5 times since Minton's (1966) paper. However, many important aspects of the taxonomy and distribution of amphibians, especially in the southern and eastern parts of the Palearctic Realm, are not yet resolved. The fossil record should be crucial to the understanding of the origin and dispersal of Recent groups; however, in many cases, the determination of fossil material is questionable.

4. There is a lack of a tradition in herpetology to analyze the distribution of amphibians throughout the Palearctic as a whole. Although there exists an enormous number of articles and books devoted to amphibian faunas of particular areas, countries, or their segments, a significantly fewer number of works focus on amphibian composition (and distribution) of large zoogeographic subdivisions of the Palearctic. The global distribution of amphibian families were treated by a few authors (e.g., Palacký, 1898; Darlington, 1957; Savage, 1973; Laurent, 1975; Duellman and Trueb, 1986). Some global zoogeographers (e.g., Wallace, 1876) used amphibians, as well as other groups of vertebrates, to illustrate the most characteristic features of the Palearctic and other realms. However, I am aware of only one publication devoted to the entire Palearctic herpetofauna—Shcherbak (1982), which treated the distribution patterns and regional compositions of amphibians and reptiles in the Palearctic. Because of their dependence on water, amphibians could be expected to have distribution patterns that do not coincide with those of reptiles, birds, or mammals (Savage, 1973). For example, quantitative analyses of distributions of amphibians, lizards, and all reptiles using river basins as geographic units have resulted in quite discordant schemes of biotic regions in Europe, as shown by comparing the results of Ramírez et al. (1992) with those of Real et al. (1992). Thus, the "Palearctic of amphibians" could be different from that of other vertebrates, insects, or plants.

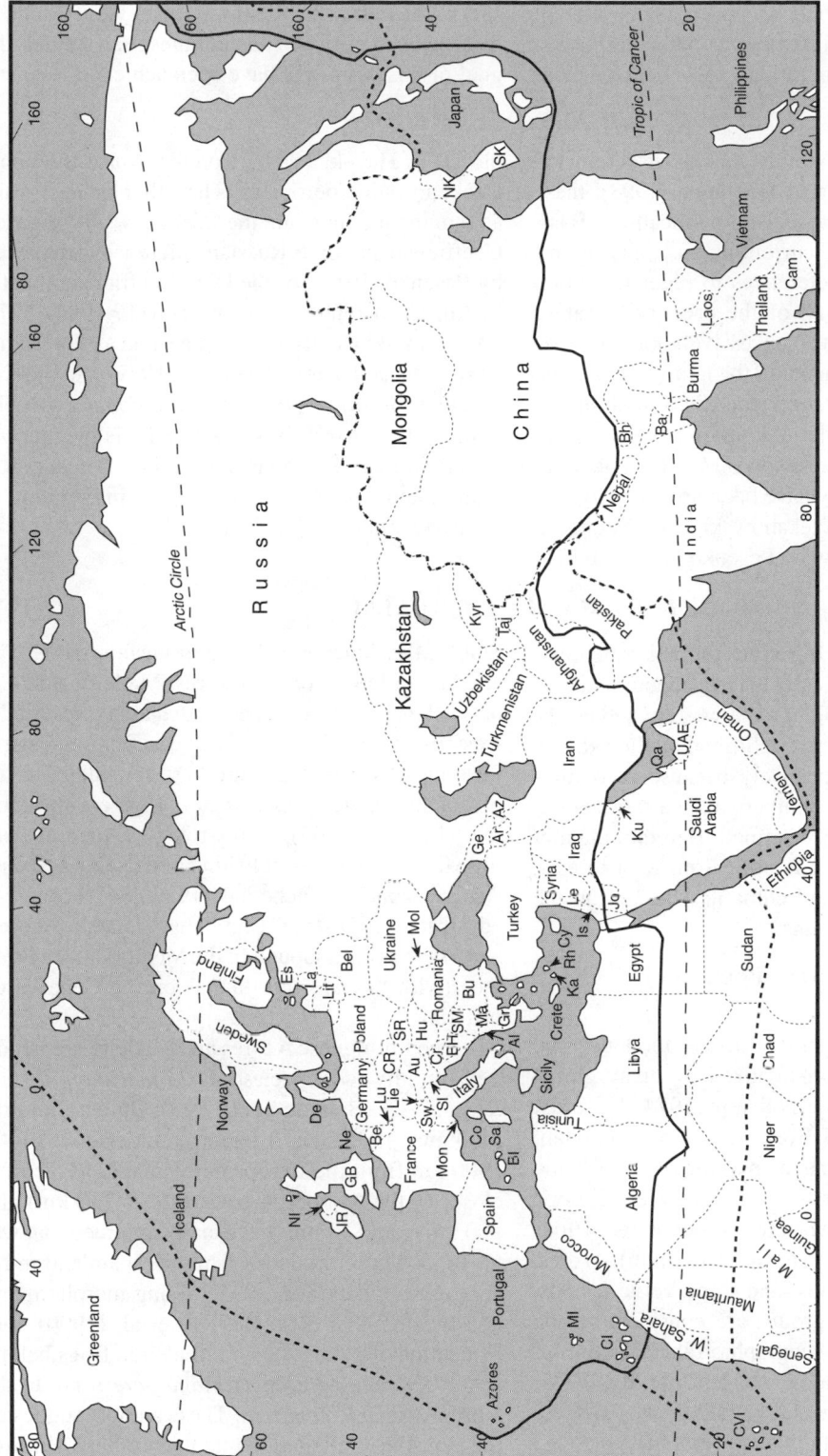

Fig. 6.1. The Palearctic Realm (solid line) and the area covered by this chapter (dashed line). The two lines coincide in some western and eastern places. Abbreviations for countries and place names: Al = Albania, Ar = Armenia, Au = Austria, Az = Azerbaijan, Ba = Bangladesh, Be = Belgium, Bel = Belorussia, BH = Bosnia and Herzegovina, Bh = Bhutan, BI = Balearic Islands, Bu = Bulgaria, Cam = Cambodia, CI = Canary Islands, Co = Corsica, CR = Czech Republic, Cr = Croatia, CVI = Cape Verde Islands, Cy = Cyprus, De = Denmark, Es = Estonia, GB = Great Britain, Ge = Georgia, Gr = Greece, IR = Ireland, Is = Israel, Jo = Jordan, Ka = Karpathos, Ku = Kuwait, Kyr = Kyrgyzstan, La = Latvia, Le = Lebanon, Lie = Liechtenstein, Lit = Lithuania, Lu = Luxembourg, Ma = Macedonia, MI = Madeira Island, Mol = Moldavia, Mon = Monaco, Ne = Netherlands, NI = Northern Ireland, NK = North Korea, Qa = Qatar, Rh = Rhodes, Sa = Sardinia, Sl = Slovenia, SK = South Korea, SM = Serbia and Montenegro, SR = Slovak Republic, Sw = Switzerland, Taj = Tadjikistan, UAE = United Arab Emirates.

In consideration of the foregoing statements, I view this chapter not only as a preliminary attempt to define amphibian distributions but as an invitation to students of amphibians to undertake more comprehensive zoogeographic analyses (and syntheses) of the Palearctic Realm. If this review results in discussion or criticism that initiates such studies, the goal of this work will have been achieved.

MATERIALS AND METHODS

The primary taxonomic lists used herein are those of Frost (1985) and Duellman (1993); these were updated from more recent publications. Basically, I tried to include current publications and reviews that included references to older literature, because inclusion of all of the primary literature would have resulted in an excessively long bibliography. I also tried to maintain the nomenclature and classification used by Inger and Zhao (both in this volume). Faunal similarities among regions were calculated using the Czekanowski Coefficient. Measurements of similarity have been proposed by various authors, such as Czekanowski, Dice, Sørensen, and others. (See review by Pesenko [1982] and Table 20 in Hayek, 1994.) I prefer to use the name Czekanowski Coefficient. The other names available in the literature are the Czekanowski-Sørensen Coefficient in some Russian references (reviewed by Pesenko, 1982), or the Dice Coefficient, mostly in American references (e.g., Hayek, 1994). The Czekanowski Coefficient is the same as the Faunal Resemblance Factor (Duellman, 1979) and the Coefficient of Biogeographical Resemblance (Duellman, 1990): $2C/(N_1 + N_2)$, where N_1 is the number of taxa in Region 1, N_2 is the number of taxa in Region 2, and C is the number of taxa in common to the two regions.

THE PALEARCTIC REALM

As noted in the Introduction, the Palearctic Realm is biogeographically heterogeneous and thus difficult to define. Although the western and northern boundaries are clearly delimited by the Atlantic and Arctic oceans, respectively, because of continuity of land masses, the southern and eastern boundaries cannot be so easily defined. Herein, I attempt to define the Palearctic Realm, from west to east, based on geological and climatic data and on the distribution of amphibians.

SOUTHWESTERN BOUNDARY

Macaronesia

Macaronesia is a subdivision of the Palearctic (or Holarctic) Realm recognized by many global biogeographers (e.g., Emeljanov, 1974; Udvardy, 1975; Bănărescu and Boşcaiu, 1978; Takhtajan, 1978). This region is made up of the North Atlantic volcanic oceanic islands off southwestern Europe and northern Africa. Macaronesia consists of four main island groups (from the north to south): Azores, Madeira, Canary Islands, and Cape Verde Islands. Commonly, all these groups are considered to be subunits of equal zoogeographic rank. Many biogeographers (Emeljanov, 1974; Udvardy, 1975; Bănărescu and Boşcaiu, 1978; Takhtajan, 1978) accorded Macaronesia a high biogeographic rank (e.g., province, subregion, or region), which is the same rank as the Saharan or Mediterranean regions. The amphibian faunas of the Azores, Madeira, and Canary Islands consist of species that are typical of the western part of the Mediterranean Region. Therefore, these archipelagos are best considered to be part of that zoogeographic region. I agree with Klemmer (1976:452–453), who wrote: "… there is very little reasoning for establishing a special Macaronesian sub-zone…" Shcherbak (1982:234) recognized "the Canarian District" (the Canary Islands, Azores, and Madeira) as a subunit of the Mediterranean Province (with "a transition fauna" on the Cape Verde Islands).

The Portuguese Azores and Madeira are inhabited by only two species—*Hyla meridionalis* and *Rana perezi* (Balletto et al., 1990; Balletto, in litt.; Malkmus, 1995). The Azorean archipelago is 1400–1900 km from the European mainland. Madeira is closer to the Moroccan coast (about 730 km) and 900 km from Lisbon. It is usually assumed that water frogs were introduced from the mainland in recent times (Malkmus, 1995). Using morphometric and biochemical data, Balletto et al. (1990) confirmed that the Azorean and Madeiran frogs belong to the southwest European *Rana perezi,* not to the North African *R. saharica.* These authors suggested the plausible explanation that the northern Macaronesian water frogs may have been derived from a single original event of natural colonization at about 400,000 years ago and that the Azores were reached via Madeira. Apart from frogs, a newt (*Triturus carnifex*) has existed in the Azores since its introduction in 1922 (Malkmus, 1995).

The Selvages Archipelago (Portugal) is located southeast of Madeira at approximately 30° N lat. No amphibians exist on these small islands (U. Joger, in litt.).

The Canary Islands are about 100 km west of southern Morocco. These islands also are inhabited by two anurans. *Hyla meridionalis,* which was described from Tenerife Island, probably is the only autochthonous amphibian species. *Rana perezi* seems to have been introduced from continental Spain (Klemmer, 1976), but status of the *Rana* needs further examination in the light of the hypothesis proposed by Balletto et al. (1990).

Located approximately 500 km west of Senegal, the Cape Verde Islands are the southernmost archipelago of Macaronesia and have a dry, tropical climate. The only amphibian is an Afrotropical toad, *Bufo regularis,* which was introduced (Schleich, 1987; Joger, 1993).

Sahara Desert

The Sahara, the largest desert in the world, occupies most of North Africa between the Atlantic Ocean and Red Sea north of approximately 15–17° N, except for a mountanous region in the extreme northwestern part of the continent. The Sahara is shared by 11 countries (Table 6:1); its area is about equal to that of the United States or Australia. Owing to its vast area, the Sahara is heterogeneous in both physical geographic (climate and landscape) and biological parameters. This heterogeneity is evident from north to south, and from west to east. Schiffers (1971) and Cloudsley-Thompson (1984) edited useful books reviewing climate, geology, soils, and plant and animal life of the Sahara Desert.

The limits of Sahara have been argued by biogeographers (e.g., Schiffers, 1971; Bons, 1973). Herein the delimitation of the Sahara Desert is based on the 100 mm isohyetal line (line depicting the area having an annual precipitation up to 100 mm [Fig. 6:2]). This parameter was used by biologists in the book edited by Cloudsley-Thompson (1984). Other authors analyzing the distribution of birds used the 200 mm isohyet. Le Berre (1989), who reviewed Saharan fishes, amphibians, and reptiles, used 150

mm isohyet. In all of these cases, climatic (rainfall) parameters have been used as biogeographic boundaries. Although at first glance the differences in the sizes of areas contoured by various isohyetal lines are small (Fig. 6:2), they resulted in different lists of species, because the northern or southern distributional limits of some amphibians coincide with the intervening areas.

To the north, the Sahara Desert is replaced by a narrow transition zone, which is bordered on the north by the Mediterranean Province. A particular geographic zone, the Sahel, is south of the Sahara. Climatically, the Sahel includes an area with annual precipitation up to approximately 800 mm. The Sahel is a broad transition between the Sahara and the African savannas.

Many publications during the past century have dealt with amphibians of the Sahara; the most important ones are referenced in Table 6:1. However, some records are taxonomically problematic or contradictory (e.g., zoogeographically key localities like the mountainous Ahaggar [Hoggar] area in southern Algeria and the Ghat Oasis in southwestern Libya). In the way of comparison, Gruber (1971) mentioned only three Saharan species of amphibians—one frog and two toads (plus three other anurans from the Nile Valley). Lambert (1984), who utilized the 100 mm isohyet in defining the Sahara, listed nine species. According to Le Berre (1989), based on the 150 mm isohyet, the Saharan amphibian fauna consists of 14 species. My list (Table 6:1) consists of 13 species and includes most species listed by Le Berre, except *Pleurodeles poireti* (Salamandridae) and *Discoglossus pictus* (Discoglossidae), the southern distributional limits of which are slightly north of the 100 mm isohyet. Also, contrary to *Rana ridibunda* in the lists of Lambert and Le Berre, the green frogs (*Rana esculenta* group) in the Sahara are allocated to two species—*Rana bedriagae* (= *R. levantina*) in the lower Nile Valley Egypt and *Rana saharica* in the remaining part of North Africa (Fig. 6:2). *Bufo vittatus* of the Nile Delta was referred to a separate species, *B. kassasii,* by Baha el Din (1993), which, by implication, should belong to the Afrotropical *Bufo funereus* group (Tandy and Tandy, 1976).

As shown in Table 6:1, North Africa (Mediterranean and Saharan Africa north of the Sahel) is inhabited by 22 species, including three salamanders (14%) belonging to two salamandrid genera and 19 anurans (86%) belonging to nine genera and five

Table 6:1. Distribution of amphibians in the Saharan (S) and Mediterranean (M) parts of North Africa. Abbreviations: AL = Algeria, CH = Chad, EG = Egypt, LI = Libya, ML = Mali, MO = Morocco, MR = Mauritania, NI = Niger, SU = Sudan, TU = Tunisia, WS = Western Sahara. Affinities to realms are abbreviated: A = Afrotropical (Ethiopian), P = Palearctic. Asterisks denote species (*) or subspecies (**) endemic to the region. Sources: Angel and Guibé (1948), Angel and Lhote (1938), Baha el Din (1993), Böhme (1978), Bons (1972, 1973), Flower (1933), Hemmer et al. (1980), Hulselmans (1977), Joger (1981), Lambert (1984), Le Berre (1989), Marx (1968), Mosauer (1934), Pasteur and Bons (1959), Rehák and Osborn (1988), Roussel and Amar (1985), Salvador (1996), Schleich et al. (1996), B.Schneider (1974), Scortecci (1936), Steinwarz and Schneider (1991), Tandy et al. (1976, 1985), Wake and Kluge (1961).

| Species | Country | | | | | | | | | | | Region | | Realm |
	WS	MO	AL	TU	LI	EG	MR	ML	NI	CH	SU	M	S	
CAUDATA: SALAMANDRIDAE:														
*Pleurodeles poireti**	−	−	+	+	−	−	−	−	−	−	−	+	−	P
Pleurodeles waltl	−	+	−	−	−	−	−	−	−	−	−	+	−	P
*Salamandra algira**	−	+	+	+	−	−	−	−	−	−	−	+	−	P
Total Caudata	0	2	2	2	0	0	0	0	0	0	0	3	0	3
ANURA: BUFONIDAE:														
*Bufo brongersmai**[1]	+	+	−	−	−	−	−	−	−	−	−	+	+	P
Bufo bufo	−	+	+	+	−	−	−	−	−	−	−	+	−	P
Bufo dodsoni	−	−	−	−	−	+	−	−	−	−	+	−	+	A
*Bufo kassasii**	−	−	−	−	−	+	−	−	−	−	−	+	−	A
Bufo mauritanicus[1]	+	+	+	+	+	−	+	+	+	−	−	+	+	A
Bufo pentoni	−	−	−	−	−	−	+	−	+	−	+	−	+	A
Bufo regularis[1]	−	−	?	−	?	+	+	+	+	+	+	+	+	A
*Bufo viridis***	+	+	+	+	+	+	−	−	−	−	−	+	+	P
Bufo xeros[1]	−	−	?	−	?	−	+	−	+	−	−	−	+	A
ANURA: DISCOGLOSSIDAE:														
*Alytes obstetricans***	−	+	−	−	−	−	−	−	−	−	−	+	−	P
*Discoglossus pictus***	−	+	+	+	−	−	−	−	−	−	−	+	−	P
ANURA: HYLIDAE:														
Hyla meridionalis	−	+	+	+	−	−	−	−	−	−	−	+	−	P
ANURA: PELOBATIDAE:														
*Pelobates varaldii**	−	+	−	−	−	−	−	−	−	−	−	+	−	P
ANURA: PIPIDAE:														
Xenopus muelleri	−	−	−	−	−	−	−	−	−	+	−	−	+	A
ANURA: RANIDAE:														
Hoplobatrachus occipitalis	−	−	−	−	+	−	+	?	+	+	−	−	+	A
Ptychadena mascareniensis	−	−	−	−	−	+	−	−	+	−	+	+	+	A
Rana bedriagae	−	−	−	−	−	+	−	−	−	−	−	−	+	P
Rana saharica[1]	+	+	+	+	+	?	−	−	−	−	−	+	+	P
Tomopterna cryptotis	−	−	−	−	−	−	+	−	+	−	−	−	+	A
Total Anura	4	9	8	6	5	6	6	2	7	3	4	11	13	19
Total Amphibia	4	11	10	8	5	6	6	2	7	3	4	15	13	22
Palearctic species	3	10	7	7	2	2	0	0	0	0	0	11	4	12
Percent Palearctic	75	91	70	88	40	33	0	0	0	0	0	73	31	55
Afrotropical species	1	1	3	1	3	4	6	2	7	3	4	4	9	10
Percent Afrotropical	25	9	30	12	60	67	100	100	100	100	100	27	69	45

[1]Schleich et al. (1996) mentioned *Rana saharica* from northwestern Egypt but gave no localities. *Bufo brongersmai* and *B. mauritanicus* were recorded from Western Sahara by Le Berre (1989) but not by Salvador (1996) and Schleich et al. (1996). There is no consensus on the identification of toads from the central and southern Sahara. Records of *B. regularis* from Algeria, Libya, Mauritania, Mali, and Nigeria, as well as records of *B. mauritanicus* from Mauritania, Mali, and Niger, were reassigned to *B. xeros* by Salvador (1996). Although Schleich et al. (1996) also considered *B. regularis* from Ghat (Libya) and neighboring areas most probably to be *B. xeros,* they recognized the validity of southern Saharan localities of *B. mauritanicus.* It should be mentioned that toads fron the southern parts of Algeria and Libya were referred to *B. regularis,* not *B. xeros,* by Tandy et al. (1985).

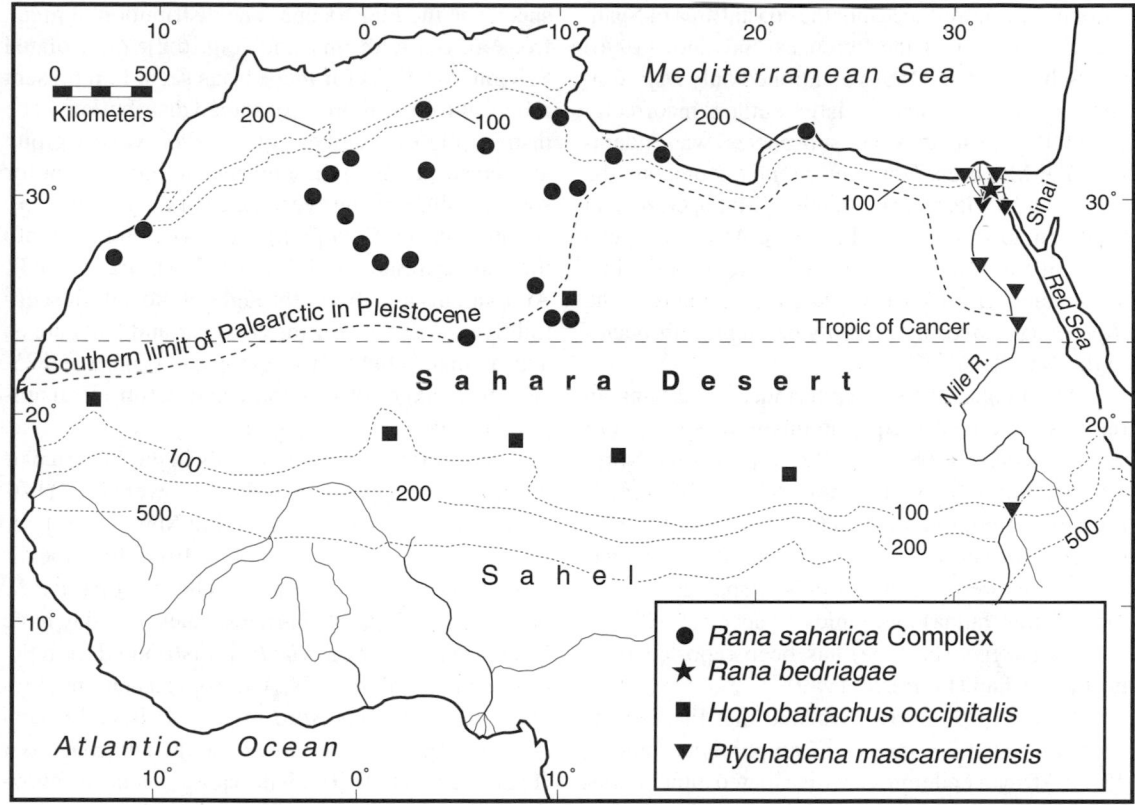

Fig. 6:2. Distribution of Palearctic (dots and asterisk) and Afrotropical (squares and triangles) ranids in the Sahara Desert. References are listed in the Table 6:1; only some peripheral Mediterranean records are shown.

families. Among North African amphibians, bufonids (9 species of *Bufo;* 41% of the amphibian fauna) are the most diverse, followed by ranids (5 species in 3 genera; 23%). Salamandrids (14%) and discoglossids (14%) have 3 species each, whereas hylids (4%) and pipids (4%) each are represented by one species.

Six species (28%) are endemic to North Africa (Table 6:1). Three of these (*Pleurodeles poireti, Salamandra algira,* and *Pelobates varaldii*) are endemic to the Maghreb area (namely Morocco and northern Algeria and Tunisia) of the Mediterranean Province. Two other species (*Bufo brongersmai* and *Rana saharica*) are distributed in both the Maghreb and the Sahara, whereas *Bufo kassasii* is endemic to the Nile Delta. *Bufo mauritanicus* may be considered a near-endemic to the region because of its isolated populations in the Sahel of Mali—Bourem on the Niger River (Angel and Lhote, 1938) and at

Burkina Faso (Tandy and Keith, 1972). However, according to Salvador (1996), but not to Schleich et al. (1996), all records of the toad from the southern Sahara probably are of *B. xeros;* if this suggestion is confirmed, *B. mauritanicus* should be classified as an endemic of Palearctic North Africa. The green frogs, united here under the name *Rana saharica,* perhaps consist of two or more species (Hemmer et al., 1980; Steinwarz and Schneider, 1991; Dubois and Ohler, 1995a).

Two other species not endemic to North Africa are represented there by endemic subspecies. These are *Alytes obstetricans maurus,* a taxon that is not a distinct species (Arntzen and Szymura, 1984; Arntzen and García-París, 1995), and *Discoglossus pictus* with two endemic subspecies—*D. p. scovazzii* in Morocco and *D. p. algirus* (= *D. p. auritus*) in northern Algeria and Tunisia (Lanza et al., 1986; Capula and Corti, 1993). The latter subspecies apparently was introduced into the coastal area of Spain and France east of the Pyrenees. Salvador (1996) and Schleich et al. (1996) mentioned only *D. p. scovazzii* in North Africa; the latter authors incorrectly stated that the species (not subspecies) was endemic to the Maghreb. However, *D. pictus* inhabits the European Mediterranean islands of Malta, Gozo, and Sicily (map in Gasc et al., 1997). Also, the green toads of the Maghreb seem to represent a distinct subspecies, *Bufo viridis boulengeri,* whereas the Egyptian populations possibly belong to subspecies in the Near East.

The faunas of the Mediterranean and Saharan regions have nearly equal numbers of species (15 vs. 13 species, or 68% vs. 59% of all known North African amphibians), but they are very different in taxonomic composition. The Sahara lacks salamanders, discoglossids, and hylids; it is characterized by the presence of bufonids (7 species; 54% of the Saharan fauna) and ranids (5 species; 38%). A larval *Xenopus* (Pipidae) has been reported from northern Chad (Le Berre, 1989).

Bons (1973) correctly stated that there exists no "true Saharan" species of amphibians. Indeed, North African endemics are distributed either mainly in the Maghreb or in the Maghreb and in the Sahara. Saharan amphibians fall into two categories, the Palearctic (Mediterranean) and Afrotropical; clear differences exist in their distributional patterns.

Böhme (1978) reported that there is an almost 800 km (airline) gap between the southern known range of "*Rana ridibunda*" and the northern range limit of *Dicroglossus* (now *Hoplobatrachus*) *occipitalis* in the western and central parts of the Sahara Desert. However, he rejected as a dubious record of green frogs from Ahaggar (Hoggar) in southern Algeria (Pellegrin, 1931) and did not mention Scortecci's (1936) record of "*Rana*" *occipitalis* from the Ghat Oasis in southwestern Libya. Confirmed records show that the Palearctic green frogs (*Rana saharica*) are discontinuously distributed in Saharan oases west of approximately 10° E and extend southward to approximately 23° N (Fig. 6:2). Hot deserts, of course, are not favorable habitat for water frogs. Therefore, the present distribution seems to indicate previous connections of water systems, although some introductions may have occurred.

In their analysis of distributions of freshwater and land molluscs in the Sahara, Sparks and Grove (1961) outlined the southern limit of the Palearctic

species in the Pleistocene. The distribution of green frogs agrees with this line (Fig. 6:2); the isolated Saharan localities for these frogs seem to represent relicts of former more widespread distributions. The distribution of members of the *Bufo viridis* group also corresponds to the Pleistocene line, except for some localities in the Fezzan area of Libya (Fig. 6:3). Afrotropical (= Ethiopian) species extend only into the southern Sahara (Table 6:1, Figs. 6:2 and 6:3). Amphibians are absent throughout most of the central Sahara, except for the Ahaggar and Ghat areas, approximately between the Tropic of Cancer and 25° N, where ranges of Palearctic and Afrotropical amphibians probably are in contact.

A different situation prevails in the eastern part of the Sahara (Egypt and Sudan). Flower (1933) listed no green frogs from Egypt, but Marx (1968) reported them, as *Rana ridibunda,* from three localities in the Cairo-Giza area. These frogs are *R. bedriagae* (= *R. levantina*), a species that may have been introduced. *Bufo viridis* is distributed from the coast southward to 25° N. Afrotropical anurans (*Ptychadena mascareniensis, Bufo regularis,* and *B. kassasii*) occupy the Nile Valley and the adjacent coast (Figs. 6:2 and 6:3). The Nile Valley is a mesic corridor which apparently is more favorable for dispersal of Afrotropical amphibians than Palearctic ones. The amphibians in the Nile Valley contradict the suggestion by Braestrup (1948) that Palearctic amphibians penetrated into tropical West Africa during cooler past periods. More likely, Saharan river and oasis systems during wetter periods might have provided more suitable conditions for the dispersal of Afrotropical groups northward.

Remarkably, *Bufo mauritanicus* may be an unusual case of the northernmost distribution of Afrotropical groups. This toad occurs in the Maghreb (B. Schneider, 1978) and has disjunct populations as far south as Burkina Faso (Tandy and Keith,1972). Such a distribution might be associated with a Palearctic lineage, and according to Siboulet (1972), *B. mauritanicus* is serologically closer to the European *B. bufo* than to the Afrotropical *B. regularis*. However, the species was treated as the sole member of the *Bufo mauritanicus* group, which is related to the Afrotropical *Bufo pentoni* and *Bufo regularis* groups (Tandy and Keith, 1972; Tandy and Tandy, 1976; Sherif et al., 1990) or to the South African *B. rangeri* (Maxson, 1981). Therefore, *B. mauritanicus* most likely belongs to a lineage of Afrotropical origin.

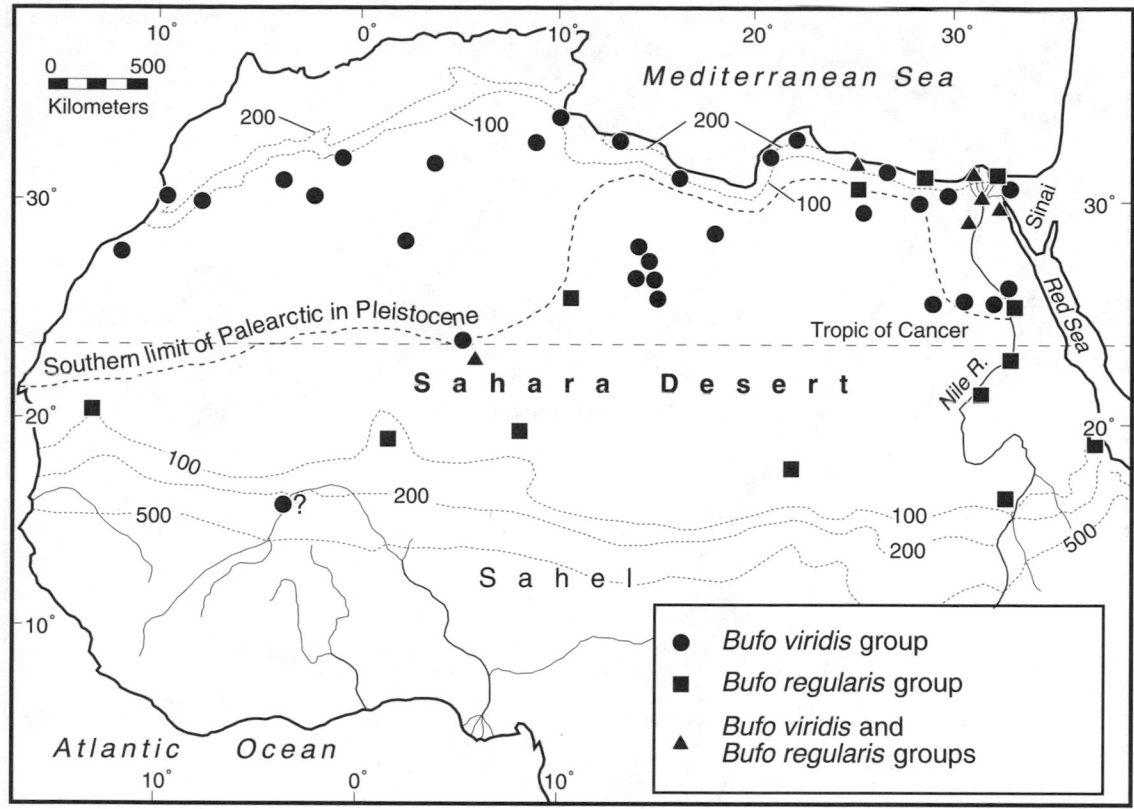

Fig. 6:3.   Distribution of toads in the Sahara Desert.   Records for Palearctic toads of the *Bufo viridis* group (*B. brongersmai* and *B. viridis*) are shown by dots; records for Afrotropical toads of the *Bufo regularis* group (*B. regularis* and *B. xeros*) are shown by squares; sympatric occurrence of toads of both groups are indicated by triangles. References are listed in Table 6:1; only some peripheral Mediterranean records are shown.

The Sahara (together or not with the Arabian Peninsula) is recognized traditionally as a province of the Palearctic (or Holarctic) Realm (e.g., Udvardy, 1975; Takhtajan, 1978; Shcherbak, 1982). Other authors (e.g., De Lattin, 1967) suggested that the southern border of the Palearctic coincides with the southern limit of the Sahara. Sclater (1858) attached only the Mediterranean part of northwestern Africa (north of the Atlas Mountains) to the Palearctic; thus, the Sahara was considered to be an area of the Ethiopian (Afrotropical) Region. Darlington (1957) and P. Müller (1986) regarded the Sahara and most of the Arabian Peninsula as a transition area between the Palearctic and Afrotropical realms.

Based on amphibian distribution, the Sahara may be divided into two parts. The northern part inhabited by green frogs and green toads is a portion of the Palearctic Realm. The southern part, which is inhabited by *Hoplobatrachus occipitalis*, *Bufo regularis*, *B. xeros,* and other apparently Afro-

tropical species (Table 6:1), should be placed in the Afrotropical Realm. The 25° N latitude with a southern loop encompassing the Ahaggar area may be used as a convenient boundary line between the two realms. This is a modification of Wallace's (1876) proposal that the Tropic of Cancer was the boundary. Alternatively, the Palearctic includes the western part of the Sahara, and the eastern Sahara (Nile area) is in the Afrotropical Realm.

Sinai Peninsula

This relatively small triangular piece of Asia is a land link between northeastern Africa and southwestern Asia. However, the intermediate geographic position of the Sinai is not reflected in the amphibian fauna. There are at least four species of amphibians in the Nile Delta (Flower, 1933; Marx, 1968; Akef and Schneider, 1989; Baha el Din, 1993), seven species in Israel (Werner, 1988), and nine in the Arabian Peninsula (Balletto et al., 1985), but only

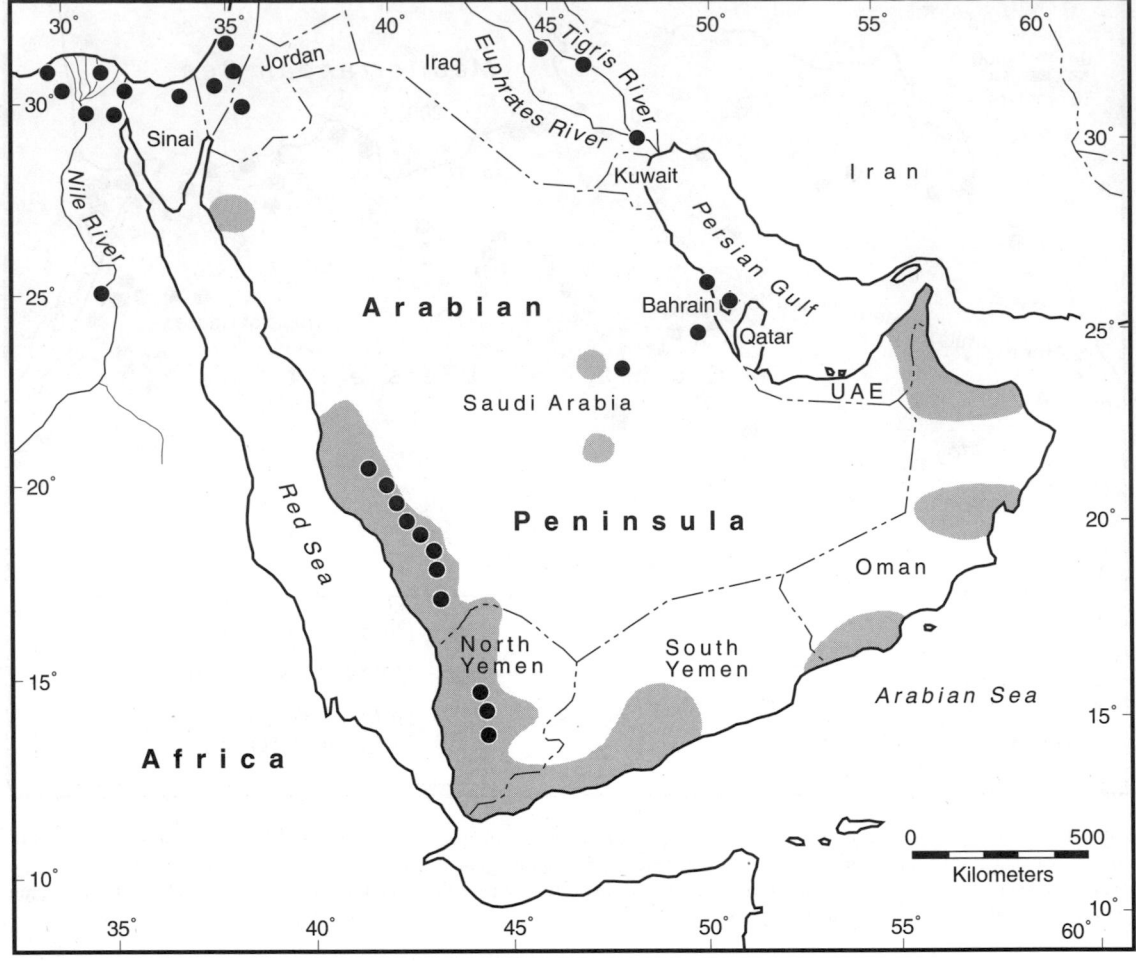

Fig. 6:4. Distribution of Palearctic amphibians (dots) and non-Palearctic amphibians (shaded areas) in the Arabian Peninsula. References are listed in the text. Dots outside the peninsula are records of *Bufo viridis* (based on various references).

two species have been reported from the Sinai Peninsula—*Bufo viridis* from the northeastern part of the Sinai (Flower, 1933) and *Ptychadena mascareniensis* from the southern part of the peninsula (Wadi Feiran) (Schmidt and Marx, 1956). Nonetheless, according to Werner (1982:155), "Of the species known to inhabit the Sinai . . . the only amphibian represented is *Bufo viridis* and that only from the northeast corner of Sinai. Here, as in the Negev of Israel, this Palearctic toad reaches the southern (desertic) limit of its distribution . . . Specific searches and inquiries for other frogs, especially in the larger oases, remained fruitless." Recently, two more anurans were found in the northeastern Sinai—*Rana bedriagae* and *Hyla savignyi;* these were reported as *R. ridibunda* from the vicinity

of Rafah and *H. arborea* from the vicinity of Sheikh Swaied and El Arish, respectively, by Baha el Din (1992). Both species had been reported previously from the Gaza area, Israel, by Mendelssohn and Steinitz (1944).

The Sinai Peninsula is in fact "an empty" territory for amphibians. *Bufo regularis* and *Ptychadena mascareniensis* occur on the western (African) side of the Suez Canal (Flower, 1933; Marx, 1968; Werner, 1987). The northernmost locality for *Bufo arabicus,* an endemic of the Arabian Peninsula, is southeast of the Sinai (Balletto et al., 1985; Fig. 6:4). The southernmost coastal record of *Pelobates syriacus* is not far north of the Sinai-Palestine border (Mendelssohn and Steinitz, 1944). The widespread *Bufo viridis* and *Rana bedriagae* occur in the north-

Table 6:2. Distribution of amphibians in the Arabian Peninsula. Abbreviations of countries: BA = Bahrain, KU = Kuwait, NY = North Yemen, OM = Oman, QA = Qatar, SA = Saudi Arabia, SY = South Yemen, UA = United Arab Emirates. Realms are abbreviated: A = Afrotropical, O = Oriental, P = Palearctic. Asterisks (*) denote species endemic to the region. Sources: Balletto et al. (1985), Briggs and Ault (1985), Leviton et al. (1992), Schätti (1989), Schätti and Gasperetti (1994).

| Species | Country | | | | | | | | Realm |
	SA	NY	SY	OM	UA	QA	BA	KU	
ANURA: BUFONIDAE:									
Bufo arabicus*	+	+	+	+	+	–	–	–	A
Bufo dhufarensis*	+	+	+	+	+	–	–	–	A
Bufo hadramautinus*	–	–	+	–	–	–	–	–	A
Bufo scorteccii*	–	+	–	–	–	–	–	–	A
Bufo tihamicus*	+	+	+	–	–	–	–	–	A
Bufo viridis	+	–	–	–	–	–	–	–	P
ANURA: HYLIDAE:									
Hyla savignyi	+	+	–	–	–	–	–	–	P
ANURA: RANIDAE:									
Euphlyctis ehrenbergii*	+	+	+	–	–	–	–	–	O
Rana "ridibunda"	+	+	–	–	–	–	+	–	P
Total species	7	7	5	2	2	0	1	0	9
Afrotropical species	3	4	4	2	2	0	0	0	5
Percent Afrotropical	43	57	80	100	100	0	0	0	56
Oriental species	1	1	1	0	0	0	0	0	1
Percent Oriental	14	14	20	0	0	0	0	0	11
Palearctic species	3	2	0	0	0	0	1	0	3
Percent Palearctic	43	29	0	0	0	0	100	0	33

eastern part of the Sinai and the African part of Egypt, with a gap in the main part of the Sinai (Figs. 6:2 and 6:4). Therefore, the Sinai probably is a major barrier to amphibian dispersal between Africa and Asia.

Arabian Peninsula

Werner (1988) mentioned the range extensions of *Bufo viridis*, *Rana bedriagae* (as *R. ridibunda*), and *Hyla savignyi* into southern Israel. An invasion of African *Bufo regularis* and *Ptychadena mascareniensis* into Sinai may be expected as a result of large agricultural projects associated with waters from the Nile (Baha el Din, 1992).

Balletto et al. (1985) provided a comprehensive survey of the amphibians of the Arabian Peninsula; subsequent publications are by Briggs and Ault (1985), Schätti (1989), Leviton et al. (1992), and Schätti and Gasperetti (1994). The peninsular amphibian fauna consists of nine anurans, six of which (67%) are endemic (Table 6:2).

The three (33%) nonendemic species (*Bufo viridis*, *Hyla savignyi*, and *Rana "ridibunda"*) are typical Palearctic elements. The geographically nearest records of water frogs are *Rana bedriagae* (= *R. levantina*) from Egypt and the Near East and *R. ridibunda susana* from southern Iran; Dubois and Ohler

(1995a) suggested that the latter should be recognized at the specific level. Endemic toads belong to two or three species groups. *Bufo tihamicus* seems to be allied to the African *B. pentoni* (*Bufo regularis* group); Leviton and Aldrich (1984) listed it as *B. pentoni tihamicus*. *Bufo arabicus* (= *B. orientalis* of various authors), *hadramautinus*, and *scorteccii* are related to the African *B. dodsoni*; thus, they are referred to the *Bufo orientalis* group (Balletto et al., 1985). *Bufo dhufarensis* has been assigned either to the *Bufo orientalis* group (Tandy and Keith, 1972) or to the *Bufo stomaticus* group in the Oriental Realm (Inger, 1972). Balletto et al. (1985) mentioned that *B. dhufarensis* is similar to the Indian *B. stomaticus* and *B. dodsoni* of Somalia. According to Schätti and Gasperetti (1994), most species of Arabian toads, except *B. tihamicus*, belong to a mainly Palearctic group; however, these authors offered no evidence for such relationships. The only nonbufonid endemic, *Euphlyctis ehrenbergii*, is related to the Oriental *E. cyanophlyctis* (Balletto et al., 1985); some authors (e.g., Leviton and Aldrich, 1984; Briggs and Ault, 1985) considered the peninsular frog to be a subspecies of the latter. Of the four species in the genus, only *E. cornii* occurs in Africa.

Like the Sahara Desert, the hot Arabia Peninsula with few bodies of water, is hostile to amphibians. The distributions of all species principally are coastal; only a few localities are inland (Fig. 6:4); perhaps these are partly the result of introductions (Balletto et al., 1985). Three Palearctic anurans have populations that are widely disjunct from their main ranges. Except for records of green frogs (*Rana "ridibunda"*) in Bahrain, all known localities of Palearctic species are confined by the southwestern coastal area by the Red Sea in Saudi Arabia (Asir) and North Yemen between approximately 21°30' N and 14°30' N. These species occur at elevations of 1500–2900 m, whereas the non-Palearctic endemic anurans occur at elevations between sea level and 2500 m.

Balletto et al. (1985) pointed out that the peninsular amphibian fauna is mostly of south-Tethyian origin and that Arabia once was a land bridge for faunal interchange between Asia and Africa. Toads endemic to the Arabian Peninsula are closely related to African species, especially species in the Horn of Africa. The separation of these taxa presumably occurred no later than the time of the opening of the Red Sea 2–5 million years ago. Reptiles endemic to the southwestern part of the Arabian Peninsula and closely related to East African species were interpreted as evidence of a probable Miocene land connection (Joger, 1987). Unlike reptiles, only rare or poorly preserved remains of unidentified anurans ("bufonoids indet." and "ranoids indet.") are known from the lower Miocene, 15–17 million years ago, of Saudi Arabia (Rage in Thomas et al., 1982). Indeed, the Arabian Peninsula was a part of Africa before its collision with Eurasia in the Oligocene; the first exchanges probably occurred 18–20 million years ago (Rage, 1995). Palearctic species possibly dispersed to Arabia during the Würm glaciation or the late pluvials (Balletto et al., 1985). The Bahrain oases inhabited by *Rana "ridibunda"* are watered by karst springs dating at least to the early Pleistocene (Briggs and Ault, 1985). There is no evidence of faunal exchange between North Africa and the Arabian Peninsula via a northern route around the Red Sea through the Sinai Peninsula; Joger (1987) rejected a possibility of the northern route for reptiles during the Quaternary.

Taking into consideration the relationships and distribution patterns of amphibians, I suggest that the Arabian Peninsula belongs to the Afrotropical (Ethiopian) Realm. The 30° N latitude may be used as a provisional boundary between the Afrotropical and Palearctic Realms. This line lies north of the limits proposed by some authors (e.g., Ripley, 1954).

SOUTHERN BOUNDARY

Iran

Unfortunately, Iran, a zoogeographically important country, lacks any recent, detailed survey of amphibian fauna, except for the short review by S. Anderson (1985), who also provided a description of physiographic regions of Iran with an analysis of the lizard fauna (S. Anderson, 1968). Eighteen species of amphibians are known from Iran; these include seven salamanders (39%) and 11 anurans (61%) (Table 6:3). Salamanders are represented by two hynobiids (11%) of the same genus and five salamandrids (28%) of two genera. Anurans include a pelobatid (6%), a hylid (6%), six toads (33%) from three species groups, and three ranids (17%) from two genera.

The taxonomy of several groups is still unclear. In the *Bufo viridis* group, *B. luristanicus* is treated either as a subspecies of *B. surdus* (Schmidtler and Schmidtler, 1969; Eiselt and Schmidtler, 1973; Leviton et al., 1992) or as a distinct species (Mertens, 1971; Inger, 1972; S. Anderson, 1985). Populations of green toads east of the Caucasus and Zagros mountains were considered to be several subspecies of *B. viridis* (Eiselt and Schmidtler,1973) or of *B. latastii* (Hemmer et al., 1978); also, a new species (*Bufo kavirensis*) was described from central Iran by Andrén and Nilson (1979). However, karyological evidence demonstrated the occurrence of both diploid (*B. viridis*) and tetraploid ("*B. danatensis*") species in the former Soviet Central Asia (i.e., Turkmenistan, Uzbekistan, Tadjikistan, Kyrgyzstan, and Kazakhstan). The latter species is widely distributed in mountainous areas east of the Caspian Sea (Borkin et al., 1986; Borkin, unpublished data) and certainly should occur in northeastern Iran (Fig. 6:5). The brown frogs (*Rana temporaria* group) of Iran, like the Caucasian and Turkish ones, are recognized either as two species (*R. camerani* and *R. macrocnemis*), or as the same species under the name *R. macrocnemis* (Eiselt and Schmidtler, 1973; Borkin, 1978; S. Anderson, 1985; Baran and Atatür, 1986). The green frogs (*Rana esculenta* group) are represented by *R. ridibunda;* however, southern populations of that species were proposed to be a distinct species, *R. susana* (Dubois and Ohler, 1995a), and the western ones may belong to *R. bedriagae.*

Table 6:3. Distribution of amphibians in administrative provinces (ostans) of Iran. Abbreviations of provinces are: BA = Saheli-ye Jazaryer va Banader-e Khalije Fars va Darya-ye'Oman (included in Kerman on some maps), BU = Bushehr-e Khalij-e Fars (included in Fars on some maps), EA = East Azarbaijan, ES = Esfahan, FA = Fars, GI = Gilan, HA = Hamadan, IL = Ilam, KE = Kerman, KH = Khorasan, KO = Kordestan, KS = Kermanshahan, KU = Khuzestan, LO = Lorestan, MA = Mazandaran, SB = Sistan and Baluchestan, SE = Semnan, TE = Tehran, WA = West Azarbaijan, ZA = Zanjan. No records for amphibians exist for two provinces (Bakhtiari va Chahar Mahall and Boyer Ahmadi-Ye Sardsir va Kohgiluyeh); they are not listed. Abbreviations for realms are: O = Oriental, P = Palearctic. Asterisks denote species (*) or subspecies (**) endemic to the country; some brown frogs recorded as "Rana camerani" are noted by (c). Sources: S. Anderson (1963), Andrén and Nilson (1979), Borkin (1978), Clergue-Gazeau and Thorn (1978), Eiselt and Schmidtler (1973), Eiselt and Steiner (1970), Guibé (1957), Mertens (1957), C. Müller (1985), N. Rastegar-Pouyani (in litt.), Schmidtler and Schmidtler (1969), Steiner (1973), Tuck (1971, 1979).

Species	WA	EA	GI	MA	KO	ZA	KS	HA	TE	IL	LO	KU	ES	FA	BU	SE	KH	KE	BA	SB	Realm
CAUDATA: HYNOBIIDAE:																					
Batrachuperus gorganensis*	–	–	–	+	–	–	–	–	–	–	–	–	–	–	–	–	–	–	–	–	P
Batrachuperus persicus*	–	–	+	+	–	–	–	–	–	–	–	–	–	–	–	–	–	–	–	–	P
CAUDATA: SALAMANDRIDAE:																					
Neurergus crocatus	+	–	–	–	–	–	–	–	–	–	–	–	–	–	–	–	–	–	–	–	P
Neurergus kaiseri*	–	–	–	–	–	–	–	–	–	–	+	–	–	–	–	–	–	–	–	–	P
Neurergus microspilotus	–	–	–	–	+	–	–	–	–	–	–	–	–	–	–	–	–	–	–	–	P
Salamandra infraimmaculata	–	–	–	–	+	–	–	–	–	–	–	–	–	–	–	–	–	–	–	–	P
Triturus karelinii	?	–	+	+	–	–	–	–	–	–	–	–	–	–	–	–	–	–	–	–	P
Total Caudata	1	0	2	3	2	0	0	0	0	0	1	0	0	0	0	0	0	0	0	0	7
ANURA: BUFONIDAE:																					
Bufo kavirensis*	–	–	–	–	–	–	–	–	+	–	–	–	–	–	–	–	–	–	–	–	P
Bufo olivaceus	–	–	–	–	–	–	–	–	–	–	–	–	–	–	–	–	–	+	+	+	O
Bufo stomaticus	–	–	–	–	–	–	–	–	–	–	–	–	–	–	–	–	+	+	–	+	O
Bufo surdus**	–	–	–	–	–	–	–	–	–	–	–	–	–	+	+	–	–	–	–	–	P
Bufo verrucosissimus	–	–	+	+	–	–	–	–	–	–	–	–	–	–	–	–	–	–	–	–	P
Bufo viridis**	+	+	+	+	+	+	+	+	+	+	+	+	+	+	–	+	+	+	+	+	P
ANURA: HYLIDAE:																					
Hyla savignyi	+	+	+	–	+	+	+	–	+	–	+	–	–	+	–	–	–	–	–	–	P
ANURA: PELOBATIDAE:																					
Pelobates syriacus	+	+	+	–	–	–	+	–	+	–	–	–	–	–	–	–	–	–	–	–	P
ANURA: RANIDAE:																					
Euphlyctis cyanophlyctis	–	–	–	–	–	–	–	–	–	–	–	–	–	–	–	–	–	–	+	+	O
Rana macrocnemis	c	c	c	c	c	c	–	–	–	–	–	–	–	–	–	–	–	–	–	–	P
Rana ridibunda complex**	+	+	+	+	+	+	+	+	+	+	+	+	+	+	+	–	+	–	+	+	P
Total Anura	5	5	6	4	4	4	4	2	5	2	3	2	2	4	2	1	3	3	4	5	11
Total Amphibia	6	5	8	7	6	4	4	2	5	2	4	2	2	4	2	1	3	3	4	5	18
Oriental species	0	0	0	0	0	0	0	0	0	0	0	0	0	0	0	0	1	2	2	3	3
Percent Oriental	0	0	0	0	0	0	0	0	0	0	0	0	0	0	0	0	33	67	50	60	17
Palearctic species	6	5	8	7	6	4	4	2	5	2	4	2	2	4	2	1	2	1	2	2	15
Percent Palearctic	100	100	100	100	100	100	100	100	100	100	100	100	100	100	100	100	67	33	50	40	83

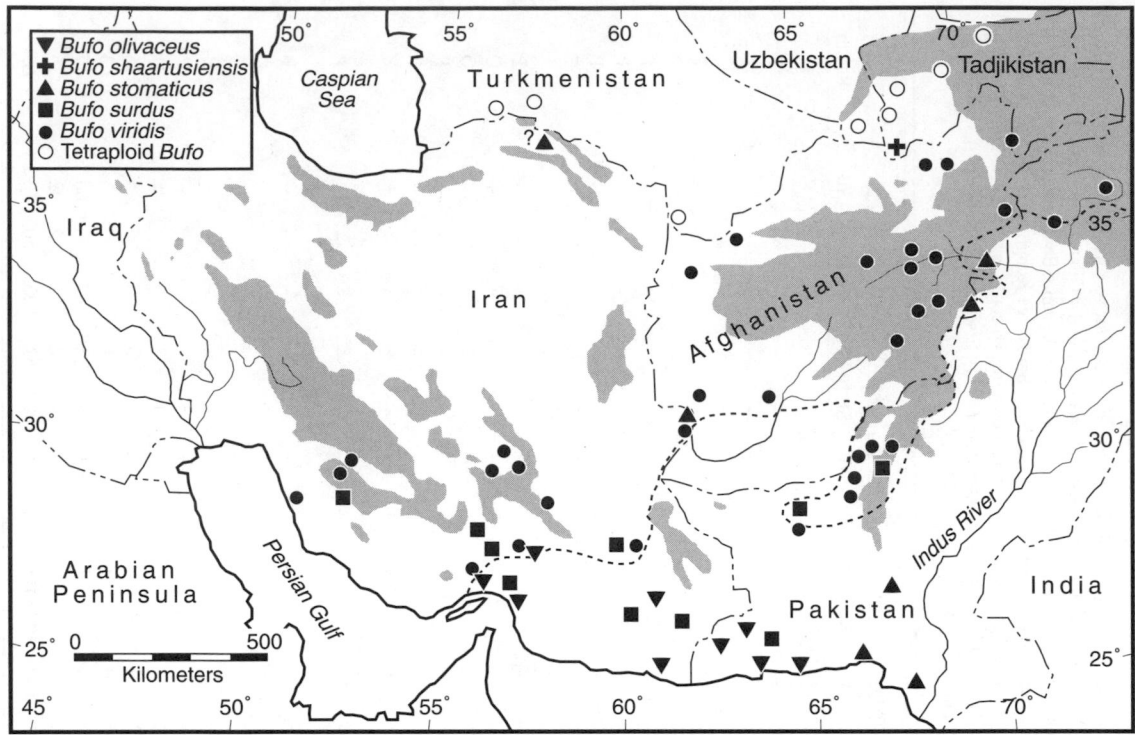

Fig. 6:5.   Distribution of toads of the Palearctic *Bufo viridis* group and Oriental *Bufo stomaticus* group in southern Iran, Afghanistan and western Pakistan. References are listed in Tables 6:3–6:6. Areas above 2000 m are shaded. The provisionary Palearctic-Oriental boundary is shown as a dashed line.

Four species (*Batrachuperus gorganensis, B. persicus, Neurergus kaiseri,* and *Bufo kavirensis*), 22% of the amphibian fauna, are endemic to Iran.

The majority (15 species, 83%) of Iranian amphibians is composed of Palearctic elements. Only three anurans (*Bufo olivaceus, B. stomaticus,* and *Euphlyctis cyanophlyctis*) are Oriental elements. The distribution of Iranian amphibians in 12 physiographic regions proposed by S. Anderson (1968) is given in Table 6:4. (Also, see Tuck, 1971.) The northern (south of the Caspian Sea) and western (Urmia-Zagros) parts of the country have a richer fauna than the central, eastern, or southern parts. These differences might be explained by climatic or historical factors. Palearctic amphibians prevail throughout most of Iran, except the Sistan Basin and the Baluchistan and Makran coast. Oriental species are largely distributed in the southeastern part of Iran. A strange record of *Bufo stomaticus* exists for the Bodjnurd area in the Kopet Dagh region in northern Khorasan Province. Eiselt and Schmidtler (1973) noted geographic replacement of the Palearctic species by Oriental ones. *Bufo viridis* is replaced by *B. oliva-*

ceus, and *Rana ridibunda* is replaced by *Euphlyctis cyanophlyctis*. However, the ranges of *Bufo surdus* and *B. olivaceus* overlap in southeastern Iran and southwestern Pakistan (Fig. 6:5). The Oriental species are primarily confined to lowlands in the Makran region and extend westward to the Bandar Abbas area. Perhaps the geological history of the area has some significance to these distribution patterns. Southern Iran (south of the Zagros Mountains) is the northeastern part of the Arabian-African plate that prior to the mid-Neogene was separated from the rest of what is now Iran by the Tethys Sea (Rage, 1995).

Afghanistan

Afghanistan is characterized by a depauperate amphibian fauna. Only six species are known from this country covered mostly by deserts and mountains—one hynobiid (17%), two toads (33%), and three ranids (50%) (Table 6:5). Toads belong to different species groups and include *Bufo stomaticus* (= *B. andersoni*) and three subspecies of *Bufo viridis* (*B. v. turanensis, oblongus,* and *pseudoraddei*).

Table 6:4. Distribution of amphibians in physiographic regions in Iran (S. Anderson, 1968; Tuck, 1971). Abbreviations: BAL = Baluchistan and Makran Coast. CAS = Caspian Coast, CEP = Central Plateau (interior desert basins), ELB = Elburz (= Alburz) Mountains, KOP = Kopet Dagh Mountains, KHP = Khuzistan Plain and Persian Gulf Coast (extension of the Mesopotamian Plain), MUG = Mughan Steppe, SIS = Sistan Basin, TUR = Turkmen Steppe, URM = Urmia (Rezaiyeh) Lake Basin, WZM = western foothills of Zagros Mountains, ZAG = Zagros Mountains. For sources, see Table 6:3.

Species	Regions											
	MUG	CAS	ELB	URM	ZAG	WZM	KHP	CEP	TUR	KOP	SIS	BAL
CAUDATA: HYNOBIIDAE:												
Batrachuperus gorganensis	–	–	+	–	–	–	–	–	–	–	–	–
Batrachuperus persicus	–	–	+	–	–	–	–	–	–	–	–	–
CAUDATA: SALAMANDRIDAE:												
Neurergus crocatus	–	–	–	+	+	–	–	–	–	–	–	–
Neurergus kaiseri	–	–	–	–	–	+	–	–	–	–	–	–
Neurergus microspilotus	–	–	–	–	+	–	–	–	–	–	–	–
Salamandra infraimmaculata	–	–	–	–	+	–	–	–	–	–	–	–
Triturus kareliniii	–	+	+	–	–	–	–	–	–	–	–	–
ANURA: BUFONIDAE:												
Bufo kavirensis	–	–	–	–	–	–	–	+	–	–	–	–
Bufo olivaceus	–	–	–	–	–	–	–	–	–	–	–	+
Bufo stomaticus	–	–	–	–	–	–	–	–	–	?	+	?
Bufo surdus	–	–	–	–	+	+	–	–	–	–	–	+
Bufo verrucosissimus	–	+	?	–	–	–	–	–	–	–	–	–
Bufo viridis	+	+	+	+	+	+	+	+	+	+	+	+
ANURA: HYLIDAE:												
Hyla savignyi	?	+	+	+	+	+	+	–	–	–	–	–
ANURA: PELOBATIDAE:												
Pelobates syriacus	?	+	?	+	–	–	–	–	–	–	–	–
ANURA: RANIDAE:												
Euphlyctis cyanophlyctis	–	–	–	–	–	–	–	–	–	–	+	+
Rana macrocnemis	–	+	+	+	+	+	–	–	–	+	–	–
Rana ridibunda complex	+	+	+	+	+	+	+	+	+	+	–	–
Total Amphibia	2	7	7	6	8	6	3	3	2	3	3	4

Table 6:5. Distribution of amphibians in geographic regions in Afghanistan. Abbreviations for geographic regions (after Humlum, 1959) are: BA = Badakhshan, CE = Central, EA = East, MO = three monsoonal areas, NO = North, NU = Nurestan, NW = Northwest, SO = South Afghanistan, VA = Vakhan, WE = West. Abbreviations of realms are: O = Oriental, P = Palearctic. Asterisk (*) denotes a species endemic to the country. Sources: Clark (1990), Clark et al. (1969), Eiselt and Schmidtler (1973), Hemmer et al. (1978), Leviton and Anderson (1961, 1963, 1970), Reilly (1983).

Species	Geographic region										Realm
	NO	BA	VA	NW	WE	CE	NU	MO	EA	SO	
CAUDATA: HYNOBIIDAE:											
*Batrachuperus mustersi**	–	–	–	–	–	–	–	–	+	–	P
ANURA: BUFONIDAE:											
Bufo stomaticus	–	–	–	–	–	–	–	+	+	?	O
Bufo viridis	+	+	?	?	+	+	+	–	+	+	P
ANURA: RANIDAE:											
Euphlyctis cyanophlyctis	–	–	–	–	–	–	–	+	–	+	O
Paa sternosignata	–	–	–	–	–	–	–	+	+	+	O
Rana ridibunda	+	+	–	?	+	+	–	–	+	–	P
Total Amphibia	2	2	?	?	2	2	1	3	5	3	6
Oriental species	0	0	0	0	0	0	0	3	2	2	3
Percent Oriental	0	0	0	0	0	0	0	100	40	67	50
Palearctic species	2	2	0	0	2	2	1	0	3	1	3
Percent Palearctic	100	100	0	0	100	100	100	0	60	33	50

At least two more species probably occur in the northern part of the country—the tetraploid *Bufo danatensis* (Fig. 6:5) and *Rana terentievi*, a *Rana ridibunda*–like frog described from southern Tadjikistan (Mezhzherin, 1992). An additional green toad, *Bufo shaartusiensis*, is known only from near the Afghanistan border in Tadjikistan (Pisanetz et al., 1996). Two other toads (*Bufo surdus* and *B. viridis zugmayeri*) are known from Baluchistan Province in Pakistan, not far from the southern border of Afghanistan.

Only one species (*Batrachuperus mustersi*) is endemic to Afghanistan. Four species (*Bufo stomaticus, B. viridis, Euphlyctis cyanophlyctis,* and *Rana ridibunda*) are shared with Iran.

The confirmed amphibian fauna consists of equal numbers of Palearctic and Oriental components. However, the former prevails throughout most of the country, except in some southern and eastern areas. Some of the eastern valleys are part of the Indus Basin. As such, they are occupied by the Oriental *Bufo stomaticus* (Fig. 6:5) and *Euphlyctis cyanophlyctis* (in the Kabul River Valley and the Khost area). Part of the area encompassing Baluchistan and Sistan is another southwestern gateway of penetration of Oriental amphibians into Afghanistan. *Bufo stomaticus* occurs in the Sistan Depression near the Iran-Afghanistan border (Fig. 6:5), and *Euphlyctis cyanophlyctis* extends into the southern Afghan Desert along the Helmand River Valley. *Paa sternosignata,* the westernmost member of this mostly Himalayan genus, occurs in southern and eastern mountainous parts of Afghanistan (Kandahar, Khost, Kabul, Paghman); this frog (as *Rana sternosignata*) was erroneously classified as a Palearctic, Central Asian species by Khan (1980). The Palearctic water frog (*Rana ridibunda*) is known from only north of the Hindukush Mountains, and green toads (*Bufo viridis*) are unknown south of approximately 31° N, where the Oriental *Euphlyctis cyanophlyctis* and *Paa sternosignata* occur (Clark, 1990).

Combining features of climate and topography, Humlum (1959) divided Afghanistan into ten "natural regions" (Table 6:5). So-called "Monsoon Afghanistan" consists of three small areas in the east (the Gumal and Kundar valleys, Khost area, and the Kabul River area) that are influenced by the Indian monsoons, as well as South Afghanistan that also has a subtropical climate; these four regions differ from Humlum's other eight regions having temperate climates. There is a good concordance between distributional patterns of the Oriental amphibians and these subtropical areas. Curiously, these areas are geologically parts of the so-called Kabul Block, which was a portion of the Indian Plate (Dercourt et al., 1986).

The Palearctic-Oriental border in southwestern Asia is sinuous. Unfortunately, the amphibian fauna in this zoogeographically important region has been poorly sampled, and taxonomic problems exist. The provisionary line between the realms (Fig. 6:5) needs to be refined by many more precise localities. Based on current knowledge, southeastern Iran and southern and easternmost Afghanistan seem to be assignable to the Oriental Realm.

Pakistan

The amphibians of Pakistan are not yet adequately known in terms of taxonomy and distribution. Apart from the basic papers by Minton (1966), who reported 11 species, and Mertens (1969), Khan (1976, 1987, 1994) provided several updated reviews based on the literature and his own data. Seventeen Pakistani species in seven genera and three families are recognized currently; these include eight toads (47%) in three species groups, one microhylid (6%), and eight ranids (47%) in seven species groups within five genera. *Rana* (*Paa*) *barmoachensis* Khan and Tasnim (1989) was synonymized with *Paa hazarensis* by Dubois (1992). *Bufo viridis*, with two subspecies described from Pakistan (*B. v. zugmayeri* Eiselt and Schmidtler [= *B. v. "arabicus"* of Khan] and *B. v. pseudoraddei* Mertens) may consist of one diploid and one tetraploid species. Recently, *Bufo siachinensis* of the *B. viridis* group has been described (Khan, 1997). The number of amphibian species in Pakistan (17; Table 6:6) is close to that of Iran (19), but the two countries share only four species (*Bufo olivaceus, B. surdus, B. viridis,* and *Euphlyctis cyanophlyctis*). Afghanistan with only six species also shares four species (*Bufo stomaticus, B. viridis, Euphlyctis cyanophlyctis,* and *Paa sternosignata*) with Pakistan. Beyond comparison, India has some 200 species of amphibians (Dutta, 1992); only 13 of these are shared with Pakistan, which has only two endemic species, *Bufo siachinensis* and *Paa hazarensis,* and one endemic subspecies, *Bufo viridis zugmayeri*.

According to Minton (1966:37), "the amphibians of … Pakistan are … derived almost equally

Table 6:6. Distribution of amphibians in Pakistan. Abbreviations for provinces are: AK = Azad Kasmir, BA = Baluchistan, GI = Gilgit (including Baltistan), NW = North West Frontier Province, PU = Punjab, SI = Sind. Abbreviations of realms are: O = Oriental, P = Palearctic. Asterisks denote species (*) or subspecies (**) endemic to the country. Sources: Dubois and Khan (1979), Dubois and Martens (1977), Eiselt and Schmidtler (1973), Khan (1976, 1987, 1994, 1997, in litt.).

Species	Province						Elevational range (m)	Realm
	BA	SI	PU	NW	GI	AK		
ANURA: BUFONIDAE:								
Bufo himalayanus	–	–	–	+	–	+	2000–3500	O
Bufo latastii	–	–	–	–	+	?	3000	P
Bufo melanostictus	–	–	+	+	–	+	1200–1400	O
Bufo olivaceus	+	–	–	–	–	–	plains	O
Bufo siachinensis*	–	–	–		+	–	5238	P
Bufo stomaticus	+	+	+	+	–	+	< 1500	O
Bufo surdus	+	–	–	–	–	–	—	P
Bufo viridis**	+	–	–	+	+	–	1500–2300	P
ANURA: MICROHYLIDAE:								
Microhyla ornata	–	–	+	+	–	+	< 1300	O
ANURA: RANIDAE:								
Euphlyctis cyanophlyctis	+	+	+	+	–	+	< 1800	O
Hoplobatrachus tigerinus	–	+	+	+	–	+	< 600	O
Limnonectes limnocharis	–	+	+	+	–	–	< 1200	O
Limnonectes syhadrensis	–	+	+	–	–	+	< 2800	O
Paa hazarensis*	–	–	–	+	–	+	1000–1500	O
Paa sternosignata	+	+	–	–	–	–	< 2100	O
Paa vicina	–	–	+	–	–	+	1100	O
Tomopterna breviceps	–	+	+	+	–	+	< 1100	O
Total Amphibia	6	7	9	10	3	10	—	17
Oriental Species	4	7	9	9	0	10	0–3500	13
Percent Oriental	67	100	100	90	0	100	—	76
Palearctic species	2	0	0	1	3	?	1500–5200	4
Percent Palearctic	33	0	0	10	100	?	—	24

from Palearctic and Oriental sources." However, the majority (76%) of the Pakistani amphibians is of Oriental origin (Table 6:6). These include the Indian *Tomopterna breviceps*, the Himalayan genus *Paa*, and widespread Indo-Malayan species such as *Bufo melanostictus, Microhyla ornata, Euphlyctis cyanophlyctis, Limnonectes limnocharis* complex, and *Hoplobatrachus tigerinus*. The Palearctic component (24%) consists only of four species of the green toad group (*Bufo latastii, B. siachinensis, B. surdus,* and *B. viridis*). Of these, *B. siachinensis* was discovered recently at an elevation of 5238 m at the foot of the Siachin Glacier in the Karakoram Mountain Range in Baltistan (Khan, 1997). This seems to be the world record for the highest distribution of a species of amphibian.

Taking into account the general topography of midwestern Asia, regional limits of the Palearctic seem to approximate the edge of the great Iranian Plateau. Khan (1980) drew the Palearctic-Oriental

dividing line passing west of the Kirthar Range and through Sulaiman and the Chitral Hills. Thus, most of Baluchistan, the western strip of Punjab, and the southwestern and western border of the North West Frontier Province lie in the Palearctic Realm. With some modifications, I support Khan's scheme.

Therefore, Pakistan, except for a few mountainous areas in Baluchistan, North West Frontier Province, and Gilgit (Fig. 6:5), should be considered as the northwestern part of the Oriental (Indo-Malayan) Realm. Unlike the Palearctic species, Oriental amphibians in Pakistan mostly inhabit lowland plains and foothills (< 1500–2000 m) rather than high mountains; an exception is *Bufo himalayanus* of the *Bufo melanostictus* group (Table 6:6).

Northwestern India

Some Palearctic species have infiltrated into northwestern India. Based on the distribution of insects and other animals, Mani (1974) recognized northwest Himalaya and the higher Himalaya in the east as two secondary centers of independent faunal differentiation; these mountains are wholly within the Palearctic. Kashmir is rich in Palearctic animals (Das, 1966). Sahi and Duda (1986) listed 12 species of amphibians from the State of Jammu and Kashmir, which incorporates the junction of the Palearctic and Oriental realms. Ten anurans include one microhylid (*Microhyla ornata*), one megophryine pelobatid (*Scutiger occidentalis*), four toads (*Bufo latastii, B. melanostictus, B. stomaticus,* and *B. viridis*), and four ranids (*Euphlyctis cyanophlyctis, Hoplobatrachus tigerinus, Limnonectes limnocharis,* and "*Rana sternosignata*" (? = *Paa vicina*). All of these species, except *Scutiger occidentalis,* also occur in Pakistan. The records of two salamanders (*Hynobius chinensis* and *Tylototriton verrucosus*) in Kashmir need confirmation. Dubois (1981) listed 11 anurans from the state, and Dutta (1992) mentioned one more species (*Scutiger nyingchiensis*). The anuran fauna of Jammu and Kashmir is nearly all Oriental, except for the *Bufo viridis* group occurring in the northern Himalayan area. Outlining the Palearctic elements in Kashmir, Das (1966) mentioned, among the amphibians, *Limnonectes limnocharis,* "*Rana pleskei,*" *Microhyla ornata, Bufo latastii,* and *Bufo viridis.* However, except for the *Bufo,* these species are obviously of Oriental origin. Sahi and Duda (1986) suspected *Bufo stomaticus* to be a Palearctic element, as did Minton (1966) and Khan (1980) for this species (as *B. andersoni*) and *B. olivaceus* in Pakistan. However, taxonomic relationships and geographical distributions of these toads are contrary to such a suggestion. Moreover, *Bufo stomaticus* is known only from elevations below 1000 m; that is normal for many Oriental species, as opposed to Palearctic anurans, which occur only in the highlands in the region.

Dubois (1981) assigned the Ladakh area northeast of Kashmir to the Tibetan zoogeographical region. Only two amphibian species are known from Ladakh—the endemic *Scutiger occidentalis* and *Bufo latastii.* The latter is known from elevations of 1600–3300 m in the Ladakh Range and in Kashmir. The records of "*Bufo viridis*" from Jammu and Kashmir mentioned by many authors seem to be misidentifications of *B. latastii* and perhaps *B. stomaticus* (Dubois and Martens, 1977; Hemmer et al., 1978). Relationships of *B. latastii* and *B. siachinensis,* recently described from Baltistan, in the north-

ern part of the State of Jammu and Kasmir (under the administration of Pakistan) need to be resolved. Based on the depauperate amphibian fauna, Ladakh could be treated as a Palearctic outpost in the western Himalaya or as a transition zone between the Palearctic and Oriental realms.

Northwestern Nepal

Another area of the western Himalaya named "the Northwest Nepal" was also recognized as a part of the Tibetan zoogeographical region by Dubois (1981). Unlike the Ladakh Range, this area is characterized by a somewhat richer amphibian fauna of seven species—*Scutiger alticola, Bufo himalayanus, Amolops formosus, Nanorana (Altirana) parkeri, Paa liebigii, P. polunini,* and *P. rostandi.* The list includes no true Palearctic species. Therefore, this local fauna cannot be considered as part of the Palearctic Realm.

SOUTHEASTERN BOUNDARY

The southeastern demarcation of the Palearctic Realm is a headache for biogeographers. Southern China is saturated with Oriental taxa. Some have penetrated quite far north; for example, the microhylid *Kaloula borealis* extends northward into Heilongjiang Province (Zhao and Adler, 1993). Contrariwise, some Palearctic groups are widespread in southern China. Some members of the Palearctic *Rana temporaria* and *Rana esculenta* groups are widely distributed in southern China. For example, *Rana japonica* penetrates into northern Vietnam, and *Rana lateralis* occurs in Burma, Thailand, and Cambodia (Dubois and Ohler, 1995a). In southwestern China, most species with Palearctic affinities inhabit mountains, whereas species with Oriental affinities inhabit the subtropical and tropical lowlands. Therefore, southwestern China, as well as parts of central and eastern China, with various combinations of the Palearctic and Oriental taxa, represent a broad transition zone between the Palearctic and Oriental realms, where the faunal replacement may be influenced by zonal changes in climate (Darlington, 1957). The distinction between the realms in this region has been the subject of a lengthy debate (e.g., P. Müller, 1986). Four main boundary lines have been suggested.

1. The "Amur Line" separates the Palearctic from the territory south of the Amur River referred to as the Amur (or Manchurian) Transition Region (Amur River Basin and Sakhalin Island, northeast-

ern China, Korean Peninsula, and main Japanese Islands) and the Sino-Indian Region (the rest of China, Ryukyu Islands, and farther south and east). This scheme is accepted for freshwater fish (e.g., Heptner, 1936). Most modern authors (e.g., Bănărescu and Boşcaiu, 1978) are inclined to include the Amur Region in the Sino-Indian Region.

2. The "Qinling [Tsinling Mountains] Line" passes along the mountain ranges of central China from the northern part of Hengduanshan Mountains in the west through the Tsinling Mountains, Funiu Shan, Dabie Shan, and along about 31° N eastward to the coast. This is the demarcation line accepted by Chinese zoogeographers (Zhao and Adler, 1993; Zhao, this volume).

3. The "Yangtze Line" is slightly south of the middle and lower Yangtze River (e.g., Heptner, 1936; Shcherbak, 1982).

4. The "South China Line" includes almost all of China (except the southernmost area and Hainan Island) in the Palearctic Realm. In this case, the southern border of the Palearctic almost coincides with the Tropic of Cancer (without Taiwan) or passes slightly to the south of it (e.g., Udvardy, 1975) and encompasses Taiwan (e.g., Emeljanov, 1974; Takhtajan, 1978).

Taking into consideration a balance between Palearctic and Oriental faunal components, I accept the Qinling Line. Thus, several zoogeographic divisions of China, where Oriental genera and species predominate, namely the Southwest China, Central China, and South China regions (Zhao and Adler, 1993; Zhao, this volume) are part of the Oriental Realm. All Japanese islands north of the Ryukyu Archipelago, the Tokara Tectonic Strait, are part of the Palearctic (Hikida et al., 1992; Fig. 6:1).

THE AMPHIBIAN FAUNA

The amphibian fauna of the Palearctic Realm is composed of 17 genera and 69 species of salamanders and 17 genera and 106 species of anurans (Table 6:7; Appendix 6:1); caecilians are absent. Among salamanders, hynobiids are characteristic of the Asian part of the realm, whereas salamandrids are dominant in the western part. Two relictual salamanders, the European *Proteus* and the Far Eastern *Andrias* (the latter with an extensive fossil record in Europe, Central Asia, and North America), as well as the Mediterranean *Hydromantes* with Nearctic affinities, round out the diverse salamander fauna. Among the eight families of anurans, some Palearctic groups are especially noteworthy. These are the spadefoots (*Pelobates*) and parsley frogs (*Pelodytes*) distributed in the western part of the realm, the so-called gray or common toads (*Bufo bufo* group), green toads (*Bufo viridis* group), treefrogs (*Hyla arborea* group), and last, but not least, the brown or grass frogs (*Rana temporaria* group) and green or water frogs (*Rana esculenta* group). It is not possible to envision the Palearctic fauna without these animals.

Although the Palearctic Realm contains a high percentage of the genera and species of several families, viz., Hynobiidae, Salamandridae, Discoglossidae, and Pelodytidae, this vast realm harbors only about 18% of the species of salamanders and 3% of the species of anurans in the world (Table 6:8).

TAXONOMIC BACKGROUND

Although some of the amphibians of the western Palearctic have been known from the time of Linnaeus, amphibian diversity in central and eastern Asia is only now being discovered adequately. Furthermore, intensive studies utilizing modern techniques of sound analysis, karyology, and biochemistry have revealed the existence of many cryptic species, even in the well-known European fauna

Table 6:7. Taxonomic composition of the amphibian fauna of the Palearctic Realm.

Family	Genera		Species	
	Number	Percent	Number	Percent
Cryptobranchidae	1	2.9	2	1.1
Hynobiidae	5	14.7	27	15.4
Plethodontidae	1	2.9	7	4.0
Proteidae	1	2.9	1	0.6
Salamandridae	9	26.5	32	18.3
Total Caudata	17	50.0	69	39.4
Bufonidae	1	2.9	24	13.7
Discoglossidae	3	8.8	13	7.4
Hylidae	1	2.9	7	4.0
Microhylidae	1	2.9	1	0.6
Pelobatidae	2	5.9	9	5.1
Pelodytidae	1	2.9	2	1.1
Ranidae	6	17.6	47	26.9
Rhacophoridae	2	5.9	3	1.7
Total Anura:	17	50.0	106	60.6
Total Amphibia	34	100.0	175	100.0

Table 6:8. Percentage of amphibians in the Palearctic Realm in relation to world fauna; numbers by families calculated from Duellman (1993: Table 1).

Family	World fauna (n)		Percent Palearctic	
	Genera	Species	Genera	Species
Cryptobranchidae	2	3	50	67
Hynobiidae	7	36	71	75
Plethodontidae	25	232	4	3
Proteidae	2	6	50	17
Salamandridae	15	55	60	58
Total Caudata	61	392	28	18
Bufonidae	31	365	3	6
Discoglossidae	4	17	75	76
Hylidae	39	719	2	1
Microhylidae	64	313	1	< 1
Pelobatidae	9	80	22	11
Pelodytidae	1	2	100	100
Ranidae	44	625	14	7
Rhacophoridae	10	204	20	1
Total Anura:	334	3967	5	3
Total Anura + Caudata	395	4360	9	4
Total Amphibia*	428	4523	8	4

* Including Gymnophiona.

(Veith, 1996a). Thus, there has been a resurgence in amphibian taxonomy and a recent increase in the number of species recognized (Fig. 6:6). Herein, I provide a brief review of the current taxonomy of Palearctic amphibians.

Cryptobranchidae

The Cryptobranchidae contains two Recent genera distributed in Eastern Asia (*Andrias*) and North America (*Cryptobranchus*), respectively. Both species of *Andrias* occur in the Palearctic, either in China or Japan (Zhao, this volume).

Hynobiidae

Four genera and six species are currently known from the territory covered by this chapter. The taxonomy of *Salamandrella keyserlingii* was reviewed by Borkin (1994), and Kuzmin (1995a) compiled data on the Far Eastern genus *Onychodactylus*. Steiner (1973) and Reilly (1983) reported on the distribution and natural history of two species of *Batrachuperus* in Iran and Afghanistan, respectively. Brushko et al. (1988) outlined the range of *Ranodon sibiricus* in Kazakhstan. Works by Thorn (1969) and Zhao et al. (1988) provided additional information on these salamanders. *Turanomolge mensbieri,* described from the former Soviet Central Asia, was synonymized with the *Triturus cristatus* complex, and the status of *Hynobius turkestanicus* remains enigmatic (Kuzmin et al., 1995).

The Hynobiidae was divided into two "natural groups" by Zhao et al. (1988)—the *Hynobius* group containing *Hynobius* (including *Satobius*), *Pachypalaminus,* and *Salamandrella,* and the *Ranodon* group containing *Batrachuperus, Liua, Onychodactylus,* and *Ranodon.* Both groups are represented in the Palearctic. Using allozyme data, Matsui et al. (1992) reexamined the generic status of some members of the *Hynobius* group. (Also see Matsui, 1987; Nishio et al., 1987; Adler and Zhao, 1990; and Kohno et al., 1991.) Currently, five groups at subgeneric and/or generic levels are recognized tentatively. (See Appendix 6:1.)

Plethodontidae

Seven (or eight) species of this New World family are currently recognized in southern Europe (Lan-

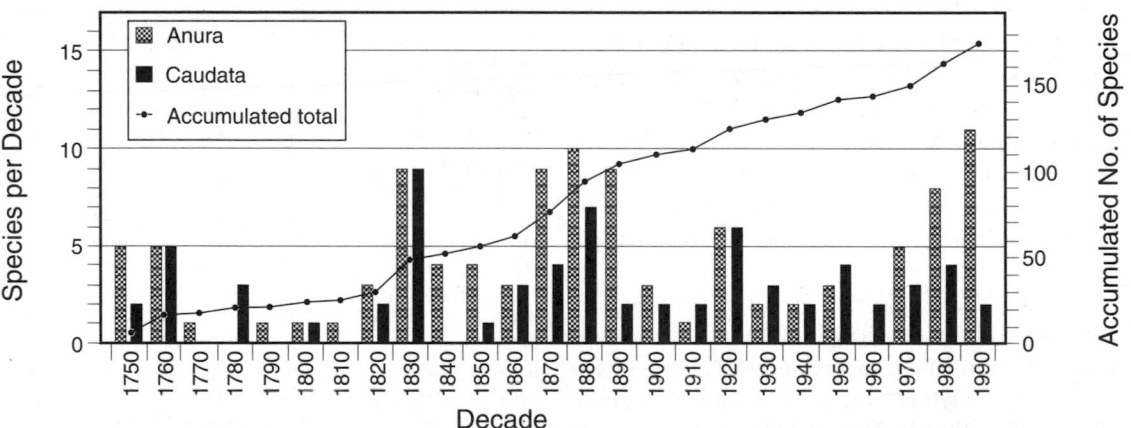

Fig. 6:6. Rates of discovery of species of anurans and salamanders and accumulated number of species in the Palearctic Realm.

za et al., 1995; Nascetti et al., 1996, and references cited therein). Lanza and Vanni (1981) split *Hydromantes* sensu lato into two genera with European (West Mediterranean) and North American species, respectively. This taxonomic arrangement and the associated nomenclatural problems of these salamanders (*Hydromantes Hydromantoides, Speleomantes*) has been debated. The name *Hydromantes* has been approved by the International Commission on Zoological Nomenclature, and the name *Speleomantes* is available for the European species if they are recognized as generically distinct (Anonymous, 1997; Lanza et al., 1995).

Proteidae

Currently two subspecies of the European paedogenetic troglobitic salamander are recognized; the black and "eyed" subspecies (*Proteus anguinus parkelj*) was described by Sket and Arntzen (1994), and its status was discussed by Grillitsch and Tiedemann (1994) and Arntzen and Sket (1997). The only other living representative of the family is *Necturus* in the Nearctic.

Salamandridae

Besides the western Palearctic, the range of this family includes the Nearctic and the northern part of the Oriental realms. Nine of the 14 genera of living salamandrids occur in the Palearctic Realm. Thorn (1969) reviewed all species and genera in the Palearctic. Genera have been clustered into three groups: (1) *Salamandra* group containing *Chioglossa, Mertensiella, Salamandra,* and *Salamandrina;* (2) *Triturus* group containing *Euproctus, Neurergus, Triturus, Cynops,* and other East Asian and North American genera; and (3) *Pleurodeles* and *Tylototriton* (Wake and Özeti, 1969; Estes, 1981; Hayashi and Matsui, 1989).

Comprehensive evolutionary analysis of the genus *Triturus* by Macgregor et al. (1990) resulted in the recognition of 12 species in two subgenera: *Triturus* and *Palaeotriton* (also, see Arntzen, 1995). The status of four members of the *Triturus cristatus* complex (*T. carnifex, cristatus, dobrogicus* and *karelinii*) of the subgenus *Triturus* also was corroborated by Litvinchuk et al. (1994). The closely related genera *Mertensiella* and *Salamandra* were reviewed by Klewen (1991) and Greven and Thiesmeier (1994). Unlike *Mertensiella* with two species, recent, mostly biochemical, studies on *Salamandra* revealed five, instead of two, species (Nascetti et al., 1988; Grossenbacher, 1994; Joger and Steinfartz, 1994, 1995; Veith, 1994, 1996b). Schmidtler (1994) reviewed new data on four allopatric species of the genus *Neurergus* in southwestern Asia. Three species of the genus *Euproctus* were assigned to two phyletic lineages—*E. montanus* and *E. platycephalus* on the western Mediterranean islands of Corsica and Sardinia that are closely related and represent a distinct cladogenetic event; the other lineage including the Pyrenean *E. asper* seems to be quite old and probably separated from the lineage to *Triturus* by a relatively short time span (Sbordoni et al., 1990; Caccone et al., 1994).

Bufonidae

Toads have a nearly cosmopolitan distribution. In the Palearctic, this family is represented only by *Bufo,* which has an intercontinental distribution and an enormous number of species. Six species groups currently recognized in the Palearctic (Inger, 1972; Tandy and Keith, 1972) contain 24 species. Two of these, the *Bufo bufo* and *Bufo viridis* groups containing 19 species (79% of Palearctic *Bufo*), occupy most of the Palearctic. Three groups have Afrotropical affinities; these are the *Bufo mauritanicus* group (1 species), *B. regularis* group (1), and *B. funereus* group (1); the *B. stomaticus* group (2 species) has Oriental affinities. These four groups occur only in peripheral parts of the Palearctic.

During the past two decades, the taxonomy of the gray toads making up the *Bufo bufo* group, has changed markedly, especially among the Far Eastern taxa. Many geographic races were elevated to full species, and some new taxa have been described (Matsui, 1980, 1984, 1986; Borkin and Roshchin, 1981; Borkin, 1984; Borkin and Matsui, 1986; Zhao and Adler, 1993). Among the western Palearctic members, the Caucasian toads seem to be a species distinct from the European *Bufo bufo* (Borkin, 1987). However, the recognition of three subspecies of *B. verrucosissimus* by Orlova and Tuniyev (1989), as well as the relationships with the Mediterranean *B. bufo spinosus,* needs further examination. According to present taxonomy, the *Bufo bufo* group contains nine species with distributions in the Palearctic.

Unlike the *Bufo bufo* group, the green toads of the *Bufo viridis* group, with ten species, primarily inhabit open landscapes rather than forested areas.

However, a thorough revision of this group is greatly needed. For example, the taxonomic position of many populations in arid parts of the Palearctic is unclear. Among Eurasian amphibians, this group displays a unique case of polyploid speciation. The tetraploid toads, arbitrarily united under the name *B. danatensis* (some senior synonyms available), range throughout the vast territory of Central Asia between the Caspian Sea and western Mongolia (Borkin et al., 1986; Borkin and Kuzmin, 1988). The distribution and taxonomy of the *Bufo viridis* group is especially interesting because these animals can be considered as an indicator of southern limits of the Palearctic in arid regions, as can the water frogs of the *Rana esculenta* group.

Discoglossidae

Traditionally, four genera have been placed in this family. However, these genera seem to belong to two monophyletic groups, which sometimes are recognized as distinct subfamilies or families—Bombinatoridae (= Bombinidae *auctorum*) and Discoglossidae (sensu stricto). Various authors have proposed different arrangements—*Alytes* + *Bombina* vs. *Discoglossus* (Lanza et al., 1976), *Alytes* + *Bombina* + *Barbourula* vs. *Discoglossus* (Dubois, 1987a), *Alytes* vs. *Discoglossus* + *Barbourula* + *Bombina* (Clarke in Evans et al., 1990), and *Bombina* + *Barbourula* vs. *Alytes* + *Discoglossus* (Ford and Cannatella, 1993). Immunological distances suggest the Cretaceous separation for *Alytes, Bombina,* and *Discoglossus,* with the latter two genera sharing a common lineage after *Alytes* diverged (Maxson and Szymura, 1984). Some fossil evidence indirectly suggests that the *Alytes* lineage and the lineage to *Discoglossus* and other genera had differentiated by 170 million years ago, i.e., in the Middle Jurassic (Evans et al., 1990). Dubois (1987c) discussed some nomenclatural aspects. Arntzen and García-París (1995) revised *Alytes* (including *Baleaphryne*), and described a new species and subspecies. The four species in this western Palearctic genus can be allocated to two (*Alytes* and *Ammoryctis*) or three (*Alytes* [*A. obstetricans*], *Ammoryctis* [*A. cisternasii*], and *Baleaphryne* [*A. muletensis* and *A. dickhilleni*]) subgenera.

Morphologically and karyologically, the genus *Bombina* can be divided into two subgenera (Tian and Hu, 1985; Dubois, 1987b; Liu and Yang, 1993). Each has a different pattern of distribution. The nom-inotypical subgenus (with three European species [*B. bombina, B. variegata,* and *B. pachypus* {Lanza and Corti, 1996}] and the Far Eastern *B. orientalis*) is endemic to the Palearctic. The other subgenus (*Grobina* [= *Glandula*] also with three species [*B. fortinuptialis, maxima,* and *microdeladigitora*]) is endemic to the northern part of the Oriental Realm (South China and northern Vietnam). There are two gaps in distribution of the genus. The first one exists among members of the subgenus *Bombina;* both European species and *B. orientalis* portray a classic case of the so-called European–Far Eastern disjunction (Borkin, 1984, 1986). Based on immunological data, the time of divergence between these Palearctic lineages was estimated to be in the late Miocene (Maxson and Szymura, 1984). Indeed, the earliest European record of *Bombina* is from the lower Miocene of Germany (Sanchiz and Schleich, 1986). Nevertheless, fertile hybrids have been obtained from laboratory crosses of both European species with *B. orientalis* (Uteshev and Borkin, 1985). The second distributional gap is between the two subgenera, i.e., between *B. (B.) orientalis* and the other Chinese species in the subgenus *Grobina.*

Biochemical (allozyme), bioacoustic, and morphometric data revealed new species and subspecies of *Discoglossus.* At the present time, this genus consists of four species distributed in the western part of the Mediterranean area (Lanza et al., 1984, 1986; Capula et al., 1985; Busack, 1986a; Capula and Corti, 1993; Fritz et al., 1994; Vences and Glaw, 1996) and an enigmatic Israeli species, *D. nigriventer,* which may be extinct (Werner, 1988, and references cited therein).

Three of the four extant genera of the Discoglossidae are distributed in the Palearctic or have Palearctic affinities, such as the subgenus *Grobina* in southern China. The only genus that is extralimital is *Barbourula* with one species in the Philippines and one in Borneo. Phenetically, *Barbourula* is most similar to *Bombina* (Clarke, 1987). Indeed, the family has a Laurasian origin that is well documented by Mesozoic remains from the Northern Hemisphere and even from India. Further resolution of relationships among discoglossids must await additional data based on study of both living and extinct taxa.

Hylidae

Treefrogs of the genus *Hyla* are widely distributed in temperate Eurasia, Mediterranean northwest

Africa, and the New World. Only 14 species (about 5% of those in the genus) are known from Eurasia. Six species are endemic to the Palearctic and seven to the northern part of the Oriental Realm (Zhao, this volume); *Hyla japonica* occurs in both realms. A few of the species are morphologically cryptic; their distinctiveness was revealed by advertisement calls, hybridization studies, or analysis of allozymes (H. Schneider, 1974; Kuramoto, 1980, 1984; Nascetti et al., 1995).

Most authors have thought that the Eurasian species formed a monophyletic *Hyla arborea* group derived from a Nearctic stock. However, I found no comprehensive analysis of the entire group. Geographically, Eurasian treefrogs belong to two assemblages; six species are in the western Palearctic, and eight species are in the eastern Palearctic and the northern periphery of the Oriental Realm. Arid Central Asia and Siberia lack hylids (Borkin, 1984, 1986); maps in Duellman and Trueb (1986), K. Anderson (1991), and some other authors are partly incorrect. Using karyological data, K. Anderson (1991) divided most species of Holarctic *Hyla* into two groups. Among Eurasian treefrogs, two Far Eastern species—the widespread *H. japonica* and the Korean sibling species *H. suweonensis*—were included in one group, and the other five Eurasian species studied (*H. arborea, chinensis, hallowelli, meridionalis,* and *savignyi*) were placed in a second group; both groups contain North American species. These data suggest at least two lineages and at least two invasions into Eurasia. As noted by K. Anderson (1991), her findings contradict some biochemical data provided by Nishioka et al. (1990). In light of these findings, it is obvious that the *Hyla arborea* group needs further study.

Microhylidae

In Eurasia, this family is typical of the Oriental Realm. Only one species, *Kaloula borealis,* occurs in the Chinese and Korean areas of the Palearctic (Zhao, this volume). Surprisingly, microhylid remains probably are known from the late Eocene of Europe (Sanchiz and Ro ek, 1996).

Pelobatidae

In the western Palearctic, this family is represented by the pelobatine *Pelobates* with four species. The North American lineage consisting of *Scaphiopus* and *Spea* is sometimes placed in its own subfamily or even family (e.g., Dubois, 1987a). The distinctive Megophryinae, which is recognized as two subfamilies (Tian and Hu, 1985; Dubois, 1987a) or a different family (Ford and Cannatella, 1993; Lathrop, 1997), contains a fascinating diversity of taxa, and certainly is an Oriental group. Only one genus (*Scutiger*) extends into the West China portion of the Palearctic (Inger, this volume; Zhao, this volume).

Pelodytidae

The family consists of only one living genus, *Pelodytes,* plus one or two fossil genera (Borkin and Anissimova, 1987; Ford and Cannatella, 1993; Henrici, 1994; and references cited therein). Two species of *Pelodytes* have disjunct distributions in the western Palearctic, namely in the Caucasus Mountains and in western Europe. However, the Oligocene-Miocene records of *Miopelodytes* and *Tephrodytes* from North America are evidence of a former broader Laurasian pattern.

Ranidae

During the last decade, the taxonomy of this almost cosmopolitan family has been transforming radically (compare lists in Frost, 1985, and Duellman, 1993). Most changes were introduced by Dubois, and his last version (Dubois, 1992) is used, in part, herein.

Among the six genera currently recognized in the Palearctic, four (*Euphlyctis* [1 species], *Limnonectes* [1], *Paa* [1], and *Nanorana* [2]) are Oriental with peripheral distributions in the southeastern Palearctic (Inger, this volume; Zhao, this volume). The same is true for *Ptychadena mascareniensis,* an Afrotropical component in the Nile Valley, Egypt. However, the genus *Rana* (sensu stricto) is characteristic of the Palearctic.

The water frogs or *Rana esculenta* group (recognized as the subgenus *Pelophylax* by Dubois [1992]) provide an example of the rapidly growing number of species recently discovered by means of various techniques; many new species have been described (some names are preoccupied). Dubois and Ohler (1995a, 1995b) recognized 26 taxa (species and subspecies). The distributions of 19 species in the *Rana esculenta* group are entirely or partly in the Palearctic. The group is known from the lower Oligocene of Germany (Sanchiz et al., 1993). Green frogs in the western Palearctic exhibit an unusual mode of speciation characterized by complicated genetic mechanisms (hybridization, so-called hemi-

Table 6:9. Zoogeographic composition of the amphibian fauna in the Palearctic Realm. Numbers in parentheses are percents.

Family	Palearctic		Afrotropical		Oriental		Nearctic	
	Genera	Species	Genera	Species	Genera	Species	Genera	Species
Cryptobranchidae	1 (100)	2 (100)	–	–	–	–	–	–
Hynobiidae	5 (100)	27 (100)	–	–	–	–	–	–
Plethodontidae	–	7 (100)	–	–	–	–	1 (100)	–
Proteidae	1 (100)	1 (100)	–	–	–	–	–	–
Salamandridae	8 (89)	31 (97)	–	–	1 (11)	1 (3)	–	–
Total Caudata	15 (88)	68 (99)	–	–	1 (6)	1 (1)	1 (6)	–
Bufonidae*	2 (33)	19 (79)	3 (50)	3 (13)	1 (17)	2 (8)	–	–
Discoglossidae	3 (100)	13 (100)	–	–	–	–	–	–
Hylidae	–	7 (100)	–	–	–	–	1? (100?)	–
Microhylidae	–	–	–	–	1 (100)	1 (100)	–	–
Pelobatidae	1 (50)	4 (44)	–	–	1 (50)	5 (56)	–	–
Pelodytidae	1 (100)	2 (100)	–	–	–	–	–	–
Ranidae	1 (17)	41 (87)	1 (17)	1 (2)	4 (67)	5 (11)	–	–
Rhacophoridae	–	–	–	–	2 (100)	3 (100)	–	–
Total Anura	6 (35)	85 (80)	1 (6)	4 (4)	8 (47)	16 (16)	1? (6?)	–
Total Amphibia	21 (65)	153 (87)	1 (3)	4 (2)	9 (26)	17 (10)	1 (3)	–

* Bufonidae, which is represented by one heterogeneous genus, is analyzed at the species group level.

clonal inheritance, polyploidy, unisexual and bisexual population systems (Graf and Polls Pelaz, 1989; Günther, 1990; Vinogradov et al., 1990). Three taxa are of hybrid origin—*R. esculenta* of central Europe, *R. grafi* of southwestern Europe, and *R. hispanica* of Italy; these taxa have been designated as kleptons by Dubois and Günther (1982) and Crochet et al. (1995). Another hybridogenic complex (*R. saharica*, with triploids) distributed in North Africa was suggested by Hemmer et al. (1980); however, some recent data do not support that condition (Buckley et al., 1995). Some sibling *Rana ridibunda*–like species were detected bioacoustically and/or electrophoretically in the eastern Mediterranean area (H. Schneider et al., 1984, 1992, 1993; Hotz et al., 1987; Beerli, 1994; Beerli et al., 1994; Sinsch and Eblenkamp, 1994) and in Central Asia (Mezhzherin, 1992).

The brown (or grass) frogs of the *Rana temporaria* group form the nominotypical subgenus *Rana*. Dubois (1992) placed the 27 species in six groups diagnosed by diploid chromosome number (24 or 26), larval denticulation, and egg size; these groupings seem to be arbitrary. For example, based on two karyological characters, the brown frogs could be divided into seven groups. Allopatric species seem to be more similar to each other than to sympatric species, and the latter were placed in different groups (Orlova et al., 1978). Mensi et al. (1992) reviewed the taxonomy of brown frogs in Europe and Asia Minor. Nishioka et al. (1992), and Green and Bor-

kin (1993) discussed evolutionary relationships (based on allozymes and karyotypes) of 12 species, mainly from the eastern Palearctic. Additional taxonomic revisions of brown frogs of some regions (e.g., *R. chensinensis* complex, *R. camerani-macrocnemis* complex, *R. arvalis-altaica* complex) are needed.

Rhacophoridae

Only a few rhacophorids have penetrated the Palearctic Realm, where the family is represented by two genera (the primitive *Buergeria* with one species, and *Rhacophorus* with two species); all three species are endemic to Japan (Wilkinson et al., 1996; Zhao, this volume). The genera belong to different tribes or subfamilies (Channing, 1989; Dubois, 1992). At least three hypotheses exist concerning the regional origins of the family—Oriental Region (Liem, 1970; Laurent, 1975), the African part of Gondwana (Savage, 1973; Duellman and Trueb, 1986), and Pangaea (Channing, 1989).

ZOOGEOGRAPHIC COMPOSITION

Zoogeographically, amphibians of the Palearctic may be divided into four main faunal categories (Table 6:9). Of course, the Palearctic faunal element prevails (65% at the generic level and 87% of the species), followed by the derivatives of Oriental groups (26% and 10%, respectively). The Afrotropical (3% and 2%) and Nearctic components (3% of genera) are minor components. Species of salamanders are less numerous than those of anurans;

Table 6:10. Endemism among amphibians in the Palearctic Realm.

Family	Endemic genera		Endemic species	
	Number	Percent	Number	Percent
Cryptobranchidae	–	–	1	50
Hynobiidae	2	40	25	93
Plethodontidae	1?	100?	7	100
Proteidae	1	100	1	100
Salamandridae	8	89	32	100
Total Caudata	11	65	66	96
Bufonidae	–	–	13	54
Discoglossidae	2	67	13	100
Hylidae	–		6	86
Microhylidae	–	–	–	–
Pelobatidae	1	50	7	78
Pelodytidae	1	100	2	100
Ranidae	–	–	37	79
Rhacophoridae	–	–	3	100
Total Anura:	4	24	79	76
Total Amphibia	15	44	145	84

the Palearctic component makes up the vast majority of genera (88%) and species (99%) of salamanders. Among anurans, Oriental genera predominate over Palearctic genera (47% vs. 35%); however, the Palearctic species are more numerous than Oriental species (80% vs. 16%). Thus, although many Oriental genera extend into the Palearctic, they are represented there by few species. The percentage of Palearctic and Oriental genera and species varies considerably by family (Table 6:9).

Numbers and percentages of amphibians endemic to the Palearctic are given in Tables 6:10 and 6:11. Endemism in salamander genera and species is higher than in anurans. In four of the five families of salamanders, more than 90% of the species are endemic. Among anurans, five of the eight families have no endemic genera. Apart from microhylids, the proportion of endemic species varies at the family level from 56% in the Bufonidae to 100% in the Discoglossidae, Pelodytidae, and Rhacophoridae.

Among the 13 families of amphibians, only one (8%), the Pelodytidae, is endemic to the Palearctic. Among the 34 genera, 15 (44%) are endemic to the Palearctic (Table 6:11); this includes 11 (32%) endemic genera of salamanders but only four (12%) endemic genera of anurans. Six families inhabiting the Palearctic have no endemic genera (Table 6:10). Generic endemism in the Plethodontidae depends on recognition of the European species in the genus *Hydromantes* (also in the Nearctic) or separate gen-

era. Nonendemic genera have high endemism at the species level. In characteristic Palearctic genera and species groups, the proportion of endemic species is 50–90%. Some Oriental genera (e.g., *Buergeria* and *Rhacophorus*) also are represented by species endemic to the Palearctic (Table 6:11).

HISTORICAL SOURCES, DISPERSAL, AND RELATIONSHIPS WITH OTHER REALMS

Savage (1973), Laurent (1975), Milner (1983), and Duellman and Trueb (1986) provided important biogeographic analyses of historical distributions of major groups of amphibians (mainly at the familial level) based on changing configuration of land masses and climates. Although the composition of the Recent fauna of the Palearctic obviously is fairly heterogeneous, it is composed largely of derivatives from Laurasian groups. Indeed, the origin of eight of the 13 families is suggested to be associated with Laurasia, the northern supercontinent that was disrupted during the early Cretaceous and evolved finally into Eurasia and North America. (For details, see above-mentioned references and Rage, 1995.) These include all five families of salamanders and three families of anurans—Discoglossidae, Pelobatidae, and Pelodytidae. Five other anuran families (Bufonidae, Hylidae, Microhylidae, Ranidae, and Rhacophoridae) seem to represent derivatives of Gondwana, the southern supercontinent fragmented into land masses in the Southern Hemisphere (South America, Africa, Madagascar, India, Australia, and Antarctica).

Paleogeographically, the Palearctic fauna has been derived from several sources and directions; these are tentatively classified as follow:

Autochthonous

These include at least five families that are derivatives of Laurasian stocks (see also under Africa, below).

Hynobiidae: A suggestion of East Asian origin of the family is generally accepted. Hynobiids may have appeared in the middle Jurassic (Milner, 1983) or early Cretaceous (Duellman and Trueb, 1986). Laurent's (1975) speculation of a Triassic origin and former widespread Holarctic distribution has no support.

Cryptobranchidae: This family possibly originated in the middle Jurassic. In the past, it was widespread in Europe, Asia, and North America, but most

Table 6:11. Numbers and percentages of all and endemic species in amphibian genera in the Palearctic Realm and their relation to the total numbers in those genera. Asterisks (*) denote genera endemic to the realm.

Genus or species group	Total world species	Palearctic species		Palearctic endemics[1]		Percent world endemics[2]
		n	%	n	%	
CAUDATA: CRYPTOBRANCHIDAE:						
Andrias	2	2	100	1	50	50
CAUDATA: HYNOBIIDAE:						
Batrachuperus	7	5	71	3	60	43
Hynobius	22	18	82	18	100	282
Onychodactylus*	2	2	100	2	100	100
Ranodon	2	1	50	1	100	50
Salamandrella*	1	1	100	1	100	100
CAUDATA: PLETHODONTIDAE:						
Hydromantes	10	7	70	7	100	70
CAUDATA: PROTEIDAE:						
Proteus*	1	1	100	1	100	100
CAUDATA: SALAMANDRIDAE:						
Chioglossa*	1	1	100	1	100	100
Cynops	8	1	12	1	100	12
Euproctus*	3	3	100	3	100	100
Mertensiella*	2	2	100	2	100	100
Neurergus*	4	4	100	4	100	100
Pleurodeles*	2	2	100	2	100	100
Salamandra*	6	6	100	6	100	100
Salamandrina*	1	1	100	1	100	100
Triturus*	12	12	100	12	100	100
ANURA: BUFONIDAE:						
Bufo	211	24	11	13	54	6
Bufo bufo group	11	9	82	5	56	45
Bufo funereus group	6	1	17	1	100	17
Bufo mauritanicus group	1	1	100	–	–	–
Bufo regularis group	9	1	11	–	–	–
Bufo stomaticus group	6	2	33	–	–	–
Bufo viridis group	10	10	100	7	70	70
ANURA: DISCOGLOSSIDAE:						
Alytes*	4	4	100	4	100	100
Bombina	7	4	57	4	100	57
Discoglossus*	5	5	100	5	100	100
ANURA: HYLIDAE:						
Hyla	281	7	2	6	86	2
ANURA: MICROHYLIDAE:						
Kaloula	9	1	11	–	–	–
ANURA: PELOBATIDAE:						
Pelobates*	4	4	100	4	100	100
Scutiger	29	5	17	3	60	10
ANURA: PELODYTIDAE:						
Pelodytes*	2	2	100	2	100	100
ANURA: RANIDAE:						
Euphlyctis	4	1	25	–	–	–
Limnonectes	62	1	2	–	–	–
Nanorana	3	2	67	–	–	–
Paa	26	1	4	–	–	–
Ptychadena	40	1	2	–	–	–
Rana	222	41	18	37	90	17
Rana esculenta group	25	19	76	17	89	68
Rana rugosa group	3	1	33	1	100	33
Rana temporaria group	26	21	81	19	90	73

Table 6:11 continued.

Genus or species group	Total world species	Palearctic species		Palearctic endemics[1]		Percent world endemics[2]
		n	%	n	%	
ANURA: RHACOPHORIDAE:						
Buergeria	4	1	25	1	100	25
Rhacophorus	57	2	4	2	100	4

[1]In relation to Palearctic congeneric species.
[2]In relation to the total number of congeneric species.

probably it originated in Asia rather than in North America (Čkhikvadze, 1982; Milner, 1983). However, Duellman and Trueb (1986) suggested that the early Cretaceous cryptobranchoid stock that remained in North America evolved to become obligate neotenes, the cryptobranchids, and these aquatic salamanders invaded Eurasia in the early Cenozoic from the west via Europe (the DeGeer Passage). The Paleocene-Eocene cryptobranchid records from Central Asia (Mongolia and eastern Kazakhstan) (Čkhikvadze, 1982; Gubin, 1991) seem to support the Asian dispersal. A new subfamily, Aviturinae with three genera, has been described from the Upper Paleocene of Mongolia (Gubin, 1991).

Proteidae: Proteus, as well as its North American relative, *Necturus,* are thought to be autochthonous relicts of a late Jurassic proteid stock (Laurent, 1975) that was distributed in the western (Euramerican) portion of the Laurasian land mass (Milner, 1983; Duellman and Trueb, 1986). In the past, proteids (Miocene *Mioproteus* and *Orthophyia*) were widespread in Europe and Central Asia (Estes, 1981; Čkhikvadze, 1984; Estes and Schleich, 1994; Roček, 1994).

Salamandridae: The late Cretaceous or early Paleocene origin of the family has been restricted to Europe with subsequent dispersal into Asia and North America (Milner, 1983; Duellman and Trueb, 1986). The earliest known European salamandrids are from the Upper Paleocene and Lower Eocene (Estes, 1981; Roček, 1994).

Discoglossidae: This family of Jurassic origin is a typical Laurasian group with Mesozoic fossils from Europe, North America, and Central Asia (i.e., Laurasia) and questionable remains from peninsular India (i.e., Gondwana), where the family is absent in the Recent (Evans et al., 1990; Evans and Milner, 1993; Prasad and Rage, 1991, 1995; Gubin, 1993; Roček and Nessov, 1993; Rage and Jaeger, 1995).

North America

In this category are three families of both Laurasian and Gondwanan origin:

Plethodontidae: These salamanders are probably North American in origin; possibly they arrived in Europe via the Greenland link in the Eocene (Lanza and Vanni, 1981) or via the Beringian land link in the Oligocene (Wake et al., 1978; Milner, 1983; Duellman and Trueb, 1986). Some other presumed ways and dates of invasion were discussed by Lanza and Vanni (1981) and Lanza et al. (1995).

Bufonidae: Savage (1973) suggested that bufonids had two centers of Cretaceous diversification (Africa and South America) as the result of the fragmentation of western Gondwana. The same has been argued for the genus *Bufo* (Tandy and Tandy, 1976; Maxson, 1981). Having entered North America in the Paleocene, *Bufo* diversified, and derivatives of temperate groups arrived in Eurasia via Beringia. Savage (1973) dated this invasion by members of the *Bufo viridis* group as apparently in the Oligocene. Since that time, the green toads spread throughout temperate Eurasia and into northern Africa. Fossils of this group are known from the middle Miocene of Europe (Sanchiz, 1977). Another Palearctic group, the *Bufo bufo* group, also seems to have originated from a North American stock, which also arrived in the Palearctic via Beringia; eastern Tibet harbors probably relict members of the *Bufo bufo* group, such as *B. tibetanus,* which might be close to the ancestral form (Borkin and Matsui, 1986). Fossils of the *Bufo bufo* group have been reported from the middle and late Miocene of Europe (Sanchiz, 1977; Hodrová, 1980).

Hylidae: The pattern of dispersal of this Gondwanan group probably closely paralleled that of bufonids (Savage, 1973). Thus, South America is suggested to be the original geographic source. A temperate-adapted North American *Hyla* crossed

Beringia apparently in the Oligocene. The earliest Eurasian fossil *Hyla* are from the Miocene of Europe (Sanchiz, 1981; Duellman and Trueb, 1986). According to K. Anderson (1991), the Eurasian species of *Hyla* may have derived from two North American stocks. A presumed hylid frog is known from the latest Cretaceous (Maastrichtian) of peninsular India (Prasad and Rage, 1995).

Africa

Savage (1973:438) wrote: "The amphibian fauna of northwest Africa is derived completely from southern Europe. All of its members belong to European species or species groups; none from tropical Africa." However, members of three Recent Afrotropical groups (*Xenopus* sp., *Ptychadena* sp., and a toad similar to *Bufo regularis*), together with a member of a Palearctic group (*Discoglossus* sp.) have been reported from the upper Miocene of Morocco (Vergnaud-Grazzini, 1966). Moreover, at least four Recent anuran species of Afrotropical affinities occur in the Palearctic part of Africa. These species belong to two families of Gondwanan origin and could be classified as autochthonous for the African part of the Palearctic. African sources include members of two families:

Bufonidae: The African branch of bufonids is represented by three lineages. *Bufo mauritanicus,* the sole member of the same species group is related to true (20-chromosome) Afrotropical species groups; it occurs in the Maghreb and western Sahara Desert. The eastern Sahara is inhabited by two other Afrotropical species, *B. regularis* of the same species group and the relictual *B. kassasii* of the *B. funereus* group. Both species are sympatric with the true Palearctic species *Bufo viridis* in the lower Nile Valley in Egypt (Fig. 6:3). Hemmer et al. (1981) assumed that an ancestral green toad stock existed in the lowermost Miocene of North Africa; they proposed that that stock gave rise to the European *Bufo calamita* lineage via a connection between Libya and Anatolia some 15–16 million years ago. The earliest fossil remains of green toads are from the Miocene of Spain (Sanchiz, 1977).

Ranidae: This family, classified by Savage (1973) under the so-called Young Tropical Unit, is clearly of African origin (Laurent, 1975). Only one species of a certainly Afrotropical component, *Ptychadena mascareniensis,* extends into the periphery of the Palearctic via the Nile Valley (Fig. 6:2).

India

After the Mesozoic breakup of Gondwana, India moved to the north and collided with Asia. This contact was suggested to have happened during the Eocene or even the Oligocene (Duellman and Trueb, 1986). However, some Laurasian fossils of late Cretaceous (Maastrichtian) age seem to indicate an earlier contact (Prasad and Rage, 1991, 1995; Rage and Jaeger, 1995). Although no confirmed Indian derivatives ranged directly into the Palearctic, three Recent species occur peripherally in the Palearctic; these are toads of the *Bufo stomaticus* group and a ranid, *Euphlyctis cyanophlyctis* (Fig. 6:5).

Southeast Asia

Liem (1970) and Laurent (1975) suggested that rhacophorids are an autochthonous group derived from a ranid stock in the Oriental Region. Contrary to previous authors, who were "dispersalists," Savage (1973) rejected any importance of southeastern Asia as a major source. He suggested that even megophryines were immigrants from the north. This area has served as a refuge for tropical groups displaced southward by late Cenozoic cooling trends and for derivatives of some ancient Indian stocks. Microhylids and rhacophorids of the African part of Gondwana possibly arrived in Asia by the drifting Indian subcontinent and dispersed eastward (Duellman and Trueb, 1986). A few representatives invaded the Palearctic (in China and Japan) from southeastern Asia. A radical interpretation has been proposed by Channing (1989) for rhacophorids. He suggested that the ancestral stock existed on Pangaea, probably about 200 million years ago. Fragmentation of Pangaea by three main steps led to vicariance, with derivatives of the treefrog stock on Africa, the Seychelles, Madagascar, and Asia. According to this hypothesis, Asian rhacophorids would be an autochthonous group. Rhacophorids, as well as microhylids, seem to have been represented in Europe prior to the so-called Grand Coupure (i.e., to the Eocene-Oligocene boundary) (Sanchiz et al., 1993; Sanchiz and Roček, 1996).

Uncertain

Conflicting hypotheses have been proposed to explain distributional patterns of at least three anuran families:

Pelobatidae: Savage (1973) assumed that this obviously Laurasian group probably originated in North America in the mid- to late Jurassic, rather than in Eurasia. This assumption seems to be supported by the earliest known fossil evidence from the Upper Jurassic of Wyoming, USA (Evans and Milner, 1993). Pelobatines might have arrived in eastern Asia in the Eocene and western Asia in the Oligocene. The earliest confirmed European fossil of the Recent genus *Pelobates* is from the basal Miocene of Germany, 24 million years ago (Böhme et al., 1982). A brief mention of an Eocene *Pelobates* from Belgium (Roček in Böhme et al., 1982), repeated by Henrici (1994), was not included in the review by Sanchiz and Roček (1996). However, European *Eopelobates* are known from the middle Eocene (Rage, 1984, Sanchiz and Roček, 1996); this genus has been grouped with *Pelobates* in the Pelobatinae (Henrici, 1994). According to estimates based on molecular data (Sage et al., 1982), the divergence between the Recent *Pelobates* and *Scaphiopus* may be dated as Cretaceous. Unfortunately, some evidence provided by fossils is problematic. Several late Cretaceous anuran genera have been described from the Soviet Central Asia and Mongolia. The species previously referred to the earliest known eopelobatines (*Eopelobates sosedkoi* Nessov, 1981, and others) are currently treated as members of another genus (*Gobiates*) and a separate family, Gobiatidae (Roček and Nessov, 1993). This family has certain similarities with the Leiopelmatidae and Discoglossidae. Two other genera (*Kizylkuma* and *Aralobatrachus*) originally described by Nessov (1981) without clear familial allocation were later assigned to eopelobatines (e.g., Duellman and Trueb, 1986). However, Roček and Nessov (1993) indentified them as valid genera of discoglossids. Therefore, there are no confirmed Cretaceous records of pelobatids from Central Asia. The early Oligocene pelobatid genus *Uldzinia* was described recently from southern Mongolia (Gubin, 1996). A pelobatid frog has been reported from the latest Cretaceous of peninsular India (Rage and Jaeger, 1995).

Pelodytidae: This group, considered by Savage (1973) as a subfamily of pelobatids, probably has been involved in the same scenario as the pelobatines. Duellman and Trueb (1986) suggested that pelodytids probably diverged from propelobatids prior to the middle Jurassic, but the earliest fossils are from the upper Eocene of Europe (Rage, 1984a; Sanchiz and Roček, 1996). Sage et al. (1982) noted that immunological data indicate divergence between

the living *Pelodytes* and pelobatines in the Cretaceous. In the Eocene, Europe was isolated by the Tethys Sea, a part of which, the Turgai Strait, separated Europe from Asia. Thus, paleontological and molecular data contradict Savage's (1973) scenario and would require at least an Eocene invasion of Europe from Asia, if the Turgai Strait was traversable by anurans (Rage, 1984a, b). Possibly the Pelobatidae and Pelodytidae are autochthonous derivatives of Laurasian stocks, like some salamanders and discoglossids.

Ranidae: Two hypotheses have been proposed. A temperate-adapted African ranine stock expanded across Eurasia, diversified into several distinct lineages, and invaded North America by the Eocene. The members of this radiation included the Old World *Rana esculenta* and *Rana temporaria* groups and several New World groups (Savage, 1973). However, according to Duellman and Trueb (1986), after the formation of the Tethys Sea in the Jurassic, Africa was not connected with Eurasia until the end of the Miocene (see maps in Dercourt et al., 1986). The earliest fossil *Rana* of Europe are from the Oligocene. Therefore, it seems more likely that the ranine stock reached Asia via drifting India. Thus, Holarctic *Rana* would have had an Asian origin (Duellman and Trueb, 1986). Nevertheless, ranids are known from the Upper Eocene of France. Having an African origan, ranids possibly invaded Europe via Asia and perhaps the partially penetrable Turgai Strait (Rage, 1984b). Other subsequent geological evidence shows that the Iberian Plate was a part of Africa from the Late Cretaceous to the mid-Eocene, and since the late Oligocene it became a part of Eurasia (Srivastava et al., 1990). Amphibians might have used such a "jumping land" for dispersal from Africa to Europe. Rage (1995) reviewed faunal exchanges and some presumed connections between Africa and Eurasia in the Mediterranean and southwestern Asia in the Late Cretaceous and Paleogene. However, because of the absence of fossil documentation, he doubted the possibility of dispersal by amphibians (and freshwater fishes) via these filters.

DOES THE HOLARCTIC EXIST FOR AMPHIBIANS?

Many authors have pointed out the faunistic similarities in the northern continents and united temperate Eurasia (Palearctic) and North America (Nearctic) as one unit, the Holarctic. Some global biogeographers (e.g., Lemée, 1967; Takhtajan, 1978) even denied the Palearctic and Nearctic as distinct regions; others recognized them as separate parts of the Holarctic (e.g., De Lattin, 1967; Emeljanov, 1974; P. Müller, 1986). The latter emphasized that the Holarctic similarity is the result of faunal exchanges between Eurasia and North America via Beringia. Among five major Recent world frog faunal regions, the Holarctic was considered by Savage (1973) to have three centers of further differentiation: (1) western Eurasia and northwestern Africa, i.e., western Palearctic; (2) eastern and southern Asia, i.e., partly eastern Palearctic; and (3) North America.

The concept of the Holarctic was based on the sharing of closely related species (or even subspecies) of some groups of plants and animals (principally birds and mammals) that inhabit tundra, taiga, and deciduous forests. Unlike these groups, amphibians exhibit another distributional pattern. First of all, there are no species or sister species shared by the Palearctic and Nearctic. Secondly, only four genera are common to both continents, these are *Hydromantes* and three nearly cosmopolitan genera of anurans—*Bufo, Hyla,* and *Rana.* Lastly, coefficients of faunal similarity reveal that the similarity at the familial level between the Palearctic and Nearctic is lower than that between two remote parts of the Palearctic (Europe and the Far East) or between the Palearctic and Oriental realms (Table 6:12 and 6:13). Numerically, the Far Eastern Palearctic fauna is more similar to the Oriental than to the Nearctic fauna (coefficients of 0.87 vs. 0.52). Therefore, I am in-

Table 6:12. Holarctic distribution of Recent amphibians. For lists of species in Palearctic and Europe, see Appendices 6:1 and 6:2.

Family	Nearctic	Palearctic	Europe	Far East
Ambystomatidae	+	–	–	–
Amphiumidae	+	–	–	–
Cryptobranchidae	+	+	–	+
Dicamptodontidae	+	–	–	–
Hynobiidae	–	+	+	+
Plethodontidae	+	+	+	–
Proteidae	+	+	+	–
Rhyacotritonidae	+	–	–	–
Salamandridae	+	+	+	+
Sirenidae	+	–	–	–
Total Caudata	9	5	4	3
Ascaphidae	+	–	–	–
Bufonidae	+	+	+	+
Discoglossidae	–	+	+	+
Hylidae	+	+	+	+
Leptodactylidae	+	–	–	–
Microhylidae	+	+	–	+
Pelobatidae	+	+	+	+
Pelodytidae	–	+	+	–
Ranidae	+	+	+	+
Rhacophoridae	–	+	–	+
Rhinophrynidae	+	–	–	–
Total Anura	8	8	6	7
Total Amphibia	17	13	10	10

clined to recognize the Palearctic and Nearctic as distinct realms in the Northern Hemisphere instead of grouping them into the Holarctic. Sometimes, the term "Holarctic distribution" is used. This term is not applicable to amphibians, because "the Holarctic amphibians" are, in fact, of different origins and ages. Thus, the Holarctic is an arbitrary term in any sense.

Table 6:13. Faunal similarities of amphibian families (as listed by Duellman, 1993). Gymnophiona of the Oriental and Afrotropical realms are included.

Regions compared	Coefficient	Common Groups	Total Groups
Palearctic-Oriental	0.77	10	26
Europe-Far East	0.70	7	20
Palearctic-Nearctic	0.60	9	30
Europe-Nearctic	0.52	7	27
Far East-Nearctic	0.52	7	27
Nearctic-Oriental	0.47	7	30
Palearctic-Afrotropical	0.31	4	26

SUBDIVISIONS OF THE PALEARCTIC REALM

Many authors have suggested various schemes for subdivisions of the Palearctic Realm. Two main approaches have been used. One is based on so-called landscape zones (geographic belts) that distinguish tundra, taiga (coniferous forests), deciduous forests, deserts, etc.; these are largely latitudinal zones, each with distinct ecological conditions. This is the basis for some floristic regions, for instance those mapped by Good (1964) (Fig. 6:7). Another approach stresses sectoral differentiation, and uses preferably historical explanations. Some authors have proposed to unite the classic zoogeographic system, faunal and floral classification, and biomes, i.e., physiognomic plant formations with the animal populations that inhabit them. The number of subdivisions and their boundaries often differ among various authors, depending on their methodology, distributional peculiarities of particular groups of organisms analyzed, and knowledge of the areas. For example, Emeljanov (1974) recognized 60 provinces in eight regions, and Takhtajan (1978) divided the Palearctic territory into six regions and 53 provinces; Udvardy (1975) established 44 biogeographic provinces of the Palearctic Realm (without regions), and Shcherbak (1982) recognized four subregions and only 13 provinces of the Palearctic Region.

Based on reptilian distributions, some herpetologists (e.g., Joger, 1987) proposed to erect the so-called Saharan-Sind Province (or Subregion), the vast arid territory in the southern part of the Palearctic from North Africa eastward to Pakistan, as a realm equivalent to the Palearctic. No consensus exists in recognizing the main subdivisions of the Palearctic. However, based on patterns of amphibian distribution, the Palearctic Realm can be divided into six regions, namely the Mediterranean, European (or Boreal), Siberian, West Asian, Central Asian, and East Asian (Fig. 6:8; Appendix 6:1). Comparative data on taxonomic composition and endemism of the amphibian faunas of these regions are given in Tables 6:14 and 6:15.

MEDITERRANEAN REGION

This region embraces the Iberian Peninsula including the northern Cantabrian Mountains and Pyrenees (see below), southwestern and southern coastal France, Italy (including the Po River area), the south Adriatic coastal area of the Balkan Peninsula, Greece, and numerous islands in the Mediterranean Sea between Africa, Europe, and Asia eastward to the islands of Samothraki and Crete. The eastern boundary is debatable. Northwestern Africa (Egypt excluded) is, undoubtedly, a part of the region. The northern Sahara and three northern "Macaronesian" archipelagos also could be included in

the Mediterranean Region (Fig. 6:8). In addition to deserts and highlands, the Mediterranean Realm has the characteristic climate of mild, rainy winters and dry, warm summers.

The Pyrenees usually are considered to be a barrier separating the Mediterranean fauna of the Iberian Peninsula from the boreal fauna of Europe. However, the ranges of many Iberian and boreal amphibians do not reach the Pyrenees (Zuiderwijk, 1980; Castanet and Guyétant, 1989; Gasc et al., 1997). Most authors (e.g., Lemée, 1967; Takhtajan, 1978; also see Fig. 6:7) join the Cantabrian and Pyrenean Mountains to the European (Boreal) Region. However, amphibian distributions do not fit to this scheme. The Pyrenees harbor 19 species of amphibians (Table 6:14). Of these, two, *Euproctus asper* and *Rana pyrenaica,* are endemic to the mountains. Sixteen species are shared with the Iberian Peninsula, and 15 are shared with the southwest of France, i.e., a lowland area approximately south of the Garonne River and Central Plateau ("Massif Central"). Only two Iberian species recorded in the Pyrenees (*Discoglossus galganoi* and *Rana iberica*) are absent in the southwestern area. Thus, the majority of amphibians in the Pyrenean region is known both to the south and to the north. Curiously, *Triturus alpestris* demonstrates a disjunction; it is widespread in Europe and represented by some isolated populations in the Iberian Peninsula. Although this species is absent in the Pyrenees, it inhabits the Cantabrian Mountains to the west.

Based on the concept of Pleistocene refugia, Zuiderwijk (1980) categorized amphibians of western Europe into two main groups: 18 species are hypothesized to have survived in, and later partly dispersed from, the western (Iberian) refuge, whereas 12 species are hypothesized to have expanded their ranges westward from the eastern (Balkans plus the Black Sea steppes) refuge. Four species (*Triturus alpestris, Salamandra salamandra, Hyla arborea,* and *Rana esculenta* complex), probably, survived the last ice age (Würm) in both refugia.

Among 15 species having distributions that pass through the Pyrenees, eight possibly have penetrated the mountains from south to north, and seven have dispersed in the opposite direction (Table 6:16). In some cases, the history of dispersal of widespread and subspecifically structured species, like *Salamandra salamandra* and *Alytes obstetricans,* has been

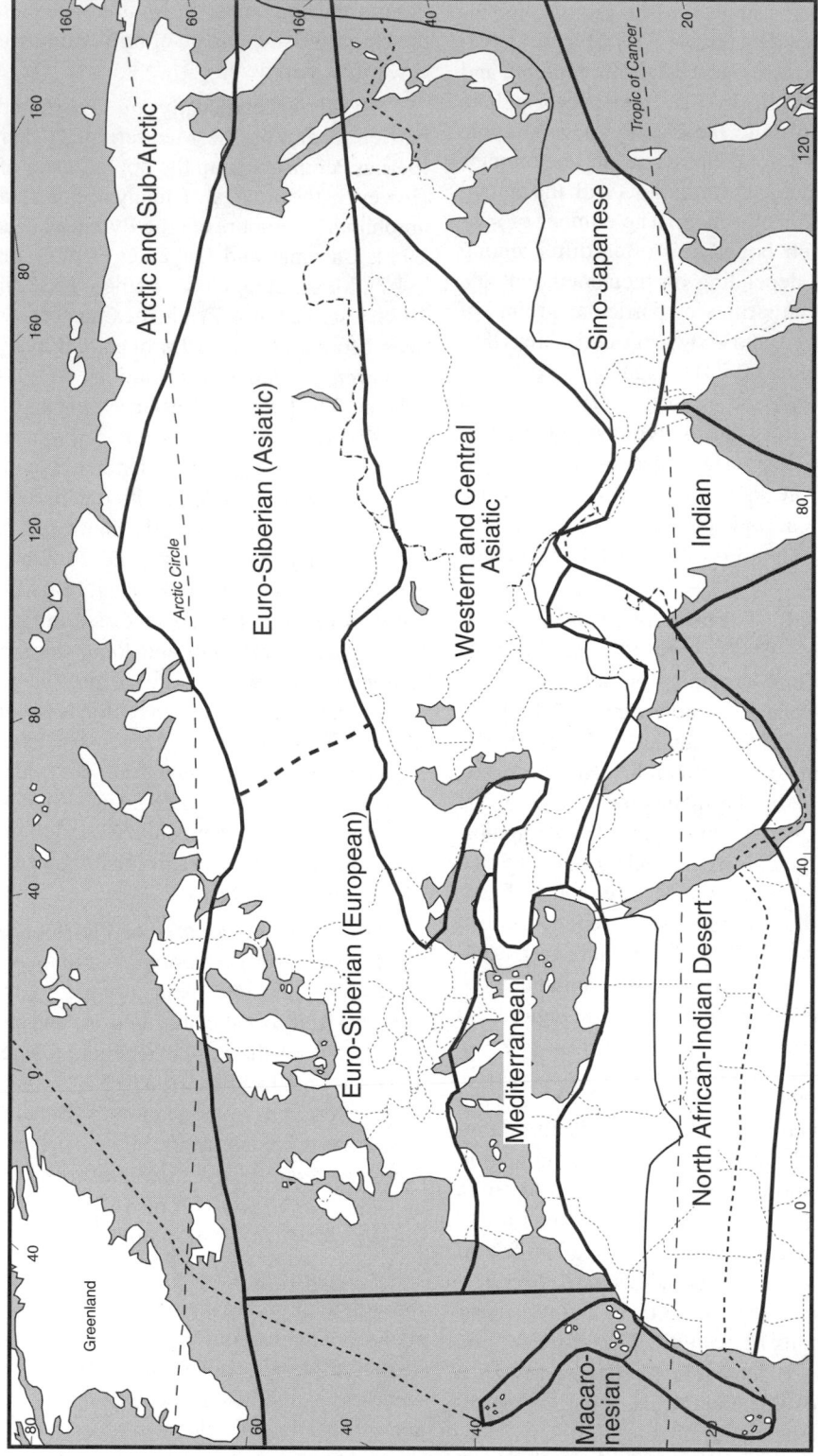

Fig. 6:7. Floristic regions of the Palearctic Realm and adjacent regions. Based on Good (1964). The border of the Palearctic Realm is the solid line; the southern limits of the area covered in this chapter are shown by the dotted line.

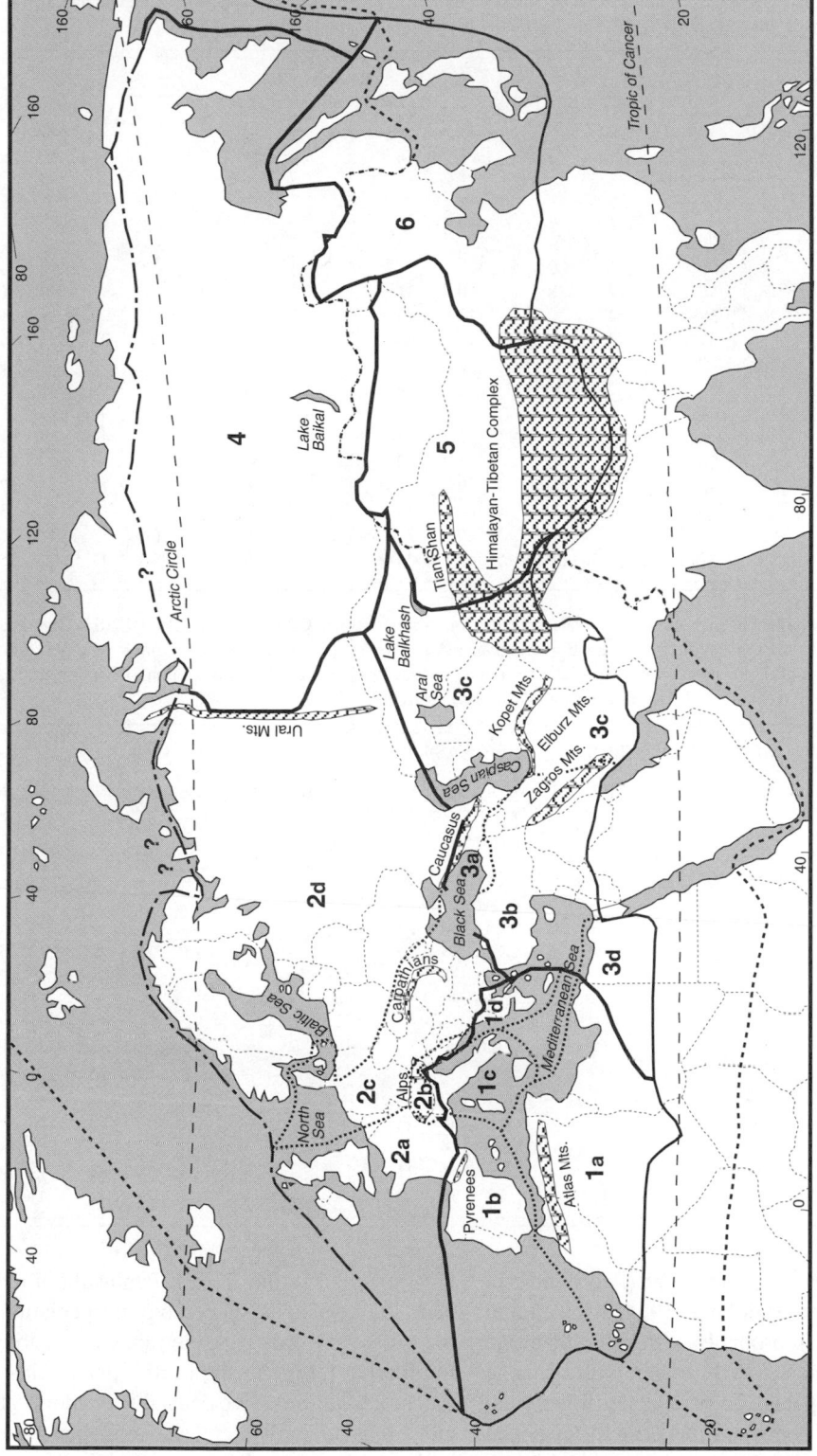

Fig. 6:8. Six major regions of amphibian distribution in the Palearctic Realm. 1 = Mediterranean Region (Provinces: 1a = Northwestern African, 1b = Iberian, 1c = Italian, 1d = Balkan). 2 = European Region (Provinces: 2a = Atlantic, 2b = Alpine-Dinaric, 2c = Central European, 2d = Eastern European). 3 = West Asian Region (Provinces: 3a = Caucasian, 3b = Near East, 3c = Iranian, 3d = Egyptian). 4 = Siberian Region. 5 = Central Asian Region. 6 = East Asian Region. Major mountain ranges are shown diagrammatically. The southern border of the Palearctic Realm is a solid line; the southern limits of the area covered in this chapter are noted by a dashed line. The northern limit of amphibian distribution is shown by a dash-dot line and is based on Anufriev and Bobretsov (1996), Borkin et al. (1984), Dokuchaev et al. (1984), Gasc et al. (1997), Kutenkov and Panarin (1995), and Kuzmin (1994).

Table 6:14. Taxonomic composition of amphibian faunas in six regions of the Palearctic Realm. Abbreviations: CAS = Central Asian, EAS = East Asian, EUR = European (or Boreal), MED = Mediterranean, SIB = Siberian, WAS = West Asian, n = number of species, % = percentage of total number of species in that family occurring in the Palearctic.

Family	MED		EUR		WAS		SIB		CAS		EAS	
	n	%	n	%	n	%	n	%	n	%	n	%
Cryptobranchidae	–	–	–	–	–	–	–	–	1	50	2	100
Hynobiidae	–	–	1	4	3	11	1	4	3	11	21	78
Plethodontidae	7	100	–	–	–	–	–	–	–	–	–	–
Proteidae	1	100	1	100	–	–	–	–	–	–	–	–
Salamandridae	18	56	12	38	10	31	1	3	–	–	1	3
Total Caudata	26	38	14	20	13	19	2	3	4	6	24	35
Bufonidae	5	21	3	12	13	54	3	12	6	25	6	25
Discoglossidae	10	77	3	23	2	15	–	–	–	–	1	8
Hylidae	4	57	2	29	2	29	1	14	–	–	2	29
Microhylidae	–	–	–	–	–	–	–	–	–	–	1	100
Pelobatidae	4	44	3	33	2	22	–	–	4	44	1	11
Pelodytidae	1	50	1	50	1	50	–	–	–	–	–	–
Ranidae	19	40	9	19	9	19	3	6	6	13	17	36
Rhacophoridae	–	–	–	–	–	–	–	–	–	–	3	100
TotalAnura	43	41	21	20	29	27	7	7	16	15	31	29
TotalAmphibia	69	39	35	20	42	24	9	5	20	11	55	31

Table 6:15. Species endemism in six regions of the Palearctic Realm. Abbreviations: CAS = Central Asian, EAS = East Asian, EUR = European (or Boreal), MED = Mediterranean, SIB = Siberian, WAS = West Asian. Within each region, n = number of endemic species, RP = percentage of regional endemism (in relation to number of species in the family in the region), PP = Palearctic percentage of species (in relation to number of species in the family in the Palearctic).

Family	MED			EUR			WAS			SIB			CAS			EAS		
	n	RP	PP	n	RP	PP	n	RP	PP	n	RP	PP	n	RP	PP	n	RP	PP
Cryptobranchidae	–	–	–	–	–	–	–	–	–	–	–	–	–	–	–	1	50	50
Hynobiidae	–	–	–	–	–	–	2	100	11	–	–	–	1	33	4	20	95	74
Plethodontidae	7	100	100	–	–	–	–	–	–	–	–	–	–	–	–	–	–	–
Proteidae	–	–	–	–	–	–	–	–	–	–	–	–	–	–	–	–	–	–
Salamandridae	11	61	34	3	33	13	8	80	25	–	–	–	–	–	–	1	100	3
Total Caudata	18	69	26	4	29	6	11	85	16	–	–	–	1	25	1	22	92	32
Bufonidae	1	20	4	–	–	–	5	38	21	–	–	–	–	–	–	3	50	12
Discoglossidae	8	80	62	–	–	–	1	50	8	–	–	–	–	–	–	1	100	8
Hylidae	3	75	43	–	–	–	–	–	–	–	–	–	–	–	–	1	50	14
Microhylidae	–	–	–	–	–	–	–	–	–	–	–	–	–	–	–	–	–	–
Pelobatidae	1	25	11	–	–	–	–	–	–	–	–	–	2	50	22	1	100	11
Pelodytidae	–	–	–	–	–	–	1	100	50	–	–	–	–	–	–	–	–	–
Ranidae	11	58	23	–	–	–	5	56	11	–	–	–	1	17	2	10	59	21
Rhacophoridae	–	–	–	–	–	–	–	–	–	–	–	–	–	–	–	3	100	100
TotalAnura	24	56	23	–	–	–	12	41	11	–	–	–	3	19	3	19	61	18
TotalAmphibia	42	61	24	4	11	2	23	55	13	–	–	–	4	20	2	41	75	23

suggested to have been complicated (Alcobendas et al., 1994; Joger and Steinfartz, 1994; Arntzen and García-París, 1995), probably with more than one invasion into the Iberian Peninsula or with a change of direction of dispersal. In such cases, it is necessary to distinguish between the primary dispersal (e.g., the invasions of *Salamandra salamandra* or *Hyla arborea* to the Iberian Peninsula in the Miocene-Pliocene, with successive subspeciation) from the secondary one (e.g., the range extensions from the Iberian refuge to the north after the last Pleistocene glaciation). Two coastal "corridors" at lower elevations probably were used as dispersal routes in both (north-south) directions. Judging from distri-

Table 6:16. Amphibians of the Iberian Peninsula, Pyrenees, and France exclusive of insular species. Asterisk (*) denotes species endemic to the Pyrenees. Sources: Arano et al. (1995), Castanet and Guyétant (1989), Crochet et al. (1995), Gasc et al. (1997), Martínez Rica (1983), Serra-Cobo (1993), and personal data.

Species	Iberian Peninsula	Pyrenees	Mainland Southwest	Central and eastern France	Primary distribution via Pyrenees
CAUDATA: SALAMANDRIDAE:					
Chioglossa lusitanica	+	–	–	–	–
Euproctus asper*	–	+	–	–	–
Pleurodeles waltl	+	–	–	–	–
Salamandra salamandra	+	+	+	+	Southward
Triturus alpestris	+	–	–	+	Southward
Triturus boscai	+	–	–	–	–
Triturus cristatus	–	–	–	+	–
Triturus helveticus	+	+	+	+	Northward
Triturus marmoratus	+	+	+	+	Northward
Triturus vulgaris	–	–	–	+	–
ANURA: BUFONIDAE:					
Bufo bufo	+	+	+	+	Southward?
Bufo calamita	+	+	+	+	Northward
Bufo viridis	–	–	–	+[1]	–
ANURA: DISCOGLOSSIDAE:					
Alytes cisternasii	+	–	–	–	–
Alytes dickhelleni	+	–	–	–	–
Alytes obstetricans	+	+	+	+	Northward
Bombina variegata	–	–	+[2]	+	–
Discoglossus galganoi	+	+	–	–	–
Discoglossus pictus[3]	+	+	+	–	Introduced
ANURA: HYLIDAE:					
Hyla arborea	+	+	+	+	Southward
Hyla meridionalis	+	+	+	+	Northward
ANURA: PELOBATIDAE:					
Pelobates cultripes	+	+	+	+	Northward
Pelobates fuscus	–	–	–	+	–
ANURA: PELODYTIDAE:					
Pelodytes punctatus	+	+	+	+	Northward
ANURA: RANIDAE:					
Rana arvalis	–	–	–	+[1]	–
Rana dalmatina	–[4]	+	+	+	Southward
Rana esculenta	–	–	–	+	–
Rana grafi	+	+	+	+	Southward
Rana iberica	+	+	–	–	–
Rana lessonae	–	–	–	+	–
Rana perezi	+	+	+	+	Northward
Rana pyrenaica*	–	+	–		–
Rana ridibunda	–	–	–	+	–
Rana temporaria	+	+	+	+	Southward
Total species	22	19	15	24	

[1]Only from the eastern part of France.
[2]An isolated population.
[3]Probably introduced (Lanza et al., 1986).
[4]Does occur in the Cantabrian Mountains.

bution patterns, most species (e.g., *Alytes obstetricans, Pelobates cultripes, Pelodytes punctatus,* and *Hyla meridionalis*) presumably went around the Pyrenees by an eastern (Mediterranean) pathway, whereas others (e.g., *Hyla arborea, Rana dalmatina*) invaded by a western (Atlantic) pathway. Some species (e.g., *Triturus helveticus, T. marmoratus,* and *Bufo bufo*) might have used both pathways. Arano et al. (1995) suggested that green frogs (*Rana perezi* and hybridogenetic *Rana grafi*) penetrated by three routes—via Euskadi (Irún) in the west, the eastern Pyrenees, and La Cerdanya in the central Pyrenees, by following the river courses.

Although the Cantabrian and Pyrenean mountains harbor relictual populations of some widespread European amphibians, like the brown frogs *Rana dalmatina* and *Rana temporaria* (latter represented by an endemic subspecies, *R. t. parvipalmata*), on the whole, the Pyrenean batrachofauna seems to be closer to that of the Iberian Peninsula than that of boreal Europe. Faunal similarities calculated by the Czekanowski Coefficient between the Iberian Peninsula and Pyrenees, southwestern France, and central and eastern France are 0.78, 0.76, and 0.61, respectively. The similarities between the Pyrenees and southwestern France, and central and eastern France are 0.88, and 0.65, respectively; that between both parts of France is only 0.72. Except *Bombina variegata* with an isolated record south of the Garonne River (Grangé in Castanet and Guyétant, 1989), all species occurring in southwestern France also are known from the Pyrenean area. Therefore, the amphibian fauna of southwestern France more closely resembles those of the Pyrenees and Iberian Peninsula than that of boreal Europe. Consequently, the northern limit of the Mediterranean Region seems to be the Garonne River and southern edge of the Central Plateau.

The Mediterranean Region has the most diverse amphibian fauna in the Palearctic; with 69 species (26 salamanders and 43 anurans) in 16 genera and 9 families; this amounts to 40% of the amphibians of the Palearctic Realm (Tables 6:14 and 6:17; Appendix 6:1). The region is characterized by the presence of plethodontids, diverse salamandrids and many discoglossids (especially *Alytes* and *Discoglossus*) as well as a pelodytid, some hylids, and many ranids. Four genera (25%) are endemic to the region; all of them are salamandrids—*Chioglossa, Euproctus, Pleurodeles,* and *Salamandrina.* Endemism at the specific level is high—61% (69% in salamanders and 56% in anurans) (Tables 6:15 and 6:17). In addition to the endemic genera, endemic species are known in *Hydromantes* (all 7 species), *Salamandra* (2), *Triturus* (2), *Bufo* (1), *Alytes* (3), *Discoglossus* (all 4 species), *Hyla* (3), *Pelobates* (1) and *Rana* (11). Many widespread species (e.g., *Salamandra salamandra, Triturus alpestris, T. vulgaris, Bufo bufo, Alytes obstetricans*) are represented by endemic subspecies.

The Mediterranean Region has many islands. Lanza and Vanni (1987) summarized data on 26 species that occur on islands; their review is updated to 38 species in Table 6:18. All islands in the Mediterranean Sea are traditionally considered to belong to the Mediterranean Region. However, based on amphibian distribution, the islands may be allocated to three regions. The easternmost islands and archipelagos like Cyprus, Lycian coastal islets, Rhodes, Karpathos, South Sporades, and some eastern North Aegean islands are inhabited by invaders (e.g., *Mertensiella luschani, Rana bedriagae*) from the West Asian Region and could be classified as a part of that region. A few of the northernmost islands in the Aegean Sea like Samothraki harbor members (e.g., *Rana ridibunda*) of the European (Boreal) Region. The numerous remaining islands do belong to the Mediterranean Region, within the limits proposed herein. The origin of islands is heterogeneous. Some of them, such as the Balearic Islands, Corsica, and Sardinia, are mostly results of (micro)plate drifting. Others were separated from the mainland more recently by eustatic oscillations (sea transgressions). These geological processes must be considered when interpreting the amphibian fauna of a particular island or archipelago (Lanza and Vanni, 1987; Sbordoni et al., 1990; Oosterbroek and Arntzen, 1992; Caccone et al., 1994). Perhaps introduction by human transportation also took place (Hemmer et al., 1981; Lanza and Vanni, 1987).

Thirty-eight amphibians (12 salamanders, 26 anurans) are known from the Mediterranean islands (Table 6:18); 13 species (34%), including 7 salamanders and 6 anurans, are endemic to islands. Besides, three other mainland species are represented by endemic subspecies on Mediterranean islands. Lanza and Vanni (1987) noted significant differences in the number of insular species between western (19 species) and eastern (11 species) parts of the Mediterranean Sea. However, more extensive sampling and new taxonomic studies demonstrate approximately a balance in faunal size of both parts—19 vs. 18 confirmed species (Table 6:18). However, the composition of the amphibian faunas in the west and the east is different. Only three widespread species (*Bufo bufo, B. viridis,* and *Rana dalmatina*) are shared by both parts. Stugren (1985) and Lanza and Vanni (1987) mentioned the absence of insular endemics in the eastern Mediterranean Sea in comparison with the west, with 11 endemics (7 salamanders and 4 anurans). Such an imbalance also is evident in Table 6:18; only two green frogs are endemic to the southern Aegean islands—Crete, Karpathos, and

Table 6:17. Distribution of amphibians in four provinces of the Mediterranean Region. Abbreviations of provinces: BAL = Balkan, IBE = Iberian, ITA = Italian, NWA = northwestern Africa. Asterisks denoted species (*) or subspecies (**) endemic to the Mediterranean Region; + = present, − = absent, E = endemic to province, I = probably introduced, P = peripheral occurrence only.

Species	NWA	IBE	ITA	BAL
CAUDATA: PLETHODONTIDAE:				
Hydromantes ambrosii*	−	−	E	−
Hydromantes flavus*	−	−	E	−
Hydromantes genei*	−	−	E	−
Hydromantes imperialis*	−	−	E	−
Hydromantes italicus*	−	−	E	−
Hydromantes strinatii*	−	−	E	−
Hydromantes supramontis*	−	−	E	−
CAUDATA: PROTEIDAE:				
Proteus anguinus	−	−	P	−
CAUDATA: SALAMANDRIDAE:				
Chioglossa lusitanica*	−	E	−	−
Euproctus asper*	−	E	−	−
Euproctus montanus*	−	−	E	−
Euproctus platycephalus*	−	−	E	−
Pleurodeles poireti*	E	−	−	−
Pleurodeles waltl*	+	+	−	−
Salamandra algira*	E	−	−	−
Salamandra atra	−	−	P?	P?
Salamandra corsica*	−	−	E	−
Salamandra salamandra**	−	+	+	+
Salamandrina terdigitata*	−	−	E	−
Triturus alpestris**	−	+	+	+
Triturus boscai*	−	E	−	−
Triturus carnifex	−	−	+	+
Triturus helveticus	−	+	−	−
Triturus italicus*	−	−	E	−
Triturus marmoratus	−	+	−	−
Triturus vulgaris**	−	−	+	+
Total Caudata	3	8	18	5
Number of endemics	2	3	12	0
Percent endemic	67	37	67	0
ANURA BUFONIDAE:				
Bufo brongersmai*	E	−	−	−
Bufo bufo**	+	+	+	+
Bufo calamita	−	+	−	−
Bufo mauritanicus	+	−	−	−
Bufo viridis**	+	I?	+	+
ANURA: DISCOGLOSSIDAE:				
Alytes cisternasii*	−	E	−	−
Alytes dickhilleni*	−	E	−	−
Alytes muletensis*	−	E	−	−
Alytes obstetricans**	+	+	−	−
Bombina pachypus*	−	−	E	−
Bombina variegata	−	−	P	+
Discoglossus galganoi*	−	E	−	−
Discoglossus montalentii*	−	−	E	−
Discoglossus pictus*	+	I	+	−
Discoglossus sardus*	−	−	E	−
ANURA: HYLIDAE:				
Hyla arborea**	−	+	−	+
Hyla intermedia*	−	−	E	−
Hyla meridionalis*	+	+	+	−

Table 6:17. Continued.

Species	NWA	IBE	ITA	BAL
Hyla sarda*	−	−	E	−
ANURA: PELOBATIDAE:				
Pelobates cultripes	−	+	−	−
Pelobates fuscus**	−	−	+	−
Pelobates syriacus	−	−	−	+
Pelobates varaldii*	E	−	−	−
ANURA: PELODYTIDAE:				
Pelodytes punctatus	−	+	+	−
ANURA: RANIDAE:				
Rana bergeri*	−	−	E	−
Rana cretensis*	−	−	−	E
Rana dalmatina	−	+	+	+
Rana epeirotica*	−	−	−	E
Rana esculenta	−	−	+	−
Rana graeca	−	−	−	+
Rana grafi	−	+	−	−
Rana hispanica*	−	−	E	−
Rana iberica*	−	E	−	−
Rana italica*	−	−	E	−
Rana kurtmuelleri*	−	−	−	E
Rana latastei*	−	−	E	−
Rana lessonae	−	−	+	+
Rana perezi	−	+	−	−
Rana pyrenaica*	−	E	−	−
Rana ridibunda	−	−	P	?
Rana saharica*	E	−	−	−
Rana shqiperica*	−	−	−	E
Rana temporaria**	−	+	+	−
Total Anura	9	17	22	12
Number of endemics	3	6	9	4
Percent endemic	33	35	41	33
Total Amphibia	12	25	40	17
Total number of endemics	5	9	21	4
Total percent endemic	42	36	53	24

Rhodes (Beerli, 1994; Beerli et al., 1994; Veith, 1996a). Also, two subspecies (of *Hyla arborea* and *Mertensiella luschani*) are endemic to Crete and the Lycian coastal islands, respectively. I am inclined to explain these strong differences in endemism by age and geological history of islands rather than by ecological reasons like the dryer climate, reduced population density in the east, or by extinction, as suggested by Stugren (1985) and Lanza and Vanni (1987). Indeed, the western Mediterranean endemics are only on the Balearic Islands, Corsica, and Sardinia, which were subjected to long microplate drifting during the Tertiary. Some of islands may have been reached during the so-called Messinian crisis, when the Mediterranean Sea dried up after the closing of the Strait of Gibraltar in the late Miocene, approximately 5–6 million years ago, or by a land bridge in the Pliocene, 2 million years ago (Lan-

Table 6:18. Distribution of amphibians on Mediterranean islands. Abbreviations: AS = Argo-Saronic Islands, BA = Balearic Islands, CO = Corsica, CP = Cyprus, CR = Crete, CY = Cyclades, DA = Dalmatian Islands, DJ = Djerba, EV = Evvoia, GA = Gallita Island, HY = Hyères Islands, IO = Ionian Islands, LY = Lycian Islands, KA = Karpathos group (Saria-Kasos), KI = Kithira, MA = Maltese Islands, NA = North Aegean Islands (Thasos-Khios), NS = North Sporades (Skiros), PE = Pelagie Islands, RH = Rhodes, SA = Sardinia, SI = Sicily, SS = South Sporades (Ikaria-Kos), TU = Tuscan Archipelago (Elba, etc.). Asterisks denote species (*) or subspecies (**) endemic to the entire island region; + = present, – = absent, E = endemic to a given island or archipelago, I = probably introduced. Sources: Beerli (1994), Beerli et al. (1994), Clark (1989, 1991), Dubois and Ohler (1995a), Eiselt (1988), Gasc et al. (1997), Grillitsch and Grillitsch (1991), Hemmer et al. (1981), Lanza and Vanni (1987), Lanza et al. (1995), Nascetti et al. (1995), Parent (1981), Sofianidou and Schneider (1989), Sofianidou et al. (1994), Sotiropoulos et al. (1995).

Species	Islands (Archipelagos)																							
	BA	HY	CO	TU	SA	GA	DJ	PE	MA	SI	DA	IO	KI	CR	AS	CY	EV	NS	NA	SS	KA	RH	LY	CP
CAUDATA: PLETHODONTIDAE:																								
Hydromantes flavus*	–	–	–	–	E	–	–	–	–	–	–	–	–	–	–	–	–	–	–	–	–	–	–	–
Hydromantes genei*	–	–	–	–	E	–	–	–	–	–	–	–	–	–	–	–	–	–	–	–	–	–	–	–
Hydromantes imperialis*	–	–	–	–	E	–	–	–	–	–	–	–	–	–	–	–	–	–	–	–	–	–	–	–
Hydromantes supramontis*	–	–	–	–	E	–	–	–	–	–	–	–	–	–	–	–	–	–	–	–	–	–	–	–
CAUDATA: PROTEIDAE:																								
Proteus anguinus	–	–	–	–	–	–	–	–	–	–	?	–	–	–	–	–	–	–	–	–	–	–	–	–
CAUDATA: SALAMANDRIDAE:																								
Euproctus montanus*	–	–	E	–	–	–	–	–	–	–	–	–	–	–	–	–	–	–	–	–	–	–	–	–
Euproctus platycephalus*	–	–	–	–	E	–	–	–	–	–	–	–	–	–	–	–	–	–	–	–	–	–	–	–
Mertensiella luschani**	–	–	–	–	–	–	–	–	–	–	–	–	–	–	–	–	–	–	–	–	+	–	+	–
Salamandra corsica*	–	–	E	–	–	–	–	–	–	–	–	–	–	–	–	–	–	–	–	–	–	–	–	–
Salamandra salamandra	–	–	–	–	–	–	–	–	–	–	–	?	–	–	–	–	+	–	–	–	–	–	–	–
Triturus carnifex	–	–	–	–	–	–	–	–	–	–	+	+	–	–	–	–	+	–	–	–	–	–	–	–
Triturus vulgaris	–	–	–	–	–	–	–	–	–	–	+	+	–	–	–	+	–	–	–	–	–	–	–	–
Total Caudata	0	0	2	–	5	0	0	0	0	0	2	2	0	0	0	1	2	0	0	0	1	0	1	0
ANURA: BUFONIDAE:																								
Bufo bufo	–	–	–	+	–	–	–	–	–	–	+	+	–	+	–	+	+	–	+	–	–	–	–	–
Bufo viridis	I	–	+	+	+	–	+	+	–	+	+	+	–	+	+	+	+	+	+	+	+	+	–	+
ANURA: DISCOGLOSSIDAE:																								
Alytes muletensis*	E	–	–	–	–	–	–	–	–	–	–	–	–	–	–	–	–	–	–	–	–	–	–	–
Bombina pachypus	–	–	–	–	–	–	–	–	–	–	–	–	–	–	–	–	–	–	–	–	–	–	–	–
Bombina variegata	–	–	–	–	–	–	–	–	–	–	+	–	–	–	–	–	–	–	–	–	–	–	–	–
Discoglossus montalentii*	–	–	E	–	–	–	–	–	–	–	–	–	–	–	–	–	–	–	–	–	–	–	–	–
Discoglossus pictus*	–	–	–	–	–	+	–	–	+	+	–	–	–	–	–	–	–	–	–	–	–	–	–	–
Discoglossus sardus*	–	–	+	+	+	–	–	–	–	–	–	–	–	–	–	–	–	–	–	–	–	–	–	–
ANURA: HYLIDAE:																								
Hyla arborea**	–	–	–	–	–	–	–	–	–	–	+	–	–	+	–	+	+	+	+	–	+	+	–	–
Hyla intermedia	–	+	–	–	–	–	–	–	–	+	–	–	–	–	–	–	–	–	–	–	–	–	–	–
Hyla meridionalis	+	–	+	–	–	–	–	–	–	–	–	–	–	–	–	–	–	–	–	–	–	–	–	–
Hyla sarda*	–	–	+	+	+	–	–	–	–	–	–	–	–	–	–	–	–	–	–	–	–	–	–	–
Hyla savignyi	–	–	–	–	–	–	–	–	–	–	–	–	–	–	–	–	–	–	–	–	–	–	–	+
ANURA: PELOBATIDAE:																								
Pelobates fuscus	–	–	–	–	–	–	–	–	–	–	–	–	–	–	–	–	–	–	+	+	–	–	–	–
Pelobates syriacus	–	–	–	–	–	–	–	–	–	–	–	–	–	–	–	–	–	–	+	–	–	–	–	–

Table 6:18 continued

Species		Islands (Archipelagos)																						
	BA	HY	CO	TU	SA	GA	DJ	PE	MA	SI	DA	IO	KI	CR	AS	CY	EV	NS	NA	SS	KA	RH	LY	CP
ANURA: RANIDAE:																								
Rana bedriagae	–	–	–	–	–	–	–	–	–	–	–	–	–	–	–	–	–	–	–	+	–	–	–	+
Rana bergeri	–	–	+	+?	+?	–	–	–	–	+	–	–	–	–	–	–	–	–	–	–	–	–	–	–
Rana cerigensis*	–	–	–	–	–	–	–	–	–	–	–	–	–	–	–	–	–	–	–	–	E	?	–	–
Rana cretensis*	–	–	–	–	–	–	–	–	–	–	–	–	–	E	–	–	–	–	–	–	–	–	–	–
Rana dalmatina	–	–	–	–	–	–	–	–	–	+	+	+	–	–	–	–	–	–	–	–	–	–	–	–
Rana epeirotica	–	–	–	–	–	–	–	–	–	–	–	+	–	–	–	–	–	–	+	–	–	–	–	–
Rana graeca	–	–	–	–	–	–	–	–	–	+	–	–	–	–	–	–	+	–	–	–	–	–	–	–
Rana hispanica	–	–	+	+?	+?	–	–	–	–	+	–	–	–	–	–	–	–	–	–	–	–	–	–	–
Rana kurtmuelleri	–	–	–	–	–	–	–	–	–	–	+	+	+?	–	+	+	+	+?	+	–	–	–	–	–
Rana perezi	I	–	–	–	–	–	–	–	–	–	–	–	–	–	–	–	–	–	–	–	–	–	–	–
Rana ridibunda	–	–	–	–	–	–	–	–	–	+	+	+	–	–	–	–	–	–	+	+	–	–	–	–
Total Anura	4	2	6	6	5	1	1	1	1	8	6	6	1	4	2	4	3	1	7	5	2	3	0?	3
Total Amphibia	4	2	8	6	10	1	1	1	1	8	8	8	1	4	2	5	5	1	7	5	3	3	1	3
Species endemic	1	0	3	0	5	0	0	0	0	0	0	0	0	1	0	0	0	0	0	0	1	0	0	0
Percentage endemics	25	0	38	0	50	0	0	0	0	0	0	0	0	25	0	0	0	0	0	0	33	0	0	0

za and Vanni, 1987; Sbordoni et al., 1990; Ooster-broek and Arntzen, 1992; Caccone et al., 1994; Lanza et al., 1995). On the other hand, the Hellenic Arc (from Crete to Rhodes), also inhabited by two endemic frogs, has been isolated from the mainland for a longer time than have other Aegean islands (Beerli, 1994).

The Mediterranean Region may be subdivided into four provinces which differ by faunal composition even at the family level (Table 6:17; Fig. 6:8). Faunal similarities among the four Mediterranean provinces shows a greater association between Northwestern Africa and Iberia than with the other provinces (Table 6:19). At the species level, the Iberian Province has nearly equivalent relationships with the other three provinces. Surprisingly, the closest similarity (0.35) exists between the Italian and Balkan provinces; this similarity can be explained by the mixture of widespread and/or boreal amphibians in those provinces. Coefficients at the genus- and species-group levels is higher than at the species level, and the relationships among provinces are modified somewhat. The Iberian Province shows a stronger association with the Italian Province than with the others. This seems to reflect the older historical connections between these parts of the western Mediterranean.

Basal lineages of many groups perhaps are restricted to the western Mediterranean, whereas more recent lineages occur in the eastern Mediterranean, including Asia Minor (Oosterbroek and Arntzen, 1992). Geologically, southwestern Europe (including the present Iberian Peninsula) has played an important role in the make up of Corsica, Sardinia, and the southern part of Italy (Dercourt et al., 1986). The historical geology seems to be mirrored in the faunal similarity above the species level (Table 6:19).

Table 6:19. Faunal similarity among four provinces of the Mediterranean Region as calculated by the Czekanowski Coefficient. Comparison by species is in the upper right; comparison by genera and species groups is in the lower left. Abbreviations of provinces in headings shown in first column. Two species probably introduced in the Iberian Province (*Bufo viridis* and *Discoglossus pictus*) are excluded.

Province	NWA	IBE	ITA	BAL
Northwestern Africa (NWA)	–	0.22	0.15	0.14
Iberian (IBE)	0.69	–	0.22	0.24
Italian (ITA)	0.56	0.77	–	0.35
Balkan (BAL)	0.60	0.69	0.80	–

Clearly, the Iberian landmass may have been an important Tertiary source for faunal dispersal, diversification, and exchanges in various directions and at different times—to the south to Africa via the Strait of Gibraltar (Bons, 1973; Busack et al., 1985; Busack, 1986a, 1986b), to the east via the microplate (island) drifting (Caccone et al., 1994), and to and from the north (Table 6:16).

Northwestern African Province

This province covers the mainland of northwestern Africa west of Egypt, the Canary Islands, the Tunisian coastal islands Gallita and Djerba, as well as Italian Pelagie Islands (Lanza and Vanni, 1987). The province has a relatively poor amphibian fauna one-sixth to one-third the size of that in the other provinces; it contains only 12 species (18% of the total Mediterranean amphibians), namely three salamanders and nine anurans. Five species in five genera (*Pleurodeles, Salamandra, Bufo, Pelobates,* and *Rana*) are endemic (Table 6:17). Although the number of species is low, the percentage (42%) of endemics is high.

Iberian Province

This province embraces the Iberian Peninsula, southwestern France northward to the Garonne River and the Central Massif, and the Mediterranean coastal zone of France (with small northern extension along the Rhone River) eastward approximately to the Var River area. The province also includes the Mediterranean Balearic Islands and probably, Madeira and Azores Islands of the Atlantic Ocean (Fig. 6:8). The fauna contains 25 species (37% of the regional list), namely eight salamanders and 17 anurans. Nine species (36%) in six genera and three families are endemic to the province (Table 6:17). The endemic genus *Chioglossa,* various salamandrids, including *Pleurodeles* and endemic species in *Euproctus* and *Triturus,* as well as the presence of all species of the genus *Alytes* (three endemic), an endemic *Discoglossus,* and two endemic brown frogs characterize the province. Some Iberian amphibians have reached the northwestern part (Ligurian) of the Italian Province (e.g., *Pelodytes punctatus*) and/or northwestern Africa (e.g., *Hyla meridionalis* and *Pleurodeles waltl*).

Italian Province

This province is situated south of Alps (Fig. 6:8); except for a few small northern areas, it corresponds to the territory outlined by Lanza and Corti (1996) as "the Italian region." Thus, the province includes the eastern part of Cote d'Azur, Mediterranean France (a transition to the Iberian Province), mainland and peninsular Italy, the southernmost corner of Switzerland (Ticino), southwestern Slovenia, and northwestern Croatia (e.g., the Trieste and Istria areas). The lowland east of the Po River is a broad transition area to the European Region. At least a part of that territory may be assigned to the European, rather than to the Mediterranean, Region. Four peripheral species are known from this transition area (Table 6:17). Apart from the mainland, the Italian Province includes the large Mediterranean islands of Sicily, Sardinia (both in Italy), and Corsica (France). Some smaller islands like the coastal Hyères islands (France) in the north, and the Maltese islands (Malta and Gozo) in the south, also are included in the Italian Province. This province has the most diversified fauna in the Mediterranean Region. It is inhabited by 40 species (57% of the regional list), namely by 18 salamanders and 22 anurans. Among them, 21 species (53%) in nine genera and five families are endemic to the province (Table 6:17). High island endemism is the characteristic feature (8 species from Corsica and Sardinia; Table 6:18). Faunistically, the province is characterized by the plethodontid *Hydromantes* (all endemics), various salamandrids including the endemic genus *Salamandrina,* one endemic species of *Salamandra,* three endemic newts (*Euproctus* and *Triturus*), various discoglossids with endemic *Discoglossus* and *Bombina* (but no *Alytes*), two endemic tree frogs, and many ranids, four of which are endemic.

Balkan Province

This province occupies the coastal Adriatic zone in Yugoslavia and northern Albania, the southern parts of Albania and Macedonia, and almost all of Greece (except the northeastern part east of approximately 25° E long.). The province also includes the Ionian Islands in the Adriatic Sea, and numerous islands in the Aegean Sea west of the arbitrary line associating Samothraki with Crete (Fig. 6:8). This province harbors 17 amphibians (25% of the regional list), namely five salamanders and 12 anurans. In the Mediterranean Region, the level of endemism in the province is lowest (24%) and least diversified taxonomically; there are only four endemic species, all *Rana* (Table 6:17).

EUROPEAN REGION

Situated north of the Mediterranean Region and the Caucasus, the European (or Boreal) Region covers all the remaining parts of Europe, except Iceland and the true Arctic zone (Fig. 6:8). In the south, the region occupies France north of the Garonne River and the Mediterranean coast, the Alps including the northernmost parts of mainland Italy, almost entirely the republics of the former Yugoslavia, the northeastern part of Greece, Turkish Thrace, and probably a small neighboring area of Asia Minor. The eastern limit roughly corresponds to the edge of the eastern slope of the Ural Mountains, with some extensions in the sub-Arctic Yamal Peninsula and the lowlands of western Siberia (west of Tyumen and Kurgan provinces in Russia). In the southeast, the border probably reaches steppe areas southward, approximately, to the Emba River and the Irghiz-Turgai area in northern Kazakhstan. The eastern range limits of *Rana temporaria* and *Pelobates fuscus* are indicators of the regional limits.

Although more than twice the size of the Mediterranean Region, the European Region has only half as many species and harbors only 20% of the species in the Palearctic Realm (Table 6:14). This difference supports the general rule of classic biogeography that more northern regions have larger territories and smaller faunas. The regional fauna contains 35 species—14 salamanders and 21 anurans in 11 genera and 9 families (Table 6:20). Taxonomically, the familial composition in the European Region is similar to that of the Mediterranean, but with the presence of a hynobiid salamander instead of plethodontids (Tables 6:14 and 6:20). Unlike the Mediterranean, the European Region is characterized by a low level of endemism (only 11%); endemics are one *Salamandra* and three *Triturus.* However, a few more species—*Proteus anguinus, Salamandra atra, Bombina bombina,* and probably *Rana esculenta*—are nearly endemic to the region, because their distributions beyond the European Region are limited and, in fact, only in transitional areas. Some species, mainly salamanders, are represented by endemic subspecies (Table 6:20).

Four provinces can be recognized in the European Region (Fig. 6:8). However, their distinctness is significantly less than the Mediterranean provinces (Table 6:20). Therefore, the proposed provisionary

Table 6:20. Distribution of amphibians in four provinces of the European Region. Abbreviations of provinces: AT = Atlantic, AD = Alpine-Dinaric, CE = Central European, EE = East European. Asterisks denote species (*) or subspecies (**) endemic to the European Region; + = present, – = absent, E = endemic to province, ? = questionable occurrence.

Species	AT	AD	CE	EE
CAUDATA: HYNOBIIDAE:				
Salamandrella keyserlingii	–	–	–	+
CAUDATA: PROTEIDAE:				
*Proteus anguinus***	–	+	–	–
CAUDATA: SALAMANDRIDAE:				
*Salamandra atra***	–	+	–	–
*Salamandra lanzai**	–	E	–	–
*Salamandra salamandra***	+	+	+	?
*Triturus alpestris***	+	+	+	–
Triturus carnifex	–	+	+	–
*Triturus cristatus**	+	+	+	+
*Triturus dobrogicus**	–	–	E	–
Triturus helveticus	+	+	–	–
Triturus karelinii	–	–	+	+
Triturus marmoratus	+	–	–	–
*Triturus montandoni**	–	–	E	–
*Triturus vulgaris***	+	+	+	+
Total Caudata	6	9	8	4
Number of endemics	0	1	2	0
Percent endemic	0	22	25	0
ANURA: BUFONIDAE:				
Bufo bufo	+	+	+	+
Bufo calamita	+	+	+	+
*Bufo viridis***	+	+	+	+
ANURA: DISCOGLOSSIDAE:				
Alytes obstetricans	+	+	+	–
Bombina bombina	–	–	+	+
Bombina variegata	+	+	+	–
ANURA: HYLIDAE:				
Hyla arborea	+	+	+	+
Hyla meridionalis	+	–	–	–
ANURA: PELOBATIDAE:				
Pelobates cultripes	+	–	–	–
Pelobates fuscus	+	–	+	+
Pelobates syriacus	–	–	+	–
ANURA: PELODYTIDAE:				
Pelodytes punctatus	+	+	–	–
ANURA: RANIDAE:				
Rana arvalis	–	–	+	+
Rana dalmatina	+	+	+	–
Rana esculenta	+	+	+	+
Rana graeca	–	–	+	–
Rana grafi	+	–	–	–
Rana lessonae	+	+	+	+
Rana perezi	+	–	–	–
Rana ridibunda	+	+	+	+
*Rana temporaria***	+	+	+	+
Total Anura	17	12	16	11
Number of endemics	0	0	0	0
Percent endemic	0	0	0	0
Total Amphibia	23	21	24	15
Total number of endemics	0	1	2	0
Total percent endemic	0	5	8	0

delimitation of provinces in the European Region is questionable and awaits more thorough study. An analysis of faunal similarity at the species level expressed by the Czekanowski Coefficient shows that the most geographically distant provinces have the lowest similarity. For example, the association of the Atlantic Province with the Alpine-Dinaric Province is 0.77 (the highest value among all pairs), with the Central European Province 0.68, and with the distant Eastern European Province only 0.59. The lowest value (0.56) is between the Alpine-Dinaric and Eastern European provinces. Essentially equal values exist between the Central European Province and the Alpine-Dinaric Province (0.71) and the Eastern European Province (0.72). The faunal similarity at the species level among the provinces in the European Region (0.56–0.77) coincides with the range of similarity at the genus- and species-group level in the Mediterranean Region (0.56–0.80), whereas the species-level similarity among the latter is only 0.13–0.36. Thus, these data support significantly weaker differentiation among subdivisions of the European Region.

Atlantic Province

This province includes the Atlantic lowlands and Central Massif of France. The provisionary border line may be drawn through Belgium and easternmost France to the Alps (Fig. 6:8). Beyond the mainland, the province also includes the British Isles with a depauperate amphibian fauna (six native species vs. 11 in the closest continental area). Actually, the province is a broad transition area between the faunas of the Iberian Province of the Mediterranean Region and the true boreal fauna of Europe (Zuiderwijk, 1980). Some Atlantic species (e.g., *Triturus helveticus*) are widespread north of the border. The Atlantic Province contains 23 species (66% of the regional list), six salamanders and 17 anurans; no species is endemic to the province (Table 6:20).

Alpine-Dinaric Province

This province embraces the Alps, as well as the Dinaric Mountains along the Adriatic side of the Balkans (Fig. 6:8). The mountains are the most characteristic feature of the province. The number of species (21) is similar to that of the Atlantic Province (Table 6:20). The fauna is characterized by the presence of a proteid, absence of pelobatids, and larger number of salamanders (in relation to anurans), including three species of *Salamandra*. Only one species (*Salamandra lanzai*) is endemic. *Proteus anguinus* and *Sala-*

mandra atra are nearly endemic; their ranges extend slightly into another region and/or province.

Central European Province

This province is situated east of the two previous provinces. The eastern boundary seems to coincide approximately with the eastern range limit of the beech, *Fagus silvatica* (Walter, 1968), so the eastern edge of the province is defined by a line passing through southern Sweden, eastern Germany (approximately the Elbe River), the Carpathians, and the Danube Delta (Fig. 6:8). The fauna consists of 24 species (69% of the regional list), namely eight salamanders (including seven *Triturus*) and 16 anurans (Table 6:20). Two newts (*Triturus dobrogicus* and *T. montandoni*) are endemic to the province.

Eastern European Province

This huge province encompasses the remaining part of the European (or Boreal) Region (Fig. 6:8). The southern limit in southern Russia is a line connecting the Kuban and Terek rivers between the Black and Caspian seas. This vast province has the poorest fauna in the European Region (Table 6:20); it is inhabited by only 15 species (43% of the regional list), namely four salamanders and 11 anurans. The presence of a hynobiid, and the absence of the proteid and pelodytids, as well as probably *Salamandra*, combine to contribute to the "picture" of the province, which has no endemics.

WEST ASIAN REGION

This vast territory is heterogeneous topographically and climatically, but most of it is arid (Fig. 6:8). The western boundary seems to coincide with a line connecting the Aegean islands of Samothraki in the north and Crete in the south. The region includes the Anatolian part of Turkey (probably exclusive of the small westernmost area), the Caucasus, Iran, and Afghanistan. The eastern edge is delimited by the western mountains in Pakistan and northwestern India (north of Jammu and Kashmir). In the north, the region covers the so-called Turan Lowland in the former Soviet Central Asia and reaches the steppe zone of northern and northeastern Kazakhstan, as well as the Tian Shan Mountains and Lake Balkhash area of eastern Kazakhstan. In the south, the region encompasses the Near East southward to the Palearctic limits north of the Arabian Peninsula. Egypt is a questionable part of the region; alternatively (and perhaps more correctly) this area may be included in the Afrotropical Realm.

Table 6:21. Distribution of amphibians in four provinces of the West Asian Region. Abbreviations of provinces: CAU = Caucasian, EGY = Egyptian, IRA = Iranian, and NE = Near East. Asterisks denote species (*) or subspecies (**) endemic to the region; + = present, – = absent, E = endemic to province, P = peripheral occurrence; ? = questionable occurrence.

Species	CAU	NE	IRA	EGY
CAUDATA: HYNOBIIDAE:				
*Batrachuperus gorganensis**	–	–	E	–
*Batrachuperus mustersi**	–	–	E	–
*Batrachuperus persicus**	–	–	E	–
CAUDATA: SALAMANDRIDAE:				
*Mertensiella caucasica**	E	–	–	–
*Mertensiella luschani**	–	E	–	–
*Neurergus crocatus**	–	E	–	–
*Neurergus kaiseri**	–	E	–	–
*Neurergus microspilotus**	–	E	–	–
*Neurergus strauchii**	–	E	–	–
*Salamandra infraimmaculata**	–	E	–	–
Triturus karelinii	+	+	+	–
*Triturus vittatus**	+	+	–	–
*Triturus vulgaris***	+	+	–	–
Total Caudata	4	9	4	0
Number of endemics	1	6	3	0
Percent endemic	25	67	75	0
ANURA: BUFONIDAE:				
Bufo bufo	–	+	–	–
Bufo danatensis	–	–	+	–
*Bufo kassasii**	–	–	–	E
*Bufo kavirensis**	–	–	E	–
Bufo latastii	–	–	+	–
Bufo olivaceus	–	–	P	–
Bufo regularis	–	–	–	+
*Bufo shaartusiensis**	–	–	E	–
*Bufo siachinensis**	–	–	E	–
Bufo stomaticus	–	–	P	–
*Bufo surdus***	–	+	+	–
*Bufo verrucosissimus**	+	?	P	–
*Bufo viridis***	+	+	+	+
ANURA: DISCOGLOSSIDAE:				
Bombina bombina	P	?	–	–
*Discoglossus nigriventer**	–	E	–	–
ANURA: HYLIDAE:				
*Hyla arborea***	+	+	–	–
Hyla savignyi	P	+	P	–
ANURA: PELOBATIDAE:				
Pelobates fuscus	P	–	–	–
Pelobates syriacus	+	+	–	–
ANURA: PELODYTIDAE:				
*Pelodytes caucasicus**	E	–	–	–
ANURA: RANIDAE:				
Euphlyctis cyanophlyctis	–	–	P	–
Ptychadena mascareniensis	–	–	–	+
Rana bedriagae	–	+	–	+
*Rana cerigensis**	–	E	–	–
Rana dalmatina	–	+	–	–
*Rana holtzi**	–	E	–	–
*Rana macrocnemis**	+	+	+	–
*Rana ridibunda***	+	?	+	–
*Rana terentievi**	–	–	E	–
Total Anura	10	12	15	5
Number of endemics	1	3	4	1
Percent endemic	10	25	27	25
Total Amphibia	14	21	19	5
Total number of endemics	2	9	7	1
Total percent endemic	14	43	37	20

The fauna consists of 42 species (13 salamanders and 29 anurans) in 14 genera and eight families (Tables 6:14 and 6:21); this amounts to 24% of the Palearctic species. The number of species is 20% higher than that in the European Region but is notably less than in the Mediterranean Region. The presence of hynobiid salamanders and an abundance of toads and salamandrids characterize the region. Except for the absence of the Proteidae, familial composition in the West Asian Region resembles that of the European Region.

Regional endemism (Table 6:15 and 6:21) is much higher (55%) than that of the European Region (11%) and only slightly lower than that in the Mediterranean Region (61%). The two endemic genera (14% of the regional list) are salamandrids—*Mertensiella* (2 species) and *Neurergus* (4 species). Seventeen other endemic species are members of seven genera (Table 6:21)—*Batrachuperus* (3), *Salamandra* (1), *Triturus* (1), *Bufo* (5), *Discoglossus* (1), *Pelodytes* (1), and *Rana* (5).

The region is subdivided into four provinces (Fig. 6:8; Table 6:21).

Caucasian Province

This province includes the Caucasus northward approximately to the Kuban and Terek rivers in southern Russia, but the Armenian Plateau is excluded. The province also includes the Black Sea coastal area in Turkey westward to the Yesilirmak Mountains at approximately 39° E lon. (Atatür and Budak, 1982; Schmidtler, 1986). In the southeast, the province extends south of the Caspian Sea eastward to the Elburz Mountains in northern Iran. The Crimea Peninsula (except the northern part), with isolated populations of *Hyla arborea* and *Triturus karelinii,* also may be assigned to the province.

The territory is inhabited by 14 species (33% of the regional list), namely four salamanders and 10 anurans (Table 6:21). Two species, *Mertensiella caucasica* and *Pelodytes caucasica,* are endemic (14% of the provincial list). Three species, *Triturus vittatus, T. vulgaris,* and *Hyla arborea,* are represented by endemic subspecies. Two boreal anurans (*Bombina bombina* and *Pelobates fuscus*), and *Hyla savignyi* have peripheral distributions in the province.

Near East Province

This province encompasses Turkey (the Pontic northeast excluded), southern part of the Transcaucasus (i.e., the northern part of the Armenian Plateau), Iran west of the Elburz and Zagros Mountains, and the Near East itself (Iraq, Syria, Lebanon, Israel, and Jordan). Also included are the Mediterranean islands of Cyprus, South Sporades, and others east of the line from Samothraki to Crete (Fig. 6:8).

The province has the richest fauna in the region, with 21 species (50%), namely nine salamanders and 12 anurans. The province is characterized by high endemism (9 species, or 43%); 67% of the salamanders and 25% of the anurans are endemic. The only endemic genus is *Neurergus* with four species. Four other genera also contain endemic species—*Mertensiella* (1), *Salamandra* (1), *Discoglossus* (1), and *Rana* (2). Four other species (*Triturus vittatus, T. vulgaris, Bufo surdus,* and *B. viridis*) are represented by endemic subspecies. Because of the arid climate, amphibians, especially salamanders, are distributed mainly in mountainous or coastal areas; they avoid hot depressions, such as inner Anatolia.

Iranian Province

This province consists of two geographically unequal parts. The southern part covers the great Iranian Plateau lying east of the Zagros Mountains and extends eastward to the western mountains of Pakistan; the Elburz Mountain Range is included; the highland node with the Karakoram Mountains and neighboring northwest Himalaya (northern Pakistan, Ladakh area) also belong to this part of the province. The larger, northern part of the province includes the Turan Lowland between the Caspian Sea and Lake Balkhash. In the south, this territory is separated by the mountain chain from the Kopet Dagh to the Pamirs and western Tian Shan. The northern limit crosses northern Kazakhstan and coincides with the border of the West Asian Region (Fig. 6:8). Thus, the province has the largest territory in the region. Geographically, this is a huge area of arid mountains and hot deserts.

The fauna consists of 19 species (or 45% of the regional list), namely four salamanders and 15 anurans. The presence of hynobiid salamanders and various toads (10 of 14 species recorded in the region, mostly in the *Bufo viridis* group), as well as peripheral occurrence of some Oriental species characterize the province, which has seven endemic species (or 37%), which belong to genera *Batrachuperus* (3 species), *Bufo* (3), and *Rana* (1). All of these endemics have fairly limited distributions (only one or a few localities). *Bufo viridis* and *Rana ridibunda* are represented by distinct subspecies.

Egyptian Province

This province is situated in northeastern Africa and covers the territory shared by northern Egypt and eastern Libya (Fig. 6:8). Therefore, this is mostly the northeastern Saharan area. The fauna is the poorest in the region, and contains five anurans only (12% of the regional list), namely three toads and two frogs (Table 6:21). Unlike members of other provinces, each species is a member of a distinct species group or genus. The Afrotropical anurans are the peculiarities of the province, including the sole endemic, *Bufo kassasii* from the Nile Delta. The Nile Valley, Mediterranean coast, and oasis system are three patterns of amphibian distribution in the province.

SIBERIAN REGION

This, the largest region, occupies the northern latitudes of Asia (Fig. 6:8). The southern border crosses northern Kazakhstan west of the Irtysh River to the Chinese part of the Altay Mountains (no data for the Mongolian Highlands). A small lowland area east of Lake Uvsunur (with a record of *Salamandrella keyserlingii*) and northern Mongolia (as indicated by the southern range limit of *Rana amurensis* [Borkin, 1988; Borkin and Kuzmin, 1988]) also are assigned to the Siberian Region. The southeast boundary almost corresponds to the Amur River (and partly coincides with the Russian-Chinese border), with a short extension northward along the Zeya River. But the Russian Far East, east of the Bureya River and south of the Amur River is a part of the East Asian Region. The Chinese part of the Altay Mountains lying in the north of the Xinjiang Province in northwestern China, is sometimes considered as a distant area of the Northeastern China Region (the Da Xingan Ling Subregion, NE-A; Zhao and Adler, 1993) or the Korean Peninsula–Northeast China Region (KN; Zhao, this volume) "because of its faunal similarities to the Greater Xingan Ranges." However, I fail to find any amphibian species shared by these areas and joined this tip of Xinjiang to the Siberian Region. The two northernmost Kurile Islands (Paramushir and Shumshu), as well as the Kamchatka Peninsula with the coastal Karaginsky Island are inhabited by *Salamandrella keyserlingii*, and, clearly, belong to the Siberian Region. Other islands in the Bering Sea have no amphibians.

Most of the region is covered by the taiga (coniferous forests) replaced by open landscapes in the north (tundra) and in the south (steppe). The ground throughout most of the territory contains permafrost. Such a severe climate does not favor high amphibian diversity. The region is inhabited by only nine species, namely two salamanders and seven anurans, belonging to five genera and five families (Appendix 6:1; Table 6:14). This is the poorest regional fauna of the Palearctic Realm (only 5%). *Salamandrella keyserlingii* and *Rana amurensis* are characteristic of the region; both are widely distributed throughout the vast territory, including the southern tundra and steppe areas. The former species reaches northward to 72° N lat. in Yakutia; it is the northernmost record for a terrestrial vertebrate ectotherm (Borkin et al., 1984). Some localities are in the coastal zone of the Arctic Ocean (Dokuchaev et al., 1984). The northern limit of amphibian distribution (Fig. 6:8) in Asia is based on the range limit of *Salamandrella keyserlingii*, a species still unreported from extreme northeastern Asia (Chukotka Peninsula), which according to some authors (e.g., Emeljanov, 1974) belongs to the Nearctic Realm.

Minor differences are observed in the southern parts of the region. *Bufo viridis* enters peripherally in the southwest. The European *Triturus vulgaris* and *Rana arvalis* are western invaders with more extended ranges in Siberia. The latter extends eastward along the Lema River to Yakutia (124° E lon.) (Borkin et al., 1984); this is the easternmost record for a European amphibian. Southeast of Yakutia is the northern limit of another brown frog, *Rana chensinensis*, which has a wide distribution in eastern and central Asia. Two other widespread anurans, *Bufo raddei* and *Hyla japonica*, are distributed along the Amur River Valley and reach northern Mongolia (via northeast China) and the area of Lake Baikal; both species prefer open habitats.

The Siberian Region has no true endemics. *Rana arvalis* is represented by a subspecies *R. a. altaica* from the Altay Mountains in southwest Siberia. This taxon is treated as a distinct species by some authors (Ye et al., 1981; Zhao and Adler, 1993; Zhao, this volume), but its status needs to be confirmed.

Taking into consideration the poverty and slight geographic differentiation of the amphibian fauna, I am not inclined to recognize any provinces in the Siberian Region based only on patterns of amphibian distribution.

CENTRAL ASIAN REGION

In the northwest, this region covers eastern Kazakhstan (east of Lake Balkhash) and most of the

Tian Shan Mountains east of the Ferghana Valley in Kyrgyzstan. The northern limit in Mongolia seems to coincide with the Hangay Mountain Range, the northern slopes of which face the steppe zone of the Siberian Region, and the southern slopes face the Gobi Desert (Borkin, 1988).

Most of the Central Asian Region is situated in southern Mongolia and western and northern China southward to the Himalayas and eastward to the Da Xingan Ling (Greater Hsingan Range). The region consists of vast depressions (to 154 m below sea level) or plateaus alternating with high mountain chains, with peaks up to 8848 m. Except for the southeastern part of the region, the drainage systems are entirely inland, not flowing into any oceans. Mostly, the climate is extremely arid, characterized by drastic diel extremes in temperature.

Only 20 amphibians, four salamanders and 16 anurans in seven genera and five families exist in the region (Table 6:14). A few hynobiid salamanders (*Batrachuperus* and *Ranodon*), some *Bufo,* the megophryine genus *Scutiger,* and ranids are typical for the Central Asian Region. Only four species (*Ranodon sibiricus, Scutiger maculatus, S. nyingchiensis,* and *Rana asiatica*) are endemic to the region (Table 6:15).

Zhao (this volume) divided this huge, arid territory into two provinces ("regions"): Mongol-Xinjiang and Qinghai-Xizang (Tibet). Zhao and Adler (1993) published a more detailed scheme with additional subdivisions (subregions). The slight discrepancies between Zhao's and my lists seems to be influenced by our different northern limits of the Central Asian Region and by some taxonomic differences. For instance, *Rana tenggerensis* was considered to be a synonym of *R. nigromaculata* by Fei et al. (1991) and Dubois and Ohler (1995a); the diploid *Bufo viridis* is replaced by a tetraploid species, *B. danatensis,* and *Hyla japonica* and *Rana amurensis* are distributed in northern, but not in southern, Mongolia. The area covered herein and situated in the former Soviet Union almost entirely could be joined with the Tian Shan Subregion (N-C) of the Nei Mongol–Xinjiang Region outlined by Zhao and Adler (1993: Fig. 37).

East Asian Region

In the north, this region includes the Russian Far East, i.e., the area south of the Amur River; Sakhalin and southern Kurile Islands (Kunashir, Shikotan, and some smaller islets) are included. Basical-ly, the territory is forested hills and mountain ridges, with some spacious flooded meadows or steppe areas (e.g., Lake Khanka area). The comparatively mild climate is significantly influenced by the Pacific Ocean. Most of the region outlined by Zhao (this volume) covers northeast China, Korean Peninsula, and Japanese Islands (exclusive of most of the Ryukyu Islands).

The amphibian fauna containing 55 species (24 salamanders and 31 anurans) in 15 genera and 10 families, is second in size only to the Mediterranean Region in the Palearctic Realm. Diverse hynobiids and ranids, some toads, and the presence of rhacophorids and a microhylid characterize the region (Table 6:14; Appendix 6:1). Zhao's list (this volume) is modified slightly, as follows: the Xinjiang *Rana altaica* is omitted, and *Hyla suweonensis,* endemic to the Korean Peninsula (Kuramoto, 1980) is added. The region has the highest level of endemism in the Palearctic Realm; 41 species (75% of the regional list) are endemic. Endemics include 92% of the salamanders and 61% of the anurans (Table 6:15; Appendix 6:1). Two hynobiid genera, *Onychodactylus* (2 species) and *Satobius* (sometimes recognized as a subgenus, 1 species), are endemic. Endemic species in other genera include *Andrias* (1), *Hynobius* (all 17), *Cynops* (1), *Bufo* (3), *Bombina* (1), *Hyla* (1), *Scutiger* (1), *Rana* (10), *Buergeria* (1), and *Rhacophorus* (2).

The East Asian Region is divided into three parts. These are the Korean Peninsula and Northeast China Region (KN, in Zhao, this volume) or Northeastern China Region (NE in Zhao and Adler, 1993), the North China Region (NC), and the Japan Proper (JP) including Tsushima Island, Osumi Group and the northern part of Tokara Group of the Ryukyu Islands (Zhao, this volume).

The Russian mainland with nine species is the northeastern part of the KN Province enclosed in the "Changbai Shan Subregion" (Zhao and Adler, 1993: Fig. 37). In Russian zoogeographical works, the fauna of this territory is often termed "Manchurian" and is represented by *Onychodactylus fischeri, Bufo gargarizans, Bombina orientalis, Hyla japonica, Rana chensinensis,* and *R. nigromaculata,* plus two Siberian species (*Salamandrella keyserlingii* and *Rana amurensis*) and the Central Asian *Bufo raddei.* Sakhalin Island with five species (*Salamandrella keyserlingii, Bufo gargarizans, Hyla japonica, Rana amurensis,* and *R. chensinensis*) also belongs to this

Table 6:22. Distribution of amphibian genera and species groups (*Bufo* and *Rana*) in six regions of the Palearctic Realm. Abbreviations of regions: CAS = Central Asian, EAS = Eastern Asian, EUR = European, MED = Mediterranean, SIB = Siberian, and WAS = Western Asian; + = presence, – = absent, P = peripheral occurrence.

Genus or Species Group	MED	EUR	WAS	SIB	CAS	EAS
CAUDATA: CRYPTOBRANCHIDAE:						
Andrias	–	–	–	–	+	+
CAUDATA: HYNOBIIDAE:						
Batrachuperus	–	–	+	–	+	–
Hynobius	–	–	–	–	–	+
Onychodactylus	–	–	–	–	–	+
Ranodon	–	–	–	–	+	–
Salamandrella	–	+	–	+	–	+
CAUDATA: PLETHODONTIDAE:						
Hydromantes	+	–	–	–	–	–
CAUDATA: PROTEIDAE:						
Proteus	P	+	–	–	–	–
CAUDATA: SALAMANDRIDAE:						
Chioglossa	+	–	–	–	–	–
Cynops	–	–	–	–	–	+
Euproctus	+	–	–	–	–	–
Mertensiella	–	–	+	–	–	–
Neurergus	–	–	+	–	–	–
Pleurodeles	+	–	–	–	–	–
Salamandra	+	+	+	–	–	–
Salamandrina	+	–	–	–	–	–
Triturus	+	+	+	+	–	–
ANURA: BUFONIDAE:						
Bufo bufo group	+	+	+	+	+	+
Bufo funereus group	–	–	+	–	–	–
Bufo mauritanicus group	+	–	–	–	–	–
Bufo regularis group	–	–	+	–	–	–
Bufo stomaticus group	–	–	P	–	–	–
Bufo viridis group	+	+	+	+	+	+
ANURA: DISCOGLOSSIDAE:						
Alytes	+	+	–	–	–	–
Bombina	+	+	P	–	–	+
Discoglossus	+	–	+	–	–	–
ANURA: HYLIDAE:						
Hyla	+	+	+	+	–	+
ANURA: MICROHYLIDAE:						
Kaloula	–	–	–	–	–	+
ANURA: PELOBATIDAE:						
Pelobates	+	+	+	–	–	–
Scutiger	–	–	–	–	+	+
ANURA: PELODYTIDAE:						
Pelodytes	+	+	+	–	–	–
ANURA: RANIDAE:						
Euphlyctis	–	–	P	–	–	–
Limnonectes	–	–	–	–	–	+
Nanorana	–	–	–	–	+	–
Paa	–	–	–	–	–	+
Ptychadena	–	–	+	–	–	–
Rana esculenta group	+	+	+	–	+	+
Rana rugosa group	–	–	–	–	–	+
Rana temporaria group	+	+	+	+	+	+
ANURA: RHACOPHORIDAE:						
Buergeria	–	–	–	–	–	+
Rhacophorus	–	–	–	–	–	+
Total Amphibia	19	13	19	6	9	18

province. The southern Kurile Islands, geologically associated with the Hokkaido mountains, have been populated from Hokkaido Island and have a depauperate amphibian fauna of only three species (*Salamandrella keyserlingii, Hyla japonica,* and *Rana chensinensis* or *R. pirica* [Borkin and Bassarukin, 1987]). The Hokkaido-Kurile area has two endemic species (and, maybe, one genus)—the hynobiid *Hynobius* (or *Satobius*) *retardatus,* and a brown frog, *Rana pirica* (plus widespread *Salamandrella keyserlingii* and *Hyla japonica*). This area seems to be closer to the Korean-Manchurian Province (Hikida et al., 1989) than to the Japan Proper Region as proposed by Zhao (this volume). The lack of consensus on the taxonomy of the Far Eastern brown frogs with 24 chromosomes (*Rana chensinensis* complex) shows the need for further study (Matsui, 1991; Matsui et al., 1993; Green and Borkin, 1993).

COMPARISON OF REGIONAL FAUNAS

The distributions of amphibian genera and species groups (only *Bufo* and *Rana*) in the six regions of the Palearctic Realm are tabulated in Table 6:22, and the faunal similarities are given in Table 6:23. The greatest similarities (0.82–0.94) at the familial level are among the three western Palearctic regions, namely the Mediterranean, European, and West Asian regions. The Central Asian Region has the weakest similarity with other regions (0.43–0.67, $\bar{x}$ = 0.58), whereas the West Asian Region has the strongest (0.62–0.94, $\bar{x}$ = 0.79). The East Asian Region also has a higher similarity with more distant regions, like the West Asian and European regions (0.78 and 0.74, respectively) than with the contiguous Central Asian and Siberian (0.67 each).

At the generic level, faunal similarities range are 0.29–0.75 ($\bar{x}$ = 0.45), but the range among four provinces of the Mediterranean Region is higher (0.56–0.80, $\bar{x}$ = 0.68). The generic level seems to be more sensitive than the familial level in evaluating faunal similarity between regions. Nevertheless, again, the western Palearctic regions have greater similarities with each other than with the eastern Palearctic regions; those two provinces, the Central and East Asian, are associated with other regions weakly (0.29–0.44, $\bar{x}$ = 0.37; and 0.32–0,45, $\bar{x}$ = 0.39, respectively). Thus, in the Palearctic, there exist two principal parts with diverse faunas, namely the western Atlantic and the eastern Pacific, whereas the central part of the Realm is characterized by a depau-

Table 6:23. Faunal similarity among the six regions in the Palearctic Realm as calculated by the Czekanowski Coefficient. Comparison by families is in the upper right; comparison by genera and species groups is in the lower left. Distributional data are in Tables 6:13 and 6:21. Abbreviations of regions in headings shown in first column.

Region	MED	EUR	WAS	SIB	CAS	EAS
Mediterranean (MED)	–	0.89	0.82	0.57	0.43	0.63
European (EUR)	0.75	–	0.94	0.71	0.57	0.74
West Asian (WAS)	0.58	0.62	–	0.77	0.62	0.78
Siberian (SIB)	0.40	0.63	0.40	–	0.60	0.67
Central Asian (CAS)	0.29	0.36	0.36	0.40	–	0.67
East Asian (EAS)	0.32	0.45	0.32	0.42	0.44	–

perate amphibian fauna. These segments are correlated with climatic and vegetation zonation (Emeljanov, 1974). The formation of such a partition most probably is result of the shaping of the post-Tethys landmass and progressive aridity and cooling of temperate Eurasia since the Oligocene.

As a rule, quantitatively, taxonomic structure of zoogeographic subdivisions is asymmetrical, i.e., two or three families are dominant in making up more than 50% of fauna. In the Palearctic Realm as a whole (Table 6:7), Ranidae (47 species, 27%), Salamandridae (32, 18%), and Hynobiidae (27, 15%) compose 60% of the species. However, the regions in the Palearctic are not a mirror of the realm and have different compositions. In the Mediterranean Region, two families, Salamandridae (18 species, 26%) and Ranidae (19 species, 28%) make up 54% of the regional fauna; discoglossids (10 species) add 15% more. The European Region is characterized by a preponderance of salamandrids (12 species, 34%) and ranids (9 species, 26%), which compose 60% of the regional species. Unlike these regions, salamandrids are replaced by bufonids in other parts of the Palearctic. Thus, bufonids and ranids (each with 3 species or 33%, together 66%) are predominant over other families in the Siberian Region. The same composition of two major families (60%) occurs in the Central Asian Region—bufonids and ranids (each 6 species or 30%). Bufonids (13 species, 31%), salamandrids (10, 24%) and ranids (9, 21%) make up 76% of the West Asian fauna. Another combination is observed in the East Asian Region where hynobiids (21 species, 38%) and ranids (17, 31%) compose 69% of the fauna, whereas bufonids (6 species) add only 11%. Thus, although bufonids are ranked fourth with 14% of the species in the Palearctic Realm, they contribute more significantly in three of the six regions.

REGIONAL DISTRIBUTION OF AMPHIBIANS

Theoretically, every species has a unique range, and the diversity of distribution patterns should approach the number of species. However, the ranges of various species have similarities that can be recognized as distribution patterns. Configurations of ranges have led some authors to suggest "centers" of origin or dispersal (or refuge), habitat, etc., and thereby offer different classifications of such patterns. However, such classifications should be viewed skeptically and used provisionally, because in most cases real evidence supporting such speculations are absent.

Based on amphibian distribution in the Palearctic, 30 patterns are proposed. Some of these partly overlap; for example, the Western Palearctic, Iberian, and Atlantic patterns overlap in the western Palearctic, and the East Asian, Manchurian, and Japanese patterns overlap in the eastern Palearctic. Among these patterns, 26 are shared by Palearctic species, three by Oriental, and one by Afrotropical (Table 6:24; Appendix 6:1).

PALEARCTIC PATTERNS

Western Palearctic

These distributions cover vast areas of Europe and western Asia as well as northwestern Africa. Such polycentric ranges of unclear origin are exhibited by *Bufo viridis* (open landscapes), *Bufo bufo* (forested areas), and possibly by *Rana ridibunda,* which is absent in southwestern Europe and Africa.

Northwestern African

Seven amphibians (e.g., *Bufo mauritanicus* and *Rana saharica*) are distributed in the Mediterranean and Saharan zones in northwestern Africa. The distribution only in Mediterranean northwestern Africa could be recognized as a particular kind (Maghrebian; *Pleurodeles poireti* and *Pelobates varaldii*). *Discoglossus pictus* is the only Maghrebian amphibian that extends beyond Africa; it is represented by a distinct subspecies in Sicily and the Maltese Islands, and by probably introduced populations in the northeastern part of the Iberian Peninsula.

Iberian

Superficially, this pattern shared by 15 amphibians can be viewed as a particular case of the Atlantic distribution. Most species are confined entirely or mainly to the Iberian Peninsula, but some amphibians extend into western Morocco (*Pleurodeles waltl*) or to southern France (*Pelobates cultripes* and *Rana perezi*). *Alytes obstetricans,* having a large range from Morocco to Germany, also seems to belong to this group. Three subdivisions of the pattern are: (1) Iberian proper (true Iberian); some species occur only south of the northern Cantabrian and Pyreneen Mountains (e.g., *Alytes cisternasii* and *A. dickhilleni*) or are essentially associated with either of these mountain ranges (e.g., *Chioglossa lusitanica* and *Rana iberica*). (2) Pyreneen; two species, *Euproctus asper* and *Rana pyrenaica,* are endemic to the Pyrenees Mountains. (3) Balearic; one species, *Alytes muletensis,* is restricted to the Balearic Islands.

Atlantic

Four species are distributed throughout most of the western part of Europe, including the Iberian Peninsula. These are *Triturus helveticus, T. marmoratus, Pelodytes punctatus,* and *Bufo calamita;* the latter extends eastward into the Baltic countries and central Belorussia.

Italian

Twenty species (e.g., *Bombina pachypus, Hyla intermedia,* and green frogs of the genus *Rana*) are entirely or essentially confined to the Italian Province. Three subdivisions are: (1) Apenninean; the ranges of 12 species (e.g., *Triturus italicus* and *Rana italica*) cover the Apennine Peninsula, and some of these (e.g., *Salamandrina terdigitata*) have a northwestern extension. (2) Ligurian; some species (e.g., *Hydromantes strinatii*) are distributed only in northwestern Italy and neighboring Mediterranean France. (3) Tyrrhenian; ten species (Table 6:18) are basically restricted to Corsica (e.g., *Euproctus montanus*) or Sardinia (e.g., *Hydromantes genei*), or to both plus a few small islands (*Discoglossus sardus* and *Hyla sarda*).

Apennine-Balkan

A newt, *Triturus carnifex,* in the *T. cristatus* group is known from the Apennine Peninsula and the western Balkans. The combined ranges of the brown frogs *Rana italica* and *R. graeca,* which may be subspecifically related also have this pattern.

Alpine-Dinaric

This pattern is shown by species that only inhabit the Alps (*Salamandra lanzai*) and/or Dinaric Alps (*S. atra*). The subterranean *Proteus anguinus* also is included.

Table 6:24. Allocation of distribution patterns in six regions of the Palearctic Realm. Abbreviation of regions: CAS = Central Asian, EAS = East Asian, EUR = European, MED = Mediterranean, SIB = Siberian, WAS = West Asian; n = number of species % = percent of species in region.

	Region												
Distribution Pattern	MED		EUR		WAS		SIB		CAS		EAS		Total	
	n	%	n	%	n	%	n	%	n	%	n	%	n	%
Palearctic Patterns:														
Western Palearctic	3	4	3	9	3	7	2	22	1	5	–	–	3	2
Northwestern African	7	10	–	–	–	–	–	–	–	–	–	–	7	4
Iberian	15	22	5	14	–	–	–	–	–	–	–	–	15	9
Atlantic	4	6	4	11	–	–	–	–	–	–	–	–	4	2
Italian	20	29	–	–	–	–	–	–	–	–	–	–	20	11
Apennine-Balkan	1	1	1	3	–	–	–	–	–	–	–	–	1	<1
Alpine-Dinaric	2	3	3	9	–	–	–	–	–	–	–	–	3	2
Adriatic	2	3	–	–	–	–	–	–	–	–	–	–	2	1
Balkan	3	4	1	3	–	–	–	–	–	–	–	–	3	2
Central European	6	9	7	20	2	5	–	–	–	–	–	–	8	5
East European	5	7	8	23	3	7	2	22	–	–	–	–	8	5
East Mediterranean	1	1	2	6	2	5	–	–	–	–	–	–	2	1
West Asian	–	–	–	–	16	38	–	–	–	–	–	–	16	9
Iranian	–	–	–	–	4	10	–	–	–	–	–	–	4	2
Hindu Kush	–	–	–	–	1	2	–	–	–	–	–	–	1	<1
Tadjik	–	–	–	–	2	5	–	–	–	–	–	–	2	1
Kashmir	–	–	–	–	2	5	–	–	–	–	–	–	2	1
Siberian	–	–	1	3	–	–	2	22	–	–	2	4	2	1
Central Asian	–	–	–	–	1	2	–	–	3	15	–	–	3	2
Mongolian	–	–	–	–	–	–	1	11	1	5	1	2	1	<1
East Asian	–	–	–	–	–	–	2	22	4	20	9	16	9	5
Manchurian	–	–	–	–	–	–	–	–	–	–	11	20	11	6
North China	–	–	–	–	–	–	–	–	–	–	1	2	1	<1
Himalayan-Tibet	–	–	–	–	–	–	–	–	3	15	–	–	3	2
Hengduanshan Palearctic	–	–	–	–	–	–	–	–	8	40	1	2	8	5
Japanese	–	–	–	–	–	–	–	–	–	–	28	51	28	16
Oriental Patterns:														
West Oriental	–	–	–	–	3	7	–	–	–	–	–	–	3	2
Hengduanshan Oriental	–	–	–	–	–	–	–	–	–	–	1	2	1	<1
Pan-Oriental	–	–	–	–	–	–	–	–	–	–	1	2	1	<1
Afrotropical Pattern:														
Nile	–	–	–	–	3	7	–	–	–	–	–	–	3	2
Total	69	100	35	100	42	100	9	100	20	100	55	100	175	100

Adriatic

Two species of green frogs (*Rana epeirotica* and *R. shqiperica*) are distributed along the Adriatic coast of the Balkan Peninsula in Yugoslavia, Albania, and Greece.

Balkan

Two species (*Rana graeca* and *R. kurtmuelleri*) are restricted to the Balkan Peninsula as well as to some neighboring Aegean islands. *Rana cretensis*, endemic to Crete, is a special case.

Central European

The so-called Nemoral pattern includes relatively widespread amphibians (e.g., *Salamandra sala-*

mandra and *Rana dalmatina*) with ranges mostly associated with broad-leaved forests. Within this broad pattern are three restricted distributions, each displayed by one species: (1) Danubian; a newt, *Triturus dobrogicus*, in the *T. cristatus* group is restricted to the Danube Basin. (2) Carpathian; another newt, *Triturus montandoni*, in the *T. vulgaris* group is endemic to the Carpathian Mountains. (3) Po; a brown frog, *Rana latastei*, is endemic to the lowlands of the Po River in northern Italy, as is *Pelobates fuscus insubricus*. The Po Valley is a transition area between the Mediterranean and European regions. *Rana latastei* is closer to *Rana dalmatina* than to *R. italica* (Mensi et al., 1992). Thus, alternatively, the Po subdivision could be ascribed to the Italian distribution pattern.

East European

This pattern includes eight widespread species that inhabit various climatic and vegetation zones, with eastern limits in eastern Europe (Russia) or in western Siberia and northern Kazakhstan. Examples are *Rana temporaria* and *R. arvalis,* which range from tundra to forest-steppe and steppe.

East Mediterranean

Two amphibians (*Triturus karelinii* and *Pelobates syriacus*) are distributed in the Balkan Peninsula and West Asia, including the Caucasus.

West Asian

Sixteen species (e.g., *Salamandra infraimmaculata, Hyla savignyi, Rana bedriagae,* and *R. macrocnemis*) are distributed over West Asia. Within this broad pattern are four kinds of more restricted ranges: (1) Caucasian; *Mertensiella caucasica* and *Pelodytes caucasicus* are confined to the Caucasus with a western extension along the Black Sea in Turkey, and *Bufo verrucosissimus* with an undetermined southern range limit also is included. (2) Near East; this includes four species that are variously distributed along the Mediterranean zone of Anatolian Turkey southward to Israel—*Mertensiella luschani* and probably *Rana holtzi* endemic to the Taurus Mountains, *Rana cerigensis* endemic to Karpathos and Rhodes, and the probably extinct *Discoglossus nigriventer* known only from Hula Lake in northeastern Israel. (3) Armenian; two newts, *Neurergus microspilotus* and *N. strauchii,* are confined, or nearly so, to the Armenian Plateau, an area between the Lesser Caucasus, Zagros and Taurus Mountains. (4) Western Zagros; a newt, *Neurergus kaiseri,* is known from the western slopes of the Zagros Mountains.

Iranian

This pattern unites amphibians distributed in the Iranian Plateau, a vast area rimmed by the Elburz and Kopet Dagh ranges in the north, by the Zagros and Makran Mountains in the west and south, and by the Hindu Kush in Afghanistan and the Sulaiman Mountains in Pakistan in the east. Three subdivisions are: (1) Elburz (= Alburz) or Northern Peripheral; two hynobiid salamanders, *Batrachuperus persicus* and *B. gorganensis,* are distributed in the area south of the Caspian Sea between the Caucasus and Kopet Dagh. (2) Iranian proper or Central; a green

toad, *Bufo kavirensis,* is endemic to the central desert part of the plateau. (3) Southern Peripheral; a green toad, *Bufo surdus,* occurs only in the southern part of the Zagros Mountains and in the Makran Mountains

Hindu Kush

The range of one species, the hynobiid *Batrachuperus mustersi,* is restricted to the southwestern spurs of the Hindu Kush Mountain Range.

Tadjik

Two anurans (*Bufo shaartusiensis* and *Rana terentievi*) are endemic to the lowlands of southern Tadjikistan.

Kashmir

Two green toads, *Bufo siachinensis* and *B. latastii,* are restricted to the highlands of the Karakoram–Ladakh–western Himalaya area.

Siberian

Two species, *Salamandrella keyserlingii* and *Rana amurensis,* are associated with taiga (coniferous forests), mostly in Siberia.

Central Asian

This pattern includes species that are distributed mostly in the lowland and/or mountain areas of the Central Asian Region. The species are *Ranodon sibiricus, Rana asiatica,* and perhaps *Bufo danatensis,* which extends through the mountain belt in the south of the former Soviet Central Asia westward to the Caspian Sea.

Mongolian

A green toad, *Bufo raddei,* is confined to open landscapes (mainly steppes and deserts) in Mongolia, northern China, southern Siberia, and the Russian Far East.

East Asian

This pattern unites nine species that are widespread in eastern Asia. For example, the range of *Hyla japonica* encompasses Japan, the Korean Peninsula, the Russian Far East, southern Siberia, and eastern China. Likewise, *Bufo gargarizans, Rana chensinensis, R. nigromaculata,* and *R. rugosa* have broad distributions. Some species are restricted to China and Korea (*Kaloula borealis*) or to China (*Andrias davidianus*). Zhao (this volume) termed this pattern as the "Pacific Palearctic" with 13 species.

Manchurian

This pattern includes 11 species distributed in the Russian Far East, probably Hokkaido Island (Japan), the Korean Peninsula and northeastern China. Three subdivisions are: (1) Two species (*Onychodactylus fischeri* and *Bombina orientalis*) with broad distributions in the continental part of the region. (2) Eight taxa restricted to the southern part of the area (e.g., the Korean Peninsula—*Hyla suweonensis, Rana chosenica, R. dybowskii* [also on Tsushima Island, Japan], and *R. amurensis coreana* [probably, a full species]), or to Manchuria, namely to the Changbai Shan Subregion (NE-B in Zhao and Adler, 1993)—*Hynobius mantchuricus* and *Rana huanrenensis,* or to both areas—*Hynobius leechii* and *Bufo stejnegeri.* (3) Two species (*Hynobius* [or *Satobius*] *retardatus* and *Rana pirica* confined to the insular Hokkaido area (Hokkaido, the southern Kurile Islands, and perhaps at least the southern part of Sakhalin Island). Zhao (this volume) allocates the first group (plus *Rana dybowskii*) to his Pacific Palearctic pattern, the second to the Korean Peninsula–Northeast China pattern, and the third (Hokkaido only) to the Japanese category (see below).

North China
This pattern is exhibited by *Scutiger liupanensis* (Zhao, this volume).

Himalayan-Tibet
Three species, *Scutiger nyingchiensis, S. boulengeri,* and *Nanorana parkeri,* exhibit this pattern (Zhao, this volume).

Hengduanshan Palearctic
This category contains 24 species that are distributed in the northern part of the Hengduanshan Mountains (Zhao, this volume). However, only eight of these species enter the Palearctic Realm.

Japanese
This pattern is characteristic of 28 species (e.g., *Hynobius naevius, Cynops pyrrhogaster, Rana ornativentris,* and *Buergeria buergeri*) distributed in the Japanese Islands between Hokkaido in the north and Ryukyu Archipelago in the south. Zhao (this volume) applies a more general name "Eastern Asia Island Palearctic." The distributions of some species are restricted to islands like Honshu (e.g., *Hynobius takedai, Rana porosa,* and *Rhacophorus arboreus*), Kyushu (e.g., *Hynobius stejnegeri*), Tsushima (*Hynobius tsuensis* and *Rana tsushimensis*), or to some smaller islands (e.g., *Hynobius okiensis*).

ORIENTAL PATTERNS

West Oriental
Three Oriental anurans, the Baluchistan *Bufo olivaceus,* as well as the widespread *Bufo stomaticus* and *Euphlyctis cyanophlyctis,* are distributed in southeastern Iran (Sistan, Baluchestan, and Makran Coast), as well as in neighboring Afghanistan and Pakistan.

Hengduanshan Oriental
Of 32 species in the southern section of the Hengduanshan Mountains (Zhao, this volume), only *Paa boulengeri* extends into the Palearctic Realm in China northward to Shanxi Province (Zhao and Adler, 1993).

Pan-Oriental
Only the widespread frog *Limnonectes limnocharis* enters northeastern China northward to the Shandong Province (Zhao and Adler, 1993).

AFROTROPICAL PATTERN

Nile
This pattern is observed in the Egyptian Province which could be assigned to the Palearctic (West Asian Region) or to the Afrotropical (Ethiopian) realms. Three species, all with Afrotropical affinities, are *Ptychadena mascareniensis, Bufo kassasii* (endemic to the Nile Delta), and *Bufo regularis.*

SUMMARY OF DISTRIBUTION PATTERNS

Among the 30 patterns, only five contain more than 10 species (Table 6:24). These are the Japanese (28 species, 16% of the Palearctic fauna), Italian (20, 11%), West Asian (16, 9%), Iberian (15, 9%), and Manchurian (11, 6%). These five patterns account for 51% of the Palearctic species; the remaining 49% are shared by 24 patterns, each with 1–9 species (0.5–5%).

The variety of distribution patterns is different in different regions. For instance, the West Asian and Mediterranean regions each have 12 patterns, and the European Region has 10 patterns, whereas the East Asian Region contains nine, the Central Asian six, and the Siberian only five. Thus, regions with fewer species are more homogeneous. The Japanese pattern constitutes 51% of the species in the East Asian Region; the Japanese Archipelago is an excellent arena for extensive insular speciation.

The arrangement of distribution patterns by family is shown in Table 6:25. Ranids are the most di-

Table 6.25. Allocation of distribution patterns in families of the Palearctic Realm. Abbreviations: BUF = Bufonidae, CRY = Cryptobranchidae, DIS = Discoglossidae, HYL = Hylidae, HYN = Hynobiidae, MIC = Microhylidae, PEB = Pelobatidae, PED = Pelodytidae, PLE = Plethodontidae, PRO = Proteidae, RAN = Ranidae, RHA = Rhacophoridae, and SAL = Salamandridae; n = number of species, % = percent of species in family in Palearctic Realm.

Distribution Pattern	CRY		HYN		PLE		PRO		SAL		BUF		DIS		HYL		MIC		PEB		PED		RAN		RHA		Total
	n	%	n	%	n	%	n	%	n	%	n	%	n	%	n	%	n	%	n	%	n	%	n	%	n	%	n
Palearctic Patterns:																											
West Palearctic	–	–	–	–	–	–	–	–	–	–	2	8	–	–	–	–	–	–	–	–	–	–	1	2	–	–	3
Northwest African	–	–	–	–	–	–	–	–	2	6	2	8	1	8	–	–	–	–	1	11	–	–	1	2	–	–	7
Iberian	–	–	–	–	–	–	–	–	4	13	–	–	5	38	1	14	–	–	1	11	–	–	4	9	–	–	15
Atlantic	–	–	–	–	–	–	–	–	2	6	1	4	–	–	–	–	–	–	–	–	1	50	–	–	–	–	4
Italian	–	–	–	–	7	100	–	–	5	16	–	–	3	23	2	29	–	–	–	–	–	–	3	6	–	–	20
Apennine-Balkan	–	–	–	–	–	–	1	100	–	–	–	–	–	–	–	–	–	–	–	–	–	–	–	–	–	–	1
Alpine-Dinaric	–	–	–	–	–	–	–	–	1	3	–	–	–	–	–	–	–	–	–	–	–	–	2	4	–	–	3
Adriatic	–	–	–	–	–	–	–	–	2	6	–	–	–	–	–	–	–	–	–	–	–	–	–	–	–	–	2
Balkan	–	–	–	–	–	–	–	–	–	–	–	–	–	–	–	–	–	–	–	–	–	–	3	6	–	–	3
Central European	–	–	–	–	–	–	–	–	4	13	–	–	1	8	1	14	–	–	–	–	–	–	2	4	–	–	8
East European	–	–	–	–	–	–	–	–	2	6	–	–	1	8	–	–	–	–	1	11	–	–	4	9	–	–	8
East Mediterranean	–	–	–	–	–	–	–	–	1	3	–	–	–	–	–	–	–	–	1	11	–	–	–	–	–	–	2
West Asian	–	–	–	–	–	–	–	–	8	25	1	4	1	8	1	14	–	–	–	–	1	50	4	9	–	–	16
Iranian	–	–	2	7	–	–	–	–	–	–	2	8	–	–	–	–	–	–	–	–	–	–	–	–	–	–	4
Hindu Kush	–	–	1	4	–	–	–	–	–	–	–	–	–	–	–	–	–	–	–	–	–	–	–	–	–	–	1
Tadjik	–	–	–	–	–	–	–	–	–	–	1	4	–	–	–	–	–	–	–	–	–	–	1	2	–	–	2
Kashmir	–	–	–	–	–	–	–	–	–	–	2	8	–	–	–	–	–	–	–	–	–	–	–	–	–	–	2
Siberian	–	–	1	4	–	–	–	–	–	–	1	4	–	–	–	–	–	–	–	–	–	–	–	–	–	–	2
Central Asian	–	–	1	4	–	–	–	–	–	–	1	4	–	–	–	–	–	–	–	–	–	–	1	2	–	–	3
Mongolian	1	50	–	–	–	–	–	–	–	–	–	–	–	–	–	–	–	–	–	–	–	–	–	–	–	–	1
East Asian	–	–	–	–	–	–	–	–	–	–	1	4	–	–	1	14	1	100	–	–	–	–	5	11	–	–	9
Manchurian	–	–	4	15	–	–	–	–	–	–	1	4	1	8	1	14	–	–	–	–	–	–	4	9	–	–	11
North China	–	–	–	–	–	–	–	–	–	–	–	–	–	–	–	–	–	–	1	11	–	–	–	–	–	–	1
Himalayan–Tibet	–	–	–	–	–	–	–	–	–	–	–	–	–	–	–	–	–	–	2	22	–	–	1	2	–	–	3
Hengduanshan Palearctic	–	–	2	7	–	–	–	–	–	–	3	13	–	–	–	–	–	–	2	22	–	–	1	2	–	–	8
Japanese	1	50	16	59	–	–	–	–	1	3	2	8	–	–	–	–	–	–	–	–	–	–	5	11	3	100	28
Oriental Patterns:																											
West Oriental	–	–	–	–	–	–	–	–	–	–	2	8	–	–	–	–	–	–	–	–	–	–	1	2	–	–	3
Hengduanshan Oriental	–	–	–	–	–	–	–	–	–	–	–	–	–	–	–	–	–	–	–	–	–	–	1	2	–	–	1
Pan-Oriental	–	–	–	–	–	–	–	–	–	–	–	–	–	–	–	–	–	–	–	–	–	–	1	2	–	–	1
Afrotropical Patterns:																											
Nile	–	–	–	–	–	–	–	–	–	–	2	8	–	–	–	–	–	–	–	–	–	–	1	2	–	–	3
Total	2	100	27	100	7	100	1	100	32	100	24	100	13	100	7	100	1	100	9	100	2	100	47	100	3	100	175

verse, with 21 of the 30 patterns, whereas bufonids have 15 and salamandrids 11. Four families exhibit only one pattern, and the remaining six families have 2–7 patterns. The diversity of patterns within a family seems to be correlated with the number of zoogeographic units (subdivisions) inhabited by the family rather than the number of species. For example, bufonids with 24 species have more than twice the number of distribution patterns (15 vs. 7) than hyno-

biids with 27 species, whereas the latter does not differ from discoglossids with only 13 species.

Among the 30 distribution patterns, the Japanese pattern predominates in hynobiids (59%), the West Asian pattern (25%) in salamandrids, and the Iberian pattern (38%) in discoglossids. The patterns in ranids and bufonids are relatively even (Table 6:25).

AREAS OF HIGH SPECIES DIVERSITY AND ENDEMISM

Traditionally, a taxon could be considered as endemic to an area if it occurs only in that area. Therefore, specifying an area (and time) is needed. Sydney Anderson (1994) proposed a classification of numerous factors influencing the number of endemics of an area. Generally, it has been assumed that a positive correlation exists between size of an area and endemism as well as diversity. For example, 84% of the amphibians (100% of salamanders, 76% of anurans) are endemic to the Palearctic Realm (Table 6:10). In Europe, a part of the realm, 70% of the amphibians (78% of salamanders, 64% of anurans) are endemic (Appendix 6:2); however, Europe and Siberia have almost the same area (ca. 10^6 km^2), but Siberia has no endemics. The fauna of Italy (ca. 0.3^6 km^2) contains 37 amphibians including 11 endemics (Appendix 6:2), whereas noticeably larger Iran (ca. 1.65^6 km^2) harbors 18 species with four endemics (Table 6:3), and Mongolia (ca. 1.57^6 km^2) with six amphibians has no endemic species or subspecies (Borkin, 1988; Borkin and Kuzmin, 1988). The fauna of the vast territory of the former Soviet Union (ca. 22.4^6 km^2) contains 41 species and has only two endemics (Appendix 6:3).

In the Palearctic as a whole, species richness tends to increase toward the southwest and toward the southeast, rather than in the central parts—cold Siberia and arid Central Asia. This latitudinal trend is shifted to the south more extensively in the east than in the west (Figs. 6:9 and 6:10). These trends are associated with the impact of the Atlantic and Pacific climates resulting in different arrangements of arid and moist zones.

At the geographic landscape level, changes in species richness are also evident. For example, in the territory of the former USSR (from north to south) tundra harbors only three species; taiga (conifer forest) is inhabited by eight amphibians, the

mixed and broad-leaved forest zone by 20, steppes by 9, deserts by five or six, and desert mountains by four species.

However, at the regional and local level, the situation is more complicated. Despite noticeable trends in biotic and abiotic factors, no clear latitudinal diversity gradient is apparent in the Iberian Peninsula. However, amphibians probably are most diverse in cooler, humid zones (Schall and Pianka, 1979; Busack and Jaksić, 1982). The species:area relationship is a continental type for anurans, and "insular" (i.e., ecologically unsaturated) type for salamanders. At the local level, Iberian amphibian diversity fails to demonstrate a significant correlation with vegetation and soil diversity (Busack and Jaksić, 1982). Huge areas of Siberia, Central Asia, and the Sahara also show no significant changes in species richness, which varies from 0 to 3 species per 100 × 100 km quadrat.

In the western Palearctic, local species richness seems to be higher in central Europe north and east of the Alps (e.g., Fig. 7 in Gasc et al., 1997), where many grids of 50 × 50 km have more than 14 species per quadrate (Fig. 6:9 and Fig. 6:10). However, continental areas with high local species richness have no or few endemics. Moreover, the European (or Boreal) Region incorporating this area has an amphibian fauna that is about one-half that of the Mediterranean Region (Table 6:14). However, the area in central Europe north and east of the Alps, situated approximately between central France and the Carpathians, is a kind of zoogeographic crossroad, where two major postglacial waves of the European fauna that dispersed from the southwest and from the southeast have met (Zuiderwijk, 1980). High species richness also occurs in northwestern Italy and in some places of the Iberian Peninsula.

In the territory of the former USSR, three areas

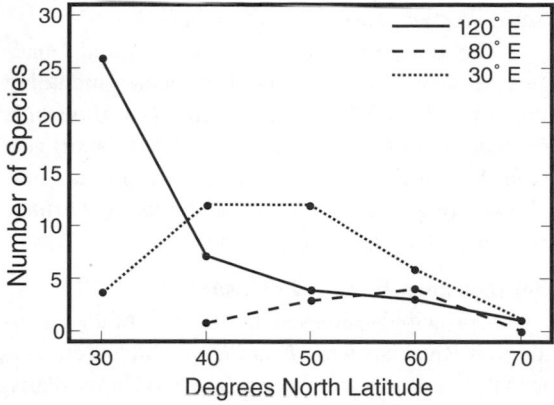

Fig. 6:9. Latitudinal changes in species richness (number of species per 100 × 100 km quadrat) in the western, central, and eastern Palearctic.

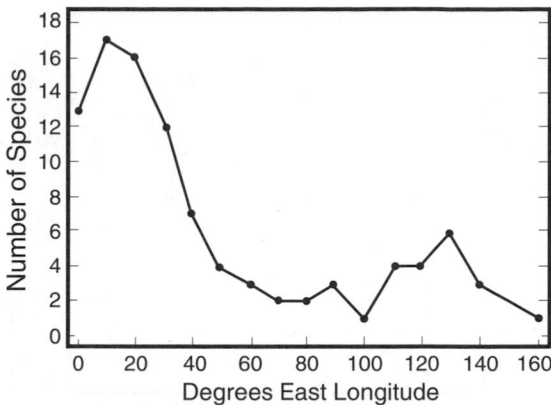

Fig. 6:10. Longitudinal changes in amphibian species richness (number of species per 100 × 100 km quadrat) at 50° N lat. in the Palearctic.

of high species richness are known. The highest number of species is from the Carpathian region (18 amphibians—6 salamanders and 12 anurans). The Caucasus harbors 14–15 species (4 salamanders, 10–11 anurans), and the southern mainland of the Russian Far East is inhabited by nine species (2 salamanders and 7 anurans). Except for the Caucasus, these regions have no endemics.

An apparent correlation exists between regional species richness and number of endemic (Tables 6:14 and 6:15), but the association between regional species richness and percentage of endemics tends to be more complicated (Fig. 6:11). Sydney Anderson (1994:461) also mentioned such a correlation with "more species in the fauna" in Australian anurans; however, "no theoretical reason has been advanced for why an increase in the number of species alone should increase endemism." Accordingly to Anderson, both parameters in Australia were positively related with area size. However, this is not true for the Palearctic, where the Mediterranean and East Asian regions with the highest endemism have total areas that do not exceed those of the other four regions (Fig. 6:8).

Fifty-one areas having high species diversity and/or endemic amphibians, or being of particular zoogeographic importance in the Palearctic Realm, except the territory of Central and Eastern Asia covered by Zhao (this volume) have been identified (Fig. 6:12). These areas include all species and even most subspecies distributed in this part of the Palearctic. The numbers in the following list of areas refer to

the location of the areas on the accompanying map (Fig. 6:12).

NORTH AFRICA

Seven areas contain all species and subspecies of amphibians endemic to the Palearctic portion of North Africa. Also see additional information in the text, Table 6:1, Pasteur and Bons (1959), Bons (1972, 1973), Le Berre (1989), Baha el Din (1993), Salvador (1996), and Schleich et al. (1996).

Mountains of northwestern Morocco (1)

The Rif and Middle Atlas Mountains harbor probably the most diverse North African amphibian fauna including regional endemics(*) at the species and subspecies level (*Pleurodeles waltl, Salamandra algira*, Alytes obstetricans maurus*, Discoglossus pictus scovazzii*, Hyla meridionalis, Bufo bufo spinosus, B. viridis boulengeri**, and *Rana saharica**).

Plains of northwestern Morocco (2)

This lowland area is inhabited by a Moroccan endemic (*Pelobates varaldii*) and other species, including regional endemics (*)—*Pleurodeles waltl, Discoglossus pictus scovazzii*, Hyla meridionalis, Bufo mauritanicus*, B. viridis boulengeri**, and *Rana saharica**.

Atlantic coast of southwestern Morocco (3)

The arid landscapes of the Sous and Draa valleys harbor the regional endemic toad *Bufo brongersmai*. A few other anurans endemic to North Africa (*Bufo mauritanicus, B.viridis boulengeri,* and *Rana saharica*) also occur there.

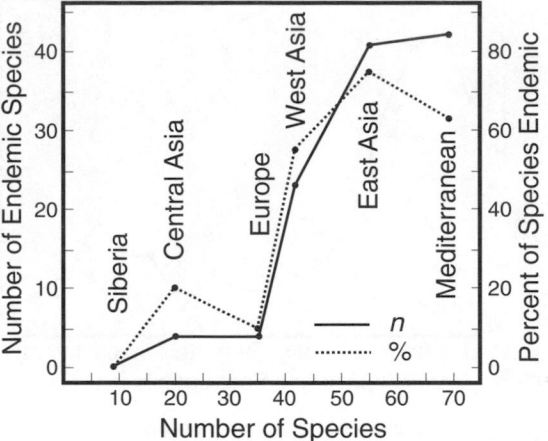

Fig. 6:11. Relationship between total number of species and number and percentage of endemic species in six regions in the Palearctic.

Tell Atlas Mountains (4)

This area is second in species richness in North Africa. Six regional endemics (*) are among the eight species recorded; these are *Pleurodeles poireti**, *Salamandra algira**, *Discoglossus pictus algirus**, *Hyla meridionalis, Bufo bufo spinosus, B. mauritanicus**, *B. viridis boulengeri**, and *Rana saharica**.

Ahaggar (=Hoggar) area, southern Algeria (5)

This mountainous area in the central Sahara is of particular zoogeographic importance because of the occurrence of both Palearctic and Afrotropical anurans (Fig.6:2 and Fig. 6:3).

Ghat Oasis, southwestern Libya (6)

This is another zoogeographically important area in the Sahara Desert with both Palearctic and Afrotropical components (Fig. 6:2 and Fig. 6:3).

Nile Delta, Egypt (7)

Four Palearctic and Afrotropical anurans (*Bufo viridis, B. regularis, Ptychadena mascareniensis,* and *Rana bedriagae*) and local endemic toad *Bufo kassasii* of the Afrotropical affinity inhabit the area.

EUROPE

The 77 European amphibians are listed by country in Appendices 6:2–6:5. The European atlas (Gasc et al., 1997), with maps of 62 species, including an introduced North American frog *Rana catesbeiana* and with a general map of amphibian species richness, is a useful basis for an evaluation of areas of high richness and endemism.

Southwestern Iberian Peninsula (8)

The area mostly between the Tajo and Guadalquivir Rivers contains twelve species including Iberian endemics (*Triturus boscai, T. marmoratus pygmaeus, Alytes cisternasii,* and *Discoglossus galganoi*), as well as some characteristic Iberian amphibians (e.g., *Pleurodeles waltl, Pelobates cultripes, Hyla meridionalis,* and *Rana perezi*).

Southeastern Iberian Peninsula (9)

A principal species in the area east of the Guadalquivir River is *Alytes dickhilleni,* which is endemic to the Betic Mountains (Arntzen and García-París, 1995). At least seven more amphibians, including both true Iberian and widespread European species, inhabit this area.

Central Spain (10)

The area north of the Tajo River is inhabited by at least 15 species, including Iberian endemic species (e.g., *Triturus boscai, Discoglossus galganoi,* and *Alytes cisternasii*) and subspecies (e.g., *Salamandra salamandra bejarae, S. s. almanzoris, Alytes obstetricans boscai,* and *Bufo bufo gredosicola*).

Northwestern Iberian Peninsula (11)

This area harbors probably the most diverse amphibian fauna of the peninsula. Among at least 17 species, Iberian endemics are dominant. The assemblage consists of western Iberian endemics (e.g., *Chioglossa lusitanica, Triturus boscai,* and *Rana iberica*), widespread Iberian species (e.g., *Discolossus galganoi*), and European species represented by endemic subspecies (e.g., *Salamandra salamandra gallaica, S. s. bernardezi, Triturus alpestris cyreni,* and *Hyla arborea molleri*). Most members of the assemblage are associated with mountainous regions. Most species are distributed north of the Douro River in Portugal, but some species extend southward approximately to the Tajo River (e.g., *Chioglossa lusitanica* and *Triturus helveticus sequeirai*), Guadiana River (*Alytes obstetricans boscai* and *Rana iberica*), or Guadalquivir River (e.g., *Discoglossus g. galganoi*).

Pyrenees (12)

Amphibians inhabiting the Pyrenees are listed in Table 6:16. Examples include the endemic *Euproctus asper, Rana pyrenaica, Salamandra salamandra fastuosa,* and *Rana temporaria parvipalmata.*

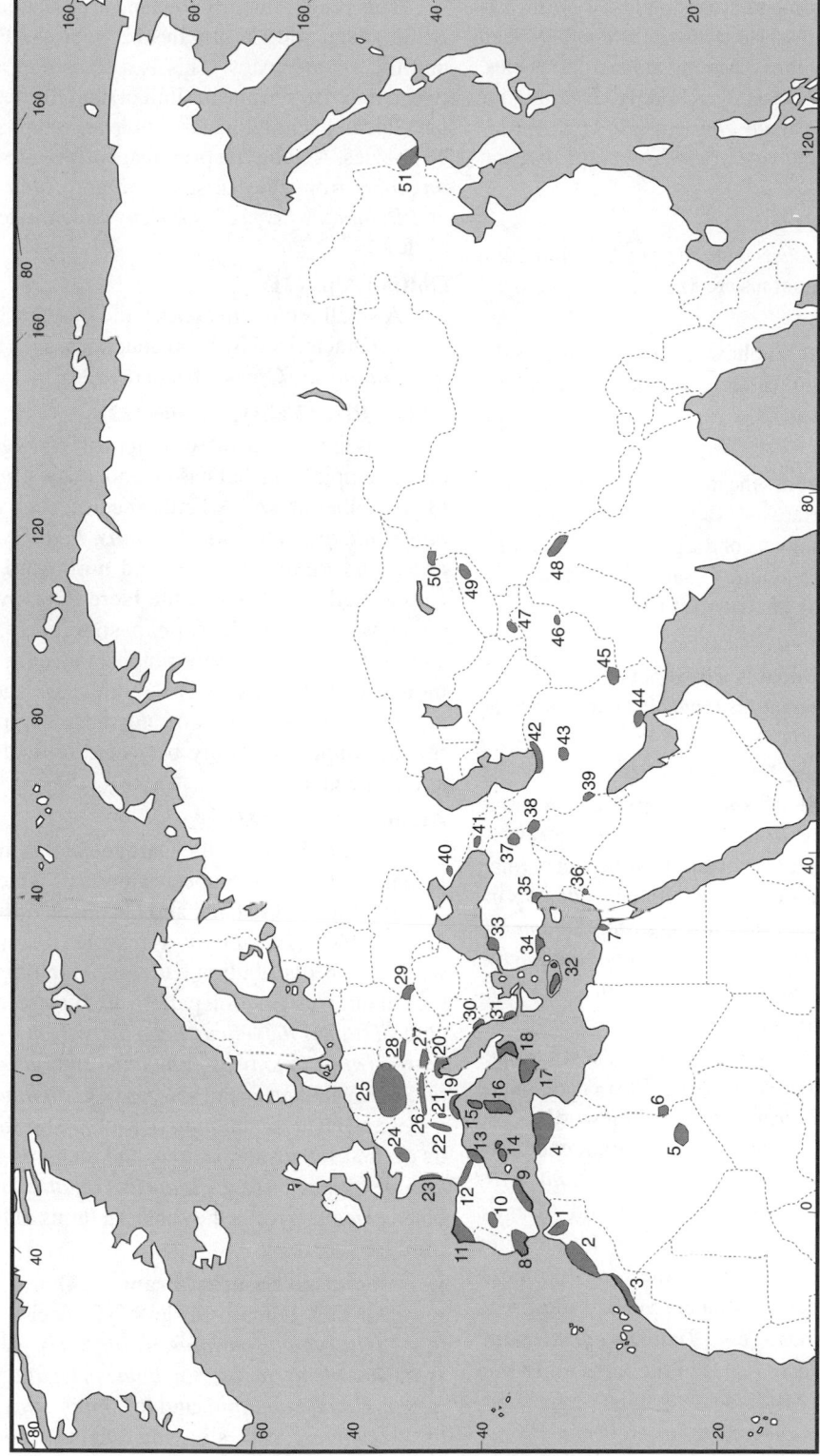

Fig. 6:12. Areas having high species diversity and/or endemic amphibians, or being of particular zoogeographic importance in the Palearctic Realm. Numbers refer to areas defined in the text.

Northeastern Iberian Peninsula (13)

This Mediterranean area shared by northeastern Spain and southwestern France is characterized by high species richness. Among at least 15 species, *Euproctus asper* occurs peripherally. Anurans include *Alytes obstetricans almogavarii, Hyla meridionalis, Pelobates cultripes, Pelodytes punctatus,* and *Rana perezi.*

Balearic Islands (14)

Alytes muletensis is endemic to the archipelago, where three more anurans are known (Table 6:18).

Corsica (15)

Three of eight amphibians are endemic to the island and two anurans are endemic to the Tyrrhenian region (Table 6:18).

Sardinia (16)

The island harbors the most diverse insular fauna in the Mediterranean (Table 6:18). Among ten species, five salamanders of the genera *Hydromantes* and *Euproctus* are endemic to Sardinia, and two Tyrrhenian anurans are shared with Corsica.

Sicily (17)

This island is inhabited by eight anurans (Table 6:18). *Discoglossus pictus pictus* is shared with the Maltese Islands.

Southern Apennine Peninsula (18)

Among 12 species known from the area, some are represented by isolated populations, a few of which are recognized as endemic subspecies (e.g., *Triturus alpestris inexpectatus*). Italian endemics include *Salamandra salamandra giglioli, Salamandrina terdigitata, Triturus italicus, Bombina pachypus, Hyla intermedia,* and *Rana italica.*

Northwestern Italy (19)

This area contains probably the richest European amphibian fauna. Among the 20 species, mainland species of *Hydromantes* are noteworthy. The northern limits of ranges of some Italian endemics (e.g., *Salamandrina terdigitata* and *Rana italica*) are in this area. A few Iberian and Atlantic species (e.g., *Hyla meridionalis* and *Pelodytes punctatus*) reach this area via the Mediterranean corridor in southeastern France. Some widespread European amphibians (e.g., *Triturus carnifex* and *Rana dalmatina*) also occur in the area, and some are represented by Italian subspecies (e.g., *Triturus alpestris apuanus* and *T. vulgaris meridionalis*).

Po-Trieste-Istria (20)

This predominantly lowland area shared by Italy, Slovenia, and Croatia lies between the Po River and the Dinaric Alps. This is a transition area between the Mediterranean (Italian) and European faunas. The rich amphibian assemblage contains at least 18 species. Among the principal species are *Proteus anguinus* from the Karst region and *Rana latastei* and *Pelobates fuscus insubricus* endemic to the Po Valley.

Cottian Alps (21)

A small mountain area shared by France and Italy is characterized by local endemics, such as *Salamandra lanzai* (Grossenbacher, 1994).

Rhone River Valley, France (22)

This area is inhabited by approximately 19 species of amphibians but has no endemics. The ranges of some Iberian and Atlantic species, such as *Pelobates cultripes, Hyla meridionalis, Pelodytes punctatus,* and *Rana perezi,* extend northward via the Rhone Valley at least to the Isere River, whereas some widespread European species reach the southern limits of their ranges via neighboring mountain landscapes. The number of species is highest in the central and northern segments of the area, approximately between the Isere River and Lake Geneva.

Atlantic France (23)

The area between the Garonne and Loire rivers is inhabited by 17–18 species, none of which is endemic (maps in Castanet and Guyétant, 1989; Gasc et al., 1997). The fauna is a mixture of amphibians with various distribution patterns. Amphibians with the Atlantic distribution pattern are typical; these include *Triturus helveticus helveticus, T. marmoratus marmoratus, Pelodytes punctatus,* and *Bufo calamita.* Some Mediterranean species (e.g., *Pelobates cultripes* and *Hyla meridionalis*) reach the northern limits of their ranges in the area, and many widespread European species (e.g., *Triturus cristatus* and *Bombina variegata*) reach the southern limits of their distributions there.

Loire-Seine rivers area, France (24)

This rich fauna of at least 17 species patterns (e.g., *Triturus marmoratus, T. alpestris, Alytes obstetricans obstetricans, Bombina variegata, Bufo calamita, Rana dalmatina,* and *R. temporaria*) consists of widespread species having Atlantic and Central European patterns.

Rhine-Elbe rivers area (25)

The large area, mainly in the hills and low mountains of western Germany, lies between the northern lowlands and the Alps, approximately in the Rhine-Maas-Mosel-Main area and west of the Elbe River; this rich fauna also extends into western Czechia and Austria (Appendix 6:2; Gasc et al., 1997). This zoogeographic crossroad has high species richness (up to 19 species) characterized by the overlap of western and eastern amphibians (e.g., *Alytes obstetricans* and *Rana arvalis*), but contains many species with the Central European distribution pattern.

Northern Alps (26)

The mountainous region shared by Switzerland, Austria, and southern Germany is inhabited by numerous populations of *Salamandra atra atra* (Grossenbacher, 1994) and some other amphibians.

Drava River area (27)

The area shared by southern Austria, Slovenia, and Croatia has up to 17 species but no endemics; widespread European amphibians with various distribution patterns are predominant.

Lower Austria (28)

The Danubian area also is characterized by up to 17 European amphibians. Among them, *Triturus dobrogicus,* a member of the superspecies *T. cristatus,* is essentially endemic to the Danube Valley.

Ukrainian Carpathian area (29)

This area includes mountains and the neighboring edge of the Pannonian lowlands. Among 18 species, *Triturus montandoni* is endemic to the Carpathian mountains. Some widespread European species (*Salamandra salamandra, Triturus alpestris,* and *Rana dalmatina*) reach the eastern limits of their ranges in this area. The species contact (and hybridization) zone is situated in the transition zone between the plain and mountains, some species pairs (e.g., *Bombina bombina* and *B. variegata*) come into contact and may hybridize.

Lake Skadar (30)

The largest lake in the Balkan Peninsula is shared by Yugoslavia (Montenegro) and Albania. The local fauna consists of 14 species with various distribution patterns including Alpine-Dinaric and Balkan patterns. Included are *Salamandra atra prenjensis, Triturus carnifex, T. alpestris montenegrinus, T.a.piperianus, T. vulgaris graecus, Bombina variegata scabra, Rana kurtmuelleri* (= *R. balcanica*), and *R. graeca* (Schneider and Haxhiu, 1994; Crnobrnja-

Isailovic and Dzukic, 1995). The area is the type locality of *Rana shqiperica* (Hotz et al., 1987).

Epeiros, western Greece (31)

At least 12 species are known from this area. Among the Balkan amphibians are *Triturus carnifex, T. vulgaris graecus, Pelobates syriacus,* and two green frogs, *Rana epeirotica* and *R. kurtmuelleri,* which generally inhabit lowlands.

Hellenic Arc (32)

The fauna of islands Crete, Karpathos, and Rhodes is listed in Table 6:18. Two endemic green frogs, *Rana cerigensis* and *R. cretensis,* as well as *Hyla arborea kretensis* and *Mertensiella luschani helverseni* are principal amphibians in the area.

WEST ASIA

Northwestern Anatolia (33)

This area is inhabited by amphibians of various origins. The local fauna (approximately 12 species) consists of some widespread European species (e.g., *Bombina bombina* and *Rana dalmatina*) that have the southeastern tips of their ranges in the area or are represented by local subspecies (*Triturus vulgaris kosswigi*). Some widespread East Mediterranean species (e.g., *Triturus karelinii* and *Pelobates syriacus*), as well as West Asian endemics (e.g., *Triturus vittatus ophryticus* and *Rana macrocnemis*) also exist there.

Lycian area, southwestern Anatolia (34)

A salamander, *Mertensiella luschani* with numerous subspecies, is characteristic of this Mediterranean area of Turkey. Other amphibians, including West Asian endemics (e.g., *Rana bedriagae*) also occur there.

Bolkar Mountains (35)

This area is within the Taurus Mountain Range in southern Turkey; it is inhabited by a fauna of at least nine species (Schmidtler et al., 1990). The subalpine brown frog *Rana holtzi* is endemic to the mountains. West Asian endemics, such as *Salamandra infraimmaculata orientalis, Triturus vittatus cilicensis, Hyla savignyi, Rana bedriagae,* and *R. macrocnemis,* are known from lower altitudes.

Hula Lake, Israel (36)

This area is the type locality of the enigmatic *Discoglossus nigriventer,* the sole discoglossid frog in the Near East. Probably the species is extinct, and only two preserved specimens are known (Werner, 1988).

Van Lake, eastern Turkey (37)

A newt, *Neurergus strauchii* described from the area, is the notable amphibian species in the area (Schmidtler and Schmidtler, 1970; Schmidtler, 1994).

Urmia Lake, Iran, and Kurdistan area (38)

This mountain area is shared by Turkey, Iraq, and Iran harbors two peculiar newts, *Neurergus crocatus* and *N. microspilotus* (Schmidtler and Schmidtler, 1970; Schmidtler, 1994), as well as *Salamandra infraimmaculata semenovi*. The six amphibians known from the Kurdistan Province and Urmia Lake area, Iran, are listed in Tables 6:3 and 6:4.

Western Zagros area, Iran (39)

The Lorestan Province is inhabited by two endemic amphibians, *Neurergus kaiseri* and *Bufo surdus luristanicus;* the latter is considered to be a full species by some authors. The provincial list of five amphibians is given in Table 6:3. (Also see Table 6:4 for physiographic regions of Iran.)

Northwestern Caucasus (40)

The Caucasian Reserve, Krasnodar Territory of Russia, protects 9 of 14 species known in the Caucasus. Among them, five amphibians (*Triturus vittatus ophryticus, T. vulgaris lantzi, Bufo verrucosissimus, Pelodytes caucasicus, Hyla arborea schelkownikowi,* and *Rana macrocnemis*) are endemic to the Caucasus or to West Asia. The East Mediterranean *Triturus karelinii* and two widespread European species (*Bufo viridis* and *Rana ridibunda*) complete the list.

Southwestern Caucasus (41)

The Trialeti Mountain Range in Georgia, south of the Great Caucasus, harbors 11 amphibians. The local fauna includes all species listed for the Caucasian Reserve, plus a salamander *Mertensiella caucasica* endemic to the southeast Black Sea zone and *Pelobates syriacus* (Tarkhnishvili and Gokhelashvili, 1995).

Elburz Mountain Range, Iran (42)

Among at least seven species (Table 6:4), two hynobiids, *Batrachuperus gorganensis* and *B. persicus,* are endemic to the area.

Dasht-i-Kavir Desert, Iran (43)

This area in the central part of Iran is inhabited by an endemic toad, *Bufo kavirensis* (Andrén and Nilson, 1979). The poor fauna of the Central Plateau, including the interior desert basin, contains only three anurans (Table 6:4).

Bandar Abbas or Kerman Area, south Iran (44)

This area is zoogeographically interesting, because the fauna includes both Palearctic components, such as *Bufo surdus, B. viridis kermanensis,* and *Rana ridibunda susana* (endemic to coastal region), and Oriental components like *Bufo olivaceus* (Fig. 6:5).

Sistan (45)

This area shared by Iran, Afghanistan, and Pakistan has an amphibian fauna which is a mixture of Palearctic (*Bufo viridis oblongus*) and Oriental anurans (*Bufo stomaticus* and *Euphlyctis cyanophlyctis*). Based on the herpetofauna as a whole and some vegetation features, Khan (1980) suggested that this area be ranked as a subregion of the Palearctic.

Paghman Mountain Range, Afghanistan (46)

This Hundu Kush area is close to Kabul. A hynobiid salamander, *Batrachuperus mustersi,* is endemic to the mountains. At lower elevations, the fauna of the East Region of Afghanistan is influenced by Oriental components, which are distributed along the Kabul River Valley (Table 6:5; Clark, 1990).

Southern Tadjikistan (47)

Recently two new species have been described from the lowlands of this republic; *Rana terentievi* and *Bufo shaartusiensis* (Mezhzherin, 1992; Pisanetz et al., 1996) probably are endemic to this area and possibly the neighboring plains of northern Afghanistan.

Baltistan, Karakoram, Ladakh Range area (48)

This mountain area shared by Pakistan and India connects the Karakoram and western Himalayan high elevations. This Palearctic "outpost" is inhabited by two Palearctic toads, *Bufo siachinensis* and *B. latastii.* A megophryine, *Scutiger occidentalis,* also is known from the Ladakh area.

CENTRAL ASIA

Issyk Kul Lake, Tian Shan Mountains (49)

This is one of the largest montane lakes in the world; it lies in Kyrgyzstan and has two peculiar anurans—the tetraploid toad *Bufo danatensis* and a brown frog, *Rana asiatica,* endemic to the northwestern part of the Central Asian Region. *Rana ridibunda* has been introduced in the eastern corner of the lake.

Dzhungarsky (Junggar) Alataw (50)

This mountain system shared by Kazakhstan and China is inhabited by an endemic hynobiid, *Ranodon sibiricus*. The tetraploid toads also occur in the area, and *Rana ridibunda* is common at lower elevations.

Ussuri River (51)

The southern part of the mainland Russian Far East harbors a mixture of the Siberian and so-called Manchurian (East Asian) components. Among nine amphibians, a hynobiid, *Onychodactylus fischeri,* listed in the Russian Red Data Book, is of a particular importance.

DISJUNCT DISTRIBUTIONS

GENERAL CLASSIFICATION

The term "discontinuous distribution" may refer to six superficially similar phenomena; these include four natural and two artificial categories, as defined below. In some cases, the first four categories can be misidentified, thereby leading to misinterpretation of the causes.

Disjunct distribution.

Discontinuity may result from the disappearance (extinction) of a species from the intermediate part of a formerly continuous range with survival in marginal parts of the range.

Disjointed distribution

If a continuous range has been separated by physical phenomena during geological time, such as land drifting, an original distribution can be fragmented.

Pseudodisjunct distribution

In the course of dispersal, species may reach some separated areas that have no connection with each other, such as the colonization of islandlike landscapes (e.g., true islands, oases in deserts, mountains separated by plains, forest fragments within steppe).

Cryptic discontinuity

An invasion of a large landmass from more than one direction also may result in an apparent discontinuity and mimic a continuous distribution. For example, green frogs and green toads are widespread in northern Africa (Fig. 6:2 and Fig. 6:3). Traditionally, the frogs were (and, from time to time, are) treated as a single species under various names, and the toads as *Bufo viridis.* However, it is clear now that African green frogs consist of at least two species, namely *Rana saharica* in the northwest and *R. bedriagae* (= *R. levantina*) in the Nile Valley. There-

fore, current distribution of green frogs in northern Africa probably is the result of two invasions, a western one from southwestern Europe, and an eastern one from the Near East. This also seems to be true for *Bufo viridis* (at the subspecific level).

Discontinuity through introduction

As a result of chance or deliberate introductions, artificial disjunctions are created. Examples include the Iberian *Discoglossus pictus* (Lanza et al., 1986), the so-called Balearic toad (Hemmer et al., 1981), and the Chinese *Rana nigromaculata* into Turkmenistan.

Cartographic discontinuity

Not too infrequently, distributions are stated incorrectly based on "disjunctions" or "isolated populations" displayed on maps showing only peripheral segments of the range of a species instead of the entire range. For example, *Pelobates syriacus* is distributed in southeastern Europe in the Balkan Peninsula and the northern Caucasus (a small extension); on European maps, this appears to be a disjunction. *Triturus karelinii* is an analogous but more complicated case and appears to have an isolated population in the Crimean Peninsula. However, both the Balkan and Caucasian, segments of these ranges are connected via West Asia (Turkey, etc.). A brown frog, *Rana macrocnemis,* of the Caucasus and Kopet Dagh Mountains, Turkmenistan, is another example on some Soviet maps. However, the "isolated" Turkmen localities are connected with the Caucasian part of the range via Iran (Borkin, 1978).

Thus, apart from cases resulting from human activity, natural discontinuous distributions can be the result of three different processes—extinction, fragmentation of ranges, and dispersal, or a combination of these processes.

Taxonomically, the discontinuity at specific

(isolated populations, subspecies), generic (various species, species groups, and subgenera), and familial (various genera, subfamilies) level can be recognized. Geographically, two kinds of discontinuities can be recognized. *Microdiscontinuity* is at the species (population) level. Many species exist within the perimeters of their overall ranges as a number of more or less contiguous isolated populations. The distribution of some anurans in the Sahara Desert (Fig. 6:2 and Fig. 6:3), the mountain brown frogs of the *Rana macrocnemis* complex in Turkey and Iran (maps in Borkin, 1978, and Baran and Atatür, 1986) are examples. As a rule, such ranges are not widely disjunct from the main range of the species and are within the same zoogeographic subdivision.

Macrodiscontinuity is associated with greater geographic distances and with other taxonomic levels (from species to family). Palearctic amphibians are involved in various patterns of the macrodiscontinuity. These include intercontinental, interrealm, interregional, and intraregional (including island-mainland) cases reviewed below.

DISJUNCT DISTRIBUTIONS IN THE PALEARCTIC

Thirty main discontinuous patterns are recognized.

Western versus Eastern Palearctic

This intracontinental pattern is classic, and numerous examples of plants, fishes, birds, amphibians, and mammals (but not reptiles) are mentioned in various textbooks on biogeography. Sometimes, the pattern is termed the European–Far Eastern disjunction, although the western Palearctic part can encompass northern Africa, western Asia, and even western Siberia. On the other hand, the Asian part of ranges of some groups include the northern areas of the Oriental Realm (e.g., southern China). The west-east disjunction is mainly associated with a huge gap in the distribution of the broad-leaved forests (the so-called Nemoral pattern) of temperate Eurasia, which are replaced by taiga (coniferous forest) northward and by desert or tropical vegetation southward. The width and a shape of the distributional gap vary greatly, depending on the animal group. Lake Baikal in south-central Siberia is a key area; some western Palearctic species (e.g., European *Bufo bufo*) reach the western shore of the lake, whereas some East Asian species (e.g., *Hyla japonica, Bufo raddei,* and probably *B. gargarizans*) occur east of the lake.

Among Recent Palearctic amphibians, six of 13 families of both Laurasian and Gondwanan origin demonstrate the west-east discontinuity. However, the cases shared by this pattern are fairly heterogeneous both geographically and taxonomically (at generic and familial levels). Examples include *Pleurodeles* in the west and *Tylototriton* in the east, *Triturus* and *Neurergus* in the west and *Cynops* and *Paramesotriton* in the east, *Bufo bufo* and *B. viridis* groups, *Bombina, Hyla,* and the *Rana esculenta* and *R. temporaria* groups.

It is commonly assumed that the gaps in distribution of closely related or allopatric pairs of species or subspecies arose in association with severe climatic changes during the Pleistocene. Commonly cited examples are *Bombina variegata* (or *B. bombina*) vs. *B. orientalis, Bufo viridis* (or *B. calamita*) vs. *B. raddei, Hyla arborea* vs. *H. japonica, Rana lessonae* vs. *R. nigromaculata,* and *R. temporaria* vs. *R. chensinensis.* However, Borkin (1984, 1986) reanalyzed these cases and rejected this common textbook explanation. He suggested that in all of these examples the species "pairs" are not closest relatives and have other sister species, and that disjunctions probably date from the Neogene. Progressive aridization and cooling of Tertiary climate, particularly in the central parts of Eurasia, may have been responsible. Borkin (1984) proposed four modes of origin of the west-east discontinuity. Three of them are true disjunctions, and one a pseudodisjunction. Some Recent East Asian genera are known as fossils from Europe. These are the salamandrid *Tylototriton* from the middle Eocene of Germany, and the cryptobranchid *Andrias* from the upper Oligocene to late Pliocene of Germany and the former Czechoslovakia (Estes, 1981; Roček, 1994).

Western Mediterranean versus Californian

This intercontinental pattern is evidenced by the plethodontid *Hydromantes.* Various possible routes and dates of invasion into Europe were discussed by Wake et al. (1978), Milner (1983), Lanza and Vanni (1981), and Lanza et al. (1995). In a broader context, North American–Eurasian discontinuities are exhibited by cryptobranchids, proteids, salamandrid genera of the *Triturus* group, and pelobatids, as well as *Bufo, Hyla,* and *Rana.* However, these cases are fairly heterogeneous and are based on different histories. (See the foregoing section on Historical Sources.)

Atlantic Europe versus the Caucasus

The herpetofauna of Europe consists of two large zoogeographic groups, the western European and the eastern European, which have different histories. The amphibain components were discussed by Zuiderwijk (1980) and the snakes by Szyndlar (1984). The origin of the amphibian groups has been explained by their survival in two refugia during Pleistocene glacial periods, followed by dispersal eastward from the western (Iberian) refuge and westward from the eastern refuge (Balkan-Pontic steppe area) during climatically more suitable times. Components of both groups made contact (parapatric or sympatric) in eastern France and central Europe. However, two cases demonstrate the Atlantic (and Iberian) versus Caucasian pattern of interregional discontinuous distribution. These are *Pelodytes punctatus* in Atlantic Europe and *P. caucasicus* in the Caucasus, and *Chioglossa lusitanica* in Iberia and *Mertensiella caucasica* in the Caucasus. The latter pair has been recognized as sister taxa based on external morphology, osteology, and behavior (Goux, 1957; Sanchiz and Młynarski, 1979; Vences, 1990), although Özeti (1967) and Wake and Özeti (1969) included *Mertensiella* in the genus *Salamandra*. The Caucasus existed as an island for a long time and was joined with Asia Minor in the Neogene, but remained separated from the eastern European mainland by a marine strait. Therefore, faunal exchanges between western Europe and Caucasus probably occurred only via West Asia (Borkin, 1987). However, more former widespread distributions are evidenced by *Mertensiella* from the lower Miocene of Germany and the upper Pliocene of Slovakia and Poland (Sanchiz and Młynarski, 1979; Hodrová, 1984) and fossil remains close to *Triturus marmoratus,* an Atlantic species, from the late Miocene of the North Caucasus and the upper Pliocene of Slovakia (Estes, 1981; Hodrová, 1984).

Western versus Eastern Mediterranean

This interregional disjunction is evidenced by the genus *Discoglossus* with four species in the western Mediterranean species and *D. nigriventer* in the Near East.

Gibraltar

This disjunction involves the Maghrebian and Iberian parts of two provinces in the Mediterranean Region (Fig. 6:8; Table 6:17). Nine cases with six families are shared by this intercontinental, but in-traregional pattern. *Pleurodeles waltl, Bufo bufo, B. viridis, Alytes obstetricans, Discoglossus pictus,* and *Hyla meridionalis* are represented in each of the two parts by isolated populations or subspecies. *Salamandra algira* and *S. salamandra, Pelobates varaldii* and *P. cultripes,* and *Rana saharica* and *R. perezi* are examples of species pairs. All these conspecific and specific pairs probably have been separated from each other for at least 5.5–7.0 million years by the opening of the Strait of Gibraltar in the late Miocene. However, some discrepancies exist between this date and estimated times of divergence based on molecular data (Bons, 1973; Arntzen and Szymura, 1984; Busack, 1986a, 1986b; Arntzen and García-París, 1995). An undetermined species of *Discoglossus* is known from the upper Miocene of Morocco (Vergnaud-Grazzini, 1966).

Sicily and Maltese Islands versus Tunisia

This intercontinental, but intraregional disjunction is exhibited by two species. *Discoglossus pictus* is shared by the mainland of northwest Africa (*D. p. algirus* [= *auritus*]) and the Mediterranean Sicily and Maltese islands (*D.p.pictus*). *Bufo viridis* has a similar disjunction, but is absent from the Maltese Islands, although Pleistocene remains are known from Malta (Sanchiz, 1979). The latter also seems to be represented by different subspecies. As in the case of the Gibraltar pattern, amphibians shared by Sicily and peninsular Italy, probably, reached the former during times of lower sea level in the Pleistocene (Lanza and Vanni, 1987).

Mainland versus Atlantic Archipelagos

This disjunction is exhibited by two anurans, *Hyla meridionalis* and *Rana perezi*. (See Macaronesia.)

Balearic

A discoglossid lineage leading to *Alytes muletensis,* which is endemic to the Balearic Islands, probably reached the archipelago during Messinian time in the late Miocene when the connection with the mainland came into existence prior to the complete desiccation of the Mediterranean Sea. The opening of the Strait of Gibraltar and the formation of the Mediterranean Sea about five million years ago separated these islands from the mainland (Arntzen and García-París, 1995). Other anurans now inhabiting the Balearic Archipelago probably were introduced by human activity (Hemmer et al., 1981). However, a Quaternary record of *Bufo viridis* exists (Sanchiz, 1979).

Tyrrhenian

The history of amphibians on two Mediterranean islands, Corsica and Sardinia (Table 6:18), involved at least three different scenarios. In the Oligocene, both islands were part of one microplate, which separated from southwestern Europe about 29 million years ago and drifted to the east. Later the microplate split into two islands; that process began 15, and was completed 9, million years ago (Sbordoni et al., 1990; Caccone et al., 1994). These orogenic events probably resulted in the isolation of *Euproctus montanus* on Corsica and *E. platycephalus* on Sardinia. The Oligocene drifting of the block also is hypothesized for an ancestral stock leading to *Hydromantes genei* (Lanza and Vanni, 1987; Lanza et al., 1995). Another scenario seems to be associated with the late Miocene Messinian crisis 5–6 million years ago when some amphibians (e.g., *Discoglossus montalentii* and three eastern Sardinian species—*Hydromantes flavus, H. imperialis,* and *H. supramontis*) probably invaded the islands (Capula et al., 1985; Lanza and Vanni, 1987; Lanza et al., 1995). A third scenario apparently occurred in the Pliocene, when *Discoglossus sardus, Hyla sarda,* and probably *Salamandra corsica* reached the islands via a land connection during one of the marine regressions about 2 million years ago (Lanza and Vanni, 1987).

Hellenic Arc

The insular Hellenic arc in the southern Aegean Sea between the Peloponnesos and Asia Minor has existed since the late Miocene. Crete, with a sole endemic frog, *Rana cretensis,* became permanently isolated from the mainlands after the Messinian crisis. Karpathos, with the endemic *Rana cerigensis,* may have separated from the Anatolian mainland in the middle or late Pliocene, and Rhodes separated in the late Pliocene or early Pleistocene (Beerli, 1994).

Coastal Mediterranean Islands

These minor insular disjunctions can be explained by changes in sea level during the Pleistocene, as has been proposed by Lanza and Vanni (1987).

Iberian

This disjunction is exhibited by *Hyla meridionalis* which has a range consisting of two isolated parts; the southern one occupies the southwest of the Iberian Peninsula, and the northern one includes the Mediterranean northeastern Spain, southern France and northwestern Italy (García París in Gasc et al., 1997). Generally, this species avoids mountainous areas. The remaining parts of the range also are fragmented—Balearic Islands (Menorca), Canary and Madeira islands, and northwestern Africa. If the fossil reported from the late Pleistocene of Britain by Holman (1992) is correctly identified, *H. meridionalis* formerly had a more extensive distribution.

Southern Mountains versus Central and Northern Europe

Apart from their primary ranges, some widespread amphibians occupy some isolated areas, mainly mountains in southern Europe. The northern Iberian populations of *Triturus alpestris* and the Pyreneen subspecies of *Rana temporaria* are examples of this interprovincial distribution. During warm interglacial phases, amphibians of colder areas may have taken refuge in mountains (Rage and Saint Girons in Castanet and Guyétant, 1989; Rage in Gasc et al., 1997).

Atlantic versus Mediterranean

An isolated part of the range of *Pelobates cultripes* is in the Loire-Garonne drainage in the Atlantic coastal area of France. These populations are separated by a wide gap from the the major part of the range in the French Mediterranean and Iberia (Castanet and Guyétant, 1989; Gasc et al., 1997) and constitute an interprovincial disjunction. Such a disjunction could be explained by species extinction in the Garonne River Valley, if the Atlantic and Mediterranean parts have been connected, or by extinction of more southern Atlantic populations, if the range expanded from the Iberian Peninsula via the Atlantic coast route. Also, *Bufo calamita* and *Bombina variegata* probably have isolated Atlantic populations (maps in Castanet and Guyétant, 1989; Gasc et al., 1997). However, these species probably have a different history.

Po Plain

The disjunct *Pelobates fuscus insubricus* represents this species on the Po Plain. Correspondingly, this is a disjunct area of the continental oak forests, the remainder of which covers southeastern Europe north of the Balkans (Fig. 2 in Gasc et al., 1997).

Southern versus Central (or Northern) Apennines

Obvious gaps of various size exist in the ranges of some amphibians between the southern (mainly in Calabria) and central (or northern) parts of peninsular Italy. This pattern of intraprovincial disjunction is exhibited by *Salamandra salamandra, Triturus alpestris, T. carnifex, Bufo bufo, B. viridis, Hyla intermedia,* and green frogs (e.g., maps in Gasc et al., 1997).

British Isles

Six native species of European amphibians occur on the British Isles—*Triturus cristatus, T. helveticus, T. vulgaris, Bufo bufo, B. calamita,* and *Rana temporaria.* Only three of these (*T. vulgaris, Bufo calamita,* and *Rana temporaria*) inhabit Ireland. This intraprovincial disjunction is the result of Quaternary events. The Pleistocene fauna was richer and contained treefrogs and green frogs (Holman, 1992).

Baltic

Some widespread species are distributed on both sides of the Danish straits in the western Baltic Sea. *Bombina bombina, Bufo calamita, B. viridis, Pelobates fuscus, Rana dalmatina, R. esculenta,* and *R. lessonae* reach the northern limits of their ranges in southern Sweden.

Balkan versus Asia Minor

Some widespread European species (e.g., *Bombina bombina* and *Rana dalmatina*) occur across the Bosphorus and Dardanelles in the western part of Anatolian Turkey. In some cases, such populations in Asia Minor are recognized subspecifically (e.g., *Triturus vulgaris kosswigi*). Two mainly East Mediterranean species, *Triturus karelinii* and *Pelobates syriacus,* also share this disjunct pattern. *Salamandra salamandra* is lacking in Thrace, the European province of Turkey, and *Bombina bombina* is known only from the western part of Thrace (Atatür and Yilmaz, 1986). The separation of Thrace Turkey from Anatolia can be dated as the early Pleistocene (ca. 1.8 million years ago).

Crimean

Among five species inhabiting the Crimea Peninsula, two (*Triturus karelinii* and *Hyla arborea*) are represented by isolated populations. Inhabitants of the Crimean forests are isolated by steppe to the north and by the Black Sea in other directions. The Crimean amphibians seem to be closer to the Caucasian than to other conspecific populations (Shcherbak, 1966; personal data). Therefore, they demonstrate the Crimea versus Caucasian intraprovincial disjunct pattern.

Caucasian versus East European

A few widespread European species preferably inhabiting forest areas are separated by a gap in the Russian steppe area from endemic subspecies in the Caucasus. This pattern is shown by *Triturus v. vulgaris* in the closest part of eastern Europe and *T. v. lantzi* in the Caucasus, and by *Hyla a. arborea* in the European part of the former USSR and *H. a. schelkownikowi* in the Caucasus (and probably also in the Crimea Peninsula). The case of *Bufo verrucosissimus,* treated by previous authors as a Caucasian subspecies of the European *B. bufo,* needs further study, because the species may be closer to the Mediterranean *B. b. spinosus* (a distinct species?) than to the boreal *B. b. bufo.*

Caucasian versus South Anatolian

Interprovincial distributions in the Caucasus (plus the Black Sea zone of Turkey) and in southern Mediterranean Turkey and neighboring countries, with an extensive gap in central Anatolia, are exhibited by two salamandrids. *Triturus vittatus ophryticus* occurs in the northern area and *T. v. vittatus* plus *T. v. cilicensis* in the southern area (Schmidtler and Schmidtler, 1967). *Mertensiella caucasica* occurs in the northern area, and *M. luschani* in the southern area.

Arabian

Three widespread anurans (*Bufo viridis, Hyla savignyi,* and *Rana "ridibunda"*) occur in the Near East and in the mountains of the Arabian Peninsula. Water frogs also occur in the eastern part of the peninsula. This is a case of an interrealm disjunction.

Sinaitic

Only two anurans, *Bufo viridis* and *Rana bedriagae* (= *R. levantina*) are shared by the Nile area in Egypt and the Near East, with a gap in the Sinai Peninsula.

Iranian versus Tibetan Plateaus

Three isolated endemic species of the hynobiid genus *Batrachuperus* occur in the Elburz (two species) and Hindu Kush (one species) mountains on the northern and northeastern edge of the Iranian Plateau. Four other species are known from the eastern border of Tibet in the Hengduanshan (Zhao, this volume). The Elburz, Kopet Dagh, and Hindu Kush have existed as a long westward extension of Asia surrounded by sea since the late Eocene-Oligocene, about 35 million years ago, when the main parts of Iran, Afghanistan, and Turan lowlands were covered by sea (maps in Dercourt et al., 1986). Therefore, hynobiids probably invaded the Iranian mountains subsequently.

Dzhungarsky (Junggar) Alataw Mountains versus Central China

This disjunct pattern is exhibited by the hynobiid genus *Ranodon,* with *R. sibiricus* in the north and *R. tsinpaensis* in the south; intervening northwestern China lacks hynobiids.

Northeastern versus Southeastern China

Bombina exhibits such an interrealm disjunct pattern. The lowland *B. orientalis* of the subgenus *Bombina* occurs in the Russian Far East, Korean Peninsula, and northeastern China. Three species of the subgenus *Grobina* are known only from mountainous southwestern China, with a wide gap in eastern China (Zhao and Adler, 1993; Zhao, this volume).

Northern China Plain

Some anurans (e.g., *Hyla japonica* and *Rana plancyi*) have a gap in their ranges between the Huang He (= Yellow River) and Huai He (Zhao, this volume).

Russian Far East Mainland versus Northern Islands

Five species of amphibians invaded Sakhalin Island via its northern part, most probably during the Pleistocene. Three of these reached the southern Kurile Islands via Hokkaido Island. This is an intraprovincial disjunct pattern.

Japanese

This important pattern is shared by 28 species distributed in the Japanese Islands between Hokkaido and the Ryukyu Archipelago (Appendix 6:1). Amphibians may have entered the Japanese main islands from the Korean Peninsula or from China directly through a wide lowland now covered by the East China Sea (Hikida et al., 1989). This pattern is an interprovincial or even an interrealm disjunction depending on taxa (e.g., *Rana rugosa* or the genus *Rhacophorus,* respectively).

CONSERVATION

Not considering the present decline in amphibian populations globally, the reality of which needs to be confirmed, the most important danger to most amphibians is habitat alteration resulting directly or indirectly from human activities. For example, habitat destruction (plus some other factors) must be responsible for the loss of many local populations of the Italian *Pelobates fuscus insubricus,* which recently had a disjunct range on the Po Plain (Corbett, 1989; Nöllert, 1990). However, in some cases, landscape modification, such as the formation of new bodies of water, resulted in increases in populations of amphibians—demographic explosion in the Moroccan mountains (Dubois, 1982) or in successful expansion of the range of a species, such as that of *Rana ridibunda* in Central Asia. Water pollution can be especially dangerous for many species because of the aquatic phase of their life cycles. High incidence of various morphological anomalies has been discovered in anurans—Sakhalin Island in the Russian Far East (Mizgireuv et al., 1984) and in industrial areas in the eastern Ukraine (Flax and Borkin, 1997).

At least in some cases, declining populations and even local extinctions can be attributed to a particular cause. For example, the introduction of the amphibian-eating snake *Natrix maura* had deleterious effects on the Mallorcan *Alytes muletensis* (Cor-

bett, 1989). Other declines are the result of introduction of fishes that feed on newts and larvae, and overcollecting.

The conservation policies and standards vary greatly among the countries covered by this chapter. Efforts for conservation of amphibians in most European countries are exemplary in contrast to those elsewhere in the world. In addition to the designation of protected areas of habitat for the natural survival of amphibians and other organisms, some countries (e.g., Denmark, France, Germany, Luxembourg, and the Netherlands) have legislation protecting all species of amphibians; however, in other countries (e.g., Turkey), no species are protected (Corbett, 1989; Engelmann et al., 1993). Traffic mortality during times of breeding migrations has been reduced, especially in England, with the construction of passages under roadways. Attempts at propogation of some rare salamanders and anurans in laboratories and zoos and repatriation of captive-bred individuals into localities where species occurred formerly have been accomplished in several countries in western Europe and the former USSR.

But populations of amphibians in many areas continue to decline because of habitat destruction, use of pesticides, and pollution. Moreover, although France has rigid protection of its native amphibians, the French (and Swiss) gastronomic predilection for consuming frog legs has resulted in decimation of populations of edible frogs of the genus *Rana* in many parts of the world. According to Honegger (1981), more than 2 million frogs were exported for food from Greece in 1975, and in the period 1968–1970 about 47 million frogs (i.e., approximately 174 tons) were exported from Italy. In the period 1971–1981, France imported 24,794 tons of frozen froglegs, of which 43% came from Indonesia, 39% from India, 7% from Greece, and 1% from eastern Europe (Dubois, 1983). Once more stringent controls were placed on depauperate populations, commerce shifted to India and Bangladesh, where populations of edible species of *Rana* have dwindled, and *Rana (Hoplobatrachus) tigerina,* once a common species in much of India, is now listed as an endangered species.

Honegger (1981) listed four European salamanders (*Salamandra salamandra almanzoris, Triturus alpestris cyreni, T. dobrogicus,* and *T. montandoni*) and seven European anurans (*Bufo bufo, Hyla meridionalis, Rana arvalis, R. esculenta, R. lessonae, R. perezi,* and *R. ridibunda*) as threatened species. He also noted that *Proteus anguinus* was threatened by pollution and collecting by farmers for pig food.

The Berne Convention on the Conservation of European Wildlife and Natural Habitats, administered by the Council on Europe (1982), has been ratified by 18 of the 21 states of western Europe. The convention concerns all European species listed in two annexes (II—strictly protected, and III—control of exploitation), except *Salamandrella keyserlingii* and *Rana shqiperica,* which are distributed in countries that are not members (Oliveira et al., 1997). Moreover, the annexes include three amphibians (*Triturus vittatus, Pelodytes caucasicus,* and *Rana macrocnemis*) in the Caucasus and Asia Minor, since Turkey ratified the convention.

A network of European Biogenetic Reserves has been proposed. These reserves should cover: (1) so-called species sites, i.e., the most important sites of critical habitats of threatened species of amphibians and reptiles, and (2) so-called species assemblage sites, i.e., centers of diversity where high numbers of species and important herpetofaunal communities exist, even if no single species is especially threatened. Some of these sites (Corbett, 1989) coincide with areas shown in Figure 6:12.

The 1996 IUCN Red List of Threatened Animals contains 17 species and two subspecies of European amphibians. Among them are *Euproctus platycephalus, Salamandra atra aurorae,* and *Alytes muletensis* classified as critically endangered; *Pelobates fuscus insubricus* as endangered; and *Chioglossa lusitanica, Hydromantes flavus, Mertensiella luschani, Proteus anguinus, Salamandra lanzai, Alytes dickhelleni,* and *Discoglossus montalentii* as vulnerable (Oliveira et al., 1997). No European species is listed on the CITES Appendices; however, European Union's Habitat and Species Directive covers most European amphibians (Oliveira et al., 1997).

The Societas Europaea Herpetologica (SEH) has prioritized three species of amphibians (*Proteus anguinus, Euproctus platycephalus,* and *Alytes muletensis*) as conservation targets and is paying special attention to four other species—*Triturus cristatus* (the superspecies), *Bombina bombina, Hyla arborea,* and *Rana holtzi* (Corbett, 1997). Furthermore, the Species Survival Group for the European Herpetofauna of the International Union for the Conservation of Nature is drawing up Action Plans for 47 species of amphibians and reptiles. Nevertheless, some European governments do not adequately fulfill their obligations under the Berne Convention, particularly with respect to protecting habitats of threatened species listed by the Convention (Corbett (1997). "However, whatever the political mechanism, we would reiterate that little will be achieved until our herpetofauna are [is] ensured a commensurate part of the nature conservation resource and its budget, and that in the interim further declines therefore seem inevitable and several extinctions now look probable" (Corbett, 1997:30).

Unlike western Europe, countries of the former Soviet Union protect fewer than 33% of species of national amphibian faunas. Of the six species listed in the Russian Red Data Book, only *Onychodactylus fischeri* is Asiatic. Amphibians are listed in the Red Data Books of nine other republics formerly part of the Soviet Union (Appendix 6:3). As many as five species are listed by Ukraine. *Bufo calamita* is listed in six Red Data Books, and *Pelobates syriacus* is listed in four books. However, some Red Data Books list species that do not occur in the country, whereas others list common species, such as *Bufo danatensis* in Kazakhstan (Appendix 6:3). Nearly all species, except *Ranodon sibiricus* in Kazakhstan and *Bufo shaartusiensis* and *Rana terentievi* in Tadjikistan, occur in protected areas in more than 150 reserves dispersed in the territory of the former USSR (Darevsky and Krever, 1987; Kuzmin, 1995b; personal data).

Apart from national and regional levels, the conservation of Palearctic amphibians needs a systems approach based on zoogeographic principles rather than on political philosophies of a divided world. Preservation of the integrity of the Palearctic herpetofauna can be accomplished by establishing appropriate reserves at least in the 51 areas that I have identified as having high species richness and/ or endemism (Fig. 6:12). Some reserves, such as the Caucasian Biosphere Reserve in the northwestern Caucasus, already exist. The suggested list of areas should be considered as a minimum. More detailed information on, and more exact limits of, areas can be found in other sources or discussed elsewhere. Suggestions of additional areas of special importance at regional, national, or local levels would be welcome.

Acknowledgments: Above all, I deeply thank William E. Duellman (USA) for his encouragement to write this chapter, for his enduring patience to cooperate with me in the course of its preparation, and for his diligent editing of the manuscript. M. Sherif Khan (Pakistan) and N. Rastegar-Pouyani (Iran) have kindly provided me some data on the distribution of Pakistani and Iranian amphibians, respectively. I also used some unpublished data on Iranian amphibians given to me many years ago by Robert G. Tuck, Jr. (USA), who was an advisor for the Tehran Museum, Iran. Emilio Balletto and Cristina Giacoma (Italy) consulted with me on amphibians of the Mediterranean area and the Arabian Peninsula. Spartak N. Litvinchuk (Russia) checked the distributions of salamandrids and some Yugoslavian amphibians. Ulrich Joger (Germany) provided data on the Atlantic archipelagos, and George D. ukic and Jelka Crnobrnja (Yugoslavia) contributed data on distributions in the former Yugoslavia. Robert F. Inger (USA) and Er-mi Zhao (China) have kindly provided me with copies of the manuscripts for their chapters contained herein. Also, I am grateful to Kraig Adler (USA), Ilya Darevsky (Russia), Notker Helfenberger and Beat Schätti (Switzerland), and A. A. Tawfik (Egypt), who sent me copies of some publications. Alain Dubois and Annemarie Ohler (France) and George R. Zug (USA) gave me access to their libraries. The Zoological Institute, Russian Academy of Sciences, and the St. Petersburg Association of Scientists and Scholars provided facilities for the preparation of the manuscript.

LITERATURE CITED

ADLER, K., AND E. ZHAO. 1990. Studies on hynobiid salamanders, with description of a new genus. Asiatic Herpetol. Res. 3:37–45.

AFRASIAB, S. R., AND H. A. ALI. 1988. A new record of toad *Bufo surdus* Boulenger (Amphibia, Bufonidae) from Iraq, with a preliminary key for Iraqi Amphibia. Bull. Iraq Nat. Hist. Mus., Univ. Baghdad, 8:115–123.

AKEF, M. S. A., AND H. SCHNEIDER. 1989. The eastern form of *Rana ridibunda* (Anura: Ranidae) inhabits the Nile Delta. Zool. Anzeiger 223 (3/4):129–138.

ALCOBENDAS, M., H. DOPAZO, AND P. ALBERCH. 1994. Genetic structure and differentiation in *Salamandra salamandra* populations from the northern Iberian Peninsula. Mertensiella 4:7–23.

ANANJEVA, N. B., L. J. BORKIN, I. S. DAREVSKY, AND N. L. ORLOV. 1998. *Amphibians and Reptiles of Russia and Adjacent Countries.* Moscow: ABF (in Russian).

ANDERSON, K. 1991. Chromosome evolution in Holarctic treefrogs. Pp. 299–331 *in* D. M.Green and S. K.Sessions (eds.), *Amphibian Cytogenetics and Evolution.* San Diego: Academic Press.

ANDERSON, S. C. 1963. Amphibians and reptiles from Iran. Proc. California Acad. Sci., Ser. 4, 31:417–498.

ANDERSON, S. C. 1968. Zoogeographic analysis of the lizard fauna of Iran. Pp. 305–371 *in* W. B. Fisher (ed.), *The Cambridge History of Iran. Volume I. The Land of Iran.* Cambridge: Cambridge Univ. Press.

ANDERSON, S. C. 1985. Amphibians. Pp. 987–990 *in* E. Yarshater (ed.), *Encyclopaedia Iranica, Vol. 1.* London: Routlege & Kegan Paul.

ANDERSON, SYDNEY. 1994. Area and endemism. Quart. Rev. Biol. 69:451–471.

ANDRÉN, C., AND G. NILSON. 1979. A new species of toad (Amphibia, Anura, Bufonidae) from the Kavir Desert, Iran. J. Herpetol. 13:93–100.

ANGEL, F., AND J. GUIBÉ. 1948. A propos d'*Arthroleptis agadesi* Angel (Batracien). Bull. Mus. Natl. Hist. Nat. Paris (2), 20: 62–63.

ANGEL, F., AND H. LHOTE. 1938. Reptiles et Amphibiens du Sahara central et du Soudan. Bull. Comité Étud. Hist. Sci. Afrique Occid. Franç., Paris, 21:345–384.

ANONYMOUS. 1997. The International Commission on Zoological Nomenclature. Opinion 1866. *Hydromantes* Gistel, 1848 (Amphibia, Caudata): *Spelerpes platycephalus* Camp, 1916 designated as the type species. Bull. Zool. Nomencl. 54:72–74.

ANUFRIEV, V. M., AND A. V. BOBRETSOV. 1996. [Amphibians and Reptiles.] St. Petersburg: Nauka(*in* Fauna of the European Northeast of Russia, Vol. 4) (in Russian).

ARANO, B., G. A. LLORENTE, P. HERRERO, AND B. SANCHÍZ. 1995 "1994." Current studies on Iberian water frogs. Zool. Poloniae 39:365–375.

ARNTZEN, J. W. 1995. European newts: a model system for evolutionary studies. Pp. 26–32 *in* G.A.Llorente, A. Montori, X.

Santos, and M.A. Carretero (eds.), *Scientia Herpetologica*. Barcelona: Soc. Española Herpetol.

ARNTZEN, J. W., AND M. GARCÍA-PARÍS. 1995. Morphological and allozyme studies of midwife toads (genus *Alytes*), including the description of two new taxa from Spain. Contrib. Zool., Amsterdam, 65: 5–34.

ARNTZEN, J. W., AND B. SKET. 1997. Morphometric analysis of black and white European cave salamander, *Proteus anguinus*. J. Zool. London 241:699-707.

ARNTZEN, J. W., AND J. M. SZYMURA. 1984. Genetic differentiation between African and European midwife toads (*Alytes, Discoglossidae*). Bijdr. Dierkunde 54:157–162.

ASTUDILLO, G., AND B. ARANO. 1995. Europa y su herpetofauna: responsibilidadesde cada pais en lo referente a su conservación. Bol. Asoc. Herpetol. Española 6:14–45.

ATATÜR, M. K., AND A. BUDAK. 1982. The present status of *Mertensiella caucasica* (Waga, 1876) (Urodela: Salamandridae) in northeastern Anatolia. Amphibia-Reptilia 4:295–301.

ATATÜR, M. K., AND I. YILMAZ. 1986. A comparison of the amphibian fauna of Turkish Thrace with that of Anatolia and Balkan states. Amphibia-Reptilia 7:135–140.

BAHA EL DIN, S. M. 1992. Notes on the herpetology of north Sinai. British Herpetol. Soc. Bull. 41:9–11.

BAHA EL DIN, S. M. 1993. A new species of toad (Anura: Bufonidae) from Egypt. J. Herpetol. Assoc. Africa 42:24–27.

BALLETTO, E., M. A. CHERCHI, AND J. GASPERETTI. 1985. Amphibians of the Arabian Peninsula. Fauna of Saudi Arabia, 7:318–392.

BALLETTO, E., C. GIACOMA, C. PALESTRINI, A. ROLANDO, M. SARÀ, A. BARBERIS, S. SALVIDIO, P. MENSI, AND L. CASSULO. 1990. On some aspects of the biogeography of northern Macaronesia. Pp. 167–199 *in* A. Azzaroli (ed.), *Biogeographical Aspects of Insularity*. Rome, Accad. Nazion. Lincei.

BĂNĂRESCU, P., AND N. BOŞCAIU. 1978. *Biogeographie. Fauna und Flora der Erde und ihre geschichtliche Entwicklung*. Jena: G.Fischer.

BARAN, I., AND M. K. ATATÜR. 1986. A taxonomic survey of the mountain frogs of Anatolia. Amphibia-Reptilia 7:115–133.

BARAN, I., AND M. ÖZ. 1994. *Salamandra salamandra* of Anatolia. Mertensiella 4:25–32.

BARUŠ, V., B. KRÁL, O. OLIVA, E. OPATRNÝ, I. REHÁK, Z. ROČEK, P. ROTH, Z. ŠPINAR, AND L. VOJTKOVÁ. 1992 *Obojživelníci - Amphibia*. Prague: Academia (Fauna ČSFR, 25).

BAŞOGLU, N., AND N. ÖZETI. 1973. The amphibians of Turkey. Ege Univ. Fen Fak. Kitaplar Ser. 50:1–155.

BEERLI, P. 1994. Genetic isolation and calibration of an average protein clock in western Palearctic water frogs of the Aegean region. Universität Zürich, Inaugural Dissertation.

BEERLI, P., H. HOTZ, H. G. TUNNER, S. HEPPICH, AND T. UZZELL. 1994. Two new water frog species from the Aegean Islands Crete and Karpathos (Amphibia, Salientia, Ranidae). Notulae Naturae 470:1–9.

BÖHME, W. 1978. Die Indentität von *Rana esculenta bilmaensis* Angel, 1936, aus der südlichen Sahara. Rev. Suisse Zool. 85:641–644.

BÖHME, W., Z. ROČEK, AND Z. V. ŠPINAR. 1982. On *Pelobates decheni* Troschel, 1861, and *Zaphrissa eurypelis* Cope,1866 (Amphibia: Salientia: Pelobatidae) from the early Miocene of Rott near Bonn,West Germany. J. Vert. Paleontol. 2:1–7.

BÖHME, W., AND H. WIEDL. 1994. Status and zoogeography of the herpetofauna of Cyprus, with taxonomic and natural history notes on selected species (genera *Rana, Coluber, Natrix,*

Vipera). Pp. 31–52 *in* R. Kinzelbach and M. Kasparek (ed.), *Zoology in the Middle East*. Vol. 10. Heidelberg: M. Kasparek Verlag.

BONS, J. 1972. Herpetologie marocaine. I. Liste commentee des Amphibiens et Reptiles du Maroc. Bull. Soc. Sci. Nat. Phys. Maroc: 52:107–126.

BONS, J. 1973. Herpetologie marocaine. II. Origines, évolution et particularités du peuplement herpétologique du Maroc. Bull. Soc. Sci. Nat. Phys. Maroc 53:63–110.

BORKIN, L. J. 1978 "1977." On a new record and taxonomic position of the brown frogs of Kopeth-Dagh, Turkmenia. Pp. 24–31 *in* N. B. Ananjeva, L. J. Borkin, and I. S. Darevsky (eds.), *Herpetological Collected Papers.*: Proc. [Trudy] Zool. Inst., USSR Acad. Sci., Leningrad 74 (in Russian, with English summary).

BORKIN, L. J. 1984. The European–Far Eastern disjunctions in distribution of amphibians: a new analysis of the problem. Pp. 55–88 *in* L. J. Borkin (ed.), *Ecology and Faunistics of Amphibians and Reptiles of the USSR and Adjacent Countries*. Proc. [Trudy] Zool. Inst., USSR Acad. Sci., Leningrad 124 (in Russian, with English summary).

BORKIN, L. J. 1986. Pleistocene glaciations and western–eastern Palearctic disjunctions in amphibian distribution. Pp. 63–66. *in* Z. Roček (ed.), *Studies in Herpetology*. Prague: Charles Univ. Press.

BORKIN, L. J. 1987 "1986." On the systematics and zoogeography of amphibians of the Caucasus. Pp. 47–58 *in* L. J. Borkin (ed.), *Herpetological Investigations in the Caucasus*. Proc. [Trudy] Zool. Inst., USSR Acad. Sci., Leningrad 158 (in Russian, with English summary).

BORKIN, L. J. 1988. General account of amphibian distribution. Pp. 213–229 *in* E. I. Vorobieva and I. S. Darevsky (eds.), *Amphibans and Reptiles of the Mongolian People's Republic. General Problems. Amphibians*. Moscow: Nauka (in Russian).

BORKIN, L. J. 1994. Systematics. Pp. 54–80 *in* E. I. Vorobyeva (ed.), *The Siberian Newt* (Salamandrella keyserlingii *Dybowski, 1870). Zoogeography, Systematics, Morphology*. Moscow: Nauka (in Russian).

BORKIN, L. J., AND E. V. ANISSIMOVA. 1987 "1986." The vertebral structure and vocal sacs in the Caucasian parsley frog (*Pelodytes caucasicus*) and its taxonomic position. Pp. 59–76 *in* L. J. Borkin (ed.), *Herpetological Investigations in the Caucasus*. Proc. [Trudy] Zool. Inst., USSR Acad. Sci., Leningrad 158 (in Russian, with English summary).

BORKIN, L. J., AND A. M. BASSARUKIN. 1987. Herpetofauna of the Kurile Reserve. Pp. 219–227 *in* I. S. Darevsky and V. G. Krever (eds.), *Amphibii i Reptilii Zapovednykh Territorii*. Moscow:Scientific-Research Laboratory of Hunting Management and Reserves (in Russian).

BORKIN, L. J., G. T. BELIMOV, AND V. T. SEDALISHCHEV. 1984. New data on distribution of amphibians and reptiles in Yakutia. Pp. 89–101 *in* L. J. Borkin (ed.), *Ecology and Faunistics of Amphibians and Reptiles of the USSR and Adjacent Countries*. Proc. [Trudy] Zool.Inst.,USSR Acad. Sci., Leningrad, 124 (in Russian, with English summary).

BORKIN, L. J., I. A. CAUNE, E. M. PISANETZ, AND Y. M. ROZANOV. 1986. Karyotype and genome size in the *Bufo viridis* group. Pp. 137–142 *in* Z.Roček (ed.), *Studies in Herpetology*. Prague: Charles Univ. Press.

BORKIN, L. J., AND I. S. DAREVSKY. 1987. A list of amphibians and reptiles of the USSR. Pp. 128–141 *in* I. S. Darevsky and V.

G. Krever (eds.), *Amphibians and Reptiles of Protected Territories*. Moscow (in Russian).

BORKIN, L. J., AND S. L. KUZMIN. 1988. Amphibians of Mongolia: species account. Pp. 30–197 *in* E. I. Vorobieva and I. S. Darevsky (eds.), *Amphibians and Reptiles of the Mongolian People's Republic. General Problems. Amphibians.* Moscow: Nauka (in Russian).

BORKIN, L. J., S. N. LITVINCHUK, AND Y. M. ROSANOV. 1997. Amphibians and reptiles of Moldavia: additions and corrections, with a list of species. Russian J. Herpetol. 4:50–62.

BORKIN, L. J., AND M. MATSUI. 1986. On systematics of two toad species of the *Bufo bufo* complex from eastern Tibet. Pp.43–54 *in* N. B. Ananjeva and L. J. Borkin (eds.), *Systematics and Ecology of Amphibians and Reptiles.* Proc. [Trudy] Zool. Inst., USSR Acad. Sci. Leningrad 157 (in Russian, with English summary).

BORKIN, L. J., AND V. V. ROSHCHIN. 1981. Electrophoretic comparison of proteins of the European and Far East toads of the *Bufo bufo* complex. Zool. Zhurnal 61:1802–1812 (in Russian, with English summary).

BRAESTRUP, F. W. 1948. Remarks on faunal exchange through the Sahara. Vidensk. Medd. Dansk Naturh. Foren. 110:1–15.

BRIGGS, J. L., AND C. R. AULT. 1985. Distribution of the genus *Rana* in southwestern Saudi Arabia. Herpetol. Rev. 16:72–75.

BRUNO, S. 1989. Introduction to a study of the herpetofauna of Albania. British Herpetol. Soc. Bull. 29:16–41.

BRUSHKO, Z. K., R. A. KUBYKIN, AND S. P. NARBAEVA. 1988. Recent distribution of Siberian salamander *Ranodon sibiricus* (Amphibia, Hynobiidae) in Dzungar Alatau. Zool. Zhurnal 67:1753–1756 (in Russian, with English summary).

BUCKLEY, D., B. ARANO, P. HERRERO, G. A. LLORENTE, AND M. ESTEBAN. 1995 "1994." Moroccan water frogs vs. *R. perezi:* allozyme studies show up their differences. Zool. Poloniae 39:377–385.

BUSACK, S. D. 1986a. Biochemical and morphological differentiation in Spanish and Moroccan populations of *Discoglossus* and the description of a new species from southern Spain (Amphibia, Anura, Discoglossidae). Ann. Carnegie Mus. 55:41–61.

BUSACK, S. D. 1986b. Biogeographic analysis of the herpetofauna separated by the formation of the Strait of Gibraltar. Natl. Geogr. Res. 2:17–36.

BUSACK, S. D., AND F. M. JAKSIĆ. 1982. Ecological and historical correlates of Iberian herpetofaunal diversity: an analysis at regional and local levels. J. Biogeogr. 9:289–302.

BUSACK, S. D., L. R. MAXON, AND M. A. WILSON. 1985. *Pelobates varaldii* (Anura: Pelobatidae): a morphologically conservative species. Copeia 1985:107–112.

CACCONE, A., M. C. MILINKOVITCH, V. SBORDONI, AND J. R. POWELL. 1994. Molecular biogeography: using the Corsica-Sardinia microplate disjunction to calibrate mitochondrial rDNA evolutionary rates in mountain newts (*Euproctus*). J. Evol. Biol. 7:227–245.

CAPULA, M., AND M. CORTI. 1993. Morphometric variation and divergence in the West Mediterranean *Discoglossus* (Amphibia: Discoglossidae). J. Zool., 231:141–156.

CAPULA, M., G. NASCETTI, B. LANZA, L. BULLINI, AND E. G. CRESPO. 1985. Morphological and genetic differentiation between the Iberian and the other West Mediterranean *Discoglossus* species(Amphibia Salientia Discoglossidae). Monit. Zool. Italiano, N.S. 19:69–90.

CASTANET, J., AND R. GUYÉTANT. 1989. *Atlas de repartition des Amphibiens et Reptiles de France.* Paris: Société Herpétologique de France.

CHANNING, A. 1989. A re-evaluation of the phylogeny of Old World treefrogs. South African J. Zool. 24:116–131.

ČKHIKVADZE, V. M. 1982. On the findings of fossil Cryptobranchidae in the USSR and Mongolia. Vertebrata Hungarica 21:63–67.

ČKHIKVADZE, V. M. 1984. Review on fossil urodelan and anuran amphibians of the USSR. Bull. [Izvestiya] Acad. Sci. Georgian [Grusin.] SSR, Ser. Biol. 10:5–13 (in Russian).

CLARK, R. 1989. A check list of the herpetofauna of the Argo-Saronic Gulf district, Greece. British Herpetol. Soc. Bull. 28:8–24.

CLARK, R. 1990. A report on herpetological observations in Afghanistan. British Herpetol. Soc. Bull. 33:20–42.

CLARK, R. 1991. A report on the herpetological investigations on the island of Samothraki, North Aegean Sea–Greece. British Herpetol. Soc. Bull. 38:3–7.

CLARK, R. J., E. D. CLARK, S. C. ANDERSON, AND A. E. LEVITON. 1969. Report on a collection of amphibians and reptiles from Afghanistan. Proc. California Acad. Sci., Ser. 4, 36:279–316.

CLARKE, B. T. 1987. A description of the skeletal morphology of *Barbourula* (Anura: Discoglossidae), with comments on its relationships. J. Nat. Hist. 21:879–891.

CLERGUE-GAZEAU, M., AND R. THORN. 1978. Une nouvelle espèce de Salamandre du genre *Batrachuperus,* en provenance de l'Iran septentrional (Amphibia, Caudata, Hynobiidae). Bull. Soc. Hist. Nat. Toulouse, 114:455–460.

CLOUDSLEY-THOMPSON, J. L. (ed.). 1984. *Sahara Desert.* Oxford: Pergamon Press [Russian version, 1990: Sahara. Moscow, Progress].

COGĂLNICEANU, D. 1991. A preliminary report on the geographical distribution of amphibians in Romania. Rev. Roumania Biol., Sér. Biol. Anim. 36(1–2):39–50.

CORBETT, K. 1989. *The Conservation of European Reptiles and Amphibians.* London: Christopher Helm.

CORBETT, K. 1997. Conservation of the European herpetofauna. Pp. 29–30 *in* Gasc, J. P., A. Cabela, J. Crnobrnja-Isailovic, D. Dolmen, K. Grossenbacher, P. Haffner, J. Lescure, H. Martens, J. P. Martinez-Rica, H. Maurin, R. Oliveiri, T. Sofianidou, M. Veith, and A. Zuiderwijk (eds.), *Atlas of Amphibians and Reptiles of Europe.* Paris: Muséum National d'Histoire Naturelle.

CRNOBRNJA-ISAILOVIC, J., AND G. DZUKIC. 1995. First report about conservation status of herpetofauna in the Skadar Lake region (Montenegro): current situation and perspectives. Pp. 373–380 *in* G. A. Llorente, A. Montori, X. Santos, and M. A. Carretero (eds.), *Scientia Herpetologica.* Barcelona: Asoc. Herpetol. Española.

CROCHET, P.-A., A. DUBOIS, A. OHLER, AND H. TUNNER. 1995. *Rana (Pelophylax) ridibunda* Pallas, 1771, *Rana (Pelophylax) perezi* Seoane, 1885, and their associated klepton (Amphibia, Anura): morphological diagnoses and description of a new taxon. Bull. Mus. Natl. Hist. Nat. Paris, Sér. 4, 17:11–30.

DAREVSKY, I. S., AND V. G. KREVER (eds.). 1987. *Amfibii i Reptilii Zapovednykh Territorii* [Amphibians and Reptiles of Protected Territories], Moscow: Central Scientific-Research Laboratory of Hunting Management and Reserves, The Main Administration of Hunting Management and Reserves at the Council of Ministers of the RSFSR (in Russian).

DARLINGTON, P. J. 1957. Zoogeography. The Geographical Distribution of Animals. New York: Wiley [Russian version: Moscow, Mir Press, 1966].

DAS, S. M. 1966. Palaearctic elements in the fauna of Kashmir. Nature 212:1327–1330.

DASZAK, P., AND S. CAWTHRAW. 1991. A review of the reptiles and amphibians of Turkey, including a literature survey and species checklist. British Herpetol. Soc. Bull. 36:14–26.

DE LATTIN, G. 1967. *Grundriss der Zoogeographie*. Jena: G. Fischer.

DERCOURT, J., L. P. ZONENSHAIN, L.-E. RICOU, V. G. KAZMIN, X. LE PICHON, A. L. KNIPPER, C. GRANDJACQUET, I. M. SBORTSCHIKOV, J. GEYSSANT, C. LEPVRIER, D. H. PECHERSKY, J. BOULIN, J.-C. SIBUET, L. A. SAVOSTIN, O. SOROKHTIN, M. WESTPHAL, M. L. BAZHENOV, J. P. LAUER, AND B. BIJU-DUVAL. 1986. Geological evolution of the Tethys belt from the Atlantic to the Pamirs since the Lias. Tectonophysics 123:241–315.

DISI, A. M. 1996. A contribution to the knowledge of the herpetofauna of Jordan. VI. The Jordan herpetofauna as a zoogeographic indicator. Herpetozoa 9:71–81.

DISI, A. M., AND W. BÖHME. 1996. Zoogeography of the amphibians and reptiles of Syria, with additional new records. Herpetozoa 9:63–70.

DOKUCHAEV, N. E., A. V. ANDREEV, AND G. I. ATRASHKEVICH. 1984. On distribution and biology of the Siberian salamander, *Hynobius keyserlingii*, in extreme north-east of Asia. Pp. 109–114 in L. J. Borkin (ed.), *Ecology and Faunistics of Amphibians and Reptiles of the USSR and adjacent Countries*. Proc. [Trudy] Zool. Inst., USSR Acad. Sci., Leningrad, 124 (in Russian, with English summary).

DUBOIS, A. 1981. Biogéographie des Amphibiens de l'Himalaya. Pp. 63-74 in Paléogéographie et Biogéographie de l'Himalaya et du Sous-Continent Indien. Paris: Centre National de la Recherche Scientifique.

DUBOIS, A. 1982. Les amphibiens de le station d'altitude d'Oukaimeden (Haut-Atlas, Maroc). Bull. Soc. Linn. Lyon 51:329–339.

DUBOIS, A. 1983. A propos de cuisses de grenouilles. Protection de amphibiens arrêtes ministériels, projets d'élevage, gestion des populations naturelles, enquêtes de répartition, production, importations et consommation: une équation difficile à résoudre. Les propositions de la Société Batrachologique de France. Alytes 2:69–111.

DUBOIS, A. 1987a "1986." Miscellanea taxinomica batrachologica (I). Alytes 5:7–95.

DUBOIS, A. 1987b "1986." Miscellanea nomenclatorica batrachologica (XII). Alytes 5:97–98.

DUBOIS, A. 1987c. Discoglossidae Günther, 1858 (Amphibia, Anura): proposed conservation. Alytes 6:56–68.

DUBOIS, A. 1992. Notes sur la classification des Ranidae (Amphibiens Anoures). Bull. Soc. Linn. Lyon 61:305–352.

DUBOIS, A., AND R. GÜNTHER. 1982. Klepton and synklepton: two new evolutionary systematics categories in zoology. Zool. Jahrb. Syst. 109:290–305.

DUBOIS, A., AND M. S. KHAN. 1979. A new species of frog (genus *Rana*, subgenus *Paa*) from northern Pakistan (Amphibia, Anura). J. Herpetol. 13:403–410.

DUBOIS, A., AND J. MARTENS. 1977. Sur les crapauds du groupe *Bufo viridis* (Amphibiens, Anoures) de l'Himalaya occidental (Cachemire et Ladakh). Bull. Soc. Zool. France 102:459–465.

DUBOIS, A., AND A. OHLER. 1995a "1994." Frogs of the subgenus *Pelophylax* (Amphibia, Anura, genus *Rana*): a catalogue of available and valid scientific names, with comments on name-bearing types, complete synonymies, proposed common names, and maps showing all type localities. Zool. Poloniae 39:139–204.

DUBOIS, A., AND A. OHLER. 1995b "1994." Catalogue of names of frogs of the subgenus *Pelophylax* (Amphibia, Anura, genus *Rana*): a few additions and corrections. Zool. Poloniae 39:205–208.

DUELLMAN, W. E. 1979. The herpetofauna of the Andes: patterns of distribution, origin, differentiation, and present communities. Pp. 371–459 in W. E. Duellman (ed.), The South American herpetofauna: its origin, evolution, and dispersal. Monogr. Mus. Nat. Hist. Univ. Kansas 7:1–485.

DUELLMAN, W. E. 1990. Herpetofaunas in neotropical rainforests: comparative composition, history and resource utilization. Pp. 455–505 in A. H. Gentry (ed.), *Four Neotropical Rainforests*. New Haven: Yale Univ. Press.

DUELLMAN, W. E. 1993. Amphibian Species of the World: Additions and Corrections. Spec. Publ. Univ. Kansas Mus. Nat. Hist. 21:1–372.

DUELLMAN, W. E., AND L. TRUEB. 1986. Biology of Amphibians. New York: McGraw-Hill Book Co.

DUTTA, S.K. 1992. Amphibians of India: updated species list with distribution record. Hamadryad 17:1–13.

EISELT, J. 1958. Der Feuersalamander *Salamandra salamandra* (L.), Beitrage zu einer taxonomischen Synthese. Abh. Ber. Naturk. Vorgeschichte, Mus. Kulturgeschichte Magdeburg 10:75–154.

EISELT, J. 1988. Krötenfrösche (*Pelobates* gen., Amphibia salientia) in Türkisch-Thrakien und Griechenland. Ann. Naturhist. Mus. Wien 90B:51–59.

EISELT, J., AND J. F. SCHMIDTLER. 1973. Froschlurche aus dem Iran unter Berücksichtigung außeniranischer Populationsgruppen. Ann. Naturhist. Mus. Wien 77:181–243.

EISELT, J., AND H. M. STEINER. 1970. Erstfund eines hynobiiden Molche in Iran. Ann. Naturhist. Mus. Wien 74:77–90.

EMELJANOV, A. F. 1974. Proposals on the classification and nomenclature of areals. Rev. Entomol. URSS 53:497–522 (in Russian, with English summary).

ENGEL, E., AND R. THORN. 1996. L'herpétofaune du Grand-Duché de Luxembourg: bilan actuel. Bull. Soc. Herpétol. France 78:61–64.

ENGELMANN, W.-E., J. FRITZSCHE, R. GÜNTHER, AND F. J. OBST. 1993. *Lurche und Kriechtiere Europas*. 2 Aufl. Radebeul: Neumann.

ESTES, R. 1981. *Gymnophiona,Caudata. Handbuch der Palaoherpetologie, 2*. Stuttgart: G. Fischer.

ESTES, R. D., AND H. H. SCHLEICH. 1994. New material of *Mioproteus caucasicus* Estes & Darevsky from South German localities (Amphibia: Caudata: Proteidae). Pp. 7–21 in H. H. Schleich (ed.), *Amphibien und Reptilien aus dem Känozoikum Eurasiens*. Frankfurt am Main: Courier Forschungs-Institut Senckenberg, 173.

EVANS, S. E., AND A. R. MILNER. 1993. Frogs and salamanders from the Upper Jurassic Morrison Formation (Quarry Nine, Como Bluff) of North America. J. Vert. Paleontol. 13:24–30.

EVANS, S. E., A. R. MILNER, AND F. MUSSETT. 1990. A discoglossid frog from the Middle Jurassic of England. Paleontology 33:299–311.

FEI, L., C.-Y. YE, AND Y.-Z. HUANG. 1991 "1990." *Key to Chinese*

Amphibia. Chongqing: Science and Technology Literature Press (in Chinese).

FLAX, N. L., AND L. J. BORKIN. 1997. High incidence of abnormalities in anurans in contaminated industrial areas (eastern Ukraine). Pp. 119–123 *in* W. Böhme, W. Bischoff, and T. Ziegler (eds.), *Herpetologia Bonensis.* Bonn, Germany: Societas Europae Herpetologicas.

FLOWER, S. S. 1933. Notes on the recent reptiles and amphibians of Egypt, with a list of the species recorded from that kingdom. Proc. Zool. Soc. London, 1933:735–851.

FORD, L. S., AND D. C. CANNATELLA. 1993. The major clades of frogs. Herpetol. Monogr. 7:94–117.

FRITZ, B., M. Vences, and F. Glaw. 1994. Comparative DNA content in *Discoglossus* (Amphibia, Anura, Discoglossidae). Zool. Anzeiger 233:135–145.

FROST, D. R. (ed.). 1985. *Amphibian Species of the World. A taxonomic and geographical reference.* Lawrence, Kansas: Allen Press and Assoc. Systematics Collections.

GASC, J. P., A. CABELA, J. CRNOBRNJA-ISAILOVIC, D. DOLMEN, K. GROSSENBACHER, P. HAFFNER, J. LESCURE, H. MARTENS, J. P. MARTINEZ-RICA, H. MAURIN, R. OLIVEIRI, T. SOFIANIDOU, M. VEITH, AND A. ZUIDERWIJK (eds.). 1997. *Atlas of Amphibians and Reptiles in Europe.* Paris: Muséum National d'Histoire Naturelle.

GOOD, R. 1964. *The Geography of Flowering Plants.* 3rd ed. London: Longmans

GOUX, L. 1957. Contribution a l'etudé écologique, biologique et biogéographique de *Chioglossa lusitanica* Barb.(Urodela Salamandridae). Bull. Soc. Zool. France 82:361–377.

GRAF, J.-D., AND M. POLLS PELAZ. 1989. Evolutionary genetics of the *Rana esculenta* complex. Pp. 289–302 *in* R. M. Dawley and J. P. Bogart (eds.), *Evolution and Ecology of Unisexual Vertebrates.* Albany, New York: New York State Mus. Bull. 466.

GREEN, D. M., AND L. J. BORKIN. 1993. Evolutionary relationships of Eastern Palearctic brown frogs, genus *Rana:* paraphyly of the 24-chromosome species group and the significance of chromosome number change. Zool. J. Linnean Soc. 109:1–25.

GREVEN, H., AND B. THIESMEIER (eds.). 1994. Biology of *Salamandra* and *Mertensiella.* Mertensiella 4:1–544.

GRILLITSCH, H., AND B. GRILLITSCH. 1991. Zur Taxonomie und Verbreitung des Feuersalamander, *Salamandra salamandra* (Linnaeus,1758) (Caudata: Salamandridae) in Griechenland. Herpetozoa 4:133–147.

GRILLITSCH, H., AND F. TIEDEMANN. 1994. Die Grottenolm-Typen Leopold Fitzingers (Caudata: Proteidae: *Proteus*). Herpetozoa 7:139–148.

GROSSENBACHER, K. 1994. Zur Systematik und Verbreitung der Alpensalamander (*Salamandra atra atra, Salamandra atra aurorae, Salamandra lanzai*). Abhand. Ber. Naturk. Magdeburg 17:75–81.

GRUBER, U. F. 1971. Reptilien. Amphibien. Pp. 533–541 *in* H. Schiffers (ed.), *Die Sahara und ihre Randgebiete. Darstellung eines Naturgroßraumes. I. Band. Physiogeographie.* München: Weltforum Verlag.

GUBIN, YU. M. 1991. Paleocene salamanders of southern Mongolia. Palaeontol. Zhurnal, Moscow 1:96–106 (in Russian).

GUBIN, YU. M. 1993. Cretaceous anurans of Mongolia. Palaeontol. Zhurnal, Moscow 1:51–56 (in Russian).

GUBIN, YU. M. 1996. The first find of pelobatids (Anura) in the Paleogene of Mongolia. Palaeontol. Zhurnal, Moscow 4:73–76 (in Russian).

GUIBÉ, J. 1957. Reptiles d'Iran récoltés par M. Francis Petter. Description d'un vipéridé nouveau: *Pseudocerastes latirostris* n. sp. Bull. Mus. Natl. Hist. Nat. Paris, Sér. 2, 29:136–142.

GÜNTHER, R. 1990. *Die Wasserfrösche Europas (Anura-Froschlurche).* Wittenberg Lutherstadt: A. Ziemsen Verlag (Die Neue Brehm-Bücherei,600).

HAXHIU, I. 1994. The herpetofauna of Albania. Amphibia: species composition, distribution, habitats. Zool. Jahrb. Syst. 121:321–334.

HAYASHI, T., AND M. MATSUI. 1989. Preliminary study of phylogeny in the family Salamandridae: allozyme data. Pp. 157–167 *in* M. Matsui, T. Hikida, and R. C. Goris (eds.), *Current Herpetology in East Asia.* Kyoto: Herpetol. Soc. Japan.

HAYEK, L.-A. C. 1994. Analysis of amphibian biodiversity data. Pp. 207–269 *in* W. R. Heyer, M. A. Donnelly, R. W. McDiarmid, L.-A.C. Hayek, and M. S. Foster (eds.), *Measuring and Monitoring Biological Diversity. Standard Methods for Amphibians.* Washington, D.C.: Smithsonian Institution Press.

HEMMER, H., B. KADEL, AND K. KADEL. 1981. The Balearic toad (*Bufo viridis balearicus* (Boettger,1881)), human bronze age culture, and Mediterranean biogeography. Amphibia-Reptilia 2:217–230.

HEMMER, H., A. KONRAD, AND K. BACHMANN. 1980. Hybridization within the *Rana ridibunda* complex of North Africa. Amphibia-Reptilia 1:41–48.

HEMMER, H., J. F. SCHMIDTLER, AND W. BÖHME. 1978. Zur Systematik zentralasiatischer Grünkröten (*Bufo viridis*-Komplex) (Amphibia, Salientia, Bufonidae). Zool. Abhand. Staat. Mus. Tierkunde Dresden 34:349–384.

HENRICI, A. C. 1994. *Tephrodytes brassicarvalis,* new genus and species (Anura: Pelodytidae),from the Arikareean Cabbage Patch beds of Montana, USA, and pelodytid-pelobatid relationships. Ann. Carnegie Mus. 63:155–183.

HEPTNER, V. G. 1936. Obshchaya Zoogeografiya [General Zoogeography]. Moscow: Biomedgiz (in Russian).

HIKIDA, T. ,H. OTA, M. KURAMOTO, AND M.TOYAMA. 1989. Zoogeography of amphibians and reptiles in East Asia. Pp.278–281 *in* M. Matsui, T. Hikida, and R. C. Goris (eds.), *Current Herpetology in East Asia.* Kyoto: Herpetol. Soc. Japan.

HIKIDA, T.,H. OTA, AND M. TOYAMA. 1992. Herpetofauna of an encounter zone of Oriental andPalearctic elements: amphibians and reptiles of the Tokara Group and adjacent islands in the northern Ryukyus, Japan. Biol. Mag. Okinawa 30:29–43.

HODROVÁ, M. 1980. A toad from the Middle Miocene at Devínska Nová Ves near Bratislava. Věstnik Ústředn. Ústavu Geol. 55: 311–316.

HODROVÁ, M. 1984. Salamandridae of the Upper Pliocene Ivanoce locality (Czechoslovakia). Acta Univ. Carolinae, Praha, Geol. 4:331–352.

HOLMAN, J. A. 1992. *Hyla meridionalis* from the Late Pleistocene (Last Interglacial Age: Ipswichian) of Britain. British Herpetol. Soc. Bull. 41:12–14.

HONEGGER, R. E. 1981. *Threatened Amphibians and Reptiles in Europe.* Wiesbaden: Akademische Verlagsgesellschaft.

HOTZ, H., T. UZZELL, R. GÜNTHER, H. G. TUNNER, AND S. HEPPICH. 1987. *Rana shqiperica,* a new European water frog species from the Adriatic Balkans (Amphibia, Salientia, Ranidae). Notulae Naturae 468:1–3.

HULSELMANS, J. L. J. 1977. Further notes on African Bufonidae, with description of new species and subspecies (Amphibia, Bufonidae). Rev. Zool. Africains 91:512–524.

HUMLUM, J. 1959. *La Géographie de l'Afghanistan. Etude d'un Pays Aride.* Copenhagen: Gyldendal.

INGER, R. F. 1972. *Bufo* of Eurasia. Pp. 102–118 *in* W. F. Blair (ed.), *Evolution in the genus* Bufo. Austin: Univ. Texas Press.

JOGER, U. 1981. Zur Herpetofaunistik Westafrikas. Bonn. Zool. Beitr., 32:297–340.

JOGER, U. 1987. An interpretation of reptile zoogeography in Arabia, with special reference to Arabian herpetofaunal relations with Africa. Pp. 257–271 *in* F. Krupp, W. Schneider, and R.Kinzelbach (eds.), Proc. Symp. Fauna and Zoogeography of the Middle East. Wiesbaden: Dr. Ludwig Reichert Verlag.

JOGER, U. 1993. On two collections of reptiles and amphibians from the Cape Verde Islands, with descriptions of three new taxa. Courier Forschungsinstitut Senckenberg 159:437–444.

JOGER, U., AND S. STEINFARTZ. 1994. Electrophoretic investigations in the evolutionary history of the West Mediterranean *Salamandra.* Mertensiella 4:241–254.

JOGER, U., AND S. STEINFARTZ. 1995. Protein electrophoretic data on taxonomic problems in East Mediterranean *Salamandra* (Urodela: Salamandridae). Pp. 33–36 *in* G. A. Llorente, A. Montori, X. Santos, and M. A. Carretero (eds.), *Scientia Herpetologica.* Barcelona: Asoc. Herpetol. Española.

KHAN, M. S. 1976. An annotated checklist and key to the amphibians of Pakistan. Biologia, Lahore, 22:201–210.

KHAN, M. S. 1980. Affinities and zoogeography of herpetiles of Pakistan. Biologia, Lahore 26(1–2):113–171.

KHAN, M. S. 1987. A field guide to the identification of herps of Pakistan. Part I: Amphibia. Biol. Soc. Pakistan Monogr. 4:1–28.

KHAN, M. S. 1994. A revised checklist and key to the amphibians of Pakistan. Hamadryad 19:11–14.

KHAN, M. S. 1997. A new toad of the genus *Bufo* from the foot of Siachin Glacier, Baltistan, northeastern Pakistan. Pakistan J. Zool. 29:43–48.

KHAN, M. S., AND R. TASNIM. 1989. A new frog of the genus *Rana,* subgenus *Paa,* from southwestern Azad Kashmir. J. Herpetol. 23:419 423.

KLEMMER, K. 1976. The Amphibia and Reptilia of the Canary Islands. Pp. 433-456 *in* G. Kunkel (ed.), *Biogeography and Ecology in the Canary Islands.* The Hague: Dr. W. Junk.

KLEWEN, R. 1991. *Die Landsalamander Europas. I. Die Gattungen* Salamandra *und* Mertensiella. Wittenberg Lutherstadt: A. Ziemsen Verlag (Die Neue Brehm-Bücherei, 584).

KOHNO, S.-I, M. KURO-O, AND C. IKEBE. 1991. Cytogenetics and evolution of hynobiid salamanders. Pp. 67–88 *in* D. M. Green and S. K. Sessions (eds.), *Amphibian Cytogenetics and Evolution.* San Diego: Academic Press.

KURAMOTO, M. 1980. Mating calls of treefrogs (genus *Hyla*) in the Far East, with description of a new species from Korea. Copeia 1980:100–108.

KURAMOTO, M. 1984. Systematic implications of hybridization experiments with some Eurasian treefrogs (genus *Hyla*). Copeia 1984:611–618.

KUTENKOV, A. P., AND A. E. PANARIN. 1995. Ecology and status of populations of the common frog (*Rana temporaria*) and the moor frog (*Rana arvalis*) in northwestern Russia with notes on their distribution in Fennoscandia. Pp. 64–70 *in* S.L. Kuzmin, C. K. Dodd, and M. M. Pikulik (eds.), *Amphibian Populations in the Commonwealth of Independent States: Current Status and Declines.* Moscow: Pensoft.

KUZMIN, S. L. 1994. Arcal. Pp. 15 53 *in* E. I. Vorobyeva (ed.), *The Siberian Newt* (Salamandrella keyserlingii *Dybowski, 1870*). *Zoogeography, Systematics, Morphology.* Moscow: Nauka (in Russian).

KUZMIN, S. L. 1995a. *The Clawed Salamanders of Asia: Genus* Onychodactylus. *Biology, Distribution, and Conservation.* Magdeburg: Westarp Wissenschaften (Die Neueu Brehm-Bücherei, 622).

KUZMIN, S. L. 1995b. Die Amphibien Russlands und angrenzender Gebiete. Magdeburg: Westarp Wissenschaften (Die Neueu Brehm-Bücherei, 627).

KUZMIN, S. L., N. S. LEBEDKINA, AND L. J. BORKIN. 1995. The taxonomic position of Central Asian urodelans *Hynobius turkestanicus* and *Turanomolge mensbieri.* Zool. Zhurnal 74:92–105 (in Russian, with English summary).

LAMBERT, M. R. K. 1984. Amphibians and reptiles. Pp. 205–227 *in* J. L. Cloudsley-Thompson (ed.), *Sahara Desert.* Oxford: Pergamon Press [Russian version, 1990: Sahara. Moscow: Progress].

LANZA, B., V. CAPUTO, G. NASCETTI, AND L. BULLINI. 1995. Morphologic and genetic studies of the European plethodontid salamanders: taxonomic inferences (genus *Hydromantes*). Mus. Reg. Sci. Natur. Torino, Monogr. 16:1–366.

LANZA, B., J. M. CEI, AND E. G. CRESPO. 1976. Further immunological evidence for the validity of the family Bombinidae (Amphibia Salientia). Monit. Zool. Italiano, N.S. 10:311–314.

LANZA, B., AND C. CORTI. 1996 "1993–1996." Evolution of knowledge on the Italian herpetofauna during the 20th Century. Boll. Mus. Civ. Stor. Nat. Verona 20:373–436.

LANZA, B., G. NASCETTI, M. CAPULA, AND L. BULLINI. 1984. Genetic relationships among West Mediterranean *Discoglossus* with the decription of a new species (Amphibia Salientia Discoglossidae). Monit. Zool. Italiano, N. S. 18:133–152.

LANZA, B., G. NASCETTI, M. CAPULA, AND L. BULLINI. 1986. Les Discoglosses de la région méditerranéenne occidentale (Amphibia; Anura; Discoglossidae). Bull. Soc. Herpetol. France 40:16–27.

LANZA, B., AND S. VANNI. 1981. On the biogeography of plethodontid salamanders (Amphibia Caudata) with a description of a new genus. Monit. Zool. Italiano, N.S. 15:117–121.

LANZA, B., AND S. VANNI. 1987. Hypotheses on the origins of the Mediterranean island batrachofauna. Bull. Soc. Zool. France 112:179–196.

LATHROP, A. 1997. Taxonomic review of the megophryid frogs (Anura: Pelobatoidea). Asiatic Herpetol. Res. 7:68–79.

LAURENT, R. F. 1975. La distribution des amphibiens et les translations continentales. Mém. Mus. Natl. Hist. Nat., Paris A88:176–191.

LE BERRE, M. 1989. *Faune du Sahara. 1. Poissons-Amphibiens-Reptiles.* Paris: Lechevalier-R.Chabaud.

LEMÉE, G. 1967. *Précis de Biogéographie.* (Russian version: 1976. *Osnovy Biogeografii.* Moscow, Progress).

LEVITON, A. E., AND M. L. ALDRICH. 1984. John Anderson (1833-1900): a zoologist in the Victorian period. Pp. v–xxxv *in* J. Anderson, *Herpetology of Arabia.* Oxford, Ohio: Soc. Study Amphibians and Reptiles (Facsimile reprint).

LEVITON, A. E., AND S. C. ANDERSON. 1961. Further remarks on the amphibians and reptiles of Afghanistan. Wasmann J. Biol. 19:269–276.

LEVITON, A. E., AND S. C. ANDERSON. 1963. Third contribution to the herpetology of Afghanistan. Proc. California Acad. Sci., Ser. 4, 31:329–339.

LEVITON, A. E., AND S. C. ANDERSON. 1970. The amphibians and reptiles of Afghanistan, a checklist and key to the herpetofauna. Proc. California Acad. Sci., Ser. 4, 38:163–206.

LEVITON, A. E., S. C. ANDERSON, K. ADLER, AND S. A. MINTON. 1992. *Handbook to Middle East Amphibians and Reptiles.* Oxford, Ohio: Society for the Study of Amphibians and Reptiles.

LIEM, S. S. 1970. The morphology, systematics, and evolution of the Old World treefrogs (Rhacophoridae and Hyperoliidae). Fieldiana: Zool. 57:i–viii + 1–145.

LITVINCHUK, S. N., T. M. SOKOLOVA, AND L. J. BORKIN. 1994. Biochemical differentiation of the crested newt (*Triturus cristatus* group) in the territory of the former USSR. Abhand. Ber. Naturk., Magdeburg 17:67–74.

LIU, C.-C., AND S. Q.HU. 1961. *Tailless Amphibians of China.* Peking: Science Press (in Chinese).

LIU, W.-Z, AND D.-T. YANG. 1993. A karyosystematic study of the genus *Bombina* from China (Amphibia: Discoglossidae). Asiatic Herpetol. Res. 5:137–142.

MACGREGOR, H. C., S. K. SESSIONS, AND J. W. ARNTZEN. 1990. An integrative analysis of phylogenetic relationships among newts of the genus *Triturus* (family Salamandridae), using comparative biochemistry, cytogenetics and reproductive interactions. J. Evol. Biol. 3:329–373.

MALKMUS, R. 1995. *Die Amphibien und Reptilien Portugals, Madeiras und der Azoren. Verbreitung, Ökologie, Schutz.* Magdeburg: Westarp Wissenschaften (Die Neue Brehm-Bücherei) 621:1–192.

MANI, M. S. (ed.). 1974. *Ecology and Biogeography in India.* The Hague: Dr. W. Junk.

MARTÍNEZ RICA, J. P. 1983. Atlas herpetológico del Pirineo. Muníbe (Soc. Cienc. Aranzadi), San Sebastián, 35:51–80.

MARX, H. 1968. Checklist of the reptiles and amphibians of Egypt. Cairo: Special Publ. U.S. Naval Medical Res. Unit Number Three. i–iv + 1-91.

MATSUI, M. 1980. The status and relationships of the Korean toad, *Bufo stejnegeri* Schmidt. Herpetologica 36:37–41.

MATSUI, M. 1984. Morphometric variation analyses and revision of the Japanese toads (genus *Bufo*, Bufonidae). Contrib. Biol. Lab. Kyoto Univ. 26:209–428.

MATSUI, M. 1986. Geographic variation in toads of the *Bufo bufo* complex from the Far East, with a description of a new subspecies. Copeia 1986:561–579.

MATSUI, M. 1987. Isozyme variation in salamanders of the *nebulosus-lichenatus* complex of the genus *Hynobius* from eastern Honshu, Japan, with a description of a new species. Japan. J. Herpetol. 12:50–64.

MATSUI, M. 1991. Original description of the brown frog from Hokkaido, Japan (genus *Rana*). Japan. J. Herpetol. 14:63–78.

MATSUI, M., T. SATO, S. TANABE, AND T. HAYASHI. 1992. Electrophoretic analyses of systematic relationships and status of two hynobiid salamanders from Hokkaido (Amphibia: Caudata). Herpetologica 48:408–416.

MATSUI, M., G.-F. WU, AND M.-T. SONG. 1993. Morphometric comparisons of *Rana chensinensis* from Shaanxi with three Japanese brown frogs (genus *Rana*). Japan. J. Herpetol. 15:29–36.

MAXSON, L. R. 1981. Albumin evolution and its phylogenetic implications in African toads of the genus *Bufo*. Herpetologica 37:96–104.

MAXSON, L. R., AND J. M. SZYMURA. 1984. Relationships among discoglossid frogs: an albumin perspective. Amphibia-Reptilia 5:245–252.

MENDELSSOHN, H., AND H. STEINITZ. 1944. Contributions to the ecological zoogeography of the amphibians in Palestine. Rev. Fac. Sci. Univ. Istanbul, Sér. B, 9:289–298.

MENSI, P., A. LATTES, B. MACARIO, S. SALVIDIO, C. GIACOMA, AND E. BALLETTO. 1992. Taxonomy and evolution of European brown frogs. Zool. J. Linnean Soc. 104:293–311.

MERTENS, R. 1957. Weitere Unterlagen zur Herpetofauna von Iran 1956. Jh. Ver. vaterl. Naturk. Württemberg, Stuttgart, 112 (1): 118-128.

MERTENS, R. 1969. Die Amphibien und Reptilien West-Pakistans. Stuttgart. Beitr. Naturk. 197:1–96.

MERTENS, R. 1971. Zur Kenntnis von *Bufo luristanicus* (Salientia, Bufonidae). Salamandra 7:83–84.

MERTENS, R., AND H. WERMUTH. 1960. *Die Amphibien und Reptilien Europas (Dritte Liste, nach dem Stand vom 1. Januar 1960).* Frankfurt am Main: Waldemar Kramer.

MEZHZHERIN, S. V. 1992. A new species of green frogs *Rana terentievi* sp. nova (Amphibia, Ranidae) from south Tadjikistan. Dopov. Akad. Nauk Ukrain. 5:154–157.

MILNER, A. R. 1983. The biogeography of salamanders in the Mesozoic and early Caenozoic: a cladistic-vicariance model. Pp. 368–399 *in* R. W. Sims, J. H. Price, and P. E. S. Whalley (eds.), *Evolution, Time and Space: the Emergence of the Biosphere.* London: Academic Press [Russian version, 1988: *Biosfera: Evolutsiya, Prostranstvo, Vremya.* Biogeograficheschkiye ocherki. Moscow: Progress].

MINTON, S. A. 1966. A contribution to the herpetology of West Pakistan. Bull. Am. Mus. Nat. Hist. 134:27–184.

MIZGIREUV, I. V., N. L. FLAX, L. J. BORKIN, AND V. V. KHUDOLEY. 1984. Dysplastic lesions and abnormalities in amphibians associated with environmental conditions. Neoplasma 31:175–181.

MORAVEC, J. (ed.).1994. *Atlas Rozšíření Obojživelníků v České republice.* Praha: Narodní Museum.

MOSAUER, W. 1934. The reptiles and amphibians of Tunisia. Publ. Univ. California Biol. Sci. 1:49–63.

MÜLLER, C. C. 1985. Der Feuersalamander (*Salamandra salamandra semenovi* Nesterov, 1916) im Iran gefunden (Caudata, Salamandridae). Zoologischer Garten, Jena, N.F., 55:348–349.

MÜLLER, P. 1986. *Biogeography.* New York: Harper & Row.

NASCETTI, G., F. ANDREONE, M. CAPULA, AND L. BULLINI. 1988. A new *Salamandra* species from southwestern Alps (Amphibia, Urodela, Salamandridae). Boll. Mus. Reg. Sci. Nat. Torino 6:617–638.

NASCETTI, G., R. CIMMARUTA, B. LANZA, AND L. BULLINI. 1996. Molecular taxonomy of European plethodontid salamanders (genus *Hydromantes*). J. Herpetol. 30:161–183.

NASCETTI, G., B. LANZA, AND L. BULLINI. 1995. Genetic data support the specific status of the Italian treefrog (Amphibia: Anura: Hylidae). Amphibia-Reptilia 16:215–227.

NESSOV, L. A. 1981. Cretaceous salamanders and frogs of Kizylkum Desert. Pp. 57–88 *in* N. B. Ananjeva and L. J. Borkin (eds.), *The Fauna and Ecology of Amphibians and Reptiles of the Palaearctic Asia.* Proc. [Trudy] Zool. Inst., USSR Acad. Sci., Leningrad 101 (in Russian, with English summary).

NISHIO, K, M. MATSUI, AND M. TASUMI. 1987. The lacrimal bone in salamanders of the genera *Hynobius* and *Pachypalami-*

nus: a reexamination of the taxonomic significance. Monit. Zool. Italiano, N. S. 21:307–315.

NISHIOKA, M., M. SUMIDA, AND L. J. BORKIN. 1990. Biochemical differentiation of the genus *Hyla* distributed in the Far East. Sci. Rep. Lab. Amphibian Biol. Hiroshima Univ. 10:93–124.

NISHIOKA, M., M. SUMIDA, L. J. BORKIN, AND Z. WU. 1992. Genetic differentiation of 30 populations of 12 brown frog species distributed in the Palearctic region elucidated by the electrophoretic method. Sci. Rep. Lab. Amphibian Biol. Hiroshima Univ. 11:109–160.

NÖLLERT, A. 1990. Die Knoblauchkröte. 2. Aufl. Wittenberg Lutherstadt: A. Ziemsen (Die Neue Brehm Bücherei, 561).

NÖLLERT, A., AND C. NÖLLERT. 1992. *Die Amphibien Europas: Bestimmung, Gefährdung, Schutz.* Stuttgart: Franckh-Kosmos.

OLIVEIRA, M. E., P. DASZKIEWICZ, AND B. GAUVRIT. 1997. Conservation of the European herpetofauna. Pp. 408–412 *in* Gasc, J. P., A. Cabela, J. Crnobrnja-Isailovic, D. Dolmen, K. Grossenbacher, P. Haffner, J. Lescure, H. Martens, J. P. Martinez-Rica, H. Maurin, R. Oliveiri, T. Sofianidou, M. Veith, and A. Zuiderwijk (eds.), *Atlas of Amphibians and Reptiles of Europe.* Paris: Muséum National d'Histoire Naturelle.

OOSTERBROEK, P., AND J. W. ARNTZEN. 1992. Area-cladograms of Circum-Mediterranean taxa in relation to Mediterranean paleogeography. J. Biogeogr. 19:3–20.

ORLOVA, V. F., V.A. BAKHAREV, AND L. J. BORKIN. 1978 "1977." Karyotypes of some brown frogs of Eurasia, and a taxonomic analysis of karyotypes of the group. Pp. 81–103 *in* N. B. Ananjeva, L. J. Borkin, and I. S. Darevsky (eds.), *Herpetological Collected Papers.* Proc. [Trudy] Zool. Inst., USSR Acad. Sci., Leningrad 74 (in Russian).

ORLOVA, V. F., AND B. S. TUNIYEV. 1989. On the taxonomy of the Caucasian common toads belonging to the group *Bufo bufo verrucosissimus* (Pallas) (Amphibia, Anura, Bufonidae). Bull. Moscow Soc. Naturalists, Ser. Biol., 94(3):13–24 (in Russian).

ÖZETI, N. 1967. The morphology of the salamander *Mertensiella luschani* (Steindachner) and the relationships of *Mertensiella* and *Salamandra.* Copeia 1967:287–298.

PALACKÝ, J. 1898. Die Verbreitung der Batrachier auf der Erde. Verhand. K. K. Zool.-Bot. Ges. Wien 48:374–382.

PARENT, G. H. 1981. Quelques observations écologiques sur l'herpétofaune de l'ile de Djerba (Tunisie méridionale). Natur Belges 62:122–149 (not seen).

PASTEUR, G., AND J. BONS. 1959. Les Batraciens du Maroc. Trav. Inst. Sci. Chérif. Rabat, Sér. Zool., 17:i–xvi + 1–241.

PELLEGRIN, J. 1931. Reptiles, Batraciens et Poissons du Sahara central recueillis par le Pr. Seurat. Bull. Mus. Natl. Hist. Nat. Paris, Sér. 2, 3:216–218.

PESENKO, YU. A. 1982. *Principles and Methods of Quantitative Analysis in Faunistic Studies.* Moscow: Nauka (in Russian).

PISANETZ, E. M., S. V. MEZHZHERIN, AND N. N. SZCZERBAK. 1996. Studies on hybridization and external morphology of Asiatic toads (Amphibia: Bufonidae) and description of a new species *Bufo shaartusiensis* sp. nov. Dopov. Nation. Akad. Nauk Ukraine. 6:147–151 (in Russian).

PRASAD, G. V. R., AND J.-C. RAGE. 1991. A discoglossid frog in the latest Cretaceous (Maastrichtian) of India. Further evidence for a terrestrial route between India and Laurasia in the latest Cretaceous. Compte Rendu. Acad. Sci. Paris, Sér. II, 313:273–278.

PRASAD, G. V. R., AND J.-C. RAGE. 1995. Amphibians and squamates from the Maastrichtian of Naskal, India. Cretaceous Res. 16:95–107.

RADOVANOVIČ, M. 1964. Die Verbreitung der Amphibien und Reptilien in Jugoslawien. Senckenberg. Biol. 45:553–561.

RAGE, J.-C. 1984a. La "Grande Coupure" éocène/oligocène et les herpétofaunes (amphibiens et reptiles): problèmes du synchronisme des événements paléobiogéographiques. Bull. Soc. Géol. France 26:1251–1257.

RAGE, J.-C. 1984b. Are the Ranidae (Anura, Amphibia) known prior to the Oligocene? Amphibia-Reptilia 5:281–288.

RAGE, J.-C. 1995. La Tethys et les dispersions transtethysiennes par voie terrestre. Biogeographica 71:109–126.

RAGE, J.-C., AND J.-J. JAEGER. 1995. The sinking Indian raft: a response to Thewissen and McKenna. Syst. Biol. 44:260–264.

RAMÍREZ, J. M., J. M. VARGAS, AND J. C. GUERRERO. 1992. Distribution patterns and species diversity in European reptiles. Pp. 371–376 *in* Z. Korsós and I. Kiss (eds.), *Proceedings of the Sixth Ordinary General Meeting Societas Europaea Herpetologica, Budapest, 1991.* Budapest: Hungarian Natural History Museum.

REAL, R., A. ANTÚNEZ, AND J. M. VARGAS. 1992. A biogeographic synthesis of European amphibians. Pp. 377–381 *in* Z. Korsós and I. Kiss (eds.), *Proceedings of the Sixth Ordinary General Meeting Societas Europaea Herpetologica, Budapest, 1991.* Budapest: Hungarian Natural History Museum.

REHÁK, I., AND D. J. OSBORN. 1988. Notes on the distribution of reptiles and amphibians in Egypt. Věstník Československ. Spolecn. Zool. 52:271–277.

REILLY, S. M. 1983. The biology of the high altitude salamander *Batrachuperus mustersi* from Afghanistan. J. Herpetol. 17:1–9.

RIPLEY, S. D. 1954. Comments on the biogeography of Arabia with particular reference to birds. J. Bombay Nat. Hist. Soc. 52:241–248.

ROČEK, Z. 1994. A review of the fossil Caudata of Europe. Abhand. Ber. Naturk. Magdeburg 17:51–56.

ROČEK, Z., AND L. A. NESSOV. 1993. Cretaceous anurans from Central Asia. Palaeontographica A, 226:1–54.

ROUSSEL, H., AND Y. AMAR. 1985. Notes de batrachologie saharienne. I. Les Amphibiens de l'Oued Saoura. Alytes 4:41–51.

SAGE, R. D., E. M. PRAGER, AND D.B. WAKE. 1982. A Cretaceous divergence time between pelobatid frogs (*Pelobates* and *Scaphiopus*): immunological studies of serum albumin. J. Zool. 198:481–494.

SAHI, D. N., AND P. L. DUDA. 1986. Affinities and distribution of amphibians and reptiles of Jammu and Kashmir State (India). Bull. Chicago Herpetol. Soc. 21:84–88.

SALVADOR, A. 1996. Amphibians of northwest Africa. Smithsonian Herpetol. Information Service 109:1–43.

SANCHIZ, F. B. 1977. La familia Bufonidae (Amphibia, Anura) en el Terciario europeo. Trab. Neógeno-Cuaternario 8:75–111.

SANCHIZ, F. B. 1979. Notas sobre la batracofauna cuaternaria de Cerdeña. Estud. Geol., Madrid, 35:437–441.

SANCHIZ, F. B. 1981. Registro fósil y antigüedad de la familia Hylidae (Amphibia, Anura) en Europa. An. II Congr. Latino-Amer. Palaeont., Porto Alegre, Brazil: 757–764.

SANCHIZ, F. B., AND M. MLYNARSKI. 1979. Pliocene salamandrids (Amphibia, Caudata) from Poland. Acta Zool. Cracoviensia 24:175–188.

SANCHIZ, B., AND Z. RO EK. 1996. An overview of the anuran fossil record. Pp. 317–328 in R. C. Tinsley, and H. R. Kobel (eds.), *The Biology of* Xenopus. Oxford: Clarendon Press.

SANCHIZ, B., AND H. H. SCHLEICH. 1986. Erstnachweis der Gattung *Bombina* (Amphibia: Anura) in Untermiozän Deutschland. Mitt. Bayer. Staatssammlung Paläont. Hist. Geol., München, 26:41–44.

SANCHIZ, B., H. H. SCHLEICH, AND M. ESTEBAN. 1993. Water frogs (Ranidae) from the Oligocene of Germany. J. Herpetology 27:486–489.

SAVAGE, J. M. 1973. The geographic distribution of frogs: patterns and predictions. Pp. 351–445 in J. L. Vial (ed.), *Evolutionary Biology of the Anurans. Contemporary Research on Major Problems.* Columbia: Univ. Missouri Press.

SBORDONI, V., A. CACCONE, G. ALLEGRUCCI, AND D. CESARONI. 1990. Molecular island biogeography. Pp. 55–83 in A. Azzaroli (ed.), *Biogeographical Aspects of Insularity.* Roma: Accad. Nazion. Lincei.

SCHALL, J. J., AND E. R. PIANKA. 1979 "1977." Species densities of reptiles and amphibians on the Iberian Peninsula. Doñana (Acta Vertebrata) 4:27–34.

SCHÄTTI, B. 1989. Amphibien und Reptilien aus der Arabischen Republik Jemen und Djibouti. Rev. Suisse Zool. 96:905–937.

SCHÄTTI, B., AND J. GASPERETTI. 1994. A contribution to the herpetofauna of southwest Arabia. Fauna of Saudi Arabia 14:348–423.

SCHIFFERS H. (ed.). 1971. *Die Sahara und ihre Randgebiete. Darstellung eines Naturgroßraumes. I. Band. Physiogeographie.* München: Weltforum Verlag.

SCHLEICH, H. H. 1987. Herpetofauna Caboverdiana. Spixiana, Suppl. 12:1–75.

SCHLEICH, H. H., W. KÄSTLE, AND K. KABISCH. 1996. *Amphibians and Reptiles of North Africa. Biology, Systematics, Field Guide.* Königstein: Koeltz Scientific Books.

SCHMIDT, K. P., AND H. MARX. 1956. The herpetology of Sinai. Fieldiana:Zool. 39:21–40.

SCHMIDTLER, J. F. 1986. Orientalische Smaragdeidechsen: 2. Über Systematik und Synökologie von *Lacerta trilineata, L. media* und *L. pamphylica* (Sauria: Lacertidae). Salamandra 22:126–146.

SCHMIDTLER, J. F. 1994. Eine Übersicht neuerer Untersuchungen und Beobachtungen an der vorderasiatischen Molchgattung *Neurergus* Cope, 1862. Abhand. Ber. Naturk. Magdeburg 17:193–198.

SCHMIDTLER, J. F., J. EISELT, AND H. SIGG. 1990. Die subalpine Herpetofauna des Bolkar-Gebirges (mittlerer Taurus, Südtürkey). Herpetofauna 12 (64):11–20.

SCHMIDTLER, J. J., AND SCHMIDTLER, J. F. 1967. Über die Verbreitung der Molchgattung *Triturus* in Kleinasien. Salamandra 3:15–36.

SCHMIDTLER, J. J., AND J. F. SCHMIDTLER. 1969. Über *Bufo surdus;* mit einem Schlüssel und Anmerkungen zu den übrigen Kröten Irans und West-Pakistans. Salamandra 5:113–123.

SCHMIDTLER, J. J., AND SCHMIDTLER, J. F. 1970. Morphologie, Biologie und Verwandtschaftsbeziehungun von *Neurergus strauchii* aus der Türkei. Senckenberg. Biol. 51:41–53.

SCHNEIDER, B. 1974. Beitrag zur Herpetofauna Tunesiens, I. *Bufo bufo spinosus.* Salamandra 10:55–60.

SCHNEIDER, B. 1978. Beitrag zur Herpetofauna Tunesiens, II. *Bufo mauritanicus.* Salamandra 14:33–40.

SCHNEIDER, H. 1974. Structure of the mating calls and relationships of the European tree frogs (Hylidae, Anura). Oecologia 14:99–110.

SCHNEIDER, H., AND I. HAXHIU. 1994. Mating-call analysis and taxonomy of the water frogs in Albania (Anura:Ranidae). Zool. Jahrb. Syst. 121:248–262.

SCHNEIDER, H., U. SINSCH, AND E. NEVO. 1992. The lake frogs in Israel represent a new species. Zool. Anzeiger 228:97–106.

SCHNEIDER, H., U. SINSCH, AND T. S. SOFIANIDOU. 1993. The water frogs of Greece: bioacoustic evidence for a new species. Zeit. Zool. Syst. Evol.-forsch. 31:47–63.

SCHNEIDER, H., T. S. SOFIANIDOU, AND P. KYRIAKOPOULOU-SKLAVOUNOU. 1984. Bioacoustic and morphometric studies in water frogs (genus *Rana*) of Lake Ioannina in Greece, and description of a new species (Anura, Amphibia). Zeit. Zool. Syst. Evol.-forsch. 22:349–366.

SCHREIBER, E. 1912. *Herpetologica Europaea. Eine systematische Bearbeitung der Amphibien und Reptilien welche bisher in Europe aufgefunden sind.* Jena: G. Fischer.

SCLATER, P. L. 1858. On the general geographic distribution of the members of the Class Aves. J. Linnean Soc., Zool. 2:130–145.

SCORTECCI, G. 1936. Gli anfibi della Tripolitania. Atti Soc. Italiano Sci. Nat. Mus. Civico Stor. Nat. Milano 75:129–226.

SERRA-COBO, J. 1993. Descripción de una nueva especie europea de rana parda (Amphibia, Anura, Ranidae). Alytes 11:1–15.

SHCHERBAK [SZCZERBAK], N. N. 1966. *Amphibians and Reptiles of the Crimea.* Herpetologica Taurica. Kiev: Naukova Dumka. (in Russian).

SHCHERBAK [SZCZERBAK], N. N. 1982. Grundzüge einer herpetogeographischen Gliederung der Paläarktis. Vertebrata Hungarica 21:227–239.

SHERIF, N., U. SINSCH, AND H. SCHNEIDER. 1990. Phylogenetic relationships in the genus *Bufo* (Amphibia, Anura). Verhand. Deutsch. Zool. Ges. 83:516–517 (abstract).

SIBOULET, R. 1972. Affinités sérologiques entre diverses espèces du genre *Bufo.* Compte Rendu Acad. Sci. Paris, Sér. D, 275:221–224.

SINSCH, U., AND B. EBLENKAMP. 1994. Allozyme variation among *Rana balcanica, R. levantina,* and *R. ridibunda* (Amphibia: Anura). Genetic differentiation corroborates the bioacoustically detected species status. Zeit. Zool. Syst. Evolut.-forsch. 32:35–43.

SKET, B., AND J. W. ARNTZEN. 1994. A black, non-troglomorphic amphibian from the karst of Slovenia: *Proteus anguinus parkelj* n. ssp. (Urodela: Proteidae). Bijdr. Dierk. 64:33–53.

SOFIANIDOU, T., AND H. SCHNEIDER. 1989. Distribution range of the Epeirus frog *Rana epeirotica* (Amphibia: Anura) and the composition of the water frog populations in western Greece. Zool. Anzeoger 223:13–25.

SOFIANIDOU, T., H. SCHNEIDER, AND U. SINSCH. 1994. Comparative electrophoretic investigation on *Rana balcanica* and *Rana ridibunda* from northern Greece. Alytes 12:93–108.

SOTIROPOULOS, K., A. LEGATIS, AND R. -M. POLYMENI. 1995. A review of the knowledge on the distribution of the genus *Triturus* (Rafinesque, 1815) in Greece (Caudata: Salamandridae). Herpetozoa 8:25–34.

SPARKS, B. W., AND A. T. GROVE. 1961. Some Quaternary fossil non-marine mollusca from the central Sahara. J. Linnean Soc. London 44:355–364.

SRIVASTAVA, S. P., H. SCHOUTEN, W. R. ROEST, K. D. KLITGORD, L.

C. Kovacs, J. Verhoef, and R. Macnab. 1990. Iberian plate kinematics: a jumping plate boundary between Eurasia and Africa. Nature 344:756–759.

Stegmann, B. 1938. Grundzüge der ornithogeographischen Gliederung des Paläarktischen Gebietes. USSR Acad. Sci., Faune de l'URSS. Oiseaux, 1 (2):1–157 (in Russian and German).

Steiner, H. M. 1973. Beiträge zur Kenntnis von Verbreitung, Ökologie und Bionomie von *Batrachuperus persicus* (Caudata, Hynobiidae). Salamandra 9:1–6.

Steinwarz, D., and H. Schneider. 1991. Distribution and bioacoustics of *Rana perezi* Seoane, 1885 (Amphibia, Anura, Ranidae) in Tunisia. Bonn. Zool. Beitr. 42: 283–297.

Stugren, B. 1985. Island biogeography from the viewpoint of Aegean herpetology. Pp. 279–283 *in* N. Coman, S. Kiss, T. Perseca, L. S. Petcrfi, and G. Racovita (eds.), *Evolution and Adaptation.* Cluj-Napoca, Romania: Univ. Babes Bolyai.

Szyndlar, Z. 1984. Fossil snakes from Poland. Acta Zool. Cracoviensia 28:3–156.

Takhtajan, A. L. 1978. *The Floristic Regions of the World.* Leningrad: Nauka (in Russian).

Tandy, M., J. P. Bogart, M. J. Largen, and D. J. Feener. 1985. Variation and evolution in *Bufo kerinyagae* Keith, *B. regularis* Reuss and *B. asmarae* Tandy et al. (Anura Bufonidae). Monit. Zool. Italiano, N.S., Suppl., 20:210–267.

Tandy, M., and R. Keith. 1972. *Bufo* of Africa. Pp. 119–170 *in* W. F. Blair (ed.), *Evolution in the genus* Bufo. Austin: Univ. Texas Press.

Tandy, M., and J. Tandy. 1976. Evolution of acoustic behaviour in African *Bufo* (Anura: Bufonidae). Zool. Africana, 11:349–368.

Tandy, M., J. Tandy, R. Keith, and A. Duff-MacKay. 1976. A new species of *Bufo* (Anura: Bufonidae) from Africa's dry savannas. Pearce-Sellards Ser. Univ. Texas 24:1–20.

Tarkhnishvili, D. N., and R. K. Gokhelashvili. 1995. Amphibian assemblages of Trialeti Ridge, Georgia, Pp. 125–135 *in* S. L. Kuzmin, C. K. Dodd, and M. M. Pikulik (eds.), *Amphibian Populations in the Commonwealth of Independent States: Current Status and Declines.* Moscow: Pensoft.

Thomas, H., S. Sen, M. Khan, B. Battail, and G. Ligabue. 1982. The Lower Miocene fauna of Al-Sarrar (Eastern Province, Saudi Arabia). Atlal, J. Saudi Arabian Archaeology 5:109–136.

Thorn, R. 1969 "1968." *Les Salamandres d'Europe, d'Asie et d'Afrique du Nord.* Paris: Paul Lechevalier.

Tian, W., and Q. Hu. 1985. Taxonomical studies on the primitive anurans of the Hengduan Mountains, with description of a new subfamily and subdivision of *Bombina*. Acta Herpetol. Sinica 4:219–224 (in Chinese, with English summary).

Tuck, R.G. 1971. Amphibians and reptiles from Iran in the United States National Museum collection. Bull. Maryland Herpetol. Soc. 7:48–86.

Tuck, R.G. 1979. Notes on the Turan Biosphere Reserve herpetofauna, northeastern Iran. Bull. Maryland Herpetol. Soc. 15:95–123.

Udvardy, M. D. F. 1975. A classification of the biogeographical provinces of the World. Occas. Pap. Internat. Union Cons. Nat. and Nat. Resources 18:1–48.

Uteshev, V., and L. J. Borkin. 1985. On interspecific hybridization of European and Far Eastern discoglossid toads of the genus *Bombina*. Zool. Anzeiger 215:355–367.

Veith, M. 1994. Morphological, molecular and life history variation in *Salamandra salamandra* (L.). Mertensiella 4:355–397.

Veith, M. 1996a. Molecular markers and species delimitation: examples from the European batrachofauna. Amphibia-Reptilia 17:303–314.

Veith, M. 1996b. Are *Salamandra atra* and *S. lanzai* sister species? Amphibia-Reptilia 17:174–177.

Vences, M. 1990. Untersuchungen zur Ökologie, Ethologie und geographischen Variation von *Chioglossa lusitanica* Bocage, 1864. Salamandra 26:267–297.

Vences, M., and F. Glaw. 1996. Further investigations on *Discoglossus* bioacoustics: relationships between *D. galganoi galganoi, D. g. jeanneae* and *D. p. scovazzi*. Amphibia-Reptilia 17:333–340.

Vergnaud-Grazzini, C. 1966. Les Amphibiens du Miocène de Beni-Mellal. Notes Service Géol. Maroc, Rabat, 27:43–74 (Notes Mém Serv. Géol. 198).

Vinogradov, A. E., L. J. Borkin, R. Günther, and J. M. Rosanov. 1990. Genome elimination in diploid and triploid *Rana esculenta* males: cytological evidence from DNA flow cytometry. Genome 33:619–627.

Vogrin, N. 1996. *Rana latastei* in Croatia and Slovenia. Froglog 18:1.

Vogrin, N. 1997. An overview of the herpetofauna of Slovenia. British Herpetol. Soc. Bull. 58:26–35.

Wake, D. B., and A. G. Kluge. 1961. The Machris expedition to Tchad, Africa. Contrib. Sci. Los Angeles Co. Mus. 40:1–12.

Wake, D. B., and N. Özeti. 1969. Evolutionary relationships in the family Salamandridae. Copeia 1969:124–137.

Wake, D. B., L. R. Maxson, and G. Z. Wurst. 1978. Genetic differentiation, albumin evolution, and their biogeographic implications in plethodontid salamanders of California and southern Europe. Evolution 32:529–539.

Wallace, A. R. 1876. *The Geographical Distribution of Animals with a Study of the Relations of Living and Extinct Faunas as Elucidating the Past Changes of the Earth's Surface.* London: Macmillan & Co.

Walter, H. 1968. *Die Vegetation der Erde in Ökophysiologischer Betrachtung. Bd II. Die Gemaßigten und Arktischen Zonen.* Jena: G. Fischer [1974, Moscow: Progress, (Russian version)].

Werner, Y. L. 1982. Herpetofaunal survey of the Sinai Peninsula (1967-1977), with emphasis on the Saharan sand community. Pp. 153–161 *in* N.J. Scott, Jr. (ed.), Herpetological communities. U.S. Fish and Wildlife Serv. Wildlife Res. Rept. 13:1–239.

Werner, Y. L. 1987. *Bufo regularis* (Amphibia: Anura) retracted from the herpetofaunas of the Negev (Israel) and Petra (Jordan). Israel J. Zool. 34:239–243.

Werner, Y. L. 1988. Herpetofaunal survey of Israel (1950-85), with comments on Sinai and Jordan and on zoogeographical heterogeneity. Pp. 355–388 *in* Y. Yom-Tov and E. Tchernov (eds.), *The Zoogeography of Israel. The Distribution and Abundance at a Zoogeographical Crossroad.* Dordrecht: Dr. W.Junk.

Wilkinson, J. A., M. Matsui, and T. Terachi. 1996. Geographic variation in a Japanese tree frog (*Rhacophorus arboreus*) revealed by PCR-aided restriction site analysis of mtDNA. J. Herpetol. 30:418–423.

Ye, C.-Y., L. Fei, and L.-G. Xiang. 1981. *Rana altaica* Kastschenko—a new record of Chinese frog from Xinjiang, China.

Acta Herpetol. Sinica 5:121–122 (in Chinese with English summary).

YILMAZ, I. 1986. On the distribution of the fire-bellied toad, *Bombina bombina,* in Turkey. Pp. 109–110 *in* R. Kinkelbach and M. Kasparek (eds.), *Zoology in the Middle East.* Vol. 1. Heidelberg: M. Kasparek Verlag.

ZAVADIL, V., J. PIÁLEK, AND L. KLEPSCH. 1994. Extension of the known range of *Triturus dobrogicus:* electrophoretic and morphological evidence for presence in the Czech Republic. Amphibia-Reptilia 15:329–335.

ZHAO E., AND K. ADLER. 1993. *Herpetology of China.* Oxford, Ohio: Soc. Study Amphibians and Reptiles.

ZHAO, E., Q. HU, Y. JIANG, AND Y. YANG. 1988. *Studies on Chinese Salamanders.* Oxford, Ohio: Soc. Study Amphibians and Reptiles.

ZUIDERWIJK, A. 1980. Amphibian distribution patterns in western Europe. Bijdr. Dierk. 50:52–72.

APPENDIX 6:1

Alphabetical list of the species of amphibians of the Palearctic Realm with a notation of Palearctic (P), Oriental (O), Afrotropical (A), or Nearctic (N) faunal affinities, as well as Palearctic endemics. Distribution of species in zoogeographic regions (Fig. 6:8) and floristic regions (Good, 1964) shown by presence (+), absence (–), or endemism (E). Abbreviations of zoogeographic regions: CAS = Central Asian, EAS = East Asian, EUR = European (Boreal), MED = Mediterranean, SIB = Siberian, WAS = West Asian. Abbreviations of floristic regions: ASA = Arctic and Sub-Arctic, ESA = Euro-Siberian (Asiatic), ESE = Euro-Siberian (European), MAC = Macaronesian (European), MED = Mediterranean, NAI = North African-Indian Desert, S-J = Sino-Japanese, WCA = Western and Central Asiatic. Abbreviations of distribution patterns: AB = Apennine-Balkan, AD = Alpine-Dinaric, Adr = Adriatic, Ape = Apenninean, Arm = Armenian, Bal = Balkan, BI = Balearic Islands, Car = Carpathian, CAs = Central Asian, Cau = Caucasian, CE = Central European, Dan = Danubian, EAs = East Asian, EE = East European, Elb = Elburz, EM = East Mediterranean, Hin = Hindu Kush, HO = Hengduanshan-Oriental, HOK = Hokkaido, HP = Hengduanshan-Palearctic, HT = Himalayan-Tibet, Ibe = Iberian, Ira = Iranian, Ita = Italian, Jap = Japanese, Kas = Kashmir, KN = Korean Peninsula and northeast China, Man = Manchurian, Mon = Mongolian, NC = North China, NE = Near East, Nil = Nile, NWA = Northwestern Africa, Pan = Pan-Oriental, Po = Po, Pyr = Pyrenean, Sib = Siberian, SP = Southern periphery (Iran), Tad = Tadjik, Tyr = Tyrrhenian, WAs = West Asian, WO = West Oriental, WP = Western Palearctic, ZAG = Western Zagros. See text for definition of patterns.

Species	Faunal affinities	Distribution pattern	Palearctic endemic	Zoogeographic Region						Floristic Region							
				MED	EUR	WAS	SIB	CAS	EAS	MAC	NAI	MED	ESE	ESA	ASA	WCA	S-J
CAUDATA: CRYPTOBRANCHIDAE:																	
Andrias davidianus 1871	P	EAs	–	–	–	–	–	+	+	–	–	–	–	–	–	+	+
Andrias japonicus 1836	P	Jap	+	–	–	–	–	–	E	–	–	–	–	–	–	–	E
CAUDATA: HYNOBIIDAE:																	
Batrachuperus gorganensis 1979	P	Elb	+	–	–	E	–	–	–	–	–	–	–	–	–	E	–
Batrachuperus karlschmidti 1950	P	HP	–	–	–	–	–	+	–	–	–	–	–	–	–	E	–
Batrachuperus mustersi 1940	P	Hin	+	–	–	E	–	+	–	–	–	–	–	–	–	E	–
Batrachuperus persicus 1970	P	Elb	+	–	–	E	–	–	–	–	–	–	E?	–	–	–	–
Batrachuperus tibetanus 1925	P	HP	–	–	–	–	–	+	–	–	–	–	–	–	–	E	–
Hynobius (Hynobius) abei 1934	P	Jap	+	–	–	–	–	–	E	–	–	–	–	–	–	–	E
Hynobius (Hynobius) dunni 1931	P	Jap	+	–	–	–	–	–	E	–	–	–	–	–	–	–	E
Hynobius (Hynobius) hidamontanus 1987	P	Jap	+	–	–	–	–	–	E	–	–	–	–	–	–	–	E
Hynobius (Hynobius) leechii 1887	P	Kn	+	–	–	–	–	–	E	–	–	–	–	–	–	–	E
Hynobius (Hynobius) lichenatus 1883	P	Jap	+	–	–	–	–	–	E	–	–	–	–	–	–	–	E
Hynobius (Hynobius) mantschuriensis 1927	P	Man	+	–	–	–	–	–	E	–	–	–	–	–	–	–	E
Hynobius (Hynobius) nebulosus 1838	P	Jap	+	–	–	–	–	–	E	–	–	–	–	–	–	–	E
Hynobius (Hynobius) nigrescens 1907	P	Jap	+	–	–	–	–	–	E	–	–	–	–	–	–	–	E
Hynobius (Hynobius) takedai 1984	P	Jap	+	–	–	–	–	–	E	–	–	–	–	–	–	–	E
Hynobius (Hynobius) tenuis 1991	P	Jap	+	–	–	–	–	–	E	–	–	–	–	–	–	–	E
Hynobius (Hynobius) tokyoensis 1931	P	Jap	+	–	–	–	–	–	E	–	–	–	–	–	–	–	E
Hynobius (Hynobius) tsuensis 1922	P	Jap	+	–	–	–	–	–	E	–	–	–	–	–	–	–	E
Hynobius (Pachypalaminus) boulengeri 1912	P	Jap	+	–	–	–	–	–	E	–	–	–	–	–	–	–	E

Appendix 6:1 Continued

Species	Faunal affinities	Distribution pattern	Palearctic endemic	Zoogeographic Region						Floristic Region							
				MED	EUR	WAS	SIB	CAS	EAS	MAC	NAI	MED	ESE	ESA	ASA	WCA	S-J
Hynobius (Pseudosalamandra) kimurae 1923	P	Jap	+	–	–	–	–	–	E	–	–	–	–	–	–	–	E
Hynobius (Pseudosalamandra) naevius 1838	P	Jap	+	–	–	–	–	–	E	–	–	–	–	–	–	–	E
Hynobius (Pseudosalamandra) okiensis 1940	P	Jap	+	–	–	–	–	–	E	–	–	–	–	–	–	–	E
Hynobius (Pseudosalamandra) stejnegeri 1923	P	Jap	+	–	–	–	–	–	E	–	–	–	–	–	–	–	E
Hynobius (Satobius) retardatus 1923	P	Hok	+	–	–	–	–	–	E	–	–	–	–	–	–	–	E
Onychodactylus fischeri 1886	P	Man	+	–	–	–	–	–	E	–	–	–	–	–	–	–	E
Onychodactylus japonicus 1782	P	Jap	+	–	–	–	–	–	E	–	–	–	–	–	–	–	E
Ranodon sibiricus 1866	P	CAs	+	–	–	–	–	E	–	–	–	–	–	–	–	E	–
Salamandrella keyserlingii 1870	P	Sib	+	–	+	–	+	–	+	–	–	+	+	+	+	+	+
Caudata: Plethodontidae:																	
Hydromantes ambrosii 1955	N	Ita	+	E	–	–	–	–	–	–	–	E	–	–	–	–	–
Hydromantes flavus 1969	N	Tyr	+	E	–	–	–	–	–	–	–	E	–	–	–	–	–
Hydromantes genei 1838	N	Tyr	+	E	–	–	–	–	–	–	–	E	–	–	–	–	–
Hydromantes imperialis 1969	N	Tyr	+	E	–	–	–	–	–	–	–	E	–	–	–	–	–
Hydromantes italicus 1923	N	Ita	+	E	–	–	–	–	–	–	–	E	–	–	–	–	–
Hydromantes strinatii 1958	N	Ita	+	E	–	–	–	–	–	–	–	+	+	–	–	–	–
Hydromantes supramontis 1986	N	Tyr	+	E	–	–	–	–	–	–	–	E	–	–	–	–	–
Caudata: Proteidae:																	
Proteus anguinus 1768	P	Adr	+	+	+	–	–	–	–	–	–	–	E	–	–	–	–
Caudata: Salamandridae:																	
Chioglossa lusitanica 1864	P	Ibe	+	E	–	–	–	–	–	–	–	+	+	–	–	–	–
Cynops pyrrhogaster 1826	O	Jap	+	–	–	–	–	–	E	–	–	–	–	–	–	–	+
Euproctus asper 1852	P	Pyr	+	E	–	–	–	–	–	–	–	+	+	–	–	–	–
Euproctus montanus 1838	P	Tyr	+	E	–	–	–	–	–	–	–	E	–	–	–	–	–
Euproctus platycephalus 1829	P	Tyr	+	E	–	–	–	–	–	–	–	E	–	–	–	–	–
Mertensiella caucasica 1876	P	Cau	+	–	–	E	–	–	–	–	–	–	E	–	–	–	–
Mertensiella luschani 1891	P	NE	+	–	–	E	–	–	–	–	–	E	–	–	–	–	–
Neurergus crocatus 1862	P	WAs	+	–	–	E	–	–	–	–	–	–	–	–	–	–	–
Neurergus kaiseri 1952	P	Zag	+	–	–	E	–	–	–	–	–	–	–	–	–	–	–
Neurergus microspilotus 1917	P	Arm	+	–	–	E	–	–	–	–	–	–	–	–	–	–	–
Neurergus strauchii 1887	P	Arm	+	–	–	E	–	–	–	–	–	E	–	–	–	–	–
Pleurodeles poireti 1835	P	NWA	+	E	–	–	–	–	–	–	–	E	–	–	–	–	–
Pleurodeles waltl 1830	P	Ibe	+	E	–	–	–	–	–	–	–	+	+	–	–	–	–
Salamandra algira 1883	P	NWA	+	E	–	–	–	–	–	–	–	E	–	–	–	–	–
Salamandra atra 1768	P	AD	+	+	+	–	–	–	–	–	–	E	–	–	–	–	–
Salamandra corsica 1838	P	Tyr	+	E	–	–	–	–	–	–	–	E	–	–	–	–	–
Salamandra infraimmaculata 1885	P	WAs	+	–	–	E	–	–	–	–	–	E	–	–	–	+	–
Salamandra lanzai 1988	P	AD	+	–	E	–	–	–	–	–	–	+	–	–	–	–	–
Salamandra salamandra 1758	P	CE	+	+	+	–	–	–	–	–	–	+	+	–	–	+	–

Appendix 6:1 Continued

Species	Faunal affinities	Distribution pattern	Palearctic endemic	Zoogeographic Region						Floristic Region							
				MED	EUR	WAS	SIB	CAS	EAS	MAC	NAI	MED	ESE	ESA	ASA	WCA	S-J
Salamandrina terdigitata 1788	P	Ita	+	E	–	–	–	–	–	–	–	+	+	–	–	–	–
Triturus (Palaeotriton) boscai 1879	P	Ibe	+	E	–	–	–	–	–	–	–	+	+	–	–	–	–
Triturus (Palaeotriton) italicus 1898	P	Ita	+	E	–	–	–	–	–	–	–	E	–	–	–	–	–
Triturus (T. vulgaris group) helveticus 1789	P	Atl	+	+	+	–	–	–	–	–	+	+	+	–	–	–	–
Triturus (T. vulgaris group) montandoni 1880	P	Car	+	–	E	+	–	–	–	–	–	–	E	–	–	–	–
Triturus (T. vulgaris group) vulgaris 1758	P	EE	+	+	+	+	+	–	–	–	+	+	+	+	+	+	–
Triturus (Triturus) alpestris 1768	P	CE	+	+	+	+	–	–	–	–	+	+	+	–	–	–	–
Triturus (Triturus) vittatus 1835	P	WAs	+	–	–	E	–	–	–	–	–	+	+	+	–	+	–
Triturus (T. cristatus group) carnifex 1768	P	AB	+	+	–	–	–	–	–	–	+	+	+	–	–	–	–
Triturus (T. cristatus group) cristatus 1768	P	EE	+	+	E	+	+	–	–	–	+	+	+	+	+	+	–
Triturus (T. cristatus group) dobrogicus 1903	P	Dan	+	–	E	E	–	–	–	–	–	+	E	–	–	–	–
Triturus (T. cristatus group) karelinii 1870	P	EM	+	+	+	+	–	–	–	–	–	+	+	–	–	+	–
Triturus (T. cristatus group) marmoratus 1800	P	Atl	+	+	+	–	–	–	–	–	+	+	+	–	–	–	–
ANURA: BUFONIDAE:																	
Bufo (B. bufo group) bufo 1758	P	WP	+	+	+	+	+	–	–	–	+	+	+	+	+	+	–
Bufo (B. bufo group) gargarizans 1842	P	EAs	–	–	–	–	?	+	E	–	–	–	–	+	+	+	E
Bufo (B. bufo group) japonicus 1838	P	Jap	+	–	–	–	–	–	E	–	–	–	–	–	–	–	E
Bufo (B. bufo group) mishanicus 1926	P	HP	–	–	–	–	–	+	E	–	–	–	–	–	+	+	E
Bufo (B. bufo group) stejnegeri 1931	P	KN	+	–	–	–	–	–	–	–	–	–	–	–	–	–	E
Bufo (B. bufo group) tibetanus 1926	P	HP	–	–	–	–	–	+	E	–	–	–	–	–	+	+	E
Bufo (B. bufo group) torrenticola 1976	P	Jap	+	–	–	–	–	–	E	–	–	–	–	–	–	–	E
Bufo (B. bufo group) tuberculatus 1926	P	HP	–	–	–	–	–	+	E	–	–	–	–	–	+	+	E
Bufo (B. bufo group) verrucosissimus 1814	P	Cau	+	–	–	E	–	–	–	–	–	E	+	–	–	+	–
Bufo (B. funereus group) kassasii 1993	A	Nil	+	–	–	E	–	–	–	–	E	+	–	–	–	+	–
Bufo (B. mauritanicus group) mauritanicus 1841	A	NWA	–	+	–	–	–	–	–	–	+	+	–	–	–	+	–
Bufo (B. regularis group) regularis 1833	A	Nil	–	+	–	+	–	–	–	–	+	+	–	–	–	+	–
Bufo (B. stomaticus group) olivaceus 1874	O	WO	–	–	–	+	–	–	–	–	+	–	–	–	–	+	–
Bufo (B. stomaticus group) stomaticus 1862	O	WO	–	–	–	+	–	+	+	–	+	–	–	+	–	+	–
Bufo (B. viridis group) brongersmai 1972	P	NWA	+	E	–	–	–	–	–	–	+	+	–	–	–	–	–
Bufo (B. viridis group) calamita 1768	P	Atl	+	+	–	–	–	–	–	–	+	+	+	+	–	–	–
Bufo (B. viridis group) danatensis 1978	P	CAs	+	–	–	E	–	+	–	–	–	–	–	?	–	E	–
Bufo (B. viridis group) kavirensis 1979	P	Ira	+	–	–	E	–	–	–	–	–	–	–	–	–	E	–
Bufo (B. viridis group) latastii 1882	P	Kas	+	–	–	E	–	–	–	–	–	E	–	+	–	+	–
Bufo (B. viridis group) raddei 1876	P	Mon	+	–	–	–	–	+	+	–	–	–	–	–	–	+	–
Bufo (B. viridis group) shaartusiensis 1996	P	Tad	+	–	–	E	–	–	–	–	–	–	+	–	–	E	–
Bufo (B. viridis group) siachinensis 1997	P	Kas	+	–	–	E	–	–	–	–	–	–	–	–	–	E	–
Bufo (B. viridis group) surdus 1891	P	SP	–	–	–	+	–	–	–	–	–	–	+	–	–	+	–
Bufo (B. viridis group) viridis 1768	P	WP	–	+	+	+	+	+	+	–	+	+	+	+	+	+	+

Appendix 6:1 Continued

Species	Faunal affinities	Distribution pattern	Palearctic endemic	Zoogeographic Region						Floristic Region							
				MED	EUR	WAS	SIB	CAS	EAS	MAC	NAI	MED	ESE	ESA	ASA	WCA	S-J
ANURA: DISCOGLOSSIDAE:																	
Alytes cisternasii 1879	P	Ibe	+	E	–	–	–	–	–	–	–	E	–	–	–	–	–
Alytes dickhilleni 1995	P	Ibe	+	E	–	–	–	–	–	–	–	E	–	–	–	–	–
Alytes muletensis 1979	P	BI	+	E	–	–	–	–	–	–	–	E	–	–	–	–	–
Alytes obstetricans 1768	P	Ibe	+	+	+	–	–	–	–	–	–	+	+	–	–	+	–
Bombina bombina 1761	P	EE	+	–	+	+	–	–	–	–	–	+	+	–	–	+	–
Bombina orientalis 1890	P	Man	+	–	–	–	–	–	E	–	–	–	–	–	–	–	E
Bombina pachypus 1838	P	Ita	+	E	–	–	–	–	–	–	–	E	+	–	–	–	–
Bombina variegata 1758	P	CE	+	+	+	–	–	–	–	–	–	+	+	–	–	–	–
Discoglossus galganoi 1985	P	Ibe	+	E	–	–	–	–	–	–	–	+	–	–	–	–	–
Discoglossus montalentii 1984	P	Tyr	+	E	–	–	–	–	–	–	–	E	–	–	–	–	–
Discoglossus nigriventer 1943	P	NE	–	–	–	E	–	–	–	–	–	E	+	–	–	–	–
Discoglossus pictus 1837	P	NWA	+	E	–	–	–	–	–	–	–	E	–	–	–	–	–
Discoglossus sardus 1837	P	Tyr	+	E	–	–	–	–	–	–	–	E	–	–	–	–	–
ANURA: HYLIDAE:																	
Hyla arborea 1758	P	CE	+	+	+	+	–	–	–	–	–	+	+	–	–	+	–
Hyla intermedia 1882	P	Ita	+	E	+	–	–	–	–	–	–	+	+	–	–	–	–
Hyla japonica 1859	P	EAs	–	–	–	–	+	–	+	–	–	–	–	+	–	–	+
Hyla meridionalis 1874	P	Ibe	+	+	+	–	–	–	–	+	–	+	+	–	–	–	–
Hyla sarda 1853	P	Tyr	+	E	–	–	–	–	–	–	–	E	–	–	–	–	–
Hyla savignyi 1827	P	WAs	+	–	–	+	–	–	–	–	+	–	+	–	–	+	–
Hyla suweonensis 1980	P	KN	+	–	–	–	–	–	E	–	–	–	–	–	–	–	E
ANURA: MICROHYLIDAE:																	
Kaloula borealis 1908	O	EAs	–	–	–	–	–	–	+	–	–	–	–	–	–	–	E
ANURA: PELOBATIDAE:																	
Pelobates cultripes 1829	P	Ibe	+	+	+	–	–	–	–	–	–	+	+	–	–	+	–
Pelobates fuscus 1768	P	EE	+	+	+	+	–	–	–	–	–	+	+	+	–	+	–
Pelobates syriacus 1889	P	EM	+	+	–	+	–	–	–	–	–	+	+	+	–	+	–
Pelobates varaldii 1959	P	NWA	+	E	–	–	–	–	–	–	–	E	–	–	–	–	–
Scutiger boulengeri 1898	O	HT	–	–	–	–	–	+	–	–	–	–	–	–	–	+	–
Scutiger liupanensis 1985	O	NC	+	–	–	–	–	E	–	–	–	–	–	–	–	+	+
Scutiger maculatus 1950	O	HP	+	–	–	–	–	E	–	–	–	–	–	–	–	+	–
Scutiger mammatus 1896	O	HP	–	–	–	–	–	+	–	–	–	–	–	–	–	+	+
Scutiger nyingchiensis 1977	O	HT	+	–	–	–	–	E	–	–	–	–	–	–	–	+	+
Scutiger occidentalis 1978	O	HT	+	–	–	–	–	E	–	–	–	–	–	–	–	+	+
ANURA: PELODYTIDAE:																	
Pelodytes caucasicus 1896	P	Cau	+	–	–	E	–	–	–	–	–	–	+	–	–	+	–
Pelodytes punctatus 1802	P	Atl	+	+	+	–	–	–	–	–	–	+	+	–	–	–	–
ANURA: RANIDAE:																	
Euphlyctis cyanophlyctis 1799	O	WO	–	–	–	+	–	–	–	–	–	–	–	–	–	+	–

Appendix 6:1 Continued

Species	Faunal affinities	Distribution pattern	Palearctic endemic	Zoogeographic Region						Floristic Region							
				MED	EUR	WAS	SIB	CAS	EAS	MAC	NAI	MED	ESE	ESA	ASA	WCA	S-J
Limnonectes limnocharis 1829	O	Pan	–	–	–	–	–	–	+	–	–	–	–	–	–	–	+
Nanorana parkeri 1927	O	HT	–	–	–	–	–	+	–	–	–	–	–	–	–	+	+
Nanorana pleskei 1896	O	HP	–	–	–	–	–	+	+	–	–	–	–	–	–	+	+
Paa boulengeri 1889	O	HO	–	–	–	–	–	–	+	–	–	–	–	–	–	+	+
Ptychadena mascareniensis 1841	A	Nil	–	–	–	E	–	–	–	+	–	–	–	–	–	–	–
Rana (*R. esculenta* group) *bedriagae* 1882	P	WAs	+	E	–	E	–	–	–	–	+	+	?	–	–	–	–
Rana (*R. esculenta* group) *bergeri* 1986	P	Ita	+	–	–	–	–	–	–	–	–	E	–	–	–	–	–
Rana (*R. esculenta* group) *cerigensis* 1994	P	NE	+	–	–	E	–	–	–	–	–	E	–	–	–	–	–
Rana (*R. esculenta* group) *chosenica* 1931	P	KN	+	–	–	–	–	–	E	–	–	–	–	–	–	–	E
Rana (*R. esculenta* group) *cretensis* 1994	P	Bal	+	E	–	–	–	–	–	–	–	E	–	–	–	–	–
Rana (*R. esculenta* group) *epeirotica* 1984	P	Adr	+	E	–	–	–	–	–	–	–	E	–	–	–	–	–
Rana (*R. esculenta* group) *esculenta* 1758	P	EE	+	+	+	–	–	–	–	–	–	–	E	–	–	–	–
Rana (*R. esculenta* group) *grafi* 1995	P	Ibe	+	+	+	–	–	–	–	–	+	+	+	–	–	–	–
Rana (*R. esculenta* group) *hispanica* 1839	P	Ita	+	E	–	–	–	–	–	–	–	E	–	–	–	–	–
Rana (*R. esculenta* group) *kurtmuelleri* 1940	P	Bal	+	E	–	–	–	–	–	–	–	E	–	–	–	–	–
Rana (*R. esculenta* group) *lessonae* 1882	P	EE	+	+	+	–	–	–	+	–	–	+	+	–	–	–	–
Rana (*R. esculenta* group) *nigromaculata* 1861	P	EAs	–	+	–	+	–	+	+	+	–	+	+	+	–	+	+
Rana (*R. esculenta* group) *perezi* 1885	P	Ibe	+	+	+	–	–	–	–	–	+	+	–	–	–	–	–
Rana (*R. esculenta* group) *plancyi* 1880	P	EAs	–	–	–	–	–	–	E	–	–	–	–	–	–	+	+
Rana (*R. esculenta* group) *porosa* 1868	P	Jap	+	–	–	–	–	–	E	–	–	–	–	–	–	–	E
Rana (*R. esculenta* group) *ridibunda* 1771	P	WP	+	+	+	+	+	+	+	+	+	+	+	+	–	+	–
Rana (*R. esculenta* group) *saharica* 1913	P	NWA	+	E	–	–	–	E	–	+	–	+	–	–	–	+	–
Rana (*R. esculenta* group) *shqiperica* 1987	P	Adr	+	E	–	E	–	+	–	–	–	E	–	–	–	–	E
Rana (*R. esculenta* group) *terentievi* 1992	P	Tad	+	–	–	E	–	–	E	–	–	–	–	–	E	–	–
Rana (*R. rugosa* group) *rugosa* 1838	P	EAs	+	–	–	–	–	–	+	–	–	+	+	+	+	+	+
Rana (*R. temporaria* group) *amurensis* 1886	P	Sib	+	–	–	–	+	–	+	–	–	+	+	+	+	+	+
Rana (*R. temporaria* group) *arvalis* 1842	P	EE	+	–	–	–	+	–	–	–	–	+	+	+	–	E	–
Rana (*R. temporaria* group) *asiatica* 1898	P	CAs	+	–	–	–	–	E	–	–	–	+	+	+	–	E	–
Rana (*R. temporaria* group) *chensinensis* 1875	P	EAs	–	–	–	–	–	+	+	–	–	+	+	+	–	+	+
Rana (*R. temporaria* group) *dalmatina* 1838	P	CE	+	+	+	–	–	–	–	–	–	+	+	–	–	+	–
Rana (*R. temporaria* group) *dybowskii* 1876	P	KN	+	–	–	E	–	–	E	–	–	–	–	–	–	E	E
Rana (*R. temporaria* group) *graeca* 1891	P	Bal	+	+	+	–	–	–	–	–	–	+	+	–	–	–	–
Rana (*R. temporaria* group) *holtzi* 1898	P	NE	+	–	–	E	–	–	–	–	–	+	–	–	–	–	–
Rana (*R. temporaria* group) *huanrenensis* 1990	P	KN	+	E	–	–	–	–	E	–	–	–	–	–	E	E	E
Rana (*R. temporaria* group) *iberica* 1879	P	Ibe	+	E	–	–	–	–	–	–	+	+	+	–	–	–	–
Rana (*R. temporaria* group) *italica* 1987	P	Ita	+	E	–	–	–	–	–	–	–	E	–	–	–	–	–
Rana (*R. temporaria* group) *japonica* 1859	P	EAs	–	–	–	–	–	–	+	–	–	+	E	–	–	+	+
Rana (*R. temporaria* group) *latastei* 1879	P	Po	+	E	–	–	–	–	–	–	–	E	E	–	–	–	–
Rana (*R. temporaria* group) *macrocnemis* 1885	P	WAs	+	–	–	E	–	–	–	–	–	+	+	–	–	+	–

Appendix 6:1 Continued

Species	Faunal affinities	Distribution pattern	Palearctic endemic	Zoogeographic Region								Floristic Region						
				MED	EUR	WAS	SIB	CAS	EAS	MAC	NAI	MED	ESE	ESA	ASA	WCA	S-J
Rana (R. temporaria group) ornativentris 1903	P	Jap	+	–	–	–	–	–	E	–	–	–	–	–	–	–	E
Rana (R. temporaria group) pirica 1991	P	Hok	+	–	–	–	–	–	E	–	–	–	–	–	–	–	E
Rana (R. temporaria group) pyrenaica 1993	P	Pyr	+	+	–	–	–	–	–	–	–	–	E	–	–	–	–
Rana (R. temporaria group) sakuraii 1990	P	Jap	+	–	–	–	–	–	E	–	–	–	–	–	–	–	E
Rana (R. temporaria group) tagoi 1928	P	Jap	+	–	–	–	–	–	E	–	–	–	–	–	–	–	E
Rana (R. temporaria group) temporaria 1758	P	EE	+	+	+	–	–	–	–	–	–	+	+	+	+	+	–
Rana (R. temporaria group) tsushimensis 1907	P	Jap	+	–	–	–	–	–	E	–	–	–	–	–	–	–	E
ANURA: RHACOPHORIDAE:																	
Buergeria buergeri 1838	O	Jap	+	–	–	–	–	–	E	–	–	–	–	–	–	–	E
Rhacophorus arboreus 1924	O	Jap	+	–	–	–	–	–	E	–	–	–	–	–	–	–	E
Rhacophorus schlegelii 1859	O	Jap	+	–	–	–	–	–	E	–	–	–	–	–	–	–	E

APPENDIX 6:2

Distribution of amphibians in Europe, which is delimited in the east by the Ural Mountains, the Ural River, and includes the northern slopes of the Greater Caucasus Mountains. Abbreviations of countries are: AL = Albania, AU = Austria, BE = Belgium, BU = Bulgaria, CZ = Czech Republic, DE = Denmark, FI = Finland, FR = France, GB = Great Britain, GE = Germany, GR = Greece, HU = Hungary, IR = Ireland, IT = Italy, LI = Liechtenstein, LU = Luxembourg, MA = Malta, MO = Monaco, NE = Netherlands, NO = Norway, PO = Poland, PR = Portugal, RO = Romania, SD = Sweden, SL = Slovakia, SP = Spain, SU = the former Soviet Union (European part; also see Appendix 6:3), SW = Switzerland, TU = Turkey (European part; also see Appendix 6:5), YU = the former Yugoslavia (also see Appendix 6:4). In the columns, + = present, – = absent, E = endemic, I = probably introduced; ? = occurrence needs to be confirmed; other letters denote endemism to particular islands in a country: B = Balearic, Co = Corsica, Cr = Crete, K = Karpathos, S = Sardinia. Asterisks after a species name denote species endemic to Europe (*) or European species with limited ranges outside of Europe (**). Sources: Astudillo and Arano (1995), Atatür and Yilmaz (1986), Baruš et al. (1992), Bruno (1989), Crochet et al. (1995), Cogălniceanu (1991), Dubois and Ohler (1995a), Eiselt (1988), Engel and Thorn (1996), Engelmann et al. (1993), Gasc et al. (1997, and references cited therein), Grossenbacher (1994), Haxhiu (1994), Lanza and Corti (1996), Lanza et al. (1995), Malkmus (1995), Moravec (1994), Nascetti et al. (1995), Nöllert and Nöllert (1992, and references cited therein), Schneider et al. (1993), Serra-Cobo (1993), Zavadil et al. (1994), and my own data.

Species	PR	SP	FR	MO	IT	MA	IR	GB	BE	LU	NE	GE	SW	LI	AU	DE	SD	NO	FI	PO	CZ	SL	HU	RO	YU	AL	GR	BU	TU	SU
CAUDATA: HYNOBIIDAE:																														
Salamandrella keyserlingii	–	–	–	–	–	–	–	–	–	–	–	–	–	–	–	–	–	–	–	–	–	–	–	–	–	–	–	–	–	+
CAUDATA: PLETHODONTIDAE:																														
Hydromantes ambrosii*	–	–	–	–	E	–	–	–	–	–	–	–	–	–	–	–	–	–	–	–	–	–	–	–	–	–	–	–	–	–
Hydromantes flavus*	–	–	–	–	S	–	–	–	–	–	–	–	–	–	–	–	–	–	–	–	–	–	–	–	–	–	–	–	–	–
Hydromantes genei*	–	–	–	–	S	–	–	–	–	–	–	–	–	–	–	–	–	–	–	–	–	–	–	–	–	–	–	–	–	–
Hydromantes imperialis*	–	–	–	–	S	–	–	–	–	–	–	–	–	–	–	–	–	–	–	–	–	–	–	–	–	–	–	–	–	–
Hydromantes italicus*	–	–	–	–	E	–	–	–	–	–	–	–	–	–	–	–	–	–	–	–	–	–	–	–	–	–	–	–	–	–
Hydromantes strinatii*	–	–	+	–	+	–	–	–	–	–	–	–	–	–	–	–	–	–	–	–	–	–	–	–	–	–	–	–	–	–
Hydromantes supramontis*	–	–	–	–	S	–	–	–	–	–	–	–	–	–	–	–	–	–	–	–	–	–	–	–	–	–	–	–	–	–
CAUDATA: PROTEIDAE:																														
Proteus anguinus*	–	–	–	–	+	–	–	–	–	–	–	–	–	–	–	–	–	–	–	–	–	–	–	–	+	–	–	–	–	–
CAUDATA: SALAMANDRIDAE:																														
Chioglossa lusitanica*	+	+	–	–	–	–	–	–	–	–	–	–	–	–	–	–	–	–	–	–	–	–	–	–	–	–	–	–	–	–
Euproctus asper*	+	+	–	–	–	–	–	–	–	–	–	–	–	–	–	–	–	–	–	–	–	–	–	–	–	–	–	–	–	–
Euproctus montanus*	–	–	Co	–	–	–	–	–	–	–	–	–	–	–	–	–	–	–	–	–	–	–	–	–	–	–	–	–	–	–
Euproctus platycephalus*	–	–	–	–	S	–	–	–	–	–	–	–	–	–	–	–	–	–	–	–	–	–	–	–	–	–	–	–	–	–
Mertensiella luschani	–	–	–	–	–	–	–	–	–	–	–	–	–	–	–	–	–	–	–	–	–	–	–	–	–	–	+	–	–	–
Pleurodeles waltl	+	+	–	–	–	–	–	–	–	–	–	–	–	–	–	–	–	–	–	–	–	–	–	–	–	–	–	–	–	–
Salamandra atra*	–	–	–	–	+	–	–	–	–	–	–	+	+	+	+	–	–	–	–	–	–	–	–	–	+	+	–	–	–	–
Salamandra corsica*	–	–	Co	–	–	–	–	–	–	–	–	–	–	–	–	–	–	–	–	–	–	–	–	–	–	–	–	–	–	–
Salamandra lanzai*	–	–	+	–	+	–	–	–	–	–	–	–	–	–	–	–	–	–	–	–	–	–	–	–	–	–	–	–	–	–
Salamandra salamandra*	+	+	+	–	+	–	–	–	+	+	+	+	+	+	+	–	–	–	–	+	+	+	+	+	+	+	+	+	+	+

Appendix 6:2 Continued

Country

Species	PR	SP	FR	MO	IT	MA	IR	GB	BE	LU	NE	GE	SW	LI	AU	DE	SD	NO	FI	PO	CZ	SL	HU	RO	YU	AL	GR	BU	TU	SU
*Salamandrina terdigitata**	–	–	–	–	E	–	–	–	–	–	–	–	–	–	–	–	–	–	–	–	–	–	–	–	–	–	–	–	–	–
*Triturus alpestris**	–	+	+	–	+	–	–	–	+	+	+	+	+	+	+	+	+	–	–	+	+	+	+	+	+	+	+	+	–	+
*Triturus boscai**	+	+	–	–	–	–	–	–	–	–	–	–	–	–	–	–	–	–	–	–	–	–	–	–	–	–	–	–	–	–
*Triturus carnifex**	–	–	+	–	+	–	–	–	–	–	–	–	+	–	+	–	–	–	–	–	–	+	+	–	+	+	+	–	–	–
*Triturus cristatus***	–	–	+	–	–	–	–	+	+	+	+	+	+	–	+	+	+	–	+	+	+	+	+	+	+	–	–	–	–	+
*Triturus dobrogicus**	–	–	–	–	–	–	–	–	–	–	–	–	–	–	+	–	–	–	–	–	–	+	+	+	+	–	–	+	–	+
*Triturus helveticus**	+	+	+	–	–	–	–	+	+	+	+	+	+	–	–	–	–	–	–	–	–	–	–	–	–	–	–	–	–	–
*Triturus italicus**	–	–	–	–	E	–	–	–	–	–	–	–	–	–	–	–	–	–	–	–	–	–	–	–	–	–	–	–	–	–
Triturus karelinii	–	–	–	–	–	–	–	–	–	–	–	–	–	–	–	–	–	–	–	–	–	–	–	–	+	–	+	+	+	+
*Triturus marmoratus**	+	+	+	–	–	–	–	–	–	–	–	–	–	–	–	–	–	–	–	–	–	–	–	–	–	–	–	–	–	–
*Triturus montandoni**	–	–	–	–	–	–	–	–	–	–	–	–	–	–	–	–	–	–	–	+	+	+	+	+	–	–	–	–	–	+
Triturus vittatus	–	–	–	–	–	–	–	–	–	–	–	–	–	–	–	–	–	–	–	–	–	–	–	–	–	–	–	–	–	+
Triturus vulgaris	–	–	+	–	+	–	+	+	+	+	+	+	+	+	+	+	+	+	+	+	+	+	+	+	+	+	+	+	+	+
ANURA: BUFONIDAE:																														
Bufo bufo	+	+	+	+	+	–	–	+	+	+	+	+	+	+	+	+	+	+	+	+	+	+	+	+	+	+	+	+	+	+
*Bufo calamita**	+	+	+	–	–	–	+	+	+	+	+	+	+	–	–	+	+	–	–	+	+	–	–	–	–	–	–	–	–	+
Bufo verrucosissimus	–	–	–	–	–	–	–	–	–	–	–	–	–	–	–	–	–	–	–	–	–	–	–	–	–	–	–	–	–	+
Bufo viridis	–	I	+	–	+	+	–	–	–	–	–	+	+	+	+	+	+	–	–	+	+	+	+	+	+	+	+	+	+	+
ANURA: DISCOGLOSSIDAE:																														
*Alytes cisternasii**	+	+	–	–	–	–	–	–	–	–	–	–	–	–	–	–	–	–	–	–	–	–	–	–	–	–	–	–	–	–
*Alytes dickhilleni**	–	E	–	–	–	–	–	–	–	–	–	–	–	–	–	–	–	–	–	–	–	–	–	–	–	–	–	–	–	–
*Alytes muletensis**	–	B	–	–	–	–	–	–	–	–	–	–	–	–	–	–	–	–	–	–	–	–	–	–	–	–	–	–	–	–
*Alytes obstetricans**	+	+	+	–	–	–	–	+	+	+	+	+	+	–	–	+	–	–	–	–	–	–	–	–	–	–	–	–	–	–
Bombina bombina	–	–	–	–	–	–	–	–	–	–	–	+	–	–	+	+	+	–	–	+	+	+	+	+	+	–	–	+	–	+
*Bombina pachypus**	–	–	–	–	E	–	–	–	–	–	–	–	–	–	–	–	–	–	–	–	–	–	–	–	–	–	–	–	–	–
*Bombina variegata**	–	–	+	–	+	–	–	–	+	+	–	+	–	–	+	+	–	–	–	+	+	+	+	+	+	+	+	+	–	+
*Discoglossus galganoi**	+	+	–	–	–	–	–	–	–	–	–	–	–	–	–	–	–	–	–	–	–	–	–	–	–	–	–	–	–	–
*Discoglossus montalentii**	–	–	Co	–	–	–	–	–	–	–	–	–	–	–	–	–	–	–	–	–	–	–	–	–	–	–	–	–	–	–
Discoglossus pictus	–	I	I	+	+	+	–	–	–	–	–	–	–	–	–	–	–	–	–	–	–	–	–	–	–	–	–	–	–	–
*Discoglossus sardus**	–	–	+	–	+	–	–	–	–	–	–	–	–	–	–	–	–	–	–	–	–	–	–	–	–	–	–	–	–	–
ANURA: HYLIDAE:																														
Hyla arborea	+	+	+	–	+	–	–	–	+	+	+	+	+	+	+	+	+	–	–	+	+	+	+	+	+	+	+	+	+	+
*Hyla intermedia**	–	–	+	–	+	–	–	–	–	–	–	–	+	–	–	–	–	–	–	–	–	–	–	–	+	–	–	–	–	–
*Hyla meridionalis**	+	+	+	+	+	+	–	–	–	–	–	–	–	–	–	–	–	–	–	–	–	–	–	–	–	–	–	–	–	–
*Hyla sarda**	–	+	+	–	+	–	–	–	–	–	–	–	–	–	–	–	–	–	–	–	–	–	–	–	–	–	–	–	–	–
ANURA: PELOBATIDAE:																														
*Pelobates cultripes**	+	+	+	–	–	–	–	–	–	–	–	–	–	–	–	–	–	–	–	–	–	–	–	–	–	–	–	–	–	–
Pelobates fuscus	–	–	+	–	+	–	–	–	+	–	+	+	–	–	+	+	+	–	–	+	+	+	+	+	+	–	–	+	+	+
Pelobates syriacus	–	–	–	–	–	–	–	–	–	–	–	–	–	–	–	–	–	–	–	–	–	–	–	+	+	?	+	+	+	+

Appendix 6:2 Continued

Species	PR	SP	FR	MO	IT	MA	IR	GB	BE	LU	NE	GE	SW	LI	AU	DE	SD	NO	FI	PO	CZ	SL	HU	RO	YU	AL	GR	BU	TU	SU
ANURA: PELODYTIDAE:																														
Pelodytes caucasicus	–	–	–	–	–	–	–	–	–	–	–	–	–	–	–	–	–	–	–	–	–	–	–	–	–	–	–	–	–	+
Pelodytes punctatus*	+	+	+	–	+	–	–	–	–	–	–	–	–	–	–	–	–	–	–	–	–	–	–	–	–	–	–	–	–	–
ANURA: RANIDAE:																														
Rana arvalis	–	–	–	–	–	–	–	–	+	–	+	+	–	–	+	+	+	+	+	+	+	+	+	+	+	–	–	–	–	+
Rana bergeri*	–	–	+	–	+	–	–	–	–	–	–	–	–	–	–	–	–	–	–	–	–	–	–	–	–	–	–	–	–	–
Rana cerigensis*	–	–	–	–	–	–	–	–	–	–	–	–	–	–	–	–	–	–	–	–	–	–	–	–	–	–	K	–	–	–
Rana cretensis*	–	–	–	–	–	–	–	–	–	–	–	–	–	–	–	–	–	–	–	–	–	–	–	–	–	–	Cr	–	–	–
Rana dalmatina	–	–	+	–	+	–	–	–	?	–	–	+	+	–	+	+	+	–	–	+	+	+	+	+	+	+	+	+	–	+
Rana epeirotica*	–	–	–	–	–	–	–	–	–	–	–	–	–	–	–	–	–	–	–	–	–	–	–	–	–	+	+	–	–	–
Rana esculenta*	–	–	+	–	+	–	–	–	+	+	+	+	+	+	+	+	+	–	–	+	+	+	+	+	+	+	–	+	–	+
Rana graeca*	–	–	–	–	–	–	–	–	–	–	–	–	–	–	–	–	–	–	–	–	–	–	–	–	+	+	+	+	–	–
Rana grafi*	–	+	+	–	–	–	–	–	–	–	–	–	–	–	–	–	–	–	–	–	–	–	–	–	–	–	–	–	–	–
Rana hispanica*	–	+	+	–	–	–	–	–	–	–	–	–	–	–	–	–	–	–	–	–	–	–	–	–	–	–	–	–	–	–
Rana iberica*	+	E	–	–	–	–	–	–	–	–	–	–	–	–	–	–	–	–	–	–	–	–	–	–	–	–	–	–	–	–
Rana italica*	–	–	–	–	E	–	–	–	–	–	–	–	–	–	–	–	–	–	–	–	–	–	–	–	–	–	–	–	–	–
Rana kurtmuelleri*	–	–	–	–	–	–	–	–	–	–	–	–	–	–	–	–	–	–	–	–	–	–	–	–	+	+	+	–	–	–
Rana latastei*	–	–	–	–	+	–	–	–	–	–	–	–	+	–	–	–	–	–	–	–	–	–	–	–	+	–	–	–	–	–
Rana lessonae*	–	–	+	–	+	–	–	+	+	–	+	+	+	+	+	–	+	–	–	+	+	+	+	+	+	–	–	+	–	+
Rana macrocnemis	–	–	–	–	–	–	–	–	–	–	–	–	–	–	–	–	–	–	–	–	–	–	–	–	–	–	–	–	+	+
Rana perezi*	+	+	+	+	–	–	–	–	–	–	–	–	–	–	–	–	–	–	–	–	–	–	–	–	–	–	–	–	–	–
Rana pyrenaica*	–	+	+	–	–	–	–	–	–	–	–	–	–	–	–	–	–	–	–	–	–	–	–	–	–	–	–	–	–	–
Rana ridibunda	–	–	+	–	–	–	–	+	+	–	+	+	–	–	+	+	–	–	–	+	+	+	+	+	+	+	+	+	+	+
Rana shqiperica*	–	–	–	–	–	–	–	–	–	–	–	–	–	–	–	–	–	–	–	–	–	–	–	–	+	+	–	–	–	–
Rana temporaria**	–	+	+	–	+	–	+	+	+	+	+	+	+	+	+	+	+	+	+	+	+	+	+	+	+	–	–	+	–	+
Total Caudata	6	8	12	0	17	0	1	3	5	5	5	7	7	4	7	3	2	2	2	5	7	6	6	6	9	5	6	5	2	9
Total Anura	11	19	24	3	20	2	2	3	11	8	12	14	11	6	13	11	11	3	3	13	13	12	13	12	18	10	14	13	9	17
Total Amphibia	17	27	36	3	37	2	3	6	16	13	17	21	18	10	20	14	13	5	5	18	20	18	19	18	27	15	20	18	11	26
Total endemics	–	3	3	–	11	–	–	–	–	–	–	–	–	–	–	–	–	–	–	–	–	–	–	–	–	–	2	–	–	–
Percent endemic	–	11	8	–	30	–	–	–	–	–	–	–	–	–	–	–	–	–	–	–	–	–	–	–	–	–	10	–	–	–

APPENDIX 6:3

Distribution of amphibians in countries of the former Soviet Union. Abbreviations of countries are: AR = Armenia, AZ = Azerbaijan, BE = Belorussia, ES = Estonia, GE = Georgia, KA = Kazakhstan, KY = Kyrgyzstan, LA = Latvia, LI = Lithuania, MO = Moldavia, TA = Tadjikistan, TU = Turkmenistan, UK = Ukraine, UZ = Uzbekistan. Within Russia: AS = Asian part of Russia (Ural Mountains and Ural River eastward), EU = European part of Russia (including the Russian Caucasus), TO = total Russia (Asian and European parts). + = present, – = absent, R = species present and listed in the Red Data Book of a particular country, ? = occurrence needs to be confirmed, * = species endemic to the former USSR as a whole, E = endemic to a given country. Sources: Darevsky and Krever (1987), Mezhzherin (1992), Kuzmin (1995a), Borkin et al. (1997), Borkin in Ananjeva et al. (1998), Pisanetz et al. (1996), my own data, and Red Data Books for particular countries. *Hynobius turkestanicus* Nikolsky 1909 has not been included, because its validity has not been confirmed (Kuzmin et al., 1995). Introductions of some species (e.g., *Rana nigro-maculata* in Turkmenistan, *Rana ridibunda* in Latvia, Siberia, etc.) have not been listed. *Salamandra salamandra, Hyla arborea,* and *Rana ridibunda* are listed in the Red Data Books of Russia, Latvia, and Estonia, respectively, but confirmed records of those species are lacking.

Species	Russia TO	EU	AS	Other Countries ES	LA	LI	BE	MO	UK	GE	AR	AZ	KA	KY	UZ	TA	TU
CAUDATA: HYNOBIIDAE:																	
Onychodactylus fischeri	R	–	+	–	–	–	–	–	–	–	–	–	–	–	–	–	–
Ranodon sibiricus	–	–	–	–	–	–	–	–	–	–	–	–	R	–	–	–	–
Salamandrella keyserlingii	+	+	+	–	–	–	–	–	–	–	–	–	+	–	–	–	–
CAUDATA: SALAMANDRIDAE:																	
Mertensiella caucasica	–	–	–	–	–	–	–	–	–	R	–	–	–	–	–	–	–
Salamandra salamandra	–	–	–	–	–	–	–	–	R	–	–	–	–	–	–	–	–
Triturus alpestris	–	–	–	–	–	–	–	–	R	–	–	–	–	–	–	–	–
Triturus cristatus	+	+	+	+	R	+	+	+	+	+	–	–	–	–	–	–	–
Triturus dobrogicus	–	–	–	–	–	–	–	+	+	–	–	–	–	–	–	–	–
Triturus karelinii	R	+	–	–	–	–	–	–	+	+	+	–	R	–	–	–	–
Triturus montandoni	–	–	–	–	–	–	–	–	R	–	–	–	–	–	–	–	–
Triturus vittatus	R	+	–	–	–	–	–	–	–	R	–	–	–	–	–	–	–
Triturus vulgaris	+	+	+	+	+	+	+	+	+	+	+	+	+	–	–	–	–
ANURA: BUFONIDAE:																	
Bufo bufo	+	+	+	+	+	+	+	+	+	–	–	–	+	–	–	–	–
Bufo calamita	R	+	–	R	R	R	R	–	R	–	–	–	–	–	–	–	–
Bufo danatensis	–	–	–	–	–	–	–	–	–	–	–	–	R	+	+	+	+
Bufo gargarizans	+	–	+	–	–	–	–	–	–	–	–	–	–	–	–	–	–
Bufo raddei	+	–	+	–	–	–	–	–	–	–	–	–	–	–	–	–	–
Bufo shaartusiensis*	–	–	–	–	–	–	–	–	–	–	–	–	–	–	–	E	–
Bufo verrucosissimus	+	+	–	–	–	–	–	–	–	+	–	R	–	–	–	–	–
Bufo viridis	+	+	+	R	+	+	+	+	+	+	+	+	+	+	+	+	+
ANURA: DISCOGLOSSIDAE:																	
Bombina bombina	+	+	–	–	R	+	+	+	+	–	–	–	+	–	–	–	–
Bombina orientalis	+	–	+	–	–	–	–	–	–	–	–	–	–	–	–	–	–
Bombina variegata	–	–	–	–	–	–	–	–	–	+	–	–	–	–	–	–	–
ANURA: HYLIDAE:																	
Hyla arborea	+	+	–	–	–	+	+	+	+	+	+	+	–	–	–	–	–
Hyla japonica	+	–	+	–	–	–	–	–	–	–	–	–	–	–	–	–	–
Hyla savignyi	–	–	–	–	–	–	–	–	–	+	+	+	–	–	–	–	–
ANURA: PELOBATIDAE:																	
Pelobates fuscus	+	+	+	R	+	+	+	+	+	–	–	–	+	–	–	–	–
Pelobates syriacus	R	+	–	–	–	–	–	–	–	R	R	R	–	–	–	–	–
ANURA: PELODYTIDAE:																	
Pelodytes caucasicus	R	+	–	–	–	–	–	–	–	R	–	R	–	–	–	–	–
ANURA: RANIDAE:																	
Rana amurensis	+	–	+	–	–	–	–	–	–	–	–	–	–	–	–	–	–

Appendix 6:3 Continued

Species	Russia			Other Countries													
	TO	EU	AS	ES	LA	LI	BE	MO	UK	GE	AR	AZ	KA	KY	UZ	TA	TU
Rana arvalis	+	+	+	+	+	+	+	+	+	–	–	–	+	–	–	–	–
Rana asiatica	–	–	–	–	–	–	–	–	–	–	–	–	R	+	–	–	–
Rana chensinensis	+	–	+	–	–	–	–	–	–	–	–	–	–	–	–	–	–
Rana dalmatina	–	–	–	–	–	–	–	–	R	–	–	–	–	–	–	–	–
Rana esculenta	+	+	–	+	+	+	+	+	+	–	–	–	–	–	–	–	–
Rana lessonae	+	+	–	+	+	+	+	?	+	–	–	–	–	–	–	–	–
Rana macrocnemis	+	+	–	–	–	–	–	–	–	+	+	+	–	–	–	–	R
Rana nigromaculata	+	–	+	–	–	–	–	–	–	–	–	–	–	–	–	–	–
Rana ridibunda	+	+	–	–	–	–	+	+	+	+	+	+	+	+	+	+	+
Rana temporaria	+	+	+	+	+	+	+	+	+	–	–	–	+	–	–	–	–
*Rana terentievi**	–	–	–	–	–	–	–	–	–	–	–	–	–	–	–	E	–
Total Caudata	6	5	4	2	2	2	2	3	7	4	1	2	3	0	0	0	0
Total Anura	22	15	12	8	9	10	11	9	13	8	6	8	9	4	3	5	4
Total Amphibia	28	20	16	10	11	12	13	12	20	12	7	10	12	4	3	5	4
Total endemics	–	–	–	–	–	–	–	–	–	–	–	–	–	–	–	2	–
Percent endemic	–	–	–	–	–	–	–	–	–	–	–	–	–	–	–	40	–

APPENDIX 6:4

Distribution of amphibians in countries of the former Yugoslavia. + = present, – = absent. Sources: Bruno (1989), Crnobrnja-Isailovic and Dzukic (1995, in litt.), Gasc et al. (1997), Grossenbacher (1994), Günther (1990), Hotz et al. (1987), S. N. Litvinchuk (in litt.), Radovanovič (1964), Schneider et al. (1993), Schneider and Haxhiu (1994), and Vogrin (1996, 1997).

	Country				
Species	Slovenia	Croatia	Bosnia and Hercegovina	Serbia and Montenegro	Macedonia
CAUDATA: PROTEIDAE:					
Proteus anguinus	+	+	+	–	–
CAUDATA: SALAMANDRIDAE:					
Salamandra atra	+	+	+	+	–
Salamandra salamandra	+	+	+	+	+
Triturus alpestris	+	+	+	+	+
Triturus carnifex	+	+	+	+	+
Triturus cristatus	–	–	–	+	–
Triturus dobrogicus	–	+	+	+	–
Triturus karelinii	–	–	–	+	+
Triturus vulgaris	+	+	+	+	+
ANURA: BUFONIDAE:					
Bufo bufo	+	+	+	+	+
Bufo viridis	+	+	+	+	+
ANURA: DISCOGLOSSIDAE:					
Bombina bombina	+	+	+	+	–
Bombina variegata	+	+	+	+	+
ANURA: HYLIDAE:					
Hyla arborea	+	+	+	+	+
Hyla intermedia	+	–	–	–	–
ANURA: PELOBATIDAE:					
Pelobates fuscus	+	+	+	+	–
Pelobates syriacus	–	–	–	+	+
ANURA: RANIDAE:					
Rana arvalis	+	+	–	+	–
Rana dalmatina	+	+	+	+	+
Rana esculenta	+	+	+	+	–
Rana graeca	–	–	+	+	+
Rana kurtmuelleri	–	–	–	+	+
Rana latastei	+	+	–	–	–
Rana lessonae	+	+	+	+	–
Rana ridibunda	+	+	+	+	+
Rana shqiperica	–	–	–	+	–
Rana temporaria	+	+	+	+	+
Total Caudata	6	7	7	8	5
Total Anura	14	13	12	16	10
Total Amphibia	20	20	19	24	15

APPENDIX 6:5

Distribution of amphibians in the Near East and Turkey (exclusive of the European part; see Appendix 6:2). + = present, – = absent, E = endemic to a country. Asterisks denote species (*) or subspecies (**) endemic to the region as a whole. Sources: Afrasiab and Ali (1988), Atatür and Yilmaz (1986), Baran and Öz (1994), Başoglu and Özeti (1973), Böhme and Wiedl (1994), Daszak and Cawthraw (1991), Disi (1996), Disi and Böhme (1996), Eiselt (1958), Joger and Steinfartz (1995), Leviton et al. (1992), Mendelssohn and Steinitz (1944), Schmidtler (1994), Schneider et al. (1992), Werner (1982, 1987, 1988), and Yilmaz (1986).

Species	Country						
	Cyprus	Turkey	Iraq	Syria	Lebanon	Israel	Jordan
CAUDATA: SALAMANDRIDAE:							
Mertensiella caucasica	–	+	–	–	–	–	–
*Mertensiella luschani***	–	+	–	–	–	–	–
Neurergus crocatus	–	+	+	–	–	–	–
Neurergus microspilotus	–	–	+	–	–	–	–
*Neurergus strauchii**	–	E	–	–	–	–	–
*Salamandra inframmaculata***	–	+	+	+	+	+	–
Salamandra salamandra	–	+	–	–	–	–	–
Triturus karelinii	–	+	–	–	–	–	–
*Triturus vittatus***	–	+	+	+	+	+	+
*Triturus vulgaris***	–	+	–	–	–	–	–
ANURA: BUFONIDAE:							
Bufo bufo	–	+	–	+	–	–	–
Bufo surdus	–	–	+	–	–	–	–
Bufo verrucosissimus	–	+	–	–	–	–	–
Bufo viridis	+	+	+	+	+	+	+
ANURA: DISCOGLOSSIDAE:							
Bombina bombina	–	+	–	–	–	–	–
*Discoglossus nigriventer**	–	–	–	–	–	E	–
ANURA: HYLIDAE:							
Hyla arborea	–	+	–	–	–	–	–
Hyla savignyi	+	+	+	+	+	+	+
ANURA: PELOBATIDAE:							
Pelobates syriacus	–	+	+	+	+	+	+
ANURA: PELODYTIDAE:							
Pelodytes caucasicus	–	+	–	–	–	–	–
ANURA: RANIDAE:							
*Rana bedriagae**	+	+	+	+	+	+	+
Rana dalmatina	–	+	–	–	–	–	–
*Rana holtzi**	–	E	–	–	–	–	–
Rana macrocnemis (+ *camerani*)		+	+	–	–	–	–
Rana ridibunda	–	+	–	–	–	–	–
Total Caudata	0	9	4	2	2	2	1
Total Anura	3	13	6	5	4	5	4
Total Amphibia	3	22	10	7	6	7	5
Total endemics	–	2	–	–	–	1	–
Percent endemic	–	9	–	–	–	14	–

ADDENDUM

Differences found by Roy and Elepfandt (1993) between the Yemenite *Euphlyctis (Rana) ehrenbergi* and *E. cyanophlyctis* from northeastern India support the specific rank of the former.

Cladistic analysis of electrophoretic data by Hedges (1986) did not show evidence of close relationships among three species of Palearctic tree frogs (*Hyla*). The three species were arranged in two phyletic lineages. The Moroccan *Hyla meridionalis* clustered with North American *Acris,* whereas *Hyla arborea* from Italy (= *H. intermedia* in current taxonomy) and *Hyla japonica* from Japan were grouped with other North American tree frogs. Formerly, Ralin (1977) suggested that the *Hyla arborea* complex was a polyphyletic assemblage with close relatives in at least two complexes of North American *Hyla.* According to some molecular estimates, Eurasian treefrogs may have diverged from North American relatives about 40 million years ago (Maxson and Wilson, 1984).

The arrangement of rhacophorid frogs by Channing (1989) has been criticized by Blommers-Schlösser (1993).

LITERATURE CITED

BLOMMERS–SCHLÖSSER, R. M. A. 1993. Systematic relationships of the Mantellinae Laurent, 1946 (Anura: Ranoidea). Ethol. Ecol. Evol. 5:199–218.

CHANNING, A. 1989. A re-evaluation of the phylogeny of Old World treefrogs. South African J. Zool. 24:116–131.

HEDGES, S. B. 1986. An electrophoretic analysis of Holarctic hylid frog evolution. Syst. Zool. 35:1–21.

MAXSON, L. R., AND A. C. WILSON. 1984. Albumin evolution and organismal evolution in tree frogs (Hylidae). Syst. Zool. 24:1–15.

RALIN, D. B. 1977. Hybridization of *Hyla cinerea* of the United States and *H. arborea savignyi* of Israel (Amphibia, Anura, Hylidae). J. Herpetol. 11:105–106.

ROY, D, AND A. ELEPFANDT. 1993. Bioacoustic analysis of frog calls from northeast India. J. Biosci. Bangalore 18:381–393.

7. Distribution Patterns of Amphibians in Temperate Eastern Asia

ZHAO ER-MI

Chengdu Institute of Biology
Academica Sinica
P. O. Box 416
Chengdu, Sichuan, 610041, China

ABSTRACT   Temperate Eastern Asia occupies a vast area of about 11,766,457 km^2 and has 316 species of amphibians belonging to three orders, 11 families, and 45 genera. The total species density is 26.9 per 10^6 km^2. Density is much greater in the southern part belonging to the Oriental Realm (89.4 per 10^6 km^2) than in the northern part belonging to the Palearctic Realm (8 per 10^6 km^2). There are no endemic families, but there are some endemic genera (7, 15.6%) and many endemic species (229, 72.5%). Southern China species (87, 27.5%) and Hengduanshan Palearctic plus Hengduanshan Oriental species (56, 17.7%) are the principal faunal elements on the mainland; these mainly range over the Central China and Southwest China regions. The Hengduanshan Mountains in southwestern China are a transition area between Palearctic and Oriental elements as well as the warm temperate lowlands to the east and the high, cold plateau to the west. Disjunct distributions at the specific and generic levels occur in the Yellow River region. Peripheral parts of Temperate Eastern Asia are invaded by species from adjacent geographic regions. A high percentage of endemism occurs on the Oriental islands—81.1% on Japan, 68.4% on the Ryukyus, 30.3% on Taiwan, and 18.4% on Hainan.

Key words: Amphibians, Distribution patterns, Temperate Eastern Asia.

INTRODUCTION

As defined here, Temperate Eastern Asia consists of the mainland of the eastern portion of Eurasia (east of about 75° E, south of about 50° N, and north of the southern border of China) and the adjacent islands. These are, from north to south, the islands of Japan, the Ryukyu Islands, Taiwan, and Hainan.

Based principally on the taxonomy and number of taxa presented by Frost (1985) and Duellman (1993), I recognize 316 species of amphibians in the region. Large portions of Temperate Eastern Asia have been poorly, or not yet, surveyed. Also, most taxa have not been studied well; some widely distributed "species" may consist of a number of cryptic species. Some species that are recognized herein actually may be conspecific with others, and a few species need to be identified precisely. Furthermore,

I have little or no personal knowledge of the landscapes and environments in much of the region. Consequently, the present survey can provide only a preliminary and very rough idea about the patterns of distribution of amphibians in the region. If my efforts are useful to future studies, I will be pleased.

Many authors have contributed to the taxonomy and distribution of amphibians in this region. The principal workers have been A. G. Bannikov (Mongolia), Alice M. Boring (northern China), Leo J. Borkin (Mongolia), Mangven L. Y. Chang (China), H. B. Ding (Fujian), S. C. Hu (China), Robert F. Inger (Ryukyu Islands; Sichuan, China), Y. S. Kang (Korea), S. L. Kuzmin (Mongolia), C.-C. Liu (China), K. Y. Lue (Taiwan), M. Matsui (Japan), K. Munkhbayar (Mongolia), Y. Okada (Japan), Jehol ([now part of Beijing and Inner Mongolia], China), C. H. Pope (Fujian), I. Sato (Japan), M. A. Smith (Hainan), L. H. Stejneger (Japan, China including Taiwan), Y. Utsunomiya (Japan, Taiwan), H. K. Won (Korea), D. T. Yang (Yunnan), H. T. Yu (Taiwan), and E.-M. Zhao (China).

MATERIALS AND METHODS

Temperate Eastern Asia is located in southeastern Eurasia and the western Pacific Ocean. It includes the whole of China (South China Sea islands not included), Japan, the Korean Peninsula, and Mongolia. It consists of mainland and islands. The mainland of Temperate Eastern Asia, as influenced by geographic position, atmospheric circulation, and topography, can be divided grossly into three natural areas—Eastern Monsoonal Area, Northwest Dry Area, and Qinghai-Xizang (Tibet) High Cold Area.

To date, 316 species of amphibians belonging to three orders, 11 families and 45 genera are known from Temperate Eastern Asia (Table 7:1). The taxonomy and numbers of taxa follow Frost (1985), as updated by Duellman (1993) with the following exceptions: *Microhyla fowleri* Taylor 1934 is regarded as a junior synonym of *Microhyla berdmorei* (Blyth, 1856) by Taylor (1962) and some other authorities. Deletion of *Rana hubeiensis*, which is a synonym of *Rana plancyi* according to Mou and Zhao (1992). Deletion of *Buergeria pollicaris*, which is a synonym of *Chirixalus eiffingeri* according to Kuramoto and Wang (1987), and deletion of *Cynops shataukokensis*, which is a synonym of *Cynops pyrrhogaster* according to Risch and Romer (1980). Addition of *Scutiger liupangensis* (Huang, 1985) and *Scutiger* (as *Oreolalax*) *multipunctatus* (Wu et al., 1993). Dubois (1992) resurrected the names *Rana caldwelli* Schmidt 1925 and *Rana emeljanovi* Nikolski 1913 without justification; they are not recognized herein. I retain the generic name *Satobius* (Adler and Zhao, 1990) for *Hynobius retardatus* of Japan, although the species is placed in *Hynobius* by Japanese workers (e.g., Matsui et al., 1992). The taxonomy of Frost (1985) and Duellman (1993) is followed to preserve uniformity among chapters and regions; it does not indicate a change in position stated by Zhao and Adler (1993).

THE AMPHIBIAN FAUNA

The amphibian fauna of Temperate Eastern Asia consists of 316 recognized species placed in 45 genera and 11 families. The largest number of species (82.3%) are anurans (Table 7:1). The total surface area of Temperate Eastern Asia is 11,766,457 km^2; the total species density is 26.9 per 10^6 km^2, and that of anurans is 22.1 per 10^6 km^2. Whereas the number of species is not as great as in tropical regions of the world, about 14% of the species of salamanders and 6.5% of the species of anurans in the world occur in Temperate Eastern Asia. Of the 316 species of amphibians in the region, 72.5% are endemic; the other species are shared with the Palearctic Region or tropical southeastern Asia.

Three globally widespread genera occur in Temperate Eastern Asia. Of these, *Rana* (sensu stricto) has 224 species, of which 63 (28.1%) occur in temperate eastern Asia. Of the 212 species of *Bufo* known worldwide, 18 (8.5%) occur in temperate eastern Asia, whereas only eight (2.8%) of the 285 species of *Hyla* occur there. Among the anurans, the Ranidae with 10 genera and 111 species is the most diverse family; the largest genera are *Rana* (63 species), *Amolops* (17 species), and *Paa* (14 species). The next largest family is the Pelobatidae (5 genera, 52 species); the largest genera are *Scutiger* (26 species) and *Megophrys* (17 species). The Rhacophoridae (6 genera, 50 species), contains the only genus (*Buergeria* with 3 species) endemic to Temperate Eastern Asia; the largest genera in the region are *Rhacophorus* with 22 species and *Philautus* with 12 species. Of the 19 species in the Bufonidae, 18 are members of the genus *Bufo*, and the other is a *Pelophryne*. The Microhylidae is represented by five genera and 16 species, of which the largest is *Microhyla* with seven species. The other two families of anurans are represented by one genus each—Discoglossidae with four species of *Bombina* and Hylidae with eight species of *Hyla*.

Salamanders are represented by three families. The most diverse is the Hynobiidae with eight genera and 33 species, 21 of which are in the genus *Hynobius*. Five genera with 20 species are in the family Salamandridae; the most speciose genera of salamandrids are *Cynops* with seven species and *Paramesotriton* with five species. The Cryptobranchidae is represented by two species of *Andrias*, and the Gymnophiona is represented by a single species of *Ichthyophis*.

Of the 30 genera of anurans, the only endemic

Table 7:1. Taxonomic composition of the amphibian fauna of Temperate Eastern Asia. Numbers in each column are total, endemics, and percent endemic.

Order	Families			Genera			Species		
Anura	7	0	0.0	30	1	3.3	260	178	68.5
Caudata	3	0	0.0	14	6	42.9	55	50	90.9
Gymnophiona	1	0	0.0	1	0	0.0	1	1	100.0
Total	11	0	0.0	45	7	15.6	316	229	72.5

genus is *Buergeria* (3 species). However, specific endemism is high, especially in the Pelobatidae, Discoglossidae, and Ranidae with 78.8, 75.0, and 70.3%, respectively, of the species endemic. Among the salamanders, *Andrias* (2 species) and three monotypic hynobiid genera (*Liua, Pachyhynobius,* and *Satobius*) are endemic to Temperate Eastern Asia,

as are two genera of salamandrids—*Cynops* (7 species) and *Pachytriton* (2 species). Specific endemism is high in the Hynobiidae (90.9%) and Salamandridae (90.0%). The single species of Gymnophiona, but no genus of caecilians, is endemic to the region.

DEFINITION OF REGIONS

As defined here, Temperate Eastern Asia consists of the mainland of eastern Asia south of 50° N and east of 75° E north of the southern boundary of mainland China and the adjacent continental islands (Fig. 7:1). Temperate Eastern Asia is divided into 11 biogeographic regions (Fig. 7:2). Seven regions are recognized on the mainland, whereas the others are constituted by individual islands or archipelagos.

Five of the regions (Korean Peninsula–Northeast China, North China, Central China, South China, and Southwest China) on the mainland can be grouped into the eastern monsoon area.

MAINLAND

This huge area lying east of the Greater Xingan Ranges and the Qinghai-Xizang (Tibet) Plateau and south of the Mongolian Plateau includes the Korean Peninsula; interior Eurasia lies to the north and west. Monsoon climate results in high temperatures and heavy rainfall during the summers, but the winters are mostly cold and dry because of cold airflow from the north. Forests were the natural vegetation, but in most places these have been destroyed by human activities during the past 1000 yr. Now the footpaths between adjacent fields join together seemingly without end, and towns and cities occur almost everywhere. An important geographical boundary on a line along the Qingling Mountains and Huai He River essentially is the demarcation between the Palearctic and Oriental realms. To the north of this line four seasons are clearly distinguishable, whereas to the south vegetation remains green throughout the year. West of the Taihang Shan–Wu Shan–Xuefeng Shan mountains (about 115° E south-southwestward to

about 110° E) the average elevation is 1000–2000 m, and the landscape consists of a series of high mountains, plateaus, and basins, including Taihang Shan (mountain), Lüliang Shan, Liupan Shan, Qinling, Dabie Shan, Loess Earth Plateau, Yunnan–Guizhou Plateau, and the Sichuan Basin. East of the Taihang Shan–Wu Shan–Xuefeng Shan mountains and the Greater Xingan ranges the average elevation is less than 500 m; the landscape consists of a series of plains and surrounding hills, including the Northeast Plain, North China Plain, Middle and Lower Yangtze River plains, and the southwestern plains of the Korean Peninsula. Among the hills are the Changbai Shan, Shandong Hills, Southeastern Coastal Hills, and the northern and eastern hills of the Korean Peninsula.

This eastern monsoon area can be divided, from north to south and from east to west, into five regions.

Korean Peninsula–Northeast China Region

The region consists of northeastern China east of the Greater Xingan Ranges and north of Daling He (about 42° N) and the Korean Peninsula. A disjunct area in the west, the Altay Mountains in the northernmost tip of Xinjiang, also is included in this region because of its faunal similarities to the Greater Xingan Ranges. This example provides evidence of "... European-Far Eastern amphiboreal disjunctions in distribution ..." which "... arose in association with severe climatic changes during the Pleistocene" (Borkin, 1986:63).

North China Region

Lying to the south of the Korean Peninsula–

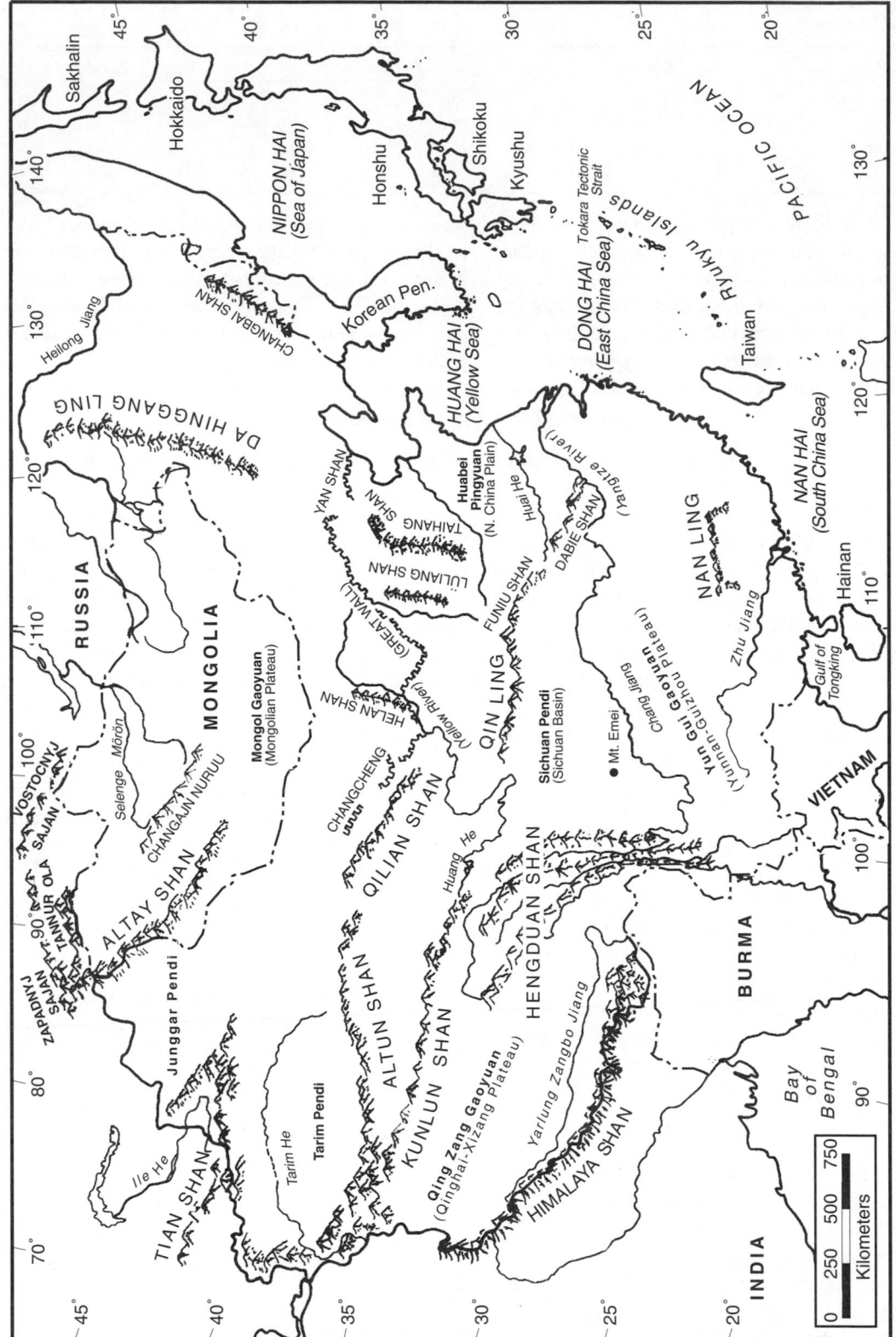

Fig. 7:1. Temperate Eastern Asia showing political and physiographic names used in text.

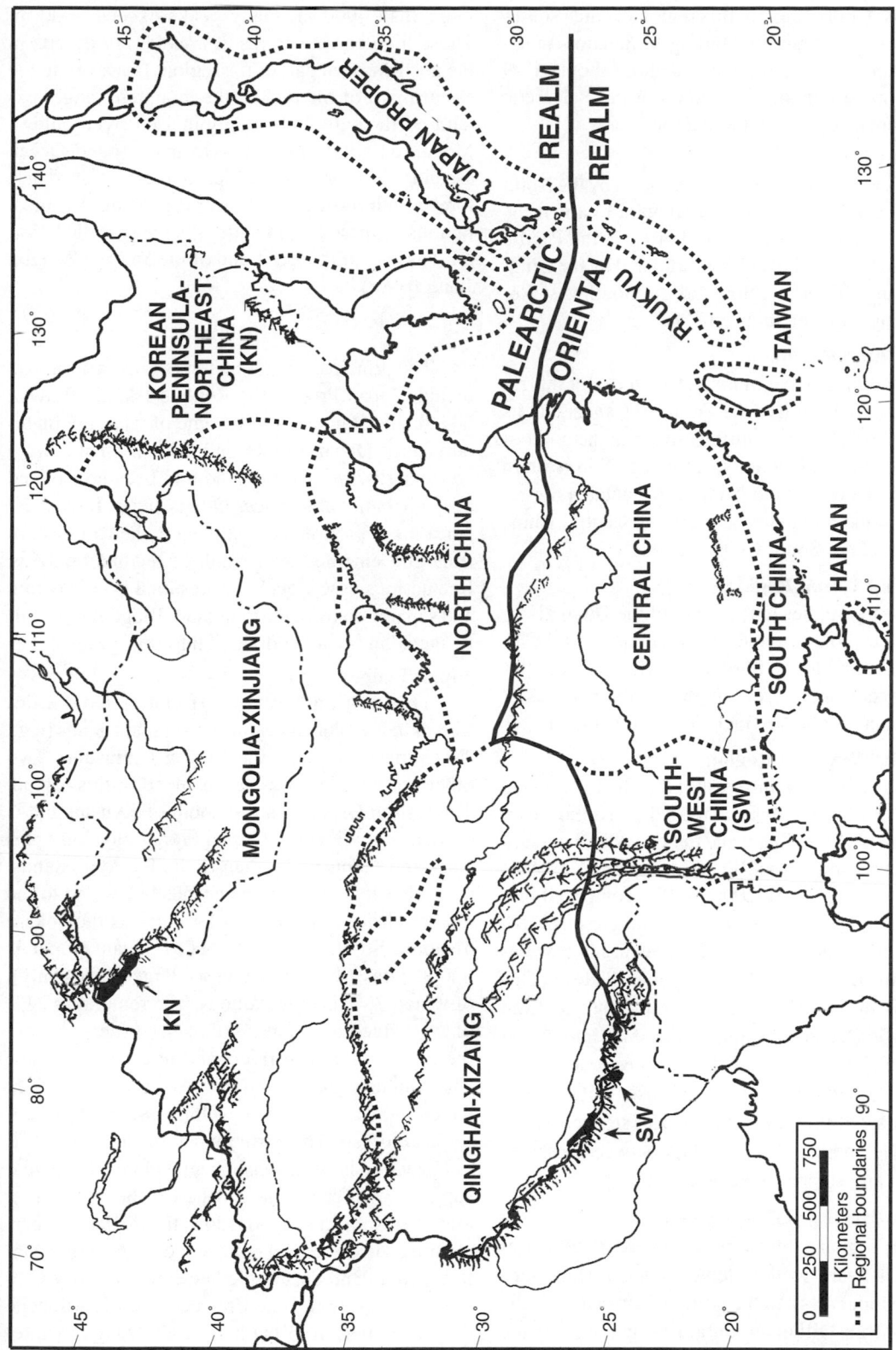

Fig. 7.2. Temperate Eastern Asia showing herpetofaunal regions.

Northeast China Region, this region extends southward to a line along the Qinling Mountains–Funiu Shan–Dabie Shan then eastward along about 31° N (approximate demarcation between the Palearctic and Oriental realms) to the Pacific Coast.

Central China Region

This region is bordered on the north by the North China Region and extends southward to about the Tropic of Cancer. The eastern border in the Pacific Coast; to the west the region is bordered by the Qinghai-Xizang (Tibet) Region and the Southwest Region along a line at about 112° E.

South China Region

Throughout most of its east-west extent this region lies to the south of the Tropic of Cancer, but its northern limit extends northward along the coast of Fujian to about 26° N and to about 25° N in western Yunnan. Generally, the Nanling Mountains are regarded as the boundary between the Central China Region and the South China Region.

Southwest China Region

This region lies to the south of the Qinghai-Xizang (Tibet) Region, to the west of the Central China Region, and to the north of the South China Region. Several small areas on the southern slopes of the Himalayas also are included in this region.

Mongolia-Xinjiang Region

This area lies to the west of the Greater Xingan Ranges, north of a line composed of Kulun Shan, Altun Shan, Qilian Shan, and the Great Wall, including the People's Republic of Mongolia. The average elevation is 1000–2000 m. It is composed of a series of high mountains, plateaus, and basins, such as Mt. Changajn, Mt. Altay, Tian Shan, Junggar Basin, Tarim Basin, and the Mongolian Plateau. This region is in the interior of Eurasia; moist air from the Pacific Ocean does not penetrate far into the region, which is mostly dry and includes vast parts of the Gobi Desert. Except for some oases that are cultivated, most of this region consists of steppe. Within China, Helan Shan is a boundary between desert to the west and steppe to the east.

Qinghai-Xizang (Tibet) Region

This region extends from the western border of China eastward to the Hengduanshan Mountains (Trans-Himalayas); to the north it is bordered by the Kulun-Altun-Qilian mountain ranges and to the south by the Himalayas. The average elevation is more than 4000 m; many peaks exceed 7000 m. These high elevations result in cold, dry deserts in the northwestern part of the region. However, in the eastern part of the region, the mountain ranges extend north-south with deeply incised river valleys. Moisture from the monsoons from the Indian Ocean ascends these valleys, which harbor moist, dense forests with vertical zonation of vegetation. The most famous of these humid valleys is the so-called "Vapor Passage of the Big Bend of the Yarlung Zangbo Jiang (river)."

ISLANDS

The principal islands of Temperate Eastern Asia include Japan Proper, the Ryukyu Islands, Taiwan Island, and Hainan Island. Some of these continental islands (Japan Proper and Hainan Island) were connected with the mainland and became isolated through fault subsidence. Others (e.g., Taiwan Island) are a part of the continental shelf of eastern Asia and emerged as a result of tectonic activity; fluctuations in sea level have resulted in connection and separation from the mainland. Biogeographically, the islands can be divided into four groups.

Japan Proper

This region includes Honshu, Hokkaido, Shikoku, Kyushu, and adjacent smaller islands (e.g., Tsushima Island near the Korean Peninsula). Two other groups of islands are included in this region. The Osumi Group lies just south of Kyushu and is more closely related to Japan Proper physiographically and faunistically than to the Ryukyu Islands. The other includes several islands belonging to the Tokara Group extending as far north as the Tokara Tectonic Strait. The total area of Japan Proper is 360,927 km^2 (the sum of the four large islands only), of which 76% is mountainous. The remaining 24% of the surface area consists of plains dispersed along the lower reaches of rivers and in coastal regions. The climate is highly variable. Areas south of 35° N have a monsoon subtropical climate supporting subtropical forests. The northern part of Honshu and all of Hokkaido have a monsoon temperate climate that supports coniferous forest, whereas the intervening area, also subject to a monsoon climate, has temperate broad-leaf forest. These islands were connected to the mainland during the late Jurassic–early Cretaceous. During the late Cretaceous–early Miocene they separated from the mainland through disintegration of the continental border.

Ryukyu Region

This region includes the Ryukyu Islands of Japan (excluding the Osumi Group and several islands of the Tokara Group) having an area of 2196 km². The northern boundary of the region is defined by the Tokara Tectonic Strait, which was recognized as "… a border zone between Palearctic and Oriental Regions" by Hikida et al. (1992). The Ryukyu Islands extend in a convex arc from a point south of Kyushu to a point northeast of Taiwan. The approximately 95 islands and islets that make up the chain have been divided into five groups by Tokunaga (1901) and subsequent biogeographers; these are (from north to south) the Tokara, Amami, Okinawa, Miyako, and Yaeyama groups. Some of the islands have little relief, and others have low hills and ridges; the greatest elevation is 690 m on Amami. The monsoon climate supports subtropical and tropical evergreen forests. (See Inger, 1947, for summary of vegetation and climate.) The Ryukyu Cordillera arose in the late Permian or early Mesozoic. The history of the islands between the Permian and the upper Eocene remains uncertain, but since the Eocene there have been three times when the Ryukyu Islands might have had direct land connections with Asia through Taiwan (Inger, 1947).

Taiwan Region

This region includes Taiwan Island (35,961 km²) and adjacent small islands (e.g., Penghu Liedao [the Pescadores]) situated southeast of mainland China and separated from it by the Taiwan Strait, 130 km in width at its narrowest point and having an average depth of about 80 m. Parallel ranges of mountains in a NNE-SSW direction on the eastern side of the island have 62 peaks higher than 3000 m; the highest peak, Yufeng of the Yu Shan, has an elevation of 3997 m and is the highest peak in eastern Asia. On the western side of the island the mountains are reduced gradually to hills, terraces, and coastal plains. Throughout the elevated parts of the island vertical zonation of vegetation is evident. Taiwan is dominated by a marine climate with strong monsoons during the summer and fall. The resulting tropical and subtropical characteristics include high temperatures and heavy rainfall; average annual temperatures in the lowlands exceed 20° C, and the annual precipitation in most places exceeds 2000 mm. Taiwan emerged in the Tertiary and has been connected with mainland Asia at various times.

Hainan Region

Hainan Island (33,900 km²) situated south of mainland China in the northwestern corner of the South China Sea is separated from the Leizhou Peninsula on the mainland by the Qiongzhou Strait only 15 km wide with an average depth of 60 m. The center of the island is mountainous; the highest peaks are Mt. Wuzhi (1867 m) and Mt. Yinggeling (1812 m). Peripherally are low mountains, hills, basins, and plains. The climate of the island is influenced by tropical monsoons. Average annual temperature is more than 23° C; annual precipitation is 1000–2600 mm with the eastern side of the island being more moist than the western side. Hainan was part of the Asian mainland; its separation by the subsidence forming the Qiongzhou Strait occurred in the Tertiary.

PATTERNS OF DISTRIBUTION

The largest number of species in a given region is 111 in Southwest China, followed by 106 in Central China, and 91 in South China; likewise, the number of endemic species is correspondingly high, but by far the highest percentage of endemic species is found in Japan Proper and the Ryukyu Islands (Table 7:2). An analysis of relationships of the 11 regions based on shared species of amphibians reveals that the highest coefficients of biogeographic resemblance (> 0.4) are between the Korean Peninsula–Northeast China Region and the Mongolia-Xinjiang Region, between the North China Region and the Mongolia-Xinjiang and Qinghai-Xizang regions, and between the China Region and the Central China and Hainan regions (Table 7:3). Of the 55 comparisons, coefficients are < 0.1 in 31 cases. This analysis shows that the regions are reasonably discrete faunistically.

The broadest distributions are exhibited by species not endemic to Temperate Eastern Asia; three such species occur in eight of the 11 regions. *Bufo gargarizans* is absent only in the South China, Japan Proper, and Hainan regions; *Limnonectes limnocharis* is absent only in the Korean Peninsula–Northeast China, Mongolia-Xinjiang, and Qinghai-Xizang regions, and *Rana nigromaculata* is absent only in three island regions—Ryukyu, Taiwan, and Hainan.

Six species shared by the Korean Peninsula–Northeast China, North China, and Mongolia-Xin-

Table 7:2. Species of amphibians in 11 regions in Temperate Eastern Asia. Numbers in each column are anurans + salamanders + caecilians = total.

Region	Total species				Shared species				Endemic species				Percent endemic			
Korean Penin.–NE China	14 +	4 +	0 =	18	10 +	2 +	0 =	12	4 +	2 +	0 =	6	28.6 +	50.0 +	00.0 =	33.3
North China	12 +	2 +	0 =	14	11 +	2 +	0 =	13	1 +	0 +	0 =	1	8.3 +	00.0 +	00.0 =	7.1
Central China	91 +	15 +	0 =	106	56 +	4 +	0 =	60	35 +	11 +	0 =	46	38.5 +	73.3 +	00.0 =	43.4
South China	83 +	7 +	1 =	91	70 +	4 +	0 =	74	13 +	3 +	1 =	17	15.7 +	42.9 +	100.0 =	18.7
Southwest China	100 +	11 +	0 =	111	55 +	4 +	0 =	59	45 +	7 +	0 =	52	45.0 +	63.6 +	00.0 =	46.8
Mongolia-Xinjiang	11 +	2 +	0 =	13	10 +	2 +	0 =	12	1 +	0 +	0 =	1	9.1 +	00.0 +	00.0 =	7.7
Qinghai-Xizang	13 +	3 +	0 =	16	11 +	3 +	0 =	14	2 +	0 +	0 =	2	15.4 +	0.00 +	00.0 =	12.5
Japan Proper	17 +	20 +	0 =	37	6 +	1 +	0 =	7	11 +	19 +	0 =	30	64.7 +	95.0 +	00.0 =	81.1
Ryukyu Islands	17 +	2 +	0 =	19	5 +	1 +	0 =	6	12 +	1 +	0 =	13	70.6 +	50.0 +	00.0 =	68.4
Taiwan	29 +	3 +	0 =	32	21 +	1 +	0 =	22	8 +	2 +	0 =	10	27.6 +	66.7 +	00.0 =	31.3
Hainan	37 +	1 +	0 =	38	30 +	1 +	0 =	31	7 +	0 +	0 =	7	18.9 +	0.00 +	0.00 =	18.4

Table 7:3. Distribution of species of amphibians in 11 regions in Temperate Eastern Asia. Abbreviations in headings to columns correspond to regions in first column. The number of species in each region is shown in boldface in common cell; the numbers of species that are in common to two regions are in the upper right, and the coefficient of biogeographic resemblance is in italics in the lower left.

Region	KN	NC	CC	SC	SW	MX	QX	JP	RK	TW	HN
Korean Peninsula–NE China (KN)	**18**	7	4	1	3	7	4	6	1	1	0
North China (NC)	*0.44*	**14**	8	2	6	7	6	3	2	2	1
Central China (CC)	*0.07*	*0.13*	**106**	41	32	3	3	4	3	16	19
South China (SC)	*0.02*	*0.04*	*0.42*	**91**	26	1	2	2	2	14	26
Southwest China (SW)	*0.05*	*0.10*	*0.30*	*0.26*	**111**	3	11	2	3	8	11
Mongolia-Xinjiang (MX)	*0.45*	*0.52*	*0.05*	*0.02*	*0.05*	**13**	3	4	1	1	0
Qinghai-Xizang (QX)	*0.24*	*0.40*	*0.05*	*0.04*	*0.17*	*0.21*	**16**	1	1	1	0
Japan Proper (JP)	*0.22*	*0.12*	*0.06*	*0.03*	*0.03*	*0.16*	*0.04*	**37**	1	1	1
Ryukyu Islands (RK)	*0.05*	*0.12*	*0.05*	*0.04*	*0.05*	*0.06*	*0.06*	*0.05*	**19**	6	2
Taiwan (TW)	*0.04*	*0.09*	*0.23*	*0.23*	*0.12*	*0.04*	*0.04*	*0.03*	*0.24*	**32**	9
Hainan (HN)	*0.00*	*0.04*	*0.26*	*0.41*	*0.15*	*0.00*	*0.00*	*0.03*	*0.07*	*0.26*	**38**

Table 7:4. Proportions of different distribution patterns in various regions. Abbreviations of regions: CC = Central China, HN = Hainan Island, JP = Japan Proper, KN = Korean Peninsula and northeastern China, MX = Mongolia-Xinjiang, NC = North China, QX = Qinghai-Xizang, RK Ryukyu islands, SC = South China, SW = Southwest China, TW = Taiwan Island.

Pattern	Total species	Regions										
		KN	NC	CC	SC	SW	MX	QX	JP	RK	TW	HN
Eastern Asian islands Palearctic	30	—	—	—	—	—	—	—	81.1	—	—	—
Pacific Palearctic	13	55.6	64.3	6.6	1.1	3.6	38.5	31.3	13.5	5.3	6.3	—
Korean Peninsula–NE China	5	27.8	—	—	—	—	—	—	—	—	—	—
Siberian Palearctic	3	16.7	7.1	—	—	—	15.4	—	2.7	—	—	—
North China	1	—	7.1	—	—	—	—	—	—	—	—	—
Central Asia	4	—	—	—	—	—	30.8	—	—	—	—	—
Mediterranean Palearctic	2	—	—	—	—	—	15.4	—	—	—	—	—
Himalayas-Tibet	3	—	—	—	—	0.9	18.8	—	—	—	—	—
Hengduanshan Palearctic	24	—	7.1	1.9	1.1	20.7	—	50.0	—	—	—	—
Himalayas-Hengduanshan	2	—	—	1.9	2.2	1.8	—	—	—	—	—	—
Himalayas	20	—	—	—	—	18.0	—	—	—	—	—	—
Hengduanshan Oriental	30	—	7.1	1.9	1.1	28.8	—	—	—	—	—	—
Southern China	87	—	—	62.3	39.3	9.0	—	—	—	—	18.8	13.2
Indochina	43	—	—	18.9	41.6	12.6	—	—	—	—	15.6	42.1
Indo-Malayan	10	—	—	3.8	9.0	0.9	—	—	—	—	9.4	15.8
Pan-Oriental	4	—	7.1	2.8	4.5	3.6	—	—	2.7	10.5	9.4	10.5
Eastern Asian islands Oriental	33	—	—	—	—	—	—	—	—	84.2	40.6	18.4
Total species per region	—	37	14	106	89	111	13	16	37	19	32	38

jiang regions are Palearctic taxa—*Bufo gargarizans, B. raddei, Hyla japonica, Rana chensinensis, R. nigromaculata,* and *Salamandrella keyserlingii.* Only one species is endemic in each of two of the regions—*Scutiger liupanensis* in North China, and *Rana tenggerensis* in Mongolia-Xinjiang. Because the Korean Peninsula–Northeast China Region has more endemics (6—*Bufo stejnegeri, Rana altaica, R. chosenica, R. huanrenensis, Hynobius leechii,* and *H. mantschuriensis*), the biotic relationship values between it and the North China and Mongolia-Xinjiang regions are lower than they are between those two regions. The amphibian fauna of the Qinghai-Xizang Region contains only two endemics (*Scutiger maculatus* and *S. nyingchiensis*) and otherwise shows a mixture of species shared with North China (e.g., *Bufo raddei* and *Rana chensinensis*) and Southwest China (e.g., *Bufo tibetanus, Scutiger boulengeri, Nanorana pleskei, Batrachuperus karlschmidti,* and *B. tibetanus*). In addition to the widespread species listed above, the North China, Southwest China, and Qinghai-Xizang regions share *Bufo minshanicus* and *Andrias davidianus.*

The only other high (> 0.4) coefficients of biogeographic resemblance is between the South China and Hainan regions, which share 26 species. Four

of these (*Rana adenopleura, R. tiannanensis, R. versabilis,* and *Polypedates megacephalus*) are endemic to Temperate Eastern Asia, whereas the rest of the species (e.g., *Bufo melanostictus, Microhyla butleri, Limnonectes cancrivora, Rana livida, Chirixalus doriae,* and *Echinotriton asperrimus*) have extensive distributions to the southwest in Indochina and Indo-Malaya.

Based on present knowledge of distribution of the 316 species of amphibians known from Temperate Eastern Asia, 17 patterns of distribution can be identified (Table 7:4). Nine of these patterns are associated with the Palearctic Realm, and eight are associated with the Oriental Realm.

PALEARCTIC REGION

Eastern Asian Islands

Of the 30 species (9.5% of the total species in Temperate Eastern Asia) that occur only on the islands of Japan Proper, 19 are salamanders (17 hynobiids) and 11 are anurans. Nine species of anurans (e.g., *Bufo torrenticola, Rana porosa, Buergeria buergeri,* and *Rhacophorus schlegelii*) and six species of salamanders (e.g., *Andrias japonicus, Hynobius nebulosus, Onychodactylus japonicus,* and *Cynops pyrrhogaster*) are widely distributed in Japan

Proper and occur on all three main islands—Honshu, Shikoku, and Kyushu. The other 15 species occur on only one island in the archipelago. These insular endemics include only two anurans—*Rana pirica* on Hokkaido Island and *Rana tsushimensis* on Tsushima Island. In contrast to the anurans, 13 of the 19 species of salamanders are endemic to particular islands. Eight species of *Hynobius* (e.g., *H. abei, lichenatus, nigrescens,* and *tenuis*) are endemic to Honshu Island, whereas other islands in the archipelago have fewer endemic salamanders—Hokkaido Island (*Satobius retardatus*), Kyushu Island (*Hynobius dunni* and *H. stejnegeri*), Oki Island (*Hynobius okiensis*), and Tsushima Island (*Hynobius tsuensis*).

Pacific Palearctic

Of 13 species (4.1% of the total species in Temperate Eastern Asia) having this distributional pattern, seven (*Bufo raddei, Kaloula borealis, Rana chensinensis, R. japonica, R. plancyi, R. rugosa,* and *Andrias davidianus*) are endemic to Temperate Eastern Asia. The other species (*Bufo gargarizans, Bombina orientalis, Hyla japonica, Rana dybowskii, R. nigromaculata,* and *Onychodactylus fischeri*) also are distributed to the north in eastern Russia. With the exception of *Rana japonica, R. plancyi,* and *Andrias davidianus,* all of the species occur in the Korean Peninsula–North China Region. Two species are extremely widespread. For example, *Bufo gargarizans* occurs in all regions except South China, Hainan, and Japan Proper, and *Rana nigromaculata* occurs throughout the mainland and Japan Proper. Some species (e.g., *Bufo raddei, Rana chensinensis,* and *Andrias davidianus*) extend westward into the Qinghai-Xizang Region. Other species (e.g., *Bombina orientalis, Rana dybowskii, R. rugosa,* and *Onychodactylus fischeri*) have more restricted distributions, principally in the Korean Peninsula–Northeast China Region; all of these, except the latter, also occur in Japan Proper.

Korean Peninsula–Northeast China

All five species (1.6% of the total) are endemic to Temperate Eastern Asia. One species (*Bufo stejnegeri*) ranges throughout the Korean Peninsula and northeast China; three species (*Hynobius leechii, H. mantschuriensis,* and *Rana huanrenensis*) occur only in northeast China, and one (*Rana chosenica*) occurs only in the Korean Peninsula.

Siberian Palearctic

None of the three species (0.9% of the total) is endemic to Temperate Eastern Asia. All of these species are distributed in Asia north of 40° N. The distribution of *Salamandrella keyserlingii* includes Siberia, the Korean Peninsula, northeastern China, northern China, the Mongolia-Xinjiang Region, and Japan Proper. *Rana amurensis* is distributed from western Siberia to Sakhalin Island and occurs in Temperate Eastern Asia only in the Korean Peninsula–Northeast China and Mongolia-Xinjiang regions, whereas *Rana altaica* enters Temperate Eastern Asia only in northern Xinjiang.

North China

One species (*Scutiger liupanensis*) is known only from Liupan Shan, Ningxia Hui Autonomous Region, in north-central China.

Central Asia

Of the four species (1.3% of the total), one (*Rana tenggerensis*) is endemic to Temperate Eastern Asia where it is restricted to the southern border of the Tengger Desert in Inner Mongolia. The other three species (*Bufo danatensis, Rana asiatica,* and *Ranodon sibiricus*) range from central Asia eastward to western Xinjiang.

Mediterranean Palearctic

Two species (*Bufo viridis* and *Rana ridibunda*) range from Europe eastward to western Xinjiang.

Himalaya-Tibet

All three species (0.9% of the total) occur on the Qinghai-Xizang Plateau in Tibet. *Scutiger nyingchiensis* is endemic to that plateau, whereas *Scutiger boulengeri* and *Nanorana parkeri* also occur at high elevations in the Himalayas.

Hengduanshan Palearctic

All 24 species (7.6% of the total) are endemic to Temperate Eastern Asia, where most species occur in the northern part of the Hengduanshan Mountains of the Southwest China Region; one species (*Nanorana ventripunctata*) is restricted to the middle part of the mountain range. This group is dominated by 10 species of *Scutiger,* four of *Batrachuperus,* and three each of *Bufo* and *Amolops.* The ranges of seven species (e.g., *Bufo tibetanus, Scutiger maculatus, Nanorana pleskei,* and *Batrachuperus karlschmidti*) extend into at least the southern part of the Qinghai-Xizang Region. *Rana chaochiaoensis* extends into western Central China and South China regions.

ORIENTAL

Himalayan-Hengduanshan

Two species (*Megophrys minor* and *M. omeimontis*) endemic to Temperate Eastern Asia occur in the lower parts of the Himalayas and extend eastward to the southern part of the Hengduanshan Mountains and their foothills.

Himalayan

Of the 20 species (6.3% of the total), all anurans, nine are endemic to Temperate Eastern Asia, where all species of this group occur on the lower part of the southern slopes of the Himalayas in the Southwest China Region. Endemics include *Megophrys kempii, Ingerana reticulatus, Paa liebigii, Philautus medogensis,* and *Rhacophorus translineatus.* Those species that are not endemic (e.g., *Bufo himalayanus, Scutiger sikimmensis, Amolops monticola, Paa blanfordi, Rana gerbillus,* and *Theloderma moloch*) have distributions westward along the front of the Himalayas.

Hengduanshan Oriental

All 30 species (9.5% of the total) are endemic to Temperate Eastern Asia, where most species are endemic to the southern part of the Hengduanshan Mountains and their foothills in the Southwest China Region. Examples are *Bufo ailaoanus, Bombina maxima, Calluella yunnanensis, Leptobrachium ailaonica, Megophrys shapingensis, Scutiger popei, Amolops lifanensis, Paa maculosa, Rana grahami, Rhacophorus gongshanensis, Cynops cyanurus,* and *Tylototriton taliangensis.* The ranges of three species in this category extend beyond the Southwest China Region. *Bufo andrewsi* ranges into Central China; *Amolops mantzorum* ranges into South China, and *Paa boulengeri* extends into Central China and North China.

South China

This is the largest distributional unit with 87 species (27.5% of the total); all are endemic to Temperate Eastern Asia, where most species are restricted to the Central China and South China regions. The one caecilian (*Ichthyophis bannanicus*) and 15 salamanders (10 salamandrids) are included in this category. Forty-six species (e.g., *Bombina fortinuptialis, Hyla zhaopingensis, Kaloula rugifera, Lep-* *tobrachium boringii, Megophrys nankiangensis, Scutiger lichuanensis, Amolops granulosus, Paa jiulongensis, Philautus albopunctatus, Rhacophorus hungfuensis, Hynobius chinensis, Cynops orientalis,* and *Paramesotriton chinensis*) are restricted to the Central China Region, and 20 species (e.g., *Kalophrynus menglienicus, Megophrys brachykolos, Ophryophryne pachyproctus, Scutiger rhodostigmatus, Amolops hongkongensis, Ingerana liui, Rana fukienensis, Philautus romeri,* and *Paramesotriton guangxiensis*) are restricted to the South China Region. Ten species (e.g., *Amolops chunganensis, Rana margaretae, Polypedates dugritei,* and *Rhacophorus nigropunctatus*) range into the Southwest China Region. The distributions of nine species include the islands of Hainan and/or Taiwan—Hainan (*Rana tiannanensis, R. versabilis,* and *Rhacophorus rhodopus*), Taiwan (*Rana fukienensis, R. latouchii, R. longicrus,* and *R. swinhoana*), Hainan and Taiwan (*Rana adenopleura* and *Polypedates megacephalus*).

Indo-Chinese

None of the 43 species (13.6% of the total) is endemic to Temperate Eastern Asia; the 41 species of anurans and two of salamanders are distributed in Indochina and extend northward to southern China. All but four species (*Amolops nasicus, Rana sauteri, Rhacophorus bipunctatus,* and *Theloderma asperum*) occur in the South China Region. These four species occur in the Southwest China Region, together with 13 other species (e.g., *Leptolalax pelodytoides, Megophrys carinensis, Chirixalus vittatus, Rhacophorus bipunctatus,* and *Tylototriton verrucosus*) that also occur in the South China Region. The ranges of 19 of the species (e.g., *Hyla chinensis, Megophrys boettgeri, Paa spinosa,* and *Polypedates dennysi*) having this pattern extend into the Central China Region. Hainan is included in the ranges of 15 of the species (e.g., *Microhyla pulchra, Phrynoglossus martensii, Rana andersonii, Chirixalus doriae,* and *Echinotriton asperrimus*), whereas three species (*Hoplobatrachus rugulosus, Rana guentheri,* and *R. taipehensis*) reach Taiwan.

Indo-Malayan

None of the 10 species (3.2% of the total) is endemic to Temperate Eastern Asia. These species are widely distributed in Indonesia, the Philippines, the Malay Peninsula, and Indo-China; their ranges extend northward into the South China Region, and one species (*Microhyla heymonsi*) ranges into the foothills of the Himalayas in the Southwest China Region, as well as onto Hainan and Taiwan. Only two of these species (*Rhacophorus reinwardtii* and *Theloderma leporosa*) are restricted to the mainland in China. *Kalophrynus pleurostigma, Kaloula pulchra, Leptobrachium hasseltii, Limnonectes cancrivora,* and *Occidozyga lima*) occur on Hainan, and *Micryletta inornata* and *Limnonectes kuhlii* occur on Taiwan.

Pan-Oriental

Four species, all anurans (1.3% of the total) have broad distributions in southern Asia; none is endemic to Temperate Eastern Asia. All four species (*Bufo melanostictus, Microhyla ornata, Limnonectes limnocharis,* and *Rhacophorus cavirostris*) occur in the South China and Southwest China regions and on Hainan. All except *Rhacophorus cavirostris* also occur in the Central China Region and on Taiwan. The range of *Microhyla ornata* also includes the Ryukyu Islands, as does that of *Limnonectes limnocharis,* which also occurs in the North China and Japan Proper regions.

Eastern Asian Island Oriental

All 33 species (10.4% of the total) are endemic to the islands off the coast of Temperate Eastern Asia. Seven species (e.g., *Pelophryne scalpta, Amolops torrentis, Philautus ocellatus,* and *Rhacophorus oxycephala*) are endemic to Hainan; 10 species (e.g., *Bufo bankorensis, Micryletta stejnegeri, Buergeria robusta,* and *Hynobius sonani*) are endemic to Taiwan, whereas 13 species (e.g., *Hyla hallowellii, Limnonectes namiyei, Rana holsti,* and *Cynops ensicauda*) are known only from the Ryukyu Islands. Three species (*Buergeria japonica, Chirixalus eiffingeri,* and *Echinotriton andersoni*) are shared by Taiwan and the Ryukyu Islands.

AREAS OF FAUNAL TRANSITION

The Hengduanshan Mountain region in southwestern China is a transition zone between Palearctic elements from the north and Oriental elements from the south. Also, it is a transition zone between the warm, temperate lowlands to the east and the high, cold plateaus to the west. This region was a sanctuary during glacial periods of the Quaternary; many primitive species remain there. Apparently because of the prominent vertical zonation of climate and vegetation, the region has been a center of speciation of some groups (e.g., *Batrachuperus,*

Amolops, and especially pelobatids). Another transition zone exists in the eastern part of the mainland where there are no significant geographic barriers. The distributions of several Oriental and Palearctic species overlap there.

DISJUNCT DISTRIBUTIONS

There are a few notable European–Far Eastern amphiboreal disjunct distributions pointed out by Borkin (1986). These include species of *Bombina* and the *Hyla arborea* complex. The viperid snake, *Vipera berus,* is an example of a species found in northern Xinjiang in the west and in northeastern China (Changbai Shan, Jilin Province) in the east but unknown in the vast area in between.

A few species (e.g., *Hyla japonica, Rana japonica, R. plancyi*) have discontinuous distributions involving the Huang (Yellow River)–Huai (river)–Hai (river) Plain or the North China Plain in the North China Region. The plain descends from Mt. Taihang and Mt. Funiu in the west to the coast of Bo Hai (sea) and Huang Hai (Yellow Sea); it extends from Mt. Yan in the north to the Huai River in the south. It is a vast low, open area with extensive agricultural development and dense human population. It also is an area in which the Huang He (Yellow River) has changed its course many times. Perhaps populations of animals within the intervening area were eliminated so often during periods of flooding that they have been unable to recover. The species mentioned above occur both to the north and to the south of, but not in, this area.

The Yellow River Plain also serves to create disjunct distributions at the generic level in *Bombina, Hyla, Cynops,* and *Hynobius.* Most species of *Hynobius* occur in Japan Proper, and two species (*H. leechii* and *H. mantschuriensis*) are in the Korean Peninsula–Northeast China Region. Two species (*H. chinensis* and *H. amjiensis*) are in the eastern part of the Central China Region, and two species (*H. formosanus* and *H. sonani*) are on Taiwan, but the genus is absent in the North China Region.

Another disjunction is associated with the Tokara Strait between Japan Proper and the Ryukyu Islands and provides the only discontinuous distribution at the familial level in Temperate Eastern Asia. Of the five genera and 50 species of rhacophorids, all of the genera and 47 (94%) of the species have ranges throughout the Southwest China, Central China, and South China regions, and on three Oriental Islands, but the remaining three species (*Buergeria buergeri, Rhacophorus arboreus,* and *R. schlegelii*) occur in Japan Proper. Thus a major discontinuity in distribution of rhacophorids exists between Japan Proper and the rest of Temperate Eastern Asia.

AREAS OF HIGH SPECIES DIVERSITY AND ENDEMISM

Herein the term "species diversity" is equivalent to beta diversity or solely the number of species in a given region (Table 7:2).

REGIONAL DIVERSITY AND ENDEMISM

On the mainland, endemism is low in the northern regions. Only one species is endemic in each of two of the regions—*Scutiger liupanensis* in North China, and *Rana tenggerensis* in Mongolia-Xinjiang. The amphibian fauna of the Qinghai-Xizang Region contains only two endemics (*Scutiger maculatus* and *S. nyingchiensis*). The Korean Peninsula–Northeast China Region has six endemics (*Bufo stejnegeri, Rana altaica, R. chosenica, R. huanrenensis, Hynobius leechii,* and *H. mantschuriensis*).

The vast Central China Region with favorable climate and diverse topography has 35 endemic anurans and 11 salamanders, including the monotypic genera *Liua* and *Pachyhynobius.* High specific endemism is found in the Hynobiidae (83.3%), Salamandridae (75.0%), and Pelobatidae (70.6%); however, the number of species of pelobatids (17) is much smaller than in the Southwest Region (28).

The highest degree of specific endemism (46.8%) is found in the Southwest Region, where parallel ranges of the southern and middle sections of the Hengduanshan Mountains are aligned in a north-south direction with high mountains alternating with deep valleys that provide a wide range of environments. This is a region of speciation of many groups of amphibians, especially anurans. For example, 21 of the 28 species of pelobatids (including 15 of 18 species of *Scutiger*) are endemic to the region, as are 17 of the 37 species of ranids. With the exception of four species of *Batrachuperus* (2 endemic), hynobiids are not represented, but two of the three species of *Tylototriton* are endemic.

Although 89 species occur in the South China Region, only 16 species are endemic; 26 species are shared with Hainan and 19 with the Central China Region. Endemics include 1 microhylid, 3 pelobatids, 6 ranids, 3 rhacophorids, 3 salamandrids, and the caecilian *Ichthyophis bannanicus.*

Among the island regions, Japan Proper with its large area and diverse topography has the highest degree of endemism; 30 of the 37 species (81.1%) are endemic. Seventeen of the 18 species of hynobiid salamanders are endemic to Japan Proper. The Ryukyu Islands have the next highest degree of insular endemism (68.4%). However, these islands differ from others by having a high percentage (88.9%) of endemic ranids (8 of 9 species). In addition, the only hylid, two of four species of rhacophorids, one of two microhylids, and one of two salamandrids are endemic.

Among island regions, Taiwan has low endemism; 10 of 32 species (31.3%) are endemic. Species restricted to Taiwan include the two species of *Hynobius,* five of eight rhacophorids, one of three bufonids, and one of four microhylids. In contrast to the Ryukyu Islands, only one of 13 ranids is endemic. Hainan has the least degree of endemism (18.4%) among island regions; the seven endemic anurans include *Pelophryne scalpta,* four species of ranids, and two of rhacophorids.

LOCAL DIVERSITY AND ENDEMISM

There are some specific areas of high endemism and/or diversity in Temperate Eastern Asia, among which Mt. Emei (previously Mt. Omei) is typical. Mt. Emei is a fault block mountain having countless sheer precipices and steep cliffs, an area of about 200 km^2, and highest elevation of 3099 m. It stands erect between the Sichuan Basin and the eastern border of the Qinghai-Xizang (Tibet) Plateau, where the climate is temperate and moist with dense forests and vertical zonation of vegetation. High diversity and endemism of amphibians, both Southern China and Hengduanshan Oriental, as well as other elements, have been found there. To date, 32 species (24 or 75% endemic to Temperate Eastern Asia) are known from Mt. Emei; the species density is as high as 16 per 100 km^2. Mt. Emei is the type locality of 13 of the 32 species; many of these are known only from the mountain. Among the species known from Mt. Emei are 30 anurans and two (both endemic) salamanders (*Batrachuperus pinchonii* and *Andrias davidianus*). The 30 species of anurans belong to six of the seven families known from Temperate Eastern Asia (no representatives of the Discoglossidae). The anuran fauna includes two (one endemic) bufonids (*Bufo andrewsi* and *B. gargarizans*), one endemic hylid (*Hyla annectans*), and two (one endemic) microhylids (*Kaloula rugifera* and *Microhyla ornata*). Of the 11 pelobatids, 10 are endemic (e.g., *Leptobrachium boringii, Megophrys omeimontis, Oreolalax multipunctatus, O. omeimontis, O. schmidti, Scutiger chintingensis*), and the other (*Leptolalax pelodytoides oshanensis*) is an endemic subspecies. The fauna is completed with 10 (six endemic) ranids (e.g., *Amolops chunganensis, Limnonectes limnocharis, Paa boulengeri, Rana chevronta, R. daunchina, R. margaretae, R. nigromaculata*) and four (three endemic) rhacophorids (e.g., *Rhacophorus chenfui, Polypedates omeimontis*).

PRESERVATION OF THE AMPHIBIAN FAUNA

There are 274 species of amphibians known from China; these include 238 anurans, 35 salamanders, and one caecilian. In the National Protective Wildlife Catalogue approved by the State Council on 10 December 1988, one anuran (*Rana tigerina* = *Hoplobatrachus rugulosus*) and six salamanders (*Andrias davidianus, Echinotriton asperrimus, E. chinhaiensis, Tylototriton kweichowensis, T. taliangensis,* and *T. verrucosus*) were listed as second-grade protected wildlife. In addition, each province has its own local protective wildlife catalogue. For example, all six salamanders and 73 anurans known from Sichuan Province are included in the provincial catalogue. To date, more than 700 nature reserves with a total of more than 50,600,000 hectares (about 5.27% of the country) have been established in China. No reserve has been established specifically to protect amphibians, and only a few reserves contain protected amphibians. For example, the giant salamander, *Andrias davidianus,* is protected, so far as I know, in the Fanjing Shan (Jiangkou, Yingjiang, and Songtao counties, Guizhou Province) and Dafengding (Ebian County, Sichuan Province) nature reserves. Some species of large pond frogs (*Rana*) regarded as a natural enemy of agricultural pests were widely protected in China, but still they were eaten by some people.

Much attention is given to protection of wildlife in Japan, where 53 species of amphibians are known. According to Kato and Ota (1993), two species (*Hynobius abei* and *H. takedai*) are endangered, four (*Hynobius hidamontanus, Rana holsti, R. ishikawae,* and *R. subaspera*) are vulnerable, and eight (*Andrias japonicus, Hynobius dunni, H. okiensis, H. stejnegeri, Salamandrella keyserlingii, Echinotriton andersoni, Limnonectes namiyei,* and *Rana porosa brevipoda*) and several local populations of other species are rare.

To date, six species of amphibians are known from the Mongolian People's Republic; these include one salamander (*Salamandrella keyserlingii*) and five anurans (*Bufo danatensis, B. raddei, Hyla japonica, Rana amurensis,* and *R. chensinensis*) (Borkin and Kuzmin, 1988). Three of these (*Salamandrella keyserlingii, Hyla japonica,* and *Rana chensinensis*) are in need of protection. Already in Mongolia there exist 13 nature reserves with an area of more than 53,633 km^2; this figure is about 3.4% of the total area of 1,566,500 km^2 in Mongolia and does not include three reserves for which the areas are unknown. Munkhbayar (cited in Munkhbayar and Semenov, 1988) proposed to establish three nature preserves especially for the protection of the three species of amphibians listed above.

DISCUSSION

Species Density

Temperate Eastern Asia crosses two zoogeographic realms. The boundary between the Palearctic and Oriental realms in this area was considered to be, from west to east, the Himalayas, Hengduanshan Mountains (at different latitudes depending on elevation), Qinling, Funiu Shan, Dabie Shan eastward at about 31° N to the coast. In the sea, the Tokara Tectonic Strait was considered to be the boundary.

Four regions on the mainland (Korean Peninsula–Northeast China, Mongolia-Xinjiang, Qinghai-Xizang [Tibet], and North China) and Japan Proper are in the Palearctic Realm and make up about 75% of the total area of Temperate Eastern Asia. The remaining 25% of the area is in the Oriental Realm made up of three regions on the mainland (Central China, South China, and Southwest China) and three insular regions (Ryukyu, Taiwan, and Hainan). Among the 316 recognized species of amphibians in Temperate Eastern Asia, 245 (77.5%) occur only in the Oriental Realm, and 53 species (16.8%) occur only in the Palearctic Realm; the remaining 18 species (5.7%) are shared by the two realms. Total species density in the Oriental Realm in Temperate Eastern Asia is 89.4 per 10^6 km², whereas that in the Palearctic part of Temperate Eastern Asia is only 8 per 10^6 km².

FAUNAL ELEMENTS

Southern China species (87; 27.5%) and Hengduanshan Palearctic plus Hengduanshan Oriental species (56; 17.7%) are the principal faunal elements on the mainland. These elements range mainly throughout the Central China and Southwest China regions and are the proper elements of Temperate Eastern Asia. The more peripheral parts of Temperate Eastern Asia commonly are invaded by various elements from neighboring geographic regions. For example, in the northwestern part of the Mongolia-Xinjiang Region, eight species (61.5% of the species in the region) are invaders from the Siberian Palearctic, Central Asia, and Mediterranean Palearctic. Likewise, in the southern part of the South China Region, 45 species (50.6% of the species in the region) are invaders from the Indo-China and Indo-Malayan regions.

INSULAR ENDEMISM

Islands commonly harbor endemic species, and the amphibian faunas on the islands in Temperate Eastern Asia are made up predominantly by endemic species. On Japan Proper there are 30 endemic species (81.1% of the total), and on the Ryukyu Islands there are 13 endemics (68.4%). The other island regions have proportionately more species shared with the mainland. Of the 32 species on Taiwan, 10 (31.3%) are endemic, 19 (59.4%) are shared with the mainland, and three (9.3%) are endemic to Taiwan and the Ryukyu Islands. Of the 38 species on Hainan, 7 (18.4%) are endemic and 29 (76.3%) are shared with mainland China; two species (5.3%) are shared one each with the Indo-China and Indo-Malayan regions and are not found on mainland China. It is clear that the number of endemic species is positively correlated with island area, diversity of topography and vegetation, climate, distance between the island and the mainland, and the length of time the island has been separated from the mainland.

DISTRIBUTIONAL ROUTES FROM THE MAINLAND TO CONTINENTAL ISLANDS

Japan Proper

Kuramoto (1979) recognized four routes by which amphibians dispersed from the mainland to Japan: (1) northern route from Amur to Sakhalin and then to Hokkaido; (2) the most important route, from Korea to Japan; (3) from south China to Taiwan and on to the Ryukyus; (4) an additional route directly from China across a former lowland now submerged by the shallow East China Sea. During my studies on the origin of the amphibian fauna of the Ryukyu Islands, I came to agree with Kuramoto's conclusions.

Ryukyu Islands

Herein, I refer only to the islands south of the Tokara Tectonic Strait. Zhao (1989) concluded that the nonendemic herpetofauna of the Ryukyus originated from the mainland by way of Taiwan. The faunal similarities between individual islands in the Ryukyus and Taiwan depend on the distance between the two (i.e., the shorter the distance, the greater the similarity). Additional evidence is provided by the colubrid snake *Opisthotropis kikuzatoi* on Kumejima Island (Toyama, 1983), although it has yet to be found on Taiwan. My conclusion is in accord with that of Hikida et al. (1992:29) that "… species belonging to Palearctic elements, supposedly having come down from Kyushu, did or did not get to the Tokara Group and seemed never to have crossed the Tokara Tectonic Strait, whereas some of those classified as the Oriental origin have extended the distribution northward across the strait."

Taiwan and Hainan Islands

Without a doubt the amphibian faunas of these two islands originated from the neighboring mainland of China; this is based on the close distances between the islands and the mainland and on the great similarity of the insular faunas with the mainland. Possibly present distributions were established when the islands were connected to the mainland during periods of lower sea level in the Pleistocene; however, some species may have reached Hainan by rafting across the narrow Qiongzhou Strait. A few species (e.g., *Leptobrachium hasseltii* and *Bufo galeatus*) are Indo-Chinese elements that may have reached Hainan during periods of low sea level, at which time the intervening Gulf of Tongking was a coastal plain (Zhao, 1989).

Acknowledgments: I am grateful to Kraig Adler, Leo J. Borkin, and Robert F. Inger for their helpful criticism on earlier drafts of this manuscript. I especially express my sincere thanks to William E. Duellman for his encouragement to prepare this paper and for many valuable comments and extensive rewriting parts of the manuscript.

LITERATURE CITED

ADLER, K., AND E. ZHAO. 1990. Studies on hynobiid salamanders, with description of a new genus. Asiatic Herpetol. Res. 3:37–45.

BORKIN, L. J. 1986. Pleistocene glaciations and western-eastern Palearctic disjunctions in amphibian distribution. Pp. 63-66 *in* Z. Rocek (ed.), *Studies in Herpetology.* Prague: Charles University.

BORKIN, L. J., AND S. L. KUZMIN. 1988. Amphibians of Mongolia: species account. Pp. 30–197 *in* E. I. Vorobieva and I. S. Darevsky (eds.), *Amphibians and Reptiles of the Mongolian People's Republic. General Problems. Amphibians.* Moscow: Nauka (in Russian).

DUBOIS, A. 1992. Notes sur la classification des Ranidae (amphibiens anoures). Bull. Mens. Soc. Linnénne Lyon 61:305–352.

DUELLMAN, W. E. 1993. Amphibian species of the world: additions and corrections. Spec. Publ. Mus. Nat. Hist. Univ. Kansas 21:1–372.

FROST, D. R. (ed.). 1985. *Amphibian Species of the World.* Lawrence, Kansas: Association of Systematics Collections and Allen Press, Inc.

HIKIDA, T., H. OTA, AND M. TOYAMA. 1992. Herpetofauna of an encounter zone of Oriental and Palearctic elements: amphibians and reptiles of the Tokara group and adjacent islands in the northern Ryukyus, Japan. Biol. Mag. Okinawa 30:29–43.

Huang, Y.-Z. 1985. A new species of pelobatid toads (Amphibia: Pelobatidae) from Ningxia Hui Autonomous Region. Acta Biol. Plateau Sinica 4:77–81 (in Chinese).

Inger, R. F. 1947. Preliminary survey of the amphibians of the Riukiu Islands. Fieldiana: Zool. 32:297–352.

KATO, T., AND H. OTA. 1993. *Endangered Wildlife of Japan.* Osaka: Hoikusha.

KURAMOTO, M. 1979. Distribution and isolation in the anurans of the Ryukyu Islands. Japan. J. Herpetol. 8:8–21 (in Japanese).

KURAMOTO, M., AND C. S. WANG. 1987. A new rhacophorid tree-frog from Taiwan, with comparisons to *Chirixalus eiffingeri* (Anura, Rhacophoridae). Copeia 1987:931–942.

MATSUI, M., T. SATO, S. TANABE, AND T. HAYASHI. 1992. Electrophoretic analysis of systematic relationships and status of two hynobiid salamanders from Hokkaido (Amphibia: Caudata). Herpetologica 48:408–416.

MOU, Y., AND E.-M. ZHAO. 1992. A study on vocalization of thirteen anuran species from China. Pp. 15-25 *in* Y.-M. Yang (ed.), *Collected Papers in Herpetology.* Chengdu: Sichuan Publ. House Science (in Chinese).

MUNKHBAYAR, K., AND D. V. SEMENOV. 1988. Significance and conservation of amphibians and reptiles. Pp. 16–29 *in* E. I. Vorobyeva and I. S. Darevsky (eds.), *Amphibians and Reptiles of Mongolian People's Republic: Amphibians.* Moscow: Science Press (in Russian).

RISCH, J.-P., AND J. D. ROMER. 1980. Origin and taxonomic status of the salamander *Cynops shataukokensis.* J. Herpetol. 14:337–341.

TAYLOR, E. H. 1962. The amphibian fauna of Thailand. Univ. Kansas Sci. Bull. 53:265–599.

TOKUNAGA, S. 1901. Notes on the raised coral reefs of the Riu kiu curve. J. Coll. Sci. Imp. Univ. Tokyo 16 (not seen).

TOYAMA, M. 1983. Taxonomic reassignment of the colubrid snake, *Opheodrys kikuzatoi,* from Jume-jima Island, Ryukyu Archipelago. Japan. J. Herpetol. 10:33–38.

WU, G.-F., E.-M. ZHAO, R. F. INGER, AND H. B. SHAFFER. 1993. A new frog of the genus *Oreolalax* (Pelobatidae) from Sichuan, China. J. Herpetol. 27:410–413.

ZHAO, E.-M. 1989. Some aspects of herpetogeography of East Asian islands. *Plenary Lectures of the First World Congress of Herpetology, Catebury, Tape 7.* London: Q. E. D. Recording Services, Ltd.

ZHAO, E.-M., AND K. ADLER. 1993. *Herpetology of China.* Soc. Stud. Amphib. Rept. and Chinese Soc. Stud. Amphib. Rept.

APPENDIX 7:1

DISTRIBUTION OF SPECIES OF AMPHIBIANS IN REGIONS IN TEMPERATE EASTERN ASIA

Abbreviations for regions: CC = Central China, HN = Hainan Island, JP = Japan Proper, KN = Korean Peninsula and northeastern China, MX = Mongolia-Xinjiang, NC = North China, QX = Qinghai-Xizang, RK Ryukyu Islands, SC = South China, SW = Southwest China, TW = Taiwan Island. Abbreviations for distribution patterns (Column DP): CA = Central Asia, EIO = Eastern Asian islands Oriental, EIP = Eastern Asian islands Palearctic, HH = Himalayas-Hengduanshan, HI = Himalayas, HO = Hengduanshan Oriental, HP = Hengduanshan Palearctic, HT = Henduanshan-Tibet, IC = Indochina, IM = Indo-Malayan, KN = Korean Peninsula and northeast China, MP = Mediterranean Palearctic, NC = North China, PO = Pan-Oriental, PP = Pacific Palearctic, SC = Southern China, SP = Siberian Palearctic. Symbols in columns: + present, – absent, • also occurring outside of temperate eastern Asia, compass direction (e.g., N = only in north), ? = questionable. Symbols in boldface indicate endemism. Symbols in column JP: HK = only on Hokkaido Island, HS = only on Honshu Island, JP = on three main islands of Japan (Honshu, Shikoku, and Kyushu), KU = only on Kyushu Island, OK = only on Oki Island; TS = only on Tsushima Island. Symbol in column KN: XJ = only in northernmost tip of Xinjiang.

Taxon	DP	KN	NC	CC	SC	SW	MX	QX	JP	RK	TW	HN
ANURA: BUFONIDAE:												
Bufo ailaoanus	HO	–	–	–	–	+	–	–	–	–	–	–
Bufo andrewsi	HO	–	–	+	–	+	–	–	–	–	–	–
Bufo bankorensis	EIO	–	–	–	–	–	–	–	–	–	+	–
Bufo burmanus	IC	–	–	–	•	–	–	–	–	–	–	–
Bufo cryptotympanicus	SC	–	–	+	+	–	–	–	–	–	–	–

Appendix 7:1 continued

Taxon	DP	KN	NC	CC	SC	SW	MX	QX	JP	RK	TW	HN
Bufo danatensis	CA	–	–	–	–	–	•	–	–	–	–	–
Bufo galeatus	IC	–	–	–	•	–	–	–	–	–	–	–
Bufo gargarizans	PP	•	•	•	–	•	•	•	–	•	•	–
Bufo himalayanus	HI	–	–	–	–	•	–	–	–	–	–	–
Bufo japonicus	EIP	–	–	–	–	–	–	–	**JP**	–	–	–
Bufo melanostictus	PO	–	?	•	•	•	–	–	–	–	•	•
Bufo minshanicus	HP	–	+	–	–	+	–	+	–	–	–	–
Bufo raddei	PP	+	+	–	–	–	+	+	–	–	–	–
Bufo stejnegeri	KN	+	–	–	–	–	–	–	–	–	–	–
Bufo tibetanus	HP	–	–	–	–	+	–	+	–	–	–	–
Bufo torrenticola	EIP	–	–	–	–	–	–	–	**JP**	–	–	–
Bufo tuberculatus	HP	–	–	–	–	+	–	+	–	–	–	–
Bufo viridis	MP	–	–	–	–	–	•	?	–	–	–	–
Pelophryne scalpta	EIO	–	–	–	–	–	–	–	–	–	–	+
ANURA: DISCOGLOSSIDAE:												
Bombina fortinuptialis	SC	–	–	+	–	–	–	–	–	–	–	–
Bombina maxima	HO	–	–	–	–	+	–	–	–	–	–	–
Bombina microdeladigitora	SC	–	–	+	+	–	–	–	–	–	–	–
Bombina orientalis	PP	•	–	–	–	–	–	–	•	–	–	–
ANURA: HYLIDAE:												
Hyla annectans	IC	–	–	•	•	•	–	–	–	–	–	–
Hyla chinensis	IC	–	–	•	•	–	–	–	–	–	•	–
Hyla hallowellii	EIO	–	–	–	–	–	–	–	–	+	–	–
Hyla japonica	PP	•	•	•	–	–	•	–	•	–	–	–
Hyla sanchiangensis	SC	–	–	+	+	–	–	–	–	–	–	–
Hyla simplex	IC	–	–	•	•	–	–	–	–	–	–	•
Hyla tsinlingensis	SC	–	–	+	–	–	–	–	–	–	–	–
Hyla zhaopingensis	SC	–	–	+	–	– .	–	–	–	–	–	–
ANURA: MICROHYLIDAE:												
Calluella yunnanensis	HO	–	–	–	–	+	–	–	–	–	–	–
Kalophrynus menglienicus	SC	–	–	–	+	–	–	–	–	–	–	–
Kalophrynus pleurostigma	IM	–	–	–	•	–	–	–	–	–	–	•
Kaloula borealis	PP	+	+	+	–	–	–	–	–	–	–	–
Kaloula pulchra	IM	–	–	–	•	–	–	–	–	–	–	•
Kaloula rugifera	SC	–	–	+	–	–	–	–	–	–	–	–
Kaloula verrucosa	HO	–	–	–	–	+	–	–	–	–	–	–
Microhyla berdmorei	IC	–	–	–	•	–	–	–	–	–	–	•
Microhyla butleri	IC	–	–	•	•	•	–	–	–	–	–	•
Microhyla heymonsi	IM	–	–	•	•	•	–	–	–	–	•	•
Microhyla mixtura	SC	–	–	+	–	–	–	–	–	–	–	–
Microhyla okinavensis	EIO	–	–	–	–	–	–	–	–	+	–	–
Microhyla ornata	PO	–	–	•	•	•	–	–	–	–	•	•
Microhyla pulchra	IC	–	–	• ·	•	–	–	–	–	–	–	•
Micryletta inornata	IM	–	–	–	•	–	–	–	–	–	•	–
Micryletta stejnegeri	EIO	–	–	–	–	–	–	–	–	–	+	–
ANURA: PELOBATIDAE:												
Leptobrachium (L.) chapaense	IC	–	–	–	•	–	–	–	–	–	–	–
Leptobrachium (L.) hasseltii	IM	–	–	–	–	–	–	–	–	–	–	•
Leptobrachium (V.) ailaonica	HO	–	–	–	–	+	–	–	–	–	–	–
Leptobrachium (V.) boringii	SC	–	–	+	–	–	–	–	–	–	–	–
Leptobrachium (V.) leishanensis	SC	–	–	+	–	–	–	–	–	–	–	–
Leptobrachium (V.) liui	SC	–	–	+	–	–	–	–	–	–	–	–
Leptolalax pelodytoides	IC	–	–	•	•	•	–	–	–	–	–	–
Megophrys boettgeri	IC	–	–	•	•	–	–	–	–	–	–	–
Megophrys brachykolos	SC	–	–	–	+	–	–	–	–	–	–	–
Megophrys carinensis	IC	–	–	•	•	•	–	–	–	–	–	–
Megophrys giganticus	HO	–	–	–	–	+	–	–	–	–	–	–
Megophrys glandulosa	HO	–	–	–	–	+	–	–	–	–	–	–

Appendix 7:1 continued

Taxon	DP	KN	NC	CC	SC	SW	MX	QX	JP	RK	TW	HN
Megophrys kempii	HI	−	−	−	−	+	−	−	−	−	−	−
Megophrys kuatunensis	SC	−	−	+	−	−	−	−	−	−	−	−
Megophrys lateralis	IC	−	−	−	•	−	−	−	−	−	−	−
Megophrys mangshanensis	SC	−	−	+	−	−	−	−	−	−	−	−
Megophrys minor	HH	−	−	+	+	+	−	−	−	−	−	−
Megophrys nankiangensis	SC	−	−	+	−	−	−	−	−	−	−	−
Megophrys omeimontis	HH	−	−	+	+	+	−	−	−	−	−	−
Megophrys pachyproctus	HI	−	−	−	−	+	−	−	−	−	−	−
Megophrys palpebralespinosa	IC	−	−	−	•	−	−	−	−	−	−	−
Megophrys parva	IC	−	−	−	•	−	−	−	−	−	−	−
Megophrys shapingensis	HO	−	−	−	−	+	−	−	−	−	−	−
Megophrys spinatus	SC	−	−	+	−	−	−	−	−	−	−	−
Ophryophryne microstoma	IC	−	−	−	•	−	−	−	−	−	−	−
Ophryophryne pachyproctus	SC	−	−	−	+	−	−	−	−	−	−	−
Scutiger (O.) chuanbeiensis	SC	−	−	+	−	−	−	−	−	−	−	−
Scutiger (O.) jingdongensis	HS	−	−	−	−	+	−	−	−	−	−	−
Scutiger (O.) lichuanensis	SC	−	−	+	−	−	−	−	−	−	−	−
Scutiger (O.) major	HS	−	−	−	−	+	−	−	−	−	−	−
Scutiger (O.) multipunctatus	HO	−	−	−	−	+	−	−	−	−	−	−
Scutiger (O.) omeimontis	HH	−	−	−	−	+	−	−	−	−	−	−
Scutiger (O.) pingii	HO	−	−	−	−	+	−	−	−	−	−	−
Scutiger (O.) popei	HO	−	−	−	−	+	−	−	−	−	−	−
Scutiger (O.) rhodostigmatus	SC	−	−	−	+	−	−	−	−	−	−	−
Scutiger (O.) rugosus	HP	−	−	−	−	+	−	−	−	−	−	−
Scutiger (O.) schmidti	HO	−	−	−	−	+	−	−	−	−	−	−
Scutiger (O.) xiangchengensis	HP	−	−	−	−	+	−	−	−	−	−	−
Scutiger (S.) boulengeri	HT	−	−	−	−	•	−	•	−	−	−	−
Scutiger (S.) chintingensis	HP	−	−	−	−	+	−	−	−	−	−	−
Scutiger (S.) glandulatus	HP	−	−	−	−	+	−	−	−	−	−	−
Scutiger (S.) gongshanensis	HP	−	−	−	−	+	−	−	−	−	−	−
Scutiger (S.) liupanensis	NC	−	+	−	−	−	−	−	−	−	−	−
Scutiger (S.) maculatus	HP	−	−	−	−	−	−	+	−	−	−	−
Scutiger (S.) mammatus	HP	−	−	−	−	+	−	+	−	−	−	−
Scutiger (S.) ningshanensis	SC	−	−	+	−	−	−	−	−	−	−	−
Scutiger (S.) nyingchiensis	HT	−	−	−	−	−	−	+	−	−	−	−
Scutiger (S.) pingwuensis	SC	−	−	+	−	−	−	−	−	−	−	−
Scutiger (S.) ruginosus	HP	−	−	−	−	+	−	−	−	−	−	−
Scutiger (S.) sikimmensis	HI	−	−	−	−	•	−	−	−	−	−	−
Scutiger (S.) tuberculatus	HP	−	−	−	−	+	−	−	−	−	−	−
Scutiger (S.) weigoldi	HO	−	−	−	−	+	−	−	−	−	−	−
ANURA: RANIDAE:												
Amolops afghanus	IC	−	−	−	•	•	−	−	−	−	−	−
Amolops chunganensis	SC	−	−	+	−	+	−	−	−	−	−	−
Amolops granulosus	SC	−	−	+	−	−	−	−	−	−	−	−
Amolops hainanensis	EIO	−	−	−	−	−	−	−	−	−	−	+
Amolops hongkongensis	SC	−	−	−	+	−	−	−	−	−	−	−
Amolops kangtingensis	HP	−	−	−	−	+	−	−	−	−	−	−
Amolops liangshanensis	HP	−	−	−	−	+	−	−	−	−	−	−
Amolops lifanensis	HO	−	−	−	−	+	−	−	−	−	−	−
Amolops loloensis	HP	−	−	−	−	+	−	−	−	−	−	−
Amolops macrorhynchus	SC	−	−	−	+	−	−	−	−	−	−	−
Amolops mantzorum	HO	−	−	−	+	+	−	−	−	−	−	−
Amolops monticola	HI	−	−	−	−	•	−	−	−	−	−	−
Amolops nasicus	IC	−	−	•	−	−	−	−	−	−	−	•
Amolops ricketti	SC	−	−	+	+	−	−	−	−	−	−	−
Amolops torrentis	EIO	−	−	−	−	−	−	−	−	−	−	+
Amolops viridimaculatus	SC	−	−	−	+	+	−	−	−	−	−	−
Amolops wuyiensis	SC	−	−	+	−	−	−	−	−	−	−	−

Appendix 7:1 continued

Taxon	DP	KN	NC	CC	SC	SW	MX	QX	JP	RK	TW	HN
Chaperina quadranus	SC	–	–	+	–	–	–	–	–	–	–	–
Chaperina unculuanus	SC	–	–	–	+	–	–	–	–	–	–	–
Hoplobatrachus rugulosus	IC	–	–	•	•	–	–	–	–	–	•	•
Ingerana liui	SC	–	–	–	+	–	–	–	–	–	–	–
Ingerana reticulatus	HI	–	–	–	–	+	–	–	–	–	–	–
Ingerana xizangensis	HI	–	–	–	–	+	–	–	–	–	–	–
Limnonectes cancrivora	IM	–	–	–	•	–	–	–	–	–	–	•
Limnonectes fragilis	EIO	–	–	–	–	–	–	–	–	–	–	+
Limnonectes kuhlii	IM	–	–	•	•	–	–	–	–	–	•	–
Limnonectes limnocharis	PO	–	•	•	•	•	–	–	•	•	•	–
Limnonectes namiyei	EIO	–	–	–	–	–	–	–	–	+	–	–
Nanorana parkeri	HT	–	–	–	–	–	–	•	–	–	–	–
Nanorana pleskei	HP	–	–	–	–	+	–	+	–	–	–	–
Nanorana ventripunctata	HP	–	–	–	–	+	–	–	–	–	–	–
Occidozyga lima	IM	–	–	•	•	–	–	–	–	–	–	•
Paa arnoldi	HI	–	–	–	–	•	–	–	–	–	–	–
Paa blanfordi	HI	–	–	–	–	•	–	–	–	–	–	–
Paa boulengeri	HO	–	+	+	–	+	–	–	–	–	–	–
Paa conaensis	HI	–	–	–	–	+	–	–	–	–	–	–
Paa exilispinosa	SC	–	–	+	+	–	–	–	–	–	–	–
Paa feae	IC	–	–	–	•	–	–	–	–	–	–	–
Paa jiulongensis	SC	–	–	+	–	–	–	–	–	–	–	–
Paa liebigii	HI	–	–	–	–	+	–	–	–	–	–	–
Paa liui	HO	–	–	–	–	+	–	–	–	–	–	–
Paa maculosa	HO	–	–	–	–	+	–	–	–	–	–	–
Paa polunini	HI	–	–	–	–	•	–	–	–	–	–	–
Paa shini	SC	–	–	+	–	–	–	–	–	–	–	–
Paa spinosa	IC	–	–	•	•	–	–	–	–	–	–	–
Paa yunnanensis	IC	–	–	•	•	•	–	–	–	–	–	–
Phrynoglossus borealis	HI	–	–	–	–	•	–	–	–	–	–	–
Phrynoglossus martensii	IC	–	–	–	•	–	–	–	–	–	–	•
Rana adenopleura	SC	–	–	+	+	+	–	–	–	–	+	+
Rana altaica	SP	XJ	–	–	–	–	–	–	–	–	–	–
Rana amurensis	SP	•	–	–	–	–	•	–	–	–	–	–
Rana andersonii	IC	–	–	•	•	–	–	–	–	–	–	•
Rana anlungensis	SC	–	–	+	–	–	–	–	–	–	–	–
Rana asiatica	CA	–	–	–	–	–	•	–	–	–	–	–
Rana chaochiaoensis	HP	–	–	+	+	+	–	–	–	–	–	–
Rana chensinensis	PP	+	+	–	–	+	+	+	–	–	–	–
Rana chevronta	HO	–	–	–	–	+	–	–	–	–	–	–
Rana chosenica	KN	+	–	–	–	–	–	–	–	–	–	–
Rana daunchina	HO	–	–	–	–	+	–	–	–	–	–	–
Rana dybowskii	PP	•	–	–	–	–	–	–	TS	–	–	–
Rana fukienensis	SC	–	–	–	+	–	–	–	–	–	+	–
Rana gerbillus	HI	–	–	–	–	•	–	–	–	–	–	–
Rana grahami	HO	–	–	–	–	+	–	–	–	–	–	–
Rana guentheri	IC	–	–	•	•	•	–	–	–	–	•	•
Rana hejiangensis	SC	–	–	+	–	–	–	–	–	–	–	–
Rana holsti	EIO	–	–	–	–	–	–	–	–	+	–	–
Rana huanrenensis	KN	+	–	–	–	–	–	–	–	–	–	–
Rana ijimae	EIO	–	–	–	–	–	–	–	–	+	–	–
Rana ishikawae	EIO	–	–	–	–	–	–	–	–	+	–	–
Rana japonica	PP	–	–	+	–	–	–	–	+	–	+	–
Rana kuangwuensis	SC	–	–	+	–	–	–	–	–	–	–	–
Rana latouchii	SC	–	–	+	+	–	–	–	–	–	+	–
Rana livida	IC	–	–	•	•	–	–	–	–	–	–	•
Rana longicrus	SC	–	–	–	+	–	–	–	–	–	+	–
Rana lungshengensis	SC	–	–	+	–	–	–	–	–	–	–	–

Appendix 7:1 continued

Taxon	DP	KN	NC	CC	SC	SW	MX	QX	JP	RK	TW	HN
Rana macrodactyla	IC	–	–	–	•	–	–	–	–	–	–	–	•
Rana margaretae	SC	–	–	+	–	+	–	–	–	–	–	–	–
Rana minima	SC	–	–	–	+	–	–	–	–	–	–	–	–
Rana narina	EIO	–	–	–	–	–	–	–	–	–	+	–	–
Rana nigrolineata	SC	–	–	–	+	–	–	–	–	–	–	–	–
Rana nigromaculata	PP	•	•	•	•	•	•	•	•	–	–	–	–
Rana nigrotympanica	SC	–	–	+	+	–	–	–	–	–	–	–	–
Rana nigrovittata	IC	–	–	–	•	–	–	–	–	–	–	–	–
Rana okinavana	EIO	–	–	–	–	–	–	–	–	–	+	–	–
Rana ornativentris	EIP	–	–	–	–	–	–	–	JP	–	–	–	–
Rana pirica	EIP	–	–	–	–	–	–	–	HK	–	–	–	–
Rana plancyi	PP	–	+	+	–	–	–	–	–	–	–	–	–
Rana pleuraden	HO	–	–	–	–	+	–	–	–	–	–	–	–
Rana porosa	EIP	–	–	–	–	–	–	–	JP	–	–	–	–
Rana psaltis	EIO	–	–	–	–	–	–	–	–	+	–	–	–
Rana ridibunda	MP	–	–	–	–	–	•	–	–	–	–	–	–
Rana rugosa	PP	+	–	–	–	–	–	–	+	–	–	–	–
Rana sakuraii	EIP	–	–	–	–	–	–	–	JP	–	–	–	–
Rana sangzhiensis	SC	–	–	+	–	–	–	–	–	–	–	–	–
Rana sauteri	IC	–	–	•	–	–	–	–	–	–	–	•	–
Rana schmackeri	SC	–	–	+	–	?	–	–	–	–	–	–	–
Rana shuchinae	HP	–	–	–	–	+	–	–	–	–	–	–	–
Rana spinulosa	EIO	–	–	–	–	–	–	–	–	–	–	–	+
Rana subaspera	EIO	–	–	–	–	–	–	–	–	+	–	–	–
Rana swinhoana	SC	–	–	+	–	–	–	–	–	–	–	+	–
Rana tagoi	EIP	–	–	–	–	–	–	–	JP	–	–	–	–
Rana taipehensis	IC	–	–	•	•	–	–	–	–	–	–	•	•
Rana taiwaniana	EIO	–	–	–	–	–	–	–	–	–	–	+	–
Rana tenggerensis	CA	–	–	–	–	–	+	–	–	–	–	–	–
Rana tiannanensis	SC	–	–	–	+	–	–	–	–	–	–	–	+
Rana tientaiensis	SC	–	–	+	–	–	–	–	–	–	–	–	–
Rana tormotus	SC	–	–	+	–	–	–	–	–	–	–	–	–
Rana tsushimensis	EIP	–	–	–	–	–	–	–	TS	–	–	–	–
Rana versabilis	SC	–	–	+	+	–	–	–	–	–	–	–	+
Rana weiningensis	HO	–	–	–	–	+	–	–	–	–	–	–	–
Rana wuchuanensis	SC	–	–	+	–	–	–	–	–	–	–	–	–
ANURA: RHACOPHORIDAE:												
Buergeria buergeri	EIP	–	–	–	–	–	–	–	JP	–	–	–	
Buergeria japonica	EIO	–	–	–	–	–	–	–	–	+	+	–	
Buergeria robusta	EIO	–	–	–	–	–	–	–	–	–	+	–	
Chirixalus doriae	IC	–	–	–	•	–	–	–	–	–	–	–	•
Chirixalus eiffingeri	EIO	–	–	–	–	–	–	–	–	+	+	–	
Chirixalus idiootocus	EIO	–	–	–	–	–	–	–	–	–	+	–	
Chirixalus vittatus	IC	–	–	–	•	•	–	–	–	–	–	–	•
Philautus albopunctatus	SC	–	–	+	–	–	–	–	–	–	–	–	–
Philautus andersoni	IC	–	–	–	•	•	–	–	–	–	–	–	–
Philautus argus	HI	–	–	–	–	•	–	–	–	–	–	–	–
Philautus gracilipes	IC	–	–	–	•	–	–	–	–	–	–	–	–
Philautus jinxiuensis	SC	–	–	+	–	–	–	–	–	–	–	–	–
Philautus longchuanensis	SC	–	–	–	+	+	–	–	–	–	–	–	–
Philautus medogensis	HI	–	–	–	–	+	–	–	–	–	–	–	–
Philautus menglaensis	SC	–	–	–	+	–	–	–	–	–	–	–	–
Philautus ocellatus	EIO	–	–	–	–	–	–	–	–	–	–	–	+
Philautus palpebralis	SC	–	–	–	+	–	–	–	–	–	–	–	–
Philautus rhododiscus	SC	–	–	+	+	–	–	–	–	–	–	–	–
Philautus romeri	SC	–	–	–	+	–	–	–	–	–	–	–	–
Polypedates dennysi	IC	–	–	•	•	–	–	–	–	–	–	–	–
Polypedates dugritei	SC	–	–	+	+	+	–	–	–	–	–	–	–

Appendix 7:1 continued

Taxon	DP	KN	NC	CC	SC	SW	MX	QX	JP	RK	TW	HN
Polypedates feae	IC	–	–	–	•	–	–	–	–	–	–	–
Polypedates megacephalus	SC	–	–	+	+	+	–	–	–	–	+	+
Polypedates mutus	IC	–	–	•	•	–	–	–	–	–	–	•
Polypedates omeimontis	SC	–	–	+	–	+	–	–	–	–	–	–
Rhacophorus arboreus	EIP	–	–	–	–	–	–	–	JP	–	–	–
Rhacophorus bipunctatus	IC	–	–	–	–	•	–	–	–	–	–	–
Rhacophorus cavirostris	PO	–	–	–	•	•	–	–	–	–	–	•
Rhacophorus chenfui	SC	–	–	+	–	+	–	–	–	–	–	–
Rhacophorus gongshanensis	HO	–	–	–	–	+	–	–	–	–	–	–
Rhacophorus hungfuensis	SC	–	–	+	–	–	–	–	–	–	–	–
Rhacophorus maximus	IC	–	–	–	•	•	–	–	–	–	–	–
Rhacophorus moltrechti	EIO	–	–	–	–	–	–	–	–	–	+	–
Rhacophorus nigropunctatus	SC	–	–	+	–	+	–	–	–	–	–	–
Rhacophorus owstoni	EIO	–	–	–	–	–	–	–	–	+	–	–
Rhacophorus oxycephala	EIO	–	–	–	–	–	–	–	–	–	–	+
Rhacophorus prasinatus	EIO	–	–	–	–	–	–	–	–	–	+	–
Rhacophorus reinwardtii	IM	–	–	–	•	–	–	–	–	–	–	–
Rhacophorus rhodopus	SC	–	–	+	–	+	–	–	–	–	–	+
Rhacophorus schlegelii	EIP	–	–	–	–	–	–	–	JP	–	–	–
Rhacophorus taipeianus	EIO	–	–	–	–	–	–	–	–	–	+	–
Rhacophorus translineatus	HI	–	–	–	–	+	–	–	–	–	–	–
Rhacophorus tuberculatus	HI	–	–	–	–	•	–	–	–	–	–	–
Rhacophorus verrucopus	HI	–	–	–	–	+	–	–	–	–	–	–
Rhacophorus verrucosus	IC	–	–	–	–	•	–	–	–	–	–	–
Rhacophorus viridis	EIO	–	–	–	–	–	–	–	–	+	–	–
Rhacophorus yaoshanensis	SC	–	–	+	–	–	–	–	–	–	–	–
Theloderma asperum	IC	–	–	–	–	•	–	–	–	–	–	–
Theloderma leporosa	IM	–	–	•	–	–	–	–	–	–	–	–
Theloderma moloch	HI	–	–	–	–	•	–	–	–	–	–	–
CAUDATA: CRYPTOBRANCHIDAE:												
Andrias davidianus	PP	–	+	+	–	+	–	+	–	–	?	–
Andrias japonicus	EIP	–	–	–	–	–	–	–	JP	–	–	–
CAUDATA: HYNOBIIDAE:												
Batrachuperus karlschmidti	HP	–	–	–	–	N	–	SE	–	–	–	–
Batrachuperus pinchonii	HP	–	–	–	–	+	–	–	–	–	–	–
Batrachuperus tibetanus	HP	–	–	–	NW	N	–	SE	–	–	–	–
Batrachuperus yenyuanensis	HP	–	–	–	–	+	–	–	–	–	–	–
Hynobius abei	EIP	–	–	–	–	–	–	–	HS	–	–	–
Hynobius amjiensis	SC	–	–	+	–	–	–	–	–	–	–	–
Hynobius boulengeri	EIP	–	–	–	–	–	–	–	JP	–	–	–
Hynobius chinensis	SC	–	–	+	–	–	–	–	–	–	–	–
Hynobius dunni	EIP	–	–	–	–	–	–	–	KS	–	–	–
Hynobius formosanus	EIO	–	–	–	–	–	–	–	–	–	+	–
Hynobius hidamontanus	EIP	–	–	–	–	–	–	–	HS	–	–	–
Hynobius kimurae	EIP	–	–	–	–	–	–	–	HS	–	–	–
Hynobius leechii	KN	+	–	–	–	–	–	–	–	–	–	–
Hynobius lichenatus	EIP	–	–	–	–	–	–	–	HS	–	–	–
Hynobius mantschuriensis	KN	+	–	–	–	–	–	–	–	–	–	–
Hynobius naevius	EIP	–	–	–	–	–	–	–	JP	–	–	–
Hynobius nebulosus	EIP	–	–	–	–	–	–	–	JP	–	–	–
Hynobius nigrescens	EIP	–	–	–	–	–	–	–	HS	–	–	–
Hynobius okiensis	EIP	–	–	–	–	–	–	–	OK	–	–	–
Hynobius sonani	EIO	–	–	–	–	–	–	–	–	–	+	–
Hynobius stejnegeri	EIP	–	–	–	–	–	–	–	KS	–	–	–
Hynobius takedai	EIP	–	–	–	–	–	–	–	HS	–	–	–
Hynobius tenuis	EIP	–	–	–	–	–	–	–	HS	–	–	–
Hynobius tokyoensis	EIP	–	–	–	–	–	–	–	HS	–	–	–
Hynobius tsuensis	EIP	–	–	–	–	–	–	–	TS	–	–	–

Appendix 7:1 continued

Taxon	DP	KN	NC	CC	SC	SW	MX	QX	JP	RK	TW	HN
Liua shihi	SC	–	–	+	–	–	–	–	–	–	–	–
Onychodactylus fischeri	PP	•	–	–	–	–	–	–	–	–	–	–
Onychodactylus japonicus	EIP	–	–	–	–	–	–	–	**JP**	–	–	–
Pachyhynobius shangchengensis	SC	–	–	+	–	–	–	–	–	–	–	–
Ranodon sibiricus	CA	–	–	–	–	–	•	–	–	–	–	–
Ranodon tsinpaensis	SC	–	–	+	–	–	–	–	–	–	–	–
Salamandrella keyserlingii	SP	•	•	–	–	–	•	–	HK	–	–	–
Satobius retardatus	EIP	–	–	–	–	–	–	•	**HK**	–	–	–
Caudata: Salamandridae:												
Cynops chenggongensis	HO	–	–	–	–	+	–	–	–	–	–	–
Cynops cyanurus	HO	–	–	–	–	+	–	–	–	–	–	–
Cynops ensicauda	EIO	–	–	–	–	–	–	–	–	+	–	–
Cynops orientalis	SC	–	–	+	–	–	–	–	–	–	–	–
Cynops orphicus	SC	–	–	–	+	–	–	–	–	–	–	–
Cynops pyrrhogaster	EIP	–	–	–	–	–	–	–	**JP**	–	–	–
Cynops wolterstorffi	HO	–	–	–	–	+	–	–	–	–	–	–
Echinotriton andersoni	EIO	–	–	–	–	–	–	–	–	+	+	–
Echinotriton asperrimus	IC	–	–	•	•	–	–	–	–	–	–	•
Echinotriton chinhaiensis	SC	–	–	+	–	–	–	–	–	–	–	–
Pachytriton brevipes	SC	–	–	+	–	–	–	–	–	–	–	–
Pachytriton labiatus	SC	–	–	+	+	–	–	–	–	–	–	–
Paramesotriton caudopunctatus	SC	–	–	+	–	–	–	–	–	–	–	–
Paramesotriton chinensis	SC	–	–	+	–	–	–	–	–	–	–	–
Paramesotriton fuzhongensis	SC	–	–	+	–	–	–	–	–	–	–	–
Paramesotriton guangxiensis	SC	–	–	–	+	–	–	–	–	–	–	–
Paramesotriton hongkongensis	SC	–	–	–	+	–	–	–	–	–	–	–
Tylototriton kweichowensis	HO	–	–	–	–	+	–	–	–	–	–	–
Tylototriton taliangensis	HO	–	–	–	–	+	–	–	–	–	–	–
Tylototriton verrucosus	IC	–	–	–	•	•	–	–	–	–	–	–
Gymnophiona: Ichthyophiidae:												
Ichthyophis bannanicus	SC	–	–	–	+	–	–	–	–	–	–	–

ADDENDUM

Since this paper was prepared, new taxa have been described, and the distributions of others have been modified. The listing of these additions and changes follows the order of taxa in Appendix 7:1.

Bufo burmanus recorded in Yunnan has been revised to include two new species in a new genus (Yang et al., 1996). The new species are *Torrentophryne aspinia* Yang and Rao and *T. tuberospinia* Yang and Liu. Both species are endemic to Yunnan, China. *Bufo burmanus* does not occur in China.

Two new species of *Bufo* (*B. kabischi* and *B. wolongensis*) are endemic to western Sichuan, China (Herrmann and Kühnel, 1997).

The records of *Amolops nasicus* from China refer to a new species, *A. mengyangensis,* that is endemic to southern Yunnan, China (Wu and Tian, 1995).

A new species, *Rana robertingeri* Wu and Zhao (1995), should be placed in the genus *Paa;* this species is endemic to southern Sichuan, China.

A new species, *Polypedates zhaojuensis,* is endemic to southern Sichuan, China (Wu and Zeng, 1994).

Two new rhacophorids (*Rhacophorus arvalis* and *R. aurantiventris*) are known only from Taiwan (Lue et al., 1994, 1995).

Literature Cited

HERRMANN, H.-J., AND K.-D. KÜHNEL. 1994. Zwei neue Arten der *Bufo bufo*-superspecies aus der Provinz Sichuan, SW-China. Sauria 19:31–40.

LUE, K. Y., J. S. LAI, AND Y. S. CHEN. 1994. A new species of *Rhacophorus* (Anura: Rhacophoridae) from Taiwan. Herpetologica 50:303–308.

LUE, K. Y., J. S. LAI, AND Y. S. CHEN. 1995. A new *Rhacophorus* (Anura: Rhacophoridae) from Taiwan. J. Herpetol. 29:338–345.

WU, G. F., AND W. S. TIAN. 1995. A new *Amolops* species from southern Yunnan. Pp. 50–52 *in* E. M. Zhao (ed.), *Amphibian Zoogeographic Division of China.* Chengdu: Herpetological Series 8 [in Chinese].

WU, G. F., AND X. M. ZENG. 1994. The karyotypic differentiation of *Polypedates dugritei* with description of a superspecies (Rhacophoridae, Anura). Sichuan J. Zool. 13:156–161 (in Chinese).

WU, G. F., AND E. M. ZHAO. 1995. A new ranid species of the *spinosae* group from Sichuan. Pp. 52–55 *in* E. M. Zhao (ed.), *Amphibian Zoogeographic Division of China.* Chengdu: Herpetological Series 8 (in Chinese).

YANG, D. T., W. Z. LIU, AND D. Q. RAO. 1996. A new toad genus of Bufonidae—*Torrentophryne* Transhimalayan Mountain of Yunnan of China with its biology. Zool. Res. Kunming 17:353–359.

8. Distribution of Amphibians in Southern Asia and Adjacent Islands

ROBERT F. INGER

Department of Zoology
Field Museum of Natural History
Chicago, Illinois 60605, USA

ABSTRACT   Approximately 650 species of amphibians are known from southern Asia. Phylogenetic relations within generally accepted families are almost unknown, which restricts the analysis of geographic distribution of this rich fauna. However, some generalizations are possible. A few species (all of them anurans) are almost ubiquitously distributed in this vast area; these are species associated with environments seriously modified by man and may be regarded as commensals of man. The remainder, more than 95%, have distributions correlated with natural environmental variations in vegetation and climate. These distributions, which also show the effects of geologic history, are clustered into the following geographic regions (defined in the text): India/Sri Lanka, the Southern Himalayan Flanks, the Northeastern Montane Region, the Southeast Asian Lowlands, the Thai-Lao Plateau, the Malay peninsular region, Sundaland, Philippine Islands, Sulawesi, and the Lesser Sundas. The fauna of India/Sri Lanka is the most distinctive, with a number of endemic lineages (as opposed to species). Differences among the other faunal regions are generally at the species level, although a few lineages are endemic to a single region. The faunas of the northern tier of regions show significant overlap with the fauna of China. Geologic evidence for the connection of India/Sri Lanka with Gondwanaland is clear and strong and probably accounts for the distinctiveness of this fauna. Two lineages with Papuan affinities accord with geologic evidence of the proximity of the eastern Philippines to New Guinea and Melanesia in the Oligocene or earlier. Pleistocene changes in climate and sea level probably account for disjuction in at least one lineage and for great overlap of species between currently separated landmasses. Present distributions also have been influenced by nonhistorical, ecological factors, such as seasonality of rainfall and topography.

Key words: Amphibians; Distribution; Southern Asia.

INTRODUCTION

Southern Asia (Fig. 8:1) as here defined is essentially coterminous with the classic biogeographer's Oriental Region—the territory south of and including the southern flanks of the Himalayas, the Indian subcontinent east of the dry Indus Valley, Sri Lanka, southeastern Asia south of the Chinese border, the Greater and Lesser Sundas, and the Philippine Islands. I am also including Sulawesi, although it lies to the east of Wallace's Line, because the amphibian fauna is so overwhelmingly related to that of the territory to the west; Whitmore (1981) noted that the Sulawesi biota, although long isolated by water, is distinctly Laurasian. The boundaries of "southern Asia." so defined, are largely natural in the sense that they coincide with pronounced physical boundaries—deep ocean basins or straits and, on the west, a semiarid area. The single largely artificial boundary is the one between China and South

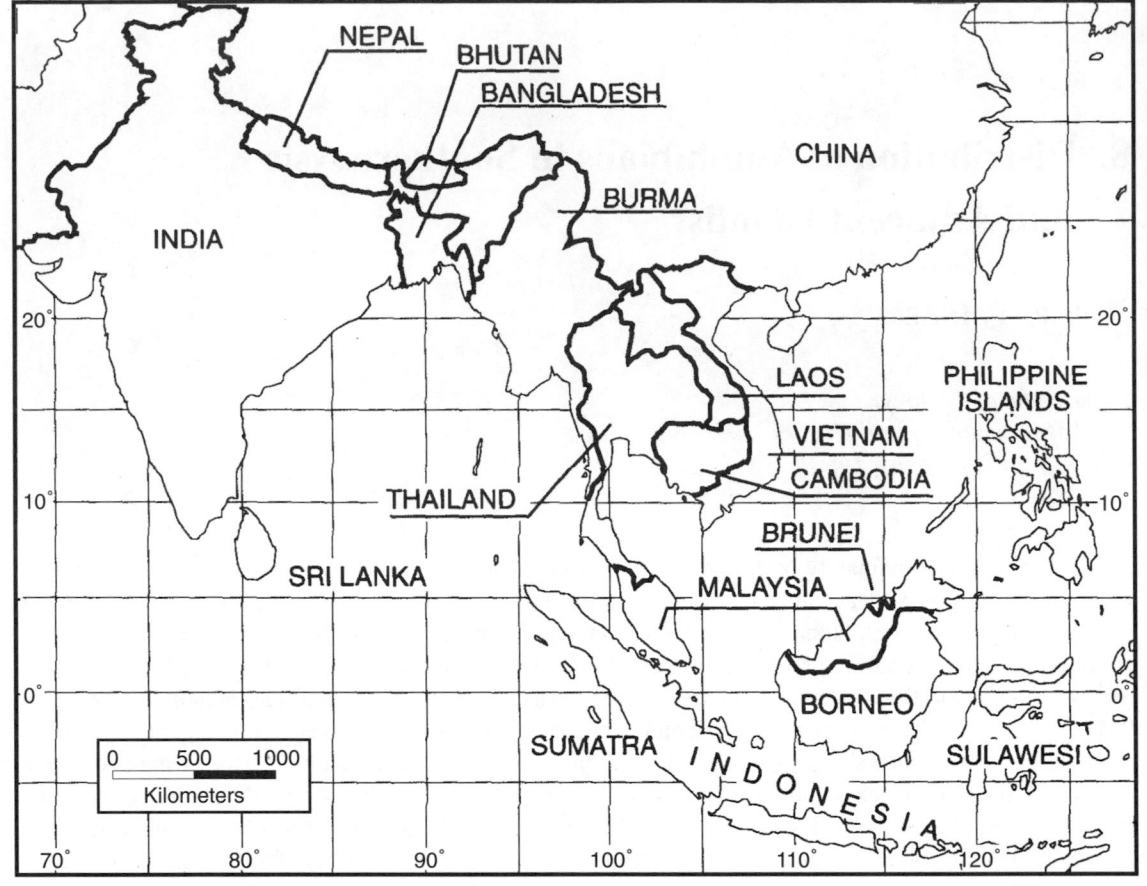

Fig. 8:1. Map of southern Asia, showing prinicipal political entities.

east Asia; although this boundary follows mountain ridges, the elevations are not high enough to constitute significant barriers to dispersal of amphibians and some of the largest rivers of Southeast Asia (e.g., Salween, Mekong) have their origins in China.

Between 600 and 650 species of amphibians are now known from southern Asia (Appendix 8:1), the uncertainty reflecting disagreements among systematists as to boundaries of species in many cases. Without a doubt, the actual number of species occurring in southern Asia is significantly higher as new species continue to be discovered. Also, as discussed below, large portions of southern Asia have been poorly sampled.

This analysis of the amphibians of southern Asia describes the broad outlines of distribution, not the details, except in unusual or illustrative cases. I analyze regional faunas and consider correlates of amphibian distribution—historical (geological), climatic and vegetational. The principal sources of distributional data used in this study are: Alcala (1986),

Berry (1975), Boulenger (1893), Bourret (1942), Dubois (1974), Inger (1966), Inger and Dutta (1986), Iskandar (ms), Kirtisinghe (1957), Nabhitabhata (ms), and Yang (1991), supplemented by Frost (1985).

As biogeographical analyses are at the mercy of taxonomic and sampling biases, I begin with those subjects.

TAXONOMIC BACKGROUND

The basic taxonomic units for this study are species and lineages of species. The principal concern here is not category level, so that genus, subgenus, and species group may be treated as equivalents. (This procedure has the not-trivial advantage of avoiding much fruitless disputation.) Distributional patterns of these units provide the basis for discussion of origins and comparisons of regional faunas. Because caecilians and salamanders form such a small portion (about 8%) of the amphibian fauna in southern Asia, limits to current understanding of the phylogenetics within these groups cause less difficulty than do anurans. The three species of sala-manders are members of three genera of Salaman-dridae. Treatment of the genera and species of cae-cilians of the area has achieved at least temporary stability through the work of Nussbaum (in Frost, 1985), and there is now reasonably good understand-ing of phylogenetic relations at the suprageneric level of the caecilian groups (Hedges et al., 1993).

"Lineages of species" implies phylogenetic re-lations, a concept posing severe problems when deal-ing with the anurans of southern Asia. Unfortunate-ly, aside from the recent phylogenetic classification of frogs as a whole (Ford and Cannatella, 1993), the phylogenies proposed for the Rhacophoridae by Liem (1970) and Channing (1989), and the revision of *Amolops* by Yang (1991), there is almost no ex-plicit phylogenetic information that might be useful in a study of the origins and distribution of anurans of southern Asia and adjacent islands.

Dubois (1987, 1992) proposed revisions of the classification of the Ranidae and Inger (1972) one for *Bufo* of Asia, but these revisions are essentially phenetic. Absence of phylogenetic background obliges me to make some arbitrary taxonomic deci-sions. As those decisions affect the patterns of re-gional and local diversity substantially, I present the classification used in this study. I adopt the propos-als of Ford and Cannatella (1993) and use the fol-lowing suprageneric levels—Bombinatoridae, Me-gophryidae, Bufonidae, Hylidae, Microhylidae, Ranidae, and Rhacophoridae.

Taxonomic decisions are not required at lower levels in two families for the purposes of this study. The Family Bombinatoridae is represented in south-ern Asia by two well-defined genera, *Bombina* with a single species and *Barbourula* with two. The Hyl-idae is represented by two species of *Hyla*.

Megophryidae

Although no phylogenetic study has been made of this family, there seems to be little disagreement on the generic boundaries, at least for taxa occur-ring in southern Asia. Here I follow Matsui (in Frost, 1985) and recognize *Leptobrachella, Leptobrachi-um, Leptolalax, Megophrys, Ophryophryne,* and *Scutiger.*

Bufonidae

Again, there is no explicit phylogenetic outline, yet there is general agreement on the genera occur-ring in southern Asia. I use the species groups of *Bufo* proposed earlier (Inger, 1972) and the generic

classification for Asian species used by Inger (in Frost, 1985).

Microhylidae

Taxonomic arrangement is as in Frost (1985).

Ranidae

Most of the taxonomic problems lie in this, the most speciose family in the region. Dubois's (1992) revision of the classification is seductive, even though it is purely phenetic, because it breaks up this very large family into a large number of "gen-era" of more manageable size. The main deficiency of his classification, besides its phenetic nature, is that many of Dubois's decisions appear to be based on skimpy evidence. Given these qualities, my use of parts of his classification is arbitrary and reflects convenience; it does not signify thoughtful agree-ment with Dubois's decisions. On the other hand, parts of Dubois's classification probably will be sup-ported by future phylogenetic studies. For example, the genus *Paa,* as used in Dubois (1992), seems to be a monophyletic unit defined by morphology of adults and tadpoles and a relatively cohesive geo-graphic distribution; the same merits are shown by other genera either redefined (e.g., *Hoplobatrachus, Nannophrys, Indirana,* and *Micrixalus*) or estab-lished (e.g., *Ingerana*) in Dubois's classification. These five are treated as lineages in this study. I also use *Limnonectes* as defined in Dubois (1987), ex-cept for those moved to *Hoplobatrachus* by Dubois (1992), although the phylogenetic basis for this tax-on is dubious; further study almost certainly will establish the paraphyly of this group. I follow Yang's (1991) lineages of the species formerly grouped under *Amolops.* Other generic or species group as-signments of ranid species follow the listing in Frost (1985) except for *Phrynoglossus* as defined by Smith (1931a; see also Dubois, 1987).

Rhacophoridae

I use Liem's arrangement, although *Philautus* presents several kinds of problems. Definition of lin-eages within this "genus" is uncertain despite the recognition of a few species groups within a restrict-ed region (Dring, 1987) or of subgenera (Dubois, 1992). *Philautus* almost certainly is polyphyletic. Nonetheless, in the absence of a comprehensive, phylogenetic analysis, I treat it as a single unit. An-other type of problem posed by *Philautus* is defini-tion of species. Individual and interspecific varia-tion are thoroughly confused in the literature, leading to uncertainties about regional diversity.

The "species boundary problem" in the case of *Philautus* has often led to the proliferation of named entities. That is not true for other anuran groups with boundary problems. It is well known that what has been called *Rana* (= *Limnonectes*) *limnocharis* consists of a number of sibling species. Dubois (1974) was able to recognize four sympatric species of this taxon in Nepal on the basis of call differentiation. Dutta (pers. comm.) is able to recognize three sympatric forms of "*limnocharis*" in Orissa, India, on the basis of calls and morphology. Other, as yet undefined, forms of "*limnocharis*" exist in Southeast Asia. The name *Rana livida* also masks several cryptic species, but these are not as well studied as *L. limnocharis*. Similarly, the name *Rana* (= *Limnonectes*) *kuhlii* applies not to a single species, but to a complex of at least two if not four species (Iskandar, pers. comm.). *Leptobrachium hasselti* is the name applied to certain populations of megophryines occurring from northeastern India to the Philippines; several geographic units have been split off as distinct species (Matsui in Frost, 1985), but the actual limits of the species remain to be determined.

Future systematic studies of the fauna of southern Asia probably will lead to the placing in synonymy of some presumed species. However, it is more likely that, quite aside from the discovery of genuine novelties, the number of sets of valid, sibling species will increase and more than offset the reductions caused by elimination of invalid species. Consequently, the numbers of species cited here for southern Asia as a whole and for its various subregions are certainly underestimates.

SAMPLING WEAKNESSES

Sampling of the amphibian fauna of southern Asia has been uneven, as in the rest of the tropics. Although new records and new species are to be expected in every part of southern Asia, some areas have been well collected. For example, the amphibians of the Philippine Islands were sampled intensively by E. H. Taylor (1917–1922), who was able to work in most of the larger and some of the smaller islands; the Philippine Zoological Expedition of the Field Museum of Natural History (1946–1947) sampled the fauna of Mindanao, Palawan, and Luzon. In the 1960s W. C. Brown and A. Alcala worked on most of the larger islands; and recently R. Crombie (1980s) has worked intensively on Luzon. We can assume, therefore, that the species of Philippine amphibians cited by Alcala (1986) is a reasonably complete list of the fauna, even though new species are being discovered (Brown and Alcala, 1994) and some nominate species appear to be clusters of two or three sibling species (Crombie, pers. comm.).

Another well-collected area is India, particularly the rich Western Ghats. Begun in the nineteenth century, sampling of the fauna of that area has continued to the present (e.g., Pillai, 1986; Inger et al., 1984; Ravichandran, 1992). Again, although new species are being found (e.g., by Ravichandran, 1992), we can safely say that the broad outlines of lineages and the magnitude of the fauna are well known.

Thailand, West Malaysia, and Borneo have been sampled rather thoroughly. The first was explored herpetologically by M. A. Smith in the 1920s, by E. H. Taylor in the 1950s, and more recently by J. Dring and D. Dammon in the 1970s. The 111 species of amphibians listed from Thailand by Nabhitbhata (ms) probably represent a very large proportion of the actual fauna. Exploration of West Malaysia begun during the British colonial period has been continued by Malaysian biologists (e.g., P. Y. Berry, B.-H. Kiew) and others. Our knowledge of the fauna of that area is good, if not complete. Borneo has attracted herpetological collectors since the mid-nineteenth century and by 1966 the number of amphibian species known from the island stood at 94 (Inger, 1966). Sampling of the fauna has intensified and the faunal list of Borneo now stands at 137 species (Inger and Tan, 1996).

At the other extreme is Burma. Intensive sampling of this large area has been carried out just once, by L. Fea in 1887–1888; based largely on Fea's collection, Boulenger (1893) listed 54 species of anurans from Burma. Since then, there has been little field work on amphibians in Burma, and the faunal list has grown to only 76 species. As a measure of the incompleteness of present knowledge of the Burmese fauna, Borneo, which is only 10% larger than Burma, has almost twice as many anurans.

Sampling of the amphibian fauna of Sulawesi is even worse. No one has sampled this fauna even as well as Fea did that of Burma a hundred years ago. The known faunal list stands at 19 species. The island is mountainous and, until recently, was covered with lowland and montane rain forests, both qualities essential to the development of a large anuran fauna. Dr. D. Iskandar of Bandung has made three short trips to Sulawesi recently and found several apparently new species (Iskandar, pers. comm.). There are undoubtedly many more.

Sumatra, in contrast to the adjacent land masses of the Malay Peninsula and Borneo, has rarely been sampled and its known amphibian fauna is only 70% and 50% that of those two areas, respectively. What is more illuminating is the complete absence of Sumatran records of the bufonid genus *Ansonia,* the megophryine *Leptobrachella,* and the ranid *Meristogenys,* which are well represented in Borneo (11, 6, and 8 species, respectively). The Pleistocene history of the Sunda Shelf (see below), of which Sumatra is a part, is such that the absence of these genera from Sumatra must be viewed as apparent rather than real.

Does the heterogeneity of sampling introduce significant bias into analysis of biogeographic patterns in southern Asia? The answer is positive only if distributions are examined in terms of political units. The answer is negative if distributions are considered in terms of geological and physiographic features. The amphibian fauna of mountainous portions of Burma, the political area with the poorest sampling history, is considered here as part of the northeastern montane fauna, which includes the faunas of parts of Thailand, Indochina, and northeastern India. Certainly, additional sampling in Burma will uncover many new records for that country, but it is not likely to change the basic known pattern of similarity of the montane fauna of Burma to those of montane northern Thailand, Assam, and Indochina. Intensive, new sampling in Sumatra will increase the list of species known from that geographic unit by perhaps 50%; again, the Pleistocene history of the Sunda Shelf ensures that the dominant relations of the Sumatran fauna will still be with those of the other Greater Sundas, Java and Borneo, and the Malay Peninsula.

Additional sampling in southern Asia will undoubtedly increase the number of species known from each area and illuminate details of the distributions of species. However, it is unlikely that new sampling will change the general biogeographic patterns now evident.

GEOLOGICAL, VEGETATIONAL, AND CLIMATOLOGICAL BACKGROUND

The major historical relations of southern Asia and adjacent islands are complex, but relatively clear in outline, although details in many areas are not. The summary that follows is based largely on two recent papers by Hall (1997, in press).

1. There is overwhelming geological evidence that India was part of Gondwanaland; its collision with Eurasia dates from the early Tertiary, perhaps the Middle Eocene.

2. Southeast Asia—that is, roughly most of Burma, all of Thailand, the Malay Peninsula, Indochina, most of Sumatra and Java, southwestern Borneo, and intervening portions of the South China Sea—has been part of the Eurasian continental mass at least since the Mesozoic, but the exact southern margins of this mass are not certain. During the Eocene, the Malay Peninsula was closer to China. The Palawan-Mindoro portion of the Philippines apparently was part of the same continental mass; however, in the early Tertiary, it was not in its present position, nor was it above sea level. West Sulawesi probably was at the margin of the continental mass.

3. Much of the Philippines Archipelago formed part of an arc with Halmahera that probably represents an early rifted (Cretaceous?) Gondwanan fragment. During the early Tertiary, this arc lay farther east and south of its present position, yet still north of the Australo-Papuan continental mass.

4. Many parts of these areas have undergone rotation during the Tertiary, especially during the Eocene-Miocene interval, some in a clockwise and other parts in a counter-clockwise direction. Changes in positions of other portions, for example, the Palawan-Mindoro piece, have involved more than angular shifts.

5. During the mid-Oligocene, large portions of Sundaland (the area encompassing the Malay Peninsula, Sumatra, Java, Borneo, and the intervening South China Sea) were exposed. At this time, most of eastern Borneo, eastern Java, the Lesser Sundas, and the Philippines probably were covered by shallow seas (Fig. 8:2A). The Malay Peninsula began to shift relative to Indochina, resulting later in their present relative positions.

6. Near the end of the Oligocene, uplift in central Borneo provided material to form deltas in northern and eastern Borneo. Northward movement of the Australian mass may have created land connections within the Philippines-Halmahera arc and between the latter and part of what is now Sulawesi. There is

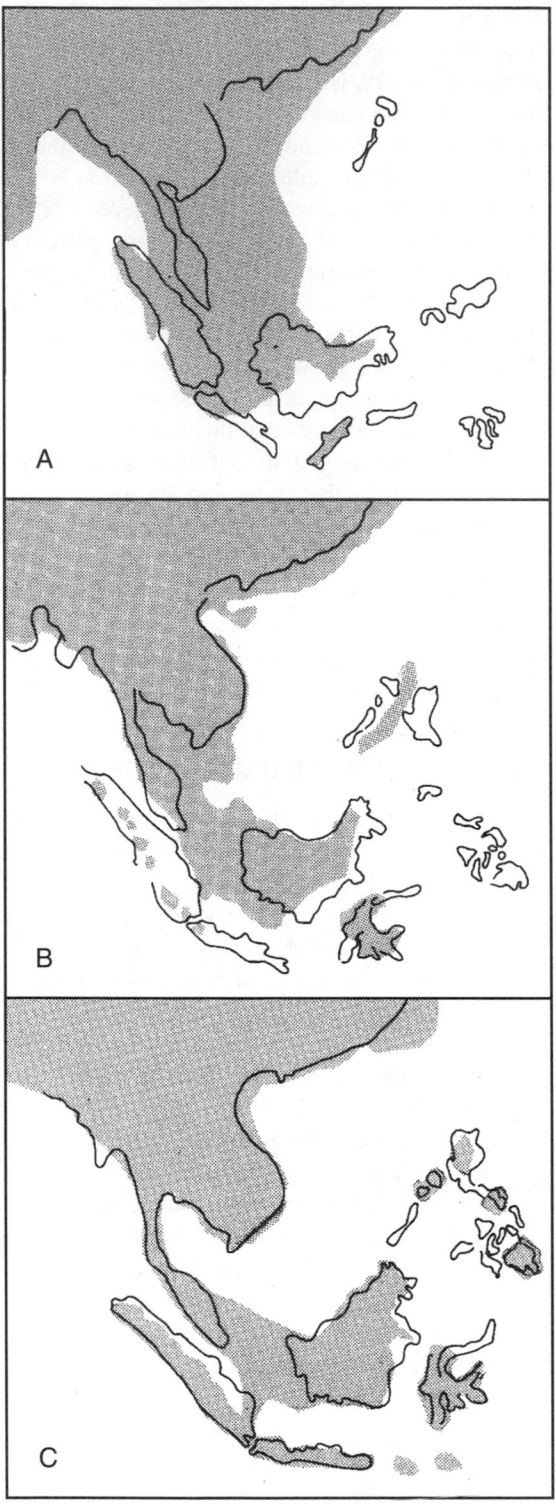

A

B

C

Fig. 8:2.   Relative land areas (shaded) during Middle Oligocene (A), Middle Miocene (B), and Early Pliocene (C). Present land masses are outlined by solid lines. Adapted from Hall (1997).

no evidence suggesting direct land connection with the Australian mass.

7. By the Late Miocene to the Early Pliocene (Fig. 8:2), the various parts of southern Asia had attained their present spatial relations; the Palawan-Mindoro piece and the rest of the Philippines had achieved their present positions, and the various parts of Sulawesi had been assembled. Much of the present area of the South China Sea was submerged, although a wide land connection between western Borneo and the Malay Peninsula may have persisted. Large portions of eastern Sumatra were inundated. Southern Asia was a land mass with possible small areas inundated. Virtually all of Borneo, much of Java, Sumatra, and Sulawesi, and parts of the Philippines were exposed land.

The most radical changes in land areas in relatively recent geological times have involved the Sunda Shelf—Borneo, Java, Sumatra, the Malay Peninsula, and the intervening South China Sea—which was alternately aggregated into one large land mass (Sundaland) and fragmented into an archipelago as glaciers grew and shrank during the Pleistocene. Current estimates of the lowering of sea level during the mid-Pleistocene are about 160 m (Gascoyne et al., 1979). That depth would expose most of the bed of the South China Sea, and now submerged Pleistocene river drainages have long been known (Heaney, 1991:Fig. 1). At that stage, the Palawan chain (Balabac, Palawan, and the Calamianes Islands) forming the southwestern portion of the Philippine Islands (Fig. 8:3) would have been part of Sundaland as the strait separating the Palawan chain from Borneo is only 140 m deep (Heaney, 1985). The remaining Philippine Islands, separated from Borneo by straits > 280 m deep, were not part of Sundaland at this time (if ever), although clearly the widths of the gaps were narrowed (see review in Heaney, 1985). Sulawesi apparently was not part of Pleistocene Sundaland; it was separated from Borneo by a relatively deep, but narrow, strait.

The physiographic picture is more complex. The area I am referring to as the northeastern montane region (Fig. 8:4) is continuous with the mountainous portion of southern and southwestern China that forms the eastern flanks of the Himalayas. The southern flanks of the Himalayan Mountains in the Nepal-Bhutan area are separated from the northeastern montane region by the lowlands of the Ganges-Brahmaputra Valley. Most of the mass of India proper

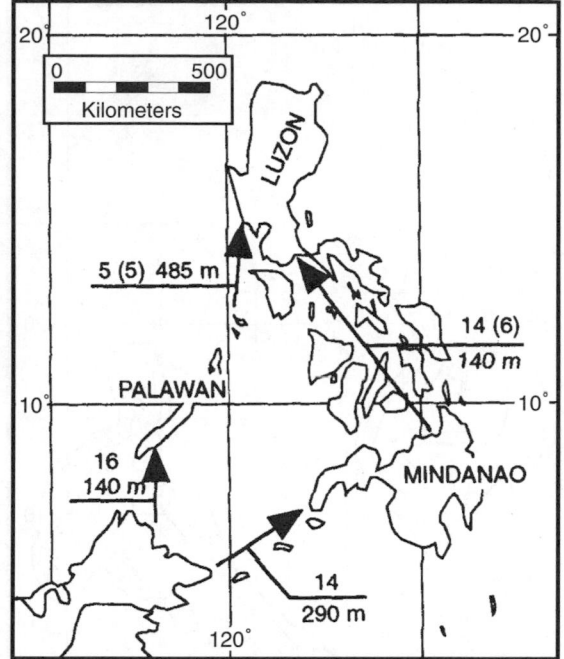

Fig. 8:3. Philippine Islands and adjacent corner of Borneo. Numbers give depths of straits (meters), species shared across straits, and (in parentheses) shared species also occurring in Borneo.

forms a plateau lying above 200 m and much of it above 500 m. Sri Lanka is part of the same continental shelf as India and is narrowly separated from India by a shallow strait. The largest areas of lowland are the Ganges-Brahmaputra and the central Thai-Cambodian area. The latter curls eastward forming a relatively narrow strip along the coast of Vietnam, and also extends southward down the Malay Peninsula. The low central portion of Burma is separated from the Thai-Cambodian lowlands by a narrow tongue of the northeastern montane region. The dry plateau of eastern Thailand and adjacent parts of Laos (Fig. 8:4) is bounded on the south by a pronounced escarpment rising out of the flat lowlands east and north of Bangkok. The escarpment is high enough (maximum elevation 1300 m) to capture water from air currents to provide (or did once) an environment that for amphibians resembled that of the montane areas to the north—moist evergreen forests and steep gradient streams. A short, isolated chain (about 250 km) of mountains with a maximum elevation of about 1650 m rises out of the lowlands of southwestern Cambodia and southeastern Thailand.

The insular portions of southern Asia include large areas below 200 m (e.g., much of Sumatra and Borneo) but also large montane areas, such as the western third of Sumatra, the spine of western Borneo, and parts of Sulawesi and the Philippines. The Greater Sundas—Borneo, Java, and Sumatra—lie on the same continental shelf as the Malay Peninsula; several small island groups (e.g., Natunas) are part of the same shelf.

Based on geologic history and physiography, southern Asia may be divided into the following 10 units (Fig. 8:4):

1. India and Sri Lanka (including all of India except for those parts noted in the next two units).

2. Southern Himalayan Flanks (Nepal, Bhutan, Sikkim, the Darjeeling district of West Bengal, northern Uttar Pradesh, Arunachal Pradesh, Himachal Pradesh, Ladakh, Jammu, and Kashmir).

3. Northeastern Montane Region (Meghalaya, Manipur, Nagaland, mountains of western, northern and eastern Burma, northern and western mountainous areas of Thailand, mountains of Annam and Tonkin in Vietnam).

4. Thai-Lao Dry Plateau.

5. Southeast Asian Lowlands (central valleys of Burma and Thailand, Cambodia, southern and coastal Vietnam, and the Red River Valley of Vietnam).

6. Tenasserim and Malay Peninsula (including peninsular parts of Burma, Thailand, and West Malaysia).

7. Sundaland (the Greater Sundas plus Natuna Islands and the Palawan chain of the Philippine Islands).

8. Sulawesi.

9. Philippine Islands (excluding the Palawan chain).

10. Lesser Sundas (from Bali on the west to Timor and Wetar on the east).

Rainfall patterns in southern Asia vary greatly (Fig. 8:5), from aseasonal to strongly seasonal and from the extraordinary total of > 10 m at Cherrapunjee, Meghalaya, to < 600 mm in parts of the Deccan of India. Annual precipitation is high (> 3000 mm) in the Western Ghats forming the western edge of the Indian peninsula, parts of the Northeast Montane Region, the Arakan Coast of western Burma, the Malay Peninsula, Borneo, Sumatra, Sulawesi, and parts of the Philippines. It is slightly less than that in montane Vietnam and Java. Rainfall is < 2000

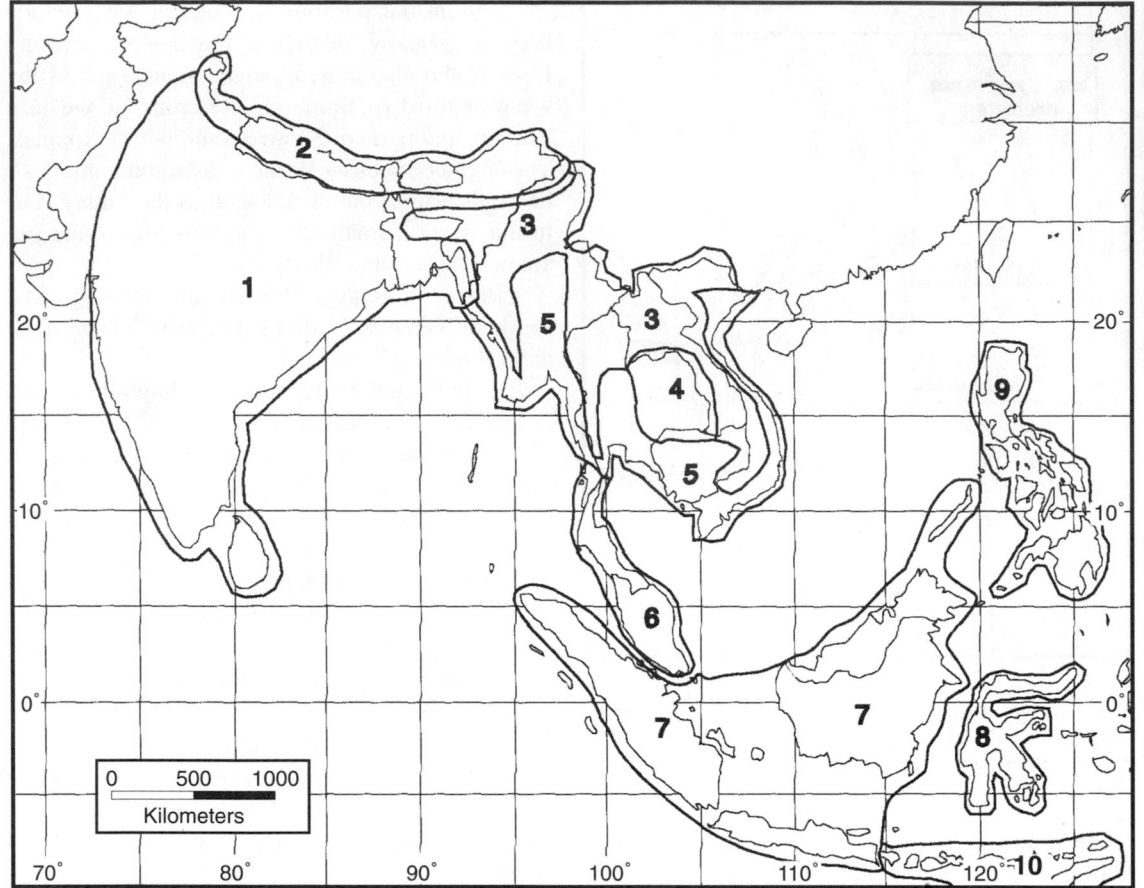

Fig. 8:4. Biogeographic regions of southern Asia based on the distribution of amphibians. 1 = India/Sri Lanka. 2 = Southern Himalayan Flanks. 3 = Northeastern Montane Region. 4 = Thai-Lao Dry Plateau. 5 = Southeast Asian Lowlands. 6 = Tenasserim and Malay Peninsula. 7 = Sundaland. 8 = Sulawesi. 9 = Philippine Islands. 10 = Lesser Sundas.

mm in the dry Thai-Lao plateau and about 1000 mm over the bulk of the Indian peninsula behind the rain shadow of the Western Ghats. There is no clear dry season in the Greater Sundas (except in eastern Java), the southern Malay Peninsula, and the Arakan and Tenasserim coasts of Burma. Dry seasons of 2–4 months characterize most of the Northeastern Montane Region, and very hot dry periods of at least four months are typical of the Thai-Lao Plateau, most of India except for the Western Ghats, and the Lesser Sundas (Anonymous, 1961).

The dominant general type of vegetation until recently over all except small portions of South Asia was forest. (See vegetation maps and discussion in Collins et al., 1991.) Evergreen rainforest was the native vegetation in the perhumid regions along the west-facing coasts of India and Burma, the Malay Peninsula, the Greater Sundas (except eastern Java),

the Philippines, and Sulawesi. Monsoon forests, largely deciduous but including areas of evergreen vegetation, characterized most of Burma, Thailand, and Indochina where precipitation is seasonal. Strongly seasonal monsoon forests also covered the Lesser Sundas. Most of India east of the Western Ghats was covered by drier scrub forest or savanna. At elevations above 1000 m, the forests of Southeast Asia change from ones in which dipterocarps dominate to montane forests in which oaks, chestnuts, and conifers are common. The last three groups of plants characterized the southern flanks of the Himalayas. Agriculture, resource extraction, and, more recently, urbanization have greatly reduced forested areas. Perhaps the greatest impact of deforestation on amphibians of southern Asia has been the siltation of formerly clear streams with resultant loss of tadpole microhabitats.

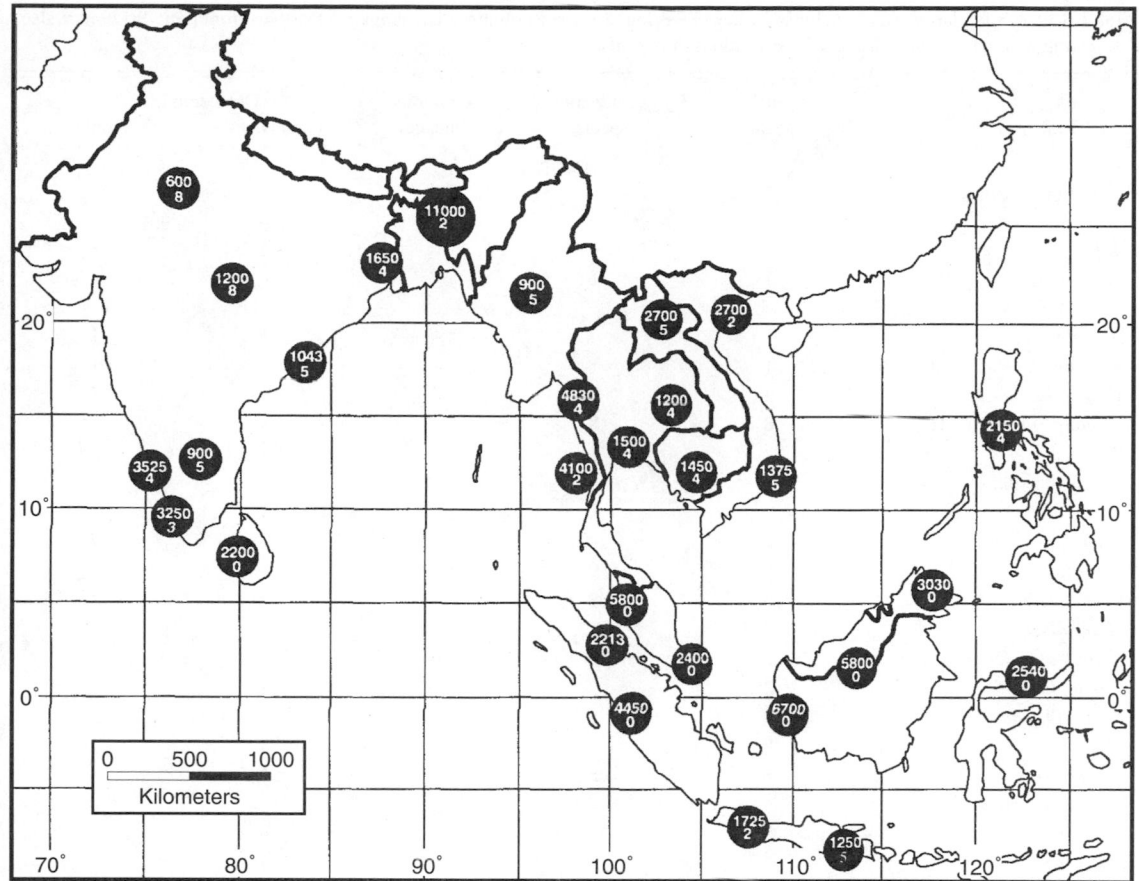

Fig. 8:5. Rainfall at selected stations in southern Asia. Upper number of each pair = total annual rainfall (mm); lower = number of months with < 50mm of rain. Sources: Anonymous (1961); Clayton (1927). Data for one locality (central spot in Borneo) based on personal data.

REGIONAL FAUNAS

Patterns of distribution are determined by historical (geological) relations of land areas and by ecological circumstances of the near past. A few exceptional species of southern Asian amphibians are virtually ubiquitous in distribution; they are mainly species that are commensals of man, following agriculture and other human activities that create severe disturbance of native vegetation and aquatic environments. However, the majority respond to variations in climate, vegetation, and topography and are rarely distributed across large areas.

The political division of southern Asia does not correspond, except occasionally, with boundaries that may influence biogeographic distributions. For example, of the 19 species of *Rana* known from Thailand, 14 occur in peninsular Thailand, which has a relatively perhumid climate and a natural cov-

er of rainforest, and only seven occur in the mountains of northern and western Thailand, where monsoon forests form the native vegetation. Most faunal reviews have been delimited by political entities (e.g., India, Thailand), and the most recent summary of the amphibian species of the world (Frost, 1985) also presents distributions in terms of political units. However, since this chapter is an attempt to understand a biological phenomenon, distribution of the amphibians of southern Asia will be presented in terms of the 10 physiographic regions listed above. A complete list of the species, showing their distributions in regions, is presented in Appendix 8:1.

INDIA AND SRI LANKA

Climates and vegetation vary greatly over this large region (see above and discussion in Das, ms),

Table 8:1. Amphibian fauna of the India/Sri Lanka Region. Last two columns give numbers of species from India/Sri Lanka also occurring in other regions. See text for definitions of regions.

Lineage	Total species	Endemic species	Endemic lineage	Occurring in	
				Himalayan	NE montane
ANURA					
BUFONIDAE:					
Ansonia	2	2	–	–	–
Bufo	11	9	–	2	1
Pedostibes	1	1	–	–	–
MICROHYLIDAE:					
Kaloula	1	–	–	1	1
Melanobatrachus	1	1	+	–	–
Microhyla	3	2	–	1	1
Ramanella	8	8	+	–	–
Uperodon	2	1	–	1	–
RANIDAE:					
Euphlyctis	2	1	–	1	1
Hoplobatrachus	2	–	–	2	1
Indirana	9	9	+	–	–
Limnonectes	10	8	–	2	1
Micrixalus	7	7	+	–	–
Nannobatrachus	2	2	+	–	–
Nannophrys	3	3	+	–	–
Nyctibatrachus	8	8	+	–	–
Rana	8	7	–	1	1
Tomopterna	6	5	–	1	–
RHACOPHORIDAE:					
Philautus	25	25	–	–	–
Polypedates	4	3	+	1	–
Rhacophorus	6	6	–	–	–
Subtotals	121	108	–	13	7
GYMNOPHIONA					
CAECILIIDAE:					
Gegeneophis	3	3	+	–	–
Indotyphlus	1	1	+	–	–
ICHTHYOPHIIDAE:					
Ichthyophis	10	10	–	–	–
URAEOTYPHLIDAE:					
Uraeotyphlus	4	4	+	–	–
Grand totals	139	126	–	13	7

but I treat it as one unit, because so few species (< 10%) are known from the humid, forested areas of the Western Ghats and Sri Lanka. Species and lineage diversity are high (Table 8:1), especially in the Western Ghats. This area of low mountains, which forms such a small proportion of the total area of India, has 114 species of anurans (Inger and Dutta, 1986; Ravichandran, 1992), and all but one of the lineages in Table 8:1 occur there. By contrast, Sri Lanka and the Eastern Ghats along India's east coast have far fewer species, 41 and 21, respectively, and

the large interior Deccan Plateau has only 18 (Inger and Dutta, 1986). The high level of endemism at both specific and generic levels (Table 8:1) in this region is striking. All 13 of the species that India/Sri Lanka share with an adjacent region are tolerant of severe anthropogenic habitat disturbance and six (e.g., *Bufo melanostictus, Microhyla ornata,* and *Hoplobatrachus tigerinus*) appear to be confined to such situations.

SOUTHERN HIMALAYAN FLANKS

The anuran fauna is rich (Table 8:2) for a rela-

Table 8:2. Amphibian fauna of the Southern Himalayan Flanks. Last two columns give numbers of species from Himalayan flanks also occurring in other regions. See text for definitions of biogeographic regions.

Lineage	Total species	Endemic species	Occurring in NE montane	China
ANURA				
MEGOPHRYIDAE:				
Megophrys	3	2	1	1
Scutiger	4	2	–	1
BUFONIDAE:				
Bufo	5	1	1	2
MICROHYLIDAE:				
Kaloula	1	–	1	1
Microhyla	1	–	1	1
Uperodon	1	–	–	–
RANIDAE:				
Altirana	1	–	–	1
Amolops	5	1	3	1
Euphlyctis	1	–	1	–
Hoplobatrachus	2	–	1	1
Limnonectes	6	3	2	1
Paa	10	6	1	3
Phrynoglossus	1	1	–	–
Rana	7	1	6	2
Tomopterna	2	1	–	–
RHACOPHORIDAE:				
Chirixalus	1	–	1	1
Philautus	4	2	2	–
Polypedates	3	1	1	1
Rhacophorus	6	4	1	2
Theloderma	2	–	2	2
CAUDATA				
SALAMANDRIDAE:				
Tylototriton	1	–	1	1
GYMNOPHIONA				
ICHTHYOPHIIDAE:				
Ichthyophis	1	1	–	–
Totals	68	26	26	22

tively small area, much of which is at high elevation. The fauna is a mixture of a few mainly Himalayan groups (*Altirana* and *Paa*), a few Indian species tolerant of severe environmental disturbances (*Uperodon systoma, Euphlyctis cyanophlyctis,* and *Tomopterna breviceps*), and one temperate taxon (*Scutiger*), but mainly tropical and subtropical groups. *Paa* reaches its maximum diversity here, where all streams are swift and provide the appropriate breeding habitat. No significant natural boundaries separate this area from China, at least along

the east, and a large portion of the species also occur in China. Taxa shared with China are not solely those with a primarily temperate range, such as *Scutiger* and *Altirana,* but include several with primarily tropical and subtropical distributions (e.g., *Chirixalus* and *Rhacophorus*). It is likely that the overlap with China will increase as the active sampling of the Yunnan fauna by the group in Kunming continues (see Yang et al., 1991), and the endemism, which is quite high now (about 40%), will probably decrease. All species shared with the Indian/Sri Lankan area are ones that thrive in areas of intense anthropogenic disturbance.

NORTHEASTERN MONTANE REGION

Endemism is high (36%) at the species level. The Ranidae dominate the fauna (Table 8:3) as they do in other regions. Many species are shared with China, perhaps reflecting the importance of the corridors formed by the headwaters of many of the largest rivers in Southeast Asia; the Mekong, Irriwaddy, Salween, and Red rivers all rise in the mountains of Yunnan where most of the shared species occur (see maps in Yang et al., 1991). Overlaps with the Southeast Asian Lowlands and Thai-Lao Plateau are smaller in absolute terms (Table 8:3) but greater relative to the total known faunas of those two regions.

This region is heterogeneous in climate and in vegetation. Evergreen and monsoon forests are mixed in complicated patterns in Burma, Meghalaya, and Nagaland in northeastern India, Indochina, and northern Thailand (Collins et al., 1991), and reflect rainfall patterns associated with prevailing winds and differences in aspects and heights of mountain ranges. These variations could result in significant geographic restriction in the distributions of species, an hypothesis that can be tested by analysis of distributions across the region. In order to examine these patterns, I have considered distributions in the four quasi-political units arranged in a roughly linear pattern from northwest to southeast: northeastern India (Meghalaya, Manipur, and Nagaland), Burma, Thailand, and Indochina (Vietnam, Laos, Cambodia). Of the 129 species in the Northeast Montane Region (Table 8:3), 83 are known from only one of the four political units. That observation plus the fact that only seven of the 48 species endemic to this region occur in more than a single political unit support the hypothesis of a high degree of geographic restriction. However, the same data

Table 8:3. Amphibian fauna of the Northeast Montane Region. See text for definitions of biogeographic regions.

Lineages	Total species	Endemic species	Species also occurring in			
			SE Asia Lowlands	Thai-Lao Plateau	Peninsula	China
ANURA						
BOMBINATORIDAE:						
Bombina	1	–	–	–	–	1
BUFONIDAE:						
Bufo	7	2	2	1	4	2
Bufoides	1	1	–	–	–	–
Pedostibes	1	1	–	–	–	–
HYLIDAE:						
Hyla	2	–	–	–	–	2
MEGOPHRYIDAE:						
Leptobrachium	2	–	–	–	1	1
Leptolalax	2	1	–	–	1	1
Megophrys	7	2	1	1	1	5
Ophryophryne	2	1	–	–	–	1
Scutiger	1	1	–	–	–	–
MICROHYLIDAE:						
Kalophrynus	1	–	1	1	1	1
Kaloula	1	–	1	1	1	1
Microhyla	5	–	5	5	5	4
Micryletta	1	–	1	1	1	1
RANIDAE:						
Amolops	6	2	–	–	1	2
Elachyglossa	1	1	–	–	–	–
Euphlyctis	1	–	–	–	–	–
Hoplobatrachus	1	–	1	1	1	–
Huia	1	–	–	–	–	1
Ingerana	1	–	–	–	1	–
Limnonectes	11	3	3	2	7	2
Occidozyga	1	–	1	1	1	1
Paa	5	1	–	–	–	4
Phrynoglossus	1	–	–	1	–	–
Pterorana	1	1	–	–	–	–
Rana	22	10	4	4	7	3
RHACOPHORIDAE:						
Chirixalus	4	1	3	3	–	2
Philautus	15	9	1	1	–	3
Polypedates	3	–	1	1	1	2
Rhacophorus	8	5	1	1	3	2
Theloderma	6	4	–	–	1	2
CAUDATA						
SALAMANDRIDAE:						
Echinotriton	1	–	–	–	–	1
Paramesotriton	1	1	–	–	–	–
Tylototriton	1	–	–	–	–	1
GYMNOPHIONA						
ICHTHYOPHIIDAE:						
Ichthyophis	4	3	1	1	1	–
Totals	129	50	26	26	39	46

Table 8:4. Amphibian fauna of the Southeast Asian Lowlands and the Thai-Lao Dry Plateau. Last column gives number of species also occurring in adjacent region. See text for definitions of biogeographic regions.

Lineages	SE Asian Lowlands		Thai-Lao Plateau		Occurring in Peninsula
	Total species	Endemic species	Total species	Endemic species	
ANURA					
BUFONIDAE:					
Bufo	5	2	1	–	3
MEGOPHRYIDAE:					
Megophrys	2	–	–	–	1
MICROHYLIDAE:					
Calluella	1	–	1	–	1
Glyphoglossus	1	–	1	–	1
Kalophrynus	1	–	1	–	1
Kaloula	2	–	2	–	2
Microhyla	8	3	5	–	5
Micryletta	1	–	1	–	1
RANIDAE:					
Hoplobatrachus	2	–	1	–	2
Limnonectes	7	3	2	–	3
Occidozyga	1	–	1	–	1
Paa	1	1	–	–	–
Phrynoglossus	1	–	2	–	1
Rana	9	3	6	–	5
RHACOPHORIDAE:					
Chirixalus	3	–	4	1	–
Philautus	1	1	1	1	–
Polypedates	1	–	1	–	1
Rhacophorus	2	2	3	1	2
Theloderma	1	–	1	–	–
GYMNOPHIONA					
ICHTHYOPHIIDAE:					
Ichthyophis	1	–	1	–	1
Totals	51	15	35	3	31

could result from inadequate sampling (see above, Sampling Weaknesses). Furthermore, as 25 species occur in three or four of these areas, environments across this region do not appear to be radically different for amphibians.

SOUTHEAST ASIAN LOWLANDS

The low diversity of this fauna (Table 8:4) probably results from two factors: (1) the almost complete conversion of the vegetation cover by agriculture; and (2) the generally flat topography, which eliminates streams with the characters—clarity, moderate to strong currents—that many of the species in the preceding region require for breeding.

There are no *Amolops* and the numbers of megophryids and species of *Paa* are greatly reduced. However, one short, isolated range of low mountains in southeastern Thailand and adjoining western Cambodia provides clear, rocky streams where species, such as *Limnonectes kuhlii* and the only species of *Megophrys* and *Paa* that occur in this region, can breed. These mountains also are the only locality for four other species (*Bufo parvus, Kalophrynus pleurostigma, Rana montivaga,* and *Theloderma stellatum*) that are restricted to humid forest.

Species and lineages of Microhylidae are relatively much more conspicuous in this region than in the preceding one, perhaps because the flat terrain and strongly seasonal rainfall provide an abundance of temporary bodies of water, the breeding habitat for all Asian species of this family. The same type of habitat is used by the lineages of Rhacophoridae, except *Philautus.* More than half (30 of 51) of the species also occur in the Peninsular Region. Of those that do not, 15 are endemic to the Southeast Asian Lowlands and are known so far from only a single locality; three are members of a lineage, *Chirixalus,* absent in the peninsula. The greatest proportional overlap is with the next region.

Thai-Lao Dry Plateau

Scarcely any of the native vegetation remains in this region. One measure of this drastic modification is the almost complete absence of forest reserves and conservation areas (see Map 27.1 in Collins et al., 1991). The small number of species known from the region probably reflects this habitat modification, as well as small area and poor sampling. The last factor may account for the small contribution of Ranidae to the total, compared to other regions. The great majority of species in this fauna are pond breeders, including all Microhylidae, all but two of the Ranidae, and all Rhacophoridae except *Philautus parvulus.*

The southwestern edge of this region is a pronounced escarpment that catches moisture in winds coming from the more humid lowlands to the south. Evergreen forests characterized this area; it is the only part of the plateau where *Rana livida* and *Rhacophorus bisacculus* have been found. Until relatively recently, small patches of seasonal evergreen forest existed on the plateau, and one such patch is the only known locality for *Rhacophorus bipunctatus* and *Theloderma stellatum* in this region. The two *Rhacophorus, Rana livida,* and *Chirixalus hansenae* are the only plateau species not shared with the lowlands.

Tenasserim-Malay Peninsular Region

Although relatively small in area, this region has a rich fauna (Table 8:5), with high diversity both at the species and lineage levels. Endemism is only moderate. The Thai and Malaysian portions of the region have been well sampled, in contrast to the Burmese Tenasserim. Only 25 of the 115 species have been reported from Tenasserim, but 73 and 98 from the Thai and Malaysian sectors, respectively. No physical characteristic can explain the low num-

Table 8:5. Amphibian fauna of the Tenasserim-Malay Peninsular Region. See text for definition of region.

Lineage	Total species	Endemic species	Also in Sundaland
ANURA			
Bufonidae:			
Ansonia	6	4	2
Bufo	6	–	5
Leptophryne	1	–	1
Pedostibes	1	–	1
Pelophryne	1	–	1
Pseudobufo	1	–	1
Megophryidae:			
Leptobrachium	3	–	2
Leptolalax	3	1	1
Megophrys	5	1	2
Microhylidae:			
Calluella	2	–	1
Chaperina	1	–	1
Glyphoglossus	1	–	–
Kalophrynus	3	2	1
Kaloula	3	–	2
Metaphrynella	1	–	1
Microhyla	9	–	7
Micryletta	2	1	1
Phrynella	1	–	1
Ranidae:			
Amolops	2	1	–
Hoplobatrachus	2	–	1
Ingerana	2	1	–
Limnonectes	14	4	8
Occidozyga	1	–	1
Phrynoglossus	2	–	1
Rana	17	2	9
Rhacophoridae:			
Nyctixalus	1	–	1
Philautus	2	1	1
Polypedates	3	–	3
Rhacophorus	10	3	5
Theloderma	3	–	2
GYMNOPHIONA			
Ichthyophiidae:			
Caudacaecilia	2	2	–
Ichthyophis	4	2	1
Totals	115	25	63

ber in Tenasserim. Compared to the preceding regions, there is a marked increase in diversity of lineages and species in the Bufonidae and Microhylidae. The dearth of species of *Philautus* (Table 8:5) is almost certainly an artifact of our ignorance of this taxonomically difficult group.

No sharp physical barrier separates this region from the Southeast Asian Lowlands and most (31 of 55) species from the latter region occur in both faunas. The peninsular region is only narrowly separated from the Northeast Montane, yet only 37 (32%) species are shared with that region (Table 8:3) and seven of those are commensals of man. The most significant overlap is with the Greater Sundas, and although the overlap varies among families, the number of species shared is directly related to the proportional contribution of each family to the peninsular fauna.

SUNDALAND

The association of the Palawan chain of the Philippines with Sundaland was based on submarine topography and geologic history, but is supported by the distribution of amphibians. Two-thirds (16 of 23) of Palawan's amphibian species are shared with Borneo (Fig. 8:3). Three Palawan endemics (*Barbourula busuangensis, Ingerana mariae,* and *Bufo philippinicus*) are closely related to Bornean species and are members of lineages not known from the rest of the Philippine Islands.

This region has the highest diversity at the species and lineage levels (Table 8:6) in southern Asia. Species diversity may be related to the often-discussed Pleistocene changes in sea level that alternately connected and separated these large islands and provided time for genetic divergence of local populations. That hypothesis could explain speciose genera with high levels of endemism, such as *Ansonia, Bufo, Kalophrynus, Limnonectes, Philautus, Rhacophorus, Meristogenys,* and *Leptobrachella.* Only a small proportion of species (50 = 27% of total) have been reported from two or more of the largest islands—Borneo, Sumatra, Java, Palawan. However, the fact that 89 of the remaining 144 species have so far been reported only from Borneo, the most thoroughly sampled island, suggests strongly that the heterogeneity of distributions within Sundaland is at least partly an effect of inadequate sampling on Sumatra, Java, and Palawan rather than a biological phenomenon. This argument is supported by the distributions of three lineages now known only from Borneo among the large islands: *Leptobrachella* (7 species) and *Meristogenys* (8), which are endemic to Sundaland, and *Ansonia* (11), which

Table 8:6. Amphibian fauna of Sundaland. See text for definitions of regions.

Lineage	Total species	Endemic species	Species also in Philippines	Sulawesi
ANURA				
BOMBINATORIDAE:				
Barbourula	2	2	–	–
BUFONIDAE:				
Ansonia	11	9	–	–
Bufo	12	6	–	–
Leptophryne	2	1	–	–
Pedostibes	4	3	–	–
Pelophryne	6	4	1	–
Pseudobufo	1	–	–	–
MEGOPHRYIDAE:				
Leptobrachella	7	7	–	–
Leptobrachium	5	1	1	–
Leptolalax	2	1	–	–
Megophrys	5	3	–	–
MICROHYLIDAE:				
Calluella	4	3	–	–
Chaperina	1	–	1	–
Gastrophrynoides	1	1	–	–
Kalophrynus	8	7	1	–
Kaloula	2	–	–	2
Metaphrynella	2	1	–	–
Microhyla	11	4	–	–
Micryletta	1	–	–	–
Phrynella	1	–	–	–
RANIDAE:				
Hoplobatrachus	1	–	–	–
Huia	3	3	–	–
Ingerana	2	2	–	–
Limnonectes	16	8	2	2
Meristogenys	8	8	–	–
Occidozyga	1	–	–	–
Phrynoglossus	2	1	1	1
Rana	13	4	3	1
Staurois	3	2	1	–
RHACOPHORIDAE:				
Nyctixalus	2	1	1	–
Philautus	17	15	–	–
Polypedates	4	1	2	–
Rhacophorus	18	13	3	–
Theloderma	2	–	–	–
GYMNOPHIONA				
ICHTHYOPHIIDAE:				
Caudacaecilia	3	3	–	–
Ichthyophis	11	10	–	–
Totals	194	124	17	6

Table 8:7. Amphibian fauna of the Philippine and Sulawesi[1] Regions. See text for definitions of regions.

| | Philippines | | Sulawesi | |
Lineage	Total species	Endemic species	Total species	Endemic species
ANURA				
BUFONIDAE:				
Ansonia	2	2	–	–
Bufo	1	–	1	1
Pelophryne	2	1	–	–
MEGOPHRYIDAE:				
Leptobrachium	1	–	–	–
Megophrys	1	1	–	–
MICROHYLIDAE:				
Chaperina	1	–	–	–
Kalophrynus	1	–	–	–
Kaloula	5	5	2	–
Oreophryne	2	2	3	3
RANIDAE:				
Limnonectes	10	8	5	3
Phrynoglossus	2	1	3	2
Platymantis	12	12	–	–
Rana	6	3	3	2
Staurois	1	–	–	–
RHACOPHORIDAE:				
Nyctixalus	2	1	–	–
Philautus	9	9	–	–
Polypedates	3	1	1	–
Rhacophorus	4	1	3	3
GYMNOPHIONA				
ICHTHYOPHIIDAE:				
Ichthyophis	2	2	–	–
Totals	61	49	21	14

[1]Recent field work by D. T. Iskandar suggests significant increases in numbers and knowledge of faunal relationships are in the offing.

Table 8:8. Amphibian fauna of the Lesser Sunda Islands.

Lineage	Total species	Endemic species	Species also in Java
ANURA			
BUFONIDAE:			
Bufo	2	–	2
MEGOPHRYIDAE:			
Leptobrachium	1	–	1
MICROHYLIDAE:			
Kaloula	1	–	1
Microhyla	1	–	1
Oreophryne	3	3	–
RANIDAE:			
Limnonectes	5	3	2
Phrynoglossus	2	1	1
Rana	3	1	2
RHACOPHORIDAE:			
Polypedates	1	–	1
Totals	18	8	10

the Peninsular Region (Table 8:5). Two lineages (*Oreophryne* and *Platymantis*) do not occur in Sundaland, but do occur in the Moluccas and the Papuan region (see distributions in Frost, 1985). On the other hand, the Philippines Region lacks the genera *Barbourula* and *Ingerana* which ally Palawan to Borneo. Another distinctive feature of the Philippine fauna is the diversity within the genus *Kaloula*. Endemism at the specific level is high; all species that are not endemic also occur on Borneo, which has long been considered the major source of the Philippine fauna (Inger, 1954). Endemism in *Limnonectes* is higher than elsewhere except India/Sri Lanka (Table 8:1), but the high rate of endemism in *Philautus* follows the pattern of this group in other regions (Tables 8:1, 8:3, 8:6).

SULAWESI

Sulawesi is almost twice the size of Java (28 species recorded), and probably has at least the same number of species. The small number reported so far (Table 8:7) is an underestimate. Dr. D. Iskandar in three brief trips has discovered several undescribed species. Despite the inadequacy of the sampling, the high rate of endemism shown in Table 8:7 is probably an accurate reflection of the fauna; the four non-endemic species (*Kaloula baleata, Limnonectes cancrivorus, Phrynoglossus laevis,* and *Rana chalconota*) thrive in disturbed forest and adjacent clear-

also occurs on the mainland. All three breed only along clear, rocky streams. The mountainous areas of Sumatra, Java, and Palawan certainly provide suitable habitats; another lineage with similar breeding requirements, *Huia,* occurs on Borneo, Sumatra, and Java. Another datum supports the hypothesis of poor sampling—only two species of *Philautus* have been recorded from Sumatra and only three from Java, in contrast to 12 from Borneo.

PHILIPPINE ISLANDS

This is a relatively small fauna (Table 8:7), considering that the land area is greater than that of

ings (pers. obs.) in Borneo. Except for the microhylid, *Oreophryne,* all lineages are Oriental.

LESSER SUNDAS

The small fauna is a mixture of Sundan and Papuan elements (Table 8:8). Eleven species have long been known from Java, which lies less than 10 km off the west end of the chain. Three of the endemics are species of the Papuan genus *Oreophryne.* The other five are ranids of the lineages *Limnonectes* (3), *Phrynoglossus,* and *Rana;* the last is related to a Papuan species (*R. papua*).

DISCUSSION

DISCONTINUITIES IN DISTRIBUTIONS

Lineages of anurans in southern Asia generally are distributed across the boundaries between adjacent regions. For example, 17 of the 19 lineages occurring in the Southeast Asian Lowlands (Table 8:4) also are present in the Northeast Montane Region (Table 8:3). Eighteen of the 22 lineages that occur in the Himalayan Flanks (Table 8:2) also are present in the Northeast Montane Region. Twenty-eight of the 31 found in the Peninsular Region (Table 8:5) also occur in Sundaland (Table 8:6).

The only exception to this pattern involves the India/Sri Lanka Region. Although it is juxtaposed to the Himalayan and Northeast Montane regions, only about half (13 and 12, respectively) of its 25 genera (Table 8:1) are shared with those regions. Ten of its genera are endemic (Table 8:1), and three others are represented elsewhere in southern Asia by only a single species. Nine of the 13 genera shared with the Himalayan region occur in at least seven of the other nine regions covered in this chapter and in China; three of the remaining six (*Uperodon, Tomopterna,* and *Euphlyctis*) extend only to the Himalayan or the adjacent Northeast Montane Region. At the species level, the isolation of India from the rest of southern Asia seems even more profound; only 13 of its 139 species are shared with other regions, 13 with the Himalayan Flanks and only seven with the Northeast Montane Region. All 13 of these shared species are tolerant of severe habitat disturbance or are confined to such situations, the kind of species that often display immunity to natural barriers to rapid dispersal. Four of these species are distributed from Sri Lanka to the Greater Sundas and two others from Sri Lanka to the Peninsular Region.

Differences between other pairs of adjacent regions are much less pronounced. Overlaps at the level of lineages is extensive. Besides those cases mentioned above, all 17 of the lineages reported from the Thai-Lao plateau also occur in the lowlands (Table 8:4), 17 of the 19 in the Southeast Asian Lowlands are found in the Peninsular Region, and 20 of the 22 in the Philippine Islands also occur in the Greater Sundas. In fact, overlap at this level is high across the entire area excluding India/Sri Lanka. The Himalayan Flanks Region shares as many lineages (13) with the Greater Sundas, 3300 km distant, as it does with the immediately adjacent India/Sri Lanka Region. The Northeast Montane Region shares almost twice as many lineages (20) with the Greater Sundas as it does with adjacent India/Sri Lanka, and 14 of those 20 occur in all regions between the montane and Sundan regions.

It is mainly at the species level that clear differences among the non-Indian regions appear. Despite the high overlap between the Himalayan Flanks and the Northeast Montane regions at the lineage level, only 41% of the Himalayan species are shared with the montane region. The Northeast Montane Region shares 52% of its 29 lineages with the Peninsular Region, but only 29% of its species (Table 8:3). The highest overlaps at the specific level between adjacent regions involve the lowland-plateau pair (27 species shared, or a 62% similarity) and the peninsular-Sundas pair (63 species shared, or a 43% similarity). The lowest overlaps involve the Sunda-Philippines pair (21 species shared) and the Sunda-Sulawesi pair (4 species), but the last number is more a measure of poor sampling than of biogeographic relations.

ORIGINS OF DISCONTINUITIES

Discontinuities in distributions may be explained by ecological or historical causes. The latter seem most obvious for the gap between the India/Sri Lanka Region and all the others. The Gondwanan origin of India is well established. Its relatively late separation from Gondwanaland and its recent connection with the rest of southern Asia could explain the evolution and isolation of nine of the 10 endemic Indian genera and the inclusion in the Indian fauna of three genera with African members (*Hoplobatrachus, Euphlyctis,* and *Tomopterna*). Two species of *Hoplobatrachus* have relatively wide distributions in southern Asia, but both are tolerant of severe anthropogenic disturbance, suggesting a capacity for rapid, recent dispersal. Indeed one, *H. rugulosus,* is known to have made the leap from Taiwan to northern Borneo in recent years as a deliberate introduction (Inger and Stuebing, 1989).

Hedges et al. (1993) suggested a different origin for the exceptional tenth genus, the caecilian *Uraeotyphlus.* They hypothesized that the uraeotyphlids and their sister lineage, the ichthyophiids (or their common antecedent), became isolated in Laurasia when that supercontinent split from Gondwanaland. Subsequently, according to this hypothesis, either proto-*Uraeotyphlus* then dispersed into India after its fusion with Asia and left no remnant survivor in what remains of Laurasia, or an early ichthyophiid dispersed into India and gave rise to *Uraeotyphlus.*

Historical factors undoubtedly have played a role in other discontinuities. The Philippine fauna includes two lineages with clear Papuan affinities, *Platymantis* and *Oreophryne.* The presence of these two genera in the Philippines (but not in Palawan) may date from either the pre-Tertiary or Oligocene (Hall, 1997, in press), when the eastern Philippines-Halmahera arc was closest to New Guinea and the Melanesian Islands. The presence of *Oreophryne* on Sulawesi may date from the Miocene when Hall (in press) postulates that a fragment of New Guinea rifted and ultimately became incorporated with Sulawesi.

The single disjunction in southern Asia at the familial level, involving the Bombinatoridae, probably is very old. The family is represented by one genus, *Barbourula,* in Borneo and the Palawan chain of the Philippine Islands and one, *Bombina,* in northern Indochina and China. Given the long history of the family, with fossils as old as the Jurassic (Estes and Sanchiz, 1982), the separation of the two genera could be traceable to a vicariant event in the late Mesozoic. However, there is no evidence that the Palawan chain included emergent portions during the early Tertiary when Palawan was closest to southern China (Hall, in press). Nor is there any evidence that Borneo was closer to Indochina or China during the Tertiary (Hall, in press), although Metcalfe (1988) suggested that southwestern Borneo was a terrane that rifted from northeastern Indochina in the late Cretaceous.

Pleistocene changes in the South China Sea during northern glacial stages united the Peninsular and Sundaland regions and affected the distributions of amphibians (Inger, 1966). They almost certainly account for the wide overlap at the specific level between these two regions. Although these changes in sea level probably did not effect juncture between Sundaland and the Philippines Region (excluding the Palawan chain), they almost certainly resulted in the narrowing of the gap between Borneo and the southeastern Philippines and in development of a land bridge between Mindanao and Luzon in the northern Philippines. Dispersal during the Pleistocene may account for the 14 species Mindanao (total fauna 40 species) shares with Borneo and the 14 species shared by Mindanao and Luzon (total fauna 23 species) (Fig. 8:3).

Pleistocene changes in climate were accompanied also by changes in altitudinal zonation of environments and, at times, lowered what is now termed the montane vegetation zone by at least 500 m in places (Holloway, 1986). A depression in the lower altitudinal limit of the montane zone would result also in great areal expansion of the zone. These changes effected well-documented alterations in the distributions of certain invertebrates (Holloway, 1986) and probably altered the distributions of anurans in southern Asia. The occurrence of *Paa fasciculispina* 1000 km south of its nearest congeners (Fig. 8:6) may have resulted from such change in climate zonation. *Leptolalax pelodytoides* with an equally wide disjunction between populations in West Malaysia and the Karen Hills of Burma (Fig. 8:7), may be another example of the same phenomenon.

ECOLOGICAL CORRELATES OF DISCONTINUITIES

Absence of a lineage or species in an area may be related to patterns of rainfall, vegetation characteristics, or structure of the environment. For most species, precise knowledge of response to quantity and seasonality of rainfall is nonexistent, obliging us to look for correspondence between ranges of species and rainfall patterns. Areas of very heavy rainfall and weak seasonality occur in many parts of southern Asia, but large areas have moderate to strong seasonality, and some areas have relatively low amounts of rain (Fig. 8:5). Much less rain falls in the part of India closest to the Himalayan and

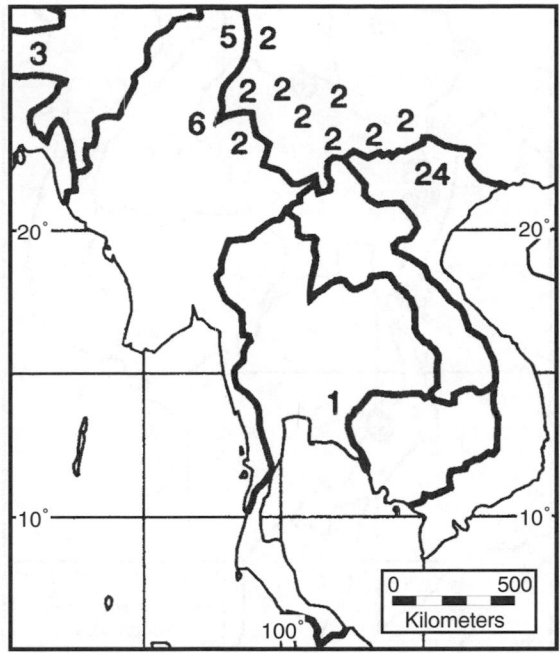

Fig. 8:6. Distribution of species of *Paa* in Southeast Asia. 1 = *P. fasciculispina*. 2 = *P. yunnanensis*. 3 = *P. blanfordi*. 4 = *P. bourreti*. 5 = *P. arnoldii*. 6 = *P. feae*.

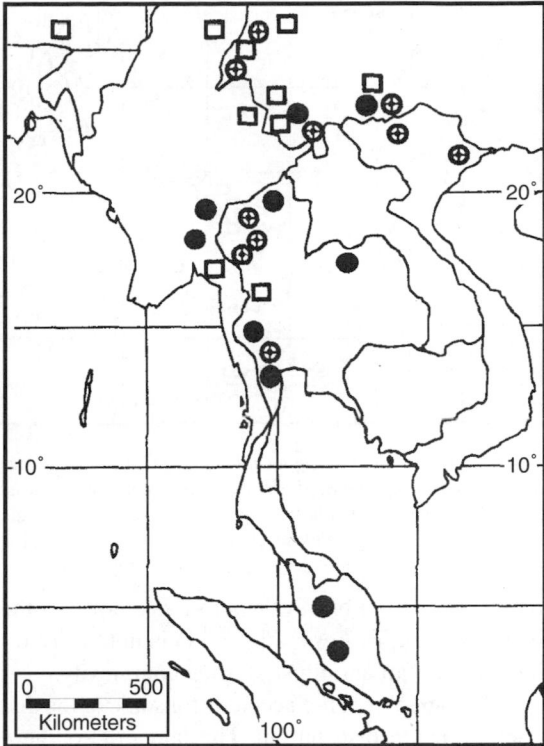

Fig. 8:7. Distribution of *Megophrys lateralis* (open square) and *Leptolalax pelodytoides* (solid circle) in Southeast Asia. Circle enclosing cross = both species.

northeastern montane regions than in those two regions. This factor may account for absence of many of the Himalayan and montane lineages from India. Probably it also is responsible for the separation of the endemic Indian lineages, which are confined to the wetter, more heavily forested areas of the Indian peninsula, from the wetter areas to the north and northeast. Had these endemic lineages (e.g., *Nyctibatrachus* and *Micrixalus*) been close to the humid areas in the northeast in the past, at least a few species should have been able to disperse into those regions.

Changes in distributions of amphibians in relation to rainfall patterns (and associated patterns of vegetation) across Southeast Asia is complex geographically because of the effects of height of land and its aspect. The relationship is most clearly shown in the Malay Peninsula. Smith (1931b) called attention to the change in herpetological faunas that occurs around 10° N (Isthmus of Kra). The Isthmus provides a rough dividing line between regions with and without a distinct, harsh dry season (Inger, 1966: Table 48). Many groups of anurans reach the northern limits of their distributions at this point (Inger, 1966; Nabhitabhata, ms) (Fig. 8:8). However, that

break in rainfall pattern applies mainly to the land east of the hills that extend north-south in the peninsula. The west-facing slopes of those hills and the mountains that continue north along the Burma-Thailand border have heavy rainfall (Fig. 8:5) and support evergreen rainforests (Map 15.1 in Collins et al., 1991). Some amphibians characteristic of the perhumid areas to the south occur in these areas north of the Isthmus of Kra (Fig. 8:9). Altogether 16 species in the families Megophryidae (e.g., *Leptolalax pelodytoides* and *Megophrys lateralis;* Fig. 8:7), Bufonidae (*Bufo asper* and *B. parvus*), Microhylidae (*Microhyla berdmorei*), Ranidae (e.g., *Ingerana tenasserimensis, Rana cubitalis,* and *R. leptoglossa,* Fig. 8:10), and Rhacophoridae (*Rhacophorus bipunctatus*) have a significant portion of their ranges in these rain-trapping north-south mountains or in similar areas at the eastern margin of the central Thai lowlands. However, the relation of distribution to patterns of rainfall (or associated type of vegetation) is not simple. The failure of some of those species to range south of the Isthmus of Kra (e.g., *Rana*

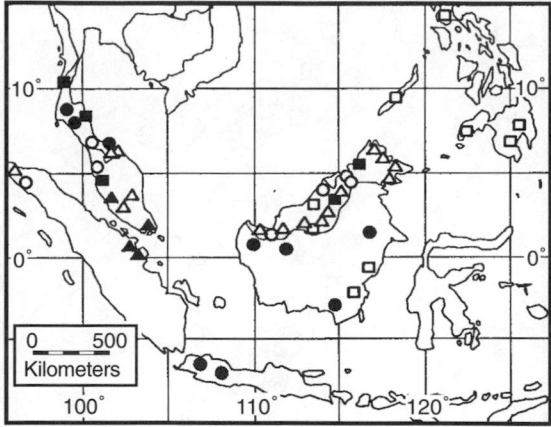

Fig.8:8. Distribution of *Rana chalconota* (solid circle), *R. signata* (open square), and *R. glandulosa* (open circle) in Southeast Asia. Open triangle = *R. chalconota* + *R. signata*, closed triangle = *R. chalconota* + *R. glandulosa*, closed square = all three species.

livida; Fig. 8:11) or of others to range north of the isthmus, (e.g., *R. hosii;* Fig. 8:11) is not easily related to rainfall and illustrates the complexity.

Topographic relief accounts for some of the discontinuities in distribution. The fauna of southern Asia includes many lineages that are obligate stream breeders. Larvae of all lineages of Megophryidae develop in streams, as do those of several bufonid, ranid, and rhacophorid lineages (Smith, 1924; Liu and Hu, 1960; Inger, 1985; Inger et al., 1986) (Ta-

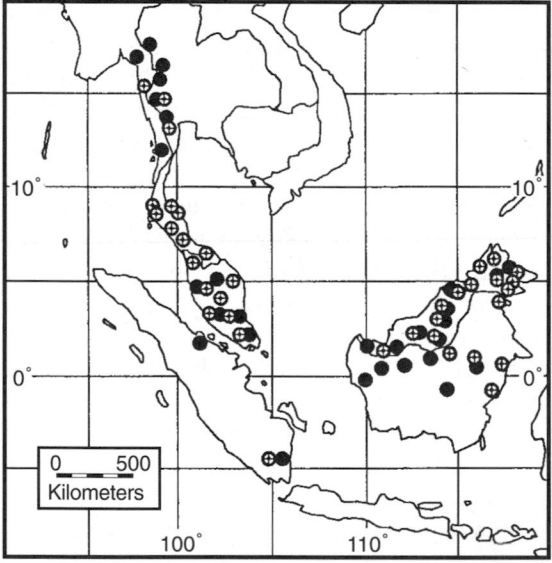

Fig. 8:9. Distribution of *Limnonectes blythi* (cross in circle) and *Bufo asper* (solid circle) in Southeast Asia and Sundaland.

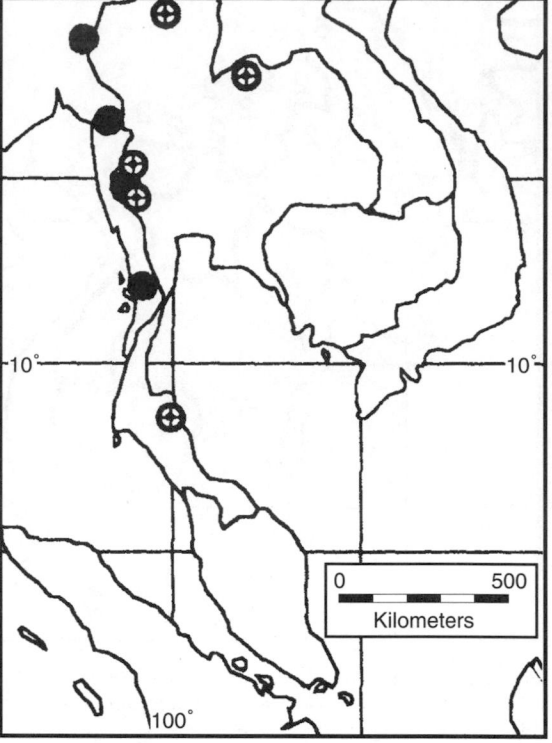

Fig. 8:10. Distribution of *Rana cubitalis* (circle enclosing cross) and *R. leptoglossa* (solid circle).

ble 8:9). These larvae vary greatly in their distributions with respect to current; some (e.g., *Limnonectes blythi;* Inger et al., 1986) live in weak currents, and others (e.g., *Amolops;* Yang, 1991) live in torrents. Lineages with swift-water tadpoles include about 18% (102) of the total species in southern Asia. Distribution of these lineages is clearly limited by topographic relief, because the clear, rocky streams they require do not occur in flat terrain. Consequently, except for two species, no species of the lineages to which these swift-water tadpoles belong occur in the Southeast Asian Lowlands. The two exceptional species, *Paa fasciculispina* and *Megophrys longipes,* live in a topographically exceptional part of those lowlands. The Western Ghats of India and the hills of Sri Lanka provide suitable habitats for swift-water tadpoles, but only one of those lineages (*Ansonia* with 2 species) occurs there.

Lineages with larvae developing in slow currents or in ponds (i.e., most lineages in southern Asia) are not limited by topography, seasonality of rainfall, or general climate (Table 8:9). They occur in all biogeographic regions covered by this chapter,

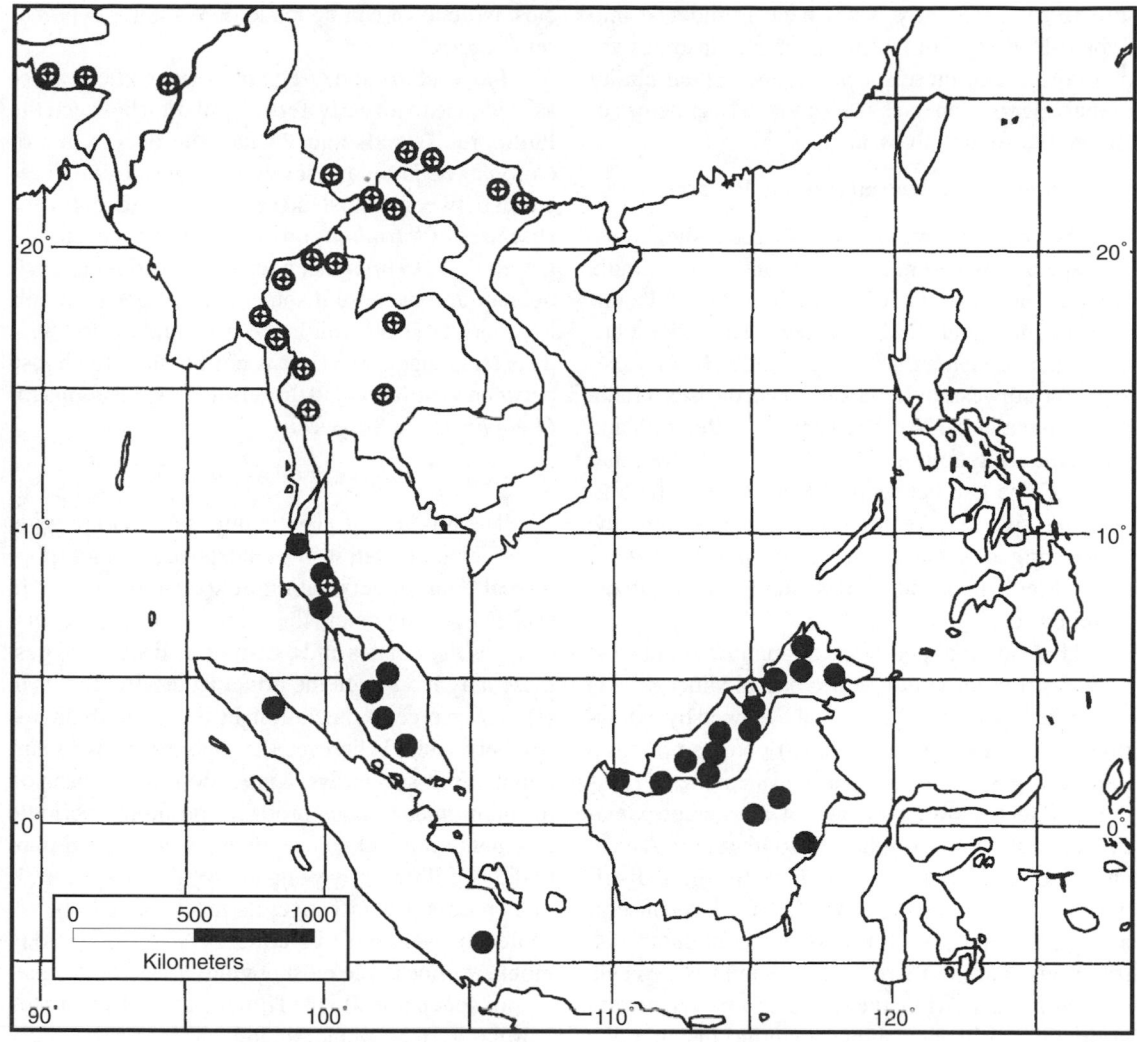

Fig. 8:11. Distribution of *Rana hosii* (solid circle) and *R. livida* (circle enclosing cross) in Southeast Asia and Sundaland.

with one exception. No lotic tadpoles have been recorded so far from Sulawesi, but no significance can be attached to this exception because of poor sampling.

COMMENSALS OF MAN

Although a number of species are able to live in severely disturbed environments in southern Asia, a few seem to be most common in such circumstances, where they form large populations, and are rarely seen in forests. The list includes *Bufo melanostictus, Kaloula pulchra, Microhyla ornata, Limnonectes limnocharis, Hoplobatrachus rugulosus, H. tigerinus, Euphlyctis cyanophlyctis, Rana erythraea, R. nicobariensis, R. taipehensis,* and

Polypedates leucomystax. These species breed in temporary pools of standing water and rarely are seen in relatively undisturbed vegetation. *Bufo melanostictus,* for example, occurs around houses, roads, and open fields, but does not penetrate forests in southern India (Inger et al., 1984) or Malaya (Berry, 1975); similarly, *Hoplobatrachus tigerinus* is common in open agricultural areas in Thailand, but does not occur in even open forests (Inger and Colwell, 1977). Furthermore, most of these species have extremely large geographic distributions. *Bufo melanostictus,* for example, occurs from Sri Lanka to southern China and southeastward to the Greater Sundas. *Hoplobatrachus rugulosus* is distributed from southern China south into the Malay Peninsula and west into Burma. Recently it has been introduced into Sabah (Borneo) from Taiwan, which illustrates another quality of these species, namely, their ability to make large dispersal strides via deliberate or accidental introduction by man.

Unresolved Distributional Patterns

Two bufonid lineages show large disjunctions. Two species of *Ansonia* occur in southwestern India and the remaining 17 in the southern Malay Peninsula, Borneo, and the Philippines. The second bufonid genus, *Pedostibes,* is represented by one species in southwestern India, one in northeastern India (Meghalaya in the Northeast Montane Region), and the remaining four in the Malay Peninsula and Borneo. The gap between Meghalaya and the Malay Peninsula may be an artifact of sampling in the intervening area, particularly in Burma, but that between Meghalaya and southwestern India is almost certainly real.

There are four possible explanations for the disjunct distributions of these two bufonid lineages: (1) a vicariance event; (2) dispersal followed by extinction in the intervening areas; (3) poor sampling in intervening areas; or (4) faulty systematics. (1) As there is no geologic evidence of close approximation of southwestern India and southeastern Asia at any time, a vicariance explanation for these distributions probably can be discarded. (2) It is conceivable that both genera formerly had continuous distributions from southern India to Southeast Asia via the Northeastern Montane Region. However, no other amphibian lineage shows a similar disjunct distribution. Although it cannot be eliminated, the dispersal hypothesis would have to be invoked for just these two lineages. (3) Poor sampling in Burma has been mentioned frequently in this chapter and may be responsible in these cases. (4) Phylogenetic analyses of southern Asian amphibians are almost nonexistent, thereby leaving most of the relations within and among lineages suspect. Although the genus *Ansonia* is defined on the basis of adult morphology, its most striking feature is a distinctive tadpole (Inger, 1960). To date, no larvae have been found in India, which raises the possibility that the Indian species are not congeneric with the southeast Asian ones but are related instead to some other bufonid lineage. *Pedostibes* also has been defined on the basis of adult morphology, but there has been no detailed study of the southern Indian and Meghalayan spe-

cies. No choice can be made now among hypotheses (2) to (4).

Platymantis and *Oreophryne* were cited earlier as evidence of an early Tertiary relation between the Philippine Islands and Papua. The occurrence of *Oreophryne* in Sulawesi might suggest a similar relation between that island and Papua, except for the absence of *Platymantis* on Sulawesi. It has been suggested that *Oreophryne,* because of its direct development and because it sometimes oviposits in hollow aerial tubers, might be a candidate for waif dispersal (Inger, 1954). It is not possible to choose between vicariance and dispersal as explanations for *Oreophryne* on Sulawesi.

Patterns of Speciation

The absence of genetic information to provide an estimate of relationships independent of morphological data hinders dating of speciation events in this fauna. However, the patterns of Pleistocene changes in climates and extent of land areas suggest these may have been the principal driving forces in relatively recent speciation in the amphibians of southern Asia. Wherever the topography is mountainous, the successive contraction and expansion of the montane zones probably promoted periodic fragmentation and genetic divergence of local populations, followed by sympatry of sibling species. If this process was effective, its results should be revealed by sympatry of congeners in present-day montane zones. The most obvious areas for this pattern of speciation are the Himalayan Flanks and the Northeastern Montane Region where *Paa* and *Amolops* have their centers of distributions and where congeners often are sympatric (Dubois, 1976; Yang, 1991). Other lineages showing the same pattern are *Megophrys* in Thailand (Nabhitabhata, ms) and Indochina (Bourret, 1942), *Ansonia* in Borneo (Inger and Dring, 1988), *Philautus* in Borneo (Dring, 1987; Inger and Stuebing, 1992), and *Leptolalax* and *Meristogenys* in Borneo (Inger, in ms).

The same phenomenon could have operated in the Western Ghats of southern India, where areas above 1000 m now form isolated patches and local endemism in montane islands seems to have been an important phenomenon (Inger et al., 1987), affecting not only the endemic lineages, such as *Nyctibatrachus* and *Indirana* (Table 8:1), but also *Philautus.*

The alternate unification and separation among the Greater Sundas and the Malay Peninsula in con-

Table 8:9. Ecological correlates of distributions of anuran lineages grouped by larval habitat requirements.

Lineage	Climate[1]	Rainfall[2]	Topography[3]	Region[4]
LINEAGES TIED TO CLEAR, ROCKY STREAMS:				
Leptobrachella	1	1	1, 2	7
Leptobrachium	1–3	1–3	1, 2	3, 6, 7, 9, 10
Leptolalax	1–3	1–3	1, 2	3, 6, 7
Megophrys	1–3	1–3	1, 2	2, 3, 5–7, 9
Ophryophryne	1, 2	2, 3	1, 2	3
Scutiger	2, 3	2	1, 2	2, 3
Ansonia	1	1, 2	1, 2	1, 6, 7, 9
Bufo asper group	1	1, 2	1, 2	3, 6, 7
Amolops	1–3	1, 2	1, 2	2, 3, 6
Huia	1, 2	1, 2	1, 2	3, 7
Meristogenys	1	1	1, 2	7
Paa	1–3	1–3	1, 2	2, 3, 5
Rana hosii-livida	1, 2	1, 2	1, 2	2, 3, 6, 7
Staurois	1	1, 2	1, 2	7, 9
Rhacophorus gauni group	1, 2	1, 2	1, 2	6, 7, 9
LINEAGES BREEDING IN STREAMS HAVING SLOW TO MODERATE CURRENT:				
Pedostibes	1	1, 2	2	1, 3, 6, 7
Limnonectes blythi group	1, 2	1, 2	1–3	3, 5–7, 9
Micrixalus	1	2	2	1
Nyctibatrachus	1	2, 3	2	1
Rana alticola	1, 2	2	2	3, 6
Rana chalconota	1	1, 2	2, 3	6, 7
Rana nigrovittata	1	1–3	1–3	3–7
Rana signata	1	1, 2	1, 2	6, 7, 9
Rana temporalis	1	2	1, 2	1
LINEAGES BREEDING IN TEMPORARY PONDS:				
Bufo (most species)	1–3	1–3	1–3	1–10
Pelophryne	1	1, 2	2	6, 7, 9
Hyla	2, 3	2, 3	2	3
Calluella	1–3	1–3	2, 3	4–7
Chaperina	1	1, 2	1–3	6, 7, 9
Glyphoglossus	1	2, 3	3	4–6
Kalophrynus	1, 2	1, 2	1–3	3–7, 9
Kaloula	1–3	1–3	1–3	1–10
Microhyla	1–3	1–3	2, 3	1–7, 10
Euphlyctis	1, 2	1–3	2, 3	1–3
Hoplobatrachus	1, 2	1–3	2, 3	1–7
Limnonectes cancrivora	1	1, 2	3	4, 6–10
Limnonectes keralensis	1	2	2	1
Limnonectes limnocharis group	1–3	1–3	1–3	1–7, 9, 10
Occidozyga	1, 2	2, 3	2, 3	3–7
Phrynoglossus	1, 2	1–3	1–3	2–4, 6–9
Rana baramica	1	1, 2	2, 3	6, 7
Rana erythraea group	1, 2	1–3	2, 3	2–7, 9
Rana luctuosa	1	1, 2	2, 3	6, 7
Chirixalus	1, 2	1–3	1–3	2–5
Polypedates	1–3	1–3	2, 3	1–7, 9, 10
Rhacophorus (most species)	1–3	1–3	1–3	1–7, 9, 10

[1] 1 = tropical, 2 = subtropical, 3 = temperate.   [2] 1 = aseasonal, 2 = moderately seasonal, 3 = strongly seasonal.
[3] 1 = steep, 2 = moderate, 3 = flat.
[4] 1 = India/Sri Lanka, 2 = Himalayan Flanks, 3 = Northeast Montane, 4 = Thai-Lao Plateau, 5 = SE Asian Lowlands, 6 = Peninsular Region, 7 = Sundaland, 8 = Sulawesi, 9 = Philippine Islands, 10 = Lesser Sundas.

cert with northern glaciations is the other Pleistocene phenomenon that probably promoted speciation. At least one species group of *Limnonectes* (*L. blythi-macrodon*) has distributional patterns suggesting this effect. Two species of that group are sympatric (and often syntopic) on Sumatra, Borneo, and the Malay Peninsula, but not the same two species in each area; the one (*L. blythi*) which has been reported from all three areas may, in fact, be a complex of cryptic sibling species. Other clusters of species in which speciation may have been affected by these Pleistocene changes include *Ansonia leptopus* and *A. penangensis* in the Malay Peninsula and Borneo; *A. mcgregori* and *A. muelleri* in the Philippines; the *Bufo biporcatus* species group (Inger, 1972) in the Greater Sundas and the Malay Peninsula; *Bufo asper* and *B. juxtasper* in Sumatra and Borneo; *Megophrys baluensis, montana,* and *nasuta* in Borneo and Sumatra; and the *Limnonectes microdiscus-palavanensis* group (Inger, 1966) in Sundaland.

COMPARISONS WITH DISTRIBUTION OF MAMMALS

Corbett and Hill (1992) reviewed the distributional relations of the mammals of southern Asia, which they refer to as the Indomalayan Region. Among their conclusions are the following: (1) The Himalayan montane forests have a distinctive fauna, which shares many species with peninsular India. (2) The Isthmus of Kra is a "… sharp faunal boundary out of proportion to the vegetational and climatic differences" on either side of it. (3) "The mammal fauna of Palawan clearly reflects its position on the Sunda shelf and distances it from the other Philippine islands …" The first conclusion partially applies to the amphibians of the Southern Himalayan Flanks Region; although the amphibian fauna of the region is distinct, it shares very few species with the Indian peninsula. The second conclusion also applies partially to amphibian distributions; a significant number of anurans reach their northern limits at the Isthmus of Kra, but a number occur also to the north of that line mainly along west-facing, perhumid slopes. However, Corbett and Hill oversimplify the climatic and vegetational changes that take place at the Isthmus. Their third conclusion matches the distributional data of the amphibians remarkably well.

The distributional divisions that Corbett and Hill constructed for mammals only partially correspond to those I suggest for amphibians. Their Indian and Himalayan subregions match the Indian and Southern Himalyan Flanks regions of this chapter. However, their Indochinese Subdivision (of their much larger Indochinese Subregion) includes the Northeast Montane, Southeast Asian Lowland, Dry Plateau, and part of the Peninsular regions of this chapter, as well as southern China. Amphibian distributions apparently show much more heterogeneity in Southeast Asia than do mammals. The differences between the two may reflect differences in environmental requirements, although differences between the sampling histories of the two groups cannot be ignored.

CONSERVATION

Two kinds of data are essential to assess the health and long-term viability of amphibian species: (1) detailed knowledge of ecological and geographic distribution, and (2) long-term information on population dynamics. Unfortunately, in tropical Asia the database is weak. General knowledge of geographic distribution is reasonably good for some species, but not for all. Detailed knowledge of distribution within historical times is weak; the geographic limits and continuity of distribution rarely are known. Knowledge of ecological distribution is patchy, whereas information on population sizes and fluctuations essentially is absent. Consequently, the conservation status of the amphibians of tropical Asia is in a general sense unknown. Thus, herein it is possible only to make general statements and to consider potential dangers. There have been no reports of extinctions or declines in amphibian populations, as have been reported from the Neotropics and Australia, and there has been only one formal attempt to resample sites after a long interval, as has been done in North America.

There are several obvious threats to the amphibian fauna of tropical Asia. The first is the great reduction in the area occupied by original vegetation. Logging, shifting agriculture, and, recently, large-scale development schemes have eliminated major portions of all types of forests over much of tropical Asia. Most amphibians in tropical Asia inhabit forests; thus, habitat loss for them has been severe. For some species with very restricted geographic ranges (e.g., *Barbourula kalimantanensis*), a modest amount of forest destruction could be catastrophic. Two other phenomena result from removal of forests—siltation of streams and fragmentation of remaining forest habitats. Because a high proportion of the amphibians in this region have stream-dwelling tadpoles, siltation of streams can be a potential threat for many species. The effects of fragmentation of environments has not been investigated, but on theoretical grounds it must be considered a hazard to the genetic structure of species.

In many parts of tropical Asia, forests have been replaced with large plantations of coffee, cacao, and oil palm, and with large-scale vegetable farming. All of these agricultural establishments rely heavily on pesticides and herbicides. No measure has yet been made of the effects of these chemicals on reproductive success of amphibians, but significant impacts are to be expected.

At the same time as these threats to amphibian diversity grow, governments in tropical Asia are establishing national parks and wildlife preserves. As these parks are distributed widely over tropical Asia and as they range from the lowlands to montane levels, they offer potential sanctuary to a wide segment of the amphibian fauna. Unfortunately, the lack of faunal inventories for most of these parks makes it difficult to assess their significance. Some governments recognize this deficiency and have instituted both inventories and monitoring programs. These new activities and concomitant education programs offer reason for cautious optimism.

CONCLUSIONS

The size of the amphibian fauna of southern Asia can be estimated only roughly as > 600 and probably < 750 species. An exact count of known species cannot be made because species boundaries are so poorly understood in certain significant lineages. There also is uncertainty about the geographic ranges of individual species, partly because of systematic problems but also because sampling of the fauna is seriously deficient in large areas. Despite these weaknesses in the data, we may be confident concerning some conclusions.

1. Perhaps the most important of these conclusions is that the amphibian fauna of southern Asia is really two distinct faunas, each with many endemic lineages and overlapping with one another at the species level only through frogs that are commensals of man.

(A) Geological evidence for the connection of India/Sri Lanka with Gondwanaland is clear and overwhelming, and the amphibian fauna shows evidence of that connection. Several ranid lineages are represented both in this region and in Africa. Lineages endemic to India/Sri Lanka include ranids and microhylids among frogs and three lineages of caecilians. Unfortunately, the phylogenetic relations of almost all of the endemics are unknown.

(B) The remainder of southern Asia and much of Sundaland have been part of the Eurasian continental mass at least since the late Mesozoic. Faunal differences among regions within this large part of southern Asia are generally at the species level, although a few lineages are endemic to single regions.

As expected, the northern tier of regions shows significant faunal overlap with southern China.

2. The disjunct distribution of the Bombinatoridae in Southeast Asia probably results from a vicariance event in the late Mesozoic.

3. Occurrence of the genera *Platymantis* and *Oreophryne* in the Philippines (except in the Palawan chain) represents Papuan elements that accord with the geological evidence of a closer spatial relationship of the Philippines with the Moluccas and New Guinea at several times in the Tertiary.

4. Pleistocene changes in climate in Southeast Asia and in sea level in the South China Sea account for disjunction in the distribution of the ranid lineage *Paa* and the great overlap at the species level among the Greater Sunda Islands and between them and the peninsula of Southeast Asia.

5. Nonhistorical, ecological factors also contribute to the patterns of present distributions and the recognition of biogeographic divisions in southern Asia.

(A) Transitions of present climatic circumstances, particularly from essentially aseasonal to strongly seasonal patterns, account for limits to the distributions of many species and, consequently, to discontinuities between certain pairs of regions.

(B) Finally, topography, by influencing the nature of streams, limits the distribution of suitable habitats for many larval anurans and so accounts for restrictions to the distributions of certain lineages.

Acknowledgments: I thank J. Nabhitabhata and D. T. Iskandar for allowing me to use their observations on the Thai and Sumatran faunas, respectively; R. Hall for discussions of geologic history and for preprints of several publications; N. Haile for guidance into the geological literature; L. R. Heaney for useful discussions and suggestions concerning the history and relations of the Philippine vertebrate fauna; E.-M. Zhao and W. R. Heyer for helpful comments on the manuscript; F. L. Tan for translations of certain Chinese language texts; and S. Hamnik for help with mapping type localities. Finally, I express my indebtedness to four deceased herpetologists: G. A. Boulenger, R. Bourret, and M. A. Smith, without whose work this study would have been impossible; and K. P. Schmidt, who introduced me to the fascinations of biogeography.

LITERATURE CITED

ALCALA, A. C. 1986. Amphibians and Reptiles. *Guide to Philippine Flora and Fauna, Vol. 10.* Quezon City: Natural Resources Management Center.

ANONYMOUS. 1961. *Temperature, Relative Humidity and Precipitation. Part V. Asia.* London: Air Ministry.

BERRY, P. Y. 1975. *The Amphibian Fauna of Peninsular Malaysia.* Kuala Lumpur: Tropical Press.

BOULENGER, G. A. 1893. Concluding report on the reptiles and batrachians obtained in Burma by Signor L. Lea, dealing with the collection made in Pegu and the Karin Hills in 1887–88. Ann. Mus. Civ. Genova (2)13:304–347.

BOULENGER, G. A. 1912. *A Vertebrate Fauna of the Malay Peninsula.* Reptilia and Batrachia. London: Taylor & Francis.

BOURRET, R. 1942. *Les Batraciens de l'Indochine.* Hanoi: Inst. Oceanographique de l'Indochine.

BROWN, W. C., AND A. C. ALCALA. 1994. Philippine frogs of the family Rhacophoridae. Proc. California Acad. Sci.48:185–220.

CHANNING, A. 1989. A re-evaluation of the phylogeny of Old World treefrogs. South African J. Sci. 24:116–131.

CLAYTON, H. H. (ed.) 1927. World weather records. Smithsonian Misc. Coll. 79:1–1198.

COLLINS, N. M., J. A. SAYER, AND T. C. WHITMORE. (eds.) 1991. *The Conservation Atlas of Tropical Forests: Asia and the Pacific.* New York: Simon and Schuster.

CORBETT, G. B., AND J. E. HILL. 1992. *The Mammals of the Indomalayan Region: A Systematic Review.* Oxford: Oxford University Press.

DAS, I. ms. Biogeography of reptiles of the Indian subcontinent.

DRING, J. 1987. Bornean frogs of the genus *Philautus* (Rhacophoridae). Amph.-Rept. 8:19–47.

DUBOIS, A. 1974. Liste commentée d'amphibiens récoltés au Népal. Bull. Mus. Natl. Hist. Nat. (3) (Zool.) 143:341–411.

DUBOIS, A. 1976. Les grenouilles du sous-genre *Paa* du Nepal, famille Ranidae, genre *Rana.* Cahiers Nepalais 6:1–275.

DUBOIS, A. 1987. Miscellanea taxinomica batrachologica (I). Alytes 5:7–95.

DUBOIS, A. 1992. Notes sur la classification des Ranidae (Amphibiens Anoures). Bull. Mens. Soc. Linn. Lyon 61:305–352.

ESTES, R., AND SANCHIZ, B. 1982. New discoglossid and palaeobatrachid frogs from the late Cretaceous of Wyoming and Montana, and a review of other frogs from the Lance and Hell Creek Formations. J. Vert. Paleo. 2:9–20.

FORD, L. S., AND D. C. CANNATELLA. 1993. The major clades of frogs. Herpetol. Monogr. 7:94–117.

FROST, D. R. (ed.). 1985. *Amphibian Species of the World.* Lawrence: Association of Systematics Collections and Allen Press, Inc..

GASCOYNE, M., G. J. BENJAMIN, AND H. P. SCHWARTZ. 1979. Sea-level lowering during the Illinoian glaciation: evidence from a Bahama 'blue hole.' Science 205:806–808.

HALL, R. 1997. Cenozoic tectonics of southeast Asia and Australasia. Pp. 47–62 *in* J. V. C. Howes and R. A. Noble (eds.), *Petroleum Systems of SE Asia and Australasia.* Jakarta, Indonesia: Indonesian Petroleum Association.

HALL, R. (in press). The plate tectonics of Cenozoic SE Asia and the distribution of land and sea. *In* R. Hall and J. D. Holloway (eds.), *Biogeography and Geological Evolution of SE Asia.* Amsterdam: Backhuys.

HEANEY, L. R. 1985. Zoogeographic evidence for Middle and Late Pleistocene land bridges to the Philippine Islands. Modern Quaternary Res. SE Asia 9:127–143.

HEANEY, L. R. 1991. A synopsis of climatic and vegetational change in Southeast Asia. Climatic Change 19:53–61.

HEDGES, S. B., R. A. NUSSBAUM, AND L. R. MAXSON. 1993. Caecilian phylogeny and biogeography inferred from mitochondrial DNA sequences of the 12S rRNA and 16S rRNA genes (Amphibia: Gymnophiona). Herpetol. Monogr. 7:64–76.

HOLLOWAY, J. D. 1986. Origins of the lepidopteran faunas in high mountains of the Indo-Australian tropics. Pp. 533-556 *in* F. Vuilleumier and M. Monasterio (eds.), *High Altitude Tropical Biogeography.* New York: Oxford Univ. Press.

HORA, S. L. 1932. Classification, bionomics and evolution of homalopterid fishes. Mem. Indian Mus. 12:263–330.

INGER, R. F. 1954. Systematics and zoogeography of Philippine Amphibia. Fieldiana: Zool. 33:183–531.

INGER, R. F. 1960. A review of the Oriental toads of the genus *Ansonia* Stoliczka. Fieldiana: Zool. 39:473–503.

INGER, R. F. 1966. The Amphibia of Borneo. Fieldiana: Zool. 52:1–402.

INGER, R. F. 1972. *Bufo* of Eurasia. Pp. 102-118, 357-360 *in* W. F. Blair (ed.), *Evolution in the genus* Bufo. Austin: Univ. Texas Press.

INGER, R. F. 1985. Tadpoles of the forested regions of Borneo. Fieldiana: Zool., n.s. 26:1–89.

INGER, R. F., AND R. K. COLWELL. 1977. Organization of contiguous communities of amphibians and reptiles in Thailand. Ecol. Monogr. 47:229–253.

INGER, R. F., AND J. DRING. 1988. Taxonomic and ecological relationships of Bornean stream toads allied to *Ansonia leptopus* (Guenther) (Anura: Bufonidae). Malayan Nat. J. 41:461–471.

INGER, R. F., AND S. DUTTA. 1986. An overview of the amphibian fauna of India. J. Bombay Nat. Hist. Soc. 83 (suppl.):135–146.

INGER, R. F., H. B. SHAFFER, M. KOSHY, AND R. BAKDE. 1984. A report on a collection of amphibians and reptiles from Ponmudi, Kerala, South India. J. Bombay Nat. Hist. Soc. 81:406–427, 551–570.

INGER, R. F., H. B. SHAFFER, M. KOSHY, AND R. BAKDE. 1987. Ecological structure of a herpetological assemblage in South India. Amph.-Rept. 8:189–202.

INGER, R. F., AND R. B. STUEBING. 1989. *Frogs of Sabah.* Sabah Parks Publication no. 10.

INGER, R. F., AND R. B. STUEBING. 1992. The montane amphibian fauna of northwestern Borneo. Malayan Nat. J. 46:41–51.

INGER, R. F., AND F.-L. TAN. 1996. Checklist of the frogs of Borneo. Raffles Bull. Zool. 44:551–574.

INGER, R. F., H. K. VORIS, AND K. J. FROGNER. 1986. Organization of a community of tadpoles in rain forest streams in Borneo. J. Tropical Ecol. 2:193–205.

JARRARD, R. D., AND S. SASAJIMA. 1980. Palaeomagnetic synthesis for Southeast Asia: constraints on plate movements. Pp. 293–316. *in* D. E. Hayes (ed.), *The Tectonic and Geologic Evolution of Southeast Asian Seas and Islands.* Am. Geophys. Union, Geophys. Monogr. 23.

KIEW, B.-H. 1984a. A new species of sticky frog (*Kalophrynus palmatissimus* n. sp.) from Peninsular Malaysia. Malayan Nat. J. 37:145–152.

KIEW, B.-H. 1984b. The conservation status of the Malaysian fauna. III. Amphibians. Malayan Nat. 37:6-10.

KIRTISINGHE, P. 1957. *The Amphibia of Ceylon.* Colombo: privately published.

LIEM, S.-S. 1970. The morphology, systematics, and evolution of the Old World treefrogs (Rhacophoridae and Hyperoliidae). Fieldiana: Zool. 57:1–145.

Liu, C.-C., and S.-C. Hu. 1960. New *Scutiger* from China with a discussion about the genus. Scientia Sinica 6:760–780.

METCALFE, I. 1988. Origin and assembly of south-east Asian continental terranes. Pp. 101–118. *in* M. G. Audley-Charles, and A. Hallam (eds.), *Gondwana and Tethys.* Geol. Soc. Am. Special Publ. 37.

NABHITABHATA, J. ms. The amphibian fauna of Thailand: a checklist with distribution maps.

PILLAI, R. S. 1986. Amphibian fauna of Silent Valley, Kerala, South India. Rec. Zool. Survey India 84:229–242.

RAVICHANDRAN, M. S. 1992. *Studies on the Amphibia of Southern Western Ghats.* Ph.D. thesis. Madras: University of Madras.

ROBERTS, T. R. 1989. The freshwater fishes of western Borneo (Kalimantan Barat, Indonesia). Mem. California Acad. Sci. 14:1–210.

SMITH, M. A. 1924. Descriptions of Indian and Indo-Chinese tadpoles. Rec. Indian Mus. 26:137–144.

SMITH, M. A. 1930. The Reptilia and Amphibia of the Malay Peninsula. Bull. Raffles Mus. 3:1–149.

SMITH, M. A. 1931a. The herpetology of Mt. Kinabalu, North Borneo, 13,455 ft. Bull. Raffles Mus. 5:3–32.

SMITH, M. A. 1931b. *The Fauna of British India. Reptilia and Amphibia. Vol. I. Loricata, Testudines.* London: Taylor & Francis.

WHITMORE, T. C. 1981. Introduction. Pp. 1–2 *in* T. C. Whitmore (ed.), *Wallace's Line and Plate Tectonics.* Oxford: Oxford Univ. Press.

YANG, D.-T. 1991. Phylogenetic systematics of the *Amolops* group of ranid frogs of southeastern Asia and the Greater Sunda Islands. Fieldiana: Zool., n.s. 63:1–42.

YANG, D.-T., LI S.-M., LIU W.-Z., LII S.-Q., AND WU B.-L. 1991. *The Amphibia Fauna of Yunnan.* Kunming: China Forestry Publishing House (in Chinese).

APPENDIX 8:1

DISTRIBUTION OF AMPHIBIANS OF SOUTHERN ASIA AND ADJACENT ISLANDS

Definition of regions are explained in the text. (Also see Fig. 8:4.) Regions are: HF = Himalayan flanks, ISL = India/Sri Lanka, LS = Lesser Sundas, NEM = Northeast Montane, PHI = Philippines, , PR = Peninsular Region, SAL = Southeast Asia Lowlands, SUL = Sulawesi, SUN = Sundaland, TLP = Thai-Lao Plateau.

Taxon	ISL	HF	NEM	TLP	SAL	PR	SUN	SUL	PHI	LS	Other
ANURA: BOMBINATORIDAE:											
Barbourula busuangensis	–	–	–	–	–	–	+	–	–	–	
Barbourula kalimantanensis	–	–	–	–	–	–	+	–	–	–	
Bombina maxima	–	–	+	–	–	–	–	–	–	–	
ANURA: BUFONIDAE:											
Ansonia albomaculata	–	–	–	–	–	–	+	–	–	–	
Ansonia fuliginea	–	–	–	–	–	–	+	–	–	–	
Ansonia guibei	–	–	–	–	–	–	+	–	–	–	
Ansonia hanitschi	–	–	–	–	–	–	+	–	–	–	
Ansonia latidisca	–	–	–	–	–	–	+	–	–	–	
Ansonia leptopus	–	–	–	–	–	+	+	–	–	–	
Ansonia longidigita	–	–	–	–	–	+	+	–	–	–	
Ansonia malayana	–	–	–	–	–	+	–	–	–	–	
Ansonia mcgregori	–	–	–	–	–	–	–	–	+	–	
Ansonia minuta	–	–	–	–	–	–	+	–	–	–	
Ansonia muelleri	–	–	–	–	–	–	–	–	+	–	
Ansonia ornata	+	–	–	–	–	–	–	–	–	–	
Ansonia penangensis	–	–	–	–	–	+	–	–	–	–	
Ansonia platysoma	–	–	–	–	–	+	–	–	–	–	
Ansonia rubigina	+	–	–	–	–	–	–	–	–	–	
Ansonia siamensis	–	–	–	–	–	+	–	–	–	–	
Ansonia spinulifer	–	–	–	–	–	–	+	–	–	–	
Ansonia tiomanicus	–	–	–	–	–	+	–	–	–	–	
Ansonia torrentis	–	–	–	–	–	–	+	–	–	–	
Bufo abatus	–	+	–	–	–	–	–	–	–	–	
Bufo asper	–	–	+	–	–	+	+	–	–	–	
Bufo beddomii	+	–	–	–	–	–	–	–	–	–	
Bufo biporcatus	–	–	–	–	–	–	+	–	–	+	
Bufo brevirostris	+	–	–	–	–	–	–	–	–	–	
Bufo burmanus	–	–	+	–	–	–	–	–	–	–	
Bufo camortensis	–	–	–	–	–	–	–	–	–	–	Andamans
Bufo celebensis	–	–	–	–	–	–	–	+	–	–	
Bufo chlorogaster	–	–	–	–	–	–	+	–	–	–	
Bufo claviger	–	–	–	–	–	–	+	–	–	–	
Bufo divergens	–	–	–	–	–	+	+	–	–	–	
Bufo fergusoni	+	–	–	–	–	–	–	–	–	–	
Bufo galeatus	–	–	–	–	+	–	–	–	–	–	
Bufo himalayanus	–	+	–	–	–	–	–	–	–	–	
Bufo hololius	+	–	–	–	–	–	–	–	–	–	
Bufo juxtasper	–	–	–	–	–	–	+	–	–	–	
Bufo kelaarti	+	–	–	–	–	–	–	–	–	–	
Bufo koynayensis	+	–	–	–	–	–	–	–	–	–	
Bufo latastii	–	+	–	–	–	–	–	–	–	–	
Bufo macrotis	–	–	+	–	+	+	–	–	–	–	
Bufo melanostictus	+	+	+	+	+	+	+	–	–	+	
Bufo microtympanum	+	–	–	–	–	–	–	–	–	–	
Bufo pageoti	–	–	+	–	–	–	–	–	–	–	
Bufo parietalis	+	–	–	–	–	–	–	–	–	–	
Bufo parvus	–	–	+	–	+	+	+	–	–	–	
Bufo philippinicus	–	–	–	–	–	–	+	–	–	–	
Bufo quadriporcatus	–	–	–	–	–	+	+	–	–	–	

Appendix 8:1 Continued

Taxon	ISL	HF	NEM	TLP	SAL	PR	SUN	SUL	PHI	LS	Other
Bufo silentvalleyensis	+	–	–	–	–	–	–	–	–	–	
Bufo stomaticus	+	+	–	–	–	–	–	–	–	–	
Bufo stuarti	–	–	+	–	–	–	–	–	–	–	
Bufo sumatrana	–	–	–	–	–	–	+	–	–	–	
Bufo tienhoensis	–	–	–	–	+	–	–	–	–	–	
Bufo valhallae	–	–	–	–	–	–	+	–	–	–	
Bufoides meghalayanus	–	–	+	–	–	–	–	–	–	–	
Leptophryne borbonica	–	–	–	–	–	+	+	–	–	–	
Leptophryne cruentata	–	–	–	–	–	–	+	–	–	–	
Pedostibes everetti	–	–	–	–	–	–	+	–	–	–	
Pedostibes hosii	–	–	–	–	–	+	+	–	–	–	
Pedostibes kempi	–	–	+	–	–	–	–	–	–	–	
Pedostibes maculatus	–	–	–	–	–	–	+	–	–	–	
Pedostibes rugosus	–	–	–	–	–	–	+	–	–	–	
Pedostibes tuberculosus	+	–	–	–	–	–	–	–	–	–	
Pelophryne albotaeniata	–	–	–	–	–	–	–	–	+	–	
Pelophryne api	–	–	–	–	–	–	+	–	–	–	
Pelophryne brevipes	–	–	–	–	–	+	+	–	+	–	
Pelophryne exigua	–	–	–	–	–	–	+	–	–	–	
Pelophryne guentheri	–	–	–	–	–	–	+	–	–	–	
Pelophryne lighti	–	–	–	–	–	–	–	–	+	–	
Pelophryne macrotis	–	–	–	–	–	–	+	–	–	–	
Pelophryne misera	–	–	–	–	–	–	+	–	–	–	
Pseudobufo subasper	–	–	–	–	–	+	+	–	–	–	
ANURA: HYLIDAE:											
Hyla annectens	–	–	+	–	–	–	–	–	–	–	
Hyla chinensis	–	–	+	–	–	–	–	–	–	–	
ANURA: MEGOPHRYIDAE:											
Leptobrachella baluensis	–	–	–	–	–	–	+	–	–	–	
Leptobrachella brevicrus	–	–	–	–	–	–	+	–	–	–	
Leptobrachella mjobergi	–	–	–	–	–	–	+	–	–	–	
Leptobrachella natunae	–	–	–	–	–	–	+	–	–	–	
Leptobrachella palmata	–	–	–	–	–	–	+	–	–	–	
Leptobrachella parva	–	–	–	–	–	–	+	–	–	–	
Leptobrachella serasanae	–	–	–	–	–	–	+	–	–	–	
Leptobrachium abbotti	–	–	–	–	–	–	+	–	+	–	
Leptobrachium chapaense	–	–	+	–	–	–	–	–	–	–	
Leptobrachium hasselti	–	–	–	–	–	–	+	–	–	+	
Leptobrachium hendricksoni	–	–	–	–	–	+	+	–	–	–	
Leptobrachium montanum	–	–	–	–	–	–	+	–	–	–	
Leptobrachium nigrops	–	–	–	–	–	+	+	–	–	–	
Leptobrachium pullus	–	–	+	–	–	+	–	–	–	–	
Leptolalax bourreti	–	–	+	–	–	–	–	–	–	–	
Leptolalax dringi	–	–	–	–	–	–	+	–	–	–	
Leptolalax gracilis	–	–	–	–	–	+	+	–	–	–	
Leptolalax heteropus	–	–	–	–	–	+	–	–	–	–	
Leptolalax pelodytoides	–	–	+	–	–	+	–	–	–	–	
Megophrys aceras	–	–	–	–	–	+	–	–	–	–	
Megophrys baluensis	–	–	–	–	–	–	+	–	–	–	
Megophrys carinensis	–	–	+	–	–	+	–	–	–	–	
Megophrys dringi	–	–	–	–	–	–	+	–	–	–	
Megophrys edwardinae	–	–	–	–	–	–	+	–	–	–	
Megophrys feae	–	–	+	–	–	–	–	–	–	–	
Megophrys intermedia	–	–	+	–	–	–	–	–	–	–	
Megophrys kempii	–	+	–	–	–	–	–	–	–	–	
Megophrys lateralis	–	–	+	–	–	–	–	–	–	–	
Megophrys longipes	–	–	–	–	+	+	–	–	–	–	
Megophrys minor	–	–	+	–	–	–	–	–	–	–	

Appendix 8:1 Continued

Taxon	ISL	HF	NEM	TLP	SAL	PR	SUN	SUL	PHI	LS	Other
Megophrys montana	–	–	–	–	–	˙+	+	–	–	–	
Megophrys nasuta	–	–	–	–	–	+	+	–	–	–	
Megophrys palpebralspinosa	–	–	+	–	–	–	–	–	–	–	
Megophrys parva	–	+	+	–	+	–	–	–	–	–	
Megophrys robusta	–	+	–	–	–	–	–	–	–	–	
Megophrys stejnegeri	–	–	–	–	–	–	–	–	+	–	
Ophryophryne microstoma	–	–	+	–	–	–	–	–	–	–	
Ophryophryne poilani	–	–	+	–	–	–	–	–	–	–	
Scutiger adungensis	–	–	+	–	–	–	–	–	–	–	
Scutiger alticola	–	+	–	–	–	–	–	–	–	–	
Scutiger nepalensis	–	+	–	–	–	–	–	–	–	–	
Scutiger occidentalis	–	+	–	–	–	–	–	–	–	–	
Scutiger sikimmensis	–	+	–	–	–	–	–	–	–	–	
ANURA: MICROHYLIDAE:											
Calluella brooksi	–	–	–	–	–	–	+	–	–	–	
Calluella flava	–	–	–	–	–	–	+	–	–	–	
Calluella guttulata	–	–	–	+	+	+	–	–	–	–	
Calluella smithi	–	–	–	–	–	–	+	–	–	–	
Calluella volzi	–	–	–	–	–	+	+	–	–	–	
Chaperina fusca	–	–	–	–	–	+	+	–	+	–	
Gastrophrynoides borneensis	–	–	–	–	–	–	+	–	–	–	
Glyphoglossus molossus	–	–	–	+	+	+	–	–	–	–	
Kalophrynus baluensis	–	–	–	–	–	–	+	–	–	–	
Kalophrynus bunguranum	–	–	–	–	–	–	+	–	–	–	
Kalophrynus heterochirus	–	–	–	–	–	–	+	–	–	–	
Kalophrynus intermedius	–	–	–	–	–	–	+	–	–	–	
Kalophrynus nubicola	–	–	–	–	–	–	+	–	–	–	
Kalophrynus palmatissimus	–	–	–	–	–	+	–	–	–	–	
Kalophrynus pleurostigma	–	–	+	+	+	+	+	–	+	–	
Kalophrynus punctatus	–	–	–	–	–	–	+	–	–	–	
Kalophrynus robinsoni	–	–	–	–	–	+	–	–	–	–	
Kalophrynus subterrestris	–	–	–	–	–	–	+	–	–	–	
Kaloula baleata	–	–	–	–	–	+	+	+	–	+	
Kaloula conjuncta	–	–	–	–	–	–	–	–	+	–	
Kaloula kalingensis	–	–	–	–	–	–	–	–	+	–	
Kaloula kokacii	–	–	–	–	–	–	–	–	+	–	
Kaloula mediolineata	–	–	–	+	+	+	–	–	–	–	
Kaloula picta	–	–	–	–	–	–	–	–	+	–	
Kaloula pulchra	+	+	+	+	+	+	+	+	–	–	
Kaloula rigida	–	–	–	–	–	–	–	–	+	–	
Melanobatrachus indicus	+	–	–	–	–	–	–	–	–	–	
Metaphrynella pollicaris	–	–	–	–	–	+	+	–	–	–	
Metaphrynella sundana	–	–	–	–	–	–	+	–	–	–	
Microhyla achatina	–	–	–	–	–	–	+	–	–	–	
Microhyla annamensis	–	–	–	–	+	–	–	–	–	–	
Microhyla annectans	–	–	–	–	–	+	+	–	–	–	
Microhyla berdmorei	–	–	+	+	+	+	+	–	–	–	
Microhyla borneensis	–	–	–	–	–	+	+	–	–	–	
Microhyla butleri	–	–	+	+	+	+	–	–	–	–	
Microhyla chakrapani	–	–	–	–	–	–	–	–	–	–	Andamans
Microhyla fusca	–	–	–	–	+	–	–	–	–	–	
Microhyla heymonsi	–	–	+	+	+	+	+	–	–	–	
Microhyla maculifera	–	–	–	–	–	–	+	–	–	–	
Microhyla ornata	+	+	+	+	+	+	+	–	–	–	
Microhyla palmipes	–	–	–	–	–	+	+	–	–	+	
Microhyla perparva	–	–	–	–	–	–	+	–	–	–	
Microhyla petrigena	–	–	–	–	–	–	+	–	–	–	
Microhyla picta	–	–	–	–	+	–	–	–	–	–	

Appendix 8:1 Continued

Taxon	ISL	HF	NEM	TLP	SAL	PR	SUN	SUL	PHI	LS	Other
Microhyla pulchra	–	–	+	+	+	+	–	–	–	–	
Microhyla rubra	+	–	–	–	–	–	–	–	–	–	
Microhyla superciliaris	–	–	–	–	–	+	+	–	–	–	
Microhyla zeylanica	+	–	–	–	–	–	–	–	–	–	
Micryletta inornata	–	–	+	+	+	+	+	–	–	–	
Micryletta stejnegeri	–	–	–	–	–	+	–	–	–	–	
Oreophryne annulatus	–	–	–	–	–	–	–	–	+	–	
Oreophryne celebensis	–	–	–	–	–	–	–	+	–	–	
Oreophryne jeffersonianus	–	–	–	–	–	–	–	–	–	+	
Oreophryne monticola	–	–	–	–	–	–	–	–	–	+	
Oreophryne nana	–	–	–	–	–	–	–	–	+	–	
Oreophryne rookmaakeri	–	–	–	–	–	–	–	–	–	+	
Oreophryne variabilis	–	–	–	–	–	–	–	+	–	–	
Oreophryne zimmeri	–	–	–	–	–	–	–	+	–	–	
Phrynella pulchra	–	–	–	–	–	+	+	–	–	–	
Ramanella anamallaiensis	+	–	–	–	–	–	–	–	–	–	
Ramanella minor	+	–	–	–	–	–	–	–	–	–	
Ramanella montana	+	–	–	–	–	–	–	–	–	–	
Ramanella mormorata	+	–	–	–	–	–	–	–	–	–	
Ramanella obscura	+	–	–	–	–	–	–	–	–	–	
Ramanella palmata	+	–	–	–	–	–	–	–	–	–	
Ramanella triangularis	+	–	–	–	–	–	–	–	–	–	
Ramanella variegata	+	–	–	–	–	–	–	–	–	–	
Uperodon globosa	+	–	–	–	–	–	–	–	–	–	
Uperodon systoma	+	+	–	–	–	–	–	–	–	–	
ANURA: RANIDAE:											
Altirana parkeri	–	+	–	–	–	–	–	–	–	–	
Amolops afghanus	–	+	+	–	–	+	–	–	–	–	
Amolops chapaensis	–	–	+	–	–	–	–	–	–	–	
Amolops formosus	–	+	+	–	–	–	–	–	–	–	
Amolops himalayanus	–	+	+	–	–	–	–	–	–	–	
Amolops kaulbacki	–	–	+	–	–	–	–	–	–	–	
Amolops larutensis	–	–	–	–	–	+	–	–	–	–	
Amolops monticola	–	+	–	–	–	–	–	–	–	–	
Amolops nepalicus	–	+	–	–	–	–	–	–	–	–	
Amolops ricketti	–	–	+	–	–	–	–	–	–	–	
Elachyglossa gyldenstolpei	–	–	+	–	–	–	–	–	–	–	
Euphlyctis cyanophlyctis	+	+	+	–	–	–	–	–	–	–	
Euphlyctis hexadactyla	+	–	–	–	–	–	–	–	–	–	
Hoplobatrachus crassus	+	+	–	–	–	–	–	–	–	–	
Hoplobatrachus rugulosus	–	–	–	–	+	+	+	–	–	–	
Hoplobatrachus tigerinus	+	+	+	+	+	+	–	–	–	–	
Iluia cavitympanum	–	–	–	–	–	–	+	–	–	–	
Huia javana	–	–	–	–	–	–	+	–	–	–	
Huia nasica	–	–	+	–	–	–	–	–	–	–	
Huia sumatrana	–	–	–	–	–	–	+	–	–	–	
Indirana beddomii	+	–	–	–	–	–	–	–	–	–	
Indirana brachytarsus	+	–	–	–	–	–	–	–	–	–	
Indirana diplosticta	+	–	–	–	–	–	–	–	–	–	
Indirana gundia	+	–	–	–	–	–	–	–	–	–	
Indirana leithi	+	–	–	–	–	–	–	–	–	–	
Indirana leptodactyla	+	–	–	–	–	–	–	–	–	–	
Indirana phrynoderma	+	–	–	–	–	–	–	–	–	–	
Indirana semipalmata	+	–	–	–	–	–	–	–	–	–	
Indirana tenuilingua	+	–	–	–	–	–	–	–	–	–	
Ingerana baluensis	–	–	–	–	–	–	+	–	–	–	
Ingerana mariae	–	–	–	–	–	–	+	–	–	–	
Ingerana tasanae	–	–	–	–	–	+	–	–	–	–	

Appendix 8:1 Continued

Taxon	ISL	HF	NEM	TLP	SAL	PR	SUN	SUL	PHI	LS	Other
Ingerana tenasserimensis	−	−	+	−	−	+	−	−	−	−	
Limnonectes acanthi	−	−	−	−	−	−	+	−	−	−	
Limnonectes andamanensis	−	−	−	−	−	−	−	−	−	−	Andamans
Limnonectes arathooni	−	−	−	−	−	−	−	+	−	−	
Limnonectes blythi	−	−	−	−	−	+	+	−	−	−	
Limnonectes brevipalmatus	+	−	−	−	−	−	−	−	−	−	
Limnonectes cancrivorus	−	−	−	−	+	+	+	+	+	+	
Limnonectes corrugatus	+	−	−	−	−	−	−	−	−	−	
Limnonectes dabanus	−	−	−	−	+	−	−	−	−	−	
Limnonectes dammermanni	−	−	−	−	−	−	−	−	−	+	
Limnonectes diuata	−	−	−	−	−	−	−	−	+	−	
Limnonectes doriae	−	−	+	−	−	+	−	−	−	−	
Limnonectes finchi	−	−	−	−	−	−	+	−	−	−	
Limnonectes gracilis	+	−	−	−	−	−	−	−	−	−	
Limnonectes greeni	+	−	−	−	−	−	−	−	−	−	
Limnonectes hascheanus	−	−	+	−	−	+	+	−	−	−	
Limnonectes heinrichi	−	−	−	−	−	−	−	+	−	−	
Limnonectes ibanorum	−	−	−	−	−	−	+	−	−	−	
Limnonectes ingeri	−	−	−	−	−	−	+	−	−	−	
Limnonectes kenepaiensis	−	−	−	−	−	−	+	−	−	−	
Limnonectes keralensis	+	−	−	−	−	−	−	−	−	−	
Limnonectes khammonensis	−	−	+	−	−	−	−	−	−	−	
Limnonectes khasianus	−	−	+	−	−	−	−	−	−	−	
Limnonectes kohchangae	−	−	−	−	+	−	−	−	−	−	
Limnonectes kuhlii	−	−	+	−	+	+	+	−	−	−	
Limnonectes laticeps	−	−	+	−	−	+	+	−	−	−	
Limnonectes leytensis	−	−	−	−	−	−	−	−	+	−	
Limnonectes limnocharis	+	+	+	+	+	+	+	−	+	+	
Limnonectes macrocephalus	−	−	−	−	−	−	−	−	+	−	
Limnonectes macrodon	−	−	−	−	−	−	+	−	−	−	
Limnonectes macrognathus	−	−	+	−	−	+	−	−	−	−	
Limnonectes magnus	−	−	−	−	−	−	−	−	+	−	
Limnonectes malesianus	−	−	−	−	−	+	+	−	−	−	
Limnonectes mawlyndipi	−	−	+	−	−	−	−	−	−	−	
Limnonectes mawphlangensis	−	+	+	−	−	−	−	−	−	−	
Limnonectes micrixalus	−	−	−	−	−	−	−	−	+	−	
Limnonectes microdiscus	−	−	−	−	−	+	−	−	−	−	
Limnonectes microtympanum	−	−	−	−	−	−	−	+	−	−	
Limnonectes modestus	−	−	−	−	−	−	−	+	−	−	
Limnonectes murthi	+	−	−	−	−	−	−	−	−	−	
Limnonectes nepalensis	−	+	−	−	−	−	−	−	−	−	
Limnonectes nilagirica	+	−	−	−	−	−	−	−	−	−	
Limnonectes nitidus	−	−	−	−	−	+	−	−	−	−	
Limnonectes palavanensis	−	−	−	−	−	−	+	−	−	−	
Limnonectes paramacrodon	−	−	−	−	−	+	+	−	−	−	
Limnonectes parvus	−	−	−	−	−	−	−	−	+	−	
Limnonectes pierrei	−	+	−	−	−	−	−	−	−	−	
Limnonectes pileatus	−	−	+	+	+	−	−	−	−	−	
Limnonectes plicatellus	−	−	−	−	−	+	−	−	−	−	
Limnonectes raja	−	−	−	−	−	+	−	−	−	−	
Limnonectes sauriceps	+	−	−	−	−	−	−	−	−	−	
Limnonectes syhadrensis	+	+	−	−	−	−	−	−	−	−	
Limnonectes teraiensis	−	+	−	−	−	−	−	−	−	−	
Limnonectes timorensis	−	−	−	−	−	−	−	−	−	+	
Limnonectes toumanoffi	−	−	−	−	+	−	−	−	−	−	
Limnonectes tweediei	−	−	−	−	−	+	−	−	−	−	
Limnonectes verruculosus	−	−	−	−	−	−	−	−	−	+	
Limnonectes visayanus	−	−	−	−	−	−	−	−	+	−	

Appendix 8:1 Continued

Taxon	ISL	HF	NEM	TLP	SAL	PR	SUN	SUL	PHI	LS	Other
Limnonectes woodworthi	−	−	−	−	−	−	−	−	+	−	
Meristogenys amoropalamus	−	−	−	−	−	−	+	−	−	−	
Meristogenys jerboa	−	−	−	−	−	−	+	−	−	−	
Meristogenys kinabaluensis	−	−	−	−	−	−	+	−	−	−	
Meristogenys macrophthalmus	−	−	−	−	−	−	+	−	−	−	
Meristogenys orphnocnemis	−	−	−	−	−	−	+	−	−	−	
Meristogenys phaeomerus	−	−	−	−	−	−	+	−	−	−	
Meristogenys poecilus	−	−	−	−	−	−	+	−	−	−	
Meristogenys whiteheadi	−	−	−	−	−	−	+	−	−	−	
Micrixalus fuscus	+	−	−	−	−	−	−	−	−	−	
Micrixalus gadgili	+	−	−	−	−	−	−	−	−	−	
Micrixalus nudis	+	−	−	−	−	−	−	−	−	−	
Micrixalus opisthorhodus	+	−	−	−	−	−	−	−	−	−	
Micrixalus saxicolus	+	−	−	−	−	−	−	−	−	−	
Micrixalus silvaticus	+	−	−	−	−	−	−	−	−	−	
Micrixalus thampi	+	−	−	−	−	−	−	−	−	−	
Nannobatrachus beddomii	+	−	−	−	−	−	−	−	−	−	
Nannobatrachus kempholeyensis	+	−	−	−	−	−	−	−	−	−	
Nannophrys ceylonensis	+	−	−	−	−	−	−	−	−	−	
Nannophrys guentheri	+	−	−	...	−	−	−	−	−	−	
Nannophrys marmoratus	+	−	−	−	−	−	−	−	−	−	
Nyctibatrachus aliciae	+	−	−	−	−	−	−	−	−	−	
Nyctibatrachus deccanensis	+	−	−	−	−	−	−	−	−	−	
Nyctibatrachus humayuni	+	−	−	−	−	−	−	−	−	−	
Nyctibatrachus major	+	−	−	−	−	−	−	−	−	−	
Nyctibatrachus minor	+	−	−	−	−	−	−	−	−	−	
Nyctibatrachus sanctipalustris	+	−	−	−	−	−	−	−	−	−	
Nyctibatrachus sylvaticus	+	−	−	−	−	−	−	−	−	−	
Nyctibatrachus vasanthi	+	−	−	−	−	−	−	−	−	−	
Occidozyga lima	−	−	+	+	+	+	+	−	−	−	
Paa annandalei	−	+	−	−	−	−	−	−	−	−	
Paa arnoldii	−	−	+	−	−	−	−	−	−	−	
Paa blanfordi	−	+	+	−	−	−	−	−	−	−	
Paa bourreti	−	−	+	−	−	−	−	−	−	−	
Paa ercepeae	−	+	−	−	−	−	−	−	−	−	
Paa fasciculispina	−	−	−	−	+	−	−	−	−	−	
Paa feae	−	−	+	−	−	−	−	−	−	−	
Paa liebigii	−	+	−	−	−	−	−	−	−	−	
Paa minica	−	+	−	−	−	−	−	−	−	−	
Paa polunini	−	+	−	−	−	−	−	−	−	−	
Paa rara	−	+	−	−	−	−	−	−	−	−	
Paa rostandi	−	+	−	−	−	−	−	−	−	−	
Paa sternosignata	−	I	−	−	−	−	−	−	−	−	NW India
Paa vicina	−	+	−	−	−	−	−	−	−	−	
Paa yunnanensis	−	−	+	−	−	−	−	−	−	−	
Phrynoglossus baluensis	−	−	−	−	−	−	+	−	−	−	
Phrynoglossus borealis	−	+	−	−	−	−	−	−	−	−	
Phrynoglossus celebensis	−	−	−	−	−	−	−	+	−	−	
Phrynoglossus diminutiva	−	−	−	−	−	−	−	−	+	−	
Phrynoglossus floresianus	−	−	−	−	−	−	−	−	−	+	
Phrynoglossus laevis	−	−	−	−	−	+	+	+	+	+	
Phrynoglossus magnapustulosa	−	−	+	+	−	−	−	−	−	−	
Phrynoglossus martensii	−	−	−	+	+	+	−	−	−	−	
Phrynoglossus semipalmata	−	−	−	−	−	−	−	+	−	−	
Platymantis cornutus	−	−	−	−	−	−	−	−	+	−	
Platymantis corrugatus	−	−	−	−	−	−	−	−	+	−	
Platymantis dorsalis	−	−	−	−	−	−	−	−	+	−	
Platymantis guentheri	−	−	−	−	−	−	−	−	+	−	

Appendix 8:1 Continued

Taxon	ISL	HF	NEM	TLP	SAL	PR	SUN	SUL	PHI	LS	Other
Platymantis hazelae	–	–	–	–	–	–	–	–	+	–	
Platymantis ingeri	–	–	–	–	–	–	–	–	+	–	
Platymantis insulatus	–	–	–	–	–	–	–	–	+	–	
Platymantis lawtoni	–	–	–	–	–	–	–	–	+	–	
Platymantis levigatus	–	–	–	–	–	–	–	–	+	–	
Platymantis polillensis	–	–	–	–	–	–	–	–	+	–	
Platymantis spelaeus	–	–	–	–	–	–	–	–	+	–	
Platymantis subterrestris	–	–	–	–	–	–	–	–	+	–	
Pterorana khare	–	–	+	–	–	–	–	–	–	–	
Rana aenea	–	–	+	–	–	–	–	–	–	–	
Rana albolineata	–	–	+	–	–	–	–	–	–	–	
Rana alticola	–	–	+	–	–	+	–	–	–	–	
Rana andersonii	–	–	+	–	–	–	–	–	–	–	
Rana aurantiaca	+	–	–	–	–	–	–	–	–	–	
Rana baramica	–	–	–	–	–	+	+	–	–	–	
Rana celebensis	–	–	–	–	–	–	–	+	–	–	
Rana chalconota	–	–	–	–	–	+	+	+	–	+	
Rana crassiovis	–	–	–	–	–	–	+	–	–	–	
Rana cubitalis	–	–	+	–	–	+	–	–	–	–	
Rana curtipes	+	–	–	–	–	–	–	–	–	–	
Rana danieli	–	–	+	–	–	–	–	–	–	–	
Rana erythraea	–	+	+	+	+	+	+	–	+	–	
Rana everetti	–	–	–	–	–	–	–	–	+	–	
Rana fansipani	–	–	+	–	–	–	–	–	–	–	
Rana florensis	–	–	–	–	–	–	–	–	–	+	
Rana garoensis	–	–	+	–	–	–	–	–	–	–	
Rana gerbillus	–	+	+	–	–	–	–	–	–	–	
Rana ghoshi	–	–	+	–	–	–	–	–	–	–	
Rana glandulosa	–	–	–	–	–	+	+	–	–	–	
Rana guentheri	–	–	–	+	–	–	–	–	–	–	
Rana hosii	–	–	–	–	–	+	+	–	–	–	
Rana humeralis	–	+	+	–	–	–	–	–	–	–	
Rana intermedia	+	–	–	–	–	–	–	–	–	–	
Rana kampeni	–	–	–	–	–	–	+	–	–	–	
Rana lateralis	–	–	+	+	+	–	–	–	–	–	
Rana leptoglossa	–	–	+	–	–	+	–	–	–	–	
Rana livida	–	+	+	–	+	+	–	–	–	–	
Rana longimanus	–	–	+	–	–	–	–	–	–	–	
Rana luctuosa	–	–	–	–	–	+	+	–	–	–	
Rana luzonensis	–	–	–	–	–	–	–	–	+	–	
Rana macrodactyla	–	–	–	+	+	+	–	–	–	–	
Rana macrops	–	–	–	–	–	–	–	+	–	–	
Rana malabarica	+	–	–	–	–	–	–	–	–	–	
Rana maosonensis	–	–	+	–	–	–	–	–	–	–	
Rana margariana	–	–	+	–	–	–	–	–	–	–	
Rana melanomenta	–	–	–	–	–	–	–	–	+	–	
Rana microlineata	–	–	+	–	–	–	–	–	–	–	
Rana milleti	–	–	–	–	+	–	–	–	–	–	
Rana miopus	–	–	–	–	–	+	–	–	–	–	
Rana montivaga	–	–	–	–	+	–	–	–	–	–	
Rana nicobariensis	–	–	–	–	–	+	+	–	+	+	
Rana nigrovittata	–	–	+	+	+	+	+	–	–	–	
Rana oatesi	–	–	–	–	+	–	–	–	–	–	
Rana paludicola	–	–	–	–	–	+	–	–	–	–	
Rana sanguinea	–	–	–	–	–	–	+	–	–	–	
Rana sauteri	–	–	–	–	+	–	–	–	–	–	
Rana senchalensis	–	+	–	–	–	–	–	–	–	–	
Rana siberu	–	–	–	–	–	–	+	–	–	–	

Appendix 8:1 Continued

Taxon	ISL	HF	NEM	TLP	SAL	PR	SUN	SUL	PHI	LS	Other
Rana signata	–	–	–	–	–	+	+	–	+	–	
Rana sikimensis	–	+	+	–	–	–	–	–	–	–	
Rana taipehensis	+	+	+	+	–	+	–	–	–	–	
Rana temporalis	+	–	–	–	–	–	–	–	–	–	
Rana travancorica	+	–	–	–	–	–	–	–	–	–	
Rana tuberculata	+	–	–	–	–	–	–	–	–	–	
Staurois latopalmatus	–	–	–	–	–	–	+	–	–	–	
Staurois natator	–	–	–	–	–	–	+	–	+	–	
Staurois tuberilinguis	–	–	–	–	–	–	+	–	–	–	
Tomopterna breviceps	+	+	–	–	–	–	–	–	–	–	
Tomopterna dobsoni	+	–	–	–	–	–	–	–	–	–	
Tomopterna leucorhynchus	+	–	–	–	–	–	–	–	–	–	
Tomopterna parambikulimana	+	–	–	–	–	–	–	–	–	–	
Tomopterna rolandae	+	–	–	–	–	–	–	–	–	–	
Tomopterna rufescens	+	–	–	–	–	–	–	–	–	–	
Tomopterna swani	–	+	–	–	–	–	–	–	–	–	
ANURA: RHACOPHORIDAE:											
Chirixalus doriae	–	+	+	+	+	–	–	–	–	–	
Chirixalus hansenae	–	–	–	+	–	–	–	–	–	–	
Chirixalus nongkhorensis	–	–	+	+	+	–	–	–	–	–	
Chirixalus simus	–	–	+	–	–	–	–	–	–	–	
Chirixalus vittatus	–	–	+	+	+	–	–	–	–	–	
Nyctixalus margaritifer	–	–	–	–	–	–	+	–	–	–	
Nyctixalus pictus	–	–	–	–	–	+	+	–	+	–	
Nyctixalus spinosus	–	–	–	–	–	–	–	–	+	–	
Philautus acutirostris	–	–	–	–	–	–	–	–	+	–	
Philautus acutus	–	–	–	–	–	–	+	–	–	–	
Philautus adspersus	+	–	–	–	–	–	–	–	–	–	
Philautus alticola	–	–	–	–	–	–	–	–	+	–	
Philautus amoenus	–	–	–	–	–	–	+	–	–	–	
Philautus andersoni	–	–	+	–	–	–	–	–	–	–	
Philautus annandalii	–	+	+	–	–	–	–	–	–	–	
Philautus aurantium	–	–	–	–	–	–	+	–	–	–	
Philautus aurifasciatus	–	–	–	–	–	–	+	–	–	–	
Philautus banaensis	–	–	–	–	+	–	–	–	–	–	
Philautus beddomi	+	–	–	–	–	–	–	–	–	–	
Philautus bombayensis	+	–	–	–	–	–	–	–	–	–	
Philautus carinensis	–	–	+	–	–	–	–	–	–	–	
Philautus chalazodes	+	–	–	–	–	–	–	–	–	–	
Philautus charius	+	–	–	–	–	–	–	–	–	–	
Philautus cherrapungiae	–	+	+	–	–	–	–	–	–	–	
Philautus cornutus	–	–	–	–	–	–	+	–	–	–	
Philautus crni	+	–	–	–	–	–	–	–	–	–	
Philautus disgregus	–	–	–	–	–	–	+	–	–	–	
Philautus elegans	+	–	–	–	–	–	–	–	–	–	
Philautus emembranatus	–	–	–	–	–	–	–	–	+	–	
Philautus femoralis	+	–	–	–	–	–	–	–	–	–	
Philautus flaviventris	+	–	–	–	–	–	–	–	–	–	
Philautus garo	–	–	+	–	–	–	–	–	–	–	
Philautus glandulosus	+	–	–	–	–	–	–	–	–	–	
Philautus gracilipes	–	–	+	–	–	–	–	–	–	–	
Philautus gryllus	–	–	+	–	–	–	–	–	–	–	
Philautus hassanensis	+	–	–	–	–	–	–	–	–	–	
Philautus hosii	–	–	–	–	–	–	+	–	–	–	
Philautus ingeri	–	–	–	–	–	–	+	–	–	–	
Philautus jacobsoni	–	–	–	–	–	–	+	–	–	–	
Philautus kempii	–	–	+	–	–	–	–	–	–	–	
Philautus kerangae	–	–	–	–	–	–	+	–	–	–	

Appendix 8:1 Continued

Taxon	ISL	HF	NEM	TLP	SAL	PR	SUN	SUL	PHI	LS	Other
Philautus kottegeharensis	+	–	–	–	–	–	–	–	–	–	
Philautus laevis	–	–	+	–	–	–	–	–	–	–	
Philautus leitensis	–	–	–	–	–	–	–	–	+	–	
Philautus leucorhinus	+	–	–	–	–	–	–	–	–	–	
Philautus longicrus	–	–	–	–	–	–	+	–	–	–	
Philautus maosonensis	–	–	+	–	–	–	–	–	–	–	
Philautus melanensis	+	–	–	–	–	–	–	–	–	–	
Philautus microtympanum	+	–	–	–	–	–	–	–	–	–	
Philautus mjobergi	–	–	–	–	–	–	+	–	–	–	
Philautus namdaphaensis	–	+	–	–	–	–	–	–	–	–	
Philautus narainensis	+	–	–	–	–	–	–	–	–	–	
Philautus nasutus	+	–	–	–	–	–	–	–	–	–	
Philautus noblei	+	–	–	–	–	–	–	–	–	–	
Philautus pallidipes	–	–	–	–	–	–	+	–	–	–	
Philautus palpebralis	–	–	+	–	–	–	–	–	–	–	
Philautus parkeri	+	–	–	–	–	–	–	–	–	–	
Philautus parvulus	–	–	+	+	–	–	–	–	–	–	
Philautus petersi	–	–	–	–	–	+	+	–	–	–	
Philautus poecilus	–	–	–	–	–	–	–	–	+	–	
Philautus pulcherrimus	+	–	–	–	–	–	–	–	–	–	
Philautus schmackeri	–	–	–	–	–	–	–	–	+	–	
Philautus shillongensis	–	–	+	–	–	–	–	–	–	–	
Philautus shyamrupus	–	+	–	–	–	–	–	–	–	–	
Philautus signatus	+	–	–	–	–	–	–	–	–	–	
Philautus similis	–	–	–	–	–	–	+	–	–	–	
Philautus surdus	–	–	–	–	–	–	–	–	+	–	
Philautus surrufus	–	–	–	–	–	–	–	–	+	–	
Philautus swamianus	+	–	–	–	–	–	–	–	–	–	
Philautus tectus	–	–	–	–	–	–	+	–	–	–	
Philautus temporalis	+	–	–	–	–	–	–	–	–	–	
Philautus travancoricus	+	–	–	–	–	–	–	–	–	–	
Philautus tytthus	–	–	+	–	–	–	–	–	–	–	
Philautus umbra	–	–	–	–	–	–	+	–	–	–	
Philautus variabilis	+	–	–	–	–	–	–	–	–	–	
Philautus vermiculatus	–	–	–	–	–	+	–	–	–	–	
Philautus williamsi	–	–	–	–	–	–	–	–	+	–	
Polypedates colletti	–	–	–	–	–	+	+	–	–	–	
Polypedates cruciger	+	–	–	–	–	–	–	–	–	–	
Polypedates eques	+	–	–	–	–	–	–	–	–	–	
Polypedates feae	–	–	+	–	–	–	–	–	–	–	
Polypedates hecticus	–	–	–	–	–	–	–	–	+	–	
Polypedates leucomystax	–	+	+	+	+	+	+	+	+	+	
Polypedates longinasus	+	–	–	–	–	–	–	–	–	–	
Polypedates macrotis	–	–	–	–	–	+	+	–	+	–	
Polypedates maculatus	+	+	–	–	–	–	–	–	–	–	
Polypedates mutus	–	–	+	–	–	–	–	–	–	–	
Polypedates otilophus	–	–	–	–	–	–	+	–	–	–	
Polypedates zed	–	+	–	–	–	–	–	–	–	–	
Rhacophorus angulirostris	–	–	–	–	–	–	+	–	–	–	
Rhacophorus annamensis	–	–	+	–	–	–	–	–	–	–	
Rhacophorus appendiculatus	–	–	+	+	–	+	+	–	+	–	
Rhacophorus baluensis	–	–	–	–	–	–	+	–	–	–	
Rhacophorus bimaculatus	–	–	–	–	–	+	+	–	+	–	
Rhacophorus bipunctatus	–	–	+	+	–	+	–	–	–	–	
Rhacophorus bisacculus	–	–	–	+	–	–	–	–	–	–	
Rhacophorus calcadensis	+	–	–	–	–	–	–	–	–	–	
Rhacophorus calcaneus	–	–	–	–	+	–	–	–	–	–	
Rhacophorus chapaensis	–	–	+	–	–	–	–	–	–	–	

Appendix 8:1 Continued

Taxon	ISL	HF	NEM	TLP	SAL	PR	SUN	SUL	PHI	LS	Other
Rhacophorus depressus	−	−	−	−	−	−	+	−	−	−	
Rhacophorus dorsoviridis	−	−	+	−	−	−	−	−	−	−	
Rhacophorus dubius	−	+	−	−	−	−	−	−	−	−	
Rhacophorus dulitensis	−	−	−	−	−	−	+	−	−	−	
Rhacophorus edentulus	−	−	−	−	−	−	−	+	−	−	
Rhacophorus everetti	−	−	−	−	−	−	+	−	−	−	
Rhacophorus fasciatus	−	−	−	−	−	−	+	−	−	−	
Rhacophorus gauni	−	−	−	−	−	−	+	−	−	−	
Rhacophorus georgi	−	−	−	−	−	−	−	+	−	−	
Rhacophorus harrissoni	−	−	−	−	−	−	+	−	−	−	
Rhacophorus javanus	−	−	−	−	−	−	+	−	−	−	
Rhacophorus jerdoni	−	+	−	−	−	−	−	−	−	−	
Rhacophorus kajau	−	−	−	−	−	−	+	−	−	−	
Rhacophorus lateralis	+	−	−	−	−	−	−	−	−	−	
Rhacophorus longinasus	+	−	−	−	−	−	−	−	−	−	
Rhacophorus malabaricus	+	−	−	−	−	−	−	−	−	−	
Rhacophorus maximus	−	+	+	−	−	+	−	−	−	−	
Rhacophorus modestus	−	−	−	−	−	−	+	−	−	−	
Rhacophorus monticola	−	−	−	−	−	−	−	+	−	−	
Rhacophorus namdophaensis	−	+	−	−	−	−	−	−	−	−	
Rhacophorus naso	−	+	−	−	−	−	−	−	−	−	
Rhacophorus nigropalmatus	−	−	−	−	−	+	+	−	−	−	
Rhacophorus notater	−	−	−	−	+	−	−	−	−	−	
Rhacophorus pardalis	−	−	−	−	−	+	+	−	+	−	
Rhacophorus pleurostictus	+	−	−	−	−	−	−	−	−	−	
Rhacophorus poecilonotus	−	−	−	−	−	−	+	−	−	−	
Rhacophorus prominanus	−	−	−	−	−	+	−	−	−	−	
Rhacophorus reinwardti	−	−	−	−	−	+	+	−	−	−	
Rhacophorus robinsoni	−	−	−	−	−	+	−	−	−	−	
Rhacophorus rufipes	−	−	−	−	−	−	+	−	−	−	
Rhacophorus taeniatus	+	−	−	−	−	−	−	−	−	−	
Rhacophorus taroensis	−	−	+	−	−	−	−	−	−	−	
Rhacophorus tuberculatus	−	+	−	−	−	−	−	−	−	−	
Rhacophorus tunkui	−	−	−	−	−	+	−	−	−	−	
Rhacophorus turpes	−	−	+	−	−	−	−	−	−	−	
Rhacophorus zamboangensis	−	−	−	−	−	−	−	−	+	−	
Theloderma asper	−	+	+	−	−	+	−	−	−	−	
Theloderma bicolor	−	−	+	−	−	−	−	−	−	−	
Theloderma corticalis	−	−	+	−	−	−	−	−	−	−	
Theloderma gordoni	−	−	+	−	−	−	−	−	−	−	
Theloderma horridum	−	−	−	−	−	+	+	−	−	−	
Theloderma leprosa	−	−	−	−	−	+	+	−	−	−	
Theloderma moloch	−	+	+	−	−	−	−	−	−	−	
Theloderma phrynoderma	−	−	+	−	−	−	−	−	−	−	
Theloderma stellatum	−	−	−	+	+	−	−	−	−	−	
CAUDATA: SALAMANDRIDAE:											
Echinotriton asperrimus	−	−	+	−	−	−	−	−	−	−	
Paramesotriton deloustali	−	−	+	−	−	−	−	−	−	−	
Tylototriton verrucosus	−	+	+	−	−	−	−	−	−	−	
GYMNOPHIONA: CAECILIIDAE:											
Gegeneophis carnosus	+	−	−	−	−	−	−	−	−	−	
Gegeneophis fulleri	+	−	−	−	−	−	−	−	−	−	
Gegeneophis ramaswamii	+	−	−	−	−	−	−	−	−	−	
Indotyphlus battersbyi	+	−	−	−	−	−	−	−	−	−	
GYMNOPHIONA: ICHTHYOPHIIDAE:											
Caudacaecilia asplenia	−	−	−	−	−	−	+	−	−	−	
Caudacaecilia larutensis	−	−	−	−	−	+	−	−	−	−	
Caudacaecilia nigroflavus	−	−	−	−	−	+	−	−	−	−	

Appendix 8:1 Continued

Taxon	ISL	HF	NEM	TLP	SAL	PR	SUN	SUL	PHI	LS	Other
Caudacaecilia paucidentula	–	–	–	–	–	–	+	–	–	–	
Caudacaecilia weberi	–	–	–	–	–	–	+	–	–	–	
Ichthyophis acuminatus	–	–	+	–	–	–	–	–	–	–	
Ichthyophis atricollaris	–	–	–	–	–	–	+	–	–	–	
Ichthyophis beddomii	+	–	–	–	–	–	–	–	–	–	
Ichthyophis bernisi	–	–	–	–	–	–	+	–	–	–	
Ichthyophis biangularis	–	–	–	–	–	–	+	–	–	–	
Ichthyophis billitonensis	–	–	–	–	–	–	+	–	–	–	
Ichthyophis bombayensis	+	–	–	–	–	–	–	–	–	–	
Ichthyophis dulitensis	–	–	–	–	–	–	+	–	–	–	
Ichthyophis elongatus	–	–	–	–	–	–	+	–	–	–	
Ichthyophis glandulosus	–	–	–	–	–	–	–	–	+	–	
Ichthyophis glutinosus	+	–	–	–	–	–	–	–	–	–	
Ichthyophis hypocyanea	–	–	–	–	–	–	+	–	–	–	
Ichthyophis javanicus	–	–	–	–	–	–	+	–	–	–	
Ichthyophis kohtaoensis	–	–	+	+	+	+	–	–	–	–	
Ichthyophis laosensis	–	–	+	–	–	–	–	–	–	–	
Ichthyophis longicephalus	+	–	–	–	–	–	–	–	–	–	
Ichthyophis malabarensis	+	–	–	–	–	–	–	–	–	–	
Ichthyophis mindanaoensis	–	–	–	–	–	–	–	–	+	–	
Ichthyophis monochrous	–	–	–	–	–	+	+	–	–	–	
Ichthyophis orthoplicatus	+	–	–	–	–	–	–	–	–	–	
Ichthyophis paucisulcus	–	–	–	–	–	–	+	–	–	–	
Ichthyophis peninsularis	+	–	–	–	–	–	–	–	–	–	
Ichthyophis pseudangularis	+	–	–	–	–	–	–	–	–	–	
Ichthyophis sikkimensis	–	+	–	–	–	–	–	–	–	–	
Ichthyophis singaporensis	–	–	–	–	–	+	–	–	–	–	
Ichthyophis subterrestris	+	–	–	–	–	–	–	–	–	–	
Ichthyophis sumatranus	–	–	–	–	–	–	+	–	–	–	
Ichthyophis supachaii	–	–	–	–	–	+	–	–	–	–	
Ichthyophis tricolor	+	–	–	–	–	–	–	–	–	–	
Ichthyophis youngorum	–	–	+	–	–	–	–	–	–	–	
GYMNOPHIONA: URAEOTYPHLIDAE:											
Uraeotyphlus malabaricus	+	–	–	–	–	–	–	–	–	–	
Uraeotyphlus menoni	+	–	–	–	–	–	–	–	–	–	
Uraeotyphlus narayani	+	–	–	–	–	–	–	–	–	–	
Uraeotyphlus oxyurus	+	–	–	–	–	–	–	–	–	–	

ADDENDUM

In the several years that this manuscript has been in press, new species of amphibians have been described from various parts of tropical Asia, and much additional field work has been carried out by a number of researchers. Most of the additions to the faunal list involve the Ranidae and Rhacophoridae, although new species of Bufonidae and Megophryidae also have been described. The rate at which new species have been discovered in various parts of tropical Asia during the last four years suggests that future discoveries may increase the total number by at least 15% and perhaps as much as 20%. The most conspicuous additions to our geographic information have been made in Vietnam and have revealed that this fauna was more poorly known than this manuscript assumed. Although our knowledge both of species richness and geographic distribution has grown in this short interval, the general patterns described in the paper remain unchanged. The area in which we may expect significant changes in the near future is geological history, particularly of the present insular regions.

9. Distribution of Amphibians in Sub-Saharan Africa, Madagascar, and Seychelles

J. C. POYNTON

Department of Zoology
Natural History Museum
London SW7 5BD
United Kingdom

ABSTRACT   Geographical sampling and taxonomic analysis of Sub-Saharan amphibians is less than adequate; the situation is better in Madagascar and the Seychelles. The amphibian faunas of Sub-Saharan Africa (± 622 species), Madagascar (156 species), and Seychelles (12 species) each show a high level of endemism and differ markedly from each other. Fourteen regions are distinguished in continental Sub-Saharan Africa, none in Madagascar and Seychelles. High-ranking Sub-Saharan regions in respect of species diversity are Intertropical Montane (255 species), and West Equatorial Lowland (179 species). Percentage species endemism ranks as West Equatorial (79%), Intertropical Montane (70%), South Temperate (64%). Percentage generic endemism ranks as Intertropical Montane (44%), West Equatorial (31%), South Temperate (30%). Savanna regions have high species diversity but relatively fewer endemic genera and species than more moist, dry or cooler regions, which presumably experienced more disruption during climatic cycles. Species diversity does not correlate significantly with endemism.

The Intertropical Montane and South Temperate regions together carry the larger part of what is regarded as an Afrotemperate fauna, which comprises half the total number of Sub-Saharan species. Use of the term "Afrotropical" for the whole of Sub-Saharan Africa is therefore criticized. It is proposed that the term "Afrotropical" is employed only in its climatic sense, the name "Sub-Saharan" being applied to the subcontinent as a whole. The Sub-Saharan fauna is thus seen to consist of two main components, Afrotropical and Afrotemperate. Gondwana-derived elements are evident in the Afrotemperate fauna, and also are indicated in the faunas of Madagascar and Seychelles. Correlations of distributional patterns with temperature and rainfall are evident, but particular environmental factors or complexes involved in causal relationships have not been identified. Effects of habitat disturbance on species diversity have not yet been comprehensively studied; human-induced disturbance needs to be restricted at least until its effects are better understood.

Key words: Zoogeography, Amphibia; Africa; Sub-Sahara; Madagascar; Seychelles; Diversity; Endemism.

INTRODUCTION

Sub-Saharan Africa, Madagascar, and the Seychelles are three fragments of the ancient supercontinent of Gondwana. They have a certain unity in being populated by the distinctive treefrog family, the Hyperoliidae (Drewes, 1984; Channing, 1989). A general coverage of the amphibians of these continental fragments has been given by Duellman (1993a); more detailed treatment has been hampered by the lack of a recent comprehensive analysis of amphibian distribution in Africa. The Madagascan fauna has been reviewed by Guibé (1978) and Blommers-Schlösser and Blanc (1991, 1993);

amphibians of the Seychelles have been treated by Nussbaum (1984). In Africa, the genus *Bufo* has been reviewed across the continent (Tandy and Keith, 1972), but, although there are several Sub-Saharan regional studies (e.g., Amiet, 1983; Channing and van Dijk, 1976; Inger, 1968; Lanza, 1981; Laurent, 1964, 1972; Lawson, 1993; Loumont, 1992; Mertens, 1965; Perret, 1966; Poynton, 1964a, 1990, 1991; Poynton and Broadley, 1985a, 1985b, 1987, 1988, 1991; Poynton and Haacke, 1993; Schiøtz, 1967, 1975; Schmidt and Inger, 1959; Stewart, 1967), the last attempt at a comprehensive treatment of the African amphibian fauna was by Noble in 1924. Space therefore has to be given in this chapter to a new attempt to investigate amphibian distribution in Sub-Saharan Africa.

The preliminary nature of this treatment needs to be emphasized at the outset. Kingdon (1990:23) asserted recently that "the flora and fauna of Africa are still largely unexplored" This is supported by a scrutiny of the intensity of collecting of amphibians in two areas of southern Africa that are thought to be relatively well collected (Poynton, 1992; Poynton and Broadley, 1991). In two transects covering 41 quarter-degree intervals, only five intervals (12%) were considered to have received seemingly comprehensive collecting. In Cameroon, which is regarded as another relatively well-studied area, collections of amphibians were shown by Amiet (1983) to exist in only 16% of the squares in a 20 x 20 km grid map. Added to a paucity of data is the undeveloped state of identifying distribution patterns, and, from there, to classifying areas of land into regions of some kind. Literature on African animal distribution has tended to give only limited attention to matters of underlying method and approach (Poynton, 1986a), which makes it seem necessary to discuss the issues at some length in the following section.

MATERIALS AND METHODS

Rosen (1988:26) has cautioned that "... in biogeography we should be alert to the hidden implications in choice and method of approach." He argued that biogeographical patterns are "perceptual, not factual" (Rosen, 1988:27), and stated that problems can arise in biogeography "... from misunderstandings about the perceptual, or hypothetical, nature of patterns" (Rosen, 1988:26). In agreement with these views, biogeographical patterns are here regarded as perceptions, which then requires accepting the inevitability of bias in these perceptions. Moreover, in agreement with views recently expressed by Huntley and Webb (1989), Levin (1989), and Colinvaux (1993), such perceived patterns are considered to reflect range groupings that are heterogeneous and ephemeral to an unknown degree (Poynton, 1994).

In attempting to define regions in Africa, authors have tended not to be explicit about whether their treatment is range-based (proceeding from a study of perceived clustering among species or other taxon ranges), or district-based (in which taxa are sorted and grouped according to preconceived biochoria, bioclimatic, or political/administrative divisions) (Poynton, 1986a). The result has been the production of noncomparable biogeographical classifications. A range-based approach generally provides more useful biogeographical classifications, but its success depends on the density of locality records, particularly near the perceived margins of a region. This is complicated by problems regarding the definition of margins, because they tend to be irreducibly fuzzy in the sense discussed by Sattler (1986). Many geographical features affect the fuzziness of margins. Those occurring in areas of high topographical relief tend to appear relatively sharper than those occurring on a plain where populations are widely scattered, and elements from different regions intermingle (Poynton and Broadley, 1991). Even where there is high relief, margins tend to be much more blurred south of the tropics than within the tropics on account of intermingling (White, 1978; Poynton, 1992). Moreover, different groups may be perceived to have different boundaries, as in the situation analyzed by Schiøtz (1967) in West Africa, where a border between west and central amphibian faunas follows the Benin (or Dahomey) Gap in the case of high forest species, but occurs along the Cross River (650 km to the east), in the case of farmbush species.

In a range-based approach, the difficulty of identifying and plotting locality records directly according to degree coordinates usually leads to partitioning a study area by means of a grid. By scoring each quadrat in some way, this method has the potential for allowing a quantitative approach to the analysis of distribution and diversity. Yet the usefulness of this method depends primarily on the grid size, and here there is an uneasy compromise: the finer the grid, the finer the resolution of the patterning; yet the method becomes more dependent on close and uniform sampling across the study area. Because collecting in Africa has been largely unsystematic, and dependent mainly upon where collectors have resided and where their favored collecting sites and routes to them have been, existing locality data are not suited to the employment of quantitative procedures over large areas, which require uniformity of sampling. To smooth out the patchiness of data, Crowe (1990) used a more than two-degree square grid in his study of amphibian distribution and diversity in southern Africa; but a grid of this size yielded an analysis that lacks close correspondence with the results of an analysis applying quarter-degree grids to more localized areas (Poynton, 1992; Poynton and Broadley, 1991).

If the distributional data are too incomplete to support quantitative procedures over large areas, the only alternative—if a range-based approach is to be followed—is to group ranges by visual sorting and matching. This procedure has been explicitly used by Poynton (1964a) and Poynton and Broadley (1991), with testable results (Poynton, 1992; Poynton and Broadley, 1991). The same procedure is followed in this study, bearing in mind that, because any agglomerative procedure is an inductive process, its validity is ultimately to be assessed by the testability and utility of the final product, not so much by the techniques carried out to achieve it (Poynton and Broadley, 1991). Biogeographical classifications, or definitions of regions, are, as Hengenveld (1990:63) pointed out, "... not right or wrong, only useful or not." The usefulness of range classification in the present context is the potential for it to identify (to the extent that limited knowledge permits) areas of particular diversity, endemism, and need for conservation attention. It can only be hoped that future work will allow greater precision in carrying out this task.

Although precision should be enhanced by a more complete database, there are conceptual difficulties that mitigate against precision however complete the data. Notable is the conflict arising from taking both diversity and endemism into account. For instance, the highland plain of Tanzania apparently has only two endemic species of amphibians, and it lacks a distinct geographical assemblage, because the other species occurring there are widespread mostly to the east and southwest. Therefore the area does not fit the "province concepts" as discussed by Rosen (1988:43), which rest on identifying regions of endemism or distinct geographical assemblages. As a result, the highland plains may not seem classifiable as a "province." Yet the area shows substantial species diversity, which in turn links it to conservation concerns, as in the case of the Serengeti National Park. Highland plains in Zimbabwe, South Africa, and Namibia present similar difficulties. This situation contrasts with the East African Lowlands where, although endemism is relatively low, a fairly distinct geographical assemblage includes high species diversity (Poynton, 1990). It also contrasts with areas of relatively high endemism but low diversity, such as montane forests or oceanic islands.

It is therefore accepted in this study that little precision is attainable in defining biogeographical units when diversity and endemism are combined. The usefulness of a classification is naturally affected by this lack of coincidence, a lack which is one aspect of the irreducible heterogeneity of biogeographical and ecological units.

In determining African phytochoria, White (1983:42) distinguished "regional centers of endemism" from "regional transition zones" on the basis of the degree of endemism shown in the chosen areas, in association with the absolute number of endemics present. A regional center of endemism was identified as having more than 50% of its species confined to it, and on this basis, the East African Highlands could be classified as a transition zone, effectively separating it from regions of high endemicity. Yet, as can be seen in Table 9:5 below, there is a wide range in percentage endemicity within the regions, without any clear break in the range. Even an arbitrary division into regions and transition zones cannot readily be made on the basis of this feature. However, it may be seen from Table 9:6 below that 67% of amphibian species are restricted to one division only, which suggests that the units arrived at do reflect clusters of ranges.

In contrast to White's chorology, and the treatment of southern African amphibians by Poynton (1964a), this study does not classify biogeographical units into regions and transition zones; and it is hierarchical in that three regions are divided into provinces. A region is here taken to be an area covered by a perceived set of common species ranges; but, in the case of the archipelago of montane areas extending west and south from Ethiopia, a single intertropical region is recognized, even though it is composed of subsets of species ranges that are generally nonoverlapping. This region corresponds with White's (1983) Afromontane regional center of endemism, excluding the area south of the Limpopo Basin. Croizat's conception of a "generalized track" (Croizat, 1968; Craw, 1983) applies to this archipelago, representing connections that are shown by many taxa along a line of disjunct areas. These areas are classified here as provinces within the region. To a lesser extent, separate groupings of species ranges are also perceptible in two other regions, namely in West Africa and in South Africa, and provinces are accordingly recognized also in these regions.

Estimation of species ranges is here based mainly on several regional studies now available, some of which are listed in the Introduction, aided by the entries in Frost (1985) and Duellman (1993b). In the case of species of the Bufonidae, Heleophrynidae, and Microhylidae, localities have been plotted in a project recently commenced to determine the known ranges of all African amphibian species. The mapped distributions of bufonids, heleophrynids, and microhylids were used in an initial sorting and matching of species ranges; this sorting was tested and modified by referring to studies covering other families, and to material in the Natural History Museum, London, where the project is centered.

The regions are not all thought to be clearly demarcated from each other, and they are presented in agreement with the view of Hengenveld (1990:49) that "... classifications are heuristic and that they should be used only as starting points for further research, rather than descriptions of objective, natural units, being ends in themselves." Figure 9:3, which portrays these divisions, could have adopted a caption that was perspicaciously used by Noble (1924:152) in a map that subdivided Africa into "... convenient areas for distributional discussion." The regions shown in Figure 9:3 are not considered to have definite boundaries, especially over plains where no boundaries may be drawn. This situation is represented in Figure 9:3 by blank areas in eastern and southern Africa.

Similarities (or dissimilarities) between species composition in some regions are quantified by use of a coefficient of community similarity that is variously attributed to Dice or to Sørensen, namely, twice the number of taxa common to both areas divided by the sum of totals of taxa present in each area, expressed as a percentage (Poynton and Broadley, 1991). For its restricted purpose, this index has been found to give a more convenient set of values than use of the Jaccard Coefficient or more elaborate techniques.

Estimation of the area of each region in Africa was made by marking out the regions on a 1:10,000,000 Lambert zenithal equal areas projection map of Africa, printed by Bartholomew Harper Collins, 1992. The outlines were transferred to graph paper with a 2.5 mm square grid and the enclosed areas estimated by counting squares. The estimates were checked against various sources such as White (1983) and Stuart et al. (1990). These sources provided area estimates for Madagascar and the Seychelles. The areas estimated for each African region should be viewed as approximate, and useful only for broad comparisons, because, as just discussed, clear boundaries for the divisions are lacking for the most part.

Climatic data have been taken mainly from Jackson (1961) and Thompson (1965). Use has been made of White's (1983) valuable summaries of climate in the regions of his *Vegetation of Africa*.

DESCRIPTION OF REGIONS

The African continent has straddled the equator since the end of the Paleogene, but a peeling away of southern landmasses during the breakup of Gondwana during the early Cretaceous (Pitman et al., 1993) left a much smaller area of African land south of the equator than north of it. Consequently, although both ends of the continent extend into a zone of Mediterranean-type climate, the southern end experiencing this climate is only about 13% of the area at the northern end.

A vast expanse of desert makes about one-half of the northern part of the continent largely uninhabitable for amphibians at present. Le Berre (1989) recorded in the Sahara presumably relict populations of the following essentially Sub-Saharan species: *Xenopus muelleri, Bufo regularis, Hoplobatrachus occipitalis,* and *Ptychadena mascareniensis.* Siboulet (1969) recorded *P. mascareniensis* and *B. regularis* in southern Algeria. Relict populations of three species with Mediterranean distributions also were recorded by Le Berre (1989) in the Sahara. One of these, *Bufo mauritanicus,* has a population recorded on the northern bend of the River Niger. This suggests that at least the western Sahara and Sahel were at one time a transition zone between what are now the Palearctic and Sub-Saharan areas. The transition probably involved many kinds of plants and animals; Moreau (1966:76) wrote of the likelihood of a "... pleasing conjunction of such birds as blue tits with elephants and magpies with white rhinos."

The Nile Delta still retains this transitional character; *Bufo regularis* and *Ptychadena mascareniensis* have been recorded there (Siboulet, 1969; El Din, 1993). Egyptian material conventionally referred to *P. mascareniensis* shows some variation (Akef, 1993), but it is not readily distinguishable from material assigned to this species from the rest of Africa. *Bufo kassasii* (El Din, 1993; *B. vittatus* of Akef and Schneider, 1993), a species that shows close affinities with *B. steindachneri* from farther south, appears to be endemic to the delta.

Excluding the Nile from Khartoum northward (15° 35' N), the present northern limit of Sub-Saharan Africa may be taken to lie south of the depauperate transition zone, a subtraction-transition zone in Darlington's (1957) sense. This limit lies more or less at the 15° N latitude. South of this is an area of about 17,900,000 km², including islands on the continental shelf. The oceanic, volcanic Gulf of Guinea islands, with a distinctive amphibian fauna, have a total area of 964 km². This island chain is of Tertiary origin and has probably been isolated since that period (Schätti and Loumont, 1992). Madagascar has an area of 594,180 km². It may have commenced separation from Africa in the Late Jurassic (Pitman et al., 1993), but connections with Africa may have remained until the mid-Cretaceous, as demonstrated by distribution patterns of archaic angiosperms and austral gymnosperms (Coetzee, 1993). The Seychelles Islands currently have an area of about 453 km². They belong to an ancient, partially submerged, continental vestige of the Gondwana mass, left in the track of northward-moving India.

Apparently this continental mass was most recently in contact with India in the early Paleocene (references in Nussbaum, 1984). The existing islands apparently have been emergent throughout the Cenozoic. At present the Seychelles microcontinent evidently is submerged to nearly its maximum extent (Nussbaum, 1984), but a drop in sea level of little over 50 m would have united most of the islands as mountains on a landmass of some 125,000 km² (Vine, 1989). Presumably, this was the condition at the commencement of the Holocene.

TOPOGRAPHY

Although an outline of Africa shows more land surface north of the equator than south of it, the reverse is true of land surface above 1000 m (Fig. 9:1). Most of Africa south of 10° S is above 1000 m; only a narrow lowland margin exists along the west, south, and southeast coast. The margin is variably wider along the eastern side of the continent north of 28° S. The upper basins of the Orange, Limpopo, and Zambesi rivers lie to some extent below 1000 m, but the overall effect of the extensive southern highlands is a lowering of surface temperature, which tends to increase the area of cool land surface in the southern part of the continent compared to the north, despite a southward tapering of the continental outline.

On the eastern side of the continent, highland is dissected by the lower Limpopo and Zambesi basins, but the highland continues north of 10° S to the edge of the Red Sea, with an interruption only in the saddle containing Lake Turkana. This highland archipelago of bevelled plains and mountains of various kinds that extends from the southern highlands across the equator to the Horn of Africa is of special biogeographical importance. Its features are complicated by an associated system of rift valleys, which have marked biogeographical effects, such as separating the mountainous eastern rim of the Zaire Basin from East African mountains. The crystalline mountains of eastern Tanzania form an interrupted side chain.

West of the eastern highlands is a scattering of isolated highland areas, the largest of which is in western Cameroon, but lowland predominates in this northwestern sector of Sub-Saharan Africa; it may be termed "Low Africa," in contrast to "High Africa" to the east and south (White, 1983). The Zaire or Congo Basin is an extensive interior lowland area in the central part of the continent, which is separated

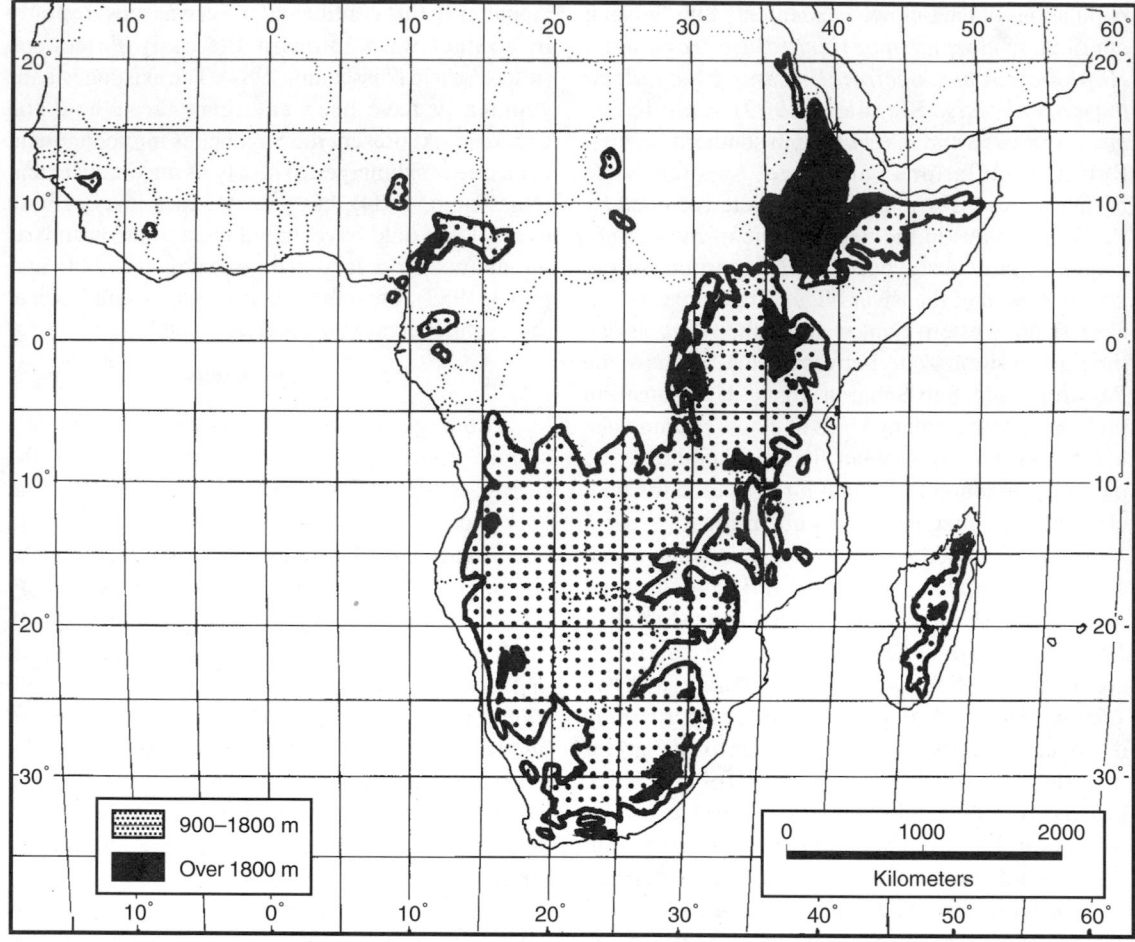

Fig. 9:1.   Simplified topographic map of Sub-Saharan Africa and Madagascar.

to a large extent by highlands in western Cameroon from the very extensive lowlands of West Africa, that spread northward into the Sahara. Although limited in area, the highland patches in Low Africa, notably the Guinean and Cameroonian highlands, nevertheless are of great biogeographical interest. The Cameroonian Highlands are complex, formed of ancient crystalline rocks and a somewhat separated Mount Cameroon, which is an active volcano. The line of volcanic mountains continues out into the Gulf of Guinea. Bioko (formerly Fernando Po) is the largest of these islands. It is situated on the continental shelf some 35 km from the mainland, and was linked to the mainland during the Pleistocene (Schätti and Loumont, 1992). The other islands of this chain are oceanic. The largest, São Tomé, is formed around a volcanic mountain 2,024 m in height.

Madagascar is relatively simple topographically. A highland spine with a narrow ridge over 2,000 m extends along its axis. Extensive coastal lowlands are present only on the western side of the island; these extend inland up to 150 km in the northwest. The Seychelles are a heterogeneous cluster of islands. The larger, granitic, islands are mountainous; a height of 917 m is reached on Mahé Island. The rugged topography provides a diversity of habitats, despite the relatively small area of the islands.

CLIMATE

Several accounts of the climate of Africa from a biological viewpoint have been written. Notable are those of Moreau (1966), White (1983), and Kingdon (1990). In the southern half of Africa, the cold Benguela Current off the west coast, together with an elevated interior, have the effect of pushing the latitudinal thermal zones toward the equator, even though parts of the interior experience great daytime heating during summer. In contrast, the east coast is washed by the warm South Equatorial and Agulhas Currents, which, in combination with mainly onshore winds, result in a southward extension of warm, relatively moist conditions. Therefore, thermal zonation south of the Tropic of Capricorn tends to be longitudinal rather than latitudinal, especially in the eastern part of the subcontinent, where a "tropical peninsula" extends down the eastern lowlands (Poynton, 1961, 1964a).

Rainfall zonation in southern Africa also is greatly influenced by ocean currents and topography, to the extent that, south of the Tropic of Capricorn, zonation is again mainly longitudinal rather than latitudinal. The west coast is largely desert, whereas most of the east coast has an annual rainfall in excess of 800 mm; rain falls in summer and peaks around January. The southern extreme of the continent experiences a Mediterranean-type climate with rain falling mostly in the southern winter months of May to August. Annual rainfall over most of the southern half of Africa is less than 1000 mm, usually considerably less. Even in the winter rainfall zone, most areas receive less than 800 mm of rain. Most of the southern part of Africa can be characterized as arid when compared with most of central Africa, and cool to cold especially in winter. Ground frost is widespread over the interior in winter, and snow would be a common occurrence if conditions were not so dry at this time.

The tendency in southern Africa toward longitudinal rather than latitudinal zonation of climate continues in East Africa, as a result of the northward extension of High Africa in this area. Mean monthly temperatures tend to remain below 18° C in the East African and Ethiopian highlands. Arid conditions prevail over the Horn of Africa, southern Sudan, and much of Kenya and northern Tanzania. It is likely that an "arid corridor" linked northeastern Africa with southwestern Africa during dry phases of the Pleistocene (Balinsky, 1962). In the East African highlands, which were once on the path of this arid corridor, annual rainfall still tends to be under 1000 mm.

In Low Africa, temperature and rainfall zonation tends to be arranged in regular latitudinal bands. This zonation is disrupted by a relatively dry interval in the region of the Bight of Benin (the Benin or Dahomey Gap, less than 1800 mm per annum). Annual rainfall is highest on the Guinea-Liberia and Nigeria-Cameroon coasts, reaching over 3000 mm, which constitute the wettest part of Africa. The restricted highland areas in Gabon, Cameroon, and Guinea are cooler; mean monthly temperatures tend to remain below 18° C. Annual rainfall on the outer Gulf of Guinea islands ranges from 600 mm to over 5000 mm. The surrounding ocean prevents temperatures exceeding 25° C.

Allowing for the irregularity in the Benin Gap, the overall climatic pattern in Africa is in the form of a warm, moist core situated around the Gulf of Guinea, with concentric zones of increasingly drier and cooler climate extending to the north, east, and south. This zonation is most compressed to the north of the Gulf of Guinea, and most spread out in the southeastern part of the continent. This overall climatic pattern is complicated by a northward swing of higher temperature and rainfall zones during the northern summer, and a southward swing during the southern summer.

Several attempts have been made to integrate patterns of rainfall and temperature in Africa into an overall bioclimatic scheme. In devising such schemes, specific reference has been made to amphibian distribution in southern Africa by Poynton (1964a, 1969), Stuckenberg (1969), and van Dijk (1971, 1982) (reviewed by Poynton and Broadley, 1978). Whatever scheme is used, the longitudinal rather than latitudinal zonation of climate in southern Africa stands out as a major feature. This tendency persists through the eastern part of Africa (Poynton, 1962).

Late-Quaternary changes in African climate were reviewed by Livingstone (1993). He emphasized the aridity of the continent compared to Madagascar and South America, with a result that "… any reduction in its already low rainfall has a more noticeable effect" (Livingstone, 1993:468). Regional variations in climatic changes have been detected, but in general the last glacial maximum was a dry period over most of Africa, with wetter conditions in the Holocene. It can be accepted that rainfall varied in the region between 50% and 150% of the present level (Livingstone, 1993). Estimates of temperatures during the last glacial maximum put the decrease from the present in the range of 6–10° C (Livingstone, 1993). Evidence from montane birds and grasses indicate range continuities between Mt.

Cameroon and East African mountains and a general reduction in the extent of forest trees (Livingstone, 1993). The southern end of the continent, influenced by the temperate westerlies, did not follow this pattern; ice-age aridity was not as marked (Livingstone, 1993).

The climate of Madagascar, like that of the adjacent African coast, shows longitudinal rather than latitudinal zonation. This is a result of the longitudinally positioned highland spine and the South East Trade Winds, which deposit heavy rain throughout the year on the eastern coast and slopes. Rainfall decreases, and becomes seasonal, to the west. Apart from the southwestern corner, where annual rainfall is less than 400 mm, the island is substantially more moist than the adjacent part of Africa; annual rainfall ranges from 350 to nearly 4000 mm (Blommers-Schlösser and Blanc, 1993). Highland temperatures fall below a minimum monthly mean of 10° C, but maximum monthly mean temperatures exceed 20°C over the whole island and reach 40° C in the southwest (Blommers-Schlösser and Blanc, 1993).

The Seychelles lie in the path of the northwest monsoons during the southern summer. These create a wet season with 3000–3800 mm of rain falling in mountainous areas. The mean temperature on the coast exceeds 26° C, and highland temperatures do not fall below 12° C.

VEGETATION ZONES

Vegetation maps of Sub-Saharan Africa reflect the tendency shown by the climatic pattern toward longitudinal zonation in the south and east, and latitudinal zonation elsewhere (Figure 9:2). Vegetation zones are accordingly arranged concentrically around a Gulf of Guinea core. Descriptions and illustrations of the vegetation have been provided notably by Moreau (1966), White (1983), and Kingdon (1990).

Humid forest predominates in the Zaire Basin in a band between about 5 degrees north and south of the equator, with a westward extension to Sierra Leone that is interrupted by the Benin Gap. North of the Zaire and Guinea blocks of forest, latitudinal zonation of vegetation passes through savanna to desert within a distance of as little as 1000 km. The northern savanna belt is connected through East Africa to vegetation types south of the Zaire forests, thereby forming a huge girdle of woodland (with occasional dry forest), savanna, and grassland around the humid forests of the Zaire Basin and Gulf of Guinea. Phytogeographers usually recognize the core of humid forest as the Guineo-Congo Region and the surrounding girdle as the Sudano-Zambesian Region (Brenan, 1978), although differences between Sudanian (Low African) and Zambezian (High African) vegetation may be reflected in a separate regional treatment (White, 1983). The concentric zones eventually deteriorate into semidesert and desert in the Sahara, in the Horn of Africa south to central Tanzania, and in the southwest of the continent, until different weather systems favor macchia at the continent's southern and northern ends.

The southern macchia has little in common with the northern macchia, having a greater diversity of shrubby forms, and a rich diversity of bulbous plants; sclerophyllous shrubland (fynbos) is the dominant vegetation type, with forest occurring only in moister enclaves, whereas in the Mediterranean Basin, forest is the regional climax (White, 1983). Floristically, fynbos is of Afromontane affinity (White, 1983; Linder, 1990). Fynbos replaced the former subtropical rainforest that occurred in the area around the early Pliocene, when winter cyclonic rain became established in the southwestern Cape (Coetzee, 1993).

East African montane areas have mixed grassland to moist woodland and forest, with floristic elements associated with Cape fynbos especially in higher areas. The currently isolated patches of montane vegetation have an archipelago-like configuration (White, 1978, 1983). A dry type of forest predominates in some patches of the eastern lowland.

During the last glacial maximum, a pattern of concentric zonation still appeared to exist, but with the core vegetation of the Gulf of Guinea consisting mainly of arid grassland and savanna, leading to desert and semidesert in North Africa, the Horn of Africa, and southwest to southcentral Africa (Coetzee, 1993, Fig. 3.3). Forest was greatly reduced in West and East Africa (Coetzee, 1993; Livingstone, 1993), but in South Africa, grassland and forest may have spread during the last glacial maximum (Coetzee, 1993). During the earlier part of the Holocene (14,000–12,000 B.P.) there is evidence of a spread of montane forest and a uniting of Afromontane vegetation between parts of the now isolated islands (Coetzee, 1993). During the Holocene climatic optimum (9,000–6,000 B.P.), there

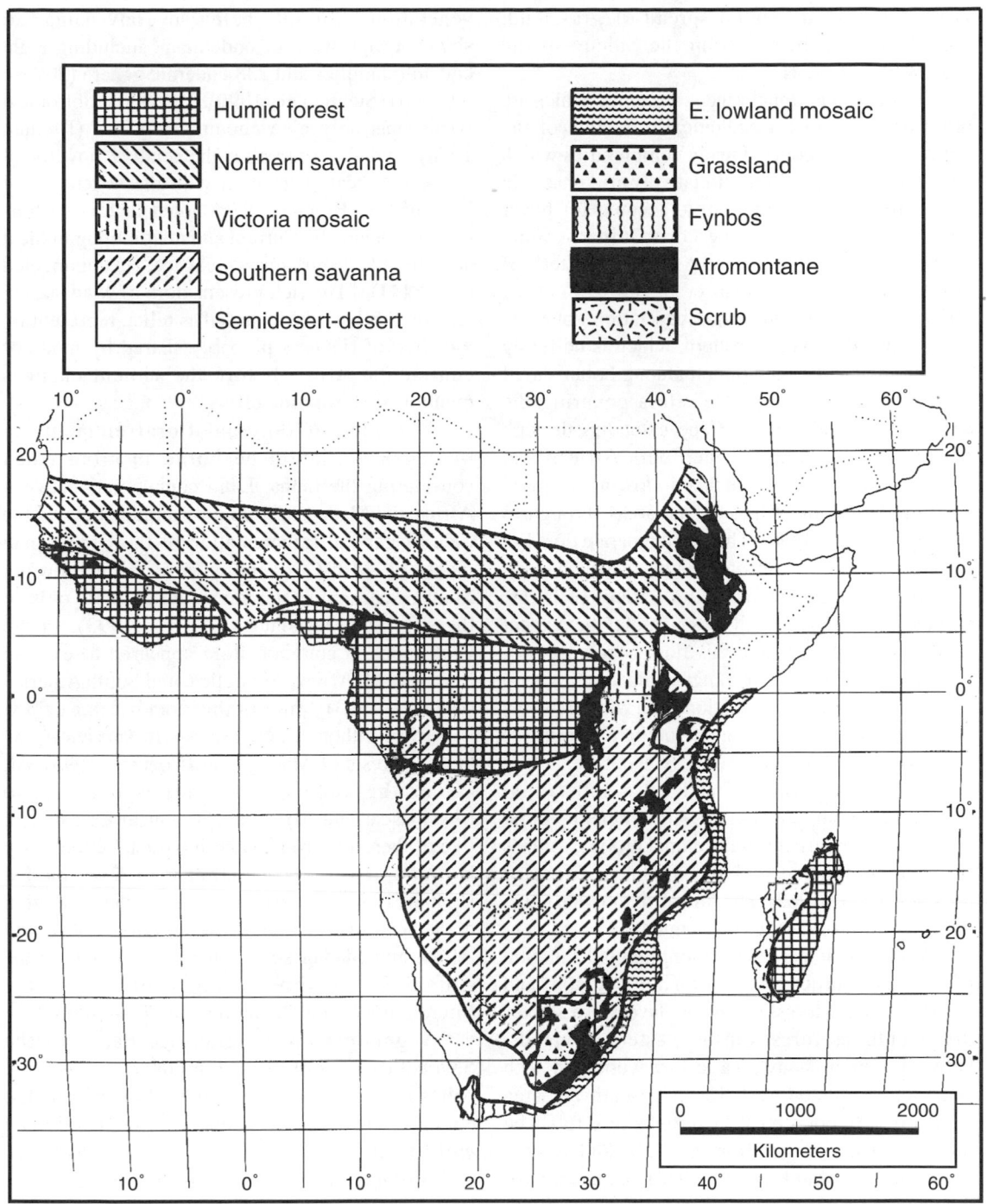

Fig. 9:2. Simplified vegetation zones in Sub-Saharan Africa and Madagascar.

is evidence of a general spread of grassland, woodland, and forest, from the Sahara to the southwest arid areas.

These conventional vegetation zones, which are more or less a direct ecological reflection of the climatic pattern, conceal many complexities which stem from past conditions. Relatively arid phases in the central Zaire Basin seem to account for fewer species in that part of the core Guineo-Congo Region, despite the currently uniform pattern of climate in the basin (Stuart et al., 1990; Coetzee, 1993). Also, the conventional vegetation zones do not adequately reveal a pattern, which is currently gaining increasing recognition among botanists, of a complex "Afrotemperate" flora covering the southwestern and southern Cape, extending through the eastern highlands of the South African and Zimbabwean plateaus, and then following the more easily recognizable Afromontane areas through to Ethiopia (Linder, 1990). The Afrotemperate flora has many components—proteaceous, restionaceous, ericaceous, graminaeous, and a variety of herbs, shrubs, and trees—not all of which may be present in any one region where this flora is represented, but nearly all of which range from the Cape to Ethiopia. Phylogenetic relationships with elements in the vegetation of the other southern continents (e.g., Axelrod and Raven, 1978; Coetzee, 1993) suggest the presence of at least a substantial Gondwanan component in this Afrotemperate flora, and invite comparisons with animal groups, such as the southern montane palaeogenic fauna discussed by Stuckenberg (1962). The African amphibian fauna will be scrutinized with this in mind.

Madagascar has a mainly longitudinal zonation of vegetation, in accordance with its topography and climate. Humid forest occurs on the eastern slopes, dry deciduous forest on the western lowlands, between which is savanna to low woodland with some high altitude forest. The drier southwest area is covered by open bush to xerophytic scrub. The most extensive vegetation on the island is now secondary grassland (White, 1983). An account of the vegetation zones from a herpetological point of interest was given by Lang (1990); Blommers-Schlösser (1993) outlined the vegetation from the aspect of amphibian habitats. Madagascar's natural vegetation (currently being severely damaged) shows a high level of endemism, including eight endemic families and 238 endemic genera (20% of the total) (Stuart et al., 1990). The whole of tropical Africa has only nine endemic families (Brenan, 1978), and, in comparing the relative poverty of tropical African vegetation with the vegetation of Madagascar, Brenan (1978) noted factors such as the cold Benguella Current and major topographical and climatic disturbances in Africa. He commented (1978:441), "The rich present flora of Madagascar may on this basis be regarded as relict, representing a degree of richness probably shared by much of continental Africa before the violent changes mentioned above took effect."

The idea of relictual floral elements in Madagascar should be borne in mind when considering the fauna. Land connections between Africa and Madagascar do not continue beyond the time of general fragmentation of the Gondwanan mass; by Mid-Cretaceous, migration between Madagascar and Africa remained possible only in the south of the island (Coetzee, 1993). In the Cretaceous, a common flora appeared to exist in Madagascar, Africa, Antarctica, and South America (Coetzee, 1993). Much of this flora became extinct in Africa, although derivatives still survive in the humid forests of Madagascar (Coetzee, 1993). All six of the angiosperm families endemic to Madagascar and Africa are present in East Africa; five of these occur in southern Africa, but only one is found in West Africa (Brenan, 1978); this led Brenan (1978:448) to remark on "… the separation of the West African rain-forest flora from both South Africa and Madagascar." This has parallels in the Amphibia, notably among members of the Microhylidae, and the ranid genus *Tomopterna*.

Vegetation on the granitic islands of the Seychelles is rainforest, dominated by endemic palms in drier areas (Vine, 1989). Several peculiar endemics such as the Coco de Mer (*Lodoicea maldivica*) and Jellyfish Tree (*Medusagyne oppositifolia*) suggest long isolation of the islands (Vine, 1989; Kingdon, 1990). The vegetation has been altered radically since settlement of the islands (White, 1983).

Table 9:1. Numbers of families, genera, and species of amphibians in the study area, together with numbers of endemics and percentages.

Taxa	Sub-Saharan Africa		Madagascar		Seychelles		Whole Area	
	Total	Endemic	Total	Endemic	Total	Endemic	Total	Endemic
Families	11	4 (36%)	5	0 (0%)	4	1 (25%)	13	1 (8%)
Genera	86	79 (92%)	18	16 (89%)	7	6 (86%)	108	1 (0.9%)
Species	622	621	156	155	12	11	788	1

DEFINITION OF THE AMPHIBIAN FAUNA

In the taxonomy of Sub-Saharan amphibians, there is still uncertainty and lack of agreement among taxonomists about the delineation and identification of many taxonomic categories. Exploration of the fauna also is far from complete. Therefore, the following analysis must be viewed as only a rough indication of the degree of evolutionary differentiation that has taken place among amphibians in some parts of Africa. There has been only one rigorous cladistic analysis at the species level, that of Gauld and Underwood (1986) on members of the *Nectophryne* group of bufonids. However, Clarke (1988) showed that the resulting classification is not sufficiently robust to accommodate easily a newly discovered species. Phylogenetic analyses have been carried out at the generic level and above by Clarke (1981), Drewes (1984), Cannatella and Trueb (1988), Channing (1989), and Blommers-Schlösser (1993), but the African amphibian fauna as a whole has yet to be treated with a rigor sufficient to inspire broad understanding and agreement among herpetologists. Therefore the number of species given here can at best be considered a rough approximation, and the numbers of genera and even families may be considered largely a reflection of impressions of individual workers.

The required taxonomic source for this review is Frost (1985), with its most recent update (Duellman, 1993b). Some species appearing in these sources have been excluded, notably several species of *Hyperolius* described by Ahl and questioned in entries in Frost. Species appearing in the literature since the publication of the update by Duellman (1993b) are included. The update is heavily affected by the classification of the Ranidae by Dubois (1992), which Dubois himself stated to be

provisional and tentative. His inclusion of many groups in the genus *Rana* is not followed here, because his selection and analyses of characters is unclear. Genera recognized here that are included in the comprehensive *Rana* of Dubois are *Hylarana* (section *Hylarana* including the subgenera *Amnirana* and *Hydrophylax* of Dubois), *Rana* (subgenera *Afrana* and *Amietia* of Dubois), and *Strongylopus* (subgenus *Strongylopus* of Dubois). Dubois's (1992) proposal to separate the "Phrynobatrachidae" from the remaining Ranidae (perhaps still a wastebasket category) is weakened by a recent osteological analysis by Blommers-Schlösser (1993), which suggests that Dubois's "Phrynobatrachidae" is not monophyletic, a view also taken by Duellman (1993a). The Sub-Saharan species included here and their provisional "higher classification" are listed in Appendix 9:1. The Appendix does not include the seven species occurring on the Gulf of Guinea islands on account of the geographical and taxonomic distinctness of the fauna.

Africa and the Seychelles have two recognized amphibian orders, the Anura and Caudata. The Anura is the only order represented on Madagascar. Below this level of classification, the faunas of Sub-Saharan Africa, Madagascar, and the Seychelles differ so much that they are here treated separately. Using the classification in Frost (1985) and Duellman (1993b) (with some modification noted above and below), Table 9:1 shows the number of families, genera, and species in each region and the number of taxa endemic to each.

One family, the Hyperoliidae, occurs over all major areas and is endemic to it (Drewes, 1984; Channing, 1989), as does one ranid genus, *Ptychadena*. According to current taxonomic assessment, a single species, *P. mascareniensis*,

Table 9:2. Numbers of species in families represented in Sub-Saharan Africa (including Gulf of Guinea islands), Madagascar, and Seychelles.

Sub-Saharan Africa		Madagascar		Seychelles	
Hyperoliidae	199	Mantellidae	68	Caeciliidae	7
Ranidae	182	Microhylidae	50	Sooglossidae	3
Bufonidae	82	Rhacophoridae	28	Hyperoliidae	1
Arthroleptidae	73	Hyperoliidae	8	Ranidae	1
Microhylidae	27	Ranidae	2		
Pipidae	21				
Caeciliidae	17				
Hemisotidae	8				
Scolecomorphidae	5				
Heleophrynidae	5				
Rhacophoridae	3				
Total	622		156		12

Table 9:3. Numbers of species in the larger Sub-Saharan genera (including Gulf of Guinea islands), and genera in Madagascar and Seychelles.

Sub-Saharan Africa		Madagascar		Seychelles	
Hyperolius	90	*Mantidactylus*	58	*Grandisonia*	5
Phrynobatrachus	65	*Boophis*	27	*Sooglossus*	2
Bufo	56	*Plethodontohyla*	13	*Tachycnemis*	1
Leptopelis	50	*Platypelis*	10	*Ptychadena*	1
Ptychadena	41	*Stumpffia*	8	*Nesomantis*	1
Arthroleptis	33	*Scaphiophryne*	8	*Hypogeophis*	1
Afrixalus	27	*Heterixalus*	8	*Prasalinia*	1
Cardioglossa	15	*Mantella*	7		
Xenopus	14	*Anodonthyla*	4		
Breviceps	13	*Dyscophus*	3		
Kassina	13	*Laurentomantis*	3		
Astylosternus	11	*Cophyla*	1		
Leptodactylodon	11	*Madecassophryne*	1		
Hylarana	11	*Rhombophryne*	1		
Rana	9	*Paradoxophyla*	1		
Hemisus	8	*Ptychadena*	1		
Cacosternum	7	*Tomopterna*	1		
Petropedetes	7	*Aglyptodactylus*	1		
Conraua	6				
Strongylopus	6				
Tomopterna	6				

covers Africa, Madagascar, the Seychelles, and other islands in the Indian Ocean. A comprehensive taxonomic investigation of this supposed species is one of the outstanding herpetological needs of the region; the uniquely wide distribution of this frog is at present unaccounted for. With the exception of this supposed species, endemism at the species level is complete within the three areas of continental Sub-Saharan Africa, Madagascar, and the Seychelles, and is high at the generic level, at least 86% (Table 9:1). Apart from the shared *Ptychadena* species or species complex, only one other genus is shared between two regions, namely the ranid *Tomopterna* (according to the current taxonomy), which occurs in Africa and Madagascar; the genus also occurs in India. Other genera shared between Sub-Saharan Africa and landmasses outside the region are *Bufo* (almost worldwide but not in Madagascar or Seychelles) and a few ranids still of uncertain composition, and referred by Dubois (1992) to the genera *Euphlyctis* (Africa and Asia), *Hoplobatrachus* (Africa and Asia), and *Rana*, including the "section" *Hylarana* (Africa and Asia). *Rana* is here considered to be mainly an Afrotemperate and Holarctic group.

Following the classification in Frost (1985) and Duellman (1993b) (with a few modifications), the numbers of genera and species recognized here in the families represented in Sub-Saharan Africa, Madagascar, and Seychelles are given in Tables 9:2 and 9:3. To avoid excessive complication, the small Gulf of Guinea fauna is included in the Sub-Saharan fauna in these tables.

The large number of species of Sub-Saharan *Hyperolius, Phrynobatrachus,* and *Bufo* could in part be a reflection of inadequate taxonomic analysis, particularly in the case of small species of *Bufo* (Poynton, 1991). On the other hand, no less than 35 of the Sub-Saharan genera that are currently recognized are monotypic. This 41% monotypy may to some extent be an artifact of overzealous taxonomic separation, but it does suggest long-standing evolutionary activity.

According to the recent analysis of Blommers-Schlösser (1993), the Mantellidae and Rhacophoridae listed in Table 9:2 should be included in the Ranidae, which would reduce the number of Madagascan families from five to three. If her analysis is correct in including the Brevicipitinae in the Hemisotidae, then this family would become the fifth largest in Sub-Saharan Africa, with 21 species, and the number of Sub-Saharan families would be reduced from eleven to nine. Of these, the Ranidae and Hyperoliidae would be represented in Africa and Madagascar, but not her Microhylidae (which excludes the Brevicipitinae). This would remove one family common to Madagascar and East Africa. *Tomopterna,* common to Madagascar and Africa, is mainly east and southern African in distribution.

In view of the evident age of at least much of the Madagascan amphibian and other faunas, it is notable that no caecilians or bufonids have been found on the island. Blommers-Schlösser and Blanc (1993) consider the Mantellidae, Microhylidae, Rhacophoridae, and *Tomopterna* among the Ranidae to have a Gondwanan origin with affinities in Africa and India. *Ptychadena mascareniensis,* widely spread over the island, is considered by them to have a more recent African origin, as has the hyperoliid genus *Heterixalus. Hoplobatrachus tigerinus* (= *Rana tigerina*) was introduced from India at the beginning of this century (Blommers-Schlösser and Blanc, 1993), and is not included in this study. More than one-third of the Madagascan species belong to the genus *Mantidactylus* of the family Mantellidae (subfamily Mantellinae of Blommers-Schlösser, 1993), a family of unsettled relationships with the Rhacophoridae (Channing, 1989; Duellman, 1993b; Blommers-Schlösser, 1993).

The Seychelles is the only region in which caecilians constitute the majority of species: seven of the 12 species belong to three caeciliid genera— *Grandisonia, Hypogeophis,* and *Praslinia.* The endemic anuran family Sooglossidae includes two genera, *Nesomantis* and *Sooglossus,* with three species. The Hyperoliidae is represented by a divergent and endemic subfamily with a monotypic genus *Tachycnemis* (Channing, 1989). Among this distinctly non-African and non-Madagascan fauna occurs, according to current taxonomic assessment, *Ptychadena mascareniensis,* on five of the eight islands (Nussbaum, 1984). If its introduction to the Seychelles and Madagascar was by man, this evidently occurred far back in time "... when travel to these islands was both difficult and infrequent" (Nussbaum, 1984:398).

In a table listing the amphibian fauna of islands in the Indian Ocean, Blommers-Schlösser and Blanc (1993:Table 2) recorded two Madagascan species on the Comores Islands (*Mantidactylus granulatus* and *Boophis opisthodon*), and the Madagascan *Rhombophryne testudo* on Réunion Island, together with the African *Bufo gutturalis* and the supposedly ubiquitous *Ptychadena mascareniensis.* The latter two species also occur on Mauritius Island, together with the American but widely introduced *Bufo marinus* (Owadally and Lambert, 1988), and an apparently introduced species of *Mantella* (Stuart et al., 1990). Bearing in mind the extinct dodo, a giant rail, and two boas on Mauritius, amphibian endemics may have passed unnoticed to extinction there.

The noncontinental islands in the Gulf of Guinea, São Tomé and Principe, have two endemic species of the caeciliid *Schistometopum,* and five endemic anuran species, two of them assigned to the endemic hyperoliid genus *Nesionixalus* (Loumont, 1992). This fauna is described more fully in the following section.

PATTERNS OF AMPHIBIAN DISTRIBUTION

DEFINITION OF REGIONS

As discussed in the section on materials and methods, biogeographical regions are regarded in this chapter as being inductive, and not clearly demarcated from each other. They are presented in agreement with Hengenveld's (1990:49) view that they should not be regarded as "... descriptions of objective natural units, being ends in themselves." As with Noble (1924:152), they are intended to be "... convenient areas for distributional discussion."

In the following list of Sub-Saharan regions, the code number refers to the tabular listing in Appendix 9:1 of species and their occurrence in the various regions. The classification into divisions adopted in this study does not wholly correspond with classifications of eastern and southern African distributions adopted in earlier studies (Poynton, 1964a; Poynton and Broadley, 1991), because of different geographical scales, area, and purpose. Such differences may be expected; Hengeveld (1990:48) went so far as to conclude that "... classifications of the same general area, but with dissimilar limits, will not be comparable." This may overstate the position. The present classification is in broad agreement with the comprehensive study of African centers of endemism by Kingdon (1990).

Sub-Saharan Africa

The following analysis covers the African mainland and islands on the continental shelf. The small amphibian fauna of the outer Gulf of Guinea islands is exclusive to them (Loumont, 1992; Schätti and Loumont, 1992). The peculiarity of these islands,

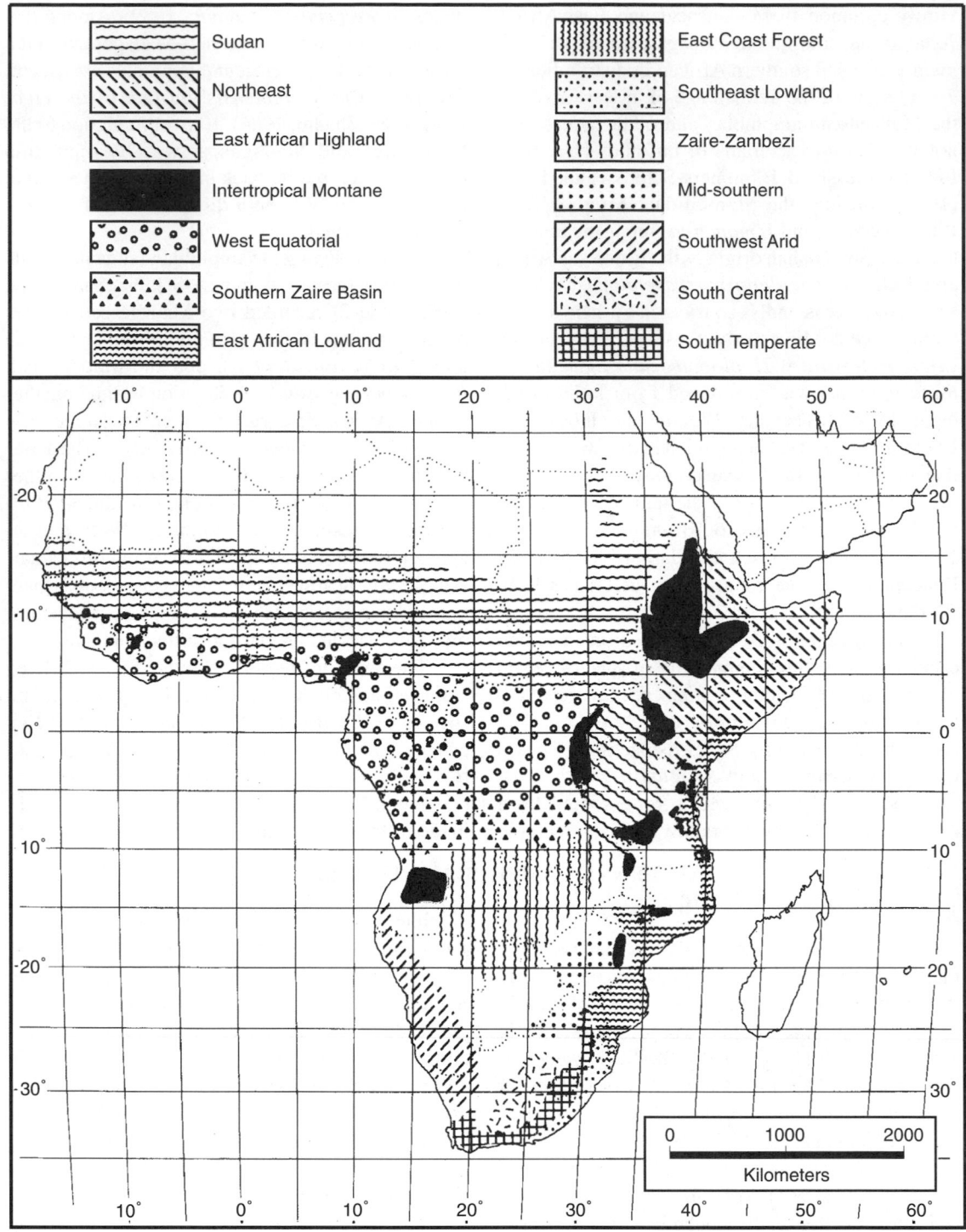

Fig. 9:3.   Biogeographic regions of Sub-Saharan Africa recognized in this study as "convenient areas for distributional discussion" (Noble, 1924:152). Unmarked areas represent extensive transitions, as discussed in the text.

with their complete isolation from the continent even during glacial periods and the 100% endemism shown by the amphibian fauna, places them outside the field of analysis of the continental fauna. The regions are indicated in Figure 9:3.

Sudan (SUD): The term "Sudan" is used in its geographical sense, not its political sense referring only to the Republic of Sudan (or Soudan). This region of 7,022,500 km² extends across Low Africa between about 8° and 15°N from Senegal to Ethiopia, mostly below an altitude of 700 m. There is a southward extension onto higher ground in northeastern Zaire and Uganda, where there is a mingling with eastern and western species (Lake Victoria regional mosaic of White, 1983, and Stuart et al., 1990). This region corresponds with the area of Schiøtz's (1976) second element of savanna treefrogs, and also with the Sudanian Region and southern part of the Sahel zone recognized by the IUCN (Stuart et al., 1990). It covers White's (1983) Sudanian regional center of endemism and also the transition zone to the south, which includes the Benin Gap that reaches the coast at the Bight of Benin. It also includes a margin of White's Sahel regional transition zone to the north. Some species are limited to the eastern or western section of this region; the entries in Appendix 9:1 accordingly are shown as E or W. The monotypic genus *Euphlyctis* (Ranidae) is endemic to the eastern section.

Rain falls during the (northern) summer; the winters are very dry. Mean range of annual precipitation is 500–1200 mm; mean temperatures are 24–28°C. Woodland is the main vegetation type, with some grassland (White, 1983).

Northeast (NEA): This region of 1,094,800 km² includes plains of the Horn of Africa, and eastern parts of Kenya to northern Tanzania, but not the coastal strip of Tanzania, Kenya, and southern Somalia south of about 2° N. It corresponds with Schiøtz's (1976) fifth element of savanna treefrogs, but not entirely with the Somalia-Masai Region of White (1983) and the IUCN (Stuart et al., 1990), because the great East African upland plains lack several Somalian species. These uplands are included in the following region. There is one endemic (monotypic) genus, *Lanzarana* (Ranidae).

The region lies mostly below 900 m. Summer rainfall is experienced at midyear north of the equator but tends to occur at the turn of the year south of the equator. Annual precipitation is less than 500 mm almost throughout the region. Mean annual temperature is above 22°C in most areas. Vegetation ranges from desert to grassland to bushland and thicket (White, 1983).

East African Highland (EAH): The highland plain region from the environs of Lake Victoria to the interval between lakes Tanganyika and Nyasa constitutes an area of 363,000 km². The region corresponds roughly to the upland savanna zone of Loveridge (1937), and includes the Lake Victoria regional mosaic, southwestern part of the Somalia-Masai region, and a northeastern portion of the Zambezian region of White (1983). The region has only two recorded endemic species, *Bufo vittatus* and *Hemisus brachydactylus,* but it has a fairly high species diversity. There are no endemic genera.

The region is composed largely of the Lake Victoria Basin, lying above 1000 m. Annual precipitation is about 900–1500 mm, confined almost entirely to summer. Mean annual temperatures are 21–24° C. The vegetation varies from wooded grassland through woodland to dry forest (White, 1983).

Intertropical Montane: The disjunct mountainous areas from Sierra Leone to Ethiopia, and south to the tip of the continent, show "a remarkable continuity and uniformity" in vegetation (White, 1983:162). This is also evident in animal distribution, which currently leads most authors to include the areas in a single archipelago-like region, called the Afromontane Region following White (1978, 1983). This complex is a "generalized track" according to Croizat's (1968) biogeographical system. A fairly typical Afromontane distribution is shown among amphibians by the *Nectophryne* group of bufonids (Grandison, 1981; Gauld and Underwood, 1986), members of which occur in mountains of Guinea-Liberia, Cameroon, eastern Zaire, Ethiopia, Tanzania, and the southwestern Cape (although *Nectophryne* of Cameroon is not itself montane). The distribution of this group leads to the inclusion of the west and east African mountainous areas in an Intertropical Montane Region. This inclusion has to be a qualified one, because, as discussed later, a marked east-west divide in the amphibian fauna is evident, whose significance has not yet been properly assessed. The matter is "still problematic," to quote

White's (1983:161) comment on the question of including West African mountains in the Afromontane floristic region.

The Cape member of the *Nectophryne* group, *Capensibufo,* appears in the Gauld-Underwood dendrogram as a sister group to the rest. Its relationship to the rest, and to other bufonids, still requires clarification. This, together with other considerations discussed below, makes it seem useful to separate Afromontane areas north and south of the Limpopo Basin. Therefore, the Intertropical Montane Region recognized here as lying north of the Limpopo Basin is not to be confused with the whole Afromontane Region as currently recognized. The Intertropical Montane Region broadly corresponds with the Afromontane Region of White (1978, 1983) and of the IUCN (Stuart et al., 1990), but it excludes the South African areas (South Temperate region in this chapter), which White himself initially excluded from his Afromontane region. A calculation of amphibian species similarity (Dice/Sørensen index) between the Afromontane areas of eastern Zimbabwe and northern Malawi gives a value of 79%, but the value obtained from comparing Afromontane eastern Zimbabwe and Eastern Transvaal–Swaziland faunas is only 40%. Nevertheless, as discussed below, there are clear faunal similarities between the Zimbabwean and South African Afromontane areas, and the assignment of South African areas to a separate South Temperate Region is provisional, awaiting further comparisons between Afromontane islands throughout the archipelago. Both regions, Intertropical Montane and South Temperate, will be discussed below as the major part of an Afrotemperate realm.

The Intertropical Montane Region, of some 755,700 km², is here divided into a number of provinces, which could be subdivided still further in more localized studies (e.g., Howell, 1993). The region has 23 endemic genera, the highest of any African region. However, unlike other regions, no single amphibian species inhabits the entire region. No equivalent to the Afroalpine floristic region in the highest areas can be distinguished in amphibian distribution; instead, these heights are areas of extreme impoverishment.

Many of the mountain areas are volcanic in origin, but overall there is great physiographic diversity. Afromontane plant communities are usually found above 2000 m in the tropics, but on seaward-facing slopes the communities occur as low as 1200 m (White, 1983). Annual precipitation is generally well above 1000 mm and thus allows for the development of forest, but many other combinations of environmental factors lead to the development of grassland, ericoid, and other communities (Meadows and Linder, 1993). Mean midwinter month temperatures are typically below 18° C, and frost occurs at higher elevations, leading on some mountains to very low biotic diversity. The name "Intertropical Montane" in this chapter uses the term "intertropical" in a purely cartographic sense; the areas are not climatically classifiable as "tropical" in any scheme that adopts Köppen's climatic classification, which defines a tropical climate as one in which the coldest month has a mean temperature of over 18° C (Poynton, 1964a). Pros and cons of this measure in relation to southern African amphibians were discussed by Poynton (1964b), Stuckenberg (1969), and Poynton and Broadley (1978).

The following provinces of the Intertropical Montane region are recognized.

1. Ethiopian (M.E.).—The fractured dome of the Ethiopian highlands forms the largest Afromontane area, of some 414,300 km². It ranks highest among the provinces in respect both of percentage endemic genera and endemic species. As pointed out by Largen and Drewes (1989), a distributional gap is shown by brevicipitines (Microhylidae) and nectophrynoids (Bufonidae) between Afromontane areas of Ethiopia and Tanzania, despite an abundance of suitable habitat in Kenya. The "Kenyan interval" is discussed below. Endemic genera are *Altiphrynoides* and *Spinophrynoides* (Bufonidae; genera of Dubois, 1987, that are still to receive general acceptance), *Paracassina* (Hyperoliidae), *Balebreviceps* (Microhylidae), *Ericabatrachus* (Ranidae), and *Sylvacaecilia* (Caeciliidae). All but *Paracassina* are monotypic.

2. Kenyan (M.K.).—This area of some 93,400 km² has no endemic genera, and only 23% of its species are endemic. It has been a center of relatively recent volcanic activity, which might account for low endemism and the existence of the "Kenyan interval" between the Ethiopian and Tanzanian montane areas, but the Aberdare Mountains in Kenya are ancient. Like the "Malawi interval" noted below, this Kenyan interval deserves study.

3. Central or western rift (M.R.).—The eastern highland rim of the Zaire Basin, with an area of some 47,800 km², has a fauna that merges into the lowland fauna. Yet endemism is relatively high; 42% of the species are endemic, and there are three monotypic endemic genera—*Laurentophryne* (Bufonidae), *Callixalus,* and *Chrysobatrachus* (Hyperoliidae).

4. Eastern (M.T.).—This province of 91,300 km² includes the Eastern Arc mountains from Taita Hills in southeast Kenya to the Uzungwa and Mahenge Mountains in southern Tanzania, which were called the Tanzanian basement mountains by Schiøtz (1976, 1981); but the province also includes the more southerly Rungwe complex, as treated by Howell (1993). It is second to the Ethiopian Province in generic richness and contains the endemic genera *Nectophrynoides* (Bufonidae), *Callulina, Hoplophryne,* and *Parhoplophryne* (Microhylidae). *Nectophrynoides* has five species, and *Hoplophryne* has two; the remainder are monotypic. *Boulengerula* (Caeciliidae) and *Scolecomorphus* (Scolecomorphidae) are centered there.

5. Malawi (M.M.).—The province of 15,500 km² includes Malawi and the adjacent Mozambique highland and eastern Zambian center of Poynton and Broadley (1991). There is some range overlap with the Eastern Province, but some distinctive genera, noted below, occur to the north and south but not in the province itself. This "Malawi interval" was noted by White (1978) to occur in the distribution of several Afromontane plants. There is a single endemic genus, *Nothophryne,* which possibly includes two species. Its apparent "Cacosternine" features indicate southern affinities (Poynton and Broadley, 1991).

6. Eastern Zimbabwe (M.Z.).—This province of 7,000 km² is the Eastern Zimbabwe center of Poynton and Broadley (1991). The southernmost species of *Stephopaedes* and *Probreviceps* occur in the province; other species of these genera occur in Tanzania, thereby exhibiting the "Malawi interval." This interval is unaccounted for (Poynton, 1991). The Chirinda Forest, where *Stephopaedes* occurs, was considered by White (1978) to be transitional between Afromontane and lowland forest. While the presence of *Stephopaedes* and *Probreviceps* indicates northern affinities of this province, the occurrence of a member of the southern *Bufo angusticeps* Group, *B. inyangae,* shows southern affinities (Poynton, 1964a), as do two species of the southern ranid genus *Strongylopus* (Poynton, 1964a; Poynton and

Broadley, 1985b, 1991). There are no endemic genera.

7. Angolan (M.A.).—The difficulty in classifying the Angolan highlands that have an area of about 63,900 km² was pointed out by Stuart et al. (1990). The highlands are given an ambiguous status by Kingdon (1990). The amphibian fauna is not clearly demarcated from that of the extensive Zambezi-Zaire divide. There are no endemic genera.

8. Cameroonian (M.C.).—Separability of the Cameroonian and Guinean highland amphibian fauna from the lowland West African fauna is discussed under the West Equatorial Subregion. Endemic genera of the Cameroonian Province with an area of 19,000 km² are the torrent-adapted *Werneria* (four species) (Bufonidae), and the monotypic genera *Alexteroon* and *Arlequinus* (Hyperoliidae), and *Idiocranium* (Caeciliidae). These genera do not appear on the island of Bioko.

9. Guinean (M.G.).—In this province of 3500 km², the only endemic genus is *Nimbaphrynoides* (Bufonidae) with two species, separated by Dubois (1987) from *Nectophrynoides.*

West Equatorial (WEQ): This region, covering 1,900,000 km², includes the lowlands bordering the Gulf of Guinea, including the island of Bioko, and northern part of the Zaire Basin. A break occurs in the Benin Gap. Forest species continue in suitable habitat through Uganda to extreme western Kenya, constituting the first forest faunal element of Schiøtz (1976) in the area of the Lake Victoria regional mosaic of White (1983) and the IUCN (Stuart et al., 1990). The region also merges into the Central Province of the Intertropical Montane Region, and the Southern Zaire Basin Region.

The eastern part of the region appears to have a lower amphibian diversity, and, as noted by Schiøtz (1976), it could be regarded as a subtraction margin of the fauna of Cameroon; but some species appear to be restricted to the eastern margin of the basin, which suggests provincial status. Areas within the arc of the Zaire River seem to be relatively depauperate and lack endemics, evidently on account of the climatic disturbances noted earlier.

Most of the region lies below 500 m. Although rainforest covers (or did cover) the greater part of the region (White, 1983), rainfall is relatively low compared with areas of rainforest in other continents, and shows marked seasonality. Summer rainfall is experienced, for the most part exceeding 1800 mm per year. It is in excess of 3000 mm on the coast of Guinea-Liberia and Nigeria-Cameroon.

The region shows notable amphibian diversity. It includes the endemic genera *Silurana, Hymenochirus,* and *Pseudhymenochirus* (Pipidae); *Didynamipus* and *Nectophryne* (Bufonidae); *Acanthixalus, Cryptothylax,* and *Opisthothylax* (Hyperoliidae); *Dimorphognathus, Phrynodon,* and *Aubria* (Ranidae); *Geotrypetes* and *Herpele* (Caeciliidae), although none of these genera occupies the whole region. Several species cover virtually the whole region, but more limited ranges tend to be grouped and suggest four centers or provinces within the total range.

1. Western (W.W.).—This province of 206,000 km², which is the Guinea Block of Schiøtz (1967), is west of the Benin Gap. The monotypic *Pseudhymenochirus* is endemic.

2. Central (W.C.).—This large province of 442,000 km² includes southeastern Nigeria, southern Cameroon, and Central African Republic to northern Gabon and Congo, and the island of Bioko (2033 km²). Two species are endemic to Bioko, *Leptopelis brevipes* and (if recognizable) *Schistometopum garzonheydti,* although other endemic subspecies may be recognized (Schätti and Loumont, 1992); otherwise the amphibian species there also occur on the adjacent mainland. Two other species of *Schistometopum* occur on outer Gulf of Guinea islands, discussed below, and one in East Africa, according to current taxonomy. Genera endemic to the province are *Didynamipus* (Bufonidae) and *Phrynodon* (Ranidae), both monotypic. Both of these occur on Bioko and the mainland.

3. Eastern (W.E.).—The eastern margin of the basin (534,000 km²) is the Ituri-Maniema forest refuge area of Kingdon (1990). There are no endemic genera.

4. Southern (W.S.).—This province of 179,000 km² includes the southern Gabon–western Congo area. *Cryptothylax* (Hyperoliidae) with two species is endemic.

Recognizing four divisions still oversimplifies distributional complexity in the region. For example, Amiet (1983) showed that species ranges in Cameroon separate according to east-west, north-south, forest-savanna, upland-lowland, and other groupings. Inclusion of the Cameroonian and Guinean highland fauna in the Intertropical Montane Region rather than in this region may be criticized on the grounds that there are marked faunal

similarities between the West African lowlands and highlands. Yet a substantial turnover of species takes place from the West African lowlands to highlands. Using data given by Amiet (1975:Table 2), calculation of the Dice/Sørensen similarity index gives a value of 27% when comparing the fauna of three stations below 750 m and the fauna of Mt. Manengouba, above 1200 m. As indicated below, this is a relatively low degree of similarity, which justifies regional separation. Lawson (1993:81), in his study of the Korup National Park in southwestern Cameroon, confirmed that the montane areas there "share with each other and the neighbouring Cameroon highlands a fauna distinctive from the adjacent lowlands."

Southern Zaire Basin (SZB): This region of 835,700 km² includes the southern Zaire Basin slopes in Zaire and northern Angola. The Lake Upemba Basin (but not the surrounding highland) is included in this region. It merges into the West Equatorial Region and Central Highland Province. As noted under the section on vegetation, this area has a history of substantial disturbance during the Pleistocene, and it could be regarded as lying in subtraction margins of the West Equatorial and Zaire-Zambesi regions, rather than being a region in its own right. It occupies the same general area as White's (1983) Guinea-Congolia/Zambezia regional transition zone. It has no endemic genera.

East African Lowlands (EAL): With an area of 546,500 km² mainly below 500 m, this region extends from southern Somalia to northeastern South Africa; it reaches inland along river basins, notably the Limpopo and the Zambesi, and marginally includes Lake Malawi. Pemba, Zanzibar, and Mafia islands are included in the region. The fauna of this region, which includes Schiøtz's (1976) coastal "farmbush element," was detailed by Poynton (1990), but some taxa included there are assigned to the following region. There are no endemic genera.

Annual rainfall is 800–1300 mm, increasing to the south; it is restricted to summer although becoming more evenly spread in the south. Relative humidity remains fairly high throughout the year. The mean midwinter monthly temperature is 18° C or more. The region includes White's (1983) Zanzibar-Inhambane regional mosaic, excluding the Usambara Mountains, and also excluding patches of lowland forest assigned to the following region. The region also includes the lower-lying strip of

White's (1983) Tongaland-Pondoland regional mosaic south to the region of Lake St. Lucia (28°S), where precipitous subtraction of the fauna occurs with replacement by southern species (Poynton, 1964a, 1990).

There are several islands off the East African coast. The only conspicuous biogeographical anomaly that is evident is the presence of *Xenopus laevis* on Inhaca Island off southern Mozambique. This species has a south temperate distribution, and is replaced by *X. muelleri* on the Mozambique Plain, apart from its southern margins (Poynton and Broadley, 1991). The population of *X. laevis* on Inhaca Island is evidently a relict of a former distribution of the species over the southern part of the plain, which could be expected to have occurred during a glacial maximum, when cooler conditions would have been likely to have favored the spread of this species, and when Inhaca Island would have been continuous with the mainland on account of lowered sea level (Poynton, 1964a).

East Coast Forest (ECF): Disjunct patches of lowland forest having a combined area of about 12,000 km² in Kenya, Tanzania, and possibly northern Mozambique make up this region, which merges with the Eastern Arc mountain forest. A patch on Mafia Island probably is now lost. The region is distinguished mainly by the presence of two genera of small bufonids, *Mertensophryne* and *Stephopaedes* (Poynton, 1991). The mid-altitude Chirinda Forest of southeastern Zimbabwe and adjacent Mozambique, where *Stephopaedes* also occurs, could be regarded as an outlier, but this forest is included here in the Intertropical Montane Eastern Zimbabwe Province, with a result that *Stephopaedes* is excluded as an east coast forest endemic. *Mertensophryne* is regarded as an endemic, although earlier records (Loveridge, 1925; Barbour and Loveridge, 1928) are from more inland areas.

Southeast Lowlands (SEL): This region, with an area of about 54,100 km², commences in the south as a narrow coastal strip at about 33° S, widens northward to include the lower plateau slopes to about 1200 m, and around 29°S leaves the coast to continue at increasing altitude along the plateau slopes to Swaziland, lying at about 1000–1500 m. The monotypic genus *Natalobatrachus* is endemic to it and occurs in sheltered, wooded ravines, which form a part of White's (1983) Tongaland-Pondoland regional mosaic. Inland, the mosaic intermingles

extensively with Afromontane vegetation (White, 1978, 1983) in the south, and with the highveld transition zone (White, 1983) farther north. In the north, it lies above the lowland strip of Tongaland mosaic. Annual rainfall is 700–1200 mm; winters are dry. Temperature is strongly influenced by altitude and latitude; mean annual temperatures are 20–17° C, and occasional frost occurs especially in river valleys, where cold air drainage takes place.

Zaire-Zambesi (ZAZ): The extensive watershed of the upper Zaire and Zambesi river systems, with an area of some 1,750,000 km², is included in this region which lies at elevations over 1000 m. It corresponds to some extent with the area of Schiøtz's (1976) third element of savanna treefrogs, but it does not include the highland area south of Lake Victoria (East African Highland Region), which has eight species not found in Zaire-Zambesi. The region includes the northwestern, north-central, and central centers of Poynton and Broadley (1991), but it is smaller than the Zambesian Region recognized by White (1983) and the IUCN (Stuart et al., 1990) in that it does not include the East African, Zimbabwean, and (in part) Angolan highlands. The monotypic *Kassinula* is the only endemic genus.

Rain occurs in summer, with an annual range of about 500–1400 mm. Mean annual temperature does not exceed 24° C, and light frost occasionally occurs in depressions. Vegetation ranges from dry evergreen forest to dry deciduous forest and woodland to bushland to grassland (White, 1983).

Mid-southern (MST): This region is partially divided by the Limpopo Basin into highland plains with an area of 239,500 km². It includes the southeastern portion of White's (1983) Zambesian regional center, but it is not included in the Zaire-Zambesi Region of the present study because its diversity is lower, and a few species, notably *Bufo fenoulheti*, have ranges centered in Zimbabwe and the northern Transvaal. The region has no endemics.

Southwest Arid (SWA): This region of 528,100 km² extends from southwestern Angola to Namaqualand in South Africa. It broadly corresponds to the Karoo-Namib regional center of White (1983) and the IUCN (Stuart et al., 1990), but excludes most of the Karoo. As with other plateau areas, there are few endemics to characterize the region, but there is a fairly high diversity of widespread species, centered to the east and south. The name "Southwest

Arid" is adopted without intending to introduce circularity into causal analysis; the name "Southwest Region" (usefully suggesting a complementary relationship with the Northeast Region) could lead to confusion with the southwestern Cape. There are no endemic genera.

Annual rainfall is generally much less than 250 mm, with a tendency to shift from summer rainfall in the north to winter rainfall in the south. Mean annual temperatures generally do not exceed 18°C, and in winter frost occurs inland. Vegetation types range from desert to shrubland (White, 1983).

South Central (SCE): The central South African plateau, with an area of some 207,200 km², can be recognized as another highland plateau region, which is poor in endemics but has a fairly high diversity of widespread species. The area includes the southeastern part of White's (1983) Kalahari-Highveld regional transition zone, most of which is shown as a blank area in Figure 9:3. The presence of temperate species and the endemic *Bufo vertebralis* make the area distinctive. There are no endemic genera.

Annual rainfall is about 400–700 mm; winters are dry and very cold with widespread frost. Vegetation types include grassland, wooded grassland, and scrub (White, 1983).

South Temperate (STE): Encompassing the eastern margin of the South African Plateau to the southern tip of the continent, this region has an area of about 200,300 km². This and the South Central Region correspond to Crowe's (1990) "9-shaped configuration," tailing down into the southwestern Cape; the South Central Region is separated from the plateau margin in this study on account of substantial faunal turnover (Poynton, 1992). The South Temperate Region corresponds with the southern end of White's (1978, 1983) Afromontane Region and the Cape Region, or the Drakensberg Mountains Subregion and the Cape Fynbos Region of the IUCN (Stuart et al., 1990). Although eight species are distributed through at least a large part of the area, and some endemic genera are widespread, a somewhat different taxonomic composition of the fauna on either side of the Great Fish River Basin (Poynton, 1988) suggests recognition of two provinces. This pattern has wide applicability among invertebrate and plant groups (Stuckenberg, 1962). Two genera are endemic to the whole region, *Heleophryne* (Heleophrynidae) with five species and *Arthroleptella* (Ranidae) with three.

1. Southeast Highland (S.H.).—This province includes the eastern plateau rim, escarpment and upper plateau slopes of South Africa. The Drakensberg and its foothills, together with the Lesotho mountains, form the main features of this Afromontane province. The general land surface with an area of 97,100 km² lies at 1200–2200 m. Temperature is governed mainly by altitude and ranges from an annual mean of about 14°C to 17°C. Mean midwinter month (July) temperatures range from less than 10°C to less than 6°C. Frost is common in winter, which is the dry season. Mean annual rainfall is about 900–>2000 mm, generally increasing with greater altitude. Snow may be widespread in spring and early summer. Vegetation is Afromontane, and includes forest (dominated by *Podocarpus* especially at higher altitudes), ericaceous and proteaceous types, and grassland (White, 1978).

Apart from the monotypic *Anhydrophryne* (Ranidae) in the extreme southern forest there are no endemic genera, but the province is characterized by a concentration of species of *Rana* (four species) and the barely distinguishable *Strongylopus* (four species) (Channing, 1979, 1981).

2. Cape (S.C.).—This province of 103,200 km² includes the Cape folded mountains and southern to southwestern forelands. Topography contrasts strongly between level forelands and parallel folded mountain ranges, with an average altitude of 1000–1500 m. Rain falls mainly, but not exclusively, in winter, with an annual mean generally between 300 and 2500 mm; the amount is strongly influenced by aspect. Temperature is strongly affected by altitude. On the forelands, the mean annual temperature is about 15–17°C, and frosts do not occur. Snow falls on mountains during winter and spring. The prevalent vegetation is sclerophyllous shrubland or fynbos. Where aspect is suitable, there is the development of forest with a mixture of Afromontane species and Cape endemics (White, 1983).

In this province, amphibian species tend to occur either on the lowlands or on mountains, making a division into lowland and montane areas useful in a more localized study. However, not all species that occur in montane areas are torrent-adapted, but may inhabit marshy patches or shallow depressions in sandstone. Endemic genera are *Capensibufo* (Bufonidae) with two species, and the monotypic *Poyntonia* and *Microbatrachella* (Ranidae). The first two genera occur in montane marshes or rock depressions (Channing and Boycott, 1989), the latter in inundated patches on the forelands.

Testing the regions: A classification of this sort can be tested by setting up transects through adjacent regions and calculating species turnover. Transects divided into quarter-degree intervals have been run between Durban and Bloemfontein in South Africa, and between Beira and Harare in Mozambique-Zimbabwe (Poynton, 1992; Poynton and Broadley, 1991). The South African transect passes through three regions recognized above, the Southeast Lowlands, South Temperate, and South Central. Using the Dice/Sørensen similarity index, the similarity between quarter-degree intervals on the coast (Southeast Lowland), the Drakensberg (South Temperate), and Bloemfontein (South Central) can be quantified. Similarity between Durban and the Drakensberg is 42%; between the Drakensberg and Bloemfontein, 38%. Similarity between Durban and Bloemfontein is only 27% (Poynton, 1992). The Mozambique-Zimbabwe transect included the East African Lowland Region, the Eastern Zimbabwe Province of the Intertropical Montane Region, and the Mid-southern Region. Similarity between the transect intervals Beira (coast) and Mutare-Inyanga (Eastern Highland) was 54%. The western end of the transect was on the central Zimbabwe Plateau (Chinhoyi). Similarity between Chinhoyi and Mutare-Inyanga is 81%, but only 43% between Chinhoyi and Beira (Poynton and Broadley, 1991).

A transect along the eastern coastal lowlands of Africa (Poynton, 1990) passes through the Northeast, East African Lowlands, and Southeastern Lowlands regions. Similarity between the area of the Kenya-Tanzania border and the tip of Somalia north of 8°N (i.e., between the Northeast and East African Lowlands regions) is 10%, and is 31% between northernmost and southernmost Natal (between the East African Lowlands and Southeastern Lowlands Regions). Similarity between the Southeastern Lowlands Region and the southwestern Cape is 18%. Similarity between the East African Lowlands and the southwestern Cape is 0% (Poynton, 1988).

With so few transects available, it is not possible to evaluate precisely the significance of these figures.

Similarity between the regions identified here is about 50% or less. If considerably more transects are run across relatively well-collected parts of Africa, it may be possible to set a criterion for regional separation according to some similarity value.

Gulf of Guinea Islands

The islands of Principe (136 km^2) and São Tomé (1269 km^2), but not Annobón (18 km^2), are occupied by an amphibian fauna peculiar to the islands (Loumont, 1992; Schätti and Loumont, 1992). This includes the endemic hyperoliid genus *Nesionixalus* with two species, debatably two species of the cacciliid genus *Schistometopum* (otherwise known only from a single species on Bioko and one in East Africa, according to current taxonomy), and the endemic *Phrynobatrachus dispar, Ptychadena newtoni,* and *Leptopelis palmatus* (Loumont, 1992; Schätti and Loumont, 1992).

Four of the seven species are restricted to São Tomé; one is restricted to Principe, and two, including one of the species of *Nesionixalus,* are common to both.

Madagascar

The island has an area of 594,180 km^2. Guibé (1978) recognized eastern and western regions. Blommers-Schlösser and Blanc (1993), using a similarity coefficient, recognized three domains— Southern, Western, and Eastern. The latter includes the central highland and montane areas. They did not assign species or genera to these domains; indeed, no such partitioning of the fauna is evident. The pattern essentially is one of differing degrees of diversity, from high diversity (97% of the total number of genera) in the Eastern Domain, fairly low (41%) in the Western Domain, and low (19%) in the Southern Domain (Blommers-Schlösser and Blanc, 1993:Table 12). According to the maps of Blommers-Schlösser and Blanc (1991), the southern area has only one species *(Scaphiophryne brevis)* with a range mainly restricted to it; consequently this area is regarded here as a depauperate zone rather than a separate region. More distinctive as an area of endemism is the region of the central highland, mountains, and slopes within the Eastern Domain, on account of 11 species of *Boophis* being restricted to it. Nevertheless, this region is swamped by the ranges of every genus, apart from the one species of *Tomopterna.* Therefore, no Madagascan

Table 9:4. Genera in continental Sub-Saharan regions and provinces: diversity and endemism (ranks in parentheses).

Region	Province	Number of genera		Number of endemic genera		Percent endemic genera	
Sudan		19	(10)	1	(9)	5	(14)
Northeast		11	(20)	1	(9)	9	(10)
East African Highland		21	(7)	0	(19)	0	(19)
Intertropical Montane	Ethiopian	15	(16)	6	(1)	40	(1)
	Kenyan	10	(25)	0	(19)	0	(19)
	Central	17	(13)	3	(4)	18	(3)
	Eastern	22	(3)	4	(2)	18	(3)
	Malawian	12	(19)	1	(9)	8	(11)
	E. Zimbabwe	11	(20)	0	(19)	0	(19)
	Angolan	8	(26)	0	(19)	0	(19)
	Cameroonian	22	(3)	4	(2)	18	(3)
	Guinean	8	(26)	1	(9)	13	(8)
	Total	52		23		44	
West Equatorial	All	15	(16)	2	(6)	13	(8)
	Western	22	(3)	1	(9)	5	(14)
	Central	37	(1)	2	(6)	5	(14)
	Eastern	20	(8)	0	(19)	0	(19)
	Southern	26	(2)	1	(9)	4	(18)
	Total	39		12		31	
S. Zaire Basin		16	(14)	0	(19)	0	(19)
E. African Lowland		20	(8)	0	(19)	0	(19)
East Coast Forest		7	(28)	1	(9)	14	(7)
Southeast Lowland		18	(12)	1	(9)	6	(12)
Zaire-Zambesi		22	(3)	1	(9)	5	(14)
Mid-southern		19	(10)	0	(19)	0	(19)
Southwest Arid		11	(20)	0	(19)	0	(19)
South Central		11	(20)	0	(19)	0	(19)
South Temperate	All	11	(20)	2	(6)	18	(3)
	SE Highland	16	(14)	1	(9)	6	(12)
	Cape	15	(16)	3	(4)	20	(2)
	Total	20		6		30	

biogeographical divisions are recognized in this study.

Seychelles

The distribution of species on these various islands with an area of 453 km^2 has been discussed by Nussbaum (1984:Table 3). Island endemism is low. One species, the caeciliid *Hypogeophis rostratus,* is represented on all eight islands listed by Nussbaum. All amphibian species, except the caeciliid *Grandisonia diminutiva,* occur on Mahé. The low endemism on individual islands led Nussbaum (1984:392) to conclude that "… the distributional pattern of amphibians in the Seychelles reflects relatively recent partitioning of a once wide-ranging fauna into island refugia through submergence of the Seychelles Microcontinent."

NUMBERS OF GENERA AND SPECIES IN EACH REGION

Sub-Saharan Africa

Bearing in mind the difficulties involved in determining numbers of taxa and biogeographical divisions, this section deals with the number of taxa believed to occur in the Sub-Saharan regions just described. Appendix 9:1 shows individual Sub-Saharan species assigned to the various divisions recognized herein and forms the basis for the following analysis. Assignment of species to divisions is conservative, in that assignments are made only on

Table 9:5. Species in continental Sub-Saharan regions and provinces: diversity and endemism (ranks in parentheses).

Region	Province	Number of species		Number of endemic species		Percent endemic species	
Sudan		66	(5)	35	(4)	53	(9)
Northeast		20	(20)	11	(14)	55	(7)
East African Highland		37	(13)	2	(25)	5	(27)
Intertropical Montane	Ethiopian	32	(15)	25	(7)	78	(1)
	Kenyan	18	(23)	5	(19)	28	(18)
	Central	69	(4)	50	(1)	72	(2)
	Eastern	57	(8)	32	(6)	56	(6)
	Malawian	32	(15)	5	(19)	16	(22)
	E. Zimbabwe	19	(22)	4	(23)	21	(19)
	Angolan	12	(26)	7	(18)	58	(5)
	Cameroonian	62	(7)	39	(3)	63	(4)
	Guinean	15	(25)	5	(19)	33	(14)
	Total	255		177		70	
West Equatorial	All	26	(19)	14	(11)	54	(8)
	Western	88	(2)	43	(2)	49	(10)
	Central	116	(1)	34	(5)	29	(15)
	Eastern	48	(11)	5	(19)	10	(25)
	Southern	63	(6)	10	(16)	16	(22)
	Total	179		142		79	
S. Zaire Basin		56	(9)	16	(9)	29	(15)
E. African Lowland		56	(9)	12	(12)	21	(19)
East Coast Forest		10	(27)	4	(23)	40	(11)
Southeast Lowland		38	(12)	11	(14)	29	(15)
Zaire-Zambesi		77	(3)	15	(10)	19	(21)
Mid-southern		36	(14)	0	(28)	0	(28)
Southwest Arid		20	(20)	8	(17)	40	(11)
South Central		17	(24)	1	(26)	6	(26)
South Temperate	All	9	(28)	1	(26)	11	(24)
	SE Highland	32	(15)	12	(12)	38	(13)
	Cape	30	(18)	21	(8)	70	(3)
	Total	53		34		64	

the grounds of well-established records. Future collecting may well show that many species are more widespread than indicated in the appendix.

The data are summarized in Tables 9:4 and 9:5, in which, All = number of taxa that range throughout the entire region (zero in the case of Intertropical Montane species); Total = total number of taxa in the region (not a simple sum of taxa in the constituent provinces in the West Equatorial region, because many occur in more than one province).

The great majority of species are restricted to only one region or province; thus, they are classifiable as endemics. As shown in Table 9:6, there is almost an exponential decline in the number of species occupying an increasing number of regions or provinces (jointly called divisions). The slight rise in the number of species that occupy four divisions is the result of several species occurring in all four of the West Equatorial provinces. A count at regional level would eliminate this. Nearly all genera have a majority of species restricted to only one division. Exceptions are *Tomopterna,* two species of which occur in three regions, whereas only one species, *T. delalandii,* seems (from uncertain data) to be restricted to a single division; and *Chiromantis,* in which no species is restricted to a single province, although *C. rufescens* seems to be restricted to the Western Equatorial Region.

Table 9:6. Number of continental Sub-Saharan divisions (regions or, if relevant, provinces) occupied by each species.

Number of divisions	Number of species	Percent of total species
1	411	66.9
2	103	16.8
3	28	4.5
4	31	5.0
5	16	2.6
6	6	1.0
7	4	0.6
8	7	1.1
9	2	0.3
10	0	0.0
11	1	0.2
12	0	0.0
13	1	0.2
14	1	0.2
15	1	0.2
16	0	0.0
17	1	0.2

These figures do not support Kingdon's (1990:22) view which at first impression seems correct, that "Africa's fauna and flora have tended to diverge into at least two major divisions — widely distributed, common and adaptable species, and more restricted species living in narrowly defined habitats or geographically small regions." If "major division" is taken to be measured by numbers of species, Kingdon's view predicts a substantial number and grouping of widespread species, which is not evident from Table 9:6. Also, there is no bimodality in a plot of single-division species against area. The impression that "widely distributed, common and adaptable species" constitute a "major division" of the amphibian fauna probably arises from the fact that, although such species are few in number, they are conspicuous because of high population densities in open country.

Oceanic Islands

As discussed earlier, the present study does not recognize divisions in Madagascar and the Seychelles, at least from the point of view of assigning taxa to each division. Diversity gradients and other differences over the regions are treated in the next section. The two Gulf of Guinea islands— São Tomé and Principe—also do not constitute two clear divisions.

AREAS OF HIGH SPECIES DIVERSITY AND ENDEMISM

In agreement with Brown (1988), species diversity is defined as the number of species in a given area. A comprehensive treatment of biodiversity needs to take into account the diversity shown at generic and family levels. This presupposes a suitable degree of phylogenetic analysis, which is largely lacking in the case of African amphibians. Nevertheless, some preliminary treatment of generic diversity is given in this section.

This section compares the number of taxa recorded in each region. An effective comparison presupposes an equal intensity of collecting in each region, but this has not yet been achieved. Therefore, the following rankings according to recorded diversity should be regarded as preliminary. An interpretation of the diversity values would be assisted by ranking the regions according to thoroughness of investigation, but this is an unknown factor; the recent discovery of a new genus in the southwestern Cape (Channing and Boycott, 1989), generally presumed to be a well-studied area, serves as a caution.

SUB-SAHARAN AFRICA

Using the data in Table 9:4, the regions are ranked in the following order according to the number of genera:

1. Intertropical Montane (total) (52)
2. West Equatorial (total) (39)
3. Zaire-Zambesi (22)
4. East African Highland (21)
5. East African Lowland (20)
5. South Temperate (total) (20)
7. Mid-southern (19)
7. Sudan (19)
9. Southeast Lowland (18)
10. Southern Zaire Basin (16)
11. Northeast (11)
11. Southwest Arid (11)
11. South Central (11)
14. East Coast Forest (7)

Although the whole Intertropical Montane Region has the greatest number of genera, no province in that region surpasses the number in the Central West Equatorial Province (37 genera, Table

9:4) and the Southern West Equatorial Province (26 genera). The provinces of the Intertropical Montane Region have only pan-African genera in common, in contrast to the occurrence of some genera that are restricted to all of the West Equatorial Region, namely *Silurana, Hymenochirus,* and *Geotrypetes.* According to number of endemic genera in the regions the ranking is:

1. Intertropical Montane (total) (23)
2. West Equatorial (total) (12)
3. South Temperate (total) (6)
4. Sudan (1)
4. Northeast (1)
4. East Coast Forest (1)
4. Southeast Lowland (1)
4. Zaire-Zambesi (1)

The remaining regions lack endemic genera. The preponderance of endemic genera in each region is shown in Table 9:4; ranking according to this value results in the following order:

1. Intertropical Montane (total) (44%)
2. West Equatorial (total) (31%)
3. South Temperate (total) (30%)
4. East Coast Forest (total) (14%)
5. Northeast (9%)
6. Southeast Lowland (6%)
7. Sudan (5%)
7. Zaire-Zambesi (5%)

This identifies three principal areas of generic diversity. The Intertropical Montane and South Temperate Regions are the major components of an Afrotemperate realm, discussed below. Together with the fourth-ranking region, the East Coast Forest, it may be noted that the environment of these regions is essentially mesic, and more likely to be subject to geographical disturbance during climatic cycles than savanna areas, which have many widespread taxa but relatively few endemics. In fifth position is a xeric region, the Northeast, at the other end of the environmental spectrum and again more likely to be subject to climatic disturbance. This trend also is evident at the species level (Table 9:5).

Noteworthy in Table 9:4 is the high ranking of most of the provinces in the percentage of endemic genera. Although generic diversity may be relatively small, the proportion of endemic genera can be high:

1. Ethiopian (Intertropical Montane) 40%
2. Cape (South Temperate) 20%
3. Central (Intertropical Montane) 18%
3. Eastern (Intertropical Montane) 18%
3. Cameroonian (West Equatorial) 18%
6. Guinean (West Equatorial) 13%
7. Malawian (Intertropical Montane) 8%
8. Southeast Highland (South Temperate) 6%
9. Western (West Equatorial) 5%
9. Central (West Equatorial) 5%
11. Southern (West Equatorial) 4%

A number of provinces have no endemic genera. The distinctiveness of the Intertropical Montane and South Temperate regions is again evident. This has implications regarding conservation which will be discussed later.

The first-ranking Ethiopian Province also takes first rank in respect of percentage endemic species (Table 9:5). The regions show the following ranking according to species number:

1. Intertropical Montane (total) (255)
2. West Equatorial (total) (179)
3. Zaire-Zambesi (77)
4. Sudan (66)
5. Southern Zaire Basin (56)
5. East African Lowland (56)
7. South Temperate (total) (53)
8. Southeast Lowland (38)
9. East African Highland (37)
10. Mid-southern (36)
11. Northeast (20)
11. Southwest Arid (20)
13. South Central (17)
14. East Coast Forest (10)

The whole Intertropical Montane Region includes the greatest number of species, but none of its provinces exceeds the 116 species in the West Equatorial Central Province, or the 88 species in the West Equatorial Western Province. Twenty-six species are distributed throughout the West Equatorial Region, whereas no Intertropical Montane species occurs in all the provinces included in the Intertropical Montane Region. This reflects the continuity of the physical geography of the West African and Zairean lowlands, in contrast to the disrupted land surface of highland areas.

Ranking regions according to number of endemic species gives the following order:

1. Intertropical Montane (total) (177)
2. West Equatorial (total) (142)
3. Sudan (35)
4. South Temperate (total) (34)

5. Southern Zaire Basin (16)
6. Zaire-Zambesi (15)
7. East African Lowland (12)
8. Northeast (11)
8. Southeast Lowland (11)
10. Southwest Arid (8)
11. East Coast Forest (4)
12. East African Highland (2)
13. South Central (1)
14. Mid-southern (0).

No particular province has an outstandingly high number of endemics. The ranking (Table 9:5) is Central Intertropical Montane (50 spp), Western West Equatorial (43 spp), Cameroonian Intertropical Montane (39 spp), and Central West Equatorial (34 spp). Apart from the Western West Equatorial Province, these provinces are either in, or circle, the Zaire Basin.

The preponderance of endemic species in each region is shown in Table 9:5. Ranking according to this feature gives the following order:

1. West Equatorial (total) (79%)
2. Intertropical Montane (total) (70%)
3. South Temperate (total) (64%)
4. Northeast (55%)
5. Sudan (53%)
6. East Coast Forest (40%)
6. Southwest Arid (40%)
8. Southern Zaire Basin (29%)
8. Southeast Lowland (29%)
10. East African Lowland (21%)
11. Zaire-Zambesi (19%)
12. South Central (6%)
13. East African Highland (5%)
14. Mid-southern (0%)

The three major areas of percentage generic diversity, the Intertropical Montane, West Equatorial, and South Temperate regions, again appear as the three main areas of percentage species diversity, although, compared to percentage generic diversity, the East Coast Forest Region loses rank to two relatively xeric regions, the Sudan and Northeast. The East Coast Forest Region shares rank with another xeric region, the Southwest Arid. This continues the tendency, noted in the genera, for xeric areas to have relatively high percentage endemism.

Table 9:5 shows that four provinces rank highest in the percentage of endemic species: Ethiopian (78%), Central Intertropical Montane (72%), Cape

(70%), and Cameroonian (63%). As noted earlier, these provinces rank high in percent endemic genera.

When the rankings for species diversity and percentage species endemics in the regions are compared, the rank correlation r' is 0.41, which is not significant ($n = 14$). This confirms the poor coincidence between diversity and endemism noted earlier. Nevertheless, it is striking that the Intertropical Montane and the West Equatorial Regions rank high both in diversity and in endemism at species and generic levels. The southwestern Cape shows a high percentage of endemism but a relatively low number of species. This feature was detected in Crowe's (1990) quantitative analysis of distribution patterns, and was commented on by Poynton (1989a).

It is generally thought that a positive relationship exists between area and taxonomic diversity, or area and percentage endemism, but differences in physical geography (primarily topography) between the Intertropical Montane and West Equatorial regions complicate area/diversity comparisons. The value, percentage endemic species/area x 10,000, is 0.4 for the West Equatorial Region, 0.9 for the Intertropical Montane Region, and 3.2 for the South Temperate Region. The relatively low degree of endemism in the West Equatorial Region compared with the Intertropical Montane Region may be the result of relatively low levels of Pleistocene fragmentation in the West Equatorial lowlands compared with disruptions in the upland areas of the Intertropical Montane Region which, for purely topographical reasons, were likely to have experienced greater fragmentation. This high degree of fragmentation most likely accounts for extensive speciation, as Amiet (1975), among others, pointed out. The cause of a high percentage endemics/area value in the South Temperate Region is less clear, because both lowland and highland species are involved. Evolutionary activity in this southern part of Africa merits close study (Poynton, 1989a).

MADAGASCAR

As noted earlier, a marked diversity gradient exists on the island. This was shown at the generic level by Blommers-Schlösser and Blanc (1993: Table 12), who gave the values of the number and percentage of genera as follows: South 6 = 19%; West 13 = 41%; Central highland 21 = 66%; Montane 23 = 72%; Eastern slopes 27 = 84%; East coast 22 =

69%. The south and particularly the southwest are depauperate areas.

SEYCHELLES

The diversity of Seychellian amphibians was detailed by Nussbaum (1984: Tables 2 and 3). The three larger islands have the highest species diversity—Mahé 11 (92%), Silhouette 9 (75%), and Praslin 8 (67%). The smaller La Digue has five species (42% diversity); even smaller islands have fewer species. If Nussbaum (1984) is correct in concluding that amphibian distribution on the Seychelles reflects recent partitioning of a once wide-ranging fauna over the Seychelles microcontinent, then the degree of extinction suffered on the existing islands is broadly related to the area of each island, although relief and habitat diversity seem to have modified a simple species-area relationship (Nussbaum, 1984). As Nussbaum observed, there is no reason to suppose that the presently recorded diversity is at a stabilized or equilibrium level.

Like the faunas of the montane areas of Africa, which for purely topographical reasons would be likely to experience fragmentation with resulting speciation, periodic submergence of the Seychelles microcontinent would, as Nussbaum (1984) pointed out, fragment populations on separate islands. Divergence in allopatry and subsequent overlapping of ranges during periods of emergence could explain why several closely related species are found on the same island at the present time of partial submergence (Nussbaum, 1984). A similar model was used by Amiet (1975) and other writers to explain the large number of related species occurring on montane islands in Africa.

GULF OF GUINEA ISLANDS

The greater area of São Tomé may be a factor contributing to its having four of the seven species, even though it is slightly farther from the mainland than Principe.

MAJOR GROUPINGS IN SUB-SAHARAN AFRICA

In the previous section, a relatively high degree of endemism was noted in the Intertropical Montane Region of Africa compared with the West Equatorial Region. This east-west difference can be related to topography. More difficult to account for are marked taxonomic differences between East and West African amphibian faunas which have been noted by several authors (Schiøtz, 1981; Howell, 1993; Nussbaum and Hinkel, 1994). As with the floristic pattern noted earlier, the taxonomic differences between East and West African faunas need to be evaluated in a broader context of eastern and much of southern Africa possessing a common fauna. Notable among Southern–East African groups is the large contingent of microhylids (allowing for the Sudanian *Phrynomantis microps*), the widespread bufonid *Schismaderma*, and the widespread ranids *Hoplobatrachus* (allowing for a westward Sudanian extension), *Cacosternum, Hildebrandtia, Pyxicephalus, Rana, Strongylopus,* and *Tomopterna*. In addition, there are geographically restricted eastern genera: the ranids *Arthroleptides, Ericabatrachus, Nothophryne, Euphlyctis, Lanzarana,* the caeciliid genera *Boulengerula* and *Sylvacaecilia,* and the scolecomorphid *Scolecomorphus*. The *Nectophryne* group of mainly montane dwarf bufonids (Gauld and Underwood,

1986) also is separated into a southern-eastern and a western division; the southern-eastern division comprises the genera *Capensibufo, Nectophrynoides, Altiphrynoides,* and *Spinophrynoides*. Other eastern dwarf bufonid genera are *Mertensophryne* and *Stephopaedes,* which are disjunctly distributed from Kenya to Zimbabwe. On account of their specialized features (Poynton, 1991), they may be regarded as the remains of an ancient group, whose taxonomic and geographical affinities have yet to be clarified.

The eastern-southern genera contrast with an assemblage of western genera (including the Central, Cameroonian, and Guinean provinces), which includes *Cardioglossa, Astylosternus, Leptodactylodon, Nyctibates, Scotobleps, Trichobatrachus* (Arthroleptidae), *Alexteroon, Arlequinus, Chlorolius, Chrysobatrachus, Cryptothylax, Opisthothylax* (Hyperoliidae), *Silurana, Hymenochirus, Pseudhymenochirus* (Pipidae), *Dimorphognathus, Petropedetes, Phrynodon, Aubria* (Ranidae), *Geotrypetes, Herpele,* and *Idiocranium* (Caeciliidae), and *Crotaphatrema* (Scolecomorphidae). Added to these genera are members of the western division of the *Nectophryne* group (Gauld and Underwood, 1986), namely *Laurentophryne, Nectophryne, Wolterstorffina, Werneria, Didynamipus,* and *Nimbaphrynoides*.

There are in fact far more genera centered in the eastern (29) or the western (29) sectors than genera common to both (14), namely *Arthroleptis, Bufo* (although showing internal east-west divisions), *Hemisus, Afrixalus, Hyperolius, Kassina, Leptopelis, Xenopus, Conraua, Phrynobatrachus, Ptychadena, Hylarana, Chiromantis,* and *Schistometopum.* This gives a similarity value of only 48%, which falls slightly below the rough 50% similarity limit between regions discussed under Definition of Regions.

The east-west divide seems so pronounced that there could be justification for dividing the Intertropical Montane Region into a western region that includes the Central, Cameroonian, and Guinean Provinces, and another region including the eastern provinces. As noted earlier, the reason for not doing so in the present study is the overall unity of intertropical montane areas suggested by the *Nectophryne* group (Gauld and Underwood, 1986), excluding *Capensibufo.* The matter needs fuller investigation.

The east-west divide at generic level also appears at species group level. This is particularly evident where there is controversy or uncertainty regarding identification of savanna taxa with eastern and western components, as in the pairs *Bufo gutturalis* and *B. regularis,* and *Phrynomantis bifasciatus* and *P. microps.* The zone of transition from one member of the pair to the other has not yet been clearly determined, but Uganda appears to be involved in both cases. The geographical transition from High Africa to Low Africa occurs in this general area, and its relation to faunal turnover deserves closer study. White (1983) drew attention to the more diversified physiography and climate in High Africa in relation to the wider range of vegetation types occurring in his Zambesian Region (High Africa) compared to his Sudanian Region (Low Africa). But the overall east-west transition is evidently complex and involves more than just the area of Uganda. In a study of African snake faunas, Hughes (1983) found a faunal similarity of 72% (common species as percent of the smaller total) between inland Kenya and Sudan and recognized coastal Kenya as being the "watershed" between faunas of the southern and northern savannas. This area was identified by Poynton (1990) as a replacement-transition zone in Darlington's (1957) sense among savanna amphibian species ranges.

In addition to an east-west divide in the amphibian fauna, a geographically complex north-south divide is evident, various aspects of which have been described and discussed by Poynton (1961, 1962, 1964a, 1982, 1986a, 1988, 1989a). The north-south divide seems to be associated largely with a tropical-nontropical differentiation of the fauna; the terms "tropical" and "nontropical" are used here in a climatological, not cartographical, sense. This differentiation apparently is associated with latitudinal and altitudinal zonation of climate. Thus, in accordance with the longitudinal trend in climate described above, the East African fauna is well represented in a "tropical peninsula" extending into northeastern South Africa, but only a small fraction of this fauna occurs in the same latitude on the highland to the west (Poynton, 1964a, 1990; Poynton and Bass, 1970; Poynton and Broadley, 1991).

The southern temperate part of the continent gives the appearance of being an independent theater of evolution (Poynton, 1989a). As Crowe (1990:149) observed from his analysis, "… there is a progressive increase in percentage endemism from Mozambique to the Cape." Eleven of the 28 genera currently recognized south of the Tropic of Capricorn seem to have had an initial range in South Africa, the term "initial range" being used in Brundin's (1981) sense to mean the limitation of a taxon to the area, or else to mean the area of greatest diversity of the taxon within its total range, including the seemingly more primitive members. The genera with an initial range apparently in the south are *Capensibufo, Heleophryne, Semnodactylus, Breviceps, Anhydrophryne, Arthroleptella, Cacosternum, Microbatrachella, Natalobatrachus, Poyntonia,* and *Strongylopus* (Poynton, 1989a). It may be noted that those genera with extensive ranges to the north, namely *Breviceps, Cacosternum,* and *Strongylopus,* avoid the west equatorial lowlands, and approach the equator only on the eastern highlands in accordance with the apparent temperate preferences of this fauna. But the historical course and processes involved in the development of this fauna have not yet been discovered, least of all why nearly two-thirds of the southern endemic genera are ranids in a broad sense, and over half are phrynobatrachines, which distinguishes the south from the tropical eastern and western divisions of the continent.

At species level, a tropical-nontropical differentiation is suggested in southern Africa by a number of species pairs, one member centered in the eastern "tropical peninsula" and the other occurring in the climatically nontropical west and south (Poynton, 1982). Like the members of the east-west pairs noted earlier, their similarities are sufficient to have caused taxonomic confusion, and so suggest relatively recent separation events.

Whereas discussion has focused on tropical-nontropical differentiation in southern Africa, this differentiation is also evident in the contrast between highland and lowland areas within the cartographical tropics; such highland areas are understood to be nontropical in a climatological sense. Species turnover from lowland to adjacent highland in the tropics is marked; in Cameroon, similarity between lowland and highland faunas is only 27% (data from Amiet, 1975). This is less even than the 43% similarity between Mozambique and Zimbabwe (Poynton and Broadley, 1991), and the 42% similarity between lowland and upland KwaZulu/ Natal (Poynton, 1992).

Given that the tropical-nontropical differentiation covers a taxonomic range from family level to the slight differentiation evident in species pairs, it is inviting to conceive of two contrasting faunal groups—one cool-adapted and presently showing fragmented or relict patterns in areas of high altitude and latitude and the other group more warm-adapted, with continuous ranges centered in the lowlands to midlands. The high altitude–high latitude group is considered in this study to comprise the characteristic fauna of the Intertropical Montane Region, the southern plateau regions (Mid-southern, South Central, and Southwest Arid) and the South Temperate Region. This fauna can largely be regarded as biogeographically equivalent to the Afrotemperate flora (e.g., Linder, 1990). The Afrotemperate amphibian fauna includes no less than half the number of Sub-Saharan species. In showing such a large nontropical component (in a climatological sense), amphibians resemble some invertebrate groups rather than other terrestrial vertebrate groups (Poynton, 1964a). This is in keeping with the development of a large north-temperate amphibian fauna (Darlington, 1957).

The existence of a large, evidently cool-adapted fauna in Africa makes the prevalent use of the name "Afrotropical" for the whole of Sub-Saharan Africa seem inappropriate. Although the name has become standard usage, it could lead to an unconscious (if not conscious) deemphasis of the large nontropical element in the Sub-Saharan fauna (the word "tropical" being taken in a climatological, not cartographic sense). It is suggested here that the term "Afrotropical" is employed only in its climatic sense, the name "Sub-Saharan" being applied to the subcontinent as a whole (if the older name "Ethiopian" is considered inappropriate). The Sub-Saharan fauna is then seen to be composed of two main components, Afrotropical and Afrotemperate. The Afrotropical component shows eastern and western subcomponents, discussed above. The Afrotemperate component appears to show a major division across the Limpopo Basin, and another seemingly lesser division between the East African and West African mountain systems.

This largely temperature-based proposal is not intended to deemphasize the importance of rainfall-related factors. Marked similarities between the fauna and flora of the arid southwest and northeast areas of Africa have been commented on by several authors (e.g., Balinsky, 1962; Kingdon, 1990). A strip of relatively arid country still connects the two areas through western Tanzania, and during dry phases of the Quaternary there probably existed a substantial arid corridor allowing faunal and floral interchange between the southwest and northeast. This corridor does not appear to be reflected in the distribution of any amphibian species, although it is indicated in disjunct distributions of members of the *Bufo garmani* and *B. vertebralis* complexes, occurring in the northeast and mainly the more arid parts of southern Africa (Poynton, 1995).

The proposal of Afrotemperate and Afrotropical divisions also does not carry the implication that vegetation patterns have any direct causal bearing on the distribution of amphibian species, despite the name "Afrotemperate" being adopted from the botanical literature. As discussed below, amphibian distributions show little direct association with the distribution of vegetation types. Nevertheless, similar environmental factors, past and present, could be expected to have similar effects on the ranges of various plants and animals. The idea of cool-adapted and warm-adapted groups leads to a dynamic model that will be discussed in the section on historical phenomena.

CORRELATIONS OF DISTRIBUTIONAL PATTERNS

TOPOGRAPHIC FEATURES

As a primary cause of climatic and distributional discontinuities, the importance of broken topography can hardly be overstated, whether the highland areas are separated by lowlands or by transgressing ocean. The differences between salient features of the Intertropical Montane Region and the West Equatorial Region have already been discussed as arising primarily from the different topographies of the two areas. Influences of High African and Low African topography on climate and distribution also have been noted. Yet, because of the obvious importance of topography, other dependent factors affecting distribution more directly may be underemphasized. Stuckenberg (1969) demonstrated a close correlation between the number of anuran species and the width of the KwaZulu/Natal coastal plain, which tapers markedly in the region of heaviest subtraction of the East African fauna. Yet, to the north of KwaZulu/Natal, the same tropical species may range widely to the west over all kinds of terrain, showing, as Stuckenberg pointed out, that topography cannot provide a simple, direct explanation for the pattern in KwaZulu/Natal. A correlation between topography and distribution in one area need not necessarily indicate a direct causal relationship (Poynton and Broadley, 1978). In cases such as genera that have torrent-dwelling tadpoles, notably *Heleophryne* of South Africa and *Werneria* of Cameroon, the relationship between topography and distribution is evidently a close one. However, specific adaptations such as these are uncommon, and it seems that topography most usually affects amphibian distribution indirectly through its effect on other environmental factors. As a result, an African-wide division into "lowland" and "montane" faunas, used most notably by Moreau (1966) in his study of African bird faunas, runs into difficulties; for example, a faunal group that is montane in the tropics becomes lowland in the cooler Cape, thereby suggesting that a climate-based terminology, rather than one based on topography, might be closer to immediate determinants of distribution (Poynton, 1967).

CLIMATE

The most elaborate attempt to correlate amphibian distribution in Sub-Saharan Africa with environmental features was presented by Crowe (1990) in his analysis of southern African vertebrate distribution, diversity, and endemism. The analysis was range-based, and use was made of grid-quadrats. The grid was more than two-degrees square, which precludes a detection of faunal changes that are reflected in some divisions recognized in the present study. As noted above, these divisions emerge when a quarter-degree grid is applied. Crowe's main finding was a longitudinal zonation of diversity in southern Africa; species richness in the west was found to be less than 20% the richness in the east, apart from the southern Cape where diversity remained 50% of that of the east. The species richness vs. environment regression results suggested "… that a combination of mean annual rainfall (57%; positive relation) and mean July (winter) temperature (22%; positive relation) are the key factors which influence species richness, explaining 79% of the geographical variation in species richness" (Crowe, 1990:149).

Crowe noted that these results are consistent with results of an earlier study of southern African amphibian distribution (Poynton 1964a, b). In this earlier work, more emphasis was placed on temperature; "… thermal conditions appear to act directly on amphibian distribution, while rainfall appears to act indirectly through its effect on the habitats of the amphibians" (Poynton, 1964b:213). Emphasis on temperature was prompted by the evidence that "Some major features of the amphibian distribution pattern do not show any conformity with the biome pattern, but coincide very closely with the general thermal pattern" (Poynton, 1964b:216). The direct effect of temperature was thought to operate particularly through breeding success, such as the annual number of nights favoring breeding behavior (Poynton, 1969).

The high value for the rainfall relationship that Crowe (1990) found is not surprising, when it is considered that his study included the arid west, with its low species diversity. If eastern areas of the subcontinent were to be examined on their own, a gain in the value for temperature could be expected, because broad conformities between distribution patterns and thermal patterns have been noted in that region (Stuckenberg, 1969; Poynton and Broadley, 1978; 1991). Crowe (1990:149) stated that the "best evidence" for accepting what he perceived to be a "rainfall-induced distribution pattern" was a supposed eastward bulge in his depauperate "western zone"; but the position of this area shown on his map for amphibians (1990:Fig. 3a), as far east as Lake Kariba, is reached by 79% of the East African lowland species of adjoining Mozambique (Poynton and Broadley, 1991). In this instance a temperature-induced pattern is indicated, because the fauna of this area shows greater similarity with the fauna around Beira than does the fauna of the intervening Zimbabwean highland (Poynton and Broadley, 1991).

Relationship between rainfall patterns and amphibian distribution in southern Africa also was investigated by van Dijk (1971, 1982), who found some correspondence, but stated (1971:94) that "… it is not thereby implied that rainfall directly determines distribution." No individual climatic factors have been identified that seem directly to determine distribution. No clear conclusions were reached in an attempt to identify particular environmental factors that may be involved in the dramatic transition from East African to South African amphibian faunas in northeastern KwaZulu/Natal (Poynton and Bass, 1970); the study merely revealed the complexity of relationships between environment and species distributions. Clearly, environmental factors do not vary alone and independently; they tend to form complexes that present extreme difficulties in identification and characterization. Grimsdell and Raw (1984) compared the distribution of amphibian species in KwaZulu/Natal with Phillips's bioclimatic regions. Some 70% of species were found to occur in the moist, warm, coastal lowlands. Diversity declined progressively into the interior, and was lowest in the highlands, where the mean midwinter month temperature is about 10° C lower than it is on the coast, and where the rainfall is more seasonal although in the same general annual range as on the coast (900–1300 mm), the high Drakensberg excluded. The number of species endemic or partially endemic to KwaZulu/Natal was more evenly spread in highlands and lowlands; this suggests a greater degree of adaptation of the endemic fauna to the more temperate highlands of Natal. However, no particular climatic factors or climatic complexes affecting the general species distribution or the distribution of endemics were identified.

A close relationship between distribution and contemporary environmental factors need not be expected; environmental tracking of ranges of species in accordance with climatic changes is likely to involve complex lagging, governed by the spreading and retracting rate of each species (Poynton, 1990). Data relevant to this are lacking.

VEGETATION

A broad correlation between vegetation patterns in Sub-Saharan Africa and the regions recognized here should be obvious, at least with respect to the primary divisions into forest, savanna, and semidesert. All writers on the subject have observed that a distinct amphibian fauna is found in forests; but, when an attempt is made to identify particular factors in the vegetation types that directly affect amphibian distribution, the matter becomes as complex and obscure as the identification of particular climatic factors, despite broad correlations between climate and distribution. For example, it might seem that a study of amphibian distribution in the *Flora Zambesiaca* area (Poynton and Broadley, 1991) would have benefited from the detailed chorological analysis presented by Wild and Barborosa (1967), yet there seemed to be little or no evidence that amphibians respond to vegetational physiognomy as identified by Wild and Barborosa, or even to the broader physiognomy used by White (1983) in his classification.

White (1983:29) observed that the ranges of animals "either greatly transgress the boundaries of vegetation types or, alternatively, when they are confined to an individual type, they usually occupy only part of it." This seems to be particularly true of amphibian species, even of those customarily grouped together as "forest species." Species adapted primarily to living in fast-flowing streams, such as those of *Heleophryne,* normally occur in forest, which tends to develop in association with such streams; but the species may be found at high altitudes where forest is replaced by grassland (Poynton, 1985; Poynton and Broadley, 1991). Other species, such as those which lay eggs in seepages, as in the case of *Arthroleptella,* take advantage of the protection provided by tree cover against (presumably) heat and desiccation; but their occurrence in grass-covered highlands shows that they are not wholly reliant on tree cover (Poynton, 1964a). Other species lay eggs in humus or in shaded soft soil, which is a characteristic substrate of forests; but where forest has been cleared, such species can continue to breed if there is scrub or dense grass cover, as in the case of *Arthroleptis* and *Breviceps*

(Poynton, 1964a; Poynton and Pritchard, 1976; Poynton and Broadley, 1991), and caecilians (Nussbaum and Hinkel, 1994). For such species, the presence of trees in closed formation seems to be of secondary importance, and the term "forest species" should not properly be applied (Poynton, 1989b; Poynton and Broadley, 1991).

It is in the African lowlands that limitation of "forest" animals to a closed-canopy habitat appears to be more marked, presumably because of the risk of desiccation in hotter low-lying areas (Poynton and Broadley, 1991). In the case of amphibians and many kinds of forest invertebrates, particular susceptibility to desiccation during the terrestrial egg and younger stages of the life history seems to be especially important regarding the selection of breeding sites (Lawrence, 1953). Where a protected environment can be actively created, for example by deep burrowing, expansion into the savannas is possible, as evidenced by more derived members of the genus *Breviceps*, savanna species of which lay eggs in deep burrows (Poynton and Pritchard, 1976; Poynton and Broadley, 1985a). In members of the *Nectophryne* group of bufonids there is a trend toward viviparity. As Wake (1980:203) pointed out, viviparity in *Nimbaphrynoides* could allow embryonic development to proceed in a "… gestation period in which the adult frogs spend the cold dry months of the year burrowed under ground." Ovoviviparity in *Nectophrynoides* may in part be a similar strategy. Nevertheless, other small bufonid genera are particularly closely associated with forest, and, especially in the case of *Mertensophryne* and *Stephopaedes*, oviposition in water-filled holes between buttress roots, or in treeholes or shells of the forest snail *Achatina* (references in Poynton, 1991), may restrict breeding populations to closed-canopy forests. Yet, a specimen of *Mertensophryne* was collected from an area where forest had been cleared (Barbour and Loveridge, 1928); thus, total dependence on forest cannot be assumed.

In the case of montane small mammals in Malawi, Happold and Happold (1989:363) found that "… forest species can survive in grasslands, but grassland species cannot survive in forest." Generally, this holds true for amphibians, especially in moist or upland areas. The tendency for "savanna" amphibians not to occur in dense forest may largely be due to a lack of suitable breeding sites and water of suitable productivity for tadpoles. The first, but by no means always the second, of these factors also applies to semideserts.

In Madagascar, the presence of digital intercalary cartilages in 60% of the species (members of the Mantellidae, Rhacophoridae, and Hyperoliidae) suggests, as Blommers-Schlösser and Blanc (1993:432) pointed out, "… une adaptation à la vie arboricole." The climate favors tree cover as the most extensive habitat type on the island, and the prevalence of this biome may largely account for the uniformity of the Madagascan amphibian fauna (although not for the absence of caeciliids and bufonids). But, as in Africa, there seems to be lack of success in identifying particular factors in the various types of tree cover that directly affect amphibian distribution.

HISTORICAL PHENOMENA

In the absence of adequate fossil material, reconstruction of the history of a fauna has to rely on phylogenetic analyses and whatever can be suggested by present distribution. Regrettably, the situation regarding Cameroon highland amphibians noted by Lawson (1993:84) is general for African amphibians, that "… many likely biogeographically informative taxa remain to be described, let alone subjected to phylogenetic analysis." The only cladistic study on African amphibians at the species level is that by Gauld and Underwood (1986) on the *Nectophryne* group of mainly montane bufonids, based on previous work by Grandison (1981). The study shows a striking agreement between phylogenetic affinities and geographical distribution. At the base of Grandison's (1981) proposed "*Nectophryne* line" is *Capensibufo* of the southwestern Cape mountains; she suggested that the affinities of *Capensibufo* were as much with the Australian Myobatrachidae as with the Bufonidae. Gauld and Underwood's dendrogram (1986: Fig. 4) shows *Capensibufo* as a sister group to the remaining members of the *Nectophryne* line, composed of six genera distributed in provinces of the Intertropical Montane Region, namely the Eastern, Ethiopian, Central, Cameroonian, and Guinean, and also the central and eastern provinces of the West Equatorial Region.

Too much may be read out of this single case, for which monophyly has not been clearly demonstrated. Nevertheless, the *Nectophryne* group may be regarded as an Afromontane group (apart from some lowland Cameroon species) which is of particular interest because the possible myobatrachid (i.e., Australian) affinities suggest a Gondwanan origin. The seemingly most primitive member of the group, *Capensibufo* of the southwestern Cape, could be regarded as a member of the montane palaeogenic fauna defined by Stuckenberg (1962) as an ancient, montane assemblage with Gondwanan affinities. The distribution of the more derived members of the *Nectophryne* group conforms with the Hennig-Brundin progression rule (Brundin, 1981), which predicts peripheral apomorphy.

The genus *Breviceps* also shows a pattern suggesting a southern initial range in Brundin's (1981) sense, as discussed earlier. But it evidently is not Gondwanan in origin; it appears to have been derived from sylvicolous East African *Probreviceps* stock (Poynton and Broadley, 1985a). The seemingly most derived species of *Breviceps* are adapted for burrowing in savannas, and range northward as far as Angola, southern Zaire, and Tanzania (Poynton and Broadley, 1985a). The same phenomenon of geographical and phylogenetic progression from the south of the continent northward may have occurred also in the case of *Cacosternum,* centered in South Africa, with one species complex spreading disjunctly northward over highland areas as far as Somalia. Because of taxonomic uncertainties, the picture in this genus is not as clear as in *Breviceps. Strongylopus* is another genus that has an evident southern initial range (but probably post-Gondwanan) with a subtraction margin extending disjunctly over highlands to Tanzania (Poynton, 1964c; Poynton and Broadley, 1985b). The species extending the farthest north, *S. fasciatus,* seems to be the most derived (Poynton, 1964c), and, like *Cacosternum,* the genus evidently followed a highland pathway, as probably did an early stock of the *Nectophryne* group.

The torrent-dwelling genus *Heleophryne* is generally regarded as an African Gondwana-derived group of species, with possibly myobatrachid affinities (Frost, 1985); therefore, it could be included in Stuckenberg's (1962) montane palaeogenic fauna. Neither this genus nor anything remotely related to it is known north of the Limpopo Basin, where suitable habitat exists. Ecologically, it is more specialized than most members of the

Nectophryne group in having torrent-adapted tadpoles. The absence of torrential streams in the Limpopo Basin may have restricted the genus to the region south of the basin.

These southern-based groups are adapted to cool conditions, judging from their distribution, which corresponds with the latitudinal and altitudinal zonation of temperature. As discussed earlier, a large part of amphibian distribution in Sub-Saharan Africa can be resolved into two contrasting faunal groups, the one cool-adapted and presently showing relict distribution in areas of high altitude and latitude, while the other group is more warm-adapted, with continuous ranges centered in the lowlands to midlands. The isolated population of *Xenopus laevis* on Inhaca Island, described earlier, provides a different case of evident temperature-related relict distribution. The overall pattern suggests a dynamic model involving reciprocal spreading and withdrawing of the interlocking ranges of cool-adapted and warm-adapted groups of species, following cyclic changes in Quaternary climate and perhaps before (Poynton, 1986a). In accordance with the cyclic changes in climate, one group may be expected to expand, showing a radial pattern, a pattern of dispersal; while contemporaneously, contracting ranges of the other group will tend to show relict patterns, influenced by isolating topographical features, which may result in vicariation. Radial and relict patterns, dispersal and vicariant events, may thus be seen to be the two sides of the same coin, as several authors have pointed out (Poynton, 1983).

Yet still to be answered are the questions of what historical-environmental factors lead to such a grouping of ranges. Some historical factors may extend back to Gondwanan times, when southern and northern components of the supercontinent existed (Coetzee, 1993). The fauna of Southern Gondwana (including southern South America, southern Africa, and Madagascar) was presumably more cool-adapted than that of Northern Gondwana, and *Heleophryne* and the *Nectophryne* group may provide examples of derivation from this fauna, in contrast to the pipids and *Bufo,* perhaps derived from the warmer Northern Gondwana (Poynton and Broadley, 1978; Cannatella and Trueb, 1988; Duellman, 1993a). But tropical-nontropical groupings of ranges are also evident in amphibians of likely post-Gondwanan distribution, notably most ranids and southern microhylids. Whatever the factors are that have led to this grouping of ranges, at least some appear to have been operating up to the present time.

In a similar vein, Blommers-Schlösser and Blanc (1993) discerned two sources of the Madagascan amphibian fauna, Gondwanan (Mantellidae, *Tomopterna* of the Ranidae, Microhylidae and Rhacophoridae) and African (Hyperoliidae and *Ptychadena*). These authors pointed out that this dichotomy reflects Madagascar's two-part history, Gondwanan and insular. The same two-part history seems to hold for the Seychelles. A more detailed history cannot yet be given. The phylogenetic and continental relationships of the Seychellian family Sooglossidae are, in Nussbaum's (1984) opinion, completely unknown, although he gave some evidence to suspect that affinities may lie with myobatrachids, as did Tyler (1985). This could indicate some remote affinity with the *Nectophryne* group of African toads as well as with Australian groups, and suggests an ancient Gondwanan connection. On the other hand, Tyler (1985) concluded that affinities of the sooglossids may be found with ranoids. He assumed African connections in view of proximity with the African continent, but on palaeogeographical grounds, a link with Indian ranoids might seem more likely.

The Seychellian caecilians seem to form a monophyletic unit (Nussbaum and Ducey, 1988). The source of this stock is obscure, although Nussbaum and Ducey (1988) reported chromosomal similarities with Central American caecilians. Nussbaum (1984) considered *Tachycnemis* to be a hyperoliid derived from Africa or Madagascar, although so distinctive as to suggest that its presence on the Seychellian microcontinent predates the arrival of *Heterixalus* on Madagascar.

Loumont (1992) noted chromosomal differentiation in the Gulf of Guinea island fauna, but considered that the evolutionary divergences shown there are recent. No suggestions emerge as to how amphibians reached the islands, apart from the possibility of island hopping (Schätti and Loumont, 1992), or on driftwood (Kingdon, 1990).

PRESERVATION OF AMPHIBIAN FAUNA

In a major IUCN report on biodiversity in Sub-Saharan Africa, Madagascar, and the Seychelles, Stuart et al. (1990:14) stated regarding Africa: "In general, the continent's protected area system is least complete in the lowland rainforests of West and Central Africa, and in the mountain regions throughout the continent." The present study confirms the common perception that both diversity and endemism are highest in these very areas, thereby suggesting that much of the amphibian fauna is at risk. The same situation was reported in Madagascar by Blommers-Schlösser and Blanc (1993), who commented on the human demographic explosion that has occurred on Madagascar during the course of this century with a resulting "savanisation" of forest habitats.

Yet the response of the amphibian fauna is complex. Andreone (1994) found that in the area of the Ranomafana rain forest reserve of Madagascar, arboreal species were better able to tolerate human disturbance than terrestrial species which require leaf litter. Species restricted to, or found more abundantly in, unaltered habitat constituted 62.5% of the fauna; but 22.5% were found only in disturbed environments, and a further 15% were found more abundantly in altered than unaltered habitats. Similarly, in the Korup National Park, Cameroon, Lawson (1993:83) observed that some of the "... most conspicuous species are not indigenous to the forest, but are savanna or forest edge forms which have followed paths of human disturbance into the dense forest." The number of amphibian species that exploit such disturbed patches is large enough to constitute what Schiøtz (1967) has termed a "farmbush fauna."

The amphibian fauna of the smaller islands may be less resilient. Settlement of the Seychelles commenced as late as 1770, but was followed by devastation of areas easily inhabitable by man (Stoddart, 1984). Nussbaum (1984) pointed out that the threat to habitat brought about by forest clearing may be compounded by the invasion of introduced plant and animal species. Alien invasion could be especially damaging on small islands, as indicated by the appalling extinction suffered by endemics on Mauritius (Owadally and Lambert, 1988; Stuart et al., 1990). Vegetation of the outer Gulf of Guinea islands has been profoundly altered by human settlement since their discovery in 1470-71 (White, 1983; Stuart et al. 1990; Loumont, 1992). Despite having some of the highest levels of species endemism in the world, no protected areas have been established (Stuart et al., 1990).

Adult frogs and also tadpoles are exploited as food by Africans (e.g., Poynton, 1964a; Lawson, 1993), but no evidence exists that this exploitation is damaging. Blommers-Schlösser and Blanc (1993) reported that amphibians are exploited in Madagascar for food and for terraria. The imported *Hoplobatrachus tigerinus* is cropped for export, which evidently relieves some pressure on the indigenous fauna (Blommers-Schlösser and Blanc, 1993).

The situation regarding conservation efforts in Sub-Saharan Africa was treated in detail by Stuart et al. (1990). They concluded that the effectiveness of the effort is patchy. Areas of southern Africa are deemed to be receiving "reasonably good conservation attention" (Stuart et al., 1990:39) although the situation in the southwestern Cape gives cause for concern; three amphibian species are endangered there (Branch, 1988). The Cape species evidently originated there, like the bulk of the present Capensis flora (Poynton, 1983; White, 1983), and cannot be replaced from any other source. Nevertheless, in an investigation of species-area relationships in southwestern Cape fynbos, Cowling and Bond (1991) found an "encouraging" prospect for the conservation of at least plant diversity.

Biotic diversity and conservation in southern Africa has been written about in some depth, notably in Huntley (1989), where it is reported that 92% of amphibian species are represented in nature reserves. Of particular conservation concern in Sub-Saharan Africa are the montane areas and patches of coastal forest of East Africa, where protected area coverage is poor to very poor (Stuart et al., 1990), and where natural areas are restricted and fragmented. The rich endemic amphibian faunas of these geographically small to minute areas seem to be at risk to a degree which still has to be determined, because it is not clear to what extent the species associated with tree cover or montane grassland can survive and breed in areas cleared and planted with exotics. As noted above, it is certainly not unknown for forest-associated amphibian populations to survive disturbed conditions.

Although the number of amphibian species suffering local extinction in urban areas is variable (e.g., Poynton, 1986b), the potential for conservation in towns and cities should not be overlooked. Active and directed management, that aims at open space systems of core areas and interlinking corridors, can lead to the recovery or reestablishment within the city of natural sites that are both ecologically resilient and retain the biogeographical characteristics of the region (Poynton and Roberts, 1985; Roberts and Poynton, 1985). As conservationists tend to treat urban areas with some aloofness, a need exists to emphasize the potential that well-managed open space systems in cities have as effective retreats for the local indigenous fauna and flora.

Blommers-Schlösser and Blanc (1993) listed species recorded in twelve "réserves naturelles intégrales" in Madagascar (ranging in area from 520 to 152,000 ha), together with some forestry stations. Twenty species (13% of the total) are listed as not being included within a protected area; these include the peculiar *Rhombophryne testudo,* which occurs on the northern end of the island. In the Seychelles, under an Ordinance of 1969, National Parks have been declared on Mahé, Praslin, and Curieuse, and Special Reserves have been set up elsewhere, with the result that "… the prospects for the proper management and conservation of the remarkable habitats, fauna and flora of this extraordinary archipelago now seem to be assured" (Stoddart, 1984:653). Nevertheless, it remains to be seen whether the areas under conservation are large enough to support the currently surviving species diversity.

It is hardly possible to anticipate the effects of conservation efforts such as these, when so little is known about the effects of habitat disturbance on species diversity, whether the disturbance is caused by human settlement or whether it occurs naturally. Natural disturbance can lead to an increase in species diversity, as with tree-falls in forest that cause open patches. Humans magnify disturbance, such as by fire in wooded areas that results in extended open habitats. The available evidence suggests that the response of species to habitat disturbance is a matter of great complexity, in which major unknowns are the relationships between disturbance and breeding failure, particularly the involvement of population and range size in breeding failure. Time is needed to allow a less crisis-ridden study of these relationships. This is perhaps the best argument for restricting human-induced habitat disturbance until its effects are better understood.

Beyond that is the fact, noted in the Introduction, that exploration of the amphibian fauna, particularly of Africa, is far from complete. The present generation of herpetologists has a responsibility to its successors in striving to preserve the faunas of Africa and its islands.

Acknowledgments: This chapter is an early product of a project on the biogeography of African amphibians assisted by funding through the World Wide Fund for Nature and centered in the Natural History Museum, London. I am grateful for the support given by the WWF and its affiliate, the Southern African Nature Foundation, and for the friendly interest and help given in many different ways by staff and visitors to the Reptiles and Amphibians Section of the Natural History Museum. In particular I am grateful to Dr. B. T. Clarke and Dr. B. Hughes for many discussions and for critically reading a draft of this chapter. I am also grateful to Mrs. M. E. du Plessis of the University of Natal, Durban, for assistance in setting up the tables, and to the university for providing facilities. Facilities also were provided by the Natal Museum, Pietermaritzburg, and I am particularly grateful to Mr. P. Croeser for assistance with computer programming in setting up the appendix. It is a pleasure to acknowledge the fruitful exchange of ideas with staff and research students of the University of Natal and the Natal Museum, particularly Dr. D. C. Roberts and Mr. M. Mattson regarding ecological and biogeographical principles. Acknowledgment should also be given to the enormous amount of cooperative work that went into the preparation and editing of *Amphibian Species of the World* and its addenda and corrigenda, without which this study would hardly have been possible.

LITERATURE CITED

AKEF, M. S. A. 1993. Morphometric studies in Water Frog *Rana mascareniensis* in Egypt (Anura: Ranidae). Proc. Egyptian Acad. Sci. 43:71–79.

AKEF, M. S. A., AND H. SCHNEIDER. 1993. Reproductive behavior and mating call pattern in Degen's toad, *Bufo vittatus,* in Egypt (Bufonidae, Amphibia). J. African Zool. 107:97–104.

AMIET, J.-L. 1975. Ecologie et distribution des Amphibiens anoures de la région de Nkongsamba (Cameroun). Ann. Fac. Sci Yaoundé 20:33–107.

AMIET, J.-L. 1983. Un essai de cartographie des anoures du Cameroun. Alytes 2:124–146.

ANDREONE, F. 1994. The amphibians of Ranomafana rain forest, Madagascar—preliminary community analysis and conservation considerations. Oryx 28:207–214.

AXELROD, D. I., AND P. H. RAVEN. 1978. Late Cretaceous and Tertiary vegetation history of Africa. Pp. 77–130 *in* M. J. A. Werger (ed.), *Biogeography and Ecology of Southern Africa.* The Hague: W. Junk.

BALINSKY, B. I. 1962. Patterns of animal distribution on the African continent. Ann. Cape Prov. Mus. 2:299–310.

BARBOUR, T., AND A. LOVERIDGE. 1928. A comparative study of the herpetological faunae of the Uluguru and Usambara Mountains, Tanganyika Territory. Mem. Mus. Comp. Zool. 50:87-265.

BLOMMERS-SCHLÖSSER, R. M. A. 1993. Systematic relationships of the Mantellinae Laurent 1946 (Anura Ranoidea). Ethol. Ecol. Evol. 5:199–218.

BLOMMERS-SCHLÖSSER, R. M. A., AND C. P. BLANC. 1991. Amphibiens (première partie). Faune de Madagascar 75(1):1–379. Paris: Mus. Natl. d'Hist. Nat.

BLOMMERS-SCHLÖSSER, R. M. A., AND C. P. BLANC. 1993. Amphibiens (deuxième partie). Faune de Madagascar 75(2):385–530. Paris: Mus. Natl. d'Hist. Nat.

BRANCH, W. R. (ed.). 1988. *South African Red Data Book—Reptiles and Amphibians.* Pretoria: Council for Scientific and Industrial Research.

BRENAN, J. P. M. 1978. Some aspects of the phytogeography of tropical Africa. Ann. Missouri Bot. Gard. 65:437–478.

BROWN, J. H. 1988. Species diversity. Pp. 57–89 *in* A. A. Myers and P. S. Giller (eds.), *Analytical Biogeography: An Integrated Approach to the Study of Animal and Plant Distributions.* London: Chapman and Hall.

BRUNDIN, L. Z. 1981. Croizat's panbiogeography versus phylogenetic biogeography. Pp. 94–138 *in* G. Nelson and D. E. Rosen (eds.), *Vicariance Biogeography: A Critique.* New York: Columbia University Press.

CANNATELLA, D. C., AND L. TRUEB. 1988. Evolution of pipoid frogs: morphology and phylogenetic relationships of *Pseudhymenochirus.* J. Herpetol. 22:439–456.

CHANNING, A. 1979. Ecological and systematic relationships of *Rana* and *Strongylopus* in southern Natal (Amphibia: Anura). Ann. Natal Mus. 23:797–831.

CHANNING, A. 1981. Southern origin of the African genus *Strongylopus* Tschudi, 1838 (Amphibia Ranidae). Monit. Zool. Italiano N.S. Suppl. 15:333–336.

CHANNING, A. 1989. A re-evaluation of the phylogeny of Old World treefrogs. S. African J. Zool 24:116–131.

CHANNING, A., AND R. C. BOYCOTT. 1989. A new frog genus and species from the mountains of the southwestern Cape, South Africa (Anura: Ranidae). Copeia 1989:467–471.

CHANNING, A., AND D. E. VAN DIJK. 1976. *A Guide to the Frogs of South West Africa.* Durban: University of Durban-Westville Press.

CLARKE, B. T. 1981. Comparative osteology and evolutionary relationships in the African Raninae (Anura: Ranidae). Monit. Zool. Italiano N.S. Suppl. 15:285–331.

CLARKE, B. T. 1988. The amphibian fauna of the East African rainforests, including the description of a new species of toad, genus *Nectophrynoides* Noble 1926 (Anura Bufonidae). Tropical Zool. 1:169–177.

COETZEE, J. A. 1993. African flora since the terminal Jurassic. Pp. 37–61 *in* P. Goldblatt (ed.), *Biological Relationships between Africa and South America.* New Haven: Yale University Press.

COLINVAUX, P. 1993. *Ecology 2.* New York: John Wiley & Sons.

COWLING, R. M. AND W. J. BOND. 1991. How small can reserves be? An empirical approach in Cape fynbos, South Africa. Biol. Conserv. 58:243–256.

CRAW, R. 1983. Panbiogeography: method and synthesis in biogeography. Pp. 405–435 *in* A. A. Myers and P. S. Giller (eds.) *Analytical Biogeography: An Integrated Approach to the Study of Animal and Plant Distributions.* London: Chapman and Hall.

CROIZAT, L. 1968. Introduction raisonée à la biogéographie de l'Afrique. Mems Soc. Broteriana 20:1–451.

CROWE, T. 1990. A quantitative analysis of patterns of distribution, species richness and endemism in southern African vertebrates. Pp. 145–160 *in* G. Peters and R. Hutterer (eds.), *Vertebrates in the Tropics.* Bonn: Alexander Koenig Zoological Research Institute and Zoological Museum.

DARLINGTON, P. J. 1957. *Zoogeography: The Geographical Distribution of Animals.* New York: Wiley and Sons.

DREWES, R. C. 1984. A phylogenetic analysis of the Hyperoliidae

(Anura): treefrogs of Africa, Madagascar, and the Seychelles Islands. Occas. Pap. California Acad. Sci. 139:1–70.

DUBOIS, A. 1987 "1986." Miscellanea taxinomica batrachologica (1). Alytes 5:7-95.

DUBOIS, A. 1992. Notes sur la classification des Ranidae (Amphibiens Anoures). Bull. Mens. Soc. Linn. Lyon 61:305–352.

DUELLMAN, W. E. 1993a. Amphibians in Africa and South America: evolutionary history and ecological comparisons. Pp. 200–243 in P. Goldblatt (ed.) *Biological Relationships between Africa and South America*. New Haven & London: Yale University Press.

DUELLMAN, W. E. 1993b. Amphibian Species of the World: Additions and Corrections. Spec. Publ. Mus. Nat. Hist. Univ. Kansas 21:1–372.

EL DIN, S. M. B. 1993. A new species of toad (Anura: Bufonidae) from Egypt. J. Herpetol. Assoc. Africa 42:24–27.

FROST, D. R. (ed.). 1985. *Amphibian Species of the World*. Lawrence, Kansas: Association of Systematic Collections and Allen Press Inc.

GAULD, I., AND G. UNDERWOOD. 1986. Some applications of the LeQuesne compatibility test. Biol. J. Linnean Soc. 29:191–222.

GRANDISON, A. C. G. 1981. Morphology and phylogenetic position of the West African *Didynamipus sjoestedti* Anderson, 1903 (Anura Bufonidae). Monit. Zool. Italiano N. S. Suppl. 15:187–215.

GRIMSDELL, J. J. R., AND L. R. G. RAW. 1984. Frog species diversity in relation to bioclimatic regions and conservation areas in Natal. Lammergeyer 33:21–29.

GUIBÉ, J. 1978. Les batrachiens de Madagascar. Bonner Zool. Monogr. 11:1–148.

HAPPOLD, D. C. D., AND M. HAPPOLD. 1989. Biogeography of montane small mammals in Malawi, central Africa. J. Biogeogr. 16:353–367.

HENGENVELD, R. 1990. *Dynamic Biogeography*. Cambridge: Cambridge University Press.

HOWELL K. M. 1993. Herpetofauna of eastern African forests. Pp. 173–210 in J. C. Lovett and S. K. Wasser (eds.) *Biogeography and Ecology of the Rain Forests of Eastern Africa*. Cambridge: Cambridge University Press.

HUGHES, B. 1983. African snake faunas. Bonn. Zool. Beitr. 34:311–356.

HUNTLEY, B. J. (ed.). 1989. *Biotic Diversity in Southern Africa: Concepts and Conservation*. Cape Town: Oxford University Press.

HUNTLEY, B. J., AND T. WEBB. 1989. Migration; species' response to climatic variations caused by changes in the earth's orbit. J. Biogeogr. 16:5–19.

INGER, R. F. 1968. Amphibia. Expl. Parc Natl. Garamba 52:1–190.

JACKSON, S. P. 1961. *Climatological atlas of Africa*. Lagos & Nairobi: CCTA.

KINGDON, J. 1990. *Island Africa: The Evolution of Africa's Rare Animals and Plants*. London: Collins.

LANG, M. 1990. Phylogenetic analysis of the genus group *Tracheloptychus-Zonosaurus* (Reptilia: Gherrosauridae), with a hypothesis of biogeographical unit relationships in Madagascar. Pp.261–274 in G. Peters and R. Hutterer (eds), *Vertebrates in the Tropics*. Bonn: Alexander Koenig Zoological Research Institute and Zoological Museum.

LANZA, B. 1981. A check-list of the Somali amphibians. Monit. Zool. Italiano N.S. Suppl. 15:151–186.

LARGEN, M. J., AND R. C. DREWES. 1989. A new genus and species of brevicipitine frog (Amphibia Anura Microhylidae) from high altitude in the mountains of Ethiopia. Trop. Zool. 2:13–30.

LAURENT, R. 1964. Reptiles et amphibiens de l'Angola. Publ. Cult. Comp. Diam. Angola 67:1–165.

LAURENT, R. 1972. Amphibiens. Expl. Parc. Natn. Virunga, 2nd Ser. 22:1–125.

LAWRENCE, R. F. 1953. *The Biology of the Cryptic Fauna of Forests with Special Reference to the Indigenous Forests of South Africa*. Cape Town: Balkema.

LAWSON, D. P. 1993. The reptiles and amphibians of the Korup National Park project, Cameroon. Herpetol. nat. Hist. 1:27–90.

LE BERRE, M. 1989. *Faune du Sahara 1 Poissons Amphibiens Reptiles*. Paris: R. Chabaud.

LEVIN, S. A. 1989. Challenges in the development of a theory of community and ecosystem structure and function. Pp. 242–255 in J. Roughgarden, R. M. May, and S. A. Levin (eds.), *Perspectives in Ecological Theory*. Princeton: Princeton Univ. Press.

LINDER, H. P. 1990. On the relationship between the vegetation and floras of the Afromontane and the Cape Regions of Africa. Mitt. Inst. Allg. Bot. Hamburg 23:777–790.

LIVINGSTONE, D. A. 1993. Evolution of African climate. Pp. 455–472 in P. Goldblatt (ed.) *Biological Relationships between Africa and South America*. New Haven & London: Yale University Press.

LOUMONT, C. 1992. Les amphibiens de São Tomé et Principe: révision systématique, cris nuptiaux et caryotypes. Alytes 10:37–62.

LOVERIDGE, A. 1925. Notes on East African batrachians, collected 1920–1923, with the descriptions of four new species. Proc. Zool. Soc. London 1925:763–791.

LOVERIDGE, A. 1937. Scientific results of an expedition to rain forest regions in eastern Africa. IX. Zoogeography and itinerary. Bull. Mus. Comp. Zool. 79:479–541.

MEADOWS, M. E., AND H. P. LINDER. 1993. A palaeoecological perspective on the origin of Afromontane grasslands. J. Biogeogr. 20:345–355.

MERTENS, R. 1965. Die amphibien von Fernando Poo. Bonn. Zool. Beitr. 16:14–29.

MOREAU, R. E. 1966. *The Bird Faunas of Africa and Its Islands*. London: Academic Press.

NOBLE, G. K. 1924. Contributions to the herpetology of the Belgian Congo based on the collection of the American Museum Congo expedition 1909–1915. Bull. Am. Mus. Nat. Hist. 49:147–347.

NUSSBAUM, R. A. 1984. Amphibians of the Seychelles. Pp. 379–415 in D. R. Stoddart (ed.), *Biogeography and Ecology of the Seychelles Islands*. Monogr. Biol. 55. The Hague: W. Junk.

NUSSBAUM, R. A., AND P. K. DUCEY. 1988. Cytological evidence for monophyly of the caecilians (Amphibia: Gymnophiona) of the Seychelles archipelago. Herpetologica 44:290–296.

NUSSBAUM, R. A., AND H. HINKEL. 1994. Revision of East African caecilians of the genera *Afrocaecilia* Taylor and *Boulengerula* Tornier (Amphibia: Gymnophiona: Caeciliaidae). Copeia 1994:760–766.

OWADALLY, A. W., AND M. LAMBERT. 1988. Herpetology in Mauritius: a history of extinction, future hope for conservation. British Herp. Soc. Bull. 23:11–20.

PERRET, J.-L. 1966. Les amphibiens du Cameroun. Zool. Jb. Syst. 93:289–464.

PITMAN, W. C., S. CANDE, J. LABREQUE, AND J. PINDELL. 1993. Fragmentation of Gondwana: The separation of Africa from South America. Pp. 15–34 in P. Goldblatt (ed.) *Biological Relationships between Africa and South America*. New Haven & London: Yale University Press.

POYNTON, J. C. 1961. Biogeography of south-east Africa. Nature 189:801–803.

POYNTON, J. C. 1962. Zoogeography of eastern Africa: an outline based on anuran distribution. Nature 194:1217–1219.

POYNTON, J. C. 1964a. The Amphibia of southern Africa: a faunal study. Ann. Natal Mus. 17:1–334.

POYNTON, J. C. 1964b. The biotic divisions of southern Africa, as shown by the Amphibia. Pp. 206–218 in D. H. S. Davis (ed.), *Ecological Studies in Southern Africa*. The Hague: W. Junk.

POYNTON, J. C. 1964c. Amphibia of the Nyasa-Luangwa region of Africa. Senck. Biol. 45:193–225.

POYNTON, J. C. 1967. Santa Rosalia in Africa or why are there so many African birds? South African J. Sci. 63:471–497.

POYNTON, J. C. 1969. Optimum temperatures and amphibian distribution in south–east Africa. Palaeoecol. Africa 4:161–162.

POYNTON, J. C. 1982. On species pairs among southern African amphibians. South African J. Zool. 17:67–74.

POYNTON, J. C. 1983. The dispersal versus vicariance debate in biogeography. Bothalia 14:455–460.

POYNTON, J. C. 1986a. Historical biogeography: theme and South African variations. Palaeoecol. Africa 17:139–154.

POYNTON, J. C. 1986b. *Hyperolius argus* (Anura) in Natal: taxonomy, biogeography and conservation. S. African J. Sci. 81:466–468.

POYNTON, J. C. 1988. Amphibian biogeography: facts in search of Quaternary theory. Palaeoecol. Africa 19:327–333.

POYNTON, J. C. 1989a. Evolutionary activity in the southern part of Africa: evidence from the Amphibia. J. Herp. Assoc. Africa 36:2–6.

POYNTON, J. C. 1989b. Biogeography of forest Amphibia. Pp. 41–47 *in* C. J. Geldenhuys (ed.), *Biogeography of the Mixed Evergreen Forests of Southern Africa.* Pretoria: Council for Scientific and Industrial Research.

POYNTON, J. C. 1990. Composition and subtraction margins of the East African lowland amphibian fauna. Pp. 285–296 *in* G. Peters and R. Hutterer (eds), *Vertebrates in the Tropics.* Bonn: Alexander Koenig Research Institute and Zoological Museum.

POYNTON, J. C. 1991. Amphibians of southeastern Tanzania, with special reference to *Stephopaedes* and *Mertensophryne* (Bufonidae). Bull. Mus. Comp. Zool. 152:451–473.

POYNTON, J. C. 1992. Amphibian diversity and species turnover in southern Africa: investigation by means of a Bloemfontein–Durban transect. J. Herpetol. Assoc. Africa 40:2–8.

POYNTON, J. C. 1994. Investigating biogeographical patterns: small steps between the obvious and the obscure. J. Herpetol. Assoc. Africa 43:1–5.

POYNTON, J. C. 1995. The "arid corridor" distribution in Africa: a search for instances among the Amphibia. Madoqua 19:45–48.

POYNTON, J. C., AND A. J. BASS. 1970. Environment and amphibian distribution in Zululand. Zool. Africana 5:41–48.

POYNTON, J. C., AND D. G. BROADLEY. 1978. The herpetofauna. Pp. 927–948 *in* M. J. A. Werger (ed.), *Biogeography and Ecology of Southern Africa.* The Hague: W. Junk.

POYNTON, J. C., AND D. G. BROADLEY. 1985a. Amphibia Zambesiaca 1. Scolecomorphidae, Pipidae, Microhylidae, Hemisidae, Arthroleptidae. Ann. Natal Mus. 26:503–553.

POYNTON, J. C., AND D. G. BROADLEY. 1985b. Amphibia Zambesiaca 2. Ranidae. Ann. Natal Mus. 27:115–181.

POYNTON, J. C., AND D. G. BROADLEY. 1987. Amphibia Zambesiaca 3. Rhacophoridae and Hyperoliidae. Ann. Natal Mus. 28:161–229.

POYNTON, J. C., AND D. G. BROADLEY. 1988. Amphibia Zambesiaca 4. Bufonidae. Ann. Natal Mus. 29:447–490.

POYNTON, J. C., AND D. G. BROADLEY. 1991. Amphibia Zambesiaca 5. Zoogeography. Ann. Natal Mus. 32:221–277.

POYNTON, J. C., AND W. D. HAACKE. 1993. On a collection of amphibians from Angola, including a new species of *Bufo* Laurenti. Ann. Transvaal Mus. 36:9–16.

POYNTON, J. C., AND S. PRITCHARD. 1976. Notes on the biology of *Breviceps* (Anura: Microhylidae). Zool. Africana 11:313–318.

POYNTON, J. C., AND D. C. ROBERTS. 1985. Urban open space planning in South Africa: a biogeographical perspective. S. African J. Sci. 81:33–37.

ROBERTS, D. C., AND J. C. POYNTON. 1985. Central and peripheral urban open spaces: need for biological evaluation. S. African J. Sci. 81:33–37.

ROSEN, B. R. 1988. Biogeographical patterns: a perceptual overview. Pp. 23–55 *in* A. A. Myers and P. S. Giller (eds.), *Analytical Biogeography: An Integrated Approach to the Study of Animal and Plant Distributions.* London: Chapman and Hall.

SATTLER, R. 1986. *Biophilosophy: Analytic and Holistic Perspectives.* Berlin: Springer–Verlag.

SCHÄTTI, B., AND C. LOUMONT. 1992. Ein beitrag zur herpetofauna von São Tomé (Golf von Guinea) (Amphibia et Reptilia). Zool. Abh. Mus. Tierkund. Dresden 47:23–36.

SCHIØTZ, A. 1967. The treefrogs (Rhacophoridae) of West Africa. Spolia Zool. Mus. Haun. 25:1–346.

SCHIØTZ, A. 1975. *The treefrogs of eastern Africa.* Copenhagen: Steenstrupia.

SCHIØTZ, A. 1976. Zoogeographical patterns in the distribution of East African treefrogs (Anura: Ranidae). Zool. Africana 11:335–338.

SCHIØTZ, A. 1981. The Amphibia of the forested basement hills of Tanzania: a biogeographical indicator group. African J. Ecol. 19:205–207.

SCHMIDT, K. P., AND R. F. INGER. 1959. Exploration du Parc National de l'Upemba. Fasc. 67 Amphibians. Bruxelles: Inst. Parcs Nation. Congo Belge.

SIBOULET, R. 1969. Amphibiens d'Algerie. Bull. Soc. Hist. Nat. Afrique du Nord 59:1–4.

STEWART, M. M.. 1967. *Amphibians of Malawi.* Albany: State Univ. New York Press.

STODDART, D. R. 1984. Impact of man in the Seychelles. Pp. 641–654 *in* D. R. Stoddart (ed.), *Biogeography and Ecology of the Seychelles Islands.* The Hague: W. Junk.

STUART, S. N., R. J. ADAMS, AND M. D. JENKINS. 1990. Biodiversity in Sub-Saharan Africa and it's islands: conservation, management, and sustainable use. Occas. Pap. IUCN Species Survival Commission 6:1–242.

STUCKENBERG, B. R. 1962. The distribution of the montane palaeogenic element in the South African invertebrate fauna. Ann. Cape Prov. Mus. 2:190–205.

STUCKENBERG, B. R. 1969. Effective temperature as an ecological factor in southern Africa. Zool. Africana 4:145–197.

TANDY, M., AND R. KEITH. 1972. *Bufo* of Africa. Pp. 119–170 *in* W. F. Blair (ed.), *Evolution in the Genus* Bufo. Austin: University of Texas Press.

THOMPSON, B. W. 1965. *The Climate of Africa.* Oxford: Oxford University Press.

TYLER, M. 1985. Phylogenetic significance of the superficial mandibular musculature and vocal sac structure of sooglossid frogs. Herpetologica 41:173–176.

VAN DIJK, D. E. 1971. The zoocartographic approach to anuran ecology. Zool. Africana 6:85–117.

VAN DIJK, D. E. 1982. Anuran distribution, rainfall and soils in southern Africa. S. African J. Sci. 78:401–406.

VINE, P. 1989. *Seychelles.* London: Immel Publishing.

WAKE, M. H. 1980. The reproductive biology of *Nectophrynoides malcolmi* (Amphibia: Bufonidae), with comments on the evolution of reproductive modes in the genus *Nectophrynoides.* Copeia 1980:193–209.

WHITE, F. 1978. The Afromontane Region. Pp.463–513 *in* M. J. A. Werger (ed.), *Biogeography and Ecology of Southern Africa.* The Hague: W. Junk.

WHITE, F. 1983. *The Vegetation of Africa.* Paris: UNESCO.

WILD, H., AND L. A. G. BARBOROSA. 1967. *Vegetation Map of the Flora Zambesiaca Area.* Harare: M. O. Collins.

APPENDIX 9:1

DISTRIBUTION OF AMPHIBIAN SPECIES IN CONTINENTAL SUB-SAHARAN REGIONS

Symbols in columns: + present, – absent. Abbreviations for regions: SUD Sudan (W = western part, E = eastern part), NEA Northeast, EAH East Africa Highland, M.E Ethiopian province of Intertropical Montane, M.K Kenyan province of Intertropical Montane, M.R central province of Intertropical Montane, M.T eastern province of Intertropical Montane, M.M Malawian province of Intertropical Montane, M.Z Zimbabwean province of Intertropical Montane, M.A Angolan province of Intertropical Montane, M.C Cameroonian province of Intertropical Montane, M.G Guinean province of Intertropical Montane, WEQ Western Equatorial, W.W western province of Western Equatorial, W.C central province of Western Equatorial, W.E eastern province of Western Equatorial, W.S southern province of Western Equatorial, SZB Southern Zaire Basin, EAL East African Lowland, ECF East Coast Forest, SEL Southeast Lowland, ZAZ Zaire-Zambesi, MST Mid-southern, SWA Southwest Arid, SCE South Central, STE South Temperate, S.H southeast highland province of South Temperate, S.C Cape province of South Temperate.

Taxon	SUD	NEA	EAH	M.E	M.K	M.R	M.T	M.M	M.Z	M.A	M.C	M.G	WEQ	W.W	W.C	W.E	W.S	SZB	EAL	ECF	SEL	ZAZ	MST	SWA	SCE	STE	S.H	S.C
ANURA: ARTHROLEPTIDAE:																												
Arthroleptis adelphus	–	–	–	–	–	–	–	–	–	–	–	–	+	–	–	–	–	–	–	–	–	–	–	–	–	–	–	–
Arthroleptis adolffriderici	–	–	–	+	+	+	–	–	–	+	+	–	+	–	–	–	–	–	–	–	–	–	–	–	–	–	–	–
Arthroleptis affinis	–	–	–	–	–	+	–	–	–	–	–	–	+	–	–	–	–	–	–	–	–	–	–	–	–	–	–	–
Arthroleptis bivittatus	–	–	–	–	–	–	–	–	–	–	–	–	+	+	+	+	+	–	–	–	–	–	–	–	–	–	–	–
Arthroleptis brevipes	–	–	–	–	–	–	–	–	–	–	–	–	+	+	+	–	–	–	–	–	–	–	–	–	–	–	–	–
Arthroleptis carquejai	–	–	–	–	–	–	–	–	–	–	–	–	–	–	–	–	–	+	–	–	–	–	–	–	–	–	–	–
Arthroleptis crusculus	–	–	–	–	–	–	–	–	–	–	+	–	–	–	–	–	–	–	–	–	–	–	–	–	–	–	–	–
Arthroleptis francei	–	–	–	–	–	+	–	–	–	–	–	–	–	–	–	–	–	–	–	–	–	–	–	–	–	–	–	–
Arthroleptis hematogaster	–	–	–	–	+	–	–	–	–	–	–	–	–	–	–	+	–	–	+	–	–	–	–	–	–	–	–	–
Arthroleptis lameeri	–	–	–	–	–	–	–	–	–	–	–	–	–	–	–	–	–	–	+	–	–	–	–	–	+	–	–	–
Arthroleptis loveridgei	–	–	–	–	+	–	–	–	–	–	–	–	–	–	–	–	–	–	–	–	–	–	–	–	–	–	–	–
Arthroleptis milletihorsini	+	–	–	–	+	–	–	–	–	–	–	–	–	–	–	–	–	–	–	–	–	–	–	–	–	–	–	–
Arthroleptis mossoensis	–	–	–	–	+	–	–	–	–	–	–	–	+	–	–	–	–	–	–	–	–	–	–	–	–	–	–	–
Arthroleptis nimbaensis	–	–	–	–	–	–	–	–	–	–	–	+	–	–	–	–	–	–	–	–	–	–	–	–	–	–	–	–
Arthroleptis phrynoides	–	–	–	–	–	–	–	–	–	–	–	–	–	–	–	–	–	–	+	–	–	–	–	–	–	–	–	–
Arthroleptis poecilonotus	+	–	–	–	+	–	–	–	–	–	–	–	+	+	+	+	+	–	+	–	–	–	–	–	–	–	–	–
Arthroleptis pyrrhoscelis	–	–	–	–	+	–	–	–	–	–	–	–	–	–	–	+	+	–	–	–	–	–	–	–	–	–	–	–
Arthroleptis reichei	–	–	–	–	+	–	+	–	–	–	–	–	–	–	–	–	–	–	–	–	–	–	–	–	–	–	–	–
Arthroleptis schubotzi	–	–	–	–	+	–	–	–	–	–	–	–	–	–	–	–	–	–	–	–	–	–	–	–	–	–	–	–
Arthroleptis spinalis	–	–	–	–	+	–	+	–	–	–	–	–	–	–	–	–	–	–	–	–	–	–	–	–	–	–	–	–
Arthroleptis stenodactylus	–	+	–	–	–	–	+	–	–	–	–	–	+	+	+	+	–	–	+	+	+	+	–	–	–	–	–	–
Arthroleptis sylvatica	–	–	–	–	–	–	–	–	–	–	–	–	–	+	+	+	–	–	–	–	–	–	–	–	–	–	–	–
Arthroleptis taeniatus	–	–	–	–	–	–	–	–	–	–	–	–	+	+	+	+	–	–	–	–	–	–	–	–	–	–	–	–
Arthroleptis tanneri	–	–	–	–	–	–	–	+	–	–	–	–	–	–	–	–	–	–	–	–	–	–	–	–	–	–	–	–
Arthroleptis troglodytes	–	–	–	–	–	–	–	–	–	–	–	–	–	–	–	–	–	–	–	–	–	+	–	–	–	–	–	–
Arthroleptis tuberosus	–	–	–	–	–	–	–	–	–	–	–	–	+	+	+	+	–	–	–	–	+	+	–	–	–	–	–	–
Arthroleptis variabilis	–	–	–	–	–	–	–	–	–	–	–	–	+	+	+	+	+	–	–	–	–	–	–	+	+	–	–	–
Arthroleptis vercammeni	–	–	–	+	–	–	–	–	–	–	–	–	–	–	–	–	–	–	–	–	–	–	–	–	–	–	–	–
Arthroleptis wahlbergii	–	–	–	–	–	–	–	–	–	–	–	–	–	–	–	–	–	–	–	–	–	–	–	–	+	–	–	–

Appendix 9:1 continued

Taxon	SUD	NEA	EAH	M.E	M.K	M.R	M.T	M.M	M.Z	M.A	M.C	M.G	WEQ	W.W	W.C	W.E	W.S	SZB	EAL	ECF	SEL	ZAZ	MST	SWA	SCE	STE	S.H	S.C
Arthroleptis xenochirus	–	–	–	–	–	–	–	–	–	–	–	–	–	–	–	–	–	–	–	–	+	–	–	–	–	–	–	–
Arthroleptis xenodactyloides	–	–	–	–	+	+	+	+	–	–	–	–	–	–	–	–	–	+	+	+	+	–	–	–	–	–	–	–
Arthroleptis xenodactylus	–	–	–	–	–	+	+	–	–	–	–	–	–	–	–	–	–	–	+	–	–	–	–	–	–	–	–	–
Arthroleptis zimmeri	–	–	–	–	–	–	–	–	–	–	–	+	+	–	–	–	–	–	–	–	–	–	–	–	–	–	–	–
Astylosternus batesi	–	–	–	–	–	–	–	–	–	+	–	+	+	+	–	–	–	–	–	–	–	–	–	–	–	–	–	–
Astylosternus diadematus	–	–	–	–	–	–	–	–	–	–	+	+	–	+	+	–	–	–	–	–	–	–	–	–	–	–	–	–
Astylosternus fallax	–	–	–	–	–	–	–	–	–	–	–	+	–	+	–	–	–	–	–	–	–	–	–	–	–	–	–	–
Astylosternus laurenti	–	–	–	–	–	–	–	–	–	–	+	+	–	+	–	–	–	–	–	–	–	–	–	–	–	–	–	–
Astylosternus montanus	–	–	–	–	–	–	–	–	–	+	–	–	–	–	–	–	–	–	–	–	–	–	–	–	–	–	–	–
Astylosternus nganhanus	–	–	–	–	–	–	–	–	–	+	–	–	–	–	–	–	–	–	–	–	–	–	–	–	–	–	–	–
Astylosternus occidentalis	–	–	–	–	–	–	–	–	–	–	–	+	+	–	–	–	–	–	–	–	–	–	–	–	–	–	–	–
Astylosternus perreti	–	–	–	–	–	–	–	–	–	+	–	–	–	–	–	–	–	–	–	–	–	–	–	–	–	–	–	–
Astylosternus ranoides	–	–	–	–	–	–	–	–	–	+	–	+	–	–	–	–	–	–	–	–	–	–	–	–	–	–	–	–
Astylosternus rheophilus	–	–	–	–	–	–	–	–	–	+	–	+	–	–	–	–	–	–	–	–	–	–	–	–	–	–	–	–
Astylosternus schioetzi	–	–	–	–	–	–	–	–	–	–	–	–	+	–	–	–	–	–	–	–	–	–	–	–	–	–	–	–
Cardioglossa aureoli	–	–	–	–	–	–	–	–	–	–	–	+	+	–	–	–	–	–	–	–	–	–	–	–	–	–	–	–
Cardioglossa cyaneospila	–	–	–	+	–	–	–	–	–	–	–	–	–	–	–	–	–	–	–	–	–	–	–	–	–	–	–	–
Cardioglossa elegans	–	–	–	–	–	–	–	–	–	+	–	+	+	–	+	–	–	–	–	–	–	–	–	–	–	–	–	–
Cardioglossa escalerae	–	–	–	–	–	–	–	–	–	–	–	+	+	+	+	+	+	–	–	–	–	–	–	–	–	–	–	–
Cardioglossa gracilis	–	–	–	–	–	–	–	–	–	+	–	+	+	+	+	+	+	–	–	–	–	–	–	–	–	–	–	–
Cardioglossa gratiosa	–	–	–	–	–	–	–	–	–	–	–	+	–	+	–	–	–	–	–	–	–	–	–	–	–	–	–	–
Cardioglossa leucomystax	–	–	–	–	–	–	–	–	–	+	–	+	+	+	+	+	+	–	–	–	–	–	–	–	–	–	–	–
Cardioglossa liberiensis	–	–	–	–	–	–	–	–	–	–	–	+	+	–	+	–	–	–	–	–	–	–	–	–	–	–	–	–
Cardioglossa melanogaster	–	–	–	–	–	–	–	–	–	+	–	+	–	–	+	–	–	–	–	–	–	–	–	–	–	–	–	–
Cardioglossa nigromaculata	–	–	–	–	–	–	–	–	–	+	–	–	–	–	–	–	–	–	–	–	–	–	–	–	–	–	–	–
Cardioglossa oreas	–	–	–	–	–	–	–	–	–	+	–	+	–	–	–	–	–	–	–	–	–	–	–	–	–	–	–	–
Cardioglossa pulchra	–	–	–	–	–	–	–	–	–	+	–	+	–	–	–	–	–	–	–	–	–	–	–	–	–	–	–	–
Cardioglossa schioetzi	–	–	–	–	–	–	–	–	–	+	–	+	–	–	+	–	–	–	–	–	–	–	–	–	–	–	–	–
Cardioglossa trifasciata	–	–	–	–	–	–	–	–	–	+	–	+	–	–	–	–	–	–	–	–	–	–	–	–	–	–	–	–
Cardioglossa venusta	–	–	–	–	–	–	–	–	–	+	–	+	–	–	–	–	–	–	–	–	–	–	–	–	–	–	–	–
Leptodactylodon albiventris	–	–	–	–	–	–	–	–	–	+	–	+	–	–	+	–	–	–	–	–	–	–	–	–	–	–	–	–
Leptodactylodon axillaris	–	–	–	–	–	–	–	–	–	+	–	+	–	–	–	–	–	–	–	–	–	–	–	–	–	–	–	–
Leptodactylodon bicolor	–	–	–	–	–	–	–	–	–	+	–	+	–	+	+	–	–	–	–	–	–	–	–	–	–	–	–	–
Leptodactylodon boulengeri	–	–	–	–	–	–	–	–	–	–	–	+	–	+	+	–	–	–	–	–	–	–	–	–	–	–	–	–
Leptodact. erythrogaster	–	–	–	–	–	–	–	–	–	+	–	+	–	–	–	–	–	–	–	–	–	–	–	–	–	–	–	–
Leptodactylodon mertensi	–	–	–	–	–	–	–	–	–	+	–	+	–	–	–	–	–	–	–	–	–	–	–	–	–	–	–	–
Leptodactylodon ornatus	–	–	–	–	–	–	–	–	–	+	–	–	–	–	–	–	–	–	–	–	–	–	–	–	–	–	–	–
Leptodactylodon ovatus	–	–	–	–	–	–	–	–	–	+	–	+	–	–	–	–	–	–	–	–	–	–	–	–	–	–	–	–
Leptodactylodon perreti	–	–	–	–	–	–	–	–	–	+	–	+	–	–	–	–	–	–	–	–	–	–	–	–	–	–	–	–
Leptodact. polyacanthus	–	–	–	–	–	–	–	–	–	+	–	+	–	–	–	–	–	–	–	–	–	–	–	–	–	–	–	–

Appendix 9:1 continued

Taxon	SUD	NEA	EAH	M.E	M.K	M.R	M.T	M.M	M.Z	M.A	M.C	M.G	WEQ	W.W	W.C	W.E.	W.S	SZB	EAL	ECF	SEL	ZAZ	MST	SWA	SCE	STE	S.H	S.C
Leptodact. ventrimarmoratus	–	–	–	–	–	–	–	–	–	–	–	–	–	–	–	–	–	–	–	–	–	–	–	–	–	–	–	–
Nyctibates corrugatus	–	–	–	–	–	–	–	–	–	–	+	–	–	+	+	–	–	–	–	–	–	–	–	–	–	–	–	–
Scotobleps gabonicus	–	–	–	–	–	–	–	–	–	–	–	–	–	+	+	–	–	–	–	–	–	–	–	–	–	–	–	–
Trichobatrachus robustus	–	–	–	–	–	–	–	–	–	+	+	–	–	+	+	–	–	–	–	–	–	–	–	–	–	–	–	–
ANURA: BUFONIDAE:																												
Altiphrynoides malcolmi	–	–	+	–	–	–	–	–	–	–	–	–	–	–	–	–	–	–	–	–	–	–	–	–	–	–	–	–
Bufo amatolicus	–	–	+	–	–	–	–	–	–	–	–	–	–	–	–	–	–	–	–	–	–	–	–	–	–	–	+	–
Bufo angusticeps	–	–	–	–	–	–	–	–	–	–	–	–	–	–	–	–	–	–	–	–	–	–	–	–	–	–	–	+
Bufo asmarae	–	–	+	–	–	–	–	–	–	–	–	–	–	–	–	–	–	–	–	–	–	–	–	–	–	–	–	–
Bufo beiranus	–	–	–	–	–	–	–	–	–	–	–	–	–	–	–	–	–	–	+	–	+	–	–	–	–	–	–	–
Bufo blanfordii	–	+	–	–	–	–	–	–	–	–	–	–	–	–	–	–	–	–	–	–	–	–	–	–	–	–	–	–
Bufo brauni	–	–	–	–	–	–	+	–	–	–	–	–	–	–	–	–	–	–	–	–	–	–	–	–	–	–	–	–
Bufo camerunensis	–	–	–	–	–	+	–	–	–	–	–	–	–	–	+	+	–	–	–	–	–	–	–	–	–	–	–	–
Bufo chappuisi	–	+	–	–	–	–	–	–	–	–	–	+	–	–	–	–	–	–	–	–	–	–	–	–	–	–	–	–
Bufo chudeaui	W	–	–	–	–	–	–	–	–	–	–	–	–	–	–	–	–	–	–	–	–	–	–	–	–	–	–	–
Bufo cristiglans	–	–	–	–	–	–	–	–	–	–	–	–	–	–	–	–	+	–	–	–	–	–	–	–	–	–	–	–
Bufo cruciger	–	–	–	–	–	–	–	–	–	–	–	–	–	+	–	–	–	–	–	–	–	–	–	–	–	–	–	+
Bufo danielae	–	+	–	–	–	–	–	–	–	–	–	–	–	+	–	–	–	–	–	–	–	–	–	–	–	–	–	–
Bufo dodsoni	–	+	–	–	–	–	–	–	–	–	–	–	–	–	–	–	–	–	–	–	–	–	–	–	–	–	–	–
Bufo dombensis	–	–	–	–	–	–	–	+	–	–	–	–	–	–	–	–	–	–	–	–	–	–	–	+	–	–	–	–
Bufo fenoulheti	–	–	–	–	–	+	–	–	–	–	–	–	–	–	–	–	–	–	+	+	–	–	+	–	–	–	–	–
Bufo fuliginatus	–	–	–	–	–	–	–	–	–	–	–	–	–	–	–	–	–	+	–	+	+	–	–	–	–	–	–	–
Bufo funereus	–	–	–	–	–	–	–	–	–	–	–	–	–	–	+	–	+	+	–	+	+	–	–	–	–	–	–	–
Bufo gariepensis	–	?	–	–	–	–	–	–	–	–	–	–	–	+	–	–	+	–	+	+	+	–	+	+	+	+	+	+
Bufo garmani	–	+	?	–	–	–	–	–	–	–	–	+	–	–	–	–	–	–	–	–	–	–	–	–	–	–	–	–
Bufo gracilipes	–	–	–	–	–	–	–	–	–	–	–	–	–	–	–	–	+	–	–	–	–	–	–	–	–	–	–	–
Bufo grandisonae	–	–	–	–	–	–	–	–	–	–	–	–	–	–	–	–	–	–	–	–	–	–	–	+	–	–	–	–
Bufo gutturalis	–	+	–	–	–	–	+	–	–	+	–	–	–	–	–	–	–	–	+	+	–	–	+	–	+	–	+	–
Bufo hoeschi	–	–	–	–	–	–	–	–	–	–	–	–	–	–	–	–	–	–	–	–	–	–	–	+	–	–	–	–
Bufo inyangae	–	–	–	–	–	–	–	–	+	–	–	–	–	–	–	–	–	–	–	–	–	–	–	–	–	–	–	–
Bufo kavanagesis	–	–	–	–	–	–	–	–	–	–	–	–	–	–	–	–	–	–	+	–	–	–	–	+	–	–	–	–
Bufo keringyagae	–	–	–	+	+	–	–	–	–	–	–	–	–	–	–	–	–	–	–	–	–	–	–	–	–	–	–	–
Bufo kisoloensis	–	–	–	+	–	–	–	–	–	–	–	–	–	–	–	–	–	–	–	–	–	–	–	–	–	–	–	–
Bufo langanoensis	–	–	–	+	+	–	–	–	–	–	–	–	–	–	–	–	–	–	–	–	–	–	–	–	–	–	–	–
Bufo latifrons	–	–	–	–	–	–	–	–	–	–	–	–	–	–	+	+	+	–	–	–	–	–	–	–	–	–	+	–
Bufo lemairii	–	–	–	–	–	–	–	–	–	–	–	–	–	–	–	+	–	–	–	–	–	–	–	–	–	+	–	–
Bufo lindneri	–	–	–	–	–	+	–	–	–	–	–	–	–	–	–	–	–	–	+	+	–	–	–	–	–	–	–	–
Bufo lonnbergi	–	–	–	–	–	–	–	–	–	–	–	–	–	–	–	–	+	+	–	–	+	–	–	–	–	–	+	–
Bufo lughensis	+	–	–	+	–	–	–	–	–	–	–	–	–	–	–	–	–	–	+	–	–	–	–	–	–	–	–	–
Bufo maculatus	+	–	?	+	–	–	–	–	–	–	–	–	–	+	+	–	–	–	+	+	+	–	+	–	–	–	+	–
Bufo melanopleura	–	–	–	–	–	–	–	–	–	–	–	–	+	–	–	–	–	+	+	–	+	–	–	+	–	+	–	+

Appendix 9:1 continued

Taxon	SUD	NEA	EAH	M.E	M.K	M.R	M.T	M.M	M.Z	M.A	M.C	M.G	WEQ	W.W	W.C	W.E	W.S	SZB	EAL	ECF	SEL	ZAZ	MST	SWA	SCE	STE	S.H	S.C
Bufo pardalis	–	–	–	–	–	–	–	–	–	–	–	–	–	–	–	–	–	–	–	–	+	–	–	–	–	–	–	–
Bufo parkeri	–	+	–	–	–	–	–	–	–	–	–	–	–	–	–	–	–	–	–	–	–	–	–	–	–	–	–	–
Bufo pentoni	+	–	–	–	–	–	–	–	–	–	–	–	–	–	–	–	–	–	–	–	–	–	–	–	–	–	–	–
Bufo perreti	–	–	–	–	–	–	–	–	–	–	–	–	–	–	–	–	–	–	–	–	–	–	–	+	+	–	–	–
Bufo poweri	–	–	–	–	–	–	–	–	–	–	–	–	–	–	–	–	–	–	–	–	+	–	+	+	+	–	+	+
Bufo rangeri	–	–	–	–	–	–	–	–	–	–	–	–	–	–	–	–	–	–	–	–	+	–	+	–	–	+	+	+
Bufo reesi	–	–	–	–	–	–	–	–	–	–	–	–	–	–	–	–	–	–	–	–	+	–	–	–	–	–	–	–
Bufo regularis	+	–	–	+	–	–	–	–	–	–	–	+	+	+	+	+	+	+	–	–	+	–	–	–	–	–	–	–
Bufo schmidti	–	–	–	–	–	–	–	–	–	–	–	–	–	–	–	–	–	+	–	–	–	–	–	–	–	–	–	–
Bufo steindachneri	+	+	–	–	–	–	–	–	–	–	–	–	–	–	–	–	–	–	–	–	–	–	–	–	–	–	–	–
Bufo superciliaris	–	–	–	–	–	–	–	–	–	–	–	–	+	+	–	–	+	–	–	–	–	–	–	–	–	–	–	–
Bufo taitanus	–	–	+	–	–	–	–	–	–	–	–	–	–	–	–	–	–	–	–	–	–	+	–	–	–	–	–	–
Bufo togoensis	–	–	–	–	–	–	–	–	–	–	–	+	+	+	–	–	–	–	–	–	–	–	–	–	–	–	–	–
Bufo tuberosus	–	–	–	–	–	–	–	–	–	–	–	–	–	+	–	–	+	–	–	–	–	–	–	–	–	–	–	–
Bufo turkanae	–	+	–	–	–	–	–	–	–	–	–	–	–	–	–	–	–	–	–	–	–	–	–	–	–	–	–	–
Bufo urunguensis	–	–	+	–	–	–	+	–	–	–	–	–	–	–	–	–	–	–	–	–	–	+	–	–	–	–	–	–
Bufo uzunguensis	–	–	+	–	–	–	–	–	–	–	–	–	–	–	–	–	–	–	–	–	–	–	–	–	–	–	–	–
Bufo vertebralis	–	–	–	–	–	–	–	–	–	–	–	–	–	–	–	–	–	–	–	–	–	–	–	–	+	–	–	–
Bufo villiersi	–	–	+	–	–	–	–	–	–	–	–	+	–	–	–	–	–	–	–	–	–	–	–	–	–	–	–	–
Bufo vittatus	–	+	–	–	–	–	–	–	–	–	–	–	–	–	–	–	–	–	–	–	–	–	–	–	–	–	–	–
Bufo xeros	+	+	–	–	–	–	–	–	–	–	–	–	–	–	–	–	–	–	–	–	–	–	–	–	–	–	–	–
Capensibufo rosei	–	–	–	–	–	–	–	–	–	–	–	–	–	–	–	–	–	–	–	–	–	–	–	–	–	–	–	–
Capensibufo tradouwi	–	–	–	–	–	–	–	–	–	–	–	–	–	–	–	–	–	–	–	–	–	–	–	–	–	–	–	–
Didynamipus sjostedti	–	–	–	+	–	–	–	–	–	–	–	–	–	–	–	–	–	–	–	–	–	–	–	–	–	–	–	–
Laurentophryne parkeri	–	–	–	–	–	+	–	–	–	–	–	–	–	–	–	–	–	–	–	–	–	–	–	–	–	–	–	–
Mertensophryne micranotis	–	–	–	–	–	–	–	–	–	–	–	–	–	–	–	–	–	–	–	+	–	–	–	–	–	–	–	–
Nectophryne afra	–	–	–	–	–	–	–	–	–	–	–	–	–	+	–	–	–	–	–	–	–	–	–	–	–	–	–	–
Nectophryne batesii	–	–	–	–	–	–	–	–	–	–	–	–	–	+	+	–	–	–	–	–	–	–	–	–	–	–	–	–
Nectophrynoides cryptus	–	–	–	–	–	+	–	–	–	–	–	–	–	–	–	–	–	–	–	–	–	–	–	–	–	–	–	–
Nectophrynoides minutus	–	–	–	–	–	+	–	–	–	–	–	–	–	–	–	–	–	–	–	–	–	–	–	–	–	–	–	–
Nectophrynoides tornieri	–	–	–	–	–	+	–	–	–	–	–	–	–	–	–	–	–	–	–	–	–	–	–	–	–	–	–	–
Nectophrynoides viviparus	–	–	–	–	–	+	–	–	–	–	–	–	–	–	–	–	–	–	–	–	–	–	–	–	–	–	–	–
Nectophrynoides wendyae	–	–	–	–	–	+	–	–	–	–	–	–	–	–	–	–	–	–	–	–	–	–	–	–	–	–	–	–
Nimbaphrynoides liberiensis	–	–	–	–	–	–	–	–	–	–	–	–	–	–	+	–	–	–	–	–	–	–	–	–	–	–	–	–
Nimbaphrynoides occidentalis	–	–	–	–	–	–	–	–	–	–	–	–	–	–	+	–	–	–	–	–	–	–	–	–	–	–	–	–
Schismaderma carens	–	–	+	–	–	–	–	–	–	–	–	–	–	–	–	–	–	–	–	–	–	–	+	–	+	–	–	–
Spinophrynoides osgoodi	–	–	–	–	+	–	–	–	–	–	–	–	–	–	–	–	–	–	–	–	–	–	–	–	–	–	–	–
Stephopaedes anotis	–	–	–	–	–	–	–	–	+	–	–	–	–	–	–	–	–	–	–	–	–	–	–	–	–	–	–	–
Stephopaedes loveridgei	–	–	–	–	–	–	–	–	–	–	–	–	–	–	–	–	–	–	–	–	–	–	–	–	–	–	+	+
Stephopaedes sp.	–	–	–	–	–	–	–	–	–	–	–	–	–	–	–	–	–	–	–	–	–	–	–	–	–	–	+	+

Appendix 9:1 continued

Taxon	SUD	NEA	EAH	M.E	M.K	M.R	M.T	M.M	M.Z	M.A	M.C	M.G	WEQ	W.W	W.C	W.E.	W.S	SZB	EAL	ECF	SEL	ZAZ	MST	SWA	SCE	STE	S.H	S.C
Werneria bambutensis	–	–	–	–	–	–	–	–	–	+	–	–	–	–	–	–	–	–	–	–	–	–	–	–	–	–	–	–
Werneria mertensiana	–	–	–	–	–	–	–	–	–	+	–	–	–	–	–	–	–	–	–	–	–	–	–	–	–	–	–	–
Werneria preussi	–	–	–	–	–	–	–	–	–	+	–	–	–	–	–	–	–	–	–	–	–	–	–	–	–	–	–	–
Werneria tandyi	–	–	–	–	–	–	–	–	–	+	–	–	–	–	–	–	–	–	–	–	–	–	–	–	–	–	–	–
Wolterstorffina mirei	–	–	–	–	–	–	–	–	–	+	–	–	–	–	–	–	–	–	–	–	–	–	–	–	–	–	–	–
Wolterstorffina parvipalmata	–	–	–	–	–	–	–	–	–	+	–	–	–	–	–	+	–	–	–	–	–	–	–	–	–	–	–	–
ANURA: HELEOPHRYNIDAE:																												
Heleophryne hewitti	–	–	–	–	–	–	–	–	–	–	–	–	–	–	–	–	–	–	–	–	–	–	–	–	–	–	–	+
Heleophryne natalensis	–	–	–	–	–	–	–	–	–	–	–	–	–	–	–	–	–	–	–	–	–	–	–	–	–	–	+	–
Heleophryne purcelli	–	–	–	–	–	–	–	–	–	–	–	–	–	–	–	–	–	–	–	–	–	–	–	–	–	–	–	+
Heleophryne regis	–	–	–	–	–	–	–	–	–	–	–	–	–	–	–	–	–	–	–	–	–	–	–	–	–	–	–	+
Heleophryne rosei	–	–	–	–	–	–	–	–	–	–	–	–	–	–	–	–	–	–	–	–	–	–	–	–	–	–	–	+
ANURA: HEMISOTIDAE:																												
Hemisus brachydactylus	–	–	+	–	–	–	–	–	–	–	–	–	–	–	–	–	–	–	–	–	–	–	–	–	–	–	–	–
Hemisus guineensis	–	–	+	–	–	–	–	–	–	–	–	–	+	+	+	+	+	+	+	–	–	–	+	–	–	–	–	–
Hemisus guttatus	–	–	–	–	–	–	–	–	–	–	–	–	–	–	–	–	–	–	–	–	+	–	–	–	–	–	–	–
Hemisus marmoratus	–	+	+	–	–	–	–	–	–	–	–	–	+	+	–	+	+	+	+	–	–	+	+	–	–	–	–	–
Hemisus microscaphus	–	–	–	+	–	–	–	–	–	–	–	–	–	–	–	–	–	–	–	–	–	–	–	–	–	–	–	–
Hemisus olivaceus	–	–	–	–	+	–	–	–	–	–	–	–	–	–	–	–	–	–	–	–	–	–	–	–	–	–	–	–
Hemisus perreti	–	–	–	–	–	–	–	–	–	–	–	–	–	–	–	–	+	–	–	–	–	–	–	–	–	–	–	–
Hemisus wittei	–	–	–	–	–	–	–	–	–	–	–	–	–	–	–	–	–	–	+	–	–	–	–	–	–	–	–	–
ANURA: HYPEROLIIDAE:																												
Acanthixalus spinosus	–	–	–	–	–	–	–	–	–	–	–	–	–	+	–	+	–	–	–	–	–	–	–	–	–	–	–	–
Afrixalus aureus	–	+	–	–	–	–	–	–	–	–	–	–	–	–	–	–	–	–	–	–	+	–	–	–	–	–	–	–
Afrixalus brachycnemis	–	–	–	–	–	–	–	–	–	–	–	–	–	–	–	–	–	–	+	–	–	–	–	–	–	–	–	–
Afrixalus clarkei	–	–	–	+	–	–	–	–	–	–	–	–	–	–	–	–	–	–	–	–	–	–	–	–	–	–	–	–
Afrixalus crotalus	–	–	–	–	–	–	–	–	–	–	–	–	–	–	–	–	–	–	+	–	–	–	+	–	–	–	–	–
Afrixalus delicatus	–	–	–	–	–	–	–	–	–	–	–	–	–	–	–	–	–	–	+	–	+	–	–	–	–	–	–	–
Afrixalus dorsalis	–	–	–	–	–	–	–	–	–	–	–	+	–	+	+	+	+	–	+	–	+	–	–	–	–	–	–	–
Afrixalus enseticola	–	–	–	+	–	–	–	–	–	–	–	+	–	–	–	–	–	–	–	–	–	–	–	–	–	–	–	–
Afrixalus equatorialis	–	–	–	–	–	–	–	–	–	–	–	–	–	–	–	–	–	–	+	–	+	–	–	–	–	–	–	–
Afrixalus fornasini	–	+	+	–	–	–	–	–	–	–	–	–	–	–	–	–	–	+	+	–	+	–	–	–	–	–	–	–
Afrixalus fulvovittatus	–	+	+	–	+	–	–	–	–	–	–	–	–	–	–	–	–	+	–	–	–	–	–	–	–	–	–	–
Afrixalus knysnae	–	–	–	–	–	–	–	–	–	–	–	–	–	–	–	–	–	–	–	–	–	–	–	–	–	+	+	+
Afrixalus lacteus	–	–	–	–	–	–	–	–	–	–	–	–	–	–	–	+	–	–	–	–	–	–	–	–	–	–	–	–
Afrixalus laevis	–	–	–	–	–	–	–	–	–	–	+	–	+	+	+	+	–	–	–	–	+	–	–	–	–	–	–	–
Afrixalus leucostictus	–	–	–	–	–	–	–	–	–	–	–	–	+	+	+	+	–	–	–	–	–	–	–	–	–	–	–	–
Afrixalus nigeriensis	–	–	–	–	–	–	–	–	–	–	–	–	+	–	+	–	–	–	–	–	–	–	–	–	–	–	–	–
Afrixalus orophilus	–	–	–	–	–	+	–	–	–	–	–	–	–	–	–	–	–	–	–	–	–	–	–	–	–	–	–	–
Afrixalus osorioi	–	–	–	–	–	+	–	–	–	–	–	–	–	–	–	–	–	+	–	–	–	–	–	–	–	–	–	–

Appendix 9:1 continued

Taxon	SUD	NEA	EAH	M.E	M.K	M.R	M.T	M.M	M.Z	M.A	M.C	M.G	WEQ	W.W	W.C	W.E	W.S	SZB	EAL	ECF	SEL	ZAZ	MST	SWA	SCE	STE	S.H	S.C
Afrixalus paradorsalis	–	–	–	–	–	–	–	–	–	–	–	–	–	+	+	+	–	–	–	–	–	–	–	–	–	–	–	–
Afrixalus schneideri	–	–	–	–	–	–	–	–	–	–	–	–	–	+	+	–	–	–	–	–	–	–	–	–	–	–	–	–
Afrixalus sp. P.& B.	–	–	–	–	–	–	–	–	–	–	–	–	–	–	–	–	–	–	+	–	–	–	–	–	–	–	–	–
Afrixalus spinifrons	–	–	–	–	–	–	–	–	–	–	–	–	–	–	–	–	–	–	+	+	–	–	–	–	–	–	–	–
Afrixalus stuhlmanni	–	–	–	–	–	–	–	–	–	–	–	–	–	–	–	–	–	–	+	–	–	–	–	+	–	–	–	–
Afrixalus sylvaticus	–	–	–	–	–	–	–	–	–	–	–	–	–	–	–	–	–	–	+	–	+	–	–	–	–	–	–	–
Afrixalus uluguruensis	–	–	–	–	–	+	–	–	–	–	–	–	–	–	–	–	–	–	–	–	–	–	–	–	–	–	–	–
Afrixalus vittiger	+	–	–	–	–	–	–	–	–	–	–	–	–	–	–	–	–	–	–	–	–	–	–	–	–	–	–	–
Afrixalus weidholzi	+	–	–	–	–	–	–	–	–	–	–	–	–	–	–	–	–	–	–	–	–	–	–	–	–	–	–	–
Afrixalus wittei	–	–	–	–	–	–	–	–	–	–	–	–	–	–	–	–	+	–	–	–	+	–	–	–	–	–	–	–
Alexteroon obstetricans	–	–	–	–	–	–	–	–	–	–	+	–	–	–	–	–	–	–	–	–	–	–	–	–	–	–	–	–
Arlequinus krebsi	–	–	–	–	–	–	–	–	–	–	+	–	–	–	–	–	–	–	–	–	–	–	–	–	–	–	–	–
Callixalus pictus	–	–	–	–	+	–	–	–	–	–	–	–	–	–	–	–	–	–	–	–	–	–	–	–	–	–	–	–
Chlorolius koehleri	–	–	–	–	+	–	–	–	–	–	+	–	–	–	+	–	–	–	–	–	–	–	–	–	–	–	–	–
Chrysobatrachus cupreonitens	–	–	–	–	+	–	–	–	–	–	–	–	–	–	–	–	–	–	–	–	–	–	–	–	–	–	–	–
Cryptothylax greshoffi	–	–	–	–	–	–	–	–	–	–	–	–	–	+	–	–	–	–	–	–	–	–	–	–	–	–	–	–
Cryptothylax minutus	–	–	–	–	–	–	–	–	–	–	–	–	–	+	–	–	–	–	–	–	–	–	–	–	–	–	–	–
Hyperolius acutirostris	–	–	–	–	–	–	–	–	–	–	+	–	–	+	–	–	–	–	–	–	–	–	–	–	–	–	–	–
Hyperolius adametzi	–	–	–	–	–	–	–	–	–	–	–	–	–	–	–	–	+	–	–	–	–	–	–	–	–	–	–	–
Hyperolius alticola	–	–	–	–	+	–	–	–	–	–	–	–	–	–	–	–	–	–	+	–	–	–	–	–	–	–	–	–
Hyperolius argus	–	–	–	–	–	–	–	–	–	–	–	–	–	–	–	–	–	–	–	+	–	–	–	–	–	–	–	–
Hyperolius atrigularis	–	–	–	–	+	–	–	–	–	–	–	–	–	–	–	–	–	–	–	–	–	–	–	–	–	–	–	–
Hyperolius balfouri	+	–	–	–	–	–	–	–	–	–	–	–	–	–	–	–	–	–	–	–	–	–	–	–	–	–	–	–
Hyperolius baumanni	–	–	–	–	–	–	–	–	–	–	–	–	+	–	–	–	–	–	–	–	–	–	–	–	–	–	–	–
Hyperolius benguellensis	–	–	–	–	–	–	–	+	–	–	–	–	+	–	–	–	–	–	–	–	–	+	–	–	–	–	–	–
Hyperolius bobirensis	–	–	–	–	–	–	–	–	–	–	–	–	+	–	+	–	–	–	–	–	–	–	–	–	–	–	–	–
Hyperolius bocagei	–	–	–	–	–	–	–	–	–	–	–	–	–	–	–	–	–	–	–	–	–	+	–	–	–	–	–	–
Hyperolius bolifambae	–	–	–	–	–	–	–	–	–	–	–	–	–	–	–	+	–	–	–	–	–	–	–	–	–	–	–	–
Hyperolius bopeleti	–	–	–	–	–	–	–	–	–	–	–	–	–	–	–	+	–	–	–	–	–	–	–	–	–	–	–	–
Hyperolius brachiofasciatus	+	–	–	–	–	–	–	–	–	–	–	–	–	–	–	+	–	–	–	–	–	–	–	–	–	–	–	–
Hyperolius castaneus	–	–	–	–	+	–	–	–	–	–	–	–	+	–	–	–	–	–	–	–	–	–	–	–	–	–	–	–
Hyperolius chlorosteus	–	–	–	–	+	–	–	–	–	–	–	–	–	–	+	–	–	–	–	–	–	–	–	–	–	–	–	–
Hyperolius chrysogaster	–	–	–	–	+	–	–	–	–	–	–	–	–	–	–	–	–	–	–	–	–	–	–	–	–	–	–	–
Hyperolius cinereus	–	–	–	–	–	–	–	–	–	+	–	–	–	–	–	–	–	–	–	–	–	–	–	–	–	–	–	–
Hyper. cinnamomeoventris	–	–	–	–	+	–	–	–	–	–	–	–	–	+	–	+	–	+	–	–	–	+	–	–	–	–	–	–
Hyperolius concolor	–	–	–	–	–	–	–	–	–	–	–	–	–	+	–	+	–	–	–	–	–	–	–	–	–	–	–	–
Hyperolius cystocandicans	–	–	–	–	–	–	–	–	–	–	–	–	–	–	–	–	+	–	–	–	–	–	–	–	–	–	–	–
Hyperolius destefanii	+	–	–	+	–	–	–	–	–	–	–	–	–	–	–	–	–	–	–	–	–	–	–	–	–	–	–	–
Hyperolius diaphanus	–	–	–	–	+	–	–	–	–	–	–	–	–	–	–	–	–	–	–	–	–	–	–	–	–	–	–	–
Hyperolius discodactylus	–	–	–	–	+	–	–	–	–	–	–	–	–	–	–	–	–	–	–	–	–	–	–	+	–	–	–	–

Appendix 9:1 continued

Taxon	SUD	NEA	EAH	M.E	M.K	M.R	M.T	M.M	M.Z	M.A	M.C	M.G	WEQ	W.W	W.C	W.E	W.S	SZB	EAL	ECF	SEL	ZAZ	MST	SWA	SCE	STE	S.H	S.C
Hyperolius endjami	−	−	−	−	−	−	−	−	−	−	−	−	+	−	−	−	−	−	−	−	−	−	−	−	−	−	−	−
Hyperolius erythromelanus	−	−	−	−	−	−	−	−	−	+	−	−	−	−	−	−	−	−	−	−	−	−	−	−	−	−	−	−
Hyperolius ferrugineus	−	−	−	−	+	−	−	−	−	−	−	−	−	−	−	−	−	−	−	−	−	−	−	−	−	−	−	−
Hyperolius frontalis	−	−	−	−	+	−	−	−	−	−	−	−	−	−	−	−	−	−	−	−	−	−	−	−	−	−	−	−
Hyperolius fusciventris	−	−	−	−	−	−	−	−	−	−	−	−	+	+	−	−	−	+	−	−	−	−	−	−	−	−	−	−
Hyperolius ghesquieri	−	−	−	−	−	−	−	−	−	−	−	−	−	−	−	−	−	+	−	−	−	−	−	−	−	−	−	−
Hyperolius guttulatus	−	−	−	−	−	−	−	−	−	−	−	−	+	+	−	−	−	−	−	−	−	−	−	−	−	−	−	−
Hyperolius horstockii	−	−	−	−	−	−	−	−	−	−	−	−	−	−	−	−	−	−	−	−	−	−	−	−	−	−	−	+
Hyperolius inornatus	−	−	−	−	−	−	−	−	−	−	−	−	−	−	−	−	+	−	−	−	−	−	−	−	−	−	−	−
Hyperolius kachalolae	−	−	−	−	−	−	−	−	−	−	−	−	−	−	−	−	−	−	−	−	−	+	−	−	−	−	−	−
Hyperolius kibarae	−	−	−	−	−	−	−	−	−	−	−	−	−	−	−	−	−	+	−	−	−	+	−	−	−	−	−	−
Hyperolius kivuensis	−	−	−	−	+	−	−	−	−	−	−	−	−	−	−	−	−	−	−	−	+	−	−	−	−	−	−	−
Hyperolius kuligae	−	−	−	−	−	−	−	−	−	−	−	−	+	−	−	+	−	−	−	−	−	−	−	−	−	−	−	−
Hyperolius lamottei	W	−	−	−	−	−	−	−	−	−	−	−	−	−	−	−	−	−	−	−	−	−	−	−	−	−	−	−
Hyperolius lateralis	−	−	−	−	+	−	−	−	−	−	−	−	−	−	−	−	−	−	−	−	+	−	−	−	−	−	−	−
Hyperolius laticeps	−	−	−	−	−	−	−	−	−	−	−	−	−	+	−	−	−	−	−	−	−	−	−	−	−	−	−	−
Hyperolius laurenti	−	−	−	−	−	−	−	−	−	−	−	−	−	+	−	−	−	−	−	−	−	−	−	−	−	−	−	−
Hyperolius leleupi	−	−	−	−	+	−	−	−	−	−	−	−	−	−	−	−	−	−	−	−	−	−	−	−	−	−	−	−
Hyperolius leucotaenius	−	−	−	−	+	−	−	−	−	−	−	−	+	−	−	−	−	+	+	−	−	−	−	−	−	−	−	−
Hyperolius marmoratus	−	−	−	−	−	+	+	+	+	−	−	−	−	−	−	−	−	−	+	+	−	−	+	−	−	−	−	−
Hyperolius minutissimus	−	−	−	−	−	−	+	−	−	+	−	−	−	−	−	−	−	−	−	−	−	−	−	−	−	−	−	−
Hyperolius mitchelli	−	−	−	−	−	−	+	−	−	−	−	−	−	−	−	−	−	−	+	−	−	−	−	−	−	−	−	−
Hyperolius montanus	−	−	−	+	−	−	−	+	−	−	−	−	−	−	−	−	−	−	−	−	−	−	−	−	−	−	−	−
Hyperolius mosaicus	−	−	+	−	−	−	−	−	−	−	−	−	+	−	−	−	−	−	−	−	−	−	−	−	−	−	−	−
Hyperolius nasutus	+	−	−	−	−	+	−	−	−	−	−	−	−	−	−	−	−	+	+	+	+	−	+	−	−	−	−	−
Hyperolius obscurus	−	−	−	−	−	−	−	−	−	−	−	−	−	−	−	−	−	+	−	−	−	−	−	−	−	−	−	−
Hyperolius occidentalis	−	−	−	−	−	−	−	−	−	−	−	−	+	+	−	+	+	−	−	−	−	−	−	−	−	−	−	−
Hyperolius ocellatus	−	−	−	−	−	−	−	−	−	−	−	−	+	+	−	+	+	+	−	−	−	−	−	−	−	−	−	−
Hyperolius pardalis	−	−	−	−	−	−	−	−	−	−	−	−	+	+	−	+	+	−	−	−	−	−	−	−	−	−	−	−
Hyperolius parkeri	−	−	−	−	−	−	−	−	−	−	−	−	−	−	+	−	+	−	+	−	−	−	−	−	−	−	−	−
Hyperolius phantasticus	−	−	−	−	−	−	−	−	−	−	−	−	−	+	−	+	+	−	−	−	−	−	−	−	−	−	−	−
Hyperolius pickersgilli	−	−	−	−	−	+	−	−	−	−	−	−	−	−	−	−	−	−	−	−	−	+	−	−	−	−	−	−
Hyperolius picturatus	−	−	−	−	−	−	+	−	−	−	−	−	+	−	−	−	−	−	−	+	−	−	−	−	−	−	−	−
Hyperolius pictus	−	−	−	−	+	−	−	−	−	−	−	−	+	+	−	+	+	−	−	−	−	−	−	−	−	−	−	−
Hyperolius platyceps	−	−	−	−	−	−	−	−	−	−	−	−	+	+	−	+	+	−	−	−	−	+	−	−	−	−	−	−
Hyperolius polli	−	−	−	−	−	−	−	−	−	−	−	−	−	−	−	−	−	+	−	−	−	−	−	+	−	−	−	−
Hyperolius polystictus	−	−	−	−	−	−	−	−	−	−	−	−	−	−	−	−	−	+	−	−	−	−	−	+	−	−	−	−
Hyperolius puncticulatus	−	−	−	−	+	+	−	−	−	−	−	−	−	−	−	−	−	−	−	−	−	−	−	−	−	−	+	−
Hyperolius punctulatus	−	−	−	+	−	−	−	−	−	+	−	−	−	−	−	−	−	−	−	−	−	−	−	−	−	−	−	−
Hyperolius pusillus	−	−	−	−	−	−	−	−	−	−	−	−	−	−	−	−	−	−	+	+	−	−	−	−	−	−	−	−

Appendix 9:1 continued

Taxon	SUD	NEA	EAH	M.E	M.K	M.R	M.T	M.M	M.Z	M.A	M.C	M.G	WEQ	W.W	W.C	W.E.	W.S	SZB	EAL	ECF	SEL	ZAZ	MST	SWA	SCE	STE	S.H	S.C
Hyperolius pustulifer	–	–	–	–	+	–	–	–	–	–	–	–	–	–	–	–	–	–	–	–	–	–	–	–	–	–	–	–
Hyperolius quinquevittatus	–	–	–	–	–	–	+	–	–	–	–	–	–	–	–	–	–	–	–	–	–	+	–	–	–	–	–	–
Hyperolius reesi	–	–	–	–	–	–	–	–	–	–	–	–	–	–	–	–	–	+	–	–	–	–	–	–	–	–	–	–
Hyperolius riggenbachi	–	–	–	–	–	–	–	–	–	–	+	–	–	–	+	–	–	–	–	–	–	–	–	–	–	–	–	–
Hyperolius robustus	–	–	–	–	–	–	–	–	–	–	–	–	–	–	–	–	+	–	–	–	–	–	–	–	–	–	–	–
Hyperolius rubrovermiculatus	–	–	–	–	–	–	–	–	–	–	–	–	–	–	–	–	–	–	–	+	–	–	–	–	–	–	–	–
Hyperolius sankuruensis	–	–	–	–	–	–	–	–	–	–	–	–	–	–	+	–	–	–	–	–	–	–	–	–	–	–	–	–
Hyperolius schouteni	+	–	–	–	+	–	+	–	–	–	–	–	–	–	–	–	–	–	–	–	–	–	–	–	–	–	–	–
Hyperolius semidiscus	–	–	–	–	–	–	–	–	–	–	–	–	–	–	–	–	–	–	–	–	+	–	–	–	–	–	–	–
Hyperolius sheldrecki	–	–	–	–	–	–	–	–	–	–	–	–	–	–	–	–	–	+	–	–	–	–	–	–	–	–	–	–
Hyperolius soror	–	–	–	–	–	–	–	–	–	–	–	–	+	–	–	–	–	–	–	–	–	–	–	–	–	–	–	–
Hyperolius spinigularis	–	–	+	–	+	–	+	+	–	–	–	–	–	–	–	–	–	–	–	–	–	–	–	–	–	–	–	–
Hyperolius steindachneri	–	–	–	–	–	–	–	–	–	–	–	–	+	+	–	–	–	+	–	–	–	–	–	–	–	–	–	–
Hyperolius sylvaticus	–	–	–	–	–	–	–	–	–	–	+	+	+	–	–	–	–	–	–	–	–	–	–	–	–	–	–	–
Hyperolius tanneri	–	–	–	–	–	–	+	–	–	–	–	–	–	–	–	–	–	–	–	–	–	–	–	–	–	–	–	–
Hyperolius tornieri	–	–	–	–	–	–	+	–	–	–	–	–	–	–	–	–	–	–	–	–	–	–	–	–	–	–	–	–
Hyperolius torrentis	–	–	–	–	–	–	–	–	–	–	+	+	+	+	+	+	–	–	–	–	–	–	–	–	–	–	–	–
Hyperolius tuberculatus	–	–	–	–	–	–	–	–	–	–	+	+	+	+	+	+	–	–	+	–	–	–	–	–	–	–	–	–
Hyperolius tuberilinguis	–	–	–	–	–	–	–	–	–	–	–	–	–	–	–	–	–	+	+	–	+	–	–	–	–	–	–	–
Hyperolius vilhenai	–	–	–	–	–	–	–	–	–	–	–	–	–	–	–	–	–	+	–	–	–	–	–	–	–	–	–	–
Hyperolius viridiflavus	+	+	–	–	+	–	+	+	–	–	–	+	–	–	–	–	–	–	–	–	–	–	–	–	–	–	–	–
Hyperolius viridigulosus	–	–	–	–	–	–	–	–	–	–	–	+	–	–	–	–	–	–	–	–	–	–	–	–	–	–	–	–
Hyperolius viridis	–	–	–	–	+	–	+	+	–	–	–	+	–	–	–	–	–	–	–	–	–	–	–	–	–	–	–	–
Hyperolius wermuthi	–	–	–	–	–	–	–	–	–	–	–	+	–	–	–	–	–	–	–	–	–	–	–	–	–	–	–	–
Hyperolius xenorhinus	+	–	–	–	–	–	–	–	–	–	–	–	+	–	–	–	–	–	–	–	–	–	–	–	–	–	–	–
Hyperolius zavattarii	+	–	–	–	–	–	–	–	–	–	–	+	–	–	–	+	–	–	–	–	–	–	–	–	–	–	–	–
Hyperolius zonatus	–	–	–	–	–	–	–	–	–	–	–	–	+	+	–	–	–	–	–	–	–	–	–	–	–	–	–	–
Kassina arboricola	+	–	–	–	–	–	–	–	–	–	–	–	–	–	+	–	–	–	–	–	–	–	–	–	–	–	–	–
Kassina cassinoides	–	–	–	–	–	–	–	–	–	–	–	+	+	+	+	+	–	–	+	–	–	–	+	–	–	–	–	–
Kassina cochranae	–	–	–	–	–	+	–	–	–	–	–	–	–	–	–	–	–	–	–	–	–	–	–	–	–	–	–	–
Kassina decorata	W	–	–	–	–	–	–	–	–	–	–	–	–	–	–	–	–	–	–	–	–	–	–	–	–	–	–	–
Kassina fusca	–	+	–	–	–	–	–	–	–	–	–	–	–	–	–	–	–	–	–	–	+	–	–	–	–	–	–	–
Kassina kuvangensis	–	–	–	–	–	–	–	–	–	–	–	–	–	–	–	–	–	–	–	–	+	–	+	–	–	–	–	–
Kassina lamottei	–	–	–	–	–	–	–	–	–	–	–	–	–	–	–	–	–	–	+	–	–	–	–	–	–	–	+	+
Kassina maculata	–	–	–	–	–	–	–	–	–	–	–	–	–	–	–	–	+	–	–	–	–	–	–	–	–	–	+	–
Kassina maculosa	+	–	–	–	–	+	–	–	–	–	–	–	–	–	–	–	–	–	–	–	–	–	–	–	–	–	+	–
Kassina mertensi	–	+	–	–	–	–	–	–	–	–	–	–	–	–	–	–	–	–	–	–	–	–	–	–	–	–	+	–
Kassina parkeri	+	+	–	–	–	–	–	–	–	–	–	–	–	–	–	–	–	–	–	–	–	–	–	–	–	–	–	–
Kassina senegalensis	+	+	–	–	–	–	+	–	–	–	–	–	–	–	–	–	+	+	+	+	+	+	+	+	+	+	+	+
Kassina somalica	–	+	–	–	–	–	–	–	–	–	–	–	–	–	–	–	+	–	+	–	–	–	–	–	–	–	–	–

Appendix 9:1 continued

Taxon	SUD	NEA	EAH	M.E	M.K	M.R	M.T	M.M	M.Z	M.A	M.C	M.G	WEQ	W.W	W.C	W.E.	W.S	SZB	EAL	ECF	SEL	ZAZ	MST	SWA	SCE	STE	S.H	S.C
Kassinula wittei	–	–	–	–	–	–	–	–	–	–	–	–	–	–	–	–	–	–	–	–	–	–	–	–	–	–	–	–
Leptopelis anchietae	–	–	–	–	–	–	–	–	+	–	–	–	–	–	–	–	–	–	–	–	+	–	–	–	–	–	–	–
Leptopelis argenteus	–	–	–	–	–	–	–	–	–	–	–	–	–	–	–	–	–	–	+	–	–	–	+	–	–	–	–	–
Leptopelis aubryi	–	–	–	–	–	–	–	–	–	–	–	–	–	–	–	+	–	–	–	–	–	–	–	–	–	–	–	–
Leptopelis barbouri	–	–	–	–	–	+	–	–	–	–	–	–	–	–	–	–	–	–	–	–	–	–	–	–	–	–	–	–
Leptopelis bequaerti	+	+	+	–	–	–	–	–	–	–	–	–	–	–	–	–	–	–	–	–	–	–	+	–	–	–	–	–
Leptopelis bocagii	–	–	–	+	–	–	–	–	–	–	–	–	–	+	+	–	–	–	+	–	+	–	+	–	–	–	–	–
Leptopelis boulengeri	–	–	–	–	–	–	–	–	–	–	–	–	–	–	+	–	–	–	–	–	–	–	–	–	–	–	–	–
Leptopelis brevipes	–	–	–	–	–	–	–	–	–	–	–	–	–	–	+	–	–	–	–	–	–	–	–	–	–	–	–	–
Leptopelis brevirostris	–	–	–	–	–	–	–	–	–	+	–	–	–	–	+	+	+	–	–	–	–	–	–	–	–	–	–	–
Leptopelis broadleyi	W	–	–	–	–	–	–	–	–	–	–	–	–	–	–	–	–	–	–	–	–	–	–	–	–	–	–	–
Leptopelis bufonides	–	–	–	–	+	–	–	–	–	+	–	–	–	+	+	–	–	–	+	–	–	–	–	–	–	–	–	–
Leptopelis calcaratus	–	–	–	–	–	–	–	–	–	–	–	–	–	–	–	–	–	–	+	–	–	–	–	–	–	–	–	–
Leptopelis christyi	–	–	–	–	+	–	–	–	–	–	–	–	–	–	–	–	–	–	–	–	–	–	–	–	–	–	–	–
Leptopelis concolor	–	–	–	–	–	–	–	–	–	–	–	–	–	–	–	–	–	–	+	–	+	–	–	–	–	–	–	–
Leptopelis cynnamomeus	–	–	–	–	–	–	–	–	–	–	–	–	–	–	–	–	–	–	–	–	–	–	–	–	–	–	–	–
Leptopelis fenestratus	–	–	–	–	+	–	–	–	–	–	–	–	–	–	–	–	–	–	–	+	–	–	–	–	–	–	–	–
Leptopelis fiziensis	–	–	–	–	+	–	–	–	–	–	–	–	–	–	–	–	–	–	–	–	–	–	–	–	–	–	–	–
Leptopelis flavomaculatus	–	–	–	–	+	–	–	+	–	–	–	–	–	–	+	–	–	–	+	–	–	–	–	–	–	–	–	–
Leptopelis gramineus	–	+	–	–	–	–	–	–	–	–	–	–	–	–	–	–	–	–	+	–	–	–	–	–	–	–	–	–
Leptopelis hyloides	–	–	–	–	–	–	–	–	–	–	–	–	–	+	–	–	–	–	–	–	–	–	–	–	–	–	–	–
Leptopelis jordani	–	–	–	–	–	–	–	–	–	–	–	–	–	–	–	–	–	–	–	–	–	–	–	–	–	–	–	–
Leptopelis karissimbensis	–	–	–	–	+	–	–	–	–	–	–	–	–	+	–	–	–	–	–	–	–	–	–	–	–	–	–	–
Leptopelis kivuensis	–	–	–	–	+	–	–	–	–	–	–	–	–	–	–	–	–	–	–	–	–	–	–	–	–	–	–	–
Leptopelis lebeaui	–	–	–	–	–	–	–	–	–	–	–	–	–	–	–	–	–	+	–	–	–	–	–	–	–	–	–	–
Leptopelis macrotis	–	–	–	–	–	–	–	–	–	+	–	–	–	+	–	–	–	–	–	–	–	–	–	–	–	–	–	–
Leptopelis marginatus	–	–	–	–	–	–	–	–	–	–	–	–	–	–	–	+	–	–	–	–	–	–	–	–	–	–	–	–
Leptopelis millsoni	–	–	–	–	–	–	–	–	–	–	–	–	–	+	+	–	–	–	–	–	–	–	–	–	–	–	–	–
Leptopelis modestus	–	–	–	–	–	–	–	–	–	+	–	–	–	+	+	–	–	–	+	–	–	–	–	–	–	–	–	–
Leptopelis mossambicus	–	–	–	–	–	–	–	–	–	–	–	–	–	–	–	–	–	–	–	–	–	–	–	+	–	–	–	–
Leptopelis natalensis	–	–	–	–	–	–	–	–	–	+	–	–	–	–	+	–	–	–	–	–	–	–	–	–	–	–	–	–
Leptopelis nordequatorialis	–	–	–	–	–	–	–	–	–	–	–	–	–	–	–	–	–	–	–	–	–	–	–	–	–	–	–	–
Leptopelis notatus	–	–	–	–	–	–	–	–	–	–	–	–	–	–	+	–	–	–	–	–	–	–	–	–	–	–	–	–
Leptopelis occidentalis	–	–	–	–	–	–	–	–	–	–	–	–	–	+	+	–	–	–	–	–	–	–	–	–	–	–	–	–
Leptopelis ocellatus	–	–	–	–	–	–	–	–	–	–	–	–	–	–	–	+	–	–	–	–	–	–	–	–	–	–	+	–
Leptopelis omissus	–	–	–	–	–	–	–	–	–	–	–	–	–	–	–	–	–	–	–	–	–	–	–	–	–	–	–	–
Leptopelis oryi	E	–	–	–	–	–	–	–	–	–	–	–	–	–	–	–	–	–	–	–	–	–	–	–	–	–	–	–
Leptopelis parbocagii	–	–	–	–	–	–	–	–	–	–	–	–	–	–	–	–	–	+	–	–	–	+	–	–	–	–	–	–
Leptopelis parkeri	–	–	–	–	–	+	–	–	–	–	–	–	–	–	–	–	–	–	–	–	–	–	–	–	–	–	+	–
Leptopelis parvus	–	–	–	–	–	–	–	–	–	–	–	–	–	–	–	–	–	+	–	–	–	–	–	–	–	–	–	+

Appendix 9:1 continued

Taxon	SUD	NEA	EAH	M.E	M.K	M.R	M.T	M.M	M.Z	M.A	M.C	M.G	WEQ	W.W	W.C	W.E	W.S	SZB	EAL	ECF	SEL	ZAZ	MST	SWA	SCE	STE	S.H	S.C
Leptopelis ragazzii	–	+	–	–	–	–	–	–	–	–	–	–	–	–	–	–	–	–	–	–	–	–	–	–	–	–	–	–
Leptopelis rufus	–	–	–	–	–	–	–	–	–	–	–	+	+	+	+	–	–	–	–	–	–	–	–	–	–	–	–	–
Leptopelis sp. P. & B.	–	–	–	–	–	–	+	–	–	–	–	–	+	+	+	–	–	–	–	–	–	–	–	–	–	–	–	–
Leptopelis susanae	–	–	+	–	–	–	–	–	–	–	–	–	–	–	–	–	–	–	–	–	–	–	–	–	–	–	–	–
Leptopelis uluguruensis	–	–	–	–	+	–	–	–	–	–	–	–	–	–	–	–	–	–	–	–	–	–	–	–	–	–	–	–
Leptopelis vannutellii	–	–	+	–	–	–	–	–	–	–	–	–	–	–	–	–	–	–	–	–	–	–	–	–	–	–	–	–
Leptopelis vermiculatus	–	–	–	–	+	–	–	–	–	–	–	–	–	–	–	–	–	–	–	–	–	–	–	–	–	–	–	–
Leptopelis viridis	+	–	–	–	–	–	–	–	–	–	–	–	–	–	–	–	–	–	–	–	–	–	–	–	–	–	–	–
Leptopelis xenodactylus	–	–	–	–	–	–	–	–	–	–	–	–	–	–	–	–	–	–	–	–	–	–	–	–	–	+	+	–
Leptopelis yaldeni	–	–	+	–	–	–	–	–	–	–	–	–	–	–	–	–	–	–	–	–	–	–	–	–	–	–	–	–
Opisthothylax immaculatus	–	–	–	–	–	–	–	–	–	–	–	–	+	+	+	–	–	–	–	–	–	–	–	–	–	–	–	–
Paracassina kounhiensis	–	–	+	–	–	–	–	–	–	–	–	–	–	–	–	–	–	–	–	–	–	–	–	–	–	–	–	–
Paracassina obscura	–	–	+	–	–	–	–	–	–	–	–	–	–	–	–	–	–	–	–	–	–	–	–	–	–	–	–	–
Phlyctimantis boulengeri	–	–	–	–	–	–	–	–	–	–	–	–	+	+	+	–	–	–	–	–	–	–	–	–	–	–	–	–
Phlyctimantis keithae	–	–	–	–	+	–	–	–	–	–	–	–	–	–	–	–	–	–	–	–	–	–	–	–	–	–	–	–
Phlyctimantis leonardi	–	–	–	–	–	–	–	–	–	–	–	–	+	+	+	–	–	–	–	–	–	–	–	–	–	–	–	–
Phlyctimantis verrucosus	–	–	–	–	–	+	–	–	–	–	–	–	–	–	–	–	–	–	–	–	–	–	–	–	–	–	–	–
Semnodactylus wealii	–	–	–	–	–	–	–	–	–	–	–	–	–	–	–	–	–	–	–	–	+	+	+	+	+	+	+	+
ANURA: MICROHYLIDAE:																												
Balebreviceps hillmani	–	–	+	–	–	–	–	–	–	–	–	–	–	–	–	–	–	–	–	–	–	–	–	–	–	–	–	–
Breviceps acutirostris	–	–	–	–	–	–	–	–	–	–	–	–	–	–	–	–	–	–	–	–	–	–	–	–	–	–	–	+
Breviceps adspersus	–	–	–	–	–	–	–	–	–	–	–	–	–	–	–	–	–	–	–	+	+	+	+	+	+	+	+	+
Breviceps fuscus	–	–	–	–	–	–	–	–	–	–	–	–	–	–	–	–	–	–	–	–	–	–	–	–	–	–	–	+
Breviceps gibbosus	–	–	–	–	–	–	–	–	–	–	–	–	–	–	–	–	–	–	–	–	–	–	–	–	+	–	–	+
Breviceps macrops	–	–	–	–	–	–	–	–	–	–	–	–	–	–	–	–	–	–	–	–	–	–	–	–	+	–	–	–
Breviceps maculatus	–	–	–	–	–	–	–	+	–	–	–	–	–	–	–	–	–	–	–	–	–	–	–	–	–	–	–	–
Breviceps montanus	–	–	–	–	–	–	–	–	–	–	–	–	–	–	–	–	–	–	–	–	–	–	–	–	–	+	+	+
Breviceps mossambicus	–	+	–	–	–	+	–	–	–	–	–	–	–	–	–	–	–	–	+	+	+	+	+	–	–	+	–	–
Breviceps namaquensis	–	–	–	–	–	–	–	–	–	–	–	–	–	–	–	–	–	–	–	+	+	–	–	+	–	–	–	–
Breviceps poweri	–	–	–	–	–	–	–	–	–	–	–	–	–	–	–	–	–	–	–	–	+	–	–	–	–	–	–	–
Breviceps rosei	–	–	–	–	–	–	–	–	–	–	–	–	–	–	–	–	–	–	–	–	–	–	–	–	–	–	+	+
Breviceps sylvestris	–	–	–	–	–	–	–	–	–	–	–	–	–	–	–	–	–	–	–	–	–	–	–	–	–	+	+	+
Breviceps verrucosus	–	–	–	–	–	–	–	–	–	–	–	–	–	–	–	–	–	–	–	–	+	–	–	–	–	–	+	–
Callulina kreffti	–	–	–	–	+	+	–	–	–	–	–	–	–	–	–	–	–	–	–	–	–	–	–	–	–	–	–	–
Probreviceps macrodactylus	–	–	–	–	+	+	–	–	–	–	–	–	–	–	–	–	–	–	–	–	–	–	–	–	–	–	–	–
Probreviceps rhodesianus	–	–	–	–	–	–	–	–	+	–	–	–	–	–	–	–	–	–	–	–	–	–	–	–	–	–	–	–
Probreviceps uluguruensis	–	–	–	–	+	–	–	–	–	–	–	–	–	–	–	–	–	–	–	–	–	–	–	–	–	–	–	–
Spelaeophryne methneri	–	–	–	+	–	–	–	–	–	–	–	–	–	–	–	–	–	–	+	+	–	–	–	–	–	–	–	–
Hoplophryne rogersi	–	–	–	–	+	–	–	–	–	–	–	–	–	–	–	–	–	–	–	–	–	–	–	–	–	–	–	–
Hoplophryne uluguruensis	–	–	–	–	+	+	–	–	–	–	–	–	–	–	–	–	–	–	–	–	–	–	–	–	–	–	–	–

Appendix 9:1 continued

Taxon	SUD	NEA	EAH	M.E	M.K	M.R	M.T	M.M	M.Z	M.A	M.C	M.G	WEQ	W.W	W.C	W.E	W.S	SZB	EAL	ECF	SEL	ZAZ	MST	SWA	SCE	STE	S.H	S.C
Parhoplophryne usambarica	–	–	–	–	–	–	–	–	–	–	–	–	–	–	–	–	–	–	–	–	–	–	–	–	–	–	–	–
Phrynomantis affinis	–	–	–	–	–	–	–	–	–	–	–	–	–	–	–	–	–	–	–	–	–	+	–	–	–	–	–	–
Phrynomantis annectens	–	–	+	–	–	–	–	–	–	–	–	–	–	–	–	–	–	–	–	–	–	–	–	+	–	–	–	–
Phrynomantis bifasciatus	–	–	+	+	–	+	–	–	–	–	–	–	–	–	–	–	–	–	+	+	+	+	+	+	–	–	–	–
Phrynomantis microps	+	–	–	–	–	–	–	–	–	–	–	–	–	–	–	–	–	–	–	–	–	–	–	–	–	–	–	–
Phrynomantis somalicus	–	+	–	–	–	–	–	–	–	–	–	–	–	–	–	–	–	–	–	–	–	–	–	–	–	–	–	–
ANURA: PIPIDAE:																												
Hymenochirus boettgeri	–	–	–	–	–	–	–	–	–	–	–	+	+	+	+	–	–	–	–	+	–	–	–	–	–	–	–	–
Hymenochirus boulengeri	–	–	–	–	–	–	–	–	–	–	–	+	+	+	+	–	–	–	–	–	–	–	–	–	–	–	–	–
Hymenochirus curtipes	–	–	–	–	–	–	–	–	–	–	–	–	–	–	–	+	+	–	–	–	–	–	–	–	–	–	–	–
Hymenochirus feae	–	–	–	–	–	–	–	–	–	–	–	–	–	–	–	+	+	–	–	–	–	–	–	–	–	–	–	–
Pseudhymenochirus merlini	–	–	–	–	–	–	–	–	–	–	–	+	–	–	–	–	–	–	–	–	–	–	–	–	–	–	–	–
Silurana epitropicalis	–	–	–	–	–	–	–	–	–	–	–	+	+	+	+	+	–	–	–	–	–	–	–	–	–	–	–	–
Silurana tropicalis	–	–	–	–	–	–	–	–	–	–	–	+	+	+	+	+	–	–	–	–	–	–	–	–	–	–	–	–
Xenopus amieti	–	–	–	–	–	–	–	–	–	–	+	–	–	–	–	–	–	–	–	–	–	–	–	–	–	–	–	–
Xenopus andrei	–	–	–	–	–	–	–	–	–	–	–	–	–	–	+	–	–	–	–	–	–	–	–	–	–	–	–	–
Xenopus borealis	–	+	+	–	–	–	–	–	–	–	–	–	–	–	–	–	–	–	–	–	–	–	–	–	–	–	–	–
Xenopus boumbaensis	–	–	–	–	–	–	–	–	–	–	+	–	–	–	–	–	–	–	–	–	–	–	–	–	–	–	–	–
Xenopus clivii	–	–	–	+	–	–	–	–	–	–	–	–	–	–	–	–	–	–	–	–	–	–	–	–	–	–	–	–
Xenopus fraseri	–	–	–	–	–	–	–	–	–	–	+	–	+	–	–	–	–	+	–	–	–	–	–	–	–	–	–	–
Xenopus gilli	+	–	–	–	–	–	–	–	–	–	–	–	–	–	–	–	–	–	–	–	–	–	–	–	–	–	–	+
Xenopus laevis	+	–	+	–	–	+	+	+	–	+	–	+	–	–	–	–	–	+	+	+	+	+	+	+	+	+	+	+
Xenopus longipes	–	–	–	–	–	–	–	–	–	–	–	–	–	–	–	–	–	–	–	–	–	–	–	–	–	–	–	–
Xenopus muelleri	+	+	+	–	–	–	–	–	–	–	–	–	+	–	–	–	–	–	+	–	–	–	–	–	–	–	–	–
Xenopus pygmaeus	–	–	–	–	–	–	–	–	–	–	+	–	–	–	–	–	–	–	–	–	–	–	–	–	–	–	–	–
Xenopus ruwenzoriensis	–	–	–	–	+	–	–	–	–	–	–	–	–	–	–	–	–	–	–	–	–	–	–	–	–	–	–	–
Xenopus vestitus	+	–	–	–	+	–	–	–	–	–	–	–	–	–	–	–	–	–	–	–	–	–	–	–	–	–	–	–
Xenopus wittei	–	–	–	–	+	–	–	–	–	–	–	–	–	–	–	–	–	–	–	–	–	–	–	–	–	–	–	–
ANURA: RANIDAE:																												
Anhydrophryne rattrayi	–	–	–	–	–	–	–	–	–	–	–	–	–	–	–	–	–	–	–	–	–	–	–	–	–	–	+	–
Arthroleptella hewitti	–	–	–	–	–	–	–	–	–	–	–	–	–	–	–	–	–	–	–	–	–	–	–	–	–	–	+	–
Arthroleptella lightfooti	–	–	–	–	–	–	–	–	–	–	–	–	–	–	–	–	–	–	–	–	–	–	–	–	–	–	+	+
Arthroleptella ngongoniensis	–	–	–	–	–	–	–	–	–	–	–	–	–	–	–	–	–	–	–	–	–	–	–	+	–	–	+	–
Arthroleptides dutoiti	–	–	–	–	+	–	–	–	–	–	–	–	–	–	–	–	–	–	–	–	–	–	–	–	–	–	–	–
Arthroleptides martiensseni	–	–	–	–	–	–	+	–	–	–	–	–	–	–	–	–	–	–	–	–	–	–	–	–	–	–	–	–
Aubria masako	–	–	–	–	–	–	–	–	–	–	–	–	–	–	–	+	–	–	–	–	–	–	–	–	–	–	–	–
Aubria subsigillata	–	–	–	–	–	–	–	–	–	–	–	–	+	+	+	–	–	–	–	–	–	–	–	–	–	–	–	–
Cacosternum boettgeri	–	+	–	–	–	–	–	–	–	–	–	–	–	–	–	–	–	–	–	+	+	+	+	+	+	+	+	+
Cacosternum capense	–	–	–	–	–	–	–	–	–	–	–	–	–	–	–	–	–	–	–	–	–	–	–	–	–	–	–	+
Cacosternum leleupi	–	–	–	–	–	–	–	–	–	–	–	–	–	–	–	–	–	–	–	–	–	–	+	–	–	–	–	–

Appendix 9:1 continued

Taxon	SUD	NEA	EAH	M.E	M.K	M.R	M.T	M.M	M.Z	M.A	M.C	M.G	WEQ	W.W	W.C	W.E.	W.S	SZB	EAL	ECF	SEL	ZAZ	MST	SWA	SCE	STE	S.H	S.C
Cacosternum namaquense	–	–	–	–	–	–	–	–	–	–	–	–	–	–	–	–	–	–	–	–	–	–	–	+	–	–	–	–
Cacosternum nanum	–	+	–	–	–	–	–	–	–	–	–	–	–	–	–	–	–	–	–	+	+	–	–	–	–	+	–	–
Cacosternum poyntoni	–	–	–	–	–	–	–	–	–	–	–	–	–	–	–	–	–	–	–	+	+	–	–	–	–	–	–	–
Cacosternum striatus	–	–	–	–	–	–	–	–	–	–	–	–	–	–	–	–	–	–	–	+	+	–	–	–	–	+	–	–
Conraua alleni	–	–	–	–	–	–	–	–	–	–	+	–	+	–	–	–	–	–	–	–	–	–	–	–	–	–	–	–
Conraua beccarii	E	–	–	–	–	–	–	–	–	–	–	–	–	–	–	–	–	–	–	–	–	–	–	–	–	–	–	–
Conraua crassipes	–	–	–	–	–	–	–	–	–	–	+	–	+	+	–	–	–	–	–	–	–	–	–	–	–	–	–	–
Conraua derooi	–	–	–	–	–	–	–	–	–	–	–	–	–	+	–	–	–	–	–	–	–	–	–	–	–	–	–	–
Conraua goliath	–	–	–	–	–	–	–	–	–	–	–	–	+	–	+	–	–	–	–	–	–	–	–	–	–	–	–	–
Conraua robusta	–	–	–	–	–	–	–	–	–	–	–	–	+	+	+	–	–	–	–	–	–	–	–	–	–	–	–	–
Dimorphognathus africanus	–	–	–	–	–	–	–	–	–	–	–	–	+	+	+	–	–	–	–	–	–	–	–	–	–	–	–	–
Ericabatrachus baleensis	–	–	+	+	–	–	–	–	–	–	–	–	–	–	–	–	–	–	–	–	–	–	–	–	–	–	–	–
Euphlyctis cornii	E	–	–	–	–	–	–	–	–	–	–	–	–	–	–	–	–	–	–	–	–	–	–	–	–	–	–	–
Hildebrand. macrotympanum	–	+	+	+	–	–	–	–	–	–	–	–	–	–	–	–	–	–	–	–	–	–	–	–	–	–	–	–
Hildebrandtia ornata	+	+	–	–	–	–	–	–	–	–	–	–	–	–	–	–	–	–	+	–	–	+	+	+	–	–	–	–
Hildebrandtia ornatissima	–	–	–	–	–	–	–	–	+	–	–	–	–	–	–	–	–	–	–	–	–	–	–	–	–	–	–	–
Hoplobatrachus demarchii	E	+	+	–	–	–	–	–	–	–	–	–	–	–	–	–	–	–	–	–	–	–	–	–	–	–	–	–
Hoplobatrachus occipitalis	+	+	–	–	–	–	–	–	–	–	–	–	+	+	+	–	–	–	+	+	+	–	–	–	–	–	–	–
Hylarana albolabris	–	–	–	–	–	–	–	–	–	–	–	–	+	+	+	+	–	–	–	–	–	–	–	–	–	–	–	–
Hylarana amnicola	–	–	–	–	–	–	–	–	–	–	–	–	+	+	+	–	–	–	–	–	–	–	–	–	–	–	–	–
Hylarana asperrima	–	–	–	–	–	–	–	–	–	–	–	–	–	–	+	–	–	–	–	–	–	–	–	–	–	–	–	–
Hylarana darlingi	–	–	–	–	–	–	–	–	–	–	–	–	–	–	+	–	–	–	–	–	–	–	+	+	–	–	–	–
Hylarana galamensis	+	+	+	–	–	–	–	–	–	–	–	–	–	+	–	–	–	–	+	–	–	–	–	–	–	–	–	–
Hylarana lemairei	–	–	–	–	–	–	–	–	–	–	–	–	+	–	–	–	–	+	+	–	+	–	–	–	–	–	–	–
Hylarana lepus	–	–	–	–	–	–	–	–	–	–	–	–	+	+	+	–	–	–	–	–	–	–	–	–	–	–	–	–
Hylarana longipes	–	–	–	–	–	–	–	–	–	–	–	–	–	–	+	–	–	–	–	–	–	–	–	–	–	–	–	–
Hylarana occidentalis	–	–	–	–	–	–	–	–	–	–	–	–	+	+	–	–	–	–	–	–	–	–	–	–	–	–	–	–
Hylarana parkeriana	–	–	–	–	–	–	–	–	–	–	–	–	–	–	–	–	–	–	–	–	–	–	–	–	–	–	–	–
Lanzarana largeni	+	–	–	–	–	–	–	–	–	–	–	–	–	–	–	–	–	–	–	–	–	–	–	–	–	–	–	–
Microbatrachella capensis	–	–	–	–	–	–	–	–	–	–	–	–	–	–	–	–	–	–	–	–	–	–	–	–	–	–	+	–
Natalobatrachus bonebergi	–	–	–	–	–	–	–	–	–	–	–	–	–	–	–	–	–	–	–	+	–	–	–	–	–	–	–	–
Nothophryne broadleyi	–	–	–	–	–	–	–	+	–	–	–	–	–	–	–	–	–	–	–	–	–	–	–	–	–	–	–	–
Petropedetes cameronensis	–	–	–	–	–	–	–	–	–	–	+	–	+	–	+	–	–	–	–	–	–	–	–	–	–	–	–	–
Petropedetes johnstoni	–	–	–	–	–	–	–	–	–	–	+	–	+	–	+	–	–	–	–	–	–	–	–	–	–	–	–	–
Petropedetes natator	–	–	–	–	–	–	–	–	–	–	+	–	+	–	+	–	–	–	–	–	–	–	–	–	–	–	–	–
Petropedetes newtoni	–	–	–	–	–	–	–	–	–	–	+	–	–	–	–	–	–	–	–	–	–	–	–	–	–	–	–	–
Petropedetes palmipes	–	–	–	–	–	–	–	–	–	–	+	–	–	–	+	–	–	–	–	–	–	–	–	–	–	–	–	–
Petropedetes parkeri	–	–	–	–	–	–	–	–	–	–	+	–	–	–	–	–	–	–	–	–	–	–	–	–	–	–	–	–
Petropedetes perreti	–	–	–	–	–	–	–	–	–	–	+	–	–	–	–	–	–	–	–	–	–	–	–	–	–	–	–	–
Phrynobatrachus accraensis	W	–	–	–	–	–	–	–	–	–	–	–	+	–	–	–	–	–	–	–	–	–	–	–	–	–	–	–

Appendix 9:1 continued

Taxon	SUD	NEA	EAH	M.E	M.K	M.R	M.T	M.M	M.Z	M.A	M.C	M.G	WEQ	W.W	W.C	W.E.	W.S	SZB	EAL	ECF	SEL	ZAZ	MST	SWA	SCE	STE	S.H	S.C
Phrynobatrachus acridoides	–	+	–	–	–	–	–	–	–	–	–	–	–	–	–	–	–	–	–	–	–	–	–	–	–	–	–	–
Phrynobatrachus acutirostris	–	–	–	–	+	–	–	–	–	–	–	–	–	–	–	–	–	–	–	–	–	–	–	–	–	–	–	–
Phrynobat. albomarginatus	–	–	–	–	–	–	–	–	–	–	–	–	–	+	–	–	–	–	–	–	–	–	–	–	–	–	–	–
Phrynobatrachus alleni	–	–	–	–	–	–	–	–	–	–	+	–	+	–	+	–	–	–	–	–	–	–	–	–	+	–	–	–
Phrynobatrachus alticola	–	–	–	–	–	–	–	–	–	–	+	+	+	+	+	–	–	–	–	–	–	–	–	–	–	–	–	–
Phrynobatrachus annulatus	–	–	–	–	–	–	–	–	–	–	+	–	+	–	+	–	–	–	–	–	–	–	–	–	+	–	–	–
Phrynobatrachus anotis	–	–	+	–	–	–	–	–	–	–	–	–	–	–	–	–	–	–	–	–	–	–	–	–	–	–	–	–
Phrynobatrachus asper	–	–	–	–	–	+	–	–	–	–	+	–	–	–	+	+	+	–	–	–	–	–	–	–	–	–	–	–
Phrynobatrachus auritus	–	–	–	–	–	–	–	–	–	–	+	–	+	+	+	+	–	–	–	–	–	–	–	–	–	–	–	–
Phrynobatrachus batesii	–	–	–	–	–	–	–	–	–	–	–	–	+	+	+	–	–	–	–	–	–	–	–	–	–	–	–	–
Phrynobatrachus bequaerti	–	–	–	–	–	+	–	–	–	–	+	–	–	–	–	–	–	–	–	–	–	–	–	–	–	–	–	–
Phrynobatrachus bottegi	–	–	–	+	–	–	–	–	–	–	–	–	–	–	–	–	–	–	–	–	–	–	–	–	–	–	–	–
Phrynobatrachus calcaratus	–	–	–	–	–	–	–	–	–	–	+	–	+	+	+	+	–	–	–	–	–	–	–	–	+	–	+	–
Phrynobatrachus congicus	–	–	–	–	–	–	–	–	–	–	–	–	+	+	–	–	–	–	–	–	–	–	–	–	–	–	–	–
Phrynobatrachus cornutus	–	–	–	–	–	–	–	–	–	–	+	–	+	+	+	–	–	–	–	–	–	–	–	–	–	–	+	–
Phrynobatrachus cricogaster	–	–	–	–	–	–	–	–	–	+	–	–	–	–	–	+	–	–	–	–	–	–	–	–	–	–	–	–
Phrynobatrachus cryptotis	–	–	–	–	–	–	–	–	–	–	–	–	–	–	–	–	+	–	+	–	–	–	–	–	–	–	–	–
Phrynobatrachus dalcqi	–	–	–	–	–	+	–	–	–	–	+	–	–	–	–	–	–	–	–	–	–	–	–	–	–	–	–	–
Phrynobatrachus dendrobates	–	–	–	–	–	+	–	–	–	–	+	–	–	–	–	–	–	–	–	–	–	–	–	–	–	–	–	–
Phrynobatrachus elberti	+	–	–	–	–	–	–	–	–	–	–	–	–	–	–	–	–	–	–	–	–	–	–	–	–	–	–	–
Phrynobatrachus francisci	+	–	–	–	–	–	–	–	–	–	–	–	–	–	–	–	–	–	–	–	–	–	–	–	–	–	–	–
Phrynobatrachus fraterculus	–	–	–	–	–	–	–	–	–	–	–	–	+	–	+	+	–	–	–	–	–	–	–	–	–	–	–	–
Phrynobatrachus gastoni	–	–	–	–	–	–	–	–	–	–	–	–	+	–	–	–	–	–	–	–	–	–	–	–	–	–	–	–
Phrynobatrachus ghanensis	–	–	–	–	–	–	–	–	–	–	–	–	+	+	+	–	–	–	–	–	–	–	–	–	–	–	–	–
Phrynobatrachus giorgii	–	–	–	–	–	–	–	–	–	–	–	–	–	–	+	–	–	–	–	–	–	–	–	–	–	–	–	–
Phrynobatrachus graueri	–	–	–	–	–	+	–	–	–	–	+	–	+	–	+	–	–	–	–	–	–	–	–	–	–	–	–	–
Phrynobatrachus guineensis	–	–	–	–	–	–	–	–	–	–	–	–	+	–	+	–	–	–	–	–	–	–	–	–	–	–	–	–
Phrynobatrachus gutturosus	–	–	–	–	–	–	–	–	–	–	–	–	+	–	+	–	–	–	–	–	–	–	–	–	–	–	–	–
Phrynobatrachus hylaios	–	–	–	–	–	–	–	–	–	–	–	–	–	+	–	–	–	–	–	–	–	–	–	–	–	–	–	–
Phrynobatrachus keniensis	–	–	–	+	–	–	–	–	–	–	–	–	–	–	–	–	–	–	–	–	–	–	–	–	–	–	–	–
Phrynobatr. kinangopensis	–	–	–	+	–	–	–	–	–	–	–	–	–	–	–	–	–	–	–	–	–	–	–	–	–	–	–	–
Phrynobatrachus krefftii	–	–	–	–	–	–	+	–	–	–	–	–	–	–	–	–	–	–	–	–	–	–	–	–	–	–	–	–
Phrynobatrachus liberiensis	–	+	–	–	–	–	–	–	–	–	–	–	+	–	+	–	–	–	–	–	–	–	–	–	–	–	–	–
Phrynobatrachus mababiensis	–	–	–	–	–	–	–	–	–	–	–	–	–	–	–	–	–	–	+	–	+	+	–	–	+	–	–	–
Phrynobat. manengoubensis	–	–	–	–	–	–	–	–	–	+	–	–	–	–	–	–	–	–	–	–	–	–	–	–	–	–	–	–
Phrynobatrachus minutus	–	–	+	–	–	–	–	–	–	–	–	–	–	–	–	–	–	–	–	–	–	–	–	–	–	–	–	–
Phrynobatrachus nanus	+	–	–	–	–	–	–	–	–	–	–	–	–	–	–	–	–	–	–	–	–	–	–	–	–	–	–	–
Phrynobatrachus natalensis	+	–	–	–	+	–	–	+	+	–	–	–	–	+	–	–	+	+	+	–	+	+	+	–	+	+	+	–
Phrynobatrachus ogoensis	–	–	–	–	–	–	–	–	–	–	–	–	–	–	–	–	–	–	–	–	–	–	–	+	–	–	–	–
Phrynobatrachus packenhami	–	–	–	–	–	–	–	–	–	–	–	–	–	–	–	–	–	–	–	+	–	–	–	–	–	–	–	–

Appendix 9:1 continued

Taxon	SUD	NEA	EAH	M.E	M.K	M.R	M.T	M.M	M.Z	M.A	M.C	M.G	WEQ	W.W	W.C	W.E	W.S	SZB	EAL	ECF	SEL	ZAZ	MST	SWA	SCE	STE	S.H	S.C
Phrynobatrachus parkeri	E	–	–	–	–	–	–	–	–	–	–	–	–	–	–	–	–	–	–	–	–	–	–	–	–	–	–	–
Phrynobatrachus parvulus	–	–	–	–	–	+	+	+	+	+	–	–	–	–	–	–	–	–	–	–	–	+	+	–	–	–	–	–
Phrynobat. perpalmatus	+	–	–	–	–	–	–	–	–	–	–	–	–	–	–	–	–	–	–	–	+	+	–	–	–	–	–	–
Phrynobat. petropedetoides	–	–	–	–	+	–	–	–	–	–	–	–	–	–	–	–	–	–	–	–	–	–	–	–	–	–	–	–
Phrynobatrachus plicatus	–	–	–	–	–	–	–	–	–	–	–	+	–	–	–	–	–	–	–	–	–	–	–	–	–	–	–	–
Phrynobatrachus pygmaeus	+	–	–	–	–	+	–	–	–	–	–	–	–	–	–	–	–	–	–	–	+	–	–	–	–	–	–	–
Phrynobatrachus rouxi	–	–	–	–	–	+	–	–	–	–	–	–	–	–	–	–	–	–	–	–	–	–	–	–	–	–	–	–
Phrynobatrachus rungwensis	–	–	–	–	–	–	–	+	–	–	–	–	–	–	–	–	–	–	–	+	+	+	+	–	–	–	–	–
Phrynobatrachus scapularis	E	–	–	–	–	–	–	–	–	–	–	–	–	–	–	–	–	–	–	–	–	–	+	+	–	–	–	–
Phrynobat. sciangallarum	–	–	–	+	–	–	–	–	–	–	–	–	–	–	–	–	–	–	–	–	–	–	–	–	–	–	–	–
Phrynobat. steindachneri	–	–	–	–	–	–	–	–	–	–	–	–	–	+	–	–	–	–	–	–	–	–	–	–	–	–	–	–
Phrynobat. stewartae	–	–	–	–	–	–	+	–	–	–	–	–	–	–	–	–	–	–	–	–	–	–	–	–	–	–	–	–
Phrynobat. sulfureogularis	–	–	–	–	–	+	–	–	–	–	–	–	–	–	–	–	–	–	–	–	–	–	–	–	–	–	–	–
Phrynobatrachus taiensis	–	–	–	–	–	–	–	–	–	–	–	–	+	–	–	–	–	–	–	–	–	–	–	–	–	–	–	–
Phrynobatrachus tellinii	E	–	–	–	–	–	–	–	–	–	–	–	–	–	–	–	–	–	–	–	–	–	–	–	–	–	–	–
Phrynobatrachus tobka	–	–	–	–	–	–	–	–	–	–	–	–	+	–	–	–	–	–	–	–	–	–	–	–	–	–	–	–
Phrynobatrachus ukingensis	–	–	–	–	–	+	+	–	–	–	–	–	–	–	–	–	–	–	–	–	–	–	–	–	–	–	–	–
Phrynobat. uzungwensis	–	–	–	–	–	+	–	–	–	–	–	–	–	–	–	–	–	–	–	–	–	–	–	–	–	–	–	–
Phrynobatrachus versicolor	–	–	–	–	–	+	+	–	–	–	–	–	–	–	–	–	–	–	–	–	–	–	–	–	–	–	–	–
Phrynobatrachus villiersi	–	–	–	–	–	–	–	–	–	–	–	–	+	–	+	–	–	–	–	–	–	–	–	–	–	–	–	–
Phrynobatrachus vogti	–	–	–	–	–	–	–	–	–	–	–	–	+	–	+	–	–	–	–	–	–	–	–	–	–	–	–	–
Phrynobatrachus werneri	–	–	–	–	–	–	–	–	–	–	+	–	–	–	–	–	–	–	–	–	–	–	–	–	–	–	–	–
Phrynobatrachus zavattarii	–	–	–	+	–	–	–	–	–	–	–	–	–	–	–	–	–	–	–	–	–	–	–	–	–	–	–	–
Phrynodon sandersoni	–	–	–	–	–	–	–	–	–	–	–	–	–	+	–	–	–	–	–	–	–	–	–	–	–	–	–	+
Poyntonia paludicola	–	–	–	–	–	–	–	–	–	–	–	–	–	–	–	+	+	–	–	–	–	–	–	–	–	–	–	–
Ptychadena aequiplicata	–	+	–	–	–	–	–	–	–	–	–	–	+	+	–	+	–	–	–	–	–	–	+	–	–	–	–	–
Ptychadena anchietae	E	–	+	–	–	–	–	–	–	–	–	–	–	–	–	+	–	–	+	–	+	+	+	+	–	–	–	–
Ptychadena ansorgii	–	+	–	–	–	–	–	–	–	–	–	–	–	–	–	–	–	–	–	–	–	–	+	–	–	–	–	–
Ptychadena broadleyi	–	–	–	–	–	–	+	–	–	–	–	–	–	–	–	–	–	–	–	–	–	–	–	–	–	–	–	–
Ptychadena bunoderma	–	–	–	–	–	–	–	–	–	–	–	–	–	–	–	–	–	–	–	–	+	–	–	–	–	–	–	–
Ptychadena christyi	–	–	–	–	–	–	–	–	–	–	–	–	–	–	+	–	–	–	–	–	–	–	–	–	–	–	–	–
Ptychadena chrysogaster	–	–	–	–	–	+	–	–	–	–	–	–	–	–	–	–	–	–	–	–	–	–	–	–	–	–	–	–
Ptychadena cooperi	–	–	–	+	–	–	–	–	–	–	–	–	–	–	–	–	–	–	–	–	–	–	–	–	–	–	–	–
Ptychadena erlangi	–	–	–	+	–	–	–	–	–	–	–	–	–	–	–	–	–	–	–	–	–	–	–	–	–	–	–	–
Ptychadena floweri	+	–	–	–	–	+	–	–	–	–	–	–	–	–	–	–	–	–	–	–	–	–	–	–	–	–	–	–
Ptychadena grandisonae	–	–	–	–	–	–	–	–	–	–	–	–	–	–	–	–	–	–	–	–	–	+	+	–	–	–	–	–
Ptychadena guibei	E	–	–	–	–	–	–	–	–	–	–	–	–	–	–	–	–	–	+	–	–	+	+	+	–	–	–	–
Ptychadena ingeri	–	–	–	–	–	–	–	–	–	–	–	–	–	–	–	–	–	–	–	–	–	–	–	–	–	–	–	–
Ptychadena keilingi	–	–	–	–	–	–	–	–	–	–	–	–	–	–	–	–	–	–	–	–	–	+	+	–	–	–	–	–
Ptychadena longirostris	W	–	–	–	–	–	–	–	–	–	–	–	+	–	–	–	–	–	–	–	–	–	–	–	–	–	–	–

Appendix 9:1 continued

Taxon	SUD	NEA	EAH	M.E	M.K	M.R	M.T	M.M	M.Z	M.A	M.C	M.G	WEQ	W.W	W.C	W.E	W.S	SZB	EAL	ECF	SEL	ZAZ	MST	SWA	SCE	STE	S.H	S.C
Ptychadena maccarthyensis	+	–	–	–	–	–	–	–	–	–	–	–	+	–	–	–	–	–	–	–	–	–	–	–	–	–	–	–
Ptychadena mapacha	–	–	–	–	–	–	–	–	–	–	–	–	–	–	–	–	–	–	–	–	–	+	–	–	–	–	–	–
Ptychadena mascareniensis	+	–	+	–	–	–	–	–	–	–	–	–	+	–	–	–	+	–	+	–	–	+	–	–	–	–	–	–
Ptychadena mossambica	–	–	–	+	–	–	–	–	–	–	–	–	+	–	–	–	–	+	+	–	+	–	–	–	–	+	–	–
Ptychadena nana	–	+	–	–	–	–	–	–	–	–	–	–	–	–	–	–	–	–	–	–	–	–	–	–	–	–	–	–
Ptychadena neumanni	–	+	–	–	–	–	–	–	–	–	–	–	–	–	–	–	–	–	–	–	–	–	–	–	–	–	–	–
Ptychadena obscura	–	–	–	–	–	–	–	–	–	–	–	–	–	–	+	–	–	+	–	+	+	+	–	–	–	–	–	–
Ptychadena oxyrhynchus	+	–	+	+	+	+	+	+	+	+	+	+	+	+	+	+	+	+	+	+	+	+	–	–	–	–	–	–
Ptychadena perplicata	–	–	–	–	–	–	–	–	+	–	–	–	–	–	+	–	–	–	–	–	–	–	–	–	–	–	–	–
Ptychadena perreti	–	–	–	–	–	–	–	–	–	–	–	–	–	+	+	+	+	–	–	+	–	–	–	–	–	–	–	–
Ptychadena porosissima	–	–	–	+	+	+	+	+	+	–	–	–	–	–	–	–	+	+	+	+	+	+	–	–	–	+	+	+
Ptychadena pumilo	+	–	–	–	–	–	–	–	–	–	–	–	–	–	–	–	–	–	–	–	–	–	–	–	–	–	–	–
Ptychadena retropunctata	–	–	+	–	–	–	–	–	–	–	–	+	–	–	–	–	–	–	–	–	–	–	–	–	–	–	–	–
Ptychadena schillukorum	+	–	+	–	–	–	–	–	–	–	–	–	+	–	–	–	+	–	+	–	–	–	–	–	–	–	–	–
Ptychadena schubotzi	+	–	–	–	–	–	–	–	–	–	–	–	–	–	–	–	–	–	–	–	–	–	–	–	–	–	–	–
Ptychadena stenocephala	+	–	–	–	–	–	–	–	–	–	–	–	–	–	–	–	–	–	–	–	–	–	–	–	–	–	–	–
Ptychadena straeleni	+	–	–	–	–	–	–	–	–	–	–	–	–	–	–	–	–	–	–	–	–	–	–	–	–	–	–	–
Ptycha. submascareniensis	W	–	–	–	–	–	–	–	–	+	–	+	–	–	–	–	–	+	–	+	+	–	–	–	–	–	–	–
Ptychadena subpunctata	–	–	–	–	–	–	–	–	–	–	–	–	–	–	–	–	–	+	–	–	+	–	–	–	–	–	–	–
Ptychadena superciliaris	–	–	–	–	–	–	–	–	–	–	–	–	+	–	–	–	–	–	–	+	–	–	–	–	–	–	–	–
Ptychadena taenioscelis	+	–	–	–	–	–	–	–	–	–	–	–	–	–	–	–	–	+	–	+	+	+	–	–	–	–	–	–
Ptychadena tournieri	W	–	–	–	–	–	–	–	+	–	–	+	–	–	–	–	–	–	–	+	–	–	–	–	–	–	–	–
Ptychadena trinodis	+	–	–	–	–	–	–	–	–	–	–	–	–	–	–	–	–	–	–	–	–	–	–	–	–	–	–	–
Ptychadena upembae	–	–	–	–	–	–	–	–	–	–	–	–	–	–	–	–	–	+	–	+	+	+	–	–	–	–	–	–
Ptychadena uzungwensis	+	–	+	–	–	+	+	+	–	–	–	–	–	–	–	–	–	+	+	+	+	+	+	–	+	–	+	–
Pyxicephalus adspersus	–	–	–	–	–	–	–	–	–	–	–	–	–	–	–	–	–	+	+	+	+	+	+	+	+	–	+	–
Pyxicephalus edulis	–	–	–	–	–	–	–	–	–	–	–	–	–	–	–	–	–	–	–	–	–	–	–	–	–	–	–	–
Pyxicephalus obbianus	+	+	–	–	–	–	–	–	–	–	–	–	–	–	–	–	–	–	–	–	–	–	–	–	–	–	–	–
Rana amieti	–	–	–	–	–	–	–	–	–	–	–	–	–	–	–	–	–	+	–	+	+	+	–	–	+	–	+	–
Rana angolensis	+	+	+	+	+	+	+	+	+	+	–	–	–	–	–	–	–	+	–	+	+	+	–	–	+	–	+	–
Rana desaegeri	–	–	–	–	–	–	–	–	–	–	–	–	–	–	–	–	–	–	–	–	–	–	–	–	–	–	–	–
Rana dracomontana	–	–	–	–	–	–	–	–	–	–	–	–	–	–	–	–	–	–	–	–	–	–	–	–	+	+	+	+
Rana fuscigula	+	–	–	–	+	+	+	+	–	–	–	+	–	–	–	–	–	–	–	–	–	–	–	–	+	+	+	+
Rana johnstoni	–	–	–	–	+	+	+	+	+	–	–	–	–	–	–	–	–	–	–	–	–	–	–	–	+	–	–	–
Rana ruwenzorica	–	–	–	+	–	–	–	–	–	–	–	–	–	–	–	–	–	–	–	–	–	–	–	–	–	–	+	–
Rana vertebralis	–	–	–	–	–	–	–	–	–	–	–	–	–	–	–	–	–	–	–	–	–	–	–	–	–	–	+	+
Rana wittei	–	–	–	+	+	+	+	+	+	–	–	–	–	–	–	–	–	–	–	–	–	–	–	–	–	–	–	–
Strongylopus bonaespei	–	–	–	–	–	–	–	–	–	–	–	–	–	–	–	–	–	–	–	–	–	–	–	–	–	–	+	+
Strongylopus fasciatus	–	–	–	–	+	+	+	+	+	–	–	–	–	–	–	–	–	–	+	+	+	+	+	–	+	+	+	+
Strongylopus grayii	–	–	–	–	–	–	–	+	+	–	–	–	–	–	–	–	–	–	+	+	+	–	+	+	+	+	+	+

Appendix 9:1 continued

Taxon	SUD	NEA	EAH	M.E	M.K	M.R	M.T	M.M	M.Z	M.A	M.C	M.G	WEQ	W.W	W.C	W.E	W.S	SZB	EAL	ECF	SEL	ZAZ	MST	SWA	SCE	STE	S.H	S.C
Strongylopus hymenopus	–	–	–	–	–	–	–	–	–	–	–	–	–	–	–	–	–	–	–	–	–	–	–	–	–	–	–	–
Strongylopus springbokensis	–	–	–	–	–	–	–	–	–	–	–	–	–	–	–	–	–	–	–	–	–	–	–	+	–	–	+	–
Strongylopus wageri	+	+	–	–	–	–	–	–	–	–	–	–	–	–	–	–	–	–	–	–	–	–	–	–	+	–	+	–
Tomopterna cryptotis	+	+	+	–	–	–	–	–	–	–	–	–	–	–	–	–	–	–	–	–	+	+	+	+	+	–	+	–
Tomopterna delalandii	–	–	–	–	–	–	–	–	–	–	–	–	–	–	–	–	–	–	–	–	–	–	–	–	–	–	–	+
Tomopterna krugerensis	–	–	–	–	–	–	–	–	–	–	–	–	–	–	–	–	–	–	+	–	+	–	+	+	–	–	–	–
Tomopterna marmorata	–	+	–	–	–	–	–	–	–	–	–	–	–	–	–	–	–	–	+	–	+	+	+	–	+	–	+	–
Tomopterna natalensis	–	–	–	–	–	–	–	–	–	–	–	–	–	–	–	–	–	–	–	+	+	–	+	–	–	–	–	–
Tomopterna tuberculosa	–	–	–	–	–	–	–	–	–	–	–	–	–	–	–	–	–	–	–	–	+	+	+	–	–	–	–	–
ANURA: RHACOPHORIDAE:																												
Chiromantis petersii	–	+	–	–	–	–	–	–	–	–	–	–	–	–	–	–	–	–	–	–	–	–	–	–	–	–	–	–
Chiromantis rufescens	–	–	–	–	–	–	–	–	–	–	–	–	–	–	+	+	–	–	–	–	–	–	–	–	–	–	–	–
Chiromantis xerampelina	–	+	+	–	–	–	–	–	–	–	–	–	–	–	–	–	–	–	+	–	+	+	–	–	–	–	–	–
GYMNOPHIONA: CAECILIIDAE:																												
Boulengerula boulengeri	–	–	–	–	–	+	–	–	–	–	–	–	–	–	–	–	–	–	–	–	–	–	–	–	–	–	–	–
Boulengerula changamwensis	–	–	–	–	–	–	+	–	–	–	–	–	–	–	–	–	–	–	+	–	–	–	–	–	–	–	–	–
Boulengerula fischeri	–	–	–	–	–	+	–	–	–	–	–	–	–	–	–	–	–	–	–	–	–	–	–	–	–	–	–	–
Boulengerula taitanus	–	–	–	–	–	+	–	–	–	–	–	–	–	–	–	–	–	–	–	–	–	–	–	–	–	–	–	–
Boulengerula uluguruensis	–	–	–	–	–	+	–	–	–	–	–	–	–	–	–	–	–	–	–	–	–	–	–	–	–	–	–	–
Geotrypetes angeli	W	–	–	–	–	–	–	–	–	–	–	–	–	–	–	–	–	–	–	–	–	–	–	–	–	–	–	–
Geotrypetes pseudoangeli	–	–	–	–	–	–	–	–	–	–	–	–	–	–	–	+	–	–	–	–	–	–	–	–	–	–	–	–
Geotrypetes seraphini	–	–	–	–	–	–	–	–	–	–	–	–	–	–	+	+	–	–	–	–	–	–	–	–	–	–	–	–
Herpele multiplicata	–	–	–	–	–	–	–	–	–	–	–	–	–	+	+	–	–	–	–	–	–	–	–	–	–	–	–	–
Herpele squalostoma	–	–	–	–	–	–	–	–	–	–	–	–	–	+	+	+	+	–	–	–	–	–	–	–	–	–	–	–
Idiocranium russelli	–	+	–	–	–	–	–	–	–	–	+	–	–	–	–	–	–	–	–	–	–	–	–	–	–	–	–	–
Schistometopum garzonheydti	–	–	–	–	–	–	–	–	–	–	–	–	–	–	+	–	–	–	–	–	–	–	–	–	–	–	–	–
Schistometopum gregorii	–	–	–	–	–	–	–	–	–	–	–	–	–	–	–	–	–	–	+	–	–	–	–	–	–	–	–	–
Sylvacaecilia grandisonae	–	–	+	–	–	–	–	–	–	–	–	–	–	–	–	–	–	–	–	–	–	–	–	–	–	–	–	–
GYMNOPHIONA: SCOLECOMORPHIDAE:																												
Crotaphatrema bornmuelleri	–	–	–	–	–	–	–	–	–	–	–	–	–	–	+	–	–	–	–	–	–	–	–	–	–	–	–	–
Crotaphatrema lamottei	–	–	–	–	–	–	–	–	–	–	+	–	–	–	–	–	–	–	–	–	–	–	–	–	–	–	–	–
Scolecomorphus kirkii	–	–	–	–	–	+	–	–	–	–	–	–	–	–	–	–	–	–	–	–	–	–	–	–	–	–	–	–
Scolecomorphus uluguruensis	–	–	–	–	–	+	–	–	–	–	–	–	–	–	–	–	–	–	–	–	–	–	–	–	–	–	–	–
Scolecomorphus vittatus	–	–	–	–	–	+	–	–	–	–	–	–	–	–	–	–	–	–	–	–	–	–	–	–	–	–	–	–

ADDENDUM

With anuran higher classification continuing to be in a chaotic state, the definition of the amphibian fauna remains blurred, and many of the perceived patterns of distribution remain unclear or unsure. Uncertainty prevails especially with the systematics of microhylids, rhacophorids, and ranids.

As noted in the text (p. 494), Blommers-Schlösser proposed excluding the Brevicipitinae from the Microhylidae and placing it with *Hemisus* in a newly defined Hemisotidae. Channing (1995) reanalyzed Blommers-Schlösser's data in a comparison of *Hemisus* and *Breviceps;* after listing several points of difference between these two genera, he suggested that placing them in the same family was "premature." As the Phrynomerinae (retained in the Microhylidae by Blommers-Schlösser) was not considered, the significance of the differences listed by Channing can hardly be evaluated. Some are questionable when the condition in *Probreviceps* is reinvestigated (in preparation); for example the structure of the snout in *P. rungwensis* indicates forward burrowing (Channing's Character 20) at least for foraging in forest litter. This situation reinforces Channing's conclusion that, "The determination of the relationships between these two enigmatic genera must await a thorough analysis of the Microhylidae and related frogs." In the meantime, biogeographical aspects relating to Madagascar and Africa must remain in suspense.

A similar conclusion was reached by Inger (1996) in a commentary on Dubois's classification of the Ranidae (discussed on p. 493). Herein, Dubois's classification was adopted partially with reservations; Inger (1996:245) believed that acceptance of the classification "even on a provisional basis is likely to lead to confusion rather than understanding." He advocated a broader and more thorough analysis of the Ranidae.

The constitution of the Ranidae vis-à-vis the Rhacophoridae is in an unsettled state in the recent literature. Apart from the problematic case of *Ptychadena, Tomopterna* was treated in the foregoing text as the only genus to be shared between Africa and Madagascar. In a phylogenetic study using mitochondrial DNA sequences, Richards and Moore (1998) cast doubt on the conventional treatment of the Madagascan *Tomopterna labrosa* as a ranid, and provided evidence of relationship with the rhacophorids *Aglyptodactylus* and *Boophis.* Using a wide range of characters, Glaw et al. (1998) found that *T. labrosa* is not closely related to the African and Asian *Tomopterna,* and provisionally placed it in a separate, new subgenus, *Sphaerotheca.* These authors found *Aglyptodactylus* to be closely related to *T. labrosa,* but less so to *Boophis.* Whether *Aglyptodactylus* should be included in the Ranidae or *T. labrosa* be included in the Rhacophoridae seems at the moment to be an open question that can be resolved only when the current classification has been reformed (M. Vences, pers. comm.). In the meantime, it seems doubtful whether any genus apart from *Ptychadena* can be regarded as common to Madagascar and Africa.

In a morphometric study of the Ptychadenini, Laurent (1997) found no significant subspecific differences between African and non-African populations of *Ptychadena mascareniensis,* despite some expectations to the contrary. However, Glaw and Vences (1994:186) reported that the calls of the Madagascan specimens "differ partly" from those recorded in South Africa. They found it to be "probably the most abundant Malagasy frog, present in every field and everywhere near water outside the forest." Showing all the characteristics of a weed, it is still not clear how this frog became so widely distributed over the Indian Ocean. Laurent identified a plesiomorphic group within *Ptychadena* composed of *P. newtoni* from São Tomé Island and five species from northeastern Africa; *P. newtoni* was considered by him to be an isolated relict and the northeastern species to be "stragglers left behind at the beginning of the African invasion from Arabie" (Laurent, 1997:183).

As noted at the beginning of this chapter, the family Hyperoliidae gives Sub-Saharan Africa, Madagascar, and the Seychelles a certain unity; this unity is confirmed by Richards and Moore's (1996) phylogeny for the family based on mitochondrial rDNA. *Heterixalus* of Madagascar and *Tachycnemis* of the Seychelles are grouped in one clade; the sister group is *Afrixalus,* and the sister group of these is *Hyperolius.* Richards and Moore's (1998) phylogenetic tree confirms earlier work that the African rhacophorid *Chiromantis* tends to group with the Asian rhacophorids, whereas the Madagascan rhacophorids form a separate clade.

The work on Madagascan amphibians by Blommers-Schlösser and Blanc (1991, 1993) has been followed by more popular yet comprehensive coverages by Glaw and Vences (1994) and Henkel and Schmidt (1995). Recognition of several new species raises the number of species listed by Glaw and Vences to 168 (given as 156 on p. 494). In a treatment of the west African savanna fauna, Rödel (1996) recognized 51 species. He listed the bullfrog as *Pyxicephalus edulis,* not *P. adspersus* as in the foregoing appendix; his description indicates that his identification is correct. Considering the inadequate understanding of the difficult genus *Phrynobatrachus,* there might be some relief that Rödel found a need to recognize only two undescribed species and to resurrect only one name (with some uncertainty), *P. latifrons* Ahl as a savanna form of *P. accraensis. Hyperolius nitidulus* Peters was treated as a full species, not counted as a subspecies of *viridiflavus* as in the foregoing appendix.

The number of species recognized in the revised edition of *South African Frogs* (Passmore and Carruthers, 1995) has risen to 98 from 84 in the 16-year interval since the publication of the first edition. This trend seems likely to continue and indicates the incomplete state of exploration and revision, even in this relatively well-studied part of Africa. Complex nomenclatural problems still remain, such as the name of the southwestern Cape toad listed as *Bufo cruciger* herein; the correct name seems to be *B. pantherinus* A. Smith, 1828 (Poynton and Lambiris, in press).

The genus *Xenopus* has received comprehensive treatment in a volume edited by Tinsley and Kobel (1996). *Silurana* is treated there as a subgenus. A marked altitudinal separation is shown by the 17 recognized species of *Xenopus;* about half are upland and half lowland (Kobel et al., 1996). On the basis of mtDNA restriction analysis, Grohovaz et al. (1996) found that populations of *X. laevis* from the southwestern Cape differed markedly from populations from other parts of South Africa. They noted that the dichotomy shown by types of mtDNA between the southwestern Cape and the rest of the subcontinent is "consistent with the broader categorization of two faunal groups of amphibians in southern Africa" (Grohovaz et al., 1996:247). The displaced breeding season of *X. laevis* in the southwestern Cape, caused by climatic differences, was seen to be involved in this dichotomy.

Of the 98 species listed by Passmore and Carruthers (1995), 50 are endemic to South Africa. Endemicity is evident in all nine families represented, except Rhacophoridae; the distinctiveness of the South African fauna evidently is related to the large area of nontropical climate, as suggested by the fact that no endemics occur in the northeastern corner of the country where a tropical climate prevails.

Faunal differences between the East African and West African savannas (see p. 510) have been emphasized further by Kobel et al. (1996), who separated *Xenopus muelleri* into an eastern form from southeastern Kenya southward and a western form from the southern Sudan westward. However, a study of the Tanzanian torrent frog *Arthroleptides martiensseni* by Klemens (1998) revealed similarities with West African *Petropedetes* that makes it highly doubtful whether two genera can be accepted. This lessens the apparent difference between East and West African amphibian forest faunas discussed herein.

Nevertheless, the existence of distinct tropical and temperate assemblages in the African amphibian fauna continues to emerge in biogeographical analyses. A transect taken from the east coast in the region of Maputo, southern Mozambique, to the inland plateau (Poynton and Boycott, 1996) revealed the same general pattern shown by two previous transects (see p. 503), namely a major east to west species turnover. Species similarity between the two ends of the transect was only 17%. The coherent pattern of lowland to highland transition confirms the definition and demarcation of a lowland Afrotropical assemblage and a highland Afrotemperate assemblage in southeastern Africa.

A distributional analysis of the amphibian fauna in the Democratic Republic of the Congo (formerly Zaire) (Poynton, in press) supports the perception of a massive east-to-west faunal turnover across the continent. A comparison of species composition in the lower Congo area and the eastern highland border gave a similarity value of only 7%. Comparison of this highland fauna with the East African lowland fauna gave a similarity value of a mere 3.3%. These values emphasize the distinctness of the highland Afrotemperate fauna from the Afrotropical faunas on the lowlands to either side. Differences in the highland and lowland amphibian faunas of Ethiopia were outlined by Largen (1998).

All six genera endemic to Ethiopia are montane, as are some 19 of the 23 endemic species.

The unity of the Seychelles amphibian fauna was emphasized in a study of *Tachycnemis seychellensis* by Nussbaum and Sheng Hai Wu (1995). Material was examined from four granitic islands, representing populations separated by marine transgressions dating from 10,000 yr B.P. Despite some complex divergent trends that were evident, the authors considered the material to represent a single species.

Amphibian species and subspecies that have been described between 1993 (the time of publication of the revised edition of *Amphibian Species of the World* [Duellman, 1993b]) and December 1997 were listed by Glaw et al. (1998). Fifteen full species (more than half of them ranids) were listed for Sub-Saharan Africa and 23 for Madagascar (more than half of them rhacophorids as currently understood). These authors believe that "the biggest burst of amphibian species discoveries is not yet reached" (Glaw et al., 1998:2). This could well be true of the situation in Sub-Saharan Africa. The incompleteness of taxonomic and distributional work in Africa was highlighted in a preliminary paper on Tanzanian bufonid diversity (Poynton, 1998), which suggested the presence in Tanzania of an undescribed genus and five (18%) undescribed species of the 28 species listed.

LITERATURE CITED

CHANNING, A. 1995. The relationship between *Breviceps* (Anura: Microhylidae) and *Hemisus* (Hemisotidae) remains equivocal. J. Herpetol. Assoc. Africa 44:55–57.

GLAW, F., J. KÖHLER, S. LOTTERS, AND M. VENCES. 1998. Preliminary list and references of newly described amphibian species and subspecies between 1993 and 1997. Elaphe 6:1–24.

GLAW, F., AND M. VENCES. 1994. *A Fieldguide to the Amphibians and Reptile of Madagascar* (2nd ed.). Bonn: Zoologisches Forschungsinstitut und Museum Alexander Koenig.

GLAW, F., M. VENCES, AND W. BÖHME. 1998. Systematic revision of the genus *Aglyptodactylus* Boulenger, 1919 (Amphibia: Ranidae), and analysis of its phylogenetic relationships to other Madagascan ranid genera (*Tomopterna, Boophis, Mantidactylus*, and *Mantella*). J. Zool. Syst. Evol. Res. 36:17–37.

GROHOVAZ, G. S., E. HARLEY, AND B. FABIAN. 1996. Significant mitochondrial DNA sequence divergence in natural populations of *Xenopus laevis* (Pipidae) from southern Africa. Herpetologica 52:247–253.

HENKEL, F.-W., AND W. SCHMIDT. 1995. *Amphibien und Reptilien Madagaskars der Maskarenen, Seychellen und Komoren.* Stuttgart: Verlag Eugen Ulmer.

INGER, R. F. 1996. Commentary on a proposed classification of the family Ranidae. Herpetologica 52:241–246.

KLEMENS, M. K. 1998. The male nuptial characteristics of *Arthroleptides martiensseni* Nieden, an endemic torrent frog from Tanzania's eastern arc mountains. Herpetol. J. 8:35–40.

KOBEL, H. R., C. LOUMONT, AND R. C. TINSLEY. 1996. The extant species. Pp. 9–33 *in* R. C. Tinsley & H. R. Kobel (eds.), *The Biology of* Xenopus. Oxford: Zoological Society of London.

LARGEN, M. J. 1998. A preliminary review of the amphibians of Ethiopia. Herpetol. J. 8:7–12.

LAURENT, R. F. 1997. Morphometric approach of the evolution of the tribe Ptychadenini. Pp. 183–187 *in* J. H. van Wyk (ed.), *Proceedings of the Fitzsimons Commemorative Symposium and Third H.A.A. Symposium on African Herpetology.* Cape Town: Herpetological Association of Africa.

NUSSBAUM, R. A., AND SHENG HAI WU. 1995. Distribution, variation, and systematics of the Seychelles treefrog, *Tachycnemis seyschellensis* (Amphibia: Anura: Hyperoliidae). J. Zool., Lond. 236:383–406.

PASSMORE, N. I., AND V. C. CARRUTHERS. 1995. *South African Frogs: A Complete Guide.* Revised edition. Johannesburg: Southern Book Publishers and Witwatersrand University Press.

POYNTON, J. C. 1998. Tanzanian bufonid diversity: preliminary findings. Herpetol. J. 8:3–6.

POYNTON, J. C. Gladwyn Kingsley Noble and the study of Africa amphibians. *In* Karl P. Schmidt and G. Kingsley Noble, *Contributions to the Herpetology of the Belgian Congo.* Ithaca, New York: Society for the Study of Amphibians and Reptiles (in press).

POYNTON, J. C., AND R. C. BOYCOTT. 1996. Species turnover between Afromontane and eastern African lowland faunas: patterns shown by amphibians. J. Biogeogr. 23:669–680.

POYNTON, J. C., AND A. J. L. LAMBIRIS. On *Bufo pantherinus* A. Smith, 1828 (Anura: Bufonidae), the Leopard Toad of the southwestern Cape, with the designation of a neotype. African J. Zool. (in press).

RICHARDS, C. M., AND W. S. MOORE. 1996. A phylogeny for the African treefrog family Hyperoliidae based on mitochondrial rDNA. Molec. Phylog. Evol. 5:522–532.

RICHARDS, C. M., AND W. S. MOORE. 1998. A molecular phylogenetic study of the Old World treefrog family Rhacophoridae. Herpetol. J. 8:41–46.

RÖDEL, M.-O. 1996. *Amphibien der westafrikanischen Savanne.* Frankfurt am Main: Edition Chimaira.

TINSLEY, R. C., AND H. R. KOBEL. 1996. *The Biology of* Xenopus. Oxford: Zoological Society of London.

10. Distribution Patterns of Amphibians in the Australo-Papuan Region

MICHAEL J. TYLER

Department of Zoology
University of Adelaide
Adelaide, South Australia 5005
Australia

ABSTRACT   The Australo-Papuan Region contains three distinct herpetofaunal units—Australia, New Guinea and the adjacent islands, and New Zealand. These units were created by different historical biogeographic events. In Australia, subdivision into provinces is largely of dubious scientific value. In particular, the largest area, the vast central Eyrean Province, is a biological dustbin. Patterns of distribution of frogs is unclear, beyond the evidence for montane and lowland elements. Anuran diversity is highest in New Guinea and in northern and southeastern Australia, but within Australia endemism is highest in southern Western Australia. Perhaps only 50% of the fauna has been described, and the situation may be much clearer when knowledge is more advanced. New Zealand has suffered a major and modern extinction of its fauna; this has constrained amphibians to narrow latitudes within the North and South islands. Within Australia, frog diversity is closely correlated with rainfall; only four species are confined to the arid zone. Even though most species of anurans in Australia are protected by regulations, some species have disappeared within the past decade or so.

Key words: Anurans, Distribution patterns, Australo-Papuan Region.

INTRODUCTION

Alfred Russel Wallace (1823–1913) would probably turn in his grave at the prospect of the recognition of a biogeographical term, such as the "Australo-Papuan Region." Nevertheless, his concept of an Australian "Region" comprising an Australian and a Papuan subregion does not address adequately the extensive, and faunally mixed, area of the southwest Pacific including New Zealand in the south and the Bismarck Archipelago, the Solomon Islands, and Fiji to the east (Fig. 10:1). Collectively, these geographic elements are linked somewhat tenuously.

Historically, the events that shaped the frog faunas differ markedly. New Zealand retained its presumed Gondwanan fauna at least since the Cretaceous; this fauna included several more native species until their extinction in the Holocene. New Guinea includes several elements—an early invasion of ranids that ranged east into the Solomon Islands and Fiji, an Australian element derived from the south, and an Oriental one from the west (Tyler, 1979). The Oriental and ranid stocks made minimal impact on Australia by remaining within the northern periphery, thereby leaving the Gondwanan components to speciate and dominate the continent.

The amphibian fauna of the entire region is poorly known. Northwestern Australia is an area that includes many localities not visited in the wet season when frogs are active, whereas New Guinea remains largely unknown. I suspect that the total fauna of New Guinea includes about 400 species, about twice the number known today. It follows that the present analysis is likely to be more deficient than I anticipate.

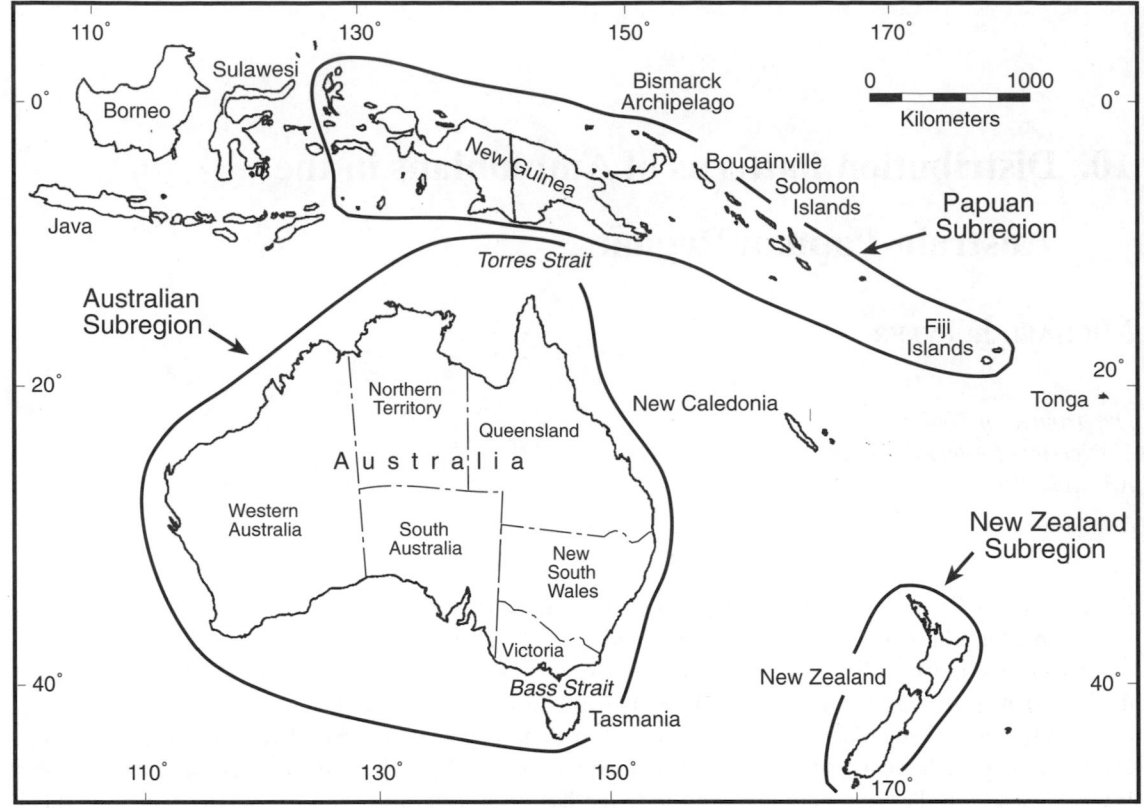

Fig. 10:1.   Geographic area of the Australo-Papuan Region as recognized herein; the subregions are encircled in heavy lines. Islands to the west of the Papuan Subregion are not part of the Australo-Papuan Region.

MATERIALS AND METHODS

The Australo-Papuan Region is defined herein to include the entire area of the southwest Pacific made up of the major landmasses of Australia and New Guinea, and islands associated with them. Farther east the Bismarck Archipelago is included and recognized as a unit separate from Bougainville and the Solomon Islands, which form an archipelago trending northwest-southeast. Other islands in the southwestern portion of the Pacific Ocean have been included if they sustain endemic or introduced species of amphibians.

The inclusion of New Zealand within an Australian-Papuan biogeographic unit presently cannot be argued on the basis of any similarity between its native anuran fauna and those of any other geographic area within the southwest Pacific. Nevertheless, it is reasonable to assume a former affinity, which a more extensive fossil record is likely to confirm. Semantic arguments may persist, but it is convenient

to recognize that this chapter is concerned with a geographic part of the world for which there is no acceptable descriptive name.

Subdivision of the region into meaningful units (subregions) is simplest if they are defined according to shared common features as outlined below and physically delimited in Figure 10:1.

• The Australian Subregion is composed of Australia, Tasmania, and all of the islands in the Torres Strait. The latter are included because their anuran faunas are almost entirely derived from the Australian mainland (Tyler, 1972).

• The Papuan Subregion is delimited on the west near Wallace's Line; this is the distributional limit of the Hylidae and Myobatrachidae, which are the Gondwanan elements of the New Guinean anuran fauna. The Fiji Islands are the eastern outpost of frogs in the southwest Pacific; two species of *Platymantis* (a speciose genus extending as far as the Philippines) occur there. This fact associates the Fiji Islands with the landmasses to the west.

• The New Zealand Subregion clearly is distinct from the other two subregions by virtue of its native anuran fauna consisting of only one endemic family (Leiopelmatidae). Several Australian species have been liberated there with various degrees of success in terms of establishment of successful breeding populations.

The sources of data on distribution include a wide variety of published and unpublished sources. Only those of major significance have been acknowledged. All species in the region are listed in Appendices 10:1 and 10:2.

PHYSICAL COMPONENTS

AUSTRALIA

Although often described as an island continent, Australia is composed of two units: mainland Australia and the associated island of Tasmania. The total land area is 7,682,300 km². The northern and southern boundaries of the continent extend farther than a cursory glance at a map would suggest, reaching north to the New Guinea coastline at 10°41' S and south of Tasmania to 43°39' S.

What sets Australia apart from other continents is not its antiquity but its stability. For most of the past 200 million years, natural forces have worked to erode the land surface, leading to its predominantly flat surface on which elevated areas are highly localized. The overall range of altitude is from −16 m at Lake Eyre to 2228 m at the summit of Mount Kosciusko. It has been calculated that 99% of Australia is below 1000 m (Galloway and Kemp, 1981). The principal zones of elevation are the Kimberley Ranges in the north of Western Australia, Arn-

hem Land in the north of the Northern Territory, the Pilbara in central Western Australia, the Macdonnell Range in Central Australia, the Flinders Ranges in South Australia, and the Great Dividing Range extending along the entire east coast and reaching into Tasmania (Fig. 10:2).

A computer-generated image enhancing the generally low relief of the continent reveals three distinct physiographic features: the Western Plateau, the Interior Lowlands, and the Eastern Uplands (Fig. 10:3). The Western Plateau consists of the Archean and Proterozoic structures, which are now dominated by sand plains and low tablelands. The Interior Lowlands are flat and low in elevation and most commonly made up of Mesozoic and Cenozoic sediments that have been deeply weathered. Here there are vast floodplains. In contrast, the extreme eastern continental component is the younger and elevated Eastern Uplands, which includes the island of Tasmania. The mountains there slope gently to the west

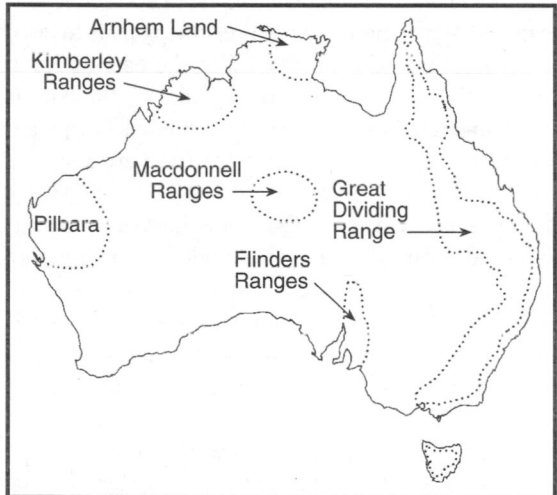

Fig. 10:2. Principal zones of elevation in Australia. Although spatially separated from the Australian mainland, the highlands constituting the greater part of Tasmania are considered to be an isolate of the Great Dividing Range.

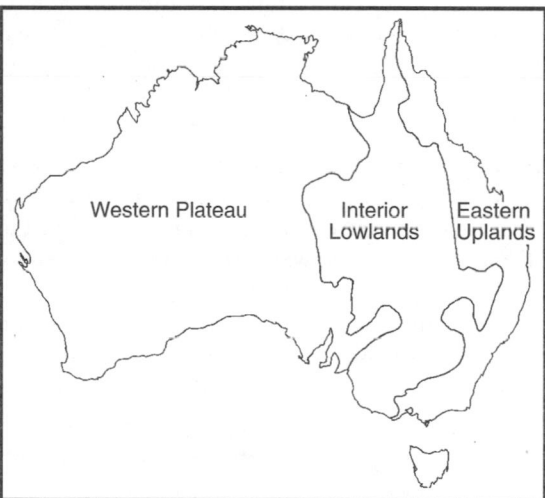

Fig. 10:3. Computer-generated image enhancing the generally low relief of the Australian mainland so as to define three primary physiographic regions (after Macquarie Library, 1996).

and more abruptly to the east where they are bordered by a narrow coastal plain.

Tasmania is the southern extremity of the eastern uplands and involves 67,800 km^2 of the total surface area of Australia. The depth of the Bass Strait separating Tasmania from the mainland has an average depth of about 70 m, and the period of isolation of the island from the mainland is approximately 50,000 yr.

In terms of climate, Australia experiences a variety of regimes. The southern two-thirds of the continent are broadly described as temperate, whereas the northern one-third (north of the Tropic of Capricorn) is tropical. The north of the continent is exposed to heavy rain in summer (a "summer wet") and the extreme south to winter rains involving cold fronts predominantly located south of the landmass at latitude 40° S. The consequence of the peripheral impact of these rains is that the greater portion of the continent at midlatitudes is dry, and the frequency and timing of any rainfall is erratic and unreliable. An exception to this generality is the east coast, which receives moisture derived from airflows from the Pacific Ocean.

There are major seasonal variations in monthly temperature maxima and minima. The hottest part of the continent is the vicinity of Marble Bar in central Western Australia, where daily maxima can reach a minimum of 38° C for several weeks, and during 1923–24 was so maintained for 161 days. The maximum recorded temperature for Australia is 52.8° C at Bourke, New South Wales, in January 1877, and the minimum is –22.2° C at Charlotte Pass in the Southern Alps in 1945 and again in 1947 (Bureau of Meteorology, 1989). By combining data from records of rainfall and temperature, it is possible to arrive at a description of climatic types (Köppen, 1932). This produces a range from hot and moist in the north to cool and wet in the south, separated by a vast area that is hot and dry.

A minimum of seven vegetation formations are detectable in a broad analysis of the natural vegetation of Australia. In terms of dominance by surface area, the most conspicuous components are low forest, scrub, hummock and tussock grasses, and graminoids.

ISLANDS

New Guinea

New Guinea is the second largest island in the

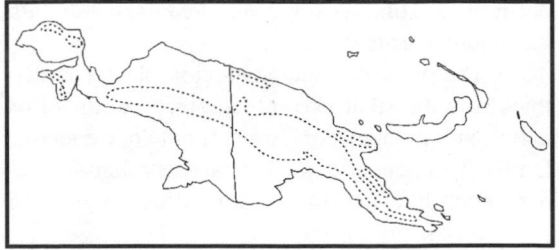

Fig. 10:4. New Guinea showing principal montane areas (enclosed by dotted lines).

world. Separated from Australia by the shallow Torres Strait (commonly no more than 10 m deep), it is physically diverse, young, and tectonically active. It is 2385 km long, up to 720 km wide and has a surface area of approximately 800,000 km^2. The central cordillera has numerous mountain peaks rising above 3000 m (Fig. 10:4). The highest peak is Mt. Jaya (5000 m). Such peaks are capped with ice and snow. The cordillera is flanked on the south and north by extensive lowland plains and swamps bisected by major river systems. Politically, the island of New Guinea is divided into two halves: the western province of Indonesia named West Irian, Irian Barat, or Irian Jaya, and the eastern Papua New Guinea. The latter comprises 462,840 km^2. Physically, New Guinea bears little resemblance to Australia, with the exception of the savanna of the extreme southern lowland portion which is the leading margin of the Australian continental plate. Tectonic activity takes the form of earthquakes and active volcanoes. A particularly active seismic zone is located directly south of New Britain to the east of Huon Peninsula on the mainland. A trench at that site is 9140 m deep. Several hundred earthquakes equal to or greater than 5.5 on the Richter Scale are associated with this New Britain island arc (Denham, 1970).

Climatically, New Guinea is best described as a humid, tropical island with moderate to very high rainfall and minimal seasonal variability (Gressitt, 1982). Its proximity to the Equator is such that winds from both hemispheres converge in the Intertropical Convergence Zone. In particular, winds coming off the Coral Sea rise at the southern flank of the central cordillera and precipitate upon the mountain summits and upper slopes. Despite high sea-surface temperatures in the vicinity, cyclones occur infrequently. There is a high variability in annual rainfall in Papua New Guinea; the annual means are 6076 mm at Gasmata and 995 mm at Port Moresby (Hart,

1970). Rainfall is generally high, with that in more than half of Papua New Guinea exceeding 2450 mm. Hart (1970:42) noted that the highest rainfall areas are "along the southern edge of the central cordillera, from the Gulf of Papua into Irian Barat and along the southern coast of New Britain." A severe drought in 1997 dried moss in rainforest to dust when handled (J. I. Menzies, pers. comm.); the impact on the frog fauna may have been disastrous.

Given the physical structure of New Guinea, it is not surprising that the simplest classification of the vegetation is linked to the vertical distribution of the various associations. At sea level, mangroves give way to dense herbaceous swamps and swamp woodland dominated by sago palm. Savanna in southern Papua merging with the swamp tree (paperbark, *Melaleuca*) and open grassland is remarkably similar to the savanna of adjacent northern Australia (F. Parker, pers. comm.). Rainforest of various types extends to approximately 3000 m, where it is replaced by grassland and, at 3500 m, by mosses, lichens, and stunted grasses (Robbins, 1970; Johns, 1982).

Associated with New Guinea is a large number of islands, many of which are faunistically significant. By far, the most significant of these are the Bismarck Archipelago and, to its southeast, Bougainville and the Solomon Islands.

Bismarck Archipelago

The Bismarck Archipelago consists of an apparently distorted semicircular arc extending from the east of the Huon Peninsula on the mainland and trending initially northeast (New Britain) and then distinctly northwest (New Ireland). Each of these islands is more than 400 km in length. The close proximity of the western end of New Britain to the Huon Peninsula suggests a direct link between the two during periods of lower sea levels; however, the great depth of the Vitiaz Strait separating them precludes this possibility.

New Britain includes most of the active volcanoes in Papua New Guinea; with the latter, these form the "Bismarck Volcanic Belt," which extends along the northern face of the cordillera. This volcanic activity has led to elevation of coral reefs to terraces more than 600 m inland (Brown, 1972). The overall structure of the island is a central mountainous spine surrounded by low-lying areas where there are few places where water can accumulate. The soils are predominantly derived from volcanic material with inherently low water-retention capacity.

Climatically there is a major difference within the 100 km or so distance from the north of New Britain to the south. At the southern margin, the annual mean precipitation is greater than 5080 mm, whereas on the north it is 1500 mm less.

Solomon Islands

The Solomon Islands trend in a northwest-southeast direction from New Ireland and decrease progressively in size. Bougainville is excluded politically (being part of Papua New Guinea), but physically it is the most northern element of that chain and is considered to be part of the Solomon Islands. The total surface area of the group is 28,446 km².

New Caledonia

The island of New Caledonia is by far the largest among several groups of islands located between latitudes 18° and 23° S and longitudes 163° and 129° E. Located approximately 1500 km east of Australia, New Caledonia has a surface area of 19,000 km². A physical environment of rugged mountains and a subtropical climate with annual rainfall up to 3000 mm is ideal for frogs, but there are no native species. The island is included only because *Litoria aurea* has been introduced there.

Fiji Islands

The Fiji Islands form an archipelago surrounding the Koro Sea about 2100 km north of New Zealand. The archipelago consists of about 300 islands scattered over 3 million km²; the total land area is only 18,272 km². Derived from volcanic activity, coral formation, and sedimentation, the largest island (Viti Levu) has several peaks exceeding 1000 m, numerous rivers, and average annual rainfall of 1500–3000 mm.

New Zealand

New Zealand consists of two large islands and a number of small islands; the total surface area is 269,057 km². Topographically, the land is a consequence of rapid and major geological change derived from its location at the site of intersection and interaction of two major continental plates. The Southern Alps, which extend for a distance of 500 km, are a direct consequence of uplift along the Alpine Fault, situated to the east of the plate junction. The earthquakes, volcanoes, and geothermal activity that characterize the area also are attributable to this same tectonic activity (Macquarie Library, 1996).

Table 10:1. Distribution of families of anurans in the Australo-Papuan Region; + = present, − = absent, I = introduced.

Family	Australia	New Guinea	Bismarck Archipelago	Solomon Islands	Fiji Islands	New Caledonia	New Zealand
Bufonidae	I	I	I	I	I	−	−
Hylidae	+	+	+	+	−	I	I
Leiopelmatidae	−	−	−	−	−	−	+
Microhylidae	+	+	+	−	−	−	−
Myobatrachidae	+	+	−	−	−	−	−
Ranidae	+	+	+	+	+	−	−

DEFINITION OF AMPHIBIAN FAUNA

Anura is the only amphibian order represented in the southwest Pacific. In a global sense, the region is low in endemic families, for there are only two—Leiopelmatidae in New Zealand and Myobatrachidae in Australia and New Guinea. The remaining families (Hylidae, Microhylidae, and Ranidae) are all shared with other continents (Table 10:1). The presumed presence of Rhacophoridae in New Guinea is based on the existence of the holotype of *Hyla wirzi* Roux (1927) referred to *Rhacophorus* by Forcart (1946). Despite reasonably extensive collecting, it has not been found subsequently, and it is considered that the reported type locality (Lake Sentani) is incorrect (J. I. Menzies, pers. comm.).

Generically, the fauna is extremely distinctive, including many genera unique to the area and only four genera (*Limnonectes, Litoria, Platymantis,* and *Rana*) shared with the Oriental Region. Of these, *Litoria* is included only because its range includes Timor and the Lesser Sunda Islands of Indonesia to the northwest of Australia—a minuscule extension.

Within the total area there are major differences in dominance of families and genera in terms of species richness. In Australia, Hylidae and Myobatrachidae are clearly the major elements (92%) of the fauna, whereas in New Guinea, Hylidae, Microhylidae, and Ranidae dominate the frog fauna. There is an abrupt diminution in family representation farther east (Fig. 10.5)

At the generic level, the number of species per genus is as much a reflection of generic definition as a genuine demonstration that a few genera are remarkably speciose. This phenomenon is particularly evident in the hylid genus *Litoria,* which currently includes 103 described species; a figure that is at least 50 short of the actual number. However, as presently defined, the morphological, biochemi-

cal, and biological attributes of *Litoria* are too extensive for a single genus. Tyler and Davies (1978) recognized and defined 37 species groups, and Savage (1986) resurrected *Pelodryas* for one of them. A multidisciplinary study of Australo-Papuan hylids in collaboration with M. Davies, S. Donnellan, and G. F. Watson that involves hundreds of characters is underway; I anticipate that its completion probably will result in the recognition of at least a dozen genera.

The only fossorial hylid genus (*Cyclorana*) contains 13 described species, but there is confusion about the identity of individuals at several central Australian localities. Either two or three species will have to be redefined to accommodate them, or further species described. Biological information to clarify the matter is lacking. *Nyctimystes* currently includes 23 species, only one of which occurs in Australia (Frost, 1985). It is a discrete genus characterized by two synapomorphies.

Within Myobatrachidae, *Limnodynastes, Neobatrachus,* and *Crinia* contain the largest numbers of species. In the case of *Limnodynastes,* there is a situation resembling that of *Litoria;* possibly the discrete morphotypes within the genus indicate the existence of at least three units best considered as distinct genera—the specialized "bullfrogs," such as *L. dorsalis* sharing the dermal calf gland; *L. ornatus* and *L. spenceri* with polymorphic skin patterns and behavioral synapomorphies; and a conglomeration of nonfossorial taxa highly successful in radiating throughout vast areas.

Neobatrachus has undergone extensive review, particularly in Western Australia, by M. Mahony and J. D. Roberts (Mahony and Roberts, 1986; Mahony et al., 1996; Roberts et al., 1991). Currently 10 species are recognized, but the possibility of there being more is very real in those areas not subjected to comparable study. *Crinia* (sensu Heyer et al., 1982)

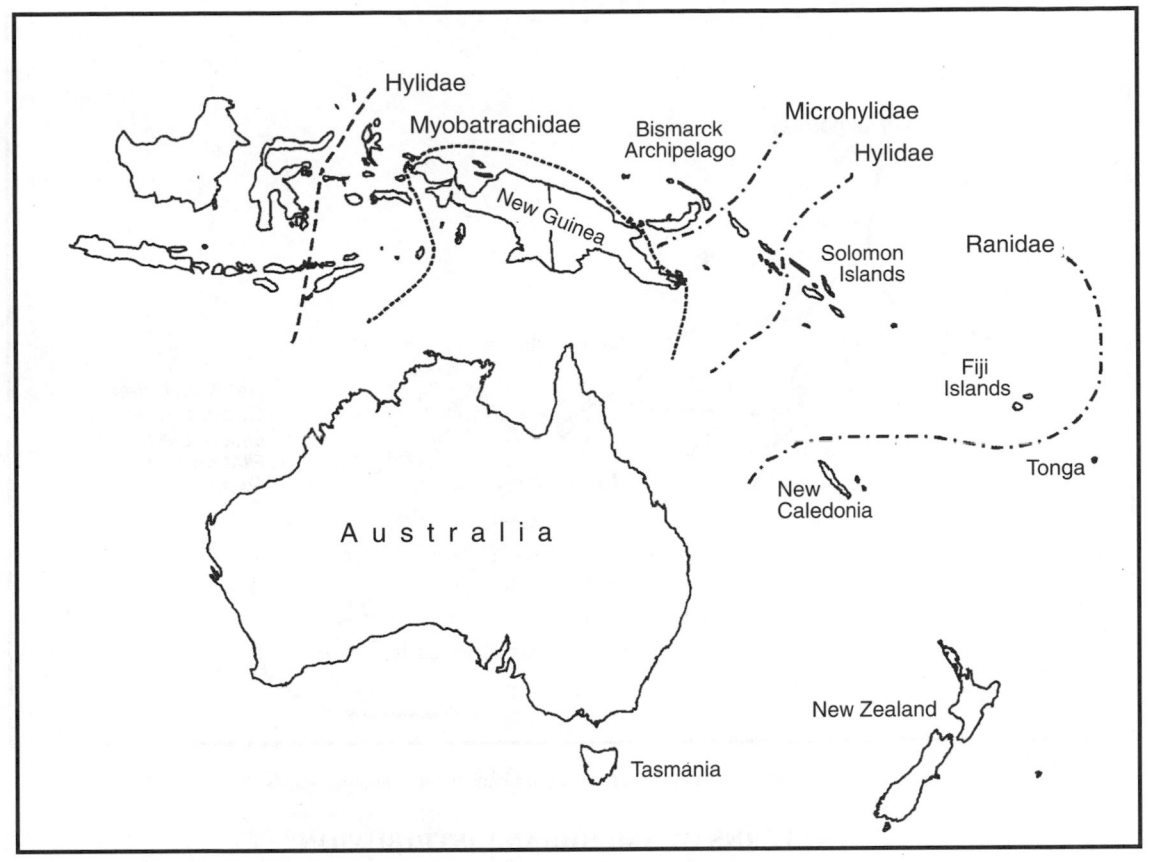

Fig. 10:5. Geographic limits of families of anurans in the Papuan Subregion.

includes 14 species. The type species (*C. georgiana*) is divergent morphologically from the other 13 species, which have been variously referred to *Ranidella*. For the purposes of this review, there is a speciose unit that will either include or exclude the single species "*Crinia*" *georgiana*. Thus, the generic association may change, but the change in generic content is minimal.

Within the Microhylidae there are several examples of species richness within genera that cannot be attributed to inadequate understanding of generic boundaries. In terms of the state of knowledge, the title of the paper by Blum and Menzies (1988) says it all: "Notes on *Xenobatrachus* and *Xenorhina* (Amphibia: Microhylidae) from New Guinea with Descriptions of Nine New Species." The extent of speciation in genera such as *Cophixalus* and *Sphenophryne* seems almost limitless. The small terrestrial species are ecologically equivalents of the myobatrachid *Crinia (Ranidella)* in Australia. These two genera are the only microhylids in Australia. Since the latest review of Zweifel (1985), at least two un-

described species of *Cophixalus* have been discovered (Davies and McDonald, in press; Richards et al., 1994), thereby bringing the current known total in Australia to 12.

Among the Ranidae, *Rana* and *Platymantis* are dominant in New Guinea, the Bismarck Archipelago, and the Solomon Islands. Although the only possible entry for ranids into this area has been from the west, the greatest extent of generic differentiation is at the eastern extremity of the family in the Solomon Islands (Fig. 10:6). There one finds two monotypic genera—the arboreal *Palmatorappia* and the forest floor–dweller *Ceratobatrachus*—and the speciose *Batrachylodes*, as well as *Rana, Platymantis*, and *Discodeles*.

Until recently, Leiopelmatidae, confined to New Zealand, was considered to be represented by three species of *Leiopelma*, of which the last (*L. archeyi* Turbot [1942]) was described more than 50 years ago. However, recent molecular studies have established that one of the small insular populations is in reality a fourth cryptic species (Bell et al., 1998).

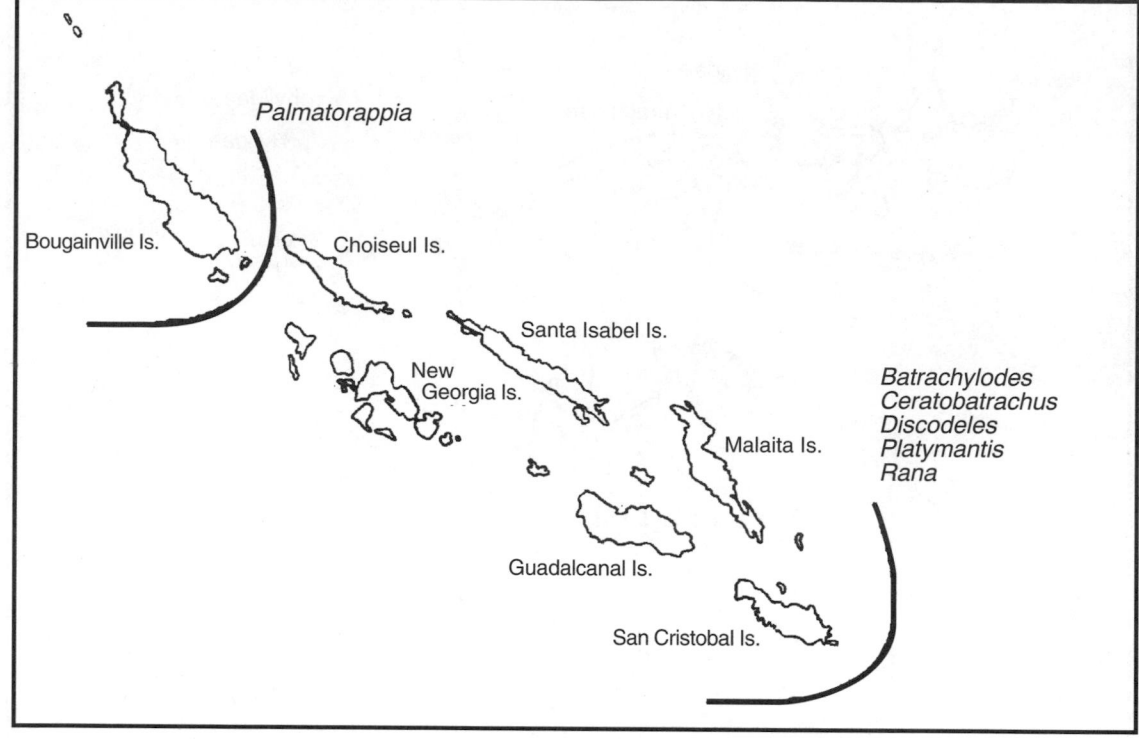

Palmatorappia

Bougainville Is.

Choiseul Is.

Santa Isabel Is.

New
Georgia Is.

Malaita Is.

Batrachylodes
Ceratobatrachus
Discodeles
Platymantis
Rana

Guadalcanal Is.

San Cristobal Is.

Fig. 10:6. Extent of distributions of genera of ranids in the Solomon Islands.

PATTERNS OF AMPHIBIAN DISTRIBUTION

Within the region as a whole, the fauna of each of the physical components regarded here is distinctive (Table 10:2), and there is minimal overlap at generic and species levels from one area to another. The exceptions are the sharing of 10 genera and 14 species between Australia and New Guinea, four genera but only one species between New Britain and the Solomon Islands, and one genus (*Platymantis*) between Fiji, the Solomon Islands, and New Guinea (Appendix 10:2).

Biogeographic provinces of Australia were proposed by Spencer (1896) and subsequently modified by various authors. The current scheme used and presumably accepted by most contributors is shown in Figure 10:7. The first contributions to an

understanding of patterns of distribution of the frog fauna were by Fletcher (1890, 1891, 1892, 1894), who gathered data in New South Wales. The first modern synthesis of the frog fauna was by Moore (1961:364–365), who clearly defined the criteria that he employed to identify the zoogeographic units that he recognized as follows:

1. A zoogeographic region [provinces of some authors] must have boundaries, and it should be possible to determine these boundaries with a moderate degree of precision...

2. A zoogeographic region should have a fauna markedly different from that of the adjacent zoogeographic regions...

3. A significant proportion of the fauna characteristic of a zoogeographic region should have ranges approximately coextensive with the region...

Table 10:2. Numbers of native genera and species in components of the Australo-Papuan Region.

Taxa	Australia	New Guinea	Bismarck Archipelago	Solomon Islands	Fiji Islands	New Caledonia	New Zealand
Genera	30	26	6	7	1	0	1
Species	211	203	14	26	2	0	4

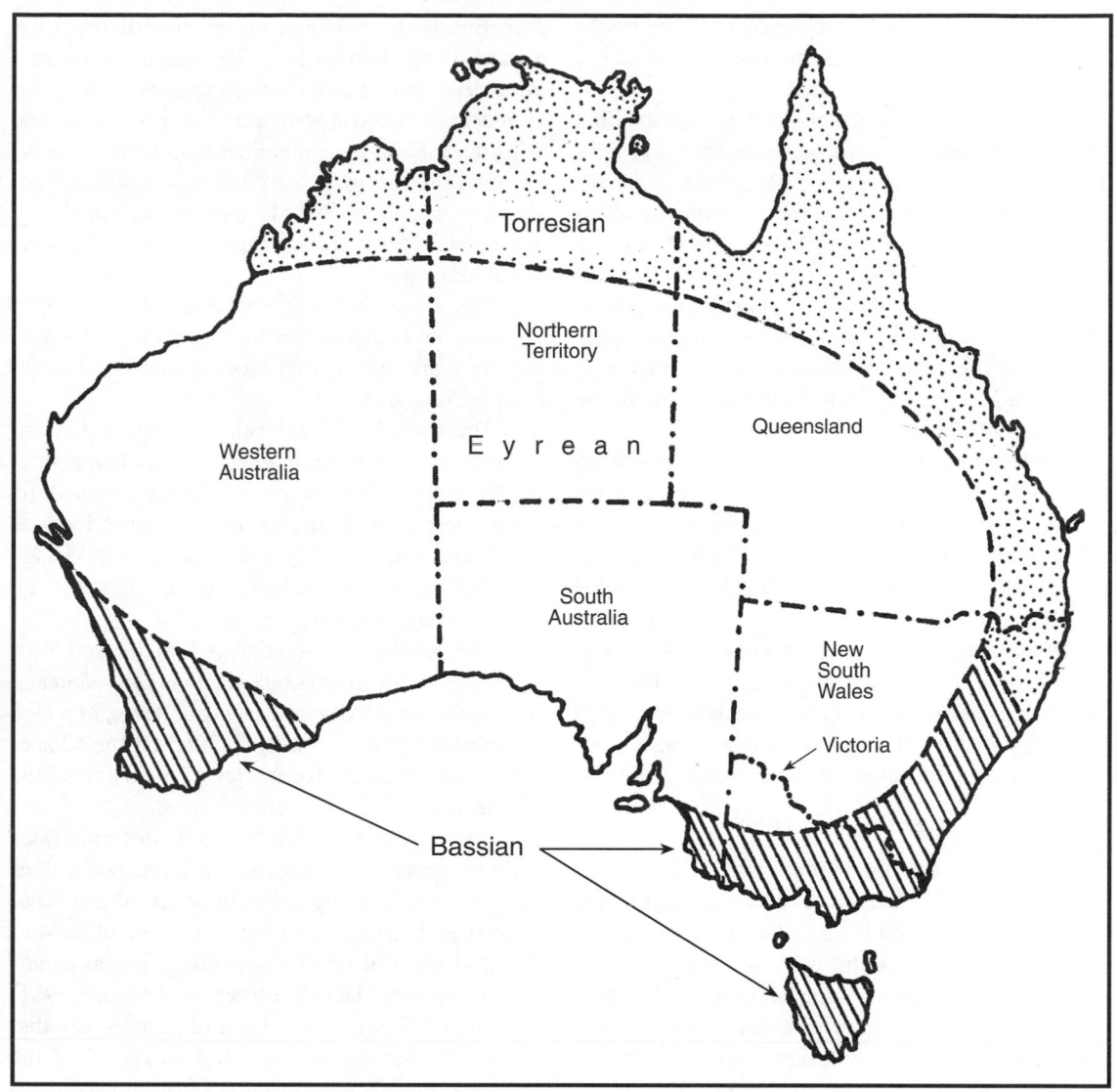

Fig. 10:7. Biogeographic provinces of Australia.

4. A zoogeographic region should be an area isolated, at the present time or during its past, for a sufficient period of time to have allowed the faunas to differentiate...

The only major flaw in his scheme was what Moore termed the "north-east crescent." This was defined as a pattern of species with continuous distributions along the east coast and then, in an arc, across the north of the continent. He noted a progressive reduction in the abundance of species from east to west. This phenomenon has been shown to be an artifact created by inadequate collecting across the north of the continent and also by compounding cryptic species. In fact, until 1974 there was no spe-

cific attempt to collect frogs in that portion of the Northern Territory or the northwest of Western Australia in the wet season (the only season when frogs are active). It follows that very little was known of the fossorial species that would be encountered only in that period. This is well-demonstrated by the fact that 13 of the 18 fossorial species in the Kimberley Division and northern part of the Northern Territory were first encountered and described in the past 25 years. In reality, the northwestern part of the continent has a rich fauna that includes broad-ranging tropical species, as well as a distinctive endemic component of 13 species (Tyler et al., 1994). *Litoria nasuta* is the only remaining species in the northeast crescent unit, and it has not been the subject of study throughout its geographic range.

More recent contributors to the understanding of Australian distribution patterns are Savage (1973), Littlejohn (1981), and Tyler et al. (1981). The quality and extent of distribution records have improved considerably over the past decade and details for most states and territories are available, the most detailed in a published format being that for Queensland by Ingram and Longmore (1991). Simultaneously the number of species known to exist in Australia has risen steadily from the 91 of Moore (1961) to 211 today.

Details of distribution in New Guinea have been provided mainly at a generic level in systematic reviews; examples are *Litoria* (as *Hyla*) by Tyler (1968), *Lechriodus* by Zweifel (1972), and *Rana* by Menzies (1987). Zweifel and Tyler (1982) provided an overview. Brown (1952) analyzed distribution of the frogs of the Solomon Islands, and there have been various presentations of the distribution of the various species of *Leiopelma* in New Zealand (e.g., Bell et al., 1985, 1998). I am unaware of any analyses of distribution in the Bismarck Archipelago or Fiji.

Australian Patterns

The major centers of abundance and diversity are the southwest, the southeastern portion of the continent east of South Australia, the boundary of southeastern Queensland with northeastern New South Wales, northern Queensland, and the Kimberley Division of northwestern Western Australia. Each area has a high degree of endemism (Fig. 10:8).

The southwestern part of Western Australia is isolated by sea to the west and south and by aridity to the north and east. This area is the major center of speciation of *Heleioporus, Neobatrachus,* and *Geocrinia.* Unique elements include the bizarre *Myobatrachus* and *Spicospina.* In contrast, *Litoria* is poorly represented. Twenty-nine of the 32 species are endemic; the other three are shared only with Central Australia.

The boundary of southeastern Queensland with New South Wales is an area of isolated rainforest. More than 50 species of anurans occur there; *Assa* and *Kyarranus* are particularly noteworthy.

Southeastern Australia is most closely linked to the southwestern part of the continent and is commonly considered to be united in a single province (the Bassian). However, no species of frogs are shared by the two regions. Tasmania has a somewhat depauperate frog fauna (6 species), but the recent discovery and description of *Bryobatrachus* and consideration of the endemic *Ranidella tasmaniensis* and *Litoria burrowsae* produce a significant endemic component. Nearly half of the species in southeastern Australia are shared with coastal Queensland to the north.

The Great Dividing Range has 40 species of anurans, only three of which are endemic. Most species are shared with coastal Queensland and/or southeastern Australia.

The Cape York Peninsula in northern Queensland is the area of the highest Papuan component on the continent. It is the only place in Australia where the Papuan genus *Nyctimystes* occurs. It shares 13 species with New Guinea and 12 with coastal Queensland; six of the latter also are shared with southeastern Australia.

The Kimberley Division is a major center of speciation of the myobatrachid genus *Uperoleia* and includes several striking species of *Litoria,* of which the most significant is *L. splendida.* Of the 42 species known from the Kimberleys, 25 are among the 33 species shared with Arnhem Land.

The Eyrean Province (Central Australia) does include a number of endemic species, but it does not have the biogeographic integrity of the other provinces. It is simply a conglomeration of seasonally arid areas of unreliable rainfall (no seasonality). The province lacks the criterion of Moore (1961) of being inhabited by a number of species with distributions contiguous with the boundaries of the Province. In reality, no species of frog approximates this criterion. Most of the species of anurans in Central Australia are shared with regions to the north. Only two species of anurans (the widespread *Litoria caerulea* and *L. rubella*) have been reported from the poorly known Pilbara Region in Western Australia

New Guinea Patterns

With so few active contributors to the study of New Guinean frogs, it is not surprising that the trend of research upon this herpetofaunally rich area is largely focused on the description of new species. Accordingly, there has been limited synthesis to identify patterns of geographic distribution. Tyler (1968) distinguished "montane" and "lowland" components within *Litoria* delimited at 3500 feet. Menzies (in

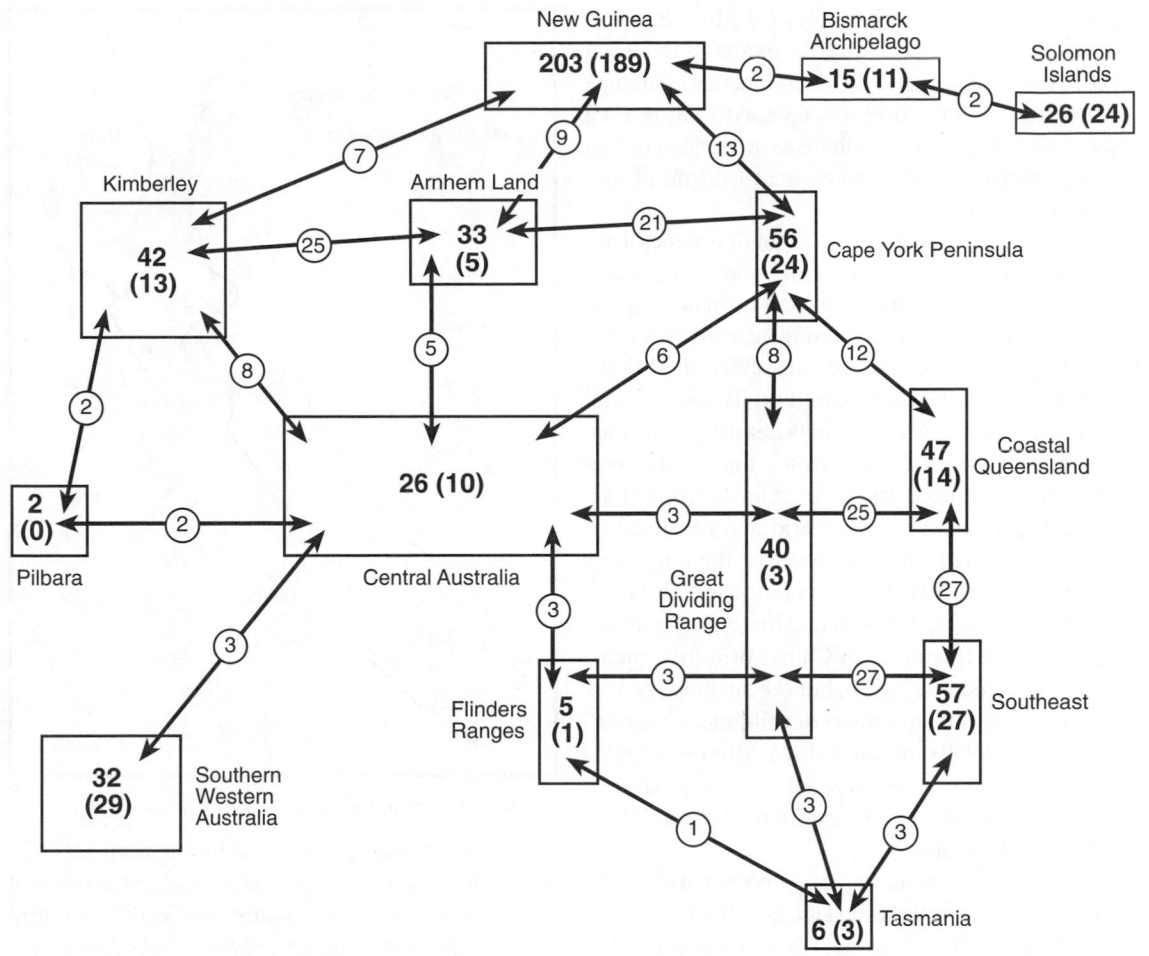

Fig. 10:8. Diagrammatic representation of the Australo-Papuan Region. Numbers in boxes are the number of native species (with number of endemics in parentheses) in each section; numbers in circles indicate species shared with different regions. No native species in New Zealand or on the Fiji Islands is shared with any other part of the Australo-Papuan Region.

litt.) drew attention to some species extending from sea level to elevations above that of the minimum of the montane zone. Nevertheless there is a distinct altitudinal distribution constraint for many species that hinges principally on the extent of reproductive dependence on free water (Zweifel and Tyler, 1982). Thus, because of the absence of free water, direct development is the only option.

Of the more than 200 species known from New Guinea, only 16, including 12 *Litoria,* are shared with Australia. Two additional species (*Litoria infrafrenata* and *Sphenophryne mehelyi*) are shared with the Bismarck Archipelago, and only two species (*Litoria thesaurensis* and *Rana kreffti*) are shared by the Bismarck Archipelago and the Solomon Islands. More than 1000 species of anurans inhabit the Oriental and Australo-Papuan regions. Only two species (members of the ranid genus *Limnonectes*) are shared by New Guinea and the Lesser Sunda Islands in the Oriental Realm.

CORRELATIONS OF DISTRIBUTION PATTERNS

HISTORICAL PHENOMENA

Savage (1973) produced a detailed account of the historical events that shaped the nature of the Australian amphibian fauna. His interpretation relied on understanding the extent and distribution of the Gondwanan fauna and climatic and other factors that influenced Australia during its northern drift to its present position. Savage's contribution antedates by one year the discovery of the first frog fossil in Australia (Tyler, 1974); the extent of the fossil record (now 33 species) provides data that challenge a few of his stimulating concepts. At long last we have a few data that contribute to an understanding of the sources and minimum age of origin of elements of the fauna.

In terms of sources of stock there is general acceptance that the Hylidae (Pelodryadidae of Savage, 1973) and Myobatrachidae are Gondwanan elements and that the Ranidae is derived from a western (Oriental Region) source. Savage perceived the Microhylidae to be a Gondwanan "Old Tropical Unit" present in tropical Australia in the early Cretaceous. He envisaged a scenario in which the family "was gradually eliminated from Australia as tropical areas on the continent became arid and remained so from the middle Cenozoic onwards; the remaining stock became established and radiated in New Guinea in middle Cenozoic, with one line moving northward through Indonesia to China; primitive members of this line had larvae, but the speciose genera in New Guinea all have direct development." In contrast, Tyler (1979) interpreted the Microhylidae as an Oriental component that entered New Guinea when that landmass collided with the Oriental Plate in the mid-Miocene.

Current understanding of paleoclimatic conditions suggests that the extensive aridity of Australia (largely the Eyrean Province) is a Pleistocene phenomenon. Data remain sketchy but do not support Savage's middle Cenozoic date of onset (Galloway and Kemp, 1981).

The fossil record is far from complete. To date, 12 species have been identified from three localities in northern Australia—Riversleigh and Floraville in northern Queensland and Bullock Creek in the Northern Territory (Tyler, 1994). These Cenozoic faunas (Oligocene-Pleistocene) do not include Microhylids. Inadequate sample size is not a factor at Riversleigh, where the disarticulated bones of at least 1000 individuals have been recovered. There remains no evidence to deduce other than that the Microhylidae entered Australia from the Oriental Region via New Guinea.

The age of the onset of aridity is relevant in another sense, namely the antiquity of the Eyrean frog fauna and its radiation. The most fundamental issue is whether the arid-adapted fauna evolved and colonized the area in response to aridity or whether

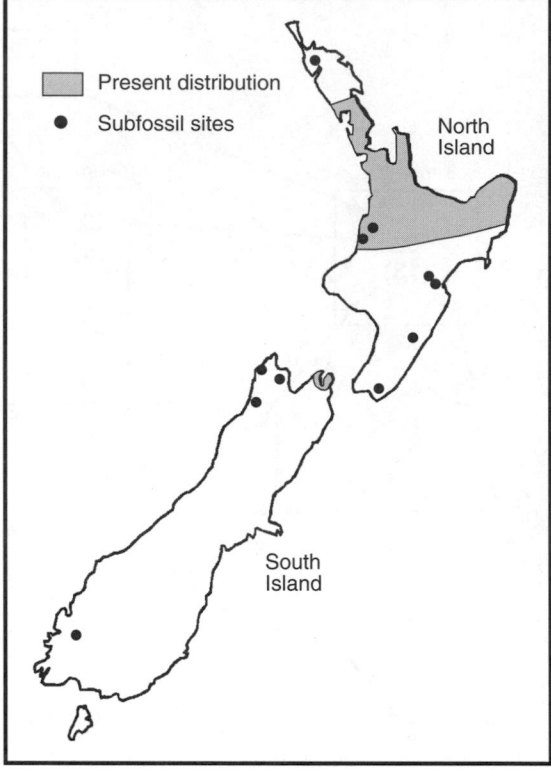

Fig. 10:9. Distribution of *Leiopelma* in New Zealand.

those species there now are surviving remnants of a much richer fauna that persisted despite the onset of aridity. Only four species are confined to the arid zone—*Cyclorana maini, C. platycephala, Litoria gilleni,* and *Uperoleia micromeles.* All others extend significantly into adjacent areas having higher rainfall.

Within New Zealand, the four extant species of *Leiopelma* collectively are concentrated in the mid-latitudes of the North Island and the northern tip of the South Island (Fig. 10:9). One species, *L. hamiltoni* (as redefined by Bell et al., 1998), is confined to Maud Island off the northern coast of South Island. However, the subfossil record based on Worthy (1987) indicates the existence of a richer Holocene fauna with a more extensive latitudinal range on both islands (Fig. 10:9).

CLIMATE

Within Australia rainfall is by far the most significant factor associated with frog diversity. This phenomenon has been well demonstrated in the Northern Territory where the isohyets follow a roughly latitudinal pattern from the high rainfall along the coast to the arid center (Fig. 10:10). Using

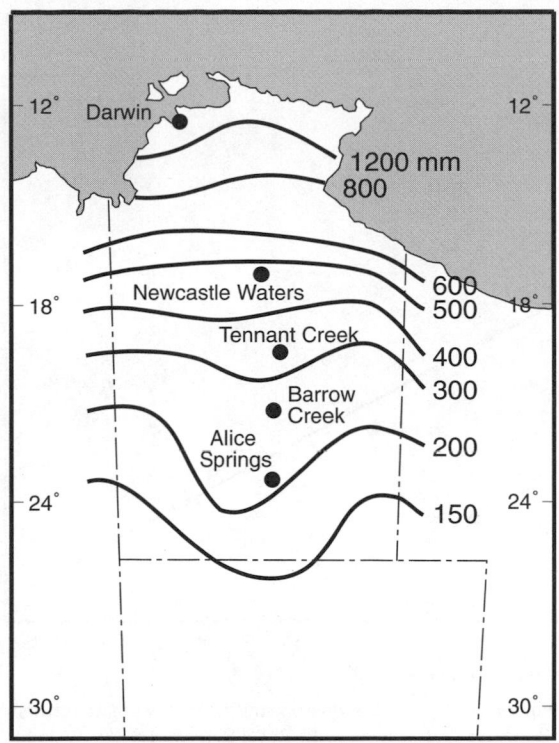

Fig. 10:10. Northern Territory of Australia showing the approximate latitudinal correspondence of isohyets. (From the Transactions of the Royal Society of South Australia; reprinted with permission of the Society.)

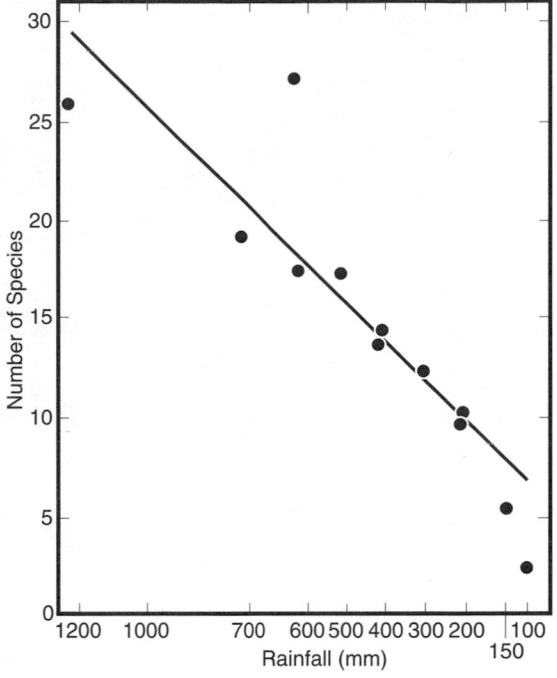

Fig. 10:11. Number of species of frogs within the Northern Territory of Australia plotted against the geographic areas defined by isohyets. (From the Transactions of the Royal Society of South Australia; reprinted with permission of the Society.)

the Northern Territory distribution data of Tyler and Davies (1986), Tyler (1994) plotted the number of species per latitudinal degree and body size against rainfall (Fig. 10:11). As indicated, the association is extremely close in each of these comparisons; the largest number of species exists in the wettest areas (Fig. 10:12), and species of small body size are most common in areas of reliable rainfall.

The inverse ratio between surface area and volume is insignificant in influencing a number of physiological parameters, but it is clear that body size (however it is expressed) influences survival in areas of unreliable rainfall. Fine details of the picture can be blurred by the construction of dams and other artificial water bodies, which create aquatic environments in areas where they would not otherwise be.

PRESERVATION OF THE AMPHIBIAN FAUNA

Within Australia, frogs are totally protected in all states and territories except South Australia. Each of the states and territories formulates and administers its own Act and Regulations, whereas the Australian Government provides support, coordinates the preparation of Action Plans (Tyler, 1997), recovery plans, etc., and has responsibility for the control and administration of national matters such as export and import.

The rigor of enforcement of regulations (as perceived by herpetologists) varies from state to state. National Parks provide the highest level of protection. In New South Wales, the presence of a species on Schedule 1 (Endangered) or Schedule 2 (Vulner-

able) of the Threatened Species Conservation Act of 1995 has significant impact on the likelihood of development of proposals gaining approval if they are found on the subject land, or considered likely to occur there. Unfortunately there is no uniform treatment of endangered or vulnerable species under the Wildlife Conservation Act. In the past, in various Australian states, decisions have been made without data to support them. For example, within months of its description in 1977, the Sandhill Frog, *Arenophryne rotunda,* was protected by legislative action, because of the belief that it was endangered. Happily this is not the case (Roberts, 1985). Irre-

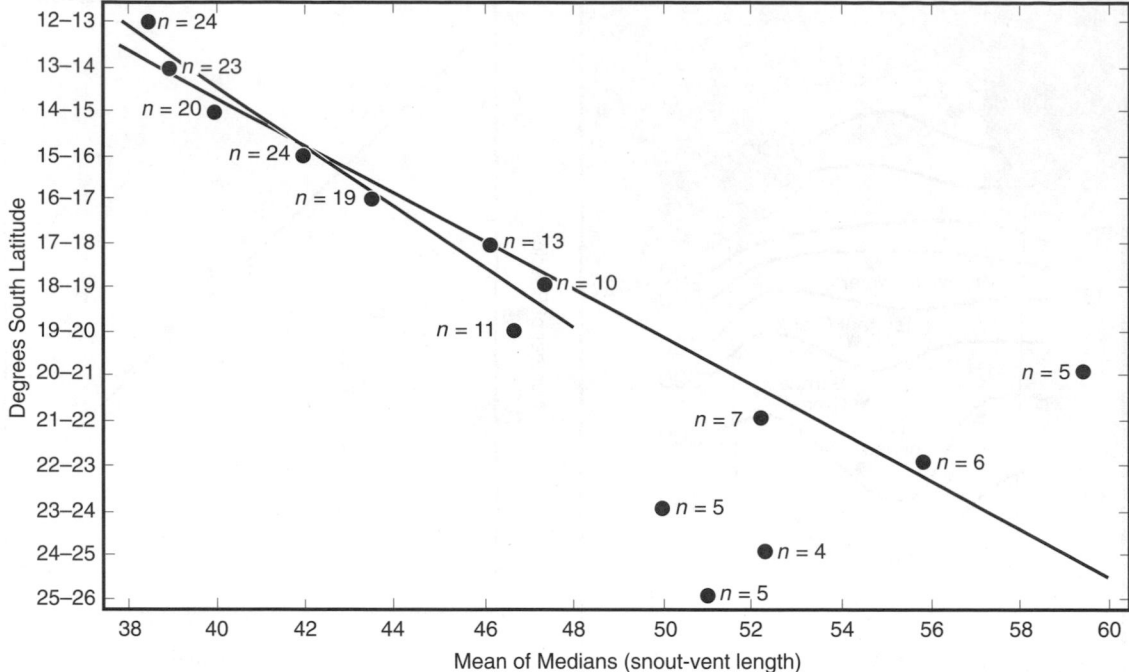

Fig. 10:12. Number of species of frogs occuring within particular latitudes (each at single-degree positions) in the Northern Territory of Australia plotted against the mean of median snout-vent lengths. (From the Transactions of the Royal Society of South Australia; reprinted with permission of the Society.)

spective of the merit of the former decisions, the overall protection of amphibians and the recognition of particular conservation needs being linked to concepts of scarcity and vulnerability, result in few species lacking adequate protection.

Legislative protection in no way protects species against many significant threatening processes responsible for the decline and even possible extinction. The best-known example is the loss of the Gastric-brooding Frog, *Rheobatrachus silus* in southeastern Queensland; simultaneously, the sympatric *Taudactylus diurnus* disappeared.

The New Guinean legislation (also operative for the Bismarck Archipelago and Bougainville) provides strict guidelines for conducting research and collecting, without inhibiting the activities of the few resident researchers or visitors to the area. Exploitation of the herpetofauna of the Solomon Islands for commercial trade is resulting in the exportation of thousands of *Ceratobatrachus;* the matter is causing considerable concern.

In New Zealand, the endemic *Leiopelma* are protected under the New Zealand Wildlife Act of 1953. Detailed study has led to exhaustive investigation of their geographic distribution and an intimate understanding of their needs (Bell, 1985).

Acknowledgments: My most significant acknowledgment is to my good friend and editor of this volume, Bill Duellman, for his understanding, support, and sustained friendship throughout the unreasonable delay in the completion of my manuscript. Despite other pressures, Bill understood the problems that slowed progress during a long period of less than perfect health. I admit to being wholly responsible for the delayed publication of this volume, and I thank my fellow authors for not resorting to physical violence as a quite reasonable expression of their frustration. Several persons assisted me in other ways. Bonnie Laucke prepared most of the figures, and Danaë Wade converted my imperfect handwriting into a carbon form. Stephen Richards provided the list of New Guinean frog species for Appendix 10:1; I thank him for maintaining this list. Jim Menzies reassured me that seeking further detail of patterns of distribution of frogs in New Guinea was an exercise in speculation. Never have I tried to write a meaningful manuscript under such trying circumstances. Thus, to Bill and others who have "gone bananas" during the process, I acknowledge my indebtedness. I thank you all.

LITERATURE CITED

BELL, B. D. 1985. Conservation status of the endemic New Zealand frogs. Pp. 449–458 *in* G. Grigg, R. Shine, and H. Ehmann (eds.), *Biology of Australasian Frogs and Reptiles.* Chipping Norton, New South Wales: Surrey Beatty.

BELL, B. D., C. H. DAUGHERTY, AND J. M. HAYS. 1998. *Leiopelma pakeka,* n. sp. (Anura: Leiopelmatidae), a cryptic species of frog from Maud Island, New Zealand, and a reassessment of the conservation status of *L. hamiltoni* from Stephens Island. J. Roy. Soc. New Zealand 28:39–54.

BELL, B. D., D. G. NEWMAN, AND C. H. DAUGHERTY. 1985. The ecological biogeography of the archaic New Zealand herpetofauna. Pp. 99–106 *in* G. Grigg, R. Shine, and H. Ehmann (eds.), *Biology of Australasian Frogs and Reptiles.* Chipping Norton, New South Wales: Surrey Beatty.

BLUM, J. P., AND J. I. Menzies. 1988. Notes on *Xenobatrachus* and *Xenorhina* (Amphibia: Microhylidae) from New Guinea with descriptions of nine new species. Alytes 7(4):125–163.

BROWN, W. C. 1952. The amphibians of the Solomon Islands. Bull. Mus. Comp. Zool. 107:1–64

BROWN, M. J. F. 1972. Landforms. Pp. 38–39 *in* R. G. Ward and D. A. M. Lea (eds.), *An Atlas of Papua and New Guinea.* Glascow: Collins and Longman.

BUREAU OF METEOROLOGY. 1989. *Climate of Australia.* Canberra: Australian Government Publishing Service.

DAVIES, M, AND K. R. McDONALD. In press. A new species of frog (Microhylidae) from C̃ape Melville, Queensland. Trans. Roy. Soc. South Australia 122.

DENHAM, D. 1970. Earthquakes. Pp. 36–37 *in* R. G. Ward and D. A. M. Lea (eds.), *An Atlas of Papua and New Guinea.* Glasgow: Collins and Longman.

FLETCHER, J. J. 1890. Contributions to a more exact knowledge of the geographical distribution of Australian Batrachia. No.1. Proc. Linnean. Soc. New South Wales Ser. 2, 5:667–676.

FLETCHER, J. J. 1891. Contributions to a more exact knowledge of the geographical distribution of Australian Batrachia. No.2. Proc. Linnean. Soc. New South Wales Ser. 2, 6: 263–274.

FLETCHER, J. J. 1892. Contributions to a more exact knowledge of the geographic distribution of Australian Batrachia. No.3. Proc. Linnean. Soc. New South Wales Ser. 2, 7: 7–19.

FLETCHER, J. J. 1894. Contributions to a more exact knowledge of the geographic distribution of Australian Batrachia. No.4. Proc. Linnean. Soc. New South Wales Ser. 2, 8: 524–533.

FORCART, L. 1946. Katalog des types Exemplare in der Amphibien Sammlung des Naturhistorisches Museums zu Basel. Verhan. Naturfor. Gesell. Basel 57:118–142.

FROST, D. R. 1985. *Amphibian Species of the World.* Lawrence, Kansas: Allen Press and Assoc. Systematics Collections.

GALLOWAY, R. W., AND E. M. KEMP. 1981. Late Caenozoic environments in Australia. Pp. 51–80 *in* A. Keast (ed.), *Ecological Biogeography of Australia.* The Hague: W. Junk.

GRESSITT, J. L. 1982. Zoogeographical summary. Pp. 897–918 *in* J. L. Gressitt (ed.), *Biogeography and Ecology of New Guinea.* The Hague:W. Junk.

HART, D. 1970. Rainfall. Pp. 42–45 *in* R. G. Ward and D.A.M. Lea (eds.), *An Atlas of Papua and New Guinea.* Glasgow: Collins and Longman.

HEYER, W. R., C. H. DAUGHERTY, AND L. R. MAXSON. 1982. Systematic resolution of the genera of the *Crinia* complex (Amphibia: Anura: Myobatrachidae). Proc. Biol. Soc. Washington 95:423–427.

INGRAM, G. J., AND N. W. LONGMORE. 1991. The frog records. Pp. 16–44 *in* G. J. Ingram and R. J. Raven (eds.), *An Atlas of Queensland's Frogs, Reptiles, Birds and Mammals.* Brisbane: Queensland Museum.

JOHNS, R. J. 1982. Plant zonation. Pp. 309–330 *in* J. L. Gressitt (ed.), *Biogeography and Ecology in New Guinea.* The Hague: W. Junk.

KÖPPEN,W. 1932. *Handbuch der Klimatologie.* Berlin: Borntrager.

LITTLEJOHN, M. J. 1981. The Amphibia of mesic Southern Australia: a zoogeographic perspective. Pp. 1305–1330 *in* A. Keast (ed.), *Ecological Biogeography of Australia.* The Hague: W. Junk.

MACQUARIE LIBRARY. 1996. *The Macquarie World Atlas.* Sydney: Macquarie.

MAHONY, M. J., S. C. DONNELLAN, AND J. D. ROBERTS. 1996. An electrophoretic investigation of relationships of diploid and tetraploid species of Australian desert frogs *Neobatrachus* (Anura: Myobatrachidae). Australian J. Zool. 44:639–650.

MAHONY, M. J., AND J. D. ROBERTS. 1986. Two new species of desert burrowing frogs of the genus *Neobatrachus* (Anura: Myobatrachidae) from Western Australia. Rec. Western Australian Mus. 13:155–170.

MENZIES, J. I. 1987. A taxonomic revision of Papuan *Rana* (Amphibia: Ranidae). Australian J. Zool. 35:373–418.

MOORE, J. A. 1961. Frogs of eastern New South Wales. Bull. Am. Mus. Nat. Hist. 121:149–386.

RICHARDS, S. J., A. J. DENNIS, M. P. TRENERRY, AND G. L. WERREN. 1994. A new species of *Cophixalus* (Anura: Microhylidae) from northern Queensland. Mem. Queensland Mus. 37:307–310.

ROBBINS, R. G. 1970. Vegetation. Pp. 46–47 *in* R. G. Ward and D. A. M. Lea (eds.), *An Atlas of Papua and New Guinea.* Glasgow: Collins and Longman.

ROBERTS, J. D. 1985. Population density estimates for *Arenophryne rotunda* : is the Round Frog rare? Pp. 463–467 *in* G. Grigg, R. Shine, and H. Ehmann (eds.), *Biology of Australasian Frogs and Reptiles.* Chipping Norton, New South Wales: Surrey Beatty.

ROBERTS, J. D., M. J. MAHONY, P. KENDRICK, AND C. M. MAJORS. 1991. A new species of burrowing frog, *Neobatrachus* (Anura: Myobatrachidae), from the eastern wheatbelt of Western Australia. Records Western Australian Mus. 15:23–32.

ROUX, J. 1927. Addition à la faune erpétologique de la Nouvelle-Guinée. Rev. Suisse Zool. 34:119–125.

SAVAGE, J. M. 1973. The geographic distribution of frogs: patterns and predictions. Pp. 351–445 *in* J. L. Vial (ed.), *Evolutionary Biology of the Anurans. Contemporary Research on Major Problems.* Columbia, Missouri: Univ. Missouri Press.

SAVAGE, J. M. 1986. Nomenclatural notes on the Anura (Amphibia). Proc. Biol. Soc. Washington 99:42–45.

SPENCER, B. 1896. *Report on the Work of the Horn Scientific Expedition to Central Australia.* Part 1. London: Dulau.

TURBOT , E.G. 1942. The distribution of the genus *Leiopelma* in New Zealand, with a description of a new species. Trans. Proc. Roy. Soc. New Zealand 71:247–253.

TYLER, M. J. 1968. Papuan hylid frogs of the genus *Hyla.* Zool. Verhand. Rijksmus. Nat. Hist. Leiden 96:1–203.

TYLER, M. J. 1972. An analysis of the lower vertebrate faunal relationships of Australia and New Guinea. Pp. 231–256 *in*

D. Walker (ed.), *Bridge and Barrier: The Natural and Cultural History of the Torres Strait.* Canberra: Research School of Pacific Studies, Dept. Biogeography and Geomorphology, Australian National University.

TYLER, M. J. 1974. First frog fossils from Australia. Nature 248:711–712.

TYLER, M. J. 1979. Herpetological relationships of South America with Australia. Pp. 73–106 *in* W. E. Duellman (ed.), The South American herpetofauna: its origin, evolution and dispersal. Monogr. Mus. Nat. Hist. Univ. Kansas 7:1–485.

TYLER, M. J. 1994. Climatic change and its implications for the amphibian fauna. Trans. Roy. Soc. S.Australia 118:53–57.

TYLER, M. J. 1997. *The Action Plan for Australian Frogs.* Canberra: Wildlife Australia.

TYLER, M. J., AND M. DAVIES. 1978. Species-groups within the Australopapuan hylid frog genus *Litoria* Tschudi. Australian J. Zool., Suppl. 63:1–47.

TYLER, M. J., AND M. DAVIES. 1986. *Frogs of the Northern Territory.* Alice Springs, Australia: Conservation Commission of the Northern Territory.

TYLER, M. J., L. A. SMITH, AND R. E. JOHNSTONE. 1994. *Frogs of Western Australia.* Perth: Western Australian Museum.

TYLER, M. J., G. F. WATSON, AND A. A. MARTIN. 1981. The Amphibia: diversity and distribution. Pp. 1272–1301 *in* A. Keast (ed.), *Ecological Biogeography of Australia.* The Hague: W. Junk.

WORTHY, T. H. 1987. Osteology of *Leiopelma* (Amphibia: Leiopelmatidae) and description of three new subfossil *Leiopelma* species. J. Roy. Soc. New Zealand 17:201–251.

ZWEIFEL, R. G. 1972. A review of the frog genus *Lechriodus* (Leptodactylidae) of New Guinea and Australia. Am. Mus. Novit. 2507:1–41.

ZWEIFEL, R. G. 1985. Australian frogs of the family Microhylidae. Bull. Am. Mus. Nat. Hist. 182:265–388.

ZWEIFEL, R. G., AND M. J. TYLER. 1982. Amphibia of New Guinea. Pp. 759–801 *in* J. L. Gressitt (ed.), *Biogeography and Ecology of New Guinea.* The Hague: W. Junk.

APPENDIX 10:1

DISTRIBUTION OF ANURANS IN GEOGRAPHIC SEGMENTS OF AUSTRALIA

In the following list, + = present, − = absent, I = introduced. Species noted by an asterisk (*) also occur in New Guinea. Abbreviations for regions are: ARN = Arnhem Land, CAU = Central Australia, COQ = Coastal Queensland, CYP = Cape York Peninsula, FLI = Flinders Ranges, GDR = Great Dividing Range, KIM = Kimberley, PIL = Pilbara, SOE = Southeast, SWA = Southwest of Western Australia, TAS = Tasmania.

Species	Geographic Segment										
	KIM	ARN	CYP	COQ	GDR	CAU	SOE	FLI	SWA	PIL	TAS
ANURA: BUFONIDAE:											
Bufo marinus	−	I	I	I	I	I	−	−	−	−	−
ANURA: HYLIDAE:											
Cyclorana australis	+	+	+	−	−	+	−	−	−	−	−
Cyclorana brevipes	−	−	−	+	+	−	−	−	−	−	−
Cyclorana cryptotis	+	+	+	−	−	−	−	−	−	−	−
Cyclorana cultripes	+	+	+	−	−	+	−	−	−	−	−
Cyclorana longipes	+	+	+	−	−	−	−	−	−	−	−
Cyclorana maculosus	+	+	−	−	−	−	−	−	−	−	−
Cyclorana maini	−	−	−	−	−	+	−	−	−	−	−
Cyclorana manya	−	−	+	−	−	−	−	−	−	−	−
Cyclorana novaehollandiae	−	−	−	+	−	+	−	−	−	−	−
Cyclorana platycephala	−	−	−	−	−	+	−	−	−	−	−
Cyclorana vagitus	+	+	−	−	−	−	−	−	−	−	−
Cyclorana verrucosus	−	−	−	+	−	+	−	−	−	−	−
Litoria adelaidensis	−	−	−	−	−	−	−	−	+	−	−
Litoria alboguttata	−	+	+	+	−	−	−	−	−	−	−
Litoria andiirrmalin	−	−	+	−	−	−	−	−	−	−	−
Litoria aurea	−	−	−	−	+	−	+	−	−	−	−
Litoria bicolor*	+	+	+	−	−	−	−	−	−	−	−
Litoria booroolongensis	−	−	−	−	+	−	+	−	−	−	−
Litoria brevipalmata	−	−	−	+	+	−	−	−	−	−	−
Litoria burrowsae	−	−	−	−	−	−	−	−	−	−	+
Litoria caerulea*	+	+	+	+	+	+	−	−	−	+	−
Litoria cavernicola	+	−	−	−	−	−	−	−	−	−	−
Litoria chloris	−	−	−	+	+	−	−	−	−	−	−
Litoria citropa	−	−	−	−	+	−	+	−	−	−	−
Litoria cooloolensis	−	−	−	+	−	−	−	−	−	−	−
Litoria coplandi	+	+	−	−	−	−	−	−	−	−	−
Litoria cyclorhynchus	−	−	−	−	−	−	−	−	+	−	−
Litoria dahlii*	+	+	+	−	−	−	−	−	−	−	−
Litoria dentata	−	−	−	−	−	−	+	−	−	−	−
Litoria electrica	−	−	−	−	−	+	−	−	−	−	−
Litoria eucnemis*	−	−	+	−	−	−	−	−	−	−	−
Litoria ewingii	−	−	−	−	+	−	+	−	−	−	+
Litoria fallax	−	−	−	−	+	−	−	−	−	−	−
Litoria flavipunctata	−	−	−	−	+	−	+	−	−	−	−
Litoria freycineti	−	−	−	−	−	−	+	−	−	−	−
Litoria genimaculata*	−	−	+	−	−	−	−	−	−	−	−
Litoria gilleni	−	−	−	−	−	+	−	−	−	−	−
Litoria gracilenta*	−	−	−	−	+	−	+	−	−	−	−
Litoria inermis	+	+	+	−	−	−	−	−	−	−	−
Litoria infrafrenata*	−	−	+	−	−	−	−	−	−	−	−
Litoria jervisiensis	−	−	−	−	−	−	+	−	−	−	−
Litoria latopalmata	−	−	+	+	+	+	+	−	−	−	−
Litoria lesueuri	−	−	+	+	+	−	+	−	−	−	−
Litoria littlejohni	−	−	−	−	+	−	+	−	−	−	−
Litoria longirostris	−	−	+	−	−	−	−	−	−	−	−
Litoria lorica	−	−	+	−	−	−	−	−	−	−	−

Appendix 10:1 continued

Species	Geographic Segment										
	KIM	ARN	CYP	COQ	GDR	CAU	SOE	FLI	SWA	PIL	TAS
Litoria meiriana	+	+	–	–	–	–	–	–	–	–	–
Litoria microbelos	+	+	+	–	–	–	–	–	–	–	–
Litoria moorei	–	–	–	–	–	–	–	–	+	–	–
Litoria nannotis	–	–	+	–	–	–	–	–	–	–	–
*Litoria nasuta**	+	+	+	+	+	–	+	–	–	–	–
*Litoria nigrofrenata**	–	–	+	–	–	–	–	–	–	–	–
Litoria nyakalensis	–	–	+	–	–	–	–	–	–	–	–
Litoria olongburensis	–	–	–	+	+	–	–	–	–	–	–
Litoria pallida	+	+	+	+	+	–	–	–	–	–	–
Litoria paraewingi	–	–	–	–	–	–	+	–	–	–	–
Litoria pearsoni	–	–	–	–	+	–	+	–	–	–	–
Litoria peronii	–	–	–	+	+	–	+	–	–	–	–
Litoria personata	–	+	–	–	–	–	–	–	–	–	–
Litoria phyllochroa	–	–	–	+	+	–	–	–	–	–	–
Litoria piperata	–	–	–	–	+	–	–	–	–	–	–
Litoria raniformis	–	–	–	+	+	–	+	–	–	–	+
Litoria revelata	–	–	–	–	+	–	+	–	–	–	–
Litoria rheocola	–	–	+	–	–	–	–	–	–	–	–
*Litoria rothii**	+	+	+	+	–	–	–	–	–	–	–
*Litoria rubella**	+	+	+	+	+	+	+	–	–	+	–
Litoria spenceri	–	–	–	–	+	–	–	–	–	–	–
Litoria splendida	+	–	–	–	–	–	–	–	–	–	–
Litoria subglandulosa	–	–	–	+	+	–	–	–	–	–	–
Litoria tornieri	+	+	+	–	–	–	–	–	–	–	–
Litoria tyleri	–	–	–	+	+	–	+	–	–	–	–
Litoria verreauxii	–	–	–	–	+	–	+	–	–	–	–
Litoria wotjulumensis	+	+	–	–	–	–	–	–	–	–	–
Litoria xanthomera	–	–	+	–	–	–	–	–	–	–	–
Nyctimystes dayi	–	–	+	–	–	–	–	–	–	–	–
ANURA: MICROHYLIDAE:											
Cophixalus bombiens	–	–	+	–	–	–	–	–	–	–	–
Cophixalus concinnus	–	–	+	–	–	–	–	–	–	–	–
Cophixalus crepitans	–	–	+	–	–	–	–	–	–	–	–
Cophixalus exiguus	–	–	+	–	–	–	–	–	–	–	–
Cophixalus hosmeri	–	–	+	–	–	–	–	–	–	–	–
Cophixalus infacetus	–	–	+	–	–	–	–	–	–	–	–
Cophixalus mcdonaldi	–	–	+	–	–	–	–	–	–	–	–
Cophixalus monticola	–	–	+	–	–	–	–	–	–	–	–
Cophixalus neglectus	–	–	+	–	–	–	–	–	–	–	–
Cophixalus ornatus	–	–	+	–	–	–	–	–	–	–	–
Cophixalus peninsularis	–	–	+	–	–	–	–	–	–	–	–
Cophixalus saxatilis	–	–	+	–	–	–	–	–	–	–	–
Sphenophryne adelphe	–	+	–	–	–	–	–	–	–	–	–
Sphenophryne fryi	–	–	+	–	–	–	–	–	–	–	–
*Sphenophryne gracilipes**	–	–	+	–	–	–	–	–	–	–	–
Sphenophryne pluvialis	–	–	+	–	–	–	–	–	–	–	–
Sphenophryne robusta	–	–	+	–	–	–	–	–	–	–	–
ANURA: MYOBATRACHIDAE:											
Adelotus brevis	–	–	–	+	+	–	+	–	–	–	–
Arenophryne rotunda	–	–	–	–	–	–	–	–	+	–	–
Assa darlingtoni	–	–	–	–	–	–	+	–	–	–	–
Bryobatrachus nimbus	–	–	–	–	–	–	–	–	–	–	+
Crinia georgiana	–	–	–	–	–	–	+	–	–	–	–
Geocrinia alba	–	–	–	–	–	–	–	–	+	–	–
Geocrinia laevis	–	–	–	–	–	–	+	–	–	–	–
Geocrinia leai	–	–	–	–	–	–	–	–	+	–	–
Geocrinia lutea	–	–	–	–	–	–	–	–	+	–	–

Appendix 10:1 continued

Species	Geographic Segment										
	KIM	ARN	CYP	COQ	GDR	CAU	SOE	FLI	SWA	PIL	TAS
Geocrinia rosea	−	−	−	−	−	−	−	−	+	−	−
Geocrinia victoriana	−	−	−	−	−	−	+	−	−	−	−
Geocrinia vitellina	−	−	−	−	−	−	−	−	+	−	−
Heleioporus albopunctatus	−	−	−	−	−	−	−	−	+	−	−
Heleioporus australiacus	−	−	−	−	−	−	+	−	−	−	−
Heleioporus barycragus	−	−	−	−	−	−	−	−	+	−	−
Heleioporus eyrei	−	−	−	−	−	−	−	−	+	−	−
Heleioporus inornatus	−	−	−	−	−	−	−	−	+	−	−
Heleioporus psammophilus	−	−	−	−	−	−	−	−	+	−	−
Kyarranus kundagungan	−	−	−	−	−	+	−	−	−	−	−
Kyarranus loveridgei	−	−	−	−	−	+	−	−	−	−	−
Kyarranus sphagnicolus	−	−	−	−	−	+	−	−	−	−	−
Lechriodus fletcheri	−	−	−	−	−	+	−	+	−	−	−
*Limnodynastes convexiusculus**	+	+	+	+	−	−	−	−	−	−	−
Limnodynastes depressus	+	−	−	−	−	−	−	−	−	−	−
Limnodynastes dorsalis	−	−	−	−	−	−	−	−	+	−	−
Limnodynastes dumerilii	−	−	−	−	−	−	+	−	−	−	−
Limnodynastes fletcheri	−	−	−	−	−	−	+	−	−	−	−
Limnodynastes interioris	−	−	−	−	−	−	+	−	−	−	−
Limnodynastes ornatus	+	+	+	+	−	−	−	−	−	−	−
Limnodynastes peroni	−	−	+	+	+	−	+	−	−	−	−
Limnodynastes salmini	−	−	−	+	+	−	+	−	−	−	−
Limnodynastes spenceri	−	−	−	−	−	+	−	−	−	−	−
Limnodynastes tasmaniensis	I	−	+	+	+	+	+	+	−	−	+
Limnodynastes terraereginae	−	−	−	+	+	−	+	−	−	−	−
Megistolotis lignarius	+	+	−	−	−	−	−	−	−	−	−
Metacrinia nichollsi	−	−	−	−	−	−	−	−	+	−	−
Mixophyes balbus	−	−	−	+	+	−	+	−	−	−	−
Mixophyes fasciolatus	−	−	−	+	+	−	+	−	−	−	−
Mixophyes fleayi	−	−	−	+	−	−	−	−	−	−	−
Mixophyes iteratus	−	−	−	+	−	−	−	−	−	−	−
Mixophyes schevilli	−	−	−	+	−	−	−	−	−	−	−
Myobatrachus gouldii	−	−	−	−	−	−	−	−	+	−	−
Neobatrachus albipes	−	−	−	−	−	−	−	−	+	−	−
Neobatrachus aquilonius	+	−	−	−	−	+	−	−	−	−	−
Neobatrachus centralis	−	−	−	−	−	+	−	−	+	−	−
Neobatrachus fulvus	−	−	−	−	−	−	−	−	+	−	−
Neobatrachus kunapalari	−	−	−	−	−	−	−	−	+	−	−
Neobatrachus pelobatoides	−	−	−	−	−	−	−	−	+	−	−
Neobatrachus pictus	−	−	−	−	−	−	+	+	−	−	−
Neobatrachus sudelli	−	−	−	−	−	−	+	−	−	−	−
Neobatrachus sutor	−	−	−	−	−	+	−	−	+	−	−
Neobatrachus wilsmorei	−	−	−	−	−	−	−	−	+	−	−
Notaden benneti	−	−	−	−	−	−	+	−	−	−	−
Notaden melanoscaphus	+	+	+	−	−	−	−	−	−	−	−
Notaden nichollsi	+	−	−	−	−	+	−	−	−	−	−
Notaden weigeli	+	−	−	−	−	−	−	−	−	−	−
Paracrinia haswelli	−	−	−	−	−	−	+	−	−	−	−
Philoria frosti	−	−	−	−	−	−	+	−	−	−	−
Pseudophryne australis	−	−	−	−	−	−	+	−	−	−	−
Pseudophryne bibroni	−	−	−	+	+	−	+	+	−	−	−
Pseudophryne coriacea	−	−	−	+	−	−	+	−	−	−	−
Pseudophryne corroboree	−	−	−	−	−	−	+	−	−	−	−
Pseudophryne covacevichae	−	−	−	+	−	−	−	−	−	−	−
Pseudophryne dendyi	−	−	−	−	−	−	+	−	−	−	−
Pseudophryne douglasi	−	−	−	−	−	−	−	−	+	−	−
Pseudophryne guentheri	−	−	−	−	−	−	−	−	+	−	−

Appendix 10:1 continued

Species	Geographic Segment										
	KIM	ARN	CYP	COQ	GDR	CAU	SOE	FLI	SWA	PIL	TAS
Pseudophryne major	−	−	−	+	−	−	−	−	−	−	−
Pseudophryne occidentalis	−	−	−	−	−	+	−	−	+	−	−
Pseudophryne pengilleyi	−	−	−	−	−	−	+	−	−	−	−
Pseudophryne raveni	−	−	−	+	−	−	−	−	−	−	−
Pseudophryne semimarmorata	−	−	−	−	−	−	+	−	−	−	−
Ranidella bilingua	+	−	−	−	−	−	−	−	−	−	−
Ranidella deserticola	+	+	−	+	−	+	−	−	−	−	−
Ranidella glauerti	−	−	−	−	−	−	−	−	+	−	−
Ranidella insignifera	−	−	−	−	−	−	−	−	+	−	−
Ranidella parinsignifera	−	−	−	−	−	−	+	−	−	−	−
Ranidella pseudinsignifera	−	−	−	−	−	−	−	−	+	−	−
Ranidella remota*	−	+	+	−	−	−	−	−	−	−	−
Ranidella riparia	−	−	−	−	−	−	−	+	−	−	−
Ranidella signifera	−	−	−	−	+	−	+	−	−	−	−
Ranidella sloanei	−	−	−	−	−	−	+	−	−	−	−
Ranidella subinsignifera	−	−	−	−	−	−	−	−	+	−	−
Ranidella tasmaniensis	−	−	−	−	−	−	−	−	−	−	+
Ranidella tinnula	−	−	−	+	+	−	−	−	−	−	−
Rheobatrachus silus	−	−	−	+	−	−	−	−	−	−	−
Rheobatrachus vitellinus	−	−	−	−	+	−	−	−	−	−	−
Spicospina flammocaerulea	−	−	−	−	−	−	−	−	+	−	−
Taudactylus acutirostris	−	−	−	−	−	−	+	−	−	−	−
Taudactylus diurnus	−	−	−	−	−	−	+	−	−	−	−
Taudactylus eungellensis	−	−	−	+	−	−	−	−	−	−	−
Taudactylus liemi	−	−	−	+	−	−	−	−	−	−	−
Taudactylus pleione	−	−	−	+	−	−	−	−	−	−	−
Taudactylus rheophilus	−	−	−	+	−	−	−	−	−	−	−
Uperoleia altissima	−	−	+	−	−	−	−	−	−	−	−
Uperoleia arenicola	−	+	−	−	−	−	−	−	−	−	−
Uperoleia aspera	+	−	−	−	−	−	−	−	−	−	−
Uperoleia borealis	+	−	−	−	−	−	−	−	−	−	−
Uperoleia capitulata	−	−	−	−	−	+	−	−	−	−	−
Uperoleia crassa	+	−	−	−	−	−	−	−	−	−	−
Uperoleia fusca	−	−	−	+	−	−	−	−	−	−	−
Uperoleia glandulosa	+	−	−	−	−	−	−	−	−	−	−
Uperoleia inundata	−	+	−	−	−	−	−	−	−	−	−
Uperoleia laevigata	−	−	−	−	−	−	+	−	−	−	−
Uperoleia lithomoda	+	+	+	−	−	−	−	−	−	−	−
Uperoleia littlejohni	−	−	−	+	−	−	−	−	−	−	−
Uperoleia marmorata	+	−	−	−	−	−	−	−	−	−	−
Uperoleia martini	−	−	−	−	−	−	+	−	−	−	−
Uperoleia micromeles	−	−	−	−	−	+	−	−	−	−	−
Uperoleia mimula	−	−	+	−	−	−	−	−	−	−	−
Uperoleia minima	+	−	−	−	−	−	−	−	−	−	−
Uperoleia mjobergi	+	−	−	−	−	−	−	−	−	−	−
Uperoleia orientalis	−	+	−	−	−	−	−	−	−	−	−
Uperoleia rugosa	−	−	−	+	+	−	+	−	−	−	−
Uperoleia russelli	+	−	−	−	−	−	−	−	−	−	−
Uperoleia talpa	+	−	−	−	−	−	−	−	−	−	−
Uperoleia trachyderma	+	−	−	−	−	+	−	−	−	−	−
Uperoleia tyleri	−	−	−	−	−	−	+	−	−	−	−
ANURA:RANIDAE:											
Rana daemeli*	−	+	+	−	−	−	−	−	−	−	−

APPENDIX 10:2

DISTRIBUTION OF ANURANS IN NEW GUINEA AND ON ARCHIPELAGOS AND ISLANDS IN THE AUSTRALO-PAPUAN REGION

In the following list, + = present, – = absent, I = introduced. Species noted by an asterisk (*) also occur in Australia. Abbreviations for regions are: BA = Bismarck Archipelago, FI = Fiji Islands, NG = New Guinea, NZ = New Zealand, SI = Solomon Islands. *Litoria aurea* has been introduced on New Caledonia.

Species	Region					Species	Region				
	NG	BA	SI	FI	NZ		NG	BA	SI	FI	NZ
ANURA: BUFONIDAE:						*Litoria pratti*	+	–	–	–	–
Bufo marinus	I	I	I	I	–	*Litoria pronimia*	+	–	–	–	–
ANURA: HYLIDAE:						*Litoria prora*	+	–	–	–	–
Litoria albolabris	+	–	–	–	–	*Litoria pygmaea*	+	–	–	–	–
Litoria amboinensis	+	–	–	–	–	*Litoria quadrilineata*	+	–	–	–	–
Litoria angiana	+	–	–	–	–	*Litoria raniformis**	–	–	–	–	I
Litoria arfakiana	+	–	–	–	–	*Litoria rothii**	+	–	–	–	–
Litoria aruensis	+	–	–	–	–	*Litoria rubella**	+	–	–	–	–
*Litoria aurea**	–	–	–	–	I	*Litoria sanguinolenta*	+	–	–	–	–
Litoria becki	+	–	–	–	–	*Litoria spinifera*	+	–	–	–	–
*Litoria bicolor**	+	–	–	–	–	*Litoria thesaurensis*	–	+	+	–	–
Litoria brongersmai	+	–	–	–	–	*Litoria timida*	+	–	–	–	–
Litoria bulmeri	+	–	–	–	–	*Litoria umbonata*	+	–	–	–	–
*Litoria caerulea**	+	–	–	–	I	*Litoria vagabunda*	+	–	–	–	–
Litoria chloronota	+	–	–	–	–	*Litoria vocivincens*	+	–	–	–	–
Litoria congenita	+	–	–	–	–	*Litoria wisselensis*	+	–	–	–	–
Litoria contrastens	+	–	–	–	–	*Litoria wollastoni*	+	–	–	–	–
*Litoria dahlii**	+	–	–	–	–	*Nyctimystes avocalis*	+	–	–	–	–
Litoria darlingtoni	+	–	–	–	–	*Nyctimystes cheesmanae*	+	–	–	–	–
Litoria dorsalis	+	–	–	–	–	*Nyctimystes daymani*	+	–	–	–	–
Litoria dorsivena	+	–	–	–	–	*Nyctimystes disruptus*	+	–	–	–	–
*Litoria eucnemis**	+	–	–	–	–	*Nyctimystes fluviatilis*	+	–	–	–	–
Litoria everetti	+	–	–	–	–	*Nyctimystes foricula*	+	–	–	–	–
*Litoria ewingii**	–	–	–	–	I	*Nyctimystes granti*	+	–	–	–	–
Litoria exophthalmus	+	–	–	–	–	*Nyctimystes gularis*	+	–	–	–	–
*Litoria genimaculata**	+	–	–	–	–	*Nyctimystes humeralis*	+	–	–	–	–
*Litoria gracilenta**	+	–	–	–	–	*Nyctimystes kubori*	+	–	–	–	–
Litoria graminea	+	–	–	–	–	*Nyctimystes montanus*	+	–	–	–	–
Litoria havina	+	–	–	–	–	*Nyctimystes narinosus*	+	–	–	–	–
*Litoria infrafrenata**	+	+	–	–	–	*Nyctimystes obsoletus*	+	–	–	–	–
Litoria iris	+	–	–	–	–	*Nyctimystes oktediensis*	+	–	–	–	–
Litoria jeudii	+	–	–	–	–	*Nyctimystes papua*	+	–	–	–	–
Litoria leucova	+	–	–	–	–	*Nyctimystes perimetri*	+	–	–	–	–
Litoria longicrus	+	–	–	–	–	*Nyctimystes persimilis*	+	–	–	–	–
Litoria louisiadensis	+	–	–	–	–	*Nyctimystes pulchra*	+	–	–	–	–
Litoria lutea	–	–	+	–	–	*Nyctimystes semipalmatus*	+	–	–	–	–
Litoria majikihise	+	–	–	–	–	*Nyctimystes trachydermis*	+	–	–	–	–
Litoria micromembrana	+	–	–	–	–	*Nyctimystes tyleri*	+	–	–	–	–
Litoria modica	+	–	–	–	–	*Nyctimystes zweifeli*	+	–	–	–	–
Litoria mucro	+	–	–	–	–	ANURA: LEIOPELMATIDAE:					
Litoria multiplica	+	–	–	–	–	*Leiopelma archeyi*	–	–	–	–	+
Litoria mystax	+	–	–	–	–	*Leiopelma hamiltoni*	–	–	–	–	+
Litoria napaea	+	–	–	–	–	*Leiopelma hochstetteri*	–	–	–	–	+
*Litoria nasuta**	+	–	–	–	–	*Leiopelma pakeka*	–	–	–	–	+
*Litoria nigrofrenata**	+	–	–	–	–	ANURA: MICROHYLIDAE:					
Litoria nigropunctata	+	–	–	–	–	*Albericus darlingtoni*	+	–	–	–	–
Litoria obtusirostris	+	–	–	–	–	*Albericus tuberculus*	+	–	–	–	–
Litoria oenicolen	+	–	–	–	–	*Albericus variegatus*	+	–	–	–	–
Litoria ollauro	+	–	–	–	–	*Aphantophryne minuta*	+	–	–	–	–

Appendix 10:2 continued

Species	NG	BA	SI	FI	NZ	Species	NG	BA	SI	FI	NZ
Aphantophryne pansa	+	–	–	–	–	*Oreophryne kampeni*	+	–	–	–	–
Aphantophryne sabini	+	–	–	–	–	*Oreophryne parkeri*	+	–	–	–	–
Asterophrys leucopus	+	–	–	–	–	*Oreophryne wolterstorffi*	+	–	–	–	–
Asterophrys turpicola	+	–	–	–	–	*Pherohapsis menziesi*	+	–	–	–	–
Barygenys atra	+	–	–	–	–	*Sphenophryne brevicrus*	+	–	–	–	–
Barygenys cheesmanae	+	–	–	–	–	*Sphenophryne brevipes*	+	–	–	–	–
Barygenys exsul	+	–	–	–	–	*Sphenophryne cornuta*	+	–	–	–	–
Barygenys flavigularis	+	–	–	–	–	*Sphenophryne crassa*	+	–	–	–	–
Barygenys maculata	+	–	–	–	–	*Sphenophryne dentata*	+	–	–	–	–
Barygenys nana	+	–	–	–	–	*Sphenophryne gracilipes**	+	–	–	–	–
Barygenys parvula	+	–	–	–	–	*Sphenophryne hooglandi*	+	–	–	–	–
Callulops comptus	+	–	–	–	–	*Sphenophryne macrorhyncha*	+	–	–	–	–
Callulops doriae	+	–	–	–	–	*Sphenophryne mehelyi*	+	+	–	–	–
Callulops eurydactylus	+	–	–	–	–	*Sphenophryne palmipes*	+	–	–	–	–
Callulops glandulosus	+	–	–	–	–	*Sphenophryne polysticta*	+	–	–	–	–
Callulops humicola	+	–	–	–	–	*Sphenophryne pusilla*	+	–	–	–	–
Callulops personatus	+	–	–	–	–	*Sphenophryne rhododactyla*	+	–	–	–	–
Callulops robustus	+	–	–	–	–	*Sphenophryne schlaginhaufeni*	+	–	–	–	–
Callulops sagittatus	+	–	–	–	–	*Xenobatrachus anornis*	+	–	–	–	–
Callulops slateri	+	–	–	–	–	*Xenobatrachus arfakianus*	+	–	–	–	–
Callulops stictogaster	+	–	–	–	–	*Xenobatrachus bidens*	+	–	–	–	–
Callulops wilhelmanus	+	–	–	–	–	*Xenobatrachus fuscigula*	+	–	–	–	–
Choerophryne rostellifer	+	–	–	–	–	*Xenobatrachus giganteus*	+	–	–	–	–
Cophixalus ateles	+	–	–	–	–	*Xenobatrachus huaon*	+	–	–	–	–
Cophixalus biroi	+	–	–	–	–	*Xenobatrachus macrops*	+	–	–	–	–
Cophixalus cheesmanae	+	–	–	–	–	*Xenobatrachus mehelyi*	+	–	–	–	–
Cophixalus cryptotympanum	+	–	–	–	–	*Xenobatrachus multisica*	+	–	–	–	–
Cophixalus daymani	+	–	–	–	–	*Xenobatrachus obesus*	+	–	–	–	–
Cophixalus kaindiensis	+	–	–	–	–	*Xenobatrachus ocellatus*	+	–	–	–	–
Cophixalus nubicola	+	–	–	–	–	*Xenobatrachus ophiodon*	+	–	–	–	–
Cophixalus parkeri	+	–	–	–	–	*Xenobatrachus rostratus*	+	–	–	–	–
Cophixalus pipilans	+	–	–	–	–	*Xenobatrachus scheepstrai*	+	–	–	–	–
Cophixalus riparius	+	–	–	–	–	*Xenobatrachus schiefenhoeveli*	+	–	–	–	–
Cophixalus shellyi	+	–	–	–	–	*Xenobatrachus subcroceus*	+	–	–	–	–
Cophixalus sphagnicola	+	–	–	–	–	*Xenobatrachus tumulus*	+	–	–	–	–
Cophixalus tagulensis	+	–	–	–	–	*Xenorhina bouwensi*	+	–	–	–	–
Cophixalus verrucosus	+	–	–	–	–	*Xenorhina eiponis*	+	–	–	–	–
Copiula fistulans	+	–	–	–	–	*Xenorhina minima*	+	–	–	–	–
Copiula minor	+	–	–	–	–	*Xenorhina oxycephala*	+	–	–	–	–
Copiula oxyrhina	+	–	–	–	–	*Xenorhina parkerorum*	+	–	–	–	–
Copiula pipiens	+	–	–	–	–	*Xenorhina similis*	+	–	–	–	–
Copiula tyleri	+	–	–	–	–	ANURA: MYOBATRACHIDAE:					
Genyophryne thomsoni	+	–	–	–	–	*Lechriodus aganoposis*	+	–	–	–	–
Hylophorbus rufescens	+	–	–	–	–	*Lechriodus melanopyga*	+	–	–	–	–
Mantophryne infulata	+	–	–	–	–	*Lechriodus platyceps*	+	–	–	–	–
Mantophryne louisadensus	+	–	–	–	–	*Limnodynastes convexiusculus**	+	–	–	–	–
Oreophryne albopunctata	+	–	–	–	–	*Mixophyes hihihorlo*	+	–	–	–	–
Oreophryne anthonyi	+	–	–	–	–	*Ranidella remota**	+	–	–	–	–
Oreophryne biroi	+	–	–	–	–	*Uperoleia minima*	+	–	–	–	–
Oreophryne brachypus	–	+	–	–	–	ANURA: RANIDAE:					
Oreophryne brevicrus	+	–	–	–	–	*Batrachylodes elegans*	–	–	+	–	–
Oreophryne crucifera	+	–	–	–	–	*Batrachylodes gigas*	–	–	+	–	–
Oreophryne flava	+	–	–	–	–	*Batrachylodes mediodiscus*	–	–	+	–	–
Oreophryne geislerorum	+	–	–	–	–	*Batrachylodes minutus*	–	–	+	–	–
Oreophryne idenburgensis	+	–	–	–	–	*Batrachylodes montanus*	–	–	+	–	–
Oreophryne inornata	+	–	–	–	–	*Batrachylodes trossulus*	–	–	+	–	–
Oreophryne insulana	+	–	–	–	–	*Batrachylodes vertebralis*	–	–	+	–	–

Appendix 10:2 continued

Species	Region					Species	Region				
	NG	BA	SI	FI	NZ		NG	BA	SI	FI	NZ
Batrachylodes wolfi	−	−	+	−	−	*Platymantis myersi*	−	−	+	−	−
Ceratobatrachus guentheri	−	−	+	−	−	*Platymantis neckeri*	−	−	+	−	−
Discodeles bufoniformis	−	−	+	−	−	*Platymantis nexipus*	−	+	−	−	−
Discodeles guppyi	−	−	+	−	−	*Platymantis papuensis*	+	−	−	−	−
Discodeles malukuna	−	−	+	−	−	*Platymantis parkeri*	−	−	+	−	−
Discodeles opisthodon	−	−	+	−	−	*Platymantis punctata*	+	−	−	−	−
Discodeles ventricosus	−	+	−	−	−	*Platymantis rhipiphalcus*	−	+	−	−	−
Limnonectes cancrivora	+	−	−	−	−	*Platymantis schmidti*	−	+	−	−	−
Limnonectes verruculosus	+	−	−	−	−	*Platymantis solomonis*	−	−	+	−	−
Palmatorappia solomonis	−	−	+	−	−	*Platymantis vitiensis*	−	−	−	+	−
Platymantis acrochordus	−	−	+	−	−	*Platymantis vituanus*	−	−	−	+	−
Platymantis aculeodactylus	−	−	+	−	−	*Platymantis weberi*	−	−	+	−	−
Platymantis akarithymus	−	+	−	−	−	*Rana arfaki*	+	−	−	−	−
Platymantis batantae	+	−	−	−	−	*Rana daemeli**	+	−	−	−	−
Platymantis boulengeri	−	+	−	−	−	*Rana garritor*	+	−	−	−	−
Platymantis cheesmanae	+	−	−	−	−	*Rana grisea*	+	−	−	−	−
Platymantis gilliardi	−	+	−	−	−	*Rana grunniens*	+	−	−	−	−
Platymantis guppyi	−	−	+	−	−	*Rana jimiensis*	+	−	−	−	−
Platymantis macrops	−	−	+	−	−	*Rana kreffti*	−	+	+	−	−
Platymantis macrosceles	−	+	−	−	−	*Rana novaebrittaniae*	+	−	−	−	−
Platymantis magnus	−	+	−	−	−	*Rana papua*	+	−	−	−	−
Platymantis mimicus	−	+	−	−	−	*Rana supragrisea*	+	−	−	−	−

Index

If no subject is noted under a scientific name, the reference is to distribution. Numbers in boldface refer to maps.

A

N